Advanced Calculus

Textbooks in Mathematics

Advanced Calculus
Theory and Practice

By John Srdjan Petrovic

CRC Press
Taylor & Francis Group
Boca Raton London New York

CRC Press is an imprint of the
Taylor & Francis Group, an **informa** business

A CHAPMAN & HALL BOOK

CRC Press
Taylor & Francis Group
6000 Broken Sound Parkway NW, Suite 300
Boca Raton, FL 33487-2742

First issued in paperback 2022

© 2021 by Taylor & Francis Group, LLC

CRC Press is an imprint of Taylor & Francis Group, an Informa business

No claim to original U.S. Government works

ISBN 13: 978-1-03-247433-5 (pbk)
ISBN 13: 978-1-138-56821-1 (hbk)
ISBN 13: 978-0-203-70514-8 (ebk)

DOI: 10.1201/9780203705148

Publisher's Note
The publisher has gone to great lengths to ensure the quality of this reprint but points out that some imperfections in the original copies may be apparent.

Visit the Taylor & Francis Web site at
http://www.taylorandfrancis.com

and the CRC Press Web site at
http://www.crcpress.com

Contents

Preface

This text was written for a one-semester or a two-semester course in advanced calculus. It has several goals: to expand the material covered in elementary calculus, and to present it in a rigorous fashion; to improve the problem-solving and proof-writing skills of students; to make the reader aware of the historical development of calculus concepts and the reasons behind them; and to point out the connection between various topics. In writing this book, I was guided by my long-standing experience and interest in teaching calculus at a variety of levels, from advanced high-school students to real and functional analysis.

For the student, my purpose was to lay out material in a precise, readable manner, with the concepts and techniques of advanced calculus clearly presented. I wanted to give students an understanding of higher mathematical theories together with a sense of why these concepts matter and where they come from. My goal was to ensure that students acquire good mathematical skills, both in proving theorems and doing careful calculations.

For the instructor, my purpose was to create a flexible, comprehensive teaching tool using proven pedagogical techniques in mathematics. I wanted to provide instructors with a text that they could use to teach advanced calculus effectively and efficiently in the most appropriate manner for their particular set of students.

This book is designed to meet the needs of a one- or two-term advanced calculus course. It is highly flexible and extremely comprehensive. It is intended not only as a successful textbook but also as a valuable resource students can consult throughout their studies and professional life.

Goals of an Advanced Calculus Course

An advanced calculus course has several purposes:

- Students should learn a specific set of mathematical objects, ideas and techniques, and the role they play in calculus.

- They should become proficient in establishing further properties of these objects and their relationships.

- They should understand the origins of particular concepts.

- They should learn to find analogies and similarities between different branches of calculus.

Changes in the Second Edition

I was lucky to get the input from several excellent reviewers and colleagues who have used the first edition. I have invested a considerable amount of time and energy to follow their suggestions and make the book more effective and student friendly. Here is the list of changes, together with the benefits they provide:

- **Improved Organization.**

  - Chapters 1 and 2 have traded places. That eliminates the need to use results before they are proved.

  - Historical facts are placed in sidebars. That way the flow of the mathematical text is not disturbed.

- **Enhanced Coverage of Axiomatic Systems.** A section is added to include Peano's system of axioms for the set of natural numbers and their use in developing the well-known properties of the set $\mathbb{N}$. This gets students in the right frame of mind to continue with the more complicated axiomatic approach to the set of real numbers.

- **Expanded and Organized Exercise Collection.**

  - There are close to 1,000 new exercises, many of them with solutions or hints.

  - Exercises are classified based on the level of difficulty. Those that are somewhat more difficult than average are marked with a single star *; much more challenging exercises are marked with two stars **.

  - Computation-oriented exercises are paired and solutions or hints provided for the odd-numbered questions.

- **Enrichment Material.**

  - Biographies of 60 mathematicians were added as sidebars.

  - Historical information has been expanded.

- **Illustrations.** About 35 new illustrations were added in order to guide students through examples or proofs.

Features of the Book

- **Motivation of Theorems and Definitions.** Main results are stated only after examples that point out the correct formulation. (See the Extreme Value Theorem on page 99 and the examples preceding it.) When a new concept is introduced, there is a justification for doing so and in such a manner. (See the topological definition of compactness in terms of open covers and the analysis on pages 332–333.) To a student, such insights can help make the theory less abstract.

- **Selection of Proofs.** Rather than aiming for the shortest or the most elegant proof, the preferred choice is to start with an idea that should look reasonable to a student and pursue it. An effort is made to supply the motivation for a particular line of reasoning. (The Mean Value Theorem is proved by noticing that the picture is a slanted version of the one in the proof of Rolle's Theorem and using a rotation.)

- **Historical Information.** A significant effort was made for the book to include a historical prospective. A reader will get a glimpse into the development of calculus and its ideas from the age of Newton and Leibniz, all the way into the twentieth century. Some of the benefits of such an approach include deeper understanding of:

  - *A concept.* Continuity of a function became a hot topic only after Arbogast showed that solutions of partial differential equations should be sought among piecewise continuous functions. The proper definition was then given by Cauchy, but even he struggled until the German school introduced the $\varepsilon - \delta$ symbolism, which allowed the understanding of the uniform continuity, and made it possible to create continuous but nowhere differentiable functions.

– *The need for some concepts.* Piecewise-defined functions make a novice uncomfortable, so it is refreshing to find out that they had the same effect on the best minds of the eighteenth century, and they were not considered to be functions. Nevertheless, they could not be outlawed once it was discovered that they often appear as a sum of a Fourier series.

– *The unity of calculus.* To a student, various topics in calculus may seem unrelated, but in reality many of them have a common root because they have been developed through attempts to solve some specific problems. For example, Cantor studied the sets on the real line because he was interested in Fourier series, and their sets of convergence. Weierstrass discovered that the uniform convergence was the reason why the sum of a power series is a continuous function and the sum of a trigonometric series need not be.

– *Mathematics as a developing science.* It is helpful to know that some of the "natural" ideas were accepted until they proved to be flawed, and that it took a long time and several tries to come up with a proper formulation. It also helps the student understand that mathematics has had (and still has) its share of disagreements and controversies, and that it is a lively area full of opportunities to contribute to its advancement. There is a long list of references which will allow the instructor to direct students to further research the history of a concept or a result.

- **Worked Examples.** The book contains close to 400 worked out examples. In most cases, a question is first asked, then its solution is presented. To signal the end of the example, the symbol ♦ is used. Examples have several different aims: to help a student develop their problem-solving skills; to motivate and lead students to important theorems and definitions; and to help students acquire a habit of scrutinizing theorems. Does the converse hold? Will it remain true if one of the hypotheses is relaxed? If not, what is the counterexample?

- **Exercises.** The book contains more than 2,000 exercises that will help students develop their problem-solving skills. Whenever appropriate, exercises are paired and solutions or hints provided for the odd-numbered questions. An effort was made to include exercises that lead the student through the original proofs of important results.

- **Key Terms and Authors.** To facilitate the use of the book, there is a comprehensive subject index and a separate author index.

- **The Philosophy.** The exposition follows the idea that learning goes from specific to general. Perhaps it was best formulated by Ralph Boas in [7]:

> *Suppose that you want to teach the 'cat' concept to a very young child. Do you explain that a cat is a relatively small, primarily carnivorous mammal with retractible claws, a distinctive sonic output, etc.? I'll bet not. You probably show the kid a lot of different cats, saying 'kitty' each time, until it gets the idea. To put it more generally, generalizations are best made by abstraction from experience.*

John B. Conway echoes the same idea in [20]:

> *To many, mathematics is a collection of theorems. For me, mathematics is a collection of examples; a theorem is a statement about a collection of examples and the purpose of proving theorems is to classify and explain the examples...*

- **Review Material.** The book contains a number of review sections and examples. They serve as refreshers, ensuring that the reader restores the knowledge accumulated through the earlier courses. In addition, they point out specific results that were previously taken for granted, and that will be proved, thereby making a connection with the previously acquired knowledge and providing motivation for the rest of the chapter.

- **Illustrations.** The book contains 126 illustrations that will help the student follow the reasoning in examples or proofs.

- **Accessibility.** The prerequisites for a course based on this book are a standard calculus sequence, linear algebra, and discrete mathematics or a "proofs" course, but these requirements can be relaxed. The first nine chapters require little more than two semesters of calculus—a moderate use of quantifiers and an understanding of mathematical induction, while some exposure to $\varepsilon - \delta$ arguments and maturity with other proofs techniques are a plus. Beyond that, a semester of multivariable calculus is needed, including some linear algebra that is used in such a course. The only exception is Chapter 12, where a solid foundation in linear algebra is necessary.

- **Breadth.** In addition to the majority of the standard topics, the book contains material that is less frequently present in textbooks. Much of this is related to the word *practice* in the title of the book: various techniques of integration (Sections 5.2.1–5.2.4), some lesser-known tests for convergence of series (Section 7.4.1), and finding limits of sequences without resorting to L'Hôpital's Rule (Section 2.9). Chapter 13 (Integrals Depending on a Parameter), in addition to possessing intrinsic beauty, shows in a very obvious way some of the shortcomings of the Riemann integral, and more than justifies the transition to the Lebesgue theory of integration. Finally, Chapter 9 contains the rudiments of Fourier series. The importance, applications, and the volume of this area of mathematics often cause a creation of a separate course. Nevertheless, its significance for the development of calculus cannot be overstated. Section 9.6 is written as an attempt to justify the interest in the previously studied material, and to explain the birth of some modern mathematical disciplines.

- **Mathematical Rigor and Precision.** All definitions and theorems in this text are stated extremely carefully so that students will appreciate the precision of language and rigor needed in calculus. Every step of a proof is either carefully justified or stated as an exercise.

- **Flexibility.** An effort was made to present the material in a unified way. Nevertheless, much of the connection rests on results that should be a part of the student's background, and the dependence of chapters on the previous material is minimized. Most sections allow the instructor to cover the material in a 50-minute class, assuming that some of the proofs are left for students to read on their own.

- **Writing style.** The writing style in this book is direct and pragmatic. Precise mathematical language is used without rhetorical flourishes. Care has been taken to make things simple and clear.

- **Writing projects.** Throughout the text, efforts were made to include the historical development of important theorems. These can be expanded by students by supplying either the proofs of original (frequently weaker) results or finding in the literature examples that contradicted them. The book contains numerous references for students to conduct this research.

- **Suggested reading.** While the book is trying to be comprehensive, it is impossible to cover everything. Whenever there is more to read about a topic, a reference is given. These include classical textbooks, expository articles, and research publications.

How to Use This Book

The book has been carefully written and constructed to support advanced calculus courses at different levels and with differing goals. It can be used either for a year-long course or for the study of single- or multi-variable calculus in a one-semester course. Tables 0.1 and 0.2 identify the core and optional sections. The symbol R denotes a review of material that should be familiar to students from elementary calculus.

Chapter	Core	Optional
1	1.2–1.3	1.1, 1.4
2	2.2–2.8	2.1^R, 2.9
3	3.3–3.6, 3.8–3.9	3.1^R, 3.2^R, 3.7
4	4.2–4.6	4.1^R
5	5.2	5.1^R, 5.2.1–5.2.4
6	6.2–6.6	6.1^R, 6.7
7	7.2–7.5	7.1^R, 7.4.1, 7.5.1, 7.5.2
8	8.1–8.4	8.3.1, 8.3.2, 8.5
9		9.1–9.6

Table 0.1: Semester 1: Functions of One Variable.

A few caveats: Section 8.5 builds upon the material from Section 3.7, and Chapter 9 on Sections 7.5.1 and 7.5.2.

Chapter	Core	Optional
10	10.1–10.6	10.7
11	11.2, 11.3, 11.6	11.1^R, 11.4, 11.5
12		12.1–12.6
13		13.1–13.5
14	14.1–14.5, 14.7	14.6
15	15.1–15.6	15.7, 15.8

Table 0.2: Semester 2: Functions of Several Variables.

A few caveats: Chapter 12 requires a solid background in linear algebra; Sections 13.3–13.5 and Section 14.6 build upon Section 6.7.

Second Edition Reviewers

Yi Hu, *Georgia Southern University*
M. Randall Holmes, *Boise State University*

Jim Lawrence, *George Mason University*
Jeffery J. Leader, *Rose-Hulman Institute of Technology*

Acknowledgments

It is my pleasant duty to thank several people who have made significant contributions to the quality of the book. Shelley Speiss is a rare combination of an artist and a computer whiz. She has made all the illustrations in the book, on some occasions making an immeasurable improvement of my original sketches. Daniel Sievewright has read the whole text (both for the first and the second edition) and even checked the calculations in many examples and exercises (I know that for a fact because he discovered numerous errors, both of a typographical and computational nature). Nathan Poirier had also discovered numerous errors in the first edition of the book. Dr. Dennis Pence has been a gracious audience during the academic year that the book was brewing. I often exchanged ideas with him about how to approach a particular issue, and his enthusiasm for the history of mathematics was contagious.

Western Michigan University has put its trust in me and allowed me to take a sabbatical leave to work on this text, and another one to produce the second edition. This is my opportunity to thank them. I hope that they will feel their expectations fulfilled.

I am grateful to Messrs. Robert Stern and Robert E. Ross, senior editors at CRC Press, who have shown a lot of patience and willingness to help throughout the publishing process. Several anonymous reviewers have made valuable suggestions, and my thanks goes to them.

My family has been, as always, extremely supportive, and helped me sustain the energy level needed to finish the project.

This book is dedicated to all my calculus teachers, and in particular to my first one, Mihail Arsenović, who hooked me on the subject.

John Srdjan Petrovic
Western Michigan University

1

Real Numbers

Important theorems of calculus often rely on the properties of real numbers. In the course of the nineteenth century it became clear that these needed to be proved as well, and for that it was necessary to make a precise definition of real numbers. This task was accomplished around 1872 by the independent efforts of Dedekind, Cantor, Heine, and Méray.

1.1 Axiomatic Systems

In any area of mathematics, statements need to be proved, and this always involves the use of previously established results. This approach, the *deductive method*, has as its foundation a set of *axioms*, from which other assertions can be derived. The first known text that followed this method is "Elements", written by a Greek mathematician Euclid (c. 300 BC). Across 13 books, Euclid covered plane and solid geometry, and elementary number theory. Many centuries later, it served as an inspiration to develop axiomatic systems in other areas of mathematics. The nineteenth century saw the emergence of such systems. As an example, we mention group theory.

An important example is the set of positive integers (also known as *natural numbers*) $1, 2, 3, \dots$. It is useful to have a symbol for this set and we will use $\mathbb{N}$. The fact that a number N is a positive integer can be written as $N \in \mathbb{N}$. Occasionally, we will be dealing with the set of non-negative integers. We will denote it by $\mathbb{N}_0$. The question whether the set $\mathbb{N}$ can be defined axiomatically became a hot topic towards the end of the nineteenth century. The best known system of axioms is due to Peano. Let us look at the full list.

Axiom 1. 1 is a natural number.

This axiom guarantees that there is at least one number in $\mathbb{N}$, i.e., that it is not the empty set. The next 4 axioms describe the equality.

Axiom 2. For every natural number x, $x = x$.

Axiom 3. For all natural numbers x and y, if $x = y$ then $y = x$.

Axiom 4. For all natural numbers x, y and z, if $x = y$ and $y = z$ then $x = z$.

Axiom 5. For all x and y, if y is a natural number and $x = y$, then x is also a natural number.

Axioms 2–4 assert that the equality is an equivalence relation. The next group of axioms describe the **successor function** S.

Axiom 6. For every natural number n, $S(n)$ is a natural number.

Axiom 7. For all natural numbers m and n, $m = n$ if and only if $S(m) = S(n)$.

Axiom 8. For every natural number n, $S(n) = 1$ is false.

We always think about $S(n)$ as $n + 1$. Of course, that means that 2 is the symbol for $S(1)$, 3 stands for $S(2)$, etc. There is one more axiom.

Axiom 9. If K is a set such that:

- 1 is in K,
- for every positive integer n, if n is in K, then $S(n)$ is in K,

then K contains every positive integer.

This is probably the most famous of his axioms. It is the foundation of the Mathematical Induction.

The axioms of $\mathbb{N}$ allow us to establish new facts about this set. Of course, we are familiar with natural numbers, so the following result is well known.

Theorem 1.1.1. *If n is a natural number different from 1, then there exists $m \in \mathbb{N}$ such that $S(m) = n$.*

Proof. We will consider the set K, defined by

$$K = \{n \in \mathbb{N} : n = 1 \text{ or } n = S(m), \text{ for some } m \in \mathbb{N}\}.$$

We want to use Axiom 9. First, by definition of the set K, we have that 1 is in K. Next, suppose that $n \in K$. Once again, by definition of K, $S(n)$ is in K. Therefore, Axiom 9 guarantees that K contains every positive integer. On the other hand, the definition of K shows that K is a subset of $\mathbb{N}$. Thus $K = \mathbb{N}$ and the theorem is proved. $\qquad\square$

GIUSEPPE PEANO (1858–1932) was an Italian mathematician and linguist. His parents were farmers and he was born in a farmhouse. His uncle was a priest in Turin and, realizing that Giuseppe was a very talented child, he took him to Turin where he attended high school and the University of Turin. He received his doctorate in 1880 and joined the university the same year. The first part of his career was brilliant. Already in 1882 he discovered that the definition of the surface area in the standard textbook was incorrect, although Schwarz (page 352) made the same discovery a year earlier (see page 500). In 1886 he proved the existence of a solution of $dy/dx = f(x, y)$ when f is continuous. Following the ideas of Grassmann (page 303) he presented the first axiomatic definition of a real linear space in [85]. In that work he introduced the modern symbols for the union and intersection of sets. Peano continued his work on the axiomatic systems and his axioms for positive integers appeared in 1889, in the book [86]. In 1890 he gave an example of a "space filling curve" (see page 482). Around that time he immersed himself in a new and extremely ambitious project known as *Formulario Mathematico*. The goal was to present the whole mathematics using the notation of mathematical logic. Although five thoroughly revised editions appeared between 1895 and 1908, the effort is nowadays considered misdirected. He also wrote an international language *Latino sine flexione* (Latin without inflections), which is a simplified version of classical Latin. Most of his books and papers are in this language.

Many statements about natural numbers involve the operation of **addition**. In order to use it properly, we must define it precisely first. Following Peano's exposition, we will require that

Axiom 10. For all $n \in \mathbb{N}$, $S(n) = n + 1$.

Axiom 11. For any $n, m \in \mathbb{N}$, $n + S(m) = S(n + m)$.

Notice that Axiom 10 simply enacts what our intuition was telling us anyway. Also, it is worth pointing out that the definition covers the addition of *any* two natural numbers. Indeed, we had proved Theorem 1.1.1 which says that any natural number is either 1 or of the form $S(m)$, for some $m \in \mathbb{N}$.

Now we can prove another well known property of natural numbers.

Theorem 1.1.2 (Associativity.). *If x, y, z are natural numbers, then $x+(y+z) = (x+y)+z$.*

Proof. Let us define a set K, a subset of $\mathbb{N}$, by

$$K = \{z \in \mathbb{N} : \text{ for all } x, y \in \mathbb{N}, x + (y + z) = (x + y) + z\}.$$

We will use Axiom 9 to prove that $K = \mathbb{N}$. First we need to show that $1 \in K$. Let $x, y \in \mathbb{N}$. Then

$$
\begin{aligned}
x + (y + 1) &= x + S(y) && \text{(by Axiom 10)} \\
&= S(x + y) && \text{(by Axiom 11)} \\
&= (x + y) + 1 && \text{(by Axiom 10).}
\end{aligned}
$$

We conclude that $1 \in K$.

Next, we want to prove that, if $z \in K$ then $S(z) \in K$. Let $z \in K$ and let $x, y \in \mathbb{N}$. Then

$$
\begin{aligned}
x + (y + S(z)) &= x + S(y + z) && \text{(by Axiom 11)} \\
&= S(x + (y + z)) && \text{(by Axiom 11)} \\
&= S((x + y) + z) && \text{(because } z \in K) \\
&= (x + y) + S(z) && \text{(by Axiom 11)}
\end{aligned}
$$

We conclude that $S(z) \in K$, and the theorem is proved. $\square$

Another important operation is **multiplication**. We define it the following way:

Axiom 12. For all $n \in \mathbb{N}$, $n \cdot 1 = n$.

Axiom 13. For any $n, m \in \mathbb{N}$, $n \cdot S(m) = n \cdot m + n$.

Now, we can establish another useful rule.

Theorem 1.1.3 (Distributive Law). *If x, y, z are natural numbers, then $x(y+z) = xy+xz$.*

Proof. Let us define a set K, a subset of $\mathbb{N}$, by

$$K = \{z \in \mathbb{N} : \text{ for all } x, y \in \mathbb{N}, x(y + z) = xy + xz\}.$$

First we will establish that $1 \in K$. Let $x, y \in \mathbb{N}$. Then

$$
\begin{aligned}
x \cdot (y + 1) &= x \cdot S(y) && \text{(by Axiom 10)} \\
&= xy + x && \text{(by Axiom 13)} \\
&= xy + x \cdot 1 && \text{(by Axiom 12).}
\end{aligned}
$$

Next, we want to prove that, if $z \in K$ then $S(z) \in K$. Let $z \in K$ and let $x, y \in \mathbb{N}$. Then

$$
\begin{aligned}
x \cdot (y + S(z)) &= x \cdot S(y + z) && \text{(by Axiom 11)} \\
&= x(y + z) + x && \text{(by Axiom 13)} \\
&= (xy + xz) + x && \text{(because } z \in K) \\
&= xy + (xz + x) && \text{(by the associativity)} \\
&= xy + x \cdot S(z) && \text{(by Axiom 13)}
\end{aligned}
$$

We conclude that $S(z) \in K$, and the theorem is proved. $\square$

Exercises

1.1.1. Prove that, if the natural numbers x, y, z are such that $x = y$ and $y \neq z$, then $x \neq z$.

1.1.2. Prove that, if x is a natural number different from 1, then there exists a unique $y \in \mathbb{N}$ such that $x = S(y)$.

1.1.3. Prove that $2 + 2 = 4$.

1.1.4. Prove that $2 \cdot 3 = 6$.

1.1.5. Prove the associativity of multiplication in $\mathbb{N}$: For any $x, y, z \in \mathbb{N}$, $x \cdot (y \cdot z) = (x \cdot y) \cdot z$.

1.1.6.* Prove the commutativity of addition in $\mathbb{N}$: For any $x, y \in \mathbb{N}$, $x + y = y + x$.

1.1.7. Use Exercise 1.1.6 to prove that, if x is a natural number, then $x \neq x + 1$.

1.1.8.* Prove that, if the natural numbers x, y, z are such that $x + z = y + z$, then $x = y$.

1.1.9.* Prove that, if the natural numbers x, y, z, w are such that $x = z$ and $y = w$, then $x + y = z + w$.

1.1.10.* Prove that, if the natural numbers x, y, z, w are such that $x = z$ and $y = w$, then $x \cdot y = z \cdot w$.

1.1.11.* Prove the commutativity of multiplication in $\mathbb{N}$: For any $x, y \in \mathbb{N}$, $x \cdot y = y \cdot x$.

1.1.12.* Prove that Axioms 1–9 define $\mathbb{N}$ uniquely up to an isomorphism. Namely, if $\mathbb{N}'$ is another set satisfying Axioms 1–9 with $1'$ instead of 1 and S' instead of S, then there exists a bijection $\varphi : \mathbb{N} \to \mathbb{N}'$ satisfying $\varphi(1) = 1'$ and $\varphi(S(x)) = S'(\varphi(x))$, for all $x \in \mathbb{N}$.

1.2   Axioms of the Set $\mathbb{R}$

The natural numbers are not suitable for the development of calculus. For example, we cannot do the usual arithmetic operations of subtraction and division. In order to remedy this deficiency, mathematicians have been using rational numbers since the earliest times. Peano's system of axioms can be expanded to justify the use of rational numbers, but we will skip the details and recommend [80] to the interested reader.

However, already in ancient Greece, it became clear that even the rational numbers are not sufficient so real numbers became more and more a part of the mainstream mathematics. In spite of their long history, the problem of finding a set of axioms that would capture their essence is relatively new in comparison. Even in the nineteenth century, mathematicians were more concerned with the construction of real numbers from the rationals. These were important efforts and we will return to them later. Right now, we will fast forward to the year 1900, and a paper [66] written by German mathematician David Hilbert (page 6). In this paper he gave a list of axioms that characterize the real numbers.

What were Hilbert's axioms for $\mathbb{R}$? The first group of axioms took care of the usual operations on real numbers and their properties. These *algebraic properties* mean that the set of real numbers $\mathbb{R}$, together with the operations of addition and multiplication, is a field. We list them in Table 1.1. While the first group of axioms deals with equalities, the second group is all about inequalities. The rules concerning the relation $\leq$ are listed in Table 1.2. Together, Field Axioms and Order Axioms make the set $\mathbb{R}$ an **ordered field**. However, they are not sufficient to describe $\mathbb{R}$, and Hilbert was forced to include a third group. The problem is that the set $\mathbb{Q}$ (rational numbers) satisfies the same axioms (Tables 1.1 and 1.2), so it is also an ordered field. In order to present that final axiom we need to introduce some new terminology.

Name	Addition	Multiplication
Closure	$a, b \in \mathbb{R} \Rightarrow a + b \in \mathbb{R}$	$a, b \in \mathbb{R} \Rightarrow ab \in \mathbb{R}$
Associativity	$(a + b) + c = a + (b + c)$	$(ab)c = a(bc)$
Identity	$a + 0 = 0 + a = a$	$a1 = 1a = a$
Inverse	$a + (-a) = (-a) + a = 0$	$aa^{-1} = a^{-1}a = 1$, if $a \neq 0$
Commutativity	$a + b = b + a$	$ab = ba$
Distributivity	$a(b + c) = ab + ac$	

Table 1.1: Field Axioms

Reflexivity	$a \leq a$
Antisymmetry	$a \leq b$ and $b \leq a \Rightarrow a = b$
Transitivity	$a \leq b$ and $b \leq c \Rightarrow a \leq c$
Trichotomy	Either $a < b$ or $a = b$ or $a > b$
	$a \leq b \Rightarrow a + c \leq b + c$; $\quad a \leq b$ and $c \geq 0 \Rightarrow ac \leq bc$

Table 1.2: Order Axioms.

Definition 1.2.1. A set A of real numbers is **bounded above** (by M) if there is a real number M such that $a \leq M$, for all $a \in A$. The number M is an **upper bound** for the set A.

Similarly we can define a property of being **bounded below** and a **lower bound** of a set.

Example 1.2.2. Let $A = (-\infty, 3)$. The set A is bounded above by 5, and it is not bounded below. ◆

Some examples may require a little work.

Example 1.2.3. Let $A = \{x \in \mathbb{Q} : x^2 < 2\}$. The set A is bounded above by 2.
Solution. Let us split the members of A into two subsets. Those with the property that $|x| \leq 1$, all the more satisfy $x \leq 2$. For the others we have $|x| > 1$, which implies that $x < x^2$, so $x < 2$. ◆

In the next example we will use the "floor" function $\lfloor x \rfloor$ (also called the greatest integer function) which gives the largest integer less than or equal to x. For example, $\lfloor 3.2 \rfloor = 3$, $\lfloor -3.2 \rfloor = -4$, $\lfloor 6 \rfloor = 6$.

Example 1.2.4. $A = \{x \in \mathbb{R} : \sin x < 1/2\}$. The set A is not bounded above.
Solution. In order to establish this assertion, we look at the definition of being bounded above, and take its negative. Using quantifiers, Definition 1.2.1 can be written as: a set A is bounded above if

$$(\exists M)(\forall a \in A) \, a \leq M.$$

Remember that, when taking the negative, the *universal* quantifier $\forall$ needs to be replaced by the *existential* quantifier $\exists$, and vice versa. Of course, the negative of $a \leq M$ is $a > M$. Thus, a set A is not bounded above if

$$(\forall M)(\exists a \in A) \, a > M,$$

and this is what we need to show. Since $a \in A$ means that $\sin a < 1/2$, this will be true, in particular, if $\sin a = 0$, i.e., if a is an integer multiple of π.

Now we can write a precise proof.

Proof. Let M be a positive real number and define $n = \lfloor M/\pi \rfloor + 1$. Then $n \in \mathbb{N}$ and $n > M/\pi$, which implies that the number $a = n\pi > M$. On the other hand, $a \in A$ because $\sin a = \sin n\pi = 0 < 1/2$.   ♦

> The greatest integer function was used by Gauss (page 196) in his 1808 work [49]. His notation was [x] and it was only in 1962 that a Canadian computer scientist Kenneth E. Iverson (1920–2004) introduced the symbol $\lfloor x \rfloor$ and the name *floor function*.

In Example 1.2.3 we have demonstrated that the set A is bounded above by 2. In fact, it is also bounded above by 9/5. Indeed, if this were not true, there would be $a \in A$ such that $a > 9/5$. However the last inequality would then imply (because both sides are positive) that $a^2 > 81/25 = 3.24 > 2$, contradicting the assumption that $a^2 < 2$. Can we do better than 9/5? It is easy to verify that 8/5 is also an upper bound of A. Is there a **least upper bound**?

The answer depends once again on the set in which we operate. If we allow real numbers, then the answer is in the affirmative, and we can even pinpoint the least upper bound: it is $\sqrt{2}$. (The proof will come later.) Clearly, if we restrict ourselves to the set of rational numbers, then the answer is in the negative. It is this advantage of the set $\mathbb{R}$ that we will enact.

The Completeness Axiom. *Every set of real numbers that is bounded above possesses a least upper bound.*

Example 1.2.5. A least upper bound and a greatest lower bound.

Let A be the open interval $(0, 2)$. A least upper bound is 2. What about a lower bound? Any negative number is a lower bound for A, and so is 0. We see that the set A has a greatest lower bound and it is 0.   ♦

Remark 1.2.6. This axiom did not appear in Hilbert's original list. In its place he had two axioms that we will quote later.

DAVID HILBERT (1862–1943) was a German mathematician. He was born and raised in Königsberg, a former German city that is now Kaliningrad, Russia. He was home-schooled until the age of 8 and he did not excel in school until the last year of high school when his mathematical talent was noticed. He enrolled at the University of Königsberg in 1880 and received his doctorate in 1884. He remained at the university until 1895 when he moved to the University of Göttingen, were he remained for the rest of his career. He is considered to have been one of the most influential and universal mathematicians of the nineteenth and early twentieth centuries. His work before 1900 was of the highest quality. In addition to being the world's leading expert in invariant theory (abstract algebra), he produced brilliant work in number theory and his book *Grundlagen der Geometrie* (The Foundations of Geometry) contains the axiomatization of geometry. At the International Congress of Mathematicians in Paris in 1900 he presented 23 unsolved problems. To this day it is generally accepted as the most influential collection of open problems ever to be produced by an individual mathematician. It appears that it inspired Hilbert himself, because his later work outshone the earlier. He turned his back on algebra and number theory to take up entirely new problems in geometry, integral equations, the calculus of variations, and mathematical physics. At the same time, Hilbert set forth a vision for a new axiomatic approach to mathematics and physics. In 1930 Hilbert retired and the city of Königsberg made him an honorary citizen of the city. He gave an address which ended with six famous words showing his enthusiasm for mathematics and his life devoted to solving mathematical problems: *We must know, we shall know.*

Judging by Example 1.2.5, a set that is bounded *below* should have a greatest lower bound. One way to guarantee such a rule would be to postulate it as an axiom. This would turn out to be really embarrassing, if we were to discover that it could be derived from the already listed axioms. So, let us try to prove it first.

The Completeness Axiom guarantees the existence of a least upper bound. The mirror image of the set $A = (0, 2)$ in Example 1.2.5 is the interval $(-2, 0)$ and all these negative lower bounds of A become positive *upper* bounds of $(-2, 0)$. Consequently, the Completeness

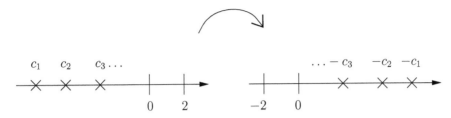

Figure 1.1: Lower bounds turn into upper bounds.

Axiom applies. This is the idea of the proof, but we need to be more precise.

Theorem 1.2.7. *Every set of real numbers that is bounded below possesses a greatest lower bound.*

Proof. Let B be a set that is bounded below by m, and consider the set $A = -B = \{-x : x \in B\}$. This set is bounded above by $-m$. Indeed, if $a \in A$, then $a = -b$, for some $b \in B$.

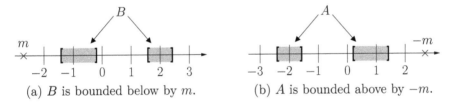

(a) B is bounded below by m.   (b) A is bounded above by $-m$.

Figure 1.2: The existence of a greatest lower bound.

By assumption, $b \geq m$ so $a = -b \leq -m$. Now the Completeness Axiom implies that A has a least upper bound c. We will complete the proof by showing that $-c$ is a greatest lower bound of B.

First, $-c$ is a lower bound of B, because, if $b \in B$, then $-b \in A$, so $-b \leq c$ and $b \geq -c$. On the other hand, $-c$ is a greatest lower bound of B. Indeed, if r were another lower bound of B and $r > -c$, then we would have that $-r < c$, so $-r$ could not be an upper bound for A. That means that there would be $a \in A$ such that $a > -r$. Since $a \in A$, then $a = -b$, for some $b \in B$, and, it would follow that $b = -a < r$. This would contradict the assumption that r is a lower bound of B. Therefore, $-c$ is a greatest lower bound of B. $\square$

Definition 1.2.8. If A is a set and M is a least upper bound of A, then we say that M is a **supremum** of A, and we write $M = \sup A$. If m is a greatest lower bound of A, then we say that m is an **infimum** of A, and we write $m = \inf A$.

Example 1.2.9. Let $A = [0, 1)$. Then $\sup A = 1$ and $\inf A = 0$. $\blacklozenge$

Example 1.2.10. Let $A = \{x : x^2 < 4\}$. Then $\sup A = 2$ and $\inf A = -2$. $\blacklozenge$

Remark 1.2.11. Notice that a supremum (or an infimum) may belong to the set but it does not have to.

Remark 1.2.12. Although the Completeness Axiom guarantees the existence of "a" least upper bound, it is not hard to see that it is "the" least upper bound. That is, there cannot be two distinct least upper bounds of a set.

The significance of the completeness of the real numbers was first recognized by Bolzano, a Bohemian mathematician, philosopher, and a Catholic priest (page 40) around 1817. For the next 50 years, many properties of real numbers were taken for granted. In the second half of the nineteenth century, it was generally accepted that the problem of giving a definition of the set $\mathbb{R}$ was quite important. The first precise formulation of any axiom related to completeness is due to German mathematician Dedekind in [26] in 1872. (See Section 1.4).

Exercises

1.2.1. If A and B are non-empty bounded subsets of $\mathbb{R}$ and $A \subset B$, prove that $\sup A \leq \sup B$.

1.2.2. Let A be a non-empty subset of $\mathbb{R}$ with the property that $\sup A = \inf A$. Prove that the set A has precisely one point.

In Exercises 1.2.3–1.2.6 determine whether the set A is bounded and find $\sup A$ and $\inf A$ if they exist.

1.2.3. $A = \{x : x^2 < 3x\}$.

1.2.4. (a) $A = \{x : 2x^2 < x^3 + x\}$. (b) $A = \{x : 4x^2 > x^3 + x\}$.

1.2.5.* $A = \left\{ \dfrac{mn}{1 + m + n} : m, n \in \mathbb{N} \right\}$.

1.2.6.* (a) $A = \left\{ \dfrac{m}{n} + \dfrac{4n}{m} : m, n \in \mathbb{N} \right\}$. (b) $A = \left\{ \dfrac{m}{m + n} : m, n \in \mathbb{N} \right\}$.

In Exercises 1.2.7–1.2.8 A and B are non-empty bounded subsets of $\mathbb{R}$.

1.2.7. If $C = \{x + y : x \in A, \, y \in B\}$, prove that C is bounded above and that $\sup C = \sup A + \sup B$.

1.2.8. (a) If $D = \{x - y : x \in A, \, y \in B\}$, prove that D is bounded above and that $\sup D = \sup A - \inf B$. (b) If $\alpha \in \mathbb{R}$, $\alpha > 0$, then $\sup\{\alpha x : x \in A\} = \alpha \sup A$.

1.2.9. Prove that a set A can have at most one supremum.

1.2.10.* Prove that, for a set $A \subset \mathbb{R}$, $s = \sup A$ if and only if

    (i) $a \leq s$, for all $a \in A$;

    (ii) for any $\varepsilon > 0$ there exists $a \in A$ such that $a > s - \varepsilon$.

In Exercises 1.2.11–1.2.12 A is a non-empty subset of $\mathbb{R}$ and f, g are functions defined on A.

1.2.11. Prove that $\sup\{f(x) + g(x) : x \in A\} \leq \sup\{f(x) : x \in A\} + \sup\{g(x) : x \in A\}$.

1.2.12.* (a) Prove that $\inf\{f(x) + g(x) : x \in A\} \geq \inf\{f(x) : x \in A\} + \inf\{g(x) : x \in A\}$. (b) Give examples to show that each of the inequalities in (a) and in Exercise 1.2.11 can be strict.

1.3 Some Consequences of the Completeness Axiom

Now that we have the Completeness Axiom we can establish several important properties of the set $\mathbb{R}$. Remember that one of the issues was the presence of gaps in the set of rational numbers. A famous example is $\sqrt{2}$.

Theorem 1.3.1. *There is no rational number a such that $a^2 = 2$.*

Proof. The proof goes all the way back to Pythagoras. We assume that such a rational number exists, so $a = p/q$. If we reduce the fraction, at least one of the integers p, q will have to be odd. Now

$$2 = a^2 = \left(\frac{p}{q}\right)^2 = \frac{p^2}{q^2}.$$

It follows that $p^2 = 2q^2$. Since the right side is an even integer, p cannot be odd. Thus, there exists an integer k such that $p = 2k$. Then $p^2 = 4k^2$, so

$$4k^2 = 2q^2.$$

This implies that $q^2 = 2k^2$, so the right side is an even integer and q cannot be odd. We have concluded that both p and q must be even, contrary to our assumption that at least one of them is odd. This contradiction shows that a cannot be a rational number. $\qquad\square$

A striking consequence of the Completeness Axiom is that it guarantees the existence of a square root of 2.

Theorem 1.3.2. *There exists a real number a such that $a^2 = 2$.*

Proof. Let $A = \{x \in \mathbb{Q} : x^2 < 2\}$. We have seen in Example 1.2.3 that the set A is bounded above by 2. By the Completeness Axiom, there exists $a = \sup A$. Since $1 \in A$, $a \geq 1$ so $a > 0$. We will establish that $a^2 = 2$. Notice that $a \leq 2$ because if $a > 2$ then $a^2 > 4$.

Suppose, to the contrary, that $a^2 \neq 2$. By the Trichotomy Axiom of an ordered field (page 5), either $a^2 > 2$ or $a^2 < 2$. We will prove that either of these two inequalities leads to a contradiction.

First we consider the possibility that $a^2 < 2$. We will show that a is "too small" and that it cannot be an upper bound for A. To accomplish this we will find ε such that $a + \varepsilon \in A$ and $0 < \varepsilon < 1$. The latter implies that $\varepsilon^2 < \varepsilon$. Since $a \leq 2$ we have

$$(a + \varepsilon)^2 = a^2 + 2\varepsilon a + \varepsilon^2 \leq a^2 + 4\varepsilon + \varepsilon = a^2 + 5\varepsilon.$$

Let us choose

$$\varepsilon = \frac{2 - a^2}{6}.$$

Then,

$$(a + \varepsilon)^2 \leq a^2 + 5\varepsilon = a^2 + 5\,\frac{2 - a^2}{6} < a^2 + (2 - a^2) = 2,$$

so $a + \varepsilon \in A$. Since $\varepsilon > 0$, we see that $a + \varepsilon > a$, contradicting the assumption that a is an upper bound of A.

Now we turn our attention to the other possibility: $a^2 > 2$. We will show that a is "too big," meaning that we can find an upper bound of A that is smaller than a. More precisely, we will find a positive number ε such that $a - \varepsilon$ is an upper bound of A.

We have

$$(a - \varepsilon)^2 = a^2 - 2\varepsilon a + \varepsilon^2 \geq a^2 - 2\varepsilon a \geq a^2 - 4\varepsilon$$

since $a \leq 2$. If we choose

$$\varepsilon = \frac{a^2 - 2}{5},$$

then

$$(a - \varepsilon)^2 \geq a^2 - 4\varepsilon = a^2 - 4\,\frac{a^2 - 2}{5} > a^2 - (a^2 - 2) = 2.$$

Since every element $x \in A$ satisfies $x^2 < 2$, then it will all the more satisfy $x^2 < (a - \varepsilon)^2$.

Thus $a - \varepsilon$ is an upper bound for A, which contradicts the fact that a is the least upper bound of A. From these two contradictions we see that $a^2 = 2$, as desired. $\quad\square$

Remark 1.3.3. Theorem 1.3.2 is merely an illustration of the power of the Completeness Axiom. It does not make it clear whether *every* gap has been patched. We will show later that this is indeed so.

One of the disadvantages of the deductive method is that every assertion needs to be proved. We are not going to aim that high, and we will be occasionally guilty of taking some facts as obvious. Sometimes, though, the omission will be only temporary. One such issue appeared in Section 1.2, and it concerns the definition of the floor function. Recall that, given $x \in \mathbb{R}$, $\lfloor x \rfloor$ is defined to be the largest integer less than or equal to x. This raises the question: How do we know that such an integer exists?

Theorem 1.3.4. *Let $a \in \mathbb{R}$ and let $A = \{n \in \mathbb{Z} : n \le a\}$. Then the set A is bounded above and $\sup A$ is an integer.*

Proof. It is obvious that A is bounded above by a, so it has the least upper bound s. Suppose, to the contrary, that s is not an integer. Since s is a least upper bound, the number $s - 1$ cannot be an upper bound of A. Therefore, there exists $N \in A$ such that $N > s - 1$. It follows that $N + 1 > s$ so $N + 1 \notin A$. Therefore, N is the largest integer in A, and we conclude that N is an upper bound of A smaller than s. This contradiction shows

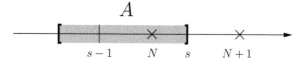

Figure 1.3: $\sup A$ must be an integer.

that s must be an integer. $\quad\square$

Next, we turn our attention to the following question: if b is a fixed real number, is there a positive integer N such that $N > b$? Now we can answer not only that, but even a more general form of the question: if a, b are positive real numbers and $a < b$, is there a positive integer N such that $aN > b$? In fact, the generalization can go one step further. The question can be asked in any ordered field, and when the answer is in the affirmative, we say that such a field is **Archimedean**, or that it has the **Archimedean property**.

Theorem 1.3.5. *The set of real numbers is an Archimedean field.*

Proof. We will first prove the case $a = 1$. Let b be a positive real number and let $B = \{n \in \mathbb{Z} : n \le b\}$. By Theorem 1.3.4, the set B is bounded above and $\sup B = \lfloor b \rfloor$. If $N = \lfloor b \rfloor + 1$ then $N \in \mathbb{N}$ and $N > b$.

The general case $a \ne 1$ follows easily now. Since $a, b > 0$, the number b/a is defined and positive. By the first part of the proof, there exists $N \in \mathbb{N}$ such that $N > b/a$. It follows that $aN > b$ and the proof is complete. $\quad\square$

Historically, the question answered by Theorem 1.3.5 intrigued ancient Greek mathematicians, except that a, b were line segments. Eudoxus of Cnidus (c.408 BC–c.355 BC), a student of Plato, postulated this fact as an axiom: *For any two line segments, there is a multiple of one whose length exceeds the other.* Euclid included this axiom in his book *Elements*, and he ascribed it to another Greek mathematician, Archimedes of Syracuse (c.287 BC–c.212 BC). Although Archimedes himself credited Eudoxus, this property and its generalizations carry his name. It appears that the term "Archimedes' axiom" was coined by Stolz (page 347) in a 1882 article.

Remark 1.3.6. Earlier, when we talked about Hilbert's axioms of $\mathbb{R}$, we only specified two groups of axioms (Field and Order Axioms). Now we can state the axioms of the third group (that Hilbert presented):

(i) $\mathbb{R}$ is an Archimedean field.

(ii) There is no ordered Archimedean field that would contain $\mathbb{R}$.

Another interesting consequence of the Completeness Axiom: the density of real numbers.

Theorem 1.3.7. *If a and b are two distinct real numbers, then there exist a rational number r and an irrational number ρ between a and b.*

Proof. Let us assume that $0 < a < b$. The idea that we will use is the following: if $b - a > 1$ and we move along the x-axis in steps of size less than 1, we must hit a point between a and b. If $b - a \leq 1$, we can stretch the interval (a, b), by multiplying both a and b by a positive integer.

First we will show that there is a rational number $r \in (a, b)$. By the Archimedean Property, there exists a positive integer q such that $q(b - a) > 1$. This implies that the distance between numbers qa and qb is bigger than 1, so there must be an integer p between them. Let $r = p/q$. Since $qa < p < qb$, we obtain that $a < r < b$.

In order to construct ρ, we consider the interval $(\frac{\sqrt{2}}{b}, \frac{\sqrt{2}}{a})$. Once again, the Archimedean Property yields a positive integer p such that

$$p\left(\frac{\sqrt{2}}{a} - \frac{\sqrt{2}}{b}\right) > 1.$$

Therefore, there exists a positive integer q between $p\frac{\sqrt{2}}{a}$ and $p\frac{\sqrt{2}}{b}$:

$$p\frac{\sqrt{2}}{b} < q < p\frac{\sqrt{2}}{a}.$$

These inequalities imply that the irrational number $\rho = \frac{p\sqrt{2}}{q}$ belongs to (a, b).

At the beginning of the proof we have made the assumption that $0 < a < b$. If $\tilde{a} < \tilde{b}$ and

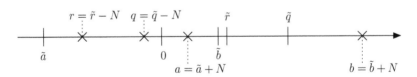

Figure 1.4: The case $\tilde{a} \leq 0$.

$\tilde{a} \leq 0$ we will proceed in the following way. The Archimedean Property yields a positive integer N such that $N > -\tilde{a}$ (if $\tilde{a} = 0$ we can just take $N = 1$), and hence $\tilde{a} + N > 0$. If we now denote $a = \tilde{a} + N$, $b = \tilde{b} + N$, then $0 < a < b$, so the previous proof applies and we get $r, \rho \in (a, b)$. Finally, we define $\tilde{r} = r - N$ and $\tilde{\rho} = \rho - N$. Then $\tilde{r} \in \mathbb{Q}$, $\tilde{\rho} \notin \mathbb{Q}$, and $\tilde{a} < \tilde{r}, \tilde{\rho} < \tilde{b}$. □

Remark 1.3.8. The irrational number $\sqrt{2}$ can be replaced by any other positive irrational number. In effect, we have a stronger result: for any irrational number σ there exists a rational number r so that $r\sigma \in (a, b)$.

Remark 1.3.9. Based on Theorem 1.3.7, we often say that the set of rational numbers is **dense** in $\mathbb{R}$. Clearly, the set of irrational numbers is also dense in $\mathbb{R}$.

Exercises

1.3.1. Let a be a real number and let $S = \{x \in \mathbb{Q} : x < a\}$. Prove that $a = \sup S$.

1.3.2. Prove that the number $\sqrt[3]{2}$ is irrational.

1.3.3.* Prove that $\sqrt{n-1} + \sqrt{n+1}$ is irrational for each $n \in \mathbb{N}$.

1.3.4. Prove that the number $\log_{10} 2$ is irrational.

1.3.5. Prove that the field of complex numbers is not an ordered field.

1.3.6. If a and b are two distinct real numbers and α is an irrational number, prove that there exists a rational number r such that $r\alpha$ lies between a and b.

1.3.7. Let $\alpha \in \mathbb{R}$ and $\varepsilon > 0$. Prove that there exist rational numbers r_1, r_2 such that $r_1 < \alpha < r_2$ and $r_2 - r_1 < \varepsilon$.

1.3.8.* Prove that there exist two irrational numbers α and β such that α^β is rational.

1.3.9.* Prove that there exists a real number a such that $a^3 = 2$.

1.3.10.* Prove that, for any irrational number α and every $n \in \mathbb{N}$, there exist a positive integer q_n and an integer p_n such that

$$\left| \alpha - \frac{p_n}{q_n} \right| < \frac{1}{n q_n}.$$

1.4   Some Thoughts about $\mathbb{R}$

We close this chapter with a brief discussion of several issues regarding the deductive approach that we have used. On a positive side, once a result is proved, there is no dilemma that it is true. However, when an object is defined in an abstract way, two questions that become relevant are the existence and the uniqueness of it.

The first issue is whether the object that we have defined exists. Why wouldn't it?

Example 1.4.1. Contradicting axioms.
    Suppose that we defined $\mathbb{R}$ as a set satisfying: Field Axioms, Order Axioms and (instead of the Completeness Axiom) the Strong Completeness Axiom:

(SCA) Every set of real numbers that is bounded above *contains* the least upper bound.

What would go wrong? We have proved in Theorem 1.3.2 that the set $A = \{x \in \mathbb{Q} : x^2 < 2\}$ has the least upper bound a that satisfies $a^2 = 2$, so (SCA) would imply that $\sqrt{2}$ belongs to A and, consequently, that $\sqrt{2}$ is a rational number. On the other hand, Theorem 1.3.1 established that $\sqrt{2}$ is not a rational number. Therefore, the axiomatic system consisting of field axioms, order axioms, and (SCA) would allow us to prove both the statement $\sqrt{2} \in \mathbb{Q}$ and its negative $\sqrt{2} \notin \mathbb{Q}$.   ◆

    When a system of axioms has the property that one can prove both a statement and its negative, we say that it is **inconsistent**. Otherwise, it is **consistent**. It is not particularly interesting to deal with an inconsistent system, because in such a system every assertion is true (and false). In disgust, we wave our hands and claim that an object satisfying Field Axioms, Order Axioms, and (SCA) *does not exist*. (At least not as a reasonable mathematical entity.) We see that the axioms in Example 1.4.1 are inconsistent. In general, in order to show that a system of axioms is inconsistent, it is sufficient to find a statement which can be proved but its negative can also be proved.

How does one verify that a system *is* consistent? How do we know that the system consisting of Field Axioms, Order Axioms, and the Completeness Axiom is consistent? The standard method is to exhibit a *model*—a concrete theory that satisfies all the axioms. For example, the Hyperbolic Geometry (a.k.a. the Geometry of Lobachevsky) was shown to be consistent using the Poincaré Model (see, e.g., [22]). What model will work for the set $\mathbb{R}$? The plan is to make the assumption that $\mathbb{Q}$ is an ordered field, and construct an extension of it that satisfies the desired axioms. We will talk about two such constructions.

Dedekind started with the idea that every point on the line should correspond to a number. It was clear that every rational number could be represented by a point on the line. Of course, we could call the other points "numbers," but we wanted to have this new collection of "numbers" satisfy axioms of an ordered field. Dedekind introduced the notion of a **cut**. A cut is a partition of the rational numbers into two non-empty parts A and B, such that:

(i) Every rational number belongs to one and only one of the sets A and B;

(ii) Every number $a \in A$ is smaller than every number $b \in B$.

We denote the cut by $A|B$. When the set A has a largest number, or when B contains a least number, such a number is rational.

Example 1.4.2. Let $A = (-\infty, 1) \cap \mathbb{Q}$ and $B = [1, \infty) \cap \mathbb{Q}$. A does not have a largest number, but B has a least number: $\inf B = 1 \in B$. So, the cut $A|B$ corresponds to 1. ◆

Example 1.4.3. Let $A = \{x \in \mathbb{Q} : x^2 < 2 \text{ or } x < 0\}$ and $B = \{x \in \mathbb{Q} : x^2 > 2 \text{ and } x > 0\}$. Again, it is easy to see that this is a cut, but this time neither A has a largest, nor does B have a least number. ◆

In a situation described in Example 1.4.3, that is, when A does not have a largest number, and when B does not have a least number, we say that a cut determines an **irrational number**. The union of all the cuts is the set of real numbers $\mathbb{R}$. It is obvious that now every real number can be identified with a point on the line and vice versa.

We said earlier that Dedekind was the first one to formulate precisely an axiom dealing with the completeness of real numbers. Now we can state his axiom:

Dedekind's Axiom: For every partition of all the points on a line into two nonempty sets such that no point of either lies between two points of the other, there is a point of one set which lies between every other point of that set and every point of the other set.

Remark 1.4.4. Although the original statement refers to "a line" and "points," it can be easily restated in terms of "the set $\mathbb{R}$" and "real numbers." In that version it says that every cut determines a real number. Although its formulation sounds complicated, it is equivalent to the Completeness Axiom (see Exercise 1.4.1).

In Dedekind's construction, each member of $\mathbb{R}$ is a cut, and the next step is to define the algebraic operations with cuts. It gets messy. As an illustration, if $A_1|B_1$ and $A_2|B_2$ are two cuts, we take the set $A = \{a_1 + a_2 : a_1 \in A_1, a_2 \in A_2\}$ and $B = \mathbb{Q} \setminus A$, and we define the "addition" of cuts as $A_1|B_1 + A_2|B_2 = A|B$. The complete details can be found in [90]. The bottom line is that when the smoke clears, we will have on our hands a genuine ordered field in which the Completeness Axiom holds. Consequently, the system of axioms consisting of Field Axioms, Order Axioms, and the Completeness Axiom must be consistent.

A different construction has been done independently by German mathematicians Georg Cantor (page 42) and Eduard Heine (page 92), and French mathematician Charles Méray (1835–1911), all about the same time as Dedekind's. The idea was to associate to every point on a line a Cauchy sequence of rational numbers. (See Section 2.6 for the definition of a Cauchy sequence.) For example, if $a = 1$ we could associate the sequence $a_n = 1$. If

$a = e$ we could associate the sequence $a_n = \left(1 + \frac{1}{n}\right)^n$. (See Section 2.5 for more about this sequence.) We will call $\mathbb{R}$ the collection of all rational Cauchy sequences. Actually, this is not quite right, because different sequences may correspond to the same number. Example: if $a = 0$ we could use $a_n = 1/n$ and $b_n = -1/n^2$. Thus, we need to identify two such sequences: $\{a_n\} \sim \{b_n\}$ if $(a_n - b_n) \to 0$. It can be verified that this is an equivalence relation, and thus $\mathbb{R}$ is really the set of equivalence classes. Now, all that remains is to define the operations and the order on $\mathbb{R}$ and verify that the axioms hold. Once again, we will bow out and direct the reader to the classic text [80] that contains all details.

While the consistency guarantees that there exists at least one object that satisfies the axioms, it is also important to determine whether such an object is *unique*. In order to clarify this issue, let us for the moment think about a geometric problem: how many different squares are there with the side of length 1? The correct answer is: only one. The fact that we may draw two such squares R_1 and R_2 on two pieces of paper does not bother us, because we can move the papers until R_1 and R_2 occupy the same position. More formally, we would say that such two figures are *congruent* because there is an *isometric transformation* F that maps the first square to the second: $F(R_1) = R_2$. In the same spirit, we would identify two different objects $\mathbb{R}_1$ and $\mathbb{R}_2$ that satisfy all the axioms for real numbers, provided that there is a "nice" map F such that $F(\mathbb{R}_1) = \mathbb{R}_2$. What would constitute a "nice" map? Since both $\mathbb{R}_1$ and $\mathbb{R}_2$ are fields, the map F needs to be a field isomorphism. In addition, it needs to preserve the ordering. Let us make this more precise. We will denote the operations on $\mathbb{R}_1$ by $+_1$ and $\cdot_1$, and the order $\leq_1$; similarly, on $\mathbb{R}_2$ we will use $+_2$, $\cdot_2$, and $\leq_2$. Then, we will identify $\mathbb{R}_1$ and $\mathbb{R}_2$ if F is a bijection such that $F(\mathbb{R}_1) = \mathbb{R}_2$ and:

$$F(x +_1 y) = F(x) +_2 F(y), \quad F(x \cdot_1 y) = F(x) \cdot_2 F(y), \quad x \leq_1 y \Leftrightarrow F(x) \leq_2 F(y).$$

It can be shown that such a map exists, so there is only one complete ordered field. The proof can be found in [64].

 RICHARD DEDEKIND (1831–1916) was a German mathematician. He grew up in Brunswick and at the age of 16 enrolled in the Collegium Carolinum where his father and his maternal grandfather were teachers. This prepared him well for the University of Göttingen, where he did his doctorate under Gauss (page 196) on Eulerian integrals. Feeling inadequately prepared he spent his next two years studying the latest developments in mathematics. After defending his Habilitation Thesis in 1854 he started teaching probability and geometry classes. When Gauss died in 1855 Dirichlet (page 64) replaced him and Dedekind learned a lot from him. He also attended lectures by Riemann (page 169). In 1858 he took a position at the Polytechnikum in Zürich. When the Collegium Carolinum became the Brunswick Polytechnikum Dedekind moved there and remained a professor at the Polytechnikum for the rest of his life. He taught even after he retired in 1894. Never married, he lived with one of his sisters. As we have seen, Dedekind worked on the foundations of mathematics: the "cuts", the construction of real numbers from rationals (see page 13). He edited and published the collected works of Dirichlet, Gauss and Riemann. The work on Dirichlet's notes inspired him to study algebraic number fields and he introduced the concept of an ideal. Among his many honors we mention the membership in the Göttingen Academy, the Berlin Academy, and the French Academy of Sciences.

Exercises

1.4.1. Suppose that $\mathbb{F}$ is an ordered field. Prove that the validity of Completeness Axiom in $\mathbb{F}$ is equivalent to Dedekind's Axiom.

2

Sequences and Their Limits

The intuitive idea of a limit is quite old. In ancient Greece, the so-called method of exhaustion was used by Archimedes, around 225 BC, to calculate the area under a parabola. In the seventeenth century, Newton and Leibniz based much of the calculus they developed on the idea of taking limits. In spite of the success that calculus brought to the natural sciences, it also drew heavy criticism due to the use of *infinitesimals* (infinitely small numbers). It was not until 1821, when Cauchy published *Cours d'Analyse*, probably the most influential textbook in the history of analysis, that calculus achieved a rigor that is quite close to modern standards.

2.1   Computing the Limits: Review

In this section we will review some of the rules for evaluating limits. Some of the easiest problems occur when a_n is a rational function, i.e., a quotient (or a *ratio*) of two polynomials.

Example 2.1.1. $a_n = \dfrac{2n - 3}{5n + 1}$.

Solution. When both polynomials have the same degree, we divide both the numerator and the denominator by the highest power of n. Here, it is just n. We obtain

$$a_n = \frac{2 - \dfrac{3}{n}}{5 + \dfrac{1}{n}}.$$

Now we take the limit, and use the rules for limits (such as "the limits of the sum equals the sum of the limits"):

$$\lim a_n = \lim \frac{2 - 3/n}{5 + 1/n} = \frac{\lim(2 - 3/n)}{\lim(5 + 1/n)} = \frac{\lim 2 - \lim 3/n}{\lim 5 + \lim 1/n}.$$

Since $\lim 2 = 2$, $\lim 5 = 5$, $\lim 1/n = 0$, and $\lim 3/n = 3 \lim 1/n = 0$, we obtain that $\lim a_n = 2/5$. ♦

Rule. *When both polynomials have the same degree, the limit is the ratio of the leading coefficients.*

Example 2.1.2. $a_n = \dfrac{3n + 5}{n^2 - 4n + 7}$.

Solution. Again, we divide both the numerator and the denominator by the highest power of n:

$$\lim a_n = \lim \frac{\dfrac{3}{n} + \dfrac{5}{n^2}}{1 - \dfrac{4}{n} + \dfrac{7}{n^2}}$$

$$= \frac{\lim\left(\dfrac{3}{n} + \dfrac{5}{n^2}\right)}{\lim\left(1 - \dfrac{4}{n} + \dfrac{7}{n^2}\right)}$$

$$= \frac{\lim\dfrac{3}{n} + \lim\dfrac{5}{n^2}}{\lim 1 - \lim\dfrac{4}{n} + \lim\dfrac{7}{n^2}}$$

$$= \frac{3\lim\dfrac{1}{n} + 5\lim\dfrac{1}{n^2}}{\lim 1 - 4\lim\dfrac{1}{n} + 7\lim\dfrac{1}{n^2}}.$$

This time, we obtain $0/1 = 0$, so $\lim a_n = 0$.   ◆

Rule. *When the degree of the denominator is higher than the degree of the numerator, the limit is 0.*

Example 2.1.3. $a_n = \dfrac{n^3 + n^2}{n^2 + 2n}$.

Solution. The highest power of n is n^3. However, we have to be careful—it would be incorrect to use the rule "the limits of the quotient equals the quotient of the limits". The reason is that, after dividing by n^3, the denominator $1/n + 2/n^2$ has limit 0. Nevertheless, the numerator is now $1 + 1/n$, which has limit 1, and we can conclude that the sequence a_n diverges to infinity.   ◆

Rule. *When the degree of the denominator is lower than the degree of the numerator, the limit is infinite.*

Frequently it is helpful to use an algebraic identity before computing the limit.

Example 2.1.4. $a_n = \sqrt{n+1} - \sqrt{n}$.

Solution. Here we will use $a^2 - b^2 = (a - b)(a + b)$. If we multiply and divide a_n by $\sqrt{n+1} + \sqrt{n}$, this identity gives that

$$a_n = \frac{(\sqrt{n+1} - \sqrt{n})(\sqrt{n+1} + \sqrt{n})}{\sqrt{n+1} + \sqrt{n}} = \frac{(n+1) - n}{\sqrt{n+1} + \sqrt{n}} = \frac{1}{\sqrt{n+1} + \sqrt{n}}.$$

Since the denominator of the last fraction diverges to infinity, we see that $\lim a_n = 0$.   ◆

Example 2.1.5. $a_n = \ln(n + 1) - \ln n$.

Solution. We will use the rule $\ln b - \ln a = \ln(b/a)$. Thus

$$a_n = \ln \frac{n+1}{n} = \ln\left(1 + \frac{1}{n}\right)$$

and $\lim a_n = \ln 1 = 0$.   ◆

Remark 2.1.6. We have used the fact that $\lim \ln\left(1 + \frac{1}{n}\right) = \ln \lim\left(1 + \frac{1}{n}\right)$. Later we will discuss whether $\lim f(x_n) = f(\lim x_n)$ for every function f.

Our next problem will use the fact that if $a_n = a^n$, for some constant $a \geq 0$, then the limit of a_n is either 0 (if $a < 1$), or 1 (if $a = 1$), or it is infinite (if $a > 1$).

Example 2.1.7. $a_n = \dfrac{3^n + 4^{n+1}}{2 \cdot 3^n - 4^n}$.

Solution. Dividing by 4^n yields

$$\frac{3^n/4^n + 4^{n+1}/4^n}{2 \cdot 3^n/4^n - 4^n/4^n} = \frac{(3/4)^n + 4}{2(3/4)^n - 1}.$$

Since $\lim(3/4)^n = 0$ we see that $\lim a_n = -4$.   ◆

Sometimes, we will need to use the Squeeze Theorem: if $a_n \le b_n \le c_n$ for each $n \in \mathbb{N}$ and if $\lim a_n = \lim c_n = L$, then $\lim b_n = L$.

Example 2.1.8. $a_n = \dfrac{\sin(n^2)}{\sqrt{n}}$.

Solution. For any $x \in \mathbb{R}$, $-1 \le \sin x \le 1$. This can be written as $|\sin x| \le 1$. Consequently, $|\sin(n^2)| \le 1$ for all $n \in \mathbb{N}$ and

$$0 \le \left| \frac{\sin(n^2)}{\sqrt{n}} \right| \le \frac{1}{\sqrt{n}}.$$

Since $\lim 1/\sqrt{n} = 0$ we obtain that $\lim a_n = 0$.   ◆

Example 2.1.9. $a_n = \dfrac{2^n}{n!}$.

Solution. Notice that

$$\frac{2^n}{n!} = \frac{2 \cdot 2 \cdot \cdots \cdot 2}{1 \cdot 2 \cdot \cdots \cdot n} = \left(\frac{2}{1}\right) \cdot \left(\frac{2}{2}\right) \cdot \left(\frac{2}{3}\right) \cdot \cdots \cdot \left(\frac{2}{n}\right) < \left(\frac{2}{1}\right) \cdot \left(\frac{2}{n}\right) = \left(\frac{4}{n}\right).$$

Therefore $0 < 2^n/n! < 4/n$, and $\lim 4/n = 0$, so we conclude that $\lim a_n = 0$.   ◆

KARL WEIERSTRASS (1815–1897) was a German mathematician, who is often cited as the "father of modern analysis". He started reading mathematical research journals while still in high school. Nevertheless, he studied finance at the University of Bohn, according to his father's wishes. Predictably, he spent time studying mathematics, so he did not get a degree. He became a high school teacher, which meant teaching many classes including physics, history, geography, even gymnastics. He continued his research, but published mostly in the school paper. However, in 1854 he published a paper in the prestigious Crelle's Journal and this led to an honorary doctorate at the University of Königsberg and a position at the University of Berlin. He started teaching and his lectures, based in large part on his own research, attracted students from all over the world. He taught a four-semester course until 1890 and the approach he used still dominates today's academia. He continued doing significant research but published little and many of his results appear in the form of notes taken by his students. (See pages 44, 99, 227, 236, 244.) Also, he demonstrated that the complex numbers are the only commutative algebraic extension of the real numbers. Weierstrass had health problems much of his adult life and in December 1861 he collapsed completely. It took him about a year to recover and afterwards he lectured sitting down while a student wrote on the blackboard. He spent his last 3 years in a wheelchair and he died of pneumonia.

Exercises

Find the following limits:

2.1.1. $\lim \dfrac{n+1}{3n-1}$.

2.1.2. $\lim \dfrac{3n^2 - 4n + 7}{2n^3 - 5n^2 + 8n - 11}$.

2.1.3. $\lim \dfrac{2n^3 + 4n^2 - 1}{3n^2 - 5n - 8}$.

2.1.4. $\lim \dfrac{4 + 5n^2}{n + n^2}$.

2.1.5. $\lim \dfrac{n - n^4}{n^3 + n^5 - 1}$.

2.1.6. $\lim \dfrac{n^5 + n^6 - 2n - 1}{n^5 + 2n^4 + 3n^3 + 4n^2 + 5n + 6}$.

2.1.7.* $\lim \left(\sqrt{n^2 + n} - n\right)$.

2.1.8.* $\lim \left(\sqrt{n^2 + 1} - n\right)$.   **2.1.9.*** $\lim \left(\sqrt[3]{n^2 + 1} - \sqrt[3]{n^2 + n}\right)$.   **2.1.10.** $\lim \dfrac{\sqrt{n+1} + 3}{\sqrt{n+2} - 4}$.

2.1.11.* $\lim \left(\sqrt{2}\sqrt[4]{2}\sqrt[8]{2}\cdots\sqrt[2^n]{2}\right)$.   **2.1.12.*** $\lim \dfrac{\sqrt[3]{n^2}\sin n!}{n+1}$.   **2.1.13.*** $\lim \sin^2\left(\pi\sqrt{n^2 + n}\right)$.

2.1.14.* $\lim \left(\cos \dfrac{x}{2} \cos \dfrac{x}{4} \cdots \cos \dfrac{x}{2^n}\right)$.   **2.1.15.** $\lim \left(\dfrac{1}{n^2} + \dfrac{2}{n^2} + \cdots + \dfrac{n-1}{n^2}\right)$.

2.1.16. $\lim \left(\dfrac{1^2}{n^3} + \dfrac{2^2}{n^3} + \cdots + \dfrac{(n-1)^2}{n^3}\right)$.   **2.1.17.*** $\lim \left(\dfrac{1}{2} \cdot \dfrac{3}{4} \cdots\cdots \dfrac{2n-1}{2n}\right)$.

2.1.18.* $\lim \left(\dfrac{1}{1 \cdot 2 \cdot 3} + \dfrac{1}{2 \cdot 3 \cdot 4} + \cdots + \dfrac{1}{n(n+1)(n+2)}\right)$.

2.2 Definition of the Limit

In the previous section we relied on some rules, such as that $\lim 1/n = 0$, or the "Squeeze Theorem". They are all intuitively clear, but we will soon encounter some that are not. For example, in Section 2.5 we will study the sequence $\{a_n\}$ defined by $a_n = \left(1 + \frac{1}{n}\right)^n$. Two most common errors are to conclude that its limit is 1 or ∞. The first (erroneous) argument typically notices that $1 + \frac{1}{n}$ has limit 1, and 1 raised to any power is 1. The second argues that since the exponent goes to infinity, and the base is bigger than 1, the limit must be infinite. In fact, the limit is neither 1 nor is it infinite. As we find ourselves in more and more complicated situations, our intuition becomes less and less reliable. The only way to ensure that our results are correct is to define precisely every concept that we use, and to furnish a proof for every assertion that we make.

Having made this commitment, let us start by defining a sequence.

Definition 2.2.1. A sequence of real numbers $\{a_n\}$ is a function a, defined on the set $\mathbb{N}$ with values in $\mathbb{R}$. Instead of $a(n)$ we almost always write a_n.

Now, we want to take a careful look at the idea of a limit. We have said that the sequence $a_n = 1/n$ has the limit 0. What does it really mean? We might say that, as n increases (without bound), a_n *is getting closer and closer to 0*. This last part is true, but it would remain true if 0 were replaced by -1: a_n *is getting closer and closer to* -1. Of course, -1 would be a poor choice for the limit since a_n will never be really close to -1. For example, none of the members of the sequence $\{a_n\}$ will fall into the interval $(-1.5, -0.5)$, with center at -1 and radius 0.5.

Figure 2.1: -1 is not the limit of $1/n$.

What about -0.001 as a limit? Although closer than -1, it suffers from the same flaw. Namely, we can find an interval around -0.001 that contains no member of the sequence $\{a_n\}$. (Example: $(-0.0015, -0.0005)$). We see that the crucial property of the limit is that there cannot be such an interval around it. More precisely, if L is the limit of $\{a_n\}$, then any attempt to select such an interval centered at L must fail. In other words, regardless of how small positive number ε is, the interval $(L-\varepsilon, L+\varepsilon)$ must contain at least one member of the sequence $\{a_n\}$. The following examples illustrate this phenomenon.

Example 2.2.2. A small interval around L contains a member of $\{a_n\}$.
Let $a_n = 1/n$, $\varepsilon = 0.02$. Here $L = 0$, so we consider the interval $(-0.02, 0.02)$, and it is

Figure 2.2: $a_{100} = 0.01 \in (-0.02, 0.02)$.

easy to see that $a_{100} = 0.01 \in (-0.02, 0.02)$. ◆

You may have noticed that $a_{80} \in (-0.02, 0.02)$ as well. Nevertheless, we are interested in limits (as n increases without bound), so there is no urgency to choose the first member of the sequence that belongs to this interval.

Example 2.2.3. A small interval around L contains a member of $\{a_n\}$.
Let $a_n = (2n-3)/(5n+1)$, $\varepsilon = 10^{-4}$. Now $L = 2/5$, so we are looking for a_n that would be on a distance from $2/5$ less than 10^{-4}. Such as, for example, $a_{68000} = (2 \cdot 68000 - 3)/(5 \cdot 68000 + 1)$ because rounded off to 5 decimals $a_{68000} \approx 0.39999$, so $|a_{68000} - \frac{2}{5}| \approx 0.00001 < 10^{-4}$. ◆

These examples illustrate the requirement that, for any $\varepsilon > 0$, the interval $(L-\varepsilon, L+\varepsilon)$ must contain at least one member of the sequence a_n. However, this is not sufficient. Going back to our original example $a_n = 1/n$, we see that this condition is satisfied by 1. Indeed, for any $\varepsilon > 0$, the interval $(1-\varepsilon, 1+\varepsilon)$ contains a_1. Of course, for small values of ε (such as 0.1) the interval $(1-\varepsilon, 1+\varepsilon)$ contains *only* a_1. What we really want is that it contains almost all a_n.

Example 2.2.4. A small interval around L contains almost all members of $\{a_n\}$.
Let $a_n = 1/n$, $\varepsilon = 0.02$. We have already concluded that $a_{100} \in (-0.02, 0.02)$. Even

Figure 2.3: $a_n \in (-0.02, 0.02)$ for $n \geq 100$.

more significant is the fact that, for any $n \geq 100$, $a_n \in (-0.02, 0.02)$. ◆

Example 2.2.5. A small interval around L contains almost all members of $\{a_n\}$.
Let $a_n = (2n-3)/(5n+1)$, $\varepsilon = 10^{-4}$. Here $|a_{68000} - \frac{2}{5}| = 0.00001 < 10^{-4}$. If $n \geq 68000$, then

$$\left| \frac{2n-3}{5n+1} - \frac{2}{5} \right| = \frac{17}{5(5n+1)} \leq \frac{17}{5(5 \cdot 68000 + 1)} \approx 0.00001 < 10^{-4}.$$ ◆

In view of these examples, we will require that, starting with some member of the sequence $\{a_n\}$, all coming after it must belong to this interval. Here is the official definition:

Definition 2.2.6. We say that a real number L is a **limit** of a sequence $\{a_n\}$, and we write $\lim a_n = L$ (or $a_n \to L$), if for any $\varepsilon > 0$ there exists a positive integer N such that, for all $n \geq N$, $|a_n - L| < \varepsilon$. In such a situation we say that the sequence $\{a_n\}$ is **convergent**. Otherwise, it is **divergent**.

Remark 2.2.7. We will mostly avoid the more cumbersome notation that includes $n \to \infty$. So, instead of $\lim_{n \to \infty} a_n$, we will write $\lim a_n$; also, we will write $a_n \to L$, as $n \to \infty$. The longer version will be used only for emphasis.

Definition 2.2.6 appeared for the first time in 1821, in a textbook *Cours d'Analyse* (*A Course of Analysis*) by French mathematician Cauchy. It completely changed mathematics by introducing the rigor that was lacking in the work of his contemporaries. We celebrate Newton (page 27) and Leibniz (page 31) for inventing calculus, but Cauchy took a giant step toward making it a rigorous discipline. In fact, pretty much everything that we will present (and prove) in this chapter can be found in his textbook.

AUGUSTIN-LOUIS CAUCHY (1789–1857) was a French mathematician. His talent for mathematics was noticed early by Lagrange (page 398) who was a family friend. Cauchy studied classical languages 1802–1804, but continued with Ecole Polytechnique 1805–1807 and Ecole des Ponts et Chaussées (School of Bridges and Roads) 1807–1810. Upon graduation he wanted an academic position, but had to settle for the job in Cherbourg port facilities and as an engineer on the Ourcq Canal project. That did not stop him from doing mathematical research. In 1814 he wrote a memoir on definite integrals that later became the basis of his theory of complex functions. Finally, in 1815, he became an assistant professor of analysis at Ecole Polytechnique. In 1816 he won the Grand Prix of the French Academy of Sciences for a work on waves. The solution of one of Fermat's claims on polygonal numbers that same year brought him fame and the membership in the Academy of Sciences. His text *Cours d'Analyse* in 1821 was designed for students at Ecole Polytechnique. It is one of the most important and influential mathematics books ever written. Much of the precision and rigor that is nowadays an essential part of mathematics dates to this book. During the 1820s he began a study of the calculus of residues and in 1829 he defined for the first time a complex function of a complex variable. The revolution of 1830 was a turning point in his life. As a devout Catholic and a royalist he developed a deep hatred towards liberals, which he maintained for the rest of his life. Because he refused to swear an oath of allegiance to the new king he lost his job and left France. He spent two years in Turin, Italy, and five years in Prague. He returned to Paris but was able to get a position only at a Jesuit school, Ecole Normale Ecclésiastique. Between 1840 and 1847 he published the 4-volume text *Exercices d'analyse et de physique mathématique* (Exercises in Analysis and Mathematical Physics) which dealt with the applications of differential equations in physics. When Louis Philippe was overthrown in 1848 Cauchy was allowed to return to Ecole Polytechnique, even though he was in a bad relationship with almost every mathematician. Cauchy is one of the most prolific writers with approximately eight hundred research articles and five complete textbooks. His collected works were published in 27 volumes. His writings cover the entire range of mathematics and mathematical physics. It is believed that more concepts and theorems have been named for Cauchy than for any other mathematician. (See also pages 37, 40, 68, 71, 96, 123, 174, 175, 198, 203, 212, 249, 257, 295, 416.)

Now we can indeed prove that $\lim 1/n = 0$. We will do this in two stages. Our first task is, given ε, to come up with N; the second part consists of proving that this choice of N indeed works. In order to find N we focus on the inequality $|a_n - L| < \varepsilon$. In our situation $L = 0$, $a_n = 1/n$ (so $|a_n| = 1/n$), and the inequality is $1/n < \varepsilon$. Since this is the same as $n > 1/\varepsilon$, we will select N so that $N > 1/\varepsilon$. (Reason: any n that satisfies $n \geq N$, will automatically satisfy $n > 1/\varepsilon$.) Since $1/\varepsilon$ need not be an integer, we use $N = \lfloor 1/\varepsilon \rfloor + 1$. This is an integer that is bigger than $1/\varepsilon$. Now that we have N, we can write the proof.

Proof. Let $\varepsilon > 0$ and define $N = \lfloor 1/\varepsilon \rfloor + 1$. Suppose that $n \geq N$. Since $N > 1/\varepsilon$, we have that $n > 1/\varepsilon$, so $1/n < \varepsilon$ and $|a_n - 0| = |\frac{1}{n}| = \frac{1}{n} < \varepsilon$. $\qquad\square$

Notice the structure of the proof: it is the same whenever we want to establish that the suspected number is indeed the limit of the sequence. Also, in the proof we do not need to explain how we found N, only to demonstrate that it works. This makes proofs shorter but hides the motivation. In the proof above (unless you can go behind the scenes), it is not immediately clear why we defined $N = \lfloor 1/\varepsilon \rfloor + 1$. In general, it is always a good idea when reading a proof to try to see where a particular choice came from. (Be warned, though: it is far from easy!)

The task of finding N can be quite complicated.

Example 2.2.8. Let $a_n = \dfrac{n^2 + 3n - 2}{2n^2 - 1}$. Prove that $\{a_n\}$ is convergent.

Solution. We "know" that the limit is $1/2$, so we focus on $|a_n - 1/2|$.

$$a_n - \frac{1}{2} = \frac{n^2 + 3n - 2}{2n^2 - 1} - \frac{1}{2} = \frac{2(n^2 + 3n - 2) - (2n^2 - 1)}{2(2n^2 - 1)} = \frac{3(2n - 1)}{2(2n^2 - 1)}.$$

Since both the numerator and the denominator of the last fraction are positive, for any $n \in \mathbb{N}$, we see that the inequality $|a_n - 1/2| < \varepsilon$ becomes

$$\frac{3(2n - 1)}{2(2n^2 - 1)} < \varepsilon.$$

In the previous example, at this point we "solved" the inequality for n. Here, it would be very hard, and in some examples it might be impossible. A better plan is to try to find a simpler expression that is bigger than (or equal to) the left side. For example, $2n - 1 < 2n$, and $2n^2 - 1 \geq 2n^2 - n^2 = n^2$ so

$$\frac{3(2n - 1)}{2(2n^2 - 1)} < \frac{3 \cdot 2n}{2n^2} = \frac{3}{n}.$$

Now, we require that $3/n < \varepsilon$, which leads to $n > 3/\varepsilon$ and $N = \lfloor 3/\varepsilon \rfloor + 1$.

Proof. Let $\varepsilon > 0$ and define $N = \lfloor 3/\varepsilon \rfloor + 1$. Suppose that $n \geq N$. Since $N > 3/\varepsilon$, we have that $n > 3/\varepsilon$ so $3/n < \varepsilon$. Now

$$\left| a_n - \frac{1}{2} \right| = \left| \frac{n^2 + 3n - 2}{2n^2 - 1} - \frac{1}{2} \right| = \left| \frac{3(2n - 1)}{2(2n^2 - 1)} \right| = \frac{3(2n - 1)}{2(2n^2 - 1)} < \frac{3}{n} < \varepsilon. \qquad \blacklozenge$$

Example 2.2.9. Let $a_n = (1/3)^n$. Prove that $\{a_n\}$ is convergent.

Solution. Since the limit is 0, and $a_n > 0$ (so that $|a_n| = a_n$), we focus on $(1/3)^n < \varepsilon$. By taking the natural logarithm from both sides, and taking advantage of the formula $\ln a^p = p \ln a$, we obtain $n \ln(1/3) < \ln \varepsilon$. Notice that $\ln(1/3) = \ln 3^{-1} = -\ln 3$, so if we divide by the *negative* number $-\ln 3$, we arrive at $n > -\ln \varepsilon / \ln 3$. Now, if $\varepsilon \geq 1$, then $\ln \varepsilon > 0$, so $-\ln \varepsilon / \ln 3$ is a negative number and any $n \in \mathbb{N}$ will satisfy the inequality $n > -\ln \varepsilon / \ln 3$. Therefore, the case of interest is when $\varepsilon < 1$. Of course, whenever $|a_n - L|$ is smaller than such an ε, it is all the more smaller than any $\varepsilon \geq 1$.

Proof. Let $\varepsilon > 0$ and, without loss of generality, suppose that $\varepsilon < 1$. Let $N = \lfloor -\ln \varepsilon / \ln 3 \rfloor + 1$, and suppose that $n \geq N$. Since $N > -\ln \varepsilon / \ln 3$, we have that $n > -\ln \varepsilon / \ln 3$ so $n \ln 3 > -\ln \varepsilon$. The last inequality can be written as $\ln 3^n > \ln \varepsilon^{-1}$, which implies that $3^n > 1/\varepsilon$ and, hence, that $1/3^n < \varepsilon$. Now

$$|a_n| = \left(\frac{1}{3} \right)^n = \frac{1}{3^n} < \varepsilon. \qquad \blacklozenge$$

Remark 2.2.10. A similar proof can be used to show that $\lim a^n = 0$, for any $a \in (0, 1)$.

> The first use of the abbreviation *lim.* (with a period at the end) was by a Swiss mathematician, Simon L' Huilier (1750–1840), in 1786. Weierstrass used it (without a period) as early as 1841, but it did not appear in print until 1894. In the 1850s, he began to write $\lim_{x=c}$, and it appears that we owe the arrow (instead of the equality) to two English mathematicians. John Gaston Leathem (1871–1923) pioneered its use in 1905, and Godfrey Harold Hardy (1877–1947) made it popular through his 1908 textbook *A Course of Pure Mathematics*.

Definition 2.2.6 applies only to the case when the limit L is a real number. It can be adjusted to the case when the limit is infinite.

Definition 2.2.11. We say that a sequence $\{a_n\}$ **diverges to** $+\infty$, and we write $\lim a_n = +\infty$ (or $a_n \to +\infty$), if for any $M > 0$ there exists a positive integer N such that, for all $n \geq N$, $a_n > M$. A sequence $\{a_n\}$ **diverges to** $-\infty$, and we write $\lim a_n = -\infty$ (or $a_n \to -\infty$), if for any $M < 0$ there exists a positive integer N such that, for all $n \geq N$, $a_n < M$. In either of the two cases, we say that $\{a_n\}$ has an infinite limit.

Remark 2.2.12. We will often omit the plus sign and write ∞ instead of $+\infty$.

Example 2.2.13. Let $a_n = 2^n$. Prove that $\{a_n\}$ diverges to $+\infty$.

Solution. We consider the inequality $a_n > M$. By taking the natural logarithm from both sides of $2^n > M$, we obtain $n \ln 2 > \ln M$ or, since $\ln 2 > 0$, $n > \ln M / \ln 2$.

Proof. Let $M > 0$ and define $N = \lfloor \ln M / \ln 2 \rfloor + 1$. Suppose that $n \geq N$. Since $N > \ln M / \ln 2$, we have that $n > \ln M / \ln 2$. It follows that $\ln 2^n = n \ln 2 > \ln M$ which implies that $2^n > M$. ♦

Exercises

In Exercises 2.2.1–2.2.8 find the limit and prove that the result is correct:

2.2.1. $\lim \dfrac{n+1}{3n-1}$.

2.2.2. $\lim \dfrac{\sqrt{n}}{1+\sqrt{n}}$.

2.2.3. $\lim \dfrac{2n^2-1}{3n^2+2}$.

2.2.4. $\lim \dfrac{3n^2+1}{2n^2+5n}$.

2.2.5. $\lim \dfrac{2n^4-1}{n^3-n^2}$.

2.2.6. $\lim \dfrac{n^3+2}{n^2+3}$.

2.2.7. $\lim \dfrac{n^2 \sin n}{n^3+1}$.

2.2.8. $\lim \dfrac{3^n+2}{5^n-3}$.

2.2.9. If a is a real number and $a_n = a$ for all $n \in \mathbb{N}$, prove that $\lim a_n = a$.

2.2.10.* Let $k \in \mathbb{N}$. If the sequence $\{b_n\}$ is obtained by deleting the first k members of the sequence $\{a_n\}$, prove that $\{b_n\}$ is convergent if and only if $\{a_n\}$ is convergent, in which case $\lim a_n = \lim b_n$.

2.2.11. Prove that $\lim a_n = -\infty$ if and only if $\lim(-a_n) = +\infty$.

2.2.12.* Prove that, for every real number c, there exists a sequence $\{a_n\}$ of rational numbers and a sequence $\{b_n\}$ of irrational numbers, such that $\lim a_n = \lim b_n = c$.

2.2.13.* Prove that $\lim \sin n$ does not exist.

2.3   Properties of Limits

In this section we will continue our study by establishing various properties of limits. Once again, much of the material can be found in Cauchy's *Cours d'Analyse*.

It is always useful to be able to replace a complicated expression with a simpler one. Usually, this strategy requires that the expressions are equal. As we have seen in Exercise 2.2.8, when proving results about limits, inequalities become as important. Some appear so frequently in the proofs that they deserve to be singled out.

Theorem 2.3.1 (The Triangle Inequality). *Let a, b be real numbers. Then:*

   (a) $|a + b| \leq |a| + |b|$;

   (b) $|a - b| \geq \big||a| - |b|\big|$;

(c) $|a - b| \geq |a| - |b|$.

Proof. (a) We will use the fact that, for any $x \in \mathbb{R}$, $x \leq |x|$. Therefore,

$$2ab \leq 2|a||b|. \tag{2.1}$$

Further, for any $x \in \mathbb{R}$, $x^2 = |x|^2$, so $a^2 + 2ab + b^2 \leq |a|^2 + 2|a||b| + |b|^2$. It follows that

$$(a + b)^2 \leq (|a| + |b|)^2.$$

Now we take the square root, and it helps to remember that, for any $x \in \mathbb{R}$, $\sqrt{x^2} = |x|$.

(b) We start by multiplying (2.1) by -1 (which yields $-2ab \geq 2|a||b|$), and add $a^2 + b^2$ to both sides. We obtain that

$$(a - b)^2 = a^2 - 2ab + b^2 \geq |a|^2 - 2|a||b| + |b|^2 = (|a| - |b|)^2,$$

and taking the square roots gives the desired inequality.

(c) This follows from (b) because $||a| - |b|| \geq |a| - |b|$. □

Remark 2.3.2. The name (Triangle Inequality) comes from similar inequalities which hold for vectors a, b. In that case a, b, and $a + b$ are the sides of a triangle and the inequality (a) of Theorem 2.3.1 expresses a geometric fact that the length of one side must be smaller

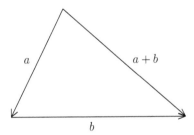

Figure 2.4: The length of $a + b$ cannot exceed the sum of the lengths of a and b.

than the sum of the lengths of the other two sides.

An essential step in many proofs will be to select the largest of several numbers. We will use the following notation: for two numbers x, y, the symbol $\mathbf{max}\{x, y\}$ will stand for the larger of these two; more generally, for n numbers $x_1, x_2, \ldots, x_n$, the symbol $\mathbf{max}\{x_1, x_2, \ldots, x_n\}$ will denote the largest of these n numbers. In the latter situation we may also write $\max\{x_k : 1 \leq k \leq n\}$.

Since we are developing mathematics in a formal way, even the most obvious things have to be proved. For example, it is intuitively clear that, if a sequence converges, it can have only one limit. Nevertheless, we will write a proof.

Theorem 2.3.3. *Every sequence can have at most one limit.*

Proof. Suppose to the contrary that there exists a sequence $\{a_n\}$ with more than one limit, and let us denote two such limits by L_1 and L_2. Let $\varepsilon = |L_1 - L_2|/3$.

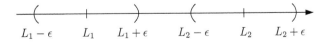

Figure 2.5: If a_n had two limits.

Since $\{a_n\}$ converges to L_1, there exists $N_1 \in \mathbb{N}$ so that

$$n \geq N_1 \quad \Rightarrow \quad |a_n - L_1| < \varepsilon.$$

Also, $\{a_n\}$ converges to L_2, so there exists $N_2 \in \mathbb{N}$ with the property that

$$n \geq N_2 \quad \Rightarrow \quad |a_n - L_2| < \varepsilon.$$

Let $N = \max\{N_1, N_2\}$ and suppose that $n \geq N$. Then

$$|L_1 - L_2| = |L_1 - a_n + a_n - L_2| \leq |a_n - L_1| + |a_n - L_2| < \varepsilon + \varepsilon = 2\varepsilon = \frac{2}{3}|L_1 - L_2|$$
$$< |L_1 - L_2|.$$

We have obtained that $|L_1 - L_2| < |L_1 - L_2|$, which is impossible, so there can be no such sequence $\{a_n\}$. □

When evaluating limits in Section 2.1 we have used several rules. Now that we have a strict definition of the limit, we can prove them.

Theorem 2.3.4. *Let $\{a_n\}$, $\{b_n\}$ be convergent sequences with $\lim a_n = a$, $\lim b_n = b$, and let α be a real number. Then the sequences $\{\alpha a_n\}$ and $\{a_n + b_n\}$ are also convergent and:*

(a) $\lim(\alpha a_n) = \alpha \lim a_n$;

(b) $\lim(a_n + b_n) = \lim a_n + \lim b_n$.

Proof. (a) We consider separately the cases $\alpha = 0$ and $\alpha \neq 0$. When $\alpha = 0$, the sequence αa_n is the zero sequence (each member of the sequence is 0), so it converges to 0 $(= \alpha \lim a_n)$. So we focus on the case when $\alpha \neq 0$. Let $\varepsilon > 0$ and select N such that, for $n \geq N$, $|a_n - a| < \varepsilon/|\alpha|$. (This is why we needed $\alpha \neq 0$!) Then, for $n \geq N$,

$$|\alpha a_n - \alpha a| = |\alpha||a_n - a| < |\alpha|\frac{\varepsilon}{|\alpha|} = \varepsilon$$

so $\lim(\alpha a_n) = \alpha a = \alpha \lim a_n$.

(b) Notice that, in order to obtain the inequality $|\alpha a_n - \alpha a| < \varepsilon$, it was not sufficient to choose n such that $|a_n - a| < \varepsilon$. We really needed $|a_n - a| < \varepsilon/|\alpha|$. A similar strategy is useful in part (b). Namely, given $\varepsilon > 0$, we are looking for n such that $|(a_n + b_n) - (a + b)| < \varepsilon$. Since

$$|(a_n + b_n) - (a + b)| = |(a_n - a) + (b_n - b)| \leq |a_n - a| + |b_n - b|,$$

one might settle for n such that $|a_n - a| < \varepsilon$ and $|b_n - b| < \varepsilon$. However, this would yield $|(a_n + b_n) - (a + b)| < 2\varepsilon$. A better idea is to try for $|a_n - a| < \varepsilon/2$ and $|b_n - b| < \varepsilon/2$.

Let $\varepsilon > 0$. We will select positive integers N_1 and N_2 with the following properties: if $n \geq N_1$ then $|a_n - a| < \varepsilon/2$, and if $n \geq N_2$ then $|b_n - b| < \varepsilon/2$. Let $N = \max\{N_1, N_2\}$ and suppose that $n \geq N$. Then

$$|(a_n + b_n) - (a + b)| = |(a_n - a) + (b_n - b)| \leq |a_n - a| + |b_n - b| < \varepsilon/2 + \varepsilon/2 = \varepsilon.$$

Therefore, $\lim(a_n + b_n) = a + b = \lim a_n + \lim b_n$. □

Theorem 2.3.4 did not address the rule for "the limit of the product." The reason is that its proof will rely on another property of convergent sequences. Let us make an observation that all members of the sequence $a_n = 1/n$ lie between -1 and 1. This can be written as $|a_n| \leq 1$ for all $n \in \mathbb{N}$. We say that the sequence $\{a_n\}$ is *bounded by 1*. More generally, we have this definition.

Definition 2.3.5. A sequence $\{a_n\}$ is **bounded (by** M) if there exists M such that $|a_n| \leq M$ for all $n \in \mathbb{N}$. If no such M exists, we say that the sequence is **unbounded**.

Example 2.3.6. An unbounded sequence.
    Let $a_n = n$. The sequence $\{a_n\}$ does not satisfy a condition of the form $|a_n| \leq M$ (for all $n \in \mathbb{N}$) regardless of the choice of M. Therefore, it is unbounded. ◆

The following theorem establishes a relationship between bounded and convergent sequences.

Theorem 2.3.7. *Every convergent sequence is bounded.*

Proof. Let $\{a_n\}$ be a convergent sequence with $\lim a_n = L$. We will show that the sequence $\{a_n\}$ is bounded. Let $\varepsilon = 1$ and choose $N \in \mathbb{N}$ so that, if $n \geq N$, $|a_n - L| < \varepsilon = 1$. The last inequality is equivalent to $-1 < a_n - L < 1$ and, hence, to

$$L - 1 < a_n < L + 1. \tag{2.2}$$

Next we consider $N + 1$ positive numbers:

$$|a_1|, |a_2|, \ldots |a_{N-1}|, |L - 1|, |L + 1|, \tag{2.3}$$

and let M be the largest among them. Then $|a_n| \leq M$ for all $n \in \mathbb{N}$. Indeed, if $1 \leq n \leq N-1$, then $|a_n|$ is one of the first $N - 1$ numbers in (2.3) and cannot be bigger than M. If $n \geq N$, there are two possibilities: $a_n \geq 0$ and $a_n < 0$. In the former, the second inequality in (2.2) shows that $|a_n| < |L + 1|$; in the latter, the first inequality in (2.2) implies that $L - 1 < a_n < 0$, so $|a_n| < |L - 1|$. Consequently, $|a_n| \leq M$. $\square$

Whenever we encounter an implication, such as "if $\{a_n\}$ is convergent then $\{a_n\}$ is bounded," it is of interest to examine whether the converse is true. Here, this would mean that "if $\{a_n\}$ is bounded then $\{a_n\}$ is convergent." However, this statement is false. The simplest method to establish that an assertion is not always true consists of exhibiting a counterexample, and we will do just that.

Example 2.3.8. A bounded divergent sequence.
    Let $a_n = (-1)^n$. It is easy to see that $|a_n| = 1$, because $(-1)^n$ is either 1 or -1. Therefore, the sequence $\{a_n\}$ is bounded by 1. On the other hand, it is not convergent. Indeed, suppose that the sequence $\{a_n\}$ is convergent and let $\lim a_n = L$. Now, consider the interval $(L - 1, L + 1)$. If $L < 0$, this interval does not contain 1, so $a_{2n} \notin (L - 1, L + 1)$ for any $n \in \mathbb{N}$. Similarly, if $L \geq 0$, $a_{2n-1} \notin (L - 1, L + 1)$ for any $n \in \mathbb{N}$. Thus, regardless of the choice of L, there are infinitely many members of the sequence outside of this interval. Therefore, the sequence $\{a_n\}$ is not convergent. ◆

Now we will prove "the limit of the product" rule and "the limit of the quotient" rule.

Theorem 2.3.9. *Let $\{a_n\}$, $\{b_n\}$ be convergent sequences with $\lim a_n = a$, $\lim b_n = b$. Then:*

(a) The sequence $\{a_n b_n\}$ is also convergent and $\lim(a_n b_n) = \lim a_n \lim b_n$.

(b) If, in addition, $b_n \neq 0$ for all $n \in \mathbb{N}$ and if $b \neq 0$, then the sequence $\{a_n/b_n\}$ is also convergent and $\lim(a_n/b_n) = \lim a_n / \lim b_n$.

Proof. Since the sequence $\{a_n\}$ is convergent, it is bounded, so there exists $M > 0$ such that $|a_n| \leq M$ for all $n \in \mathbb{N}$.

(a) Let $\varepsilon > 0$. Suppose first that $b = 0$. Since $\lim b_n = 0$, there exists $N \in \mathbb{N}$ such that, for $n \geq N$, $|b_n| < \varepsilon/M$. So, let $n \geq N$. Then

$$|a_n b_n| = |a_n||b_n| < M \frac{\varepsilon}{M} = \varepsilon,$$

which shows that $\lim(a_n b_n) = 0 = ab = \lim a_n \lim b_n$.

Next, we consider the case when $b \neq 0$. Since $\lim a_n = a$, there exists $N_1 \in \mathbb{N}$ such that

$$n \geq N_1 \quad \Rightarrow \quad |a_n - a| < \frac{\varepsilon}{2|b|}.$$

Also, $\lim b_n = b$, so there exists $N_2 \in \mathbb{N}$ such that

$$n \geq N_2 \quad \Rightarrow \quad |b_n - b| < \frac{\varepsilon}{2M}.$$

Let $N = \max\{N_1, N_2\}$ and suppose that $n \geq N$. Then

$$\begin{aligned}
|a_n b_n - ab| &= |a_n b_n - a_n b + a_n b - ab| \\
&\leq |a_n b_n - a_n b| + |a_n b - ab| = |a_n||b_n - b| + |b||a_n - a| \\
&\leq M \cdot \frac{\varepsilon}{2M} + |b| \cdot \frac{\varepsilon}{2|b|} = \frac{\varepsilon}{2} + \frac{\varepsilon}{2} = \varepsilon.
\end{aligned}$$

(b) We will show that $\lim 1/b_n = 1/b$. Since $a_n/b_n = a_n \cdot (1/b_n)$, the result will then follow from (a). Let $\varepsilon > 0$. The assumption that $b \neq 0$ implies that $\varepsilon|b|^2/2 > 0$, so there exists $N_1 \in \mathbb{N}$ such that

$$n \geq N_1 \quad \Rightarrow \quad |b_n - b| < \frac{\varepsilon|b|^2}{2}.$$

Also, let N_2 be a positive integer such that

$$n \geq N_2 \quad \Rightarrow \quad |b_n - b| < \frac{|b|}{2}.$$

It follows that, for $n \geq N_2$,

$$|b_n| = |b + b_n - b| \geq |b| - |b_n - b| > |b| - \frac{|b|}{2} = \frac{|b|}{2}.$$

Let $N = \max\{N_1, N_2\}$ and suppose that $n \geq N$. Then

$$\left| \frac{1}{b_n} - \frac{1}{b} \right| = \frac{|b_n - b|}{|b_n b|} < \frac{\varepsilon|b|^2/2}{|b| \cdot |b|/2} = \varepsilon. \qquad \square$$

Our next goal is to establish the Squeeze Theorem.

Theorem 2.3.10. *Let $\{a_n\}, \{b_n\}, \{c_n\}$ be sequences such that $\lim a_n = \lim c_n = L$ and suppose that, for all $n \in \mathbb{N}$, $a_n \leq b_n \leq c_n$. Then $\{b_n\}$ is a convergent sequence and $\lim b_n = L$.*

Proof. Let $\varepsilon > 0$. We select N_1 and N_2 so that, if $n \geq N_1$ then $|a_n - L| < \varepsilon$, and if $n \geq N_2$ then $|c_n - L| < \varepsilon$. Let $N = \max\{N_1, N_2\}$ and suppose that $n \geq N$. Then $-\varepsilon < a_n - L < \varepsilon$ and $-\varepsilon < c_n - L < \varepsilon$. Therefore,

$$-\varepsilon < a_n - L \leq b_n - L \leq c_n - L < \varepsilon.$$

Consequently, $-\varepsilon < b_n - L < \varepsilon$ or, equivalently, $|b_n - L| < \varepsilon$, and $\lim b_n = L$. $\qquad \square$

We mention here a few useful results.

Proposition 2.3.11. *Let $\{a_n\}$ be a convergent sequence with $\lim a_n = a$, and suppose that $a_n \geq 0$ for all $n \in \mathbb{N}$. Then $a \geq 0$.*

Proof. Suppose to the contrary that $a < 0$, and let $\varepsilon = |a|/2$. The reason that a cannot be the limit of $\{a_n\}$ is that the interval $(a - \varepsilon, a + \varepsilon)$ contains no a_n. Indeed, $a < 0$ implies that $|a| = -a$ so

$$a + \varepsilon = a + \frac{|a|}{2} = a - \frac{a}{2} = \frac{a}{2} < 0,$$

so if $a_n \in (a - \varepsilon, a + \varepsilon)$ then $a_n < a + \varepsilon < 0$ which is a contradiction. So, no member of the sequence $\{a_n\}$ is between $a - \varepsilon$ and $a + \varepsilon$ and a cannot be the limit of $\{a_n\}$. $\square$

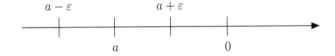

Figure 2.6: $a < 0$ cannot be the limit.

From this result we obtain an easy consequence.

Corollary 2.3.12. *Let $\{a_n\}$, $\{b_n\}$ be two convergent sequences, let $\lim a_n = a$, $\lim b_n = b$, and suppose that $a_n \leq b_n$ for all $n \in \mathbb{N}$. Then $a \leq b$.*

Proof. Let $c_n = b_n - a_n$. Then $c_n \geq 0$ and $\lim c_n = \lim b_n - \lim a_n = b - a$. By Proposition 2.3.11, $\lim c_n \geq 0$ so $b \geq a$. $\square$

Remark 2.3.13. If we replace the hypotheses $\lim a_n = a$, $\lim b_n = b$ with $\lim a_n = \infty$ in Corollary 2.3.12, then it follows from Definition 2.2.11 that $\lim b_n = \infty$.

ISAAC NEWTON (1643–1727) was an English physicist, mathematician, and astronomer. He has been considered by many to be the greatest scientist ever. His father died before he was born and his mother remarried, so Newton grew up with his grandmother. He enrolled in Trinity College in Cambridge in 1661, where he studied law and learned philosophy and astronomy, but it appears that he taught himself mathematics. He graduated in 1665 and spent 2 years at home while the university was closed because of the plague. This is the period when Newton laid the foundations of calculus. When the university reopened, he returned to Cambridge where he attended the lectures of Isaac Barrow (1630–1677) and had a number of discussions with him. When Barrow left to devote himself to divinity, Newton filled the position of the Lucasian Professor. The next 18 years was the highly productive period with important work in optics, physics and celestial mechanics, which culminated in the theory of universal gravitation. He discovered that white light is a mixture of different types of rays but advocated the corpuscular theory of light. He was always very sensitive to criticism, so he delayed further publication on this subject for more than 30 years until Hook (one of the critics of his theory of light) died. On the other hand, the astronomer Halley convinced him to publish his work in physics resulting in his 1687 book *Philosophiae naturalis principia mathematica* (Mathematical Principles of Natural Philosophy). Such is the fame of the book, that it is often referred to as *Principia*. In 1689 he was elected to Parliament and moved to London. He took over several administrative duties (he was the Master of the Mint) and stopped research. He was the president of the Royal Society from 1703 until his death. In 1705 he was knighted, the first scientist to be so honored for his work, although it is likely that it was politics, rather than science that motivated this act. The last portion of his life was dominated by the controversy with Leibniz (page 31) over which of them had invented the calculus. The Royal Society produced a study which "showed" that Newton was the true discoverer and labeled Leibniz a fraud. This study was cast into doubt when it was later found that Newton himself wrote the concluding remarks on Leibniz. It should be said that Newton was relatively modest about his achievements, writing in 1676: *If I have seen further it is by standing on the shoulders of giants.*

Exercises

2.3.1. Suppose that $\lim a_n = a$. Prove that $\lim |a_n| = |a|$. Is the converse true?

2.3.2. Suppose that $\{a_n\}, \{b_n\}$ are sequences satisfying $0 \le a_n \le b_n$ for all $n \in \mathbb{N}$. Prove that if $\{b_n\}$ is bounded then so is $\{a_n\}$.

2.3.3. Suppose that $\{a_n\}$ is a convergent sequence. Prove that $\lim(a_{n+1} - a_n) = 0$. Is the converse true?

2.3.4. If $\lim a_n = 0$ and if $\{b_n\}$ is a bounded sequence, prove that $\lim a_n b_n = 0$.

2.3.5. Suppose that $\{a_n\}$ is a sequence of positive numbers such that $\lim \dfrac{a_{n+1}}{a_n} = L < 1$. Prove that $\lim a_n = 0$.

2.3.6. If $\lim a_n = L$ and $|b_n - a_n| \le \dfrac{1}{n}$ for all $n \in \mathbb{N}$, prove that $\lim b_n = L$.

2.3.7. Suppose that $\lim a_n = 0$. Find $\lim a_n^n$.

2.3.8. Let $\{a_n\}, \{b_n\}$ be two sequences of positive numbers, and suppose that $\lim(a_n/b_n) = L$. Prove or disprove: (a) If $\{a_n\}$ is a convergent sequence, then so is $\{b_n\}$. (b) If $\{b_n\}$ is a convergent sequence, then so is $\{a_n\}$.

2.3.9.* Suppose that $\lim a_n = a$ and $\lim b_n = b$. Prove that

$$\lim \max\{a_n, b_n\} = \max\{a, b\}.$$

2.3.10.* Prove or disprove: if $\{b_n\}$ is a convergent sequence and $c_n = n(b_n - b_{n-1})$, then $\{c_n\}$ is a bounded sequence.

2.4   Monotone Sequences

Ideally, we would like to find the exact limit of a convergent sequence. Unfortunately, this is often impossible, so we are forced to estimate the limit only approximately. A very important example of this phenomenon are the infinite series, and we will talk about them in Chapter 7. For now, let us look at $a_n = (1 + \frac{1}{n})^n$. The best we can do is approximate the limit by replacing n with a large positive integer. However, this strategy works only when the limit *does* exist. Therefore, even when we cannot find the limit, it is useful if we can establish that the sequence converges. In this section we will learn about one situation where we can draw such a conclusion.

Definition 2.4.1. We say that a sequence $\{a_n\}$ is **monotone increasing** if $a_{n+1} > a_n$ for all $n \in \mathbb{N}$, and it is **monotone decreasing** if $a_{n+1} < a_n$ for all $n \in \mathbb{N}$. We often omit the word *monotone* and say just "increasing" or just "decreasing."

Example 2.4.2. An increasing sequence.

Let $a_n = (n-1)/n$. Then $\{a_n\}$ is an increasing sequence. Indeed,

$$a_{n+1} - a_n = \frac{(n+1)-1}{n+1} - \frac{n-1}{n} = \frac{n \cdot n - (n+1)(n-1)}{n(n+1)} = \frac{1}{n(n+1)} > 0.$$

So, $a_{n+1} > a_n$.   ◆

Example 2.4.3. A sequence that is neither increasing nor decreasing.

Let $a_n = \dfrac{1 + (-1)^n}{n}$. This sequence is neither increasing nor decreasing. Reason: $a_1 = 0$, $a_2 = 1$, $a_3 = 0$.   ◆

Remark 2.4.4. If the condition $a_{n+1} > a_n$ is replaced by $a_{n+1} \geq a_n$, then the sequence $\{a_n\}$ is **non-decreasing**; similarly, if instead of $a_{n+1} < a_n$ we use $a_{n+1} \leq a_n$, then the sequence $\{a_n\}$ is **non-increasing**. Although this is more precise, the distinction will seldom play any role, and we will refer to sequence $\{a_n\}$ that satisfies $a_{n+1} \geq a_n$ as **increasing**.

Recall that a sequence $\{a_n\}$ is bounded if there is a number M such that $|a_n| \leq M$, for all $n \in \mathbb{N}$.

Definition 2.4.5. We say that a sequence $\{a_n\}$ is **bounded above** if there is a number M such that $a_n \leq M$, for all $n \in \mathbb{N}$; it is **bounded below** if there is a number m such that $a_n \geq m$, for all $n \in \mathbb{N}$.

Clearly, a sequence is bounded if it is bounded both above and below.

Example 2.4.6. A sequence bounded below but not above.

Let $a_n = n^2$. It is easy to see that the sequence $\{a_n\}$ is bounded below, because $a_n \geq 0$, for all $n \in \mathbb{N}$. We will show that it is not bounded above.

Solution. This is harder to prove, because we need the negative of the statement that $\{a_n\}$ is bounded above: there exists M such that, for all n, $a_n \leq M$. Using the quantifiers, this can be written as

$$(\exists M)(\forall n)\, a_n \leq M,$$

and the negative is

$$(\forall M)(\exists n)\, a_n > M,$$

and this is what we need to show. In other words, we need to prove that, for every M, there exists n, such that $a_n > M$.

Proof. Let $M > 0$, and define $n = \lfloor \sqrt{M} \rfloor + 1$. Then $n > \sqrt{M}$ so $a_n = n^2 > M$. ♦

Our interest in sequences that are bounded above (or below) is justified by the following result.

Theorem 2.4.7 (Monotone Convergence Theorem). *If a sequence is increasing and bounded above, then it is convergent.*

Proof. Let $\{a_n\}$ be a sequence and M a number such that, for all $n \in \mathbb{N}$, $a_n \leq M$. Let A be the set of all the values that the sequence a_n takes on:

$$A = \{a_n : n \in \mathbb{N}\}.$$

Then, the set A is bounded above by M and, using the Completeness Axiom, A has the least upper bound L. We will show that $\lim a_n = L$.

Let $\varepsilon > 0$. Since $L = \sup A$, it follows that $L - \varepsilon$ is not an upper bound. Consequently, there is a positive integer N such that $a_N > L - \varepsilon$. We will show that, for $n \geq N$, $|a_n - L| < \varepsilon$.

Figure 2.7: There exists $N \in \mathbb{N}$ such that $a_N > L - \varepsilon$.

First, the sequence a_n is increasing so, if $n \geq N$,

$$a_n \geq a_N > L - \varepsilon. \tag{2.4}$$

On the other hand, L is an upper bound for A, so $a_n \leq L$ for all $n \in \mathbb{N}$. Thus, if $n \geq N$,

$$L - \varepsilon < a_n \leq L < L + \varepsilon. \tag{2.5}$$

We conclude that $|a_n - L| < \varepsilon$, and the theorem is proved. $\square$

Remark 2.4.8. There is an equivalent formulation of this theorem, namely: if a sequence is *decreasing* and bounded *below*, then it is convergent. It can be easily derived from Theorem 2.4.7 by considering the sequence $\{-a_n\}$. Indeed, if $\{a_n\}$ is decreasing and bounded below, then $\{-a_n\}$ is increasing and bounded above, and hence convergent. Of course, if $\{-a_n\}$ is convergent with limit L, then $\{a_n\}$ converges to $-L$.

Let us look at some applications of this result.

Example 2.4.9. Using Monotone Convergence Theorem.

Let $a_1 = \sqrt{2}$, $a_2 = \sqrt{2 + \sqrt{2}}$, $a_3 = \sqrt{2 + \sqrt{2 + \sqrt{2}}}$, ... Find $\lim a_n$.

Solution. We notice that if the innermost $\sqrt{2}$ is deleted we obtain the previous member of the sequence. For example, if in a_3 the innermost $\sqrt{2}$ is replaced by 0 we get $\sqrt{2 + \sqrt{2 + 0}} = \sqrt{2 + \sqrt{2}} = a_2$. Since $\sqrt{2} > 0$, the sequence $\{a_n\}$ is increasing.

In order to establish that the sequence $\{a_n\}$ is bounded above, we notice that $a_{n+1} = \sqrt{2 + a_n}$, for $n \in \mathbb{N}$. We will prove by induction that $a_n \leq 2$, for all $n \in \mathbb{N}$. It is obvious that $a_1 \leq 2$. If $a_n \leq 2$, then $a_{n+1} = \sqrt{2 + a_n} \leq \sqrt{2 + 2} = 2$. Therefore, the sequence $\{a_n\}$ is bounded and increasing and, by Theorem 2.4.7, it must be convergent.

In fact, we can now calculate the limit $L = \lim a_n$. To that end, we take the limit of both sides of the equality $a_{n+1} = \sqrt{2 + a_n}$, and we obtain that $L = \sqrt{2 + L}$. Therefore, $L^2 = 2 + L$ and $(L + 1)(L - 2) = 0$. Since $a_n \geq 0$, Proposition 2.3.11 shows that $L \neq -1$, and we conclude that $L = 2$. $\blacklozenge$

A careful reader will have noticed that we have made a leap of faith (similar to the one in Exercise 2.1.5) by assuming that $\lim \sqrt{2 + a_n} = \sqrt{2 + \lim a_n}$. As promised, we will return to this issue later in the book.

Before we solve another problem, we will need to establish an inequality.

Lemma 2.4.10. *Let a, b be distinct real numbers. Then $(a + b)^2 < 2(a^2 + b^2)$.*

Proof. We notice that, if $a \neq b$ then

$$a^2 - 2ab + b^2 = (a - b)^2 > 0. \tag{2.6}$$

(Of course, $(a - b)^2$ cannot be negative, and the assumption that $a \neq b$ guarantees that it cannot be equal to 0.) If we add $a^2 + 2ab + b^2$ to both sides of the inequality (2.6), we get that $2(a^2 + b^2) > a^2 + 2ab + b^2 = (a + b)^2$. $\square$

Example 2.4.11. Using The Monotone Convergence Theorem.

Let $a_n = 1/\sqrt{n(n + 1)} + 1/\sqrt{(n + 1)(n + 2)} + \cdots + 1/\sqrt{(2n - 1)2n}$. Prove that $\{a_n\}$ is a convergent sequence.

Solution. We consider the difference

$$a_{n+1} - a_n = -\frac{1}{\sqrt{n(n + 1)}} + \frac{1}{\sqrt{2n(2n + 1)}} + \frac{1}{\sqrt{(2n + 1)(2n + 2)}}$$

and we will show that it is a negative number for any $n \in \mathbb{N}$. If we multiply the right-hand side by $\sqrt{n(n + 1)2(2n + 1)}$ (which is a positive number), we obtain

$$-\sqrt{2(2n + 1)} + \sqrt{n + 1} + \sqrt{n}. \tag{2.7}$$

If we apply Lemma 2.4.10 with $a = \sqrt{n}$ and $b = \sqrt{n+1}$, we have that

$$\left(\sqrt{n} + \sqrt{n+1}\right)^2 < 2(2n+1)$$

and, taking the square root of both sides shows that the quantity in (2.7) is negative. It follows that $a_{n+1} - a_n < 0$ so $\{a_n\}$ is a decreasing sequence, bounded below by 0, hence it must be convergent.

What is $\lim a_n$? It is a hard question, but if we use a computer algebra system we can obtain that, for example, $a_{100} = 0.6931487434$ and $a_{1000} = 0.6931471953$. In fact the limit is $\ln 2 \approx 0.6931471806$.   ◆

Example 2.4.12. An increasing sequence.

Let $a_1 = 1$, $a_{n+1} = \dfrac{2(2a_n + 1)}{a_n + 3}$. Prove that the sequence $\{a_n\}$ is increasing.

Solution. We say that such a sequence is defined **recursively**. First we will establish, using induction, that the sequence is bounded: $0 < a_n < 2$. Clearly, this is true for $n = 1$, so we assume that it is true for some positive integer n, and we will prove that $0 < a_{n+1} < 2$. Since $a_n > 0$, both the numerator $2(2a_n + 1)$ and the denominator $a_n + 3$ are positive, so $a_{n+1} > 0$. Next we notice that, since $a_n < 2$,

$$a_{n+1} = \frac{4a_n + 12 - 10}{a_n + 3} = 4 - \frac{10}{a_n + 3} < 4 - \frac{10}{2 + 3} = 2$$

so $0 < a_{n+1} < 2$.

On the other hand,

$$
\begin{aligned}
a_{n+1} - a_n &= \frac{2(2a_n + 1)}{a_n + 3} - a_n \\
&= \frac{2(2a_n + 1) - a_n(a_n + 3)}{a_n + 3} \\
&= \frac{-a_n^2 + a_n + 2}{a_n + 3} \\
&= \frac{(a_n + 1)(2 - a_n)}{a_n + 3} \\
&> 0
\end{aligned}
$$

so the sequence $\{a_n\}$ is increasing. We conclude that it converges, and by passing to the limit we obtain that its limit a satisfies the equation $a = 2(2a + 1)/(a + 3)$. This leads to $a(a + 3) = 4a + 2$ and thus to $a^2 - a - 2 = 0$. Out of the two solutions $a = -1$ and $a = 2$ it is clear that the limit is $a = 2$.   ◆

GOTTFRIED LEIBNIZ (1646–1716) was a German philosopher and mathematician. He belongs to the Pantheon of Mathematics for inventing calculus (independently of Newton). As a high school student in Leipzig he read books on metaphysics and theology. He entered the University of Leipzig at the age of 14, graduated 2 yeas later, and continued working towards a doctorate in law. Perhaps because of his youth he could not receive his doctorate in Leipzig, so he did it at the University of Altdorf. He was then employed by baron von Boineburg and moved to Frankfurt. His position required him to be involved in projects that were scientific, literary and political. In 1672 he was sent to Paris, to the court of Louis XIV, to discuss the political situation and divert a possible attack by France. Leibniz used the opportunity to study mathematics and physics with Dutch physicist and mathematician Christiaan Huygens. As a part of his peace mission he visited London where he met with the leading scientists and was elected into the Royal Society, largely based on a calculating machine that he had designed. Leibniz remained in Paris for 2 years developing calculus. Throughout his life he kept a vast correspondence with over 600 scientists, and some of his results appear in these letters. He discussed infinite series with Grandi (see page 281) and logarithms of negative numbers with Johann Bernoulli (page 62). In an unpublished manuscript in 1675, he used for the first time notation $\int f(x)\,dx$ and stated the product rule for differentiation. The next year, he discovered the formula for $d(x^n)$, for a rational n. In 1676 he accepted the position of librarian and of Court Councillor at Hanover where he remained for the rest of his life. He devised a cataloging system that would serve as a guide for many of Europe's largest libraries. In addition, he designed windmills and pumps, studied geology, developed the binary system of numbers and in 1684 published details of his differential calculus in *Nova Methodus pro Maximis et Minimis* (New method for maxima and minima), but without proofs. An article on integral calculus followed in 1686. During the last part of his life he was involved in the dispute with Newton about who invented calculus. Leibniz was one of the most prolific inventors in the field of mechanical calculators. He made major contributions to physics and technology, and anticipated notions that surfaced much later in philosophy, probability theory, biology, medicine, geology, psychology, linguistics, and computer science. He wrote works on philosophy, politics, law, ethics, theology, history, and philology. (See also pages 34, 63, 114, 175, 211, 256, 281, 414.)

Exercises

In Exercises 2.4.1–2.4.8 determine whether the sequence $\{a_n\}$ is increasing, decreasing, or not monotone at all.

2.4.1. $a_n = \dfrac{n}{n^2+1}$.

2.4.2. $a_n = \dfrac{1}{3n+5}$.

2.4.3. $a_n = \dfrac{n!}{(2n+1)!!}$.

2.4.4. $a_n = \dfrac{1}{2^n}$.

2.4.5. $a_n = \dfrac{n}{2^n}$.

2.4.6. $a_n = \dfrac{n^2}{2^n}$.

2.4.7. $a_n = \dfrac{n-1}{n}$.

2.4.8.* $a_n = \dfrac{n}{\sqrt{n}+2}$.

2.4.9.* Let $c > 0$ and $a_1 = \dfrac{c}{2}$, $a_{n+1} = \dfrac{1}{2}(c + a_n^2)$. Determine all c for which the sequence converges. For such c find $\lim a_n$.

2.4.10.* Let $A > 0$, $a_1 > 0$, and $a_{n+1} = \dfrac{1}{2}\left(a_n + \dfrac{A}{a_n}\right)$. Prove that $\{a_n\}$ converges to $\sqrt{A}$.

In Exercises 2.4.11–2.4.13 show that the sequence $\{a_n\}$ converges and find its limit.

2.4.11.* $a_1 = 2$, $a_{n+1} = 2 + \dfrac{1}{3 + \frac{1}{a_n}}$.   2.4.12.* $a_1 = \dfrac{3}{2}$, $a_{n+1} = \sqrt{3a_n - 2}$.

2.4.13.* $a_1 = 0$, $a_2 = \dfrac{1}{2}$, $a_{n+2} = \dfrac{1}{3}(1 + a_{n+1} + a_n^3)$.

2.5   Number e

In this section we will consider a very important sequence. For that we will need some additional tools. The first one is an inequality.

Theorem 2.5.1 (Bernoulli's Inequality). *If $x > -1$ and $n \in \mathbb{N}$ then*

$$(1+x)^n \geq 1 + nx.$$

Proof. We will use mathematical induction. When $n = 1$, both sides of the inequality are $1+x$. So, we assume that the inequality is true for a positive integer n, i.e., $(1+x)^n \geq 1+nx$. We will show that $(1+x)^{n+1} \geq 1 + (n+1)x$. If we use the hypothesis, we see that

$$(1+x)^{n+1} = (1+x)^n(1+x)$$

$$\geq (1 + nx)(1 + x) = 1 + x + nx + nx^2$$
$$\geq 1 + x + nx = 1 + (n + 1)x$$

and the proof is complete. □

The inequality carries the name of a Swiss mathematician Jacob Bernoulli, because it appeared in his work [4] in 1689. Historians of mathematics point out at exactly the same result in [2] by Isaac Barrow (one of Newton's teachers), except that the latter publication appeared almost 20 years earlier, in 1670. In both, a more general result is proved (see Exercise 2.5.10).

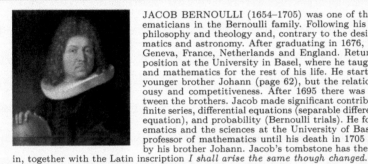

JACOB BERNOULLI (1654–1705) was one of the many prominent mathematicians in the Bernoulli family. Following his father's wish, he studied philosophy and theology and, contrary to the desires of his parents, mathematics and astronomy. After graduating in 1676, he spent 7 years living in Geneva, France, Netherlands and England. Returning to Basel, he took a position at the University in Basel, where he taught mechanics (until 1687) and mathematics for the rest of his life. He started collaborating with his younger brother Johann (page 62), but the relationship soured due to jealousy and competitiveness. After 1695 there was little communication between the brothers. Jacob made significant contributions in the theory of infinite series, differential equations (separable differential equations, Bernoulli equation), and probability (Bernoulli trials). He founded a school for mathematics and the sciences at the University of Basel and worked there as a professor of mathematics until his death in 1705 when the chair was filled in by his brother Johann. Jacob's tombstone has the logarithmic spiral carved in, together with the Latin inscription *I shall arise the same though changed.*

Now we can consider a very important sequence which is the central topic of this section.

Theorem 2.5.2. *The sequence* $a_n = \left(1 + \dfrac{1}{n}\right)^n$ *is convergent.*

Proof. First we will show that $\{a_n\}$ is an increasing sequence. Instead of $a_{n+1} - a_n > 0$, we will establish that $a_{n+1}/a_n > 1$. (Since $a_n > 0$, the two conditions are equivalent.) We notice that

$$\frac{a_{n+1}}{a_n} = \frac{\left(1 + \dfrac{1}{n+1}\right)^{n+1}}{\left(1 + \dfrac{1}{n}\right)^n} = \left(\frac{1 + \dfrac{1}{n+1}}{1 + \dfrac{1}{n}}\right)^{n+1}\left(1 + \frac{1}{n}\right).$$

Further,

$$\frac{1 + \dfrac{1}{n+1}}{1 + \dfrac{1}{n}} = \frac{\dfrac{n+2}{n+1}}{\dfrac{n+1}{n}} = \frac{n(n+2)}{(n+1)^2} = \frac{n^2 + 2n}{n^2 + 2n + 1} = 1 - \frac{1}{n^2 + 2n + 1}.$$

Therefore, using Bernoulli's Inequality,

$$\left(\frac{1 + \dfrac{1}{n+1}}{1 + \dfrac{1}{n}}\right)^{n+1} \geq 1 - (n+1) \cdot \frac{1}{n^2 + 2n + 1} = 1 - \frac{1}{n+1} = \frac{n}{n+1}.$$

It follows that

$$\frac{a_{n+1}}{a_n} \geq \frac{n}{n+1}\left(1 + \frac{1}{n}\right) = \frac{n}{n+1}\frac{n+1}{n} = 1,$$

and the sequence $\{a_n\}$ is increasing.

In order to show that it is bounded above, we will consider a sequence $b_n = (1+1/n)^{n+1}$.

An argument, similar to the one above, can be used to establish that $b_{n+1}/b_n \leq 1$ so $\{b_n\}$ is a decreasing sequence. Further, $a_n \leq b_n \leq b_1$ for any $n \in \mathbb{N}$, so the sequence $\{a_n\}$ is bounded above by $b_1 = 4$. Therefore, it is convergent, and the proof is complete. $\square$

Remark 2.5.3. The limit of the sequence $\{a_n\}$ lies between $a_1 = 2$ and $b_1 = 4$. This number is the well-known constant e, and it can be calculated to any number of decimals. For example, $a_{100} \approx 2.704813829$. Typically, we approximate e by 2.7.

> The sequence $\{a_n\}$ appeared for the first time in 1683, in the work of Jacob Bernoulli on compound interest. However, he only obtained that its limit lies between 2 and 3. Prior to that, e was present through the use of natural logarithms, but the number itself was never explicitly given. It is considered that the first time it appears in its own right is in 1690, in a letter from Leibniz to Huygens, who used the notation b. The letter e was introduced by a Swiss mathematician Leonhard Euler (page 49) in a letter to German mathematician Christian Goldbach (1690–1764) in 1731.

Although we can use the sequence $\{a_n\}$ for approximating e, there are more efficient ways. Before we get to that, we will need a way to expand the expression $\left(1 + \frac{1}{n}\right)^n$. Recall that $(a+b)^2 = a^2 + 2ab + b^2$. Another useful formula is $(a+b)^3 = a^3 + 3a^2b + 3ab^2 + b^3$. What about the higher powers? One might make a guess that $(a+b)^4$ will have the terms a^4, a^3b, a^2b^2, ab^3, and b^4, but it is not clear how to determine the coefficients. In order to describe them, we need to introduce the **binomial coefficients**. For positive integers n and k that satisfy $n \geq k$, we define the number

$$\binom{n}{k} = \frac{n!}{k!(n-k)!} = \frac{n(n-1)(n-2)\ldots(n-k+1)}{1 \cdot 2 \cdot 3 \cdot \ldots \cdot k}.$$

In addition, if $n \in \mathbb{N}$, we define $\binom{n}{0} = 1$. Now we can get the formula for $(a+b)^n$ for any $n \in \mathbb{N}$.

Theorem 2.5.4 (The Binomial Formula). *For any $n \in \mathbb{N}$,*

$$(a+b)^n = \sum_{k=0}^{n} \binom{n}{k} a^{n-k} b^k. \tag{2.8}$$

Proof. We will use mathematical induction. When $n = 1$, we obtain

$$(a+b)^1 = \binom{1}{0} a^1 b^0 + \binom{1}{1} a^0 b^1.$$

Since $\binom{1}{0} = \binom{1}{1} = 1$, we see that the equality for $n = 1$ is correct. Thus we assume that formula (2.8) is correct for some positive integer n, and we establish its validity for $n + 1$. Namely, we will prove that

$$(a+b)^{n+1} = \sum_{k=0}^{n+1} \binom{n+1}{k} a^{n+1-k} b^k. \tag{2.9}$$

Now,

$$(a+b)^{n+1} = (a+b)(a+b)^n$$

$$= (a+b) \sum_{k=0}^{n} \binom{n}{k} a^{n-k} b^k$$

$$= \sum_{k=0}^{n} \binom{n}{k} a^{n-k+1} b^k + \sum_{k=0}^{n} \binom{n}{k} a^{n-k} b^{k+1}.$$

Here we notice that a substitution $m = k + 1$ transforms the second sum into

$$\sum_{m=1}^{n+1} \binom{n}{m-1} a^{n-m+1} b^m$$

and, since m is the index of summation, it can be replaced by any other letter (e.g., k):

$$\sum_{k=1}^{n+1} \binom{n}{k-1} a^{n-k+1} b^k.$$

Thus, we obtain that

$$(a + b)^{n+1} = \sum_{k=0}^{n} \binom{n}{k} a^{n-k+1} b^k + \sum_{k=1}^{n+1} \binom{n}{k-1} a^{n-k+1} b^k. \qquad (2.10)$$

The first term in the first sum (using $k = 0$) equals a^{n+1}, which is the same as the first term in (2.9). Similarly, the last term in the second sum (using $k = n + 1$) equals b^{n+1}, which is the same as the last term in (2.9). The remaining terms in (2.10) can be put together as

$$\sum_{k=1}^{n} \left[\binom{n}{k} + \binom{n}{k-1} \right] a^{n-k+1} b^k.$$

Comparing this with the terms in (2.9), we see that it remains to prove that $\binom{n}{k} + \binom{n}{k-1} = \binom{n+1}{k}$, for $1 \le k \le n$. [For those familiar with the Pascal triangle, this is well known!] This requires just some algebra:

$$\begin{aligned}
\binom{n}{k} + \binom{n}{k-1} &= \frac{n!}{k!(n-k)!} + \frac{n!}{(k-1)!(n-k+1)!} \\
&= \frac{n!(n-k+1) + n!k}{k!(n-k+1)!} \\
&= \frac{(n+1)!}{k!(n-k+1)!} \\
&= \binom{n+1}{k},
\end{aligned}$$

and the proof is complete. $\qquad \square$

Special cases of the Binomial Formula were known in the ancient world: Euclid (about 300 BC) for $n = 2$ in Greece, Aryabhata (476–550 AD) in India for $n = 3$. The first proof (using induction) was given by a Persian mathematician Al-Karaji (953–1029). The "Pascal triangle" was also known to Zhu Shijie (1270–1330) in China. In the Western world, it appears in 1544 [97] by Michael Stifel (1486 or 1487–1567). Blaise Pascal (1623–1662), a French mathematician, was the first one to organize all the information together and add many applications of the triangle in [83] in 1653.

Bernoulli discovered the sequence $a_n = \left(1 + \frac{1}{n}\right)^n$, and we know that it converges to e. Euler found another one.

Theorem 2.5.5. *The sequence* $c_n = 1 + 1 + \dfrac{1}{2!} + \dfrac{1}{3!} + \cdots + \dfrac{1}{n!}$ *converges to* e.

Proof. We will first establish the inequality $a_n \le c_n \le e$, for all $n \in \mathbb{N}$, then use the Squeeze Theorem. As before, $a_n = (1 + 1/n)^n$.

If we apply Binomial Formula to $a_n = (1 + 1/n)^n$ we have that

$$a_n = \sum_{k=0}^{n} \binom{n}{k} \frac{1}{n^k}.$$

It is easy to see that the terms obtained for $k = 0$ and $k = 1$ are both equal to 1. For $k \geq 2$,

$$\binom{n}{k} \frac{1}{n^k} = \frac{n(n-1)\ldots(n-k+1)}{1 \cdot 2 \cdot \ldots \cdot k} \cdot \frac{1}{n^k}$$

$$= \frac{1}{k!} \cdot \frac{n}{n} \cdot \frac{n-1}{n} \cdot \ldots \cdot \frac{n-k+1}{n}$$

$$= \frac{1}{k!} \left(1 - \frac{1}{n}\right) \left(1 - \frac{2}{n}\right) \ldots \left(1 - \frac{k-1}{n}\right).$$

Let N be a positive integer greater than 2, and notice that, if $n \geq N$, then

$$a_n \geq \sum_{k=0}^{N} \binom{n}{k} \frac{1}{n^k} = 1 + 1 + \sum_{k=2}^{N} \frac{1}{k!} \left(1 - \frac{1}{n}\right) \left(1 - \frac{2}{n}\right) \ldots \left(1 - \frac{k-1}{n}\right).$$

If we take the limit as $n \to \infty$ of both sides we obtain that

$$e = \lim a_n \geq 1 + 1 + \sum_{k=2}^{N} \frac{1}{k!} = c_N.$$

Thus $c_n \leq e$, for all $n \in \mathbb{N}$.

On the other hand,

$$a_n = 1 + 1 + \sum_{k=2}^{n} \frac{1}{k!} \left(1 - \frac{1}{n}\right) \left(1 - \frac{2}{n}\right) \ldots \left(1 - \frac{k-1}{n}\right) \leq 1 + 1 + \sum_{k=2}^{n} 1/k! = c_n.$$

Therefore, $a_n \leq c_n \leq e$ for all $n \in \mathbb{N}$ and it follows from the Squeeze Theorem that c_n is a convergent sequence with limit e. $\square$

> Theorem 2.5.5 is due to Euler (page 49). It appeared in his book [41] published in 1748. He also proved there that $\lim a_n = e$, calculated e to 18 decimal places, and gave an incomplete argument that e is not a rational number.

Both sequences $\{a_n\}$ and $\{c_n\}$ have limit e, but the inequality $a_n \leq c_n \leq e$ shows that $\{c_n\}$ converges faster than $\{a_n\}$. Actually, the convergence is *much* faster. For example, $a_{100} \approx 2.704813829$, which differs from the correct value ($2.71828\ldots$) by more than 0.01. On the other hand, $c_4 \approx 2.708333333$, which has the accuracy better than 0.01. In fact, we have the following estimate. We leave the proof to the reader (Exercise 2.5.9).

Lemma 2.5.6. *Let $n \in \mathbb{N}$, let c_n be as in Theorem 2.5.5, and define $\theta_n = (e - c_n) \, n!$. Then $0 < \theta_n < 1$.*

Lemma 2.5.6 allows us to estimate the error of approximating e by c_n. Namely, it shows that the (positive) quantity $e - c_n$ is smaller than $1/n!$. For example, if we require that the error does not exceed 10^{-4}, it suffice to take $n = 8$: $1/8! \approx 2.48 \times 10^{-5} < 10^{-4}$.

Notice that both sequences $\{a_n\}$ and $\{c_n\}$ consist of rational numbers. Nevertheless, e is an irrational number (Exercise 2.5.11).

Exercises

2.5.1.* Prove that $b_n = (1 + 1/n)^{n+1}$ is a decreasing sequence.

2.5.2. Let $a_n = \left(1 + \dfrac{x}{n}\right)^n$. Show that the sequence is bounded and increasing for $n > -x$.

2.5.3.* Find $\lim n \left(\sqrt[n]{e} - 1\right)$.

2.5.4. Find $\lim \dfrac{e^{\frac{1}{n}} + e^{\frac{2}{n}} + \cdots + e^{\frac{n}{n}}}{n}$.

2.5.5. Let $a_1 = 0$, $a_2 = 1$, and $a_{n+2} = \dfrac{(n+2)a_{n+1} - a_n}{n+1}$. Prove that $\lim a_n = e$.

2.5.6. Find $\lim \left(1 - \frac{1}{n}\right)^n$.

2.5.7.* Prove that the sequence $a_n = 1 + \dfrac{1}{2} + \dfrac{1}{3} + \cdots + \dfrac{1}{n} - \ln n$ is decreasing and bounded below. Conclude that it is convergent.

Remark 2.5.7. The limit of this sequence is known as *Euler's constant*, and it is approximately equal to 0.5772.

2.5.8.* Let $a_n = 3 - \displaystyle\sum_{k=1}^{n} \dfrac{1}{k(k+1)(k+1)!}$. Prove that $\lim a_n = e$.

2.5.9.* Prove Lemma 2.5.6.

2.5.10.* A sequence $\{a_n\}$ is a *geometric* sequence if there exists $q > 0$ such that $a_{n+1}/a_n = q$, $n \in \mathbb{N}$. A sequence $\{b_n\}$ is an *arithmetic* sequence if there exists $d > 0$ such that $a_{n+1} - a_n = d$, $n \in \mathbb{N}$. If $\{a_n\}$ and $\{b_n\}$ are such sequences, and if $a_1 = b_1 > 0$, $a_2 = b_2 > 0$, prove that $a_n > b_n$ for $n \geq 3$. Use this result to derive the Bernoulli's Inequality.

2.5.11.* Prove that e is not a rational number.

2.6   Cauchy Sequences

In Section 2.4 we first encountered the situation where our main goal was to prove that a sequence converges. The method we used required that a sequence is monotone increasing or decreasing. In this section we will add to our bag of tricks another tool that can be used to detect whether a sequence is convergent (without calculating its limit). This one, however, can be applied to sequences that are not necessarily monotone. Like much else in this chapter, it was introduced by Cauchy and we honor him by calling this property that a sequence may have by his name.

Definition 2.6.1. A sequence $\{a_n\}$ is a **Cauchy sequence** if, for any $\varepsilon > 0$, there exists a positive integer N such that, if $m \geq n \geq N$, then $|a_m - a_n| < \varepsilon$.

Example 2.6.2. A Cauchy sequence.
   Prove that $a_n = 1 + \dfrac{1}{2^2} + \dfrac{1}{3^2} + \cdots + \dfrac{1}{n^2}$ is a Cauchy sequence.

Solution. This will require a similar strategy as with limits, i.e., we start by investigating the expression $a_m - a_n$. If $m \geq n$ then

$$a_m - a_n = \frac{1}{(n+1)^2} + \frac{1}{(n+2)^2} + \cdots + \frac{1}{m^2}$$

$$< \frac{1}{(n+1)n} + \frac{1}{(n+2)(n+1)} + \cdots + \frac{1}{m(m-1)}.$$

Notice that, for any x,

$$\frac{1}{x} - \frac{1}{x+1} = \frac{1}{x(x+1)}.$$

Therefore,

$$a_m - a_n < \left(\frac{1}{n} - \frac{1}{n+1}\right) + \left(\frac{1}{n+1} - \frac{1}{n+2}\right) + \cdots + \left(\frac{1}{m-1} - \frac{1}{m}\right) = \frac{1}{n} - \frac{1}{m} < \frac{1}{n}.$$

Clearly, $a_m - a_n > 0$, so $|a_m - a_n| < 1/n$. Thus, it suffices to choose N so that $1/N < \varepsilon$.

Proof. Let $\varepsilon > 0$ and let $N = \lfloor 1/\varepsilon \rfloor + 1$. Then $N > 1/\varepsilon$ and $1/N < \varepsilon$. When $m \geq n \geq N$,

$$|a_m - a_n| = a_m - a_n = \frac{1}{(n+1)^2} + \frac{1}{(n+2)^2} + \cdots + \frac{1}{m^2}$$

$$< \frac{1}{(n+1)n} + \frac{1}{(n+2)(n+1)} + \cdots + \frac{1}{m(m-1)}$$

$$= \left(\frac{1}{n} - \frac{1}{n+1}\right) + \left(\frac{1}{n+1} - \frac{1}{n+2}\right) + \cdots + \left(\frac{1}{m-1} - \frac{1}{m}\right)$$

$$= \frac{1}{n} - \frac{1}{m} < \frac{1}{n} \leq \frac{1}{N} < \varepsilon.$$

Therefore, $\{a_n\}$ is a Cauchy sequence.   ◆

Before we look at another example, we will establish a summation formula.

Theorem 2.6.3. *Let x be a real number different from 1, and $S_n = 1 + x + x^2 + \cdots + x^n$. Then $S_n = (x^{n+1} - 1)/(x - 1)$.*

Proof. Notice that

$$xS_n = x + x^2 + \cdots + x^{n+1} = S_n + x^{n+1} - 1,$$

so $(x - 1)S_n = xS_n - S_n = x^{n+1} - 1$ and the result follows.   □

Example 2.6.4. A Cauchy sequence.

Prove that $a_n = \frac{\sin 1}{2} + \frac{\sin 2}{2^2} + \cdots + \frac{\sin n}{2^n}$ is a Cauchy sequence.

Solution. If $m \geq n$ then

$$|a_m - a_n| = \left| \frac{\sin(n+1)}{2^{n+1}} + \frac{\sin(n+2)}{2^{n+2}} + \cdots + \frac{\sin m}{2^m} \right|$$

$$\leq \left| \frac{\sin(n+1)}{2^{n+1}} \right| + \left| \frac{\sin(n+2)}{2^{n+2}} \right| + \cdots + \left| \frac{\sin m}{2^m} \right|$$

$$\leq \frac{1}{2^{n+1}} + \frac{1}{2^{n+2}} + \cdots + \frac{1}{2^m} \qquad (2.11)$$

$$= \frac{1}{2^{n+1}} \cdot \frac{1 - 1/2^{m-n}}{1 - 1/2}$$

$$< \frac{1}{2^{n+1}} \cdot \frac{1}{1 - 1/2} = \frac{1}{2^n}.$$

Given $\varepsilon > 0$ we will require that $1/2^N < \varepsilon$. This leads to $2^N > 1/\varepsilon$. By taking the natural logarithm of both sides, we have that $N \ln 2 > \ln(1/\varepsilon)$, hence $N > \ln(1/\varepsilon)/\ln 2$. In other words,

$$N > \frac{\ln \frac{1}{\varepsilon}}{\ln 2} \quad \text{if and only if} \quad \frac{1}{2^N} < \varepsilon. \tag{2.12}$$

Proof. Let $\varepsilon > 0$ and let $N = \max\{\lfloor \ln(1/\varepsilon)/\ln 2 \rfloor + 1, 1\}$. Then $N > \ln \frac{1}{\varepsilon}/\ln 2$ so (2.12) implies that $1/2^N < \varepsilon$. Now, if $m \geq n \geq N$ then, using (2.11),

$$|a_m - a_n| < \frac{1}{2^n} \leq \frac{1}{2^N} < \varepsilon.$$

Therefore, the sequence $\{a_n\}$ is a Cauchy sequence. $\blacklozenge$

Example 2.6.5. A sequence that is not a Cauchy sequence.

Prove that $a_n = 1 + \frac{1}{2} + \frac{1}{3} + \cdots + \frac{1}{n}$ is *not* a Cauchy sequence.

Solution. In order to articulate the negative, it is usually helpful to write the statement in a more formal way, using quantifiers. Definition 2.6.1 can be formulated as: $\{a_n\}$ is a Cauchy sequence if

$$(\forall \varepsilon)(\exists N)(\forall m, n) \, m \geq n \geq N \Rightarrow |a_m - a_n| < \varepsilon. \tag{2.13}$$

The appearance of $(\forall m, n)$ is perhaps unexpected, but makes sense because the inequality $|a_m - a_n| < \varepsilon$ must be true for *all* m, n that satisfy $m \geq n \geq N$.

If we take the negative of (2.13), in addition to changing the quantifiers, we need to write the negative of the implication

$$m \geq n \geq N \Rightarrow |a_m - a_n| < \varepsilon.$$

The general rule is that, whenever we have an implication $p \Rightarrow q$, its negative is $p \wedge \neg q$, meaning "p and the negative of q." Here, this means "$m \geq n \geq N$ and $|a_m - a_n| \geq \varepsilon$." Thus, the negative of (2.13) is

$$(\exists \varepsilon)(\forall N)(\exists m, n) \quad m \geq n \geq N \quad \text{and} \quad |a_m - a_n| \geq \varepsilon. \tag{2.14}$$

As usual, we focus on $|a_m - a_n| \geq \varepsilon$. Let $m \geq n$. Then

$$a_m - a_n = \frac{1}{n+1} + \frac{1}{n+2} + \cdots + \frac{1}{m} \geq \frac{1}{m} + \frac{1}{m} + \cdots + \frac{1}{m} = (m-n)\frac{1}{m}.$$

Now, the inequality to consider is $(m-n)/m \geq \varepsilon$. Since $(m-n)/m = 1 - n/m$ this leads to $n/m \leq 1 - \varepsilon$. For example, let $\varepsilon = 1/2$. For this to work, we need to show that, for any $N \in \mathbb{N}$, we can find $m \geq n \geq N$ such that $|a_m - a_n| \geq 1/2$, which is the same as $n/m \leq 1 - 1/2$. It is not hard to see that the last inequality is satisfied if we take $m = 2n$.

Finally, we can write a careful proof of (2.14).

Proof. Let $\varepsilon = 1/2$ and let $N \in \mathbb{N}$. If we define $n = N$ and $m = 2N$, then $m \geq n \geq N$ and

$$|a_m - a_n| = \frac{1}{n+1} + \frac{1}{n+2} + \cdots + \frac{1}{m}$$

$$\geq \frac{1}{m} + \frac{1}{m} + \cdots + \frac{1}{m} = (m-n)\frac{1}{m} = (2N-N)\frac{1}{2N} = \frac{1}{2}.$$

Therefore $\{a_n\}$ is not a Cauchy sequence. $\blacklozenge$

Notice that the definition of a Cauchy sequence (Definition 2.6.1 above) is similar to the definition of a convergent sequence (Definition 2.2.6). It turns out that the two concepts are equivalent.

Theorem 2.6.6 (Cauchy's Test). *A sequence is a Cauchy sequence if and only if it is convergent.*

We will postpone the proof until the next section. Here, we will make a few observations.

Remark 2.6.7. The main difference between the two definitions is that, for the Cauchy sequence, there is no mention of the suspected limit L. This makes it very suitable for establishing the convergence of a sequence when we don't know what the limit is. Such is the situation in Examples 2.6.2 and 2.6.4.

Remark 2.6.8. You may have noticed that all examples in this section were of a similar form: the general term a_n was a sum of n terms. Another way of saying the same thing is that $\{a_n\}$ is the sequence of *partial sums* of an infinite series. The convergence (or divergence) of such a sequence means, by definition, that the series converges (or diverges). Therefore the series $\sum_{n=1}^{\infty} \frac{\sin n}{2^n}$ and $\sum_{n=1}^{\infty} \frac{1}{n^2}$ are convergent, while the series $\sum_{n=1}^{\infty} \frac{1}{n}$ diverges. As we will see later, one of the major roles of sequences will be in the study of infinite series.

Theorem 2.6.6 was proved by Cauchy in *Cours d'Analyse* in 1821. Four years earlier, Bolzano explicitly stated the same result and gave an incomplete proof. It is very likely that Cauchy was aware of this work.

BERNARD BOLZANO (1781–1848) was a Bohemian mathematician, philosopher, and a Catholic priest. His parents were devout Roman Catholics, his mother a German speaking Czech and his father an Italian immigrant. Bolzano was of delicate health, suffering from lung problems throughout his life. From his parents he inherited a strong desire to do what "will most further the good of the totality". He studied philosophy, physics, and mathematics at the Charles University of Prague. He was influenced by a book by German mathematician Kaestner, *Mathematical Foundations*. From Kaestner Bolzano took the idea to prove results which seem obvious. In 1804 he defended his doctoral thesis on geometry and, in succession, became an ordained priest, a professor of religion at the Charles University, a member of the Royal Bohemian Society of Sciences, and the Dean of the Faculty of Philosophy at Charles University. His anti-militarism and concerns about the domination of the German speaking Bohemians over the Czech speaking ones, brought him a suspension in 1819 and he was tried by the Church 1821-1825. Refusing to recant his views, he resigned from the University and during the next 20 years lived mostly in the countryside with friends. In 1841 he moved back to Prague and became active in the Royal Bohemian Society of Sciences, becoming their president 1842-43. His poor health caught up with him and he died in 1848. Bolzano published very little: a book on the foundations of mathematics in 1810 and a few papers. The one in 1817 contained, in addition to the proof of the Intermediate Value Theorem (see page 44), the definition of a Cauchy sequence and the definition of the derivative without the use of infinitesimals. In *Paradoxes of Infinity* that appeared in print three years after his death, the word *set* appeared for the first time. At the end of his life he entrusted his notes to a student of his, who passed it on to the Austrian Academy of Sciences. It was not until 1930 that *Functionenlehre* (Function theory) appeared, with 3 more publications in the next 5 years. In the 1970s this work was continued, and it is estimated that, out of the projected 120 volumes of Bolzano's work, about one half had been published by the year 2000.

Exercises

In Exercises 2.6.1–2.6.6 use the definition to determine which of the following sequences is a Cauchy sequence:

2.6.1.* $a_n = 1 + \frac{1}{4} + \frac{2^2}{4^2} \cdots + \frac{n^2}{4^n}.$ 2.6.2. $a_n = \frac{\arctan 1}{2} + \frac{\arctan 2}{2^2} + \cdots + \frac{\arctan n}{2^n}.$

2.6.3.* $a_n = \frac{1}{2^2} + \frac{2}{3^2} + \cdots + \frac{n}{(n+1)^2}.$ 2.6.4. $a_n = \sqrt{n}.$ 2.6.5. $a_n = \frac{n+1}{n}.$

2.6.6. $a_n = \frac{1}{\ln 2} + \frac{1}{\ln 3} \cdots + \frac{1}{\ln(n+1)}.$

2.6.7. Let $\{a_n\}$ be a sequence such that $|a_{n+1} - a_n| \to 0$. Prove or disprove: $\{a_n\}$ is a Cauchy sequence.

2.6.8.* Let $0 < r < 1$, $M > 0$, and suppose that $\{a_n\}$ is a sequence such that, for all $n \in \mathbb{N}$, $|a_{n+1} - a_n| \le Mr^n$. Prove that $\{a_n\}$ is a Cauchy sequence.

2.6.9.* Suppose that $\{a_n\}$ is a sequence such that, for all $n \in \mathbb{N}$,

$$|a_n| < 2, \quad \text{and} \quad |a_{n+2} - a_{n+1}| \le \frac{1}{8}|a_{n+1}^2 - a_n^2|.$$

Prove that $\{a_n\}$ is a Cauchy sequence.

2.7 Bolzano–Weierstrass Theorem

We have an outstanding debt from Section 2.6: a proof of Cauchy's Test (Theorem 2.6.6). In this section we will settle the score by deriving it from a very important result that is due to Bolzano and Weierstrass.

We will talk first about *nested* intervals.

Example 2.7.1. A sequence of intervals.
Let $n \in \mathbb{N}$, and let $J_n = (-1/n, 1/n)$. Since $1/(n+1) < 1/n$ we have that

$$-\frac{1}{n} < -\frac{1}{n+1} < \frac{1}{n+1} < \frac{1}{n},$$

so $J_{n+1} \subset J_n$, for each $n \in \mathbb{N}$. We make an observation that 0 belongs to each of the intervals J_n. ◆

In the situation like in Example 2.7.1, when $J_{n+1} \subset J_n$ for all $n \in \mathbb{N}$, we say that $\{J_n\}$ is a sequence of **nested intervals**. (Here, as well as everywhere in the text, the symbol $\subset$ allows the possibility that the sets are equal.) In Example 2.7.1 we noticed that $0 \in J_n$, for all $n \in \mathbb{N}$. We will be interested whether, assuming that $\{J_n\}$ is a sequence of non-empty, nested intervals, there exists a number that belongs to all of them.

Example 2.7.2. A sequence of nested intervals with a common point.
Let $J_n = [0, 1/n]$. Since $[0, \frac{1}{n+1}] \subset [0, \frac{1}{n}]$, the intervals are nested. Again, 0 belongs to each of the intervals J_n. ◆

Example 2.7.3. A sequence of nested intervals without a common point.
Let $J_n = (0, 1/n)$. Although intervals are nested, there is no number that belongs to each interval J_n. ◆

It is a good idea to look at Examples 2.7.2 and 2.7.3 and try to analyze why two very similar collections of intervals produce a different result. It looks as if the existence of a common point had to do with intervals being closed. Before we jump to any conclusions, let us look at another example.

Example 2.7.4. A sequence of nested intervals without a common point.
Let $J_n = [n, \infty)$. It is not hard to see that we have a sequence of nested intervals, and that there is no number that would belong to all intervals. ◆

Notice that the intervals are closed (they contain their endpoints). So, either we have to change our idea of a closed interval, or (better!) avoid intervals that stretch to infinity. In other words, we will require that our intervals are closed and finite.

Theorem 2.7.5. *Every sequence of nested non-empty, closed, and finite intervals has a non-empty intersection.*

Proof. Let us denote these intervals by $[a_n, b_n]$. The assumption that they are nested means that

$$a_1 \le a_2 \le \cdots \le a_n \le \cdots \le b_n \le \cdots \le b_2 \le b_1.$$

Therefore, the sequence $\{a_n\}$ is an increasing sequence that is bounded above by b_k, for each $k \in \mathbb{N}$. This implies that there exists $\sup a_n$ and, being the least upper bound, $\sup a_n \le b_k$, for any $k \in \mathbb{N}$. Of course, that implies that the sequence $\{b_n\}$ is bounded below by $\sup a_n$, so there exists $\inf b_n$ and $\sup a_n \le \inf b_n$. It follows that every number x such that $\sup a_n \le x \le \inf b_n$ belongs to each interval $[a_n, b_n]$. $\qquad\square$

Remark 2.7.6. The question whether Theorem 2.7.5 is true can be asked in any ordered field. If the answer is yes and if the field is, in addition, an Archimedean field, then the Completeness Axiom must hold (see Exercise 2.7.9).

If we strengthen the hypotheses of Theorem 2.7.5, we can guarantee the uniqueness of the common point.

Theorem 2.7.7. *Let $\{[a_n, b_n]\}$ be a sequence of nested non-empty, closed, and finite intervals. If, in addition, $\lim(b_n - a_n) = 0$, then there is a unique point that belongs to all intervals.*

Proof. It is helpful to recall the fact established in the proof of the Monotone Convergence Theorem (Theorem 2.4.7): an increasing sequence that is bounded above converges to its least upper bound. Since the sequence $\{a_n\}$ is a sequence with these properties, it must converge to $\sup a_n$. Now, $b_n = a_n + (b_n - a_n)$, so $\{b_n\}$ is a convergent sequence, and $\lim b_n = \lim a_n + 0 = \sup a_n$. Of course, $\{b_n\}$ is a decreasing sequence, so $\lim b_n = \inf b_n$. Consequently, $\sup a_n = \inf b_n$. Since $x \in [a_n, b_n]$, for each $n \in \mathbb{N}$, if and only if $\sup a_n \le x \le \inf b_n$, we see that the only common point is $x = \sup a_n = \inf b_n$. $\qquad\square$

The property of real numbers expressed in Theorem 2.7.7 was used by many mathematicians (Cauchy, Bolzano, etc.) without proof. It is sometimes referred to as the Bolzano–Weierstrass Property because Weierstrass was among the first to realize that a proof was needed but he attributed it to Bolzano. Theorem 2.7.7 is also known under the name Cantor's Intersection Theorem.

GEORG CANTOR (1845–1918) was a German mathematician, best known as the inventor of set theory. He was born in Saint Petersburg, Russia but his family moved to Germany when he was eleven. He studied mathematics at the Polytechnic of Zürich and the University of Berlin. In 1869 he got a position at the University of Halle, where he remained for the rest of his career. As a student his area of research was number theory, but in Halle he became interested in analysis under the influence of Heine (page 92), and he proved the uniqueness of the representation by a trigonometric series. He became a lifelong friend with Dedekind, and has likely inspired "Dedekind cuts". In 1873 he proved that rational numbers were countable, and a year later, in [13], using the nested intervals, that real numbers were not. This established that there exist infinite sets of different sizes. Cantor introduced the concept of a cardinal number and the power set of A (the set of all possible subsets of A). He proved that the cardinal number of the power set of A is strictly larger than the cardinal number of A. His notation for the cardinal numbers was the Hebrew letter $\aleph$ (aleph) with a natural number subscript. The Continuum hypothesis, introduced by Cantor, was presented by Hilbert as the first of his twenty-three open problems at the 1900 International Congress of Mathematicians in Paris. It conjectures that there is no cardinal number between those of the natural numbers ($\aleph_0$) and real numbers (continuum). Cantor also proved that there exists a 1-1 correspondence of points on the interval $[0, 1]$ and points in p-dimensional space. Surprised by his own discovery he wrote *I see it, but I don't believe it!* From 1884 until the end of his life, Cantor suffered from depression and mental illness. He continued doing research but his best years were behind him. He retired from teaching in 1913 and died in 1918 from a heart attack.

It is often the case that we are dealing with a divergent sequence—yet, we would like to be able to make a statement about its long-term behavior. For example, it was established in Example 2.3.8 that the sequence $a_n = (-1)^n$ is divergent. Nevertheless, we have a perfect understanding of its *asymptotic* behavior (meaning: as $n \to \infty$). When n is an even number, say $n = 2k$ with $k \in \mathbb{N}$, then $a_n = 1$. On the other hand, when n is an odd number, say $n = 2k - 1$ with $k \in \mathbb{N}$, then $a_n = -1$. In order to describe a situation like this we will develop some terminology first.

Notice that the sequence $a_1, a_3, a_5, \ldots$ is convergent. This is not the whole sequence $\{a_n\}$.

Definition 2.7.8. Let $\{a_n\}$ be a sequence, and let $n_1 < n_2 < n_3 < \ldots$ be an infinite, strictly increasing sequence of positive integers. We say that the sequence $a_{n_1}, a_{n_2}, a_{n_3}, \ldots$ is a **subsequence** of $\{a_n\}$.

Example 2.7.9. A subsequence.

Let $a_n = 1/(n+3)$, $n_k = k^2$. What subsequence is $\{a_{n_k}\}$?

Here we have $n_1 = 1$, $n_2 = 4$, $n_3 = 9$, $n_4 = 16$, etc. The subsequence is $a_1 = \frac{1}{4}$, $a_4 = \frac{1}{7}$, $a_9 = \frac{1}{12}$, $a_{16} = \frac{1}{19}$, etc. ◆

Remark 2.7.10. When $n_k = k$, i.e., when $n_1 = 1$, $n_2 = 2$, etc., we obtain the whole sequence $\{a_n\}$. So, every sequence is a subsequence of itself.

The names of Bolzano and Weierstrass are also associated with the following result.

Theorem 2.7.11 (Bolzano–Weierstrass Theorem). *Every bounded sequence has a convergent subsequence.*

Proof. Let $\{a_n\}$ be a sequence that is bounded by M: $|a_n| \le M$. We define the interval $J_1 = [-M, M]$ and $n_1 = 1$. We split the interval J_1 into two closed intervals of equal length, and we notice that at the same time this partitions the sequence $\{a_n\}$ in two: those that belong to the left half of J_1 and those that belong to the right half of J_1. At least one of these two collections must have infinitely many members of $\{a_n\}$ and we will denote by J_2 the portion of J_1 that does. (If both do, either choice will do.) Among those that belong

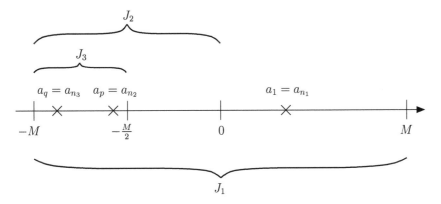

Figure 2.8: Sequences $\{J_n\}$ and $\{a_{n_k}\}$.

to J_2, we pick one different from $a_{n_1}(= a_1)$, say a_p, and we define $n_2 = p$. Now we repeat the same procedure for J_2. We split J_2 into two closed intervals of equal length, and we denote by J_3 the one that contains infinitely many members. Then we select $a_q \in J_3$ so that $q > n_2 > n_1$, and we define $n_3 = q$. We continue this process and obtain a nested sequence of intervals $\{J_n\}$ and a strictly increasing sequence of positive integers $n_1 < n_2 < n_3 < \ldots$ (Figure 2.8). Due to the construction, the length of J_2, $\ell(J_2) = \ell(J_1)/2$, $\ell(J_3) = \ell(J_1)/2^2$, and in general, $\ell(J_n) = \ell(J_1)/2^{n-1} \to 0$. By Theorem 2.7.7, there is a unique point a that belongs to all intervals. We will show that the subsequence $\{a_{n_k}\}$ converges to a.

Let $\varepsilon > 0$. One way to ensure that $|a_{n_k} - a| < \varepsilon$ is to notice that both a_{n_k} and a belong to J_k, and choose k so that $\ell(J_k) < \varepsilon$. Since $\lim \ell(J_k) = 0$, there exists $K \in \mathbb{N}$ such that, if $k \geq K$, $\ell(J_k) < \varepsilon$. For any such k, both a_{n_k} and a belong to J_k, so $|a_{n_k} - a| < \varepsilon$ and the proof is complete. $\qquad\square$

Remark 2.7.12. The question whether every bounded sequence has a convergent subsequence is meaningful in every ordered field. It turns out that the answer is in the affirmative if and only if the Completeness Axiom holds (see Exercise 2.7.9).

> Weierstrass proved Theorem 2.7.11 and presented it in a lecture in 1874. The result was initially called the Weierstrass Theorem until it was discovered that it was a part of Bolzano's proof of the Intermediate Value Theorem (Theorem 3.9.1), published in 1817.

The bisection method used in the proof of the Bolzano–Weierstrass Theorem can be traced all the way back to Euclid. It implies that, unlike $\mathbb{Q}$, the set $\mathbb{R}$ has no gaps in it. Suppose that there exists a point P on the number line that does not represent a real number. We select real numbers a_1 on the left and b_1 on the right of P, and then construct inductively a sequence of intervals $J_n = [a_n, b_n]$ that all contain P. By Theorem 2.7.7, there exists a unique real number z that belongs to all the intervals J_n, and it follows that P is precisely the representation of z.

Finally, we return to the Cauchy Test. We will prove that a sequence is convergent if and only if it is a Cauchy sequence. Let us first look at the implication: if $\{a_n\}$ is a Cauchy sequence, then it converges. A typical proof that a sequence is convergent consists of establishing the inequality $|a_n - L| < \varepsilon$, where L is the suspected limit. An obvious obstruction to this strategy is that, if a sequence $\{a_n\}$ is Cauchy, we have no idea what it might converge to. We might try to use as L the limit of some convergent subsequence $\{a_{n_k}\}$. The Bolzano–Weierstrass Theorem will guarantee the existence of such a subsequence, as long as we can be assured that $\{a_n\}$ is bounded. So, this is where we start.

Theorem 2.7.13. *Every Cauchy sequence is bounded.*

Proof. Let $\{a_n\}$ be a Cauchy sequence and let $\varepsilon = 1$. Then there exists N such that, if $m \geq n \geq N$, $|a_m - a_n| < \varepsilon = 1$. The last inequality is equivalent to $-1 < a_m - a_n < 1$ and, hence, to $a_n - 1 < a_m < a_n + 1$. In particular, when $n = N$, we have that for all $m \geq N$, $a_N - 1 < a_m < a_N + 1$. Next we consider $N + 1$ positive numbers: $|a_1|, |a_2|, \ldots |a_{N-1}|$, $|a_N - 1|$, and $|a_N + 1|$, and let

$$M = \max\{|a_1|, |a_2|, \ldots |a_{N-1}|, |a_N - 1|, |a_N + 1|\}.$$

Then $|a_n| \leq M$ for all $n \in \mathbb{N}$. Indeed, if $1 \leq n \leq N - 1$, then $|a_n|$ is one of the first $N - 1$ listed numbers and cannot be bigger than M. If $n \geq N$, then $a_N - 1 < a_n < a_N + 1$ so the distance between a_n and the origin cannot exceed the bigger of the numbers $|a_N - 1|$, and $|a_N + 1|$. Consequently, $|a_n| \leq M$. $\qquad\square$

Now we can prove Cauchy's Test. For convenience, we restate it here.

Theorem 2.7.14 (Cauchy's Test). *A sequence is a Cauchy sequence if and only if it is convergent.*

Proof. Suppose first that $\{a_n\}$ is a convergent sequence with $\lim a_n = L$, and let $\varepsilon > 0$. By definition, there exists a positive integer N such that, if $n \geq N$, $|a_n - L| < \varepsilon/2$. If $m \geq n \geq N$, then

$$|a_m - a_n| = |a_m - L + L - a_n| \leq |a_m - L| + |a_n - L| < \frac{\varepsilon}{2} + \frac{\varepsilon}{2} = \varepsilon,$$

so $\{a_n\}$ is a Cauchy sequence.

In order to prove the converse, we will assume that $\{a_n\}$ is a Cauchy sequence. By Theorem 2.7.13 this sequence is bounded. The Bolzano–Weierstrass Theorem implies that it has a convergent subsequence $\{a_{n_k}\}$. Let $\lim a_{n_k} = L$. We will show that the sequence $\{a_n\}$ is convergent and that $\lim a_n = L$.

Let $\varepsilon > 0$. First we select a positive integer N so that

$$|a_m - a_n| < \varepsilon/2, \text{ for } m \geq n \geq N. \tag{2.15}$$

Then we select a positive integer K so that

$$|a_{n_k} - L| < \varepsilon/2, \text{ for } k \geq K. \tag{2.16}$$

Let $n \geq N$, let $i = \max\{K, n\}$, and let $m = n_i$. Since $i \geq n$, we have that $m = n_i \geq n_n \geq n \geq N$. Consequently, inequality (2.15) holds. Also, $m = n_i$ and $i \geq K$, so inequality (2.16) implies that $|a_m - L| < \varepsilon/2$. It follows that

$$|a_n - L| = |a_n - a_m + a_m - L| \leq |a_n - a_m| + |a_m - L| < \frac{\varepsilon}{2} + \frac{\varepsilon}{2} = \varepsilon. \qquad \square$$

Exercises

2.7.1. Let $\{a_n\}$ be a convergent sequence and let $\lim a_n = L$. Prove that every subsequence of $\{a_n\}$ converges to L.

2.7.2. Let $\{a_n\}$ be an increasing sequence that has a bounded subsequence. Prove that the sequence $\{a_n\}$ is convergent.

2.7.3. Suppose that $\{a_n\}$ is a bounded sequence, and that all its convergent subsequences have the same limit L. Prove that $\{a_n\}$ converges to L.

2.7.4. Suppose that the subsequences $\{a_{2n}\}$ and $\{a_{2n+1}\}$ of $\{a_n\}$ both have limit L. Prove that $\{a_n\}$ converges to L.

2.7.5.* Prove that every sequence $\{a_n\}$ has a monotone subsequence.

2.7.6.* Prove that a sequence $\{a_n\}$ converges to a real number L if and only if every subsequence of $\{a_n\}$ has a subsequence converging to L.

2.7.7. Prove that a sequence that is not bounded above has a subsequence converging to $+\infty$.

2.7.8.* Find one convergent subsequence of the sequence $a_n = \sin n$.

2.7.9.* Suppose that $\mathbb{F}$ is an ordered Archimedean field. Prove that the following are equivalent:

(a) Completeness Axiom holds in $\mathbb{F}$.

(b) Every non-decreasing sequence in $\mathbb{F}$ that is bounded above must be convergent.

(c) Every sequence of nested, non-empty, closed, and finite intervals in $\mathbb{F}$ has a non-empty intersection.

(d) Every bounded sequence in $\mathbb{F}$ has a convergent subsequence.

(e) Every Cauchy sequence in $\mathbb{F}$ must be convergent.

2.8   Limit Superior and Limit Inferior

In the previous section we dealt with subsequences. Here, we will be especially interested in *convergent* subsequences of a given sequence.

Example 2.8.1. Convergent subsequences.

Let $a_n = 1 + \dfrac{n}{n+1} \cos \dfrac{n\pi}{2}$. What are the convergent subsequences of $\{a_n\}$?
It is helpful to consider $b_n = \cos \frac{n\pi}{2}$ for several values of n. We calculate

$$b_1 = \cos \frac{\pi}{2} = 0, \quad b_2 = \cos \pi = -1, \quad b_3 = \cos \frac{3\pi}{2} = 0, \quad b_4 = \cos 2\pi = 1,$$

$$b_5 = \cos \frac{5\pi}{2} = 0, \quad b_6 = \cos 3\pi = -1, \text{ etc.}$$

We see that the sequence $\{b_n\}$ is periodic, i.e., $b_{n+4} = b_n$, for all $n \in \mathbb{N}$. Trigonometric considerations confirm this:

$$b_{n+4} = \cos \frac{(n+4)\pi}{2} = \cos \frac{n\pi + 4\pi}{2} = \cos \left(\frac{n\pi}{2} + 2\pi \right) = \cos \frac{n\pi}{2} = b_n.$$

Since $b_1 = b_3 = 0$, we have that $b_{2k-1} = 0$, for all $k \in \mathbb{N}$. It follows that $a_{2k-1} = 1$, and it is obvious that the limit of this subsequence is 1. Further, $b_2 = -1$, so $b_{4k+2} = -1$, and

$$a_{4k+2} = 1 - \frac{4k+2}{4k+3} = \frac{1}{4k+3}.$$

The limit of $\{a_{4k+2}\}$ is 0. Finally, $b_4 = 1$, so $b_{4k} = 1$, and $a_{4k} = 1 + \frac{4k}{4k+1}$ which converges to 2.   ◆

Numbers like 0, 1, and 2 in Example 2.8.1 are of significance for the sequence $\{a_n\}$.

Definition 2.8.2. Let $\{a_n\}$ be a sequence, and let $\{a_{n_k}\}$ be a convergent subsequence of $\{a_n\}$, with $\lim a_{n_k} = c$. We say that c is an **accumulation point** of the sequence $\{a_n\}$.

Example 2.8.3. Accumulation points.

The sequence $a_n = (-1)^n$ has 2 accumulation points: 1 and -1.

Example 2.8.4. Accumulation points.

The sequence in Example 2.8.1 has 3 accumulation points: 0, 1, and 2.

Remark 2.8.5. When $\{a_n\}$ is convergent itself, with $\lim a_n = L$, we can take the subsequence to be the whole sequence, so L is an accumulation point. In other words, the limit is also an accumulation point. However, as the examples above show, the converse is not true.

We see that, when the sequence is convergent, it has only one accumulation point (its limit). Can there be a sequence without any accumulation points? Recall that, by the Bolzano–Weierstrass Theorem, if a sequence is bounded, it must have an accumulation point. So, we had better look at an unbounded sequence.

Example 2.8.6. A sequence without an accumulation point.

The sequence $a_n = n$ has no accumulation points.

Suppose, to the contrary, that c is an accumulation point of $\{a_n\}$, and that a subsequence $\{a_{n_k}\}$ converges to c. Notice that $a_{n_k} = n_k$, so it is an increasing sequence of positive integers. Let K be a positive integer with the property that, for $k \geq K$, $n_k > c$. Then, for $k \geq K + 1$, $n_k > c + 1$, so there are infinitely many members of $\{n_k\}$ outside of the interval $(c - 1, c + 1)$. (See Figure 2.9.) Consequently, c cannot be the limit of $\{a_{n_k}\}$.   ◆

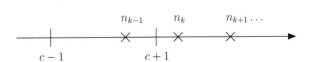

Figure 2.9: Infinitely many members of $\{n_k\}$ outside of $(c-1, c+1)$.

At the other end of the spectrum, one may ask whether it is possible that a sequence has infinitely many accumulation points.

Example 2.8.7. A sequence with infinitely many accumulation points.
Find the accumulation points of $1, \frac{1}{2}, 1, \frac{1}{2}, \frac{1}{3}, 1, \frac{1}{2}, \frac{1}{3}, \frac{1}{4}, 1, \frac{1}{2}, \frac{1}{3}, \frac{1}{4}, \frac{1}{5}, \ldots$.

Solution. We notice that $a_1 = a_3 = a_6 = a_{10} = \cdots = 1$, so we have that 1 is an accumulation point. How about $1/2$? Again, $a_2 = a_4 = a_7 = a_{11} = \cdots = 1/2$, so $1/2$ is also an accumulation point. In fact, the same is true of $1/3$, $1/4$, etc. So we can say that, for every $k \in \mathbb{N}$, the number $1/k$ is an accumulation point of a_n. Have we missed something? Clearly, we can select a subsequence $1, 1/2, 1/3, 1/4, \ldots$ that converges to 0. Therefore, the set of accumulation points includes 0 as well. ♦

If we denote by A the set of accumulation points in Example 2.8.7, then $\sup A = 1$ and $\inf A = 0$. We make an observation that both of these numbers are accumulation points. In other words, the sequence $\{a_n\}$ has both the largest and the smallest accumulation points. Will this always be the case?

Theorem 2.8.8. *For a bounded sequence $\{a_n\}$, there exist the largest and the smallest accumulation points.*

Proof. Let $V(a_n)$ denote the set of accumulation points of $\{a_n\}$. First we notice that, by Bolzano–Weierstrass Theorem, the set $V(a_n)$ cannot be empty. Next, we will show that $V(a_n)$ is a bounded set. We will do that by contrapositive: we will assume that $V(a_n)$ is not bounded, and we will prove that $\{a_n\}$ is not bounded. Let $M > 0$. We will show that there exists $n \in \mathbb{N}$ such that $a_n > M$. Since $V(a_n)$ is not bounded, there exists $a \in V(a_n)$ such that $a > M + 1$. The fact that a is an accumulation point of $\{a_n\}$ implies that there

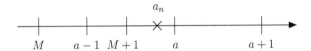

Figure 2.10: $a > M + 1$ implies $a_n > M$.

exists $n \in \mathbb{N}$ such that $|a_n - a| < 1$ or, equivalently, $a - 1 < a_n < a + 1$. Therefore,

$$a_n > a - 1 > (M+1) - 1 = M.$$

Thus, the sequence $\{a_n\}$ is not bounded. We conclude that $V(a_n)$ is a bounded set.

Now, we can use the Completeness Axiom. Let $L = \sup V(a_n)$. We will show that L is an accumulation point (hence the largest one). Let $\varepsilon > 0$. Then $L - \varepsilon/2$ is not an upper bound for $V(a_n)$, so there exists $a \in V(a_n)$ such that

$$L - \frac{\varepsilon}{2} < a \leq L.$$

Thus, $|L - a| < \varepsilon/2$. Since a is an accumulation point of $\{a_n\}$, there exists $n \in \mathbb{N}$ such that

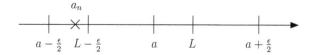

$$\text{Figure 2.11: } |L - a_n| < \varepsilon.$$

$|a_n - a| < \varepsilon/2$. Now, for that n,

$$|L - a_n| \le |L - a| + |a - a_n| < \frac{\varepsilon}{2} + \frac{\varepsilon}{2} = \varepsilon.$$

We conclude that L is an accumulation point of $\{a_n\}$, and we will leave the existence of the smallest accumulation point as an exercise. □

Cauchy used this result in *Cours d'Analyse*, but he never proved it.

Theorem 2.8.8 tells us that, if a sequence is bounded, then it has both a largest and a smallest accumulation point.

Definition 2.8.9. If $\{a_n\}$ is a bounded sequence, we call its largest accumulation point **the limit superior** of the sequence $\{a_n\}$, and we write $\limsup a_n$ (or $\overline{\lim} a_n$). We denote the smallest accumulation point of $\{a_n\}$ by $\liminf a_n$ (or $\underline{\lim} a_n$), and we call it **the limit inferior** of the sequence $\{a_n\}$. When $\{a_n\}$ is not bounded above, we will say that $\limsup a_n = +\infty$. When $\{a_n\}$ is not bounded below, we will say that $\liminf a_n = -\infty$.

Example 2.8.10. Limit superior and limit inferior.
    Let $\{a_n\}$ be the sequence
$$1, 2, 1, 3, 1, 4, 1, 5, \dots.$$
Then $\limsup a_n = +\infty$ and $\liminf a_n = 1$.   ◆

Limit superior (as well as limit inferior) share many properties with the usual limit. The following theorem shows some of these.

Theorem 2.8.11. *Let $\{a_n\}$, $\{b_n\}$ be bounded sequences, and let $\alpha \ge 0$, $\beta \le 0$. Then:*

(a) $\limsup(\alpha a_n) = \alpha \limsup a_n$;

(b) $\limsup(\beta a_n) = \beta \liminf a_n$;

(c) $\limsup(a_n + b_n) \le \limsup a_n + \limsup b_n$.

Proof. Both (a) and (b) are the consequence of the following statement: if γ is a non-zero real number, then a is an accumulation point of $\{a_n\}$ if and only if γa is an accumulation point of $\{\gamma a_n\}$. The statement is correct since a is an accumulation point of $\{a_n\}$ if and only if there exists a subsequence $\{a_{n_k}\}$ converging to a, which is equivalent to $\{\gamma a_{n_k}\}$ converging to γa. Of course, this is the case if and only if γa is an accumulation point of $\{\gamma a_n\}$. If we denote by $V(a_n)$, the set of accumulation points of a sequence $\{a_n\}$, then we have just proved that
$$\gamma V(a_n) = V(\gamma a_n),$$
if $\gamma \neq 0$. It is easy to see that the equality is also true when $\gamma = 0$, because both sides are the singleton $\{0\}$. Let w be the largest accumulation point of $\{a_n\}$. If $\gamma \ge 0$, then γw is the largest element of $\gamma V(a_n)$, hence of $V(\gamma a_n)$. In other words, $\gamma w = \limsup(\gamma a_n)$,

which is (a). If $\gamma \leq 0$, then γw is the *smallest* element of $\gamma V(a_n)$, and it follows that $\gamma w = \liminf(\gamma a_n)$. This settles (b).

In order to prove (c), let $\varepsilon > 0$, and consider the number $z = \limsup a_n + \varepsilon/2$. It cannot be an accumulation point of the sequence $\{a_n\}$, so there exists $\delta > 0$ such that the interval $(z - \delta, z + \delta)$ contains at most a finite number of elements of $\{a_n\}$. Therefore, there exists

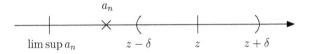

Figure 2.12: $a_n \leq z - \delta$ for $n \geq N_1$.

$N_1 \in \mathbb{N}$ such that

$$n \geq N_1 \quad \Rightarrow \quad a_n \leq z - \delta < z = \limsup a_n + \frac{\varepsilon}{2}.$$

Similarly, there exists $N_2 \in \mathbb{N}$ such that

$$n \geq N_2 \quad \Rightarrow \quad b_n < \limsup b_n + \frac{\varepsilon}{2}.$$

Let $N = \max\{N_1, N_2\}$ and let $n \geq N$. Then $a_n + b_n < \limsup a_n + \limsup b_n + \varepsilon$, so any accumulation point of $\{a_n + b_n\}$ (including the largest) cannot exceed $\limsup a_n + \limsup b_n + \varepsilon$. Consequently,

$$\limsup(a_n + b_n) \leq \limsup a_n + \limsup b_n + \varepsilon.$$

Since ε is arbitrary, (c) is proved. □

Parts (a) and (b) show that the rules for the limit superior and limit inferior are not the same as for the limits. Part (c) is even worse: where we were hoping for an equality, we have an inequality. Now, is it because proving the equality might be too hard, or could it be that the left and the right sides are not always equal?

Example 2.8.12. A strict inequality in Theorem 2.8.11 (c).
Let $a_n = (-1)^n$, $b_n = (-1)^{n+1}$. Both sequences consist of alternating 1 and -1, so $\limsup a_n = \limsup b_n = 1$. It follows that $\limsup a_n + \limsup b_n = 2$. However, $a_n + b_n = 0$, for every $n \in \mathbb{N}$. Thus, $\limsup(a_n + b_n) = 0$. ◆

LEONHARD EULER (1707–1783) was a Swiss mathematician, probably the best mathematician in the eighteenth century and one of the greatest of all time. He inherited the interest for mathematics from his father who, while studying to be a minister, lived in the house of Jacob Bernoulli and attended his lectures. Euler learned mathematics from his father and later on his own, even taking private lessons. He enrolled at the University of Basel to study theology following his father's wishes. On the recommendation of the family friend Johann Bernoulli, he transitioned into mathematics graduating in 1726. Already in 1727 he placed second for the Grand Prize of the Paris Academy. The topic was the best arrangement of masts on a ship. He accepted a position at the University of Saint Petersburg where he taught physics and, after Daniel Bernoulli (page 262) left, mathematics. This was the time when he established himself as an extraordinary mathematician, winning the Grand Prize of the Paris Academy in 1738 and 1740. He was offered a position of director of mathematics at the Berlin Academy which he accepted in 1741 when the changes at the Russian throne made the future uncertain for foreigners. Euler spent 25 years in Berlin producing about 380 articles. In addition, he supervised the observatory and the botanical gardens; selected the personnel; oversaw finances; managed the publication of various calendars and maps, etc. His relationship with king Frederick II had over time cooled down. When the position of the president of the Academy became vacant Euler was not promoted. Even worse, it was offered to d'Alembert (page 201) who declined, so Euler returned to Saint Petersburg in 1766. His eyesight was already bad by that time, and he became completely blind in 1771. Nevertheless, he produced almost a half of his work after that. He was helped by a group of scientists, including his two sons, who developed Euler's ideas, did calculations, and compiled examples. When he died, the Saint Petersburg Academy continued to publish his unpublished work for nearly 50 more years. He made important contributions in physics, astronomy, music, logic, graph theory and number theory. He brought together the work of Leibniz and Newton creating mathematical analysis. He introduced the concept of a function and the notation $f(x)$, and he was the first to consider sin and cos as functions. We owe to him Euler's constant (Remark 2.5.7), Euler-MacLaurin summation formula, Beta and Gamma functions (Section 13.5), the symbols e, i, $\sum$, Δy, and he popularized the use of π. In differential equations he introduced the integrating factor, the variation of parameters, and power series solutions. Laplace is credited with the statement: *Read Euler, read Euler, he is the master of us all.*

Exercises

2.8.1. Let $\{a_n\}$ be a bounded sequence such that $\limsup a_n = \liminf a_n$. Prove that $\{a_n\}$ is a convergent sequence.

2.8.2. Let $\{a_n\}$ be a sequence of positive numbers. Prove that $\limsup\left(\dfrac{1}{a_n}\right) = \dfrac{1}{\liminf a_n}$.

2.8.3.* Let $\{a_n\}$ be a bounded sequence. Prove that $L = \limsup a_n$ if and only if:

(a) For every $\varepsilon > 0$ there exists $N \in \mathbb{N}$ such that $a_n < L + \varepsilon$, for all $n \geq N$, and

(b) For every $\varepsilon > 0$ and every $N \in \mathbb{N}$ there exists $n > N$ such that $L - \varepsilon < a_n$.

2.8.4.* Let $\{a_n\}$ be a bounded sequence. Prove that $L = \liminf a_n$ if and only if:

(a) For every $\varepsilon > 0$ there exists $N \in \mathbb{N}$ such that $a_n > L - \varepsilon$, for all $n \geq N$, and

(b) For every $\varepsilon > 0$ and every $N \in \mathbb{N}$ there exists $n > N$ such that $a_n < L + \varepsilon$.

In Exercises 2.8.5–2.8.7 determine the set of accumulation points of the sequence $\{a_n\}$.

2.8.5.* $a_n = \dfrac{2n^2}{7} - \left\lfloor \dfrac{2n^2}{7} \right\rfloor$. **2.8.6.** $a_n = n^{(-1)^n n}$.

2.8.7.* $a_{n+1} = \begin{cases} \dfrac{a_n}{2}, & \text{if } n \text{ is even} \\ \dfrac{1 + a_n}{2}, & \text{if } n \text{ is odd.} \end{cases}$

2.8.8. Let $\{a_n\}$, $\{b_n\}$ be two bounded sequences, and suppose that $a_n \leq b_n$ for all $n \in \mathbb{N}$. Prove that $\limsup a_n \leq \limsup b_n$, and $\liminf a_n \leq \liminf b_n$.

2.8.9. Let $a_n, b_n \geq 0$ for all $n \in \mathbb{N}$. Prove that:

(a) $\limsup(a_n b_n) \leq \limsup a_n \limsup b_n$.

(b) $\liminf a_n \liminf b_n \leq \liminf(a_n b_n)$.

(c) $\liminf(a_n b_n) \leq \liminf a_n \limsup b_n$.

(d) $\liminf a_n \limsup b_n \leq \limsup(a_n b_n)$.

Give examples to show that each inequality may be strict. Give examples to show that the assumption $a_n, b_n \geq 0$ cannot be relaxed.

2.8.10.* Prove that every uncountable subset of $\mathbb{R}$ must have an accumulation point.

2.8.11.* Let $a_n > 0$ for all $n \in \mathbb{N}$. Prove that

$$\liminf \frac{a_{n+1}}{a_n} \le \liminf \sqrt[n]{a_n} \le \limsup \sqrt[n]{a_n} \le \limsup \frac{a_{n+1}}{a_n}.$$

Give examples to show that each inequality may be strict.

2.8.12. Prove that a bounded sequence has the smallest accumulation point.

2.8.13.* Suppose that $\{a_n\}$ is a sequence of positive numbers. Then

$$\limsup(a_n b_n) = \limsup a_n \limsup b_n$$

for each positive sequence $\{b_n\}$ if and only if the sequence $\{a_n\}$ is convergent.

2.8.14.* Prove the identities: (a) $\liminf x_n = \lim_{n \to \infty} \left(\inf_{m \ge n} x_m \right)$; (b) $\limsup x_n = \lim_{n \to \infty} \left(\sup_{m \ge n} x_m \right)$.

2.9 Some Interesting Limits

In this section we will consider some harder problems concerning limits. Admittedly, some of them become easy when derivatives, and in particular L'Hôpital's Rule are used. In this text, we will avoid such tools until they are firmly established (with a proof).

Example 2.9.1. $a_n = \sqrt[n]{a}$, $a > 0$.

Solution. In order to compute this limit we consider first the case $a \ge 1$. We will apply Bernoulli's Inequality (Theorem 2.5.1) with $x = \sqrt[n]{a} - 1$. This yields the inequality

$$a = (1 + x)^n \ge 1 + nx = 1 + n(\sqrt[n]{a} - 1),$$

which can be combined with the fact that $\sqrt[n]{a} - 1 \ge 0$ to obtain

$$0 \le \sqrt[n]{a} - 1 \le \frac{a - 1}{n}. \tag{2.17}$$

Now the Squeeze Theorem implies that $\lim a_n = 1$. When $0 < a < 1$, we have $1/a > 1$ so $\lim \sqrt[n]{1/a} = 1$, and Theorem 2.3.9 implies that, once again, $\lim \sqrt[n]{a} = 1$. ◆

Example 2.9.2. $a_n = \dfrac{n^k}{a^n}$, $a > 1$, $k \in \mathbb{N}$.

Solution. Let $b = \sqrt[k]{a}$. Clearly, $b > 1$ and

$$0 < \frac{n^k}{a^n} = \left(\frac{n}{\sqrt[k]{a^n}} \right)^k = \left(\frac{n}{b^n} \right)^k.$$

We will show that $n/b^n \to 0$. If we write $b = 1 + (b - 1)$, then

$$b^n = \sum_{i=0}^{n} \binom{n}{i} (b - 1)^i.$$

Since $b > 1$, all terms in this sum are positive, so

$$b^n > \binom{n}{2} (b - 1)^2 = \frac{n(n-1)}{2} (b - 1)^2. \tag{2.18}$$

Now

$$0 < \frac{n}{b^n} < \frac{2n}{n(n-1)(b-1)^2} = \frac{2}{(n-1)(b-1)^2} \to 0.$$

Thus, $n/b^n \to 0$ which implies that $a_n \to 0$. ◆

Example 2.9.3. $a_n = \dfrac{\ln n}{n}$.

Solution. We have established in the previous exercise that, when $b > 1$, $n/b^n \to 0$. Therefore, when n is large enough, we have

$$\frac{1}{b^n} < \frac{n}{b^n} < 1.$$

Let $\varepsilon > 0$, and let $b = e^\varepsilon$. Then

$$\frac{1}{e^{n\varepsilon}} < \frac{n}{e^{n\varepsilon}} < 1$$

which, after multiplying by $e^{n\varepsilon}$, becomes $1 < n < e^{n\varepsilon}$. If we apply the natural logarithm to this inequality we obtain that $0 < \ln n < \ln e^{n\varepsilon} = n\varepsilon \ln e = n\varepsilon$. Dividing by n yields

$$0 < \frac{\ln n}{n} < \varepsilon.$$

Since ε was arbitrary, we conclude that $\lim a_n = 0$.   ◆

Example 2.9.4. $a_n = \sqrt[n]{n}$.

Solution. If we take $b = \sqrt[n]{n}$ and apply (2.18), we obtain

$$n > \frac{n(n-1)}{2} \left(\sqrt[n]{n} - 1 \right)^2.$$

It follows that

$$0 < \sqrt[n]{n} - 1 < \sqrt{\frac{2}{n-1}} \to 0,$$

so $\lim \sqrt[n]{n} = 1$.   ◆

Example 2.9.5. $a_n = \dfrac{1}{\sqrt[n]{n!}}$.

Solution. We will show first that $n! > (n/3)^n$, using induction. When $n = 1$ we have $1 > 1/3$, so suppose that it is true for n, and let us establish this inequality for $n+1$. Using the hypothesis,

$$(n+1)! = (n+1)n! > (n+1) \left(\frac{n}{3} \right)^n = \left(\frac{n+1}{3} \right)^{n+1} \frac{3}{\left(1 + \frac{1}{n} \right)^n} > \left(\frac{n+1}{3} \right)^{n+1}.$$

Therefore, $n! > (n/3)^n$ for all $n \in \mathbb{N}$. Consequently,

$$0 < \frac{1}{\sqrt[n]{n!}} < \frac{3}{n} \to 0$$

and $\lim a_n = 0$.   ◆

Example 2.9.6. Let $\lim a_n = a$ and $b_n = \dfrac{a_1 + a_2 + \cdots + a_n}{n}$. Prove that the sequence $\{b_n\}$ is convergent and that $\lim b_n = a$.

Solution. Let $\varepsilon > 0$. Then, there exists $N_1 \in \mathbb{N}$, such that for any $n \geq N_1$, we have

$$|a_n - a| < \frac{\varepsilon}{2}.$$

Further, $\lim 1/n = 0$ so there exists $N_2 \in \mathbb{N}$, such that for any $n \geq N_2$, we have

$$\frac{1}{n} < \frac{\varepsilon}{2 \left(|a_1 - a| + |a_2 - a| + \cdots + |a_{N_1} - a| \right)}.$$

Now, let $n \geq N = \max\{N_1, N_2\}$. Then

$$
\begin{aligned}
|b_n - a| &= \left| \frac{a_1 + a_2 + \cdots + a_n}{n} - a \right| \\
&= \left| \frac{(a_1 - a) + (a_2 - a) + \cdots + (a_{N_1} - a) + (a_{N_1+1} - a) + \cdots + (a_n - a)}{n} \right| \\
&\leq \frac{|a_1 - a| + |a_2 - a| + \cdots + |a_{N_1} - a|}{n} + \frac{|a_{N_1+1} - a| + \cdots + |a_n - a|}{n} \\
&< \frac{\varepsilon}{2} + \frac{\frac{\varepsilon}{2} + \frac{\varepsilon}{2} + \cdots + \frac{\varepsilon}{2}}{n} = \frac{\varepsilon}{2} + \frac{\varepsilon}{2} = \varepsilon. \qquad \blacklozenge
\end{aligned}
$$

Example 2.9.7. Prove that the sequence $a_n = \sin n$ is not convergent.

Solution. Suppose, to the contrary, that the $\lim \sin n = L$ exists. We will use the trigonometric formula

$$
\sin a - \sin b = 2 \cos \frac{a+b}{2} \sin \frac{a-b}{2}.
$$

If we take $a = n + 2$ and $b = n$, then

$$
0 = \lim \sin(n + 2) - \lim \sin n = \lim (\sin(n + 2) - \sin n) = \lim 2 \cos(n + 1) \sin 1,
$$

so it follows that $\lim \cos n = 0$. Another trigonometric formula

$$
\cos a - \cos b = -2 \sin \frac{a+b}{2} \sin \frac{a-b}{2}
$$

applied to $a = n + 2$ and $b = n$ yields

$$
0 = \lim \cos(n + 2) - \lim \cos n = \lim (\cos(n + 2) - \cos n) = -2 \lim \sin(n + 1) \sin 1.
$$

From here we see that $\lim \sin n = 0$. This is a contradiction because $\sin^2 n + \cos^2 n = 1$ implies that

$$
1 = \lim \left(\sin^2 n + \cos^2 n \right) = \lim \sin^2 n + \lim \cos^2 n = 0.
$$

Therefore, the sequence $a_n = \sin n$ does not have a limit. $\qquad \blacklozenge$

Example 2.9.8. Let $\lim a_n = +\infty$. Prove that $\lim \left(1 + \dfrac{1}{a_n} \right)^{a_n} = e$.

Solution. First we make the observation that, if n_k is any increasing sequence of positive integers then

$$
\lim_{k \to \infty} \left(1 + \frac{1}{n_k + 1} \right)^{n_k} = \lim_{k \to \infty} \left(1 + \frac{1}{n_k} \right)^{n_k + 1} = \lim_{k \to \infty} \left(1 + \frac{1}{n_k} \right)^{n_k} = e. \qquad (2.19)
$$

Next, we will establish the inequality

$$
\left(1 + \frac{1}{\lfloor a_n \rfloor + 1} \right)^{\lfloor a_n \rfloor} \leq \left(1 + \frac{1}{a_n} \right)^{a_n} \leq \left(1 + \frac{1}{\lfloor a_n \rfloor} \right)^{\lfloor a_n \rfloor + 1}
$$

and the result will then follow from (2.19) and the Squeeze Theorem. Notice that $\lfloor a_n \rfloor + 1 \geq a_n$ and $\lfloor a_n \rfloor \leq a_n$, so

$$
\left(1 + \frac{1}{\lfloor a_n \rfloor + 1} \right)^{\lfloor a_n \rfloor} \leq \left(1 + \frac{1}{a_n} \right)^{\lfloor a_n \rfloor} \leq \left(1 + \frac{1}{a_n} \right)^{a_n}.
$$

Also, $a_n \geq \lfloor a_n \rfloor$ and $a_n \leq \lfloor a_n \rfloor + 1$, so

$$
\left(1 + \frac{1}{a_n} \right)^{a_n} \leq \left(1 + \frac{1}{\lfloor a_n \rfloor} \right)^{a_n} \leq \left(1 + \frac{1}{\lfloor a_n \rfloor} \right)^{\lfloor a_n \rfloor + 1}. \qquad \blacklozenge
$$

Example 2.9.9. Find $\lim n \sin(2\pi e n!)$.

Solution. By Lemma 2.5.6, $e = c_n + \theta_n/n!$, with $c_n = 1 + 1 + \frac{1}{2!} + \frac{1}{3!} + \cdots + \frac{1}{n!}$ and $0 < \theta_n < 1$. Therefore,

$$\sin(2\pi e n!) = \sin\left[2\pi(c_n n! + \theta_n)\right] = \sin(2\pi\theta_n),$$

because $c_n n!$ is an integer. Further, the proof of Lemma 2.5.6 (see Exercise 2.5.9) shows that $\theta_n < \frac{1}{n+1}$, so $\theta_n \to 0$. As a consequence,

$$\lim \frac{\sin(2\pi\theta_n)}{2\pi\theta_n} = 1.$$

Finally,

$$\theta_n = (e - c_n)n! \geq (c_{n+1} - c_n)n! = \frac{1}{(n+1)!}\, n! = \frac{1}{n+1}, \quad \text{so}$$

$$\frac{n}{n+1} \leq n\theta_n \leq 1,$$

and we conclude that

$$n \sin(2\pi e n!) = n\, \frac{\sin(2\pi\theta_n)}{2\pi\theta_n}\, (2\pi\theta_n) \to 2\pi. \qquad \blacklozenge$$

Example 2.9.10. Suppose that $\{a_n\}$ is a sequence such that $0 \leq a_{n+m} \leq a_n a_m$, for all $m, n \in \mathbb{N}$. Prove that the sequence $\{\sqrt[n]{a_n}\}$ is convergent.

Solution. Using induction on n, we will prove that $0 \leq a_n \leq a_1^n$. The case $n = 1$ is obvious and so is the non-negativity of each a_n. Suppose that $a_n \leq a_1^n$, for some $n \in \mathbb{N}$. Then, substituting $m = 1$ in $a_{m+n} \leq a_m a_n$, we have that

$$a_{n+1} \leq a_n a_1 \leq a_1 \cdot a_1^n = a_1^{n+1}.$$

It follows that $0 \leq \sqrt[n]{a_n} \leq a_1$, so the sequence $\{a_n\}$ is bounded. Let $L = \limsup \sqrt[n]{a_n}$. We will show that, for each $n \in \mathbb{N}$, $\sqrt[n]{a_n} \geq L$. It will follow that $\liminf \sqrt[n]{a_n} \geq \limsup \sqrt[n]{a_n}$, i.e., that $\sqrt[n]{a_n}$ is convergent.

So, let $n \in \mathbb{N}$ be arbitrary, and let $\{a_{n_k}\}$ be a subsequence of $\{a_n\}$ such that $\{\sqrt[n_k]{a_{n_k}}\}$ converges to L. For each $k \in \mathbb{N}$, we write

$$n_k = q_k n + r_k, \quad \text{with } q_k \in \mathbb{N}_0 \quad \text{and} \quad r_k \in \{0, 1, 2, \ldots, n-1\}.$$

Then

$$a_{n_k} = a_{q_k n + r_k} \leq a_{q_k n} a_{r_k} \leq a_n^{q_k} a_{r_k}, \quad \text{so}$$

$$\sqrt[n_k]{a_{n_k}} \leq a_n^{q_k/n_k} \sqrt[n_k]{a_{r_k}}.$$

For any $k \in \mathbb{N}$, a_{r_k} lies between the smallest and the largest of n numbers $a_{r_0}, a_{r_1}, \ldots, a_{r_{n-1}}$, so $\sqrt[n_k]{a_{r_k}} \to 1$. Also, $n_k/q_k = n + r_k/q_k \to n$, so we obtain that $L \leq \sqrt[n]{a_n}$. $\qquad \blacklozenge$

Exercises

In Exercises 2.9.1–2.9.6 find the limit:

2.9.1. $\lim \sqrt[n]{1^7 + 2^7 + \cdots + n^7}$.   2.9.2. $\lim(2^n + 3^n)^{1/n}$.

2.9.3.* $\lim\{(2 + \sqrt{3})^n\}$, where $\{x\} = x - \lfloor x \rfloor$.

2.9.4. $\lim n q^n$, if $|q| < 1$.   2.9.5. $\lim(1 + 3n)^{1/n}$.   2.9.6. $\lim 2^{-1/\sqrt{n}}$.

2.9.7.* Given $x \geq 1$, show that $\lim(2\sqrt[n]{x} - 1)^n = x^2$.

2.9.8.* Let $\{a_n\}$ be a sequence of positive numbers. Prove that

$$\limsup \left(\frac{a_1 + a_{n+1}}{a_n} \right)^n \geq e.$$

2.9.9.* Suppose that the terms of the sequence $\{a_n\}$ satisfy the inequalities $0 \leq a_{n+m} \leq a_n + a_m$. Prove that the sequence $\{a_n/n\}$ converges.

2.9.10.* (Stolz) Let $\{b_n\}$ be a strictly increasing sequence satisfying $\lim b_n = +\infty$, and let $\{a_n\}$ be a sequence of real numbers such that $\lim(a_{n+1} - a_n)/(b_{n+1} - b_n) = L$. Prove that $\lim a_n/b_n = L$.

In Exercises 2.9.11–2.9.17 find the limits.

2.9.11. $\lim \dfrac{1}{\sqrt{n}} \left(1 + \dfrac{1}{\sqrt{2}} + \cdots + \dfrac{1}{\sqrt{n}} \right).$ 2.9.12. $\lim \dfrac{n}{a^{n+1}} \left(a + \dfrac{a^2}{2} + \cdots + \dfrac{a^n}{n} \right), a > 1.$

2.9.13.* $\lim \dfrac{1^k + 2^k + \cdots + n^k}{n^{k+1}}, k \in \mathbb{N}.$ 2.9.14.* $\lim \dfrac{1}{\sqrt{n}} \left(\dfrac{1}{\sqrt{n}} + \dfrac{1}{\sqrt{n+1}} + \cdots + \dfrac{1}{\sqrt{2n}} \right).$

2.9.15.* $\lim \dfrac{1}{n^{k+1}} \left(k! + \dfrac{(k+1)!}{1!} + \cdots + \dfrac{(k+n)!}{n!} \right), k \in \mathbb{N}.$

2.9.16. $\lim \dfrac{1 + 1 \cdot a + 2 \cdot a^2 + \cdots + n \cdot a^n}{n \cdot a^{n+1}}, a > 1.$

2.9.17.* $\lim \left[\dfrac{1}{n^k} \left(1^k + 2^k + \cdots + n^k \right) - \dfrac{n}{k+1} \right], k \in \mathbb{N}.$

2.9.18. Let $\lim a_n = L$. Find $\lim \dfrac{1}{\sqrt{n}} \left(a_1 + \dfrac{a_2}{\sqrt{2}} + \dfrac{a_3}{\sqrt{3}} + \cdots + \dfrac{a_n}{\sqrt{n}} \right).$

2.9.19. Let $\{a_n\}$ be a sequence such that $\lim(a_{n+1} - a_n) = L$. Prove that $\lim a_n/n = L$.

3

Continuity

Until the end of the eighteenth century the continuity was built in the concept of a function. The work of Fourier on problems in thermodynamics brought forward functions that were not continuous. The nineteenth century saw plenty of results about continuity by the leading mathematicians of the period: Cauchy, Bolzano, Weierstrass, and others. They were all based on the solid understanding of the limit.

3.1   Computing Limits of Functions: Review

We will review some of the rules of calculating limits of functions.

Example 3.1.1. $\lim_{x \to 2} (3x - 1)$.

Solution. We use several rules: "the limit of the sum/difference equals the sum/difference of the limits" and "the limit of a product equals the product of limits":

$$\lim_{x \to 2} (3x - 1) = \lim_{x \to 2} (3x) - \lim_{x \to 2} 1 = \lim_{x \to 2} 3 \lim_{x \to 2} x - \lim_{x \to 2} 1.$$

Further, we use the rules that, if c is a constant, then $\lim_{x \to a} c = c$, and $\lim_{x \to a} x = a$. Therefore,

$$\lim_{x \to 2} (3x - 1) = 3 \cdot 2 - 1 = 5.$$ ◆

Example 3.1.2. $\lim_{x \to 2} 2^{3x-1}$.

Solution. We have already calculated that $\lim_{x \to 2} (3x - 1) = 5$, so $\lim_{x \to 2} 2^{3x-1} = 2^5 = 32$. ◆

Remark 3.1.3. We have here used the rule that $\lim_{x \to 2} 2^{f(x)} = 2^{\lim_{x \to 2} f(x)}$. It is a nice feature of the exponential function 2^x that allows such a rule and it has a name: *continuity*. We have made a similar observation for the logarithms in Remark 2.1.6 and for the square roots in Example 2.4.9.

Example 3.1.4. $\lim_{x \to -1} \dfrac{x^3 - 5x + 7}{x^2 + 3x - 4}$.

Solution. In addition to already mentioned rules, we will use "the limit of a quotient equals the quotient of limits". Thus,

$$\lim_{x \to -1} \frac{x^3 - 5x + 7}{x^2 + 3x - 4} = \frac{\lim_{x \to -1} (x^3 - 5x + 7)}{\lim_{x \to -1} (x^2 + 3x - 4)}.$$

Now,

$$\lim_{x \to -1} (x^3 - 5x + 7) = \lim_{x \to -1} x^3 - \lim_{x \to -1} 5x + \lim_{x \to -1} 7 = (-1)^3 - 5(-1) + 7 = 11, \text{ and}$$

$$\lim_{x \to -1} (x^2 + 3x - 4) = \lim_{x \to -1} x^2 + \lim_{x \to -1} 3x - \lim_{x \to -1} 4 = (-1)^2 + 3(-1) - 4 = -6, \text{ so}$$

$$\lim_{x \to -1} \frac{x^3 - 5x + 7}{x^2 + 3x - 4} = -\frac{11}{6}. \qquad \blacklozenge$$

Example 3.1.5. $\lim_{x \to 1} \dfrac{x^3 - 5x + 7}{x^2 + 3x - 4}$.

Solution. Now we cannot use the rule for quotients, because the limit of the denominator is 0:

$$\lim_{x \to 1} (x^2 + 3x - 4) = \lim_{x \to 1} x^2 + \lim_{x \to 1} 3x - \lim_{x \to 1} 4 = 1^2 + 3 \cdot 1 - 4 = 0.$$

On the other hand,

$$\lim_{x \to 1} (x^3 - 5x + 7) = \lim_{x \to 1} x^3 - \lim_{x \to 1} 5x + \lim_{x \to 1} 7 = 1^3 - 5 \cdot 1 + 7 = 3.$$

Therefore the limit does not exist. $\qquad \blacklozenge$

Rule. *When the limit of the denominator is 0 and the limit of the numerator is not 0, the limit does not exist.*

Example 3.1.6. $\lim_{x \to 1} \dfrac{x^3 - 8x + 7}{x^2 + 3x - 4}$.

Solution. The difference between this limit and the one in Example 3.1.5, is that now the numerator has also the limit 0:

$$\lim_{x \to 1} (x^3 - 8x + 7) = \lim_{x \to 1} x^3 - \lim_{x \to 1} 8x + \lim_{x \to 1} 71^3 - 8 \cdot 1 + 7 = 0.$$

Here, we will use the L'Hôpital's Rule.

$$\lim_{x \to 1} \frac{x^3 - 8x + 7}{x^2 + 3x - 4} = \lim_{x \to 1} \frac{(x^3 - 8x + 7)'}{(x^2 + 3x - 4)'} = \lim_{x \to 1} \frac{3x^2 - 8}{2x + 3}.$$

Now we can use the rule for quotients, and we obtain

$$\lim_{x \to 1} \frac{x^3 - 8x + 7}{x^2 + 3x - 4} = \frac{\lim_{x \to 1} (3x^2 - 8)}{\lim_{x \to 1} (2x + 3)} = \frac{3 \cdot 1^2 - 8}{5} = \frac{-5}{5} = -1. \qquad \blacklozenge$$

Rule. *When the limits of both the denominator and the numerator are 0, we can use the L'Hôpital's Rule.*

Remark 3.1.7. Although the stated rule is correct, strictly speaking we should not use it, because we are far from having it proved. In fact, in the long chain of results that will be used in the proof of the L'Hôpital's Rule, some of the results will be about limits. More significantly, the L'Hôpital's Rule is not very intuitive, and does not help us understand the nature of the limit. Until we prove it, we will refrain from using it.

Example 3.1.8 (Example 3.1.6 redone). $\lim_{x \to 1} \dfrac{x^3 - 8x + 7}{x^2 + 3x - 4}$.

Solution. We will factor the numerator and the denominator. The fact that both take the value 0 when $x = 1$ implies that one of the factors must be $x - 1$:

$$x^3 - 8x + 7 = (x - 1)(x^2 + x - 7) \text{ and } x^2 + 3x - 4 = (x - 1)(x + 4).$$

Further, the fact that $x \to 1$ means that $x \neq 1$, so

$$\frac{x^3 - 8x + 7}{x^2 + 3x - 4} = \frac{x^2 + x - 7}{x + 4}.$$

We obtain

$$\lim_{x \to 1} \frac{x^3 - 8x + 7}{x^2 + 3x - 4} = \lim_{x \to 1} \frac{x^2 + x - 7}{x + 4} = \frac{1^2 + 1 - 7}{1 + 4} = \frac{-5}{5} = -1.$$

Example 3.1.9. $\lim\limits_{x \to \infty} \left(1 + \dfrac{1}{x}\right)^x = e.$

Solution. By Example 2.9.8,

$$\lim_{n \to \infty} \left(1 + \frac{1}{a_n}\right)^{a_n} = e,$$

for *any* sequence a_n such that $a_n \to \infty$. Theorem 3.4.9 will show that this implies that $\lim_{x \to \infty} \left(1 + \frac{1}{x}\right)^x = e$.

Example 3.1.10. $\lim\limits_{x \to -\infty} \left(1 + \dfrac{1}{x}\right)^x = e.$

Solution. If we use a substitution $x = -t$, then $t \to +\infty$ and the desired equality becomes

$$\lim_{t \to \infty} \left(1 + \frac{1}{-t}\right)^{-t} = e.$$

Now $1 + \frac{1}{-t} = \frac{t-1}{t}$ so

$$\left(1 + \frac{1}{-t}\right)^{-t} = \left(\frac{t-1}{t}\right)^{-t} = \left(\frac{t}{t-1}\right)^{t} = \left(1 + \frac{1}{t-1}\right)^{t} = \left(1 + \frac{1}{t-1}\right)^{t-1} \left(1 + \frac{1}{t-1}\right).$$

If $t - 1 = s$, then $s \to \infty$, so it remains to prove that

$$\lim_{s \to \infty} \left(1 + \frac{1}{s}\right)^{s} \left(1 + \frac{1}{s}\right) = e.$$

This follows directly from Example 3.1.9.

Example 3.1.11. $\lim\limits_{x \to 0} (1 + x)^{1/x} = e.$

Solution. In order to prove this equality, we will consider separately the left limit and the right limit. In both we will use the substitution $x = 1/t$. When $x \to 0+$, we have that $t \to +\infty$ and the desired equality becomes

$$\lim_{t \to +\infty} \left(1 + \frac{1}{t}\right)^{t} = e$$

which was established in Example 3.1.9. Similarly, When $x \to 0-$, we have that $t \to -\infty$, so we obtain

$$\lim_{t \to -\infty} \left(1 + \frac{1}{t}\right)^{t} = e$$

which was established in Example 3.1.10.

Example 3.1.12. $\lim\limits_{x \to 0} \dfrac{\log_a(1+x)}{x} = \log_a e.$

Solution. It is easy to see that

$$\frac{\log_a(1+x)}{x} = \log_a(1+x)^{1/x}$$

and the result now follows from Example 3.1.11 and the continuity of $\log_a x$.   ◆

Example 3.1.13. $\lim\limits_{x \to 0} \dfrac{\ln(1+x)}{x} = 1.$

Solution. The result is an immediate application of Example 3.1.12 with $a = e$.   ◆

Example 3.1.14. $\lim\limits_{x \to 0} \dfrac{a^x - 1}{x} = \ln a.$

Solution. If we introduce the substitution $u = a^x - 1$, then $x = \log_a(u+1)$ and $u \to 0$, so the equality to prove becomes

$$\lim_{u \to 0} \frac{u}{\log_a(u+1)} = \ln a.$$

By Example 3.1.12, this limit equals $1/\log_a e = \ln a$. (The last equality is a consequence of the identity $\log_a b \log_b a = 1$, which is true for all $a, b > 0$.)   ◆

Example 3.1.15. $\lim\limits_{x \to 0} \dfrac{e^x - 1}{x} = 1.$

Solution. This follows from the result of Example 3.1.14 with $a = e$.   ◆

Example 3.1.16. $\lim\limits_{x \to 0} \dfrac{(1+x)^\alpha - 1}{x} = \alpha.$

Solution. The trick is to write

$$\frac{(1+x)^\alpha - 1}{x} = \alpha \, \frac{\ln(1+x)}{x} \, \frac{(1+x)^\alpha - 1}{\ln(1+x)^\alpha}.$$

Then Example 3.1.13 shows that the first fraction has limit 1. For the second fraction, the substitution $u = (1+x)^\alpha - 1$ yields $u \to 0$ and

$$\lim_{u \to 0} \frac{u}{\ln(u+1)} = 1$$

again by Example 3.1.13.   ◆

Example 3.1.17. $\lim\limits_{x \to 0} \dfrac{\sin x}{x} = 1.$

Solution. Both the numerator and the denominator have limit 0, so we cannot apply the rule for quotients. Our goal is to establish that the inequality

$$\cos x \le \frac{\sin x}{x} \le 1 \tag{3.1}$$

holds for $x \in (-\pi/2, \pi/2)$ and $x \neq 0$. Once this is done, the result will follow from the Squeeze Theorem. Further, it is enough to establish (3.1) for $x \in (0, \pi/2)$. If $x \in (-\pi/2, 0)$ then $x = -t$ for some $t \in (0, \pi/2)$, and (3.1) becomes

$$\cos(-t) \le \frac{\sin(-t)}{-t} \le 1. \tag{3.2}$$

However, $\cos(-t) = \cos t$, and $\sin(-t) = -\sin t$ so (3.2) is the same as (3.1).

So, let $x \in (0, \pi/2)$. If we take reciprocal values in (3.1) we get an equivalent inequality

$$1 \le \frac{x}{\sin x} \le \frac{1}{\cos x}.$$

If we now multiply by (the positive quantity) $\sin x$ we obtain that (3.1) is equivalent to

$$\sin x \le x \le \tan x.$$

These inequalities are obvious when one looks at Figure 3.1. ◆

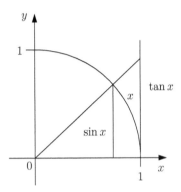

Figure 3.1: $\sin x \le x \le \tan x$.

Example 3.1.18. $f(x) = \begin{cases} -1, & \text{if } x < 3 \\ 1, & \text{if } x \ge 3, \end{cases}$ $\lim\limits_{x \to 3} f(x)$.

Solution. The limit does not exist. The one-sided limits are

$$\lim_{x \to 3^-} f(x) = -1, \quad \lim_{x \to 3^+} f(x) = 1. \quad ◆$$

Example 3.1.19. $\lim\limits_{x \to 1} \dfrac{x^2 - 5x + 7}{x^2 + 3x - 4}.$

Solution. Here the numerator has the limit 3, the denominator has the limit 0, so the limit does not exist. More precisely, the limit is infinite. From the left, when $x < 1$, $x^2 + 3x - 4 = (x-1)(x+4) < 0$, and the numerator is close to 3 (hence positive), so

$$\lim_{x \to 1^-} \frac{x^2 - 5x + 7}{x^2 + 3x - 4} = -\infty.$$

From the right, when $x > 1$, $x^2 + 3x - 4 = (x-1)(x+4) > 0$, so

$$\lim_{x \to 1^+} \frac{x^2 - 5x + 7}{x^2 + 3x - 4} = +\infty. \quad ◆$$

JOHANN BERNOULLI (1667–1748) was a younger brother of Jacob Bernoulli and the father of Daniel Bernoulli (page 262). At Basel University Johann took courses in medicine but he studied mathematics with Jacob. In 1691 Johann went to Geneva where he lectured on the differential calculus. After that he went to Paris, where he met Guillaume de l'Hôpital (page 127), perhaps the best mathematician in Paris at that time. Nevertheless, l'Hôpital needed Johann to teach him calculus. For that service Bernoulli was paid a handsome sum. The lessons continued after Johann returned to Basel. In 1696 l'Hôpital published a book based on these lectures but did not acknowledge that it was based on Johann's lectures. The truth came out in 1922 when a copy of his course, made by his nephew Nicolaus, was discovered. In 1691 Johann solved the *catenary problem* (finding the shape of a hanging rope). His doctoral dissertation was on the muscular movement, an application of mathematics to medical science. In 1695 he moved to Groningen, Netherlands, to be a professor there. After 10 years he moved back to Basel, where he was supposed to teach Greek. However, during his trip back, his brother Jacob passed away, so Johann became a professor of mathematics. He made many significant contributions to mathematics. His 1718 solution of the *isoperimetric problem* (finding a curve of a given length that maximizes the enclosed area) led to the development of the calculus of variations. In the Newton-Leibniz debate over who deserved credit for the discovery of calculus, Johann Bernoulli showed that there were problems which could be solved using Leibniz methods, while Newton had failed to solve them.

Exercises

In Exercises 3.1.1–3.1.13 find the limit without the use of the L'Hôpital's Rule:

3.1.1. $\displaystyle\lim_{x\to 2} \frac{x^3 - 2x^2 - 4x + 8}{x^4 - 8x^2 + 16}$.

3.1.2. $\displaystyle\lim_{x\to 3} \frac{x^2 - 5x + 6}{x^2 - 8x + 15}$.

3.1.3.* $\displaystyle\lim_{x\to 16} \frac{\sqrt[4]{x} - 2}{\sqrt{x} - 4}$.

3.1.4. $\displaystyle\lim_{x\to 4} \frac{\sqrt{1 + 2x} - 3}{\sqrt{x} - 2}$.

3.1.5.* $\displaystyle\lim_{x\to 1} \frac{(1 - \sqrt{x})(1 - \sqrt[3]{x})\ldots(1 - \sqrt[n]{x})}{(1 - x)^{n-1}}$.

3.1.6.* $\displaystyle\lim_{x\to 7} \frac{\sqrt{x + 2} - \sqrt[3]{x + 20}}{\sqrt[4]{x + 9} - 2}$.

3.1.7.* $\displaystyle\lim_{x\to 0} \frac{1 - \cos x}{x^2}$.

3.1.8. $\displaystyle\lim_{x\to 0} \frac{\sin 5x - \sin 3x}{\sin x}$.

3.1.9.* $\displaystyle\lim_{x\to 0} \frac{1 - \cos x \cos 2x}{1 - \cos x}$.

3.1.10. $\displaystyle\lim_{x\to 0} \frac{1 - \cos^\alpha x}{x^2}$, $\alpha \in \mathbb{R}$.

3.1.11.* $\displaystyle\lim_{x\to 0} \frac{\ln \cos 3x}{\ln \cos 2x}$.

3.1.12.* $\displaystyle\lim_{x\to \infty} \frac{\ln(x^2 - x + 1)}{\ln(x^{10} + x + 1)}$.

3.1.13. $\displaystyle\lim_{x\to 0} \frac{a^{x^2} - b^{x^2}}{(a^x - b^x)^2}$, $a, b > 0$, $a \neq b$.

3.2 Review of Functions

Although the concept of a function should be familiar to a student from the elementary calculus, we will review it here. When presenting a function it is necessary to point out at two sets A and B and a rule that assigns to each number in A a *unique* number in B. In other words, we need to verify that:

(a) we can assign a number from B to *each* element of A;

(b) we can assign *only one* number to each element of A.

The words *mapping* and *transformation* are often used instead of function.

Example 3.2.1. A function.

$f(x) = x^2$, $A = (-\infty, \infty)$, $B = (-\infty, \infty)$.

This is a well known function. It is easy to see that every element in A can be used as an input for f; also for any input, the output is just one number. ◆

Remark 3.2.2. Notice that not every number in B can be an output. Even though it may look as a flaw, it is not required for a function to "hit" every member of the set B. Functions that go this extra step are called **surjective** or **onto**. Also, notice that some numbers in A share the same number in B. For example, $f(-3) = f(3) = 9$. Again, neither (a) nor (b) requires that each number in A has its own number in B. When that happens we say that f is **injective** or **one-to-one**.

Example 3.2.3. Same formula, different function.
  $g(x) = x^2$, $A = (-\infty, \infty)$, $B = [0, \infty)$.

Do not dismiss this as the same function as in Example 3.2.1! We have changed the set B, and that makes a big difference. The function g is surjective, while f is not. ◆

Example 3.2.4. Not a function.
  $h(x) = x^2$, $A = (-\infty, \infty)$, $B = (0, \infty)$.

One more occurrence of the same rule, yet this is not a function. Reason: 0 is an element of A but $h(0)$ is not an element of B. In other words, condition (a) is violated. ◆

The functional notation like $f(x)$ is very useful when we have more than one function, so that we can make a distinction between them. When there is only one function in sight, we will use the letter y.

> The word "function" was used first by Leibniz in 1673, although not quite in the present day meaning. In 1698, in a letter to Leibniz, Johann Bernoulli narrowed the meaning closer to what we accept today. The notation $f(x)$ was first used by Euler in 1734. During the eighteenth century, the notion evolved to describe an expression or formula involving variables and constants. It took a joint work of many mathematicians throughout the nineteenth century to hammer down the definition that we use nowadays.

Example 3.2.5. A function.
  $y = 3x + 2$, $A = (0, 1)$, $B = (0, \infty)$.

It is only necessary to check that, if $0 < x < 1$ then $y > 0$. Of course, if $x > 0$, then $3x + 2 > 0$, so this is indeed a function. ◆

Example 3.2.6. A function.
  $x = (y - 2)/3$, $A = (0, 1)$, $B = (0, \infty)$.

A closer look reveals that this is the same function as the one in Example 3.2.5. Just solve it for y! ◆

Example 3.2.7. A function.
  $y = \sqrt{x}$, $A = (0, 10)$, $B = (-\infty, \infty)$.

Is this a function? What if $x = 4$? Then $y = \sqrt{4}$ and what is this number: 2 or -2? If we want this to be a function we had better stick to rule (b). Therefore $\sqrt{4}$ cannot be *both* 2 and -2. Notice that, for each $x \in A$, there are two "candidates" for $\sqrt{x}$: one positive, the other negative. It has been agreed upon that $\sqrt{x}$ will denote the former. So, $\sqrt{4} = 2$. ◆

Example 3.2.8. Not a function.
  $y = \sqrt{x}$, $A = (-1, 1)$, $B = (-\infty, \infty)$.

This is not a function: it violates rule (a), because $\sqrt{-1/2}$ is undefined. ◆

Example 3.2.9. Not a function.
  $x = y^2$, $A = (0, 10)$, $B = (-\infty, \infty)$.

Another "small" change from Example 3.2.7. However, this is not a function, because it violates rule (b). For example, if $x = 4$, there are two numbers in B that satisfy equation $4 = y^2$, namely 2 and -2. This can be seen also if we solve the equation $x = y^2$ for y, because we get $y = \pm\sqrt{x}$. ◆

Example 3.2.10. $x = y^2$, $A = (0, 10)$, $B = (0, \infty)$.

The change in the set B now makes this a function. Although the rule can be written as $y = \pm\sqrt{x}$, it does not lead to two numbers in B. The reason is that, as explained in Example 3.2.7, $\sqrt{x} \geq 0$ for any $x \in A$, and $-\sqrt{x} \leq 0$ for any $x \in A$. However, there are no negative numbers in B so there can be no more than one number associated to $x \in A$. ◆

The set A is called the **domain** of f and it is, unless specified, taken to be as large as possible. That is, it should include all numbers to which a number in B can be assigned.

Example 3.2.11. The domain of a function.
$y = \sqrt{x}$, $B = (-\infty, \infty)$. What is the domain of the function?

Now, it is clear that x must be a non-negative number ($x \geq 0$), so we would assume that $A = [0, \infty)$. ◆

Example 3.2.12. The domain of a function. $y = \sqrt{x}$, $B = [3, \infty)$. What is the domain of the function?

Although the formula $y = \sqrt{x}$ excludes only negative numbers, we can see that, if $x = 1$, there is no number in B that would correspond to it. A closer look reveals that the set A should be $[9, \infty)$. ◆

The set B is called a **codomain** of f, and unless specified differently, we will assume it is $(-\infty, \infty)$. As we have seen in Example 3.2.11, when f is not surjective there are numbers in B that do not correspond to any numbers in A. The **range** of f is the set of those numbers in B that *do* correspond to numbers in A.

Example 3.2.13. The domain, the codomain, and the range of a function. $y = \sqrt{x - 3}$.

The domain $A = [3, \infty)$, the codomain is $B = (-\infty, \infty)$, and the range is $[0, \infty)$. ◆

PETER GUSTAV LEJEUNE DIRICHLET (1805–1859) was a German mathematician. He was born in Düren, France, but the city became a part of Prussia in 1815. His family last name is corrupted *le jeune de Richellet* (a youth from Richellet) and it reflects the fact that his grandfather came from Richellet in Belgium. Although he excelled in mathematics, Dirichlet was unable to graduate high school due to the lack of fluency in Latin. He went to Paris to study at Collège de France. At the age of 20 he produced a proof of Fermat's last theorem for the case $n = 5$, which was the first advance in the theorem since Fermat's own proof of the case $n = 4$ and Euler's proof for $n = 3$. His lecture at the French Academy of Sciences in 1825 put him in touch with the leading French mathematicians, who raised his interest in theoretical physics, especially Fourier's analytic theory of heat. Even Gauss (page 196) was impressed so his recommendation was instrumental in getting a position for Dirichlet at the University of Breslau in Prussia. Since an advanced degree was required for the job, the University of Bonn awarded him an honorary doctorate. He finally completed all formal requirements for a full professor only in 1851. In Breslau Dirichlet did research in number theory and on the strength of his publications he was granted a transfer to Berlin in 1826, but only to teach at the Prussian Military Academy. In 1832 he became the youngest member of the Prussian Academy of Sciences at the age of 27. Nevertheless, he still had to teach both at the Military Academy and the University of Berlin without being paid a professor's salary. When Gauss (page 196) died in 1855 Dirichlet got the deserved appointment at the University of Göttingen. Sadly, only 3 years later he had a heart attack and he passed away in 1859. Dirichlet is credited with being one of the first to give the modern formal definition of a function. Throughout the text we will encounter his many contributions, especially in the theory of infinite series, most notably the Fourier series, about which he learned as a student in Paris.

Exercises

In Exercises 3.2.1–3.2.4 determine the domain of f:

3.2.1. $f(x) = \sqrt{\sin(\sqrt{x})}$.

3.2.2. $f(x) = \sqrt{3x - x^3}$.

3.2.3. $f(x) = \arccos(2\sin x)$.

3.2.4. $f(x) = \log_2 \log_3 \log_4 x$.

In Exercises 3.2.5–3.2.8 determine the domain and the range of f:

3.2.5. $f(x) = \sqrt{2 + x - x^2}$.   3.2.6. $f(x) = \ln(1 - 2\cos x)$.

3.2.7. $f(x) = \arccos \dfrac{2x}{1 + x^2}$.   3.2.8. $f(x) = \arcsin \ln \dfrac{x}{10}$.

In Exercises 3.2.9–3.2.12 find the set E_y to which f maps the set E_x:

3.2.9. $f(x) = x^2$, $E_x = [-1, 2]$.   3.2.10. $f(x) = \cot \dfrac{\pi x}{4}$, $E_x = (0, 1]$.

3.2.11. $f(x) = \dfrac{1}{1 - x}$, $E_x = (0, 1)$.   3.2.12. $f(x) = \sqrt{x - x^2}$, $E_x = (0, 1)$.

In Exercises 3.2.13–3.2.16 find $f(x)$ given that:

3.2.13. $f(x + 1) = x^2 - 3x + 2$.   3.2.14. $f\left(x + \dfrac{1}{x}\right) = x^2 + \dfrac{1}{x^2}$, $x \geq 2$.

3.2.15. $f\left(\dfrac{1}{x}\right) = x + \sqrt{1 + x^2}$, $x \geq 1$.   3.2.16. $f\left(\dfrac{x}{x + 1}\right) = x^2$.

A function f defined on a symmetric interval $[-a, a]$ is **even** if $f(-x) = f(x)$ for all $x \in [-a, a]$, and f is **odd** if $f(-x) = -f(x)$ for all $x \in [-a, a]$. In Exercises 3.2.17–3.2.22 determine if f is an even or an odd function:

3.2.17. $f(x) = \ln \dfrac{1 - x}{1 + x}$.   3.2.18. $f(x) = x^n$, $(n \in \mathbb{N})$.   3.2.19. $f(x) = \cos x$.

3.2.20. $f(x) = \sin x$.   3.2.21. $f(x) = a^x + a^{-x}$, $(a > 0)$.

3.2.22. $f(x) = \ln(x + \sqrt{1 + x^2})$.

3.3   Continuous Functions: A Geometric Viewpoint

All throughout calculus, there are many nice results that require that a particular function be continuous. In order to use such theorems it will be necessary to tell whether a function is continuous or not. One possible approach is geometric: just look at the graph and see whether there are any holes in it. Now, a look at the graph reveals that $y = x^2$ is a continuous function, and so is $y = 3x$. What about $y = x^2 + 3x$? Of course, we can look at the graph, but it would be nice if we knew that "a sum of continuous functions is a continuous function". (This is especially true when the sum is infinite, so graphing may not be easy.) In order to prove a result like this, we will need a more rigorous definition of continuity. Let us start with a few examples.

Example 3.3.1. A continuous function.
     $f_1(x) = -1$, $A_1 = (-\infty, -3)$.
It is not hard to see that this function, defined on the set A_1, is continuous.   ◆

Example 3.3.2. A continuous function.
     $f_2(x) = 1$, $A_2 = [-3, \infty)$.
Again, this function, defined on the set A_2, is continuous.   ◆

Example 3.3.3. A function that is not continuous.
$$f(x) = \begin{cases} -1, & \text{if } x < -3 \\ 1, & \text{if } x \geq -3. \end{cases}$$

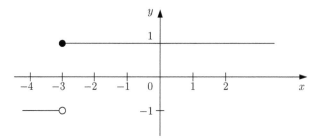

Figure 3.2: The graph of a discontinuous function.

A look at the graph reveals that this is not a continuous function, in spite of the fact that it combines functions f_1 and f_2 from Examples 3.3.1 and 3.3.2. We can see that there is a "jump" at $x = -3$.   ◆

So, we cannot declare that the function f in Example 3.3.3 is continuous. It seems a pity, because it has only one "flaw". Rather than dismiss such a function for not being continuous, we single out the culprit. We say that f is **discontinuous** at $x = -3$ and continuous at every other point. In general, the question whether a function is continuous or not is answered for each point of the domain separately. Because of that, we say that the continuity is a *local* property of a function. Almost all functions that we will work with will be continuous with at most a few exceptions, so it will be easier to list the points of discontinuity (if there are any).

Example 3.3.4. A continuous function with a hole in the graph.
$$f(x) = \begin{cases} -1, & \text{if } x < 3 \\ 1, & \text{if } x > 3. \end{cases}$$
Surprise! This function is continuous at every point. Of course, "every point" really means "every point of its domain". Since 3 does not belong to the domain of f, we must conclude that there are no discontinuities.   ◆

Let us look closely at the discontinuity in Example 3.3.3. The jump occurs because, coming along the graph from the left, we arrive at the point with coordinates $(3, -1)$ while coming from the right takes us to $(3, 1)$. For a function to be continuous, we would like to be at the same point. One way to formulate this is to require that, as x approaches to 3, the left limit of f (which equals -1) and the right limit of f (which equals 1) be the same. Notice that this is indeed the case if we consider $x = 2$: both the left and the right limit are equal to -1. When both limits exist and are the same, it means that there exists the limit of the function. So, $\lim_{x \to 2} f(x) = -1$ but $\lim_{x \to 3} f(x)$ does not exist. Therefore, in order for any function f (not just the one in Example 3.3.3) to be continuous at a point a, we will require that $\lim_{x \to a} f(x)$ exists.

Example 3.3.5. Continuity at a point.
$f(x) = 2x - 1$, $a = 4$
The graph of f (Figure 3.3) shows that it is continuous at $a = 4$. Also, $\lim_{x \to 4}(2x - 1)$ exists and equals 7.   ◆

Example 3.3.6. Discontinuity at a point where the limit exists.
$$f(x) = \begin{cases} x^2, & \text{if } x \neq 3 \\ 2, & \text{if } x = 3, \end{cases} \quad a = 3$$

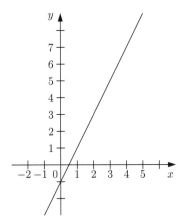

Figure 3.3: f is continuous at $a = 4$.

It is easy to see that this function is not continuous at $a = 3$. Geometrically, the graph has a "hiccup" at that point (Figure 3.4(a)). Yet, $\lim_{x \to 3} f(x)$ exists and equals 9. ◆

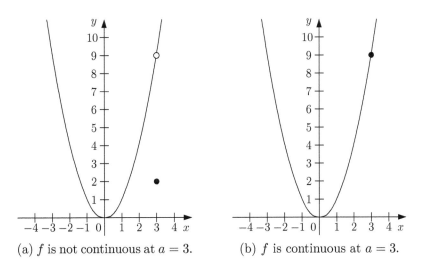

(a) f is not continuous at $a = 3$.　(b) f is continuous at $a = 3$.

Figure 3.4: Making a function continuous at $a = 3$.

The last example shows that, while a natural condition, the existence of the limit at a point is not sufficient to guarantee the continuity at that point. A closer look reveals that f would have been continuous at 3 if, in the definition of f we replaced 2 with 9.

Example 3.3.7. Continuity at a point where the limit exists.

$$f(x) = \begin{cases} x^2, & \text{if } x \neq 3 \\ 9, & \text{if } x = 3, \end{cases} \quad a = 3.$$

Of course, now the formula $f(x) = x^2$ holds not only when $x \neq 3$ but when $x = 3$ as well. Therefore, f is continuous at $x = 3$ (Figure 3.4(b)). ◆

The last two examples point out towards the proper list of requirements for a function to be continuous at a point.

Definition 3.3.8. Let f be a function and let a be a real number in the domain of f. Then f is **continuous at a** if $\lim_{x\to a} f(x)$ exists and equals $f(a)$.

As we mentioned before, most of the functions that we will encounter will be either continuous at every point or have a few discontinuities. If a function is continuous at every point of a set A, we will say that it is **continuous on** A. When A is the whole domain, we will say that f is continuous.

> Definition 3.3.8 is in Cauchy's *Cours d'Analyse*, although it is stated in a somewhat informal way there.

Our definition of continuity makes it clear that we have to understand limits better. For example, the function $f(x) = x^2$ is continuous at $a = 2$ if $\lim_{x\to 2} x^2 = 2^2$. Of course, we "know" that this is true, but we need to be able to prove it. Just like in the case of sequences, proving an equality like this will require a careful definition of a limit, and we will focus on that in the next section.

Exercises

Let $c \in (a,b)$ and let f be defined on $[a,c) \cup (c,b]$. We say that f has a **removable discontinuity** at $x = c$ if $\lim_{x\to c} f(x)$ exists. We say that f has a **jump discontinuity** at $x = c$ if both $\lim_{x\to c^-} f(x)$ and $\lim_{x\to c^+} f(x)$ exist and are finite, but not equal to each other. We say that f has an **essential discontinuity** at $x = c$ if one or both of $\lim_{x\to c^-} f(x)$ and $\lim_{x\to c^+} f(x)$ does not exist or is infinite.

In Exercises 3.3.1–3.3.8 find the points at which f is not continuous and determine their type:

3.3.1. $f(x) = \dfrac{x^2 - 1}{x^2 - 3x + 2}$.			3.3.2. $f(x) = \dfrac{x}{(1+x)^2}$.			3.3.3. $f(x) = \dfrac{\frac{1}{x} - \frac{1}{x+1}}{\frac{1}{x-1} - \frac{1}{x}}$.

3.3.4.* $f(x) = \dfrac{x}{\sin x}$.			3.3.5. $f(x) = \arctan \dfrac{1}{x}$.			3.3.6. $f(x) = \sqrt{x}\arctan \dfrac{1}{x}$.

3.3.7.* $f(x) = e^{x + \frac{1}{x}}$.			3.3.8.* $f(x) = \dfrac{1}{1 - e^{\frac{x}{1-x}}}$.

In Exercises 3.3.9–3.3.14 determine the value(s) of c so that f is continuous on $\mathbb{R}$:

3.3.9. $f(x) = \begin{cases} 2x + c & \text{if } x \le 1 \\ x^2 + 3 & \text{if } x > 1. \end{cases}$			3.3.10.* $f(x) = \begin{cases} |2x + c| & \text{if } x < -3 \\ 1 - x^2 & \text{if } x \ge -3. \end{cases}$

3.3.11. $f(x) = \begin{cases} \sin(x + c) & \text{if } x < 1 \\ x^2 - 1 & \text{if } x \ge 1. \end{cases}$			3.3.12. $f(x) = \begin{cases} ce^{3x} & \text{if } x \ge -1 \\ x^3 + x + 1 & \text{if } x < -1. \end{cases}$

3.3.13. $f(x) = \begin{cases} (x - 2c)^2 & \text{if } x > 4 \\ 2x - c & \text{if } x \le 4. \end{cases}$			3.3.14. $f(x) = \begin{cases} \frac{x}{x-1} & \text{if } x < 0 \\ e^{-x} + c & \text{if } x \ge 0. \end{cases}$

3.3.15.* Determine the values of a and b so that f is continuous on $\mathbb{R}$ if

$$f(x) = \begin{cases} x + 4 & \text{if } x < 1 \\ ax^2 + bx + 2 & \text{if } 1 \le x < 3 \\ 6x + a - b & \text{if } x \ge 3. \end{cases}$$

3.4 Limits of Functions

Our goal is to state a precise definition of the limit. It will be helpful to remember the definition we stated for sequences (Definition 2.2.6). In that scenario L is a limit of a sequence a_n if, for any $\varepsilon > 0$, the interval $(L - \varepsilon, L + \varepsilon)$ contains all members of the sequence starting with a_N. Therefore, we will once again consider an arbitrary $\varepsilon > 0$, and the interval $(L - \varepsilon, L + \varepsilon)$. The difference was that, with sequences we were interested in having the values of the sequence fall in that interval, whereas here we will ask for the values of the function to do so.

Example 3.4.1. For what x is $f(x)$ in $(L - \epsilon, L + \epsilon)$?
$f(x) = x^2$, $a = 3$, $\varepsilon = 1$.

When x approaches 3, we expect that x^2 will approach 9. So, we assume that $L = 9$ and consider the interval $(8, 10)$. We are hoping that, as x gets closer to 3, the values of f will belong to the interval $(8, 10)$. Where is that magic line after which x is close enough? Notice that, if $2.9 < x < 3.1$ then $x^2 > 2.9^2 = 8.41$ and $x^2 < 3.1^2 = 9.61$, so $x^2 \in (8, 10)$. We make an observation that the inequalities $2.9 < x < 3.1$ and $8 < x^2 < 10$ can be written as $|x - 3| < 0.1$ and $|x^2 - 1| < 9$. Thus, if $|x - 3| < 0.1$, then $|x^2 - 1| < 9$ (Figure 3.5(a)). ♦

Example 3.4.2. For what x is $f(x)$ in $(L - \epsilon, L + \epsilon)$?
$f(x) = x^2$, $a = 3$, $\varepsilon = 0.1$.

Now we consider the interval $(8.9, 9.1)$ and we notice that, if $2.9 < x < 3.1$ it may not be close enough. For example, $2.95^2 = 8.7025 \notin (8.9, 9.1)$. So we need to get closer to 9. How about $2.99 < x < 3.01$, i.e. $|x - 3| < 0.01$ (Figure 3.5(b))? Now $x^2 > 2.99^2 = 8.9401$ and $x^2 < 3.01^2 = 9.0601$, so $x^2 \in (8.9, 9.1)$. ♦

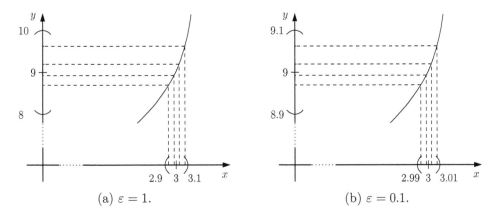

(a) $\varepsilon = 1$. (b) $\varepsilon = 0.1$.

Figure 3.5: $f(x) = x^2$, $a = 3$.

Let us summarize our experience from these two examples. In both we were given a positive number ε and we came up with a positive number: it was 0.1 in Example 3.4.1, and 0.01 in Example 3.4.2. This new number (usually denoted by δ) had a role similar to N for sequences. The number N measured how far ("towards infinity") we needed to go so that each a_n belongs to $(L - \varepsilon, L + \varepsilon)$. Here, for the limit of a function, the number δ measures how close we need to get to a so that each $f(x)$ belongs to $(L - \varepsilon, L + \varepsilon)$. Instead of intervals, it is customary to use inequalities with absolute values (just like for sequences). In the previous example we would say that, $|x - 3| < 0.01$ implies that $|x^2 - 9| < 0.1$.

In both examples, the number $a = 3$ belonged to the domain of the function. This is not always the case.

Example 3.4.3. For what x is $f(x)$ in $(L - \epsilon, L + \epsilon)$?
$$f(x) = \frac{x^2 - 1}{x - 1}, \ a = 1, \ \varepsilon = 0.1.$$

The limit is 2 and we want $f(x)$ to be between 1.9 and 2.1. It is not hard to see that the

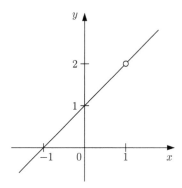

Figure 3.6: $f(x) = (x^2 - 1)/(x - 1)$, $a = 1$, $\varepsilon = 0.1$.

formula for $f(x)$ can be simplified to $f(x) = x + 1$, so if we want to have $1.9 < x + 1 < 2.1$, we should take $0.9 < x < 1.1$. In other words, $|x - 1| < 0.1$ implies that $|f(x) - 2| < 0.1$. The problem with the last statement is that the inequality $|x - 1| < 0.1$ would allow $x = 1$, and this number is not in the domain of f. Nevertheless, the implication is correct for any $x \neq 1$, and in order to accommodate for that, we will use the inequalities $0 < |x - 1| < 0.1$ instead. ◆

Example 3.4.3 reminds us that, when looking for a limit as $x \to a$, the number a need not be in the domain of f. On the other hand, it is essential that x can be taken close to a. Therefore, we will always assume that this is the case. We say that a number a is a **cluster point** of a set A, if every interval $(a - \delta, a + \delta)$ contains at least one point of A, not counting a.

Example 3.4.4. Cluster points of a set.
The set of cluster points of $A = (0, 1)$ is $[0, 1]$.

Solution. Indeed, if we take any $a \in [0, 1]$, and any $\delta > 0$, then the interval $(a - \delta, a + \delta)$ contains at least one point of $(0, 1)$, not counting a. On the other hand, if we take any $a \notin [0, 1]$, then it is not a cluster point of A. For example, $a = 1.1$ is not a cluster point because, if we take $\delta = 0.05$, then $a - \delta = 1.05$, $a + \delta = 1.15$, and the interval $(1.05, 1.15)$

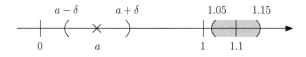

Figure 3.7: a is a cluster point of $[0, 1]$ but 1.1 is not.

contains no points of A. ◆

Cluster points were used by Weierstrass in his lectures but were never given a name. Cantor introduced them in an article in 1872 and called them "Grenzpunkt". In English, in addition to cluster point, they are often called limit points or accumulation points.

Now we are ready for the formal definition of the limit.

Definition 3.4.5. Let f be a function defined on a set A and let a be a cluster point of A. We say that L is the **limit** of f as x approaches to a, and we write $\lim_{x \to a} f(x) = L$, if for any $\varepsilon > 0$ there exists $\delta > 0$ such that $|f(x) - L| < \varepsilon$, whenever $0 < |x - a| < \delta$ and $x \in A$.

This definition appears in Cauchy's *Cours d'Analyse*, although it is stated with a minimal use of symbols. The symbol ε appears elsewhere in the book and is considered to be the first letter of the French word "erreur" (error). He also used the letter δ to denote a small quantity. It is believed that it comes from the word "différence" (French for difference).

With Definition 3.4.5 in hand, we can start proving results about limits.

Example 3.4.6. Proof that $\lim_{x \to a} f(x) = L$.
$f(x) = 3x - 2$, $a = 4$. We will show that $\lim_{x \to a} f(x) = 10$.

Solution. Our strategy is the same as with sequences: we consider the inequality $|f(x) - L| < \varepsilon$, i.e., $|(3x - 2) - 10| < \varepsilon$. Notice that

$$|(3x - 2) - 10| = |3x - 12| = 3|x - 4|$$

so we obtain $3|x - 4| < \varepsilon$ or $|x - 4| < \varepsilon/3$. This shows that all it takes is to select $\delta \leq \varepsilon/3$. With such a δ, whenever $|x - 4| < \delta$ we will have $|x - 4| < \varepsilon/3$.

Proof. Let $\varepsilon > 0$ and select $\delta = \varepsilon/3$. Suppose that $0 < |x - 4| < \delta$. Then $|x - 4| < \varepsilon/3$ which implies that

$$|f(x) - L| = |(3x - 2) - 10| = 3\,|x - 4| < 3 \cdot \frac{\varepsilon}{3} = \varepsilon. \qquad \blacklozenge$$

It happened frequently in the proofs about the limits of sequences that we needed to define N as the largest of several numbers. Here, we will often need to pick δ as the smallest of several numbers. We will use the following notation: for two numbers x, y, the symbol $\min\{x, y\}$ will stand for the smaller of these two; more generally, for n numbers $x_1, x_2, \ldots, x_n$, the symbol $\min\{x_1, x_2, \ldots, x_n\}$ will denote the smallest of these n numbers.

Example 3.4.7. Proof that $\lim_{x \to a} f(x) = L$.
$f(x) = x^2$, $a = 3$. We will show that $\lim_{x \to a} f(x) = 9$.

Solution. Since the limit is 9, we focus on the inequality $|x^2 - 9| < \varepsilon$. In the previous example it helped that $|x - 4|$ appeared as a factor. Here, we are hoping for $|x - 3|$. The good news is that $|x^2 - 9| = |x - 3||x + 3|$. The bad news is that the other factor $|x + 3|$ is a variable quantity, so it would be a mistake to declare $\delta = \varepsilon/|x + 3|$. Remember, δ cannot depend on x—only on ε. We will use a strategy that is similar to the one that was successful with sequences. Namely, we will look for an expression that is bigger than $|x + 3|$ but simpler. (Ideally, it would be a constant.)

The role of δ is to measure how close x is to 3 and the only restriction is that it has to be a positive number. If one δ works, any smaller number will do as well. In particular, we can decide that, for example, $\delta < 1$. In that case $2 < x < 4$ so $|x| = x < 4$. Consequently, $|x + 3| \leq |x| + 3 < 4 + 3 = 7$ and, assuming that $|x - 3| < \delta$, we have $|x^2 - 9| \leq 7|x - 3|$. Since we would like to have $7|x - 3| < \varepsilon$, we need $\delta \leq \varepsilon/7$.

Proof. Let $\varepsilon > 0$ and select $\delta = \min\{1, \varepsilon/7\}$. Suppose that $0 < |x - 3| < \delta$. Then $|x - 3| < 1$ so $2 < x < 4$ and $|x| = x < 4$. Further $|x - 3| < \varepsilon/7$ which implies that

$$|x^2 - 9| = |(x - 3)(x + 3)| \leq |(x - 3)|\,(|x| + 3) < 7\,|(x - 3)| < 7 \cdot \frac{\varepsilon}{7} = \varepsilon. \qquad \blacklozenge$$

As you can see, proving that a function has a particular limit can get very messy. However, if we knew for a fact that the function $f(x) = x^2$ is continuous at $a = 3$, we would conclude that $\lim_{x \to 3} f(x) = f(3)$, which means that $\lim_{x \to 3} x^2 = 3^2 = 9$, and we would be done. Let us think ahead, and ask ourselves how to prove that $f(x) = x^2$ (or a more complicated function) is continuous at $a = 3$. We will do that in the coming section and it will pay off to establish more general theorems about continuity. For example, we will prove that $f(x) = x$ is a continuous function and that the product of continuous functions is continuous. If $\lim_{x \to a} f(x) = f(a)$ and $\lim_{x \to a} g(x) = g(a)$, we will claim that $\lim_{x \to a} f(x)g(x) = f(a)g(a)$, so we will need to establish theorems of the type "the limit of the product equals the product of the limits". We have proved similar theorems for limits of sequences, so it would be nice if we could somehow take advantage of these results. Let us look again at Example 3.4.6. The assumption is that $a = 4$, so we will take a sequence $\{a_n\}$ that converges to 4. Such is, for example, $a_n = 4 + 1/n$. Now, let us compute the limit of $\{3a_n - 2\}$, as $n \to \infty$.

$$\lim_{n \to \infty} (3a_n - 2) = \lim_{n \to \infty} \left[3 \left(4 + \frac{1}{n} \right) - 2 \right] = 10.$$

So, we got the correct result, but the question remains: if we replace a_n by a different sequence (still converging to 4) would the limit still be 10? If you are convinced that the answer is in the affirmative, you had better look at the next example.

Example 3.4.8. $\lim f(a_n) \neq \lim f(b_n)$.

$f(x) = \sin \left(\dfrac{1}{x} \right)$, $a = 0$. What is $\lim_{x \to 0} f(x)$?

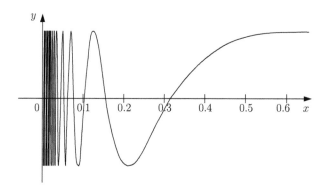

Figure 3.8: The graph of $y = \sin \frac{1}{x}$ for $x > 0$.

A look at the graph reveals that it has an x-intercept between 0.3 and 0.4, then two more between 0.1 and 0.2, etc. Solving the equation $\sin \frac{1}{x} = 0$ yields $1/x = n\pi$ so $x = 1/(n\pi)$, and the observed x-intercepts are $1/\pi \approx 0.3183098861$, $1/(2\pi) \approx 0.1591549430$, and $1/(3\pi) \approx 0.1061032954$. If we define a sequence $a_n = 1/(n\pi)$, then $\lim a_n = 0$ and

$$\lim_{n \to \infty} \sin \left(\frac{1}{a_n} \right) = \lim_{n \to \infty} \sin(n\pi) = \lim_{n \to \infty} 0 = 0.$$

On the other hand, the graph attains its maximum at a value of x close to 0.6, then between 0.1 and 0.2, etc. The maximum value of the sine function is 1, so we are looking for x such that $\sin \frac{1}{x} = 1$. We know that $\sin \frac{\pi}{2} = 1$, $\sin \frac{5\pi}{2} = 1$, $\sin \frac{9\pi}{2} = 1$ and, in general,

$\sin\left(2n\pi + \frac{\pi}{2}\right) = 1$. Thus, $\frac{1}{x} = 2n\pi + \frac{\pi}{2}$, so $x = 1/(2n\pi + \pi/2)$. The observed points are $1/(0 \cdot \pi + \pi/2) \approx 0.6366197722$ and $1/(2\pi + \pi/2) \approx 0.1273239544$. We will define a sequence $b_n = 1/(2n\pi + \pi/2)$. Then $\lim b_n = 0$, and

$$\lim_{n\to\infty} \sin\left(\frac{1}{b_n}\right) = \lim_{n\to\infty} \sin\left(2n\pi + \frac{\pi}{2}\right) = \lim_{n\to\infty} 1 = 1.$$

Now, two different sequences gave us two different results, so we should conclude that the limit $\lim_{x\to 0} \sin\frac{1}{x}$ does not exist. ♦

This example shows that, in order for the limit of a function to exist, we must get the same result for *every* sequence.

Theorem 3.4.9. *Let f be a function with domain A and let a be a cluster point of A. Then $\lim_{x\to a} f(x) = L$ if and only if, for every sequence $\{a_n\} \subset A$ converging to a, with $a_n \neq a$, $\lim_{n\to\infty} f(a_n) = L$.*

Proof. Suppose first that $\lim_{x\to a} f(x) = L$, and let $\{a_n\}$ be an arbitrary sequence in A converging to a. We will show that $\lim_{n\to\infty} f(a_n) = L$. So, let $\varepsilon > 0$. Since $\lim_{x\to a} f(x) = L$, there exists $\delta > 0$ such that

$$0 < |x - a| < \delta \Rightarrow |f(x) - L| < \varepsilon. \tag{3.3}$$

Further, $\lim a_n = a$ and $a_n \neq a$ so there exists $N \in \mathbb{N}$ such that, if $n \geq N$ then $0 < |a_n - a| < \delta$. By (3.3), if $n \geq N$, we have that $|f(a_n) - L| < \varepsilon$.

Now we will prove the converse. Namely, we will assume that, for every sequence $\{a_n\}$ converging to a and satisfying $a_n \neq a$, $\lim_{n\to\infty} f(a_n) = L$, and we will establish that $\lim_{x\to a} f(x) = L$. Suppose, to the contrary, that $\lim_{x\to a} f(x) \neq L$. By carefully applying the negative in the definition of the limit (Definition 3.4.5), we see that this means:

$$(\exists \varepsilon)(\forall \delta)(\exists x) \, 0 < |x - a| < \delta \text{ and } |f(x) - L| > \varepsilon.$$

Let ε_0 be such a number, and let $\delta_n = 1/n$, $n = 1, 2, \ldots$. Then, for each $n \in \mathbb{N}$, there exists x_n such that $0 < |x_n - a| < 1/n$ and $|f(x_n) - L| > \varepsilon_0$. Since $|x_n - a| < 1/n$, for each $n \in \mathbb{N}$, the Squeeze Theorem implies that $\lim(x_n - a) = 0$, so $\lim x_n = a$. Also, $0 < |x_n - a|$ shows that $x_n \neq a$. Our assumption now allows us to conclude that $\lim_{n\to\infty} f(x_n) = L$ which contradicts the inequality $|f(x_n) - L| > \varepsilon_0$. Therefore, $\lim_{x\to a} f(x) = L$. □

Theorem 3.4.9 allows us to translate questions about the limits of functions to the language of sequences. The proof of the following result gives a blueprint for such approach.

Theorem 3.4.10. *Let f, g be two functions with a domain A and let a be a cluster point of A. Also, let α be a real number. If $\lim_{x\to a} f(x) = L_1$ and $\lim_{x\to a} g(x) = L_2$ then:*

(a) $\lim_{x\to a}(\alpha f(x)) = \alpha \lim_{x\to a} f(x);$

(b) $\lim_{x\to a}(f(x) + g(x)) = \lim_{x\to a} f(x) + \lim_{x\to a} g(x);$

(c) $\lim_{x\to a}(f(x)g(x)) = (\lim_{x\to a} f(x))(\lim_{x\to a} g(x));$

(d) $\lim_{x\to a}(f(x)/g(x)) = \lim_{x\to a} f(x)/\lim_{x\to a} g(x)$ *if, in addition $L_2 \neq 0$.*

Proof. Let $\{a_n\}$ be an arbitrary sequence in A converging to a. The hypotheses of the theorem can be written as

$$\lim_{n\to\infty} f(a_n) = L_1 \quad \text{and} \quad \lim_{n\to\infty} g(a_n) = L_2,$$

and the assertion in (a) as

$$\lim_{n\to\infty} (\alpha f(a_n)) = \alpha \lim_{n\to\infty} f(a_n). \tag{3.4}$$

If we think of $\{f(a_n)\}$ as a convergent sequence $\{b_n\}$, then (3.4) is $\lim(\alpha b_n) = \alpha \lim b_n$ which is precisely Theorem 2.3.4 (a). Similarly, part (b) is a consequence of Theorem 2.3.4 (b), while parts (c) and (d) follow from Theorem 2.3.9. $\qquad \square$

Exercises

In Exercises 3.4.1–3.4.6 find the limit and prove that the result is correct using the definition of the limit:

3.4.1. $\lim\limits_{x\to 2} (3x^2 + 8x)$.

3.4.2. $\lim\limits_{x\to 4} \dfrac{1}{x^2 + 5x - 24}$.

3.4.3. $\lim\limits_{x\to 1} \dfrac{x}{x + 2}$.

3.4.4. $\lim\limits_{x\to 0} x \sin \dfrac{1}{x}$.

3.4.5. $\lim\limits_{x\to 1} \dfrac{x^2 - 1}{x - 1}$.

3.4.6. $\lim\limits_{x\to 2} \dfrac{x^2 + 4x - 12}{x^2 - 2x}$.

In Exercises 3.4.7–3.4.8 prove that the limit does not exist:

3.4.7. $\lim\limits_{x\to 0} \dfrac{|x|}{x}$.

3.4.8. $\lim\limits_{x\to 3} \begin{cases} 5 & \text{if } x < 3 \\ 7 & \text{if } x \geq 3. \end{cases}$

In Exercises 3.4.9–3.4.12 find the set B of all cluster points of A:

3.4.9. $A = (0,1) \cup \{2\}$.   3.4.10. $A = \mathbb{Q}$.

3.4.11. $A = \mathbb{R} \setminus \{0\}$.   3.4.12. $A = \{\frac{1}{n} : n \in \mathbb{N}\}$.

In Exercises 3.4.13–3.4.15 suppose that $\lim_{x\to a} f(x) = L$ and prove:

3.4.13. $\lim_{x\to a} |f(x)| = |L|$.   3.4.14. $\lim_{x\to a} \sqrt{f(x)} = \sqrt{L}$.

3.4.15. If $a \neq 0$ then $\lim_{x\to 1/a} f(1/x) = L$.

3.4.16. Prove that there can be at most one number L satisfying Definition 3.4.5.

3.4.17. Prove that a number a is a cluster point of a set A if and only if for any $\delta > 0$ the interval $(a - \delta, a + \delta)$ contains infinitely many points of A.

3.4.18. Prove that a finite set can have no cluster points.

3.4.19.* Prove that if A is a bounded set and sup A is not in A, then sup A is a cluster point of A. What if sup $A \in A$?

3.4.20.* Find two functions f and g such that $\lim_{x\to a} f(x) = b$ and $\lim_{x\to b} g(x) = c$, but $\lim_{x\to a} g(f(x)) \neq c$.

3.5   Other Limits

In this section we will consider some situations that have not been covered by Definition 3.4.5. These are one-sided limits, limits at infinity, and infinite limits.

3.5.1 One-Sided Limits

We will start the discussion of the one-sided limits with the following function that had already appeared in Example 3.3.3 (see Figure 3.2). Let

$$f(x) = \begin{cases} -1, & \text{if } x < -3 \\ 1, & \text{if } x \geq -3. \end{cases} \tag{3.5}$$

What is $\lim_{x \to -3} f(x)$?

If we apply Theorem 3.4.9 with $a_n = -3 - \frac{1}{n}$ then $a_n < -3$, for all $n \in \mathbb{N}$, so $f(a_n) = -1$ and $\lim f(a_n) = -1$. On the other hand, if we define $b_n = -3 + \frac{1}{n}$ then $b_n > -3$, for all $n \in \mathbb{N}$, so $f(b_n) = 1$ and $\lim f(b_n) = 1$. We conclude that the limit does not exist.

Nevertheless, there are some facts worth noting. If instead of f we consider the function $f_1(x) = -1$ on the domain $A_1 = (-\infty, -3)$, we can easily calculate $\lim_{x \to -3} f_1(x) = -1$. Also, the function $f_2(x) = 1$ on the domain $A_2 = [-3, \infty)$ has $\lim_{x \to -3} f_2(x) = 1$. This leads to the following definition.

Definition 3.5.1. Let f be a function defined on A and let a be a cluster point of A. We say that L is the **left limit** of f as x approaches to a, and we write $\lim_{x \to a^-} f(x) = L$, if for any $\varepsilon > 0$ there exists $\delta > 0$ such that $|f(x) - L| < \varepsilon$, whenever $a - \delta < x < a$ and $x \in A$.

Remark 3.5.2. An analogous definition can be stated for the right limits. We will leave it to the reader to formulate such a statement.

Remark 3.5.3. If f_1 is defined by $f_1(x) = -1$, for $x \in (-\infty, -3)$, Definition 3.5.1 shows that the limit of f_1 as $x \to -3$ is at the same time the left limit. We will typically ignore the distinction in a case like that, and we will emphasize the fact that the limit is one-sided only when the other one-sided limit is also considered.

Just like for the (two-sided) limits, there is a strong connection with limits of sequences. The following is an analogue of Theorem 3.4.9 for the left limits. A similar result holds for the right limits.

Theorem 3.5.4. *Let f be a function defined on A and let a be a cluster point of A. Then $\lim_{x \to a^-} f(x) = L$ if and only if, for any sequence $\{a_n\} \subset A$ converging to a from the left, $\lim_{n \to \infty} f(a_n) = L$.*

Going back to the function f defined by (3.5), we can ask whether it is continuous. This question has already been asked and answered in Example 3.3.3: f is continuous at every point except at $x = -3$. However, at that time our argument was visual. Now we can give a formal proof that f is not continuous at $x = -3$. Let $\varepsilon_0 = 1$, and let $\delta > 0$. Then, the interval $(-3 - \delta, -3 + \delta)$ contains $c = -3 - \delta/2$ and it is easy to see that $f(c) = -1$, so $|f(c) - f(-3)| = 2 > \varepsilon_0$.

What about the functions f_1 and f_2? For the function f_2, we see that

$$\lim_{x \to -3} f_2(x) = 1 = f_2(-3),$$

so f_2 is continuous at $x = -3$. For f_1 this question is meaningless, because f_1 is not defined at $x = -3$. If we were to redefine f_1 so that its domain is $(-\infty, -3]$ and f_1 coincides with f on $(-\infty, -3]$, then we would have $f_1(-3) = f(-3) = 1$, and we would conclude that f_1 is not continuous at $x = -3$, because

$$\lim_{x \to -3} f_1(x) = -1 \neq f_1(-3).$$

Since f_2 and (the redefined) f_1 are just restrictions of f to $[-3, +\infty)$ and $(-\infty, -3]$, we say that f is continuous at $x = -3$ *from the right* but it is not continuous *from the left*. In general we have the following definition.

Definition 3.5.5. Let f be a function defined on A and let a be a cluster point of A. Then f is **continuous from the left** (respectively, **from the right**) at a if $\lim_{x \to a^-} f(x)$ (respectively, $\lim_{x \to a^+} f(x)$) exists and equals $f(a)$.

Exercises

In Exercises 3.5.1–3.5.14 find the limit and give a strict "$\varepsilon - \delta$" proof that the result is correct:

3.5.1. $\lim\limits_{x \to 0^+} \dfrac{1}{1 + e^{1/x}}$.

3.5.2. $\lim\limits_{x \to 0^-} \dfrac{1}{1 + e^{1/x}}$.

3.5.3. $\lim\limits_{x \to 8^-} \left\lfloor \dfrac{x}{2} \right\rfloor$.

3.5.4. $\lim\limits_{x \to 8^+} \left\lfloor \dfrac{x}{2} \right\rfloor$.

3.5.5. $\lim\limits_{x \to 0^+} \dfrac{|x|}{x}$.

3.5.6. $\lim\limits_{x \to 0^-} \dfrac{|x|}{x}$.

3.5.7. $\lim\limits_{x \to 0^-} \dfrac{x^2}{|x|}$.

3.5.8. $\lim\limits_{x \to 0^+} \dfrac{x^2}{|x|}$.

3.5.9. $\lim\limits_{x \to 0^+} x \lfloor x \rfloor$.

3.5.10. $\lim\limits_{x \to 0^-} x \lfloor x \rfloor$.

3.5.11. $\lim\limits_{x \to 1^+} x \lfloor x \rfloor$.

3.5.12. $\lim\limits_{x \to 1^-} x \lfloor x \rfloor$.

3.5.13. $\lim\limits_{x \to 1^-} \arctan \dfrac{1}{1-x}$.

3.5.14. $\lim\limits_{x \to 1^+} \arctan \dfrac{1}{1-x}$.

3.5.15. Prove Theorem 3.5.4.

3.5.16.* Let f be a function defined on a domain A and let a be a cluster point of A. Suppose that f has both the right and the left limit at $x = a$. Then f has the limit at $x = a$ if and only if the left and the right limit are equal.

3.5.17.* Let f be a bounded function on $[a, b]$. (This means that its range is a bounded set.) Define $m(x) = \inf\{f(t) : t \in [a, x]\}$ and $M(x) = \sup\{f(t) : t \in [a, x]\}$. Prove that m and M are continuous from the left on (a, b).

3.5.18.* Let f be an increasing function on (a, b) and let $c \in (a, b)$. Prove that:

(a) $\lim_{x \to c^-} f(x) = \sup_{x < c} f(x)$.

(b) $\lim_{x \to c^+} f(x) = \inf_{x > c} f(x)$.

3.5.2   Limits at Infinity

Now we will discuss the limits at infinity. Once again, we start with an example. Let $f(x) = 1/x$. We "know" that $\lim\limits_{x \to +\infty} f(x) = 0$, but what does that mean?

If we make a goofy (and incorrect) attempt to apply Definition 3.4.5, we would say that for any $\varepsilon > 0$ there exists $\delta > 0$ such that $|f(x) - L| < \varepsilon$, whenever $|x - a| < \delta$ and $x \in A$. Clearly, $f(x) = 1/x$ and $L = 0$. The problem is that a is $+\infty$, so $|x - a| < \delta$ makes no sense. Of course, the reason for the last inequality in the definition is that it is a precise way of saying that $x \to a$. So, in our case we want to make more precise the statement $x \to +\infty$. We had the same situation with sequences, where we asked for $N \in \mathbb{N}$ and $n \geq N$. Here, we will ask for $M > 0$ (it does not have to be an integer) and $x > M$.

Definition 3.5.6. Let f be a function defined on a domain $(a, +\infty)$, for some $a \in \mathbb{R}$. We say that L is the **limit** of f as x approaches to $+\infty$, and we write $\lim_{x \to +\infty} f(x) = L$, if for any $\varepsilon > 0$ there exists $M > 0$ such that $|f(x) - L| < \varepsilon$, whenever $x > M$ and $x > a$.

Example 3.5.7. Using Definition 3.5.6.

Prove that $\lim_{x \to +\infty} 1/x = 0$.

Solution. As usual, we look at $|f(x) - L| < \varepsilon$, which is here $|1/x| < \varepsilon$. It can be written as $|x| > 1/\varepsilon$, so we should take $M \geq 1/\varepsilon$.

Proof. Let $\varepsilon > 0$, and select $M = 1/\varepsilon$. Suppose now that $x > M$, so that $x > 0$ and $1/x < 1/M$. Then

$$\left| \frac{1}{x} - 0 \right| = \frac{1}{|x|} = \frac{1}{x} < \frac{1}{M} = \varepsilon. \qquad \blacklozenge$$

Remark 3.5.8. Geometrically, we have just established that the function $f(x) = 1/x$ has a horizontal asymptote $y = 0$ (Figure 3.9).

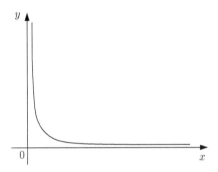

Figure 3.9: $f(x) = 1/x$ has a horizontal asymptote $y = 0$.

Example 3.5.9. Proof that $\lim_{x \to +\infty} f(x) = L$.

$f(x) = \arctan x$. We will show that $\lim_{x \to +\infty} f(x) = \pi/2$.

Solution. We consider $|\arctan x - \pi/2| < \varepsilon$, and we write it as

$$\frac{\pi}{2} - \varepsilon < \arctan x < \frac{\pi}{2} + \varepsilon.$$

The second inequality is automatically satisfied because the range of $\arctan x$ is $(-\pi/2, \pi/2)$. Since $\tan x$ is an increasing function, the first inequality leads to

$$\tan\left(\frac{\pi}{2} - \varepsilon \right) < x.$$

It is a slight cause for concern that, if $\varepsilon = \pi$, the left hand side of the last inequality is not defined.

Proof. Let $\varepsilon > 0$, and let $\eta = \min\{\varepsilon, \pi/4\}$. We define $M = \tan(\pi/2 - \eta)$. Then $M > 0$. Suppose that $x > M$. This implies that $\arctan x > \arctan M$. Therefore,

$$\left| \arctan x - \frac{\pi}{2} \right| = \frac{\pi}{2} - \arctan x$$
$$< \frac{\pi}{2} - \arctan M$$
$$\leq \frac{\pi}{2} - \arctan \left(\tan \left(\frac{\pi}{2} - \eta \right) \right) = \frac{\pi}{2} - \left(\frac{\pi}{2} - \eta \right) = \eta \leq \varepsilon. \qquad \blacklozenge$$

Once again, there is a strong connection between the limits at infinity and the limits of sequences. We leave the proof as an exercise.

Theorem 3.5.10. *Let f be a function defined on a domain $(a, +\infty)$, for some $a \in \mathbb{R}$. Then $\lim_{x \to +\infty} f(x) = L$ if and only if, for any sequence $\{a_n\}$ converging to $+\infty$, $\lim_{n \to \infty} f(a_n) = L$.*

Example 3.5.11. The limit does not exist.

$f(x) = \sin x$. We will show that $\lim_{x \to +\infty} f(x)$ does not exist.

Solution. In order to achieve this, we will use Theorem 3.5.10. Thus, it suffices to find two sequences $\{a_n\}$ and $\{b_n\}$ converging to $+\infty$ such that $\lim_{n \to \infty} \sin(a_n) \neq \lim_{n \to \infty} \sin(b_n)$. Such sequences are, for example, $a_n = n\pi$ and $b_n = \pi/2 + 2n\pi$. It is easy to see that $\lim a_n = \lim b_n = +\infty$. However, $\sin a_n = \sin(n\pi) = 0$, so $\lim \sin a_n = 0$. On the other hand, $\sin b_n = \sin(\pi/2 + 2n\pi) = 1$, so $\lim \sin b_n = 1$. ◆

Exercises

In Exercises 3.5.19–3.5.27 find the limit and give a strict "$\varepsilon - \delta$" proof that the result is correct:

3.5.19. $\lim\limits_{x \to \infty} \dfrac{2x + 3}{x - 4}$.

3.5.20. $\lim\limits_{x \to \infty} \dfrac{x}{3x - 2}$.

3.5.21. $\lim\limits_{x \to \infty} \dfrac{2x + 3}{x^2 - 4}$.

3.5.22. $\lim\limits_{x \to \infty} \dfrac{3x^2 + x + 1}{2x^2 - 3x + 4}$.

3.5.23. $\lim\limits_{x \to \infty} \dfrac{1}{e^x - 1}$.

3.5.24. $\lim\limits_{x \to -\infty} \dfrac{1}{e^x - 1}$.

3.5.25. $\lim\limits_{x \to \infty} \dfrac{\sqrt{x + \sqrt{x + \sqrt{x}}}}{\sqrt{x + 1}}$.

3.5.26.* $\lim\limits_{x \to \infty} (\sin \sqrt{x + 1} - \sin \sqrt{x})$.

3.5.27.* $\lim\limits_{x \to \infty} \dfrac{\ln(1 + \sqrt{x} + \sqrt[3]{x})}{\ln(1 + \sqrt[3]{x} + \sqrt[4]{x})}$.

3.5.28.* Prove that, if $\alpha > 0$, then $\lim\limits_{x \to \infty} \dfrac{\ln x}{x^\alpha} = 0$.

In Exercises 3.5.29–3.5.30 find constants a and b such that $\lim_{x \to \infty} (f(x) - ax - b) = 0$.

3.5.29. $f(x) = \dfrac{x^2 + 1}{x + 1}$.

3.5.30. $f(x) = \dfrac{x^3 - 2x + 1}{2x^2 - 5}$.

In Exercises 3.5.31–3.5.35 find the limits:

3.5.31. $\lim\limits_{x \to \infty} \left(\dfrac{x^2 - 1}{x^2 + 1} \right)^{x^2}$.

3.5.32. $\lim\limits_{x \to \infty} x^{1/3} \left((x + 1)^{2/3} - (x - 1)^{2/3} \right)$.

3.5.33. $\lim\limits_{x \to \infty} \dfrac{\ln(x^2 - x + 1)}{\ln(x^{10} + x + 1)}$.

3.5.34. $\lim\limits_{x \to -\infty} \dfrac{\ln(1 + 3^x)}{\ln(1 + 2^x)}$.

3.5.35. $\lim\limits_{x \to +\infty} \dfrac{\ln(1 + 3^x)}{\ln(1 + 2^x)}$.

3.5.36. Prove Theorem 3.5.10.

3.5.37. Suppose that f is a function and L is a real number such that $\lim_{x \to \infty} xf(x) = L$. Prove that $\lim_{x \to \infty} f(x) = 0$.

3.5.38. State and prove the analogue of Theorem 3.4.10 for the case when x approaches to ∞.

3.5.3 Infinite Limits

We will consider the situation when $f(x) \to \infty$. Let us start with an example.

Suppose that $f(x) = 1/x^2$. As $x \to 0$, x^2 is also approaching 0, so its reciprocal value increases to $+\infty$. We would like to make a precise description of a situation like this. It is helpful to remember Definition 2.2.11, which says that $\lim a_n = +\infty$ if for any $M > 0$ there exists a positive integer N such that, for $n \geq N$, $a_n > M$. The difference is that, at present, we are not interested in $n \to \infty$ (this is where N plays role) but $x \to 0$. Actually, infinite limits usually involve one-sided limits, i.e., $x \to a^+$ or $x \to a^-$. We will state our definition for the case $x \to a^+$ and $f(x) \to +\infty$.

Definition 3.5.12. Let f be a function defined on A and let a be a cluster point of A. We say that f has an **infinite limit** as x approaches to a from the right, and we write $\lim_{x \to a^+} f(x) = +\infty$, if for any $M > 0$ there exists $\delta > 0$ such that $f(x) > M$, whenever $a < x < a + \delta$ and $x \in A$.

The remaining variations ($x \to a^-$ and $f(x) \to -\infty$) can be formulated similarly. We will leave them to the reader, and instead we will look at a few examples.

Example 3.5.13. Proof that $\lim_{x \to a} f(x) = +\infty$.

$f(x) = 1/x^2$, $a = 0$.

Solution. We will prove that $\lim_{x \to 0} 1/x^2 = +\infty$. The inequality $f(x) > M$ translates to $1/x^2 > M$ or $x^2 < 1/M$. This leads to $|x| < 1/\sqrt{M}$, so it suffices to take $\delta = 1/\sqrt{M}$.
Proof. Let $M > 0$ and let $\delta = 1/\sqrt{M}$. If $0 < |x| < \delta$ then $0 < x^2 < 1/M$ so $1/x^2 > M$. ◆

Remark 3.5.14. Geometrically, we have just established that the function $f(x) = 1/x^2$ has a vertical asymptote $x = 0$ (Figure 3.10).

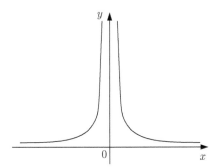

Figure 3.10: $f(x) = 1/x^2$ has a vertical asymptote $x = 0$.

In our next example we will look for the right limit.

Example 3.5.15. Proof that $\lim_{x \to a^+} f(x) = +\infty$.

$f(x) = e^{1/x}$, $a = 0$.

Solution. We will prove that $\lim_{x \to 0^+} f(x) = +\infty$. The inequality $f(x) > M$ is $e^{1/x} > M$ which leads to $1/x > \ln M$ and, therefore, to $x < 1/\ln M$ (assuming that $M > 1$). Thus, it suffices to take $\delta \leq 1/\ln M$, and notice that $M > 1$ if and only if $\ln M > 0$.
Proof. Let $M > 0$ and take

$$\delta = \begin{cases} 1/\ln M, & \text{if } \ln M > 0 \\ 1, & \text{if } \ln M \leq 0. \end{cases}$$

Suppose now that $0 < x < \delta$. Then $1/x > 1/\delta$, so

$$f(x) = e^{1/x} > e^{1/\delta} = \begin{cases} e^{\ln M}, & \text{if } \ln M > 0 \\ e^1, & \text{if } \ln M \leq 0 \end{cases} = \begin{cases} M, & \text{if } M > 1 \\ e, & \text{if } M \leq 1 \end{cases} \geq M. \qquad ◆$$

As one might expect, there is a connection with limits of sequences. We will leave the proof as an exercise.

Theorem 3.5.16. *Let f be a function defined on A and let a be a cluster point of A. Then f has an infinite limit as x approaches to a from the right if and only if, for any sequence $\{a_n\} \subset A$, $a_n > a$, converging to a, $\lim_{n\to\infty} f(a_n) = +\infty$.*

In the remaining portion of this section we will consider the infinite limits *at infinity*. For example, let $f(x) = 2^x$. What happens when $x \to +\infty$? What does it mean that this limit is infinite?

Just like in Definition 3.5.12 we will require that, for any $M > 0$, $f(x) > M$ or $f(x) < -M$ (if the limit is $-\infty$). Of course, such an inequality need not be true for all x, only when "x is large enough". A look at Definition 3.5.6 shows that a formal way to state that is to require that $f(x) > M$ when $x > K$. This leads to the following definition.

Definition 3.5.17. Let f be a function defined on a domain $(a, +\infty)$, for some $a \in \mathbb{R}$. We say that f has the **limit** $+\infty$ as x approaches to $+\infty$, and we write $\lim_{x\to+\infty} f(x) = +\infty$, if for any $M > 0$ there exists $K > 0$ such that $f(x) > M$, whenever $x > K$ and $x > a$.

Remark 3.5.18. Similar definitions can be stated when $x \to -\infty$ as well as when f has the limit $-\infty$. We will leave this task to the reader. Also, it is not hard to formulate and prove a theorem that establishes that such definitions are equivalent to those that use sequences.

Example 3.5.19. Proof that $\lim_{x\to+\infty} f(x) = +\infty$.

$f(x) = 2^x$. We will show that $\displaystyle\lim_{x\to+\infty} f(x) = +\infty$.

Solution. We need to demonstrate that, for any $M > 0$ there exists $K > 0$ such that $f(x) > M$, whenever $x > K$. If we look at the inequality $f(x) > M$, we see $2^x > M$, and taking the natural logarithm we get $\ln 2^x = x \ln 2 > \ln M$. Solving for x yields $x > \ln M/\ln 2$, so a good candidate for K is $\ln M/\ln 2$, provided that it is positive. Of course, if $\ln M \le 0$ then $M \le 1$, so $2^x > M$ as soon as $x > 0$.

Proof. Let $M > 0$ and choose $K = \max\{\ln M/\ln 2, 1\}$. Notice that $2^{\ln M/\ln 2} = M$ because both sides have the same logarithm:

$$\ln 2^{\ln M/\ln 2} = \frac{\ln M}{\ln 2} \ln 2 = \ln M.$$

Now, if $x > K$ then

$$f(x) = 2^x > 2^K = \begin{cases} 2^{\ln M/\ln 2}, & \text{if } M > 1 \\ 2^1, & \text{if } M \le 1 \end{cases} = \begin{cases} M, & \text{if } M > 1 \\ 2, & \text{if } M \le 1 \end{cases} \ge M. \qquad \blacklozenge$$

Exercises

In Exercises 3.5.39–3.5.48 give a strict proof of the asserted equalities:

3.5.39. $\displaystyle\lim_{x\to2^+} \frac{x+3}{x-2} = +\infty$.   3.5.40. $\displaystyle\lim_{x\to2^-} \frac{x+3}{x-2} = -\infty$.   3.5.41. $\displaystyle\lim_{x\to\frac{\pi}{2}^-} \tan x = +\infty$.

3.5.42. $\displaystyle\lim_{x\to\frac{\pi}{2}^+} \tan x = -\infty$.   3.5.43. $\displaystyle\lim_{x\to1^-} \frac{x}{x^2-1} = -\infty$.   3.5.44. $\displaystyle\lim_{x\to1^+} \frac{x}{x^2-1} = +\infty$.

3.5.45. $\displaystyle\lim_{x\to+\infty} \sqrt{x} = +\infty$.   3.5.46. $\displaystyle\lim_{x\to+\infty} \ln x = +\infty$.

3.5.47. $\displaystyle\lim_{x\to+\infty} \ln(1 + e^x) = +\infty$.   3.5.48. $\displaystyle\lim_{x\to-\infty} \frac{x^2+1}{x+1} = -\infty$.

3.5.49. Prove Theorem 3.5.16.

3.5.50. State and prove the analogue of Theorem 3.5.16 for the case when x approaches to ∞.

3.5.51. Let $a \in \mathbb{R}$ and suppose that f is a function that is positive on an open interval containing a. Prove that $\lim_{x \to a} f(x) = +\infty$ if and only if $\lim_{x \to a} 1/f(x) = 0$.

3.6 Properties of Continuous Functions

In this section we return to continuous functions. We have defined in Section 3.3 (Definition 3.3.8) that a function f is continuous at $x = a$ if $\lim_{x \to a} f(x) = f(a)$. In Section 3.4 we have learned the precise definition of the limit (Definition 3.4.5), as well as the more practical one using sequences (Theorem 3.4.9). The latter allows us to formulate an equivalent characterization of continuity in terms of sequences.

Corollary 3.6.1. *A function f is continuous at $x = a$ if and only if, for every sequence $\{a_n\}$ converging to a, $\lim_{n \to \infty} f(a_n) = f(a)$.*

> The alternative definition is often called the Heine definition of continuity, because Heine (page 92) was the first to write about it in [63] in 1872. It should be said that in this article Heine credits Cantor for the idea.

Now we can put all these to work and prove that some functions are continuous.

Example 3.6.2. Proof that f is continuous.
   $f(x) = C$, with C a real number.

Solution. Let a be a real number, and let us prove that f is continuous at $x = a$. We will show that $\lim_{x \to a} f(x) = f(a)$. By Theorem 3.4.9, we need to prove that, if $\lim a_n = a$ then $\lim_{n \to \infty} f(a_n) = f(a)$. Since $f(x) = C$, the last limit is $\lim_{n \to \infty} C = C$, which is obvious. Therefore, f is continuous at every point of the real line. ◆

Example 3.6.3. Proof that f is continuous.
   $f(x) = x$.

Solution. Let a be a real number, and let $\{a_n\}$ be a sequence that converges to a. The equality to show is $\lim_{n \to \infty} f(a_n) = f(a)$. Since $f(x) = x$, this comes down to $\lim_{n \to \infty} a_n = a$. Therefore, f is continuous at every $a \in \mathbb{R}$. ◆

In order to prove that more complicated functions are continuous we will need results about combinations of continuous functions.

Theorem 3.6.4. *Let f, g be two functions with a domain A and let $a \in A$. Also, let α be a real number. If f and g are continuous at $x = a$ then the same is true for: (a) αf; (b) $f + g$; (c) fg; (d) f/g if, in addition $g(a) \neq 0$. (e) If g is continuous at $f(a)$ then the composition $g \circ f$ is continuous at $x = a$.*

Proof. (a) Notice that αf is a function that assigns to each x the number $\alpha f(x)$. Therefore $(\alpha f)(a) = \alpha f(a)$, and we need to show that $\lim_{x \to a} (\alpha f(x)) = \alpha f(a)$. Combining Theorem 3.4.10 (a) and the fact that f is continuous we obtain that

$$\lim_{x \to a} (\alpha f(x)) = \alpha \lim_{x \to a} f(x) = \alpha f(a).$$

The proofs of assertions (b), (c), and (d) are similar and we leave them as an exercise. Finally, we need to show that $\lim_{x \to a} g(f(x)) = g(f(a))$. By Theorem 3.4.9 this is equivalent to establishing that, if a_n is a sequence converging to a, then $\lim g(f(a_n)) = g(f(a))$. Since f is continuous at $x = a$, and $\lim a_n = a$, we have that $\lim f(a_n) = f(a)$. Now, the continuity of g at $f(a)$ implies that $\lim g(f(a_n)) = g(f(a))$. $\qquad\square$

Remember, f is a **rational function** if it can be represented as a quotient of two polynomials. It is a direct consequence of Examples 3.6.2 and 3.6.3, and Theorem 3.6.4 that rational functions are continuous.

Theorem 3.6.5. *Every rational function is continuous at every point of its domain.*

Now we turn our attention to some discontinuous functions.

Example 3.6.6. A function with a discontinuity at every integer.
  $f(x) = \lfloor x \rfloor$. This function has a discontinuity at every integer, and it is continuous at any other point.

Solution. Let us start with the case when a is not an integer. Let n be an integer such that $n < a < n + 1$. Let $d = \min\{a - n, n + 1 - a\}$. Then $d > 0$. For any $\varepsilon > 0$ we can

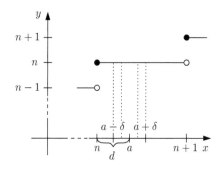

Figure 3.11: When a is not an integer.

take $\delta = d/2$. Indeed, if $0 < |x - a| < \delta$, then $n < x < n + 1$. Therefore, $|f(x) - f(a)| = \left| \lfloor x \rfloor - \lfloor a \rfloor \right| = |n - n| = 0 < \varepsilon$. Thus, f is continuous at $x = a$.
  Suppose now that $a \in \mathbb{Z}$, and let $\{a_n\}, \{b_n\}$ be two sequences converging to a such that, for all $n \in \mathbb{N}$,
$$a - 1 < a_n < a < b_n < a + 1.$$
Then $f(a_n) = a - 1$ and $f(b_n) = a$, for all $n \in \mathbb{N}$, so $\lim f(a_n) = a - 1$ and $\lim f(b_n) = a$. It follows that the limit $\lim_{x \to a} f(x)$ does not exist and f is not continuous at $x = a$. $\blacklozenge$

Example 3.6.7. A function with a discontinuity at every real number.
  The Dirichlet function
$$f(x) = \begin{cases} 1, & \text{if } x \in \mathbb{Q} \\ 0, & \text{if } x \notin \mathbb{Q} \end{cases}$$
has a discontinuity at every point on the real line.

Solution. Let a be a rational number, so that $f(a) = 1$. We will show that there exists a sequence a_n converging to a such that $\lim f(a_n) \neq 1$. Clearly, this goal will be accomplished if we select each a_n to be an irrational number, because then $f(a_n) = 0$. In order to obtain such a sequence we notice that, if $n \in \mathbb{N}$, the interval $(a - \frac{1}{n}, a + \frac{1}{n})$ contains an irrational number a_n (Theorem 1.3.7). Then $|a_n - a| < 1/n$ so the Squeeze Theorem implies that $a_n \to a$. We leave the case when a is an irrational number for a reader to prove. $\blacklozenge$

The Dirichlet function forced mathematicians to modify the concept of a function. Until then they were studying only "nice" ones and the one above would not even qualify as a function. Trouble is, it can be defined as the limit of "acceptable" functions: $f(x) = \lim_{k \to \infty} \left(\lim_{j \to \infty} \left(\cos(k!\pi x)^{2j} \right) \right)$. Dirichlet presented this function in [33] in 1829.

Example 3.6.8. A function with a discontinuity at every nonzero rational number.

The Thomae function

$$f(x) = \begin{cases} \dfrac{1}{q}, & \text{if } x = \dfrac{p}{q}, \, p \in \mathbb{Z}, \, q \in \mathbb{N}, \text{ and } p, q \text{ are mutually prime} \\ 0, & \text{if } x \text{ is irrational or } x = 0 \end{cases}$$

is continuous at a if and only if a is either irrational or 0.

Solution. The fact that f is discontinuous at every rational point can be proved just like in Example 3.6.7. Therefore, we consider only the case when a is an irrational number, so that $f(a) = 0$. Further, we will assume that $a \in (0, 1)$. Let $\varepsilon > 0$ and let $n = \lfloor 1/\varepsilon \rfloor + 1$. The largest value that f takes is $f(1/2) = 1/2$. The next two largest are $f(1/3) = f(2/3) = 1/3$, followed by $f(1/4) = f(3/4) = 1/4$, etc. That means that there is only a finite number of values of f that are bigger than $1/n$, and they are all attained at numbers of the form p/q with $q < n$. For example, if $n = 5$, then f takes values bigger than $1/5$ precisely at $1/2$, $1/3, 2/3, 1/4$, and $3/4$. Let δ denote the distance from a to the nearest fraction of the form p/q with $q < n$, and let $|x - a| < \delta$. Then x is not one of these fractions, so $f(x) \leq 1/n$. Since $n > 1/\varepsilon$ we have that $1/n < \varepsilon$ and $f(x) < \varepsilon$. The fact that $f(a) = 0$ now implies that $|f(x) - f(a)| < \varepsilon$ and we conclude that f is continuous at $x = a$. ♦

CARL JOHANNES THOMAE (1840–1921) was a German mathematician. As a high school student he excelled in mathematics. In 1861 he enrolled in the University of Halle. Heine (page 92) who taught there was a big influence on Thomae, giving him a love for function theory which he pursued for the rest of his career. In 1862 he moved to Göttingen, hoping to take courses by Riemann (page 169), but Riemann became ill of tuberculosis. In 1864, upon graduation, he studied elliptic functions with Weierstrass in Berlin. In 1866 he got a teaching position at the University of Göttingen. In 1867, Thomae was appointed at Halle where he became a colleague of Heine and Cantor. In 1874 he moved to the University of Freiburg to replace Paul du Bois-Reymond (page 177). Finally, in 1879 he moved to Jena, and remained there for the rest of his life. Thomae retired in 1914 but continued to publish papers up to 1919. His best work was in the theory of elliptic functions. The function given in Example 3.6.8 is in his book [103], published in 1875. Also, his 1870 textbook on complex analysis contains the first example of a function $f : \mathbb{R}^n \to \mathbb{R}$ that is separately continuous but not jointly continuous, although it appears that the example is due to Heine. (Example 10.3.9 gives another instance of the same phenomenon.)

Remember that, if f is an injective function, there exists its inverse function f^{-1}. Also, the graphs of f and f^{-1} are symmetric with respect to the graph of $y = x$, so if f is continuous, we expect the same for f^{-1}.

Theorem 3.6.9. *Let f be a strictly increasing, continuous function. Then there exists a function f^{-1} and it is continuous.*

Proof. First we notice that f^{-1} is also strictly increasing. Indeed, let $y_1 < y_2$ and suppose to the contrary that $f^{-1}(y_1) \geq f^{-1}(y_2)$. Since f is strictly increasing, it would follow that $f(f^{-1}(y_1)) \geq f(f^{-1}(y_2))$ which is the same as $y_1 \geq y_2$. This contradiction shows that f^{-1} is strictly increasing.

Let $\varepsilon > 0$, and let $b \in \mathbb{R}$. We will show that f^{-1} is continuous at $y = b$. Let $f^{-1}(b) = a$ and define $\delta = \frac{1}{2} \min\{f(a+\varepsilon) - f(a), f(a) - f(a-\varepsilon)\}$. Both $f(a+\varepsilon) - f(a)$ and $f(a) - f(a-\varepsilon)$

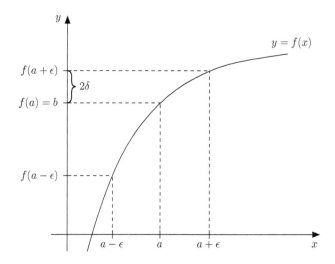

Figure 3.12: f^{-1} is continuous at $b = f(a)$.

are positive because f is strictly increasing, so $\delta > 0$. Further, $\delta < f(a + \varepsilon) - f(a)$ implies that $f(a) + \delta < f(a + \varepsilon)$ and, hence, that $f^{-1}(f(a) + \delta) < a + \varepsilon$. Since $f(a) = b$ and $b = f^{-1}(a)$ we obtain that

$$f^{-1}(b + \delta) - f^{-1}(b) < \varepsilon. \tag{3.6}$$

Similarly, $\delta < f(a) - f(a - \varepsilon)$ so $f(a) - \delta > f(a - \varepsilon)$ and $f^{-1}(f(a) - \delta) > a - \varepsilon$. Thus,

$$f^{-1}(b - \delta) - f^{-1}(b) > -\varepsilon. \tag{3.7}$$

Suppose now that $0 < |y - b| < \delta$ or, equivalently, $b - \delta < y < b + \delta$, $y \neq b$. Using (3.6) and (3.7) it follows that

$$-\varepsilon < f^{-1}(b - \delta) - f^{-1}(b) < f^{-1}(y) - f^{-1}(b) < f^{-1}(b + \delta) - f^{-1}(b) < \varepsilon,$$

so $|f^{-1}(y) - f^{-1}(b)| < \varepsilon$. $\qquad\qquad\qquad\qquad\qquad\qquad\qquad\qquad\qquad\qquad\square$

Remark 3.6.10. The case when f is a strictly decreasing, continuous function can be proved in an analogous fashion.

Theorem 3.6.9 may seem to have left out a large class of continuous functions: those that do have an inverse function but are not strictly increasing (or decreasing). Actually, there are no such functions. As soon as a function is injective (otherwise there is no inverse function) and continuous, it has to be strictly monotone (increasing or decreasing). We will state the theorem now but we will postpone the proof until Section 3.9.

Theorem 3.6.11. *Let f be a continuous injective function defined on (a, b). Then f is either strictly increasing or strictly decreasing on (a, b).*

There is a very useful property of continuous functions that we will use later.

Theorem 3.6.12. *Let f be a continuous function on an interval (a, b) and let $c \in (a, b)$. If $f(c) > 0$ then there exists $\delta > 0$ such that $f(x) > 0$ for $x \in (c - \delta, c + \delta)$.*

Proof. Suppose to the contrary that no such δ exists. Then, for every $\delta > 0$, there exists $x \in (c - \delta, c + \delta)$ such that $f(x) \leq 0$. If we take $\delta = 1/n$, we obtain a sequence $\{x_n\}$ in $(c - 1/n, c + 1/n)$ with $f(x_n) \leq 0$. The inequality $|x_n - c| < 1/n$ shows that the sequence

$\{x_n\}$ converges to c, and the continuity of f, using Theorem 3.4.9, implies that the sequence $\{f(x_n)\}$ converges to $f(c)$. Since $f(x_n) \leq 0$, Corollary 2.3.12 implies that $f(c) \leq 0$, which contradicts the assumption that $f(c) > 0$. □

Remark 3.6.13. An analogous theorem holds when $f(c) < 0$.

The converse of Theorem 3.6.12 is almost true. When $f(x) > 0$ for all x, $0 < |x - c| < \delta$, it does not follow that $f(c) > 0$. Example: $f(x) = x^2$, $c = 0$. However, it does follow that $f(c) \geq 0$.

Theorem 3.6.14. *Let f be a continuous function at a point c in (a, b). If $f(x) \geq 0$ on $(a, c) \cup (c, b)$ then $f(c) \geq 0$.*

Proof. Let $\{a_n\}$ be a sequence in (a, b) converging to c, and $a_n \neq c$. Then $f(a_n) \geq 0$, and Proposition 2.3.11 implies that $\lim f(a_n) \geq 0$. Since f is continuous at c, this means that $f(c) \geq 0$. □

Another important result is the Squeeze Theorem. We leave its proof as an exercise.

Theorem 3.6.15. *Let f, g, h be functions defined on A such that, for all $x \in A$, $f(x) \leq g(x) \leq h(x)$, and let c be a cluster point of A. If $\lim_{x \to c} f(x) = \lim_{x \to c} h(x) = L$, then $\lim_{x \to c} g(x)$ exists and equals L.*

Exercises

In Exercises 3.6.1–3.6.4 give a strict "$\varepsilon - \delta$" proof that f is continuous at a:

3.6.1. $f(x) = \sqrt{x}$, $a = 4$.

3.6.2. $f(x) = x^3$, $a = 2$.

3.6.3. $f(x) = \ln x$, $a = 3$.

3.6.4. $f(x) = \cos x$, $a = \pi/3$.

In Exercises 3.6.5–3.6.8 give a strict "$\varepsilon - \delta$" proof that f is continuous at every point of its domain:

3.6.5. $f(x) = 3x - 2$.

3.6.6. $f(x) = \sqrt[3]{x}$.

3.6.7. $f(x) = |x|$.

3.6.8. $f(x) = \dfrac{x}{x + 1}$.

3.6.9. Let $a \in \mathbb{R}$. Prove or disprove that $f + g$ has a discontinuity at a if: (a) Both f and g have a discontinuity at a; (b) f is continuous at a but g is not.

3.6.10. Let $a \in \mathbb{R}$. Prove or disprove that fg has a discontinuity at a if: (a) Both f and g have a discontinuity at a; (b) f is continuous at a but g is not.

3.6.11. Let $a \in \mathbb{R}$ and suppose that f is not continuous at a. Prove or disprove: f^2 is not continuous at a.

3.6.12. Let f be a continuous function on $\mathbb{R}$ and let $\{a_n\}$ be a bounded sequence. Prove or disprove: $\limsup f(a_n) = f(\limsup a_n)$.

3.6.13. Prove that the Dirichlet function has a discontinuity at every irrational number.

3.6.14.* Let
$$f(x) = \begin{cases} \sin |x|, & \text{if } x \text{ is rational} \\ 0, & \text{if } x \text{ is irrational.} \end{cases}$$

Prove that f is continuous at a if and only if $a = k\pi$ for some $k \in \mathbb{Z}$.

3.6.15.* Prove that if f and g are continuous functions then so are $\min\{f(x), g(x)\}$ and $\max\{f(x), g(x)\}$.

In Exercises 3.6.16–3.6.19, $a \in \mathbb{R}$ and f is a function on $\mathbb{R}$. Determine whether the given condition implies that f is continuous at a and, if not, what property of f does it describe:

3.6.16.* For any $\delta > 0$ there exists $\varepsilon > 0$ such that, if $|x - a| < \delta$ then $|f(x) - f(a)| < \varepsilon$

3.6.17.* For any $\varepsilon > 0$ there exists $\delta > 0$ such that, if $|f(x) - f(a)| < \varepsilon$ then $|x - a| < \delta$.

3.6.18.* For any $\delta > 0$ there exists $\varepsilon > 0$ such that, if $|f(x) - f(a)| < \varepsilon$ then $|x - a| < \delta$.

3.6.19. Let f, g be two functions defined and continuous on $[a, b]$ and suppose that $f(x) = g(x)$ for all rational $x \in [a, b]$. Prove that $f(x) = g(x)$ for all $x \in [a, b]$.

3.6.20. Prove the Squeeze Theorem 3.6.15.

3.6.21.* A function f is convex on $\mathbb{R}$ if for any two points $x, y \in \mathbb{R}$ and any $t \in [0, 1]$,

$$f(tx + (1 - t)y) \leq tf(x) + (1 - t)f(y).$$

Suppose that f is convex on $\mathbb{R}$.

(a) Prove that, if $r < s < t$, then $\dfrac{f(s) - f(r)}{s - r} \leq \dfrac{f(t) - f(r)}{t - r}$.

(b) Prove that, if $r < s < t$, then $\dfrac{f(s) - f(r)}{s - r} \leq \dfrac{f(t) - f(s)}{t - s}$.

(c) Prove that f is continuous.

3.6.22.* A set A is *open* if, for every $a \in A$, there exists $\varepsilon > 0$ such that $(a - \varepsilon, a + \varepsilon) \subset A$. By definition, if $f : \mathbb{R} \to \mathbb{R}$, the *inverse image* (or the *preimage*) of a set A is the set $f^{-1}(A) = \{x \in \mathbb{R} : f(x) \in A\}$. Prove that a function f is continuous on $\mathbb{R}$ if and only if, for any open set A, the set $f^{-1}(A)$ is open.

3.7 Continuity of Elementary Functions

This section is devoted to a large class of functions that we call *elementary functions*. These include: polynomials, rational functions, exponential and logarithmic functions, power functions, trigonometric functions and their inverses, as well as all functions obtained from the already listed through composition and combinations using the four arithmetic operations. They have been around for a long time, and they date back to the period when the continuity was an integral part of the notion of a function. In this section we will verify that they are indeed all continuous (within their domains).

Rational functions. By Theorem 3.6.5, every rational function is continuous at every point of its domain.

Trigonometric functions. At present, we will stick with their geometric definitions (the ratio of the appropriate sides of a right triangle). We will be able to derive a purely analytic definition of these functions in Chapter 8.

We will need a few more or less known trigonometric facts. We list them in the following proposition and we leave them as an exercise.

Proposition 3.7.1. *(a)* $\sin(\frac{\pi}{2} - x) = \cos x$; *(b)* $\sin x - \sin y = 2\cos\frac{x+y}{2}\sin\frac{x-y}{2}$; *(c) if* $|x| \leq \frac{\pi}{2}$ *then* $|\sin x| \leq |x|$.

Now we can establish that trigonometric functions are continuous.

Theorem 3.7.2. *Trigonometric functions* $\sin x$, $\cos x$, $\tan x$, $\cot x$, $\sec x$, *and* $\csc x$ *are continuous at every point of their domains.*

Proof. In view of Theorem 3.6.4 and Proposition 3.7.1 (a), it suffices to establish the continuity of the function $f(x) = \sin x$. Let $a \in \mathbb{R}$. We will show that

$$\lim_{x \to a} \sin x = \sin a. \tag{3.8}$$

By Theorem 3.4.9, we need to demonstrate that, if $\lim a_n = a$ then

$$\lim_{n \to \infty} \sin a_n = \sin a. \tag{3.9}$$

To that end, we apply Proposition 3.7.1 (b), together with the fact that $|\cos x| \leq 1$ for all $x \in \mathbb{R}$, and we obtain that

$$|\sin a_n - \sin a| = \left| 2 \cos \frac{a_n + a}{2} \sin \frac{a_n - a}{2} \right| \leq 2 \left| \sin \frac{a_n - a}{2} \right|.$$

Now Proposition 3.7.1 (c) shows that

$$|\sin a_n - \sin a| \leq 2 \left| \frac{a_n - a}{2} \right| = |a_n - a|$$

and the result follows from the Squeeze Theorem. $\qquad\square$

> The functions sine and cosine were used by Indian astronomers around 400 AD. All six trigonometric functions (that we use today) were known to Arab mathematicians by the nineth century. Where did the name "sine" come from? The Sanskrit word for half the chord, *jya-ardha*, was abbreviated to *jiva*. In Arabic this became *jiba* but, since there are no symbols for vowels in Arabic, it was written *jb*. Thinking that this stood for *jaib* (Arabic for *bay*) it was translated into Latin as *sinus*. The modern abbreviations sin, cos, tan appear for the first time in 1626, in the work *Trigonométrie* by French mathematician Albert Girard (1595–1632).

Exponential functions. Any discussion of exponential functions must start with the understanding of what they mean. Let $a > 0$. We define $a^1 = a$ and, inductively, $a^{n+1} = a^n \cdot a$. It is not hard to establish that, if $x, y \in \mathbb{N}$,

$$a^{x+y} = a^x a^y, \quad \text{and} \quad (a^x)^y = a^{xy}. \tag{3.10}$$

> The notation a^x (with $x \in \mathbb{N}$) was used first in 1637 by Descartes (page 89). Negative integers and fractions as exponents were first used with the modern notation by Newton in 1676 in a letter to Henry Oldenburg, secretary of the Royal Society.

Next, we define $a^0 = 1$, $a^{-1} = 1/a$, and $a^{-n} = (a^{-1})^n$, for $n \in \mathbb{N}$. Finally, for a rational number p/q, we define $a^{p/q} = \sqrt[q]{a^p}$. It is quite straightforward to prove that equalities (3.10) hold for $x, y \in \mathbb{Q}$. Now we come to the real problem: how to define a^x when x is an irrational number. For example: what is $2^{\sqrt{3}}$? Once again, we need to take advantage of the Completeness Axiom. Let $a > 1$, and let γ be an irrational number. In order to define a^γ we consider two sets: $A = \{a^r : r \in \mathbb{Q} \text{ and } r < \gamma\}$, $B = \{a^r : r \in \mathbb{Q} \text{ and } r > \gamma\}$. Then the set A is bounded above (by any number in B) and B is bounded below. Let $\alpha = \sup A$, $\beta = \inf B$. It is not hard to see that $\alpha \leq \beta$. We will show that $\alpha = \beta$, and we will define $a^\gamma = \alpha$.

Let $\varepsilon > 0$. In order to prove that $\alpha = \beta$, we will establish that $|\beta - \alpha| < \varepsilon$. Since ε is arbitrary, the desired equality will follow. Of course, $\beta - \alpha \geq 0$ so it suffices to show that $\beta - \alpha < \varepsilon$. By Example 2.9.1, $\lim(a^{1/n} - 1) = 0$, so there exists $N \in \mathbb{N}$ so that, for $n \geq N$, $|a^{1/n} - 1| < \varepsilon/\alpha$. (Remark: $\alpha \neq 0$ because all numbers in A are positive.) For such N, we select two rational numbers r_1, r_2 so that $r_1 < \gamma < r_2$ and $r_2 - r_1 < 1/N$. Then $a^{r_1} \in A$ and $a^{r_2} \in B$, so

$$a^{r_1} \leq \alpha \leq \beta \leq a^{r_2}.$$

It follows that

$$\beta - \alpha \le a^{r_2} - a^{r_1} = a^{r_2-r_1}a^{r_1} - a^{r_1} = a^{r_1}\left(a^{r_2-r_1} - 1\right) < \alpha\left(a^{1/N} - 1\right) < \alpha \cdot \frac{\varepsilon}{\alpha} = \varepsilon,$$

so $\alpha = \beta$.

The case $a < 1$ can be reduced to the case above; in this situation we have that $1/a > 1$ and we define $a^\gamma = (1/a)^{-\gamma}$. It is not hard to see that, if $\gamma = p/q$ is a rational number, the "old" definition $a^\gamma = \sqrt[q]{p}$ and the "new" one $a^\gamma = \alpha$ coincide. It is much more tedious to show that the usual rules of exponentiation (3.10) hold. We will skip this and focus on the continuity.

Theorem 3.7.3. *Let $a > 0$. The exponential function $f(x) = a^x$ is continuous at every point $c \in \mathbb{R}$.*

Proof. We will prove the case $a > 1$. Let $\varepsilon > 0$ and let $c \in \mathbb{R}$. Since $\lim a^{1/n} = 1$, there exists $N \in \mathbb{N}$ so that, for $n \ge N$, $|a^{1/n} - 1| < \varepsilon/a^c$. Let $\delta = 1/N$, and suppose that $|x - c| < \delta$. Then

$$|a^x - a^c| = \left|a^c\left(a^{x-c} - 1\right)\right| \le a^c\left|a^{x-c} - 1\right|. \tag{3.11}$$

Since $|x - c| < \delta$, we have that either $0 \le x - c < \delta$ or $-\delta < x - c < 0$. In the former case,

$$\left|a^{x-c} - 1\right| = a^{x-c} - 1 < a^\delta - 1 = \left|a^\delta - 1\right|,$$

while in the latter case,

$$\left|a^{x-c} - 1\right| = 1 - a^{x-c} < 1 - a^{-\delta} = a^{-\delta}\left(a^\delta - 1\right) \le a^\delta - 1 = \left|a^\delta - 1\right|.$$

Either way, $|a^{x-c} - 1| \le |a^\delta - 1|$ which, combined with (3.11), yields

$$|a^x - a^c| \le a^c\left|a^\delta - 1\right| = a^c\left|a^{1/N} - 1\right| < a^c \cdot \frac{\varepsilon}{a^c} = \varepsilon.$$

Thus, $f(x) = a^x$ is continuous if $a > 1$. When $a < 1$ we can write $f(x) = 1/(1/a)^x$ and the continuity of f now follows from the first part of the proof, Example 3.6.2, and Theorem 3.6.4 (d). $\qquad\square$

> It seems that Johann Bernoulli was the first to study the exponential function. In 1697 he published [5] where he obtained various exponential series (such as (8.35)). These ideas flourished in the work of Euler.

Logarithmic functions. Remember that $y = \log_a x$ is another way of saying that $a^y = x$. In other words, the logarithmic functions are inverse functions of the exponential functions. This allows us to conclude about its continuity.

Theorem 3.7.4. *Let $a > 0$ and $a \ne 1$. The logarithmic function $f(x) = \log_a x$ is continuous at every point $c > 0$.*

Proof. Since the (continuous) function $y = a^x$ is strictly increasing or strictly decreasing, the continuity of its inverse function follows from Theorem 3.6.9. $\qquad\square$

> Logarithms were introduced by John Napier (1550–1617), a Scottish mathematician, in 1614, in a book [82]. The name *logarithmus* is his invention. It is a hybrid of two Greek words: *logos* (meaning *ratio*) and *arithmos* (meaning *number*). Logarithms are useful in calculations, because the formula $\log(ab) = \log a + \log b$ allows us to replace multiplication by addition. Rather than calculate the logarithm of a number every time, it was much easier to have them in the form of tables. The first such table was compiled by Henry Briggs (1561–1630), an English mathematician, who changed the original Napier's logarithms into base 10 logarithms. The tables of logarithms were in use well into the second half of

the twentieth century. The logarithmic functions were introduced by Euler in the eighteenth century. He defined the exponential function and the natural logarithm by $e^x = \lim(1+x/n)^n$, $\ln(x) = \lim n(x^{1/n}-1)$, and he proved that the two functions are inverse to one another.

Inverse trigonometric functions. They are continuous for exactly the same reason as logarithms. One only needs to restrict the trigonometric functions to domains which would guarantee that they are strictly increasing or strictly decreasing. For example, for $f(x) = \sin x$ we take as its domain $[-\frac{\pi}{2}, \frac{\pi}{2}]$. Then it has an inverse function known as arcsin x (or $\sin^{-1}(x)$), with domain $[-1,1]$. Similarly, we restrict $\tan x$ to $(-\frac{\pi}{2}, \frac{\pi}{2})$, $\cos x$ to $[0, \pi]$, while for $\cot x$ it is $(0, \pi)$, and we obtain arccos x (a.k.a. $\cos^{-1} x$) defined on $[-1,1]$, arctan x (a.k.a. $\tan^{-1} x$), and arccot x (a.k.a. $\cot^{-1} x$) both defined for all $x \in \mathbb{R}$.

Theorem 3.7.5. *The inverse trigonometric functions are continuous at every point of their domains.*

The idea behind the symbol arcsin x can be seen in the early versions. For example, French mathematician Marquis de Condorcet (1743–1794) wrote arc(sin. $= x$) in 1769. It seems that the present day notation solidified during the nineteenth century. The use of the exponent -1 (as in $\sin^{-1}(x)$) was introduced in 1813 by John Herschel (1792–1871), an English astronomer who has named seven moons of Saturn and four moons of Uranus. In this text we will avoid this type of notation.

Power functions. These are functions of the form $f(x) = x^\alpha$, where α is an arbitrary real constant. Such functions are defined for $x > 0$. Using the fact that, for $x > 0$, $x = e^{\ln x}$ we can write $f(x) = e^{\ln x^\alpha} = e^{\alpha \ln x}$.

Theorem 3.7.6. *Let $\alpha \in \mathbb{R}$. The function $f(x) = x^\alpha$ is continuous for $x > 0$.*

RENÉ DESCARTES (1596–1650) was a French philosopher, a mathematician and a scientist. His mother died when he was a year old, so he grew up with his grandmother. He was of poor health from the start. For example, while attending the Jesuit Collège Royal Henry-Le-Grand at La Flèche he was permitted to stay in bed until 11 am (instead of getting up at 5 am as other students), a practice he maintained all his life. He studied law at the University of Poitiers, receiving the degree in 1616, but never practiced law. In 1618 he enlisted in the military school at Breda in the Dutch Republic. In addition to military engineering he studied mathematics and mechanics. In 1619 he joined the army of Duke Maximilian of Bavaria for a year. On the night of 10–11 November 1619, Descartes had three dreams and believed that a divine spirit revealed to him a new philosophy. Upon exiting, he had formulated analytical geometry and the idea of applying the mathematical method to philosophy. He concluded that the pursuit of science should be a central part of his life's work. In 1620 Descartes left the army, and visited various countries before returning to Paris where he composed his first essay on method: *Regulae ad Directionem Ingenii* (*Rules for the Direction of the Mind*). He sold all of his property to invest in bonds, which provided a comfortable income for the rest of his life. For the next 20 years he lived in the Dutch Republic. Despite frequent moves, he wrote all his major work during this period. In 1633, Galileo was condemned by the Catholic Church, and Descartes abandoned plans to publish *Treatise on the World*, his work of the previous four years. Nevertheless, in 1637 he published part of this work in three essays: *Les Météores* (*The Meteors*), *La Dioptrique* (*Dioptrics*) and *La Géométrie* (*Geometry*), preceded by an introduction, his famous *Discours de la méthode* (*Discourse on the Method*). In it, Descartes lays out four rules of thought, meant to ensure that our knowledge rests upon a firm foundation. In 1649 he went to Stockholm on the invitation by Queen Christina of Sweden. However, she wanted to study mathematics at 5 am, next to an open window. After only a few months he contracted a pneumonia and died. While some of his scientific work has been proven wrong, his mathematical legacy is huge. By creating analytical geometry (a.k.a. "Cartesian geometry", from his Latinized name Cartesius), he provided the basis for the calculus of Newton and Leibniz. In addition, this brought forward algebra as a fundamental branch of mathematics. He invented the convention of representing unknowns in equations by x, y, and z, and knowns by a, b, and c. He is often called the father of the modern Western philosophy and his *Meditationes de Prima Philosophia* (*Meditations on First Philosophy*) continues to be a standard text at most university philosophy departments. His best known philosophical statement is *Cogito ergo sum* (I think, therefore I am).

Exercises

3.7.1. Prove that $f(x) = |x|$ is an elementary function.

3.7.2.* Give an example of a function that is not an elementary function.

3.7.3. Prove that if f is a rational function it cannot have a jump discontinuity.

3.7.4. Prove Proposition 3.7.1.

3.7.5. Simplify $\arcsin(\cos 2x)$, $0 \leq x \leq \dfrac{\pi}{2}$.

3.7.6. Simplify $\arcsin(\cos 2x)$, $\dfrac{\pi}{2} \leq x \leq \dfrac{3\pi}{2}$.

3.7.7.* Prove that $f(x) = \arcsin x$ is continuous directly, without the use of Theorem 3.6.9.

3.7.8.* Prove that $f(x) = \arccos x$ is continuous directly, without the use of Theorem 3.6.9.

3.7.9. Let $x \in \mathbb{R}$ and $a > 1$. Prove that there exists an increasing sequence of rational numbers $\{r_n\}$ and a decreasing sequence of rational numbers $\{s_n\}$ such that $\lim r_n = \lim s_n = x$ and $\lim a^{r_n} = \lim a^{s_n} = a^x$.

3.7.10. Let $a > 1$. Prove that the exponential function $y = a^x$ is monotone.

3.7.11.* Let $x \in \mathbb{R}$ and $a > 1$. Prove that if $\{x_n\}$ is any sequence of real numbers converging to x then $\lim a^{x_n} = a^x$.

3.7.12.* Let $a > 1$. Prove that, for any real numbers x, y, $a^{x+y} = a^x a^y$

3.7.13.* Let $a > 1$. Prove that, for any real numbers x, y, $(a^x)^y = a^{xy}$.

3.7.14. The **hyperbolic functions** are: **hyperbolic sine** $f(x) = \sinh x$, and the **hyperbolic cosine** $f(x) = \cosh x$, defined by

$$\sinh x = \frac{e^x - e^{-x}}{2}, \quad \cosh x = \frac{e^x + e^{-x}}{2}.$$

Prove that:

(a) $\sinh x$ is an odd function, and $\cosh x$ is an even function;

(b) $\cosh^2 x - \sinh^2 x = 1$;

(c) $\sinh x$ and $\cosh x$ are both continuous;

(d) $y = \sinh x$ has an inverse function $y = \operatorname{arsinh} x$ (called *area hyperbolic sine*), given by the formula $\operatorname{arsinh} x = \ln\left(x + \sqrt{x^2 + 1}\right)$

(e) The restriction of $y = \cosh x$ to $[0, +\infty)$ has an inverse function $y = \operatorname{arcosh} x$ (called *area hyperbolic cosine*), given by the formula $\operatorname{arcosh} x = \ln\left(x + \sqrt{x^2 - 1}\right)$ for $x \geq 1$;

(f) $y = \operatorname{arsinh} x$ and $y = \operatorname{arcosh} x$ are continuous.

Remark 3.7.7. The word *area* in the names of the inverse hyperbolic function can be traced to the following fact: the area of the shaded region in Figure 3.13 equals $\operatorname{arcosh} a$.

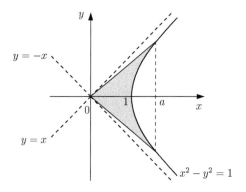

Figure 3.13: The area equals $\operatorname{arcosh} a$.

3.7.15. Prove that the shaded area in Figure 3.13 equals arcosh a.

3.7.16.* Let $f : (-a, a)\backslash\{0\} \to \mathbb{R}$. Show that $\lim_{x \to 0} f(x) = L$ if and only if $\lim_{x \to 0} f(\sin x) = L$.

3.7.17. Prove Theorem 3.7.4 directly, without the use of Theorem 3.6.9.

3.8   Uniform Continuity

Let us look closely at Definition 3.3.8 and Example 3.4.7. It was shown that $\lim_{x \to 3} x^2 = 9$ which means that $f(x) = x^2$ is continuous at $a = 3$. In particular, it was demonstrated that, given $\varepsilon > 0$, one can take $\delta = \min\{1, \varepsilon/7\}$. What if we change a?

Example 3.8.1. A different point requires a different δ.
  Let $f(x) = x^2$, $a = 40$. We will show that $\delta = \min\{1, \varepsilon/7\}$ does not work any more.

Solution. Let $\varepsilon = 0.7$. If we, once again, take $\delta = \min\{1, \varepsilon/7\}$, we obtain that $\delta = 0.1$. Does $|x - 40| < 0.1$ imply that $|x^2 - 40^2| < 0.7$? Certainly $x = 40.01$ satisfies the inequality $|40.01 - 40| < 0.1$. However, $|40.01^2 - 40^2| = 0.8001$ so it is not less than 0.7. Therefore, the suggested δ does not work.
  The solution is to choose a different δ. An analysis similar to the one in Example 3.4.7 shows that a winning choice is $\delta = \min\{1, \varepsilon/81\}$.

Proof. Let $\varepsilon > 0$ and select $\delta = \min\{1, \varepsilon/81\}$. Suppose that $|x - 40| < \delta$. Then $|x - 40| < 1$ so $39 < x < 41$ and $|x| = x < 41$. Further $|x - 40| < \varepsilon/81$ which implies that

$$|x^2 - 40^2| = |(x - 40)(x + 40)| \le |x - 40|\,(|x| + 40) < 81\,|x - 40| < 81 \cdot \frac{\varepsilon}{81} = \varepsilon. \quad \blacklozenge$$

  This example shows that, while f is continuous at both $a = 3$ and $a = 40$, when $\varepsilon > 0$ is given, the choice of δ need not be the same. Of course, the smaller δ (in this case $\min\{1, \varepsilon/81\}$) would work for both $a = 3$ and $a = 40$, but might be inadequate for another value of a.
  Let us now revisit Example 3.4.6. We have established that $\lim_{x \to 4}(3x - 2) = 10$, so $f(x) = 3x - 2$ is continuous at $a = 4$. Further, it was shown that, given $\varepsilon > 0$, we can take $\delta = \varepsilon/3$. So, we ask the same question: what if we select a different point a?

Example 3.8.2. The same δ works for a different point.

Let $f(x) = 3x - 2$, $a = 40$. We will show that $\delta = \varepsilon/3$ still works.

Solution. Suppose that $\varepsilon > 0$ is given, that $\delta = \varepsilon/3$, and that x satisfies the condition $|x - 40| < \varepsilon/3$. Does that imply that $|f(x) - f(40)| < \varepsilon$? We will show that the answer is in the affirmative.

Proof. Let $\varepsilon > 0$ and select $\delta = \varepsilon/3$. Suppose that $|x - 40| < \delta$. Then

$$|f(x) - f(40)| = |(3x - 2) - 118| = |3x - 120| = 3\,|x - 40| < 3\,\frac{\varepsilon}{3} = \varepsilon. \qquad \blacklozenge$$

Examples 3.4.7 and 3.4.6, together with Examples 3.8.1 and 3.8.2, show that functions $f(x) = x^2$ and $g(x) = 3x - 2$ behave differently when it comes to continuity. They are both continuous, but it appears that for g, given $\varepsilon > 0$, the same δ will work, regardless of the point $x = a$. This motivates the following definition.

Definition 3.8.3. Let f be a function defined on a domain A. We say that f is **uniformly continuous** on A if for any $\varepsilon > 0$ there exists $\delta > 0$ such that, for any $x, a \in A$,

$$|f(x) - f(a)| < \varepsilon \quad \text{whenever} \quad |x - a| < \delta.$$

The difference between the usual continuity and the uniform continuity is quite subtle. It even escaped the great Cauchy. The awareness of the distinction came through the lectures of Dirichlet (c. 1854) and Weierstrass (c. 1861). The first to publish the definition of the uniform continuity was Heine in [63] in 1872.

EDUARD HEINE (1821–1981), was a German mathematician. Born and raised in Berlin, he studied at the University of Berlin, but spent 3 semesters at the University of Göttingen where he attended lectures by Gauss (page 196). His 1842 dissertation was in the area of differential equations. His advisors were Dirksen and Ohm (a younger brother of physicist Ohm), but Heine dedicated the work to Dirichlet. After graduating he spent a year in Königsberg attending lectures by Jacobi (page 357) and the physicist F. Neumann (1798-1895), one of the pioneers of theoretical physics. Heine also participated in the work of the famous Königsberg mathematical seminar. In 1844 he passed his habilitation at the University of Bonn, where he taught until 1856, when he moved to the University of Halle, where he remained for the rest of his life. In 1863 he became a member of the Berlin Academy and the Göttingen Academy of Sciences. He served as the rector of the university 1864-65. In 1877 he was awarded the Gauss medal for research. Throughout his career he maintained a close contact with the Berlin mathematicians Weierstrass, Kummer (page 234), and Leopold Kronecker (1823–1891), through numerous trips to Berlin. His most celebrated article is [63] from 1872. It is the first systematic exposition of the real analysis in the sense of Weierstrass. The work ends with Theorem 3.8.7. It also contains the Heine definition of continuity (Corollary 3.6.1).

One way to pinpoint the difference between the two types of continuity is to write both definitions using quantifiers. If f is continuous on A, it means that

$$(\forall \varepsilon)(\forall a)(\exists \delta)(\forall x) \quad |x - a| < \delta \Rightarrow |f(x) - f(a)| < \varepsilon; \qquad (3.12)$$

if f is uniformly continuous on A, it means that

$$(\forall \varepsilon)(\exists \delta)(\forall a)(\forall x) \quad |x - a| < \delta \Rightarrow |f(x) - f(a)| < \varepsilon. \qquad (3.13)$$

In the former case a is given before we need to find δ, while in the latter, we need to find δ before x and a are selected. In other words, in the case of uniform continuity, δ depends only on ε and not on the choice of a. We will see later that this seemingly minor difference will be of help when proving various theorems. At present we will be interested in the relationship between the continuity and the uniform continuity. It is clear that (3.13) implies (3.12), and Example 3.8.1 seems to be saying that the converse need not be true. So let us prove it.

Theorem 3.8.4. *The function $f(x) = x^2$ is not uniformly continuous on $\mathbb{R}$.*

Proof. Suppose, to the contrary, that f is uniformly continuous on $\mathbb{R}$. Let $\varepsilon > 0$. By definition, there exists $\delta > 0$ that can be used for any $a \in \mathbb{R}$. Namely, regardless of the choice of a, as soon as $|x - a| < \delta$ we should have $|f(x) - f(a)| < \varepsilon$. It is easy to see that $x = a + \delta/2$ satisfies $|x - a| < \delta$. Therefore, we should have $|f(a + \delta/2) - f(a)| < \varepsilon$, and this should be true for any $a \in \mathbb{R}$. However,

$$\left| f\left(a + \frac{\delta}{2}\right) - f(a) \right| = \left| \left(a + \frac{\delta}{2}\right)^2 - a^2 \right| = \left| \left(2a + \frac{\delta}{2}\right) \frac{\delta}{2} \right|$$

and the last expression can be made arbitrarily large when a increases without a bound. Thus, we cannot have $|f(a+\delta/2) - f(a)| < \varepsilon$ for all $a \in \mathbb{R}$, and f is not uniformly continuous on $\mathbb{R}$. $\square$

Now we see that the set of uniformly continuous functions on $\mathbb{R}$ is a proper subset of the set of all continuous functions on $\mathbb{R}$. For example, the function $f(x) = x^2$ is continuous but not uniformly continuous on $\mathbb{R}$. What happens if we replace $\mathbb{R}$ by a different set A?

Example 3.8.5. A uniformly continuous function.
Let $f(x) = x^2$, $A = [0, 1]$. We will show that f is uniformly continuous on A.

Solution. Let $\varepsilon > 0$. If we examine the inequality $|f(x) - f(a)| < \varepsilon$, we see that

$$|f(x) - f(a)| = |x^2 - a^2| = |x - a||x + a| \leq |x - a|(|x| + |a|).$$

Since $x, a \in A = [0, 1]$, it follows that $|x| + |a| \leq 2$. Therefore, $|f(x) - f(a)| \leq 2|x - a|$, and if we take $\delta = \varepsilon/2$ it will work regardless of a. In other words, f is uniformly continuous on $[0, 1]$.

Proof. Let $\varepsilon > 0$, and take $\delta = \varepsilon/2$. If $x, a \in [0, 1]$, and if $|x - a| < \delta$, then

$$|f(x) - f(a)| = |x^2 - a^2| = |x - a||x + a| \leq |x - a|(|x| + |a|) \leq 2|x - a| < 2\delta = 2\frac{\varepsilon}{2} = \varepsilon. \; \blacklozenge$$

The proof seems to be telling us that the crucial difference between Theorem 3.8.4 and Example 3.8.5 was that in the latter the domain A was bounded. However, this is not enough.

Example 3.8.6. A continuous function that is not uniformly continuous.
Let $f(x) = 1/x$, $A = (0, 1)$. We will show that, even though the set A is bounded, f is *not* uniformly continuous on A.

Solution. Indeed,

$$|f(x) - f(a)| = \left| \frac{1}{x} - \frac{1}{a} \right| = \left| \frac{x - a}{xa} \right|.$$

If $x = a + \delta/2$ then $|x - a| < \delta$. However,

$$|f(x) - f(a)| = \left| \frac{\frac{\delta}{2}}{a\left(a + \frac{\delta}{2}\right)} \right| \to \infty,$$

as $a \to 0$. So, f is not uniformly continuous on $(0, 1)$.

Proof. Suppose, to the contrary, that f is uniformly continuous on $(0, 1)$. Let $\varepsilon > 0$. By definition, there exists $\delta > 0$ that can be used for any $x, a \in (0, 1)$. Namely, regardless of the choice of x and a, as soon as $|x - a| < \delta$ we should have $|f(x) - f(a)| < \varepsilon$. It is easy to see that $x = a + \delta/2$ satisfies $|x - a| < \delta$. Therefore, we should have $|f(a + \delta/2) - f(a)| < \varepsilon$, and this should be true for any $a \in (0, 1)$. However,

$$\left| f\left(a + \frac{\delta}{2}\right) - f(a) \right| = \left| \frac{\frac{\delta}{2}}{a\left(a + \frac{\delta}{2}\right)} \right|$$

and the last expression can be made arbitrarily large when $a \to 0$. Thus, we cannot have $|f(a + \delta/2) - f(a)| < \varepsilon$ for all $a \in (0, 1)$, and f is not uniformly continuous on $(0, 1)$. ♦

In this case, the interval A was bounded, but it was not closed. It is precisely the combination of these two conditions that guarantees the uniform continuity.

Theorem 3.8.7. *A continuous function on an interval $[a, b]$ is uniformly continuous.*

Proof. Let f be a function that is defined and continuous on $[a, b]$ and suppose, to the contrary, that it is not uniformly continuous. Taking the negative of (3.13) yields

$$(\exists \varepsilon_0)(\forall \delta)(\exists a)(\exists x) \quad |x - a| < \delta \text{ and } |f(x) - f(a)| \geq \varepsilon_0.$$

Let ε_0 be as above and, for every $n \in \mathbb{N}$, let $\delta = 1/n$. We obtain sequences $\{a_n\}$ and $\{x_n\}$, such that for all $n \in \mathbb{N}$,

$$|x_n - a_n| < \frac{1}{n} \text{ and } |f(x_n) - f(a_n)| \geq \varepsilon_0. \tag{3.14}$$

For each $n \in \mathbb{N}$, $x_n \in [a, b]$, so the sequence $\{x_n\}$ is bounded. By the Bolzano–Weierstrass Theorem, it has a convergent subsequence $\{x_{n_k}\}$, converging to a real number x. The inequalities $a \leq x_{n_k} \leq b$ and Corollary 2.3.12 imply that $x \in [a, b]$, so f is continuous at x. Since (3.14) holds for every $n \in \mathbb{N}$, it holds for n_k, so we have, for all $k \in \mathbb{N}$,

$$|x_{n_k} - a_{n_k}| < \frac{1}{n_k} \text{ and } |f(x_{n_k}) - f(a_{n_k})| \geq \varepsilon_0. \tag{3.15}$$

It follows that $a_{n_k} \to x$. Indeed,

$$|a_{n_k} - x| = |a_{n_k} - x_{n_k} + x_{n_k} - x| \leq |a_{n_k} - x_{n_k}| + |x_{n_k} - x| < \frac{1}{n_k} + |x_{n_k} - x| \to 0$$

so the Squeeze Theorem shows that $a_{n_k} \to x$. Since f is continuous at x, Corollary 3.6.1 allows us to conclude that both $\lim f(x_{n_k}) = f(x)$ and $\lim f(a_{n_k}) = f(x)$. However, this contradicts the inequality $|f(x_{n_k}) - f(a_{n_k})| \geq \varepsilon_0$ in (3.15). □

The first published version of Theorem 3.8.7 is in Heine's 1872 paper [63]. In 1878, Dini (page 374) published a book [32]. He included the statement and the proof of this theorem but he credited Cantor for the proof. He claimed that he had learned of it through Schwarz (page 352), who was a student of Weierstrass (just like Cantor). Not surprisingly, this result is now referred to as the Heine–Cantor Theorem, or sometimes the Cantor Theorem. In 1904, however, G. Arendt (1832–1915), a former student of Dirichlet's, published his teacher's lectures from 1854, and the theorem, as well as its proof is there. Arendt has put a star wherever he was not completely sure that he was following the lectures faithfully, but there is no star next to the result in question.

Exercises

In Exercises 3.8.1–3.8.8 determine whether f is uniformly continuous on the given interval:

3.8.1. $f(x) = \dfrac{1}{x}$, $[1, +\infty)$.

3.8.2. $f(x) = \dfrac{x}{4 - x^2}$, $[-1, 1]$.

3.8.3. $f(x) = e^x \cos(1/x)$, $(0, 1)$.

3.8.4. $f(x) = \ln x$, $(0, 1)$.

3.8.5. $f(x) = \sqrt{x}$, $[1, +\infty)$.

3.8.6. $f(x) = e^x$, $[0, +\infty)$.

3.8.7. $f(x) = \tan x$, $(-\pi/2, \pi/2)$.

3.8.8. $f(x) = \tan x$, $(-\pi/4, \pi/4)$.

3.8.9. Show that the function $f(x) = |\sin x|/x$ is uniformly continuous on each of the intervals $(-1, 0)$ and $(0, 1)$, but not on their union.

3.8.10. Suppose that f and g are uniformly continuous on a finite interval (a, b). Show that the same is true of their sum and product.

3.8.11.* Let $A \subset \mathbb{R}$ be a bounded set, and let f be a uniformly continuous function on A. Prove that f is a bounded function.

3.8.12.* Give an example of functions f and g that are uniformly continuous but their product is not.

3.8.13. Prove that if f is defined and continuous in $[a, +\infty)$ and if there exists a finite limit $\lim_{x \to +\infty} f(x)$, then f is uniformly continuous in $[a, +\infty)$.

3.8.14.* Give an example of a function f that is continuous and bounded but not uniformly continuous.

3.8.15. Suppose that f is defined and continuous on $[a, b]$. Prove that f is uniformly continuous on (a, b) if and only if it is uniformly continuous on $[a, b]$.

3.8.16. Suppose that f is uniformly continuous on $[a, b]$ and that g is uniformly continuous on the set $f([a, b])$. Then $g \circ f$ is uniformly continuous on $[a, b]$.

3.8.17. Suppose that f is uniformly continuous on (a, b) and that $\{x_n\}$ is a Cauchy sequence in (a, b). Prove that $\{f(x_n)\}$ is a Cauchy sequence.

3.8.18.* A function f is **periodic** with period $T \neq 0$ if $f(x + T) = f(x)$ for all $x \in \mathbb{R}$. Prove that if f is continuous and periodic, then it is uniformly continuous.

3.8.19.* Suppose that f satisfies the Lipschitz condition on (a, b): there exists $M > 0$ such that, for any $x, y \in (a, b)$, $|f(x) - f(y)| \leq M|x - y|$. Prove that f is uniformly continuous on (a, b). Give an example of a function f and an interval (a, b) so that f is uniformly continuous on (a, b) but does not satisfy the Lipschitz condition on (a, b).

3.9   Two Properties of Continuous Functions

In this section we will look at two very useful properties of functions, both consequences of the continuity. The first one is known as the *intermediate value property*, the second one as the *extreme value property*.

We will start with the former. It owes its name to the geometric observation: if two points A and B are on the opposite sides of the x-axis, and if we connect them without lifting a pen, the connecting curve has to cross the x-axis (Figure 3.14(a)).

Theorem 3.9.1. *Let f be a continuous function on $[a, b]$. If $f(a) < 0$ and $f(b) > 0$ then there exists $c \in (a, b)$ such that $f(c) = 0$.*

Proof. Let us denote $J_1 = [a, b]$, and consider the midpoint $c_1 = (a + b)/2$. If $f(c_1) = 0$ we will take $c = c_1$. Otherwise, $f(c_1) > 0$ or $f(c_1) < 0$. In the former case, we see that $f(a) < 0$ and $f(c_1) > 0$; in the latter $f(c_1) < 0$ and $f(b) > 0$. Either way, f will take values of the opposite sign at the endpoints of the interval $[a, c_1]$ or $[c_1, b]$. Let us denote that interval by J_2, its endpoints by a_2, b_2, and consider its midpoint $c_2 = (a_2 + b_2)/2$ (see Figure 3.14(b)). If $f(c_2) = 0$ we will take $c = c_2$. Otherwise, we obtain the interval J_3 with endpoints a_3, b_3 such that $f(a_3) < 0$ and $f(b_3) > 0$ (Figure 3.14(b)). Continuing the process we obtain a sequence of closed nested intervals $J_n = [a_n, b_n]$ such that $f(a_n) < 0$ and $f(b_n) > 0$, and

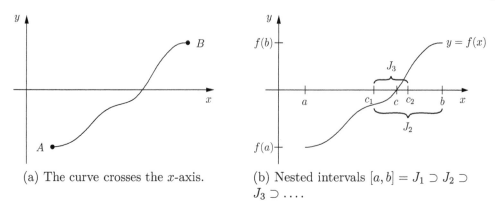

(a) The curve crosses the x-axis.

(b) Nested intervals $[a,b] = J_1 \supset J_2 \supset J_3 \supset \dots.$

Figure 3.14: The intermediate value property.

$\ell(J_n) = (b-a)/2^{n-1}$. By Theorem 2.7.7, taking into account that $\ell(J_n) \to 0$, there exists a unique point c that belongs to all intervals J_n. It remains to show that $f(c) = 0$.

Since $c \in J_n$ for all $n \in \mathbb{N}$, it follows that its distance from each endpoint a_n or b_n cannot exceed the length of the interval $(b-a)/2^{n-1}$. Therefore we have that $\lim a_n = \lim b_n = c$. The continuity of f now implies that $f(c) = \lim f(a_n)$, and since $f(a_n) < 0$ for all $n \in \mathbb{N}$, Corollary 2.3.12 shows that $f(c) \leq 0$. On the other hand, $f(c) = \lim f(b_n)$ and $f(b_n) > 0$ for all $n \in \mathbb{N}$, so we see that $f(c) \geq 0$. Consequently, $f(c) = 0$. $\qquad\square$

Remark 3.9.2. If the hypotheses $f(a) < 0$ and $f(b) > 0$ are replaced by $f(a) > 0$ and $f(b) < 0$, the conclusion still holds. To see this, let g be a function defined by $g(x) = -f(x)$. Then g satisfies all the hypotheses of Theorem 3.9.1, so there exists $c \in (a,b)$ such that $g(c) = 0$. Clearly, $f(c) = 0$ as well.

> Theorem 3.9.1 is known as Bolzano's theorem. He proved it in 1817. The algorithm used in the proof (splitting intervals in 2 equal parts) was used throughout the eighteenth century to approximate the solution c. In those days, the existence of a solution was taken for granted. Bolzano was among the first to understand the need for a proof.

The geometric motivation for Theorem 3.9.1 can be applied not only to the x-axis but to any horizontal line.

Theorem 3.9.3 (The Intermediate Value Theorem). *Let f be a continuous function on the interval $[a,b]$. If C is any number between $f(a)$ and $f(b)$ then there exists $c \in (a,b)$ such that $f(c) = C$.*

Proof. Suppose that $f(a) < C < f(b)$. The other possibility $f(a) > C > f(b)$ will follow from the one we are considering, just like in Remark 3.9.2. Notice that in this theorem, we are proving that any curve connecting $f(a)$ and $f(b)$ must cross the horizontal line $y = C$. We will use a geometric idea: translate the picture so that the horizontal line coincides with the x-axis (see Figure 3.15). Algebraically, we will introduce a new function g defined by $g(x) = f(x) - C$. Now $g(a) = f(a) - C < 0$, $g(b) = f(b) - C > 0$, g is continuous, so by Theorem 3.9.1 there exists $c \in (a,b)$ such that $g(c) = 0$. This implies that $f(c) - C = 0$, and $f(c) = C$. $\qquad\square$

> Cauchy proved Theorem 3.9.3 in *Cours d'Analyse* and derived Theorem 3.9.1 as a consequence.

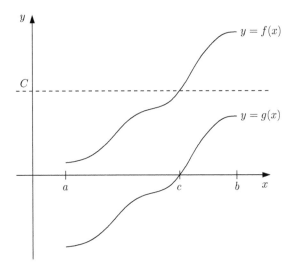

Figure 3.15: The translated graph of f.

Remark 3.9.4. Theorems 3.9.1 and 3.9.3 stop being true if the interval $[a, b]$ is replaced by any domain A. For example, if $A = [0,1] \cup [2,3]$ and if $f(x) = -1$ for $x \in [0,1]$ and $f(x) = 1$ for $x \in [2,3]$, then f is a continuous function on A, and $f(0) < 0$, $f(2) > 0$. However, there is no point c such that $f(c) = 0$.

Remark 3.9.5. The intermediate value property was in the eighteenth century deemed so important, that it served as a definition of a continuous function. Theorem 3.9.3 shows that every continuous function has this property, but the converse is not true (see Exercise 3.9.7).

Now we can prove the result that we have stated earlier (Theorem 3.6.11): if a continuous function is injective, it has to be strictly increasing or decreasing.

Proof of Theorem 3.6.11. Suppose, to the contrary, that the continuous and injective function f is not monotone. Then we can find 3 points $a < b < c$ such that either

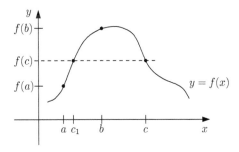

Figure 3.16: If f is not monotone.

$$f(a) < f(b) \text{ and } f(b) > f(c), \quad \text{or} \quad f(a) > f(b) \text{ and } f(b) < f(c).$$

We will prove the former case, the latter can be established by considering the function $g = -f$. Next, we compare $f(a)$ and $f(c)$. We will consider the case $f(a) < f(c)$, and leave the other as an exercise. So, suppose that $f(a) < f(c) < f(b)$. By Theorem 3.9.3, there exists $c_1 \in (a, b)$ such that $f(c_1) = f(c)$. Clearly $c_1 \neq c$ but $f(c_1) = f(c)$, contradicting the assumption that f is injective. □

Now, we will talk about the extreme value property. Let us start with an example. Let $f(x) = x^2$ and $A = [-2, 2]$. We can see that the range B of f is the interval $[0, 4]$. Therefore $\sup B = 4$ and $\inf B = 0$. In a situation like this we should write $\sup\{f(x) : x \in A\} = 4$, but we almost always use a shorter $\sup_{x \in A} f(x) = 4$, or even $\sup f(x) = 4$ when no ambiguity can occur. Similarly, we write $\inf_{x \in A} f(x) = 0$, or $\inf f(x) = 0$.

Notice that both of these numbers belong to the range of f. When the supremum of a set B belongs to B, then we call it the *maximum* of B. Similarly, if the infimum of a set B belongs to B, then we call it the *minimum* of B. So, while it is true that $\sup B = 4$ and $\inf B = 0$, it is more precise to say that $\max B = 4$ and $\min B = 0$. In other words, $\max_{x \in A} f = 4$, $\min_{x \in A} f = 0$. We say that f *attains* its (absolute) maximum and minimum values.

Example 3.9.6. A function that does not attain its minimum on a given set.
Let $f(x) = \begin{cases} x^2, & \text{if } x \in [-2, 0) \cup (0, 2] \\ 1, & \text{if } x = 0. \end{cases}$ We will show that f does not attain its minimum.

Solution. The domain of f is again $A = [-2, 2]$, but the range $B = (0, 4]$. Here, f attains

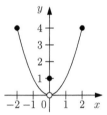

Figure 3.17: f does not attain its minimum.

its maximum value at $x = 2$ (and $x = -2$): $\max_{x \in A} f(x) = 4$. However, $\inf_{x \in A} f(x) = 0$ and 0 does not belong to the range of f. So, f does not attain its minimum value. ◆

Example 3.9.7. A function that does not attain its maximum on a given set.
Let $f(x) = \begin{cases} x^2, & \text{if } x \in [-2, 0] \\ 1/x, & \text{if } x \in (0, 2]. \end{cases}$ We will show that f does not attain its maximum.

Solution. The domain of f is again $A = [-2, 2]$, but the range $B = [0, \infty)$. Thus f attains its minimum value at $x = 0$: $\min_{x \in A} f(x) = 0$. However, the set B is not bounded above so

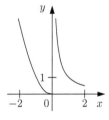

Figure 3.18: f does not attain its maximum.

$\sup B = \infty$ and f does not have a maximum value (let alone attain it). ◆

An analysis of these examples shows that the reason that the function in Example 3.9.6 does not attain its minimum value is the fact that it is discontinuous at $x = 0$. Similarly, the function in Example 3.9.7 is not even bounded. (Should we blame the discontinuity at $x = 0$?) Clearly, if f is discontinuous, there is no guarantee that it will be bounded and, even if it is bounded like the function in Example 3.9.6, it need not attain its minimum value. What if f *is* continuous? Can we find a function with the domain $[-2, 2]$ that is continuous but fails to be bounded or to attain its maximum (or minimum) value? The answer is: "No!" and it has been supplied by Weierstrass.

Theorem 3.9.8 (The Extreme Value Theorem). *Let f be a continuous function defined on a closed interval $[a, b]$. Then f is bounded and it attains both its minimum and its maximum value.*

Proof. First we will prove that f is bounded on $A = [a, b]$. By definition, this means that

$$(\exists M)(\forall x \in A)|f(x)| \leq M.$$

We will argue by contradiction, so we will assume that

$$(\forall M)(\exists x \in A)|f(x)| > M.$$

If we take $M = n$, then for every $n \in \mathbb{N}$ there exists $x_n \in A$ such that $|f(x_n)| > n$. In other words, we obtain a sequence $\{x_n\}$ in A, and it is clear that this is a bounded sequence. By the Bolzano–Weierstrass Theorem, it has a convergent subsequence $\{x_{n_k}\}$ converging to a limit L. Since $a \leq x_{n_k} \leq b$, Corollary 2.3.12 implies that $a \leq L \leq b$. Further, by the continuity of f, $f(L) = \lim f(x_{n_k})$. However, the inequality $|f(x_{n_k})| > n_k$ shows that $f(x_{n_k})$ is not a bounded sequence, so it cannot be convergent by Theorem 2.3.7. This contradiction shows that f must be a bounded function on $A = [a, b]$.

Since f is bounded, its range B is a bounded set. Let $M = \sup B$ and $m = \inf B$. We will show that there exists a point $c \in [a, b]$ such that $f(c) = M$. Let n be a positive integer. Since M is the least upper bound of B, the number $M - 1/n$ cannot be an upper bound of B. Consequently, there exists $y_n \in [a, b]$ such that

$$M - \frac{1}{n} < f(y_n) < M. \tag{3.16}$$

Once again, the Bolzano–Weierstrass Theorem guarantees the existence of a subsequence $\{y_{n_k}\}$ and a number $c \in [a, b]$ to which the said subsequence converges. Further, the Squeeze Theorem applied to (3.16) shows that, for each $k \in \mathbb{N}$,

$$\lim \left(M - \frac{1}{n_k}\right) \leq \lim f(y_{n_k}) \leq M$$

so $\lim f(y_{n_k}) = M$. By the continuity of f, $M = f(\lim y_{n_k}) = f(c)$. So, f attains its maximum value.

The fact that f attains its minimum value can now be established by considering the function $g = -f$. $\qquad \square$

Theorem 3.9.8 is known as "The Maximum Theorem", "The Extreme Value Theorem", and Weierstrass called it "The Principal Theorem" in his lectures in 1861. The result was originally proved by Bolzano, but his proof was not published until 1930! The first publication was by Cantor in 1870 in [10].

Exercises

3.9.1. Suppose that f is continuous on $[0, 2]$ and $f(0) = f(2)$. Prove that there exist $x_1, x_2 \in [0, 2]$ such that $x_2 - x_1 = 1$ and $f(x_1) = f(x_2)$.

3.9.2.* Suppose that f is continuous on $[0,2]$ and $f(0) = f(2)$. Prove that for any $\varepsilon > 0$ there exist $x_1, x_2 \in [0,2]$ such that $|x_2 - x_1| < \varepsilon$ and $f(x_1) = f(x_2)$.

3.9.3.* Give an example of a continuous function $f : \mathbb{R} \to \mathbb{R}$ that attains each of its values exactly 3 times. Is there a continuous function $f : \mathbb{R} \to \mathbb{R}$ that attains each of its values exactly 2 times?

3.9.4. Let f be a continuous function on (a,b), and let $x_1, x_2, \ldots, x_n \in (a,b)$. Then, there exists a point $x \in (a,b)$ such that

$$f(x) = \frac{f(x_1) + f(x_2) + \cdots + f(x_n)}{n}.$$

3.9.5. Let $f : [0,1] \to [0,1]$ be continuous. Prove that there exists $c \in [0,1]$ such that $f(c) = c$.

3.9.6.* A function f has the *intermediate value property* on the interval $[a,b]$ if for any $u, v \in [a,b]$, and any y between $f(u)$ and $f(v)$, there exists c between u and v, such that $f(c) = y$. Prove that, if f is a strictly increasing function on $[a,b]$ that has the intermediate value property, then it is continuous.

3.9.7. Let

$$f(x) = \begin{cases} \sin \dfrac{1}{x}, & \text{if } x \in (0,1] \\ 0, & \text{if } x = 0. \end{cases}$$

Prove that f is not continuous but has the intermediate value property on $[0,1]$.

3.9.8.* Let $a \in \mathbb{R}$ and let f be a function that is continuous and bounded on $(a, +\infty)$. Prove that, for every $T \in \mathbb{R}$, there exists a sequence x_n such that $\lim x_n = +\infty$ and $\lim (f(x_n + T) - f(x_n)) = 0$.

3.9.9. Give an example of a bounded function on $[0,1]$ that attains neither an infimum nor a supremum.

3.9.10. Suppose that f is a continuous function on $[a, +\infty)$ and that there exists a finite limit $\lim_{x \to +\infty} f(x)$. Show that f is bounded.

3.9.11. Show that if f is continuous and periodic, then f attains both its minimum and its maximum.

3.9.12. Let f be a continuous function defined on $[a,b]$. Then there exist real numbers c, d such that $f([a,b]) = [c,d]$.

3.9.13.* Let f be a continuous function on $[a,b]$ and define $M(x) = \sup\{f(t) : t \in [a,x]\}$. Prove that M is continuous on $[a,b]$.

3.9.14.* Let f be a continuous function on $[a,b]$ and define $m(x) = \inf\{f(t) : t \in [a,x]\}$. Prove that m is continuous on $[a,b]$.

4

Derivative

We live in a world that is full of changes and it is to our advantage to keep track of them. Unfortunately, many of them are quite complicated, and it is often practical to approximate the non-linear dependence between variable quantities by the linear ones. This task consists of two parts. First, we need to find an appropriate linear approximation. This is where the derivatives are indispensable. Then, it is important to prove that this approximation is indeed a good one. The limits of functions studied in Chapter 3 are used to accomplish this goal.

4.1 Computing the Derivatives: A Review

Example 4.1.1. Find the derivative of $y = x^3 + 2x - 5$.

Solution. We use the rules for the sum/difference and for the multiplication by a constant c

$$(f + g)' = f' + g' \quad \text{and} \quad (cf)' = cf',$$

and we obtain $y' = (x^3)' + (2x)' - (5)' = (x^3)' + 2(x)' - (5)'$. The power rule states that, for any $\alpha \in \mathbb{R}$,

$$(x^\alpha)' = \alpha x^{\alpha-1},$$

and the derivative of any constant function (such as $y = 5$) equals 0. Thus, $y' = 3x^2 + 2.$ ◆

Example 4.1.2. Find the derivative of $y = (2x^2 - 5x + 1)e^x$.

Solution. We use the product rule for derivatives:

$$(f \cdot g)' = f' \cdot g + f \cdot g',$$

as well as the fact that the derivative of e^x is e^x to obtain that

$$
\begin{aligned}
y' &= (2x^2 - 5x + 1)'e^x + (2x^2 - 5x + 1)(e^x)' \\
&= (4x - 5)e^x + (2x^2 - 5x + 1)(e^x) \\
&= (2x^2 - x - 4)e^x.
\end{aligned}
$$
◆

Example 4.1.3. Find the derivative of $y = \dfrac{x \sin x + \cos x}{x \cos x - \sin x}$.

Solution. We use the quotient rule for derivatives:

$$\left(\frac{f}{g}\right)' = \frac{f' \cdot g - f \cdot g'}{g^2},$$

and we obtain that

$$y' = \frac{(x \sin x + \cos x)'(x \cos x - \sin x) - (x \sin x + \cos x)(x \cos x - \sin x)'}{(x \cos x - \sin x)^2}.$$

The product rule and the fact that $(\sin x)' = \cos x$, $(\cos x)' = -\sin x$ gives

$$(x\sin x + \cos x)' = (x)'\sin x + x(\sin x)' + (\cos x)' = \sin x + x\cos x - \sin x = x\cos x, \quad \text{and}$$
$$(x\cos x - \sin x)' = (x)'\cos x + x(\cos x)' - (\sin x)' = \cos x - x\sin x - \cos x = -x\sin x.$$

It follows that

$$\begin{aligned}
y' &= \frac{(x\cos x)(x\cos x - \sin x) - (x\sin x + \cos x)(-x\sin x)}{(x\cos x - \sin x)^2} \\[2mm]
&= \frac{x^2\cos^2 x - x\cos x\sin x + x^2\sin^2 x + x\cos x\sin x}{(x\cos x - \sin x)^2} \\[2mm]
&= \frac{x^2}{(x\cos x - \sin x)^2}.
\end{aligned}$$

◆

Example 4.1.4. Find the derivative of $y = \ln\sin x$.

Solution. Now we need the chain rule:

$$[f(g(x))]' = f'(g(x))\cdot g'(x).$$

We will apply it to $f(x) = \ln x$ and $g(x) = \sin x$. We need to know that $(\ln x)' = 1/x$, so we have that

$$y' = \frac{1}{\sin x}\cdot\cos x = \cot x.$$

◆

Example 4.1.5. Find the derivative of $y = \arctan(1/x)$.

Solution. In addition to the chain rule, we need the fact that

$$\left(\frac{1}{x}\right)' = (x^{-1})' = (-1)x^{-2} = -\frac{1}{x^2}$$

and that $(\arctan x)' = 1/(1 + x^2)$. Then

$$y' = \frac{1}{1 + \left(\frac{1}{x}\right)^2}\cdot\left(\frac{1}{x}\right)' = \frac{1}{1 + \frac{1}{x^2}}\left(-\frac{1}{x^2}\right) = \frac{-1}{x^2 + 1}.$$

◆

Example 4.1.6. Find the derivative of $y = \sqrt{\tan\left(\frac{1}{2}x\right)}$

Solution. We will apply the chain rule with $f(x) = \sqrt{x}$ and $g(x) = \tan\left(\frac{1}{2}x\right)$. First,

$$(\sqrt{x})' = \left(x^{1/2}\right)' = \frac{1}{2}x^{-\frac{1}{2}} = \frac{1}{2}\cdot\frac{1}{x^{1/2}} = \frac{1}{2\sqrt{x}}.$$

Now we have that

$$y' = \frac{1}{2\sqrt{\tan\left(\frac{1}{2}x\right)}}\left(\tan\left(\frac{1}{2}x\right)\right)'.$$

In order to calculate the derivative of $\tan\left(\frac{1}{2}x\right)$ we will apply the chain rule again, this time with $f(x) = \tan x$ and $g(x) = \frac{1}{2}x$. Since $(\tan x)' = \sec^2 x$ we obtain

$$y' = \frac{1}{2\sqrt{\tan\left(\frac{1}{2}x\right)}}\sec^2\left(\frac{1}{2}x\right)\left(\frac{1}{2}x\right)' = \frac{1}{2\sqrt{\tan\left(\frac{1}{2}x\right)}}\sec^2\left(\frac{1}{2}x\right)\frac{1}{2}.$$

◆

Example 4.1.7. Find the derivative of $y = x^{\sin x}$

Solution. Here we use the fact that $\ln y = \ln x^{\sin x} = \sin x \ln x$ so $y = e^{\sin x \ln x}$. Now we can apply the chain rule, with $f(x) = e^x$ and $g(x) = \sin x \ln x$, to obtain

$$y' = e^{\sin x \ln x} (\sin x \ln x)' = e^{\sin x \ln x} [(\sin x)' \ln x + \sin x (\ln x)']$$

$$= e^{\sin x \ln x} \left(\cos x \ln x + (\sin x)\frac{1}{x} \right) = x^{\sin x} \left(\cos x \ln x + (\sin x)\frac{1}{x} \right). \qquad \blacklozenge$$

Exercises

Find the derivatives of the following functions:

4.1.1. $y = \dfrac{1 + x - x^2}{1 - x + x^2}$.

4.1.2. $y = \sqrt{x + \sqrt{x + \sqrt{x}}}$.

4.1.3. $y = \dfrac{x}{\sqrt{1 - x^2}}$.

4.1.4. $y = \sqrt[3]{\dfrac{1 + x^3}{1 - x^3}}$.

4.1.5. $y = \sin(\sin(\sin x))$.

4.1.6. $y = 2^{\tan \frac{1}{x}}$.

4.1.7. $y = x^{a^a} + a^{x^a} + a^{a^x}$.

4.1.8. $y = \ln^3 x^2$.

4.1.9. $y = \ln \dfrac{x^2 - 1}{x^2 + 1}$.

4.1.10. $y = \arccos \sqrt{1 - x^2}$.

4.1.11. $y = \arctan \dfrac{1 + x}{1 - x}$.

4.2 Derivative

We would like to establish the properties that we have used in Section 4.1. In order to do that, we will need a precise definition of the derivative. Let us first look at an example that will motivate the definition.

Example 4.2.1. Calculate $\sqrt[3]{9}$ (without a calculator).

Solution. Since $\sqrt[3]{8} = 2$, we know that the result is bigger than 2. Certainly, it is less than 3, because $3^3 = 27$, and it appears that it is closer to 2, than to 3. The problem is that, using paper and pencil only, we can do only the four arithmetic operations. Thus, it would be nice if we could replace $y = \sqrt[3]{x}$ with a linear function. The usual formula $y = mx + b$ is not useful here, because we do not have information about the y-intercept b. It is more practical to use the "point-slope" formula:

$$y - y_0 = m(x - x_0)$$

where m is the slope, and (x_0, y_0) is a point on the graph of the linear function. This is because we know $\sqrt[3]{8}$, and we can use $x_0 = 8$ and $y_0 = 2$. So, our linear function will have an equation

$$y = 2 + m(x - 8)$$

and its graph will have a common point $(8, 2)$ with the graph of $y = \sqrt[3]{x}$. It remains to select the slope m. Of course, we are free to choose m, but we want it to yield the best possible approximation to $y = \sqrt[3]{x}$. In other words we are hoping to have

$$\sqrt[3]{x} \approx 2 + m(x - 8).$$

What that means is that

$$\sqrt[3]{x} = 2 + m(x - 8) + r(x), \qquad (4.1)$$

where the error $r(x) \to 0$, when $x \to 8$. One measure of the quality of approximation is the rate at which the error term goes to 0. We will ask that $r(x) \to 0$ very fast, faster than $x - 8$. This means that we want

$$\lim_{x \to 8} \frac{r(x)}{x - 8} = 0.$$

Now we can write (4.1) as

$$\frac{\sqrt[3]{x} - 2}{x - 8} = m + \frac{r(x)}{x - 8}.$$

If we let $x \to 8$, the right hand side has limit m, so it remains to calculate

$$\lim_{x \to 8} \frac{\sqrt[3]{x} - 2}{x - 8}.$$

We will postpone the evaluation of this limit until the next section, and we will state that it equals $1/12$. We conclude that the best linear approximation at $x = 8$ is $y = 2 + \frac{1}{12}(x - 8) = \frac{1}{12}x + \frac{4}{3}$. Geometrically, this is the equation of the tangent line to the graph of $y = \sqrt[3]{x}$ at $x = 8$.

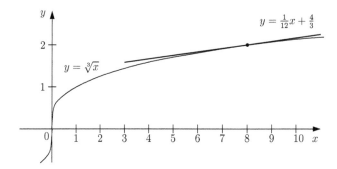

Figure 4.1: $y = \frac{1}{12}x + \frac{4}{3}$ is the best linear approximation at $x = 8$.

Consequently, we can approximate

$$\sqrt[3]{9} \approx \frac{1}{12} \cdot 9 + \frac{4}{3} = \frac{3}{4} + \frac{4}{3} = \frac{25}{12} \approx 2.083333\ldots.$$

This is not too bad, since the calculator gives 2.080083823 so our result differs only starting with the 3rd decimal, and the difference is less than 0.5×10^{-2}. ♦

We can try the same approach with any function $y = f(x)$, and a point $(x_0, f(x_0))$ on its graph. Then the linear approximation is $y = f(x_0) + m(x - x_0)$ and the optimal m would be

$$\lim_{x \to x_0} \frac{f(x) - f(x_0)}{x - x_0}.$$

In order for this method to work, the limit above must exist. This leads us to the definition of the derivative.

Definition 4.2.2. Let f be a function defined on (a, b) and let $c \in (a, b)$. If the limit

$$\lim_{x \to c} \frac{f(x) - f(c)}{x - c} \tag{4.2}$$

exists, we say that f is **differentiable at** $x = c$ and we call this limit the **derivative of** f **at** c and denote it by $f'(c)$ or $\frac{df}{dx}(c)$. If f is differentiable at every point of a set A, we say that it is differentiable on A.

It is a good moment to say a word or two about the notation and terminology. The fraction in (4.2) is called the **difference quotient**, its numerator is typically denoted by Δf or Δy, and the denominator by Δx. On the other hand, we have developed a linear approximation

$$y = y_0 + f'(x_0)(x - x_0),\tag{4.3}$$

and it is customary to write dy for the difference $y - y_0$ and dx for $x - x_0$. Notice that $dx = \Delta x$ but dy need not be equal to Δy. This is because Δy represents the change in y along the graph of $y = f(x)$, while dy is the change in y along the tangent line. The quantity dy is the **differential** of $y = f(x)$. Equation (4.3) can be written as $dy = y'\,dx$ and it holds whenever we have a differentiable function. In other words, if a function f is differentiable, then $df = f'dx$.

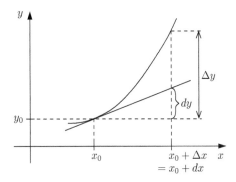

Figure 4.2: dy is the change in y along the tangent line.

The definition of the derivative was first given by Bolzano in 1817, and then popularized by Cauchy. We owe the word "derivative" and the "prime" symbol to denote the derivative to Lagrange (page 398). They appear in his article [76] in 1770. He also wrote $du = u'\,dx$ in 1772 in [77], but the symbols dx, dy, and dy/dx were invented by Leibniz in 1675.

We notice that, just like continuity, differentiability is a *local* property of a function, i.e., the statements about the differentiability of a functions are made for one point at a time. Of course, if a function is differentiable at every point of its domain, we will simply say that it is differentiable. So, it is interesting to compare the differentiability and continuity.

Theorem 4.2.3. *If a function f is differentiable at $x = c$, then it is continuous at $x = c$.*

Proof. Let $\varepsilon > 0$. By the definition of differentiability, there exists δ_1 such that

$$0 < |x - c| < \delta_1 \quad \Rightarrow \quad \left| \frac{f(x) - f(c)}{x - c} - f'(c) \right| < 1.$$

For such x,

$$|f(x) - f(c) - f'(c)(x - c)| < |x - c|,$$

and the triangle inequality implies that

$$\begin{aligned}
|f(x) - f(c)| &= |f'(c)(x - c) + (f(x) - f(c) - f'(c)(x - c))| \\
&\leq |f'(c)(x - c)| + |f(x) - f(c) - f'(c)(x - c)| \\
&< |f'(c)|\,|x - c| + |x - c| \\
&= |x - c|\,(|f'(c)| + 1).
\end{aligned}\tag{4.4}$$

Let

$$\delta = \min\left\{\delta_1, \frac{\varepsilon}{|f'(c)| + 1}\right\}.$$

Now, if $|x - c| < \delta$ then $|x - c| < \delta_1$ and, using (4.4),

$$|f(x) - f(c)| < |x - c|\,(|f'(c)| + 1) \leq \frac{\varepsilon}{|f'(c)| + 1}\,(|f'(c)| + 1) = \varepsilon. \qquad \square$$

So, differentiability implies continuity. Is the converse true? The answer is no, and the standard example is the function $f(x) = |x|$ and $c = 0$. Since $|x| = \sqrt{x^2}$, it is an elementary

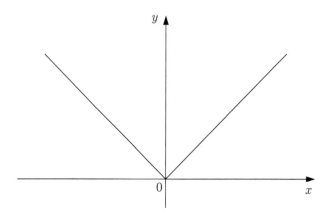

Figure 4.3: $y = |x|$ is not differentiable at 0.

function, so it is continuous at every point, including $c = 0$. However, it is not differentiable at 0. The limit

$$\lim_{x \to 0} \frac{|x| - |0|}{x - 0} \qquad\qquad (4.5)$$

does not exist, because the left limit is -1 and the right limit is 1. Geometrically, the culprit is a "sharp corner" of the graph of f at $x = 0$. Everywhere else, f either has equation $f(x) = x$ or $f(x) = -x$, and we will soon prove that f is differentiable at any $x \neq 0$.

> The fact that differentiability implies continuity is probably to blame for the relatively belated interest in the latter property. While the derivative has been a central object since the days of Newton and Leibniz, a serious study of continuity began only in the nineteenth century. Until then, the concept of a function included differentiability, so there was no need to think about their continuity. This changed with the discovery that many important functions (such as sums of some Fourier series) were not differentiable.

Based on the function $y = |x|$, one might be tempted to conclude that continuous functions are "almost" differentiable. Certainly, there can be more than one "corner" on the graph. For example, the function $y = ||x| - 1|$ has 3 such points (-1, 0, and 1) where it is not differentiable (Figure 4.4(a)). While the equations may be harder to come up with, geometrically it is easy to visualize a zig-zag graph with any number of "sharp corners". At all these points, such a function will be continuous but not differentiable. Figure 4.4(b) presents one example.

We see that a continuous function may have infinitely many points where it fails to be differentiable. How far can this go? The answer is very counter-intuitive: there exists a function that is continuous at every point of the real line, yet it is not differentiable at any point. The first such function was produced by Weierstrass, but many others have been

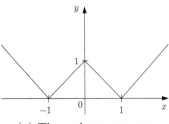
(a) Three sharp corners.

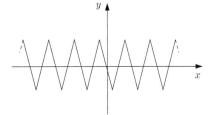
(b) Infinitely many sharp corners.

Figure 4.4: The graph has sharp corners.

exhibited later. We will present one in Section 8.3. While the existence of such a function may seem like some kind of a bizarre accident, it is quite the opposite. In a more advanced course it can be shown that, in some sense, among all continuous functions, those that have a derivative at even one point represent a collection of a negligible size.

Let us now return to the function $f(x) = |x|$, and the limit (4.5) which does not exist, but the one-sided limits do. In such a situation, we say that f is both **right differentiable** and **left differentiable**, and that its **left derivative** $f'_-(0) = -1$ and the **right derivative** $f'_+(0) = 1$.

Example 4.2.4. A function that is right differentiable but not left differentiable at a point.

Let $f = \chi_{[0,1]}$, the *characteristic function* of the interval $[0,1]$, i.e.

$$f(x) = \begin{cases} 1, & \text{if } x \in [0,1] \\ 0, & \text{if } x \notin [0,1]. \end{cases}$$

We will show that f is right differentiable but not left differentiable at $x = 0$.

Solution. Indeed,

$$\lim_{h \to 0^+} \frac{f(0+h) - f(0)}{h} = \lim_{h \to 0^+} \frac{1-1}{h} = 0,$$

so $f'_+(0) = 0$. On the other hand

$$\lim_{h \to 0^-} \frac{f(0+h) - f(0)}{h} = \lim_{h \to 0^-} \frac{0-1}{h},$$

so the limit does not exist, and f is not left differentiable at $x = 0$. ♦

One-sided derivatives are quite handy when a function is defined on a closed interval $[a, b]$. When the right derivative at $x = a$ exists, we will often say that f is differentiable at $x = a$. Similarly, if f is left differentiable at $x = b$, we will say that f is differentiable at $x = b$. Therefore, the statement that f is differentiable in $[a, b]$ will mean that it is differentiable in (a, b) and that it is right differentiable at $x = a$ and left differentiable at $x = b$.

Example 4.2.5. A function that is differentiable on $[a, b]$.

Let $f(x) = \begin{cases} x^2 \ln x, & \text{if } x \in (0,1] \\ 0, & \text{if } x = 0, \end{cases}$. We will show that f is differentiable at $a = 0$.

Solution. Indeed,

$$\lim_{h \to 0^+} \frac{f(0+h) - f(0)}{h} = \lim_{h \to 0^+} \frac{h^2 \ln h}{h} = \lim_{h \to 0^+} h \ln h = 0,$$

so f is right differentiable at $x = 0$ and $f'_+(0) = 0$. We leave the proof that f is differentiable at $b = 1$ to the reader. ◆

Exercises

In Exercises 4.2.1–4.2.4 use the definition to find the derivatives of the following functions:

4.2.1. $f(x) = x^2 - 2x + 3$.

4.2.2. $f(x) = \dfrac{x+1}{x-1}$, $x \neq 1$.

4.2.3. $f(x) = \sqrt{3x - 2}$, $x > 2/3$.

4.2.4. $f(x) = x|x|$.

In Exercises 4.2.5–4.2.8 use the linear approximation to calculate:

4.2.5. $\ln 2$.

4.2.6. $e^{0.2}$.

4.2.7. $\arctan 0.9$.

4.2.8. $\sin(\pi/7)$.

4.2.9. Prove that the function

$$f(x) = \begin{cases} xe^{-1/x^2}, & \text{if } x \neq 0 \\ 0, & \text{if } x = 0 \end{cases}$$

is differentiable at $x = 0$ and find $f'(0)$.

4.2.10. Prove that the function

$$f(x) = \begin{cases} x\sin(1/x), & \text{if } x \neq 0 \\ 0, & \text{if } x = 0 \end{cases}$$

is continuous but not differentiable at $x = 0$.

4.2.11. Let

$$f(x) = \begin{cases} x + x^2, & \text{if } x \text{ is rational} \\ x, & \text{if } x \text{ is irrational.} \end{cases}$$

Prove that f is differentiable at $x = 0$.

4.2.12. Suppose that the function f satisfies $-x^2 \leq f(x) \leq x^2$ for all $x \in \mathbb{R}$. Prove that f is differentiable at $x = 0$ and find $f'(0)$.

4.2.13. Find

$$\lim_{x \to a} \frac{xf(a) - af(x)}{x - a}.$$

4.2.14. Find

$$\lim_{x \to a} \frac{f(x)g(a) - f(a)g(x)}{x - a}.$$

4.2.15. Suppose that f is continuous at $x = 0$. Prove that $g(x) = xf(x)$ is differentiable at $x = 0$. Is the converse true?

4.2.16. Suppose that f is differentiable at $x = c$. Prove that the limit

$$\lim_{h \to 0} \frac{f(c + h) - f(c - h)}{2h}$$

exists and equals $f'(c)$.

4.2.17.* Give an example to show that the limit in Exercise 4.2.16 may exist even when f is not differentiable at $x = c$.

4.2.18. Suppose that f satisfies the Hölder condition on (a,b): there exists $M > 0$ and $\alpha \geq 0$ such that, for any $x, y \in (a,b)$, $|f(x) - f(y)| \leq M|x - y|^\alpha$. Prove that if $\alpha > 1$ then f is differentiable on (a,b). Give an example to show that this need not be true if $\alpha = 1$.

4.2.19. Suppose that the function f is defined on $[a,b]$ and that it is right differentiable at $x = a$. Prove that it is continuous from the right at $x = a$.

4.2.20. Prove that the characteristic function of $[0,1]$ is left differentiable but not right differentiable at $x = 1$.

4.2.21. Suppose that the function f is defined on $[a,b]$ and that it is both left differentiable and right differentiable at $c \in (a,b)$. Prove that f is continuous at $x = c$.

4.2.22.* Give an example of a function f such that the limits $\lim_{x\to 0-} f'(x)$ and $\lim_{x\to 0+} f'(x)$ exist and are equal, but f has neither the left nor the right derivative at $x = 0$.

4.3 Rules of Differentiation

Although Definition 4.2.2 can be occasionally useful to calculate a derivative, we will use it to establish much more practical rules of differentiation.

Theorem 4.3.1. *Let f, g be two functions defined on (a,b) and let $c \in (a,b)$. Also, let α be a real number. If f and g are differentiable at $x = c$ then the same is true for $f + g$ and αf and:*

(a) $(\alpha f)'(c) = \alpha f'(c)$;

(b) $(f + g)'(c) = f'(c) + g'(c)$.

Proof. We notice that, for $x \in (a,b)$ and $x \neq c$,

$$\frac{(\alpha f)(x) - (\alpha f)(c)}{x - c} = \frac{\alpha f(x) - \alpha f(c)}{x - c} = \alpha \frac{f(x) - f(c)}{x - c}.$$

and

$$\frac{(f + g)(x) - (f + g)(c)}{x - c} = \frac{(f(x) + g(x)) - (f(c) + g(c))}{x - c} = \frac{f(x) - f(c)}{x - c} + \frac{g(x) - g(c)}{x - c}.$$

The assertions (a) and (b) now follow from Theorem 3.4.10 (a) and (b). □

Next we will prove the "product rule" and the "quotient rule".

Theorem 4.3.2. *Let f, g be two functions defined on (a,b) and let $c \in (a,b)$. If f and g are differentiable at $x = c$ then the same is true for $f \cdot g$ and*

$$(f \cdot g)'(c) = f'(c)g(c) + f(c)g'(c).$$

If, in addition, $g(c) \neq 0$, then the function f/g is differentiable at c and

$$\left(\frac{f}{g}\right)'(c) = \frac{f'(c)g(c) - f(c)g'(c)}{g(c)^2}.$$

Proof. Once again, let $x \in (a, b)$ and $x \neq c$. We consider the difference quotient

$$\frac{(f \cdot g)(x) - (f \cdot g)(c)}{x - c} = \frac{f(x)g(x) - f(c)g(c)}{x - c}$$

and we use some algebra:

$$\frac{f(x)g(x) - f(c)g(c)}{x - c} = \frac{f(x)g(x) - f(c)g(x) + f(c)g(x) - f(c)g(c)}{x - c}$$
$$= \frac{f(x) - f(c)}{x - c}g(x) + f(c)\frac{g(x) - g(c)}{x - c}. \tag{4.6}$$

Similarly,

$$\left(\frac{f}{g}\right)(x) - \left(\frac{f}{g}\right)(c) = \frac{f(x)}{g(x)} - \frac{f(c)}{g(c)} = \frac{f(x)g(c) - f(c)g(x)}{g(x)g(c)}$$
$$= \frac{f(x)g(c) - f(c)g(c) + f(c)g(c) - f(c)g(x)}{g(x)g(c)}$$
$$= \frac{[f(x) - f(c)]g(c) - f(c)[g(x) - g(c)]}{g(x)g(c)},$$

and it follows that

$$\frac{\left(\frac{f}{g}\right)(x) - \left(\frac{f}{g}\right)(c)}{x - c} = \frac{\frac{f(x)-f(c)}{x-c}g(c) - f(c)\frac{g(x)-g(c)}{x-c}}{g(x)g(c)}. \tag{4.7}$$

If we now take the limits as $x \to c$ in (4.6) and (4.7), the result follows from the continuity of g and Theorem 3.4.10. $\square$

Our next target is the Chain Rule.

Theorem 4.3.3. *Let f be a function defined on (a_1, b_1), let $c \in (a_1, b_1)$, and let $f(c) \in (a_2, b_2)$. Suppose that f is differentiable at $x = c$ and that g is defined on (a_2, b_2) and differentiable at $f(c)$. Then the composition $g \circ f$ is differentiable at $x = c$, and*

$$(g \circ f)'(c) = g'(f(c))f'(c).$$

Proof. Again we start with the difference quotient

$$\frac{(g \circ f)(x) - (g \circ f)(c)}{x - c} = \frac{g(f(x)) - g(f(c))}{x - c} \tag{4.8}$$

We would like to multiply and divide (4.8) by $f(x) - f(c)$, but we can do that only if $f(x) \neq f(c)$. The good news is that, if $f(x) = f(c)$, the expression (4.8) equals 0, so we can write it as $g'(f(c)) \cdot \frac{f(x)-f(c)}{x-c}$ because the latter is also equal to 0. Thus,

$$\frac{(g \circ f)(x) - (g \circ f)(c)}{x - c} = \begin{cases} \dfrac{g(f(x)) - g(f(c))}{f(x) - f(c)} \cdot \dfrac{f(x) - f(c)}{x - c}, & \text{if } f(x) \neq f(c) \\ g'(f(c)) \cdot \dfrac{f(x) - f(c)}{x - c}, & \text{if } f(x) = f(c). \end{cases}$$

It is not hard to see that, as $x \to c$, the fraction $\frac{f(x)-f(c)}{x-c}$ has the limit $f'(c)$. Thus it remains to consider the case when $f(x) \neq f(c)$ and

$$\lim_{x \to c} \frac{g(f(x)) - g(f(c))}{f(x) - f(c)}. \tag{4.9}$$

If we denote $y = f(x)$ and $d = f(c)$, then the continuity of f at $x = c$ implies that, as $x \to c$, $y \to d$. Therefore the limit (4.9) equals

$$\lim_{y \to d} \frac{g(y) - g(d))}{y - d} = g'(d) = g'(f(c)). \qquad \square$$

Our last task concerns the differentiability of the inverse function (when it exists).

Theorem 4.3.4. *Let f be an injective function defined on (a, b) and let $c \in (a, b)$. If f is differentiable at $x = c$, with $f'(c) \neq 0$, then its inverse function f^{-1} is differentiable at $d = f(c)$, and*

$$\left(f^{-1}\right)'(d) = \frac{1}{f'(c)}.$$

Proof. Let $y \neq d$ be in the domain of f^{-1}. If $x = f^{-1}(y)$, then $x \neq c$. Now

$$\frac{f^{-1}(y) - f^{-1}(d)}{y - d} = \frac{x - c}{f(x) - f(c)} = \frac{1}{\frac{f(x)-f(c)}{x-c}}.$$

Further, when $y \to d$, the continuity of f^{-1} at $y = d$ implies that $f^{-1}(y) \to f^{-1}(d)$, i.e., $x \to c$. It follows that

$$\lim_{y \to d} \frac{f^{-1}(y) - f^{-1}(d)}{y - d} = \lim_{x \to c} \frac{1}{\frac{f(x)-f(c)}{x-c}} = \frac{1}{f'(c)}. \qquad \square$$

With rules of differentiation in hand, we can establish that all elementary functions are differentiable wherever they are defined.

Theorem 4.3.5. *Every rational function is differentiable at every point of its domain.*

Proof. In view of the rules of differentiation (Theorems 4.3.1 and 4.3.2) it suffices to establish the differentiability of the functions $f(x) = x$ and $g(x) = 1$. So, let $c \in \mathbb{R}$. Since

$$\frac{f(x) - f(c)}{x - c} = \frac{x - c}{x - c} = 1 \quad \text{and} \quad \frac{g(x) - g(c)}{x - c} = \frac{1 - 1}{x - c} = 0$$

it is clear that both f and g are differentiable at $x = c$, and that $f'(c) = 1$, $g'(c) = 0$. $\square$

Theorem 4.3.6. *Trigonometric functions $\sin x$, $\cos x$, $\tan x$, $\cot x$, $\sec x$, and $\csc x$ are differentiable at every point of their domains.*

Proof. We will show that the functions $f(x) = \sin x$ and $g(x) = \cos x$ are differentiable, and the rest will follow by the Quotient Rule (Theorem 4.3.2). Let $c \in \mathbb{R}$. Using Proposition 3.7.1 (b), we have that

$$\frac{f(x) - f(c)}{x - c} = \frac{\sin x - \sin c}{x - c} = 2 \cos \frac{x + c}{2} \frac{\sin \frac{x-c}{2}}{x - c} = \cos \frac{x + c}{2} \frac{\sin \frac{x-c}{2}}{\frac{x-c}{2}}.$$

If we take the limit as $x \to c$, using the continuity of $\cos x$ and Example 3.1.17, we obtain that $f(x) = \sin x$ is differentiable at $x = c$ and $f'(c) = \cos c$. The differentiability of $\cos x$ now follows from the equality $\cos x = \sin(\frac{\pi}{2} - x)$ and the Chain Rule. $\square$

Next, we will prove that the exponential functions are differentiable.

Theorem 4.3.7. *Let $a > 0$. The exponential function $f(x) = a^x$ is differentiable at every point $c \in \mathbb{R}$.*

Proof. Using a substitution $u = x - c$, the difference quotient is

$$\frac{a^x - a^c}{x - c} = a^c \frac{a^{x-c} - 1}{x - c} = a^c \frac{a^u - 1}{u}.$$

Since $u \to 0$ when $x \to c$, by Example 3.1.14,

$$f'(c) = \lim_{x \to c} \frac{a^x - a^c}{x - c} = \lim_{u \to 0} a^c \frac{a^u - 1}{u} = a^c \ln a. \qquad \square$$

Now, we can prove a theorem about logarithms. Remember that the logarithmic functions are inverse functions of the exponential functions.

Theorem 4.3.8. *Let $a > 0$ and $a \neq 1$. The logarithmic function $f(x) = \log_a x$ is differentiable at every point $c > 0$.*

Proof. Since $f^{-1}(x) = a^x$ which is differentiable at $d = \log_a c$, Theorem 4.3.4 implies that f is differentiable at c. Further,

$$f'(c) = \frac{1}{(f^{-1})'(d)} = \frac{1}{a^d \ln a} = \frac{1}{c \ln a}.$$

In the special case when $a = e$ we obtain that the derivative of $g(x) = \ln x$ at $x = c$ equals $1/c$. $\qquad \square$

The inverse trigonometric functions are differentiable for exactly the same reason as logarithms.

Theorem 4.3.9. *The inverse trigonometric functions are differentiable at every point of their domains.*

Proof. We will show that $f(x) = \arcsin x$ is differentiable at $x = c$ (with $|c| \leq 1$) and leave the rest as an exercise. Since $f^{-1}(x) = \sin x$ and $\sin x$ is differentiable at $d = \arcsin c$, it follows that f is differentiable at $x = c$. Further, $(\sin x)' = \cos x$ so

$$f'(c) = \frac{1}{(f^{-1})'(d)} = \frac{1}{\cos d} = \frac{1}{\cos(\arcsin c)}.$$

Although this is a correct answer, we should simplify it. By the Fundamental Trigonometric Identity, $\cos^2(\arcsin c) = 1 - \sin^2(\arcsin c) = 1 - c^2$. Further, for any $c \in [-1, 1]$, $\arcsin c \in [-\pi/2, \pi/2]$, so $\cos(\arcsin c) \geq 0$. Consequently, $\cos(\arcsin c) = \sqrt{1 - c^2}$ and

$$f'(c) = \frac{1}{\sqrt{1 - c^2}}. \qquad \square$$

Finally, we look at the power functions $f(x) = x^\alpha$, with $\alpha \in \mathbb{R}$, and $x > 0$.

Theorem 4.3.10. *Let $\alpha \in \mathbb{R}$ and $c > 0$. Then the function $f(x) = x^\alpha$ is differentiable at $x = c$, and $f'(c) = \alpha c^{\alpha - 1}$.*

Proof. Since $f(x) = e^{\alpha \ln x}$ the differentiability of f follows from the differentiability of the exponential function $g(x) = e^x$, the logarithmic function $h(x) = \ln x$, and the Chain Rule. Further, $f(x) = g(\alpha h(x))$ so, by the Chain Rule

$$f'(c) = g'(\alpha h(c))(\alpha h)'(c) = e^{\alpha h(c)} \alpha h'(c) = e^{\alpha \ln c} \alpha \frac{1}{c} = c^\alpha \alpha \frac{1}{c} = \alpha c^{\alpha - 1}. \qquad \square$$

PIERRE de FERMAT (1601–1665) was a French lawyer and an amateur mathematician. There is no certainty about his date of birth. It is possible that he had an older brother (also named Pierre) who died young. He studied at the University of Toulouse and in the second half of the 1620s he lived in Bordeaux, where he began his mathematical research on maxima and minima. He then went to Orléans to study law. By 1631 he was back in Toulouse as a lawyer and a government official which allowed him to add "de" in front of Fermat. In 1636 he got in touch with leading Paris mathematicians. He sent them some of his results in the form of a challenge. Since they could not solve the problems posed, he was asked to explain his methods. He quickly gained the reputation as one of the best mathematicians in the world. Throughout his life, he was not interested in publishing his results, so his method came to us through a textbook by French mathematician Pierre Hérigone (1580–1643). His standings in mathematics suffered because of his feud with Descartes, probably instigated when Fermat wrote a bad review of Descartes' book *La Dioptrique* (*Dioptrics*). Fermat disagreed with the ideas in the book and proposed the principle that light always follows the shortest path. In part because of the feud, and in part due to an epidemic of plague, he was without a contact with Paris until 1654 when he communicated his ideas on probability to Pascal. During this period he worked on number theory. He continued posing problems but never published any results, communicating most of his work in letters to friends, often with little or no proof of his theorems. The most famous of them is the so-called "Fermat's Last Theorem" which was discovered by his son in the margin of Diophantus's *Arithmetica* and became widely known in 1670 when the son published this book with his father's notes. Incidentally, the problem was solved only in 1993, using methods that were not available to Fermat, so it is likely that his claim to have solved the problem was incorrect. In the next section we will look at his calculus contribution.

Exercises

4.3.1. Suppose that f and g are defined on $\mathbb{R}$ and that f is differentiable at $x = a$ but g is not. Prove or disprove: $f + g$ is not differentiable at $x = a$.

4.3.2. Suppose that f and g are defined on $\mathbb{R}$ and that neither f nor g is differentiable at $x = a$. Prove or disprove: $f + g$ is not differentiable at $x = a$.

4.3.3.* Let f be a function continuous on (a, b). Show that f is differentiable at $c \in (a, b)$ if and only if $\lim\limits_{\substack{x \to c^+ \\ y \to c^-}} \dfrac{f(x) - f(y)}{x - y}$ exists.

4.3.4. Let f be a function defined on $\mathbb{R}$ and suppose that there exists $M > 0$ such that, for any $x, y \in \mathbb{R}$, $|f(x) - f(y)| \leq M|x - y|^2$. Prove that f is a constant function.

4.3.5. Suppose that f is differentiable at $x = 0$ and that $f'(0) \neq 0$. Find the limit

$$\lim_{x \to 0} \frac{f(x)e^x - f(0)}{f(x)\cos x - f(0)}.$$

4.3.6. Suppose that f is differentiable at $x = a$, and let $n \in \mathbb{N}$. Find the limit

$$\lim_{x \to a} \frac{a^n f(x) - x^n f(a)}{x - a}.$$

4.3.7. Find the derivative of the function $f(x) = \sqrt[3]{x}|x|$.

4.3.8. Find $f'(0)$ if $f(x) = x(x - 1)(x - 2) \ldots (x - 1000)$.

4.3.9. Show that the function

$$f(x) = \begin{cases} x^2 \sin \dfrac{1}{x}, & \text{if } x \neq 0 \\ 0, & \text{if } x = 0 \end{cases}$$

is differentiable but its derivative is not continuous. [The example appears in [105] by Volterra (page 158) in 1881.]

4.3.10. Find the derivative of $f(x) = \log_x 2$.

4.3.11. Prove that the function $f(x) = \arctan x$ is differentiable and find its derivative.

4.3.12. Prove that the function $f(x) = \arccos x$ is differentiable and find its derivative.

4.3.13. Find the derivatives of $f(x) = \operatorname{arsinh} x$ and $g(x) = \operatorname{arcosh} x$.

4.3.14. Find the derivatives of hyperbolic functions $f(x) = \sinh x$ and $g(x) = \cosh x$.

4.3.15.* Prove that the nth derivative of $f(x) = \sin x$ is $f^{(n)}(x) = \sin(x + n\pi/2)$.

4.3.16.* Prove that the nth derivative of $f(x) = \cos x$ is $f^{(n)}(x) = \cos(x + n\pi/2)$.

In Exercises 4.3.17–4.3.18 suppose that $f(a) > 0$ and that f is differentiable at $x = a$.

4.3.17.* Find $\displaystyle\lim_{n\to\infty} \left(\frac{f(a + 1/n)}{f(a)} \right)^{1/n}$.   **4.3.18.*** Find $\displaystyle\lim_{x\to a} \left(\frac{f(x)}{f(a)} \right)^{\frac{1}{\ln x - \ln a}}$.

4.3.19.* Prove the Leibniz Formula: If f and g are n times differentiable, then

$$(fg)^{(n)} = \sum_{k=0}^{n} \binom{n}{k} f^{(k)} g^{(n-k)}.$$

4.4   Monotonicity: Local Extrema

In elementary calculus, we use the derivative of a function to determine intervals on which it is increasing/decreasing, as well as points of local maximum/minimum (i.e., local extrema).

Example 4.4.1. Finding local extrema of a function.
   Let $f(x) = 3x - x^3$. We will find the local minimum and the local maximum of f.

Solution. The derivative is $f'(x) = 3 - 3x^2 = 3(1 - x)(1 + x)$ and it is easy to see that $f'(x) > 0$ when either both $1 + x > 0$, $1 - x > 0$, or both $1 + x < 0$, $1 - x < 0$. The former two inequalities yield $-1 < x < 1$ while the latter two yield $x < -1$ and $x > 1$ which cannot hold simultaneously. Thus, $f'(x) > 0$ when $-1 < x < 1$, and $f'(x) < 0$ when $x < -1$ or $x > 1$. This means that f is an increasing function for $-1 < x < 1$, and decreasing for $x < -1$ and $x > 1$. It follows that at $x = -1$, f has a local minimum and at $x = 1$ a local maximum. Figure 4.5 represents a summary of our investigation. ◆

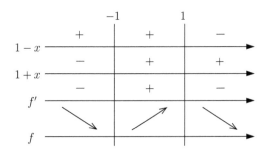

Figure 4.5: Using the derivative to find local extrema.

In this section we will work on justifying the reasoning that allowed us to make all these conclusions about the function f. We will start with a simple result that goes all the way back to Fermat.

Theorem 4.4.2 (Fermat's Theorem). *Let f be defined on an interval $[a,b]$ and suppose that it attains its greatest or its smallest value at a point $c \in (a,b)$. If f is differentiable at c, then $f'(c) = 0$.*

Proof. We will assume that f attains its greatest value at $c \in (a,b)$, i.e., that $f(x) \le f(c)$ for all $x \in [a,b]$. If $x < c$ then
$$\frac{f(x) - f(c)}{x - c} \ge 0$$
and Theorem 3.6.14 implies that $f'(c) \ge 0$. On the other hand, if $x > c$ then
$$\frac{f(x) - f(c)}{x - c} \le 0$$
so $f'(c) \le 0$. Combining these two inequalities ($f'(c) \ge 0$ and $f'(c) \le 0$), we obtain that $f'(c) = 0$. □

It follows from Theorem 4.4.2 that if a function f attains its extreme value at a point $c \in (a,b)$, then either $f'(c) = 0$ or f is not differentiable at c. A point c with either of the two properties is called a **critical point** of f.

Fermat's Theorem shows that, if at some point c, $f'(c)$ exists and is different from 0, then f cannot attain its extreme value at c. However, this result does not shed any light on the issue whether f is increasing or decreasing. For that we will need another theorem. Here is a stepping stone in that direction.

Theorem 4.4.3 (Rolle's Theorem). *Suppose that f is a function defined and continuous on an interval $[a,b]$, that it is differentiable in (a,b), and that $f(a) = f(b)$. Then there exists $c \in (a,b)$ such that $f'(c) = 0$.*

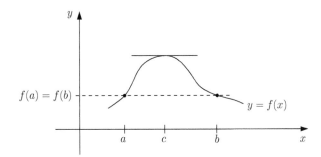

Figure 4.6: The tangent line must be horizontal somewhere.

Proof. We start with the Weierstrass Theorem (Theorem 3.9.8) which guarantees that f attains its largest value M and its smallest value m on $[a,b]$. There are two possibilities: either $M = m$ or $M > m$. In the former case, the inequality $m \le f(x) \le M$ implies that f is constant on $[a,b]$, so $f'(x) = 0$ for all $x \in (a,b)$ and we can take for c any point in (a,b). If $M > m$, the assumption that $f(a) = f(b)$ shows that at least one of M and m is attained at a point $c \in (a,b)$. By Fermat's Theorem, $f'(c) = 0$. □

Geometrically, Rolle's Theorem says that, under the listed assumptions, if f takes the same value at the endpoints, then somewhere in between the tangent line to the graph of f is horizontal (Figure 4.6). In other words, somewhere in between there is a point at which the tangent line is parallel to the (horizontal) line connecting $f(a)$ and $f(b)$. What if the latter line is not horizontal?

MICHEL ROLLE (1652–1719) was a French mathematician. He had some elementary school education, but was mostly self-educated. In 1675 he went to Paris to work as a scribe, and studied mathematics on his own. In 1682 he solved a problem posed by French mathematician Ozanam: *Find four numbers the difference of any two being a perfect square, in addition the sum of the first three numbers being a perfect square.* He was rewarded by a small pension and the Secretary of State for War Louvois hired him as a tutor for his son. Louvois helped Rolle become a member of the Académie Royale des Sciences in 1685. He published his most important work *Traité d'algèbre* (*Treatise on Algebra*) in 1690 on the theory of equations. He introduced the method of cascades without the proof. The proof appeared in *Démonstration d'une Méthode pour resoudre les Egalitez de tous les degrez* (*A proof of a method to solve equations of any degree*) in 1691. In the same work he gave the first known formal proof of Theorem 4.4.3. Rolle is also remembered for popularizing the symbol for equality =, which had been invented by a Welsh doctor and mathematician Robert Recorde, and the symbol for the nth root $\sqrt[n]{}$, although it had been suggested (for the cube root) by Albert Girard. Rolle was an outspoken critic of calculus, and his opposition had a positive effect on the new discipline. Eventually, he formally recognized its value by 1706. In 1708 he suffered a stroke and stopped doing mathematics. He died from the second stroke in 1719. The name "Rolle's theorem" was first used by Moritz Wilhelm Drobisch (1802–1896), a German mathematician, in 1834.

Example 4.4.4. The existence of a tangent line parallel to a chord.

Let $f(x) = (x^2 + 3)/4$ and $A = [-1, 3]$. We will show that there is a tangent line parallel to a chord.

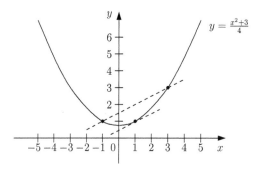

Figure 4.7: The tangent line is parallel to the chord.

Solution. The values at endpoints are $f(-1) = 1$ and $f(3) = 3$, so the line connecting them has the slope

$$m = \frac{3-1}{3-(-1)} = \frac{2}{4} = \frac{1}{2}.$$

The derivative $f'(x) = x/2$, and at $x = 1$ (which belongs to $(-1, 3)$), $f'(1) = 1/2$. In other words, the tangent line at $(1, 1)$ is parallel to the chord connecting $(-1, 1)$ and $(3, 4)$ (see Figure 4.7).   ♦

The observation made in Example 4.4.4 is in fact universally true. The result is most often referred to as the Mean Value Theorem or Lagrange's Theorem.

Theorem 4.4.5 (Mean Value Theorem). *Suppose that f is a function defined and continuous on an interval $[a, b]$, and that it is differentiable in (a, b). Then there exists $c \in (a, b)$ such that*

$$f'(c) = \frac{f(b) - f(a)}{b - a}.$$

Proof. Since the picture is essentially the same as in Rolle's Theorem, we will try to rotate the graph. To do that, we will subtract from f a linear function $y = Ax + B$, and we will select the coefficients A and B so that the resulting functions satisfies the hypotheses of Rolle's

Theorem. If $F(x) = f(x) - Ax - B$, we will ask that $F(a) = F(b)$. Then $f(a) - Aa - B = f(b) - Ab - B$ and it follows that

$$A = \frac{f(b) - f(a)}{b - a},$$

while B is arbitrary, so we will take $B = 0$. Therefore, the function

$$F(x) = f(x) - \frac{f(b) - f(a)}{b - a} x$$

satisfies $F(a) = F(b)$. Since linear functions are differentiable (and, hence, continuous), F satisfies all the hypotheses of Rolle's Theorem. It follows that there exists $c \in (a, b)$ such that $F'(c) = 0$. Clearly,

$$F'(x) = f'(x) - \frac{f(b) - f(a)}{b - a}$$

so

$$0 = F'(c) = f'(c) - \frac{f(b) - f(a)}{b - a}. \qquad \square$$

This result can be found in [18], a book by Bonaventura Cavalieri (1598–1647)—an Italian mathematician and a Jesuate—in 1635. Also, it appeared in the work of Lagrange (page 398) [79] in 1797, although it lacked the rigor and precision of the Cauchy's exposition in *Cours d'Analyse*. The name "Mean value theorem" was used in 1899 in a paper by J. K. Whittemore, a Yale professor. The elegant proof that we have presented is due to French mathematician Pierre Ossian Bonnet (1819–1892), and it was first published in 1868, in a calculus textbook *Cours de calcul différentiel et intégral* (*Course of differential and integral calculus*) by French mathematician Joseph Serret (1819–1885) who was at one time a professor of differential and integral calculus at the Sorbonne.

Now we return to our initial goal: establishing the connection between the sign of the derivative and the monotonicity of a function. This relationship needs to be more precisely formulated, because the differentiability of a function is a local property, while the monotonicity is definitely not. We say that it is a *global* property because it describes the behavior of a function on an interval. Therefore, we need to consider the differentiability as a global property as well, i.e., we need to look for the sign of the derivative at every point of an interval.

Theorem 4.4.6. *Let f be a function that is differentiable on an open interval (a, b), and suppose that $f'(x) > 0$ for all $x \in (a, b)$. Then f is strictly increasing on (a, b).*

Proof. Let $x_1 < x_2$ be any two points in (a, b). We will show that $f(x_1) < f(x_2)$. The function f satisfies the hypotheses of Lagrange's Theorem on the closed interval $[x_1, x_2]$ so there exists $c \in (x_1, x_2)$ such that

$$f'(c) = \frac{f(x_2) - f(x_1)}{x_2 - x_1}.$$

Consequently, $f(x_2) - f(x_1) = f'(c)(x_2 - x_1) > 0$. $\qquad \square$

Remark 4.4.7. The same proof shows that, if $f'(c) < 0$ for all $c \in (a, b)$, then f is strictly decreasing on (a, b).

Theorem 4.4.6 (together with Remark 4.4.7) now justifies the conclusions that we made in Example 4.4.1: since $f'(x) > 0$ when $-1 < x < 1$ we conclude that f is increasing on the interval $(-1, 1)$.

It is important to notice that the implication in Theorem 4.4.6 goes only one way. That is, if a function is strictly increasing on an interval (a, b), it does not follow that $f'(x) > 0$ for all $x \in (a, b)$.

Example 4.4.8. A strictly increasing function whose derivative is not always positive.

Let $f(x) = x^3$. We will show that f is strictly increasing on $(-1, 1)$ but $f'(0) = 0$.

Solution. It is easy to see that $x_1 < x_2$ implies $f(x_1) < f(x_2)$, so f is strictly increasing on $(-1, 1)$. However, $f'(x) = 3x^2$, and $f'(0) = 0$. Thus, although f is strictly increasing, it is not true that $f'(x) > 0$ for all $x \in (-1, 1)$. $\blacklozenge$

Going back to Example 4.4.4 we see that Theorem 4.4.6 does not provide information about the points where f' equals 0. Fermat's Theorem asserts that these are the only points where f can attain its greatest/smallest value but, as Example 4.4.8 shows, there need not be a local extremum there. The way we resolved this issue was by looking at the behavior of f on the left and on the right of the point under scrutiny. For example, when considering $x = 1$, we noticed that f is increasing if $x \in (0, 1)$ and decreasing if $x \in (1, 2)$. This allowed us to conclude that f has a local maximum at $x = 1$.

Another way to obtain the same answer is by the **Second Derivative Test**. Since $f''(x) = -6x$, we see that $f''(1) = -6 < 0$, so f has a local maximum at $x = 1$. Also, $f''(-1) = 6 > 0$, so f has a local minimum at $x = -1$. While this reasoning is correct, it needs to be justified. Since we are interested in the second derivative, we will say that a function f is **twice differentiable** on (a, b), if f is differentiable on (a, b), and its derivative f' is also differentiable on (a, b).

Theorem 4.4.9 (Second Derivative Test). *Let f be a function that is twice differentiable on an open interval (a, b), let $c \in (a, b)$, and suppose that $f'(c) = 0$. If $f''(c) < 0$ then f has a local maximum at $x = c$.*

Proof. By definition,

$$f''(c) = \lim_{x \to c} \frac{f'(x) - f'(c)}{x - c}.$$

Since $f'(c) = 0$, we have that

$$f''(c) = \lim_{x \to c} \frac{f'(x)}{x - c}.$$

If we define a function F on (a, b) by

$$F(x) = \begin{cases} \dfrac{f'(x)}{x - c}, & \text{if } x \neq c \\ f''(c), & \text{if } x = c \end{cases}$$

then F is continuous at c and $F(c) < 0$. By Theorem 3.6.12, there exists $\delta > 0$ so that $F(x) < 0$ for $x \in (c - \delta, c + \delta)$. It follows that, for $|x - c| < \delta$ and $x \neq c$,

$$\frac{f'(x)}{x - c} < 0.$$

Thus, $f'(x) > 0$ if $c - \delta < x < c$, and $f'(x) < 0$ if $c < x < c + \delta$. Now, Theorem 4.4.6 (together with Remark 4.4.7) implies that f has a local maximum at $x = c$. $\square$

Remark 4.4.10. If the condition $f''(c) < 0$ is replaced with $f''(c) > 0$, the same proof shows that f has a local minimum at $x = c$.

Exercises

In Exercises 4.4.1–4.4.4 determine the intervals on which f is monotone:

4.4.1. $f(x) = 2 + x - x^2$.

4.4.2. $f(x) = \dfrac{2x}{1+x^2}$.

4.4.3. $f(x) = \dfrac{\sqrt{x}}{x+100}$.

4.4.4. $f(x) = \dfrac{x^2}{2^x}$.

In Exercises 4.4.5–4.4.8 find the local extrema of f and determine whether they represent a local maximum or a local minimum:

4.4.5. $f(x) = x^3 - 6x^2 + 9x - 4$.

4.4.6. $f(x) = x + \dfrac{1}{x}$.

4.4.7. $f(x) = \sqrt{2x - x^2}$.

4.4.8. $f(x) = xe^{-x}$.

4.4.9.* Suppose that f has continuous derivative on (a,b), and let $c \in (a,b)$. Prove or disprove: there exist points $x_1, x_2 \in (a,b)$ such that

$$\frac{f(x_2) - f(x_1)}{x_2 - x_1} = f'(c).$$

4.4.10.* Suppose that f and g are both n times differentiable, that $f^{(k)}(a) = g^{(k)}(a)$, for $0 \le k \le n-1$, and that $f^{(n)}(x) > g^{(n)}(x)$, for all $x > a$. Prove that $f(x) > g(x)$, for all $x > a$.

In Exercises 4.4.11–4.4.14 establish the inequalities:

4.4.11. $e^x > 1 + x$ for $x \ne 0$.

4.4.12. $\sin x > x - x^3/6$ for $x > 0$.

4.4.13. $x^\alpha - 1 > \alpha(x-1)$ for $x > 1$, $\alpha \ge 2$.

4.4.14. $x - x^2/2 < \ln(1+x) < x$ for $x > 0$.

4.4.15. Suppose that f is differentiable at every point of $\mathbb{R}$, and that $f'(x) = 0$ for all $x \in \mathbb{R}$. Prove that f is a constant function.

4.4.16.* Suppose that f is continuous and right differentiable at every point of $\mathbb{R}$, and that $f'_+(x) = 0$ for all $x \in \mathbb{R}$. Prove that f is a constant function.

4.4.17.* Prove the Cauchy's Mean Value Theorem: If f and g are continuous in $[a,b]$ and differentiable in (a,b), and if g' does not vanish in (a,b), then there exists $c \in (a,b)$ such that

$$\frac{f(b) - f(a)}{g(b) - g(a)} = \frac{f'(c)}{g'(c)}.$$

4.4.18. Prove that the equation $7x^3 - 5x^2 + 4x - 10 = 0$ has exactly one real root.

4.4.19. Prove that the equation $3x^4 - 8x^3 + 7x^2 - 45 = 0$ has exactly two real roots.

4.4.20.* Prove that between any two real roots of $e^x \sin x = 1$ there is at least one real root of $e^x \cos x = -1$.

4.4.21.* Suppose that f is differentiable in (a,b) and that it has one-sided derivatives $f'_+(a) \ne f'_-(b)$ at the endpoints. If C is a real number between $f'_+(a)$ and $f'_-(b)$, prove that there exists $c \in (a,b)$ such that $f'(c) = C$. [This is known as Darboux's Theorem. It appeared in the paper of French mathematician Darboux [25]. (See page 156.)]

4.4.22. Use Exercise 4.4.21 to prove that, if a function f has a jump at an interior point of the interval $[a,b]$, then it cannot be the derivative of any function.

4.4.23. Let

$$f(x) = \begin{cases} x + x^2 \sin \dfrac{2}{x}, & \text{if } x \neq 0 \\ 0, & \text{if } x = 0. \end{cases}$$

Prove that f is differentiable at 0 and that $f'(0) > 0$ but that f is not increasing in any interval $(-a, a)$.

4.4.24. Let

$$f(x) = \begin{cases} x^4 \left(2 + \sin \dfrac{1}{x} \right), & \text{if } x \neq 0 \\ 0, & \text{if } x = 0. \end{cases}$$

Show that f has a relative minimum at 0, but $f'(x)$ takes both positive and negative values in every interval $(0, a)$.

4.4.25.* Let

$$f(x) = \begin{cases} e^{-3x} x^2 \sin \dfrac{1}{x}, & \text{if } 0 < x \leq \dfrac{1}{3} \\ 0, & \text{if } x = 0. \end{cases}$$

Show that the derivative of f does not satisfy the Extreme Value Theorem. That is, find $m = \inf\{f'(x) : x \in [0, 1/3]\}$ and show that there is no $c \in [0, 1/3]$ such that $f'(c) = m$.

4.4.26.* A function f is convex on $[a, b]$ if for any two points $x, y \in [a, b]$ and any $t \in [0, 1]$,

$$f(tx + (1 - t)y) \leq tf(x) + (1 - t)f(y).$$

Prove that, if f is twice differentiable in (a, b) and $f''(x) > 0$ for $x \in (a, b)$, then f is convex.

4.4.27. Suppose that f has a bounded derivative on (a, b). Prove that f is uniformly continuous on (a, b).

4.5 Taylor's Formula

One of the major themes in calculus is approximation. We have seen in Example 4.2.1 that we can approximate $f(x) = \sqrt[3]{x}$ by $y = 2 + \frac{1}{12}(x - 8) = \frac{1}{12}x + \frac{4}{3}$ in the vicinity of $x = 8$. In the process, we took advantage of the differentiability of f at $x = 8$. When a function is twice differentiable at a point, we can improve the approximation by using a quadratic function instead of linear. The more derivatives we can take at the point, the higher degree polynomial we can build to serve as an approximation to the given function.

Let us try this on a very simple function. Suppose that $f(x) = 3x^3 - 6x^2 - 8x + 5$, $a = 2$, and that we want to represent f as a cubic polynomial

$$p(x) = a_3(x - 2)^3 + a_2(x - 2)^2 + a_1(x - 2) + a_0.$$

The equality $p(2) = f(2)$ yields

$$a_0 = 3 \cdot 2^3 - 6 \cdot 2^2 - 8 \cdot 2 + 5 = -11.$$

Next we compare the derivatives:

$$p'(x) = 3a_3(x - 2)^2 + 2a_2(x - 2) + a_1, \quad \text{and} \quad f'(x) = 9x^2 - 12x - 8.$$

The equality $p'(2) = f'(2)$ implies that

$$a_1 = 9 \cdot 2^2 - 12 \cdot 2 - 8 = 4.$$

Since

$$p''(x) = 6a_3(x - 2) + 2a_2, \quad \text{and} \quad f''(x) = 18x - 12,$$

by equating $p''(2) = 2a_2$ and $f''(2) = 24$, we obtain that $a_2 = 12$. Notice that $a_2 = p''(2)/2$. Finally, $p'''(x) = 6a_3$ so $a_3 = p'''(2)/6$. At the same time, $f'''(x) = 18$, so $f'''(2)/6 = 3$ and $a_3 = 3$. The polynomial we get is

$$p(x) = 3(x - 2)^3 + 12(x - 2)^2 + 4(x - 2) - 11.$$

It is not hard to verify that $p(x) = f(x)$.

There is a lesson to be learned from this example. The coefficients of the "approximating" polynomial were computed using the formulas:

$$a_0 = f(2), \quad a_1 = f'(2), \quad a_2 = \frac{f''(2)}{2} = \frac{f''(2)}{2!}, \quad a_3 = \frac{f'''(2)}{6} = \frac{f'''(2)}{3!}.$$

We are inclined to believe that, if f is a polynomial of degree n, and we want to represent it as a polynomial

$$p(x) = a_0 + a_1(x - c) + a_2(x - c)^2 + \cdots + a_n(x - c)^n, \tag{4.10}$$

then

$$a_0 = f(c), \, a_1 = f'(c), \, a_2 = \frac{f''(c)}{2!}, \ldots, a_n = \frac{f^{(n)}(c)}{n!}. \tag{4.11}$$

This may serve as an inspiration to try to approximate any function f by a polynomial (4.10) with coefficients as in (4.11). (Such a polynomial is called the **Taylor polynomial**.) The big question is: how accurate is this approximation? Is there a way to get a hold on the difference between f and p?

Theorem 4.5.1 (Taylor's Formula). *Let $n \in \mathbb{N}_0$ and suppose that a function f is $(n + 1)$ times differentiable in $[a, b]$. If $c \in [a, b]$, then for any $x \in [a, b]$,*

$$f(x) = f(c) + f'(c)(x - c) + \frac{f''(c)}{2!}(x - c)^2 + \cdots + \frac{f^{(n)}(c)}{n!}(x - c)^n + r_n(x),$$

where $r_n(x)$ has the following properties:

(a) there exists x_0 between x and c such that

$$r_n(x) = \frac{f^{(n+1)}(x_0)}{(n + 1)!}(x - c)^{n+1};$$

(b) there exists θ, $0 < \theta < 1$, such that

$$r_n(x) = \frac{f^{(n+1)}((1 - \theta)c + \theta x)}{n!}(1 - \theta)^n(x - c)^{n+1}.$$

(c) If, in addition, $f^{(n+1)}$ is continuous in $[a, b]$,

$$r_n(x) = \int_c^x \frac{f^{(n+1)}(t)}{n!}(x - t)^n \, dt.$$

Proof. Let c and $x \neq c$ be fixed numbers in $[a,b]$. (If $x = c$ there is nothing to prove.) Clearly we can write

$$f(x) = f(c) + f'(c)(x-c) + \frac{f''(c)}{2!}(x-c)^2 + \frac{f^{(n)}(c)}{n!}(x-c)^n + r_n(x), \qquad (4.12)$$

and the task is to show that $r_n(x)$ can be represented in the forms (a)–(c). Let u be any differentiable function on $[a,b]$ with the property that $u(x) = 0$ and $u(c) = 1$. We consider the function

$$F(t) = -f(x) + f(t) + f'(t)(x-t) + \cdots + \frac{f^{(n)}(t)}{n!}(x-t)^n + u(t)r_n(x). \qquad (4.13)$$

Let us make it clear that F is a function of t (and x is fixed). Since $u(c) = 1$, (4.12) and (4.13) imply that $F(c) = 0$. On the other hand, $u(x) = 0$ so $F(x) = 0$. The hypotheses of the theorem, together with the assumption that u is a differentiable function, guarantee that F is differentiable on the interval $[x,c]$ (or $[c,x]$, depending whether $x < c$ or $x > c$). Therefore, Rolle's Theorem implies that there exists x_0 between x and c such that $F'(x_0) = 0$. Now we calculate $F'(t)$. Of course, $-f(x)$ is a constant, so its derivative is 0. Notice that, for any k, $1 \leq k \leq n$,

$$\left(\frac{f^{(k)}(t)}{k!}(x-t)^k\right)' = \frac{f^{(k+1)}(t)}{k!}(x-t)^k - \frac{f^{(k)}(t)}{(k-1)!}(x-t)^{k-1}.$$

Thus,

$$F'(t) = f'(t) + \left(\frac{f^{(2)}(t)}{1!}(x-t) - \frac{f^{(1)}(t)}{(0)!}(x-t)^0\right)$$
$$+ \left(\frac{f^{(3)}(t)}{2!}(x-t)^2 - \frac{f^{(2)}(t)}{(1)!}(x-t)^1\right)$$
$$+ \left(\frac{f^{(4)}(t)}{3!}(x-t)^3 - \frac{f^{(3)}(t)}{(2)!}(x-t)^2\right) + \cdots$$
$$\cdots + \left(\frac{f^{(n+1)}(t)}{n!}(x-t)^n - \frac{f^{(n)}(t)}{((n-1))!}(x-t)^{n-1}\right) + u'(t)r_n(x)$$
$$= \frac{f^{(n+1)}(t)}{n!}(x-t)^n + u'(t)r_n(x). \qquad (4.14)$$

The fact that $F'(x_0) = 0$ now implies that

$$\frac{f^{(n+1)}(x_0)}{n!}(x-x_0)^n = -u'(x_0)r_n(x). \qquad (4.15)$$

Let us summarize what we have established so far. If u is a differentiable function, such that $u(x) = 0$, and $u(c) = 1$, then formula (4.15) holds. In order to prove (a) we take

$$u(t) = \frac{(x-t)^{n+1}}{(x-c)^{n+1}}.$$

Then u satisfies the required conditions and $u'(t) = -(n+1)(x-t)^n/(x-c)^{n+1}$ so

$$\frac{f^{(n+1)}(x_0)}{n!}(x-x_0)^n = \frac{(n+1)(x-x_0)^n}{(x-c)^{n+1}}r_n(x),$$

whence (a) follows.

In order to establish (b) we take

$$u(t) = \frac{x - t}{x - c}. \tag{4.16}$$

Now u has the needed properties and $u'(t) = -1/(x - c)$. Therefore,

$$\frac{f^{(n+1)}(x_0)}{n!}(x - x_0)^n = \frac{1}{x - c}\, r_n(x),$$

and it follows that

$$r_n(x) = \frac{f^{(n+1)}(x_0)}{n!}(x - x_0)^n(x - c).$$

Since x_0 is between x and c, there exists θ between 0 and 1, such that $x_0 = (1 - \theta)c + \theta x$. Now

$$x - x_0 = x - (1 - \theta)c - \theta x = (1 - \theta)(x - c),$$

and we obtain (b).

Finally, let u be as in (4.16) and let us integrate (4.14). Then implies that

$$\int_c^x F'(t)\, dt = \int_c^x \frac{f^{(n+1)}(t)}{n!}(x - t)^n\, dt + \int_c^x u'(t) r_n(x)\, dt.$$

The fact that $F(x) = F(c) = 0$ implies that the left side equals 0. Similarly,

$$\int_c^x u'(t) r_n(x)\, dt = r_n(x) \int_c^x u'(t)\, dt = r_n(x)\, u(t)\Big|_c^x = -r_n(x),$$

whence (c) follows. □

Remark 4.5.2. When $n = 0$, assertion (a) is just the Mean Value Theorem.

BROOK TAYLOR (1685–1731) was an English mathematician who is best known for "Taylor's theorem" and the "Taylor series". He came from a wealthy family and he received a solid education (at home) that included music and painting. From 1703 to 1709 he studied at St. John's College in Cambridge. Based on the knowledge he had shown he was elected to the Royal Society in 1712. For example, that same year he solved a problem concerning Kepler's second law. From 1714 to 1718 he was the Secretary of the Royal Society. That period was also his most productive. Taylor's most important work is in two books. [101] contains what is now known as *Taylor series, calculus of finite differences, integration by parts*, and the formula for the change of variables. It should be said that special cases of Taylor's Theorem were discovered independently by Newton, Leibniz, Johann Bernoulli and French mathematician Abraham de Moivre (1667–1754). Be as it may, its significance was not recognized until Lagrange (page 398) obtained the explicit formula for the error in the revised edition of [79] in 1813. In the second book *Linear Perspective* Taylor laid the foundation for descriptive and projective geometry. His personal life was not happy. He was married twice and both wives died at childbirth. He died at the age of 46.

Taylor's Formula establishes a very compact form for the error of approximating f with the Taylor's polynomial. The form in (a) is often called the Lagrange form, while the form in (b) is the Cauchy form. Notice that x_0 depends on x and c. Although Lagrange's form is easier to use, there are situations in which it does not furnish the complete information, and the Cauchy form is needed.

Remark 4.5.3. By considering the Lagrange form of the remainder $r_n(x)$, we see that, as $x \to c$, $r_n(x)$ goes to 0 faster than $(x - c)^n$.

Now we can look at a familiar Taylor polynomial and estimate the error of approximation.

Example 4.5.4. Estimating the error of approximation.
Let $f(x) = e^x$, $c = 0$, $x = 1$ and $n = 4$. We will show that the error of approximation of $f(1)$ by the Taylor polynomial of degree 4 is no bigger than 0.075.
Solution. The Taylor polynomial of degree n, for $c = 0$, is

$$p_n(x) = 1 + x + \frac{x^2}{2!} + \frac{x^3}{3!} + \cdots + \frac{x^n}{n!}$$

and the remainder in the Lagrange form is

$$\frac{f^{(n+1)}(x_0)}{(n+1)!} x^{n+1} = \frac{e^{x_0}}{(n+1)!} x^{n+1}$$

for some x_0 between 0 and x. For $x = 1$, we get that

$$e \approx 1 + 1 + \frac{1}{2!} + \frac{1}{3!} + \cdots + \frac{1}{n!}$$

and the error can be estimated:

$$\left| \frac{e^{x_0}}{(n+1)!} \right| \leq \frac{e}{(n+1)!} < \frac{3}{(n+1)!}.$$

When $n = 4$, the Taylor's Formula gives that the error is no bigger than $3/5! = 0.075$. Actually, as Lemma 2.5.6 shows, the accuracy is better than $1/4! \approx 0.042$.   ◆

Exercises

In Exercises 4.5.1–4.5.6 write the Taylor polynomial of degree n for $f(x)$ at $x = c$:

4.5.1. $f(x) = \dfrac{1 + x + x^2}{1 - x + x^2}$, $n = 4$, $c = 0$.   4.5.2. $f(x) = e^{2x - x^2}$, $n = 5$, $c = 0$.

4.5.3. $f(x) = \ln \cos x$, $n = 6$, $c = 0$.   4.5.4. $f(x) = \tan x$, $n = 5$, $c = 0$.

4.5.5. $f(x) = \sqrt{x}$, $n = 3$, $c = 1$.   4.5.6. $f(x) = x^x - 1$, $n = 3$, $c = 1$.

4.5.7.* Let f be twice differentiable function on (a, b) and let $c \in (a, b)$. Prove that

$$f''(c) = \lim_{h \to 0} \frac{f(c + h) + f(c - h) - 2f(c)}{h^2}.$$

4.5.8.* Suppose that $a > 0$, $x > 0$ and n is a positive integer bigger than 1. Prove that

$$\sqrt[n]{a^n + x} = a + \frac{x}{na^{n-1}} - r, \quad \text{where} \quad 0 < r < \frac{(n-1)x^2}{2n^2 a^{2n-1}}.$$

In Exercises 4.5.9–4.5.12 estimate the error of approximation:

4.5.9. $\tan x \approx x + \dfrac{x^3}{3}$, for $|x| \leq 0.1$.   4.5.10. $\ln(1+x) \approx x - \dfrac{x^2}{2} + \dfrac{x^3}{3} - \dfrac{x^4}{4}$, for $|x| \leq 0.5$.

4.5.11. $\cos x \approx 1 - \dfrac{x^2}{2} + \dfrac{x^4}{24}$, for $|x| \leq 0.5$.

4.5.12. $\sqrt{x} \approx 1 + \dfrac{x-1}{2} - \dfrac{(x-1)^2}{8}$, for $|x - 1| \leq 0.5$.

In Exercises 4.5.13–4.5.16 use Taylor's Formula to evaluate a with accuracy of ε.

4.5.13. $a = \sin 1°$, $\varepsilon = 10^{-8}$.

4.5.14. $a = \sqrt{5}$, $\varepsilon = 10^{-4}$.

4.5.15. $a = \log_{10} 11$, $\varepsilon = 10^{-5}$.

4.5.16. $a = e$, $\varepsilon = 10^{-9}$.

In Exercises 4.5.17–4.5.20 use Taylor's Formula to evaluate the limits:

4.5.17. $\displaystyle \lim_{x \to 0} \frac{\sin(\sin x) - x\sqrt[3]{1 - x^2}}{x^5}$.

4.5.18. $\displaystyle \lim_{x \to 0} \frac{e^x \sin x - x(1 + x)}{x^3}$.

4.5.19. $\displaystyle \lim_{x \to \infty} x - x^2 \ln(1 + 1/x)$.

4.5.20.* $\displaystyle \lim_{x \to 0} \frac{1 - (\cos x)^{\sin x}}{x^2}$.

4.6 L'Hôpital's Rule

When computing limits, it is often very convenient to use the L'Hôpital's Rule. In reality, there is not one L'Hôpital's Rule, i.e., there are several related results that go under this common name. We start with the most straightforward.

Theorem 4.6.1 (L'Hôpital's Rule). *Let f and g be functions continuous and differentiable on (a, b), and suppose that $\lim_{x \to a^+} f(x) = \lim_{x \to a^+} g(x) = 0$. If $g'(x) \neq 0$ in (a, b), and if $\lim_{x \to a^+} f'(x)/g'(x)$ exists (finite or infinite), then so does $\lim_{x \to a^+} f(x)/g(x)$ and*

$$\lim_{x \to a^+} \frac{f(x)}{g(x)} = \lim_{x \to a^+} \frac{f'(x)}{g'(x)}. \tag{4.17}$$

Proof. We define $f(a) = g(a) = 0$, so f and g are both continuous in $[a, b_0]$, for any $b_0 < b$. Let $x \in (a, b_0)$. We apply Cauchy's Mean Value Theorem (Exercise 4.4.17) to conclude that there exists $c \in (a, x)$ such that

$$\frac{f'(c)}{g'(c)} = \frac{f(x) - f(a)}{g(x) - g(a)} = \frac{f(x)}{g(x)}.$$

Let us denote by $L = \lim_{x \to a^+} f'(x)/g'(x)$, and let $\varepsilon > 0$. There exists $\delta > 0$ such that

$$a < x < a + \delta \quad \Rightarrow \quad \left| \frac{f'(x)}{g'(x)} - L \right| < \varepsilon.$$

For such x, $a < c < a + \delta$ as well, so

$$\left| \frac{f(x)}{g(x)} - L \right| = \left| \frac{f'(c)}{g'(c)} - L \right| < \varepsilon,$$

and we see that (4.17) holds.

If $\lim_{x \to a^+} f'(x)/g'(x) = \infty$, and if $M > 0$, there exists $\delta > 0$ such that

$$a < x < a + \delta \quad \Rightarrow \quad \frac{f'(x)}{g'(x)} > M.$$

For such x, $a < c < a + \delta$ as well, so $f'(c)/g'(c) > M$, hence $f(x)/g(x) > M$, and (4.17) holds once again. $\square$

The first to use the name L'Hôpital's Rule was French mathematician Edouard Goursat (1858–1936) in his three-volume book *Cours d'Analyse Mathématique (A Course in Mathematical Analysis)*, published between 1902 and 1913. Goursat is now remembered principally as an expositor for this highly acclaimed text, which was translated to English in 1904. It set a standard for the high-level teaching of mathematical analysis, especially complex analysis.

We have formulated L'Hôpital's Rule for the case when a is a (finite) number. It is often useful to consider the situation when a is replaced by ∞.

Theorem 4.6.2. *Let f and g be two functions continuous and differentiable on $(a, +\infty)$ and suppose that*

$$\lim_{x \to \infty} f(x) = \lim_{x \to \infty} g(x) = 0.$$

If $g'(x) \neq 0$ in $(a, +\infty)$, and if $\lim_{x \to \infty} f'(x)/g'(x)$ exists (finite or infinite), then so does $\lim_{x \to \infty} f(x)/g(x)$ and

$$\lim_{x \to \infty} \frac{f(x)}{g(x)} = \lim_{x \to \infty} \frac{f'(x)}{g'(x)}.$$

Proof. The assumptions of the theorem are that the functions F and G, defined by $F(x) = f(1/x)$ and $G(x) = g(1/x)$, are continuous and differentiable on $(0, 1/a)$ and that they satisfy $\lim_{x \to 0} F(x) = \lim_{x \to 0} G(x) = 0$. Thus, with $x = 1/t$, and applying Theorem 4.6.1

$$\lim_{x \to \infty} \frac{f(x)}{g(x)} = \lim_{t \to 0^+} \frac{F(t)}{G(t)} = \lim_{t \to 0^+} \frac{F'(t)}{G'(t)}.$$

If we denote $h(x) = 1/x$, then $F = f \circ h$ and, by the Chain Rule,

$$F'(t) = f'(h(t))h'(t) = f'\left(\frac{1}{t}\right)\frac{-1}{t^2}.$$

Now, the substitution $x = 1/t$ yields $F'(t) = -x^2 f'(x)$, and similarly $G'(t) = -x^2 g'(x)$. Therefore,

$$\lim_{t \to 0^+} \frac{F'(t)}{G'(t)} = \lim_{x \to \infty} \frac{-x^2 f'(x)}{-x^2 g'(x)} = \lim_{x \to \infty} \frac{f'(x)}{g'(x)}. \qquad \square$$

Both theorems that we have established so far have dealt with the indeterminate form $\left(\frac{0}{0}\right)$. Another situation in which the L'Hôpital's Rule can be useful is when the indeterminate form is $\left(\frac{\infty}{\infty}\right)$.

Theorem 4.6.3. *Let f and g be functions continuous and differentiable on (a, b) and suppose that*

$$\lim_{x \to a^+} f(x) = \lim_{x \to a^+} g(x) = \infty.$$

If $g'(x) \neq 0$, and if $\lim_{x \to a^+} f'(x)/g'(x)$ exists (finite or infinite), then

$$\lim_{x \to a^+} \frac{f(x)}{g(x)} = \lim_{x \to a^+} \frac{f'(x)}{g'(x)}.$$

Proof. Suppose that $\lim_{x \to a^+} f'(x)/g'(x) = L$ and let $\varepsilon > 0$. Then there exists $\eta > 0$ such that $a + \eta < b$ and

$$a < x < a + \eta \quad \Rightarrow \quad \left|\frac{f'(x)}{g'(x)} - L\right| < \frac{\varepsilon}{4}. \tag{4.18}$$

Let $x_0 = a + \eta \in (a, b)$. We will use the following identity

$$\frac{f(x)}{g(x)} - L = \frac{f(x_0) - Lg(x_0)}{g(x)} + \left(1 - \frac{g(x_0)}{g(x)}\right)\left(\frac{f(x) - f(x_0)}{g(x) - g(x_0)} - L\right) \tag{4.19}$$

which holds for all $x \in (a, x_0)$. By the Cauchy's Mean Value Theorem (Exercise 4.4.17), there exists $c \in (x, x_0)$ such that

$$\frac{f(x) - f(x_0)}{g(x) - g(x_0)} = \frac{f'(c)}{g'(c)},$$

whence (4.18) implies that

$$\left| \frac{f(x) - f(x_0)}{g(x) - g(x_0)} - L \right| < \frac{\varepsilon}{4}. \tag{4.20}$$

The inequality (4.20) is true for all $x \in (a, x_0)$, and we will now make a choice of x sufficiently close to a. Since $\lim_{x \to a^+} g(x) = \infty$, there exists $\delta_1 > 0$ such that

$$a < x < a + \delta_1 \text{ and } x \in [a, x_0] \quad \Rightarrow \quad \left| \frac{f(x_0) - Lg(x_0)}{g(x)} \right| < \frac{\varepsilon}{2}. \tag{4.21}$$

Also, there exists $\delta_2 > 0$ such that, if $a < x < a + \delta_2$ and $x \in [a, x_0]$, then $g(x) > g(x_0)$ and $g(x) > 0$ so

$$\left| 1 - \frac{g(x_0)}{g(x)} \right| \le 2. \tag{4.22}$$

Let $\delta = \min\{\delta_1, \delta_2, x_0 - a\}$, and let $a < x < a + \delta$. Then $x \in (a, x_0)$ and inequalities (4.20)–(4.22) hold. It follows from (4.19) that

$$\left| \frac{f(x)}{g(x)} - L \right| \le \left| \frac{f(x_0) - Lg(x_0)}{g(x)} \right| + \left| 1 - \frac{g(x_0)}{g(x)} \right| \left| \frac{f(x) - f(x_0)}{g(x) - g(x_0)} - L \right|$$

$$< \frac{\varepsilon}{2} + 2\frac{\varepsilon}{4} = \varepsilon.$$

This settles the case when the limit $\lim_{x \to a^+} f'(x)/g'(x)$ is finite. If this limit is infinite the proof is similar and it is left as an exercise. $\qquad \square$

GUILLAUME de L'HÔPITAL (1661–1704) was a French mathematician. He came from a wealthy family. His father was a Lieutenant-general in the King's Army. Young Guillaume showed mathematical talent when, at the age of 15, he solved a problem proposed by Pascal. By the unspoken rule of the day, he was predestined for a military career, which he combined with the study of mathematics. He eventually resigned citing near-sightedness, but he was not heartbroken and devoted himself to his passion. He became a member of a circle that contained the most respected mathematicians and scientists of the day. In 1691, Johann Bernoulli was visiting Paris, and he agreed to lecture to the group for 6 months. After that, L'Hôpital hired him for private instructions. One of the solutions that Bernoulli showed him was to the problem of finding a curve whose subtangent has a fixed length. L'Hôpital then wrote about it to Huygens, who assumed it was L'Hôpital's. This did not sit well with Bernoulli, so L'Hôpital offered him a bigger sum to keep the results coming, and the contract stipulated that the material was not to be shared with anybody else. In 1696, L'Hôpital's famous book [81] was published—the first textbook on the differential calculus, used throughout the eighteenth century. In the introduction he acknowledged using the discoveries of the Bernoullis and of Leibniz. Chapter 9 contains what we now call L'Hôpital's Rule. When the manuscript of Bernoulli's course from 1691-92 was discovered in the Basel University library, it became obvious how closely the book followed the course. To be fair to L'Hôpital, he was a nobleman, accustomed to paying for the services of professionals, and the concept of intellectual property was yet to be invented. His letters to Leibniz and Huygens show that he did not lie about the authorship but referred to Bernoulli in a condescending way. In 1693, L'Hôpital was elected to the French academy of sciences and even served twice as its vice-president. Among his accomplishments were the determination of the arc length of the logarithmic graph, one of the solutions to the brachistochrone problem, and the discovery of a turning point singularity on the involute of a plane curve near an inflection point.

The L'Hôpital's Rule can be used for the indeterminate forms $\left(\frac{0}{0}\right)$ and $\left(\frac{\infty}{\infty}\right)$. When the indeterminate form is of a different type, such as $\left(0^0\right)$, $\left(0^\infty\right)$, $\left(1^\infty\right)$, etc., we need to use algebraic manipulations first.

Example 4.6.4. Indeterminate form 1^∞.

Find $\lim\limits_{x \to 1} x^{\frac{1}{1-x}}$.

Solution. This limit is of the form (1^∞), so we cannot apply the L'Hôpital's Rule directly. Therefore, we will first use some algebra. Since $x = e^{\ln x}$ when $x > 0$, and since $x \to 1$ means that we can assume that $x > 0$, we have

$$x^{\frac{1}{1-x}} = e^{\ln x \frac{1}{1-x}} = e^{\frac{\ln x}{1-x}}.$$

When $x \to 1$, the exponent $\ln x/(1-x)$ is of the form $\left(\frac{0}{0}\right)$, so we can apply the L'Hôpital's Rule. Further, $(1-x)' = -1 \neq 0$ so

$$\lim_{x \to 1} \frac{\ln x}{1-x} = \lim_{x \to 1} \frac{\frac{1}{x}}{-1} = -1,$$

and we obtain that $\lim\limits_{x \to 1} x^{\frac{1}{1-x}} = e^{-1}$.   ◆

Example 4.6.5. Indeterminate form $\infty - \infty$.

Find $\lim\limits_{x \to 0^+} \left(\cot x - \frac{1}{x} \right)$.

Solution. When $x \to 0^+$, both terms in parentheses go to $+\infty$, so we are looking at the indeterminate form $\infty - \infty$. However,

$$\cot x - \frac{1}{x} = \frac{\cos x}{\sin x} - \frac{1}{x} = \frac{x \cos x - \sin x}{x \sin x}$$

and the last fraction has both the numerator and the denominator go to 0, as $x \to 0^+$. This means that we can use the L'Hôpital's Rule:

$$\lim_{x \to 0^+} \frac{x \cos x - \sin x}{x \sin x} = \lim_{x \to 0^+} \frac{\cos x - x \sin x - \cos x}{\sin x + x \cos x} = \lim_{x \to 0^+} \frac{-x \sin x}{\sin x + x \cos x}.$$

The new fraction is of the form $\left(\frac{0}{0}\right)$, so we are allowed to use the L'Hôpital's Rule again. However, that will not lead to the solution. Instead, we will rely on the known result $\lim_{x \to 0} \sin x/x = 1$:

$$\lim_{x \to 0^+} \frac{-x \sin x}{\sin x + x \cos x} = \lim_{x \to 0^+} \frac{-\sin x}{\frac{\sin x}{x} + \cos x} = \frac{0}{1+1} = 0.$$ ◆

Exercises

In Exercises 4.6.1–4.6.23 find the limits:

4.6.1. $\lim\limits_{x \to 0} \dfrac{\tan x - x}{x - \sin x}$

4.6.2. $\lim\limits_{x \to 0} \dfrac{x \cot x - 1}{x^2}$

4.6.3. $\lim\limits_{x \to \pi/4} \dfrac{\sqrt[3]{\tan x} - 1}{2 \sin^2 x - 1}$

4.6.4. $\lim\limits_{x \to 0} \dfrac{a^x - a^{\sin x}}{x^3}$, $a > 0$

4.6.5. $\lim\limits_{x \to +\infty} \dfrac{\ln x}{x^2}$

4.6.6. $\lim\limits_{x \to +\infty} \dfrac{x^n}{a^x}$, $a > 0, n > 0$

4.6.7. $\lim\limits_{x \to \pi/2^-} \dfrac{3 + 4 \sec x}{2 + \tan x}$

4.6.8. $\lim\limits_{x \to 0^+} \dfrac{\ln \tan 2x}{\ln \tan 3x}$

4.6.9. $\lim\limits_{x \to 0} (\cot x)^{\sin x}$

4.6.10. $\lim\limits_{x \to 1^-} \ln x \ln(1-x)$

4.6.11. $\lim\limits_{x \to 0^+} x^x$

4.6.12. $\lim\limits_{x \to 0^+} x^{\frac{1}{1+\ln x}}$

4.6.13. $\lim\limits_{x \to 0} \dfrac{(1+x)^{1/x} - e}{x}$

4.6.14.* $\lim\limits_{x \to 0} e^{-1/x^2} x^{-100}$

4.6.15. $\lim\limits_{x \to 0} (\cos x)^{1/x^2}$

4.6.16. $\lim\limits_{x\to 0}\left(\dfrac{\arcsin x}{x}\right)^{1/x^2}$

4.6.17. $\lim\limits_{x\to +\infty}\dfrac{x^{\ln x}}{(\ln x)^x}$.

4.6.18. $\lim\limits_{x\to 0}\dfrac{\sin x - \arctan x}{x^2\ln x}$

4.6.19.* $\lim\limits_{x\to 0}\left(\dfrac{a^x - x\ln a}{b^x - x\ln b}\right)^{1/x^2}$

4.6.20. $\lim\limits_{x\to 0^+} x^{\sin x}$

4.6.21. $\lim\limits_{x\to 0}\left(\dfrac{1}{\sin^2 x} - \dfrac{1}{x^2}\right)$.

4.6.22.* $\lim\limits_{x\to 0}\dfrac{1}{x}\left(\dfrac{1}{\tanh x} - \dfrac{1}{\tan x}\right)$

4.6.23. $\lim\limits_{x\to 0}\left(\dfrac{1}{\ln(x + \sqrt{1+x^2})} - \dfrac{1}{\ln(1+x)}\right)$

4.6.24. Let f and g be functions continuous and differentiable in $(a, +\infty)$, let $\lim_{x\to\infty} f(x) = \lim_{x\to\infty} g(x) = \infty$, let $g'(x) \neq 0$ in $(a, +\infty)$, and let the limit $\lim_{x\to\infty} f'(x)/g'(x)$ exist (finite or infinite). Prove that the limit $\lim_{x\to\infty} f(x)/g(x)$ exists and that

$$\lim_{x\to\infty}\frac{f(x)}{g(x)} = \lim_{x\to\infty}\frac{f'(x)}{g'(x)}.$$

5

Indefinite Integral

In Chapter 4 we have looked at the problem of finding the derivative f of a given function F. Now, we are interested in the opposite: given f, find the function F. This practical problem was explored in the seventeenth century by the early masters of calculus: Newton, Leibniz, Johann Bernoulli, etc. Many powerful techniques were introduced by Euler in the eighteenth century.

5.1 Computing Indefinite Integrals: A Review

Example 5.1.1. $\int (6x^2 - 3x + 5)\, dx$.

Solution. We use the rules for the sum/difference

$$\int [f(x) + g(x)]\, dx = \int f(x)\, dx + \int g(x)\, dx,$$

the rule for the multiplication by a constant c

$$\int cf(x)\, dx = c \int f(x)\, dx,$$

and we obtain

$$\int (6x^2 - 3x + 5)\, dx = \int 6x^2\, dx - \int 3x\, dx + \int 5\, dx$$
$$= 6 \int x^2\, dx - 3 \int x\, dx + 5 \int dx$$
$$= 2x^3 - \frac{3}{2}x^2 + 5x + C. \qquad \blacklozenge$$

Example 5.1.2. $\int (1 + \sqrt{x})^4\, dx$.

Solution. The integrals are simpler to calculate when the integrand is written as a sum (rather than as a product):

$$\int (1 + \sqrt{x})^4\, dx = \int (1 + 4\sqrt{x} + 6x + 4x\sqrt{x} + x^2)\, dx$$
$$= \int dx + 4 \int x^{1/2}\, dx + 6 \int x\, dx + 4 \int x^{3/2}\, dx + \int x^2\, dx$$
$$= x + \frac{8}{3}x^{3/2} + 3x^2 + \frac{8}{5}x^{5/2} + \frac{1}{3}x^3 + C. \qquad \blacklozenge$$

Example 5.1.3. $\int \dfrac{dx}{x - a}$.

Solution. It is easy to see that the result is

$$\int \frac{dx}{x - a} = \ln |x - a| + C. \tag{5.1}$$

It is worth remembering that we really have two different rules in action here: if $x > a$ then $\ln(x - a)$ has the derivative $1/(x - a)$, so

$$\int \frac{dx}{x - a} = \ln(x - a) + C, \quad \text{if } x > a.$$

On the other hand if $x < a$, then $\ln(x - a)$ is not defined. However, $\ln(a - x)$ is defined and differentiable, and its derivative is also $1/(x - a)$, so

$$\int \frac{dx}{x - a} = \ln(a - x) + C, \quad \text{if } x < a.$$

These two formulas are usually combined into (5.1). ◆

Example 5.1.4. $\int \cos^2 3x \, dx$.

Solution. Here we use a trigonometric formula: $\cos^2 x = (1 + \cos 2x)/2$. Therefore,

$$\int \cos^2 3x \, dx = \int \frac{1 + \cos 6x}{2} \, dx$$

$$= \frac{1}{2} \left(\int dx + \int \cos 6x \, dx \right)$$

$$= \frac{1}{2} \left(x + \frac{\sin 6x}{6} \right) + C$$

$$= \frac{1}{2} x + \frac{1}{12} \sin 6x + C. \qquad ◆$$

One of the most effective techniques for the computation of an integral is the *substitution* method.

Example 5.1.5. $\int \frac{x \, dx}{1 + x^4}$.

Solution. We use the substitution $u = x^2$. Then $du = 2x \, dx$ and we obtain

$$\int \frac{x \, dx}{1 + x^4} = \int \frac{\frac{1}{2} \, du}{1 + u^2} = \frac{1}{2} \arctan u + C = \frac{1}{2} \arctan x^2 + C. \qquad ◆$$

Example 5.1.6. $\int \frac{dx}{x \ln x}$.

Solution. Here we use $u = \ln x$, so $du = \frac{1}{x} dx$. Now

$$\int \frac{dx}{x \ln x} = \int \frac{du}{u} = \ln |u| + C = \ln |\ln x| + C. \qquad ◆$$

In Examples 5.1.5 and 5.1.6 we have used substitutions of the form $u = \varphi(x)$. Sometimes it is more useful to make a substitution of the form $x = \varphi(t)$.

Example 5.1.7. $\int \sqrt{a^2 - x^2} \, dx$, $a > 0$.

Solution. We will use the substitution $x = a \sin t$. Notice that $a^2 - x^2 \geq 0$ so $|x| \leq a$ and $|a \sin t| \leq a$. This shows that this substitution is well defined (and, we hope, useful). We assume that $t \in [-\pi/2, \pi/2]$ so that the inverse function exists and $t = \arcsin(x/a)$. Since $x = a \sin t$, we have that $dx = a \cos t \, dt$, so

$$\int \sqrt{a^2 - x^2} \, dx = \int \sqrt{a^2 - (a \sin t)^2} \, a \cos t \, dt$$

$$= a \int \sqrt{a^2 - a^2 \sin^2 t} \cos t \, dt$$

$$= a \int \sqrt{a^2(1 - \sin^2 t)} \cos t \, dt$$

$$= a \int |a| \sqrt{\cos^2 t} \cos t \, dt$$

$$= a^2 \int |\cos t| \cos t \, dt.$$

The assumption that $t \in [-\pi/2, \pi/2]$ implies that $|\cos t| = \cos t$, so the integrand is $\cos^2 t = (1 + \cos 2t)/2$. Now

$$\int \sqrt{a^2 - x^2} \, dx = a^2 \int \frac{1 + \cos 2t}{2} \, dx$$

$$= \frac{a^2}{2} \left(t + \frac{\sin 2t}{2} \right) + C$$

$$= \frac{a^2}{2} t + \frac{a^2}{4} \sin 2t + C.$$

Notice that $\sin 2t = 2 \sin t \cos t = 2 \sin t \sqrt{1 - \sin^2 t}$ (because $\cos t \geq 0$), and it follows that $\sin 2t = 2 \frac{x}{a} \sqrt{1 - \left(\frac{x}{a} \right)^2}$. Therefore,

$$\int \sqrt{a^2 - x^2} \, dx = \frac{a^2}{2} \arcsin \frac{x}{a} + \frac{a^2}{4} 2 \frac{x}{a} \sqrt{1 - \frac{x^2}{a^2}} + C$$

$$= \frac{a^2}{2} \arcsin \frac{x}{a} + \frac{x}{2} \sqrt{a^2 - x^2} + C. \qquad \blacklozenge$$

Example 5.1.8. $\int \sqrt{x^2 + a^2} \, dx$, $a > 0$.

Solution. We will use the hyperbolic substitution $x = a \sinh t$. Then, $dx = a \cosh t \, dt$, and $t = \operatorname{arsinh} \left(\frac{x}{a} \right)$. By Exercise 3.7.14, $\operatorname{arsinh} x = \ln \left(x + \sqrt{x^2 + 1} \right)$. It follows that $\operatorname{arsinh} \left(\frac{x}{a} \right) = \ln \left(\left(\frac{x}{a} \right) + \sqrt{\left(\frac{x^2}{a^2} \right) + 1} \right)$. Now

$$\int \sqrt{x^2 + a^2} \, dx = \int \sqrt{a^2 \sinh^2 t + a^2} \, a \cosh t \, dt$$

$$= \int \sqrt{a^2 \cosh^2 t} \, a \cosh t \, dt$$

$$= a^2 \int \cosh^2 t \, dt$$

$$= a^2 \int \frac{1 + \cosh 2t}{2} \, dt$$

$$= \frac{a^2}{2} \left(t + \frac{\sinh 2t}{2} \right).$$

Further,

$$\sinh 2t = 2 \sinh t \cosh t = 2 \frac{x}{a} \sqrt{1 + \frac{x^2}{a^2}}, \quad \text{and}$$

$$t = \operatorname{arsinh} \left(\frac{x}{a} \right) = \ln \left(\left(\frac{x}{a} \right) + \sqrt{\left(\frac{x^2}{a^2} \right) + 1} \right).$$

It follows that

$$\int \sqrt{x^2 + a^2} \, dx = \frac{a^2}{2} \left(\ln \left(\left(\frac{x}{a} \right) + \sqrt{\left(\frac{x^2}{a^2} \right) + 1} \right) + \frac{x}{a} \sqrt{1 + \frac{x^2}{a^2}} \right)$$

$$= \frac{x}{2}\sqrt{x^2+a^2} + \frac{a^2}{2}\ln\frac{x+\sqrt{x^2+a^2}}{a}. \qquad \blacklozenge$$

Another useful technique is the Integration by Parts formula

$$\int u\,dv = uv - \int v\,du.$$

Example 5.1.9. $\int x^3 \ln x\,dx.$

Solution. We take $u = \ln x$, $dv = x^3\,dx$. It follows that $du = \frac{1}{x}\,dx$ and $v = \frac{x^4}{4}$. Then

$$\int x^3 \ln x\,dx = \ln x\,\frac{x^4}{4} - \int \frac{x^4}{4}\frac{1}{x}\,dx$$

$$= \ln x\,\frac{x^4}{4} - \frac{1}{4}\int x^3\,dx$$

$$= \ln x\,\frac{x^4}{4} - \frac{1}{4}\frac{x^4}{4} + C$$

$$= \ln x\,\frac{x^4}{4} - \frac{x^4}{16} + C. \qquad \blacklozenge$$

An interesting trick can be used in the following example.

Example 5.1.10. $I_1 = \int e^{2x}\sin 3x\,dx$, $I_2 = \int e^{2x}\cos 3x\,dx.$

Solution. We will compute both I_1 and I_2 at the same time, using the integration by parts. Further, in both integrals we will set $dv = e^{2x}\,dx$ so that $v = \frac{1}{2}e^{2x}$. In I_1, though, we will use $u_1 = \sin 3x$ and in I_2, $u_2 = \cos 3x$, so $du_1 = 3\cos 3x\,dx$ and $du_2 = -3\sin 3x\,dx$. Then

$$I_1 = \sin 3x\,\frac{1}{2}e^{2x} - \int \frac{1}{2}e^{2x}\,3\cos 3x\,dx = \frac{1}{2}\sin 3x\,e^{2x} - \frac{3}{2}I_2, \quad \text{and,}$$

$$I_2 = \cos 3x\,\frac{1}{2}e^{2x} - \int \frac{1}{2}e^{2x}(-3\sin 3x)\,dx = \frac{1}{2}\cos 3x\,e^{2x} + \frac{3}{2}I_1.$$

If we substitute the second equation in the first one we obtain

$$I_1 = \frac{1}{2}\sin 3x\,e^{2x} - \frac{3}{2}\left(\frac{1}{2}\cos 3x\,e^{2x} + \frac{3}{2}I_1\right) = \frac{1}{2}\sin 3x\,e^{2x} - \frac{3}{4}\cos 3x\,e^{2x} - \frac{9}{4}I_1,$$

and solving for I_1 yields

$$I_1 = \frac{2}{13}\sin 3x\,e^{2x} - \frac{3}{13}\cos 3x\,e^{2x}.$$

A similar calculation yields

$$I_2 = \frac{2}{13}\cos 3x\,e^{2x} + \frac{3}{13}\sin 3x\,e^{2x}. \qquad \blacklozenge$$

The following example shows yet another trick in action.

Example 5.1.11. $I_n = \int \frac{1}{(x^2+1)^n}\,dx.$

Solution. We apply the integration by parts formula with $u = 1/(x^2+1)^n$, $dv = dx$, so that $du = -n(x^2+1)^{-n-1}\,2x\,dx$ and $v = x$. Now

$$I_n = \frac{1}{(x^2+1)^n}\,x - \int -n(x^2+1)^{-n-1}\,2x \cdot x\,dx$$

$$= \frac{x}{(x^2 + 1)^n} + 2n \int \frac{x^2}{(x^2 + 1)^{n+1}} \, dx.$$

Notice that

$$\frac{x^2}{(x^2 + 1)^{n+1}} = \frac{x^2 + 1 - 1}{(x^2 + 1)^{n+1}} = \frac{1}{(x^2 + 1)^n} - \frac{1}{(x^2 + 1)^{n+1}},$$

so

$$I_n = \frac{x}{(x^2 + 1)^n} + 2n \int \frac{1}{(x^2 + 1)^n} \, dx - 2n \int \frac{1}{(x^2 + 1)^{n+1}} \, dx$$

$$= \frac{x}{(x^2 + 1)^n} + 2nI_n - 2nI_{n+1}.$$

Solving for I_{n+1} yields

$$I_{n+1} = \frac{2n - 1}{2n} I_n + \frac{x}{2n(x^2 + 1)^n}.$$

Now that we have this *recursive* formula, we can calculate I_n for any $n \in \mathbb{N}$, so long as we know I_1. But

$$I_1 = \int \frac{dx}{x^2 + 1} = \arctan x + C$$

so, for example,

$$I_2 = \frac{1}{2} I_1 + \frac{x}{2(x^2 + 1)} = \frac{1}{2} \arctan x + \frac{x}{2(x^2 + 1)} + C. \qquad \blacklozenge$$

Exercises

In Exercises 5.1.1–5.1.19 evaluate the integrals:

5.1.1. $\int \dfrac{x + 1}{\sqrt{x}}$

5.1.2. $\int \sqrt[3]{1 - 3x} \, dx$

5.1.3. $\int \dfrac{dx}{2 + 3x^2}$

5.1.4. $\int \dfrac{dx}{2 - 3x^2}$

5.1.5. $\int \dfrac{dx}{x\sqrt{x^2 + 1}}$

5.1.6. $\int \dfrac{x}{4 + x^4} \, dx$

5.1.7. $\int \dfrac{dx}{\sin^2 x + 2\cos^2 x}$

5.1.8. $\int \dfrac{dx}{e^x + e^{-x}}$

5.1.9. $\int \dfrac{6^x}{9^x - 4^x} \, dx$

5.1.10. $\int \dfrac{dx}{\sin x}$

5.1.11. $\int \sin^4 x \, dx$

5.1.12. $\int \dfrac{dx}{\sqrt{x + 1} + \sqrt{x - 1}}$

5.1.13. $\int \dfrac{dx}{1 + e^x}$

5.1.14. $\int \dfrac{x^5}{\sqrt{1 - x^2}} \, dx$

5.1.15. $\int \dfrac{dx}{\sqrt{(1 - x^2)^3}}$

5.1.16. $\int 3^{\sqrt{2x+1}} \, dx.$

5.1.17. $\int x^2 \arccos x \, dx.$

5.1.18. $\int x^n \ln x \, dx.$

5.1.19. $\int \dfrac{x \ln(x + \sqrt{1 + x^2})}{\sqrt{1 + x^2}} \, dx.$

5.2   Antiderivative

In this section we will establish the rules that we have used in the previous section. We start with the definition of the antiderivative.

Definition 5.2.1. Let f be a function defined on (a, b), and suppose that there exists a function F defined on (a, b), such that $F'(x) = f(x)$, for all $x \in (a, b)$. We say that F is an **antiderivative** of f, or a **primitive function** of f.

The term "primitive function" was introduced by Lagrange (page 398) in 1797. We owe the "antideriva-tive" to French mathematician Sylvestre-François Lacroix (1765–1843). He was known for a number of textbooks held in high esteem.

Clearly, an antiderivative is not unique. Namely, if F is an antiderivative of f, then so is $F + C$, where C is any constant. For example, if $f(x) = x^3$, then $F(x) = x^4/4$ is an antiderivative of f, and so is every function of the form $x^4/4 + C$. Could there be other antiderivatives of f?

Theorem 5.2.2. *Let F be a function that is differentiable on an open interval (a, b), and suppose that $F'(x) = 0$ for all $x \in (a, b)$. Then $F(x)$ is constant on (a, b).*

Proof. Suppose to the contrary that there exist two points $x_1 < x_2$ in (a, b) such that $F(x_1) \neq F(x_2)$. Applying the Mean Value Theorem (Theorem 4.4.5) we deduce that there exists a point $c \in (x_1, x_2)$ such that

$$F'(c) = \frac{F(x_2) - F(x_1)}{x_2 - x_1}.$$

This is a contradiction since the left side is 0 (by assumption) and the right side is not. □

From here we deduce an easy corollary.

Corollary 5.2.3. *Let F and G be two functions that are differentiable on an open interval (a, b), and suppose that $F'(x) = G'(x)$ for all $x \in (a, b)$. Then $G(x) = F(x) + C$ on (a, b).*

Proof. Let H be the function on (a, b), defined by $H(x) = F(x) - G(x)$. By assumption, H is differentiable on (a, b) and $H'(x) = F'(x) - G'(x) = 0$. It follows from Theorem 5.2.2 (applied to H instead of F) that $H(x) = C$ on (a, b), so $G(x) = F(x) + C$. □

Now we know that every antiderivative of x^3 must differ from $F(x) = x^4/4$ by a constant. This leads us to a definition.

Definition 5.2.4. Let f be a function defined on a set A, and suppose that F is an antiderivative of f. We call the collection of all functions of the form $F + C$ the **indefinite integral** of f and we write $\int f(x)\,dx = F(x) + C$.

Having defined the indefinite integral we can now derive some of its properties.

Theorem 5.2.5. *Let f, g be two functions defined on a set A, and let $\alpha \in \mathbb{R}$, $\alpha \neq 0$. Suppose that F is an antiderivative of f and that G is an antiderivative of g. Then the functions αf and $f + g$ have antiderivatives as well and:*

(a) $\int \alpha f(x)\,dx = \alpha \int f(x)\,dx$;

(b) $\int (f(x) + g(x))\,dx = \int f(x)\,dx + \int g(x)\,dx$.

Proof. We will prove only part (a) and leave (b) as an exercise. Notice that both the left and the right side describe a collection of functions. A function F belongs to the left side if and only if $F'(x) = \alpha f(x)$. A function G belongs to the right side if and only if $(G(x)/\alpha)' = f(x)$. Now Theorem 4.3.1 shows that this is equivalent to $G'(x) = \alpha f(x)$, and the assertion is proved. □

Remark 5.2.6. Notice that the assertion (a) is not true if $\alpha = 0$. In that case the left side contains all the functions F such that $F' = 0$, i.e., all constant functions. However, the right side contains only the zero function, because every function from the collection $\int f(x)\,dx$ is multiplied by $\alpha = 0$.

The first appearance of the integral symbol was in an unpublished paper by Leibniz in 1675. The integral symbol was actually a long letter S for "summa" (Latin for a sum), because Leibniz wrote about "Calculus Summatorius". At the same time, Johann Bernoulli called it "Calculus Integralis" and advocated the use of the letter I for integrals. It appears that there was a gentleman's agreement: today we use the symbol that is due to Leibniz and the name that comes from Bernoulli.

Next we address the substitution method. As we have seen in the previous section, there are two different types of substitution.

Theorem 5.2.7. *Suppose that F is an antiderivative of f, and $u = \varphi(x)$ is a differentiable function. Then*

$$\int f(\varphi(x))\varphi'(x)\, dx = \int f(u)\, du. \tag{5.2}$$

Proof. An antiderivative of the left side is $F(\varphi(x))$ and an antiderivative of the right side is $F(u)$, and they are equal since $u = \varphi(x)$. □

We notice that both types of substitution use the same formula (5.2). In Example 5.1.5, we are given the integral on the left side, and using $\varphi(x) = x^2$ we transform it into the integral on the right side. On the other hand, in Example 5.1.7 we start with the integral on the right side of (5.2), and we use the substitution $u = \varphi(x) = a \sin x$ to transform it into the integral on the left.

The substitution was extensively used by Newton in his *Principia* in 1687.

Theorem 5.2.8 (Integration by Parts)**.** *Let f, g be two differentiable functions defined on a set A, and suppose that fg' has an antiderivative. Then so does gf' and*

$$\int fg'\, dx = fg - \int gf'\, dx.$$

Proof. The derivative of the left side is fg' while on the right side we obtain

$$(fg)' - gf'.$$

The result now follows from the Product Rule for derivatives. □

Newton used integration by parts in *Principia*, although he did not formulate it as a rule. As a new method to calculate integrals, it was introduced by Taylor in [101] in 1715. The first use of the name Integration by Parts comes in 1828 in [107] by George Walker (1793–1830), a minister and the headmaster of the Leeds Free Grammar School.

We have seen that all elementary functions (provided that they are differentiable) have derivatives that are also elementary functions. The situation with antiderivatives is very different. There are numerous examples of elementary functions with antiderivatives not being in this class of functions. We list some of these functions:

$$e^{-x^2},\ \sin(x^2),\ \frac{\sin x}{x},\ \frac{1}{\ln x}, \ldots.$$

In the remainder of this section we will consider several classes of functions for which elementary antiderivatives exist, and look at some of the methods that lead to the results. With the advent of technology, some of the integration techniques may seem outdated. Yet, even a sophisticated computer algebra system could not handle Example 5.2.12.

Exercises

In Exercises 5.2.1–5.2.8 evaluate the integrals using an appropriate substitution:

5.2.1. $\int \cos^5 x \sqrt{\sin x}\, dx$ 5.2.2. $\int \dfrac{\ln x}{x\sqrt{1+\ln x}}\, dx$

5.2.3. $\int \dfrac{\arctan \sqrt{x}\, dx}{\sqrt{x}(1+x)}$ 5.2.4. $\int \dfrac{1}{\sqrt{a^2+x^2}}\, dx,\ a>0$

5.2.5. $\int \dfrac{1}{\sqrt{a^2-x^2}}\, dx$ 5.2.6. $\int \dfrac{1}{\sqrt{x^2-a^2}}\, dx,\ a>0$

5.2.7. $\int \sqrt{x^2-a^2}\, dx$ 5.2.8. $\int \dfrac{x^2}{\sqrt{a^2+x^2}}\, dx,\ a>0$

In Exercises 5.2.9–5.2.12 evaluate the integrals using integration by parts:

5.2.9. $\int \arcsin x\, dx$ 5.2.10. $\int x \arctan x\, dx$

5.2.11. $\int x \cos x\, dx$ 5.2.12. $\int x \sinh x\, dx$

5.2.13. Calculate $\int x f''(x)\, dx$.

5.2.14. Calculate $\int f'(2x)\, dx$.

5.2.15. Find $f(x)$ if $f'(x^2) = 1/x,\ x>0$.

5.2.16. Find $f(x)$ if $f'(\sin^2 x) = \cos^2 x$.

5.2.17. Let f be a monotone continuous function and let F be an antiderivative of f. Prove that $\int f^{-1}(x)\, dx = x f^{-1}(x) - F(f^{-1}(x)) + C$.

5.2.18. Prove part (b) of Theorem 5.2.5.

5.2.1   Rational Functions

If $f(x)$ is a rational function, that means that it can be written as a quotient of two polynomials P and Q. The standard technique to find an antiderivative of f is to factor the denominator Q. The factorization can be (in the absence of complex numbers) performed up to quadratic factors. In other words every polynomial Q can be factored into factors that fall into one of the 4 groups:

(i) $x - a$, for some $a \in \mathbb{R}$;

(ii) $(x - a)^k$, for some $a \in \mathbb{R}$ and $k \in \mathbb{N},\ k \geq 2$;

(iii) $x^2 + px + q$, for some $p, q \in \mathbb{R},\ p^2 - 4q < 0$;

(iv) $(x^2 + px + q)^k$, for some $p, q \in \mathbb{R},\ p^2 - 4q < 0$, and $k \in \mathbb{N},\ k \geq 2$.

The fact that $p^2 - 4q < 0$ in factors of type (iii) and (iv), means that they have no real roots, so that they cannot be factored any further. (Those with the knowledge of complex numbers will know that if we allow polynomials with complex coefficients, then every polynomial of degree > 1 can be factored.)

The next step is to use the **Partial Fractions Decomposition** to write f as a sum of rational functions,

$$\frac{P(x)}{Q(x)} = P_0(x) + \frac{P_1(x)}{Q_1(x)} + \frac{P_2(x)}{Q_2(x)} + \cdots + \frac{P_n(x)}{Q_n(x)}$$

where each of the polynomials $Q_1, Q_2, \ldots, Q_n$ has the form of one of the 4 types above. Furthermore, if $Q_i(x)$ is of type (i) or (ii), $P_i(x)$ is a constant, while for types (iii) and (iv) it is a linear function of the form $Mx + N$. Thus, assuming we have been able to factor the denominator, all we need to know is how to find an antiderivative for rational functions of type:

(i) $\dfrac{1}{x-a}$, for some $a \in \mathbb{R}$;

(ii) $\dfrac{1}{(x-a)^k}$, for some $a \in \mathbb{R}$ and $k \in \mathbb{N}$, $k \geq 2$;

(iii) $\dfrac{Mx + N}{x^2 + px + q}$, for some $M, N, p, q \in \mathbb{R}$, $p^2 - 4q < 0$;

(iv) $\dfrac{Mx + N}{(x^2 + px + q)^k}$, for some $M, N, p, q \in \mathbb{R}$, $p^2 - 4q < 0$, and $k \in \mathbb{N}$, $k \geq 2$.

In elementary calculus, you have learned that

$$\int \frac{1}{x-a} = \ln|x-a| + C, \quad \text{and} \quad \int \frac{1}{(x-a)^k} = \frac{-1}{(k-1)(x-a)^{k-1}}, \quad k \geq 2.$$

Therefore, we focus on integrands of types (iii) and (iv). In both of these it is useful to "complete the square" in the denominator:

$$x^2 + px + q = \left(x + \frac{p}{2}\right)^2 + q - \frac{p^2}{4}.$$

We have made the assumption that $q - p^2/4 > 0$, so there exists $a \in \mathbb{R}$ such that $q - p^2/4 = a^2$. If we now introduce a substitution $u = x + p/2$ then we obtain integrals of the form

(iii)$'$ $\displaystyle\int \frac{M'u + N'}{u^2 + a^2}\, du$, for some $M', N', a \in \mathbb{R}$;

(iv)$'$ $\displaystyle\int \frac{M'u + N'}{(u^2 + a^2)^k}\, du$, for some $M', N', a \in \mathbb{R}$, and $k \in \mathbb{N}$, $k \geq 2$.

Example 5.2.9. Reducing to type (iv).

Reduce $\int \dfrac{2x - 3}{(x^2 + 2x + 6)^3}\, dx$ to an integral of type (iv).

Solution. We see that this is an integral of type (iv), with $M = 2$, $N = -3$, $p = 2$, $q = 6$ (so that $q - p^2/4 = 5 > 0$), and $k = 3$. Now

$$x^2 + 2x + 6 = (x+1)^2 + 5 = (x+1)^2 + (\sqrt{5})^2,$$

and we use the substitution $u = x + 1$. Then $du = dx$ so we obtain

$$\int \frac{2x - 3}{(x^2 + 2x + 6)^3}\, dx = \int \frac{2(u-1) - 3}{(u^2 + 5)^3}\, du = \int \frac{2u - 5}{(u^2 + 5)^3}\, du$$

which is an integral of type (iv)$'$, with $M' = 2$, $N' = -5$, $a = \sqrt{5}$, $k = 3$. $\blacklozenge$

Integrals of type (iii)$'$ and (iv)$'$ can be further broken into two somewhat familiar integrals:

$$\int \frac{M'u + N'}{(u^2 + a^2)^k}\, du = M' \int \frac{u}{(u^2 + a^2)^k}\, du + N' \int \frac{1}{(u^2 + a^2)^k}\, du.$$

These can be calculated using the substitution $v = u^2$ in the first, and $u = aw$ in the second one.

Example 5.2.10. An integral of type (iii)′.

Find $\int \dfrac{2u-5}{u^2+5}\,du.$

Solution. As suggested

$$\int \frac{2u-5}{u^2+5}\,dx = 2\int \frac{u}{u^2+5}\,du - 5\int \frac{1}{u^2+5}\,du$$

and we use the substitution $v = u^2$ in the first (so that $dv = 2u\,du$), and $u = \sqrt{5}w$ in the second one. We obtain

$$2\int \frac{1}{v+5}\frac{1}{2}\,dv - 5\int \frac{1}{5w^2+5}\sqrt{5}\,dw = \ln|v+5| - \sqrt{5}\arctan w + C$$

$$= \ln|u^2+5| - \sqrt{5}\arctan \frac{u}{\sqrt{5}} + C. \qquad \blacklozenge$$

Example 5.2.11. An integral of type (iv)′.

Find $\int \dfrac{2u-5}{(u^2+5)^3}\,du.$

Solution. Once again we write

$$\int \frac{2u-5}{(u^2+5)^3}\,du = 2\int \frac{u}{(u^2+5)^3}\,du - 5\int \frac{1}{(u^2+5)^3}\,du$$

and we use $v = u^2$ in the first and $u = \sqrt{5}w$ in the second integral. Thus, we have

$$2\int \frac{1}{(v+5)^3}\frac{1}{2}\,dv - 5\int \frac{1}{(5w^2+5)^3}\sqrt{5}\,dw = \frac{-1}{2(v+5)^2} - \frac{\sqrt{5}}{25}\int \frac{1}{(w^2+1)^3}\,dw.$$

The last integral is I_3 in Example 5.1.11, and it equals

$$\frac{1}{4}\frac{w}{(w^2+1)^2} + \frac{3}{8}\frac{w}{w^2+1} + \frac{3}{8}\arctan w.$$

Therefore,

$$\int \frac{2u-5}{(u^2+5)^3}\,du = \frac{-1}{2(u^2+5)^2} - \frac{1}{4}\frac{u}{(u^2+5)^2} - \frac{3}{40}\frac{u}{u^2+5} - \frac{3\sqrt{5}}{200}\arctan \frac{u}{\sqrt{5}} + C. \qquad \blacklozenge$$

> In his book [40], Euler made a statement that, in order to calculate the integral of any rational function, it is sufficient to be able to integrate functions of the form (i)–(iv). For integrals of type (iv), he did some special cases, but not the most general case. It should be noted that, at that time, there was no proof that every polynomial with real coefficients can be factored into polynomials of degree up to 2. That factorization result is a consequence of the Fundamental Theorem of Algebra: every polynomial with complex coefficients can be written as the product of linear factors. Euler wrote his book in 1748 and it was published in 1755, way before the proof of FTA was found. Several mathematicians made unsuccessful attempts to prove it, including Euler himself in 1749, Lagrange (page 398) in 1772, and Gauss (page 196) in 1799. However, the first correct proof was published in 1806 by Jean-Robert Argand (1768–1822), a gifted French amateur mathematician. The first textbook containing a proof of the theorem was Cauchy's *Cours d'Analyse*.

Exercises

In Exercises 5.2.19–5.2.30 evaluate the integrals:

5.2.19. $\int \dfrac{x}{(x+1)(x+2)(x+3)}\,dx.$ 5.2.20. $\int \dfrac{x^3+1}{x^3-5x^2+6}\,dx.$ 5.2.21. $\int \dfrac{x^{10}}{x^2+x-2}\,dx.$

5.2.22. $\int \left(\dfrac{x}{x^2 - 3x + 2}\right)^2 dx.$ **5.2.23.** $\int \dfrac{dx}{x^3 + 1}.$ **5.2.24.** $\int \dfrac{x^4}{x^4 + 5x^2 + 4} dx.$

5.2.25. $\int \dfrac{x^2 + 1}{x^4 + x^2 + 1} dx.$ **5.2.26.*** $\int \dfrac{dx}{(x^3 + 1)^2}.$ **5.2.27.** $\int \dfrac{x^2 \, dx}{(x^2 + 2x + 2)^2}.$

5.2.28. $\int \dfrac{x \, dx}{(x - 1)^2(x^2 + 2x + 2)}.$ **5.2.29.** $\int \dfrac{dx}{x^4 + 1}.$ **5.2.30.*** $\int \dfrac{dx}{x^{2n} + 1}.$

In Exercises 5.2.31–5.2.32 determine the necessary and sufficient condition that the coefficients have to satisfy in order for the integral to represent a rational function:

5.2.31. $\int \dfrac{ax^2 + bx + c}{x^3(x - 1)^2} dx.$ **5.2.32.*** $\int \dfrac{\alpha x^2 + \beta x + \gamma}{(ax^2 + 2bx + c)^2} dx.$

5.2.33.* Derive a recursive formula for

$$I_n = \int \frac{dx}{(ax^2 + bx + c)^n}, \quad a \neq 0.$$

Use it to find $\int (x^2 + x + 1)^{-3} \, dx.$

5.2.2   Irrational Functions

Many of these functions do not have elementary antiderivatives, so we will take a look at some that do. Here, we will be interested in functions of the form

$$R\left(x, \sqrt[m]{\frac{\alpha x + \beta}{\gamma x + \delta}}\right), \tag{5.3}$$

where R is a rational function, and m is a positive integer.

Example 5.2.12. An integral of type as in (5.3).

Verify that $\int \dfrac{dx}{\sqrt[3]{(x - 1)(x + 1)^2}}$ is of type as in (5.3).

Remark 5.2.13. Even a sophisticated computer algebra system could not handle the integral in Example 5.2.12.

Solution. The integrand can be written as

$$\frac{1}{\sqrt[3]{(x - 1)(x + 1)^2}} = \sqrt[3]{\frac{x + 1}{(x - 1)(x + 1)^3}} = \sqrt[3]{\frac{x + 1}{x - 1}} \, \frac{1}{x + 1}. \qquad \blacklozenge$$

When the integrand is as in (5.3), the recommended substitution is

$$t = \sqrt[m]{\frac{\alpha x + \beta}{\gamma x + \delta}}.$$

If we solve for x, we get that $t^m = \frac{\alpha x + \beta}{\gamma x + \delta}$ so $t^m(\gamma x + \delta) = \alpha x + \beta$ and, hence,

$$x = \frac{\delta t^m - \beta}{\alpha - \gamma t^m}.$$

Consequently, x and dx are rational functions of t, and the whole integrand becomes a rational function.

Example 5.2.14. Calculating an integral of type as in (5.3).

Find $\int \sqrt[3]{\dfrac{x+1}{x-1}}\,\dfrac{1}{x+1}\,dx$.

Solution. We have $\alpha = \beta = \gamma = 1$, $\delta = -1$, $m = 3$. Thus,

$$t = \sqrt[3]{\frac{x+1}{x-1}}, \quad x = \frac{t^3+1}{t^3-1}, \quad \text{and} \quad dx = \frac{-6t^2}{(t^3-1)^2}\,dt.$$

We obtain

$$\int t\,\frac{1}{\frac{t^3+1}{t^3-1}+1}\,\frac{-6t^2}{(t^3-1)^2}\,dt = \int t\,\frac{t^3-1}{2t^3}\,\frac{-6t^2}{(t^3-1)^2}\,dt = \int \frac{-3}{t^3-1}\,dt$$

$$= -\ln|t-1| + \frac{1}{2}\ln\left(t^2+t+1\right) + \sqrt{3}\arctan\frac{(2t+1)\sqrt{3}}{3} + C.$$

Consequently,

$$\int \sqrt[3]{\frac{x+1}{x-1}}\,\frac{1}{x+1}\,dx = -\ln\left|\sqrt[3]{\frac{x+1}{x-1}}-1\right|$$

$$+ \frac{1}{2}\ln\left(\sqrt[3]{\left(\frac{x+1}{x-1}\right)^2}+\sqrt[3]{\frac{x+1}{x-1}}+1\right) + \sqrt{3}\arctan\frac{\left(2\sqrt[3]{\dfrac{x+1}{x-1}}+1\right)\sqrt{3}}{3} + C. \ \blacklozenge$$

Example 5.2.15. Calculating an integral of type as in (5.3).

Find $\int \dfrac{\sqrt{x+1}+2}{(x+1)^2 - \sqrt{x+1}}\,dx$.

Solution. In this example $\alpha = \beta = \delta = 1$, $\gamma = 0$, and $m = 2$. Therefore,

$$t = \sqrt{x+1}, \quad x = t^2 - 1, \quad dx = 2t\,dt$$

so we obtain

$$\int \frac{t+2}{t^4-t}\,2t\,dt = \int \frac{2(t+2)}{t^3-1}\,dt = 2\ln|t-1| - \ln\left(t^2+t+1\right) - \frac{2\sqrt{3}}{3}\arctan\frac{(2t+1)\sqrt{3}}{3} + C$$

$$= 2\ln\left|\sqrt{x+1}-1\right| - \ln\left(x+2+\sqrt{x+1}\right) - \frac{2\sqrt{3}}{3}\arctan\frac{(2\sqrt{x+1}+1)\sqrt{3}}{3} + C. \ \blacklozenge$$

> This type of substitution, applied to the case of the square root ($n = 2$), is due to Euler, and can be found in his book [80].

Exercises

In Exercises 5.2.34–5.2.49 evaluate the integrals by reducing them to the integrals of rational functions:

5.2.34. $\displaystyle\int \frac{1-\sqrt[3]{x-1}}{1+\sqrt[4]{x-1}}\,dx$.

5.2.35. $\displaystyle\int \frac{dx}{x(1+2\sqrt{x}+\sqrt[3]{x})}$.

5.2.36. $\displaystyle\int \frac{x\sqrt[3]{2+x}}{x+\sqrt[3]{2+x}}\,dx$.

5.2.37. $\displaystyle\int \frac{dx}{\left(1+\sqrt[4]{x}\right)^2\sqrt{x}}$.

5.2.38. $\int \dfrac{\sqrt{x+1}-\sqrt{x-1}}{\sqrt{x+1}+\sqrt{x-1}}\,dx.$

5.2.39.* $\int \dfrac{x}{\sqrt[4]{x^3(1-x)}}\,dx.$

5.2.40. $\int x\sqrt{\dfrac{x+1}{x-3}}\,dx.$

5.2.41.* $\int \sqrt[3]{1-\dfrac{1}{x-2}}\,dx.$

5.2.42. $\int \dfrac{dx}{x\sqrt{x^2-1}}.$

5.2.43. $\int \dfrac{dx}{x^2\sqrt{x^2+1}}.$

5.2.44.* $\int \dfrac{\sqrt{x^2-1}}{x}\,dx.$

5.2.45. $\int \dfrac{x^2}{\sqrt{x^2+1}}\,dx.$

5.2.46. $\int \dfrac{dx}{(1-x^2)^{3/2}}.$

5.2.47. $\int \dfrac{x^2}{(1-x^2)^{3/2}}\,dx.$

5.2.48. $\int \dfrac{dx}{\sqrt{2x^2+x-1}}.$

5.2.49.* $\int \dfrac{dx}{x+\sqrt{x^2-x+1}}.$

5.2.50.* Let $R(x,y,z)$ be a rational function of 3 variables. Prove that evaluation of the integral

$$\int R\left(x,\ \sqrt{ax+b},\ \sqrt{cx+d}\right)\,dx$$

can be reduced to the evaluation of an integral with a rational integrand.

5.2.3 Binomial Differentials

The title refers to the integrands of the form

$$x^m(a+bx^n)^p\,dx \tag{5.4}$$

where $a,b \in \mathbb{R}$ and the exponents $m,n,p \in \mathbb{Q}$. We will assume that at least one of the numbers

$$p,\ \frac{m+1}{n},\ \frac{m+1}{n}+p \tag{5.5}$$

is an integer and we will show how to reduce such an integral to the integration of a rational function.

> Euler knew about the three conditions (5.5) for the integrability of binomial differentials. Chebyshev (page 145) demonstrated in 1853 that in all other cases the integral of a binomial differential cannot be expressed in finite form through elementary functions.

Case 1: p is an integer. Let λ be the smallest common multiple of the denominators of the rational numbers m and n. Then, the substitution $x = t^\lambda$ transforms the expression (5.4) into a rational function of x.

Example 5.2.16. The integrand of type as in (5.4) with p an integer.

Reduce $\int \dfrac{\sqrt{x}}{(1+\sqrt[3]{x})^2}\,dx$ to an integral of a rational function.

Solution. This is a binomial differential with $m = 1/2$, $n = 1/3$, and $p = -2$. It is easy to see that $\lambda = 6$, so we use the substitution $x = t^6$. Then $dx = 6t^5\,dt$, so the integral becomes

$$\int \frac{t^3}{(1+t^2)^2}\,6t^5\,dt$$

which is a rational function. $\qquad\blacklozenge$

In order to consider the remaining two cases, we introduce a substitution $x = t^{1/n}$. Then $dx = \frac{1}{n} t^{\frac{1}{n}-1} \, dt$ and (5.4) becomes

$$t^{\frac{m}{n}} (a + bt)^p \frac{t^{\frac{1}{n}-1}}{n} \, dt = \frac{1}{n} t^{\frac{m+1}{n}-1} (a + bt)^p \, dt.$$

To simplify the writing, we will denote $q = \frac{m+1}{n} - 1$ and notice that the two remaining cases are when either q or $q + p$ is an integer, and the integral is of the form

$$\int t^q (a + bt)^p \, dt. \tag{5.6}$$

Case 2: q is an integer. Then we can use the substitution $u = \sqrt[\nu]{a + bt}$, where $p = \mu/\nu$. Indeed,

$$t = \frac{u^\nu - a}{b}, \quad \text{and} \quad dt = \frac{\nu u^{\nu-1}}{b} \, du,$$

so the integral (5.6) becomes

$$\int \frac{(u^\nu - a)^q}{b^q} u^\mu \, \nu \frac{u^{\nu-1}}{b} \, du.$$

Since q, μ, ν are all integers, this is again a rational function.

Example 5.2.17. An integral of type as in (5.6) with q an integer.

Find $\int \dfrac{\sqrt[3]{1 + \sqrt[4]{x}}}{\sqrt{x}} \, dx$.

Solution. It is easy to see that $m = -1/2$, $n = 1/4$, and $p = 1/3$, so $q = (m+1)/n = 2$. Further, $\nu = 3$ and we use the substitution $u = \sqrt[3]{1 + \sqrt[4]{x}}$. Then $x = (u^3 - 1)^4$ so $dx = 4(u^3 - 1)^3 3u^2 \, du$ and we obtain

$$\int \frac{u}{(u^3 - 1)^2} \, 12(u^3 - 1)^3 u^2 \, du = 12 \int u^3 (u^3 - 1) \, du = 12 \frac{u^7}{7} - 12 \frac{u^4}{4} + C$$

$$= \frac{12}{7} \left(\sqrt[3]{1 + \sqrt[4]{x}} \right)^7 - 3 \left(\sqrt[3]{1 + \sqrt[4]{x}} \right)^4 + C. \qquad \blacklozenge$$

Case 3: $q + p$ is an integer. We will rewrite the integral (5.6) as

$$\int t^{q+p} \left(\frac{a + bt}{t} \right)^p \, dt. \tag{5.7}$$

and use the substitution $u = \sqrt[\nu]{(a + bt)/t}$, where again $p = \mu/\nu$. This time, $u^\nu = (a + bt)/t$ and, solving for t yields

$$t = \frac{a}{u^\nu - b}, \quad \text{and} \quad dt = \frac{-a\nu u^{\nu-1}}{(u^\nu - b)^2} \, du.$$

Now the integral (5.7) becomes

$$\int \left(\frac{a}{u^\nu - b} \right)^{q+p} u^\mu \frac{-a\nu u^{\nu-1}}{(u^\nu - b)^2} \, du = -a^{q+p+1} \nu \int \frac{u^{\mu+\nu-1}}{(u^\nu - b)^{q+p+2}} \, du.$$

Since $q + p$ is an integer, the integrand is a rational function of u.

Example 5.2.18. An integral of type as in (5.6) with $q + p$ an integer.

Find $\int \dfrac{1}{\sqrt[4]{1 + x^4}}\, dx$.

Solution. Here $m = 0$, $n = 4$, $p = -1/4$, so $(m+1)/n = 1/4$ and

$$q + p = \frac{m+1}{n} + p = 0$$

which is an integer. Since $\nu = 4$ we will use the substitution $u = \sqrt[4]{(1 + x^4)/x^4}$. Solving for x yields

$$x = (u^4 - 1)^{-1/4} \quad \text{and} \quad dx = -u^3(u^4 - 1)^{-5/4}\, du.$$

Further,

$$1 + x^4 = 1 + (u^4 - 1)^{-1} = \frac{u^4}{u^4 - 1} \quad \text{and} \quad \sqrt[4]{1 + x^4} = \frac{u}{\sqrt[4]{u^4 - 1}}.$$

Finally, we obtain

$$\int \frac{\sqrt[4]{u^4 - 1}}{u}(-u^3)(u^4 - 1)^{-5/4}\, du = -\int \frac{u^2}{u^4 - 1}\, du$$

$$= -\int \frac{u^2}{(u-1)(u+1)(u^2+1)}\, du.$$

Using partial fraction decomposition we obtain that

$$-\frac{u^2}{(u-1)(u+1)(u^2+1)} = -\frac{1}{4}\frac{1}{u-1} + \frac{1}{4}\frac{1}{u+1} - \frac{1}{2}\frac{1}{u^2+1}$$

and, therefore, the antiderivative is

$$-\frac{1}{4}\ln|u - 1| + \frac{1}{4}\ln|u + 1| - \frac{1}{2}\arctan u + C$$

and the result is obtained by replacing u with $\sqrt[4]{(1 + x^4)/x^4}$. ◆

PAFNUTY CHEBYSHEV (1821–1894) was a Russian mathematician. He grew up in a wealthy family, and was educated at home. Having one leg shorter prevented him from taking part in many of the normal childhood activities. When the family moved to Moscow in 1832, he was tutored by Pogorelski who was considered one of the best teachers in Moscow. In 1837 Chebyshev enrolled in Moscow University. During his student years, a big influence on him was Nikolai Brashman (1796–1866), a professor of applied mathematics. His masters thesis was on probability, his thesis at the University of Saint Petersburg (allowing him to teach there) was on some special integrals, and his doctoral dissertation was on number theory. He remained at Saint Petersburg for the rest of his life, retiring in 1882, always praised as a teacher. Chebyshev never married and lived alone in a large house with ten rooms. He had a daughter whom he refused to officially acknowledge, but supported her nevertheless. He is considered a founding father of Russian mathematics, although he wanted to be considered as a "world-wide mathematician". His papers were written in French and published in the most prestigious journals in France and Germany. He traveled to Western Europe almost every summer in order to maintain contact with leading mathematicians. His many honors include the memberships in the Saint Petersburg Academy of Sciences, the Berlin Academy of Sciences, and the Royal Society of London. He was also awarded the French Légion d'Honneur. His mathematical contributions were numerous. In Number Theory he proved that between n and $2n$ there is always a prime number, and he came very close to proving the Prime Number Theorem. In Probability he proved what is today called the Chebyshev's Inequality, generalized the Central Limit Theorem of de Moivre and Laplace, and introduced the concept of a random variable and the expectation. He constructed a calculating machine, and seven of his mechanical inventions were exhibited at the World's Exposition in Chicago in 1893.

Exercises

In Exercises 5.2.51–5.2.65 evaluate the integrals:

5.2.51. $\int \dfrac{\sqrt[4]{x}}{(1+\sqrt{x})^3}\,dx.$　　　5.2.52.* $\int \dfrac{dx}{\sqrt[3]{x}(1-\sqrt{x})^2}.$　　　5.2.53. $\int \dfrac{dx}{x\sqrt[3]{1+x^5}}.$

5.2.54. $\int \dfrac{x^5}{\sqrt{1-x^2}}\,dx.$　　　5.2.55. $\int \dfrac{x\,dx}{\sqrt{1+\sqrt[3]{x}}}.$　　　5.2.56. $\int \dfrac{dx}{x^3\sqrt[5]{1+1/x}}.$

5.2.57. $\int \dfrac{dx}{x\sqrt{2+x^2}}.$　　　5.2.58. $\int \dfrac{dx}{x\sqrt{2-x^2}}.$　　　5.2.59. $\int \dfrac{x^5}{\sqrt{1-x^2}}\,dx.$

5.2.60.* $\int \sqrt{x^3+x^4}\,dx.$　　　5.2.61. $\int \dfrac{x\,dx}{\sqrt{1+\sqrt[3]{x^2}}}.$　　　5.2.62.* $\int x\sqrt[3]{1+x^3}\,dx.$

5.2.63.* $\int \sqrt[3]{3x-x^3}.$　　　5.2.64. $\int \dfrac{dx}{\sqrt[3]{1+x^3}}.$　　　5.2.65. $\int \dfrac{x\,dx}{\sqrt[3]{1+x^6}}.$

5.2.66. Suppose that m is a rational number. In which cases is the integral $\int \sqrt{1+x^m}\,dx$ an elementary function?

5.2.4　Some Trigonometric Integrals

Let $R(x,y)$ be a rational function of 2 variables, and suppose that we are evaluating

$$\int R(\sin x, \cos x)\,dx. \tag{5.8}$$

If we use the substitution $u = \tan(x/2)$ (assuming $-\pi < x < \pi$) the integrand becomes a rational function of u. Let us verify this:

$$u^2 = \tan^2 \frac{x}{2} = \frac{\sin^2(x/2)}{\cos^2(x/2)} = \frac{1-\cos x}{1+\cos x},$$

so if we solve this equation for $\cos x$ we get

$$\cos x = \frac{1-u^2}{1+u^2}.$$

Further,

$$\sin x = 2\sin\frac{x}{2}\cos\frac{x}{2} = 2\tan\frac{x}{2}\cos^2\frac{x}{2} = 2\tan\frac{x}{2}\frac{1+\cos x}{2}$$

$$= u\left(1+\frac{1-u^2}{1+u^2}\right) = \frac{2u}{1+u^2}.$$

Finally, $u = \tan(x/2)$ implies that $x = 2\arctan u$, so $dx = 2\,du/(1+u^2)$. Therefore

$$\int R(\sin x, \cos x)\,dx = \int R\left(\frac{2u}{1+u^2}, \frac{1-u^2}{1+u^2}\right)\frac{2}{1+u^2}\,du$$

and it is not hard to see that the integrand is a rational function of u.

Example 5.2.19. An integral of type as in (5.8).

Find $\int \dfrac{1}{\cos x - 3\sin x + 3}\,dx.$

Solution. Let $u = \tan(x/2)$. Using the formulas

$$\cos x = \frac{1-u^2}{1+u^2}, \quad \sin x = \frac{2u}{1+u^2}, \quad dx = \frac{2}{1+u^2}\,du,$$

we obtain

$$\cos x - 3\sin x + 3 = \frac{1-u^2}{1+u^2} - 3\frac{2u}{1+u^2} + 3 = \frac{1-u^2-6u+3+3u^2}{1+u^2} = \frac{2u^2-6u+4}{1+u^2}.$$

Therefore,

$$\int \frac{1}{\cos x - 3\sin x + 3}\,dx = \int \frac{1+u^2}{2u^2-6u+4}\frac{2}{1+u^2}\,du = \int \frac{1}{u^2-3u+2}\,du$$

$$= \int \left(\frac{1}{u-2} - \frac{1}{u-1} \right) du = \ln|u-2| - \ln|u-1| + C$$

$$= \ln\left|\tan\frac{x}{2} - 2\right| - \ln\left|\tan\frac{x}{2} - 1\right| + C. \qquad \blacklozenge$$

The substitution $u = \tan(x/2)$ was one of the topics in Euler's [40]. It will transform any integral of the form (5.8) to an integral of a rational function. The downside is that, quite often, the resulting rational integrand will be quite complicated. There are, in fact, several situations where a simpler substitution can be more effective.

The first such case occurs when the rational function R satisfies the equality

$$R(-x, y) = -R(x, y).$$

Now, the recommended substitution is $u = \cos x$.

Example 5.2.20. Substitution $u = \cos x$.

Find $\int \dfrac{1}{\sin x\,(2\cos^2 x - 1)}\,dx$.

Solution. It is not hard to see that, if $\sin x$ is replaced by $-\sin x$, the integrand changes the sign. We will use $u = \cos x$, so $du = -\sin x\,dx$. We obtain

$$\int \frac{\sin x}{\sin^2 x(2\cos^2 x - 1)}\,dx = \int \frac{-1}{(1-u^2)(2u^2-1)}\,du$$

$$= \int \left(\frac{1}{2}\frac{1}{u-1} - \frac{1}{2}\frac{1}{u+1} - \frac{1}{u\sqrt{2}-1} + \frac{1}{u\sqrt{2}+1} \right) du$$

$$= \frac{1}{2}\ln|u-1| - \frac{1}{2}\ln|u+1| - \frac{1}{\sqrt{2}}\ln|u\sqrt{2}-1| + \frac{1}{\sqrt{2}}\ln|u\sqrt{2}+1| + C$$

$$= \frac{1}{2}\ln\left|\frac{\cos x - 1}{\cos x + 1}\right| - \frac{1}{\sqrt{2}}\ln\left|\frac{\sqrt{2}\cos x - 1}{\sqrt{2}\cos x + 1}\right| + C.$$

It is a good idea to try to use the substitution $u = \tan(x/2)$ and convince yourself of the advantage of $u = \cos x$. $\qquad \blacklozenge$

The second case occurs when

$$R(x, -y) = -R(x, y).$$

The recommended substitution is now $u = \sin x$.

Example 5.2.21. Substitution $u = \sin x$.

Find $\int \sin^2 x \cos^3 x \, dx$.

Solution. With $u = \sin x$ (and $du = \cos x \, dx$) we obtain

$$\int \sin^2 x \cos^2 x \cos x \, dx = \int u^2 (1 - u^2) \, du = \frac{1}{3} u^3 - \frac{1}{5} u^5 + C = \frac{1}{3} \sin^3 x - \frac{1}{5} u \sin^5 x + C.$$

Once again, the substitution $u = \tan(x/2)$ would have resulted in a much more complicated rational function. ◆

Finally, if

$$R(-x, -y) = R(x, y),$$

the recommended substitution is $u = \tan x$. Here we assume that $-\pi/2 < x < \pi/2$. Since $\sin x = u \cos x$ we have that

$$u^2 \cos^2 x = \sin^2 x = 1 - \cos^2 x \quad \text{and} \quad \cos^2 x = \frac{1}{1 + u^2}.$$

Notice that $\cos x \geq 0$ for $x \in (-\pi/2, \pi/2)$, so

$$\cos x = \frac{1}{\sqrt{1 + u^2}} = \frac{1}{\sqrt{1 + \tan^2 x}}.$$

Also, $du = \sec^2 x \, dx = (1 + \tan^2 x) \, dx$, and it follows that

$$dx = \frac{du}{1 + u^2}.$$

Finally,

$$\sin x = u \cos x = \frac{u}{\sqrt{1 + u^2}},$$

and the integral (5.8) becomes

$$\int R \left(\frac{u}{\sqrt{1 + u^2}}, \frac{1}{\sqrt{1 + u^2}} \right) \frac{1}{1 + u^2} \, du.$$

Example 5.2.22. Substitution $u = \tan x$.

Find $\int \dfrac{1}{\sin^4 x \cos^2 x} \, dx$.

Solution. We will use $u = \tan x$. As we have seen,

$$\cos x = \frac{1}{\sqrt{1 + u^2}}, \quad \sin x = \frac{u}{\sqrt{1 + u^2}}, \quad dx = \frac{du}{1 + u^2}.$$

Therefore,

$$\int \frac{1}{\sin^4 x \cos^2 x} \, dx = \int \left(\frac{\sqrt{1 + u^2}}{u} \right)^4 (\sqrt{1 + u^2})^2 \frac{du}{1 + u^2} = \int \frac{(1 + u^2)^2}{u^4} \, du$$

$$= \int (u^{-4} + 2u^{-2} + 1) \, du = -\frac{1}{3} u^{-3} - 2u^{-1} + u + C$$

$$= -\frac{1}{3 \tan^3 x} - \frac{2}{\tan x} + \tan x + C. \qquad ◆$$

Exercises

In Exercises 5.2.67–5.2.81 evaluate the integrals:

5.2.67. $\displaystyle\int \frac{2\sin x - \cos x}{3\sin^2 x + 4\cos^2 x}\, dx.$

5.2.68. $\displaystyle\int \frac{dx}{2\sin x - \cos x + 5}.$

5.2.69.* $\displaystyle\int \frac{dx}{a + b\tan x}, \; a, b \in \mathbb{R}.$

5.2.70. $\displaystyle\int \frac{\sin^2 x \cos x}{\sin x + \cos x}\, dx.$

5.2.71. $\displaystyle\int \frac{\sin^5 x}{\cos^4 x}\, dx.$

5.2.72. $\displaystyle\int \frac{\cos^4 x}{\sin^3 x}\, dx.$

5.2.73. $\displaystyle\int \sin^2 x \cos^3 x\, dx.$

5.2.74. $\displaystyle\int \frac{1}{\sin^2 x \cos x}\, dx.$

5.2.75. $\displaystyle\int \frac{dx}{\cos^5 x}.$

5.2.76. $\displaystyle\int \sin^2 x \cos^4 x\, dx.$

5.2.77.* $\displaystyle\int \frac{\sin x}{a\cos x + b\sin x}\, dx, \; a, b \in \mathbb{R}.$

5.2.78.* $\displaystyle\int \frac{\cos x}{a\cos x + b\sin x}\, dx, \; a, b \in \mathbb{R}.$

5.2.79. $\displaystyle\int \frac{\sin x \cos x}{\cos x + \sin x}\, dx.$

5.2.80. $\displaystyle\int \frac{dx}{a + b\cos x}, \; a, b \in \mathbb{R}.$

5.2.81. $\displaystyle\int \frac{1 - r^2}{1 - 2r\cos x + r^2}\, dx, \; 0 < r < 1, \; -\pi < x < \pi.$

6

Definite Integral

When Newton and Leibniz defined the integral, it was as a sum. In fact, the symbol for integral is a stylized letter S (as in "summa"—the Latin for a sum). Soon thereafter, they discovered that the same result can be obtained using antiderivatives. Since their definitions involved the controversial infinitesimals (infinitely small numbers), the mathematical community preferred the idea of the integral as the antiderivative, and in the eighteenth century that was the prevailing viewpoint. It was only in the nineteenth century, when the need for a rigorous approach became obvious, that Cauchy and Riemann reverted to the old definition (this time without infinitesimals). Although they made significant progress, there was still room for improvement, which came through the work of Lebesgue in the twentieth century.

6.1 Computing Definite Integrals: A Review

Example 6.1.1. $\int_0^\pi \cos x\, dx$.

Solution. An antiderivative of $f(x) = \cos x$ is $F(x) = \sin x$, so the result is, by the Fundamental Theorem of Calculus, $F(\pi) - F(0)$. We obtain

$$\int_0^\pi \cos x\, dx = \sin x \Big|_0^\pi = \sin \pi - \sin 0 = 0 - 0 = 0. \qquad \blacklozenge$$

Example 6.1.2. $\int_{-2}^3 2x\, e^{x^2}\, dx$.

Solution. We will use a substitution $u = x^2$, because $du = 2x\, dx$. In order to use this substitution, the limits of the integral become $(-2)^2 = 4$ and $3^2 = 9$. It follows that

$$\int_{-2}^3 2x\, e^{x^2}\, dx = \int_4^9 e^u\, du = e^u \Big|_4^9 = e^9 - e^4 \approx 8048.485778. \qquad \blacklozenge$$

Example 6.1.3. $\int_{\sinh 1}^{\sinh 2} \dfrac{1}{\sqrt{1+x^2}}\, dx$.

Solution. We will use a substitution $x = \sinh t$. This has several implications. First, $dx = \cosh t\, dt$. Second, the limits of integration will change: when $x = \sinh 1$, $t = 1$, and when $x = \sinh 2$, $t = 2$. Finally, $1 + x^2 = 1 + \sinh^2 t = \cosh^2 t$. Actually, it was this fundamental identity

$$\cosh^2 t - \sinh^2 t = 1$$

that prompted us to use this particular substitution. Further, for any $t \in \mathbb{R}$, $\cosh t = (e^t + e^{-t})/2 > 0$, so $\sqrt{\cosh^2 t} = \cosh t$. Now,

$$\int_{\sinh 1}^{\sinh 2} \frac{1}{\sqrt{1+x^2}}\, dx = \int_1^2 \frac{1}{\sqrt{\cosh^2 t}} \cosh t\, dt = \int_1^2 dt = t \Big|_1^2 = 2 - 1 = 1. \qquad \blacklozenge$$

Example 6.1.4. $\int_0^2 |1 - x|\, dx.$

Solution. The function $|1 - x|$ is awkward, but we notice that, for $0 \leq x \leq 1$, $1 - x \geq 0$, so $|1 - x| = 1 - x$. On the other hand, if $1 \leq x \leq 2$, then $1 - x \leq 0$ so $|1 - x| = -(1 - x) = x - 1$. Therefore, we will split the domain of integration $0 \leq x \leq 2$ into $0 \leq x \leq 1$ and $1 \leq x \leq 2$. We obtain

$$\int_0^2 |1 - x|\, dx = \int_0^1 |1 - x|\, dx + \int_1^2 |1 - x|\, dx = \int_0^1 (1 - x)\, dx + \int_1^2 (x - 1)\, dx$$

$$= \left(x - \frac{1}{2} x^2 \right) \Big|_0^1 + \left(\frac{1}{2} x^2 - x \right) \Big|_1^2 = \left(1 - \frac{1}{2} \right) - 0 + \left(\frac{1}{2} 2^2 - 2 \right) - \left(\frac{1}{2} - 1 \right)$$

$$= \frac{1}{2} - \left(-\frac{1}{2} \right) = 1. \qquad \blacklozenge$$

Example 6.1.5. $\int_{-\pi}^{\pi} \cos nx \cos mx\, dx,\ m, n \in \mathbb{N}.$

Solution. We will use the trigonometric formula

$$\cos nx \cos mx = \frac{1}{2} \left(\cos(n + m)x + \cos(n - m)x \right).$$

Assuming for a moment that $m \neq n$,

$$\int_{-\pi}^{\pi} \cos nx \cos mx\, dx = \frac{1}{2} \left(\frac{\sin(n + m)x}{n + m} + \frac{\sin(n - m)x}{n - m} \right) \Big|_{-\pi}^{\pi} = 0,$$

because each of the functions $\sin(n + m)x$ and $\sin(n - m)x$ vanishes at both π and $-\pi$. When $m = n$, we need the formula

$$\cos^2 x = \frac{1 + \cos 2x}{2},$$

which implies that

$$\int_{-\pi}^{\pi} \cos mx \cos mx\, dx = \int_{-\pi}^{\pi} \frac{1 + \cos(2m)x}{2}\, dx = \frac{1}{2} \left(\frac{\sin(2m)x}{2m} + x \right) \Big|_{-\pi}^{\pi} = \pi.$$

Thus,

$$\int_{-\pi}^{\pi} \cos nx \cos mx\, dx = \begin{cases} 0, & \text{if } m \neq n \\ \pi, & \text{if } m = n. \end{cases} \qquad \blacklozenge$$

Example 6.1.6. $\int_{-\pi}^{\pi} x \cos nx\, dx,\ n \in \mathbb{N}.$

Solution. We will use integration by parts, with $u = x$, $dv = \cos nx\, dx$. Then $du = dx$ and $v = \sin nx / n$, so

$$\int_{-\pi}^{\pi} x \cos nx\, dx = x\, \frac{\sin nx}{n} \Big|_{-\pi}^{\pi} - \int_{-\pi}^{\pi} \frac{\sin nx}{n}\, dx = \frac{\cos nx}{n^2} \Big|_{-\pi}^{\pi} = \frac{\cos n\pi - \cos(-n\pi)}{n^2} = 0. \quad \blacklozenge$$

Example 6.1.7. $\int_1^{\infty} \dfrac{dx}{(2x + 3)^2}.$

Solution. We will first calculate $\int_1^b \frac{dx}{(2x+3)^2}$, where b is a real number such that $b > 1$. Using the substitution $u = 2x + 3$, we obtain

$$\int_5^{2b+3} \frac{\frac{1}{2}\, du}{u^2} = \frac{1}{2} \left(-\frac{1}{u} \right) \Big|_5^{2b+3} = \frac{1}{2} \left(-\frac{1}{2b + 3} + \frac{1}{5} \right).$$

Now we take the limit as $b \to +\infty$, and we obtain $1/10$. $\qquad \blacklozenge$

Example 6.1.8. $\int_0^\infty x e^{-x}\, dx$.

Solution. Again, we start with $\int_0^b x e^{-x}\, dx$. Using Integration by Parts with $u = x$ and $dv = e^{-x}\, dx$, so that $du = dx$ and $v = -e^{-x}$, we obtain

$$x\left(-e^{-x}\right)\Big|_0^b - \int_0^b -e^{-x}\, dx = b\left(-e^{-b}\right) - \left(e^{-x}\right)\Big|_0^b = b\left(-e^{-b}\right) - \left(e^{-b} - 1\right).$$

It is obvious that, as $b \to +\infty$, $e^{-b} \to 0$. In fact, the same is true for be^{-b}. This can be seen using the L'Hôpital's Rule:

$$\lim_{b\to\infty} be^{-b} = \lim_{b\to\infty} \frac{b}{e^b} = \lim_{b\to\infty} \frac{1}{e^b} = 0.$$

It follows that the result is 1. ◆

Example 6.1.9. $\int_0^\infty \cos x\, dx$.

Solution. Integration with limits 0 and b yields

$$\int_0^b \cos x\, dx = \sin x\Big|_0^b = \sin b.$$

However, the limit $\lim_{b\to\infty} \sin b$ does not exist (Example 3.5.11). Thus, the integral $\int_0^\infty \cos x\, dx$ does not converge. ◆

Example 6.1.10. $I_n = \int_0^\infty x^n e^{-x}\, dx$, $n \in \mathbb{N}$.

Solution. Using Integration by Parts with $u = x^n$ and $dv = e^{-x}\, dx$, so that $du = nx^{n-1}\, dx$ and $v = -e^{-x}$, we obtain

$$I_n = \left(-x^n e^{-x}\right)\Big|_0^\infty - \int_0^\infty -nx^{n-1}e^{-x}\, dx = \lim_{x\to\infty}\left(-x^n e^{-x}\right) + nI_{n-1} = nI_{n-1}.$$

It follows that $I_n = n!I_0$. Since

$$I_0 = \int_0^\infty e^{-x}\, dx - e^{-x}\Big|_0^\infty = 1,$$

we obtain that $I_n = n!$. ◆

Example 6.1.11. $\int_0^{\pi/2} \frac{\cos x}{\sqrt[3]{\sin x}}\, dx$.

Solution. The integrand is undefined when $x = 0$, so we select an arbitrary $a \in (0, \pi/2)$, and we calculate $\int_a^{\pi/2} \frac{\cos x}{\sqrt[3]{\sin x}}\, dx$. The substitution $u = \sin x$ implies that $du = \cos x\, dx$, so we obtain

$$\int_{\sin a}^1 \frac{du}{\sqrt[3]{u}} = \int_{\sin a}^1 u^{-1/3}\, du = \frac{3}{2} u^{2/3}\Big|_{\sin a}^1 = \frac{3}{2}\left(1 - (\sin a)^{2/3}\right).$$

Now we take the limit as $a \to 0^+$, and we obtain $3/2$. ◆

Example 6.1.12. $\int_0^1 \ln x\, dx$.

Solution. The integrand is undefined when $x = 0$, so we select an arbitrary $a \in (0, 1)$, and we calculate $\int_a^1 \ln x\, dx$. Using Integration by Parts with $u = \ln x$ and $dv = dx$, so that $du = \frac{1}{x}\, dx$ and $v = x$, we obtain

$$(x\ln x)\Big|_a^1 - \int_0^1 \frac{1}{x} x\, dx = -a\ln a - x\Big|_a^1 = -a\ln a - (1 - a).$$

Now we take the limit as $a \to 0^+$. Since $a \ln a = \frac{\ln a}{1/a}$ and the latter is an indeterminate form $\left(\frac{\infty}{\infty}\right)$, we can use the L'Hôpital's Rule, leading to $\frac{1/a}{-1/a^2} = -a$. Since $\lim_{a \to 0^+}(-a) = 0$ we see that the result is -1.   ◆

Example 6.1.13. $\int_1^2 \frac{dx}{x^2 - 3x + 2}$.

Solution. Since $x^2 - 3x + 2 = (x-1)(x-2)$ we see that the integrand is undefined at both endpoints of the interval $[1, 2]$. Therefore, we will select real numbers a, b so that $1 < a < b < 2$, and we will calculate $\int_a^b \frac{dx}{x^2 - 3x + 2}$. Using Partial Fraction Decomposition,

$$\int_a^b \frac{dx}{x^2 - 3x + 2} = \int_a^b \frac{dx}{(x-1)(x-2)} = \int_a^b \frac{dx}{x-2} - \int_a^b \frac{dx}{x-1}$$

$$= \ln|x-2|\Big|_a^b - \ln|x-1|\Big|_a^b = \ln|b-2| - \ln|a-2| - \ln|b-1| + \ln|a-1|$$

$$= \ln\left|\frac{b-2}{b-1}\right| + \ln\left|\frac{a-1}{a-2}\right|.$$

Now we take the limits as $a \to 1^+$ and $b \to 2^-$. It turns out that neither one exists. For example, when $a \to 1^+$, $\frac{a-1}{a-2} \to 0$, so $\ln\left|\frac{a-1}{a-2}\right| \to -\infty$ and, similarly, $\lim_{b \to 2^-} \ln\left|\frac{b-2}{b-1}\right| = -\infty$. Thus, the integral $\int_1^2 \frac{dx}{x^2 - 3x + 2}$ does not converge.   ◆

Exercises

In Exercises 6.1.1–6.1.16 find the integral. (m, n are natural numbers.)

6.1.1. $\int_{-1}^1 \frac{x}{x^2 + x + 1}\, dx$.
6.1.2. $\int_1^9 x\sqrt[3]{1-x}\, dx$.
6.1.3.* $\int_0^3 \arcsin\sqrt{\frac{x}{1+x}}\, dx$.

6.1.4.* $\int_1^e (x \ln x)^2\, dx$.
6.1.5. $\int_1^{e^{2n\pi}} \left|(\cos \ln x)'\right|\, dx$.
6.1.6. $\int_{1/e}^e |\ln x|\, dx$.

6.1.7. $\int_2^\infty \frac{dx}{x^2 + x - 2}$.
6.1.8. $\int_0^\infty \frac{x \ln x}{(1 + x^2)^2}\, dx$.
6.1.9. $\int_{-\pi}^\pi \cos nx \sin mx\, dx$.

6.1.10. $\int_{-\pi}^\pi \sin nx \sin mx\, dx$.
6.1.11. $\int_{-1}^1 \frac{dx}{\sqrt{1-x^2}}$.
6.1.12. $\int_0^1 \frac{dx}{(2-x)\sqrt{1-x}}$.

6.1.13.* $\int_0^\pi \frac{x \sin x}{1 + \cos^2 x}\, dx$.
6.1.14. $\int_{-\pi}^\pi x \sin nx\, dx$.
6.1.15.* $\int_0^{\pi/2} \sin^n x\, dx$.

6.1.16.* $\int_0^{\pi/2} \cos^n x\, dx$.

6.2   Definite Integral

In the previous section we have repeatedly taken advantage of the Fundamental Theorem of Calculus. It is, clearly, a very important result, and we will set as our target to prove it. In order to accomplish this goal, we need to come up with a rigorous definition of the definite integral. We will start with an example.

Example 6.2.1. Approximating the area under a graph.
    Approximate the area under the graph of $f(x) = x^2$ on $A = [1, 2]$.

Solution. Let us select, for example, the points $x_0 = 1$, $x_1 = 1.1$, $x_2 = 1.35$, $x_3 = 1.5$,

$x_4 = 1.7$, $x_5 = 2$. We consider inscribed rectangles, with one side the interval $[x_{k-1}, x_k]$ and the length of the other side $m_k = \inf\{f(x) : x \in [x_{k-1}, x_k]\}$ (as in Figure 6.1). Here,

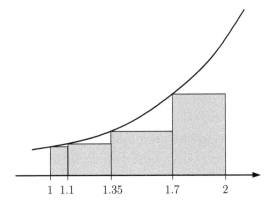

Figure 6.1: Approximating the area by inscribed rectangles.

$$m_1 = 1^2 = 1, \quad m_2 = 1.1^2 = 1.21, \quad m_3 = 1.35^2 = 1.8225,$$
$$m_4 = 1.5^2 = 2.25, \quad m_5 = 1.7^2 = 2.89.$$

Therefore, the total area of the inscribed rectangles is

$$1(1.1 - 1) + 1.21(1.35 - 1.1) + 1.8225(1.5 - 1.35) + 2.25(1.7 - 1.5) + 2.89(2 - 1.7) = 1.992875.$$

Similarly, we construct the circumscribed rectangles, of height $M_k = \sup\{f(x) : x \in [x_{k-1}, x_k]\}$. Since

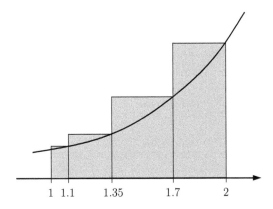

Figure 6.2: Approximating the area by circumscribed rectangles.

$$M_1 = 1.21, \quad M_2 = 1.8225, \quad M_3 = 2.25, \quad M_4 = 2.89, \quad M_5 = 4,$$

the total area of the circumscribed rectangles equals

$$1.21(1.1 - 1) + 1.8225(1.35 - 1.1) + 2.25(1.5 - 1.35) + 2.89(1.7 - 1.5) + 4(2 - 1.7) = 2.684125.$$

The picture suggests that the exact area is somewhere between 1.992875 and 2.684125. For example, we can approximate it by the average of these two numbers: $(1.992875 + 2.684125)/2 = 2.3385$. ♦

In general, let f be a bounded function defined on the interval $[a, b]$. First we create a **partition** P of the interval $[a, b]$: we select a positive integer n and **partition points**

$$x_0 = a < x_1 < x_2 < \cdots < x_{n-1} < x_n = b.$$

We write $P = \{x_0, x_1, x_2, \ldots, x_n\}$, and we denote by Δx_1, Δx_2, $\ldots$, Δx_n the lengths of these intervals: $\Delta x_1 = x_1 - x_0$, $\Delta x_2 = x_2 - x_1$, etc. Although it is sometimes beneficial to have all these intervals of the same length, we will not make such an assumption here.

If P_1 and P_2 are two partitions of $[a, b]$ we say that P_2 is **finer** than P_1, or that P_2 is a *refinement* of P_1, if $P_1 \subset P_2$. In other words, to make a partition finer, we need to include additional partition points. Usually, this leads to a better approximation of the area.

Example 6.2.2. Approximating the area under a graph.

Use the partition $P_2 = \{1, 1.1, 1.35, 1.5, 1.6, 1.7, 1.9, 2\}$ to approximate the area under the graph of $f(x) = x^2$ on $A = [1, 2]$.

Solution. Partition P_2 has all the partition points as partition P_1 from Example 6.2.1, plus two additional: 1.6 and 1.9. A calculation shows that

$$m_1 = 1, \ m_2 = 1.21, \ m_3 = 1.8225, \ m_4 = 2.25, \ m_5 = 2.56, \ m_6 = 2.89, \ m_7 = 3.61,$$
$$M_1 = 1.21, \ M_2 = 1.8225, \ M_3 = 2.25, \ M_4 = 2.56, \ M_5 = 2.89, \ M_6 = 3.61, \ M_7 = 4.$$

Therefore, the total area of the inscribed rectangles is now 2.095875, and the total area of the circumscribed rectangles is 2.581125. These numbers should be compared with those from Example 6.2.1: 1.992875 and 2.684125. Although we did not obtain the exact value, the finer partition P_2 has narrowed down the interval in which this number has to be. ♦

In order to improve the approximation it is not sufficient to increase the number of partition points. It is also important to make sure that, as partitions are being refined, the lengths of *all* the intervals converge to 0. To make the last requirement more precise it is helpful to introduce the **norm of a partition**: the length of the largest subinterval. We write $\|P\| = \max\{\Delta x_k : 1 \le k \le n\}$, and we will require that $\|P\| \to 0$. In Example 6.2.1 we had $\|P_1\| = 0.3 \, (= \Delta x_5)$, and in Example 6.2.2, $\|P_2\| = 0.2$.

Since f is a bounded function, its range is a bounded set. Thus the numbers $M_i = \sup\{f(x) : x \in [x_{i-1}, x_i]\}$ and $m_i = \inf\{f(x) : x \in [x_{i-1}, x_i]\}$ are well defined. The sums

$$L(f, P) = \sum_{i=1}^{n} m_i \Delta x_i \ \text{ and } \ U(f, P) = \sum_{i=1}^{n} M_i \Delta x_i,$$

are the **lower** and the **upper** Darboux sums.

Darboux sums were introduced by Riemann (page 169) in his Habilitation Thesis [89] in 1854.

JEAN-GASTON DARBOUX (1842–1917) was a French mathematician. He attended high school in Nimes and Montpellier and in 1861 came to Paris to study at the Ecole Polytechnique and then the Ecole Normale Supérieure. Already as a student he published his first paper. After receiving his doctorate in 1866 he taught at a high school for 5 years, then at the Ecole Normale Supérieure for 9 years. During that time he started teaching at Sorbonne, where he remained until his death. From 1889 to 1893 he served as dean of the Faculty of Science. He was considered an exceptional teacher, writer, and administrator. Among many honors bestowed on him are the membership in the Royal Society of London and their Sylvester Medal. In 1884 he was elected to the Académie des Sciences, becoming its secretary in 1900. He was a plenary speaker in the International Congress of Mathematicians 1908 in Rome. His main contributions are in the fields of differential geometry and analysis, especially partial differential equations, and he edited the Selected Works of Fourier (page 239).

An important property of Darboux sums is that they are "monotone" relative to the choice of a partition.

Lemma 6.2.3. *Let f be a bounded function on $[a,b]$ that satisfies $|f(x)| \leq M$ for all $x \in [a,b]$. Let P_1 and P_2 be two partitions of $[a,b]$ such that $P_1 \subset P_2$ and P_2 has n more points than P_1. Then $U(f,P_2) \leq U(f,P_1) \leq U(f,P_2)+3nM\|P_1\|$ and $L(f,P_1) \leq L(f,P_2) \leq L(f,P_1) + 3nM\|P_1\|$.*

Proof. We will prove the inequalities for the upper sums and leave the lower sums to the reader. Since P_2 contains all points in P_1, it is possible to select a chain of partitions $Q_1, Q_2, \ldots, Q_{n+1}$ such that

$$P_1 = Q_1 \subset Q_2 \subset \cdots \subset Q_{m-1} \subset Q_{n+1} = P_2$$

and each partition in the chain differs from the previous one by exactly one point. Clearly, it suffices to prove that

$$U(f,Q_{i+1}) \leq U(f,Q_i) \leq U(f,Q_{i+1}) + 3M\|Q_i\|,$$

i.e., the case when the finer partition has exactly one additional point. Therefore, we may assume that P_1 and P_2 differ by one point.

Let $P_1 = \{x_0, x_1, \ldots, x_n\}$, and let $P_2 = \{x_0, x_1, \ldots, x_m, y, x_{m+1}, \ldots x_n\}$. We will use the notation $M'_m = \sup\{f(x) : x \in [x_m, y]\}$ and $M''_m = \sup\{f(x) : x \in [y, x_{m+1}]\}$. Then all the terms in $U(f,P_2)$ and $U(f,P_1)$ will be equal except that $U(f,P_2)$ will have two terms

$$M'_m (y - x_m) + M''_m (x_{m+1} - y)$$

while $U(f,P_1)$ will have one term instead:

$$M_m (x_{m+1} - x_m).$$

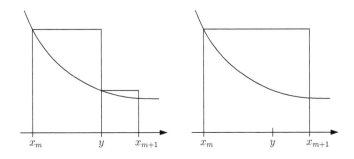

(a) $M'_m(y-x_m)+M''_m(x_{m+1}-y)$ (b) $M_m(x_{m+1} - x_m)$

Figure 6.3: With and without an additional partition point.

Since $M_m \geq M'_m$ and $M_m \geq M''_m$ we see that

$$
\begin{aligned}
M'_m (y - x_m) + M''_m (x_{m+1} - y) &\leq M_m (y - x_m) + M_m (x_{m+1} - y) \\
&= M_m (y - x_m + x_{m+1} - y) \\
&= M_m (x_{m+1} - x_m).
\end{aligned}
$$

It follows that $U(f,P_2) \leq U(f,P_1)$. On the other hand,

$$
\begin{aligned}
U(f,P_1) - U(f,P_2) &= M_m (x_{m+1} - x_m) - M'_m (y - x_m) - M''_m (x_{m+1} - y) \\
&\leq M\|P_1\| + M\|P_1\| + M\|P_1\| = 3M\|P_1\|,
\end{aligned}
$$

and the lemma is proved. $\qquad\square$

Remember that if an increasing sequence is bounded above, then it is convergent. Lemma 6.2.3 shows that the lower Darboux sums are increasing, and the upper ones are decreasing. Are they bounded? The following lemma provides the answer.

Lemma 6.2.4. *Let f be a bounded function on $[a, b]$ and let $\mathcal{P}$ be the set of all partitions of $[a, b]$. If we denote by $U = \inf\{U(f, P) : P \in \mathcal{P}\}$ and $L = \sup\{L(f, P) : P \in \mathcal{P}\}$ then $L \leq U$.*

Proof. Suppose, to the contrary, that $L > U$, and let $V = (U + L)/2$. From here we derive two conclusions. First, L is the least upper bound and $V < L$, so V is too small to be an upper bound. Therefore, there exists a partition $P_1 \in \mathcal{P}$ such that

$$V < L(f, P_1). \tag{6.1}$$

Second, U is the greatest lower bound and $V > U$, so V is too big to be a lower bound. Therefore, there exists a partition $P_2 \in \mathcal{P}$ such that

$$V > U(f, P_2). \tag{6.2}$$

Let $P = P_1 \cup P_2$, a partition that refines both P_1 and P_2. By Lemma 6.2.3,

$$U(f, P) \leq U(f, P_2) \quad \text{and} \quad L(f, P_1) \leq L(f, P). \tag{6.3}$$

Combining (6.1), (6.2), and (6.3), we obtain

$$U(f, P) \leq U(f, P_2) < V < L(f, P_1) \leq L(f, P),$$

hence $U(f, P) < L(f, P)$ which is impossible. □

Remark 6.2.5. The numbers U and L are often called the **upper (Darboux) integral** and the **lower (Darboux) integral** of f on $[a, b]$. Sometimes they are denoted by

$$\overline{\int_a^b} f(x)\, dx \quad \text{and} \quad \underline{\int_a^b} f(x)\, dx.$$

The names Upper and Lower Darboux integrals and the notation were introduced by Italian mathematician Vito Volterra (1860–1940) in [105] in 1881.

Now we can state the Darboux definition of integrability.

Definition 6.2.6. Let f be a bounded function that is defined on the interval $[a, b]$. We say that f is **(Darboux) integrable** on $[a, b]$ if $L = U$. In that case, we call this common value the **(Darboux) integral** of f over $[a, b]$ and we denote it by $\int_a^b f(x)\, dx$.

In practice, it may be hard to calculate the exact values of the upper and lower integrals. If we are merely trying to establish the integrability of a function, the following sufficient condition may be easier to verify.

Proposition 6.2.7. *A function f is Darboux integrable on $[a, b]$ if and only if for every $\varepsilon > 0$ there exists a partition P of $[a, b]$ such that $U(f, P) - L(f, P) < \varepsilon$.*

Proof. Suppose that $L = U$, and let $\varepsilon > 0$. The definition of L and U implies that there exist partitions P_1 and P_2 such that

$$L - \frac{\varepsilon}{2} < L(f, P_1) \quad \text{and} \quad U + \frac{\varepsilon}{2} > U(f, P_2).$$

If $P = P_1 \cup P_2$, then Lemma 6.2.3 implies that

$$L - \frac{\varepsilon}{2} < L(f, P) \quad \text{and} \quad U + \frac{\varepsilon}{2} > U(f, P).$$

It follows that $U(f, P) - L(f, P) < (U + \varepsilon/2) - (L - \varepsilon/2) = \varepsilon.$

In the other direction, let $\varepsilon > 0$, and let P be a partition postulated by the Proposition. By Lemma 6.2.4,

$$U < U(f, P) < L(f, P) + \varepsilon < L + \varepsilon,$$

so $U - L < \varepsilon$ and the Proposition is proved. □

In the next section we will use Definition 6.2.6 and Proposition 6.2.7 to establish the integrability, or the lack thereof, for some classes of functions.

> The notation $\int_a^b f(x)\, dx$ was introduced by Fourier (page 239) in 1822. Before him, Euler wrote the limits in brackets and used the Latin words *ab* and *ad* (*from* and *to*).

Exercises

In Exercises 6.2.1–6.2.4 find $L(f, P)$ and $U(f, P)$:

6.2.1. $f(x) = 2x^2 - x$, $P = \{0, \frac{1}{3}, 1, \frac{3}{2}\}$. 6.2.2. $f(x) = x^3$, $P = \{0, \frac{1}{3}, \frac{1}{2}, 1\}$.

6.2.3. $f(x) = 3x + 2$, $P = \{\frac{1}{3}, \frac{2}{3}, 1, \frac{3}{2}, 2\}$. 6.2.4. $f(x) = \sin x$, $P = \{\frac{\pi}{6}, \frac{\pi}{4}, \frac{\pi}{2}, \frac{3\pi}{4}, \frac{5\pi}{6}\}$.

6.2.5. In Lemma 6.2.3 prove that $L(f, P_2) \geq L(f, P_1)$.

6.2.6. Suppose that f is integrable on $[a, b]$ and that $\int_a^b f(x)\, dx > 0$. Prove that there exists $[c, d] \subset [a, b]$ and $m > 0$ such that $f(x) \geq m$ for $x \in [c, d]$.

6.2.7.* Let f and g be two bounded functions on $[a, b]$ and let P be a partition of $[a, b]$. Prove that $U(f + g, P) - L(f + g, P) \leq U(f, P) - L(f, P) + U(g, P) - L(g, P)$. Is it true that $U(f + g, P) \leq U(f, P) + U(g, P)$?

6.2.8. Suppose that f is a bounded function on $[a, b]$ and that there exists a partition P of $[a, b]$ such that $L(f, P) = U(f, P)$. Prove that f is constant on $[a, b]$.

6.2.9. Let $f(x) = x^2$. Find a partition P of the interval $[0, 3]$ such that $U(f, P) - L(f, P) < 0.01$.

6.2.10.* Suppose that $f : [a, b] \to \mathbb{R}$ is a bounded function. Prove:

(a) If there exists a sequence $\{P_n\}$ of partitions of $[a, b]$ such that $\lim L(f, P_n) = A$, then $\underline{\int_a^b} f(x)\, dx \geq A$.

(b) If there exists a sequence $\{Q_n\}$ of partitions of $[a, b]$ such that $\lim U(f, Q_n) = B$, then $\overline{\int_a^b} f(x)\, dx \leq B$.

(c) If in addition $A = B$, then f is integrable on $[a, b]$ and $\int_a^b f(x)\, dx = A$.

In Exercises 6.2.11–6.2.16 show that the given function is integrable and calculate $\int_a^b f(x)\, dx$:

6.2.11. $f(x) = x^2$, $a = 0$, $b = 1$. 6.2.12. $f(x) = 3x - 2$, $a = 0$, $b = 2$.

6.2.13. $f(x) = x^2 + 2x$, $a = 0$, $b = 1$. 6.2.14. $f(x) = 1 - x$, $a = -1$, $b = 2$.

6.2.15. $f(x) = \chi_{(a,b)}$, $a = 3$, $b = 6$. 6.2.16. $f(x) = \chi_{(a-1,b+1)}$, $a = 3$, $b = 6$.

6.2.17.* Suppose that $f : [a, b] \to \mathbb{R}$ is a bounded function and let $\{P_n\}$ be a sequence of partitions of $[a, b]$ such that $P_n \subset P_{n+1}$, for all $n \in \mathbb{N}$. Prove:

(a) The limits $A = \lim L(f, P_n)$ and $B = \lim U(f, Q_n)$ both exist and satisfy $A \le B$.

(b) If $A = B$, then f is integrable on $[a, b]$ and $\int_a^b f(x)\, dx = A$.

(c) The converse of (b) is not true.

6.2.18. Suppose that $f : [a, b] \to \mathbb{R}$ is a non-negative bounded function. Prove that $\underline{\int_a^b} f(x)\, dx \ge 0$.

6.2.19. Let f, g be two bounded functions on $[a, b]$ and let L_f, L_g, U_f, U_g denote their upper and lower integrals. If $f(x) \le g(x)$, for all $x \in [a, b]$, prove that $L_f \le L_g$ and $U_f \le U_g$.

6.2.20.* Let f, g be two bounded functions on $[a, b]$ and let L_f, L_g, L_{f+g} denote the lower integrals of $f, g, f + g$. Prove that $L_{f+g} \ge L_f + L_g$.

6.2.21.* Let f be a bounded function on $[a, b]$. Prove that f is Darboux integrable on $[a, b]$ if and only if for every $\varepsilon > 0$, there exists $\delta > 0$ such that, if P is a partition of $[a, b]$ and $\|P\| < \delta$ then $U(f, P) - L(f, P) < \varepsilon$.

6.2.22.* Let function f be integrable on $[a, b]$ and $I = \int_a^b f(x)\, dx$. Then, for any $\varepsilon > 0$, there exists a positive number δ, such that if P is any partition of $[a, b]$ and $\|P\| < \delta$, then $|L(f, P) - I| < \varepsilon$, and $|U(f, P) - I| < \varepsilon$.

6.3   Integrable Functions

In this section our goal is to get a solid grasp on the class of integrable functions. A very useful tool will be Proposition 6.2.7. We will base our approach on a strong connection between continuous and integrable functions. In one direction, this relationship is quite unambiguous.

Theorem 6.3.1. *Every continuous function on $[a, b]$ is integrable.*

Proof. Let f be a continuous function on $[a, b]$ and let $\varepsilon > 0$. We will exhibit a partition P of $[a, b]$ such that $U(f, P) - L(f, P) < \varepsilon$. By Theorem 3.8.7, f is uniformly continuous on $[a, b]$, so there exists $\delta > 0$ such that

$$x, y \in [a, b], \quad |x - y| < \delta \quad \Rightarrow \quad |f(x) - f(y)| < \frac{\varepsilon}{b - a}.$$

Let P be any partition of $[a, b]$ with norm $\|P\| < \delta$. If $P = \{x_0, x_1, \ldots, x_n\}$, and $1 \le k \le n$, then any subinterval $[x_k, x_{k+1}]$ has the length less than δ. Thus,

$$x, y \in [x_k, x_{k+1}] \Rightarrow |f(x) - f(y)| < \frac{\varepsilon}{b - a}. \tag{6.4}$$

The function f is continuous on $[x_k, x_{k+1}]$ so, by Theorem 3.9.8 it attains its maximum M_k and minimum m_k. Let $f(\xi_k') = m_k$ and $f(\xi_k'') = M_k$. It follows that

$$M_k - m_k = f(\xi_k'') - f(\xi_k') < \frac{\varepsilon}{b - a}.$$

Consequently,

$$U(f,P) = \sum_{k=1}^{n} M_k \Delta x_k < \sum_{k=1}^{n} m_k \Delta x_k + \sum_{k=1}^{n} \frac{\varepsilon}{b-a} \Delta x_k = \sum_{k=1}^{n} m_k \Delta x_k + \varepsilon$$
$$= L(f,P) + \varepsilon.$$

The result now follows from Proposition 6.2.7. $\qquad\square$

Although there are many continuous functions (all elementary functions are continuous), we want to extend the class of integrable functions beyond the continuous ones. It turns out that this is possible, to some extent.

Theorem 6.3.2. *Let f be a bounded function defined on $[a,b]$, and let $c \in [a,b]$. Suppose that f has a discontinuity at $x = c$, and that it is continuous at any other point of $[a,b]$. Then f is integrable on $[a,b]$.*

Proof. Let $\varepsilon > 0$. Since f is a bounded function, there exists M such that $|f(x)| \le M$ for all $x \in [a,b]$. Let $\varepsilon_1 = \varepsilon/(24M)$, and let $a_1, b_1 \in [a,b]$ so that

$$a \le a_1 \le c \le b_1 \le b \quad \text{and} \quad |c - a_1| < \varepsilon_1, \quad |c - b_1| < \varepsilon_1.$$

Let $\varepsilon_2 = \varepsilon/3$. The function f is uniformly continuous on $[a, a_1]$ so there exists δ_1 such that,

$$x, y \in [a, a_1] \quad \text{and} \quad |x - y| < \delta_1 \Rightarrow |f(x) - f(y)| < \frac{\varepsilon_2}{b-a}. \tag{6.5}$$

Also, f is uniformly continuous on $[b_1, b]$ so there exists δ_2 such that,

$$x, y \in [b_1, b] \quad \text{and} \quad |x - y| < \delta_2 \Rightarrow |f(x) - f(y)| < \frac{\varepsilon_2}{b-a}. \tag{6.6}$$

Let $\delta = \min\{\delta_1, \delta_2, \varepsilon_1\}$, and let $P = \{x_0, x_1, \dots, x_n\}$ be any partition of $[a,b]$ such that $\|P\| < \delta$. We will show that $U(f,P) - L(f,P) < \varepsilon$.

Let x_j be the rightmost point in P that satisfies $x_j \le a_1$, and let x_m be the leftmost point in P that satisfies $b_1 \le x_m$.

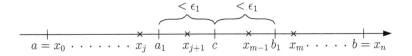

Figure 6.4: c is separated from the rest of $[a,b]$.

Now both $U(f,P)$ and $L(f,P)$ can be broken in 3 sums: in $U_1(f,P)$ and $L_1(f,P)$, the summation is over $1 \le i \le j$, in $U_2(f,P)$ and $L_2(f,P)$ it is over $j + 1 \le i \le m$, and over $m + 1 \le i \le n$ in $U_3(f,P)$ and $L_3(f,P)$. Notice that (6.5) and (6.6) are analogous to (6.4) and $\|P\| < \delta \le \delta_1$ so, just like in the proof of Theorem 6.3.1,

$$|U_1(f,P) - L_1(f,P)| < \varepsilon_2. \tag{6.7}$$

Similarly, $\|P\| < \delta \le \delta_2$ implies that

$$|U_3(f,P) - L_3(f,P)| < \varepsilon_2. \tag{6.8}$$

Finally,

$$|U_2(f,P) - L_2(f,P)| = \left| \sum_{i=j}^{m-1} M_i \Delta x_i - \sum_{i=j}^{m-1} m_i \Delta x_i \right| = \left| \sum_{i=j+1}^{m} (M_i - m_i) \Delta x_i \right|$$

$$\leq \sum_{i=j+1}^{m} |M_i - m_i| \, \Delta x_i$$

$$\leq \sum_{i=j+1}^{m} (|M_i| + |m_i|) \, \Delta x_i$$

$$\leq \sum_{i=j+1}^{m} 2M \Delta x_i = 2M(x_m - x_j).$$

Now,

$$x_m - x_j = (x_m - x_{m-1}) + (x_{m-1} - x_{j+1}) + (x_{j+1} - x_j)$$
$$\leq \delta + 2\varepsilon_1 + \delta$$
$$\leq \varepsilon_1 + 2\varepsilon_1 + \varepsilon_1 = 4\varepsilon_1,$$

so

$$|U_2(f,P) - L_2(f,P)| \leq 2M \cdot 4\varepsilon_1 = 8M\varepsilon_1.$$

Combining with the estimates (6.7) and (6.8) we obtain

$$| U(f,P) - L(f,P) |$$
$$\leq |U_1(f,P) - L_1(f,P)| + |U_2(f,P) - L_2(f,P)| + |U_3(f,P) - L_3(f,P)|$$
$$< \varepsilon_2 + 8M\varepsilon_1 + \varepsilon_2 = 2\varepsilon_2 + 8M\varepsilon_1 = 2\frac{\varepsilon}{3} + 8M\frac{\varepsilon}{24M} = \frac{2}{3}\varepsilon + \frac{1}{3}\varepsilon = \varepsilon. \qquad \square$$

If we analyze the proof of Theorem 6.3.2 we see that the crucial feature was to "isolate" the point of discontinuity. Once a small interval around it was created, the estimate $|U(f,P) - L(f,P)| < \varepsilon$ came from two sources. First, f was continuous outside of the small interval, so we were able to make the appropriate Darboux sums close by. Second, within the small interval, the difference $M_i - m_i$ was bounded by $2M$, and it was multiplied by the length of the "small" interval. This suggests that, if f had more than one discontinuity, we could use a similar strategy of "isolating" each point of discontinuity.

Corollary 6.3.3. *Let $n \in \mathbb{N}$, and let $c_1, c_2, \ldots, c_n \in [a,b]$. Suppose that f is a bounded function defined on $[a,b]$, with a discontinuity at $x = c_k$, $1 \leq k \leq n$, and that it is continuous at any other point of $[a,b]$. Then f is integrable on $[a,b]$.*

Remark 6.3.4. Functions that are continuous in $[a,b]$ except at $c_1, c_2, \ldots, c_n \in [a,b]$, and that have both the left and the right limits at these points, are called *piecewise continuous*.

An important consequence of Corollary 6.3.3 is that, if we change a value of a function at a finite number of points, it will not affect the definite integral.

Corollary 6.3.5. *Let f and g be bounded function defined on $[a,b]$, let $n \in \mathbb{N}$, and let $c_1, c_2, \ldots, c_n \in [a,b]$. Suppose that $f(x) = g(x)$ for all $x \in [a,b]$, except at $x = c_k$, $1 \leq k \leq n$. If f is integrable on $[a,b]$, then so is g, and $\int_a^b f(x)\,dx = \int_a^b g(x)\,dx$.*

For example, if a bounded function is defined only on (a,b), we can define it any way we like at the endpoints, without affecting the integrability.

The previous results show that a function with a finite number of discontinuities is integrable. Can we push this farther?

Example 6.3.6. An integrable function with infinitely many discontinuities.

We will show that the Thomae function

$$f(x) = \begin{cases} \dfrac{1}{q}, & \text{if } x = \dfrac{p}{q}, \, p \in \mathbb{Z}, \, q \in \mathbb{N}, \text{ and } p, q \text{ are mutually prime} \\ 0, & \text{if } x \text{ is irrational or } x = 0 \end{cases}$$

is integrable on $[0, 1]$ in spite of the fact that it has a discontinuity at every rational number in $[0, 1]$.

Solution. It was shown in Example 3.6.8 that f is continuous at c if and only if c is an irrational number or 0. We will show that f is integrable on $[0, 1]$. Let $\varepsilon > 0$ and let

$$n = \left\lfloor \frac{2}{\varepsilon} \right\rfloor + 1, \quad N = \frac{n(n+1)}{2} + 1, \quad P = \left\{ 0, \frac{1}{N}, \frac{2}{N}, \ldots, \frac{N-1}{N}, 1 \right\}.$$

By Theorem 1.3.7, each interval $[\frac{i-1}{N}, \frac{i}{N}]$ contains an irrational number α_j, and $f(\alpha_j) = 0$ so $\inf\{f(x) : x \in [\frac{i-1}{N}, \frac{i}{N}]\} = 0$. Therefore, $L(f, P) = 0$.

In order to evaluate $U(f, P)$ we make the following observation. The upper Darboux sum equals

$$\sum_{i=1}^{N} M_i \frac{1}{N} = \frac{1}{N} \sum_{i=1}^{N} M_i$$

and it cannot be bigger than if the terms of the sum were N largest values that f takes. The largest value of f is $f(1) = 1$, followed by $f(1/2) = 1/2$. The next two largest are $f(1/3) = f(2/3) = 1/3$, then $f(1/4) = f(3/4) = 1/4$, etc. For each $n \geq 2$, f takes up to $n - 1$ times the value $1/n$. (For $n = 3$, it is exactly 2 times; for $n = 4$, it is less than 3 times.) Thus, the sum of the largest $1 + 1 + 2 + 3 + \cdots + n$ values of f is less than

$$1 + 1 \cdot \frac{1}{2} + 2 \cdot \frac{1}{3} + 3 \cdot \frac{1}{4} + \cdots + n \cdot \frac{1}{n+1}. \tag{6.9}$$

Now we realize that the choice of N was not random. It was based on the identity

$$1 + 2 + 3 + \cdots + k = \frac{k(k+1)}{2},$$

so $N = 1 + 1 + 2 + 3 + \cdots + n$. Thus, the sum of N largest values of f cannot exceed the sum in (6.9) and, all the more, it is less than

$$1 + 2 \cdot \frac{1}{2} + 3 \cdot \frac{1}{3} + 4 \cdot \frac{1}{4} + \cdots + (n+1) \cdot \frac{1}{n+1} = n + 1.$$

It follows that

$$U(f, P) < \frac{1}{N}(n+1) < \frac{n+1}{\frac{n(n+1)}{2}} = \frac{2}{n} < \varepsilon.$$

Thus $U(f, P) - L(f, P) < \varepsilon$ and f is integrable on $[0, 1]$. ♦

The Thomae function has a discontinuity at every rational point of $[0, 1]$. Does continuity matter at all?

Example 6.3.7. A function that is not integrable.

The Dirichlet function

$$f(x) = \begin{cases} 1, & \text{if } x \in \mathbb{Q} \\ 0, & \text{if } x \notin \mathbb{Q} \end{cases}$$

is not integrable on any interval $[a, b]$.

Solution. By Proposition 6.2.7, a function f is integrable on $[a, b]$ if and only if

$$(\forall \varepsilon > 0)(\exists P) \quad U(f, P) - L(f, P) < \varepsilon.$$

Consequently, f is *not* integrable on $[a, b]$ if and only if

$$(\exists \varepsilon_0 > 0)(\forall P) \quad U(f, P) - L(f, P) \geq \varepsilon_0.$$

Let $\varepsilon_0 = (b - a)/2$ and let $P = \{x_0, x_1, \ldots, x_n\}$ be any partition of $[a, b]$. By Theorem 1.3.7, between any two partition points x_k and x_{k+1} there exists a rational number r_k and an irrational number α_k. Therefore, $m_k = f(\alpha_k) = 0$ and $M_k = f(r_k) = 1$, whence $L(f, P) = 0$ and $U(f, P) = b - a$. It follows that $U(f, P) - L(f, P) = b - a > (b - a)/2 = \varepsilon_0$, so f is not integrable on $[a, b]$. ♦

Remark 6.3.8. The Thomae example shows that a function can be integrable even if it has infinitely many discontinuities, but the Dirichlet function serves as a warning that too many points of discontinuity can preclude integrability. How many is too many? Exercise 6.3.8 gives a partial answer but not a final one. The complete characterization of integrability in terms of the size of the set of discontinuities is known as the Lebesgue Theorem and requires the knowledge of Measure Theory.

Exercises

In Exercises 6.3.1–6.3.6 determine whether f is integrable on $[a, b]$:

6.3.1. $a = -1$, $b = 1$,

$$f(x) = \begin{cases} \dfrac{\sin x}{x}, & \text{if } x \neq 0 \\ 1, & \text{if } x = 0. \end{cases}$$

6.3.2. $a = -1$, $b = 1$,

$$f(x) = \begin{cases} \dfrac{\sin x}{x}, & \text{if } x \neq 0 \\ 0, & \text{if } x = 0. \end{cases}$$

6.3.3. $a = 0$, $b = 1$,

$$f(x) = \begin{cases} \sin \dfrac{1}{x}, & \text{if } x \neq 0 \\ 0, & \text{if } x = 0. \end{cases}$$

6.3.4. $a = -1$, $b = 1$,

$$f(x) = \begin{cases} \sin \dfrac{1}{x}, & \text{if } x \neq 0 \\ 0, & \text{if } x = 0. \end{cases}$$

6.3.5. $a = 0$, $b = 3$, $f(x) = \lfloor x \rfloor$.

6.3.6. $a = 0.1$, $b = 0.2$, $f(x) = \left\lfloor \dfrac{1}{x} \right\rfloor$.

6.3.7. Prove Corollary 6.3.5.

6.3.8. A set $A \subset \mathbb{R}$ is said to have *content 0* if for every $\varepsilon > 0$, there exists a positive integer n, and intervals $[a_i, b_i]$, $1 \leq i \leq n$, of total length less than ε such that $A \subset \cup_{i=1}^{n} [a_i, b_i]$. Prove that if a bounded function f is continuous except on a set of content 0, then f is integrable.

In Exercises 6.3.9–6.3.11 determine whether f is integrable on $[a, b]$:

6.3.9. $a = 0$, $b = 1$,

$$f(x) = \begin{cases} \dfrac{1}{x} - \left\lfloor \dfrac{1}{x} \right\rfloor, & \text{if } x \neq 0 \\ 0, & \text{if } x = 0. \end{cases}$$

6.3.10. $a = 0$, $b = 1$,

$$f(x) = \begin{cases} 1, & \text{if } x = \dfrac{1}{n}, \, n \in \mathbb{N} \\ 0, & \text{otherwise.} \end{cases}$$

6.3.11. $a = 0$, $b = 1$, $f(x) = \operatorname{sgn}\left(\sin \dfrac{\pi}{x}\right)$.

6.3.12.* Given a partition $P = \{x_0, x_1, \ldots, x_n\}$ of $[a, b]$, a function $f : [a, b] \to \mathbb{R}$ is called a **step function** if it can be written as $f(x) = \sum_{k=1}^{n} \alpha_k \chi_{[x_{k-1}, x_k)}(x)$, where α_i are real numbers. Prove that every step function is integrable on $[a, b]$ and that $\int_a^b f(x)\, dx = \sum_{k=1}^{n} \alpha_k \Delta x_k$.

6.3.13.* Let f be a bounded function on $[a, b]$. Prove that f is integrable on $[a, b]$ if and only if for every $\varepsilon > 0$ there exist step functions σ and τ such that

(a) $\sigma(x) \leq f(x) \leq \tau(x)$, for all $x \in [a, b]$;

(b) $\int_a^b (\tau(x) - \sigma(x))\, dx < \varepsilon$.

6.3.14.* Let f be integrable on $[a, b]$. Prove that for every $\varepsilon > 0$ there exists a continuous function g such that $\int_a^b |f(x) - g(x)|\, dx < \varepsilon$.

6.3.15. Let f be a bounded function on $[a, b]$ and integrable on $[c, d]$, whenever $a < c < d < b$. Prove that f is integrable on $[a, b]$.

6.3.16. Suppose that a function f is continuous at every point of the open interval $(0, 1)$. Does it follow that f is integrable on $(0, 1)$?

6.3.17. Give an example of two integrable functions f and g such that $g \circ f$ is not integrable.

6.3.18.* Prove that every monotone function on $[a, b]$ is integrable.

6.3.19. Let f be integrable on $[a, b]$. Prove that $|f|$ is integrable on $[a, b]$. Is the converse true?

6.3.20.* Suppose that a function f is discontinuous at every point of the interval $[0, 1]$. Prove that f is not integrable on $[0, 1]$.

6.4 Riemann Sums

More than half a century before Darboux, Cauchy had introduced a different definition of the definite integral. Although he stated it only for continuous functions, we will not make such a restriction.

Just like in the case of the upper and lower sums, we start by partitioning the interval $[a, b]$: $P = \{x_0, x_1, \ldots, x_n\}$. Unlike the Darboux approach, we are not assuming that the function f is bounded. We select **intermediate points** $\xi_1 \in [x_0, x_1]$, $\xi_2 \in [x_1, x_2]$, etc. The collection of intermediate points is denoted by $\xi = \{\xi_k : 1 \leq k \leq n\}$. In applications, it is quite common to take ξ_k to be the left end-point or the right end-point (sometimes even a mid-point) of the interval $[x_{k-1}, x_k]$. However, at present our choice of these points within the appropriate intervals is going to be quite arbitrary.

Next, we consider rectangles R_k, with one side the interval $[x_{k-1}, x_k]$ and the length of the other side $f(\xi_k)$. The area of R_k is $f(\xi_k)\Delta x_k$, so summing up these areas we obtain

$$\sum_{k=1}^{n} f(\xi_k)\Delta x_k = f(\xi_1)\Delta x_1 + f(\xi_2)\Delta x_2 + \cdots + f(\xi_n)\Delta x_n.$$

Such a sum is usually called a **Riemann Sum** for f on $[a, b]$ and it is denoted by $S(f, P, \xi)$.

One of the earliest uses of the phrase "Riemann sum" can be found in 1935 in a book by a Harvard professor Joseph Leonard Walsh (1895–1973).

Example 6.4.1. Approximating the area under a graph.

Use a Riemann sum to approximate the area under the graph of $f(x) = x^2$ on the interval $[1, 2]$.

Solution. Let us take, for example, $P = \{1, 1.1, 1.35, 1.5, 1.7, 2\}$ and $\xi_1 = 1.05$, $\xi_2 = 1.32$, $\xi_3 = 1.5$, $\xi_4 = 1.6$, $\xi_5 = 2$.

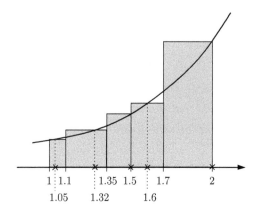

Figure 6.5: A Riemann sum for $y = x^2$ on $[1, 2]$.

Then the Riemann sum for this partition and these intermediate points is:

$$1.05^2 \cdot 0.1 + 1.32^2 \cdot 0.25 + 1.5^2 \cdot 0.15 + 1.6^2 \cdot 0.2 + 2^2 \cdot 0.3 = 2.5935.$$

It is a reasonable approximation of the area, which we "know" to be

$$\int_1^2 x^2\, dx = \frac{x^3}{3}\Big|_1^2 = \frac{8}{3} - \frac{1}{3} = \frac{7}{3} \approx 2.3333.$$

Of course, a better approximation requires a finer partition with a smaller norm.   ♦

Now we can state the definition.

Definition 6.4.2. Let f be a function that is defined on the interval $[a, b]$. We say that a real number I is the (**Riemann**) **integral** of f over $[a, b]$, if for any $\varepsilon > 0$ there exists a positive number δ, such that if P is any partition of $[a, b]$ with $\|P\| < \delta$ and ξ is any choice of intermediate points, then

$$|S(f, P, \xi) - I| < \varepsilon. \tag{6.10}$$

In this situation we say that f is (**Riemann**) **integrable** on $[a, b]$ and we write $I = \int_a^b f(x)\, dx$.

> The definition is due to Riemann (page 169), so we often use his name before the words integral and integrable. It appears in his Habilitation Thesis [89] in 1854. Riemann followed the ideas of Cauchy, but made a significant improvement. Namely, Cauchy had considered only continuous integrands and only some special choices of intermediate points. (In the presence of continuity, it makes little difference what intermediate points one selects.) As we have seen earlier, continuity is desirable but not necessary for a function to be integrable.

The two definitions of integrability (Riemann and Darboux) are equivalent. In order to see that, we look at the Darboux sums $\sum m_k \Delta x_k$ and $\sum M_k \Delta x_k$. Since $M_k = \sup\{f(\xi_k) : \xi_k \in [x_{k-1}, x_k]\}$, we should expect that, in some way, the upper Darboux sum is the supremum of Riemann sums.

Lemma 6.4.3. *Let f be a bounded function on $[a,b]$ and let P be a fixed partition of $[a,b]$. For each choice ξ of intermediate points, let $S(f,P,\xi)$ be the corresponding Riemann sum, and let $\mathcal{S}$ denote the collection of the numbers $S(f,P,\xi)$, for all possible ξ. Then $\sup \mathcal{S} = U(f,P)$ and $\inf \mathcal{S} = L(f,P)$.*

Proof. We will prove that $\sup \mathcal{S} = U(f,P)$ and leave the second equality to the reader. For any $\xi = \{\xi_i : 1 \le i \le n\}$, $f(\xi_i) \le M_i$ so

$$S(f,P,\xi) = \sum_{i=1}^{n} f(\xi_i)\Delta x_i \le \sum_{i=1}^{n} M_i \Delta x_i = U(f,P),$$

and it follows that $U(f,P)$ is an upper bound of $\mathcal{S}$. The challenge is to prove that it is the *least* upper bound.

Let $\varepsilon > 0$. We will show that $U(f,P) - \varepsilon$ is not an upper bound of $\mathcal{S}$. By definition, M_i is the least upper bound of the set $\{f(x) : x \in [x_{i-1}, x_i]\}$, so $M_i - \varepsilon/(b-a)$ is not an upper bound. Consequently, for each i, $1 \le i \le n$, there exists $\xi_i \in [x_{i-1}, x_i]$ such that $f(\xi_i) > M_i - \varepsilon/(b-a)$ or, equivalently, $M_i < f(\xi_i) + \varepsilon/(b-a)$. Now,

$$\begin{aligned}
U(f,P) &= \sum_{i=1}^{n} M_i \Delta x_i \\
&< \sum_{i=1}^{n} \left(f(\xi_i) + \frac{\varepsilon}{b-a} \right) \Delta x_i = \sum_{i=1}^{n} f(\xi_i)\Delta x_i + \frac{\varepsilon}{b-a} \sum_{i=1}^{n} \Delta x_i \\
&= S(f,P,\xi) + \frac{\varepsilon}{b-a}(b-a) = S(f,P,\xi) + \varepsilon.
\end{aligned}$$

Thus, $U(f,P) - \varepsilon < S(f,P,\xi)$ and $U(f,P) - \varepsilon$ is not an upper bound. $\qquad\square$

The Darboux definition of the integral requires the function f to be bounded, but there is no such restriction in Definition 6.4.2. Nevertheless, it is there, just not immediately obvious.

Theorem 6.4.4. *Every Riemann integrable function is bounded.*

Proof. Let f be a function defined on $[a,b]$, and suppose that it is not bounded. We will show that it is not integrable. Definition 6.4.2 can be written as

$$(\exists I)(\forall \varepsilon)(\exists \delta)(\forall P) \quad \|P\| < \delta \Rightarrow |S(f,P,\xi) - I| < \varepsilon,$$

so its negative is

$$(\forall I)(\exists \varepsilon)(\forall \delta)(\exists P) \quad \|P\| < \delta \ \text{ and } \ |S(f,P,\xi) - I| \ge \varepsilon, \tag{6.11}$$

and we will prove (6.11). So, let $I \in \mathbb{R}$ and let us take $\varepsilon = 1$. Suppose now that $\delta > 0$, and let P be any partition of $[a,b]$ that satisfies $\Delta x_i < \delta/2$ for all i. Then $\|P\| < \delta$. Since f is unbounded, there exists a subinterval $[x_{k-1}, x_k]$ such that f is unbounded on that subinterval. Let us select an intermediate point ξ_i in each of the remaining intervals and let

$$S' = \sum_{i=1}^{k-1} f(\xi_i)\Delta x_i + \sum_{i=k+1}^{n} f(\xi_i)\Delta x_i.$$

Since f is unbounded in $[x_{k-1}, x_k]$, there exists a point $\xi_k \in [x_{k-1}, x_k]$ such that

$$|f(\xi_k)| > \frac{1}{\Delta x_k} \left(1 + |S' - I| \right).$$

Now $S(f, P, \xi) = S' + f(\xi_k)\Delta x_k$ so

$$
\begin{aligned}
|S(f, P, \xi) - I| &= |S' + f(\xi_k)\Delta x_k - I| \\
&\geq |f(\xi_k)|\,\Delta x_k - |S' - I| \\
&> \frac{1}{\Delta x_k}\left(1 + |S' - I|\right)\Delta x_k - |S' - I| \\
&= (1 + |S' - I|) - |S' - I| = 1.
\end{aligned}
$$

$\square$

Theorem 6.4.4 confirms what was explicit in Definition 6.2.6: an unbounded function cannot be integrable. Therefore, we will make a standing assumption that all functions are bounded. We will return to the unbounded functions in Section 6.7.2 where we will extend the concept of integrability so that they can be included.

Now we can prove the equivalence of two definitions of the definite integral.

Theorem 6.4.5. *A function f is Riemann integrable on $[a, b]$ if and only if it is Darboux integrable on $[a, b]$. Moreover, in that case, the Darboux and the Riemann integrals of f are equal.*

Proof. Suppose first that f is Riemann integrable on $[a, b]$ and let $\varepsilon > 0$. The hypothesis implies that there exists $\delta > 0$ so that if P is any partition of $[a, b]$ satisfying $\|P\| < \delta$, and ξ is any selection of intermediate points, then $|S(f, P, \xi) - I| < \varepsilon/3$. Let P be one such partition. Using Lemma 6.4.3

$$
U(f, P) = \sup_{\xi}\{S(f, P, \xi)\} \leq I + \frac{\varepsilon}{3},
$$

and, similarly, $I - \varepsilon/3 \leq L(f, P)$. Therefore,

$$
U(f, P) - L(f, P) \leq \left(I + \frac{\varepsilon}{3}\right) - \left(I - \frac{\varepsilon}{3}\right) = \frac{2\varepsilon}{3} < \varepsilon,
$$

and the conclusion follows from Proposition 6.2.7 and Theorem 6.4.4.

Next, we concentrate on the converse. Let $\varepsilon > 0$. The definition of L and U implies that there exist partitions P_1 and P_2 such that

$$
L - \frac{\varepsilon}{2} < L(f, P_1) \quad \text{and} \quad U + \frac{\varepsilon}{2} > U(f, P_2). \tag{6.12}
$$

By definition, f is bounded, so there exists $M > 0$ such that $|f(x)| \leq M$ for $x \in [a, b]$. Let n denote the number of partition points of $P = P_1 \cup P_2$, and let

$$
\delta = \frac{\varepsilon}{6nM}.
$$

Suppose that Q is a partition satisfying $\|Q\| < \delta$. We will show that, for any selection ξ of intermediate points of Q, $|S(f, Q, \xi) - L| < \varepsilon$. Let $R = Q \cup P$. Since R has at most n more points than Q, Lemma 6.2.3 shows that

$$
U(f, Q) - U(f, R) \leq 3Mn\|Q\| < 3Mn\delta = 3Mn\frac{\varepsilon}{6nM} = \frac{\varepsilon}{2}.
$$

Since $P_2 \subset P \subset R$ it follows that $U(f, R) \leq U(f, P_2)$, so

$$
U(f, Q) - U(f, P_2) \leq U(f, Q) - U(f, R) < \frac{\varepsilon}{2}.
$$

A similar argument shows that $L(f, P_1) - L(f, Q) < \varepsilon/2$.

Let ξ be a collection of intermediate points of Q. Using Lemma 6.4.3, (6.12), and Proposition 6.2.7,

$$L - \varepsilon < L(f, P_1) - \frac{\varepsilon}{2} < L(f, Q) \leq S(f, Q, \xi) \leq U(f, Q) < U(f, P_2) + \frac{\varepsilon}{2} < U + \varepsilon.$$

Since $L = U$ it follows that $|S(f, Q, \xi) - L| < \varepsilon$, so f is Riemann integrable and the theorem is proved. $\quad\square$

Theorem 6.4.5 features in Riemann's Habilitation Thesis [89], but without proof. Darboux stated it precisely and proved it in [25] in 1875.

BERNHARD RIEMANN (1826–1866) was a German mathematician, one of the greatest of all time. He was home schooled by his father, a Lutheran minister, until the age of 14. He attended a high school in Hanover and in Lüneburg. He excelled in mathematics. On one occasion he read a 900 page book in six days. He entered the University of Göttingen in 1846 to study theology, but soon switched to mathematics. He spent two years in Berlin where he laid the foundation for his most important work on complex functions. He returned to Göttingen in 1849 and graduated in 1851. Apart from his mentor Gauss (page 196), he was influenced by a physicist Weber and by a mathematician Listing, a pioneer in topology. In his thesis, Riemann described what is now known as Riemann surfaces. His Habilitation Thesis [89] in 1854 introduced the concepts that we call Riemannian geometry and the curvature tensor. Sixty years later they were instrumental in Einstein's general theory of relativity. (The thesis was published only in 1867, a year after his death, through the effort of his friend Dedekind.) In 1857 he submitted a paper on abelian functions to Crelle's Journal. His paper appeared in the same volume with a paper of Weierstrass. Such was the strength and the originality of Riemann's ideas, that Weierstrass withdrew his paper. In 1859 he was accepted in the Berlin Academy of Sciences. His introductory speech was about the estimates for the number of prime numbers less than a given number. One of the main tools was the *zeta-function*. He proposed a hypothesis about the location of zeros of this function (the Riemann hypothesis). In 1862 he contracted tuberculosis and died 4 years later. Riemann's contributions include the integration theory (Riemann integral), infinite series (Riemann's Theorem, p. 220), Fourier series (Riemann-Lebesgue Lemma, p. 265, and Riemann's Localization Theorem, p. 269). He is one of the giants of nineteenth century mathematics.

Exercises

6.4.1. Let $f(x) = 2x - 3$, $P = \{0, \frac{1}{3}, 1, \frac{3}{2}\}$, $\xi = \{\frac{1}{7}, \frac{3}{4}, \frac{6}{5}\}$. Find $S(f, P, \xi)$.

6.4.2. Let $f(x) = 1 + \sqrt{x}$, let P be the partition of $[1, 6]$ into 10 intervals of equal lengths, and let ξ consist of the midpoints of these intervals. Find $S(f, P, \xi)$.

In Exercises 6.4.3–6.4.8 write the integral as a limit of a sum and evaluate it without using the Fundamental Theorem of Calculus:

6.4.3. $\int_0^1 x^2 \, dx$.

6.4.4. $\int_0^1 2^x \, dx$.

6.4.5.* $\int_a^b x^m \, dx$, $0 < a < b$, $m \neq -1$.

6.4.6.* $\int_a^b \frac{1}{x} \, dx$, $0 < a < b$.

6.4.7.* $\int_0^{\pi/2} \sin x \, dx$.

6.4.8.* $\int_0^{\pi/2} \cos x \, dx$.

In Exercises 6.4.9–6.4.14 find the limits by using Riemann integrals of suitably chosen functions:

6.4.9. $\lim\limits_{n \to \infty} \left(\frac{n}{n^2 + 1^2} + \frac{n}{n^2 + 2^2} + \cdots + \frac{n}{n^2 + n^2} \right)$.

6.4.10. $\lim\limits_{n \to \infty} \left(\frac{1}{n+1} + \frac{1}{n+2} + \cdots + \frac{1}{n+n} \right)$.

6.4.11. $\lim\limits_{n \to \infty} \dfrac{1^k + 2^k + \cdots + n^k}{n^{k+1}}$, $k \geq 0$.

6.4.12.* $\lim\limits_{n\to\infty} \left(\dfrac{2^{1/n}}{n+1} + \dfrac{2^{2/n}}{n+\frac{1}{2}} + \cdots + \dfrac{2^{n/n}}{n+\frac{1}{n}} \right).$

6.4.13.* $\lim\limits_{n\to\infty} \dfrac{1}{n} \sqrt[n]{(n+1)(n+2)\ldots(n+n)}.$

6.4.14. $\lim\limits_{n\to\infty} \dfrac{1}{n} \left(\sin\dfrac{\pi}{n} + \sin\dfrac{2\pi}{n} + \cdots + \sin\dfrac{n\pi}{n} \right).$

6.4.15. Without using Theorem 6.4.5, prove that there can be at most one real number I satisfying the conditions of Definition 6.4.2.

6.4.16. In Lemma 6.4.3 prove that $\inf S = L(f, P)$.

6.4.17. Suppose that f is integrable on $[a, b]$ and that $f(x) = 0$ for all rational $x \in [a, b]$. Prove that $\int_a^b f(x)\,dx = 0$.

6.4.18. Let f be an even function and suppose that f is integrable on $[0, a]$. Prove that f is integrable on $[-a, a]$ and that $\int_{-a}^0 f(x)\,dx = \int_0^a f(x)\,dx$.

6.4.19. Let f be an odd function and suppose that f is integrable on $[0, a]$. Prove that f is integrable on $[-a, a]$ and that $\int_{-a}^0 f(x)\,dx = - \int_0^a f(x)\,dx$.

6.4.20. Let

$$f(x) = \begin{cases} x, & \text{if } x \in \mathbb{Q} \\ -x, & \text{if } x \notin \mathbb{Q}. \end{cases}$$

Show that f is not integrable on $[0, 1]$.

6.4.21.* Prove that $f(x)$ is integrable on $[a, b]$ if and only if for every $\varepsilon > 0$ there exists $\delta > 0$ such that for every partition P of $[a, b]$ with $\|P\| < \delta$ and any two selections of intermediate points ξ and ξ', $|S(f, P, \xi) - S(f, P, \xi')| < \varepsilon$.

6.5 Properties of Definite Integrals

So far we have been concerned with the existence of the integral. Now we are going to start working on its evaluation. In this section we will establish some simple algebraic rules for the definite integrals that are well known from elementary calculus. In the process, we will use both the Riemann and the Darboux definitions of the definite integral.

We will start with a well known property of integrals.

Theorem 6.5.1. *Let f, g be two functions that are integrable on $[a, b]$, and let $\alpha \in \mathbb{R}$. Then the functions αf and $f + g$ are integrable on $[a, b]$ as well and:*

(a) $\int_a^b \alpha f(x)\,dx = \alpha \int_a^b f(x)\,dx$;

(b) $\int_a^b (f(x) + g(x))\,dx = \int_a^b f(x)\,dx + \int_a^b g(x)\,dx$.

Proof. (a) We notice that, if $\alpha = 0$, the right side equals 0, while the left side is a definite integral of the zero function. Since every Riemann sum of a zero function equals 0, it follows that $\int_a^b 0\,dx = 0$, and the equality is true. Therefore, we turn our attention to the case $\alpha \neq 0$. Let $I = \int_a^b f(x)\,dx$, and let $\varepsilon > 0$. We will show that $\int_a^b \alpha f(x)\,dx = \alpha I$. By definition, there exists $\delta > 0$, such that if P is any partition of $[a, b]$ and $\|P\| < \delta$, then $|S(f, P, \xi) - I| < \varepsilon/|\alpha|$. Let P be such a partition, and let us consider $S(\alpha f, P, \xi)$. If $P = \{x_0, x_1, \ldots, x_n\}$, and $\xi = \{\xi_1, \xi_2, \ldots, \xi_n\}$, then

$$S(\alpha f, P, \xi) = \sum_{k=1}^{n} (\alpha f)(\xi_k) \Delta x_k = \alpha \sum_{k=1}^{n} f(\xi_k) \Delta x_k = \alpha S(f, P, \xi).$$

It follows that

$$|S(\alpha f, P, \xi) - \alpha I| = |\alpha| \, |S(f, P, \xi) - I| < |\alpha| \frac{\varepsilon}{|\alpha|} = \varepsilon.$$

(b) Let $I_f = \int_a^b f(x) \, dx$, $I_g = \int_a^b g(x) \, dx$, and let $\varepsilon > 0$. By definition, there exists $\delta_1 > 0$, such that if P is any partition of $[a, b]$ then

$$\|P\| < \delta_1 \quad \Rightarrow \quad |S(f, P, \xi) - I_f| < \frac{\varepsilon}{2}.$$

Also, there exists $\delta_2 > 0$, such that if P is any partition of $[a, b]$ then

$$\|P\| < \delta_2 \quad \Rightarrow \quad |S(g, P, \xi) - I_g| < \frac{\varepsilon}{2}.$$

Therefore, if we define $\delta = \min\{\delta_1, \delta_2\}$, then $\|P\| < \delta$ implies that both

$$|S(f, P, \xi) - I_f| < \frac{\varepsilon}{2} \quad \text{and} \quad |S(g, P, \xi) - I_g| < \frac{\varepsilon}{2}.$$

If P is such a partition, then

$$S(f+g, P, \xi) = \sum_{k=1}^{n} (f+g)(\xi_k) \Delta x_k = \sum_{k=1}^{n} f(\xi_k) \Delta x_k + \sum_{k=1}^{n} g(\xi_k) \Delta x_k = S(f, P, \xi) + S(g, P, \xi).$$

It follows that

$$|S(f+g, P, \xi) - (I_f + I_g)| = |S(f, P, \xi) + S(g, P, \xi) - I_f - I_g|$$
$$\leq |S(f, P, \xi) - I_f| + |S(g, P, \xi) - I_g| < \frac{\varepsilon}{2} + \frac{\varepsilon}{2} = \varepsilon. \qquad \square$$

The next property is also frequently used in calculations.

Theorem 6.5.2. *Let f be a function that is integrable on $[a, b]$, and let c be a point such that $a \leq c \leq b$. Then f is integrable on $[a, c]$ and on $[c, b]$, and*

$$\int_a^b f(x) \, dx = \int_a^c f(x) \, dx + \int_c^b f(x) \, dx. \tag{6.13}$$

Proof. We will show that f is integrable on $[a, c]$ and leave the integrability on $[c, b]$ as an exercise. Let $\varepsilon > 0$. According to Proposition 6.2.7, we will accomplish our goal if we find a partition Q of $[a, c]$ such that $U(f, Q) - L(f, Q) < \varepsilon$.

By Theorem 6.4.5, there exists $\delta > 0$, such that if P is any partition of $[a, b]$ and $\|P\| < \delta$, then $U(f, P) - L(f, P) < \varepsilon$. Let P be such a partition, and let

$$P' = P \cup \{c\} = \{y_0, y_1, \ldots, y_m\},$$

with $c = y_j$. Then $\|P'\| < \delta$, so $U(f, P') - L(f, P') < \varepsilon$. Let

$$Q = \{y_0, y_1, \ldots, y_j\}, \quad Q' = \{y_j, y_{j+1}, \ldots, y_m\}.$$

It is easy to see that, $L(f, P') = L(f, Q) + L(f, Q')$ and $U(f, P') = U(f, Q) + U(f, Q')$. Further,

$$U(f, Q) - L(f, Q) \le U(f, Q) - L(f, Q) + U(f, Q') - L(f, Q') = U(f, P') - L(f, P') < \varepsilon$$

so the integrability of f on $[a, c]$ has been established.

Assuming that f is integrable on $[c, b]$, let us denote

$$I_1 = \int_a^c f(x)\, dx, \quad I_2 = \int_c^b f(x)\, dx, \quad I = \int_a^b f(x)\, dx.$$

It remains to prove the equality $I = I_1 + I_2$. Let $\varepsilon > 0$. By Exercise 6.2.22, there exists δ_1, such that if P_1 is any partition of $[a, c]$, then

$$\|P_1\| < \delta_1 \Rightarrow |L(f, P_1) - I_1| < \frac{\varepsilon}{3}. \tag{6.14}$$

Similarly, there exists δ_2, such that if P_2 is any partition of $[c, b]$, then

$$\|P_2\| < \delta_2 \Rightarrow |L(f, P_2) - I_2| < \frac{\varepsilon}{3}. \tag{6.15}$$

Finally, there exists δ_3, such that if P is any partition of $[a, b]$, then

$$\|P\| < \delta_3 \Rightarrow |L(f, P) - I| < \frac{\varepsilon}{3}. \tag{6.16}$$

Let $\delta = \min\{\delta_1, \delta_2, \delta_3\}$, let P_1 and P_2 be partitions of $[a, c]$ and $[c, b]$, respectively, such that $\|P_1\| < \delta$ and $\|P_2\| < \delta$, and let $P = P_1 \cup P_2$. Then P is a partition of $[a, b]$ and $\|P\| < \delta$, so estimates (6.14)–(6.16) hold. Finally, $L(f, P) = L(f, P_1) + L(f, P_2)$. Thus,

$$\begin{aligned} |I - (I_1 + I_2)| &= |I - L(f, P) - (I_1 - L(f, P_1)) - (I_2 - L(f, P_2))| \\ &\le |I - L(f, P)| + |I_1 - L(f, P_1)| + |I_2 - L(f, P_2)| \\ &< \frac{\varepsilon}{3} + \frac{\varepsilon}{3} + \frac{\varepsilon}{3} = \varepsilon. \end{aligned}$$

Since ε is arbitrary, we see that $I = I_1 + I_2$. $\qquad\square$

Theorem 6.5.2 shows that, if a function is integrable on $[a, b]$, then it is integrable on both $[a, c]$ and $[c, b]$. What if f is integrable on $[a, c]$ and $[c, b]$? Does it follow that it is integrable on $[a, b]$? The answer is yes, and we will leave the proof as an exercise.

Theorem 6.5.3. *Let $c \in [a, b]$, and suppose that f is integrable on $[a, c]$ and on $[c, b]$. Then f is integrable on $[a, b]$, and (6.13) holds.*

The arithmetic established in Theorem 6.5.2 shows that it makes sense to define

$$\int_a^a f(x)\, dx = 0, \quad \text{and} \quad \int_b^a f(x)\, dx = -\int_a^b f(x)\, dx.$$

Another useful property of integrals is its *positivity*. In other words, if the integrand is positive then so is the integral.

Theorem 6.5.4. *Let f be an integrable function on $[a, b]$ and suppose that $f(x) \ge 0$ for all $x \in [a, b]$. Then $\int_a^b f(t)\, dt \ge 0$.*

Proof. Since $f(x) \ge 0$ it follows that, for any partition P of $[a, b]$, each infimum $m_i \ge 0$. Consequently, every lower Darboux sum $L(f, P) \ge 0$ and, all the more, $L = \sup L(f, P) \ge 0$. The integrability of f now implies that $\int_a^b f(t)\, dt = L \ge 0$. $\qquad\square$

From here we deduce an easy corollary.

Corollary 6.5.5. *Let f, g be two integrable functions on $[a, b]$ and suppose that $f(x) \le g(x)$ for all $x \in [a, b]$. Then $\int_a^b f(t)\, dt \le \int_a^b g(t)\, dt$.*

Proof. The function $g - f$ satisfies all the hypotheses of Theorem 6.5.4, so we conclude that $\int_a^b [g(t) - f(t)]\, dt \ge 0$. Now the result follows from the additivity of the integral (Theorem 6.5.1). $\qquad \square$

Exercises

6.5.1. In Theorem 6.5.2 prove that f is integrable on $[c, b]$.

6.5.2. Prove Theorem 6.5.3.

6.5.3. Prove that, if f is integrable on $[a, b]$, then

$$\left| \int_a^b f(x)\, dx \right| \le \int_a^b |f(x)|\, dx.$$

6.5.4.* Prove that if f and g are integrable on $[a, b]$, then so is fg.

6.5.5. Prove that $\lim_{n \to \infty} \int_0^{2\pi} \dfrac{\sin nx}{x^2 + n^2}\, dx = 0$.

6.5.6. Let f and g be integrable functions on $[a, b]$. Prove that the functions $\max\{f(x),\, g(x)\}$ and $\min\{f(x),\, g(x)\}$ are also integrable on $[a, b]$.

In Exercises 6.5.7–6.5.10 calculate $\int_a^b \max\{f(x),\, g(x)\}\, dx$ and $\int_a^b \min\{f(x),\, g(x)\}\, dx$:

6.5.7. $f(x) = 3x - 2$, $g(x) = x^2$, $a = 0$, $b = 4$.

6.5.8. $f(x) = e^x$, $g(x) = e^{-x}$, $a = -1$, $b = 1$.

6.5.9. $f(x) = \sin x$, $g(x) = \cos x$, $a = 0$, $b = \pi$.

6.5.10. $f(x) = 3x - 2$, $g(x) = x^3$, $a = 0$, $b = 4$.

6.5.11. Let $f^+(x) = \max\{f(x),\, 0\}$ and $f^-(x) = \max\{-f(x),\, 0\}$. Prove that f is integrable on $[a, b]$ if and only if both f^+ and f^- are integrable on $[a, b]$.

6.5.12. Give an example to show that the integrability of f^+ on $[a, b]$ does not imply that f is integrable on $[a, b]$.

6.5.13. Let f be a positive and continuous function on $[a, b]$ and let $M = \sup\{f(x) : x \in [a, b]\}$. Prove that

$$\lim_{n \to \infty} \left(\int_a^b (f(x))^n\, dx \right)^{1/n} = M.$$

6.5.14.* If f is continuous on $[a, b]$ and if $\int_a^b f(x)g(x)\, dx = 0$ for every continuous function g, prove that $f(x) = 0$ for all $x \in [a, b]$.

6.5.15. Let f be an integrable function on $[a, b]$, and let g be a uniformly continuous function on $f([a, b])$. Prove that $g \circ f$ is integrable on $[a, b]$.

6.5.16. If f is a non-negative integrable function on $[a, b]$ and if $1/f$ is bounded, prove that $1/f$ is integrable on $[a, b]$.

6.5.17. If f is continuous on $[a, b]$ and $f(x) \ge 0$, for $x \in [a, b]$, but f is not the zero function, prove that $\int_a^b f(x)\, dx > 0$.

6.5.18. Let $f, g : [a, b] \to \mathbb{R}$ be continuous functions, and suppose that $\int_a^b f(x)\, dx = \int_a^b g(x)\, dx$. Prove that there exists $c \in [a, b]$ such that $f(c) = g(c)$.

6.6 Fundamental Theorem of Calculus

In this section we will finally address the essential question: assuming that f is integrable on $[a, b]$, how do we compute $\int_a^b f(x)\,dx$? By definition, it is the limit of Riemann sums or, more specifically, Darboux sums (Exercise 6.2.22). We will present a much more efficient method, based on finding an antiderivative (The Fundamental Theorem of Calculus).

We will start with a mean value theorem for integrals.

Theorem 6.6.1 (Mean Value Theorem). *Let f be an integrable function on $[a, b]$, and suppose that, for all $x \in [a, b]$, $m \le f(x) \le M$. Then there exists $\mu \in [m, M]$ such that*

$$\int_a^b f(t)\,dt = \mu(b - a).$$

Proof. If P is a partition of $[a, b]$, then

$$m(b - a) = m\sum_{i=1}^n \Delta x_i \le L(f, P) \le L \le U \le U(f, P) \le M\sum_{i=1}^n \Delta x_i = M(b - a).$$

Since f is integrable, $I = \int_a^b f(t)\,dt = L$. Therefore, $m(b - a) \le I \le M(b - a)$ and, if we denote $\mu = I/(b - a)$, then $\mu \in [m, M]$. $\qquad\square$

This result takes a nice form when f is continuous.

Corollary 6.6.2. *Let f be a continuous function on $[a, b]$. Then there exists $c \in [a, b]$ such that*

$$\int_a^b f(t)\,dt = f(c)(b - a).$$

Proof. Since f is continuous, by Weierstrass Theorem it is bounded, say $m \le f(x) \le M$. Further, by the Mean Value Theorem for integrals, there exists $\mu \in [m, M]$ such that $\int_a^b f(t)\,dt = \mu(b - a)$. Finally, by the Intermediate Value Theorem, there exists $c \in [a, b]$ such that $\mu = f(c)$, and the result follows. $\qquad\square$

Corollary 6.6.2 appears in Cauchy's *Cours d'Analyse*.

Now we will prove the main result of this section, and probably of the whole integral calculus.

Theorem 6.6.3 (Fundamental Theorem of Calculus). *Suppose that f is an integrable function on $[a, b]$, and let*

$$F(x) = \int_a^x f(t)\,dt.$$

Then the function F is continuous on $[a, b]$. Furthermore, if $c \in [a, b]$ and f is continuous at $x = c$, then F is differentiable at $x = c$, and $F'(c) = f(c)$.

Proof. Let $\varepsilon > 0$. By assumption, f is integrable, hence bounded. Let $|f(x)| \le M$, for all $x \in [a, b]$, and let $\delta = \varepsilon/M$. If $|x - c| < \delta$, then

$$|F(x) - F(c)| = \left| \int_a^x f(t)\,dt - \int_a^c f(t)\,dt \right| = \left| \int_c^x f(t)\,dt \right|$$

$$\leq \left| \int_c^x |f(t)|\, dt \right| \leq M \left| \int_c^x dt \right| = M|x - c| < M\delta = \varepsilon,$$

so F is continuous at $x = c$.

Supppose now that, in addition, f is continuous at $x = c$. Then there exists $\delta_1 > 0$ such that, if $|x - c| < \delta_1$, then $|f(t) - f(c)| < \varepsilon$. It follows that, for $0 < |x - c| < \delta_1$,

$$\left| \frac{F(x) - F(c)}{x - c} - f(c) \right| = \left| \frac{1}{x - c} \int_c^x f(t)\, dt - \frac{1}{x - c} \int_c^x f(c)\, dt \right| = \frac{1}{|x - c|} \left| \int_c^x [f(t) - f(c)]\, dt \right|$$

$$\leq \frac{1}{|x - c|} \left| \int_c^x |f(t) - f(c)|\, dt \right| \leq \frac{1}{|x - c|} \left| \int_c^x \varepsilon\, dt \right| = \frac{1}{|x - c|} \varepsilon |x - c| = \varepsilon.$$

Thus, F is differentiable at $x = c$ and $F'(c) = f(c)$. $\qquad\square$

The Fundamental Theorem of Calculus has an easy consequence that we use when calculating a definite integral. It is sometimes called the Second Fundamental Theorem of Calculus.

Corollary 6.6.4. *Let f be a continuous function on $[a, b]$ and let F be any primitive function of f. Then*

$$\int_a^b f(t)\, dt = F(b) - F(a). \tag{6.17}$$

Proof. Let $G(x) = \int_a^x f(t)\, dt$. By the Fundamental Theorem of Calculus, G is an antiderivative of f and

$$\int_a^b f(t)\, dt = G(b) - G(a).$$

If F is any other antiderivative of f, Corollary 5.2.3 shows that $G(x) = F(x) + C$. Consequently, $F(b) - F(a) = G(b) - G(a)$. $\qquad\square$

When f is not continuous but merely integrable, (6.17) need not be true. The problem is that f need not have a primitive function (Exercise 6.6.12). When it does, we have

$$\int_a^b F'(t)\, dt = F(b) - F(a), \tag{6.18}$$

and the equality is true whenever F' is integrable. In this form, the result was proved by Darboux in [25] in 1875 (see Exercise 6.6.16). By the way, the assumption that F' is integrable cannot be omitted. Just because F is differentiable, does not mean that F' is integrable (see Exercise 6.6.13).

> The relation between integration and antiderivation can be found in the work of Scottish mathematician and astronomer James Gregory (1638–1675) around 1668 and Isaac Barrow (1630–1677), an English Christian theologian and mathematician (c. 1670). However, they did not recognize its importance. Newton (1667) and Leibniz (1677) used it as a powerful computational tool. Because of that, the theorem is sometimes called the Newton–Leibniz Theorem. Throughout the eighteenth century the concept of the integral was synonymous with the indefinite integral. Cauchy was the first to properly define the definite integral in 1823, and give a rigorous proof of the Fundamental Theorem of Calculus for the case when f is a continuous function.

In the remaining portion of this section we will look at some consequences of the Fundamental Theorem of Calculus. The notation $f\,|_a^b$ means $f(b) - f(a)$.

Theorem 6.6.5 (Integration by Parts in Definite Integrals). *Suppose that f, g are two differentiable functions on $[a, b]$ and that their derivatives are continuous on $[a, b]$. Then*

$$\int_a^b f \, dg = fg \Big|_a^b - \int_a^b g \, df. \tag{6.19}$$

Proof. We use the Product Rule for derivatives:

$$(fg)' = f'g + fg'.$$

It can be viewed as the fact that fg is a primitive function for $f'g + fg'$. By the Fundamental Theorem of Calculus,

$$\int_a^b [f'(t)g(t) + f(t)g'(t)] \, dt = fg \Big|_a^b$$

and the result follows. $\qquad\square$

> The vertical bar to indicate evaluation of an antiderivative at the two limits of integration was used first by French mathematician Pierre Frederic Sarrus (1798–1861) in 1823. He is immortalized by the discovery of a memorization rule for computing the determinant of a 3-by-3 matrix, named *Sarrus' scheme*.

Another important consequence of the Fundamental Theorem of Calculus is that it provides a justification for the substitution method.

Theorem 6.6.6 (Substitution in Definite Integrals). *Let f be a continuous function on $[a, b]$, and suppose that φ is a function with a domain $[c, d]$ and range contained in $[a, b]$, so that $\varphi(c) = a$, $\varphi(d) = b$. Also, let φ be differentiable on $[c, d]$, and suppose that its derivative φ' is continuous on $[c, d]$. Then*

$$\int_a^b f(x) \, dx = \int_c^d f(\varphi(t)) \, \varphi'(t) \, dt. \tag{6.20}$$

Proof. Let F be any primitive function of f. By Corollary 6.6.4, the left hand side of (6.20) equals $F(b) - F(a)$. On the other hand,

$$(F \circ \varphi)'(t) = F'(\varphi(t))\varphi'(t) = f(\varphi(t)) \, \varphi'(t)$$

so $F \circ \varphi$ is a primitive function for $f(\varphi(t))\varphi'(t)$. Once again using Corollary 6.6.4, the right hand side of (6.20) equals

$$(F \circ \varphi)(d) - (F \circ \varphi)(c) = F(\varphi(d)) - F(\varphi(c)) = F(b) - F(a),$$

so (6.20) is established. $\qquad\square$

> The expression "the Fundamental Theorem of Calculus" had been sporadically used during the middle of the nineteenth century among mathematicians in Berlin. By 1876, when "Fundamentalsatz der Integral-rechnung" appeared in a paper of Du Bois-Reymond (see below), it had become part of the established terminology. It seems that it has been brought to the United States by a Canadian mathematician Daniel Alexander Murray (1862–1934), who studied in Berlin. He was an instructor at Cornell University in 1898 when his book *An Elementary Course in the Integral Calculus* was published. The word *fundamental* is well justified because the theorem establishes a relationship between two very different concepts. On one hand, the definite integral was defined with the area under the graph in mind. On the other hand, the derivative came from the need to find a linear approximation (or the slope of the tangent line, if we wish to think geometrically).

PAUL du BOIS-REYMOND (1831–1889) was born in Berlin. His father was German and his mother was French, so all 5 children were bilingual. Paul wanted to follow the example of his 13-year older brother, who was a physiologist, so he studied medicine at the University of Zürich. While taking classes in Königsberg he changed to mathematical physics. His 1853 doctoral dissertation at the University of Berlin was titled *On the equilibrium of fluids*. His first job was as a high school teacher in Berlin. In 1864 he published a paper with important results about partial differential equations. That led to a position at the University of Heidelberg in 1865. From 1870 to 1874 he taught at the University of Freiburg, and from 1874 to 1884 at the University of Tübingen, where he was a doctoral adviser of Otto Hölder (1859–1937), famous for the Hölder's Inequality (see page 295). In 1884 he took a position at the Technische Hochschule Charlottenberg in Berlin. His research was focused on partial differential equations and functions of a real variable. In 1873 he gave an example of a continuous function whose Fourier series diverges at a dense set of points (see page 268). In 1875 he published an example of a continuous function that has no derivative at any point (see page 244). It is not well known that he was the first to use, in 1875, the "diagonal argument", which Cantor would later make famous.

Exercises

In Exercises 6.6.1–6.6.4 find the derivative with respect to x:

6.6.1. $\int_a^b \sin x^2 \, dx$.   6.6.2. $\int_x^b \sin t^2 \, dt$.   6.6.3. $\int_a^{x^2} \sin t^2 \, dt$.   6.6.4. $\int_x^{x^2} \sin t^2 \, dt$.

In Exercises 6.6.5–6.6.6 find the limits:

6.6.5. $\lim_{x \to 0+} \frac{1}{x^3} \int_0^{x^2} \sin \sqrt{t} \, dt$.   6.6.6. $\lim_{x \to 0} \frac{1}{x} \int_0^x (1 + \sin t)^{1/t} \, dt$.

6.6.7. Let f be an integrable function on $[a, b]$. Prove that there exists $c \in [a, b]$ such that

$$\int_a^c f(x) \, dx = \int_c^b f(x) \, dx.$$

6.6.8.* Let f be a continuous function on $[a, b]$ and suppose that $\int_a^b f(x) \, dx = 0$. Prove that there exists $c \in (a, b)$ such that

$$\int_a^c f(x) \, dx = f(c).$$

6.6.9. Let f be a continuous function on $\mathbb{R}$ and define

$$G(x) = \int_0^x (x - t) f(t) \, dt.$$

Prove that $G''(x) = f(x)$, for all $x \in \mathbb{R}$.

6.6.10.* Let f be a continuous function on $[a, b]$ and suppose that $\int_c^d f(x) \, dx = 0$, for all $[c, d] \subset [a, b]$. Prove that $f(x) = 0$, for all $x \in [a, b]$.

6.6.11. Let $p(x) = a_1 x + a_2 x^2 + \cdots + a_n x^n$ and suppose that

$$\frac{a_1}{2} + \frac{a_2}{3} + \cdots + \frac{a_n}{n+1} = 0.$$

Prove that there exists $c \in (0, 1)$ such that $p(c) = 0$.

6.6.12.* Let f be the Thomae function (page 163), and let $g(x) = \int_0^x f(t) \, dt$. Prove that $g'(x) = f(x)$ if and only if $x \notin \mathbb{Q}$. Thus, an integrable function need not have a primitive function.

6.6.13.* Let $f(x) = \begin{cases} x^2 \sin \frac{1}{x^2}, & \text{if } x \neq 0 \\ 0, & \text{if } x = 0. \end{cases}$ Prove that f is differentiable on $[-1, 1]$, but f' is not integrable on $[-1, 1]$. Thus, a function may have a primitive function without being integrable.

6.6.14. Give an example to show that Corollary 6.6.2 is not true if f is not continuous but merely integrable.

6.6.15. Let f be a function defined and continuous on $[a, b]$ with the exception of $c \in (a, b)$ where it has a jump. Prove that the function F defined by $F(x) = \int_a^x f(t) \, dt$ has both one-sided derivatives at $x = c$. Give an example to show that these derivatives need not be equal.

6.6.16.* Prove (6.18) under the assumption that F' is integrable on $[a, b]$.

6.6.17.* Let f, g be continuous functions on $[a, b]$ and suppose that g is non-negative. Prove that there exists a point $c \in [a, b]$ such that

$$\int_a^b f(x) g(x) \, dx = f(c) \int_a^b g(x) \, dx.$$

6.6.18.* Let f, g be continuous functions on $[a, b]$ and suppose that f is non-negative and decreasing. Prove that there exists a point $c \in [a, b]$ such that

$$\int_a^b f(x) g(x) \, dx = f(a) \int_a^c g(x) \, dx.$$

6.6.19.* Let f, g be continuous functions on $[a, b]$ and suppose that f is increasing. Prove that there exists a point $c \in [a, b]$ such that

$$\int_a^b f(x) g(x) \, dx = f(a) \int_a^c g(x) \, dx + f(b) \int_c^b g(x) \, dx.$$

6.6.20.* Let f be a positive continuous function on $[a, b]$ and let $c > 0$. Suppose that

$$f(x) \le c \int_a^x f(t) \, dt$$

for all $x \in [a, b]$. Prove that $f(x) = 0$ for all $x \in [a, b]$.

In Exercises 6.6.21–6.6.22 f is continuous on $[0, 1]$. Prove that:

6.6.21. $\int_0^{\pi/2} f(\sin x) \, dx = \int_0^{\pi/2} f(\cos x) \, dx.$ **6.6.22.** $\int_0^{\pi} x f(\sin x) \, dx = \dfrac{\pi}{2} \int_0^{\pi} f(\sin x) \, dx.$

In Exercises 6.6.23–6.6.25 find the integrals using a substitution:

6.6.23. $\int_0^{\ln 2} \sqrt{e^x - 1} \, dx.$ **6.6.24.** $\int_0^1 \dfrac{\arcsin \sqrt{x}}{\sqrt{x(1-x)}} \, dx.$ **6.6.25.** $\int_0^{\pi} \dfrac{x \sin x}{1 + \cos^2 x} \, dx.$

6.6.26. Let f be a monotone continuous function on $[a, b]$, and let $f(a) = c$, $f(b) = d$. Prove that its inverse function g satisfies the relation

$$\int_a^b f(x) \, dx + \int_c^d g(x) \, dx = bd - ac.$$

In Exercises 6.6.27–6.6.30 find the integrals using integration by parts:

6.6.27. $\int_0^{\ln 2} x e^{-x} \, dx.$ **6.6.28.** $\int_0^{\pi} x \sin x \, dx.$ **6.6.29.** $\int_0^1 \arccos x \, dx.$

6.6.30. Find $\int_0^{\pi/2} \sin^n x \, dx$, if $n \in \mathbb{N}$.

6.7   Infinite and Improper Integrals

In this section we will extend the notion of the definite integral to two new situations. In the *infinite integral* the domain of a function f is not a finite interval $[a, b]$ but an infinite interval, such as $[a, +\infty)$. In the *improper integral*, the integrand f is not a bounded function.

6.7.1   Infinite Integrals

Example 6.7.1. Calculating an infinite integral.

Calculate $\int_0^\infty \dfrac{dx}{1 + x^2}$.

Solution. The standard procedure is to evaluate, if possible,

$$\int_0^b \frac{dx}{1 + x^2},$$

and then take the limit as $b \to +\infty$. Now $\int \dfrac{dx}{1 + x^2} = \arctan x + C$ so

$$\int_0^b \frac{dx}{1 + x^2} = \arctan x \Big|_0^b = \arctan b - \arctan 0 = \arctan b.$$

Therefore,

$$\int_0^\infty \frac{dx}{1 + x^2} = \lim_{b \to +\infty} \arctan b = \frac{\pi}{2}. \qquad \blacklozenge$$

In general, we have the following definition.

Definition 6.7.2. Let $a \in \mathbb{R}$, let f be a function defined on the infinite interval $[a, +\infty)$, and suppose that, for any $b > a$, f is integrable on $[a, b]$. If there exists the limit

$$\lim_{b \to +\infty} \int_a^b f(t) \, dt \qquad (6.21)$$

then we call it the **infinite integral** of f, and we write $\int_a^{+\infty} f(t) \, dt$. In this situation we say that f is **integrable on** $[a, +\infty)$.

Notice that, if we define as usual $F(x) = \int_a^x f(t) \, dt$, then the infinite integral

$$I = \lim_{x \to +\infty} F(x).$$

Therefore, if the limit exists, we say that the integral $\int_a^{+\infty} f(t) \, dt$ *converges*.

> The definition of an infinite integral presented here can be found in Cauchy's work [16] from 1823.

It might appear that there is nothing to say about infinite integrals, and that the only difference from the usual definite integrals is that, at the end, one has to take the limit. The problem is that, as we have discussed earlier, many functions do not have elementary antiderivatives, so the approach based on the Fundamental Theorem of Calculus fails. In such a situation, a different strategy is needed, and we will develop one in this section.

The approach we will use is to determine whether the limit in (6.21) exists at all. If it does, then we can approximate the exact value by replacing $+\infty$ in the upper limit by a sufficiently large positive number. Estimating the error of such an approximation is another important part of this method. We will focus on the existence of the limit, and we will present several tests.

Theorem 6.7.3 (Comparison Test). *Let f and g be two functions defined on $[a, +\infty)$ and integrable on $[a, b]$ for all $b \geq a$. Suppose that $0 \leq f(x) \leq g(x)$ for all $x \in [a, +\infty)$. If $\int_a^\infty g(x)\,dx$ converges then so does $\int_a^\infty f(x)\,dx$.*

Proof. The assumption that $0 \leq f(x) \leq g(x)$, together with Theorem 6.5.4 and Corollary 6.5.5, implies that, for all $x \in [a, +\infty)$

$$0 \leq F(x) = \int_a^x f(t)\,dt \leq \int_a^x g(t)\,dt = G(x).$$

Further, if $a \leq x_1 < x_2$, using Theorems 6.5.2 and 6.5.4,

$$G(x_2) = \int_a^{x_2} g(t)\,dt = \int_a^{x_1} g(t)\,dt + \int_{x_1}^{x_2} g(t)\,dt \geq \int_a^{x_1} g(t)\,dt = G(x_1).$$

Thus, G is an increasing function, and the same argument shows that F is an increasing function. If $\int_a^\infty g(x)\,dx$ converges then $L = \lim_{x \to +\infty} G(x)$ exists, and G is a bounded function (Exercise 6.7.17). It follows that, if x_n is an increasing sequence with $\lim x_n = +\infty$, then $F(x_n)$ is a monotone increasing sequence of real numbers, that is bounded by L, so it is convergent.

Can two different (increasing) sequences x_n and y_n, if they both converge to $+\infty$, yield different limits for $F(x_n)$ and $F(y_n)$? The answer is no! If x_n and y_n are such sequences, we will consider the sequence z_n, obtained by interlacing x_n and y_n. Namely, let $A = \{x_n, y_n : n \in \mathbb{N}\}$, and

$$z_1 = \min(A), \quad z_2 = \min\left(A \setminus \{z_1\}\right), \quad z_3 = \min\left(A \setminus \{z_1, z_2\}\right), \quad \ldots$$

Then z_n is an increasing sequence and $\lim z_n = +\infty$, so $F(z_n)$ must be a convergent sequence, say $\lim F(z_n) = w$. Therefore, each subsequence of $F(z_n)$ must converge to w. In particular, $\lim F(x_n) = w$ and $\lim F(y_n) = w$.

We conclude that, for any sequence x_n with $\lim x_n = +\infty$, $F(x_n)$ converges to the same limit. Thus, $\lim_{x \to +\infty} F(x)$ exists. □

Example 6.7.4. Establishing the convergence of an infinite integral.

Determine whether the integral $I = \int_1^\infty \dfrac{dx}{\sqrt{1 + x^4}}$ converges.

Solution. The function $f(x) = 1/\sqrt{1 + x^4}$ does not have an elementary antiderivative. Instead, we consider $g(x) = 1/x^2$. Both f and g are continuous, hence integrable, on $[1, b]$ for all $b \geq 1$. In addition, $\sqrt{1 + x^4} > \sqrt{x^4} = x^2$, so $0 < f(x) < g(x)$. Finally,

$$\int_1^b \frac{1}{x^2} = -\frac{1}{x}\Big|_1^b = -\frac{1}{b} + 1 \to 1, \quad \text{as } b \to +\infty,$$

so the integral I converges. Now we can approximate it by replacing $+\infty$ by a large number and use approximate integration. For example

$$\int_1^{1000} \frac{dx}{\sqrt{1 + x^4}} \approx 0.9260373385, \quad \text{and} \quad \int_1^{10000} \frac{dx}{\sqrt{1 + x^4}} \approx 0.9269373385. \quad \blacklozenge$$

Next, let us consider the integral $\int_2^\infty dx/\sqrt{x^4-1}$ and determine whether it converges. We would like to use an estimate similar to the one in the previous Example 6.7.4. Unfortunately, if we denote $f(x) = 1/\sqrt{x^4-1}$ and use $g(x) = 1/x^2$, then they do not satisfy $0 < f(x) < g(x)$. In fact, quite the opposite is true: $0 < g(x) < f(x)$. Yet, it is easy to see that, when x is large, $x^4 - 1$ is very close to x^4, so $f(x)$ is very close to $g(x)$. Since $\int_2^\infty g(x)\,dx$ converges, we would expect that $\int_2^\infty f(x)\,dx$ converges as well. The next result gives a sufficient condition for the integrals $\int_a^\infty f(x)\,dx$ and $\int_a^\infty g(x)\,dx$ to be *equiconvergent*, i.e., that either they both converge or they both diverge.

Theorem 6.7.5 (Limit Comparison Test). *Let f and g be two functions defined and positive on $[a, +\infty)$ and integrable on $[a, b]$ for all $b \geq a$. Suppose that*

$$\lim_{x \to +\infty} \frac{f(x)}{g(x)}$$

exists and is not equal to 0. Then the integrals $\int_a^\infty f(x)\,dx$ and $\int_a^\infty g(x)\,dx$ are equiconvergent.

Proof. Let $\lim_{x \to +\infty} f(x)/g(x) = C$. By definition, for any positive $\varepsilon < C$, there exists $M > 0$ such that, for $x > M$,

$$C - \varepsilon < \frac{f(x)}{g(x)} < C + \varepsilon.$$

Since $g(x) > 0$ we obtain that, for $x > M$,

$$(C - \varepsilon)g(x) < f(x) < (C + \varepsilon)g(x). \tag{6.22}$$

Now, if $\int_a^\infty g(x)\,dx$ converges, then so does $\int_M^\infty g(x)\,dx$ (Exercise 6.7.19), as well as

$$\int_M^\infty (C + \varepsilon)g(x)\,dx.$$

It follows from (6.22) that $\int_M^\infty f(x)\,dx$ converges and we conclude that $\int_a^\infty f(x)\,dx$ converges. On the other hand, if $\int_a^\infty f(x)\,dx$ converges, using a similar argument, the left hand inequality in (6.22) implies that so does $\int_a^\infty (C - \varepsilon)g(x)\,dx$. Consequently, $\int_a^\infty g(x)\,dx$ converges.

This shows that the convergence of either integral implies the convergence of the other one. The implications involving their divergence are just the contrapositives of those for convergence. □

Remark 6.7.6. The assumption that $C \neq 0$ was used only when showing that the convergence of $\int_a^\infty f(x)\,dx$ implies the convergence of $\int_a^\infty g(x)\,dx$. The other implication remains valid even if $C = 0$.

Remark 6.7.7. If $\lim_{x \to +\infty} f(x)/g(x) = \infty$, then we can only conclude the first inequality in (6.22). In this situation, we can only derive the divergence of $\int_a^\infty f(x)\,dx$ from the divergence of $\int_a^\infty g(x)\,dx$.

Example 6.7.8. Establishing the convergence of an infinite integral.

Use the Limit Comparison Test to show that $\displaystyle\int_2^\infty \frac{dx}{\sqrt{x^4-1}}$ converges.

Solution. We will indeed use $g(x) = 1/x^2$. Both $f(x) = 1/\sqrt{x^4-1}$ and $g(x)$ are positive on $[2, +\infty)$ and integrable on $[2, b]$ for any $b > 2$. Furthermore,

$$\frac{f(x)}{g(x)} = \frac{1/\sqrt{x^4-1}}{1/x^2} = \frac{x^2}{\sqrt{x^4-1}} = \frac{1}{\sqrt{(x^4-1)/x^4}} = \frac{1}{\sqrt{1-1/x^4}},$$

so

$$\lim_{x \to +\infty} \frac{f(x)}{g(x)} = \lim_{x \to +\infty} \frac{1}{\sqrt{1 - 1/x^4}} = 1.$$

By the Limit Comparison Test the integrals are equiconvergent. Since it was shown in Example 6.7.4 that $\int_2^\infty g(x)\, dx$ converges, then so does $\int_2^\infty f(x)\, dx$. ◆

Example 6.7.9. Establishing the convergence of an infinite integral.

Determine whether the integral $\int_1^\infty x^{a-1} e^{-x}\, dx$ converges if $a \in \mathbb{R}$.

Solution. We will use $g(x) = 1/x^2$. Both $f(x) = x^{a-1} e^{-x}$ and $g(x)$ are positive on $[1, \infty)$ and integrable on $[1, b]$ for any $b > 1$. Furthermore,

$$\lim_{x \to +\infty} \frac{f(x)}{g(x)} = \lim_{x \to +\infty} \frac{x^{a-1} e^{-x}}{1/x^2} = \lim_{x \to +\infty} \frac{x^{a+1}}{e^x} = 0.$$

Since $\int_1^\infty g(x)\, dx$ converges, the Limit Comparison Test implies that the same is true of $\int_1^\infty f(x)\, dx$. ◆

Exercises

In Exercises 6.7.1–6.7.6 evaluate the infinite integrals:

6.7.1. $\int_1^\infty \dfrac{e^{-\sqrt{x}}}{\sqrt{x}}\, dx.$

6.7.2. $\int_3^\infty \dfrac{dx}{x^2 + x - 2}.$

6.7.3. $\int_0^\infty \dfrac{dx}{x^4 + 4}.$

6.7.4.* $\int_1^\infty \dfrac{dx}{x\sqrt{1 + x^5 + x^{10}}}.$

6.7.5. $\int_1^\infty \dfrac{x \ln x}{(1 + x^2)^2}\, dx.$

6.7.6.* $\int_0^\infty \dfrac{dx}{(x^2 + 1)^n}, \; n \in \mathbb{N}.$

6.7.7. Show that the integral $\int_1^\infty \dfrac{1}{x^p}\, dx$ converges if and only if $p > 1$.

In Exercises 6.7.8–6.7.16 test the integrals for convergence:

6.7.8. $\int_1^\infty \dfrac{x\, dx}{3x^4 + 5x^2 + 1}.$

6.7.9. $\int_2^\infty \dfrac{x^2 - 1}{\sqrt{x^6 + 16}}\, dx.$

6.7.10. $\int_0^\infty e^{-x^2}\, dx.$

6.7.11. $\int_0^\infty \dfrac{x^2\, dx}{x^4 - x^2 + 1}.$

6.7.12. $\int_1^\infty \dfrac{dx}{\sqrt[3]{x^2 + 1}}\, dx.$

6.7.13. $\int_1^\infty \dfrac{\arctan x}{x}\, dx.$

6.7.14. $\int_1^\infty \dfrac{\arctan x}{x^2}\, dx.$

6.7.15. $\int_1^\infty \dfrac{\ln(1 + x)}{x}\, dx.$

6.7.16. $\int_1^\infty \dfrac{\ln(1 + x)}{x^2}\, dx.$

6.7.17.* Let f be an increasing function for $x \geq a$, and suppose that a finite limit $L = \lim_{x \to +\infty} f(x)$ exists. Prove that $f(x) \leq L$ for all $x \in [a, +\infty)$ and that f is bounded on $[a, +\infty)$.

6.7.18. Let f be a continuous function for $x \geq a$, and suppose that a finite limit $L = \lim_{x \to +\infty} f(x)$ exists. Prove that there exists $M > 0$ such that $|f(x)| \leq M$ for all $x \in [a, +\infty)$.

6.7.19. Let f be a function defined on the infinite interval $[a, +\infty)$, and suppose that, for any $b > a$, f is integrable on $[a, b]$. Prove that, if $c > a$, the integral $\int_a^\infty f(x)\, dx$ converges if and only if $\int_c^\infty f(x)\, dx$ converges.

6.7.20 (Cauchy's Test).* Let f be a function defined for $x \geq a$, and suppose that, for any $b > a$, f is integrable on $[a, b]$. Prove that the integral $\int_a^\infty f(x)\, dx$ exists if and only if, for every $\varepsilon > 0$, there exists $B > 0$ such that, for any $b_2 \geq b_1 \geq B$,

$$\left| \int_a^{b_2} f(x)\, dx - \int_a^{b_1} f(x)\, dx \right| < \varepsilon.$$

6.7.21. A function f is **absolutely integrable on** $[a, +\infty)$ if the integral $\int_a^\infty |f(x)|\, dx$ converges. (We also say that the integral $\int_a^\infty f(x)\, dx$ is *absolutely convergent.*) Prove that, if f is integrable on $[a, b]$ for every $b > a$, and if f is absolutely integrable on $[a, +\infty)$, then f is integrable on $[a, +\infty)$.

6.7.22 (Abel's Test). Let f and g be functions defined and continuous on $[a, b]$ for all $b \geq a$. Suppose that $\int_a^\infty f(x)\, dx$ converges and that g is a monotone decreasing, bounded function. Then the integral $\int_a^\infty f(x)g(x)\, dx$ converges.

6.7.23 (Dirichlet's Test). Let f and g be functions defined and continuous on $[a, b]$ for all $b \geq a$. Suppose that there exists $M > 0$ such that $|\int_a^b f(x)\, dx| \leq M$ for all $b \geq a$, and that g is a monotone decreasing function such that $\lim_{x \to \infty} g(x) = 0$. Then the integral $\int_a^\infty f(x)g(x)\, dx$ converges.

In Exercises 6.7.24–6.7.27 test the integrals for absolute and conditional convergence:

6.7.24.* $\int_1^\infty \dfrac{\sin^2 x}{x}\, dx.$ 6.7.25.* $\int_0^\infty \dfrac{\sqrt{x}\cos x}{x + 100}\, dx.$ 6.7.26.* $\int_1^\infty \dfrac{\sin x}{x}\, dx.$

6.7.27.* $\int_1^\infty \dfrac{\ln^2 x \cos x}{x}\, dx.$

6.7.2   Improper Integrals

Let us consider the problem of evaluating $\int_0^1 dx/\sqrt{x}$. It is not hard to see that the function $f(x) = 1/\sqrt{x}$ is not bounded on the interval $(0, 1)$ (Figure 6.6). Therefore, it is not integrable

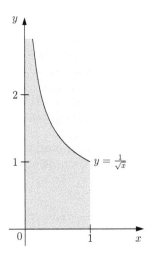

Figure 6.6: The region under the curve is unbounded.

(Theorem 6.4.4). In fact, if we think of the integral as the area under the graph, the region in question is unbounded. (The region extends along the y-axis infinitely high.) On the other hand,

$$\int \frac{dx}{\sqrt{x}} = 2\sqrt{x} + C$$

so an (unjustified) application of the Fundamental Theorem of Calculus would yield

$$\int_0^1 \frac{dx}{\sqrt{x}} = 2\sqrt{1} - 2\sqrt{0} = 2.$$

Once again, the limits come to the rescue.

Definition 6.7.10. Let $b > a$ be real numbers, let f be a function defined on the interval $(a, b]$, and suppose that, for any $c \in (a, b]$, f is integrable on $[c, b]$. If there exists the limit

$$\lim_{c \to a^+} \int_c^b f(t)\, dt$$

then we call it the **improper integral** of f, and we write $\int_a^b f(t)\, dt$.

Notice that, if we define as usual $F(x) = \int_x^b f(t)\, dt$, then the improper integral

$$I = \lim_{x \to a^+} F(x).$$

Just like for the infinite integrals, it is important to be able to tell whether such an integral converges. The similarity does not end there. The main tools are, once again, the comparison tests.

Theorem 6.7.11 (Comparison Test). *Let f and g be two functions defined on $(a, b]$ and integrable on $[c, b]$ for all $c \in (a, b]$. Suppose that $0 \le f(x) \le g(x)$ for all $x \in (a, b]$. If $\int_a^b g(x)\, dx$ converges then so does $\int_a^b f(x)\, dx$.*

Theorem 6.7.12 (Limit Comparison Test). *Let f and g be two functions defined and positive on $(a, b]$ and integrable on $[c, b]$ for all $c \in (a, b]$. Suppose that*

$$\lim_{x \to a^+} \frac{f(x)}{g(x)}$$

exists and is not equal to 0. Then integrals $\int_a^b f(x)\, dx$ and $\int_a^b g(x)\, dx$ are equiconvergent.

We will leave the proofs to the reader. Also, we will not formulate the definition of the improper integral in the case when the function f is unbounded in the vicinity of the right endpoint of the domain.

Instead we will look at examples of the application of the Limit Comparison Test. A function that is frequently used in such situations is $g(x) = 1/(x - a)^p$ so we start with it.

Example 6.7.13. An improper integral.

Determine whether the integral $\displaystyle\int_a^b \frac{dx}{(x-a)^p}$ converges.

Solution. For $c \in (a, b]$, and $p \ne 1$,

$$\int_c^b \frac{dx}{(x-a)^p} = \frac{(x-a)^{-p+1}}{-p+1}\Big|_c^b = \frac{(b-a)^{-p+1}}{-p+1} - \frac{(c-a)^{-p+1}}{-p+1}.$$

It is easy to see that, as $c \to a^+$, the limit exists if and only if $-p + 1 > 0$, i.e., if and only if $p < 1$. If $p > 1$, the integral diverges. When $p = 1$,

$$\int_c^b \frac{dx}{x-a} = \ln|x-a|\,\Big|_c^b = \ln|b-a| - \ln|c-a| \to +\infty, \quad \text{as } c \to a^+.$$

We conclude that $\displaystyle\int_a^b \frac{dx}{(x-a)^p}$ converges for $p < 1$ and diverges if $p \ge 1$. $\blacklozenge$

We can now use this result to establish the convergence (or lack thereof).

Example 6.7.14. Using the limit comparison test.

Determine whether the integral $\int_0^1 x^q \ln x \, dx$ converges.

Solution. We want to apply the Limit Comparison Test with $g(x) = 1/x^p$. Of course, $f(x) = x^q \ln x$. Now

$$\lim_{x \to 0^+} \frac{f(x)}{g(x)} = \lim_{x \to 0^+} \frac{x^q \ln x}{1/x^p} = \lim_{x \to 0^+} x^{p+q} \ln x, \tag{6.23}$$

so if $p+q > 0$, we have an indeterminate form $(0 \cdot \infty)$. With the aid of the L'Hôpital's Rule,

$$\lim_{x \to 0^+} x^{p+q} \ln x = \lim_{x \to 0^+} \frac{\ln x}{x^{-p-q}} \quad \left(= \frac{\infty}{\infty} \right)$$

$$= \lim_{x \to 0^+} \frac{1/x}{(-p-q)x^{-p-q-1}}$$

$$= \lim_{x \to 0^+} \frac{1}{(-p-q)x^{-p-q}}$$

$$= \lim_{x \to 0^+} \frac{x^{p+q}}{(-p-q)} = 0.$$

Although the limit is 0, Remark 6.7.6 shows that the convergence of $\int_0^1 1/x^p \, dx$ implies the convergence of $\int_0^1 x^q \ln x \, dx$. Further, Example 6.7.13 shows that the former integral converges when $p < 1$. Combined with $p + q > 0$ this gives $q > -p > -1$. In other words, $\int_0^1 x^q \ln x \, dx$ converges for $q > -1$.

When $p + q \leq 0$ the limit in (6.23) is infinite, so by Remark 6.7.7, the divergence of $\int_0^1 1/x^p \, dx$ implies the divergence of $\int_0^1 x^q \ln x \, dx$. Again using Example 6.7.13, the former integral diverges when $p \geq 1$. Combining with $p + q \leq 0$, we obtain $q \leq -p \leq -1$. Thus, $\int_0^1 x^q \ln x \, dx$ diverges for $q \leq -1$. ◆

Example 6.7.15. Using the limit comparison test.

Determine whether the integral $\int_0^1 \dfrac{dx}{\sqrt[3]{x(e^x - e^{-x})}}$ converges.

Solution. The integrand is unbounded when $x \to 0^+$. Notice that

$$\lim_{x \to 0^+} \frac{e^x - e^{-x}}{x} = \lim_{x \to 0^+} \frac{e^x + e^{-x}}{1} = e^0 + e^{-0} = 2.$$

Therefore, $f(x) = 1/\sqrt[3]{x(e^x - e^{-x})}$ behaves like $1/\sqrt[3]{x \cdot 2x}$. This shows that we should take $g(x) = x^{-2/3}$. Then

$$\frac{f(x)}{g(x)} = \frac{1/\sqrt[3]{x(e^x - e^{-x})}}{x^{-2/3}} = \frac{x^{2/3}}{\sqrt[3]{x(e^x - e^{-x})}} = \sqrt[3]{\frac{x^2}{x(e^x - e^{-x})}} = \sqrt[3]{\frac{x}{e^x - e^{-x}}},$$

and

$$\lim_{x \to 0^+} \frac{f(x)}{g(x)} = \sqrt[3]{\lim_{x \to 0^+} \frac{x}{e^x - e^{-x}}} = \sqrt[3]{\frac{1}{2}}.$$

By the Limit Comparison Test, the integrals $\int_0^1 \dfrac{dx}{\sqrt[3]{x(e^x - e^{-x})}}$ and $\int_0^1 \dfrac{dx}{x^{2/3}}$ are equiconvergent. The latter integral converges by Example 6.7.13, with $p = 2/3 < 1$. It follows that the former integral also converges. ◆

Example 6.7.16. Using the limit comparison test.

Determine whether the integral $\int_0^1 \frac{dx}{\sqrt[4]{1-x^4}}$ converges.

Solution. Here, the function $f(x) = 1/\sqrt[4]{1-x^4}$ is unbounded as $x \to 1^-$. Although f is unbounded at the right endpoint of the interval, we will apply the same strategy. This time, though, we will make a comparison with $g(x) = 1/(1-x)^p$. It is not hard to see that, using a substitution $u = 1 - x$,

$$\int_0^1 \frac{dx}{(1-x)^p} = \int_1^0 \frac{-du}{u^p} = \int_0^1 \frac{du}{u^p}$$

so the integral converges if and only if $p < 1$. Notice that $1 - x^4 = (1-x)(1+x+x^2+x^3)$ so, when $x \to 1^-$, $1 - x^4$ behaves like $1 - x$. (More precisely like $4(1-x)$, but we are interested in convergence, so constant factors are of no significance.) Therefore, $f(x)$ is comparable to $1/\sqrt[4]{1-x} = (1-x)^{-1/4}$. This suggests that we use $g(x) = 1/(1-x)^{1/4}$. Since $p = 1/4 < 1$, the integral $\int_0^1 \frac{dx}{(1-x)^{1/4}}$ converges. On the other hand,

$$\frac{\dfrac{1}{\sqrt[4]{1-x^4}}}{\dfrac{1}{(1-x)^{1/4}}} = \frac{(1-x)^{1/4}}{\sqrt[4]{1-x^4}} = \sqrt[4]{\frac{1-x}{1-x^4}} = \sqrt[4]{\frac{1-x}{(1-x)(1+x+x^2+x^3)}} = \sqrt[4]{\frac{1}{1+x+x^2+x^3}},$$

so

$$\lim_{x \to 1^-} \frac{f(x)}{g(x)} = \lim_{x \to 1^-} \sqrt[4]{\frac{1}{1+x+x^2+x^3}} = \sqrt[4]{\lim_{x \to 1^-} \frac{1}{1+x+x^2+x^3}} = \sqrt[4]{\frac{1}{4}},$$

which shows that the integrals are equiconvergent. Consequently, $\int_0^1 \frac{dx}{\sqrt[4]{1-x^4}}$ converges. ◆

Example 6.7.17. Using the limit comparison test.

Determine whether the integral $\int_0^1 x^{a-1}e^{-x}\,dx$ converges, if $a > 0$.

Solution. When $x \to 0^+$, $f(x) = x^{a-1}e^{-x} \sim x^{a-1}$, so we will use the Limit Comparison Test with $g(x) = x^{a-1}$. Now,

$$\lim_{x \to 0^+} \frac{f(x)}{g(x)} = \lim_{x \to 0^+} \frac{x^{a-1}e^{-x}}{x^{a-1}} = \lim_{x \to 0^+} e^{-x} = 1.$$

By the Limit Comparison Test, the integrals $\int_0^1 x^{a-1}e^{-x}\,dx$ and $\int_0^1 x^{a-1}\,dx$ are equiconvergent. The latter integral converges by Example 6.7.13, with $p = 1 - a < 1$, so the former integral also converges. ◆

Exercises

In Exercises 6.7.28–6.7.33 evaluate the integrals:

6.7.28. $\int_0^4 \frac{dx}{\sqrt{4-x}}$.

6.7.29. $\int_0^1 \frac{dx}{\sqrt{1-x^2}}$.

6.7.30. $\int_0^1 \ln x\,dx$.

6.7.31. $\int_0^1 \frac{dx}{(2-x)\sqrt{1-x}}$.

6.7.32.* $\int_0^{\pi/2} \ln \sin x\,dx$.

6.7.33.* $\int_0^1 \frac{x^n\,dx}{\sqrt{(1-x)(1+x)}}$, $n \in \mathbb{N}$.

In Exercises 6.7.34–6.7.41 test the integral for convergence:

6.7.34. $\int\limits_{2}^{3} \dfrac{dx}{x^2(x^3-8)^{2/3}}$.   6.7.35. $\int\limits_{0}^{\pi/2} \dfrac{\sin x \, dx}{x}$.   6.7.36. $\int\limits_{0}^{1} \dfrac{dx}{\ln x}$.

6.7.37. $\int\limits_{0}^{1} \dfrac{x^q}{\sqrt{1-x^4}} \, dx$.   6.7.38. $\int\limits_{0}^{1} \dfrac{\ln x}{1-x^2} \, dx$.   6.7.39. $\int\limits_{0}^{\pi/2} \dfrac{\ln \sin x \, dx}{\sqrt{x}}$.

6.7.40.* $\int\limits_{0}^{\pi/2} \dfrac{dx}{(\cos x)^{1/n}}$, $n > 1$. 6.7.41.* $\int_{0}^{1} x^q \ln^m x \, dx$, $q, m \in \mathbb{R}$.

6.7.42.* Prove Theorem 6.7.11.

6.7.43.* Prove Theorem 6.7.12.

6.7.44 (Cauchy's Test).* Let f be a function defined on the interval $(a, b]$, and suppose that, for any $c \in (a, b]$, f is integrable on $[c, b]$. Prove that the integral $\int_a^b f(x) \, dx$ exists if and only if, for every $\varepsilon > 0$, there exists $\delta > 0$ such that, for any $c_1, c_2 \in (a, a + \delta)$,

$$\left| \int\limits_{c_1}^{b} f(x) \, dx - \int\limits_{c_2}^{b} f(x) \, dx \right| < \varepsilon.$$

6.7.45 (Abel's Test).* Let f and g be functions defined and continuous on $[a, b']$ for all $a \leq b' < b$. Suppose that $\int_a^b f(x) \, dx$ converges and that g is a monotone decreasing, bounded function. Then the integral $\int_a^b f(x)g(x) \, dx$ converges.

6.7.46 (Dirichlet's Test).* Let f and g be functions defined and continuous on $[a, b']$ for all $a \leq b' < b$. Suppose that there exists $M > 0$ such that $|\int_a^{b'} f(x) \, dx| \leq M$ for all $a \leq b' < b$, and that g is a monotone decreasing function such that $\lim_{x \to b} g(x) = 0$. Then the integral $\int_a^b f(x)g(x) \, dx$ converges.

7

Infinite Series

We have seen that *finite* Riemann sums can be used to approximate definite integrals. However, they do just that—approximate. In order to obtain the exact result, we need the limits of these sums. Similarly, Taylor polynomials provide approximations of functions, but there is the error term r_n. Remark 4.5.3 reminds us that, as $n \to \infty$, this error goes to 0, which means that we would have an exact equality between the function and the Taylor polynomial if the latter had infinitely many terms. Such sums (with infinitely many terms) are the infinite series and they will be the focus of our study in this chapter.

7.1 Review of Infinite Series

Example 7.1.1. Compute the sum of the infinite series $\sum_{n=0}^{\infty} \frac{1}{2^n}$.

Solution. This is the sum

$$1 + \frac{1}{2} + \frac{1}{2^2} + \frac{1}{2^3} + \ldots$$

so it is a *geometric series* with *ratio* $r = 1/2$. The formula for such a sum is $a/(1-r)$, where a is the initial term in the sum. Here, $a = 1$ so the sum equals $1/(1 - 1/2) = 2$. ♦

Example 7.1.2. Compute the sum of the infinite series $\sum_{n=1}^{\infty} 9 \cdot 10^{-n}$.

Solution. Once again, this is a geometric series: $a_n = 9 \cdot 10^{-n}$ so

$$\frac{a_{n+1}}{a_n} = \frac{9 \cdot 10^{-(n+1)}}{9 \cdot 10^{-n}} = \frac{10^n}{10^{n+1}} = \frac{1}{10}.$$

The initial term $a = a_1 = 9 \cdot 10^{-1}$ and the ratio $r = 1/10$. It follows that the series converges to

$$\frac{9 \cdot 10^{-1}}{1 - 1/10} = \frac{9/10}{9/10} = 1.$$

Notice that the series can be written as

$$9 \cdot 10^{-1} + 9 \cdot 10^{-2} + 9 \cdot 10^{-3} + \ldots = 9 \cdot 0.1 + 9 \cdot 0.01 + 9 \cdot 0.001 + \ldots$$
$$= 0.9 + 0.09 + 0.009 + \ldots$$
$$= 0.999\ldots.$$

This confirms that the decimal number $0.999\ldots$, where 9 is repeated infinitely many times, is in fact 1. ♦

Example 7.1.3. Compute the sum of the infinite series $\sum_{n=1}^{\infty} \ln\left(1 + \frac{1}{n}\right)$.

Solution. A little algebra is helpful here:

$$\ln\left(1 + \frac{1}{n}\right) = \ln\frac{n+1}{n} = \ln(n+1) - \ln n.$$

Therefore, the sum of the first n terms

$$\begin{aligned}
s_n &= \sum_{i=1}^{n} \ln\left(1 + \frac{1}{i}\right) \\
&= \sum_{i=1}^{n} [\ln(i+1) - \ln i] \\
&= [\ln 2 - \ln 1] + [\ln 3 - \ln 2] + \cdots + [\ln(n+1) - \ln n] \\
&= \ln(n+1) - \ln 1 = \ln(n+1).
\end{aligned}$$

Since $\lim \ln(n+1) = \infty$, the series diverges. ♦

It is always useful if the series becomes "telescoping", by which we mean that we have all the intermediate terms (except the first and the last) cancel. In the last exercise it was not hard to spot this, but sometimes we need a little algebra.

Example 7.1.4. Compute the sum of the infinite series $\sum_{n=1}^{\infty} \frac{1}{n(n+1)}$.

Solution. We make observation that

$$\frac{1}{n(n+1)} = \frac{1}{n} - \frac{1}{n+1}.$$

Therefore,

$$\begin{aligned}
s_n &= \sum_{i=1}^{n} \frac{1}{i(i+1)} = \sum_{i=1}^{n} \left(\frac{1}{i} - \frac{1}{i+1}\right) \\
&= \left(1 - \frac{1}{2}\right) + \left(\frac{1}{2} - \frac{1}{3}\right) + \cdots + \left(\frac{1}{n} - \frac{1}{n+1}\right) = 1 - \frac{1}{n+1} \to 1.
\end{aligned}$$

We conclude that the sum of the series equals 1. ♦

Example 7.1.5. Compute the sum of the infinite series $\sum_{n=1}^{\infty} \cos nx$.

Solution. We will use the trigonometric formula

$$\sin a - \sin b = 2\cos\frac{a+b}{2}\sin\frac{a-b}{2}.$$

If we take $a = \left(n + \frac{1}{2}\right)x$ and $b = \left(n - \frac{1}{2}\right)x$, we obtain

$$\sin\left(n + \frac{1}{2}\right)x - \sin\left(n - \frac{1}{2}\right)x = 2\cos nx \sin\frac{x}{2}.$$

Therefore, assuming that x is not an integer multiple of 2π (otherwise $\cos nx = 1$ and each term in the series is 1, so the series diverges)

$$\sum_{i=1}^{n} \cos ix = \sum_{i=1}^{n} \frac{\sin\left(i + \frac{1}{2}\right)x - \sin\left(i - \frac{1}{2}\right)x}{2\sin\frac{x}{2}}$$

$$= \frac{\sin\left(1+\frac{1}{2}\right)x - \sin\left(1-\frac{1}{2}\right)x}{2\sin\frac{x}{2}} + \frac{\sin\left(2+\frac{1}{2}\right)x - \sin\left(2-\frac{1}{2}\right)x}{2\sin\frac{x}{2}} + \cdots$$

$$\cdots + \frac{\sin\left(n+\frac{1}{2}\right)x - \sin\left(n-\frac{1}{2}\right)x}{2\sin\frac{x}{2}}$$

$$= \frac{\sin\left(n+\frac{1}{2}\right)x - \sin\left(1-\frac{1}{2}\right)x}{2\sin\frac{x}{2}}.$$

Using the same strategy as in Example 2.9.7 we can now show that there is no limit as $n \to \infty$, so the given series diverges. ◆

Example 7.1.6. Find the sum of the series $\sum_{n=1}^{\infty} \frac{1}{\left(\sqrt{n} + \sqrt{n+1}\right)\sqrt{n(n+1)}}$.

Solution. Here, the trick is

$$\frac{1}{\sqrt{n}} - \frac{1}{\sqrt{n+1}} = \frac{\sqrt{n+1} - \sqrt{n}}{\sqrt{n}\sqrt{n+1}}$$

$$= \frac{\sqrt{n+1} - \sqrt{n}}{\sqrt{n}\sqrt{n+1}} \frac{\sqrt{n+1} + \sqrt{n}}{\sqrt{n+1} + \sqrt{n}}$$

$$= \frac{\left(\sqrt{n+1}\right)^2 - \left(\sqrt{n}\right)^2}{\sqrt{n(n+1)}(\sqrt{n+1} + \sqrt{n})}$$

$$= \frac{1}{\left(\sqrt{n} + \sqrt{n+1}\right)\sqrt{n(n+1)}}.$$

Therefore,

$$\sum_{i=1}^{n} \frac{1}{\left(\sqrt{i} + \sqrt{i+1}\right)\sqrt{i(i+1)}} = \sum_{i=1}^{n} \left(\frac{1}{\sqrt{i}} - \frac{1}{\sqrt{i+1}}\right)$$

$$= \left(\frac{1}{\sqrt{1}} - \frac{1}{\sqrt{2}}\right) + \left(\frac{1}{\sqrt{2}} - \frac{1}{\sqrt{3}}\right) + \cdots + \left(\frac{1}{\sqrt{n}} - \frac{1}{\sqrt{n+1}}\right)$$

$$= 1 - \frac{1}{\sqrt{n+1}} \to 1.$$

It follows that the sum of the series is 1. ◆

Example 7.1.7. Prove that the series $\sum_{n=1}^{\infty} \frac{1}{n2^n}$ converges.

Solution. We are not required to find the sum, just prove that the series converges. We will use the Ratio Test. Here $a_n = 1/(n2^n)$ so

$$\frac{a_{n+1}}{a_n} = \frac{\frac{1}{(n+1)2^{n+1}}}{\frac{1}{n2^n}} = \frac{n2^n}{(n+1)2^{n+1}} = \frac{n}{2(n+1)} \to \frac{1}{2} < 1.$$

It follows that the series converges. ◆

Example 7.1.8. Prove that the series $\sum_{n=1}^{\infty} \frac{n^2}{\left(3+\frac{1}{n}\right)^n}$ converges.

Solution. Since we only need to establish the convergence, we can use the Root Test:

$$\sqrt[n]{a_n} = \sqrt[n]{\frac{n^2}{\left(3+\frac{1}{n}\right)^n}} = \frac{\sqrt[n]{n^2}}{3+\frac{1}{n}} \to \frac{1}{3} < 1,$$

so the series converges. ◆

Example 7.1.9. Prove that the series $\sum\limits_{n=1}^{\infty} \dfrac{n^2-1}{2n^2-3n+5}$ diverges.

Solution. When establishing that a series is divergent, it makes sense to use Divergence Test. Therefore, we compute

$$\lim \frac{n^2-1}{2n^2-3n+5} = \frac{1}{2}.$$

Since the limit is not 0, we conclude that the series diverges. ◆

Example 7.1.10. Prove that the series $\sum\limits_{n=1}^{\infty} \dfrac{n-1}{2n^2-3n+5}$ diverges.

Solution. Divergence Test is of no use here, because $\lim \frac{n-1}{2n^2-3n+5} = 0$. Instead, we will compare the series with $\sum_{n=1}^{\infty} \frac{1}{n}$. We compute

$$\lim \frac{\frac{n-1}{2n^2-3n+5}}{\frac{1}{n}} = \lim \frac{n(n-1)}{2n^2-3n+5} = \lim \frac{n^2-n}{2n^2-3n+5} = \frac{1}{2}.$$

Since the limit is a real number different from 0, the Limit Comparison Test guarantees that either both series converge or they both diverge. The series $\sum_{n=1}^{\infty} \frac{1}{n}$ is the so-called Harmonic Series, known to be divergent. Therefore, $\sum_{n=1}^{\infty} \frac{n-1}{2n^2-3n+5}$ diverges as well. ◆

Example 7.1.11. Prove that the series $\sum\limits_{n=1}^{\infty} \dfrac{2n+1}{\sqrt{n^5+1}}$ converges.

Solution. This time we will compare the given series with $\sum_{n=1}^{\infty} \frac{1}{n^{3/2}}$.

$$\lim \frac{\frac{2n+1}{\sqrt{n^5+1}}}{\frac{1}{n^{3/2}}} = \lim \frac{n^{3/2}(2n+1)}{\sqrt{n^5+1}} = \lim \frac{n^{5/2}\left(2+\frac{1}{n}\right)}{n^{5/2}\sqrt{1+\frac{1}{n^5}}} = 2.$$

The fact that this limit exists and is not equal to 0 implies that either both series converge or they both diverge. However, the series $\sum_{n=1}^{\infty} \frac{1}{n^{3/2}}$ is of the form $\sum_{n=1}^{\infty} \frac{1}{n^p}$, and such a series converges if and only if $p > 1$. Here $p = 3/2$, so $\sum_{n=1}^{\infty} \frac{1}{n^{3/2}}$ is a convergent series, and it follows that the same is true of $\sum_{n=1}^{\infty} \frac{2n+1}{\sqrt{n^5+1}}$. ◆

Exercises

In Exercises 7.1.1–7.1.3 find the sum of the series:

7.1.1. $1 - \dfrac{1}{3} + \dfrac{1}{9} - \dfrac{1}{27} + \dots$ 7.1.2. $\dfrac{1}{2} + \dfrac{3}{2^2} + \dfrac{5}{2^3} + \dfrac{7}{2^4} + \dots$

7.1.3. $\dfrac{1}{1\cdot 4} + \dfrac{1}{4\cdot 7} + \dfrac{1}{7\cdot 10} + \dots$

In Exercises 7.1.4–7.1.9 investigate the convergence of the series $\sum\limits_{n=1}^{\infty} a_n$.

7.1.4. $a_n = \dfrac{1000^n}{n!}$.   7.1.5. $a_n = \dfrac{(n!)^2}{(2n)!}$.   7.1.6. $a_n = \dfrac{n!}{n^n}$.

7.1.7. $a_n = \left(\dfrac{n-1}{n+1}\right)^{n(n-1)}$.   7.1.8. $a_n = \dfrac{(n!)^2}{2^{n^2}}$.   7.1.9. $a_n = \dfrac{4\cdot 7\cdot 10\cdots\cdots(3n+1)}{2\cdot 6\cdot 10\cdots\cdots(4n-2)}$.

7.2 Definition of a Series

Let us revisit Example 7.1.1 where we have stated that a geometric series with ratio r and the initial term a has the sum $a/(1-r)$. Using this formula we have concluded that the infinite sum

$$1 + \frac{1}{2} + \frac{1}{2^2} + \frac{1}{2^3} + \cdots$$

equals 2. Now, where did the formula come from? What does it mean that the sum is 2?

If we calculate the sum of the first 7 terms

$$1 + \frac{1}{2} + \frac{1}{2^2} + \frac{1}{2^3} + \cdots + \frac{1}{2^6}$$

we obtain 1.984375000. We write $s_6 = 1.984375000$, and we call this number a *partial sum*. Here is a table with some calculations:

n	6	9	12	16
s_n	1.984375000	1.998046875	1.999755859	1.999984741

n	20	24	28
s_n	1.999999046	1.999999940	1.999999996

So, 2 is the limit of the sequence $\{s_n\}$. In general, we have the following definition.

Definition 7.2.1. Let $\{a_n\}_{n\in\mathbb{N}_0}$ be a sequence of real numbers, and define the sequence $\{s_n\}_{n\in\mathbb{N}_0}$ by $s_0 = a_0$ and $s_n = s_{n-1} + a_n$ for $n \in \mathbb{N}$. We say that the **infinite series** $\sum_{n=0}^{\infty} a_n$ **is convergent** if the sequence $\{s_n\}$ of partial sums is convergent. If $\lim s_n = s$ we say that the series **converges** to s and we call this number the **sum of the series** $\sum_{n=0}^{\infty} a_n$. If the sequence $\{s_n\}$ diverges, we say that the series $\sum_{n=0}^{\infty} a_n$ is **divergent**.

In the example above, $s_n = 1 + 1/2 + 1/2^2 + \cdots + 1/2^n$. Using Theorem 2.6.3,

$$s_n = \frac{(1/2)^{n+1} - 1}{1/2 - 1} = -2\left(\frac{1}{2^{n+1}} - 1\right) \to -2(-1) = 2, \quad \text{as } n \to \infty.$$

In fact, Theorem 2.6.3 allows us to find the sum of any geometric series. A geometric series is characterized by the fact that, for any $n \geq 0$, $a_{n+1}/a_n = r$. A consequence of this relation is that, if we denote $a_0 = a$, then $a_1 = a_0 r = ar$, $a_2 = a_1 r = (ar)r = ar^2$, etc. We see that $a_n = ar^n$, and

$$s_n = a + ar + ar^2 + \ldots ar^n = a(1 + r + r^2 + \cdots + r^n) = a\frac{r^{n+1} - 1}{r - 1}. \tag{7.1}$$

When calculating $\lim s_n$ we encounter 3 possibilities: $|r| < 1$, $|r| = 1$, and $|r| > 1$. When $|r| > 1$, we know that $\lim r^{n+1} = \infty$, so the series diverges. When $|r| = 1$, the sequence of partial sums is either $s_n = na$ (if $r = 1$) or $a, 0, a, 0, a, 0, \ldots$ (if $r = -1$), neither of which converges. Finally, if $|r| < 1$, $\lim r^{n+1} = 0$, so (7.1) shows that

$$\lim s_n = a\frac{-1}{r - 1} = \frac{a}{1 - r}.$$

In the eighteenth century, the distinction between convergent and divergent series was blurred. It was Fourier (page 239) who defined a convergent series first in 1811, using partial sums. Yet, he still thought that the series $1 - 1 + 1 - 1 + \dots$ had sum $1/2$.

By definition, the convergence of a series is really the convergence of the *sequence* $\{s_n\}$, and the sum of the series is simply the limit of $\{s_n\}$. This allows us to derive some results about infinite series, based on the properties of sequences.

Theorem 7.2.2. *Let $k \in \mathbb{N}$ and let $\sum_{n=k}^{\infty} a_n$ be the series obtained by deleting the first k terms in the series $\sum_{n=0}^{\infty} a_n$. Then these series either both converge or both diverge.*

Proof. If we denote by $\{s_n\}$ (respectively, $\{t_n\}$) the partial sums of $\sum_{n=0}^{\infty} a_n$ (respectively, $\sum_{n=k}^{\infty} a_n$), then for any $n \in \mathbb{N}$,

$$
\begin{aligned}
t_n &= a_k + a_{k+1} + \dots + a_{k+n-1} \\
&= s_{k+n-1} - (a_0 + a_1 + \dots + a_{k-1}) \qquad (7.2) \\
&= s_{k+n-1} - s_{k-1}.
\end{aligned}
$$

It follows that (for a fixed k) $\lim t_n = \lim s_{k+n-1} - s_{k-1}$, so either both limits exist or neither one does. $\square$

The computations performed in the proof of Theorem 7.2.2 can be used to derive another result.

Theorem 7.2.3. *Let $k \in \mathbb{N}$ and let $T_k = \sum_{n=k}^{\infty} a_n$ be the sum of the series obtained by deleting the first k terms in the convergent series $\sum_{n=0}^{\infty} a_n$. Then $\lim_{k \to \infty} T_k = 0$.*

We call T_k the kth *tail* of the series $\sum_{n=0}^{\infty} a_n$. This theorem is usually stated (in a rather informal way) as "the tail of a convergent series goes to 0".

Proof. The assumption that the series $\sum_{n=0}^{\infty} a_n$ converges means, by definition, that the sequence $\{s_n\}$ converges, say $\lim s_n = s$. Theorem 7.2.2 guarantees that the series $\sum_{n=k}^{\infty} a_n$ converges as well, so the sequence of its partial sums $\{t_n\}$ converges to T_k. If we let $n \to \infty$ in (7.2), we see that $T_k = s - s_{k-1}$ for any fixed $k \in \mathbb{N}$. If we now let $k \to \infty$ in $T_k = s - s_{k-1}$ we see that $T_k \to 0$. $\square$

Another result along these lines can be used to detect the divergence of a series.

Theorem 7.2.4 (The Divergence Test). *If $\sum_{n=0}^{\infty} a_n$ is a convergent series then $\lim a_n = 0$.*

The reason for the name is the contrapositive: If the sequence $\{a_n\}$ does not converge to 0, then the series $\sum_{n=0}^{\infty} a_n$ diverges.

Proof. If we denote by s_n the nth partial sum of the given series, then

$$
s_{n+1} = a_0 + a_1 + a_2 + \dots + a_{n+1} = s_n + a_{n+1}.
$$

If we now let $n \to \infty$, then both s_{n+1} and s_n have the same limit. Therefore, $\lim a_n = 0$. $\square$

We illustrate the Divergence Test with an example.

Example 7.2.5. Using the Divergence Test.

Determine whether the infinite series $\sum_{n=1}^{\infty} \dfrac{n}{n+1}$ converges.

Solution. Here $a_n = n/(n+1)$ and $\lim a_n = 1 \neq 0$. By the Divergence Test, the series diverges. ◆

Fourier (page 239) was among the first to stress that a necessary condition for convergence is that the terms go to 0. In 1812, Gauss (page 196) wrote one of the first serious works on series [50]. In his earlier works he had considered a series convergent if its terms go to 0, but in 1812 he realized that this condition is only necessary. It appears that Bolzano was the first one to have a clear notion of a convergence, but since his work remained unpublished the accolades go to Cauchy and his *Cours d'Analyse*.

As its name suggests, the Divergence Test can be used only to detect divergence. Namely, if $\lim a_n = 0$ we cannot conclude whether the series converges or diverges. For example, when $a_n = 1/2^n$ (which satisfies the condition $\lim a_n = 0$) we have seen that the series converges. However, as the next example will demonstrate, there are divergent series that satisfy the same condition.

Example 7.2.6. A divergent series $\sum a_n$ such that $\lim a_n = 0$.

Determine whether the infinite series $\sum\limits_{n=1}^{\infty} \dfrac{1}{n}$ converges.

This series is known as the **Harmonic series**, and it is a divergent series. Unfortunately, the Divergence Test is not of any help. Namely, $a_n = 1/n \to 0$, and the test works only when the limit is *not* equal to 0. However, we have established in Example 2.6.5 that the sequence

$$a_n = 1 + \frac{1}{2} + \frac{1}{3} + \cdots + \frac{1}{n}$$

is not a Cauchy sequence. By Theorem 2.6.6, it is not convergent, so the Harmonic series diverges. ♦

Remark 7.2.7. The name *harmonic* series is related to the concept of the *harmonic mean*. A number c is called the harmonic mean of the numbers a and b if

$$\frac{1}{c} = \frac{1}{2}\left(\frac{1}{a} + \frac{1}{b}\right). \tag{7.3}$$

Notice that, for $n \geq 2$, every term a_n in the harmonic series is the harmonic mean of the neighboring terms a_{n-1} and a_{n+1}. Indeed, with $c = 1/n$, $a = 1/(n-1)$ and $b = 1/(n+1)$ the equality (7.3) becomes

$$n = \frac{1}{2}[(n-1) + (n+1)]$$

which is easy to verify.

The fact that the harmonic series diverges was first proven in the fourteenth century by Nicole Oresme (c.1320–1382), a philosopher and mathematician in the geographic area of today's France. By the seventeenth century, this was forgotten and a fresh proof was given by Italian mathematician Pietro Mengoli (1626–1686) and, independently, 40 years later by Jacob Bernoulli.

As we have seen in Example 7.2.6, it is useful to rely on the Cauchy's Test for sequences. Here, we rephrase it for series. We will leave the proof as an exercise.

Theorem 7.2.8. *The series $\sum_{n=0}^{\infty} a_n$ converges if and only if for every $\varepsilon > 0$, there exists $N \in \mathbb{N}$ such that, if $m \geq n \geq N$ then $\left|\sum_{k=n+1}^{m} a_k\right| < \varepsilon$.*

The following result is a direct consequence of the definition of a convergent series and Theorem 2.3.4.

Theorem 7.2.9. *Let $\sum_{n=0}^{\infty} a_n$, $\sum_{n=0}^{\infty} b_n$ be convergent series and let $\alpha \in \mathbb{R}$. Then the series $\sum_{n=0}^{\infty} \alpha a_n$ and $\sum_{n=0}^{\infty} (a_n + b_n)$ are also convergent and:*

(a) $\sum_{n=0}^{\infty} (\alpha a_n) = \alpha \sum_{n=0}^{\infty} a_n$;

(b) $\sum_{n=0}^{\infty} (a_n + b_n) = \sum_{n=0}^{\infty} a_n + \sum_{n=0}^{\infty} b_n$.

The first successful summation of a series on record is due to Archimedes around 225 BC. In order to compute the area under a parabola his method required finding the sum of a geometric series. The first use of the symbol $\sum$ was by Euler in [40], although it received little attention. The modern notation started developing in the early nineteenth century in the writing of Cauchy, Fourier, etc.

CARL FRIEDRICH GAUSS (1777–1855) was a German mathematician. He is considered by many to have been the greatest mathematician ever, and is sometimes referred to as the Prince of Mathematicians. He showed a mathematical talent at an early age. A stipend from the Duke of Brunswick–Wolfenbüttel allowed him to attend Brunswick Collegium Carolinum from 1792. He continued his studies at the University of Göttingen in 1795, and got his doctorate from the University of Helmstedt in 1798. The topic of his dissertation was the fundamental theorem of algebra. For the next 10 years Gauss lived in Brunswick off his stipend, and devoted himself to research. *Disquisitiones Arithmeticae* (Investigations in Arithmetic) in 1801 established the number theory as a discipline and introduced the modular arithmetic, including the symbol $\equiv$. In the last section he presented the construction of a regular 17-gon using a compass and straight edge. That same year he successfully predicted the orbital position of Ceres, the largest object in the asteroid belt that lies between the orbits of Mars and Jupiter, using the least square approximation method. In 1807 he became the director of the Göttingen observatory. Soon afterwards, he published a two volume treatise on the motion of celestial bodies, the first part focusing on mathematics. In this period he published his work on series, approximate integration, and statistics. After 1828 he added to his list of interests what we now call differential geometry, and soon afterwards the study of the terrestrial magnetism. He specified the location for the magnetic South pole and built a primitive telegraph which could send messages over a distance of 5000 ft. By 1840 he started working on optics. Throughout his illustrious career he made numerous important discoveries, but published only a few of them. The vast majority remained in his diaries, showing that he had obtained some results decades before others. For example, he calculated the coefficients of a Fourier series 2 years before Fourier (page 239). He discussed the existence of non-Euclidean geometries at least 25 years before Lobachevsky. His reluctance to publish is likely due to his perfectionism. He wanted to present proofs completely void of the intuition behind them.

Exercises

In Exercises 7.2.1–7.2.12 find the sum of the series or determine that it diverges:

7.2.1. $\displaystyle\sum_{n=1}^{\infty} \frac{1}{(3n-2)(3n+1)}$.

7.2.2. $\displaystyle\sum_{n=1}^{\infty} \left(\frac{1}{2^n} + \frac{1}{3^n} \right)$.

7.2.3. $\displaystyle\sum_{n=1}^{\infty} \left(\sqrt{n+1} - \sqrt{n} \right)$.

7.2.4. $\displaystyle\sum_{n=1}^{\infty} \frac{n}{2n+3}$.

7.2.5. $\displaystyle\sum_{n=1}^{\infty} \frac{1}{n^2+5n+6}$.

7.2.6.* $\displaystyle\sum_{n=1}^{\infty} \sin n$.

7.2.7.* $\displaystyle\sum_{n=1}^{\infty} \frac{1}{n(n+1)(n+2)}$.

7.2.8. $\displaystyle\sum_{n=1}^{\infty} n \ln \left(1 + \frac{1}{n} \right)$.

7.2.9. $\displaystyle\sum_{n=1}^{\infty} \sin \frac{\pi n!}{5040}$.

7.2.10. $\displaystyle\sum_{n=1}^{\infty} \frac{n}{(2n+1)!!}$.

7.2.11.* $\displaystyle\sum_{n=1}^{\infty} (-1)^{n+1} \frac{2n+1}{n(n+1)}$.

7.2.12. $1 - \dfrac{1}{2} + \dfrac{1}{4} - \dfrac{1}{8} + \dfrac{1}{16} - \dots$.

7.2.13. Let s_n denote the nth partial sum of the series $\sum_{n=1}^{\infty} a_n$, for each $n \in \mathbb{N}$. Suppose that, for each $n \in \mathbb{N}$, $s_n = (n+1)/n$. Find a_n.

7.2.14. Suppose that $\{a_n\}$ is a decreasing sequence of positive numbers. Prove that $\sum_{n=0}^{\infty} a_n$ converges if and only if $\sum_{n=0}^{\infty} 2^n a_{2^n}$ converges.

7.2.15. Prove Theorem 7.2.8.

7.2.16. Prove Theorem 7.2.9.

7.2.17. Let $\sum_{n=1}^{\infty} a_n$ and $\sum_{n=1}^{\infty} b_n$ be two divergent series with non-negative terms. Determine whether the series converges: (a) $\sum_{n=1}^{\infty} \min\{a_n, b_n\}$; (b) $\sum_{n=1}^{\infty} \max\{a_n, b_n\}$.

In Exercises 7.2.18–7.2.21 use Cauchy's Test to prove that the series converges or that it diverges:

7.2.18. $\displaystyle\sum_{n=1}^{\infty} \frac{\cos(x^n)}{n^2}$.   7.2.19.* $\displaystyle\sum_{n=1}^{\infty} \frac{1}{\sqrt{n(n+1)}}$.   7.2.20.* $1 + \dfrac{1}{2} - \dfrac{1}{3} + \dfrac{1}{4} + \dfrac{1}{5} - \dfrac{1}{6} + \ldots$.

7.2.21.* $\dfrac{\cos x - \cos 2x}{1} + \dfrac{\cos 2x - \cos 3x}{2} + \cdots + \dfrac{\cos nx - \cos(n+1)x}{n} + \ldots$.

7.2.22. Let $\{a_n\}$ be a sequence of real numbers, let $\{n_k\}$ be an increasing sequence of positive integers, and let $A_k = a_{n_k} + a_{n_k+1} + \cdots + a_{n_{k+1}-1}$, for each $k \in \mathbb{N}$. Prove that, if $\sum_{n=1}^{\infty} a_n$ converges then $\sum_{k=1}^{\infty} A_k$ converges. Give an example to show that the converse is not true.

7.2.23. Suppose that the sequence $\{a_n\}$ satisfies the following condition: For every $p > 0$, $\lim(a_n + a_{n+1} + \cdots + a_{n+p}) = 0$. Prove or disprove: the series $\sum_{n=1}^{\infty} a_n$ converges.

7.3   Series with Positive Terms

One of the powerful tools to detect the convergence of a sequence was the Monotone Convergence Theorem (Theorem 2.4.7). We can apply this result to the sequence $\{s_n\}$ if it is monotone increasing, i.e. $s_{n+1} \geq s_n$, for all $n \in \mathbb{N}$. This condition implies that $a_n = s_{n+1} - s_n \geq 0$, so it is natural to consider, for the moment at least, those series that have non-negative terms. From the Monotone Convergence Theorem we immediately obtain the following result.

Theorem 7.3.1. *Let $a_n \geq 0$ for all $n \geq 0$. Then the series $\sum_{n=0}^{\infty} a_n$ converges if and only if there exists a real number M such that $s_n \leq M$, for all $n \geq 0$.*

Theorem 7.3.1 will allow us to develop some more sophisticated tests. As our first applications, we will establish the Comparison Test. Here, the idea is to compare $\{s_n\}$ and the sequence of partial sums of another series, assuming that we know whether the other series converges or diverges.

Theorem 7.3.2 (Comparison Test). *Let $\{a_n\}$, $\{b_n\}$ be two sequences of positive numbers, and suppose that there exists $N \in \mathbb{N}$ such that $a_n \leq b_n$ for all $n \geq N$. Then:*

(a) the convergence of $\sum_{n=0}^{\infty} b_n$ implies the convergence of $\sum_{n=0}^{\infty} a_n$;

(b) the divergence of $\sum_{n=0}^{\infty} a_n$ implies the divergence of $\sum_{n=0}^{\infty} b_n$.

Proof. Assertion (b) is just the contrapositive of (a), so we will prove only (a). To that end, let us denote by $\{s_n\}$ (respectively, $\{t_n\}$) the partial sums of $\sum_{n=0}^{\infty} a_n$ (respectively, $\sum_{n=0}^{\infty} b_n$). Let $M = \max\{s_n - t_n : 1 \leq n < N\}$. Notice that, if $n \geq N$,

$$s_n - t_n = s_{N-1} + \sum_{k=N}^{n} a_k - \left(t_{N-1} + \sum_{k=N}^{n} b_k\right) = s_{N-1} - t_{N-1} + \sum_{k=N}^{n} (a_k - b_k)$$

$$\leq M + \sum_{k=N}^{n} (a_k - b_k).$$

The inequality $a_n \leq b_n$ implies that $s_n \leq t_n + M$ for all $n \in \mathbb{N}$. The sequence $\{t_n\}$ is convergent, hence bounded, so $\{s_n\}$ is bounded as well. By Theorem 7.3.1, the series $\sum_{n=0}^{\infty} a_n$ is convergent. $\square$

Example 7.3.3. Using the comparison test.

Determine whether the infinite series $\sum_{n=1}^{\infty} \frac{1}{1+a^n}$, $a > 0$, converges.

Solution. If $a < 1$, $\lim 1/(1 + a^n) = 1$, and if $a = 1$, $\lim 1/(1 + a^n) = 1/2$. In both cases, the series diverges by the Divergence Test. If $a > 1$, we use the inequality

$$\frac{1}{1+a^n} < \frac{1}{a^n}.$$

The series $\sum_{n=1}^{\infty} 1/a^n$ is a geometric series with $r = 1/a < 1$, so it converges. Now the convergence of $\sum_{n=1}^{\infty} 1/(1 + a^n)$ follows from Theorem 7.3.2. ◆

Although Theorem 7.3.2 can be used to determine the convergence or divergence of a series, it requires some algebraic skills: the inequality $a_n \leq b_n$ needs to be proved. The following result is sometimes easier to apply.

Theorem 7.3.4 (Limit Comparison Test). *Let $\{a_n\}$, $\{b_n\}$ be two sequences of strictly positive numbers. Then:*

(a) if $\lim a_n/b_n$ exists, the convergence of $\sum_{n=0}^{\infty} b_n$ implies the convergence of $\sum_{n=0}^{\infty} a_n$;

(b) if $\lim a_n/b_n$ exists and it is not zero, or if $\lim a_n/b_n = \infty$, the divergence of $\sum_{n=0}^{\infty} b_n$ implies the divergence of $\sum_{n=0}^{\infty} a_n$.

Proof. (a) If the sequence $\{a_n/b_n\}$ is convergent, it is bounded, so there exists $M > 0$ such that $a_n \leq M b_n$. If $\sum_{n=0}^{\infty} b_n$ converges, then by Theorem 7.2.9 so does $\sum_{n=0}^{\infty} M b_n$, and the assertion (a) follows from Theorem 7.3.2.

(b) By Theorem 2.3.9, the sequence $\{b_n/a_n\}$ converges. Applying part (a) we see that the convergence of $\sum_{n=0}^{\infty} a_n$ implies the convergence of $\sum_{n=0}^{\infty} b_n$. By taking the contrapositive we obtain the assertion (b). □

Remark 7.3.5. The assumption that $\{a_n\}$, $\{b_n\}$ are *strictly* positive is made so that the quotient a_n/b_n would be defined. It can be relaxed by requiring that there exists $N \in \mathbb{N}$ such that $a_n, b_n > 0$, for $n \geq N$. We leave the proof to the reader.

Example 7.3.6. Using the limit comparison test.

Determine whether the infinite series $\sum_{n=1}^{\infty} \ln\left(1 + \frac{1}{n}\right)$ converges.

Solution. Here $a_n = \ln(1 + 1/n)$ and we will take $b_n = 1/n$. Then

$$\frac{a_n}{b_n} = \frac{\ln\left(1 + \frac{1}{n}\right)}{\frac{1}{n}} = n \ln\left(1 + \frac{1}{n}\right) = \ln\left(1 + \frac{1}{n}\right)^n \to \ln e = 1.$$

By Theorem 7.3.4 (b), having in mind that the Harmonic series diverges, we conclude that $\sum_{n=1}^{\infty} a_n$ diverges. ◆

> The Comparison Test (Theorem 7.3.2) was used by Gauss in 1812 in [50], but the serious treatment, together with Theorem 7.3.4 is due to Cauchy in *Cours d'Analyse*.

Another useful test for the convergence of a series is based on a very simple geometric principle. Suppose that $n \in \mathbb{N}$, and that f is a decreasing, continuous function on the interval $[n, n+1]$. Let R', R'' be the rectangles with one side the interval $[n, n+1]$, and the vertical side of length $f(n)$ for R', and $f(n+1)$ for R''. Finally, let R be the region under the graph of f, above the interval $[n, n+1]$ (see Figure 7.1). Then their areas satisfy the

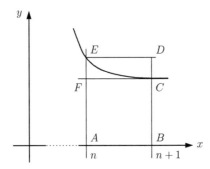

Figure 7.1: $Area(ABDE) \geq Area(ABCE) \geq Area(ABCF)$.

inequality

$$Area(R') \geq Area(R) \geq Area(R'').$$

Equivalently,

$$f(n) \geq \int_n^{n+1} f(t)\, dt \geq f(n+1) \geq 0. \tag{7.4}$$

This inequality can be used to prove the following result, that is often called the Integral Test, or the Cauchy–MacLaurin Test.

Theorem 7.3.7 (The Integral Test). *Let f be a decreasing, continuous, positive function on $[1, +\infty)$, and let $a_n = f(n)$, for every $n \in \mathbb{N}$. Then the series $\sum_{n=1}^{\infty} a_n$ converges if and only if the infinite integral $\int_1^{\infty} f(t)\, dt$ converges.*

Proof. If we denote by s_n the nth partial sum of the series $\sum_{n=1}^{\infty} a_n$, and if we sum the inequalities (7.4) for $n = 1, 2, \ldots, N$, we obtain

$$s_N \geq \int_1^{N+1} f(t)\, dt \geq s_{N+1} - a_1.$$

If the integral $\int_1^{\infty} f(t)\, dt$ converges, then the right hand inequality shows that the sequence $\{s_N\}$ is bounded and Theorem 7.3.1 implies that the series $\sum_{n=1}^{\infty} a_n$ converges.

On the other hand, suppose that the integral $\int_1^{\infty} f(t)\, dt$ diverges, and let $F(x) = \int_1^x f(t)\, dt$. We will show that the series $\sum_{n=1}^{\infty} a_n$ diverges. Let $M > 0$. Then there exists $x_0 \geq 1$ such that $F(x) = \int_1^x f(t)\, dt$ satisfies $F(x_0) \geq M$. Let $N = \lfloor x_0 \rfloor$. The inequality $N + 1 \geq x_0$ and the positivity of f imply that $F(N + 1) \geq F(x_0) \geq M$. It follows that $s_N \geq M$ and, since M was arbitrary, we have that the series $\sum_{n=1}^{\infty} a_n$ diverges. $\qquad\square$

> An early form of this test was known to Indian mathematician and astronomer Madhava (c. 1350–c. 1425). In Europe, it was later developed by Scottish mathematician Colin MacLaurin (1698–1746) and Cauchy.

As the first application of the Integral Test, we will consider the class of series that are often referred to as the "p-series".

Example 7.3.8. Convergence of p-series.

Determine whether the infinite series $\sum_{n=1}^{\infty} \dfrac{1}{n^p}$ converges.

Solution. Let $f(x) = 1/x^p$. This is a continuous function for $x \geq 1$, and it is positive.

Since $f'(x) = -px^{-p-1}$, we see that, for $p > 0$ (and $x \geq 1$), the derivative $f'(x) < 0$, so the function f is decreasing. Thus, we consider the integral $\int_1^\infty 1/x^p \, dx$. By Exercise 6.7.7 the integral converges if and only if $p > 1$. The case $p = 1$ gives the Harmonic series $\sum_{n=1}^\infty 1/n$ that diverges. Finally, if $p \leq 0$, then $\lim 1/n^p \neq 0$ so the series diverges. We conclude that the p–series converges if and only if $p > 1$. ♦

Example 7.3.9. Applying the integral test.

Determine whether the infinite series $\sum_{n=2}^\infty \dfrac{1}{n \ln n}$ converges.

Solution. We take $f(x) = 1/(x \ln x)$, and we notice that, for $x \geq 2$, f is continuous and positive. Further,

$$f'(x) = \frac{-1}{(x \ln x)^2} \left(\ln x + x \frac{1}{x} \right) = -\frac{1 + \ln x}{(x \ln x)^2} < 0,$$

so f is decreasing. Finally, in order to find an antiderivative of $1/(x \ln x)$ we use the substitution $u = \ln x$. Then $du = 1/x \, dx$ so

$$\int \frac{dx}{x \ln x} = \int \frac{du}{u} = \ln u + C = \ln(\ln x) + C.$$

Therefore,

$$\int_2^\infty \frac{1}{x \ln x} \, dx = \lim_{b \to +\infty} \ln(\ln x) \Big|_2^b = \lim_{b \to +\infty} \ln(\ln b) - \ln(\ln 2) = +\infty,$$

and the series $\sum_{n=2}^\infty \dfrac{1}{n \ln n}$ diverges. ♦

Now that we know the behavior of the p-series and the geometric series, we are in a better position to apply the Comparison Tests.

Example 7.3.10. Applying the limit comparison test.

Determine whether the infinite series $\sum_{n=1}^\infty \dfrac{1}{\sqrt{n(n^2+1)}}$ converges.

Solution. For large n, $n^2 + 1$ is "close to" n^2. (This is vague, but we are just looking for the series to compare with.) Therefore, $\sqrt{n(n^2 + 1)}$ is close to $\sqrt{n(n^2)} = n^{3/2}$. We will use the series $\sum_{n=1}^\infty \frac{1}{n^{3/2}}$.

Proof. Given $a_n = \dfrac{1}{\sqrt{n(n^2+1)}}$ we define $b_n = \dfrac{1}{n^{3/2}}$. The series $\sum_{n=1}^\infty \frac{1}{n^{3/2}}$ converges by Example 7.3.8. Since

$$\lim \frac{a_n}{b_n} = \lim \frac{\frac{1}{\sqrt{n(n^2+1)}}}{\frac{1}{n^{3/2}}} = \lim \frac{n^{3/2}}{\sqrt{n(n^2 + 1)}} = \sqrt{\lim \frac{n^3}{n(n^2 + 1)}} = \sqrt{1} = 1,$$

Theorem 7.3.4 implies the convergence of $\sum_{n=1}^\infty \frac{1}{\sqrt{n(n^2+1)}}$. ♦

Example 7.3.11. Applying the limit comparison test.

Determine whether the series $\sum_{n=1}^\infty \dfrac{1}{(\ln n)^{\ln n}}$ converges.

Solution. First we need some algebra. We would like to compare this series with the one where the term is $1/n^s$. It follows that we are comparing $(\ln n)^{\ln n}$ with n^s or, if we take logarithms,

$$\ln\left[(\ln n)^{\ln n}\right] \quad \text{with} \quad \ln n^s.$$

The first expression equals $\ln n \ln(\ln n)$ and the second one equals $s \ln n$, so after dividing by $\ln n$ we are left with $\ln(\ln n)$ and s. Clearly, the left side "wins", regardless of how large s is. However, we have taken reciprocals at the beginning of our investigation, so $1/(\ln n)^{\ln n}$ is actually smaller than $1/n^s$. So we can take, for example, $s = 2$.

Proof. Given $a_n = 1/(\ln n)^{\ln n}$ we define $b_n = 1/n^2$. The series $\sum_{n=1}^{\infty} 1/n^2$ converges by Example 7.3.8. Now

$$\lim \frac{a_n}{b_n} = \lim \frac{\frac{1}{(\ln n)^{\ln n}}}{\frac{1}{n^2}} = \lim \frac{n^2}{(\ln n)^{\ln n}} = \lim \frac{e^{\ln n^2}}{e^{\ln((\ln n)^{\ln n})}}$$

$$= \lim e^{2\ln n - \ln n \ln(\ln n)} = \lim e^{\ln n(2 - \ln(\ln n))} = 0$$

because $\lim \ln n = +\infty$ and $\lim(2 - \ln(\ln n)) = -\infty$. Theorem 7.3.4 implies the convergence of $\sum_{n=1}^{\infty} \dfrac{1}{(\ln n)^{\ln n}}$. ♦

JEAN le ROND d'ALEMBERT (1717–1783) was a French mathematician and a co-editor of *Encyclopédie*. His mother was a former nun that left the church and found her place in Paris, in the world of political intrigues and love affairs. D'Alembert was a product of such an affair with an artillery officer. When only a few days old, he was left at the steps of the church of St. Jean Le Rond. According to the tradition, he was named after the saint. His father did not acknowledge him, but arranged for d'Alembert to be in care of a woman, in whose home he stayed until well into middle age. He studied at the Collège des Quatre Nations, where he changed his name to Jean d'Alembert. He received a good mathematical training there and got acquainted with physics ideas of Descartes. Upon graduating in 1735, he spent some time studying law and medicine, but eventually turned to mathematics. He presented his first paper to the Paris Academy of Science in 1739, and by 1741 he was a member of the Academy. He was respected but he made many enemies through his unwillingness to accept other mathematicians' ideas when they clashed with his own, even when he was wrong. His theories about real world phenomena were mathematically impeccable, but often based on unrealistic assumptions that were not supported by the physical evidence. So, he was elected into the French Academy in 1754, and from 1772 till his death, he served as the secretary. On the other hand, he could not publish his results in Paris, and after a dispute with Euler, not in Berlin either. The only way was to publish his collected works, which appeared in 8 volumes, between 1761 and 1780. Outside of mathematics, he became a popular member of the social scene around 1746, attending several different "salons" (literary and artistic circles). As an atheist, he was buried in an unmarked grave. Some of his important contributions came through the work on the 28-volume Encyclopedia, for which he wrote many articles. For example, he defined the derivative as the limit of a quotient of increments, and formulated the Ratio Test that we will present in the next section. In [23] he pioneered the study of the wave equation (see page 259).

Exercises

In Exercises 7.3.1–7.3.12 use Comparison Test or Limit Comparison Test to prove that the series converges or diverges:

7.3.1. $\displaystyle\sum_{n=1}^{\infty} \left(1 - \cos\frac{1}{n}\right).$

7.3.2. $\displaystyle\sum_{n=1}^{\infty} (\sqrt[n]{a} - 1),\ a > 1.$

7.3.3. $\displaystyle\sum_{n=2}^{\infty} \frac{1}{\ln(n!)}.$

7.3.4. $\displaystyle\sum_{n=1}^{\infty} \frac{n + \sqrt{n}}{2n^3 - 1}.$

7.3.5. $\displaystyle\sum_{n=1}^{\infty} \frac{\ln n}{n^2 + 1}.$

7.3.6. $\displaystyle\sum_{n=1}^{\infty} e^{-n^2}.$

7.3.7. $\displaystyle\sum_{n=1}^{\infty} \frac{n!}{n^n}.$

7.3.8. $\displaystyle\sum_{n=1}^{\infty} \sqrt{n \arctan\frac{1}{n^3}}.$

7.3.9. $\displaystyle\sum_{n=1}^{\infty} \frac{(n!)^2}{(2n)!}.$

7.3.10. $\displaystyle\sum_{n=1}^{\infty} 2^n \sin\left(\frac{x}{3^n}\right).$

7.3.11. $\displaystyle\sum_{n=1}^{\infty} \frac{n^{n-1}}{(2n^2 + n + 1)^{(n+1)/2}}.$

7.3.12. $\dfrac{1}{3} + \left(\dfrac{2}{3}\right)^2 + \left(\dfrac{1}{3}\right)^3 + \left(\dfrac{2}{3}\right)^4 + \left(\dfrac{1}{3}\right)^5 + \left(\dfrac{2}{3}\right)^6 + \dots.$

7.3.13. Prove Theorem 7.3.1.

7.3.14. Let $a_n \geq 0$ for all $n \in \mathbb{N}$, and suppose that $\sum_{n=1}^{\infty} a_n$ converges. Prove that $\sum_{n=1}^{\infty} a_n^2$ converges.

7.3.15.* Let $\sum_{n=1}^{\infty} a_n$ and $\sum_{n=1}^{\infty} b_n$ be series of positive terms satisfying $\dfrac{a_{n+1}}{a_n} \leq \dfrac{b_{n+1}}{b_n}$. Prove that if $\sum_{n=1}^{\infty} b_n$ converges then $\sum_{n=1}^{\infty} a_n$ converges.

7.3.16. Suppose that $\sum_{n=1}^{\infty} a_n$ is a divergent series of positive numbers. Prove that the series $\sum_{n=1}^{\infty} \dfrac{a_n}{1 + a_n}$ diverges as well.

7.3.17. Suppose that $a_n = \begin{cases} \frac{1}{n}, & \text{if } n = 1, 4, 9, \dots, m^2, \dots \\ \frac{1}{n^2}, & \text{otherwise.} \end{cases}$ Determine whether the series $\sum_{n=1}^{\infty} a_n$ converges.

7.3.18.* Let $\sum_{n=1}^{\infty} a_n$ be a convergent series of nonnegative terms. Prove or disprove: (a) $\liminf na_n = 0$; (b) $\limsup na_n = 0$.

In Exercises 7.3.19–7.3.25 use Integral Test to prove that the series converges or diverges:

7.3.19. $\displaystyle\sum_{n=1}^{\infty} ne^{-n^2}.$ **7.3.20.** $\displaystyle\sum_{n=1}^{\infty} \dfrac{n}{n^2 + 1}.$ **7.3.21.*** $\displaystyle\sum_{n=1}^{\infty} \dfrac{n}{2^n}.$

7.3.22. $\displaystyle\sum_{n=1}^{\infty} \dfrac{e^{\arctan n}}{n^2 + 1}.$ **7.3.23.** $\displaystyle\sum_{n=1}^{\infty} \dfrac{1}{4n^2 + 1}.$ **7.3.24.** $\displaystyle\sum_{n=1}^{\infty} \dfrac{1}{\sqrt{n^2 + 9}}.$

7.3.25. $\displaystyle\sum_{n=1}^{\infty} \dfrac{1}{\ln(n+1) \cdot \ln(1 + n^n)}.$

In Exercises 7.3.26–7.3.28 determine all the values of p for which the series converges:

7.3.26.* $\displaystyle\sum_{n=2}^{\infty} \dfrac{1}{n(\ln n)^p}.$ **7.3.27.*** $\displaystyle\sum_{n=1}^{\infty} n\left(1 + n^2\right)^p.$ **7.3.28.*** $\displaystyle\sum_{n=1}^{\infty} \dfrac{\ln n}{n^p}.$

7.3.29.* Use the Integral Test to prove that

$$\frac{9}{8} < \sum_{n=1}^{\infty} \frac{1}{n^3} < \frac{5}{4}.$$

7.3.30.* Find a function f that is continuous, non-negative and $\int_0^{\infty} f(x)\, dx$ converges but $\sum_{n=0}^{\infty} f(n)$ diverges.

7.3.31.* Find a function f that is continuous, non-negative and $\sum_{n=0}^{\infty} f(n)$ converges but $\int_0^{\infty} f(x)\, dx$ diverges.

7.4   Root and Ratio Tests

In this section our goal is to prove that the Root Test and the Ratio Test give the correct information about the convergence of a series. Both tests are based on comparison, and just like the Comparison Test they come in two flavors (asymptotic or not). We will look at some related tests as well.

For a series

$$\sum_{n=1}^{\infty} a_n = a_1 + a_2 + a_3 + \dots \tag{7.5}$$

we will use the notation

$$\mathcal{C}_n = \sqrt[n]{a_n}, \quad \mathcal{D}_n = \frac{a_{n+1}}{a_n}$$

and we will write $\mathcal{C} = \lim \mathcal{C}_n$ and $\mathcal{D} = \lim \mathcal{D}_n$ when these limits either exist or when they are infinite.

Theorem 7.4.1 (The Root Test). *Let $\{a_n\}$ be a sequence of positive numbers and suppose that there exist $r < 1$ and $N \in \mathbb{N}$ such that $\mathcal{C}_n \leq r$ for $n \geq N$. Then the series (7.5) converges. On the other hand, if $\mathcal{C}_n \geq 1$ for $n \geq N$, then the series (7.5) diverges. In particular, it converges if $\mathcal{C} < 1$ and diverges if $\mathcal{C} > 1$.*

Proof. Suppose that $\sqrt[n]{a_n} \leq r < 1$ for $n \geq N$. For such n, $a_n \leq r^n$ and the convergence of the series (7.5) follows from the Comparison Test, since the geometric series $\sum_{n=1}^{\infty} r^n$ converges. If $\mathcal{C} < 1$, then we take a real number r satisfying $\mathcal{C} < r < 1$. Since $\mathcal{C} = \lim \sqrt[n]{a_n} < r$ there exists $N \in \mathbb{N}$, such that for $n \geq N$, $\sqrt[n]{a_n} < r$. The convergence of the series (7.5) now follows from the first part of the proof.

The inequality $\mathcal{C}_n \geq 1$ implies that $a_n \geq 1$. If this is true for $n \geq N$, then a_n cannot have limit 0, and the Divergence Test shows that the series (7.5) diverges. In particular, if $\mathcal{C} > 1$, then there exists $N \in \mathbb{N}$ such that $\mathcal{C}_n \geq 1$ for $n \geq N$, and the series (7.5) diverges. $\qquad\square$

Example 7.4.2. Applying the Root Test.

Determine whether the infinite series $\sum_{n=1}^{\infty} \dfrac{1}{(\ln n)^n}$ converges.

Solution. Here

$$\mathcal{C}_n = \sqrt[n]{a_n} = \sqrt[n]{\frac{1}{(\ln n)^n}} = \frac{1}{\ln n} \to 0.$$

Since $\mathcal{C} < 1$ we conclude that the series $\sum_{n=1}^{\infty} 1/(\ln n)^n$ converges. $\qquad\blacklozenge$

> The Root Test is often referred to as the Cauchy Test, because it comes to us from *Cours d'Analyse*.

Theorem 7.4.3 (The Ratio Test). *Let $\{a_n\}$ be a sequence of positive numbers and suppose that there exist $r < 1$ and $N \in \mathbb{N}$ such that $\mathcal{D}_n \leq r$ for $n \geq N$. Then the series (7.5) converges. On the other hand, if $\mathcal{D}_n \geq 1$ for $n \geq N$, then the series (7.5) diverges. In particular, it converges if $\mathcal{D} < 1$ and diverges if $\mathcal{D} > 1$.*

Proof. Suppose that $a_{n+1}/a_n \leq r$ for $n \geq N$. Then

$$a_{N+1} \leq r a_N, \quad a_{N+2} \leq r a_{N+1} \leq r^2 a_N$$

and, inductively,

$$a_{N+k} \leq r^k a_N,$$

for any $k \in \mathbb{N}$. The convergence of the series (7.5) now follows from the Comparison Test, because $\sum_{k=1}^{\infty} r^k a_N$ is a geometric series with the initial term $r a_N$ and ratio $r < 1$. When $\mathcal{D} < 1$ we take a real number r such that $\mathcal{D} < r < 1$. Then there exists $N \in \mathbb{N}$, such that for $n \geq N$, $\mathcal{D}_n < r$ and the problem is reduced to the previous one.

The inequality $\mathcal{D}_n \geq 1$ implies that $a_{n+1} \geq a_n$. If this is true for all $n \geq N$, then the sequence $\{a_n\}$ cannot have limit 0, and by the Divergence Test, the series (7.5) diverges. In the special case when $\mathcal{D} > 1$ (even if the limit is infinite), there exists $N \in \mathbb{N}$ such that, for $n \geq N$, $\mathcal{D}_n \geq 1$, so the series (7.5) diverges. $\qquad\square$

Example 7.4.4. Applying the Root Test.

Determine whether the infinite series $\sum\limits_{n=1}^{\infty} \dfrac{10^n}{n!}$ converges.

Solution. Here

$$\mathcal{D}_n = \frac{a_{n+1}}{a_n} = \frac{\frac{10^{n+1}}{(n+1)!}}{\frac{10^n}{n!}} = \frac{10^{n+1}}{(n+1)!}\frac{n!}{10^n} = \frac{10}{n+1}.$$

It is not hard to see that $10/(n+1) < 1$ when $n \geq 10$ so we conclude that the series $\sum_{n=1}^{\infty} 10^n/n!$ converges.   ◆

> The test was first published in [24] in 1768 by d'Alembert so it is sometimes referred to as d'Alembert's test. Truth be told, he was interested in a very specific series (the Binomial Series, see page 252), and the fashionable topic of the time: is the sequence $\{|a_n|\}$ decreasing? Cauchy gave the first serious treatment in *Cours d'Analyse*.

In most applications, the Ratio Test is easier to use, and we will always try it first. However, the Root Test is more powerful. That means that whenever we are able to decide whether a series is convergent or divergent by using the Ratio Test, the Root Test would have been applicable. (Although, perhaps, harder to use.) On the other hand, there are situations when the Ratio Test is inconclusive (the limit of $\mathcal{D}_n$ is either 1 or does not exist), but the Root Test can provide the answer. Here is a more precise statement.

Theorem 7.4.5. *Let $\{a_n\}$ be a sequence of positive numbers and suppose that $\lim \mathcal{D}_n = L$. Then the sequence $\mathcal{C}_n = \sqrt[n]{a_n}$ converges to L as well.*

Proof. Suppose first that $L > 0$ and let $\varepsilon > 0$. Without loss of generality we can choose ε so that $L - \varepsilon/2 > 0$. There exists $N \in \mathbb{N}$ such that, for $n \geq N$,

$$L - \frac{\varepsilon}{2} < \frac{a_{n+1}}{a_n} < L + \frac{\varepsilon}{2}.$$

It follows that, for any $k \in \mathbb{N}$,

$$\left(L - \frac{\varepsilon}{2}\right)^k a_N < a_{N+k} < \left(L + \frac{\varepsilon}{2}\right)^k a_N.$$

Next, by taking the $(N+k)$th roots, we obtain that, for any $k \in \mathbb{N}$,

$$\left(L - \frac{\varepsilon}{2}\right)^{\frac{k}{N+k}} a_N^{\frac{1}{N+k}} < \sqrt[N+k]{a_{N+k}} < \left(L + \frac{\varepsilon}{2}\right)^{\frac{k}{N+k}} a_N^{\frac{1}{N+k}}.$$

It is not hard to see that, when $k \to \infty$, the leftmost expression converges to $L - \varepsilon/2$ and the rightmost to $L + \varepsilon/2$. This implies that there exists $K_1, K_2 \in \mathbb{N}$ such that

$$k \geq K_1 \quad \Rightarrow \quad \left(L - \frac{\varepsilon}{2}\right)^{\frac{k}{N+k}} a_N^{\frac{1}{N+k}} > L - \varepsilon, \quad \text{and}$$

$$k \geq K_2 \quad \Rightarrow \quad \left(L + \frac{\varepsilon}{2}\right)^{\frac{k}{N+k}} a_N^{\frac{1}{N+k}} < L + \varepsilon.$$

If $K = \max\{K_1, K_2\}$ and $k \geq K$, then

$$L - \varepsilon < \sqrt[N+k]{a_{N+k}} < L + \varepsilon.$$

It follows that $\mathcal{C}_n$ converges to L.

The case $L = 0$ is proved in a similar way, except that $L - \varepsilon/2$ should be replaced by 0 throughout.   □

> Theorem 7.4.5 is due to Cauchy and can be found in *Cours d'Analyse*.

Unfortunately, the converse does not hold.

Example 7.4.6. Root Test is stronger than Ratio Test.

Determine whether the infinite series $\sum_{n=1}^{\infty} 2^{(-1)^n - n}$ converges.

Solution. First we try the Ratio Test.

$$\mathcal{D}_n = \frac{a_{n+1}}{a_n} = \frac{2^{(-1)^{n+1} - (n+1)}}{2^{(-1)^n - n}} = 2^{(-1)^{n+1} - (n+1) - (-1)^n + n} = 2^{2(-1)^{n+1} - 1},$$

so $\mathcal{D}_{2n} = 2^{-3} = 1/8$ and $\mathcal{D}_{2n-1} = 2^1 = 2$. Clearly, the sequence $\{\mathcal{D}_n\}$ does not satisfy the hypotheses of the Ratio Test, so we cannot apply it.

Let us try the Root Test.

$$\mathcal{C}_n = \sqrt[n]{a_n} = \sqrt[n]{2^{(-1)^n - n}} = 2^{\frac{(-1)^n - n}{n}} \to 2^{-1} < 1,$$

so the series converges. ◆

Exercises

In Exercises 7.4.1–7.4.9 use Root Test to determine whether the series converges.

7.4.1. $\sum_{n=1}^{\infty} n^4 e^{-n^2}$.

7.4.2. $\sum_{n=1}^{\infty} \frac{1}{n^n}$.

7.4.3. $\sum_{n=2}^{\infty} \left(\frac{n-1}{n+1}\right)^{n(n-1)}$.

7.4.4. $\sum_{n=1}^{\infty} \frac{n}{2^n}$.

7.4.5. $\sum_{n=1}^{\infty} \frac{n^2}{(2 + \frac{1}{n})^n}$.

7.4.6. $\sum_{n=2}^{\infty} (\sqrt[n]{n} - 1)^n$.

7.4.7.* $\sum_{n=1}^{\infty} \left(\frac{n^2 + 1}{n^2 + n + 1}\right)^{n^2}$.

7.4.8.* $\sum_{n=1}^{\infty} \frac{n^{n + \frac{1}{n}}}{(n + \frac{1}{n})^n}$.

7.4.9.* $\sum_{n=1}^{\infty} \left(\frac{1 + \cos n}{2 + \cos n}\right)^{2n - \ln n}$.

7.4.10. Use Root Test to determine whether the series

$$1 + 2r + r^2 + 2r^3 + r^4 + 2r^5 + \ldots$$

converges, if (a) $r = 2/3$; (b) $r = 4/3$.

In Exercises 7.4.11–7.4.17 use Ratio Test to determine whether the series converges.

7.4.11. $\sum_{n=1}^{\infty} \frac{2^n n!}{n^n}$.

7.4.12. $\sum_{n=1}^{\infty} \frac{3^n n!}{n^n}$.

7.4.13. $\sum_{n=1}^{\infty} \frac{n^2}{2^n}$.

7.4.14. $\sum_{n=1}^{\infty} \frac{n!}{e^n}$.

7.4.15. $\sum_{n=1}^{\infty} \frac{(n!)^2}{2^{n^2}}$.

7.4.16. $\sum_{n=1}^{\infty} \frac{2^n + 1}{3^n + n}$.

7.4.17. $\frac{4}{3} + \frac{4 \cdot 7}{2 \cdot 6} + \frac{4 \cdot 7 \cdot 10}{2 \cdot 6 \cdot 10} + \ldots$.

7.4.18.* Use Ratio Test to determine for what values of k will the series $\sum_{n=1}^{\infty} (n!)^2/(kn)!$ converge.

7.4.19.* The sequence a_n is defined by

$$a_1 = 2, \quad a_{n+1} = \frac{5n + 1}{4n + 3} a_n, \text{ for } n \geq 1.$$

Use Ratio Test to determine whether the series $\sum_{n=1}^{\infty} a_n$ converges.

7.4.20. Give an example of a divergent series such that $\mathcal{D}_n < 1$ for all $n \in \mathbb{N}$.

7.4.21. Prove the stronger version of Root Test: If $\{a_n\}$ is a sequence of positive numbers and $\limsup \mathcal{C}_n = r$, then the series $\sum_{n=1}^{\infty} a_n$ converges if $r < 1$ and diverges if $r > 1$.

7.4.22. Prove the stronger version of Ratio Test: If $\{a_n\}$ is a sequence of positive numbers and $\limsup \mathcal{D}_n < 1$, then the series $\sum_{n=1}^{\infty} a_n$ converges; if $\liminf \mathcal{D}_n > 1$, then the series $\sum_{n=1}^{\infty} a_n$ diverges.

In Exercises 7.4.23–7.4.24 use Exercise 7.4.21 or Exercise 7.4.22 to determine whether the series converges.

7.4.23. $\displaystyle\sum_{n=1}^{\infty} \frac{5 - (-1)^n}{3 \cdot 2^n}.$

7.4.24. $\displaystyle\sum_{n=1}^{\infty} \frac{n^3 \left(\sqrt{2} + (-1)^n\right)^n}{3^n}.$

7.4.25. Give an example of a sequence $\{a_n\}$ of positive numbers such that $\liminf \mathcal{D}_n < 1 < \limsup \mathcal{D}_n$ and such that the series $\sum_{n=1}^{\infty} a_n$ is: (a) convergent; (b) divergent.

7.4.26. Prove the stronger version of Theorem 7.4.5: Let $\{a_n\}$ be a sequence of positive numbers. If $\limsup \mathcal{D}_n < 1$ then $\limsup \mathcal{C}_n < 1$; if $\liminf \mathcal{D}_n > 1$ then $\liminf \mathcal{C}_n > 1$.

7.4.27. Prove that

$$\lim \sqrt[n]{\frac{n^n}{n!}} = e.$$

7.4.1 Additional Tests for Convergence

We will present a few more tests for convergence of a series.

Theorem 7.4.7 (Kummer's Test). *Let $\{a_n\}, \{c_n\}$ be two sequences of strictly positive numbers such that the series $\sum_{n=1}^{\infty} 1/c_n$ diverges. For $n \in \mathbb{N}$, let*

$$\mathcal{K}_n = c_n \cdot \frac{a_n}{a_{n+1}} - c_{n+1} \tag{7.6}$$

and suppose that there exist $r > 0$ and $N \in \mathbb{N}$ such that $\mathcal{K}_n \geq r$ for $n \geq N$. Then the series (7.5) converges. On the other hand, if $\mathcal{K}_n \leq 0$ for $n \geq N$, then the series (7.5) diverges. In particular, if $\lim \mathcal{K}_n = \mathcal{K}$, the series $\sum_{n=1}^{\infty} a_n$ converges if $\mathcal{K} > 0$ and diverges if $\mathcal{K} < 0$.

Proof. Suppose first that $\mathcal{K}_n \geq r > 0$ for $n \geq N$. It follows that, for $n \geq N$,

$$c_n a_n - c_{n+1} a_{n+1} \geq r a_{n+1} > 0. \tag{7.7}$$

We see that the sequence $\{c_n a_n\}$ is a decreasing sequence, and since it is bounded below by 0, it converges to a limit L. Therefore,

$$\sum_{k=1}^{n} (c_k a_k - c_{k+1} a_{k+1}) = c_1 a_1 - c_{n+1} a_{n+1} \to c_1 a_1 - L,$$

so the series $\sum_{k=1}^{\infty} (c_k a_k - c_{k+1} a_{k+1})$ converges. Now the inequality (7.7) shows that the series $\sum_{k=1}^{\infty} r a_{k+1}$ converges, whence $\sum_{k=1}^{\infty} a_k$ converges.

If $\mathcal{K}_n \leq 0$ for $n \geq N$, then

$$\frac{a_{n+1}}{a_n} \geq \frac{c_n}{c_{n+1}},$$

so we have a sequence of inequalities

$$\frac{a_{N+1}}{a_N} \geq \frac{c_N}{c_{N+1}}, \quad \frac{a_{N+2}}{a_{N+1}} \geq \frac{c_{N+1}}{c_{N+2}}, \quad \frac{a_{N+3}}{a_{N+2}} \geq \frac{c_{N+2}}{c_{N+3}}, \ldots.$$

If we multiply $k - 1$ successive inequalities we obtain

$$\frac{a_{N+1}}{a_N} \frac{a_{N+2}}{a_{N+1}} \cdots \frac{a_{N+k}}{a_{N+k-1}} \geq \frac{c_N}{c_{N+1}} \frac{c_{N+1}}{c_{N+2}} \cdots \frac{c_{N+k-1}}{c_{N+k}}$$

which yields $a_{N+k}/a_N \geq c_N/c_{N+k}$ and, hence,

$$a_{N+k} \geq a_N c_N \frac{1}{c_{N+k}}.$$

Now the divergence of $\sum_{n=1}^{\infty} a_n$ follows from the divergence of $\sum_{n=1}^{\infty} 1/c_n$.

The case when the limit $\mathcal{K}$ exists, and satisfies $\mathcal{K} > 0$ or $\mathcal{K} < 0$, follows easily from the previous and we leave it as an exercise. □

The test was published by Kummer (page 234) in 1835. In the original version there was an additional condition that $\lim a_n c_n = 0$. In 1867 Dini (page 374) proved the modern version.

Clearly, the Kummer Test covers many situations, because we have a choice of a divergent series $\sum_{n=1}^{\infty} 1/c_n$ to make. Some special cases are worth mentioning.

If we take $c_n = 1$, for all $n \in \mathbb{N}$, then $\mathcal{K}_n = a_n/a_{n+1} - 1 = 1/\mathcal{D}_n - 1$. If $\mathcal{D}_n \leq r < 1$ and if we denote $r' = \frac{1}{r} - 1$, then

$$\mathcal{K}_n = \frac{1}{\mathcal{D}_n} - 1 \geq \frac{1}{r} - 1 = r' > 0.$$

On the other hand, if $\mathcal{D}_n \geq 1$ then

$$\mathcal{K}_n = \frac{1}{\mathcal{D}_n} - 1 \leq \frac{1}{1} - 1 = 0.$$

Thus, the Ratio Test is a consequence of Kummer's Test.

Another important case is $c_n = n$. The series $\sum_{n=1}^{\infty} 1/c_n$ is the Harmonic series which diverges, so the Kummer's Test applies. We obtain that

$$\mathcal{K}_n = n \cdot \frac{a_n}{a_{n+1}} - (n + 1).$$

If we denote

$$\mathcal{R}_n = n \left(\frac{a_n}{a_{n+1}} - 1 \right)$$

we see that $\mathcal{R}_n = \mathcal{K}_n + 1$. Thus we obtain the result which had earlier been established by Raabe.

Theorem 7.4.8 (Raabe's Test). *Let $\{a_n\}$ be a sequence of strictly positive numbers and let*

$$\mathcal{R}_n = n \left(\frac{a_n}{a_{n+1}} - 1 \right).$$

Suppose that there exist $r > 1$ and $N \in \mathbb{N}$ such that $\mathcal{K}_n \geq r$ for $n \geq N$. Then the series (7.5) converges. On the other hand, if $\mathcal{R}_n \leq 1$ for $n \geq N$, then the series (7.5) diverges. In particular, if $\mathcal{R}_n \to \mathcal{R}$, the series $\sum_{n=1}^{\infty} a_n$ converges if $\mathcal{R} > 1$ and diverges if $\mathcal{R} < 1$.

Example 7.4.9. Applying Raabe's Test.

Determine whether the series $\displaystyle\sum_{n=1}^{\infty} \frac{n!}{(a+1)(a+2)\ldots(a+n)}$, $a > 0$, converges.

Solution. First we try the Ratio Test.

$$\mathcal{D}_n = \frac{a_{n+1}}{a_n} = \frac{\frac{(n+1)!}{(a+1)(a+2)\ldots(a+n+1)}}{\frac{n!}{(a+1)(a+2)\ldots(a+n)}} = \frac{n+1}{a+n+1},$$

so $\mathcal{D} = 1$, and the test is inconclusive. However,

$$\mathcal{R}_n = n\left(\frac{1}{\mathcal{D}_n} - 1\right) = n\left(\frac{a+n+1}{n+1} - 1\right) = n\frac{a+n+1-(n+1)}{n+1} = a\frac{n}{n+1}$$

so $\mathcal{R} = a$. Consequently, the series $\sum_{n=1}^{\infty} n!/[(a+1)(a+2)\ldots(a+n)]$ converges if $a > 1$ and diverges if $a < 1$. Finally, if $a = 1$,

$$a_n = \frac{n!}{(1+1)(1+2)\ldots(1+n)} = \frac{1}{n+1}$$

so the obtained series is the Harmonic series and it diverges.   ◆

JOSEPH LUDWIG RAABE (1801–1859) was a Swiss mathematician, born in the area of today's Ukraine. Coming from a poor family, he had to earn his living from a young age. He did that by teaching himself mathematics and giving private lessons throughout high school and the Vienna Polytechnikum. Inspired by his astronomy professor he started publishing papers in 1826. In the fall of 1831, a cholera epidemic reached Vienna so Raabe left for Zürich where he taught in a high school. When the University of Zürich was founded in 1833 he was hired as an assistant professor. He became a professor at the ETH Zürich in 1855 but left in 1858 and died the next year. His successor at that position was Dedekind. In addition to the Raabe Test that was in his 1834 paper, he is also known for the formula $\int_a^{a+1} \log \Gamma(t)\, dt = \frac{1}{2}\log 2\pi + a\log a - a$, $a \geq 0$.

Exercises

In Exercises 7.4.28–7.4.33 use Raabe's Test to determine whether the series converges:

7.4.28. $\left(\frac{1}{3}\right)^2 + \left(\frac{1\cdot 4}{3\cdot 6}\right)^2 + \left(\frac{1\cdot 4\cdot 7}{3\cdot 6\cdot 9}\right)^2 + \cdots + \left(\frac{1\cdot 4\cdot 7\cdot\ \cdots\ (3n-2)}{3\cdot 6\cdot 9\cdot\ \cdots\ 3n}\right)^2 + \cdots.$

7.4.29. $\dfrac{1}{3} + \dfrac{1\cdot 4}{3\cdot 6} + \dfrac{1\cdot 4\cdot 7}{3\cdot 6\cdot 9} + \cdots + \dfrac{1\cdot 4\cdot 7\cdot\ \cdots\ (3n-2)}{3\cdot 6\cdot 9\cdot\ \cdots\ 3n} + \cdots.$

7.4.30. $1 + \sum_{n=1}^{\infty} \dfrac{(2n-1)!!}{(2n)!!}\dfrac{1}{2n+1}.$ 7.4.31.* $\sum_{n=1}^{\infty} \dfrac{1}{n!}\left(\dfrac{n}{e}\right)^n.$

7.4.32. $\dfrac{2}{9} + \dfrac{2\cdot 5}{9\cdot 12} + \dfrac{2\cdot 5\cdot 8}{9\cdot 12\cdot 15} + \cdots.$ 7.4.33. $\sum_{n=1}^{\infty} \dfrac{\sqrt{n!}}{(2+\sqrt{1})(2+\sqrt{2})\ldots(2+\sqrt{n})}.$

In Exercises 7.4.34–7.4.36 use Raabe's Test to determine all values of the parameter p for which the series converges:

7.4.34. $\sum_{n=1}^{\infty} \dfrac{n!\,e^n}{n^{n+p}}.$ 7.4.35.* $\sum_{n=1}^{\infty} \dfrac{1\cdot 3\cdot\ \cdots\ (2n-1)}{p(p+2)\ldots(p+2n-2)},\ p > 0.$

7.4.36.* $\left(\dfrac{1}{2}\right)^p + \left(\dfrac{1\cdot 3}{2\cdot 4}\right)^p + \left(\dfrac{1\cdot 3\cdot 5}{2\cdot 4\cdot 6}\right)^p + \cdots.$

In Exercises 7.4.37–7.4.39 use Raabe's Test to determine the condition that positive numbers p and q, and a real number r must satisfy in order for the series to converge:

7.4.37.* $\sum_{n=1}^{\infty} \dfrac{p(p+1)\ldots(p+n-1)}{n!}\cdot\dfrac{1}{n^q}.$ 7.4.38.* $\sum_{n=1}^{\infty} \left(\dfrac{1\cdot 3\cdot\ \cdots\ (2n-1)}{2\cdot 4\cdot\ \cdots\ 2n}\right)^p\cdot\dfrac{1}{n^q}.$

7.4.39.* $\displaystyle\sum_{n=1}^{\infty}\left(\frac{p\cdot(p+1)\cdot\cdots\cdot(p+n-1)}{q\cdot(q+1)\cdot\cdots\cdot(q+n-1)}\right)^{r}.$

7.4.40.* Give an example of a series for which Raabe's Test gives $\mathcal{R}=1$ and the series is (a) convergent; (b) divergent.

7.4.41. In Kummer's Test, prove the limit case.

7.4.42. State and prove the lim sup and lim inf version of Raabe's Test.

7.4.43. State and prove the lim sup and lim inf version of Kummer's Test.

7.4.44.* The sequence a_n is defined by

$$a_1 = a_2 = 1, \quad a_{n+1} = a_n + \frac{a_{n-1}}{n^2}, \text{ for } n \geq 2.$$

Use Raabe's Test to determine whether the series $\sum_{n=1}^{\infty}\frac{1}{a_n}$ converges.

7.4.45.* Prove Bertrand's Test: if $\mathcal{B}_n = \ln n(\mathcal{R}_n - 1)$ and $\mathcal{B} = \lim \mathcal{B}_n$, then the series (7.5) converges if $\mathcal{B} > 1$ and diverges if $\mathcal{B} < 1$.

7.5 Series with Arbitrary Terms

In this section we will remove our restriction that the terms of the series must be positive. This makes the study of their convergence much harder, because the comparison is not effective any more. For example, consider the series

$$-1 + \frac{1}{2^2} - 3^2 + \frac{1}{4^2} - 5^2 + \cdots$$

Notice that $-1 < 1$, $-3^2 < \frac{1}{3^2}$, etc., so each term a_n of the given series satisfies the inequality $a_n \leq b_n$, where $b_n = 1/n^2$. In spite of the fact that $\sum_{n=1}^{\infty} b_n$ converges, the given series diverges. (For example, by the Divergence Test.)

The obvious problem here is that, although the terms a_n are dominated from the right by b_n, there is no such control from the left. One way to avoid this lack of symmetry is to consider the *absolute convergence*. Let $\sum_{n=1}^{\infty} a_n$ be a series, and consider $\sum_{n=1}^{\infty} |a_n|$. If the latter series converges, we say that the series $\sum_{n=1}^{\infty} a_n$ **converges absolutely**. Of course, the series $\sum_{n=1}^{\infty} |a_n|$ has positive terms and we can apply any of the tests studied in the previous section.

Example 7.5.1. An absolutely convergent series.

Prove that the series $\displaystyle\sum_{n=1}^{\infty} \frac{n^2}{(-2)^n}$ converges absolutely.

Solution. The series can be written as

$$-\frac{1}{2} + \frac{2^2}{2^2} - \frac{3^2}{2^3} + \frac{4^2}{2^4} - \cdots$$

so some of the terms are positive, some negative. However, we will consider the series

$$\sum_{n=1}^{\infty}\left|\frac{n^2}{(-2)^n}\right| = \sum_{n=1}^{\infty}\frac{n^2}{2^n}.$$

This is a series with positive terms, and we can use the Ratio Test. Now,

$$\mathcal{D}_n = \frac{\frac{(n+1)^2}{2^{n+1}}}{\frac{n^2}{2^n}} = \frac{(n+1)^2}{2n^2} \to \frac{1}{2}, \quad n \to \infty.$$

Thus, the series $\sum_{n=1}^{\infty} n^2/(-2)^n$ is absolutely convergent.   ♦

Suppose that we find out that a series is absolutely convergent. What does it say about the convergence of the given series?

Theorem 7.5.2. *An absolutely convergent series is convergent.*

Proof. Let $\sum_{k=1}^{n} |a_k|$ be convergent. We will use Theorem 7.2.8 to prove that $\sum_{k=1}^{n} a_k$ converges as well. Let $\varepsilon > 0$. By Theorem 7.2.8, there exists $N \in \mathbb{N}$ such that

$$m \geq n \geq N \quad \Rightarrow \quad |a_{n+1}| + |a_{n+2}| + \cdots + |a_m| < \varepsilon.$$

For such m, n,

$$|a_{n+1} + a_{n+2} + \cdots + a_m| \leq |a_{n+1}| + |a_{n+2}| + \cdots + |a_m| < \varepsilon$$

so the series $\sum_{k=1}^{n} a_k$ is convergent.   □

Now we will look at several examples, in which we will determine that the series converges absolutely and from that conclude that the series converges.

Example 7.5.3. An absolutely convergent series.

Prove that the series $\sum\limits_{n=1}^{\infty} \dfrac{a^n}{n!}$ is convergent for any $a \in \mathbb{R}$.

Solution. Notice that there are no restrictions on a, so if $a < 0$, the positive and negative terms will alternate. However, we will consider the series

$$\sum_{n=1}^{\infty} \left| \frac{a^n}{n!} \right| = \sum_{n=1}^{\infty} \frac{|a|^n}{n!}.$$

This is a series with positive terms, and we can use the Ratio Test. Now,

$$\mathcal{D}_n = \frac{\frac{|a|^{n+1}}{(n+1)!}}{\frac{|a|^n}{n!}} = \frac{|a|}{n+1} \to 0, \quad n \to \infty.$$

Thus, the series $\sum_{n=1}^{\infty} a^n/n!$ is absolutely convergent for any $a \in \mathbb{R}$ and, by Theorem 7.5.2, it is convergent.   ♦

Example 7.5.4. An absolutely convergent series.

Prove that the series $\sum\limits_{n=1}^{\infty} \dfrac{\sin n}{n^2}$ is convergent.

Solution. Once again, $a_n = \sin n/n^2$ takes both positive and negative values. For example, it is positive for $n = 1, 2, 3$, negative for $n = 4, 5, 6$, positive for $n = 7, 8, 9$, etc. However,

$$|a_n| = \frac{|\sin n|}{n^2} \leq \frac{1}{n^2},$$

for all $n \in \mathbb{N}$, and the series $\sum_{n=1}^{\infty} 1/n^2$ converges as p-series, with $p = 2$. By the Comparison Test, the series $\sum_{n=1}^{\infty} a_n$ is absolutely convergent, hence convergent.   ♦

We see that one possible strategy, when a series has terms of both signs, is to work on proving that the series is absolutely convergent. Unfortunately, this strategy does not always work because, as we will show in Example 7.5.6, there are convergent series that are not absolutely convergent. How does one establish the convergence of such a series? One special case occurs when the series is *alternating*, i.e., when the positive and negative terms alternate. More precisely, this occurs when $a_n = (-1)^{n+1}|a_n|$, for all $n \in \mathbb{N}$. For such a series we can apply the following test.

Theorem 7.5.5 (Alternating Series Test). *Let $\{c_n\}$ be a decreasing sequence of positive numbers that converges to 0. Then the alternating series $\sum_{n=1}^{\infty}(-1)^{n+1}c_n$ converges.*

Proof. Let $s_n = \sum_{k=1}^{n}(-1)^{k+1}c_k$, for $n \in \mathbb{N}$. Then

$$s_{2m} = (c_1 - c_2) + (c_3 - c_4) + \cdots + (c_{2m-1} - c_{2m}).$$

Notice that each expression in parentheses is positive, so the subsequence $\{s_{2m}\}$ is increasing. On the other hand

$$s_{2m} = c_1 - (c_2 - c_3) - (c_4 - c_5) - \cdots - (c_{2m-2} - c_{2m-1}) - c_{2m} \le c_1$$

so $\{s_{2m}\}$ is also bounded above. It follows that it is convergent, say $\lim_{m\to\infty} s_{2m} = S$.

It remains to show that $\lim_{m\to\infty} s_{2m-1} = S$. However, $s_{2m-1} = s_{2m} - c_{2m} \to S$, since $c_m \to 0$. Thus the sequence of partial sums $\{s_n\}$ converges, and the alternating series $\sum_{n=1}^{\infty}(-1)^{n+1}c_n$ converges. $\quad\square$

> Alternating Series Test is often referred to as Leibniz Test, because it appears in his letter from 1705.

Example 7.5.6. Applying Alternating Series Test.

Prove that the series $\sum_{n=1}^{\infty} \dfrac{(-1)^{n+1}}{n}$ converges.

Solution. This is an alternating series, with $c_n = 1/n$. It is often called the Alternating Harmonic Series. Clearly, $\{1/n\}$ is a decreasing sequence of positive numbers that converges to 0. By the Alternating Series Test, the series $\sum_{n=1}^{\infty}(-1)^{n+1}/n$ converges. $\quad\blacklozenge$

Notice that the series in Example 7.5.6 is not absolutely convergent: $\sum_{n=1}^{\infty} 1/n$ is the Harmonic series. We say that the Alternating Harmonic Series is **conditionally convergent**.

It is important to notice that the requirement that the sequence $\{c_n\}$ be decreasing cannot be relaxed.

Example 7.5.7. The condition that $\{c_n\}$ is decreasing cannot be removed.

We will show that the alternating series

$$\frac{1}{\sqrt{2}-1} - \frac{1}{\sqrt{2}+1} + \frac{1}{\sqrt{3}-1} - \frac{1}{\sqrt{3}+1} + \cdots + \frac{1}{\sqrt{n}-1} - \frac{1}{\sqrt{n}+1} + \ldots \qquad (7.8)$$

is divergent although $c_n \to 0$.

Solution. Indeed, the sum of $2n$ terms

$$s_{2n} = \sum_{k=2}^{n+1}\left(\frac{1}{\sqrt{k}-1} - \frac{1}{\sqrt{k}+1}\right) = \sum_{k=2}^{n+1}\frac{(\sqrt{k}+1)-(\sqrt{k}-1)}{(\sqrt{k}+1)(\sqrt{k}-1)} = \sum_{k=2}^{n+1}\frac{2}{k-1} = 2H_n$$

where H_n is the nth partial sum of the Harmonic series. Therefore, $\lim s_{2n} = \infty$ and the series (7.8) diverges. The reason that the Alternating Series Test does not apply is that the sequence $\{c_n\}$ is not decreasing. Namely, $c_{2n-1} = 1/(\sqrt{n+1}-1)$ and $c_{2n} = 1/(\sqrt{n+1}+1)$, so it is easy to see that, for each $n \in \mathbb{N}$, $c_{2n} > c_{2n-1}$. $\quad\blacklozenge$

Many eighteenth century mathematicians ignored the difference between the absolute and the conditional convergence, in spite of the evidence to the contrary. It was well known that the Harmonic Series diverges and that the Alternating Harmonic Series converges. Once again, Cauchy was the first to make this distinction.

Exercises

In Exercises 7.5.1–7.5.13 test the series for conditional and absolute convergence:

7.5.1. $\sum_{n=2}^{\infty} \dfrac{(-1)^{n-1}}{n \ln^2 n}$.

7.5.2. $\sum_{n=1}^{\infty} \dfrac{(-1)^{n-1} 2^n}{n^3}$.

7.5.3. $\sum_{n=1}^{\infty} (-1)^n \arcsin \dfrac{1}{n}$.

7.5.4. $\sum_{n=1}^{\infty} \dfrac{(-1)^n}{n^2 + n + 2}$.

7.5.5. $\sum_{n=1}^{\infty} \dfrac{(-1)^{n-1} n}{n^2 + 10}$.

7.5.6. $\sum_{n=1}^{\infty} (-1)^n \sin \dfrac{a}{n}$, $a \in \mathbb{R}$.

7.5.7. $\sum_{n=1}^{\infty} \dfrac{(-1)^{n-1} n}{3n + 1}$.

7.5.8. $\sum_{n=1}^{\infty} \dfrac{(-1)^{n+1}}{2^n}$.

7.5.9.* $\sum_{n=1}^{\infty} (-1)^n \left(\sqrt[n]{n} - 1 \right)$.

7.5.10. $\sum_{n=1}^{\infty} \dfrac{\cos n\pi}{\sqrt{n}}$.

7.5.11. $\sum_{n=1}^{\infty} \dfrac{\sin \sqrt{n}}{n^{3/2}}$.

7.5.12.* $\sum_{n=1}^{\infty} \sin \left(\pi \sqrt{n^2 + a^2} \right)$, $a \in \mathbb{R}$.

7.5.13. $1 - \dfrac{3}{2} + \dfrac{5}{4} - \dfrac{7}{8} + \dots$.

7.5.14. Prove that the series

$$1 - 2 + \frac{1}{2} - 1 + \frac{1}{3} - \frac{2}{3} + \frac{1}{4} - \frac{2}{4} + \dots + \frac{1}{n} - \frac{2}{n} + \dots$$

diverges.

7.5.15. Prove that the series

$$\frac{1}{2} - \frac{1}{2} + \frac{1}{4} - \frac{1}{4} + \frac{1}{4} - \frac{1}{4} + \frac{1}{8} - \frac{1}{8} + \frac{1}{8} - \frac{1}{8} + \frac{1}{8} - \frac{1}{8} + \frac{1}{16} - \dots$$

converges conditionally and find its sum.

In Exercises 7.5.16–7.5.18 determine all values of the parameter p for which the series converges conditionally or absolutely:

7.5.16. $\sum_{n=1}^{\infty} \dfrac{(-1)^n (\ln n)^p}{n}$.

7.5.17. $\sum_{n=1}^{\infty} (-1)^n \sin \dfrac{p}{n}$.

7.5.18.* $\sum_{n=1}^{\infty} (-1)^n \dfrac{(\ln n)^{\ln n}}{n^p}$.

7.5.19.* Determine the condition that positive numbers p and q must satisfy in order for the series to converge conditionally or absolutely:

$$\sum_{n=1}^{\infty} (-1)^n \frac{(\ln n)^p}{n^q}.$$

7.5.20. Prove or disprove: if $\sum_{n=0}^{\infty} a_n$ converges and $\lim b_n = 0$, then $\sum_{n=0}^{\infty} a_n b_n$ converges.

7.5.21. Prove that if $\sum_{n=0}^{\infty} a_n$ converges and $\sum_{n=0}^{\infty} b_n$ converges absolutely then the series $\sum_{n=0}^{\infty} a_n b_n$ converges absolutely.

7.5.22. Let $\sum_{n=0}^{\infty} a_n$ be a conditionally convergent series, and let $\sum_{n=0}^{\infty} p_n$ denote the series obtained by deleting all the negative terms in $\sum_{n=0}^{\infty} a_n$. Also, let $\sum_{n=0}^{\infty} q_n$ denote the series obtained by deleting all the positive terms in $\sum_{n=0}^{\infty} a_n$. Prove that neither of the series $\sum_{n=0}^{\infty} p_n$ and $\sum_{n=0}^{\infty} q_n$ is convergent.

7.5.23. Prove that if $\sum_{n=0}^{\infty} a_n$ converges absolutely then $\sum_{n=0}^{\infty} a_n^2$ converges. Show that the assumption about *absolute* convergence is essential.

7.5.24. Prove that the series $\sum_{n=1}^{\infty} \frac{(-1)^{\lfloor \sqrt{n} \rfloor}}{n}$ converges.

7.5.25.* Suppose that $\{a_n\}$ is a positive decreasing sequence, ε_n equal 1 or -1 for all $n \in \mathbb{N}$, and the series $\sum_{n=1}^{\infty} \varepsilon_n a_n$ converges. Prove that $\lim(\varepsilon_1 + \varepsilon_2 + \cdots + \varepsilon_n) a_n = 0$.

7.5.1 Additional Tests for Convergence

We will present a few more tests for convergence of a series with terms of both signs. For the first one we need a lemma.

Lemma 7.5.8 (Summation by Parts). *Let $\{a_n\}$ and $\{b_n\}$ be two sequences of real numbers, let $m \in \mathbb{N}$, and let $B_n = \sum_{k=1}^{n} b_k$ for all $n \in \mathbb{N}$. Then*

$$\sum_{k=n}^{m} a_k b_k = a_{m+1} B_m - a_n B_{n-1} - \sum_{k=n}^{m} (a_{k+1} - a_k) B_k. \tag{7.9}$$

Notice the similarity between the formula (7.9) and the Integration by Parts formula (6.19).

Proof. We start with the equality $b_k = B_k - B_{k-1}$, multiply it by a_k to obtain $a_k b_k = a_k B_k - a_k B_{k-1}$, and we sum up these equalities for $k = n, n+1, \ldots, m$. We obtain

$$\sum_{k=n}^{m} a_k b_k = \sum_{k=n}^{m} a_k B_k - \sum_{k=n}^{m} a_k B_{k-1}. \tag{7.10}$$

Further, if we use the substitution $k = j + 1$ in the rightmost sum,

$$\sum_{k=n}^{m} a_k B_{k-1} = \sum_{j=n-1}^{m-1} a_{j+1} B_j = \sum_{j=n}^{m} a_{j+1} B_j + a_n B_{n-1} - a_{m+1} B_m. \tag{7.11}$$

Now (7.9) follows from (7.10) and (7.11). □

With the aid of the Summation by Parts, we can establish Abel's Test.

Theorem 7.5.9 (Abel's Test). *If the series $\sum_{k=1}^{\infty} b_k$ converges, and the sequence $\{a_n\}$ is monotone and bounded, then the series $\sum_{k=1}^{\infty} a_k b_k$ converges.*

Proof. Let s_n be the nth partial sum of the series $\sum_{k=1}^{\infty} a_k b_k$. We will show that s_n is a Cauchy sequence.

Let $\varepsilon > 0$. By assumption, the sequence $\{B_n\}$ (partial sums of $\sum_{k=1}^{\infty} b_k$) is convergent, hence bounded. Therefore, there exists M_1 such that

$$|B_n| \leq M_1, \text{ for any } n \in \mathbb{N}.$$

Similarly, the sequence $\{a_n\}$ is bounded, so there exists M_2 such that

$$|a_n| \leq M_2, \text{ for any } n \in \mathbb{N}.$$

If we denote $M = \max\{M_1, M_2\}$, we have that $|a_n| \leq M$ and $|B_n| \leq M$, for all $n \in \mathbb{N}$.

Let $B = \lim B_n$. Then there exists $N_1 \in \mathbb{N}$, such that

$$|B_n - B| < \frac{\varepsilon}{4M}, \text{ for } n \geq N_1.$$

Since the sequence $\{a_n\}$ is monotone and bounded, it is convergent, hence a Cauchy sequence, so there exists $N_2 \in \mathbb{N}$, such that

$$|a_m - a_n| < \frac{\varepsilon}{4M}, \text{ for } m \geq n \geq N_2.$$

Also, there exists $N_3 \in \mathbb{N}$, such that

$$|a_m - a_n| < \frac{\varepsilon}{4|B|}, \text{ for } m \geq n \geq N_3.$$

Let $N = \max\{N_1, N_2, N_3\}$, and suppose that $m \geq n \geq N$. Then, using Lemma 7.5.8,

$$|s_m - s_{n-1}| = \left| a_{m+1}B_m - a_n B_{n-1} - \sum_{k=n}^{m} (a_{k+1} - a_k)B_k \right|$$

$$\leq |a_{m+1}B_m - a_n B_{n-1}| + \sum_{k=n}^{m} |a_{k+1} - a_k| \, |B_k|$$

$$\leq |a_{m+1}B_m - a_{m+1}B| + |a_{m+1}B - a_n B| + |a_n B - a_n B_{n-1}|$$
$$+ \sum_{k=n}^{m} |a_{k+1} - a_k| \, M$$

$$\leq |a_{m+1}| \frac{\varepsilon}{4M} + |B| \frac{\varepsilon}{4|B|} + |a_n| \frac{\varepsilon}{4M} + M \sum_{k=n}^{m} |a_{k+1} - a_k|.$$

Consider the sum

$$|a_{n+1} - a_n| + |a_{n+2} - a_{n+1}| + \cdots + |a_{m+1} - a_m|.$$

Since $\{a_n\}$ is monotone the terms are either all positive or all negative. In the former case we obtain

$$(a_{n+1} - a_n) + (a_{n+2} - a_{n+1}) + \cdots + (a_{m+1} - a_m) = a_{m+1} - a_n$$

and in the latter

$$-(a_{n+1} - a_n) - (a_{n+2} - a_{n+1}) - \cdots - (a_{m+1} - a_m) = -(a_{m+1} - a_n).$$

Either way, we obtain $|a_{m+1} - a_n|$. Therefore,

$$|s_m - s_{n-1}| \leq M \frac{\varepsilon}{4M} + \frac{\varepsilon}{4} + M \frac{\varepsilon}{4M} + M|a_{m+1} - a_n|$$
$$\leq \frac{\varepsilon}{4} + \frac{\varepsilon}{4} + \frac{\varepsilon}{4} + M \frac{\varepsilon}{4M} = \varepsilon. \qquad \square$$

Example 7.5.10. Applying Abel's Test.

Prove that the series $\sum_{n=1}^{\infty} (-1)^n \dfrac{\arctan n}{\sqrt{n}}$ converges.

Solution. The first thought might be to try the Alternating Series Test. However, that would require proving that the sequence $\{\arctan n / \sqrt{n}\}$ is monotone decreasing. It is much easier to apply the Abel's Test. In order to do that, we need to write $(-1)^n \arctan n/\sqrt{n}$ as a product of two factors: one should be monotone and convergent, the other should give rise to a convergent series. Here, this is accomplished if we take $a_n = \arctan n$ and $b_n = (-1)^n/\sqrt{n}$. The sequence $\{a_n\}$ is increasing and it converges to $\pi/2$. The series $\sum_{n=1}^{\infty} b_n$ converges by the Alternating Series Test. Therefore, by the Abel's Test, the series $\sum_{n=1}^{\infty}(-1)^n \arctan n/\sqrt{n}$ converges. ♦

NIELS HENRIK ABEL (1802–1829) was the most famous Norwegian mathematician. Norway was a part of Denmark until 1814 and a part of Sweden until 1905. Abel grew up extremely poor. He did not excel in school until 1817, when a new mathematics teacher came to his school. He started reading the works of Euler, Newton, and Lagrange. His material situation became even worse when his father died in 1820. He was given a small scholarship to finish school and enroll in the university in Oslo, where he graduated in 1822. In 1824, he proved the impossibility of solving the general equation of the fifth degree in radicals. The Norwegian government financed his travels to Berlin and Paris. In Berlin he found a warm welcome and published several papers, including [1], which contains both Summation by Parts Lemma and the Abel's Test. In Paris his work received lukewarm reception. He returned home very ill at the end of 1826. For the next 3 years he did some remarkable work on elliptic functions as his health deteriorated. His friends in Berlin were finally able to secure him a position at the University of Berlin, but by the time the news arrived, he had died. After his death, a manuscript was found where he had shown (2 years before Galois) that there is no general algebraic formula for the roots of a polynomial equation of degree greater than four.

The other important test is due to Dirichlet. It was published in 1863, 4 years after his death, when his friend Dedekind edited Dirichlet's lecture notes.

Theorem 7.5.11 (The Dirichlet's Test). *Let $\{B_n\}$ be the sequence of partial sums of the series $\sum_{k=1}^{\infty} b_k$. If the sequence $\{B_n\}$ is bounded, and the sequence $\{a_n\}$ is monotone and converges to 0, then the series $\sum_{k=1}^{\infty} a_k b_k$ converges.*

Proof. Let s_n be the nth partial sum of the series $\sum_{k=1}^{\infty} a_k b_k$. We will show that $\{s_n\}$ is a Cauchy sequence.

Let $\varepsilon > 0$. By assumption, the sequence $\{B_n\}$ is bounded, so there exists M such that

$$|B_n| \leq M, \text{ for any } n \in \mathbb{N}.$$

Since $\lim a_n = 0$, there exists $N_1 \in \mathbb{N}$ such that

$$|a_n| < \frac{\varepsilon}{3M}, \text{ for } n \geq N_1.$$

Further, $\{a_n\}$ is a Cauchy sequence, so there exists $N_2 \in \mathbb{N}$ such that

$$|a_m - a_n| < \frac{\varepsilon}{3M}, \text{ for } m \geq n \geq N_2.$$

Let $N = \max\{N_1, N_2\}$, and suppose that $m \geq n \geq N$. Then

$$|s_m - s_{n-1}| = \left| a_{m+1} B_m - a_n B_{n-1} - \sum_{k=n}^{m} (a_{k+1} - a_k) B_k \right|$$

$$\leq |a_{m+1}||B_m| + |a_n||B_{n-1}| + \sum_{k=n}^{m} |a_{k+1} - a_k||B_k|$$

$$\leq \frac{\varepsilon}{3M} M + \frac{\varepsilon}{3M} M + \sum_{k=n}^{m} |a_{k+1} - a_k| M.$$

Just like in the proof of the Abel's Test, the monotonicity of $\{a_n\}$ implies that the last sum equals $|a_{m+1} - a_n|M$, so

$$|s_m - s_{n-1}| \leq \frac{\varepsilon}{3} + \frac{\varepsilon}{3} + M|a_{m+1} - a_n| < \frac{2\varepsilon}{3} + M \frac{\varepsilon}{3M} = \varepsilon. \qquad \square$$

Example 7.5.12. Applying Dirichlet's Test.

Determine for what x does the series $\sum\limits_{n=1}^{\infty} \dfrac{\cos nx}{n}$ converge.

Solution. If $x = 2k\pi$ where $k \in \mathbb{Z}$, then $\cos nx = 1$ so we get the Harmonic series, hence a divergent series. What about $x \neq 2k\pi$? We will apply the Dirichlet's Test, with $a_n = 1/n$, $b_n = \cos nx$. Clearly, $\{a_n\}$ is monotone and converges to 0. In Example 7.1.5 we have established that, when $x \neq 2k\pi$ and $n \in \mathbb{N}$,

$$B_n = \frac{\sin\left(n + \dfrac{1}{2}\right)x - \sin\dfrac{1}{2}x}{2\sin\dfrac{x}{2}}.$$

Therefore,

$$|B_n| \leq \frac{1}{\left|\sin\dfrac{x}{2}\right|},$$

so the sequence $\{B_n\}$ is bounded, and the series $\sum_{n=1}^{\infty} \cos nx/n$ is convergent.   ◆

Example 7.5.12 is an example of a Fourier series. Such series are often only conditionally convergent, and they are almost never alternating. This motivated Abel and Dirichlet to find the tests for convergence that we have presented.

Exercises

In Exercises 7.5.26–7.5.33 test the series for convergence and absolute convergence:

7.5.26. $\displaystyle\sum_{n=1}^{\infty} \frac{\ln^{100} n}{n} \sin\frac{n\pi}{4}$.

7.5.27. * $\displaystyle\sum_{n=1}^{\infty} \frac{\sin an}{\ln n}$, $a \in \mathbb{R}$.

7.5.28. $\displaystyle\sum_{n=1}^{\infty} \frac{\cos\frac{\pi n^2}{n+1}}{\ln^2 n}$.

7.5.29. * $\displaystyle\sum_{n=1}^{\infty} \frac{(\sin n)(\sin n^2)}{n}$.

7.5.30. $\displaystyle\sum_{n=1}^{\infty} (-1)^n \frac{\sin^2 n}{n}$.

7.5.31. * $\displaystyle\sum_{n=2}^{\infty} \frac{\sin\left(n + \frac{1}{n}\right)}{\ln\ln n}$.

7.5.32. $\displaystyle\sum_{n=1}^{\infty} (-1)^n \frac{\sqrt[n]{\ln x}}{n}$, $x > 1$.

7.5.33. * $\displaystyle\sum_{n=1}^{\infty} \frac{\sin n}{n}\left(1 + \frac{1}{2} + \frac{1}{3} + \cdots + \frac{1}{n}\right)$.

7.5.34. * Determine all values of the parameter p for which the series

$$\sum_{n=1}^{\infty} \frac{\sin\frac{n\pi}{4}}{n^p + \sin\frac{n\pi}{4}}$$

converges.

In Exercises 7.5.35–7.5.38 the harmonic series is modified by introducing $+$ and $-$ signs. Determine whether the obtained series converges if the signs are:

7.5.35. * In pairs: $1 + \dfrac{1}{2} - \dfrac{1}{3} - \dfrac{1}{4} + \dfrac{1}{5} + \dfrac{1}{6} - \cdots$.

7.5.36. * In threes: $1 + \dfrac{1}{2} + \dfrac{1}{3} - \dfrac{1}{4} - \dfrac{1}{5} - \dfrac{1}{6} + \cdots$.

7.5.37. * In groups of $1, 2, 3, 4, \ldots$: $1 - \dfrac{1}{2} - \dfrac{1}{3} + \dfrac{1}{4} + \dfrac{1}{5} + \dfrac{1}{6} - \cdots$.

7.5.38. * In groups of $1, 2, 4, 8, \ldots$: $1 - \dfrac{1}{2} - \dfrac{1}{3} + \dfrac{1}{4} + \dfrac{1}{5} + \dfrac{1}{6} + \dfrac{1}{7} - \cdots$.

7.5.39. Consider the series $\dfrac{\ln 2}{2\ln 2} + \dfrac{\ln 4}{3\ln 3} - \dfrac{\ln 6}{4\ln 4} - \dfrac{\ln 8}{5\ln 5} - \dfrac{\ln 10}{6\ln 6} + \dfrac{\ln 12}{7\ln 7} + \dfrac{\ln 14}{8\ln 8} + \cdots,$
where the terms of the same sign occur in groups of $2, 3, 4, 5$, etc., and determine whether it is convergent.

7.5.40. Suppose that the series $\sum_{n=1}^{\infty} a_n$ is convergent and that $\sum_{n=1}^{\infty}(b_n - b_{n+1})$ is absolutely convergent. Prove that the series $\sum_{n=1}^{\infty} a_n b_n$ is convergent.

7.5.41. Let $\{a_n\}$ be a given sequence and $c \in \mathbb{R}$ so that the *Dirichlet series*

$$\sum_{n=1}^{\infty} \frac{a_n}{n^x}$$

converges when $x = c$. Prove that the series converges for all $x > c$.

7.5.2 Rearrangement of a Series

When dealing with finite sums, the terms can be put in any order. What about infinite sums? The answer is very different depending on whether the series in question is absolutely convergent or only conditionally convergent. We will say that the series

$$\sum_{n=1}^{\infty} b_n = b_1 + b_2 + b_3 + \cdots \tag{7.12}$$

is a **rearrangement** of a series

$$\sum_{n=1}^{\infty} a_n = a_1 + a_2 + a_3 + \cdots \tag{7.13}$$

if they contain all the same terms. More precisely, this is the case if and only if there exists a bijection between the sets $\{a_1, a_2, a_3, \dots\}$ and $\{b_1, b_2, b_3, \dots\}$.

Theorem 7.5.13. *Let (7.13) be an absolutely convergent series and let (7.12) be a rearrangement of (7.13). Then the series (7.12) converges absolutely to the same sum as the series (7.13).*

Proof. Since, by assumption, the series $\sum_{n=1}^{\infty} |a_n|$ converges, the sequence of its partial sums $\{s_n\}$ is bounded, so there exists M such that

$$s_n = \sum_{i=1}^{n} |a_i| \leq M, \text{ for any } n \in \mathbb{N}.$$

We will show that the sequence $\{t_n\}$ of the partial sums of the series $\sum_{n=1}^{\infty} |b_n|$ is also bounded. Indeed, let n be an arbitrary positive integer. Then

$$t_n = |b_1| + |b_2| + \cdots + |b_n| = |a_{k_1}| + |a_{k_2}| + \cdots + |a_{k_n}| \leq s_{k_n} \leq M.$$

Now Theorem 7.3.1 implies that the series (7.12) converges absolutely. Let A be the sum of (7.13), and let B be the sum of (7.12). We will show that $A = B$ by demonstrating that, for any $\varepsilon > 0$, $|A - B| < \varepsilon$.

So, let $\varepsilon > 0$, and let A_n and B_n denote the nth partial sums of (7.13) and (7.12). There exists $N_1, N_2 \in \mathbb{N}$ such that

$$|A_n - A| < \frac{\varepsilon}{3}, \text{ for } n \geq N_1,$$

$$|B_n - B| < \frac{\varepsilon}{3}, \text{ for } n \geq N_2.$$

Further, $\{s_n\}$ is a Cauchy sequence, so there exists $N_3 \in \mathbb{N}$ such that

$$|s_m - s_n| < \frac{\varepsilon}{3}, \text{ for } m \geq n \geq N_3.$$

Let $N = \max\{N_1, N_2, N_3\}$ and fix $n \geq N$. For any $j \in \mathbb{N}$, there exists $k_j \in \mathbb{N}$ such that $a_j = b_{k_j}$. Thus the set $\{a_1, a_2, \ldots, a_n\} = \{b_{k_1}, b_{k_2}, \ldots, b_{k_n}\}$. Let $k = \max\{k_1, k_2, \ldots, k_n\}$. Then

$$\{a_1, a_2, \ldots, a_n\} \subset \{b_1, b_2, \ldots, b_k\}. \tag{7.14}$$

On the other hand, for any $j \in \mathbb{N}$, there exists $l_j \in \mathbb{N}$ such that $b_j = a_{l_j}$. Thus, $\{b_1, b_2, \ldots, b_k\} = \{a_{l_1}, a_{l_2}, \ldots, a_{l_k}\}$. Let $m = \max\{l_1, l_2, \ldots, l_k\}$. Then

$$\{a_1, a_2, \ldots, a_n\} \subset \{b_1, b_2, \ldots, b_k\} \subset \{a_1, a_2, \ldots, a_m\}.$$

Clearly, $m \geq k \geq n \geq N$. Also,

$$|A - B| \leq |A - A_m| + |A_m - B_k| + |B_k - B| < \frac{\varepsilon}{3} + |A_m - B_k| + \frac{\varepsilon}{3}.$$

We will now show that $|A_m - B_k| < \varepsilon/3$.

Notice that

$$A_m - B_k = (a_1 + a_2 + \cdots + a_m) - (b_1 + b_2 + \cdots + b_k).$$

Since each term in the second parentheses is one of the terms in the first parentheses, they will all cancel. Thus, the remaining terms form a subset of $\{a_1, a_2, \ldots, a_m\}$. On the other hand, (7.14) shows that the canceled terms include all a_j with $1 \leq j \leq n$. In other words the remaining terms form a subset of $\{a_{n+1}, a_{n+2}, \ldots, a_m\}$. Consequently,

$$|A_m - B_k| \leq |a_{n+1}| + |a_{n+2}| + \cdots + |a_m| = s_m - s_n < \frac{\varepsilon}{3}. \qquad \square$$

The situation with conditionally convergent series is quite different. It turns out that, by rearranging such a series we can get it to converge to a different sum.

Example 7.5.14. Rearranging a series yields a different sum.

The series $1 - \dfrac{1}{2} + \dfrac{1}{3} - \dfrac{1}{4} + \cdots + \dfrac{1}{2k-1} - \dfrac{1}{2k} + \ldots$ can be rearranged so that it converges to a twice less sum.

Solution. This is a conditionally convergent series (Example 7.5.6). Let us denote its nth partial sum by A_n and its sum by A. Let us rearrange its terms in the following manner:

$$\left(1 - \frac{1}{2} - \frac{1}{4}\right) + \left(\frac{1}{3} - \frac{1}{6} - \frac{1}{8}\right) + \cdots + \left(\frac{1}{2k-1} - \frac{1}{4k-2} - \frac{1}{4k}\right) + \ldots. \tag{7.15}$$

We will show that this series converges to $A/2$. Indeed,

$$\frac{1}{2k-1} - \frac{1}{4k-2} - \frac{1}{4k} = \frac{2}{4k-2} - \frac{1}{4k-2} - \frac{1}{4k} = \frac{1}{4k-2} - \frac{1}{4k} = \frac{1}{2}\left(\frac{1}{2k-1} - \frac{1}{2k}\right),$$

so if we denote the nth partial sum of the series (7.15) by B_n, we have that

$$B_{3n} = \sum_{k=1}^{n} \frac{1}{2}\left(\frac{1}{2k-1} - \frac{1}{2k}\right) = \frac{1}{2}A_{2n} \to \frac{1}{2}A.$$

Since $B_{3n-1} = B_{3n} + \dfrac{1}{4n}$ and $B_{3n+1} = B_{3n} + \dfrac{1}{2n-1}$, it is easy to see that $\lim B_{3n-1} = \lim B_{3n+1} = A/2$. $\blacklozenge$

The series in the last example shows that, with conditionally convergent series, a rearrangement can converge to a different sum. Can it diverge?

Example 7.5.15. Rearranging a series yields a divergent series.

The series $1 - \dfrac{1}{2} + \dfrac{1}{3} - \dfrac{1}{4} + \cdots + \dfrac{1}{2k-1} - \dfrac{1}{2k} + \ldots$ can be rearranged so that it becomes divergent.

Solution. Let $S_n = 1 + 1/3 + 1/5 + \cdots + 1/(2n-1)$, and let H_n denote the nth partial sum of the Harmonic series. It is not hard to see that

$$S_n = \left(1 + \frac{1}{2} + \frac{1}{3} + \cdots + \frac{1}{2n}\right) - \left(\frac{1}{2} + \frac{1}{4} + \cdots + \frac{1}{2n}\right) = H_{2n} - \frac{1}{2}H_n.$$

Since $H_{2n} > H_n$ we see that $S_n > H_n/2 \to \infty$. Consequently, for any $M > 0$, there exists $n \in \mathbb{N}$, such that $S_n > M$. Let us define the sequence $n_1 < n_2 < n_3 < \ldots$ so that

$$S_{n_1} > \frac{1}{2} + 1,$$

$$S_{n_2} > \frac{1}{2} + \frac{1}{4} + 2,$$

$$S_{n_3} > \frac{1}{2} + \frac{1}{4} + \frac{1}{6} + 3,$$

$$\ldots$$

$$S_{n_k} > \frac{1}{2} + \frac{1}{4} + \cdots + \frac{1}{2k} + k$$

$$\ldots$$

Now we consider the rearrangement

$$\frac{1}{3} + \frac{1}{5} + \cdots + \frac{1}{2n_1 - 1} - \frac{1}{2} + \frac{1}{2n_1 + 1} + \frac{1}{2n_1 + 3} + \cdots + \frac{1}{2n_2 - 1} - \frac{1}{4}$$
$$+ \frac{1}{2n_2 + 1} + \frac{1}{2n_2 + 3} + \cdots + \frac{1}{2n_3 - 1} - \frac{1}{6} + \ldots \quad (7.16)$$

It is easy to see that, if we denote by B_n the nth partial sum of this series, we have that

$$B_{n_1+1} = S_{n_1} - \frac{1}{2} > 1,$$

$$B_{n_2+2} = S_{n_2} - \frac{1}{2} - \frac{1}{4} > 2,$$

$$\ldots$$

$$B_{n_k+k} = S_{n_k} - \frac{1}{2} - \frac{1}{4} - \cdots - \frac{1}{2k} > k,$$

$$\ldots$$

Therefore, the subsequence B_{n_k+k} has an infinite limit, and the series (7.16) diverges. ◆

The first examples that illustrate the phenomena exhibited in Examples 7.5.14 and 7.5.15 were given by Dirichlet in [34] in 1837 (see Exercises 7.5.45 and 7.5.48), who became aware of them while investigating Fourier series.

The fact that we have used the Alternating Harmonic Series in Example 7.5.15 was due to the simplicity of demonstration. In fact, the same magic can be done with any conditionally convergent series.

Theorem 7.5.16 (Riemann's Theorem). *Let $\sum_{n=1}^{\infty} a_n$ be a conditionally convergent series and let A denote any real number or infinity. There exists a rearrangement $\sum_{n=1}^{\infty} b_n$ that converges to A.*

Proof. We will prove the case when A is a finite number and leave the infinite case as an exercise. Let us denote by p_n the positive and by q_n the negative terms in the series $\sum_{n=1}^{\infty} a_n$. Since the series is not absolutely convergent, neither $\sum_{n=1}^{\infty} p_n$ nor $\sum_{n=1}^{\infty} q_n$ converges. Since both are the series with positive terms, their partial sums P_n and Q_n must have infinite limits. In particular, there exists a positive integer n such that $P_n > A$. Let n_1 be the smallest positive integer with this property. We will start the desired rearrangement with

$$p_1 + p_2 + \cdots + p_{n_1}.$$

Since we have "overshot" A, we need to take some negative terms, until we get below A. More precisely, let m_1 be the smallest positive integer such that $P_{n_1} - Q_{m_1} < A$. Thus, the desired rearrangement will start with

$$p_1 + p_2 + \cdots + p_{n_1} - q_1 - q_2 - \cdots - q_{m_1} < A.$$

Now we use positive terms again. Let n_2 be the smallest positive integer such that

$$p_1 + p_2 + \cdots + p_{n_1} - q_1 - q_2 - \cdots - q_{m_1} + p_{n_1+1} + p_{n_1+2} + \cdots + p_{n_2} > A.$$

Continuing this process, we obtain two sequences n_k and m_k of positive integers, such that

$$P_{n_k} - Q_{m_k} < A, \quad \text{and} \quad P_{n_{k+1}} - Q_{m_k} > A.$$

It remains to show that this rearrangement converges to A. To that end, the fact that $\sum_{n=1}^{\infty} a_n$ converges implies that $\lim a_n = 0$, so $\lim p_n = 0$ and $\lim q_n = 0$. Let $\varepsilon > 0$, and choose $N \in \mathbb{N}$ so that

$$p_{n_k} < \varepsilon, \quad q_{m_k} < \varepsilon, \quad \text{for} \quad k \geq N.$$

Let $n \geq N$ and let B_n be the nth partial sum of the rearrangement.

The terms of the series change sign when they get to the term number n_1, $n_1 + m_1$, $n_2 + m_1$, $n_2 + m_2$, etc. So, n must fall between two subsequent turning points. Suppose that $n_{k+1} + m_k \leq n < n_{k+1} + m_{k+1}$. Then

$$|B_n - A| = B_n - A$$
$$\leq B_{n_{k+1}+m_k} - A = (P_{n_{k+1}} - Q_{m_k}) - A$$
$$= (p_1 + p_2 + \cdots + p_{n_{k+1}-1} + p_{n_{k+1}} - Q_{m_k}) - A$$
$$= (P_{n_{k+1}-1} - Q_{m_k}) - A + p_{n_{k+1}}.$$

By definition, n_{k+1} is the *smallest* positive integer such that $P_{n_{k+1}} - Q_{m_k} > A$. Consequently,

$$P_{n_{k+1}-1} - Q_{m_k} < A$$

whence

$$|B_n - A| < p_{n_{k+1}} < \varepsilon.$$

A similar proof can be written in the case when $n_k + m_k \leq n < n_{k+1} + m_k$.   □

Riemann proved this theorem in 1854, in his Habilitation Thesis.

Exercises

7.5.42. Prove the converse of Theorem 7.5.13: If every rearrangement of a series converges to the same sum, then the series is absolutely convergent.

7.5.43. Complete the proof of Riemann's Theorem by considering the case when $n_k + m_k \leq n < n_{k+1} + m_k$.

7.5.44. Complete the proof of Riemann's Theorem by considering the case when A is $+\infty$.

7.5.45.* Show that the series $\sum_{n=1}^{\infty} \frac{(-1)^{n+1}}{\sqrt{n}}$ is convergent but $1 + \frac{1}{\sqrt{3}} - \frac{1}{\sqrt{2}} + \frac{1}{\sqrt{5}} + \frac{1}{\sqrt{7}} - \frac{1}{\sqrt{4}} + \dots$ diverges.

7.5.46. Find a rearrangement of the series $\sum_{n=1}^{\infty} \frac{(-1)^{n+1}}{\sqrt{n}}$ that converges to 0. Write the first 30 terms and justify your choice!

7.5.47. Find a rearrangement of the series $\sum_{n=1}^{\infty} \frac{(-1)^{n+1}}{\sqrt{n}}$ that converges to $+\infty$. Write the first 30 terms and justify your choice!

7.5.48. Show that the series $\sum_{n=1}^{\infty} \frac{(-1)^{n+1}}{n}$ and $1 + \frac{1}{3} - \frac{1}{2} + \frac{1}{5} + \frac{1}{7} - \frac{1}{4} + \dots$ converge to different values. [You cannot use the result of Exercise 7.5.49.]

7.5.49. Prove that if Alternating Harmonic Series is rearranged by alternating p positive and q negative terms, the sum of the rearrangement is $\ln 2 + \frac{1}{2} \ln \frac{p}{q}$.

7.5.50. Let $\{p_n\}, \{q_n\}$ be two sequences of non-negative integers such that

$$\lim \frac{p_1 + p_2 + \dots + p_n}{q_1 + q_2 + \dots + q_n} = \lim \frac{p_1 + p_2 + \dots + p_n + p_{n+1}}{q_1 + q_2 + \dots + q_n} = L.$$

Suppose that Alternating Harmonic Series is rearranged by alternating p_1 positive terms, q_1 negative terms, p_2 positive terms, q_2 negative terms, etc. Prove that the obtained series converges and that its sum is $\ln 2 + \frac{1}{2} \ln L$.

7.5.51. Use the result of Exercise 7.5.50 to find a rearrangement of Alternating Harmonic Series that converges to $\ln 2 + \frac{1}{2} \ln \pi$.

7.5.52. Let $\{a_n\}$ be a monotone sequence of positive integers such that $\lim n a_n = L$ and let $p, q \in \mathbb{N}$. Suppose that the convergent series $\sum_{n=1}^{\infty} (-1)^{n+1} a_n$ is rearranged by alternating p positive and q negative terms. Prove that the obtained series converges and that its sum is $S + \frac{1}{2} \ln \frac{p}{q}$, where S is the sum of $\sum_{n=1}^{\infty} (-1)^{n+1} a_n$.

8

Sequences and Series of Functions

In this chapter we will study the sequences of functions. Such an example is

$$\sin x, \sin 2x, \sin 3x, \dots$$

and we will write $f_n(x) = \sin nx$. The difference between the sequence above and those that we have studied in Chapter 2 is that the members are not any more real numbers but functions. Nevertheless, we will ask the same type of questions as in Chapter 2: does a given sequence converge and, if it does, to what limit? Once we get a good grasp on the convergence of sequences of functions, we will look at the series of functions.

8.1 Convergence of a Sequence of Functions

Suppose that a sequence of functions $\{f_n\}$ is defined by

$$f_n(x) = \frac{nx}{1 + n + x^2}, \tag{8.1}$$

and let us find $\lim f_n(x)$. After dividing both the numerator and the denominator by n, we get

$$\lim_{n \to \infty} \frac{x}{\frac{1}{n} + 1 + \frac{x^2}{n}}.$$

Since $\lim 1/n = 0$ and $\lim x^2/n = 0$, we see that $\lim f_n(x) = x$.

In this example, the members of the sequence were functions, so we are not surprised to discover that the limit is also a function $f(x) = x$. We would like to adapt the definition of the limit to this new situation. Let us look carefully at Definition 2.2.6. A number L qualified as a limit of the sequence $\{a_n\}$ if the members of $\{a_n\}$ were as close to L as needed, and this distance was measured on the number line. For example, the fact that $\lim 1/n = 0$ is intuitively clear: if we replace n by a large number, say $n = 1000$, then $1/1000 = 0.001$ is close to 0. If we apply the same reasoning to the sequence $\{f_n\}$, we should have f_{1000} "close to" f. However, $f_{1000}(x) = (1000x)/(1001 + x^2)$ so we need to clarify in what sense is this function close to $f(x) = x$.

It turns out that this can be (and is) done in several different ways. We will first talk about the **pointwise** convergence.

Definition 8.1.1. We say that a sequence of functions $\{f_n\}$, defined on a common domain A, converges **pointwise** to a function f on A if, for any $x \in A$, the sequence of real numbers $\{f_n(x)\}$ converges to a real number $f(x)$.

Example 8.1.2. A pointwise convergent sequence of functions.

Prove that the sequence $\{f_n\}$, defined by (8.1), converges pointwise to $f(x) = x$.

Solution. Since $1 + n + x^2 \neq 0$ for any $x \in \mathbb{R}$, the common domain of all functions f_n is the whole real line. Let $x \in \mathbb{R}$ and let $\varepsilon > 0$. We define

$$N = \max\left\{ \left\lfloor (1 + x^2) \left(\frac{|x|}{\varepsilon} - 1 \right) \right\rfloor + 1, \, 1 \right\}.$$

Now, if $n \geq N$, then

$$n > (1 + x^2) \left(\frac{|x|}{\varepsilon} - 1 \right) = |x| \frac{1 + x^2}{\varepsilon} - (1 + x^2),$$

so

$$1 + n + x^2 > \frac{|x|(1 + x^2)}{\varepsilon} = \frac{|x + x^3|}{\varepsilon},$$

and

$$\frac{|x + x^3|}{1 + n + x^2} < \varepsilon.$$

Therefore,

$$|f_n(x) - f(x)| = \left| \frac{nx}{1 + n + x^2} - x \right| = \left| \frac{nx - x(1 + n + x^2)}{1 + n + x^2} \right| = \frac{|-x - x^3|}{1 + n + x^2} < \varepsilon. \qquad \blacklozenge$$

Notice that, by selecting a value of x and keeping it fixed, we have been able to use the same techniques as for sequences of real numbers. That means that all the results from Chapter 2 carry over and it would seem that there is not much to say about sequences of functions. That might have been true if we had decided to stick with the pointwise convergence. Unfortunately, its relative simplicity comes with a price.

Example 8.1.3. A sequence of continuous functions converges to a function that is not continuous.

Let $\{f_n\}$ be defined by $f_n(x) = x^n$, $0 \leq x \leq 1$. We will show that $\lim f_n$ is not a continuous function.

Solution. If $x < 1$, then $\lim x^n = 0$ by Remark 2.2.10. However, if $x = 1$, then $f_n(1) = 1^n = 1$, so $\lim f_n(1) = 1$. It follows that the limit function is

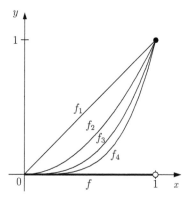

Figure 8.1: The limit function is not continuous.

$$f(x) = \begin{cases} 0, & \text{if } 0 \leq x < 1 \\ 1, & \text{if } x = 1. \end{cases}$$

We notice that f has a discontinuity at $x = 1$, in spite of the fact that, for each $n \in \mathbb{N}$, the function f_n is continuous at every point of $[0, 1]$. $\qquad \blacklozenge$

One way to avoid the situation like this is to introduce a different definition of the convergence of a sequence of functions. In order to do that, it is a good idea to see what "went wrong" in Example 8.1.3. Notice that

$$f(x) - f(1) = [f(x) - f_n(x)] + [f_n(x) - f_n(1)] + [f_n(1) - f(1)]$$

and each of the three expressions in brackets is supposed to be small. The first and the last because of the convergence of f_n to f, and the middle one for a fixed n, as $x \to 1$, because each f_n is continuous. Yet, the left hand side equals 1, no matter what $x \neq 1$ we take. The problem is that, as x gets closer to 1, we need to choose larger and larger n for which $f_n(x)$ and $f(x)$ are close. Namely, for $0 \leq x < 1$,

$$|f(x) - f_n(x)| = |0 - x^n| = x^n,$$

and

$$x^n < \varepsilon \quad \Leftrightarrow \quad n > \frac{\ln \varepsilon}{\ln x}.$$

Thus, the closer x is to 1, the larger n is. For example, if we set $\varepsilon = 0.1$, then $x = 0.99$ will require $n \geq 230$, $x = 0.999$ will need $n \geq 2302$, and for $x = 0.9999$, it takes $n \geq 23{,}025$. In other words, any fixed $n \in \mathbb{N}$ will be inadequate as $x \to 1$.

Assuming that we do not want to run into situations like this one, we had better put a more stringent condition on the sequence $\{f_n\}$. More precisely, we will require that, given ε, the positive integer N works for *all* x in the domain. Basically, this forces all sequences $\{f_n(x)\}$ (i.e., for each fixed x) to converge at a comparable (uniform) rate.

Definition 8.1.4. We say that a sequence of functions $\{f_n\}$, defined on a common domain A, converges **uniformly** to a function f on A, if for any $\varepsilon > 0$ there exists a positive integer N such that, for $n \geq N$, and $x \in A$, $|f_n(x) - f(x)| < \varepsilon$.

Example 8.1.5. A uniformly convergent sequence of functions.

The sequence $f_n(x) = \dfrac{nx^3}{1 + nx^2}$ converges uniformly on $\mathbb{R}$.

Solution. After dividing both the numerator and the denominator by n, we get

$$f_n(x) = \frac{x^3}{\frac{1}{n} + x^2}.$$

Since $\lim 1/n = 0$, we see that, if $x \neq 0$, $\lim f_n(x) = x^3/x^2 = x$. If $x = 0$, then $f_n(0) = 0 \to 0 = x$, so $f(x) = x$ is the limit. We will show that the convergence is uniform on $\mathbb{R}$. As always, we start with the inequality $|f_n(x) - f(x)| < \varepsilon$. Here,

$$\left| \frac{nx^3}{1 + nx^2} - x \right| = \left| \frac{nx^3 - x(1 + nx^2)}{1 + nx^2} \right| = \frac{|nx^3 - x - nx^3|}{1 + nx^2} = \frac{|x|}{1 + nx^2}.$$

Thus, we would like to find N such that, regardless of x, $|x|/(1 + nx^2) < \varepsilon$ as soon as $n \geq N$. It would greatly help our cause if we could find an expression that is bigger then $|x|/(1 + nx^2)$ but does not depend on x. We can use the fact that $a^2 + b^2 \geq 2ab$, for any real numbers a, b. Therefore, $1 + nx^2 \geq 2|x|\sqrt{n}$ and

$$\left| \frac{nx^3}{1 + nx^2} - x \right| = \frac{|x|}{1 + nx^2} \leq \frac{|x|}{2|x|\sqrt{n}} = \frac{1}{2\sqrt{n}}. \tag{8.2}$$

With x out of the picture, we can make $1/(2\sqrt{n}) < \varepsilon$, which happens if and only if $n > 1/(2\varepsilon)^2$.

Proof. Let $\varepsilon > 0$. We define $N = \lfloor 1/(2\varepsilon)^2 \rfloor + 1$. If $n \geq N$ then $n > 1/(2\varepsilon)^2$ which implies that $1/(2\sqrt{n}) < \varepsilon$. Now, using (8.2),

$$|f_n(x) - f(x)| = \left| \frac{nx^3}{1 + nx^2} - x \right| \leq \frac{1}{2\sqrt{n}} < \varepsilon.$$

We conclude that the sequence $\{f_n\}$ converges uniformly on $\mathbb{R}$ to $f(x) = x$. ♦

We have a strong suspicion that the sequence in Example 8.1.3 does not converge uniformly. How do we prove that? The following result is very useful when establishing uniform convergence or lack thereof.

Theorem 8.1.6. *A sequence of functions $\{f_n\}$, defined on a common domain A, converges uniformly to a function f on A if and only if $\sup_{x \in A} |f_n(x) - f(x)| \to 0$, as $n \to \infty$.*

Proof. Suppose first that $\{f_n\}$ converges uniformly to f on A, and let $\varepsilon > 0$. By definition, there exists $N \in \mathbb{N}$ such that

$$|f_n(x) - f(x)| < \frac{\varepsilon}{2}, \quad \text{for} \quad n \geq N, \quad x \in A.$$

So, for that N and $n \geq N$, we see that the set $\{|f_n(x) - f(x)| : x \in A\}$ is bounded and that $\varepsilon/2$ is an upper bound. Consequently,

$$\sup_{x \in A} |f_n(x) - f(x)| \leq \frac{\varepsilon}{2} < \varepsilon,$$

so $\sup_{x \in A} |f_n(x) - f(x)| \to 0$, as $n \to \infty$.

Suppose now that $\sup_{x \in A} |f_n(x) - f(x)| \to 0$, as $n \to \infty$, and let $\varepsilon > 0$. Notice that $\sup_{x \in A} |f_n(x) - f(x)|$ is a real number (and not a function) for any $n \in \mathbb{N}$. Thus, there exists $N \in \mathbb{N}$ such that

$$\sup_{x \in A} |f_n(x) - f(x)| < \varepsilon, \quad \text{for} \quad n \geq N.$$

For that N and $n \geq N$, let $x \in A$. Then

$$|f_n(x) - f(x)| \leq \sup_{x \in A} |f_n(x) - f(x)| < \varepsilon. \qquad \square$$

Example 8.1.7. A pointwise convergent sequence of functions that does not converge uniformly.

We return to Example 8.1.3, and we will show that $\{f_n\}$ does not converge uniformly.

Solution. Here $A = [0,1]$, $f_n(x) = x^n$, and $f(x) = 0$ if $0 \leq x < 1$ and $f(x) = 1$ if $x = 1$. Therefore

$$|f_n(x) - f(x)| = \begin{cases} |x^n|, & \text{if } 0 \leq x < 1 \\ 0, & \text{if } x = 1 \end{cases}$$

so $\sup_{x \in A} |f_n(x) - f(x)| = 1$. It follows that

$$\sup_{x \in A} |f_n(x) - f(x)| \to 1, \quad n \to \infty.$$

By Theorem 8.1.6, the sequence $\{f_n\}$ does not converge uniformly on A. ♦

The first mention of uniform convergence is in an 1838 paper by Christoph Gudermann (1798–1852), best known as a teacher of Weierstrass. The importance of this mode of convergence was fully recognized and utilized by Weierstrass, starting with an 1841 paper *Zur Theorie der Potenzreihen* (*On the Theory of Power Series*), published in 1894. In German, the expression he used is "gleichmäßig konvergent". Independently, Stokes (page 513) introduced it in 1847. For further reading on this topic we recommend [55].

Notice that Definition 8.1.4 includes the explicit mention of the set A. The reason is that the concept of uniform convergence is highly sensitive to the change of the domain. Namely, the same sequence of functions may converge uniformly on one set, but not on another.

Example 8.1.8. A change of domain makes a sequence of functions uniformly convergent.

Prove that the sequence $\{f_n\}$, defined by $f_n(x) = x^n$, converges uniformly for $0 \le x \le 1/2$.

Solution. The limit function is $f(x) = 0$, and $|f_n(x) - f(x)| = |x^n - 0| = x^n$, so

$$\sup\{|f_n(x) - f(x)| : 0 \le x \le 1/2\} = \sup\{x^n : 0 \le x \le 1/2\} = 1/2^n \to 0.$$

It follows that the convergence is uniform on $[0, 1/2]$. ◆

If $\{a_n\}$ is a sequence of real numbers, then it is a Cauchy sequence if and only if it converges. A sequence of functions $\{f_n\}$ converges pointwise if and only if $\{f_n(x)\}$ converges for every x, hence if and only if $\{f_n(x)\}$ is a Cauchy sequence for every x. Thus, we have the following result.

Theorem 8.1.9. *A sequence of functions $\{f_n\}$, defined on a common domain A, converges pointwise to a function f on A, if and only if for any $\varepsilon > 0$ and any $x \in A$, there exists a positive integer N such that, for $m \ge n \ge N$, $|f_m(x) - f_n(x)| < \varepsilon$.*

Theorem 8.1.9 states that a sequence of functions is pointwise convergent if and only if it is a pointwise Cauchy sequence. What happens if we replace the word pointwise with uniform? Will it remain true? The answer is in the affirmative and it is supplied by the next theorem.

Theorem 8.1.10. *A sequence of functions $\{f_n\}$, defined on a common domain A, converges uniformly on A, if and only if for any $\varepsilon > 0$, there exists a positive integer N such that, for $m \ge n \ge N$ and any $x \in A$, $|f_m(x) - f_n(x)| < \varepsilon$.*

Proof. Suppose first that the sequence $\{f_n\}$ converges uniformly to f on A, and let $\varepsilon > 0$. Then there exists a positive integer N such that

$$|f_n(x) - f(x)| < \frac{\varepsilon}{2}, \quad \text{for any } n \ge N, \quad x \in A.$$

Let $m \ge n \ge N$ and $x \in A$. Then

$$|f_m(x) - f_n(x)| \le |f_m(x) - f(x)| + |f(x) - f_n(x)| < \frac{\varepsilon}{2} + \frac{\varepsilon}{2} = \varepsilon.$$

Now we will prove the converse. First we will establish that the sequence $\{f_n\}$ converges pointwise. Let $\varepsilon > 0$ and let $c \in A$. By assumption, there exists $N \in \mathbb{N}$ such that

$$|f_m(x) - f_n(x)| < \frac{\varepsilon}{2}, \quad \text{for } m \ge n \ge N \text{ and } x \in A. \tag{8.3}$$

In particular, for such m, n, $|f_m(c) - f_n(c)| < \varepsilon/2$, so the sequence $f_n(c)$ is a Cauchy sequence, hence convergent. Let $f(c)$ denote its limit. Since c was arbitrary, we obtain a correspondence $c \mapsto f(c)$ for any $c \in A$. In other words, we have a function f that is a pointwise limit of f_n. The convergence is, in fact, uniform. Indeed, if in the inequality (8.3) we let $m \to \infty$, we obtain

$$|f(x) - f_n(x)| \leq \frac{\varepsilon}{2} < \varepsilon, \quad \text{for} \ \ n \geq N \ \text{ and } \ x \in A,$$

so f_n converges uniformly on A. $\square$

We can see that two modes of convergence are different. When $\{f_n\}$ converges to f pointwise, there is no guarantee that relevant properties of the functions f_n will be shared by f. In the next section we will show that the uniform convergence, although more demanding, is much better in this regard.

Exercises

In Exercises 8.1.1–8.1.6 determine whether the sequence $\{f_n\}$ converges uniformly on $[0, 1]$:

8.1.1. $f_n(x) = x^n - x^{n+1}$.   8.1.2. $f_n(x) = nxe^{-nx}$.   8.1.3. $f_n(x) = \dfrac{nx}{1 + n + x}$.

8.1.4. $f_n(x) = nx(1 - x^2)^n$.   8.1.5. $f_n(x) = \dfrac{\lfloor nf(x) \rfloor}{n}$, f is a function defined on $[0, 1]$.

8.1.6. $f_n(x) = \begin{cases} n^2 x, & \text{if } 0 \leq x \leq \dfrac{1}{n} \\ n^2 \left(\dfrac{2}{n} - x \right), & \text{if } \dfrac{1}{n} < x < \dfrac{2}{n} \\ 0, & \text{if } x \geq \dfrac{2}{n}. \end{cases}$

In Exercises 8.1.7–8.1.9 determine whether the sequence $\{f_n\}$ converges uniformly on $\mathbb{R}$:

8.1.7. $f_n(x) = \dfrac{\sin nx}{n}$.   8.1.8. $f_n(x) = \sin \dfrac{x}{n}$.

8.1.9. $f_n(x) = f(x + \frac{1}{n})$ and (a) f is continuous on $\mathbb{R}$; (b) f is uniformly continuous on $\mathbb{R}$.

In Exercises 8.1.10–8.1.11 determine whether the sequence $\{f_n\}$ converges uniformly on $(0, +\infty)$:

8.1.10. $f_n(x) = n \left(\sqrt{x + \dfrac{1}{n}} - \sqrt{x} \right)$.

8.1.11. $f_n(x) = g(nx)$, $g(x) = \begin{cases} x, & \text{if } 0 \leq x < \dfrac{1}{2} \\ 1 - x, & \text{if } \dfrac{1}{2} \leq x \leq 1 \\ 0, & \text{if } x > 1. \end{cases}$

In Exercises 8.1.12–8.1.14 determine one interval where the sequence $\{f_n\}$ converges uniformly, and one interval where it does not:

8.1.12. $f_n(x) = \dfrac{\ln(1 + nx)}{n}$.   8.1.13. $f_n(x) = \dfrac{x^n}{1 + x^n}$.   8.1.14. $f_n(x) = \dfrac{nx}{1 + n^2 x^2}$.

8.1.15.* Prove that if a sequence $\{f_n\}$ converges uniformly on a set A, then any subsequence converges uniformly on A. Give an example to show that $\{f_n\}$ may converge non-uniformly but have a uniformly convergent subsequence.

8.1.16.* Prove that if a monotonic sequence $\{f_n\}$ converges on a set A and has uniformly convergent subsequence, then $\{f_n\}$ converges uniformly.

8.1.17. Prove Dini's Theorem: Let the functions $\{f_n\}$ and f be defined and continuous on $[a,b]$, let $f_n(x) \le f_{n+1}(x)$ for all $n \in \mathbb{N}$ and all $x \in [a,b]$, and let $\{f_n\}$ converge to f pointwise. Then the convergence is uniform on $[a,b]$.

In Exercises 8.1.18–8.1.21 $\{f_n\}, \{g_n\}$ are two uniformly convergent sequences on $[a,b]$.

8.1.18. Prove that $\{f_n + g_n\}$ converges uniformly on $[a,b]$.

8.1.19. Give an example to show that $\{f_n g_n\}$ need not converge uniformly to fg on $[a,b]$.

8.1.20. Let M be such a positive number that $|f_n(x)|, |g_n(x)| \le M$ for all $n \in \mathbb{N}$ and all $x \in [a,b]$. Prove that $\{f_n g_n\}$ converges uniformly on $[a,b]$.

8.1.21.* Let M be such a positive number that $|f_n(x)| \le M$ for all $n \in \mathbb{N}$ and all $x \in [a,b]$. Prove or disprove: $\{f_n g_n\}$ converges uniformly on $[a,b]$.

8.1.22.* Give an example of 2 sequences of continuous functions $\{f_n\}, \{g_n\}$ on (a,b) such that $\{f_n\}$ converges uniformly to f and that $\{g_n\}$ converges uniformly to g on (a,b), but $\{f_n g_n\}$ does not converge uniformly to fg on (a,b).

8.1.23.* For each $n \in \mathbb{N}$, let $f_n(x) = x(1 + 1/n)$ and

$$g_n(x) = \begin{cases} \dfrac{1}{n}, & \text{if } x = 0 \text{ or } x \text{ is irrational} \\[2mm] q + \dfrac{1}{n}, & \text{if } x = \dfrac{p}{q}, \ p \in \mathbb{Z}, \ q \in \mathbb{N}, \text{ and } p, q \text{ are mutually prime.} \end{cases}$$

Prove that both $\{f_n\}$ and $\{g_n\}$ converge uniformly on any bounded interval, but $\{f_n g_n\}$ does not converge uniformly on any bounded interval.

8.2 Uniformly Convergent Sequences of Functions

One of the reasons for introducing the concept of uniform convergence was that the pointwise convergence failed to preserve continuity. We will now show that the uniform convergence accomplishes this goal.

Theorem 8.2.1. *Let $\{f_n\}$ be a sequence of functions defined on a common domain A, and suppose that, for each $n \in \mathbb{N}$, f_n is continuous at $a \in A$. If $\{f_n\}$ converges uniformly to a function f on A, then f is continuous at $x = a$.*

Proof. Let $\varepsilon > 0$. The uniform convergence implies that there exists $N \in \mathbb{N}$ such that,

$$|f_n(x) - f(x)| < \frac{\varepsilon}{3}, \quad \text{for} \ \ n \ge N, \quad x \in A.$$

In particular,

$$|f_N(x) - f(x)| < \frac{\varepsilon}{3}, \quad \text{for all} \ \ x \in A. \tag{8.4}$$

By assumption, the function f_N is continuous at $x = a$, so there exists $\delta > 0$ such that

$$|x - a| < \delta, \quad x \in A \quad \Rightarrow \quad |f_N(x) - f_N(a)| < \frac{\varepsilon}{3}. \tag{8.5}$$

Now that we have our δ, it remains to show that

$$|x - a| < \delta, \quad x \in A \quad \Rightarrow \quad |f(x) - f(a)| < \varepsilon.$$

We write

$$f(x) - f(a) = [f(x) - f_N(x)] + [f_N(x) - f_N(a)] + [f_N(a) - f(a)]$$

so (8.4) and (8.5) imply that

$$|f(x) - f(a)| \le |f(x) - f_N(x)| + |f_N(x) - f_N(a)| + |f_N(a) - f(a)| < \frac{\varepsilon}{3} + \frac{\varepsilon}{3} + \frac{\varepsilon}{3} = \varepsilon. \quad \square$$

Remark 8.2.2. As a consequence of Theorem 8.2.1 we have the equality:

$$\lim_{x \to a} \lim_{n \to \infty} f_n(x) = \lim_{n \to \infty} \lim_{x \to a} f_n(x). \tag{8.6}$$

Indeed, both sides equal $f(a)$.

Theorem 8.2.1 shows that the class of continuous functions is closed under the operation of taking uniform limits. What happens if we consider a larger class of integrable functions? Will it be closed as well?

Theorem 8.2.3. *Let $\{f_n\}$ be a sequence of functions defined and integrable on $[a, b]$. If $\{f_n\}$ converges uniformly to a function f on $[a, b]$, then f is integrable on $[a, b]$. Furthermore, if*

$$F_n(x) = \int_a^x f_n(t)\, dt \quad and \quad F(x) = \int_a^x f(t)\, dt,$$

then the sequence $\{F_n\}$ converges uniformly to F on $[a, b]$.

Proof. Let $\varepsilon > 0$. Since $\{f_n\}$ converges uniformly to f, there exists $N \in \mathbb{N}$ such that

$$|f_n(x) - f(x)| < \frac{\varepsilon}{4(b - a)}, \quad \text{for } n \ge N, \quad x \in [a, b].$$

In particular,

$$|f_N(x) - f(x)| < \frac{\varepsilon}{4(b - a)}, \quad \text{for all } x \in [a, b].$$

It follows that

$$\sup\{f_N(x) - f(x) : x \in [a, b]\} \le \frac{\varepsilon}{4(b - a)} < \frac{\varepsilon}{3(b - a)}, \quad \text{and}$$

$$\inf\{f_N(x) - f(x) : x \in [a, b]\} \ge -\frac{\varepsilon}{4(b - a)} > -\frac{\varepsilon}{3(b - a)}.$$

The function $f_N(x)$ is integrable, so there exists a partition P of $[a, b]$ such that

$$U(f_N, P) - L(f_N, P) < \frac{\varepsilon}{3}.$$

We will show that $U(f, P) - L(f, P) < \varepsilon$. Indeed, using Exercise 6.2.7,

$$U(f, P) - L(f, P) \le U(f - f_N, P) - L(f - f_N, P) + U(f_N, P) - L(f_N, P)$$
$$< |U(f - f_N, P)| + |L(f - f_N, P)| + \frac{\varepsilon}{3}. \tag{8.7}$$

By definition,

$$L(f - f_N, P) = \sum_{i=1}^{n} m_i \Delta x_i, \quad \text{and} \quad U(f - f_N, P) = \sum_{i=1}^{n} M_i \Delta x_i, \quad \text{where}$$

$$M_i = \sup\{f - f_N(x) : x \in [x_{i-1}, x_i]\}, \quad m_i = \inf\{f - f_N(x) : x \in [x_{i-1}, x_i]\}.$$

It is not hard to see that

$$-\frac{\varepsilon}{3(b-a)} < \inf\{f_N(x) - f(x) : x \in [a,b]\} \le m_i \le M_i$$

$$\le \sup\{f_N(x) - f(x) : x \in [a,b]\} < \frac{\varepsilon}{3(b-a)}.$$

Consequently,

$$|L(f - f_N, P)| \le \sum_{i=1}^{n} |m_i| \Delta x_i < \frac{\varepsilon}{3(b-a)} \sum_{i=1}^{n} \Delta x_i = \frac{\varepsilon}{3},$$

and, similarly, $|U(f - f_N, P)| < \varepsilon/3$. Combining with (8.7) we obtain that $|L(f, P) - U(f, P)| < \varepsilon$, so the function f is integrable on $[a, b]$.

Finally, we will prove that the sequence $\{F_n\}$ converges uniformly to F. Let $\varepsilon > 0$ and let $N \in \mathbb{N}$ such that

$$n \ge N, \quad t \in [a, b] \quad \Rightarrow \quad |f_n(t) - f(t)| < \frac{\varepsilon}{2(b-a)}.$$

Using the inequality established in Exercise 6.5.3, it follows that, for $n \ge N$ and $x \in [a, b]$,

$$|F_n(x) - F(x)| = \left| \int_a^x f_n(t)\, dt - \int_a^x f(t)\, dt \right| = \left| \int_a^x [f_n(t) - f(t)]\, dt \right|$$

$$\le \int_a^x |f_n(t) - f(t)|\, dt$$

$$\le \int_a^x \frac{\varepsilon}{2(b-a)}\, dt = \frac{\varepsilon}{2(b-a)}(x-a) \le \frac{\varepsilon}{2} < \varepsilon.$$

Thus, $\{F_n\}$ converges uniformly to F and the proof is complete. $\qquad\square$

The last assertion of the theorem can be written as

$$\lim_{n \to \infty} \int_a^x f_n(t)\, dt = \int_a^x \lim_{n \to \infty} f_n(t)\, dt. \tag{8.8}$$

for any $x \in [a, b]$. So, (8.8) is true when $\{f_n\}$ converges uniformly on $[a, b]$. In the absence of the uniform convergence, things can go either way. For example, it is possible that the limit function is not integrable (Exercise 8.2.22). Even when it is integrable, equality (8.8) need not be true (Exercise 8.3.36). On the other hand, it is easy to see that the sequence $\{f_n\}$ defined in Example 8.1.3 satisfies (8.8) on $[0, 1]$, in spite of the fact that it does not converge uniformly on $[0, 1]$. Another example is furnished by Exercise 8.2.23. Much better understanding of equality (8.8) requires Lebesgue theory of measure and integration.

> The incorrect statement (without the uniform convergence) can be found in Cauchy's publication [16] from 1823. The first correct statement (and the proof) is attributed to Weierstrass by his followers (see [62]).

Next, we turn our attention to differentiability. Suppose that all functions f_n are differentiable at $x = a$, and that the sequence $\{f_n\}$ converges uniformly to f. Does that imply that f is differentiable at $x = a$? This is quite different from the continuity. Intuitively, if each f_n is continuous then none of them has a "jump", and because they are close to f,

f cannot have a jump. However, the existence of the derivative translates into the visual concept of "smoothness". Thus, the question becomes: what if all functions f_n are smooth? Will that force f to be smooth? If we consider the function $f(x) = |x|$, it is not differentiable at 0, but there could be many smooth functions in its vicinity.

Example 8.2.4. A sequence of differentiable functions converges uniformly to a function that is not differentiable.
Prove that the sequence $\{f_n\}$, defined by $f_n(x) = \sqrt{x^2 + 1/n}$, consists of differentiable functions, and that it converges uniformly on $[-1, 1]$ to $f(x) = |x|$, which is not differentiable at $x = 0$.

Solution. It is not hard to see that each function f_n is differentiable on $[-1, 1]$, and that $\{f_n\}$ converges pointwise to f. Further,

$$|f_n(x) - f(x)| = \left| \sqrt{x^2 + \frac{1}{n}} - |x| \right| = \left| \frac{\left(\sqrt{x^2 + \frac{1}{n}}\right)^2 - |x|^2}{\sqrt{x^2 + \frac{1}{n}} + |x|} \right| = \left| \frac{\frac{1}{n}}{\sqrt{x^2 + \frac{1}{n}} + |x|} \right|.$$

Therefore,

$$\sup_{x \in [-1,1]} |f_n(x) - f(x)| = \sup_{x \in [-1,1]} \left| \frac{\frac{1}{n}}{\sqrt{x^2 + \frac{1}{n}} + |x|} \right| = \frac{\frac{1}{n}}{\sqrt{\frac{1}{n}}} = \frac{1}{\sqrt{n}} \to 0.$$

By Theorem 8.1.6, $\{f_n\}$ converges uniformly to f on $[-1, 1]$. Thus, each function f_n is differentiable, but f is not differentiable at 0.   ◆

What went wrong? Part of the problem is that

$$f_n'(x) = \frac{x}{\sqrt{x^2 + \frac{1}{n}}} \to \frac{x}{|x|}, \quad n \to \infty,$$

but the function $g(x) = x/|x|$ is not the derivative of any function. (See Exercise 4.4.22 on page 119.) It seems natural to impose more restrictions on the sequence $\{f_n'\}$.

Theorem 8.2.5. *Let $\{f_n\}$ be a sequence of functions defined and differentiable on $[a, b]$. Suppose that there exists a point $x_0 \in [a, b]$ such that the sequence $\{f_n(x_0)\}$ converges, and that $\{f_n'\}$ converges uniformly to a function g on $[a, b]$. Then the sequence $\{f_n\}$ converges uniformly to a function f on $[a, b]$. Furthermore, f is differentiable on $[a, b]$ and $f' = g$.*

Proof. First we will prove that the sequence $\{f_n\}$ is a uniform Cauchy sequence. Let $\varepsilon > 0$. Since the sequence $f_n(x_0)$ converges, it is a Cauchy sequence so there exists $N_1 \in \mathbb{N}$ such that

$$m \geq n \geq N_1 \quad \Rightarrow \quad |f_m(x_0) - f_n(x_0)| < \frac{\varepsilon}{2}. \tag{8.9}$$

Also, $\{f_n'\}$ is a uniform Cauchy sequence so there exist $N_2 \in \mathbb{N}$ such that

$$m \geq n \geq N_2, \quad x \in [a, b] \quad \Rightarrow \quad |f_m'(x) - f_n'(x)| < \frac{\varepsilon}{2(b - a)}. \tag{8.10}$$

If we define $N = \max\{N_1, N_2\}$, and take $m \geq n \geq N$ and $x \in [a, b]$, then applying the Mean Value Theorem to the function $f_m - f_n$,

$$[f_m(x) - f_n(x)] - [f_m(x_0) - f_n(x_0)] = [f'_m(y) - f'_n(y)](x - x_0) \tag{8.11}$$

for some y between x_0 and x, hence in $[a, b]$. Consequently, combining (8.9)–(8.11),

$$|f_m(x) - f_n(x)| \leq |f_m(x_0) - f_n(x_0)| + |f'_m(y) - f'_n(y)||x - x_0|$$
$$< \frac{\varepsilon}{2} + \frac{\varepsilon}{2(b-a)}(b-a) = \varepsilon,$$

so $\{f_n\}$ is a uniform Cauchy sequence. By Theorem 8.1.10, $\{f_n\}$ converges uniformly, and we will denote its limit by f.

It remains to show that f is differentiable on $[a, b]$ and $f' = g$. Let $\varepsilon > 0$, and let $c \in [a, b]$. We will demonstrate that there exists $\delta > 0$ such that

$$0 < |x - c| < \delta, \quad x \in [a, b] \quad \Rightarrow \quad \left| \frac{f(x) - f(c)}{x - c} - g(c) \right| < \varepsilon. \tag{8.12}$$

Let $N_3 \in \mathbb{N}$ such that

$$m \geq n \geq N_3, \quad x \in [a, b] \quad \Rightarrow \quad |f'_m(x) - f'_n(x)| < \frac{\varepsilon}{4}. \tag{8.13}$$

Also, let $N_4 \in \mathbb{N}$ such that

$$|f'_n(c) - g(c)| < \frac{\varepsilon}{3}, \quad \text{for} \quad n \geq N_4.$$

Let $M = \max\{N_3, N_4\}$. Now,

$$|f'_M(c) - g(c)| < \frac{\varepsilon}{3}. \tag{8.14}$$

Since the function f_M is differentiable at c, there exists $\delta > 0$ such that

$$0 < |x - c| < \delta, \quad x \in [a, b] \quad \Rightarrow \quad \left| \frac{f_M(x) - f_M(c)}{x - c} - f'_M(c) \right| < \frac{\varepsilon}{3}. \tag{8.15}$$

Finally, $f_n(c) \to f(c)$ so, repeating the argument above (with c instead of x_0, and M instead of n) we get, for $m \geq M$, the analogue of (8.11):

$$[f_m(x) - f_M(x)] - [f_m(c) - f_M(c)] = [f'_m(z) - f'_M(z)](x - c)$$

for some z between c and x. If we divide by $x - c$, then (8.13) implies

$$\left| \frac{f_m(x) - f_m(c)}{x - c} - \frac{f_M(x) - f_M(c)}{x - c} \right| \leq |f'_m(z) - f'_M(z)| < \frac{\varepsilon}{4}.$$

Taking the limit as $m \to \infty$, we obtain that

$$\left| \frac{f(x) - f(c)}{x - c} - \frac{f_M(x) - f_M(c)}{x - c} \right| \leq \frac{\varepsilon}{4} < \frac{\varepsilon}{3}. \tag{8.16}$$

So, for these δ and M, if $0 < |x - c| < \delta$ and $x \in [a, b]$, the estimate (8.12) follows from the inequality

$$\left| \frac{f(x) - f(c)}{x - c} - g(c) \right| \leq \left| \frac{f(x) - f(c)}{x - c} - \frac{f_M(x) - f_M(c)}{x - c} \right|$$
$$+ \left| \frac{f_M(x) - f_M(c)}{x - c} - f'_M(c) \right| + |f'_M(c) - g(c)|$$

and (8.14), (8.15), and (8.16). □

ERNST KUMMER (1810–1893) was a German mathematician. He started as a student of theology at the University of Halle, but switched to mathematics. In 1831 he won a prize for a mathematics essay and on the strength of it he was awarded a doctorate. Yet he worked as a high school teacher, first in his hometown, then for 10 years in Liegnitz in today's Poland where he had a big influence on Kronecker who was his student there. Kummer's work on hypergeometric series impressed Jacobi (page 357) and Dirichlet which led to the membership in the Berlin Academy. In 1840 he was married to Ottilie Mendelssohn, a cousin of the composer Felix Mendelssohn whose sister Rebecca was the wife of Dirichlet. In 1842 he became a professor at the University of Breslau, now Wrocław in Poland. When Dirichlet left Berlin for Göttingen, Kummer filled the position. He managed to bring in Weierstrass and Kronecker, thus creating Berlin the leading center for mathematics. In 1861 Kummer and Weierstrass established Germany's first pure mathematics seminar. He was extremely popular as a teacher. His lectures were reportedly attended by 250 students! He mentored many outstanding mathematicians, including Cantor, Frobenius (page 449), and Schwarz (page 352), who married one of Kummer's daughters. He retired in 1883 citing a failing memory. He received the Grand Prize of the Paris Academy of Sciences for his work on Fermat's Last Theorem. He received numerous other honors, including membership in the Paris Academy of Sciences and in the Royal Society of London. His main contributions are in algebra (Kummer surface, Kummer extensions of fields), function theory (the study of hypergeometric series), and in number theory (the proof of Fermat's Last Theorem for a large class of integers). He introduced the word *ideal* in connection with *ideal numbers* and inspired Dedekind's definition of ideals for rings.

Exercises

8.2.1. Suppose that the sequence $\{f_n\}$ converges to f uniformly on $\mathbb{R}$ and that each f_n is a bounded function. Prove that f is bounded.

8.2.2. Suppose that the sequence $\{f_n\}$ converges to f uniformly on $\mathbb{R}$ and that each f_n is a uniformly continuous function. Prove that f is uniformly continuous.

8.2.3.* Suppose that the sequence of polynomials $\{p_n\}$ converges to f uniformly on $\mathbb{R}$. Prove that f is a polynomial.

8.2.4. Suppose that the sequence $\{f_n\}$ converges to f uniformly on a set A and that f is a bounded function. Prove that there exist constants N and M such that $|f_n(x)| \leq M$ for all $x \in A$ and all $n \geq N$.

8.2.5. Suppose that the sequence $\{f_n\}$ converges to f uniformly on a set A and that each f_n is a bounded function. Prove that there exists a constant M such that $|f_n(x)| \leq M$ for all $x \in A$.

8.2.6.* Let f be defined on $[\frac{1}{2}, 1]$ and continuous at $x = 1$. (a) Prove that $|x^n f(x)|$ converges pointwise on $[\frac{1}{2}, 1]$; (b) Prove that this convergence is uniform if and only if f is bounded and $f(1) = 0$.

8.2.7. Prove or disprove: If $\{f_n\}$ is a sequence of differentiable functions on $[a, b]$ that converges to a differentiable function f, then $\{f_n'\}$ converges to f'.

8.2.8. Construct an example of a sequence $\{f_n\}$ of functions such that each of them is discontinuous at every point of $[a, b]$, but the sequence converges uniformly to a function f that is continuous at every point of $[a, b]$.

8.2.9. Let $f_n(x) = \frac{\cos nx}{n}$. Prove that $\{f_n\}$ is a sequence of functions that are differentiable on $[0, +\infty)$, and that converges uniformly to 0 on $[0, +\infty)$, yet $\{f_n'(x)\}$ diverges for $x \neq k\pi$, $k \in \mathbb{Z}$.

8.2.10. Let $\{f_n\}$ be a sequence of functions on $[a, b]$ that converges on $[a, b]$ and uniformly on (a, b). Prove that $\{f_n\}$ converges uniformly on $[a, b]$.

8.2.11.* Suppose that the sequence of increasing functions $\{f_n\}$ converges pointwise to a continuous function f on $[a, b]$. Prove that $\{f_n\}$ converges uniformly on $[a, b]$.

A sequence $\{f_n\}$ is *equicontinuous* on a set A if $\forall \varepsilon > 0\ \exists \delta > 0$ such that for all $n \in \mathbb{N}$ and all $x, y \in A$, $|x - y| < \delta \Rightarrow |f_n(x) - f_n(y)| < \varepsilon$.

8.2.12.* Suppose that the sequence $\{f_n\}$ converges to f uniformly on $[a, b]$ and that each f_n is a continuous function. Prove that $\{f_n\}$ is equicontinuous.

8.2.13.* Suppose that the sequence $\{f_n\}$ is equicontinuous and converges to f pointwise on $[a, b]$. Prove that $\{f_n\}$ converges uniformly on $[a, b]$.

8.2.14.* Let $\{f_n\}$ be a sequence of continuous functions on $[a, b]$ that converges uniformly on (a, b) to a function f. Prove that $\{f_n\}$ converges uniformly on $[a, b]$.

8.2.15.* Let $\{f_n\}$ be a sequence of continuous functions on $[a, b]$, and suppose that, for any $[\alpha, \beta] \subset (a, b)$, $\{f_n\}$ converges uniformly on $[\alpha, \beta]$ to a function f. Prove or disprove: (a) the sequence $\{f_n\}$ converges pointwise on $[a, b]$; (b) the sequence $\{f_n\}$ converges uniformly on $[a, b]$.

In Exercises 8.2.16–8.2.20 compute the limits and justify your computation:

8.2.16.* $\displaystyle \lim_{n \to \infty} \int_0^1 \frac{dx}{\left(1 + \frac{x}{n}\right)^n}$. 8.2.17. $\displaystyle \lim_{n \to \infty} \int_0^1 \frac{nx}{1 + n^2 x^{3/2}} \, dx$. 8.2.18.* $\displaystyle \lim_{n \to \infty} \int_0^2 \sqrt[n]{2^n + |x|^n} \, dx$.

8.2.19. $\displaystyle \lim_{n \to \infty} \int_0^{\pi/2} \cos^n x \sin^{2n} x \, dx$. 8.2.20.* $\displaystyle \lim_{n \to \infty} \int_0^\pi n \sin \sqrt{4\pi^2 n^2 + x^2} \, dx$.

8.2.21. Let $f_n(x) = n^p x e^{-nx}$, for $n \in \mathbb{N}$. Find all values of p such that: (a) $\{f_n\}$ converges on $[0, 1]$; (b) $\{f_n\}$ converges uniformly on $[0, 1]$; (c) $\int_0^1 \lim f_n(x) \, dx = \lim \int_0^1 f_n(x) \, dx$.

8.2.22.* Construct an example of a sequence $\{f_n\}$ of functions integrable on $[a, b]$ such that its pointwise limit f is not integrable on $[a, b]$.

8.2.23. Show that the sequence $\{f_n\}$, defined by $f_n(x) = nx(1 - x)^n$, converges pointwise but not uniformly on $[0, 1]$, yet

$$\lim \int_0^1 f_n(x) \, dx = \int_0^1 \lim f_n(x) \, dx. \tag{8.17}$$

Thus, the uniform convergence is a sufficient but not a necessary condition for (8.17).

8.2.24. Let $\{f_n\}$ be a sequence of functions defined and bounded on $\mathbb{R}$, and suppose that $\{f_n\}$ converges uniformly to a function f on every finite interval $[a, b]$. Does it follow that

$$\lim_{n \to \infty} \sup_{x \in \mathbb{R}} f_n(x) = \sup_{x \in \mathbb{R}} f(x)?$$

8.2.25. Let

$$f_n(x) = \begin{cases} \frac{1}{n} - \frac{x}{n^2}, & \text{if } 0 \le x \le n \\ 0, & \text{if } x > n. \end{cases}$$

Show that $\{f_n\}$ converges uniformly to $f = 0$ on $[0, +\infty)$, and that each function f_n is integrable on $[0, +\infty)$, but

$$\lim \int_0^\infty f_n(x) \, dx \ne \int_0^\infty f(x) \, dx.$$

8.3 Function Series

Some of the most important sequences of functions show up, just like with sequences of numbers, in the study of series. When $\{f_n\}$ is a sequence of functions we can consider the convergence of the series $\sum_{n=1}^{\infty} f_n$. The series $\sum_{n=1}^{\infty} f_n$ converges if the sequence of its partial sums $\{s_n\}$ converges. If $\{s_n\}$ converges pointwise, then the series **converges pointwise**. If $\{s_n\}$ converges uniformly, then the series **converges uniformly**.

When studying the pointwise convergence we effectively study one numerical series at a time. Therefore, we have all the methods of Chapter 7 at our disposal. In this section we will focus on the tests for uniform convergence. One way to decide whether a series converges uniformly is to consider the sequence $\{s_n\}$. However, sequences where the general term s_n is a sum of n terms are typically very hard. Thankfully, there are some tests especially designed for function series (rather than sequences). One of the most effective is due to Weierstrass. It can be found in his publication [110] from 1880.

Theorem 8.3.1 (The Weierstrass M-Test). *Let $\{f_n\}$ be a sequence of functions defined on a common domain A, and let $\{M_n\}$ be a sequence of positive numbers such that*

$$|f_n(x)| \leq M_n, \quad \text{for each } n \in \mathbb{N}, \text{ and any } x \in A.$$

If the series $\sum_{n=1}^{\infty} M_n$ converges then $\sum_{n=1}^{\infty} f_n$ converges uniformly.

Proof. Let $\varepsilon > 0$, and let $\{s_n\}$ be the nth partial sum of $\sum_{n=1}^{\infty} f_n$. We will use Theorem 8.1.10 to show that $\{s_n\}$ converges uniformly. Namely, we will show that there exists a positive integer N such that

$$|s_m(x) - s_n(x)| < \varepsilon, \quad \text{for } m \geq n \geq N \text{ and any } x \in A.$$

If we denote by S_n the nth partial sum of $\sum_{n=1}^{\infty} M_n$, then $\{S_n\}$ is a convergent sequence so there exists $N \in \mathbb{N}$ such that

$$|S_m - S_n| < \varepsilon, \quad \text{for } m \geq n \geq N.$$

For this N and $m \geq n \geq N$,

$$
\begin{aligned}
|s_m(x) - s_n(x)| &= |f_{n+1}(x) + f_{n+2}(x) + \cdots + f_m(x)| \\
&\leq |f_{n+1}(x)| + |f_{n+2}(x)| + \cdots + |f_m(x)| \\
&\leq M_{n+1} + M_{n+2} + \cdots + M_m = S_m - S_n \\
&< \varepsilon.
\end{aligned}
$$
$\square$

We will provide an immediate example for the application of the Weierstrass M-Test.

Example 8.3.2. Applying Weierstrass M-Test.

Prove that the series $\sum_{n=1}^{\infty} \dfrac{x^n}{n^2}$ converges uniformly for $x \in [-1, 1]$

Solution. Since $|x| \leq 1$ we have that $|x^n/n^2| \leq 1/n^2$ and we define $M_n = 1/n^2$. The series $\sum_{n=1}^{\infty} 1/n^2$ converges (a p-series, with $p = 2$), and it follows that $\sum_{n=1}^{\infty} x^n/n^2$ converges uniformly on $[-1, 1]$. $\blacklozenge$

The series $\sum_{n=1}^{\infty} x^n/n^2$ converges for each $x \in [-1, 1]$, so its limit is a function f defined on $[-1, 1]$. What can we say about f? Since $f = \lim s_n$, and each s_n is a polynomial (hence continuous), Theorem 8.2.1 implies that f is continuous on $[-1, 1]$. Is it differentiable? We would like to apply Theorem 8.2.5, so we need to check whether its hypotheses are true. Clearly, each function s_n is differentiable and $\{s_n(x_0)\}$ converges for any $x_0 \in [-1, 1]$. It remains to verify that $\{s'_n\}$ converges uniformly on $[-1, 1]$. It is easy to see that

$$s'_n = \left(\sum_{k=1}^{n} \frac{x^k}{k^2} \right)' = \sum_{k=1}^{n} \left(\frac{x^k}{k^2} \right)' = \sum_{k=1}^{n} \frac{kx^{k-1}}{k^2} = \sum_{k=1}^{n} \frac{x^{k-1}}{k}.$$

However, this series does not converge when $x = 1$. It does converge for $x = -1$. What about $|x| < 1$?

Example 8.3.3. A uniformly convergent series.

Prove that the series $\sum_{n=1}^{\infty} \dfrac{x^n}{n}$ converges uniformly on $[-r, r]$, for any $r \in [0, 1)$.

Solution. Let $0 < r < 1$, and let $|x| \le r$. Then $|x^n/n| \le r^n/n$. Further,

$$C_n = \sqrt[n]{\frac{r^n}{n}} = \frac{r}{\sqrt[n]{n}} \to r < 1, \quad n \to \infty,$$

so Cauchy's Test implies that the series $\sum_{n=1}^{\infty} r^n/n$ converges. Now, with $M_n = r^n/n$, the Weierstrass M-Test implies the uniform convergence of $\sum_{n=1}^{\infty} x^n/n$ on $[-r, r]$. ◆

It can be shown that the series $\sum_{n=1}^{\infty} x^n/n$ does *not* converge uniformly on $(-1, 1)$. However, it is not a big loss. Since it converges uniformly on $[-r, r]$, it follows that the limit f of the series $\sum_{n=1}^{\infty} x^n/n^2$ is a differentiable function on $[-r, r]$. Since r is arbitrary, we conclude that f is differentiable on $(-1, 1)$.

Although the Weierstrass M-Test is sufficient to handle the majority of the situations, it fails to distinguish between the absolute and conditional convergence. In other words, when we apply the M-Test, we establish not only the convergence of $\sum_{n=1}^{\infty} f_n$ but of $\sum_{n=1}^{\infty} |f_n|$ as well. In Section 7.5 we have encountered two tests that were able to detect conditional convergence. Here, we will establish their generalizations. The first one represents a generalization of Abel's Test for series of numbers (page 213).

Theorem 8.3.4 (Abel's Test). *Suppose that $\{a_n\}, \{b_n\}$ are two sequences of functions defined on a common domain A such that the series $\sum_{k=1}^{\infty} b_k$ converges uniformly on A, that the sequence $\{a_n(x)\}$ is monotone (for each $x \in A$), and that there exists $M > 0$ such that $|a_n(x)| \le M$ for all $x \in A$ and $n \in \mathbb{N}$. Then the series $\sum_{k=1}^{\infty} a_k b_k$ converges uniformly on A.*

Proof. Let s_n denote the nth partial sum of the series $\sum_{k=1}^{\infty} a_k b_k$. We will show that $\{s_n\}$ is a uniform Cauchy sequence. Let $\varepsilon > 0$. The assumption that the series $\sum_{k=1}^{\infty} b_k$ converges uniformly implies that the sequence of its partial sums $\{B_n\}$ converges uniformly to its sum B or, equivalently, that the sequence of its remainders $R_n = B - B_n$ converges uniformly to 0. Therefore, there exists $N \in \mathbb{N}$, such that

$$|R_n(x)| < \frac{\varepsilon}{4M} \quad \text{for} \quad n \ge N, \quad \text{and any} \quad x.$$

If we apply the Summation by Parts formula (7.5.8), we obtain that

$$s_m(x) - s_{n-1}(x) = a_{m+1}(x)B_m(x) - a_n(x)B_{n-1}(x) - \sum_{k=n}^{m} [a_{k+1}(x) - a_k(x)]B_k(x)$$

$$= a_{m+1}(x)[B(x) - R_m(x)] - a_n(x)[B(x) - R_{n-1}(x)]$$

$$- \sum_{k=n}^{m} [a_{k+1}(x) - a_k(x)][B(x) - R_k(x)]$$

$$= B(x)\left(a_{m+1}(x) - a_n(x) - \sum_{k=n}^{m} [a_{k+1}(x) - a_k(x)] \right)$$

$$- a_{m+1}(x)R_m(x) + a_n(x)R_{n-1}(x) + \sum_{k=n}^{m} [a_{k+1}(x) - a_k(x)]R_k(x)$$

$$= -a_{m+1}(x)R_m(x) + a_n(x)R_{n-1}(x) + \sum_{k=n}^{m} [a_{k+1}(x) - a_k(x)]R_k(x).$$

Therefore, for $m \geq n \geq N + 1$ and each x,

$$|s_m(x) - s_{n-1}(x)| \leq |a_n(x)||R_{n-1}(x)| + \sum_{k=n}^{m} |a_{k+1}(x) - a_k(x)| \, |R_k(x)| + |a_{m+1}(x)||R_m(x)|$$

$$\leq M \frac{\varepsilon}{4M} + \frac{\varepsilon}{4M} \sum_{k=n}^{m} |a_{k+1}(x) - a_k(x)| + M \frac{\varepsilon}{4M}.$$

Since $\{a_n(x)\}$ is monotone the terms in the last sum are all of the same sign, so after cancellation we are left with $|a_{m+1}(x) - a_n(x)|$ which is no bigger than $2M$. It follows that

$$|s_m(x) - s_{n-1}(x)| \leq \frac{\varepsilon}{4} + \frac{\varepsilon}{4M} \, 2M + \frac{\varepsilon}{4} = \varepsilon.$$

Thus, $\{s_n\}$ is a uniform Cauchy sequence and the result follows from Theorem 8.1.10. □

Remark 8.3.5. Although the statement of the theorem is in alignment with the Abel's Test for series of numbers (Theorem 7.5.9), the proof is different. If $\{B_n\}$ is a sequence of numbers then the assumption that it is convergent implies that it is bounded. However, this is not true when $B_n(x) = \sum_{k=1}^{n} b_k(x)$ is a uniformly convergent sequence of functions. For example, if $B_n(x) = x$ for each $n \in \mathbb{N}$, then $\{B_n(x)\}$ converges uniformly on $\mathbb{R}$ to $B(x) = x$ but neither B nor any of B_n is a bounded function.

We will also state a test that represents a generalization of Dirichlet's test for numerical series (Theorem 7.5.11). We will leave the proof as an exercise.

Theorem 8.3.6 (Dirichlet's Test). *Suppose that $\{a_n\}, \{b_n\}$ are two sequences of functions defined on a common domain A such that the sequence $\{a_n(x)\}$ is monotone decreasing (for each $x \in A$) and converges to 0 uniformly on A, and that there exists $M > 0$ such that $|\sum_{k=1}^{n} b_k(x)| \leq M$ for all $x \in A$ and $n \in \mathbb{N}$. Then the series $\sum_{k=1}^{\infty} a_k b_k$ converges uniformly on A.*

The tests presented in this section all give sufficient conditions for a function series to converge uniformly on a set A. What if the convergence is not uniform? How does one prove this? The main tool is Theorem 8.1.6, adapted to series of functions. Next, there is a version of the Divergence Test (Exercise 8.3.10). Namely, if the sequence $\{f_n\}$ does not converge uniformly to 0 on A, then $\sum_{n=1}^{\infty} f_n$ is not a uniformly convergent series on A.

Example 8.3.7. A series that is pointwise but not uniformly convergent on a given set.
    Prove that the series $\sum_{n=0}^{\infty} x^n$ converges pointwise but not uniformly on $(-1, 1)$.
Solution. For each $x \in (-1, 1)$ this is a geometric series, and its ratio x satisfies $|x| < 1$. Thus, the series converges to $f(x) = 1/(1 - x)$ pointwise on $(-1, 1)$. However, the sequence $\{s_n\}$, defined by $s_n(x) = \sum_{k=0}^{n} x^k$, does not converge uniformly to f on $(-1, 1)$. Indeed,

$$s_n(x) - f(x) = \frac{1 - x^{n+1}}{1 - x} - \frac{1}{1 - x} = \frac{-x^{n+1}}{1 - x},$$

and the function $|x|^{n+1}/|1 - x|$ is unbounded on $(-1, 1)$, so the conclusion follows from Theorem 8.1.6.   ♦

Another method of establishing that a series fails to converge uniformly on a set A, is to use Cauchy's Test (Theorem 8.1.10).

Example 8.3.8. A series that is pointwise but not uniformly convergent on a given set.

The series $\sum_{n=1}^{\infty} xe^{-nx^2}$ converges pointwise but not uniformly on $(0,1)$.

Solution. We need the negatives of both equivalent statements in Theorem 8.1.10: $\{f_n\}$ does not converge uniformly on A if and only if there exists $\varepsilon_0 > 0$ such that, for any $N \in \mathbb{N}$, there exist $m \geq n \geq N$ and $x_N \in A$ satisfying $|f_m(x_N) - f_n(x_N)| \geq \varepsilon_0$. We will apply this result to the sequence of partial sums $\{s_n\}$ of the given series, i.e.,

$$s_n(x) = \sum_{k=1}^{n} xe^{-kx^2}.$$

So, let $\varepsilon_0 = e^{-2}$ and let $N \in \mathbb{N}$. We will choose $m = 2N$, $n = N$, and $x_n = 1/\sqrt{N} \in (0,1)$. Then

$$|s_{2N}(x_N) - s_N(x_N)| = \left| \sum_{k=N+1}^{2N} x_N e^{-kx_N^2} \right| = \frac{1}{\sqrt{N}} \left| \sum_{k=N+1}^{2N} e^{-k/N} \right|$$

$$\geq \frac{1}{\sqrt{N}} \left(N e^{-2N/N} \right) = \sqrt{N}\, e^{-2} \geq e^{-2}.$$

Thus, the series $\sum_{n=1}^{\infty} xe^{-nx^2}$ does not converge uniformly on $(0,1)$. ♦

JOSEPH FOURIER (1768–1830) was a French mathematician and physicist. His parents died while he was a child and he attended the Benedictine college Ecole Royale Militaire of Auxerre, where he demonstrated his interest for mathematics by reading mathematical books. He spent 2 years in a Benedictine abbey preparing to be a priest, but instead became a teacher at his former school Ecole Royale Militaire. During the French revolution he was involved in politics and his life was in danger more than once. When the Ecole Normale in Paris opened he went to study there and attended classes by Lagrange (page 398) and Laplace. He began teaching at the Collège de France and doing research. From 1795 he taught at the Ecole Polytechnique. Fourier accompanied Napoleon Bonaparte on his Egyptian expedition in 1798, as scientific adviser, and was appointed secretary of the Institut d'Egypte. In 1801 he returned to Paris but soon Napoleon sent him to serve as a Prefect in Grenoble, where he remained for the next 15 years. He was charged with the publication of the Egyptian materials and he supervised several projects including the draining of swamps. In 1807 he submitted his work [46] to the Paris Institute and it raised a number of objections, including the assumed expansion of functions into trigonometric series. Even though the work won the prize of the Institute in 1811 the paper was not published until 1822. In the next chapter we will explore a special type of series of functions named after him. After the fall of Napoleon in 1815, Fourier returned to Paris where he was the director of the Statistical Bureau of the Seine. In 1822 he became Permanent Secretary of the French Academy of Sciences. Because of his work in Egyptology he was elected in 1826 to the Académie Française.

8.3.1 Applications to Differential Equations

In this part of the text we consider the initial value problem:

$$y' = F(x,y), \quad y(x_0) = y_0, \tag{8.18}$$

where F is a function of 2 variables defined on a rectangle R, and (x_0, y_0) is a point in R. We will show that very modest assumptions about the function F are sufficient to guarantee the existence and the uniqueness of the solution. Warning: the statement and the proof require some basic understanding of the multivariable calculus.

Theorem 8.3.9. *Let F be a function defined on a rectangle $R = [a,b] \times [c,d]$ and suppose that both F and its partial derivative F_y (with respect to y) are continuous in R. Let $(x_0, y_0) \in (a,b) \times (c,d)$. Then there exists a real number h such that $(x_0 - h, x_0 + h) \subset (a,b)$ and a function f defined on $(x_0 - h, x_0 + h)$ that is a solution of (8.18). Further, this solution is unique, i.e., if g is another solution of (8.18), then $f(x) = g(x)$ for all $x \in (x_0 - h, x_0 + h)$.*

Proof. Both F and F_y are continuous, hence bounded, in R so there exists $M > 0$ such that, for $(x, y) \in R$,

$$|F(x, y)| \le M, \quad |F_y(x, y)| \le M.$$

Let $h, k > 0$ so that $Mh \le k$ and that

$$R' \equiv [x_0 - h, x_0 + h] \times [y_0 - k, y_0 + k] \subset R.$$

The initial value problem $y' = F(x, y)$, $y(x_0) = y_0$ is equivalent to the *integral equation*

$$y(x) = y_0 + \int_{x_0}^{x} F(t, y(t))\, dt. \tag{8.19}$$

We will construct a sequence of functions $y_0(x) = y_0, y_1(x), y_2(x), \ldots$ by the recursive formula

$$y_{n+1}(x) = y_0 + \int_{x_0}^{x} F(t, y_n(t))\, dt. \tag{8.20}$$

Our first task is to ensure that these functions are well defined. Since F is defined on R', it suffices to demonstrate that, for every $n \in \mathbb{N}$ and every t satisfying $|t - x_0| \le h$, $|y_n(t) - y_0| \le k$. The last inequality is obvious for $n = 0$, so let us assume that it is true for n, and prove that it holds for $n + 1$. Now

$$|y_{n+1}(x) - y_0| = \left| \int_{x_0}^{x} F(t, y_n(t))\, dt \right| \le \left| \int_{x_0}^{x} |F(t, y_n(t))|\, dt \right|.$$

By induction hypothesis, $(t, y_n(t)) \in R'$ so $|F(t, y_n(t))| \le M$. This implies that

$$|y_{n+1}(x) - y_0| \le M|x - x_0| \le Mh \le k.$$

Thus, all functions $y_n(x)$ are well defined.

Next, we will show that this sequence converges uniformly for $|x - x_0| \le h$. This will follow from the inequalities

$$|y_n(x) - y_{n-1}(x)| \le M^n \frac{|x - x_0|^n}{n!} \tag{8.21}$$

which hold for all $n \in \mathbb{N}$. We will establish this by induction. It is easy to see that (8.21) is true for $n = 1$, so we will assume that it is valid for n and prove it for $n + 1$. Clearly,

$$|y_{n+1}(x) - y_n(x)| \le \left| \int_{x_0}^{x} |F(t, y_n(t)) - F(t, y_{n-1}(t))|\, dt \right|.$$

By the Mean Value Theorem, for each $(x, y_1), (x, y_2) \in R'$, there exists a real number $\xi = \xi(x)$ between y_1 and y_2, such that

$$F(x, y_1) - F(x, y_2) = (y_1 - y_2)F_y(x, \xi(x)). \tag{8.22}$$

From this we obtain that

$$|y_{n+1}(x) - y_n(x)| \le \left| \int_{x_0}^{x} |y_n(t) - y_{n-1}(t)|\, |F_y(t, \xi(t))|\, dt \right|$$

$$\leq \left| \int_{x_0}^{x} \left(M^n \frac{|t - x_0|^n}{n!} \right) M \, dt \right| = \frac{M^{n+1}}{n!} \left| \frac{|t - x_0|^{n+1}}{n+1} \right|_{x_0}^{x} = M^{n+1} \frac{|x - x_0|^{n+1}}{(n+1)!}.$$

Thus, (8.21) holds for all $n \in \mathbb{N}$. As a consequence, we obtain a weaker estimate

$$|y_n(x) - y_{n-1}(x)| \leq M^n \frac{h^n}{n!}$$

which implies, using the Weierstrass M-test, that the series

$$y_0 + \sum_{n=1}^{\infty} [y_n(x) - y_{n-1}(x)]$$

converges uniformly. (The verification that the series $\sum_{n=1}^{\infty} M^n h^n / n!$ converges is left to the reader.) However, the nth partial sum of this series is precisely $y_n(x)$, so the sequence $\{y_n\}$ converges uniformly. Let y be the limit function. If we let $n \to \infty$ in (8.20) we obtain that

$$y(x) = y_0 + \int_{x_0}^{x} F(t, y(t)) \, dt.$$

Indeed, (8.22) together with the boundedness of F_y shows that $F(t, y_n(t))$ converges uniformly to $F(t, y(t))$, and the passage of the limit into the integral is justified by (8.8).

Thus, it remains to show that the solution is unique. Suppose, to the contrary, that there exists a function g defined on $(x_0 - h, x_0 + h)$ and satisfying (8.18). Then, g satisfies the equation

$$g(x) = y_0 + \int_{x_0}^{x} F(t, g(t)) \, dt. \tag{8.23}$$

First we will show that, for $t \in (x_0 - h, x_0 + h)$, the point $(t, g(t)) \in R'$, i.e., that $|g(t) - y_0| \leq k$. We will argue by contradiction. Namely, if $|g(t) - y_0| > k$ for some $t \in (x_0 - h, x_0 + h)$, then the continuity of g implies that there exists $t \in (x_0 - h, x_0 + h)$ such that $|g(t) - y_0| = k$. Let t_0 be the closest such number to x_0,

$$t_0 = \inf\{t > x_0 : |g(t) - y_0| = k\}, \quad \text{or} \quad t_0 = \sup\{t < x_0 : |g(t) - y_0| = k\}.$$

At least one of the two sets is non-empty, and the infimum/supremum cannot be x_0, otherwise there would be a sequence $\{t_n\}$ converging to x_0, with $\{g(t_n)\}$ not converging to $g(x_0) = y_0$. Now

$$k = |g(t_0) - y_0| = |g(t_0) - g(x_0)| = |t_0 - x_0| \, |g'(\xi)| \tag{8.24}$$

for some ξ between t_0 and x_0. By assumption, $|g(\xi) - y_0| < k$, so $(\xi, g(\xi)) \in R'$, which implies that

$$|g'(\xi)| = |F(\xi, g(\xi))| \leq M. \tag{8.25}$$

Combining (8.24) and (8.25) yields $k \leq M \, |t_0 - x_0| < Mh \leq k$. This contradiction shows that, for $t \in (x_0 - h, x_0 + h)$, the point $(t, g(t)) \in R'$. Consequently,

$$|F(t, g(t))| \leq M, \text{ for } t \in (x_0 - h, x_0 + h). \tag{8.26}$$

Next, we will show that, for each $x \in (x_0 - h, x_0 + h)$, the sequence $y_n(x)$ converges to $g(x)$. This will follow from the estimate

$$|g(x) - y_n(x)| \leq M^{n+1} \frac{|x - x_0|^{n+1}}{(n+1)!}, \tag{8.27}$$

which we will prove by induction. The case $n = 0$ is an easy consequence of (8.23) and (8.26), so suppose that the estimate is true for n. Then

$$|g(x) - y_{n+1}(x)| = \left| \int_{x_0}^{x} [F(t, g(t)) - F(t, y_n(t))]\, dt \right| = \left| \int_{x_0}^{x} [g(t) - y_n(t)]\, F_y(t, \xi(t))\, dt \right|$$

$$\leq \left| \int_{x_0}^{x} |g(t) - y_n(t)|\, |F_y(t, \xi(t))|\, dt \right|$$

$$\leq M \left| \int_{x_0}^{x} M^{n+1} \frac{|t - x_0|^{n+1}}{(n+1)!}\, dt \right| = \frac{M^{n+2}}{(n+1)!} \left| \frac{|t - x_0|^{n+2}}{n+2} \right|_{x_0}^{x}$$

$$= M^{n+2} \frac{|x - x_0|^{n+2}}{(n+2)!}.$$

Thus (8.27) is established, and it implies that, for each $x \in (x_0 - h, x_0 + h)$, $|g(x) - y_n(x)| \to 0$, as $n \to \infty$. On the other hand, $\lim y_n = f$, so $g(x) = f(x)$ for all $x \in (x_0 - h, x_0 + h)$, and the theorem is proved. $\qquad\square$

8.3.2   Continuous Nowhere Differentiable Function

We close this section with an example of a continuous function that is not differentiable at any point. This example is due to Weierstrass.

Example 8.3.10. Let $0 < b < 1$, let a be an odd positive integer, and suppose that $ab > 1 + \frac{3\pi}{2}$. The function

$$f(x) = \sum_{n=0}^{\infty} b^n \cos(a^n \pi x) \tag{8.28}$$

is continuous on $\mathbb{R}$, but has no finite derivative at any point of the real line.

Since $|b^n \cos(a^n \pi x)| \leq b^n$ and $\sum_{n=0}^{\infty} b^n$ is a geometric series with ratio $b \in (0, 1)$, the Weierstrass M-test shows that the series in (8.28) converges uniformly, so the continuity of f follows from Theorem 8.2.1.

Let $c \in \mathbb{R}$. We will show that f is not differentiable at $x = c$. In fact, we will demonstrate that

$$\lim_{h \to 0} \frac{f(c + h) - f(c)}{h}$$

is infinite. Let $M > 0$. Our plan is to find $h \in (0, 3/2)$ such that

$$\left| \frac{f(c + h) - f(c)}{h} \right| > M. \tag{8.29}$$

Since $ab > 1$, and $\frac{2}{3} - \frac{\pi}{ab-1} > 0$, there exists $m \in \mathbb{N}$ such that

$$(ab)^m \left(\frac{2}{3} - \frac{\pi}{ab - 1} \right) > M.$$

Let m be such a number, and let us write $a^m c = p + r$, where $p \in \mathbb{Z}$ and $-\frac{1}{2} < r < \frac{1}{2}$. We define

$$h = \frac{1 - r}{a^m}.$$

It is not hard to see that $0 < h < \frac{3}{2a^m} < \frac{3}{2}$.

Let us now focus on the inequality (8.29). We write

$$\frac{f(c+h) - f(c)}{h} = \sum_{n=0}^{\infty} b^n \frac{\cos(a^n \pi(c+h)) - \cos(a^n \pi c)}{h}$$

$$= \sum_{n=0}^{m-1} b^n \frac{\cos(a^n \pi(c+h)) - \cos(a^n \pi c)}{h} + \sum_{n=m}^{\infty} b^n \frac{\cos(a^n \pi(c+h)) - \cos(a^n \pi c)}{h}.$$

Let us denote these two sums by S_m and R_m.

First we will estimate S_m. By the Mean Value Theorem, there exists $0 < \theta < 1$ such that

$$|\cos(a^n \pi(c+h)) - \cos(a^n \pi c)| = |a^n \pi h \sin(a^n \pi(c+\theta h))| \le a^n \pi |h|$$

so

$$|S_m| \le \sum_{n=0}^{m-1} b^n \frac{a^n \pi |h|}{|h|} = \pi \frac{(ab)^m - 1}{ab - 1} \le \pi \frac{(ab)^m}{ab - 1}.$$

Next, we will turn to R_m. If $n \in \mathbb{N}$, $n \ge m$, then

$$a^n \pi(c+h) = a^{n-m} a^m \pi(c+h) = a^{n-m} \pi(a^m c + a^m h)$$
$$= a^{n-m} \pi(p + r + 1 - r) = a^{n-m} \pi(p+1).$$

Since a is an odd integer,

$$\cos(a^n \pi(c+h)) = \cos(a^{n-m} \pi(p+1)) = (-1)^{a^{n-m}(p+1)} = (-1)^{p+1}. \tag{8.30}$$

Also, using the identity $\cos(x+y) = \cos x \cos y - \sin x \sin y$,

$$\cos(a^n \pi c) = \cos(a^{n-m} \pi a^m c) = \cos(a^{n-m} \pi(p+r))$$
$$= \cos(a^{n-m} \pi p) \cos(a^{n-m} \pi r) - \sin(a^{n-m} \pi p) \sin(a^{n-m} \pi r)$$
$$= \cos(a^{n-m} \pi p) \cos(a^{n-m} \pi r) \tag{8.31}$$
$$= (-1)^{a^{n-m} p} \cos(a^{n-m} \pi r) = (-1)^p \cos(a^{n-m} \pi r).$$

Combining (8.30) and (8.31), we have that

$$R_m = \sum_{n=m}^{\infty} \frac{b^n}{h} \left((-1)^{p+1} - (-1)^p \cos(a^{n-m} \pi r) \right)$$
$$= \frac{(-1)^{p+1}}{h} \sum_{n=m}^{\infty} b^n \left(1 + \cos(a^{n-m} \pi r) \right).$$

Since the series above has all terms positive, its sum is bigger than the first term (for $n = m$), so

$$|R_m| \ge \frac{1}{|h|} b^m \left(1 + \cos(\pi r) \right) \ge \frac{b^m}{|h|} > \frac{2(ab)^m}{3}.$$

Finally,

$$\left| \frac{f(c+h) - f(c)}{h} \right| \ge |R_m| - |S_m|$$

$$\ge \frac{2(ab)^m}{3} - \pi \frac{(ab)^m}{ab - 1} = (ab)^m \left(\frac{2}{3} - \frac{\pi}{ab - 1} \right) > M,$$

and the proof is complete. ♦

Weierstrass presented this example before the Berlin Academy in 1872, but it was published in 1875 in [36] by Du Bois-Reymond. There is some evidence that as early as 1861, Riemann had used a similar function in his lectures. They were all preceded by Bolzano who had constructed one such function in 1830, but his result was not published until 1922!

Exercises

In Exercises 8.3.1–8.3.9 determine whether the series converges uniformly on a set A:

8.3.1. $\sum_{n=1}^{\infty} \dfrac{1}{x^2 + n^2}$, $A = \mathbb{R}$.

8.3.2. $\sum_{n=1}^{\infty} \dfrac{x}{1 + n^4 x^2}$, $A = [0, +\infty)$.

8.3.3. $\sum_{n=1}^{\infty} \dfrac{n^2}{\sqrt{n!}}(x^n + x^{-n})$, $A = [\frac{1}{2}, 2]$.

8.3.4. $\sum_{n=1}^{\infty} \dfrac{\sin nx}{\sqrt[3]{n^4 + x^4}}$, $A = \mathbb{R}$.

8.3.5. $\sum_{n=1}^{\infty} \ln\left(1 + \dfrac{x^2}{n \ln^2 n}\right)$, $A = [-1, 1]$.

8.3.6.* $\sum_{n=1}^{\infty} x^2 e^{-nx}$, $A = [0, +\infty)$.

8.3.7.* $\sum_{n=1}^{\infty} x^2 (1 - x^2)^{n-1}$, $A = [-1, 1]$.

8.3.8.* $\sum_{n=1}^{\infty} n^2 x^2 e^{-n^2 |x|}$, $A = \mathbb{R}$.

8.3.9. $\sum_{n=1}^{\infty} \dfrac{\cos nx}{1 + n^4}$.

8.3.10.* Suppose that the series $\sum_{n=1}^{\infty} f_n(x)$ converges uniformly on a set A. Prove that the sequence $\{f_n\}$ converges uniformly to 0 on A.

8.3.11. Prove the following "alternating series test": If $\{f_n\}$ is a decreasing sequence of non-negative functions on a set A and $\sup_{x \in A} f_n(x) \to 0$, as $n \to \infty$, then $\sum_{n=1}^{\infty} (-1)^{n+1} f_n(x)$ converges uniformly on A.

8.3.12.* Prove Dirichlet's Test (Theorem 8.3.6).

In Exercises 8.3.13–8.3.22 determine whether the series converges uniformly on a set A:

8.3.13. $\sum_{n=1}^{\infty} \dfrac{(-1)^{n-1}}{n + x^2}$, $A = \mathbb{R}$.

8.3.14. $\sum_{n=2}^{\infty} \dfrac{(-1)^{n+1}}{\sqrt{n} + \cos x}$, $A = \mathbb{R}$.

8.3.15. $\sum_{n=1}^{\infty} (1 - x) x^n$, $A = (-1/2, 1/2)$.

8.3.16. $\sum_{n=1}^{\infty} (1 - x) x^n$, $A = (-1, 1)$.

8.3.17. $\sum_{n=1}^{\infty} \dfrac{(-1)^{n+1}}{n^x}$, $A = [a, +\infty)$, $a > 0$.

8.3.18. $\sum_{n=1}^{\infty} (-1)^{n+1} \dfrac{e^{-nx}}{\sqrt{n + x^2}}$, $A = [0, +\infty)$.

8.3.19. $\sum_{n=1}^{\infty} \dfrac{(-1)^{n+1}}{n + x^2} \arctan nx$, $A = \mathbb{R}$.

8.3.20. $\sum_{n=2}^{\infty} \dfrac{(-1)^{n+1} \cos \frac{x}{n}}{\sqrt{n} + \cos x}$, $A = [0, +\infty)$.

8.3.21.* $\sum_{n=1}^{\infty} \dfrac{\sin x \sin nx}{\sqrt{n + x}}$, $A = \mathbb{R}$.

8.3.22.* $\sum_{n=1}^{\infty} \dfrac{\sin nx}{n}$, $A = [a, 2\pi - a]$, $0 < a < \pi$.

In Exercises 8.3.23–8.3.29 determine the set A of values of x for which the series converges pointwise, and a set $B \subset A$ (as large as you can) such that the series converges uniformly on B:

8.3.23. $\sum_{n=1}^{\infty} \dfrac{1}{2n - 1}\left(\dfrac{x + 2}{x - 1}\right)^n$.

8.3.24. $\sum_{n=1}^{\infty} \dfrac{x^2}{(1 + x^2)^n}$.

8.3.25. $\sum_{n=1}^{\infty} \dfrac{\cos nx}{n}$.

8.3.26. $\sum_{n=1}^{\infty} e^{nx} \cos nx$.

8.3.27. $\sum_{n=1}^{\infty} \dfrac{1}{(x + n)(x + n - 1)}$.

8.3.28. $\sum_{n=1}^{\infty} \dfrac{1}{n(1 + x^2)^n}$.

8.3.29. $\sum_{n=1}^{\infty} \dfrac{e^{nx}}{n^2 - n + 1}$.

8.3.30. Prove that $\sum_{n=1}^{\infty} \dfrac{(-1)^{n-1}}{n+x^2}$ converges uniformly but not absolutely.

8.3.31. Give an example of a series that converges absolutely but not uniformly.

8.3.32.* Suppose that the series $\sum_{n=1}^{\infty} f_n(x)$ converges uniformly and absolutely on a set A. Prove or disprove: the series $\sum_{n=1}^{\infty} |f_n(x)|$ converges uniformly on A.

In Exercises 8.3.33–8.3.35 compute the integrals and justify your computation:

8.3.33. $\displaystyle\int_0^{\pi} \sum_{n=1}^{\infty} \frac{\cos nx}{n^2}\, dx.$
8.3.34. $\displaystyle\int_0^{1/2} \sum_{n=1}^{\infty} nx^{n-1}\, dx.$
8.3.35. $\displaystyle\int_1^{2} \sum_{n=1}^{\infty} ne^{-nx}\, dx.$

8.3.36. The purpose of this exercise is to point out at a weakness of Riemann integral: it is not compatible with pointwise limits. It is taken from Darboux's 1875 *Mémoire* [25]. Let $f_n(x) = -2n^2 x e^{-n^2 x^2} + 2(n+1)^2 x e^{-(n+1)^2 x^2}$, $n \in \mathbb{N}$.

(a) Prove that the series $\sum_{n=1}^{\infty} f_n(x)$ converges pointwise and find its sum f.

(b) Let $a > 0$, and let $a_n = \int_0^a f_n(x)\, dx$. Find $\sum_{n=1}^{\infty} a_n$.

(c) Show that $\int_0^a f(x)\, dx \neq \sum_{n=1}^{\infty} a_n$.

8.4 Power Series

Among series of functions, power series hold a very important position. As we will see they converge uniformly, and the limit function is differentiable. In fact, historically speaking, until the end of the eighteenth century, the concept of a function was almost synonymous with a convergent power series.

In the next few sections we will take a closer look at this class of series. As usual, the first order of business is to determine when they converge, and this is what we will do in this section. By a **power series** we mean a series of the form $\sum_{n=0}^{\infty} a_n(x-c)^n$, although we will be mostly interested in the case $c = 0$.

Example 8.4.1. Determine for what values of x the series $\sum_{n=0}^{\infty} \dfrac{x^n}{n!}$ converges.

Here $c = 0$ and $a_n = 1/n!$. When studying the convergence of a power series, most of the time the winning strategy is to use the Ratio Test:

$$\mathcal{D}_n = \frac{|a_{n+1} x^{n+1}|}{|a_n x^n|} = |x| \frac{\frac{1}{(n+1)!}}{\frac{1}{n!}} = \frac{|x|}{n+1} \to 0, \quad n \to \infty,$$

so the series converges absolutely for all $x \in \mathbb{R}$. Furthermore, the convergence is uniform on $[-M, M]$, for any $M > 0$. Indeed,

$$\left| \frac{x^n}{n!} \right| \leq \frac{M^n}{n!}$$

and the series $\sum_{n=0}^{\infty} \dfrac{M^n}{n!}$ converges (e.g., by the Ratio Test). Therefore, the uniform convergence follows from the Weierstrass M-Test, with $M_n = M^n/n!$. ♦

Example 8.4.2. Determine for what values of x the series $\sum_{n=0}^{\infty} \dfrac{x^n}{n^2}$ converges.

Now $c = 0$ and $a_n = 1/n^2$. Therefore,

$$\mathcal{D}_n = \frac{|a_{n+1}x^{n+1}|}{|a_n x^n|} = |x| \frac{\frac{1}{(n+1)^2}}{\frac{1}{n^2}} = \frac{|x| n^2}{(n+1)^2} \to |x|, \quad n \to \infty.$$

We conclude that the series converges when $|x| < 1$ and diverges for $|x| > 1$. ◆

In both Examples 8.4.1 and 8.4.2, the set of all x for which the series converges was an interval. In the former, this interval was infinite, while in the latter it was $(-1, 1)$. It turns out that this is not a coincidence.

Theorem 8.4.3. *Suppose that the power series converges for $x = r > 0$. Then the series converges absolutely for all $x \in (-r, r)$, and it converges uniformly in $[-r', r']$ for any $r' < r$.*

Proof. Let $\sum_{n=0}^{\infty} a_n x^n$ be a power series, let $r > 0$, and suppose that $\sum_{n=0}^{\infty} a_n r^n$ converges. Then the sequence $\{a_n r^n\}$ must converge to 0, so it is bounded: $|a_n r^n| \leq M$ for all $n \in \mathbb{N}$. If $|x| < r$, then

$$\sum_{n=0}^{\infty} |a_n x^n| \leq \sum_{n=0}^{\infty} |a_n r^n| \left|\frac{x}{r}\right|^n \leq M \sum_{n=0}^{\infty} \left|\frac{x}{r}\right|^n. \tag{8.32}$$

The last series converges as a geometric series, so $\sum_{n=0}^{\infty} a_n x^n$ converges absolutely. Further, if $|x| < r' < r$, then (8.32) shows that

$$\left|\sum_{n=0}^{\infty} a_n x^n\right| \leq \sum_{n=0}^{\infty} |a_n x^n| \leq M \sum_{n=0}^{\infty} \left|\frac{r'}{r}\right|^n.$$

Since the last series converges, the uniform convergence of $\sum_{n=0}^{\infty} a_n x^n$ follows from the Weierstrass M-Test. □

Remark 8.4.4. The center of the *interval of convergence* was 0 because we have considered a power series $\sum_{n=0}^{\infty} a_n x^n$ with $c = 0$. In general, the center of the interval of convergence for $\sum_{n=0}^{\infty} a_n (x - c)^n$ will be c.

Theorem 8.4.3 established that a power series always converges in an interval of radius R. We call them the **interval of convergence** and the **radius of convergence**. Still, some questions remain open. What about the endpoints of the interval? In Example 8.4.2, $R = 1$, so we are interested in $x = 1$ and $x = -1$. In the former case we have the series $\sum_{n=0}^{\infty} 1/n^2$ which converges as a p-series; in the latter case, we have $\sum_{n=0}^{\infty} (-1)^n / n^2$ which converges absolutely. Conclusion: the series converges for both $x = -R$ and $x = R$. Before we start believing that this is always true, let us look at a few more examples.

Example 8.4.5. A series diverging at both endpoints of the interval of convergence.

We will show that the series $\sum_{n=0}^{\infty} x^n$ diverges at both endpoints of the interval of convergence.

Solution. Here, $a_n = 1$, so

$$\mathcal{D}_n = \frac{|a_{n+1}x^{n+1}|}{|a_n x^n|} = |x|.$$

and $R = 1$. However, the situation is quite different from the one in Example 8.4.2. When $|x| = 1$ we obtain either the series $\sum_{n=0}^{\infty} 1^n$ or $\sum_{n=0}^{\infty} (-1)^n$. They both diverge by the Divergence Test. Conclusion: the series diverges for both $x = -R$ and $x = R$. ◆

Examples 8.4.2 and 8.4.5 have the common feature that the series either converges or diverges at both endpoints of the interval of convergence. Could there be a different situation at the two ends?

Example 8.4.6. A series converging at only one endpoint of the interval of convergence.

We will show that the series $\sum_{n=0}^{\infty} \frac{x^n}{n}$ converges at one endpoint of the interval of convergence and diverges at the other one.

Solution. Here

$$\mathcal{D}_n = \frac{|a_{n+1} x^{n+1}|}{|a_n x^n|} = |x| \frac{\frac{1}{n+1}}{\frac{1}{n}} = |x| \frac{n}{n+1} \to |x|, \quad n \to \infty,$$

so $R = 1$. When $x = 1$, we obtain the Harmonic series $\sum_{n=0}^{\infty} 1/n$, which diverges. When $x = -1$, we have the Alternating Harmonic series $\sum_{n=0}^{\infty} (-1)^n/n$ which converges. Conclusion: the series converges for $x = -R$ and diverges for $x = R$. ♦

We leave it as an exercise to find a power series that converges for $x = R$ and diverges for $x = -R$.

Theorem 8.4.3 shows that every power series has its radius of convergence, but does not specify how to find it. Examples 8.4.1–8.4.6 make it clear that, if $\rho = \lim |a_{n+1}/a_n|$ exists, then the series converges for $|x|\rho < 1$. In other words, if $\rho \neq 0$, then $R = 1/\rho$, and if $\rho = 0$, then R is infinite. What if ρ is infinite? We might expect R to be 0.

Example 8.4.7. A series with radius of convergence 0.

We will show that the series $\sum_{n=0}^{\infty} n! x^n$ has the radius of convergence 0.

Solution. Here, if $x \neq 0$,

$$\lim \frac{|a_{n+1} x^{n+1}|}{|a_n x^n|} = \lim |x| \frac{(n+1)!}{n!} = |x| \lim(n+1) \to \infty, \quad n \to \infty,$$

so the series diverges for any $x \neq 0$. Of course, if $x = 0$, all the terms for $n \geq 1$ are equal to 0, so the series converges. Thus, $R = 0$. ♦

It looks as if we have all angles covered, but this is not the case yet. What if $\lim |a_{n+1}/a_n|$ does not exist? We have established in Theorem 7.4.5 that the Root Test is stronger than the Ratio Test (see also Example 7.4.6). More precisely, if $\rho = \lim |a_{n+1}/a_n|$ exists, then $\rho = \lim \sqrt[n]{|a_n|}$. This means that, if the Ratio Test fails, we may still use the Root Test. What if even $\lim \sqrt[n]{|a_n|}$ does not exist?

Example 8.4.8. Determining the radius of convergence of a series.

Determine the radius of convergence of the series $\sum_{n=0}^{\infty} (2 + (-1)^n)^n x^n$.

Solution. Here, $a_{2n} = 3^{2n}$ and $a_{2n-1} = 1$. Therefore,

$$\mathcal{D}_{2n} = \left| \frac{x^{2n+1}}{3^{2n} x^{2n}} \right| = \frac{|x|}{3^{2n}}$$

$$\mathcal{D}_{2n-1} = \left| \frac{3^{2n} x^{2n}}{x^{2n-1}} \right| = 3^{2n} |x|.$$

Clearly, the sequence $\{\mathcal{D}_n\}$ is not convergent (if $x \neq 0$). Even the part of Theorem 7.4.3

that uses inequalities instead of limits is of no use, because it requires that $\mathcal{D}_n \le r < 1$ for n large enough. However, $\mathcal{D}_{2n-1} = 3^{2n}|x| \to \infty$ unless $x = 0$, so the Ratio Test would only tell us that the series converges for $x = 0$. On the other hand, it can detect the divergence only when $\mathcal{D}_n \ge 1$ for large n. Since $\mathcal{D}_{2n} \to 0$, the test is inconclusive.

Let us try the Root Test.

$$\mathcal{C}_n = \sqrt[n]{\left|(2 + (-1)^n)^n\, x^n\right|} = |2 + (-1)^n|\,|x| = \begin{cases} 3|x|, & \text{if } n \text{ is even} \\ |x|, & \text{if } n \text{ is odd.} \end{cases}$$

Again, the sequence $\{\mathcal{C}_n\}$ does not converge, so we try the part of Theorem 7.4.1 that uses inequalities. Now, we need that

$$\mathcal{C}_n \le r < 1, \quad \text{for } n \ge N.$$

Since $3|x| \ge |x|$, all we need is that $3|x| \le r < 1$, which is true whenever $|x| < 1/3$. So, the power series converges for $|x| < 1/3$.

What about $|x| \ge 1/3$? Is it possible that the radius of convergence is bigger than $1/3$? The answer is no, because already for $x = 1/3$ the series diverges. Indeed,

$$a_{2n}\left(\frac{1}{3}\right)^{2n} = \frac{\left(2 + (-1)^{2n}\right)^{2n}}{3^{2n}} = 1$$

for all $n \in \mathbb{N}$, and the Divergence Test applies.   ◆

We have seen that the sequence $\{\sqrt[n]{|a_n|}\}$ did not have a limit, because it was equal to 3 for even n, and 1 for odd n. It did have two accumulation points (1 and 3) and it turned out that the radius of convergence was determined by the larger of the two. More formally, 3 is $\limsup \sqrt[n]{|a_n|}$, and we have obtained that $R = 1/3$. It turns out that this is true in general.

Theorem 8.4.9 (Cauchy–Hadamard Theorem). *Let $\{a_n\}$ be a sequence of real numbers and let $\rho = \limsup \sqrt[n]{|a_n|}$ (finite or infinite). The radius of convergence of the power series $\sum_{n=0}^{\infty} a_n x^n$ is $R = 1/\rho$ (with obvious modifications if ρ equals 0 or ∞).*

Proof. If ρ is infinite, then we need to show that $R = 0$. Every series $\sum_{n=0}^{\infty} a_n x^n$ converges for $x = 0$, so we have to prove that it diverges for $x \ne 0$. So, let $x \ne 0$ be fixed. Since $\limsup \sqrt[n]{|a_n|}$ is infinite, there exists a subsequence $\{\sqrt[n_k]{|a_{n_k}|}\}$ such that

$$\sqrt[n_k]{|a_{n_k}|} \ge \frac{1}{|x|}, \quad \text{for all } k \in \mathbb{N}.$$

This implies that

$$|a_{n_k} x^{n_k}| \ge 1, \quad \text{for all } k \in \mathbb{N},$$

and the Divergence Test applies.

Suppose now that ρ is finite. Because of that, the inequalities

$$|x| < R \quad \text{and} \quad |x|\rho < 1$$

are equivalent. (Of course, if $\rho = 0$, the first inequality simply means that x can be any real number.) First we will show that, if x is a real number satisfying $|x|\rho < 1$, the series $\sum_{n=0}^{\infty} a_n x^n$ converges. Since the series converges at $x = 0$, we can assume that $x \ne 0$. Let r be a real number such that $|x|\rho < r < 1$, and let

$$\varepsilon = \frac{r - |x|\rho}{|x|}.$$

Clearly, $\varepsilon > 0$ so Exercise 2.8.3 shows that there exists $N \in \mathbb{N}$ such that $\sqrt[n]{|a_n|} < \rho + \varepsilon$, for $n \geq N$. For such n,

$$\sqrt[n]{|a_n x^n|} = |x|\sqrt[n]{|a_n|} < |x|(\rho + \varepsilon) = |x|\rho + |x|\frac{r - |x|\rho}{|x|} = r,$$

so $|a_n x^n| < r^n$. Since $0 \leq r < 1$, the geometric series $\sum_{n=0}^{\infty} r^n$ converges, and the convergence of $\sum_{n=0}^{\infty} a_n x^n$ follows by the Comparison Test.

So far we have established that our choice of R was not too big. Now we will show that it is not too small, meaning that if $|x| > R$ then the series $\sum_{n=0}^{\infty} a_n x^n$ diverges. Once again, the inequalities

$$|x| > R \quad \text{and} \quad |x|\rho > 1$$

are equivalent, and the second one is impossible if $\rho = 0$. So, let us assume that ρ is not zero and that $|x|\rho > 1$. It follows that $1/|x| < \rho$. By definition, ρ is the largest accumulation point of $\{\sqrt[n]{|a_n|}\}$, so there is a subsequence $\{\sqrt[n_k]{|a_{n_k}|}\}$ converging to ρ. Since $1/|x| < \rho$, there is a positive integer K such that

$$\sqrt[n_k]{|a_{n_k}|} > \frac{1}{|x|} \quad \text{for} \quad k \geq K.$$

For such k,

$$|a_{n_k} x^{n_k}| > 1$$

which implies that the sequence $\{a_n x^n\}$ cannot converge to 0. By the Divergence Test, the series $\sum_{n=0}^{\infty} a_n x^n$ diverges. $\qquad\square$

Theorem 8.4.9 gives a formula for the radius of convergence R, and it completely settles the cases $|x| < R$ and $|x| > R$. However, it does not provide an answer when $|x| = R$. Unfortunately, this has to be done on a series by series basis.

The theorem can be found in Cauchy's *Cours d'Analyse*, but it was largely forgotten at the time when Hadamard rediscovered it as a doctoral student in 1888.

JACQUES HADAMARD (1865–1963) was a French mathematician. As a child he struggled with arithmetic but by the fifth grade he was getting awards in several subjects and catching up with mathematics. He placed first on the entrance exams for both Ecole Polytechnique and Ecole Normale Supérieure, chose the latter and graduated in 1888. He taught in several high schools while working on his doctorate. At that time one of his students was Fréchet. His 1892 dissertation was a pioneering work on the general theory of analytic functions. His research on functions of a complex variable brought him the Grand Prix des Sciences Mathématiques and culminated in a proof of the Prime Number Theorem in 1896. In the same year he was awarded the Bordin Prize of the French Academy of Sciences for his work on geodesics in the differential geometry of surfaces and dynamical systems. After that he started working on mathematical physics and, in particular on partial differential equations, the calculus of variations and the foundations of functional analysis. He introduced the idea of well-posed problem and the method of descent in the theory of partial differential equations, leading up to his seminal book on the subject, based on lectures given at Yale University in 1922. Later in his life he wrote on probability theory and mathematical education. Outside of mathematics he started being politically active with the "Dreyfus affair" in 1894. Alfred Dreyfus, whose wife was a second cousin of Hadamard, was accused of selling military secrets to the Germans and he was sentenced to life in prison. The process split France (in part because Dreyfus was a Jew) and it took more than 10 years to clear his name. Hadamard lost two older sons in WWI, and the third one was killed in WWII. He immersed himself in work, writing close to 300 scientific papers and books, including books for wider audience such as *The Psychology of Invention in the Mathematical Field*.

Exercises

In Exercises 8.4.1–8.4.12 determine for what values of x the series converges:

8.4.1. $\displaystyle\sum_{n=1}^{\infty} \frac{n(x-1)^n}{2^n(3n-1)}$.

8.4.2. $\displaystyle\sum_{n=1}^{\infty} \frac{n!}{n^n} x^n$.

8.4.3. $\displaystyle\sum_{n=1}^{\infty} \frac{3^n + (-2)^n}{n} (x+1)^n$.

8.4.4. $\displaystyle\sum_{n=1}^{\infty} \left(1 + \frac{1}{n}\right)^{n^2} x^n$.

8.4.5. $\displaystyle\sum_{n=1}^{\infty} \frac{1}{2^n} x^{n^2}$.

8.4.6. $\displaystyle\sum_{n=1}^{\infty} \frac{(-1)^{\lfloor \sqrt{n} \rfloor}}{n} x^n$.

8.4.7.* $\displaystyle\sum_{n=1}^{\infty} \frac{1}{a^n + b^n} x^n$, $a, b > 0$.

8.4.8.* $\displaystyle\sum_{n=1}^{\infty} \left(1 + \frac{1}{2} + \frac{1}{3} + \cdots + \frac{1}{n}\right) x^n$.

8.4.9.* $\displaystyle\sum_{n=1}^{\infty} \frac{1}{\sin^n n} x^n$.

8.4.10.* $\displaystyle\sum_{n=1}^{\infty} x^{n!}$.

8.4.11.* $\displaystyle\sum_{n=1}^{\infty} \frac{1}{\sqrt[n]{n + n^n}} x^n$.

8.4.12. $\displaystyle\left(\frac{1}{3}\right)^2 x - \left(\frac{1 \cdot 2}{3 \cdot 5}\right)^2 x^2 + \left(\frac{1 \cdot 2 \cdot 3}{3 \cdot 5 \cdot 7}\right)^2 x^3 - \cdots$.

In Exercises 8.4.13–8.4.16 the radii of convergence of $\sum_{n=1}^{\infty} a_n x^n$ and $\sum_{n=1}^{\infty} b_n x^n$ are denoted by R_1 and R_2, respectively.

8.4.13. Suppose that $R_1 \neq R_2$ and that the radius of convergence of $\sum_{n=1}^{\infty}(a_n + b_n)x^n$ is R. Prove that $R = \min\{R_1, R_2\}$. What can be said if $R_1 = R_2$?

8.4.14.* Suppose that the radius of convergence of $\sum_{n=1}^{\infty}(a_n b_n)x^n$ is R. Prove that $R \geq R_1 R_2$. Give an example to show that the inequality may be strict.

8.4.15.* Suppose that $R_1, R_2 \in (0, \infty)$, that $b_n \neq 0$ for all $n \in \mathbb{N}$, and that the radius of convergence of $\sum_{n=1}^{\infty}(a_n/b_n)x^n$ is R. Prove that $R \geq R_1/R_2$. Give an example to show that the inequality may be strict.

8.4.16.* The product of the series $\sum_{n=1}^{\infty} a_n x^n$ and $\sum_{n=1}^{\infty} b_n x^n$ is defined by $\sum_{n=1}^{\infty} c_n x^n$, where $c_n = a_0 b_n + a_1 b_{n-1} + \cdots + a_n b_0$ for all $n \in \mathbb{N}_0$. Suppose that it has the radius of convergence R. Prove that $R \geq \min\{R_1, R_2\}$. Give an example to show that the inequality may be strict.

In Exercises 8.4.17–8.4.20 the radius of convergence of $\sum_{n=1}^{\infty} a_n x^n$ is $R \in (0, \infty)$. Find the radius of convergence of the series:

8.4.17. $\displaystyle\sum_{n=1}^{\infty} 2^n a_n x^n$.

8.4.18. $\displaystyle\sum_{n=1}^{\infty} n^n a_n x^n$.

8.4.19. $\displaystyle\sum_{n=1}^{\infty} \frac{n^n}{n!} a_n x^n$.

8.4.20. $\displaystyle\sum_{n=1}^{\infty} a_n^2 x^n$.

8.4.21.* Describe all power series that are uniformly convergent on $\mathbb{R}$.

8.5 Power Series Expansions of Elementary Functions

Taylor's Formula (page 121) shows that, assuming that an elementary function f is $n+1$ times differentiable, one can approximate it by a polynomial p_n of degree n. Since most of the elementary functions are infinitely differentiable, we can keep adding terms of higher and higher degree, and we obtain a sequence of polynomials. These polynomials are given by (4.10) and their coefficients by (4.11). If the sequence $\{p_n\}$ converges to f, we have a power series representation of f, called the **Taylor series** of f. In this section we will look at some elementary functions and demonstrate the convergence of the appropriate Taylor series. We will restrict our attention to the case when the expansion point is $c = 0$. Such series were extensively studied by MacLaurin, and they are called **MacLaurin series**. In reality, as MacLaurin himself was quick to admit, this is just a special case of Taylor series.

We will start with the following simple result.

Proposition 8.5.1. *Let f be a function defined in the interval $[0, b]$ with derivatives of all orders, and suppose that there exists $M > 0$ such that*

$$|f^{(n)}(x)| \le M, \quad x \in [0, b], \quad n \in \mathbb{N}. \tag{8.33}$$

Then, for all $x \in [0, b]$,

$$f(x) = \sum_{n=0}^{\infty} \frac{f^{(n)}(0)}{n!} x^n. \tag{8.34}$$

Proof. Let $r_n(x) = f(x) - p_n(x)$, with p_n as in (4.10). We will show that the sequence $\{r_n\}$ converges uniformly to 0. By Taylor's Formula, there exists x_0 between 0 and x such that

$$r_n(x) = \frac{f^{(n+1)}(x_0)}{(n+1)!} x^{n+1}.$$

Therefore,

$$|r_n(x)| \le M \frac{b^{n+1}}{(n+1)!} \to 0, \quad n \to \infty.$$

The last limit can be proved to be 0 in a way similar to Exercise 2.1.9. Alternatively, the Ratio Test shows that $\sum b^{n+1}/(n+1)!$ converges, so its terms must converge to 0. Either way, $\{r_n\}$ converges uniformly to 0. $\qquad\square$

Example 8.5.2. The MacLaurin series for $f(x) = e^x$ converges for each $x \in \mathbb{R}$.

Since $f^{(n)}(x) = e^x$, the estimate (8.33) holds, for any $x \in [-b, b]$ and all $n \in \mathbb{N}$, with $M = e^b$. Therefore, Proposition 8.5.1 implies that

$$e^x = \sum_{n=0}^{\infty} \frac{x^n}{n!} = 1 + x + \frac{x^2}{2!} + \frac{x^3}{3!} + \cdots \tag{8.35}$$

for all $x \in [-b, b]$. Since b can be any real number we see that this expansion is true for all $x \in \mathbb{R}$. $\qquad\blacklozenge$

Example 8.5.3. The MacLaurin series for $\sin x$ and $\cos x$.

We will write the MacLaurin series for $f(x) = \sin x$ and $g(x) = \cos x$, and we will show that they both converge for each $x \in \mathbb{R}$.

Solution. The nth derivatives are $f^{(n)}(x) = \sin(x + n\pi/2)$ and $g^{(n)}(x) = \cos(x + n\pi/2)$ (Exercises 4.3.15 and 4.3.16). Therefore, the inequality (8.33) holds with $M = 1$, and any $b \in \mathbb{R}$. It follows that

$$\sin x = \sum_{n=1}^{\infty} \frac{(-1)^{n+1}}{(2n-1)!} x^{2n-1}, \quad \cos x = \sum_{n=0}^{\infty} \frac{(-1)^n}{(2n)!} x^{2n},$$

for all $x \in \mathbb{R}$. $\qquad\blacklozenge$

Example 8.5.4. The MacLaurin series for $\sinh x$ and $\cosh x$.

We will write the MacLaurin series for $f(x) = \cosh x$ and $g(x) = \sinh x$, and we will show that they both converge for each $x \in \mathbb{R}$.

Solution. If we replace x with $-x$ in (8.35) we obtain

$$e^{-x} = \sum_{n=0}^{\infty} \frac{(-x)^n}{n!} = 1 - x + \frac{x^2}{2!} - \frac{x^3}{3!} + \dots.$$

Therefore, the Taylor series for the hyperbolic functions are

$$\cosh x = \frac{e^x + e^{-x}}{2} = 1 + \frac{x^2}{2!} + \frac{x^4}{4!} + \dots, \quad \sinh x = \frac{e^x - e^{-x}}{2} = x + \frac{x^3}{3!} + \frac{x^5}{5!} + \dots,$$

and these expansions hold for all $x \in \mathbb{R}$.   ♦

Next we turn our attention to the **Binomial Series**. These are the power series expansions for the functions of the form $f(x) = (1 + x)^\alpha$, where α is any real number. The case $\alpha = 0$ is not particularly interesting, and neither is the case when α is a positive integer which is covered by the Binomial Formula (page 34). For any other α we have that

$$f^{(n)}(x) = \alpha(\alpha - 1)(\alpha - 2)\dots(\alpha - n + 1)(1 + x)^{\alpha - n}, \quad n = 0, 1, 2, \dots.$$

We would like to prove the equality (8.34), but Proposition 8.5.1 is of little use here. The problem is that, no matter what $b > 0$ we select, $f^{(n)}(x)$ is not bounded as $n \to \infty$. In fact, the smallest value of $|f^{(n)}(x)|$ is at $x = 0$, and it equals $|\alpha(\alpha - 1)(\alpha - 2)\dots(\alpha - n + 1)| \to \infty$.

Nevertheless, the Taylor series of f converges to f. We will establish this in two steps. First, we will prove that it is a convergent series. If we use the Ratio Test, we obtain

$$\mathcal{D}_n = \left| \frac{\frac{\alpha(\alpha-1)(\alpha-2)\dots(\alpha-n)}{(n+1)!} x^{n+1}}{\frac{\alpha(\alpha-1)(\alpha-2)\dots(\alpha-n+1)}{n!} x^n} \right| = \frac{|\alpha - n|}{n + 1} |x| \to |x|$$

so the series converges for $|x| < 1$.

It remains to demonstrate that its limit is f. We will accomplish this by showing that $r_n(x) \to 0$, for $|x| < 1$. By Taylor's Formula (b) (Theorem 4.5.1, page 121), the remainder $r_n(x)$ in the Cauchy form is

$$r_n(x) = \frac{f^{(n+1)}((1 - \theta)c + \theta x)}{n!} (1 - \theta)^n (x - c)^{n+1},$$

for some $0 < \theta < 1$. Of course, we are using $c = 0$, so

$$r_n(x) = \frac{\alpha(\alpha - 1)(\alpha - 2)\dots(\alpha - n)}{n!} (1 + \theta x)^{\alpha - n - 1} (1 - \theta)^n x^{n+1}.$$

We will write it as

$$\left[\frac{(\alpha - 1)(\alpha - 2)\dots(\alpha - 1 - n + 1)}{n!} x^n \right] \left[\alpha x (1 + \theta x)^{\alpha - 1} \right] \left(\frac{1 - \theta}{1 + \theta x} \right)^n. \tag{8.36}$$

Notice that the first expression in brackets is the general term in the Taylor series for the function $(1 + x)^{\alpha - 1}$. Since this series converges, its terms must converge to 0. Thus, it suffices to establish that the remaining factors in (8.36) remain bounded as $n \to \infty$. It is easy to see that the second factor in brackets depends on n only through θ, and $1 - |x| \le 1 + \theta x \le 1 + |x|$, so it remains bounded. Finally, the last fraction is between 0 and 1. Indeed, both the numerator and the denominator are positive, and $x \ge -1$ implies that $-\theta \le \theta x$ so $1 - \theta \le 1 + \theta x$.

Example 8.5.5. The MacLaurin series for $f(x) = \sqrt{1+x}$.

Find the MacLaurin series for $f(x) = \sqrt{1+x}$.

Solution. Here, $\alpha = 1/2$, so the derivatives $f^{(n)}(0)$ form a sequence

$$1, \ \frac{1}{2}, \ \frac{1}{2}\left(-\frac{1}{2}\right), \ \frac{1}{2}\left(-\frac{1}{2}\right)\left(-\frac{3}{2}\right), \ \frac{1}{2}\left(-\frac{1}{2}\right)\left(-\frac{3}{2}\right)\left(-\frac{5}{2}\right), \ \dots.$$

We see that, for $n \geq 2$,

$$f^{(n)}(0) = (-1)^{n-1} \frac{(2n-3)!!}{2^n},$$

and the series becomes

$$\sqrt{1+x} = 1 + \frac{1}{2}x - \frac{1}{2^2(2!)}x^2 + \frac{3!!}{2^3(3!)}x^3 - \frac{5!!}{2^4(4!)}x^4 + \dots. \qquad \blacklozenge$$

Example 8.5.6. The MacLaurin series for $f(x) = \dfrac{1}{\sqrt{1+x}}$.

Find the MacLaurin series for $f(x) = \dfrac{1}{\sqrt{1+x}}$.

Solution. Now, $\alpha = -1/2$, so the derivatives $f^{(n)}(0)$ form a sequence

$$1, \ -\frac{1}{2}, \ -\frac{1}{2}\left(-\frac{3}{2}\right), \ -\frac{1}{2}\left(-\frac{3}{2}\right)\left(-\frac{5}{2}\right), \ \dots.$$

We obtain that, for $n \geq 1$,

$$f^{(n)}(0) = (-1)^n \frac{(2n-1)!!}{2^n},$$

and the series becomes

$$\frac{1}{\sqrt{1+x}} = 1 - \frac{1}{2}x + \frac{3!!}{2^2(2!)}x^2 - \frac{5!!}{2^3(3!)}x^3 + \frac{7!!}{2^4(4!)}x^4 - \dots. \qquad (8.37)$$

$$\blacklozenge$$

Example 8.5.7. The MacLaurin series for $f(x) = \dfrac{1}{1+x}$.

Find the MacLaurin series for $f(x) = \dfrac{1}{1+x}$.

Solution. Since $\alpha = -1$, the derivatives $f^{(n)}(0)$ form a sequence

$$-1, \ (-1)(-2), \ (-1)(-2)(-3), \ \dots$$

so $f^{(n)}(0) = (-1)^n n!$. The Taylor series is

$$\frac{1}{1+x} = 1 - x + x^2 - x^3 - \dots. \qquad (8.38)$$

Of course, this is a geometric series, so we have just confirmed the well known formula. $\blacklozenge$

By Theorem 8.4.3, the sequence of partial sums $\{s_n\}$ of a power series converges uniformly in $[-r, r]$, for any $r < R$. Since $s_n(x)$ is a polynomial for each $n \in \mathbb{N}$, it is continuous, hence integrable, so (8.8) shows that

$$\int_0^x f(x)\,dx = \sum_{n=0}^{\infty} a_n \frac{x^{n+1}}{n+1}, \qquad (8.39)$$

for any $x \in [0, R)$. This last property is often phrased as "a power series can be integrated term by term". (See also Exercise 8.5.26.) We will use it to derive the power series expansion for some functions.

Example 8.5.8. The MacLaurin series for $f(x) = \ln(1+x)$.

Find the MacLaurin series for $f(x) = \ln(1+x)$.

Solution. Integrating the series (8.38) term by term, over the interval $[0, x]$, with $|x| < 1$, we obtain that

$$\ln(1+x) = x - \frac{x^2}{2} + \frac{x^3}{3} - \frac{x^4}{4} + \cdots, \tag{8.40}$$

and this series converges for $|x| < 1$.

Why $\ln(1+x)$? Why not $\ln x$? The reason is that the function $f(x) = \ln x$ is not defined at $x = 0$, so there can be no MacLaurin series for this function.   ◆

Example 8.5.9. The MacLaurin series for $f(x) = \arctan x$.

Find the MacLaurin series for $f(x) = \arctan x$.

Solution. If we replace x with t^2 in (8.38), we obtain a power series expansion

$$\frac{1}{1+t^2} = 1 - t^2 + t^4 - t^6 + \cdots. \tag{8.41}$$

Next, integrating this series term by term, over the interval $[0, x]$, with $|x| < 1$, yields

$$\arctan x = x - \frac{x^3}{3} + \frac{x^5}{5} - \frac{x^7}{7} + \cdots, \tag{8.42}$$

and the series converges for $|x| < 1$.   ◆

Example 8.5.10. The MacLaurin series for $f(x) = \arcsin x$.

Find the MacLaurin series for $f(x) = \arcsin x$.

Solution. If we replace x by $-t^2$ in (8.37), we get

$$\frac{1}{\sqrt{1-t^2}} = 1 + \frac{1}{2}t^2 + \frac{3!!}{2^2(2!)}t^4 + \frac{5!!}{2^3(3!)}t^6 + \frac{7!!}{2^4(4!)}t^8 + \cdots.$$

Integrating this series term by term, over the interval $[0, x]$, with $|x| < 1$, we obtain that

$$\arcsin x = x + \frac{1}{2}\frac{x^3}{3} + \frac{3!!}{2^2(2!)}\frac{x^5}{5} + \frac{5!!}{2^3(3!)}\frac{x^7}{7} + \frac{7!!}{2^4(4!)}\frac{x^9}{9} + \cdots. \qquad ◆$$

So far, we have made quite good use of the fact that a power series can be integrated term by term. Can we differentiate it term by term? We want to apply Theorem 8.2.5, and it is not hard to see that the hypothesis that needs to be verified is the uniform convergence of the derivatives. More precisely, if s_n is the nth partial sum of the power series, then we need to establish that the sequence $\{s_n'\}$ converges uniformly.

Theorem 8.5.11. *Let $\sum_{n=0}^{\infty} a_n x^n$ be a power series with the radius of convergence R. Then the series $\sum_{n=1}^{\infty} n a_n x^{n-1}$ has the radius of convergence R.*

Proof. By Cauchy–Hadamard Theorem, the series $\sum_{n=1}^{\infty} n a_n x^{n-1}$ has the radius of convergence $R' = 1/\rho'$, where

$$\rho' = \limsup \sqrt[n]{|na_n|}.$$

By Exercise 2.8.13,

$$\rho' = \limsup \sqrt[n]{|a_n|} \lim \sqrt[n]{n} = \limsup \sqrt[n]{|a_n|} = \rho,$$

so it follows that the series $\sum_{n=0}^{\infty} a_n x^n$ and $\sum_{n=1}^{\infty} n a_n x^{n-1}$ have the same radius of convergence.   □

Corollary 8.5.12. *Let $\sum_{n=0}^{\infty} a_n x^n$ be a power series with the radius of convergence R. The series is an infinitely differentiable function in $(-R, R)$ and it can be differentiated term by term:*

$$\left(\sum_{n=0}^{\infty} a_n x^n\right)' = \sum_{n=1}^{\infty} n a_n x^{n-1}.$$

Example 8.5.13. The MacLaurin series for $f(x) = \dfrac{1}{(1-x)^2}$.

Find the MacLaurin series for $f(x) = \dfrac{1}{(1-x)^2}$.

Solution. We will replace x by $-x$ in (8.38). This yields

$$\frac{1}{1-x} = 1 + x + x^2 + x^3 + \dots$$

which holds for $|x| < 1$. If we now differentiate term by term we obtain, for $|x| < 1$,

$$\frac{1}{(1-x)^2} = 1 + 2x + 3x^2 + 4x^3 + \dots. \qquad \blacklozenge$$

Another consequence of Theorem 8.5.11 is that the sum of a power series is continuous on the interval of convergence. A much harder question concerns the continuity at the endpoints: $x = R$ and $x = -R$. Of course, if the series diverges at such a point, then f is not even defined there. What if it converges at an endpoint?

Theorem 8.5.14 (Abel's Theorem). *If a power series converges at $x = R$ then it converges uniformly in $[0, R]$.*

Proof. The idea of the proof is to write

$$\sum_{n=0}^{\infty} a_n x^n = \sum_{n=0}^{\infty} a_n R^n \left(\frac{x}{R}\right)^n$$

and then apply Abel's Test for uniform convergence (Theorem 8.3.4). The series $\sum_{n=0}^{\infty} a_n R^n$ converges and its terms do not depend on x, so the convergence is automatically uniform. The sequence $(x/R)^n$ is monotone decreasing for each $x \in [0, R]$, and $|(x/R)^n| \le 1$ for any $x \in [0, R]$ and $n \in \mathbb{N}$, so it is uniformly bounded. Thus, the conditions of the Abel's Test are satisfied, and the power series $\sum_{n=0}^{\infty} a_n x^n$ converges uniformly in $[0, R]$. $\qquad \square$

It follows from Abel's Theorem and Theorem 8.2.1 that, if a power series converges at $x = R$, its sum is continuous at $x = R$. In particular, Equation 8.6 shows that

$$\lim_{x \to R^-} \sum_{n=0}^{\infty} a_n x^n = \sum_{n=0}^{\infty} a_n R^n. \qquad (8.43)$$

Example 8.5.15. The sum of the Alternating Harmonic series.

We will show that the Alternating Harmonic series converges to $\ln 2$.

Solution. We will apply (8.43) to the function $f(x) = \ln(1+x)$. The MacLaurin series for f is

$$\ln(1+x) = x - \frac{x^2}{2} + \frac{x^3}{3} - \frac{x^4}{4} + \dots, \qquad (8.44)$$

for $-1 < x < 1$. When $x = 1$ we obtain the convergent series

$$1 - \frac{1}{2} + \frac{1}{3} - \frac{1}{4} + \dots$$

and formula (8.43) shows that its sum equals $\lim_{x \to 1^-} \ln(1+x) = \ln 2$. Thus, the Alternating Harmonic series converges to $\ln 2$. ♦

Example 8.5.16. Approximating π.

We will find an approximation for π.

Solution. We will apply (8.43) to $f(x) = \arctan x$. The MacLaurin series for f is

$$\arctan x = x - \frac{x^3}{3} + \frac{x^5}{5} - \frac{x^7}{7} + \dots,$$

and it converges for all $x \in (-1, 1)$. For $x = 1$, the series

$$1 - \frac{1}{3} + \frac{1}{5} - \frac{1}{7} + \dots$$

converges by the Alternating Series Test. It follows from Abel's Theorem that its sum is $\arctan 1 = \pi/4$. Thus we obtain that

$$\frac{\pi}{4} = 1 - \frac{1}{3} + \frac{1}{5} - \frac{1}{7} + \dots. \tag{8.45}$$

This formula can be used to approximate π. It was known to Leibniz and it is considered to be the first series of this kind. (The significance is mostly historical—the series converges much slower than some other series that were discovered later.) ♦

We will finish this section with a remark that concerns the convergence of a Taylor series. If f is a function that can be represented by a convergent Taylor series, then it must be infinitely differentiable (Corollary 8.5.12). By Exercise 8.5.9, for any function there can be at most one power series representation, and the coefficients are calculated using formulas (4.11). All that remains is to establish that the Taylor series converges to f. When this happens, we say that f is **analytic**. In this paragraph we have brought to light the fact that, if f is analytic, then it is infinitely differentiable. Unfortunately, in spite of all the examples in this section, the converse is not true.

Example 8.5.17. An infinitely differentiable function that is not analytic.

We will show that the function $f(x) = \begin{cases} e^{-1/x^2}, & \text{if } x \neq 0 \\ 0, & \text{if } x = 0 \end{cases}$ is infinitely differentiable, but not analytic.

Solution. We will demonstrate that: (a) f is infinitely differentiable, and (b) for any $n \in \mathbb{N}_0$, $f^{(n)}(0) = 0$. Thus the Taylor series of f has all coefficients 0, and its sum is 0 for any $x \in \mathbb{R}$, but it converges to $f(x)$ only at $x = 0$.

For $x \neq 0$, the function f has derivatives of all orders. Let $g(x) = -1/x^2$, and let us define a sequence of functions $g_n(x)$ by

$$g_0(x) = 1, \quad g_{n+1}(x) = g'(x)g_n(x) + g'_n(x), \text{ for } n \in \mathbb{N}_0. \tag{8.46}$$

Then, for $x \neq 0$, $f(x) = e^{g(x)}$ and we will establish that,

$$f^{(n)}(x) = f(x)g_n(x). \tag{8.47}$$

Clearly, (8.47) is true for $n = 0$. We will assume that it is true for some $n \in \mathbb{N}$, and we will prove that it holds for $n + 1$ as well. Indeed,

$$f^{(n+1)}(x) = (f(x)g_n(x))' = f'(x)g_n(x) + f(x)g_n'(x)$$
$$= f(x)g'(x)g_n(x) + f(x)g_n'(x) = f(x)\left[g'(x)g_n(x) + g_n'(x)\right]$$

so (8.47) is true for any n. Also, (8.46) can be used to deduce that $g_n(x)$ is a polynomial in $1/x$ of degree $3n$, so $g_n(x)/x$ is a polynomial in $1/x$ of degree $3n + 1$. Using the L'Hôpital's Rule (see Exercise 4.6.14),

$$\lim_{h \to 0} \frac{e^{g(h)}}{h^m} = 0, \quad \text{for all } m \in \mathbb{N}_0,$$

and it follows that

$$\lim_{h \to 0} \frac{e^{g(h)}g_n(h)}{h} = 0, \quad \text{for all } n \in \mathbb{N}_0.$$

Finally, we can show that f has derivatives of all orders at $x = 0$, and that they are all equal to 0. We see that $f(0) = 0$. If we assume that $f^{(n)}(0) = 0$, then

$$f^{(n+1)}(0) = \lim_{h \to 0} \frac{f^{(n)}(h) - f^{(n)}(0)}{h} = \lim_{h \to 0} \frac{f^{(n)}(h)}{h} = \lim_{h \to 0} \frac{e^{g(h)}g_n(h)}{h} = 0.$$

Consequently, $f^{(n)}(0) = 0$ for all $n \in \mathbb{N}_0$. ◆

Example 8.5.17 also implies that two different functions can have the same Taylor series. Cauchy showed this in 1822 and it effectively brought to an end the eighteenth century belief that calculus can be based on algebra, of which Lagrange was one of the major proponents.

Exercises

In Exercises 8.5.1–8.5.7 write the MacLaurin series for the given function and determine the set on which it converges.

8.5.1. $y = \dfrac{1}{2 - x}$.

8.5.2. $y = \operatorname{artanh} x$.

8.5.3. $y = \ln(x + \sqrt{1 + x^2})$.

8.5.4. $y = \sin^3 x$.

8.5.5. $y = \dfrac{1}{(1 - x)(2 - x)}$.

8.5.6. $y = \ln(1 + x + x^2 + x^3)$.

8.5.7. $y = \dfrac{\ln(1 + x)}{1 + x}$.

8.5.8.* Let $f(x) = \dfrac{1}{1 + x + x^2 + x^3}$. Find $f^{(1000)}(0)$.

8.5.9.* Prove that if f has a Taylor series, the expansion is unique.

8.5.10. The function f is defined by $f(x) = \sum_{n=1}^{\infty} x^n/n^2$. Prove that there exists a constant C such that

$$f(x) + f(1 - x) + \ln x \ln(1 - x) = C,$$

for all $x \in (-1, 1)$. Use the equality in Exercise 9.2.10 to find C.

In Exercises 8.5.11–8.5.14 use term by term differentiation to find the sum of the series:

8.5.11. $\displaystyle\sum_{n=1}^{\infty} (-1)^{n-1} \frac{x^{2n-1}}{2n - 1}$.

8.5.12.* $\displaystyle\sum_{n=0}^{\infty} \frac{x^{2n}}{(2n)!}$.

8.5.13.* $\displaystyle\sum_{n=1}^{\infty} \frac{x^n}{n(n + 1)}$.

8.5.14.* $1 + \dfrac{x}{2} + \dfrac{1 \cdot 3}{2 \cdot 4} x^2 + \dfrac{1 \cdot 3 \cdot 5}{2 \cdot 4 \cdot 6} x^3 + \ldots.$

In Exercises 8.5.15–8.5.17 use term by term integration to find the sum of the series:

8.5.15. $\displaystyle\sum_{n=1}^{\infty} (-1)^{n-1} n^2 x^n.$ **8.5.16.*** $\displaystyle\sum_{n=1}^{\infty} n(n+1) x^n.$ **8.5.17.** $\displaystyle\sum_{n=1}^{\infty} n x^{n-1}.$

8.5.18. Suppose that $\{a_n\}$ is a sequence of non-negative numbers such that the limit $\lim_{x \to R^-} \sum_{n=1}^{\infty} a_n x^n$ exists and equals S. Prove that $\sum_{n=1}^{\infty} a_n R^n = S$.

In Exercises 8.5.19–8.5.25 find the sum of the series:

8.5.19. $\displaystyle\sum_{n=1}^{\infty} \dfrac{n}{(n-1)!}.$ **8.5.20.** $\displaystyle\sum_{n=0}^{\infty} \dfrac{(-1)^n}{3n+1}.$ **8.5.21.** $\displaystyle\sum_{n=1}^{\infty} \dfrac{(n+1)^2}{n!}.$

8.5.22. $\displaystyle\sum_{n=1}^{\infty} \dfrac{(n+1)(n+2)}{n!}.$ **8.5.23.** $\displaystyle\sum_{n=1}^{\infty} \dfrac{n(n+1)}{2^n}.$ **8.5.24.*** $\displaystyle\sum_{n=1}^{\infty} (-1)^{n-1} \dfrac{1}{n(n+1)}.$

8.5.25.* $\dfrac{1}{1 \cdot 2} + \dfrac{1}{3 \cdot 4} + \dfrac{1}{5 \cdot 6} + \ldots.$

8.5.26. Suppose that the series $\sum_{n=1}^{\infty} a_n x^n$ converges uniformly on $[a, b]$. Prove that

$$\int_a^b \left(\sum_{n=1}^{\infty} a_n x^n \right) dx = \sum_{n=1}^{\infty} a_n \left(\dfrac{b^{n+1}}{n+1} - \dfrac{a^{n+1}}{n+1} \right).$$

In Exercises 8.5.27–8.5.30 use power series (and justify your work) to compute the integrals:

8.5.27. $\displaystyle\int_0^1 \dfrac{\ln(1+x)}{x} \, dx.$ **8.5.28.*** $\displaystyle\int_0^{\pi} \dfrac{\ln(1 + a \cos x)}{\cos x} \, dx, \ |a| < 1.$

8.5.29.* $\displaystyle\int_{-\pi}^{\pi} \dfrac{1 - r^2}{1 - 2r \cos x + r^2} \, dx, \ |r| < 1.$ **8.5.30.*** $\displaystyle\int_0^{\pi} \ln(1 - 2r \cos x + r^2) \, dx, \ |r| < 1.$

9

Fourier Series

Throughout the eighteenth century, leading mathematicians became aware that power series were insufficient to represent functions, and that a different type of series was needed. Considerations of physical problems, such as the behavior of a vibrating string, suggested that their terms should be trigonometric functions. There were serious problems with trigonometric series (e.g., the convergence and the term by term differentiation) and very little progress was made. When Fourier showed that these series can be used to solve the problems of heat flow, there was no way back. The answers were needed and a better understanding of the fundamental concepts of calculus was necessary. Much of the nineteenth century mathematics has its roots in the problems associated to Fourier series.

9.1 Introduction

Let us start by considering a simple physical problem. A flexible string is stretched along the x-axis, between $x = 0$ and $x = \pi$, and it is free to vibrate in the xy-plane. The task at hand is to come up with a function $u(x,t)$ that gives the displacement of the string at a point x at time t. It is not hard, using Newton's Second Law, to derive the equation

$$u_{tt} = c^2 u_{xx} \tag{9.1}$$

which u must satisfy. Here, u_{tt} and u_{xx} are the second-order partial derivatives of u (twice with respect to t and twice with respect to x), while c is a constant depending on the properties of the string. In 1747 d'Alembert published a general form of the solution

$$u(x,t) = F(ct+x) + G(ct-x)$$

in [23]. He then used the natural boundary conditions $u(0,t) = u(\pi,t) = 0$, for all $t \geq 0$, to conclude that $G(ct) = -F(ct)$ and that F must be periodic, with period 2π. D'Alembert also required that F be at the least twice differentiable (since u had to satisfy (9.1)).

Euler disagreed with the last condition, believing that it is too restrictive. Only a year after d'Alembert, he used a different method to obtain in [42] essentially the same solution. He set conditions $u(x,0) = f(x)$, $u_t(x,0) = g(x)$, which led to

$$F(x) + F(-x) = f(x), \quad F(x) - F(-x) = \frac{1}{c} \int_0^x g(t)\, dt.$$

Since f represents the initial shape of the string, Euler argued that it can be anything that "one can draw", including functions that do not necessarily have a derivative at every point (like the one in Figure 9.1). If one were to follow d'Alembert, the function f could be represented as a power series. Euler's argument indicated that such a representation might not be available. He considered the functions of the form

$$f(x) = \alpha \sin x + \beta \sin 2x + \ldots, \tag{9.2}$$

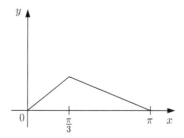

Figure 9.1: The function $y = f(x)$ is not differentiable at $x = \pi/3$.

and the corresponding solutions

$$u(x,t) = \alpha \sin x \cos ct + \beta \sin 2x \cos 2ct + \ldots \qquad (9.3)$$

but did not make it clear whether the sum was finite or infinite. He was able to calculate the coefficients although he initially missed the "easy" way. For example, to calculate β, we can multiply f by $\sin 2x$ and integrate from $-\pi$ to π. Using Exercises 6.1.9 and 6.1.10,

$$\int_{-\pi}^{\pi} f(x) \sin 2x \, dx = \alpha \int_{-\pi}^{\pi} \sin x \sin 2x \, dx + \beta \int_{-\pi}^{\pi} \sin 2x \sin 2x \, dx + \cdots = \pi \beta.$$

The next advancement was made by Daniel Bernoulli. He attacked the problem of the vibrating string from the physical viewpoint in [3], regarding functions $\sin nx \cos nct$ as simple harmonics, and claiming that infinite sums in (9.3), and thus in (9.2), encompassed all solutions.

Again, Euler disagreed. He praised the physical nature of the argument but did not accept the generality of the solution. His viewpoint, typical for the eighteenth century, was that a function must be defined by a single algebraic formula, hence its behavior on the interval $[0, \pi]$ determined its values everywhere in its domain. Since the series (9.2) is a periodic, odd function, the solution is far from general. For example, $y = x^2$ is not periodic, so Euler believed that it cannot be represented by a trigonometric series.

In 1807, Fourier published the paper [46] on the equation for heat diffusion

$$u_t = c^2 u_{xx} \qquad (9.4)$$

where u is the temperature of a thin rod at the point x $(0 \le x \le \pi)$, at time $t \ge 0$, and c is a constant. Equations (9.1) and (9.4) are very much alike, so it should not come as a surprise that Fourier's solution had much in common with the Bernoulli's series. In particular, he obtained that the function $f(x) = u(x, 0)$ can be written as

$$f(x) = \frac{1}{2\pi} \int_{-\pi}^{\pi} f(t) \, dt + \frac{1}{\pi} \sum_{n=1}^{\infty} \left(\cos nx \int_{-\pi}^{\pi} f(t) \cos nt \, dt + \sin nx \int_{-\pi}^{\pi} f(t) \sin nt \, dt \right). \qquad (9.5)$$

Unaware of Euler's work, Fourier also missed the easy way to the formula, and his derivation was very questionable. Add to that the already existing prejudice towards non-differentiable functions, and it is no wonder that his theory was rejected by the leading mathematicians of the time. It could not be totally ignored, though, because it gave correct results. It is almost impossible to accurately measure the position of a vibrating string at a specific time, but Fourier's formulas could be experimentally verified, and they worked. The French Academy found itself in the awkward position, so in 1811 Fourier was awarded the prize by the Institut de France, but (contrary to the standard practice) the prize winning paper was not published. It saw the light only in 1824 when he himself was the Secretary of the Académie des Sciences.

Mathematicians were thus left with the challenge to explain the success of the Fourier's theory. This had a profound impact on the next 100 years of mathematical research, stretching all the way to the present. Clearly, the definition of a function had to be freed from its algebraic shackles, and the modern definition of the function emerged through the work of Dirichlet and Riemann. Further, the coefficients of the series were given by integrals, and Fourier insisted that f could be given by any graph that "bounds a definite area". Today, we would say "integrable", but there was no such concept at that time. Cauchy started a serious work on defining the integral, and the improvements throughout the nineteenth century were crowned by the Lebesgue theory of integration.

Today, we call the series (9.5) the **Fourier series** of f. Fourier series are **trigonometric series**, i.e. series of the form

$$\frac{a_0}{2} + \sum_{n=1}^{\infty} (a_n \cos nx + b_n \sin nx). \tag{9.6}$$

Numbers a_n, b_n are called the Fourier coefficients of f and they are given by

$$a_0 = \frac{1}{\pi} \int_{-\pi}^{\pi} f(x)\,dx, \quad a_m = \frac{1}{\pi} \int_{-\pi}^{\pi} f(x) \cos mx\,dx, \quad b_m = \frac{1}{\pi} \int_{-\pi}^{\pi} f(x) \sin mx\,dx. \tag{9.7}$$

We have yet to discover whether f is equal to its Fourier series.

Example 9.1.1. A Fourier series for $f(x) = x$.

Write a Fourier series for $f(x) = x$ on the interval $[-\pi, \pi)$.

Solution. Clearly, f is a continuous function on $[-\pi, \pi)$, so we can use formulas (9.7).

$$a_0 = \frac{1}{\pi} \int_{-\pi}^{\pi} x\,dx = \frac{1}{\pi} \frac{x^2}{2}\Big|_{-\pi}^{\pi} = 0.$$

For a_m and b_m we use Example 6.1.6 and Exercise 6.1.14 to obtain $a_m = 0$ and $b_m = -\frac{2}{m} \cos m\pi$. We cannot put the equality between f and its Fourier series, so we write

$$x \sim \sum_{n=1}^{\infty} -\frac{2}{n} \cos n\pi \sin nx = 2\left(\sin x - \frac{1}{2} \sin 2x + \frac{1}{3} \sin 3x - \frac{1}{4} \sin 4x + \dots\right). \tag{9.8}$$

$\blacklozenge$

It turned out that each of the coefficients $a_0, a_1, a_2, \dots$ was 0, so the Fourier series was the "sine series". This was not an accident. The function $f(x) = x$ is an odd function (Exercise 3.2.18) and $\cos x$ is an even function (Exercise 3.2.19). Thus, $x \cos x$ is a product of an even and an odd function, so it is an odd function (Exercise 9.1.3). By Exercise 6.4.19,

$$\int_{-\pi}^{\pi} x \cos x\,dx = 0,$$

and the same is true for any odd function instead of $x \cos x$. (Such as, e.g., $x \cos mx$.) In a similar way, if f is an even function we will have all coefficients b_n equal to 0.

We don't know yet whether an equality must hold in (9.8). In fact, Lagrange expressed serious doubts whether the Fourier series converge at all. In the absence of general tests, Fourier established the convergence of several specific series, among them (9.8).

Example 9.1.2. A Fourier series

Write a Fourier series for $f(x) = \begin{cases} -1, & \text{if } -\pi \le x < 0 \\ 1, & \text{if } 0 \le x < \pi. \end{cases}$

Solution. Again, this is an odd function. (Actually, it is not—it fails the test at $x = 0$ but a Riemann integral is not sensitive to the change of value at one point.) So, $a_n = 0$ for all $n \in \mathbb{N}_0$ and we only need to calculate the coefficients b_n. By Exercise 6.4.19,

$$b_m = \frac{1}{\pi} \int_{-\pi}^{\pi} f(x) \sin mx \, dx = \frac{2}{\pi} \int_{0}^{\pi} f(x) \sin mx \, dx = \frac{2}{\pi} \int_{0}^{\pi} \sin mx \, dx$$

$$= \frac{2}{\pi} \left(-\frac{\cos mx}{m} \Big|_{0}^{\pi} \right) = \frac{2}{\pi} \left(-\frac{\cos m\pi}{m} + \frac{1}{m} \right)$$

$$= \frac{2}{m\pi} (1 - \cos m\pi).$$

Therefore,

$$f(x) \sim \sum_{n=1}^{\infty} \frac{2}{n\pi} (1 - \cos n\pi) \sin nx = \frac{4}{\pi} \left(\sin x + \frac{1}{3} \sin 3x + \frac{1}{5} \sin 5x + \dots \right). \qquad \blacklozenge$$

Let us denote by $S(x)$ the sum of the last series. Notice that $S(0) = 0$, but $f(0) = 1 \neq 0$. Thus, the Fourier series of f does not converge pointwise to f. What could be the reason? Is it because f has a discontinuity at $x = 0$? In the next section we will look for conditions that ensure the pointwise convergence of a Fourier series.

DANIEL BERNOULLI (1700–1782) was the middle son of Johann Bernoulli. He studied philosophy and logic at Basel University and, during the same time, he learned the methods of the calculus from his father and his older brother. He loved mathematics but was forced to study medicine. He became an expert in both: his doctoral dissertation was on the mechanics of breathing, and his most important work was the 1738 book *Hydrodynamica*, that contains the first correct analysis of water flowing from a hole in a container. He won the prize of the Paris Academy for designing an hour glass to be used at sea. On the strength of his work *Mathematical Exercises*, he was offered a position at St.Petersburg in 1725. So was his brother Nicholaus, but he died after several months. To help him cope, his father managed to send Euler, one of his best students, to St. Petersburg to work with Daniel. The time between 1727 (when Euler arrived) until 1733 was Daniel's most productive time. In 1733 Bernoulli left for the University of Basel, where he taught botany (until 1743), physiology (until 1750) and physics until his death.

Exercises

9.1.1. Prove that the function $u(x, t) = K \sin \dfrac{\pi(x + ct)}{L}$ is a solution of (9.1) that satisfies the boundary conditions $u(0, t) = u(L, t) = 0$.

9.1.2. Suppose that F and G are twice differentiable functions on $[0, L]$. Prove that $u(x, t) = F(ct + x) + G(ct - x)$ is a solution of (9.1).

9.1.3. Prove that the product of an even and an odd function is odd.

9.1.4. Suppose that f is an integrable periodic function of period $2p$. Prove that its Fourier series is

$$\frac{a_0}{2} + \sum_{n=1}^{\infty} \left(a_n \cos \frac{n\pi x}{p} + b_n \sin \frac{n\pi x}{p} \right),$$

and that the coefficients are given, for $n \in \mathbb{N}_0$, by

$$a_n = \frac{1}{p} \int\limits_{-p}^{p} f(x) \cos \frac{n\pi x}{p} \, dx, \quad b_n = \frac{1}{p} \int\limits_{-p}^{p} f(x) \sin \frac{n\pi x}{p} \, dx.$$

In Exercises 9.1.5–9.1.18 write the Fourier series for the given function on the given set.

9.1.5. $y = 1$, $-\pi \le x \le \pi$.

9.1.6. $y = \dfrac{\pi - x}{2}$, $0 \le x \le 2\pi$.

9.1.7. $y = \begin{cases} \pi + x, & \text{if } -\pi \le x \le 0 \\ \pi - x, & \text{if } 0 \le x \le \pi. \end{cases}$

9.1.8. $y = \begin{cases} 0, & \text{if } -\pi \le x \le 0 \\ \pi - x, & \text{if } 0 < x \le \pi. \end{cases}$

9.1.9. $y = |x|$, $-\pi \le x \le \pi$.

9.1.10. $y = |\sin x|$, $-\pi \le x \le \pi$.

9.1.11. $y = |\cos x|$, $-\pi \le x \le \pi$.

9.1.12. $y = x^2$, $-\pi \le x \le \pi$.

9.1.13. $y = e^x$, $-\pi \le x \le \pi$.

9.1.14. $y = x$, $0 \le x \le 2\pi$.

9.1.15. $y = x^2$, $-1 \le x \le 1$.

9.1.16. $y = x^2$, $0 \le x \le 2\pi$.

9.1.17. $y = \begin{cases} 0, & \text{if } -5 \le x \le 0 \\ 3, & \text{if } 0 < x \le 5. \end{cases}$

9.1.18. $y = \begin{cases} 3 + 2x, & \text{if } -3 \le x \le 0 \\ 3 - 2x, & \text{if } 0 \le x \le 3. \end{cases}$

9.1.19. Use the Fourier series for $y = \cos ax$, $a > 0$, on $(-\pi, \pi)$ to derive the formula

$$\frac{1}{2a} + \sum_{n=1}^{\infty} (-1)^n \frac{a}{a^2 - n^2} = \frac{\pi}{2 \sin a\pi}.$$

In Exercises 9.1.20–9.1.23 write the Fourier sine series and the Fourier cosine series for the given function on the given set.

9.1.20. $y = 1$, $0 < x \le \pi$.

9.1.21. $y = \cos x$, $0 < x \le \pi$.

9.1.22. $y = x^3$, $0 < x \le \pi$.

9.1.23. $y = \pi - x$, $0 < x \le \pi$.

9.1.24. Prove that

$$\cos x - \frac{\cos 5x}{5} + \frac{\cos 7x}{7} - \frac{\cos 11x}{11} + \cdots = \begin{cases} \pi/(2\sqrt{3}), & \text{if } 0 \le x \le \pi/3 \\ \pi/(4\sqrt{3}), & \text{if } x = \pi/3 \\ 0, & \text{if } \pi/3 < x < 2\pi/3 \\ -\pi/(4\sqrt{3}), & \text{if } x = 2\pi/3 \\ -\pi/(2\sqrt{3}), & \text{if } 2\pi/3 \le x \le \pi. \end{cases}$$

9.1.25. Express the Fourier coefficients of the function $g(x) = f(x + h)$ using the Fourier coefficients of f.

9.2 Pointwise Convergence of Fourier Series

Many mathematicians, starting with Fourier himself, tried to find sufficient conditions on f for its Fourier series to converge. Some necessary conditions are obvious. Since the sum of a Fourier series is a periodic function with period 2π, the same must be true of f. Also, the Fourier coefficients (9.7) are defined as integrals, so f must be integrable. It turns out that it is beneficial to tighten up the latter condition and require that f be *piecewise continuous*.

Definition 9.2.1. A function f defined on $[a, b]$ is **piecewise continuous** if it is continuous at every point except at $c_1, c_2, \ldots, c_n \in [a, b]$, and if it has both the left and the right limits at these points. If, in addition, f is a periodic function with period 2π, we will write $f \in PC(2\pi)$.

We will make a standing assumption that f belongs to the class $PC(2\pi)$. When dealing with such functions it is useful to have a shorthand for one-sided limits. We will use the notation pioneered by Dirichlet in 1837: $f(c-)$ for the left and $f(c+)$ for the right limit of f at $x = c$.

In Example 9.1.2 we have established that $f(0) \neq S(0)$. Notice that $f(0-) = -1$, $f(0+) = 1$, and

$$S(0) = \frac{f(0-) + f(0+)}{2}.$$

We will prove that, under appropriate conditions,

$$S(x) = \frac{f(x-) + f(x+)}{2}, \quad \text{for all } x \in \mathbb{R}.$$

Almost all the work in this direction was done by Dirichlet, and published in [33] in 1829, although some of the ideas can be traced back to Fourier.

A technical difficulty, when establishing that a series converges to a particular sum, lies in finding a compact form for the remainder of the series. For the Taylor series, we had the Lagrange, the Cauchy, and the Integral form (page 121). Here, we will use what became known as the nth **Dirichlet kernel**, defined as

$$D_n(x) = \frac{1}{2} + \sum_{k=1}^{n} \cos kx. \tag{9.9}$$

We have established in Example 7.1.5 that

$$D_n(x) = \begin{cases} \dfrac{\sin\left(n + \frac{1}{2}\right)x}{2\sin\frac{x}{2}}, & \text{if } 0 < |x| \le \pi \\ n + \frac{1}{2}, & \text{if } x = 0. \end{cases} \tag{9.10}$$

Using the functions D_n, Dirichlet obtained a nice formula for the partial sums of a Fourier series.

Lemma 9.2.2. *Let $f \in PC(2\pi)$ and let S_n denote the nth partial sum of its Fourier series (9.6). Then, for each $n \in \mathbb{N}$ and $x \in \mathbb{R}$,*

$$S_n(x) = \frac{2}{\pi} \int_0^{\pi} \frac{f(x+t) + f(x-t)}{2} D_n(t)\, dt. \tag{9.11}$$

Proof. We start with the definition of S_n and do some easy transformations:

$$S_n(x) = \frac{a_0}{2} + \sum_{k=1}^{n} (a_k \cos kx + b_k \sin kx)$$

$$= \frac{1}{2\pi} \int_{-\pi}^{\pi} f(t)\, dt + \frac{1}{\pi} \int_{-\pi}^{\pi} \sum_{k=1}^{n} f(t) (\cos kt \cos kx + \sin kt \sin kx)\, dt$$

$$= \frac{1}{\pi} \int_{-\pi}^{\pi} f(t) \left(\frac{1}{2} + \sum_{k=1}^{n} \cos k(x-t) \right) dt$$

$$= \frac{1}{\pi} \int_{-\pi}^{\pi} f(t) D_n(x - t) \, dt$$

$$= \frac{1}{\pi} \int_{x+\pi}^{x-\pi} f(x - s) D_n(s) \, (-ds) \qquad \text{[using } t = x - s]$$

$$= \frac{1}{\pi} \int_{x-\pi}^{x+\pi} f(x - s) D_n(s) \, ds$$

$$= \frac{1}{\pi} \int_{-\pi}^{\pi} f(x - s) D_n(s) \, ds \quad \text{[because } f \text{ and } D_n \text{ are periodic]}$$

$$= \frac{1}{\pi} \left(\int_{-\pi}^{0} f(x - s) D_n(s) \, ds + \int_{0}^{\pi} f(x - s) D_n(s) \, ds \right).$$

If we now substitute $s = -w$ in the first integral, and use the fact that D_n is an even function (obvious from its definition (9.9)), it becomes

$$\int_{\pi}^{0} f(x + w) D_n(-w) \, (-dw) = \int_{0}^{\pi} f(x + w) D_n(w) \, dw.$$

Thus,

$$S_n(x) = \frac{1}{\pi} \left(\int_{0}^{\pi} f(x + w) D_n(w) \, dw + \int_{0}^{\pi} f(x - w) D_n(w) \, dw \right)$$

and the result follows. $\qquad\qquad\square$

Our next goal is to see what happens in (9.11) when $n \to \infty$. Formula (9.10) suggests that it is of interest to consider the limit

$$\lim_{n \to \infty} \int_{0}^{\pi} \frac{f(x + t) + f(x - t)}{2} \frac{\sin \left(n + \frac{1}{2} \right) t}{\sin \frac{t}{2}} \, dt$$

or, more generally,

$$\lim_{n \to \infty} \int_{a}^{b} F(t) \sin \left(n + \frac{1}{2} \right) t \, dt$$

for $F \in PC(2\pi)$. Dirichlet was the first to prove that the limit is 0, albeit under some additional assumptions on F. In his dissertation, Riemann was able to prove the stronger version. Half a century later, after Lebesgue developed his theory of integration, the assertion became a simple consequence of the basic results of this theory. Nowadays it is known as the Riemann–Lebesgue Lemma.

Theorem 9.2.3 (The Riemann–Lebesgue Lemma). *If F is an integrable function on $[a, b]$ then*

$$\lim_{\lambda \to \infty} \int_{a}^{b} F(t) \cos \lambda t \, dt = \lim_{\lambda \to \infty} \int_{a}^{b} F(t) \sin \lambda t \, dt = 0.$$

Proof. We will prove only that the first limit equals 0. Let $\varepsilon > 0$. By Proposition 6.2.7 there exists a partition $P = \{t_0, t_1, \ldots, t_n\}$ of $[a, b]$ such that

$$U(F, P) - L(F, P) < \frac{\varepsilon}{2}.$$

Let

$$M_i = \sup\{F(t) : t \in [t_{i-1}, t_i]\}, \quad \text{and} \quad m_i = \inf\{F(t) : t \in [t_{i-1}, t_i]\},$$

for $1 \le i \le n$, and let

$$K = 4 \sum_{i=1}^{n} \frac{|m_i|}{\varepsilon + 1}.$$

Now, if $\lambda \ge K$ then $\lambda > 4 \sum_{i=1}^{n} |m_i|/\varepsilon$, so

$$\frac{1}{\lambda} < \frac{\varepsilon}{4 \sum_{i=1}^{n} |m_i|}.$$

Therefore,

$$\left| \int_a^b F(t) \cos \lambda t \, dt \right| = \left| \sum_{i=1}^{n} \int_{t_{i-1}}^{t_i} F(t) \cos \lambda t \, dt \right|$$

$$= \left| \sum_{i=1}^{n} \int_{t_{i-1}}^{t_i} (F(t) - m_i) \cos \lambda t \, dt + \sum_{i=1}^{n} \int_{t_{i-1}}^{t_i} m_i \cos \lambda t \, dt \right|$$

$$\le \sum_{i=1}^{n} \int_{t_{i-1}}^{t_i} |F(t) - m_i| \, dt + \sum_{i=1}^{n} |m_i| \left| \int_{t_{i-1}}^{t_i} \cos \lambda t \, dt \right|$$

$$\le \sum_{i=1}^{n} \int_{t_{i-1}}^{t_i} (M_i - m_i) \, dt + \sum_{i=1}^{n} |m_i| \left| \frac{\sin \lambda t}{\lambda} \Big|_{t_{i-1}}^{t_i} \right|$$

$$\le U(F, P) - L(F, P) + \sum_{i=1}^{n} |m_i| \frac{2}{|\lambda|}$$

$$< \frac{\varepsilon}{2} + 2 \sum_{i=1}^{n} |m_i| \frac{\varepsilon}{4 \sum_{i=1}^{n} |m_i|} = \varepsilon. \qquad \square$$

Our next step towards establishing the convergence of Fourier series is just a little algebra. Since $\int_0^\pi \cos kt \, dt = 0$, for any $k \in \mathbb{N}$, it follows that

$$\int_0^\pi D_n(t) \, dt = \frac{\pi}{2}. \tag{9.12}$$

Let $c \in \mathbb{R}$. Using (9.11) and (9.12),

$$S_n(c) - \frac{f(c+) + f(c-)}{2} = \frac{2}{\pi} \int_0^\pi \frac{f(c+t) + f(c-t)}{2} D_n(t) \, dt - \frac{2}{\pi} \int_0^\pi \frac{f(c+) + f(c-)}{2} D_n(t) \, dt$$

$$= \frac{1}{\pi} \left(\int_0^\pi (f(c+t) - f(c+)) D_n(t) \, dt + \int_0^\pi (f(c-t) - f(c-)) D_n(t) \, dt \right).$$

Since $D_n(t) = \sin(n + \frac{1}{2})t / \sin \frac{t}{2}$ except at $t = 0$, and a Riemann integral over $[0, 1]$ equals the integral over $(0, 1)$, we have that

$$S_n(c) - \frac{f(c+) + f(c-)}{2}$$

$$= \frac{1}{\pi} \int_0^\pi \frac{f(c+t) - f(c+)}{2 \sin \frac{t}{2}} \sin\left(n + \frac{1}{2}\right) t \, dt + \frac{1}{\pi} \int_0^\pi \frac{f(c-t) - f(c-)}{2 \sin \frac{t}{2}} \sin\left(n + \frac{1}{2}\right) t \, dt.$$

We would like to apply the Riemann–Lebesgue Lemma, and this requires that each of the fractions

$$\frac{f(c+t) - f(c+)}{2 \sin \frac{t}{2}}, \quad \frac{f(c-t) - f(c-)}{2 \sin \frac{t}{2}} \tag{9.13}$$

represents an integrable function. Since f is, by assumption, piecewise continuous, and $\sin \frac{t}{2}$ is continuous, each of the functions in (9.13) will be piecewise continuous (hence integrable) as long as they have right hand limits at 0. We can write

$$\frac{f(c+t) - f(c+)}{2 \sin \frac{t}{2}} = \frac{f(c+t) - f(c+)}{t} \frac{t}{2 \sin \frac{t}{2}}$$

and notice that the last factor has limit 1 (as $t \to 0^+$), so we concentrate on

$$\lim_{t \to 0^+} \frac{f(c+t) - f(c+)}{t} \quad \text{and} \quad \lim_{t \to 0^+} \frac{f(c-t) - f(c-)}{t}. \tag{9.14}$$

When f is continuous at $t = c$, $f(c-) = f(c+) = f(c)$ so the limits in (9.14) are the one-sided derivatives of f at $t = c$ (page 107). Should we assume that f has both the left-hand and the right-hand derivative at $t = c$?

The answer is a qualified yes. Exercise 4.2.21 shows that, when f has both the left-hand and the right-hand derivative at $t = c$, then f is continuous at $t = c$. Therefore, the old definition of the one-sided derivatives needs to be adapted to the situation when f may have a discontinuity at $t = c$.

It is useful to look at Example 9.1.2, and the point $x = 0$. It is not hard to see that $f'_+(0) = 0$, but the left-hand derivative

$$\lim_{h \to 0^-} \frac{f(0+h) - f(0)}{h} = \lim_{h \to 0^-} \frac{-1 - 1}{h}$$

is infinite. The problem is caused by the fact that, for negative h, $f(h)$ and $f(0)$ are far apart. We will remedy this by replacing $f(0)$ by $f(0-)$ which, in this example, equals -1. Thus, we will say that

$$f'_-(a) = \lim_{h \to 0^-} \frac{f(a+h) - f(a-)}{h}, \quad f'_+(a) = \lim_{h \to 0^+} \frac{f(a+h) - f(a+)}{h}. \tag{9.15}$$

Now, $f'_-(0)$ exists and equals 0. Of course, if f is continuous at $x = a$, then $f(a-) = f(a+) = f(a)$ and (9.15) is the same as the old definition.

With this modification, we will add to our standing assumption that $f \in PC(2\pi)$ another one.

Definition 9.2.4. A function f is **piecewise smooth** in $[-\pi, \pi]$ if both f and f' belong to $PC(2\pi)$. We will denote this class of functions by $PC^1(2\pi)$.

Remark 9.2.5. It is assumed in this definition that one-sided derivatives of f at the exceptional points are defined by (9.15).

If $f \in PC^1(2\pi)$, then the limits in (9.14) exist, and the proof of the Dirichlet's theorem about the pointwise convergence of a Fourier series is now complete.

Theorem 9.2.6. *Let f be a function in $PC^1(2\pi)$. Then the Fourier series of f converges to $(f(c+) + f(c-))/2$, for any $c \in \mathbb{R}$.*

Remark 9.2.7. If f is *continuous* at $x = c$, and if both $f'_-(c)$ and $f'_+(c)$ exist, then $S_n(c)$ converges to $f(c)$. Therefore, if f is continuous and piecewise smooth on $[-\pi, \pi]$, its Fourier series converges to f at every point.

Now we can revisit formula (9.8). The function $f(x) = x$ belongs to $PC(2\pi)$ and has a discontinuity at $x = (2k - 1)\pi$, for all $k \in \mathbb{Z}$. However, it is continuous and piecewise smooth in $(-\pi, \pi)$, so

$$x = 2 \left(\sin x - \frac{1}{2} \sin 2x + \frac{1}{3} \sin 3x - \frac{1}{4} \sin 4x + \ldots \right), \quad -\pi < x < \pi. \qquad (9.16)$$

When Dirichlet formulated Theorem 9.2.6, he did not mention one-sided derivatives. Instead, he required that the function has "a finite number of turns". He believed that this condition was not really necessary and that he could prove the theorem without this assumption. In a letter to Gauss in 1853, he even presented a sketch of a proof. It turned out that he was wrong, and a counterexample was provided in 1876 by Du Bois-Reymond. In his article [37] he found a continuous function whose Fourier series diverges at a dense set of points. Simpler examples were later provided by Schwarz (page 352) in 1880, and Lebesgue in 1906. In 1910 and 1913 Godfrey Harold Hardy (1877–1947) found easier proofs for Du Bois-Reymond's function. In fact, an even stronger theorem is true: for every countable set $A \subset [-\pi, \pi]$ there exists a continuous function whose Fourier series diverges at each point of A. The proof can be found in Chapter 8 of [114].

Henri Lebesgue

HENRI LEBESGUE (1875–1941) was a French mathematician. He attended the Ecole Normale Supérieure in Paris and received a teaching diploma in mathematics in 1897. He spent the next 2 years reading papers by French mathematician René-Louis Baire (1874–1932) on discontinuous functions, which served as an inspiration for his most important work. From 1899 to 1902 he taught in a high school in Nancy. His 1901 paper gave the definition of the integral that could be applied to a much larger class of functions than the Riemann integral. In particular, many discontinuous functions became integrable. His 1902 dissertation, with Borel (page 335) as his doctoral advisor at the Sorbonne, is considered to be one of the finest ever written by a mathematician. It contains a full account of the theory of measure and the Lebesgue integral. In 1903, he published a paper where he applied his new integral to trigonometric series. It contains the Riemann–Lebesgue Lemma together with several results which are true for the Lebesgue integral but not for the Riemann integral. Upon graduating Lebesgue taught at the University of Rennes, the University of Poitiers, the Sorbonne and, from 1921, at the Collège de France, where he lectured and did research for the rest of his life. He made major contributions in other areas of mathematics, including topology. By 1922 he had written nearly 90 books and papers. After 1922 he remained active, but his contributions were directed towards pedagogical issues, historical work, and elementary geometry. He received a number of prizes and he was honored with election to many academies, including the Académie des Sciences, and the Royal Society.

Exercises

In Exercises 9.2.1–9.2.2 find the right-hand derivative of f at $x = 0$, if it exists. It is given that $f(0) = 0$:

9.2.1. $f(x) = x \sin \dfrac{1}{x}$.   9.2.2. $f(x) = x^2 \sin \dfrac{1}{x}$.

In Exercises 9.2.3–9.2.6 determine whether f belongs to $PC(2\pi)$, $PC^1(2\pi)$, or neither. It is given that $f(0) = 0$:

9.2.3. $f(x) = \sin \dfrac{1}{x}$.   9.2.4. $f(x) = x \sin \dfrac{1}{x}$.

9.2.5. $f(x) = x^2 \sin \dfrac{1}{x}$.   9.2.6. $f(x) = x^3 \sin \dfrac{1}{x}$.

9.2.7. Prove the second equality in the Riemann–Lebesgue Lemma: If F is an integrable function on $[a, b]$ then

$$\lim_{\lambda \to \infty} \int_a^b F(t) \sin \lambda t \, dt = 0.$$

9.2.8. Let f be a continuous function in (a, b). Prove that

$$\lim_{n \to \infty} \int_a^b f(x) |\sin nx| \, dx = \lim_{n \to \infty} \int_a^b f(x) |\cos nx| \, dx = \frac{2}{\pi} \int_a^b f(x) \, dx.$$

9.2.9. Use (9.16) to prove that

$$\frac{\pi}{4} = 1 - \frac{1}{3} + \frac{1}{5} - \frac{1}{7} + \dots.$$

9.2.10. Use the Fourier series for $y = x^2$ to prove the equality

$$\sum_{n=1}^{\infty} \frac{1}{n^2} = \frac{\pi^2}{6}.$$

In Exercises 9.2.11–9.2.13 use the Fourier series of a suitable function to find the sum:

9.2.11. $\dfrac{1}{1^2} + \dfrac{1}{3^2} + \dfrac{1}{5^2} + \dfrac{1}{7^2} + \dots.$ **9.2.12.** $\dfrac{1}{1^3} - \dfrac{1}{3^3} + \dfrac{1}{5^3} - \dfrac{1}{7^3} + \dots.$

9.2.13. $\dfrac{1}{1^2} - \dfrac{1}{2^2} + \dfrac{1}{3^2} - \dfrac{1}{4^2} + \dots.$

9.2.14. Let f be a continuous function in $PC^1(2\pi)$. Prove that

$$\lim_{n \to \infty} n \int_{-\pi}^{\pi} f(t) \cos nt \, dt = \lim_{n \to \infty} n \int_{-\pi}^{\pi} f(t) \sin nt \, dt = 0.$$

9.2.15. Calculate $\displaystyle \lim_{n \to \infty} \int_0^\pi \sqrt{x} \sin^2 nx \, dx$.

9.2.16. Calculate $\displaystyle \lim_{n \to \infty} \int_0^\pi \sqrt{x} \sin^m nx \, dx$, where $m \in \mathbb{N}$.

9.2.17. Suppose that $f \in PC(2\pi)$ and let S_n denote the nth partial sum of its Fourier series. Prove that there exists $M > 0$ such that $|S_n(x)| \leq M \ln n$, for each $n \in \mathbb{N}$ and $x \in \mathbb{R}$.

9.2.18. Prove Riemann's Localization Theorem: If $f \in PC(2\pi)$, then its Fourier series will converge at $c \in [-\pi, \pi)$ if and only if there exists $\delta \in (0, \pi)$ such that the limit

$$\lim_{n \to \infty} \frac{2}{\pi} \int_0^\delta \frac{f(c+t) + f(c-t)}{2} \frac{\sin(n + \frac{1}{2})t}{t} \, dt$$

exists, in which case the limit equals the sum of the Fourier series at $x = c$.

9.2.19.* Prove the original version of Theorem 9.2.6: If f is a piecewise monotonic function in $PC(2\pi)$, then the Fourier series of f converges to $(f(c+) + f(c-))/2$, for any $c \in \mathbb{R}$.

9.2.20. Prove the following result that is due to Dini (page 374):

(a) If g is an integrable function on $[0, \pi]$ and if there exists $\delta > 0$ such that $\int_0^\delta \frac{g(t) - g(0+)}{t} \, dt$ exists, then

$$\lim_{\lambda \to \infty} \frac{2}{\pi} \int_0^\delta g(t) \frac{\sin \lambda t}{t} \, dt = g(0+).$$

(b) If $\lim S_n(x) = S(x)$ exists, if $g(t) = (f(x+t) + f(x-t))/2$, and if there exists $\delta > 0$ such that $\int_0^\delta \frac{g(t) - S(x)}{t} \, dt$ exists, then the Fourier series of f converges to $S(x)$.

9.2.21. Suppose that f is a periodic function with period 2π and that it satisfies the Lipschitz condition: there exists $M > 0$ such that, for any $x, y \in \mathbb{R}$, $|f(x) - f(y)| \le M|x - y|$. Prove that the Fourier series of f converges to $f(x)$, for all $x \in \mathbb{R}$.

9.2.22. Suppose that $f \in PC(2\pi)$ and let $x \in \mathbb{R}$. If the integral

$$\int_0^{\pi/2} \left| \frac{f(x+t) - f(x)}{t} \right| dt$$

is finite, prove that the Fourier series of f converges to $f(x)$.

9.2.23. * The purpose of this problem is to reconstruct the original Dirichlet's proof of the original version of Theorem 9.2.6, as stated in Exercise 9.2.19. The difference from Exercise 9.2.19 is that Bonnet's Mean Value Theorem (Exercise 6.6.18) was not available to Dirichlet.

(a) Write the integral

$$\int_0^{\delta} \frac{\sin \lambda t}{\sin t} g(t)\, dt = \sum_{k=1}^{n} \int_{(k-1)\pi/\lambda}^{k\pi/\lambda} \frac{\sin \lambda t}{\sin t} g(t)\, dt + \int_{n\pi/\lambda}^{\delta} \frac{\sin \lambda t}{\sin t} g(t)\, dt$$

and prove that the terms of this sum alternate in sign and decrease in absolute value.

(b) Apply Mean Value Theorem (Theorem 6.6.1) to conclude that each of the integrals above is of the form $\mu_k I_k$, where

$$g\left(\frac{(k-1)\pi}{\lambda} \right) \le \mu_k \le g\left(\frac{k\pi}{\lambda} \right).$$

(c) Prove that, as $\lambda \to \infty$,

$$I_k \to \int_{(k-1)\pi}^{k\pi} \frac{\sin t}{t}\, dt, \quad \text{and} \quad \mu_k \to g(0+).$$

(d) Use Example 13.5.1 to prove that, when $\lambda \to \infty$, the integral

$$\int_0^{\delta} \frac{\sin \lambda t}{\sin t} g(t)\, dt$$

converges to $\frac{\pi}{2} g(0+)$.

(e) Conclude that the original version of Theorem 9.2.6 is true.

9.3 Uniform Convergence of Fourier Series

Pointwise convergence was not the only hot topic in the study of Fourier series. When such a series converges it is of interest to determine the properties of the limit function such as continuity or differentiability. As we have seen in Chapter 8, the uniform convergence plays a major role in answering such questions and in this section we will establish sufficient conditions for a function f to have a uniformly convergent Fourier series. As in the previous section, we will work with functions of the class $PC(2\pi)$.

We will start with an inequality that is due to German mathematician and astronomer Friedrich Wilhelm Bessel (1784–1846). He proved the inequality in 1828. We will postpone the proof.

Theorem 9.3.1 (The Bessel's Inequality). *Let $f \in PC(2\pi)$, and let $\{a_n\}_{n=0}^{\infty}, \{b_n\}_{n=1}^{\infty}$ be its Fourier coefficients. Then*

$$\frac{1}{2} a_0^2 + \sum_{n=1}^{\infty} \left(a_n^2 + b_n^2 \right) \le \frac{1}{\pi} \int_{-\pi}^{\pi} |f(t)|^2 \, dt. \tag{9.17}$$

The following is an easy consequence of the Bessel's inequality. We leave the proof as an exercise.

Corollary 9.3.2. *If $f \in PC(2\pi)$, and $\{a_n\}, \{b_n\}$ are its Fourier coefficients, then the series $\sum_{k=0}^{\infty} a_k^2$ and $\sum_{k=1}^{\infty} b_k^2$ are convergent.*

Now we can establish a sufficient condition for a Fourier series to converge uniformly.

Theorem 9.3.3. *Let f be a continuous function in $PC^1(2\pi)$. Then the Fourier series of f converges uniformly and absolutely on $\mathbb{R}$.*

Proof. The assumption that $f' \in PC(2\pi)$ implies that f has one-sided derivatives at every point. By Remark 9.2.7, the Fourier series of f converges to f pointwise. It remains to prove that it converges uniformly. It is easy to see that, for any $n \in \mathbb{N}$ and any $x \in \mathbb{R}$,

$$|a_n \cos nx| \le |a_n|, \quad |b_n \sin nx| \le |b_n|$$

so the result will follow from the Weierstrass M-Test, provided that we can prove the convergence of $\sum_{n=1}^{\infty} (|a_n| + |b_n|)$. Let $\{a_n'\}, \{b_n'\}$ denote the Fourier coefficients of f'. Then

$$a_n' = \frac{1}{\pi} \int_{-\pi}^{\pi} f'(t) \cos nt \, dt, \quad b_n' = \frac{1}{\pi} \int_{-\pi}^{\pi} f'(t) \sin nt \, dt, \quad \text{for all } n \in \mathbb{N}.$$

If we use Integration by Parts, with $u = \cos nt$, $dv = f'(t) \, dt$, then $du = -n \sin nt \, dt$, $v = f(t)$, so

$$a_n' = \frac{1}{\pi} \left(\cos nt \, f(t) \Big|_{-\pi}^{\pi} - \int_{-\pi}^{\pi} f(t)(-n \sin nt) \, dt \right) = \frac{1}{\pi} n \pi b_n = n b_n.$$

Similarly, $b_n' = -n a_n$. Admittedly, Integration by Parts in Definite Integrals (Theorem 6.6.5) requires that the function f be differentiable, with a continuous derivative. Although this is not the case here, the formula is still valid (see Exercise 9.3.2). Next, using the inequality $ab \le (a^2 + b^2)/2$,

$$\sum_{n=1}^{\infty} |a_n| = \sum_{n=1}^{\infty} \frac{|b_n'|}{n} \le \frac{1}{2} \sum_{n=1}^{\infty} \left(|b_n'|^2 + \frac{1}{n^2} \right).$$

The series $\sum_{n=1}^{\infty} 1/n^2$ converges as a p-series, with $p = 2$. The series $\sum_{n=1}^{\infty} |b_n'|^2$ converges by Corollary 9.3.2. Thus the series $\sum_{n=1}^{\infty} |a_n|$ converges, and the same argument shows that $\sum_{n=1}^{\infty} |b_n|$ converges as well. By the Weierstrass M-Test, the Fourier series of f converges uniformly. $\qquad\square$

Example 9.3.4. A uniformly convergent Fourier series.

Prove that the Fourier series for $f(x) = x^2$ converges uniformly on $\mathbb{R}$.

Solution. We really consider the function defined by $f(x) = x^2$ for $x \in [-\pi, \pi)$ and extended periodically to $\mathbb{R}$. Such a function is, clearly, continuous for $-\pi < x < \pi$. Moreover,

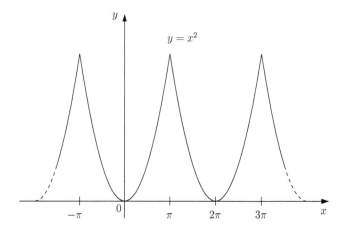

Figure 9.2: The graph of a periodic extension of $f(x) = x^2$.

$f(-\pi) = f(\pi)$ so f is continuous on $\mathbb{R}$. Also, when $-\pi < x < \pi$, $f'(x) = 2x$ which is continuous and f has one-sided derivatives at $x = -\pi$ and $x = \pi$, so f is a continuous function in $PC^1(2\pi)$. We conclude that Theorem 9.3.3 applies, whence the Fourier series of f converges uniformly. (See Exercise 9.3.3.) ♦

Theorem 9.3.3 requires that we know the function f. Often, we only have a trigonometric series, and a test for uniform convergence is needed. While the Weierstrass M-test is quite powerful, it can be used only when the series is absolutely convergent as well. Exercise 9.3.7 shows that there are uniformly convergent series that do not converge absolutely. (However, see Exercise 9.3.23, for an unexpected converse.) The tests of Abel and Dirichlet (Section 8.3) were designed specifically for such situations.

As a consequence of Theorem 9.3.3 we obtain a sufficient condition for differentiation of a Fourier series term by term.

Theorem 9.3.5. *Let f be a function in $PC^1(2\pi)$ such that $f' \in PC^1(2\pi)$. Then the Fourier series of f can be differentiated term by term. More precisely, if $f(x) = \frac{a_0}{2} + \sum_{n=1}^{\infty}(a_n \cos nx + b_n \sin nx)$, then*

$$\frac{f'(x+) + f'(x-)}{2} = \sum_{n=1}^{\infty} n\left(-a_n \sin nx + b_n \cos nx\right).$$

Proof. By Remark 9.2.7 both f and f' have a convergent Fourier series:

$$f(x) = \frac{a_0}{2} + \sum_{n=1}^{\infty}(a_n \cos nx + b_n \sin nx), \text{ and}$$

$$f'(x) \sim \frac{a_0'}{2} + \sum_{n=1}^{\infty}(a_n' \cos nx + b_n' \sin nx).$$

The hypotheses of the theorem are even more restrictive than those in Theorem 9.3.3, so

$a'_n = nb_n$ and $b'_n = -na_n$. Finally, f is periodic, so

$$a'_0 = \frac{1}{2\pi} \int_{-\pi}^{\pi} f'(x)\,dx = \frac{1}{2\pi} f(x)\Big|_{-\pi}^{\pi} = \frac{1}{2\pi}(f(\pi) - f(-\pi)) = 0.$$

Thus, the Fourier series of f' is

$$f'(x) \sim \sum_{n=1}^{\infty}(nb_n \cos nx - na_n \sin nx) = \sum_{n=1}^{\infty}(a_n(\cos nx)' + b_n(\sin nx)').$$

By Theorem 9.2.6 the last series converges to $(f'(x+) + f'(x-))/2$. $\qquad\square$

Example 9.3.6. A Fourier series that can be differentiated term by term.

Prove that the Fourier series of $f(x) = x^3 - \pi^2 x$ can be differentiated term by term.

Solution. It is easy to see that f is twice differentiable for $-\pi < x < \pi$. Further, $f(-\pi) = f(\pi)$, so f is continuous on $\mathbb{R}$. Also, $f'_+(-\pi)$ and $f'_-(\pi)$ both exist and are equal to $2\pi^2$. Consequently, f is differentiable on $\mathbb{R}$ and f' is piecewise continuous. Now Theorem 9.3.5 shows that its Fourier series can be differentiated term by term (see Exercise 9.3.4). $\qquad\blacklozenge$

Next, we turn our attention to the term by term integration of a Fourier series.

Theorem 9.3.7. *Suppose that $f \in PC(2\pi)$. Then its Fourier series can be integrated term by term.*

Proof. Since f is a piecewise continuous, it is integrable. Let a_0 and $\{a_n, b_n : n \in \mathbb{N}\}$ be its Fourier coefficients, and let

$$F(x) = \int_{-\pi}^{x} f(t)\,dt.$$

Clearly, $F(-\pi) = 0$ and to ensure that F is periodic we need to have $F(\pi) = 0$. Notice that $F(\pi) = a_0\pi$, so we will first consider the case when $a_0 = 0$. Let $\{A_n\}$, $\{B_n\}$ be the Fourier coefficients of F. Using integration by parts, just like in the proof of Theorem 9.3.3, $A_n = -b_n/n$ and $B_n = a_n/n$. Finally, since $F(-\pi) = F(\pi) = 0$,

$$A_0 = \frac{1}{\pi}\int_{-\pi}^{\pi} F(x)dx = \frac{1}{\pi}(F(x)x)\Big|_{-\pi}^{\pi} - \frac{1}{\pi}\int_{-\pi}^{\pi} F'(x)x\,dx = -\frac{1}{\pi}\int_{-\pi}^{\pi} f(x)x\,dx.$$

Thus, the Fourier series associated to F is

$$F(x) \sim \frac{A_0}{2} + \sum_{n=1}^{\infty}(A_n \cos nx + B_n \sin nx)$$

$$= -\frac{1}{2\pi}\int_{-\pi}^{\pi} f(x)x\,dx + \sum_{n=1}^{\infty}\left(\frac{-b_n}{n}\cos nx + \frac{a_n}{n}\sin nx\right).$$

By the Fundamental Theorem of Calculus, F is continuous everywhere and differentiable except at the discontinuities of f. By Exercise 6.6.15, F has one-sided derivatives at these points, so Remark 9.2.7 implies that

$$F(x) = -\frac{1}{2\pi}\int_{-\pi}^{\pi} f(x)x\,dx + \sum_{n=1}^{\infty}\left(\frac{-b_n}{n}\cos nx + \frac{a_n}{n}\sin nx\right). \qquad (9.18)$$

Further,

$$\int\limits_{-\pi}^{x} \cos nt \, dt = \frac{1}{n} \sin nt \, \Big|_{-\pi}^{x} = \frac{1}{n} \sin nx, \quad \text{and}$$

$$\int\limits_{-\pi}^{x} \sin nt \, dt = -\frac{1}{n} \cos nt \, \Big|_{-\pi}^{x} = \frac{1}{n} \left((-1)^n - \cos nx \right).$$

It follows that

$$\int\limits_{-\pi}^{x} f(t) \, dt = -\frac{1}{2\pi} \int_{-\pi}^{\pi} f(x) x \, dx + \sum_{n=1}^{\infty} \left(a_n \int\limits_{-\pi}^{x} \cos nt \, dt + b_n \int\limits_{-\pi}^{x} \sin nt \, dt - \frac{(-1)^n b_n}{n} \right).$$

However, (9.18) shows that

$$0 = F(\pi) = -\frac{1}{2\pi} \int_{-\pi}^{\pi} f(x) x \, dx + \sum_{n=1}^{\infty} \frac{b_n}{n} (-1)^n$$

and we obtain that

$$\int\limits_{-\pi}^{x} f(t) \, dt = \sum_{n=1}^{\infty} \left(a_n \int\limits_{-\pi}^{x} \cos nt \, dt + b_n \int\limits_{-\pi}^{x} \sin nt \, dt \right).$$

Finally, if $a_0 \neq 0$, then $g = f - a_0/2$ has the same Fourier coefficients as f, except that the "zero coefficient" is 0, so it can be integrated term by term. That way,

$$\int\limits_{-\pi}^{x} \left[f(t) - \frac{a_0}{2} \right] dt = \int\limits_{-\pi}^{x} f(t) \, dt - \frac{a_0}{2} \int\limits_{-\pi}^{x} dt = \sum_{n=1}^{\infty} \left(a_n \int\limits_{-\pi}^{x} \cos nt \, dt + b_n \int\limits_{-\pi}^{x} \sin nt \, dt \right). \quad \square$$

The remarkable thing about this result is that, if f is merely piecewise continuous, its Fourier series need not converge to f at every point. Nevertheless, the Fourier series of its antiderivative F *does* converge to F at *every* point.

Remark 9.3.8. Formula (9.18) can be written using indefinite integrals:

$$\int f(x) \, dx = \sum_{n=1}^{\infty} \left(a_n \int \cos nx \, dx + b_n \int \sin nx \, dx \right) + C.$$

Example 9.3.9. A Fourier series that can be integrated term by term.

Prove that the Fourier series of $f(x) = x$ can be integrated term by term.

Solution. In the previous section we have established the equality (9.16)

$$x = 2 \left(\sin x - \frac{1}{2} \sin 2x + \frac{1}{3} \sin 3x - \frac{1}{4} \sin 4x + \dots \right), \quad -\pi < x < \pi.$$

Since $f \in PC(2\pi)$ we can integrate it term by term, and obtain

$$\frac{x^2}{2} = -2 \left(\cos x - \frac{1}{4} \cos 2x + \frac{1}{9} \cos 3x - \frac{1}{16} \cos 4x + \dots \right) + C. \tag{9.19}$$

Further, equality (9.19) holds for any $x \in \mathbb{R}$. Substituting $x = \pi$ and using Exercise 9.2.10 yields $C = \pi^2/6$.   ♦

Example 9.3.10. A Fourier series that can be integrated term by term.

Prove that the Fourier series of $f(x) = \begin{cases} -1, & \text{if } -\pi \le x < 0 \\ 1, & \text{if } 0 \le x < \pi \end{cases}$ can be integrated term by term.

Solution. We have seen in Example 9.1.2 that the Fourier series for f is

$$\frac{4}{\pi}\left(\sin x + \frac{1}{3}\sin 3x + \frac{1}{5}\sin 5x + \dots\right),$$

But the equality does not hold at $x = 0$. Nevertheless, if we integrate it term by term, we obtain an equality. On one hand, f has a primitive function $|x|$. On the other hand, integrating the right side yields

$$-\frac{4}{\pi}\left(\cos x + \frac{1}{3^2}\cos 3x + \frac{1}{5^2}\cos 5x + \dots\right).$$

Thus,

$$|x| = -\frac{4}{\pi}\left(\cos x + \frac{1}{3^2}\cos 3x + \frac{1}{5^2}\cos 5x + \dots\right) + C,$$

and substituting $x = \pi/2$ allows us to conclude that $C = \pi/2$. ◆

> Theorem 9.3.7 was proved by Fatou in his 1906 article [43]. Pierre Fatou (1878–1929) was a French mathematician and astronomer. You will learn more about his contribution when you study Lebesgue's theory of integration.

Exercises

9.3.1. Prove Corollary 9.3.2.

9.3.2. Prove the following generalization of Theorem 6.6.5: let $u, v \in PC(2\pi)$, and let $U(x) = \int_{-\pi}^{x} u(t)\,dt$, $V(x) = \int_{-\pi}^{x} v(t)\,dt$. Then

$$\int_{-\pi}^{\pi} U(t)v(t)\,dt = U(t)V(t)\Big|_{-\pi}^{\pi} - \int_{-\pi}^{\pi} u(t)V(t)\,dt.$$

9.3.3. Find the Fourier series for $y = x^2$. Show directly (without using Theorem 9.3.3) that the series converges uniformly.

9.3.4. Find the Fourier series for $f(x) = x^3 - \pi^2 x$. Without using Theorem 9.3.5 show that its term by term derivative is the Fourier series for f'.

9.3.5. Suppose that a function f is a uniform limit of a trigonometric series. Prove that this series must be the Fourier series of f.

9.3.6. Prove that the series (9.16) converges uniformly.

9.3.7. Prove that the series

$$\sum_{n=2}^{\infty} \frac{\sin nx}{n \ln n}$$

converges uniformly on $\mathbb{R}$ but not absolutely.

9.3.8. Let $\{a_n\}$, $\{b_n\}$ be two sequences of real numbers such that the series $\sum_{n=1}^{\infty}(|a_n| + |b_n|)$ is convergent. Prove that the series (9.6) converges uniformly and absolutely to a continuous function f, and that (9.6) is the Fourier series of f.

9.3.9. Let f be absolutely integrable on $[A, B]$. Prove that

$$\int_a^b f(x) \sin \lambda x \, dx \quad \text{and} \quad \int_a^b f(x) \cos \lambda x \, dx$$

converge to 0, as $\lambda \to +\infty$, uniformly with respect to $[a, b] \subset [A, B]$.

9.3.10. Prove Dini's Test for uniform convergence: If f is continuous on $[a, b]$ and if there exists $\delta > 0$ such that the integral

$$\int_0^\delta \frac{f(x+t) + f(x-t) - 2f(x)}{t} \, dt$$

converges uniformly with respect to $x \in [a, b]$, then the Fourier series of f converges uniformly on $[a, b]$.

In Exercises 9.3.11–9.3.12 use the Fourier series of $f(x) = \pi - x$, $0 < x < 2\pi$ to find the sum:

9.3.11. $\displaystyle\sum_{n=1}^{\infty} \frac{\cos nx}{n^2}$.
9.3.12. $\displaystyle\sum_{n=1}^{\infty} \frac{\sin nx}{n^3}$.

9.3.13. Prove that the function

$$\sum_{n=1}^{\infty} \frac{n^3}{n^4 + 1} \sin nx$$

has derivatives of all orders.

9.3.14. Give an example of a Fourier series that cannot be differentiated term by term.

9.3.15. Suppose that $\{a_n\}, \{b_n\}$ are two bounded sequences of real numbers. Prove that the function $u(r, \theta) = \sum_{n=0}^{\infty} (a_n r^n \cos n\theta + b_n r^n \sin n\theta)$ satisfies *Laplace's equation in polar coordinates*

$$\frac{\partial^2 u}{\partial r^2} + \frac{1}{r} \frac{\partial u}{\partial r} + \frac{1}{r^2} \frac{\partial^2 u}{\partial \theta^2} = 0, \tag{9.20}$$

in the disk $0 \leq r < 1$, $-\pi \leq \theta \leq \pi$.

9.3.16. Prove that the function

$$u(x, t) = \sum_{n=1}^{\infty} \frac{4}{(2n-1)\pi} e^{-(2n-1)^2 t} \sin(2n-1)x$$

satisfies the *heat equation* $u_t = u_{xx}$ and the initial condition $u(x, 0) = 1$.

In Exercises 9.3.17–9.3.21 find all functions $y(t) \in \mathrm{PC}(2\pi)$ which satisfy the differential equation:

9.3.17. $y''(t) - 2y(t + \pi) = \cos t$.
9.3.18. $y'' + 10y = t$.

9.3.19. $y'' + 10y = f(t)$, where $f \in \mathrm{PC}(2\pi)$ is defined by $f(t) = -1$ if $-\pi < t < 0$ and $f(t) = 1$ if $0 < t < \pi$.

9.3.20. $y'' + y = \sum_{n=1}^{\infty} \frac{1}{n^2} \cos nt$.
9.3.21. $y'' + y' + y = \sum_{n=1}^{\infty} \frac{1}{n^3} \cos nt$.

9.3.22. Prove the following version of Dirichlet's Theorem: If f is a function in $\mathrm{PC}^1(2\pi)$ then its Fourier series converges to f *piecewise uniformly* on $[-\pi, \pi]$. [A series converges **piecewise uniformly** on a set A, if there exist $c_1, c_2, \ldots, c_n \in A$ such that, for any $\delta > 0$, it converges uniformly on $A \setminus \bigcup_{k=1}^{n} (c_k - \delta, c_k + \delta)$].

9.3.23. This problem will show that if a Fourier series of an integrable function f converges absolutely on $[-\pi, \pi]$ then it converges uniformly on $\mathbb{R}$.

(a) Show that there exist $\rho_n \geq 0$ and $\alpha_n \in [-\pi, \pi]$, such that $a_n \cos nx + b_n \sin nx = \rho_n \cos(nx + \alpha_n)$.

(b) Use the fact that $|\cos x| \geq \cos^2 x$ to prove that there exists $\gamma > 0$ such that

$$\int_{-\pi}^{\pi} |\cos(nx + \alpha_n)| \, dx \geq \gamma, \quad \text{for all } n \in \mathbb{N}.$$

(c) Prove that the series $\sum_{n=1}^{\infty} \rho_n$ converges.

(d) Prove that $\sum_{n=1}^{\infty}(|a_n| + |b_n|)$ converges. Conclude that the Fourier series of f converges uniformly on $\mathbb{R}$.

This is a very weak form of what is known as Denjoy–Luzin Theorem. In its full generality, this theorem requires only that the Fourier series converges absolutely on a "large" subset of $[-\pi, \pi]$. It was proved independently in 1912 by French mathematician Arnaud Denjoy (1884–1974) and Russian mathematician Nikolai Nikolaevich Luzin (1883–1950). Luzin is best known for his work in set theory and point-set topology. Denjoy worked in harmonic analysis, and he introduced an alternative concept of the integral.

9.4 Cesàro Summability

In Section 9.2 we have established some sufficient conditions for a Fourier series of a function f to converge pointwise. Unfortunately, none of these conditions fulfilled Euler's expectation that f can be anything that "one can draw", i.e., a continuous function. The reason is that, when f is merely continuous, the Fourier series can diverge at infinitely many points. In this section we will learn how to overcome this predicament.

Dealing with divergent series can be (and is) done in several different ways. We will present here the method that bears the name of the Italian mathematician Ernesto Cesàro (page 280). It is based on a simple fact, established in Example 2.9.6, that if a sequence $\{a_n\}$ is convergent, then so is the sequence $\{b_n\}$ defined by $b_n = (a_1 + a_2 + \cdots + a_n)/n$. Further, the implication is not reversible (take $a_n = (-1)^n$), so the averaging improves the chances of convergence. This is a particularly useful strategy when applied to the sequence $\{s_n\}$ of partial sums of a divergent series: $s_n = \sum_{k=0}^{n} a_k$.

Definition 9.4.1. If $\{a_n\}$ is a sequence of real numbers, the sequence of Cesàro means is defined by $c_n = \frac{1}{n+1} \sum_{k=0}^{n} a_k$. If $\sum_{n=0}^{\infty} a_n$ is a series of real numbers and $s_n = \sum_{k=0}^{n} a_k$, the Cesàro means of the series is the sequence $\{\sigma_n\}$, defined by $\sigma_n = \frac{1}{n+1} \sum_{k=0}^{n} s_k$, and we say that a series $\sum_{n=0}^{\infty} a_n$ is **Cesàro summable** and its sum is S, if

$$\lim_{n \to \infty} \sigma_n = S.$$

The main result of this section is that, if f is a continuous function, then the Cesàro means of its Fourier series converge to f uniformly. This result, published in [44] in 1904 by a Hungarian mathematician Leopold Fejér (1880–1959) is in sharp contrast with the fact that, for a "large" class of continuous functions, the Fourier series do not converge necessarily at every point.

A crucial tool in the proof of the announced result will be the so-called nth **Fejér kernel** defined by

$$F_n(x) = \frac{D_0(x) + D_1(x) + \cdots + D_n(x)}{n+1}$$

where D_k is the kth Dirichlet kernel,

$$D_k(x) = \begin{cases} \dfrac{\sin\left(k + \frac{1}{2}\right)x}{2\sin\frac{x}{2}}, & \text{if } 0 < |x| \le \pi \\ k + \frac{1}{2}, & \text{if } x = 0. \end{cases}$$

It is helpful to notice that

$$2\sin\left(k + \frac{1}{2}\right)x \sin\frac{x}{2} = \cos kx - \cos(k+1)x.$$

It follows that, if $0 < |x| \le \pi$,

$$F_n(x) = \frac{1}{n+1}\sum_{k=0}^{n} \frac{\sin\left(k+\frac{1}{2}\right)x}{2\sin\frac{x}{2}} = \frac{1}{n+1}\sum_{k=0}^{n} \frac{\cos kx - \cos(k+1)x}{4\sin^2\frac{x}{2}}$$

$$= \frac{1 - \cos(n+1)x}{4(n+1)\sin^2\frac{x}{2}} = \frac{\sin^2\frac{(n+1)x}{2}}{2(n+1)\sin^2\frac{x}{2}}.$$

Next, we notice that the function $f(x) = \sin x - \frac{2x}{\pi}$ has the derivative $f'(x) = \cos x - \frac{2}{\pi}$. Then $f'(x) = 0$ when $x = \arccos\frac{2}{\pi}$. Since $f''(x) = -\sin x$ and $\arccos\frac{2}{\pi} \in (0, \pi/2)$, it follows that f has a local maximum at $\arccos\frac{2}{\pi}$. The minimum of f has to be at one of the endpoints. Since $f(0) = f(\pi/2) = 0$, we conclude that

$$\sin x \ge \frac{2x}{\pi}, \quad \text{for all } x \in \left(0, \frac{\pi}{2}\right).$$

It is not hard to see that the opposite inequality is true if $-\pi/2 < x < 0$. This implies that, if $0 < \delta \le |x| \le \pi$, then

$$0 \le F_n(x) \le \frac{1}{2(n+1)}\left(\frac{\pi}{x}\right)^2 \le \frac{1}{2(n+1)}\left(\frac{\pi}{\delta}\right)^2.$$

Also,

$$\int_{-\pi}^{\pi} F_n(t)\,dt = \int_{-\pi}^{\pi} \frac{1}{n+1}\sum_{k=0}^{n} D_k(t)\,dt = \frac{1}{n+1}\sum_{k=0}^{n}\int_{-\pi}^{\pi} D_k(t) = \frac{1}{n+1}\sum_{k=0}^{n}\pi = \pi.$$

Now we can prove the promised result.

Theorem 9.4.2 (Fejér's Theorem). *Let f be a continuous, periodic function, with period 2π. Then the Cesàro means of its Fourier series converge to f uniformly.*

Proof. By Lemma 9.2.2,

$$S_n(x) = \frac{1}{\pi}\int_{-\pi}^{\pi} f(x-s)D_n(s)\,ds,$$

which implies that

$$\sigma_n(x) = \frac{1}{n+1}\sum_{k=0}^{n} S_k(x) = \frac{1}{n+1}\sum_{k=0}^{n}\frac{1}{\pi}\int_{-\pi}^{\pi} f(x-s)D_k(s)\,ds$$

$$= \frac{1}{\pi} \int\limits_{-\pi}^{\pi} f(x-s) \frac{1}{n+1} \sum_{k=0}^{n} D_k(s) \, ds = \frac{1}{\pi} \int\limits_{-\pi}^{\pi} f(x-s) F_n(s) \, ds.$$

Thus

$$\sigma_n(x) - f(x) = \frac{1}{\pi} \int\limits_{-\pi}^{\pi} f(x-s) F_n(s) \, ds - \frac{1}{\pi} \int\limits_{-\pi}^{\pi} f(x) F_n(s) \, ds,$$

and

$$|\sigma_n(x) - f(x)| \leq \frac{1}{\pi} \int\limits_{-\pi}^{\pi} |f(x-s) - f(x)| \, F_n(s) \, ds.$$

Let $\varepsilon > 0$. Since f is continuous and periodic, it is uniformly continuous (Exercise 3.8.18). Consequently, there exists $0 < \delta < \pi$ such that $|f(x) - f(y)| < \varepsilon/2$ whenever $|x - y| < \delta$. It follows that

$$\frac{1}{\pi} \int\limits_{-\delta}^{\delta} |f(x-s) - f(x)| \, F_n(s) \, ds < \frac{1}{\pi} \int\limits_{-\delta}^{\delta} \frac{\varepsilon}{2} F_n(s) \, ds \leq \frac{1}{\pi} \frac{\varepsilon}{2} \int\limits_{-\pi}^{\pi} F_n(s) \, ds = \frac{\varepsilon}{2}.$$

Next, f is continuous and periodic, so it is bounded (Exercise 3.9.11), say $|f(x)| \leq M$, for all $x \in \mathbb{R}$. Let $N = \lfloor \frac{4\pi^2 M}{\delta^2 \varepsilon} \rfloor$, and let $n \geq N$. Then $n+1 > \frac{4\pi^2 M}{\delta^2 \varepsilon}$ so $\frac{4\pi^2 M}{\delta^2(n+1)} < \varepsilon$. This implies that

$$\frac{1}{\pi} \int\limits_{\delta}^{\pi} |f(x-s) - f(x)| \, F_n(s) \, ds \leq \frac{1}{\pi} \int\limits_{\delta}^{\pi} 2M \frac{1}{2(n+1)} \left(\frac{\pi}{\delta}\right)^2 ds \leq \frac{\varepsilon}{4}.$$

Similarly,

$$\frac{1}{\pi} \int\limits_{-\pi}^{-\delta} |f(x-s) - f(x)| \, F_n(s) \, ds \leq \frac{\varepsilon}{4}.$$

Therefore,

$$|\sigma_n(x) - f(x)| \leq \frac{1}{\pi} \int\limits_{-\pi}^{-\delta} |f(x-s) - f(x)| \, F_n(s) \, ds$$

$$+ \frac{1}{\pi} \int\limits_{-\delta}^{\delta} |f(x-s) - f(x)| \, F_n(s) \, ds$$

$$+ \frac{1}{\pi} \int\limits_{\delta}^{\pi} |f(x-s) - f(x)| \, F_n(s) \, ds$$

$$< \frac{\varepsilon}{4} + \frac{\varepsilon}{2} + \frac{\varepsilon}{4} = \varepsilon.$$

Thus, σ_n converges uniformly to f, and the theorem is proved. $\qquad\square$

Remark 9.4.3. For every $n \in \mathbb{N}$, σ_n is of the form

$$\sigma_n(x) = A_0 + \sum_{k=1}^{n} (A_k \cos kx + B_k \sin kx). \qquad (9.21)$$

Such functions are called *trigonometric polynomials*.

ERNESTO CESÀRO (1859–1906) was an Italian mathematician. The unification of Italy in 1861 brought hard time for farmers, including Cesàro's family. After completing high school he went to study at the École des Mines in Liège, Belgium, where his older brother was a lecturer. When his father died in 1879 he returned to Italy. Eventually, he won a scholarship to continue his studies in Liège and he published his first paper (on number theory) in 1883, then nine more in 1885. In that time period he visited Paris and got inspired by Darboux's lectures on geometry. In 1884 he moved to Italy to study at the University of Rome and within 2 years published 80 articles in a variety of areas including geometry, complex analysis, and probability. He received his doctorate in 1887. He taught in a high school for a year, then at the University of Palermo for 3 years, and finally at the University of Naples for the rest of his life. He died of injuries sustained while trying to save his son from drowning. His main contributions are in differential geometry. In *Lessons of Intrinsic Geometry*, written in 1894, he explained the construction of a fractal curve. In addition, he worked on divergent series and in mathematical physics, where he popularized the ideas of James Clerk Maxwell (1831–1879), a Scottish scientist, considered to be the founder of electrical engineering.

In the next section we will look at several applications of the Fejér's Theorem. Here, we will use it to prove a very important result about the approximation of continuous functions by polynomials. This theorem was proved (in a very different way) by Weierstrass in 1885.

Theorem 9.4.4 (Weierstrass Approximation Theorem). *Let f be a continuous function on a closed interval $[a, b]$. There exists a sequence of polynomials that converges to f uniformly on $[a, b]$.*

Proof. Our plan is to apply the Fejér's Theorem, which requires that f be periodic, with period 2π, and continuous. Even if $[a, b] = [-\pi, \pi]$, and we extended f periodically beyond $[a, b]$, it might not be continuous. One way to avoid these problems is to replace f by a "better" function g (see Figure 9.3):

$$g(x) = \begin{cases} f\left(a + (x + \pi)\,\dfrac{(b - a)}{\pi}\right), & \text{if } -\pi \le x < 0 \\[3mm] f\left(a + (\pi - x)\,\dfrac{(b - a)}{\pi}\right), & \text{if } 0 \le x \le \pi. \end{cases}$$

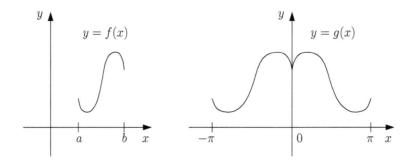

Figure 9.3: Graphs of f and g.

Notice that g is continuous on $[-\pi, \pi]$, because of the continuity of f, and the fact that at $x = 0$ both formulas have the same value $f(b)$. Further, $g(-\pi) = g(\pi) = f(a)$, so g can be periodically extended to a continuous function on $\mathbb{R}$.

Let $\varepsilon > 0$. By the Fejér's Theorem, there exists a trigonometric polynomial $\sigma_n(x)$ given by (9.21), so that

$$|\sigma(x) - g(x)| < \frac{\varepsilon}{2}, \quad \text{for } x \in [-\pi, \pi].$$

Now we can represent each function $\cos kx$ by a uniformly convergent Taylor series. Consequently, there exists a polynomial $p_k(x)$, so that

$$|A_k \cos kx - p_k(x)| < \frac{\varepsilon}{4n}, \quad \text{for } x \in [-\pi, \pi], \text{ and } 1 \le k \le n.$$

Similarly, there exists a polynomial $q_k(x)$, so that

$$|B_k \sin kx - q_k(x)| < \frac{\varepsilon}{4n}, \quad \text{for } x \in [-\pi, \pi], \text{ and } 1 \le k \le n.$$

If we denote $P(x) = A_0 + \sum_{k=1}^{n}(p_k(x) + q_k(x))$, then P is a polynomial and

$$
\begin{aligned}
|\sigma(x) - P(x)| &= \left| \sum_{k=1}^{n}(A_k \cos kx + B_k \sin kx) - \sum_{k=1}^{n}(p_k(x) + q_k(x)) \right| \\
&\le \sum_{k=1}^{n}(|A_k \cos kx - p_k(x)| + |B_k \sin kx - q_k(x)|) \\
&< \sum_{k=1}^{n}\left(\frac{\varepsilon}{4n} + \frac{\varepsilon}{4n} \right) = \frac{\varepsilon}{2}.
\end{aligned}
$$

Therefore,

$$|g(x) - P(x)| \le |g(x) - \sigma(x)| + |\sigma(x) - P(x)| < \frac{\varepsilon}{2} + \frac{\varepsilon}{2} = \varepsilon, \quad \text{for } x \in [-\pi, \pi].$$

In particular, this is true for $x \in [-\pi, 0]$, so

$$\left| f\left(a + (x + \pi)\frac{b - a}{\pi} \right) - P(x) \right| < \varepsilon. \tag{9.22}$$

Let us denote $t = a + (x + \pi)\frac{b-a}{\pi}$. Then it is not hard to see that $t \in [a, b]$ and that $x = -\pi + (t - a)\frac{\pi}{b-a}$. Further, if $Q(t) = P(-\pi + (t - a)\frac{\pi}{b-a})$, then Q is a polynomial in t and (9.22) becomes

$$|f(t) - Q(t)| < \varepsilon, \quad \text{for } t \in [a, b]. \tag{9.23}$$

It follows that, for any $\varepsilon > 0$, there exists a polynomial Q, such that (9.23) holds, and the theorem is proved. $\square$

The problem of summability was one of the central issues of the eighteenth century. In particular, there was plenty of controversy about the series

$$1 - 1 + 1 - 1 + \dots. \tag{9.24}$$

Anyone worth their salt tried to attach a number to this series. Guido Grandi (1671–1742), an Italian monk and mathematician, considered it in his 1703 book. He started with the binomial expansion

$$\frac{1}{1 + x} = 1 - x + x^2 - x^3 + \cdots$$

and substituted $x = 1$ to get $1 - 1 + 1 - 1 + \cdots = 1/2$. Grandi is quoted to have argued that since the sum was both 0 and 1/2, he had proved that the world could be created out of nothing. The series (9.24) is known today as "Grandi's series", although both Leibniz and Jacob Bernoulli had considered it earlier, and had come up with 1/2 as an acceptable sum. Euler agreed with them, but there was plenty of disagreement. Jean-Francois Callet (1744–1799), a French mathematician, pointed out that $1 - 1 + 1 - 1 + \dots$ could also be obtained from the series

$$\frac{1 + x}{1 + x + x^2} = 1 - x^2 + x^3 - x^5 + x^6 - x^8 + \cdots;$$

substituting $x = 1$ now suggests a value of 2/3. Previously, Daniel Bernoulli had made a similar claim, based on a probabilistic argument.

In the nineteenth century, a strict line between convergent and divergent series was drawn. As a consequence, the study of the latter stopped for almost a century. Nevertheless, many results about convergent series established during that period turned out to be useful when the examination of divergent series resumed. For example, Abel established the equality (8.43): if the power series converges for $x = R$ then

$$\lim_{x \to R^-} \sum_{n=0}^{\infty} a_n x^n = \sum_{n=0}^{\infty} a_n R^n.$$

When the series on the right side diverges but the limit on the left side exists, it can be used as a method to ascribe a sum to a divergent series. We call it **Abel's summability** (see Exercise 9.4.18).

A germ of the idea to use the averages can be found in an 1836 paper by Raabe. In 1880, Frobenius (page 449) proved what is now known as Frobenius' theorem (Exercise 9.4.19). In modern terms it says that if a series is Cesàro summable then it is Abel summable to the same sum. However, Frobenius did not mention divergent series in his article, and it was Cesàro who used first a systematic approach to the summability of divergent series in [19] in 1890.

Exercises

In Exercises 9.4.1–9.4.6 find the limit of Cesàro means:

9.4.1. $\displaystyle\sum_{n=1}^{\infty} \sin nx.$ 9.4.2. $\displaystyle\frac{1}{2} + \sum_{n=1}^{\infty} \cos nx.$

9.4.3. $1 + 0 - 1 + 1 + 0 - 1 + 1 + 0 - 1 + \ldots.$

9.4.4. $1 + 0 + 0 - 1 + 0 + 0 + 1 + 0 + 0 - 1 + +0 + 0 + \ldots.$

9.4.5. $1 - 2 + 3 - 4 + 5 - 6 + \ldots.$

9.4.6. $1 - 2^2 + 3^2 - 4^2 + 5^2 - 6^2 + \ldots.$

In Exercises 9.4.7–9.4.12 use Abel's method of summation (p. 282) to find the sum of the series:

9.4.7. $1 - 2 + 3 - 4 + 5 - 6 + \ldots.$

9.4.8. $p - 2p^2 + 3p^3 - 4p^4 + \ldots, |p| < 1.$

9.4.9. $1 + 0 - 1 + 1 + 0 - 1 + 1 + 0 - 1 + \ldots.$

9.4.10. $1 + 0 + 0 - 1 + 0 + 0 + 1 + 0 + 0 - 1 + +0 + 0 + \ldots.$

9.4.11. $1 - 2 + 3 - 4 + 5 - 6 + \ldots.$

9.4.12. $1 - 2^2 + 3^2 - 4^2 + 5^2 - 6^2 + \ldots.$

9.4.13. Prove the following generalization of Theorem 9.4.2: Suppose that f is a periodic function with period 2π, and integrable on $[0, 2\pi]$. Let

$$s(x) = \lim_{t \to 0^+} \frac{f(x + t) + f(x - t)}{2},$$

at the points where the limit exists. Prove that at these points $\sigma_n(x) \to s(x)$.

9.4.14. Prove that a trigonometric series is a Fourier series of a continuous function if and only if the sequence σ_n converges uniformly on $\mathbb{R}$.

9.4.15. Suppose that $f \in PC(2\pi)$ and let $M = \sup\{|f(x)| : x \in [0, 2\pi]\}$. Prove that $|\sigma_n(x)| \le M$ for all $n \in \mathbb{N}$ and $x \in [0, 2\pi]$.

9.4.16. Suppose that f is a periodic function with period 2π, and integrable on $[0, 2\pi]$. Prove that $\sigma_n(x)$ converges to $f(x)$ at every point x that satisfies

$$\lim_{t \to 0} \frac{1}{t} \int_0^t |f(x + u) - f(x)|\, du = 0.$$

9.4.17. Let $0 \le r < 1$ and $0 \le t \le 2\pi$. Prove that

$$\frac{1}{2} + \sum_{n=1}^{\infty} r^n \cos nt = \frac{1 - r^2}{1 - 2r \cos t + r^2}.$$

9.4.18. Let

$$P(r,t) = \frac{1 - r^2}{1 - 2r \cos t + r^2}, \quad A_r(f,x) = \frac{1}{2\pi} \int_{-\pi}^{\pi} P_r(t) f(x - t) \, dt$$

be the Poisson kernel and the Abel mean of the Fourier series of f. Let f be a continuous, periodic function, with period 2π, and let r_n be an increasing sequence of positive numbers converging to 1. Prove that the Abel means $A_{r_n}(f,x)$ converge to f uniformly.

9.4.19. The purpose of this problem is to reconstruct the original Frobenius' proof of his theorem from [48]: if $s_n = \sum_{k=0}^{n} a_k$, $\sigma_n = \frac{1}{n+1} \sum_{k=0}^{n} s_k$, and $\lim_{n\to\infty} \sigma_n = M$, then the series $\sum_{k=0}^{\infty} a_n x^n$ converges for $-1 < x < 1$ and $\lim_{x\to 1^-} \sum_{k=0}^{\infty} a_n x^n = M$.

Given $\varepsilon > 0$, choose $N \in \mathbb{N}$ such that the quantities $\sigma_{N+k} - M = \varepsilon_k$, $k \in \mathbb{N}_0$, all satisfy $|\varepsilon_k| < \varepsilon$.

(a) Prove that

$$a_{N+k} = (N + k + 1)\varepsilon_{k+1} - 2(N + k)\varepsilon_k + (N + k - 1)\varepsilon_{k-1}, \quad k = 1, 2, \ldots.$$

Deduce that the series $\sum_{k=0}^{\infty} a_n x^n$ converges for $-1 < x < 1$.

(b) Let

$$G(x) = M + \sum_{k=0}^{N} a_n x^n - s_N - N\varepsilon_0 \left(1 - x^{N+1}\right) + (N + 1)\varepsilon_1 \left(1 - x^N\right).$$

Prove that, for $-1 < x < 1$,

$$\sum_{k=0}^{\infty} a_n x^n - G(x) = \sum_{k=1}^{\infty} (N + k)\varepsilon_k \left(x^{N+k-1} - 2x^{N+k} + x^{N+k+1}\right).$$

(c) Show that $|\sum_{k=0}^{\infty} a_n x^n - G(x)| < \varepsilon$ for $0 < x < 1$.

(d) Conclude that $\lim_{x\to 1^-} \sum_{k=0}^{\infty} a_n x^n$ exists and equals M.

9.4.20. Exercise 9.4.19 asserts that if a series is Cesàro summable then it is Abel summable to the same sum. Show that the series $\sum_{n=0}^{\infty} (-1)^n n$ is Abel summable but not Cesàro summable.

9.4.21. Prove that if a series $\sum_{n=0}^{\infty} a_n$ is Cesàro summable to S, and if $\lim_{k\to\infty} ka_k = 0$ then $\sum_{n=0}^{\infty} a_n = S$. Give an example to show that without the condition $\lim_{k\to\infty} ka_k = 0$ the implication need not be true.

9.4.22. Prove that if a series $\sum_{n=0}^{\infty} a_n$ is Abel summable to S, and if $\lim_{k\to\infty} ka_k = 0$ then $\sum_{n=0}^{\infty} a_n = S$. Give an example to show that without the condition $\lim_{k\to\infty} ka_k = 0$ the implication need not be true.

9.4.23. Prove that if a series $\sum_{n=0}^{\infty} a_n$ is Abel summable to S, and if $a_n \ge 0$ for all $n \in \mathbb{N}$, then $\sum_{n=0}^{\infty} a_n = S$. Give an example to show that without the condition $a_n \ge 0$ the implication need not be true.

9.5   Mean Square Convergence of Fourier Series

Neither the pointwise nor the uniform convergence seems to be well suited for Fourier series. Part of the problem is that the Fourier coefficients of a function f are defined by integrals. This implies that two different functions may have the same Fourier series. Such is a situation if $f(x)$ and $g(x)$ are equal except on a "small" set (e.g., a finite set). On such a set the series cannot converge pointwise because it would have 2 different sums. Also, even if f is continuous, the example of Du Bois-Reymond shows that its Fourier series may diverge at infinitely many points. Based on these observations, we should welcome a type of convergence that would be insensitive to such exceptional sets. The good news is that, even if f is merely Riemann integrable, the set of points at which its Fourier series diverges cannot be too big (in spite of being possibly infinite). In fact, it turns out that it is small enough to be ignored in the process of integration. All of this leads to the following definition.

Definition 9.5.1. Let f be a Riemann integrable function on $(-\pi, \pi)$ and let S_n denote the nth partial sum of the Fourier series of f. We say that the Fourier series of f **converges in the mean** to f if

$$\lim_{n \to \infty} \int_{-\pi}^{\pi} |f(x) - S_n(x)|^2 \, dx = 0.$$

The main result of this section is that, if f is continuous, then its Fourier series converges in the mean to f. In order to prove this assertion, we need a technical result.

Lemma 9.5.2. *Let f be a periodic function with period 2π that is integrable on $[-\pi, \pi]$, and let*

$$T_n(t) = \frac{1}{2}\alpha_0 + \sum_{k=1}^{n} (\alpha_k \cos kt + \beta_k \sin kt)$$

be a trigonometric polynomial. Then

$$\int_{-\pi}^{\pi} |f(t) - T_n(t)|^2 \, dt = \int_{-\pi}^{\pi} |f(t)|^2 \, dt - \pi \left[\frac{1}{2}a_0^2 + \sum_{k=1}^{n} (a_k^2 + b_k^2) \right]$$

$$+ \pi \left[\frac{1}{2}(\alpha_0 - a_0)^2 + \sum_{k=1}^{n} \left((\alpha_k - a_k)^2 + (\beta_k - b_k)^2 \right) \right].$$

Proof. The idea is to simplify the quantity $|f - T_n|^2$, and then take the integral.

$$|f - T_n|^2 = |f|^2 - 2fT_n + T_n^2$$

$$= |f|^2 - 2f \left(\frac{1}{2}\alpha_0 + \sum_{k=1}^{n} (\alpha_k \cos kt + \beta_k \sin kt) \right) + \left(\frac{1}{2}\alpha_0 + \sum_{k=1}^{n} (\alpha_k \cos kt + \beta_k \sin kt) \right)^2.$$

Notice that

$$\int_{-\pi}^{\pi} 2f(t)T_n(t) \, dt = \int_{-\pi}^{\pi} 2f(t) \left(\frac{1}{2}\alpha_0 + \sum_{k=1}^{n} (\alpha_k \cos kt + \beta_k \sin kt) \right) dt$$

$$= \alpha_0 \int_{-\pi}^{\pi} f(t) \, dt + 2 \sum_{k=1}^{n} \left(\alpha_k \int_{-\pi}^{\pi} f(t) \cos kt \, dt + \beta_k \int_{-\pi}^{\pi} f(t) \sin kt \, dt \right)$$

$$= \pi \alpha_0 a_0 + 2 \sum_{k=1}^{n} \left(\pi \alpha_k a_k + \pi \beta_k b_k \right).$$

Further,

$$T_n(t)^2 = \left(\frac{1}{2} \alpha_0 + \sum_{k=1}^{n} \left(\alpha_k \cos kt + \beta_k \sin kt \right) \right)^2$$

$$= \frac{1}{4} \alpha_0^2 + \alpha_0 \sum_{k=1}^{n} \left(\alpha_k \cos kt + \beta_k \sin kt \right) + \left(\sum_{k=1}^{n} \left(\alpha_k \cos kt + \beta_k \sin kt \right) \right)^2,$$

so

$$\int_{-\pi}^{\pi} T_n(t)^2 \, dt = \frac{1}{4} \alpha_0^2 \int_{-\pi}^{\pi} dt + \alpha_0 \sum_{k=1}^{n} \int_{-\pi}^{\pi} \left(\alpha_k \cos kt + \beta_k \sin kt \right) \, dt$$

$$+ \sum_{k=1}^{n} \sum_{j=1}^{n} \int_{-\pi}^{\pi} \left(\alpha_k \cos kt + \beta_k \sin kt \right) \left(\alpha_j \cos jt + \beta_j \sin jt \right) \, dt$$

$$= \frac{\pi}{2} \alpha_0^2 + \sum_{k=1}^{n} \sum_{j=1}^{n} \int_{-\pi}^{\pi} \left(\alpha_k \alpha_j \cos kt \cos jt + \alpha_k \beta_j \cos kt \sin jt \right.$$

$$\left. + \beta_k \alpha_j \sin kt \cos jt + \beta_k \beta_j \sin kt \sin jt \right) \, dt$$

$$= \frac{\pi}{2} \alpha_0^2 + \sum_{k=1}^{n} \int_{-\pi}^{\pi} \left(\alpha_k^2 \cos^2 kt + \beta_k^2 \sin^2 kt \right) \, dt$$

$$= \frac{\pi}{2} \alpha_0^2 + \sum_{k=1}^{n} \left(\pi \alpha_k^2 + \pi \beta_k^2 \right).$$

Therefore,

$$\int_{-\pi}^{\pi} |f(t) - T_n(t)|^2 \, dt = \int_{-\pi}^{\pi} |f(t)|^2 \, dt - \pi \alpha_0 a_0 - 2 \sum_{k=1}^{n} \left(\pi \alpha_k a_k + \pi \beta_k b_k \right)$$

$$+ \frac{\pi}{2} \alpha_0^2 + \sum_{k=1}^{n} \left(\pi \alpha_k^2 + \pi \beta_k^2 \right),$$

and it remains to verify that

$$\frac{\pi}{2} \alpha_0^2 + \sum_{k=1}^{n} \left(\pi \alpha_k^2 + \pi \beta_k^2 \right) - \pi \alpha_0 a_0 - 2 \sum_{k=1}^{n} \left(\pi \alpha_k a_k + \pi \beta_k b_k \right)$$

$$= \pi \left[\frac{1}{2} (\alpha_0 - a_0)^2 + \sum_{k=1}^{n} \left((\alpha_k - a_k)^2 + (\beta_k - b_k)^2 \right) \right] - \pi \left[\frac{1}{2} a_0^2 + \sum_{k=1}^{n} \left(a_k^2 + b_k^2 \right) \right].$$

We leave this task to the reader. $\qquad \square$

Now we can prove the main result of this section.

Theorem 9.5.3. *Let f be a continuous function on $[-\pi, \pi]$, and periodic, with period 2π. Then the Fourier series of f converges to f in the mean.*

Proof. Lemma 9.5.2 shows that, for a fixed $n \in \mathbb{N}$, the quantity $\int_{-\pi}^{\pi} |f(t) - T_n(t)|^2\, dt$ is minimal when the trigonometric polynomial T_n is precisely the nth partial sum S_n of the Fourier series of f. In particular,

$$\int_{-\pi}^{\pi} |f(t) - S_n(t)|^2\, dt \leq \int_{-\pi}^{\pi} |f(t) - \sigma_n(t)|^2\, dt$$

where $\sigma_n(x) = \frac{1}{n+1} \sum_{k=0}^{n} S_k(x)$. Let $\varepsilon > 0$. By Fejér's Theorem, there exists $N \in \mathbb{N}$ such that, if $n \geq N$, $|f(x) - \sigma_n(x)| < \sqrt{\varepsilon/(2\pi)}$, for all $x \in [-\pi, \pi]$. This implies that

$$\int_{-\pi}^{\pi} |f(t) - \sigma_n(t)|^2\, dt < \int_{-\pi}^{\pi} \left(\sqrt{\frac{\varepsilon}{2\pi}} \right)^2\, dt = \frac{\varepsilon}{2\pi} \int_{-\pi}^{\pi} dt = \varepsilon.$$

It follows that

$$\int_{-\pi}^{\pi} |f(t) - S_n(t)|^2\, dt < \varepsilon, \quad \text{for } n \geq N,$$

and the theorem is proved. $\qquad\square$

The following is a nice consequence of the mean convergence.

Theorem 9.5.4 (Parseval's Identity). *Let f be a function that is continuous on $[-\pi, \pi]$, and periodic, with period 2π. If $\{a_n\}, \{b_n\}$ are the Fourier coefficients of f then*

$$\frac{1}{\pi} \int_{-\pi}^{\pi} |f(t)|^2\, dt = \frac{1}{2} a_0^2 + \sum_{n=1}^{\infty} \left(a_n^2 + b_n^2 \right). \tag{9.25}$$

Proof. We use Lemma 9.5.2 with $T_n = S_n$, the nth partial sum of the Fourier series of f, and we obtain that

$$\frac{1}{\pi} \int_{-\pi}^{\pi} |f(t) - S_n(t)|^2\, dt = \frac{1}{\pi} \int_{-\pi}^{\pi} |f(t)|^2\, dt - \left[\frac{1}{2} a_0^2 + \sum_{k=1}^{n} \left(a_k^2 + b_k^2 \right) \right].$$

The result now follows by taking the limit as $n \to \infty$. $\qquad\square$

Remark 9.5.5. The Bessel's Inequality (9.17) is just "one half" of the Parseval's Identity.

MARC-ANTOINE PARSEVAL des CHÊNES (1755–1836) was a French mathematician. He came from a family of wealthy landowners. In the years following the French revolution, as an ardent royalist, he was imprisoned in 1792. Later, he made his life even more difficult by publishing poetry against Napoleon, who ordered his arrest, so he had to flee the country. Parseval had only 5 mathematical publications, all presented to the Académie des Sciences. Theorem 9.5.4 can be found in the 1799 memoir, and in an improved form in 1801. In it he claimed, without a proof, the assertion shown in Exercise 9.5.11. In the paper, the 2 in front of $a_0 b_0$ is missing, and it is not known whether this was merely a typographical error.

The assumption that f is continuous in Parseval's Identity is too restrictive. It is not hard to extend the result to functions in $PC(2\pi)$ (Exercise 9.5.12). In fact, de la Vallée-Poussin (page 420) proved in 1893 that both Theorem 9.5.3 and Parseval's Identity are true if f is only a square-integrable function on $[-\pi, \pi]$ (see Exercise 9.5.24). (We say that f is **square integrable** if f^2 is integrable.) If the Riemann integral in (9.25) is replaced by the Lebesgue integral, the formula is again valid and expresses a basic fact in the theory of the so-called Hilbert space.

An interesting consequence of the Parseval's Identity is that not every trigonometric series is a Fourier series of a square integrable function.

Example 9.5.6. A convergent trigonometric series whose sum is not square integrable on $[-\pi, \pi]$.

We will show that the sum of the series $\sum_{n=2}^{\infty} \frac{\sin nt}{\ln n}$ is not a square integrable function.

Solution. It is not hard to establish (Exercise 7.5.27) that the series converges for any $t \in \mathbb{R}$. Let f denote its sum. The point of this example is that f is not an integrable function.

Suppose, to the contrary, that f is integrable. By Exercise 6.5.4 the function $|f|^2$ is also integrable, so the integral

$$\int_{-\pi}^{\pi} |f(t)|^2 \, dt$$

is finite. The Parseval's Identity implies that the series

$$\sum_{n=1}^{\infty} (a_n^2 + b_n^2)$$

converges. However, $a_n = 0$, and $b_n = 1/\ln n$, and it follows that the series $\sum_{n=2}^{\infty} 1/(\ln n)^2$ converges, which is obviously not true. Thus, the series $\sum_{n=1}^{\infty} \sin nt/\ln n$, in spite of being a convergent trigonometric series, is *not* a Fourier series of any integrable function. ◆

Exercises

9.5.1. Let $f_n(x) = \dfrac{n}{1 + n^2 x^2}$, for $x \in (0,1)$. Prove that $\{f_n\}$ converges pointwise on $(0,1)$ but not in the mean and not uniformly.

9.5.2. Let $f_n(x) = \begin{cases} -1 + (1+x)^n, & \text{if } x \in [-1,0] \\ 1 - (1-x)^n, & \text{if } x \in [0,1]. \end{cases}$ Prove that $\{f_n\}$ converges pointwise and in the mean, but not uniformly on $[-1,1]$.

9.5.3. Let $f_n(x) = \begin{cases} n^{1/3}, & \text{if } 0 \le x \le 1/n \\ 0, & \text{if } 1/n < x \le 1. \end{cases}$ Prove that $\{f_n\}$ converges in the mean on $(0,1)$ but not pointwise.

9.5.4. Let $f_n(x) = \dfrac{1}{\sqrt{n}} \chi_{[0,n]}(x)$, for $x \in \mathbb{R}$. Prove that $\{f_n\}$ converges uniformly on $\mathbb{R}$ but not in the mean.

9.5.5. Let $f_n(x) = \begin{cases} 0, & \text{if } 0 \le x \le 1/n \\ \ln x, & \text{if } 1/n < x \le 1. \end{cases}$ Prove that $\{f_n\}$ is a mean Cauchy sequence that converges to an unbounded function.

9.5.6. Let $\{f_n\}$ be a sequence of integrable functions on $[0,1]$ that converges in the mean to a function f and pointwise to a function g. Suppose also that f and g are continuous. Prove that $f = g$.

9.5.7. Let $\{f_n\}$ be a sequence of integrable functions on $[0,1]$ that converges in the mean to a function f. Prove or disprove: (a) for every $x \in [0,1]$, $\{f_n(x)\}$ converges to $f(x)$; (b) there exists $x \in [0,1]$, such that $\{f_n(x)\}$ converges to $f(x)$. What if we assume that each f_n and f are continuous functions on $[0,1]$?

9.5.8. Let $f \in PC(2\pi)$ and let $\varepsilon > 0$. Prove that there exists a continuous periodic function with period 2π such that

$$\int_{-\pi}^{\pi} |f(t) - g(t)|^2 \, dt < \varepsilon.$$

9.5.9. Prove that Theorem 9.5.3 remains true if f is assumed to be in $PC(2\pi)$.

9.5.10. Let f be a continuous 2π-periodic function such that f' exists in $[-\pi, \pi]$ with a possible exception of finitely many points, and suppose that f' is square integrable. Prove that the Fourier series of f converges uniformly and absolutely to $f(x)$.

9.5.11. Prove the original result of Parseval:
If $f(x) = \sum_{n=0}^{\infty} a_n \cos nx$ and $g(x) = \sum_{n=0}^{\infty} b_n \cos nx$, then

$$\frac{2}{\pi} \int_0^{\pi} f(x)g(x)\, dx = 2a_0 b_0 + \sum_{n=1}^{\infty} a_n b_n.$$

9.5.12. Prove that Parseval's Identity (Theorem 9.5.4) remains true if f is assumed to be in $PC(2\pi)$.

In Exercises 9.5.13–9.5.19 use the Parseval's Identity to prove that:

9.5.13. $\displaystyle\sum_{n=1}^{\infty} \frac{1}{n^4} = \frac{\pi^4}{90}.$ **9.5.14.** $\displaystyle\sum_{n=1}^{\infty} \frac{1}{n^6} = \frac{\pi^6}{945}.$ **9.5.15.** $\displaystyle\sum_{n=1}^{\infty} \frac{1}{(2n-1)^4} = \frac{\pi^4}{96}.$

9.5.16. $\displaystyle\sum_{n=1}^{\infty} \frac{1}{(2n-1)^2} = \frac{\pi^2}{8}.$ **9.5.17.** $\displaystyle\sum_{n=1}^{\infty} \frac{1}{n^8} = \frac{\pi^8}{9450}.$ **9.5.18.** $\displaystyle\sum_{n=1}^{\infty} \frac{1}{(2n-1)^6} = \frac{\pi^6}{960}.$

9.5.19. $\displaystyle\frac{1}{1^2 \cdot 3^2} + \frac{1}{3^2 \cdot 5^2} + \frac{1}{5^2 \cdot 7^2} + \cdots = \frac{\pi^2 - 8}{16}.$

9.5.20. Write Parseval's Identity for the function $f(x) = \begin{cases} 1, & \text{if } |x| < \alpha \\ 0, & \text{if } \alpha \le |x| < \pi. \end{cases}$ Use this to

find $\displaystyle\sum_{n=1}^{\infty} \frac{\sin^2 n\alpha}{n^2}$ and $\displaystyle\sum_{n=1}^{\infty} \frac{\cos^2 n\alpha}{n^2}$

In Exercises 9.5.21–9.5.23 use the Parseval's Identity to compute the integrals:

9.5.21. $\displaystyle\int_0^{\pi} \ln^2\left(2 \sin \frac{x}{2}\right) dx.$ **9.5.22.** $\displaystyle\int_0^{\pi} \ln^2\left(2 \cos \frac{x}{2}\right) dx.$ **9.5.23.** $\displaystyle\int_0^{\pi} \ln^2 \tan \frac{x}{2}\, dx.$

9.5.24. The purpose of this problem is to prove Theorem 9.5.3 and Parseval's Identity under the assumption that f is only a square-integrable function on $[-\pi, \pi]$. The suggested proof is a slight modification of the original de la Vallée-Poussin's proof from [29].

(a) Prove that the Poisson kernel

$$P(r, t) \equiv \frac{1 - r^2}{1 - 2r \cos t + r^2} = 1 + 2 \sum_{n=1}^{\infty} r^n \cos nt.$$

(b) Prove that the Abel mean

$$A_r(f, x) = \frac{1}{2\pi} \int_{-\pi}^{\pi} P_r(t) f(x - t)\, dt$$

is a square integrable function for $0 < r < 1$, and that

$$\lim_{r \uparrow 1} \int_{-\pi}^{\pi} A_r(f, x)\, dx = \int_{-\pi}^{\pi} f(x)\, dx.$$

(The symbol $r \uparrow 1$ means that r approaches 1 from below, in an increasing manner.)

(c) Prove that the convergence in (b) is uniform.

(d) Use (b) to prove that

$$\lim_{r \uparrow 1} \int_{-\pi}^{\pi} (A_r(f, x))^2 \, dx = \int_{-\pi}^{\pi} f^2(x) \, dx.$$

(e) Prove that

$$\frac{1}{\pi} \int_{-\pi}^{\pi} (A_r(f, x))^2 \, dx = \frac{1}{2} a_0^2 + \sum_{n=1}^{\infty} \left(a_n^2 + b_n^2 \right) r^n.$$

(f) Use (e) to prove the Parseval's Identity for f.

(g) Use (f) and Lemma 9.5.2 to prove Theorem 9.5.3 under the assumption that f is only integrable on $[-\pi, \pi]$, and periodic with period 2π.

9.6 Influence of Fourier Series

In this section we will take a look at some of the developments in mathematics that came in response to the problems raised by Fourier's pioneering essay [46] from 1807. The article [52] is a comprehensive source of information on the subject. We also recommend [104].

The Concept of a Function. In the eighteenth century, it was generally accepted that a function needed to be defined by a *single* analytic expression. There was no need for a precise definition, because the functions that were encountered were either given by simple algebraic formulas, or as sums of convergent power series. Truth be told, there were some disputes about what could be accepted as a function, primarily in response to the vibrating string problem, yet mostly of a theoretical nature.

The Fourier series dealt a blow to the algebraic nature of a function. On one hand, they were defined using "nice" functions $\cos nx$ and $\sin nx$, for $n \in \mathbb{N}$, and their (infinite) sums. On the other hand, the result of such a summation often did not look like a function to an eighteenth century mathematician. Fourier himself insisted that there need not be a "common law" by which the value of a function is calculated, although some historians claim that all he meant was that a function could have infinitely many discontinuities. In his *Cours d'Analyse* Cauchy echoed Fourier's idea, but was also guilty of implicitly assuming additional properties of functions. It seems that the modern, completely general, definition of a function is due to Dirichlet, although some historians feel that it took over only in the work of Riemann.

The Integral. The calculus of Newton and Leibniz featured *infinitesimals* (infinitely small numbers). Although the success of calculus somewhat justified their use, mathematicians felt uneasy about them. When the Fundamental Theorem of Calculus allowed them to avoid integral sums, and use antiderivatives, there was a general sigh of relief, and the integral became the indefinite integral. Since most of the functions had either a nice antiderivative or could be represented as convergent power series (that could be integrated termwise), the indefinite integral was quite sufficient.

The Fourier series brought forward functions that had neither nice antiderivatives nor power series representations. Yet, the coefficients of the series were defined as integrals, so a rigorous definition of the integral was necessary. Cauchy defined the integral of a continuous function, and Riemann and Darboux built on his work and extended the class of integrable functions. The work did not stop there. The Riemann integral was far from perfect. We have seen that a differentiable function f can have a derivative f' that is not Riemann integrable. The Riemann integral and the limit process (including the derivative and the integral) do not always commute. We will see that there are difficulties in the integration of a function depending on several variables. These problems were in part resolved by the concept of the Lebesgue integral, so the search for the ultimate concept of the integral continues.

Uniform Convergence. In 1787, Arbogast (see below) solved the prize problem posed by the Saint Petersburg Academy. The problem asked which functions could be used in solving partial differential equations, and Arbogast showed that the answer is the class of piecewise continuous functions. Since Fourier series are used to solve the Heat Equation, it was natural to try to prove that they represent continuous functions.

In *Cours d'Analyse*, Cauchy even "proved" it. Already in 1826 Abel remarked that this theorem is wrong, and then, in 1829, Dirichlet's proof of Theorem 9.2.6 in [33] settled the issue. An analysis of Cauchy's argument revealed that he had used more than just the pointwise convergence of the series. Weierstrass was among the first to study this different kind of convergence, which he called *uniform convergence* (gleichmäßige konvergenz), and he is considered to be the first to realize its significance. He defined it formally, for functions of several variables, and incorporated it in theorems on the term by term integrability and differentiability of function series.

> LOUIS ARBOGAST (1759–1803) was a French mathematician. Little is known about his childhood and education. What is known is that in 1780 he was a lawyer at the Sovereign Council of Alsace but, seven years later, he was professor of mathematics at the Collège de Colmar. That same year his essay to the Saint Petersburg Academy settled the dispute between Euler and d'Alembert about the nature of the solutions of partial differential equations. Two years later he moved to Strasbourg where he taught mathematics at the Ecole d'Artillerie. In addition, he was appointed as professor of physics at the Collège Royal, served as its rector in 1791, and after that as the rector of the University of Strasbourg. As a member of the Legislative Assembly he introduced plans for reforms in the schools at all levels. Arbogast was elected to the Académie des Sciences in 1792. In 1795 he was put in charge of planning the Ecole Centrale du Bas-Rhin and, once set up, he became the professor of mathematics there in 1796, holding the post until 1802. He was the first to write about calculus in terms of operators. Also, he was responsible for the law introducing the metric system in France.

Set Theory. One of the important questions about Fourier series is the uniqueness of representation, i.e., whether two different trigonometric series could both converge pointwise to the same function f. Equivalently, is there a nonzero trigonometric series that converges pointwise to the zero function?

In his Habilitation Thesis [89] Riemann showed that if the series converges for all values of x in $[-\pi, \pi)$, then the representation is unique. Next, Heine proved in 1870 in [62] that, if a trigonometric series converges piecewise uniformly to 0 (see Exercise 9.3.22) then all coefficients must be zeros. Cantor then took over. In 1870 he showed in [10] (see also [11]) that the same conclusion holds if a trigonometric series converges to 0 for every value of x, with a possible exception of finitely many points. So, he wanted to know whether the exceptional set can be infinite.

Cantor defined the *derivative set* A' of a set A as the set of its cluster points. Inductively, the $(n+1)$st derivative of A, is the derivative set of the nth derivative set $A^{(n)}$. In 1872 he was able to prove in [12] that, if a set A has the property that, for some $n \in \mathbb{N}$, $A^{(n)}$ is empty, and if a trigonometric series converges to 0 for every value of x except on A, then all coefficients of the series must be zeros. Clearly, such sets have a very complicated structure, and it is of interest to learn more about them. Cantor's study signifies the beginning of the set theory as an independent branch of mathematics.

Measure Theory. As we have seen, Fourier series instigated the development of the concept of the definite integral and the class of integrable functions. Already, Cauchy had established that continuous functions are integrable, but Riemann's definition of the integral allowed for many discontinuous functions to be integrable. The question was: how discontinuous could a function be so that it still is integrable?

In his dissertation in 1870, German mathematician Hermann Hankel (1839–1873) considered the set of points where a bounded function is discontinuous. As we know (page 68), at such points, the function can have only a jump. Hankel defined by S_σ the set of points where the function f has a jump greater than $\sigma > 0$. He then proved that a bounded function is integrable if and only if, for every $\sigma > 0$, the set S_σ can be enclosed in a finite collection of intervals of arbitrarily small total length. Nowadays, we would say that S_σ has *content zero*. So, the question is: what do sets with content zero look like?

Hankel himself erroneously believed that a set A has content zero if and only if it is nowhere dense (for any $x, y \in A$, there is an entire interval between them that contains no points of A). This was quickly refuted by an Oxford professor Henry Smith (1826–1883). In his 1875 article [95], he gave an example of a nowhere dense set of a positive content. Thus, in an indirect way, Fourier series are responsible for the development of the measure theory—the metric properties of sets of points.

It should be said that the study of discontinuous functions was not the only reason behind the interest in the concept of the content. For example, Peano was trying to make a rigorous definition of the area. In 1887, in [84], he defined both the inner and the outer content of a region, and the area as the common value (when these two are equal). As we will see in Chapter 14, Jordan (page 441) defined the content (measure) in the same way, and he used it for integration. This was then improved by Lebesgue and Borel, but that is the story for another course.

Those wishing to learn more about the history of calculus should consult books such as [39], [57], [73], to mention a few.

10

Functions of Several Variables

In this chapter we will start our work in the multivariable calculus. Our study will be similar to what has been done in Chapters 1–6. In particular, we will be interested in the precise understanding of the derivative. This will, of course, require some discussion of limits. Just like in the case of functions of a single variable, we will need to investigate the properties of the n dimensional domain $\mathbb{R}^n$.

10.1 Subsets of $\mathbb{R}^n$

Our plan is to study functions of several variables, so let us start with the simplest case, when the number of independent variables is 2. Now the domain of f consists of *pairs* of real numbers. Therefore, it is a subset of the Cartesian product $\mathbb{R} \times \mathbb{R}$, commonly abbreviated as $\mathbb{R}^2$. For example, $f(x, y) = x^2 + y^2$ is defined at $(x, y) = (3, -1)$ and $f(3, -1) = 3^2 + (-1)^2 = 10$. When f depends on 3 variables, we will use $\mathbb{R}^3$ to denote the set of all ordered triples (x, y, z) and, in general, $\mathbb{R}^n$ is the set of all ordered n-tuples $(x_1, x_2, \ldots, x_n)$. Notice that, beyond $n = 3$, it is more practical to switch from different letters (like x, y, z) to one letter with subscripts. Throughout the text we will use the following convention. A point in $\mathbb{R}^n$, for $n \geq 2$, will be denoted by a lowercase **bold** letter, and its *coordinates* by the same letter and subscripts $1, 2, \ldots, n$. For example, we will write $\mathbf{a} = (a_1, a_2, a_3, a_4)$ for a point in $\mathbb{R}^4$. This is especially useful when considering the general case (meaning that n is not specified) or $n \geq 4$.

Let us now think ahead. In the next section we will discuss limits of functions defined on subsets of $\mathbb{R}^n$. Therefore, we will need to adapt Definition 3.4.5 to the multivariable setting. In particular, we turn our attention to the inequality $|x - a| < \delta$. Remember, x and a were real numbers, but now they are points in $\mathbb{R}^n$. For example, in $\mathbb{R}^2$ we would replace them by $\mathbf{x} = (x_1, x_2)$ and $\mathbf{a} = (a_1, a_2)$. In order to generalize this inequality to $\mathbb{R}^n$, we are facing two obstacles. First, we need to define $\mathbf{x} - \mathbf{a}$, for $\mathbf{x}, \mathbf{a} \in \mathbb{R}^n$, and second, the absolute value of a point in $\mathbb{R}^2$ has not been defined.

The algebraic operations in $\mathbb{R}^n$ are defined coordinatewise. This means that, if $\mathbf{a}, \mathbf{b} \in \mathbb{R}^n$ and $\alpha \in \mathbb{R}$, then

$$\mathbf{a} + \mathbf{b} = (a_1 + b_1, a_2 + b_2, \ldots, a_n + b_n), \quad \text{and} \quad \alpha \mathbf{a} = (\alpha a_1, \alpha a_2, \ldots, \alpha a_n).$$

We also mention that $\mathbf{0}$ stands for a point with all coordinates equal to 0.

The absolute value had the purpose of measuring distance. The fact that $|-5 - (-2)| = 3$ simply states that the distance between -5 and -2 on the real line equals 3. Therefore, we need a way to measure the distance in $\mathbb{R}^n$.

Definition 10.1.1. If $\mathbf{a} \in \mathbb{R}^n$, we define the **Euclidean norm** of $\mathbf{a}$ by

$$\|\mathbf{a}\| = \sqrt{a_1^2 + a_2^2 + \cdots + a_n^2}. \tag{10.1}$$

In $\mathbb{R}^2$, if $\mathbf{a} = (a_1, a_2)$, the Pythagorean Theorem shows that $\|\mathbf{a}\|$ is precisely the distance of (a_1, a_2) from the origin. In $\mathbb{R}^n$, for $n \geq 4$, we cannot visualize points, so we may wonder

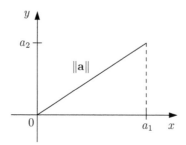

Figure 10.1: $\|\mathbf{a}\| = \sqrt{a_1^2 + a_2^2}$ is the distance from the origin to (a_1, a_2).

whether the Euclidean norm is a way to go. Namely, does it have the properties that one would expect? For example, the point $(1, 3)$ is at the distance $\sqrt{10}$ from the origin. If we multiply it by 2, the distance doubles: we get $(2, 6)$ whose distance from the origin is $\sqrt{40} = 2\sqrt{10}$. Will the same happen in $\mathbb{R}^n$? The following theorem states that the norm shares some of the most useful features of the distance in $\mathbb{R}^2$.

Theorem 10.1.2. *The Euclidean norm has the following properties:*

   (a) for any $\mathbf{a} \in \mathbb{R}^n$, $\|\mathbf{a}\| \geq 0$;

   (b) $\|\mathbf{a}\| = 0$ if and only if $\mathbf{a} = \mathbf{0}$;

   (c) if $\alpha \in \mathbb{R}$ and $\mathbf{a} \in \mathbb{R}^n$, $\|\alpha \mathbf{a}\| = |\alpha| \|\mathbf{a}\|$;

   (d) if $\mathbf{a}, \mathbf{b} \in \mathbb{R}^n$, $\|\mathbf{a} + \mathbf{b}\| \leq \|\mathbf{a}\| + \|\mathbf{b}\|$.

We will leave the proof of the properties (a)–(c) as an exercise, and we will prove only the Triangle Inequality (d). Our first step in this direction is to establish a frequently used inequality.

Theorem 10.1.3 (Cauchy–Schwarz Inequality). *Let $\mathbf{a}, \mathbf{b} \in \mathbb{R}^n$. Then*

$$\sum_{i=1}^n a_i b_i \leq \sqrt{\sum_{i=1}^n a_i^2} \sqrt{\sum_{i=1}^n b_i^2}.$$

Proof. By Theorem 10.1.2 (b), if either $\|\mathbf{a}\| = 0$ or $\|\mathbf{b}\| = 0$, then both sides of the inequality are 0. Thus, we may assume that neither one equals 0. We will apply the inequality $xy \leq (x^2 + y^2)/2$ to

$$a_i b_i = \left(a_i \sqrt{\frac{\|\mathbf{b}\|}{\|\mathbf{a}\|}} \right) \left(b_i \sqrt{\frac{\|\mathbf{a}\|}{\|\mathbf{b}\|}} \right)$$

for $1 \leq i \leq n$. We obtain n inequalities

$$a_i b_i \leq \frac{1}{2} \left(a_i \sqrt{\frac{\|\mathbf{b}\|}{\|\mathbf{a}\|}} \right)^2 + \frac{1}{2} \left(b_i \sqrt{\frac{\|\mathbf{a}\|}{\|\mathbf{b}\|}} \right)^2 = \frac{1}{2} \left(a_i^2 \frac{\|\mathbf{b}\|}{\|\mathbf{a}\|} \right) + \frac{1}{2} \left(b_i^2 \frac{\|\mathbf{a}\|}{\|\mathbf{b}\|} \right),$$

and summing them up yields

$$\sum_{i=1}^n a_i b_i \leq \frac{1}{2} \|\mathbf{a}\|^2 \left(\frac{\|\mathbf{b}\|}{\|\mathbf{a}\|} \right) + \frac{1}{2} \|\mathbf{b}\|^2 \left(\frac{\|\mathbf{a}\|}{\|\mathbf{b}\|} \right) = \|\mathbf{a}\| \|\mathbf{b}\|. \qquad \square$$

The Cauchy–Schwarz Inequality was proved by Cauchy in *Cours d'Analyse*. In 1859 Russian mathematician Viktor Bunyakovsky (1804–1889) obtained the integral form of this inequality (Exercise 10.1.3) in [9]. In 1885 Schwarz (page 352) generalized the inequality to hold for integrals over surfaces in [94]. The Cauchy–Schwarz Inequality is sometimes called the Cauchy–Bunyakovsky Inequality, and they are all frequently referred to as the Hölder's Inequality, by German mathematician Otto Hölder (1859–1937) who, in 1884, generalized further the inequality to

$$\int fg \le \left(\int |f|^p\right)^{1/p}\left(\int |g|^q\right)^{1/q},$$

where $\frac{1}{p} + \frac{1}{q} = 1$.

With the aid of the Cauchy–Schwarz Inequality, we can establish another very useful inequality.

Theorem 10.1.4 (Minkowski's Inequality). *Let $\mathbf{a}, \mathbf{b} \in \mathbb{R}^n$. Then*

$$\left(\sum_{i=1}^n |a_i + b_i|^2\right)^{1/2} \le \left(\sum_{i=1}^n |a_i|^2\right)^{1/2} + \left(\sum_{i=1}^n |b_i|^2\right)^{1/2}.$$

Proof. It is easy to see that

$$\|\mathbf{a} + \mathbf{b}\|^2 = \sum_{i=1}^n |a_i + b_i|^2 = \sum_{i=1}^n (a_i^2 + 2a_i b_i + b_i^2) = \sum_{i=1}^n a_i^2 + 2\sum_{i=1}^n a_i b_i + \sum_{i=1}^n b_i^2,$$

so the Cauchy–Schwarz Inequality implies that

$$\|\mathbf{a} + \mathbf{b}\|^2 \le \|\mathbf{a}\|^2 + 2\|\mathbf{a}\|\|\mathbf{b}\| + \|\mathbf{b}\|^2 = (\|\mathbf{a}\| + \|\mathbf{b}\|)^2.$$

The result is now obtained by taking the square roots. □

Of course, Minkowski's Inequality is precisely the Triangle Inequality asserted in Theorem 10.1.2 (d). The Euclidean norm allows us to talk about the distance between points in $\mathbb{R}^n$. If $\mathbf{a}, \mathbf{b} \in \mathbb{R}^n$, we define the **Euclidean distance** between them by $\|\mathbf{a} - \mathbf{b}\|$. For $n = 2$, $\|\mathbf{a} - \mathbf{b}\| = \sqrt{(a_1 - b_1)^2 + (a_2 - b_2)^2}$ which is the standard distance formula in $\mathbb{R}^2$.

HERMANN MINKOWSKI (1864–1909) was a German mathematician, born in Russia, in the territory of today's Lithuania. His family moved to Königsberg when he was 8. He showed a mathematical talent in high school, reading Dedekind, Dirichlet, and Gauss. He attended the University of Königsberg where he struck a friendship with Hilbert who was also a student there. While still a student, he shared the Grand Prix of the French Academy of Sciences with Henry Smith, on a problem about quadratic forms. At the age of 21 he received his doctorate with a dissertation that continued the prize winning work. Minkowski stayed in Königsberg for another 2 years doing research, then taught at the University of Bonn for 7 years, at Königsberg for 2 years, and ETF Zürich for 6 years. During the Zürich years, Einstein was a student enrolled in several courses that Minkowski taught. In 1902 he moved to the University of Göttingen where Hilbert had arranged a position for him. He died at the age of 44 from appendicitis. His main achievement was the *geometry of numbers*, a study of convex bodies and integer vectors in a space of n dimensions, that proved a valuable tool in number theory. He is best known for his pioneering work in the geometry of 4 dimensions. He astutely realized that Einstein's special theory of relativity could be best understood in a four dimensional space, using time as the fourth dimension. It is less known that he suggested to Hilbert the topic of his historic 1900 address in Paris (see page 6).

As we have said before, our interest in limits has led us to the inequality $\|\mathbf{x} - \mathbf{a}\| < \delta$ or, more precisely, the set

$$\{\mathbf{x} \in \mathbb{R}^n : \|\mathbf{x} - \mathbf{a}\| < \delta\}.$$

When $n = 2$ and $\mathbf{a} = (0,0)$, this is the set defined by $x^2 + y^2 < \delta^2$ which is the open disk with center at the origin and radius δ (Figure 10.2(a)). When $n = 3$ and $\mathbf{a} = (0,0,0)$, this

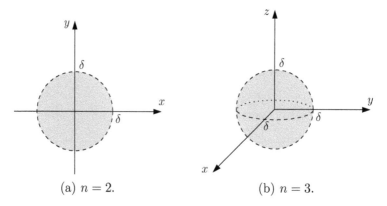

(a) $n = 2$.   (b) $n = 3$.

Figure 10.2: $\{\mathbf{x} \in \mathbb{R}^n : \|\mathbf{x} - \mathbf{a}\| < \delta\}$.

is the open ball with center at the origin and radius δ (Figure 10.2(b)). In general, a set in $\mathbb{R}^n$ defined by the inequality

$$(x_1 - c_1)^2 + (x_2 - c_2)^2 + \cdots + (x_n - c_n)^2 < r^2$$

is an **open n-ball** with center $\mathbf{c} = (c_1, c_2, \ldots, c_n)$ and radius r. We will use notation $B_r(\mathbf{c})$. If $<$ is replaced by $\leq$, then it is a **closed n-ball**. On the other hand, if $<$ is replaced by $=$, then it is a **sphere** with center $\mathbf{c}$ and radius r. For technical reasons, it will be easier to replace the inequality $\|\mathbf{x} - \mathbf{a}\| < \delta$, if possible, by the set of inequalities $|x_k - a_k| < \delta$, $1 \leq k \leq n$. This raises the question what set these inequalities define.

Example 10.1.5. A set defined by inequalities $|x - 3|, |y + 1| < \delta$.
Since x can be any number between $3 - \delta$ and $3 + \delta$, and since y can be any number between $-1 - \delta$ and $-1 + \delta$, the set in question is a rectangle $(3 - \delta, 3 + \delta) \times (-1 - \delta, -1 + \delta)$. (It is actually a square. See Figure 10.3.)   ◆

Let us compare the inequality $(x - 3)^2 + (y + 1)^2 < \delta^2$ (which describes the open disk with center $(3, -1)$ and radius δ) and $|x - 3| < \delta$, $|y + 1| < \delta$. It is easy to see that the former implies that $(x - 3)^2 < \delta^2$ and $(y + 1)^2 < \delta^2$ so, after taking square roots, we obtain that

$$|x - 3| < \delta \quad \text{and} \quad |y + 1| < \delta.$$

Geometrically, the disk $B_\delta(3, -1)$ (with center $(3, -1)$ and radius δ) is contained in the slightly bigger rectangle $(3 - \delta, 3 + \delta) \times (-1 - \delta, -1 + \delta)$.

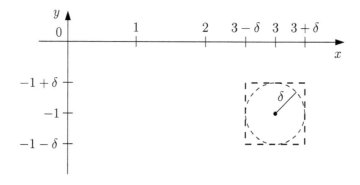

Figure 10.3: $B_\delta(3, -1)$ and the rectangle $(3 - \delta, 3 + \delta) \times (-1 - \delta, -1 + \delta)$.

In general, a set in $\mathbb{R}^n$ defined by the inequalities

$$a_1 \leq x_1 \leq b_1, \quad a_2 \leq x_2 \leq b_2, \quad \ldots, \quad a_n \leq x_n \leq b_n$$

is known as the **closed n-dimensional rectangle**. If, instead, we set

$$a_1 < x_1 < b_1, \quad a_2 < x_2 < b_2, \quad \ldots, \quad a_n < x_n < b_n$$

then it is an **open n-dimensional rectangle**.

Returning to Definition 3.4.5, we see that it involves the concept of a cluster point (in $\mathbb{R}$), so we will need to understand what it takes for a point to be a cluster point of a set in $\mathbb{R}^n$. In Section 3.4 we have defined a cluster point a of a set $A \subset \mathbb{R}$ by the condition that every interval $(a - \delta, a + \delta)$ contain at least one point of A, not counting a. In $\mathbb{R}^2$, and more generally in $\mathbb{R}^n$, an open ball is a good replacement for an interval.

Definition 10.1.6. A point $\mathbf{a}$ is a **cluster point** of a set $A \subset \mathbb{R}^n$ if, for every $\delta > 0$, the n-ball $B_\delta(\mathbf{a})$ contains at least one point of A, not counting $\mathbf{a}$.

Example 10.1.7. Cluster points of a set.

What are the cluster points of the set $A = \{(x, y) : x^2 + y^2 < 1\}$?

Solution. A point (a, b) is a cluster point of A if and only if $a^2 + b^2 \leq 1$, i.e., if and only if it belongs to the *closed unit disk* B. Indeed, if $(a, b) \in B$ then every disk $B_\delta((a, b))$ contains other points of A. On the other hand, if (a, b) lies outside of B, then $\|(a, b)\| > 1$, so any disk $B_r((a, b))$ with radius less than $\|(a, b)\| - 1$ cannot intersect A. Let us see if we can prove this.

Proof. Suppose that $a^2 + b^2 > 1$, and let $0 < r < \|(a, b)\| - 1$. Then $B_r((a, b))$ does not intersect the set A. Indeed, if $(x, y) \in B_r((a, b))$, then

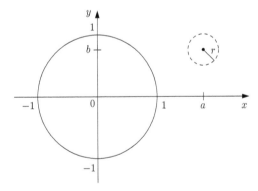

Figure 10.4: (a, b) is not a cluster point of A.

$$\|(x, y)\| \geq \|(a, b)\| - \|(x, y) - (a, b)\| > \|(a, b)\| - r > \|(a, b)\| - (\|(a, b)\| - 1) = 1,$$

so $(x, y) \notin A$.

In the other direction, suppose that $a^2 + b^2 \leq 1$. If $(a, b) = (0, 0)$ then, for any $0 < \delta < 1$,

$$\left(\frac{\delta}{2}, 0 \right) \in A \cap B_\delta((0, 0)) \setminus (0, 0),$$

so $(0, 0)$ is a cluster point of A. If $(a, b) \neq (0, 0)$, let r be a positive number satisfying

$$1 - \frac{\delta}{\|(a, b)\|} < r < 1,$$

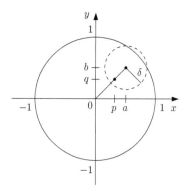

Figure 10.5: (a, b) is a cluster point of A.

and let $(p, q) = r(a, b)$. Then $\|(p, q)\| = r\|(a, b)\| \leq r < 1$, so $(p, q) \in A$. Since $r \neq 1$, $(p, q) \neq (a, b)$. Finally,

$$\|(p, q) - (a, b)\| = (1 - r)\|(a, b)\| < \frac{\delta}{\|(a, b)\|}\|(a, b)\| = \delta$$

so $(p, q) \in B_\delta((a, b))$. Consequently, (a, b) is a cluster point of A.   ◆

Now we are ready for a serious study of functions of several variables and their limits. We will do that in the next section.

Exercises

10.1.1. Prove Theorem 10.1.2 (a)–(c).

10.1.2. Prove that, if $\mathbf{a}, \mathbf{b} \in \mathbb{R}^n$, then $\|\mathbf{a} - \mathbf{b}\| \geq \|\mathbf{a}\| - \|\mathbf{b}\|$.

10.1.3. Prove the Cauchy–Schwarz–Bunyakovsky Inequality for integrals: If f, g are square-integrable functions on (a, b) then so is fg, and

$$\left| \int_a^b f(x)g(x)\, dx \right| \leq \left(\int_a^b f(x)^2\, dx \right)^{1/2} \left(\int_a^b g(x)^2\, dx \right)^{1/2}.$$

10.1.4. Prove the Cauchy–Schwarz Inequality for infinite series: If the series $\sum_{n=0}^\infty a_n^2$ and $\sum_{n=0}^\infty b_n^2$ are convergent then so is $\sum_{n=0}^\infty a_n b_n$, and

$$\left| \sum_{n=0}^\infty a_n b_n \right| \leq \left(\sum_{n=0}^\infty a_n^2 \right)^{1/2} \left(\sum_{n=0}^\infty b_n^2 \right)^{1/2}.$$

10.1.5. Prove that every n-ball $B_\delta(\mathbf{a})$ is contained in a rectangle $|x_k - a_k| < \delta$, $1 \leq k \leq n$.

10.1.6. Let $|x_k - a_k| < \delta$, $1 \leq k \leq n$, be a rectangle in $\mathbb{R}^n$. Prove that it is contained in an n-ball $B_r(\mathbf{a})$, and find a minimal r for which this is true.

10.1.7. Let $A = \{(x, y) \in \mathbb{R}^2 : x^2 + y^2 \leq 1\}$. Prove that the set of its cluster points is A.

10.1.8. Let $A = \{(x_1, x_2, \ldots, x_n) \in \mathbb{R}^n : x_1^2 + x_2^2 + \cdots + x_n^2 < 1\}$, and $B = \{(x_1, x_2, \ldots, x_n) \in \mathbb{R}^n : x_1^2 + x_2^2 + \cdots + x_n^2 \leq 1\}$. Prove that the set of cluster points of either set is B.

10.1.9. Let $A = \{(x, y) \in \mathbb{R}^2 : x^2 + y^2 = 1\}$. Determine the set of cluster points of A.

10.1.10. Let $A = \{(x,y) \in (0,1) \times (0,1) : x \in \mathbb{Q},\ y \in \mathbb{Q}\}$. Determine the set of cluster points of A.

10.1.11. Give an example of an infinite set without a cluster point.

10.1.12. Give an example of an unbounded set with exactly one cluster point.

10.1.13. Give an example of an unbounded set with exactly two cluster points.

10.1.14. Give an example of a bounded set without a cluster point.

10.1.15. Suppose that $\mathbf{a} \in \mathbb{R}^n$ is a cluster point of a set A and let $\delta > 0$. Prove that every n-ball $B_\delta(\mathbf{a})$ contains infinitely many members of A.

10.1.16. Suppose that a set $A \subset \mathbb{R}^n$ has a cluster point, and let $\delta > 0$. Prove that there exist $\mathbf{a}, \mathbf{b} \in A$ such that $\|\mathbf{a} - \mathbf{b}\| < \delta$.

10.1.17. Suppose that a set $A \subset \mathbb{R}^n$ is uncountable. Prove that it must have a cluster point.

If $A \subset \mathbb{R}^n$ the set of its cluster points is denoted by A'. In Exercises 10.1.18–10.1.20 prove or disprove that:

10.1.18. $A \subset B \Rightarrow A' \subset B'$.   **10.1.19.** $(A \cap B)' = A' \cap B'$.   **10.1.20.** $(A \cup B)' = A' \cup B'$.

10.2   Functions and Their Limits

In this section our goal is to state a precise definition of the limit, and establish its properties. As usual, we start with an example.

Let $f(x,y) = x^2 + y^2$, $\mathbf{a} = (3,-1)$, and $\varepsilon = 0.1$. Our goal is to find $\delta > 0$ such that

$$\|(x,y) - (3,-1)\| < \delta \quad \Rightarrow \quad |f(x,y) - f(3,-1)| < \varepsilon.$$

As in the case of a single variable, we investigate the inequality $|f(x,y) - f(3,-1)| < \varepsilon$. Here, this is $|x^2 + y^2 - 10| < 0.1$. The idea is to try to isolate $x - 3$ and $y + 1$. Notice that

$$x^2 + y^2 - 10 = x^2 - 9 + y^2 - 1 = (x-3)(x+3) + (y+1)(y-1).$$

Next we look at the condition $\|(x,y) - (3,-1)\| < \delta$. As we have seen in the previous section, we may replace it by $|x - 3| < \delta$, $|y + 1| < \delta$. Now we can copy the idea of Example 3.4.7: promise to have $2 < x < 4$ and $-2 < y < 0$ so that $|x| < 4$ and $|y| < 2$. Then

$$\begin{aligned}
|x^2 + y^2 - 10| &\le |x-3|(|x|+3) + |y+1|(|y|+1) \\
&\le 7|x-3| + 3|y+1| \\
&< 7\delta + 3\delta = 10\delta.
\end{aligned}$$

So, it suffices to take $\delta = \min\{\frac{\varepsilon}{10}, 1\} = 0.01$.

Now we can state the definition.

Definition 10.2.1. Let $A \subset \mathbb{R}^n$, let $f : A \to \mathbb{R}$, and let $\mathbf{a} = (a_1, a_2, \ldots, a_n)$ be a cluster point of A. We say that L is the **limit** of f as $\mathbf{x}$ approaches to $\mathbf{a}$, and we write $\lim_{\mathbf{x} \to \mathbf{a}} f(\mathbf{x}) = L$, if for any $\varepsilon > 0$ there exists $\delta > 0$ such that $|f(\mathbf{x}) - L| < \varepsilon$, whenever $|x_k - a_k| < \delta$, $1 \le k \le n$, and $\mathbf{x} \in A \setminus \{\mathbf{a}\}$.

Example 10.2.2. The limit of f as $x \to a$.

Let $f(x,y) = \dfrac{x+4y}{1-x+y}$, $\mathbf{a} = (1,-1)$. Find $\lim_{(x,y)\to(1,-1)} f(x,y)$ and prove that this is indeed the limit.

Solution. If we substitute $x=1$ and $y=-1$ in f, we obtain 3. Therefore, we consider the inequality

$$\left| \frac{x+4y}{1-x+y} - 3 \right| < \varepsilon.$$

First,

$$\frac{x+4y}{1-x+y} - 3 = \frac{x+4y-3(1-x+y)}{1-x+y} = \frac{4x+y-3}{1-x+y} = \frac{4(x-1)+(y+1)}{(1-x)+(y+1)-1}.$$

Therefore,

$$\left| \frac{x+4y}{1-x+y} - 3 \right| \le \frac{4|x-1|+|y+1|}{1-|(1-x)+(y+1)|}.$$

Now, assuming that $|x-1| < \delta$ and $|y+1| < \delta$, so that $|(1-x)+(y+1)| \le |x-1|+|y+1| < 2\delta$, and that $\delta < 1/4$,

$$\left| \frac{x+4y}{1-x+y} - 3 \right| \le \frac{4\delta+\delta}{1-2\delta} < \frac{5\delta}{1-\frac{1}{2}} = 10\delta.$$

Proof. Let $\varepsilon > 0$ and define $\delta = \min\{\varepsilon/10, 1/4\}$. If $|x-1| < \delta$ and $|y+1| < \delta$, then

$$|1-x+y| = |(1-x)+(y+1)-1|$$
$$\ge 1 - |(1-x)+(y+1)|$$
$$\ge 1 - (|x-1|+|y+1|)$$
$$> 1 - 2\delta$$
$$\ge 1 - \frac{1}{2} = \frac{1}{2},$$

and

$$|x+4y-3(1-x+y)| = |4(x-1)+(y+1)| \le 4|x-1|+|y+1| < 4\delta+\delta = 5\delta.$$

Therefore,

$$\left| \frac{x+4y}{1-x+y} - 3 \right| = \left| \frac{x+4y-3(1-x+y)}{1-x+y} \right| < \frac{5\delta}{\frac{1}{2}} = 10\delta < \varepsilon. \qquad \blacklozenge$$

Example 10.2.2 shows that the use of the $\varepsilon - \delta$ language makes proofs about limits straightforward but long. In the case of one variable, Theorem 3.4.9 allowed us to consider convergent sequences instead, and that made a whole lot of difference. If we want to do the same in the multivariable setting, we must define a sequence first.

Definition 10.2.3. A sequence $\mathbf{a}_k = (a_1^{(k)}, a_2^{(k)}, \ldots, a_n^{(k)}) \in \mathbb{R}^n$ is a mapping from $\mathbb{N}$ to $\mathbb{R}^n$. We say that $\{\mathbf{a}_k\}$ converges to $\mathbf{a} = (a_1, a_2, \ldots, a_n)$ if, for each $1 \le i \le n$, the sequence of real numbers $\{a_i^{(k)}\}$ converges to a_i.

Example 10.2.4. The limit of a sequence.

Let $\mathbf{a}_k = \left(\dfrac{1}{k}, \dfrac{k-1}{k} \right)$. We will show that $\lim \mathbf{a}_k = (0,1)$.

Solution. This is a sequence in $\mathbb{R}^2$ and it is fairly easy to see that it converges to $(0,1)$. How does one *prove* that? By Definition 10.2.3, we would need to prove that $\lim_{k\to\infty} 1/k = 0$ and $\lim_{k\to\infty}(k-1)/k = 1$, two exercises that belong to Chapter 2. $\qquad \blacklozenge$

Now we can prove the announced generalization of Theorem 3.4.9.

Theorem 10.2.5. *Let $A \subset \mathbb{R}^n$, let $f : A \to \mathbb{R}$, and let $\mathbf{a}$ be a cluster point of A. Then $\lim_{\mathbf{x} \to \mathbf{a}} f(\mathbf{x}) = L$ if and only if, for every sequence $\{\mathbf{a}_k\} \subset A$ converging to $\mathbf{a}$, $\lim_{k \to \infty} f(\mathbf{a}_k) = L$.*

Proof. Suppose first that $\lim_{\mathbf{x} \to \mathbf{a}} f(\mathbf{x}) = L$, and let $\mathbf{a}_k = (a_1^{(k)}, a_2^{(k)}, \ldots, a_n^{(k)})$ be an arbitrary sequence in A converging to $\mathbf{a} = (a_1, a_2, \ldots, a_n)$. We will show that $\lim_{k \to \infty} f(\mathbf{a}_k) = L$. So, let $\varepsilon > 0$. Since $\lim_{\mathbf{x} \to \mathbf{a}} f(\mathbf{x}) = L$, there exists $\delta > 0$ such that

$$|x_1 - a_1| < \delta, |x_2 - a_2| < \delta, \ldots, |x_n - a_n| < \delta \Rightarrow |f(\mathbf{x}) - L| < \varepsilon. \tag{10.2}$$

Further, $\lim \mathbf{a}_k = \mathbf{a}$. By definition, the sequence $a_i^{(k)} \to a_i$ for each $1 \le i \le n$. Therefore, for each i, $1 \le i \le n$, there exists $N_i \in \mathbb{N}$ such that

$$k \ge N_i \quad \Rightarrow \quad |a_i^{(k)} - a_i| < \delta.$$

Let $N = \max\{N_1, N_2, \ldots, N_n\}$, and let $k \ge N$. By (10.2), $|f(\mathbf{a}_k) - L| < \varepsilon$.

Now we will prove the converse. Namely, we will assume that, for every sequence $\{\mathbf{a}_k\}$ converging to $\mathbf{a}$, $\lim_{k \to \infty} f(\mathbf{a}_k) = L$, and we will establish that $\lim_{\mathbf{x} \to \mathbf{a}} f(\mathbf{x}) = L$. Suppose, to the contrary, that $\lim_{\mathbf{x} \to \mathbf{a}} f(\mathbf{x}) \ne L$. By carefully applying the negative in the definition of the limit (Definition 10.2.1), we see that this means:

$$(\exists \varepsilon)(\forall \delta)(\exists \mathbf{x}) \text{ so that } (\forall i)|x_i - a_i| < \delta \text{ and } |f(\mathbf{x}) - L| \ge \varepsilon. \tag{10.3}$$

Let ε_0 be such a number and, for each $k \in \mathbb{N}$, let $\delta_k = 1/k$. Then there exists a sequence $\{\mathbf{x}_k\}$ such that

$$|x_i^{(k)} - a_i| < \frac{1}{k}, \quad 1 \le i \le n, \text{ and } |f(\mathbf{x}_k) - L| \ge \varepsilon_0.$$

This implies that, for each i, $1 \le i \le n$, $x_i^{(k)} \to a_i$, as $k \to \infty$, hence $\mathbf{x}_k \to \mathbf{a}$. In view of the assumption, it follows that $\lim_{k \to \infty} f(\mathbf{x}_k) = L$ which contradicts the inequality $|f(\mathbf{x}_k) - L| \ge \varepsilon_0$. Therefore, $\lim_{\mathbf{x} \to \mathbf{a}} f(\mathbf{x}) = L$. $\square$

The use of sequences often makes it easy to establish that a limit *does not* exist.

Example 10.2.6. The limit of a function does not exist.

Prove that the limit $\lim\limits_{(x,y) \to (0,0)} \dfrac{xy}{x^2 + y^2}$ does not exist.

Solution. Let $f(x, y) = xy/(x^2 + y^2)$. By Theorem 10.2.5, if the limit L existed we would necessarily have $\lim_{k \to \infty} f(\mathbf{a}_k) = L$ for any sequence $\{\mathbf{a}_k\}$ converging to $(0, 0)$. Therefore, it will suffice to find two sequences $\{\mathbf{a}_k\}$ and $\{\mathbf{b}_k\}$ converging to $(0, 0)$, but $\lim_{k \to \infty} f(\mathbf{a}_k) \ne \lim_{k \to \infty} f(\mathbf{b}_k)$. Here we can take $\mathbf{a}_k = (1/k, 1/k)$ and $\mathbf{b}_k = (1/k, 2/k)$. Then

$$f(\mathbf{a}_k) = f\left(\frac{1}{k}, \frac{1}{k}\right) = \frac{\frac{1}{k}\frac{1}{k}}{\left(\frac{1}{k}\right)^2 + \left(\frac{1}{k}\right)^2} = \frac{\frac{1}{k^2}}{\frac{2}{k^2}} \to \frac{1}{2}, \text{ and}$$

$$f(\mathbf{b}_k) = f\left(\frac{1}{k}, \frac{2}{k}\right) = \frac{\frac{1}{k}\frac{2}{k}}{\left(\frac{1}{k}\right)^2 + \left(\frac{2}{k}\right)^2} = \frac{\frac{2}{k^2}}{\frac{5}{k^2}} \to \frac{2}{5}.$$

Thus, the limit does not exist.

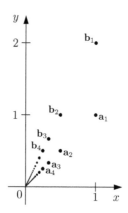

Figure 10.6: $\mathbf{a}_k = (\frac{1}{k}, \frac{1}{k})$, $\mathbf{b}_k = (\frac{1}{k}, \frac{2}{k})$.

Geometrically, each member of the sequence $\{\mathbf{a}_k\}$ lies on the line $y = x$, and $\{\mathbf{b}_k\}$ lies on the line $y = 2x$ (Figure 10.6). ◆

In many examples, all it takes is to find two sequences converging along two different straight lines. For an example where this is not sufficient and curved paths need to be considered see Exercise 10.2.28.

There is another important moral of Example 10.2.6. It is tempting, when evaluating a limit as $\mathbf{x} \to \mathbf{a}$, to do it one coordinate at a time. This is a bad idea, since it may lead to a wrong conclusion. As we have seen,

$$\lim_{(x,y)\to(0,0)} \frac{xy}{x^2 + y^2}$$

does not exist. Yet

$$\lim_{x\to 0}\lim_{y\to 0} \frac{xy}{x^2 + y^2} = \lim_{x\to 0} \frac{0}{x^2} = 0, \quad \text{and} \quad \lim_{y\to 0}\lim_{x\to 0} \frac{xy}{x^2 + y^2} = \lim_{y\to 0} \frac{0}{y^2} = 0.$$

Thus, we cannot replace the limit by the iterated limits. Example 10.2.6 shows that, in order to reduce a multivariable problem to a single variable one, switching to sequences is a way to go. Indeed, the sequences $\{\mathbf{a}_k\}$ and $\{\mathbf{b}_k\}$ are sequences of *pairs* of real numbers, but we had to compute limits of sequences $\{(1/k^2)/(2/k^2)\}$ and $\{(2/k^2)/(5/k^2)\}$ which are sequences of real numbers. This is particularly important because it allows us to extend many useful theorems to the multivariable setting. The following is an analogue of Theorem 3.4.10.

Theorem 10.2.7. *Let $A \subset \mathbb{R}^n$, let $f, g : A \to \mathbb{R}$, and let $\mathbf{a}$ be a cluster point of A. Also, let α be a real number. If $\lim_{\mathbf{x}\to\mathbf{a}} f(\mathbf{x}) = L_1$ and $\lim_{\mathbf{x}\to\mathbf{a}} g(\mathbf{x}) = L_2$ then:*

(a) $\lim_{\mathbf{x}\to\mathbf{a}} [\alpha f(\mathbf{x})] = \alpha \lim_{\mathbf{x}\to\mathbf{a}} f(\mathbf{x})$;

(b) $\lim_{\mathbf{x}\to\mathbf{a}} [f(\mathbf{x}) + g(\mathbf{x})] = \lim_{x\to\mathbf{a}} f(\mathbf{x}) + \lim_{\mathbf{x}\to\mathbf{a}} g(\mathbf{x})$;

(c) $\lim_{\mathbf{x}\to\mathbf{a}} [f(\mathbf{x})g(\mathbf{x})] = [\lim_{\mathbf{x}\to\mathbf{a}} f(\mathbf{x})] [\lim_{\mathbf{x}\to\mathbf{a}} g(\mathbf{x})]$;

(d) $\lim_{\mathbf{x}\to\mathbf{a}} [f(\mathbf{x})/g(\mathbf{x})] = \lim_{\mathbf{x}\to\mathbf{a}} f(\mathbf{x})/ \lim_{\mathbf{x}\to\mathbf{a}} g(\mathbf{x})$ *if, in addition $L_2 \neq 0$.*

Proof. We will prove (a) and leave the remaining assertions for exercise. By Theorem 10.2.5 it suffices to show that, for an arbitrary sequence $\{\mathbf{a}_k\} \subset D$ converging to $\mathbf{a}$,

$$\lim_{k\to\infty} [\alpha f(\mathbf{a}_k)] = \alpha \lim_{k\to\infty} f(\mathbf{a}_k).$$

However, $\{f(\mathbf{a}_k)\}$ is a sequence of real numbers so Theorem 2.3.4 applies. $\qquad\square$

> Functions of more than one variable were present in the days of Newton and Leibniz. The idea to consider pairs of numbers as individual objects came into prominence in the middle of the nineteenth century, in the work of Grassmann (see below), Hamilton, and Peano. Sir William Rowan Hamilton (1805–1865) was an Irish mathematician, famous for discovering quaternions.

When f is a function of one variable and the task at hand is to prove that $\lim_{x\to a} f(x) = L$, the Squeeze Theorem (Theorem 3.6.15) can be a very powerful tool. The same is true when f depends on more than one variable.

Theorem 10.2.8 (The Squeeze Theorem). *Let f, g, h be functions defined on $A \subset \mathbb{R}^n$ such that, for all $\mathbf{x} \in A$, $f(\mathbf{x}) \leq g(\mathbf{x}) \leq h(\mathbf{x})$, and let $\mathbf{c}$ be a cluster point of A. If $\lim_{\mathbf{x}\to\mathbf{c}} f(\mathbf{x}) = \lim_{\mathbf{x}\to\mathbf{c}} h(\mathbf{x}) = L$, then $\lim_{\mathbf{x}\to\mathbf{c}} g(\mathbf{x})$ exists and equals L.*

Example 10.2.9. Using The Squeeze Theorem.

Prove that $\displaystyle\lim_{(x,y)\to(0,0)} \frac{x^2 y}{x^2 + y^2} = 0$.

Solution. Since $x^2 + y^2 \geq x^2$ we have that

$$0 \leq \left| \frac{x^2 y}{x^2 + y^2} \right| = \frac{x^2 |y|}{x^2 + y^2} \leq \frac{x^2 |y|}{x^2} = |y| \to 0.$$

Therefore, using the Squeeze Theorem,

$$\lim_{(x,y)\to(0,0)} \frac{x^2 y}{x^2 + y^2} = 0. \qquad \blacklozenge$$

HERMANN GRASSMANN (1809–1877) was a German mathematician and a linguist. His father was a minister who taught mathematics and physics at a high school in Stettin (now Szczecin in Poland), and wrote several textbooks. Grassman did not excel in high school, but progressed enough to go to the University of Berlin to study theology. Upon graduating he decided that he wanted to teach mathematics and took a required exam. He did not do well and as a result he was allowed to teach only in the lower level classes. Over the next 8 years, Grassmann passed examinations enabling him to teach mathematics, physics, chemistry, and mineralogy at all secondary school levels. As a part of the certification he wrote an essay on the theory of the tides. This work contains the first appearance of Linear Algebra and the concept of a vector space. He applied his own ideas that were brewing since 1832—the use of vectors and vector-valued functions. He organized these original methods in [54]. The work was largely ignored for about 50 years and many mathematicians, including Kummer, did not understand its significance. He tried to improve the exposition in 1862 pretty much rewriting the text, but to no avail. Much later, these publications inspired Peano (see page 2) and Cartan's study of differential forms (see Section 15.7). The concept of a vector space became widely known only around 1920. Disappointed, Grassmann turned to historical linguistics and the study of Sanskrit. Ironically, he got more recognition for this work, and he was elected to the American Oriental Society and in 1876, he received an honorary doctorate from the University of Tübingen.

Exercises

In Exercises 10.2.1–10.2.4 find the limit and prove that your result is correct:

10.2.1. $\displaystyle\lim_{(x,y)\to(0,\pi/2)} \cos x \sin y$.

10.2.2. $\displaystyle\lim_{(x,y,z)\to(1,2,-3)} \arctan \frac{x+z}{y}$.

10.2.3. $\displaystyle\lim_{(x,y)\to(0,0)} \frac{x-y}{\sqrt{x}-\sqrt{y}}$.

10.2.4. $\displaystyle\lim_{(x,y,z)\to(0,\sqrt{\pi},1)} \left(e^{xz} \cos y^2 - x \right)$.

In Exercises 10.2.5–10.2.8 find the limit and use the Squeeze Theorem to prove that your result is correct:

10.2.5. $\displaystyle\lim_{(x,y,z)\to(0,0,0)} \frac{xyz}{x^2+y^2+z^2}$.

10.2.6. $\displaystyle\lim_{(x,y)\to(0,0)} \frac{x\sin(x+y)}{x+y}$.

10.2.7. $\displaystyle\lim_{(x,y)\to(0,0)} \frac{y^2}{\sqrt{x^2+y^2}}$.

10.2.8.* $\displaystyle\lim_{(x,y)\to(0,0)} \frac{x^3\cos(x-y)}{x^2+y^2}$.

In Exercises 10.2.9–10.2.12 find the limit by using an appropriate substitution that will reduce it to a limit of a single-variable function:

10.2.9. $\displaystyle\lim_{(x,y)\to(0,0)} \frac{\sin(x+y^2)}{x+y^2}$.

10.2.10. $\displaystyle\lim_{(x,y)\to(0,0)} \frac{e^{\sin(x+y)}-1}{x+y}$.

10.2.11. $\displaystyle\lim_{(x,y)\to(1/2,2)} \frac{x^2y^2-1}{x^3y^3-1}$.

10.2.12. $\displaystyle\lim_{(x,y)\to(3,-1)} \frac{1-\cos(x+3y)}{(x+3y)^2}$.

In Exercises 10.2.13–10.2.16 show that the limit does not exist:

10.2.13. $\displaystyle\lim_{(x,y)\to(0,0)} \frac{x}{x^2+y^2}$.

10.2.14. $\displaystyle\lim_{(x,y,z)\to(0,0,0)} \frac{xyz}{x^3+y^3+z^3}$.

10.2.15. $\displaystyle\lim_{(x,y)\to(0,0)} \frac{x^2y}{x^4+y^2}$.

10.2.16. $\displaystyle\lim_{(x,y)\to(0,0)} \frac{xy^2-x^2y}{(x^2+y^2)^{3/2}}$.

In Exercises 10.2.17–10.2.22 find the limit and prove that your result is correct, or show that the limit does not exist:

10.2.17. $\displaystyle\lim_{(x,y)\to(0,0)} \frac{x^3y}{x^6+y^2}$.

10.2.18. $\displaystyle\lim_{(x,y,z)\to(0,0,0)} \frac{x^3}{x^2+y^2+z^2}$.

10.2.19. $\displaystyle\lim_{(x,y)\to(0,1)} \frac{\sin xy}{x}$.

10.2.20. $\displaystyle\lim_{(x,y)\to(1,1)} \frac{3x^2-4xy+y^2}{x^2-y^2}$.

10.2.21. $\displaystyle\lim_{(x,y,z)\to(0,0,0)} \frac{xy^2z^3}{x^6+2y^6+3z^6}$.

10.2.22.* $\displaystyle\lim_{(x,y)\to(0,0)} \frac{\ln x-\ln(x+y)}{y}(x+y)$.

10.2.23. Prove that there can be at most one number L satisfying Definition 10.2.1.

10.2.24. Prove parts (b)–(d) of Theorem 10.2.7.

10.2.25. Prove Theorem 10.2.8.

10.2.26. Prove that a point $\mathbf{a}$ is a cluster point of a set $A\subset\mathbb{R}^n$ if and only if there exists a sequence $\{\mathbf{a}^{(k)}\}\subset A\setminus\{\mathbf{a}\}$ that converges to $\mathbf{a}$.

10.2.27.* Let $A\subset\mathbb{R}^2$, let $f:A\to\mathbb{R}$, and let $\mathbf{a}=(a,b)$ be a cluster point of A. Suppose that $\lim_{\mathbf{x}\to\mathbf{a}}f(\mathbf{x})=L$ along any polygonal line with vertices $\{\mathbf{a}_n\}\subset A$, where $\lim\mathbf{a}_n=\mathbf{a}$. Prove that $\lim_{\mathbf{x}\to\mathbf{a}}f(\mathbf{x})=L$.

10.2.28.* Let $p(x)$ be a polynomial of degree $n>1$ satisfying $p(0)=0$ and let

$$f(x,y)=\begin{cases}0, & \text{if } y=p(x)\\ 1, & \text{if } y\neq p(x).\end{cases}$$

Prove that: (a) the limit of f as $(x,y)\to(0,0)$ along $y=p(x)$ is zero; (b) the limit of f as $(x,y)\to(0,0)$ along any line through the origin is one; (c) if q is any non-constant polynomial such that $q(0)=0$ and $q\neq p$, then the limit of f as $(x,y)\to(0,0)$ along $y=q(x)$ is 1.

10.2.29.* Give an example of a function f and a point (a,b) such that $\lim_{x\to a}\lim_{y\to b}f(x,y)$ exists but $\lim_{y\to b}\lim_{x\to a}f(x,y)$ does not.

10.2.30.* Give an example of a function f and a point (a,b) such that $\lim_{x\to a}\lim_{y\to b}f(x,y)$ and $\lim_{y\to b}\lim_{x\to a}f(x,y)$ exist but are not equal.

10.3 Continuous Functions

The definition of a continuous function $f : \mathbb{R}^n \to \mathbb{R}$ is almost the same as in Chapter 3.

Definition 10.3.1. Let f be a function with a domain $D \subset \mathbb{R}^n$ and let $\mathbf{a} \in D$. Then f is **continuous** at $\mathbf{a}$ if $\lim_{\mathbf{x} \to \mathbf{a}} f(\mathbf{x})$ exists and equals $f(\mathbf{a})$. If f is continuous at every point of a set A, we say that it is continuous on A.

As a straightforward consequence of Theorem 10.2.5 we have an equivalent definition of continuity.

Corollary 10.3.2. *A function f with a domain $A \subset \mathbb{R}^n$ is continuous at $\mathbf{a} \in A$ if and only if, for every sequence $\{\mathbf{a}_k\} \subset A$ converging to $\mathbf{a}$, $\lim_{k \to \infty} f(\mathbf{a}_k) = f(\mathbf{a})$.*

Example 10.3.3. A function continuous at a point.

Prove that the function $f(x, y) = x^2 + y^2$ is continuous at $\mathbf{a} = (3, -1)$.

Solution. Let $\{a_n\}$ and $\{b_n\}$ be two sequences of real numbers such that $\lim a_n = 3$ and $\lim b_n = -1$. By Theorem 3.4.10,

$$\lim f(a_n, b_n) = \lim(a_n^2 + b_n^2) = (\lim a_n)^2 + (\lim b_n)^2 = 3^2 + (-1)^2 = f(3, -1),$$

so f is continuous at $\mathbf{a}$. ◆

Example 10.3.4. Points of continuity and discontinuity of a function.

Find the points of continuity of $f(x, y) = \begin{cases} -1, & \text{if } x^2 + y^2 < 1 \\ 1, & \text{if } x^2 + y^2 \geq 1. \end{cases}$

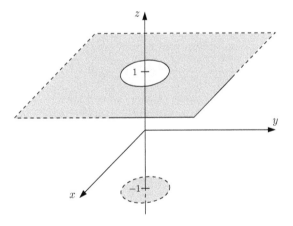

Figure 10.7: The graph of f.

Solution. The function f is continuous at every point of the open unit disk and at every point outside of the closed unit disk. It has a discontinuity at every point of the unit circle.

Proof. Let (a, b) be a point in the open unit disk $\mathbb{D}$, and let $\{(a_n, b_n)\}$ be a sequence converging to (a, b) (Figure 10.8). We will show that $\{f(a_n, b_n)\}$ converges to $f(a, b)$. Since $f(a, b) = -1$ it suffices to prove that $(a_n, b_n) \in \mathbb{D}$, at least for large values of n, because that will make $f(a_n, b_n) = -1$. Let

$$\varepsilon = \frac{1 - \|(a, b)\|}{2}.$$

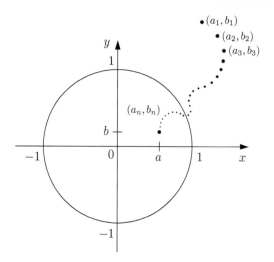

Figure 10.8: $(a_n, b_n) \in \mathbb{D}$ for large values of n.

By definition, $(a_n, b_n) \to (a, b)$ means that $a_n \to a$ and $b_n \to b$. The former implies that there exists $N_1 \in \mathbb{N}$ such that

$$n \geq N_1 \quad \Rightarrow \quad |a_n - a| < \varepsilon.$$

Similarly, $b_n \to b$ so there exists N_2 such that

$$n \geq N_2 \quad \Rightarrow \quad |b_n - b| < \varepsilon.$$

Therefore, if $n \geq N = \max\{N_1, N_2\}$, then

$$\|(a_n, b_n) - (a, b)\| = \sqrt{(a_n - a)^2 + (b_n - b)^2} < \sqrt{\varepsilon^2 + \varepsilon^2} = \varepsilon\sqrt{2} < 2\varepsilon,$$

and it follows that

$$\|(a_n, b_n)\| \leq \|(a_n, b_n) - (a, b)\| + \|(a, b)\|$$
$$< 2\varepsilon + \|(a, b)\| = 1 - \|(a, b)\| + \|(a, b)\| = 1.$$

So, $\|(a_n, b_n)\| < 1$ which implies that $f(a_n, b_n) = -1 \to -1 = f(a, b)$. Consequently, f is continuous at (a, b). We leave the case when (a, b) lies outside of the closed unit disk as an exercise.

It remains to prove that, if (a, b) lies on the unit circle, f is not continuous at (a, b). In that case, we define a sequence $\{(a_n, b_n)\}$ by $a_n = a(1 - 1/n)$ and $b_n = b$ (Figure 10.9). Then $(a_n, b_n) \to (a, b)$ and

$$a_n^2 + b_n^2 = a^2 \left(1 - \frac{1}{n}\right)^2 + b^2 < a^2 + b^2 = 1.$$

Therefore, $\|(a_n, b_n)\| < 1$ which implies that $f(a_n, b_n) = -1$. On the other hand, $f(a, b) = 1$ so the sequence $\{f(a_n, b_n)\}$ does not converge to $f(a, b)$. We conclude that f is not continuous at (a, b). ♦

Next we turn our attention to the operations that preserve continuity.

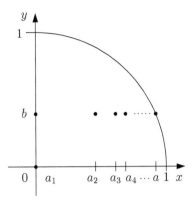

Figure 10.9: f is not continuous at (a, b).

Theorem 10.3.5. *Let f, g be two functions with a domain $A \subset \mathbb{R}^n$ and let $\mathbf{a} \in A$. Also, let α be a real number. If f and g are continuous at $\mathbf{a}$ then the same is true for: (a) αf; (b) $f + g$; (c) fg; (d) f/g if, in addition, $g(\mathbf{a}) \neq 0$.*

Proof. (a) We need to show that $\lim_{\mathbf{x} \to \mathbf{a}}[\alpha f(\mathbf{x})] = \alpha f(\mathbf{a})$. Combining Theorem 10.2.7 (a) and the fact that f is continuous we obtain that

$$\lim_{\mathbf{x} \to \mathbf{a}} [\alpha f(\mathbf{x})] = \alpha \lim_{\mathbf{x} \to \mathbf{a}} f(\mathbf{x}) = \alpha f(\mathbf{a}).$$

The proofs of assertions (b), (c), and (d) are similar and we leave them as an exercise. $\square$

Another procedure that results in a continuous function is a composition. In order to make the statement precise, we will make the following assumptions. Let $g_1, g_2, \ldots, g_n$ be functions with a domain $A \subset \mathbb{R}^p$ and let $\mathbf{a} \in A$. We will often write $\mathbf{g} = (g_1, g_2, \ldots, g_n)$ and consider $\mathbf{g}$ as a function from A to $\mathbb{R}^n$. Then $E = \mathbf{g}(A) \subset \mathbb{R}^n$ and we will assume that f is a function defined on some set E_0 that contains E. In that case we define $f \circ \mathbf{g}$ as a function defined on A, so that, for every $\mathbf{x} = (x_1, x_2, \ldots, x_p) \in A$,

$$(f \circ \mathbf{g})(\mathbf{x}) = f\left(g_1(x_1, x_2, \ldots, x_p), g_2(x_1, x_2, \ldots, x_p), \ldots, g_n(x_1, x_2, \ldots, x_p)\right).$$

Theorem 10.3.6. *Let $g_1, g_2, \ldots, g_n$ be functions with a domain $A \subset \mathbb{R}^p$ and suppose that they are all continuous at $\mathbf{a} \in A$. Further, let f be defined on some set E_0 that contains $E = \mathbf{g}(A)$, and suppose that f is continuous at $\mathbf{g}(\mathbf{a})$. Then $f \circ \mathbf{g}$ is continuous at $\mathbf{a}$.*

Proof. Let $\varepsilon > 0$. Since f is continuous at $\mathbf{g}(\mathbf{a})$, there exists $\eta > 0$ such that

$$|u_i - g_i(\mathbf{a})| < \eta, \quad 1 \le i \le n, \quad \text{and} \quad \mathbf{u} \in E_0 \quad \Rightarrow \quad |f(\mathbf{u}) - f(\mathbf{g}(\mathbf{a}))| < \varepsilon.$$

The function g_1 is continuous at $\mathbf{a}$ so there exists $\delta_1 > 0$ such that

$$|g_1(\mathbf{x}) - g_1(\mathbf{a})| < \eta, \quad \text{for } |x_k - a_k| < \delta_1, \quad 1 \le k \le p, \quad \text{and} \quad \mathbf{x} \in A.$$

Using the same argument we obtain that for each i, $1 \le i \le n$, there exists $\delta_i > 0$ such that

$$|g_i(\mathbf{x}) - g_i(\mathbf{a})| < \eta, \quad \text{for } |x_k - a_k| < \delta_i, \quad 1 \le k \le p, \quad \text{and} \quad \mathbf{x} \in A.$$

If we now define $\delta = \min\{\delta_1, \delta_2, \ldots, \delta_n\}$ and if we select $\mathbf{x} \in A$ satisfying $|x_k - a_k| < \delta$, $1 \le k \le p$, then we will have $|g_i(\mathbf{x}) - g_i(\mathbf{a})| < \eta$, for all $1 \le i \le n$. This implies that

$$|f(g_1(\mathbf{x}), g_2(\mathbf{x}), \ldots, g_n(\mathbf{x})) - f(g_1(\mathbf{a}), g_2(\mathbf{a}), \ldots, g_n(\mathbf{a}))| < \varepsilon$$

and the theorem is proved.   □

We define elementary functions in the same way as in Section 3.7, i.e., these are constants, exponential and logarithmic functions, power functions, trigonometric functions and their inverses, as well as all functions obtained from the already listed through composition and combinations using the four arithmetic operations. The difference is that, having more than one independent variable, we need to replace the identity function by the functions of the form $f(x_1, x_2, \ldots, x_n) = x_i$, $1 \le i \le n$. For example, every polynomial in two variables is an elementary function. We leave the proof of this fact to the reader.

The results of this section allow us to conclude that elementary functions of several variables are continuous in their domains.

Theorem 10.3.7. *Let A be a subset of $\mathbb{R}^n$ and let $f : A \to \mathbb{R}$ be an elementary function. Then f is continuous on A.*

Example 10.3.8. A continuous extension of a function.

The function $f(x, y) = x \sin \dfrac{y}{x}$ is undefined when $x = 0$. Can we extend it to a continuous function on $\mathbb{R}$?

Solution. Geometrically, the domain A of f is the whole plane with the y-axis deleted. By Theorem 10.3.7, if $(a, b) \in A$, f is continuous at (a, b). What if $(a, b) \notin A$? Is there a way to extend the domain of f to the whole xy-plane, so that this extension is a continuous function? If $(a, b) \notin A$ and if we denote the extension $\hat{f}$, then $\hat{f}$ will be continuous at (a, b) if $\lim_{(x,y) \to (a,b)} \hat{f}(x, y) = \hat{f}(a, b)$. In other words, the smart choice for $\hat{f}(a, b)$ will be the limit $\lim_{(x,y) \to (a,b)} f(x, y)$. Does this limit exist?

The interesting points (a, b) are those where f is not defined, i.e., when $a = 0$. We will show that $\lim_{(x,y) \to (0,b)} f(x, y) = 0$. We must make sure that $(x, y) \in A$, otherwise f would be undefined. For $(x, y) \in A$, $f(x, y) = x \sin(y/x)$, so the task is to prove that

$$\lim_{\substack{(x,y) \to (0,b) \\ x \ne 0}} x \sin \frac{y}{x} = 0.$$

This follows from the inequality $|x \sin(y/x)| \le |x|$ and the Squeeze Theorem. Thus, the function

$$\hat{f} = \begin{cases} x \sin \dfrac{y}{x}, & \text{if } x \ne 0 \\ 0, & \text{if } x = 0 \end{cases}$$

is continuous in $\mathbb{R}^2$.   ◆

We have seen in Section 10.2 (page 302) that, in general,

$$\lim_{(x,y) \to (a,b)} f(x, y) \ne \lim_{x \to a} \lim_{y \to b} f(x, y).$$

Therefore, if f is continuous in each variable (keeping the others fixed), it does not follow that f is continuous. Here is an example.

Example 10.3.9. Separate continuity does not imply continuity.

Let $f(x,y) = \begin{cases} \left(\dfrac{x^2-y^2}{x^2+y^2}\right)^2, & \text{if } (x,y) \neq (0,0) \\ 1, & \text{if } (x,y) = (0,0). \end{cases}$ Prove that f is continuous in each

variable, but not continuous at $(0,0)$.

Solution. First we will show that, for a fixed y, f is a continuous function of x. So, let y be fixed. Recall that, if $x \to 0$, it also means that $x \neq 0$, so f is computed according to the first formula. Further, if $y = 0$, then

$$f(x,y) = \left(\frac{x^2-0^2}{x^2+0^2}\right)^2 = 1.$$

Thus,

$$\lim_{x\to 0} f(x,y) = \lim_{x\to 0} \begin{cases} \left(\dfrac{x^2-y^2}{x^2+y^2}\right)^2, & \text{if } y \neq 0 \\ 1, & \text{if } y = 0 \end{cases} = \begin{cases} \left(\dfrac{0^2-y^2}{0^2+y^2}\right)^2, & \text{if } y \neq 0 \\ 1, & \text{if } y = 0 \end{cases} = 1.$$

Since $f(0,y)$ is also 1, we obtain that f is a continuous function of x. In a similar fashion, one can show that, for a fixed x, f is a continuous function of y. Nevertheless, f is not continuous at $(0,0)$. To establish this, let $(a_k, b_k) = (1/k, 1/k)$. Then $(a_k, b_k) \to (0,0)$, but

$$f(a_k, b_k) = f\left(\frac{1}{k}, \frac{1}{k}\right) = \left(\frac{\left(\frac{1}{k}\right)^2 - \left(\frac{1}{k}\right)^2}{\left(\frac{1}{k}\right)^2 + \left(\frac{1}{k}\right)^2}\right)^2 = 0 \not\to 1. \qquad \blacklozenge$$

Exercises

In Exercises 10.3.1–10.3.4 prove that the function f is continuous at $\mathbf{a}$:

10.3.1. $f(x,y) = xy + 6x$, $\mathbf{a} = (-1,2)$.   10.3.2. $f(x,y) = xe^{x+y}$, $\mathbf{a} = (2,3)$.

10.3.3. $f(x,y) = \dfrac{\sin(2y-x)}{x-y}$, $\mathbf{a} = (1,3)$.

10.3.4. $f(x,y) = \begin{cases} \dfrac{\sin(x^2+y^2)}{x^2+y^2}, & \text{if } (x,y) \neq (0,0) \\ 1, & \text{if } (x,y) = (0,0), \end{cases}$ $\mathbf{a} = (0,0)$.

In Exercises 10.3.5–10.3.8 prove that the function f is continuous at every point of its domain:

10.3.5. $f(x,y) = xy^2 - 3x$.

10.3.6. $f(x,y) = \dfrac{e^{x+y}}{x+y+1}$.

10.3.7. $f(x,y) = \dfrac{1+\sqrt{x+2y}}{x+y}$.

10.3.8. $f(x,y) = \dfrac{\ln(2y-x)}{x}$.

10.3.9. Prove that the function f in Example 10.3.4 is continuous at every point (a,b) that satisfies $\|(a,b)\| > 1$.

10.3.10. Let f be a function defined on a set $A \subset \mathbb{R}^n$ and suppose that f is continuous at $\mathbf{a} \in A$. Prove that, if $f(\mathbf{a}) > 0$ there exists an n-ball B such that $f(\mathbf{x}) > 0$ for all $\mathbf{x} \in B$.

10.3.11. Let f be a function f defined on a set $A \subset \mathbb{R}^2$, let f be a continuous function of x (with y fixed), and let f satisfy the Lipschitz condition with respect to y, i.e., there exists $M > 0$ such that

$$|f(x,y_1) - f(x,y_2)| \leq M|y_1 - y_2|$$

for all $(x,y_1),(x,y_2) \in A$. Prove that f is continuous in A.

10.3.12. Prove parts (b)–(d) of Theorem 10.3.5.

10.3.13. Prove Theorem 10.3.7.

In Exercises 10.3.14–10.3.19 determine whether the function f can be defined at $(0,0)$ so that it is continuous.

10.3.14. $f(x,y) = \dfrac{\sin(x^4 + y^2)}{x^2 + y^2}$.

10.3.15. $f(x,y) = \dfrac{e^{xy} - 1}{|x| + |y|}$.

10.3.16. $f(x,y) = \dfrac{xy}{|x| + |y|}$.

10.3.17. $f(x,y) = \dfrac{x^2 y^3}{x^4 + y^6}$.

10.3.18. $f(x,y) = xy \ln(x^2 + y^2)$.

10.3.19. $f(x,y) = \dfrac{x^p y^q + x^r y^s}{x^q y^p + x^s y^r}$, $p,q,r,s > 0$.

10.3.20. Let f be continuous at $(a,b) \in \mathbb{R}^2$. Prove that the function $g(x) = f(x,b)$ is continuous at $x = a$, and that the function $h(y) = f(a,y)$ is continuous at $y = b$.

10.3.21. In Example 10.3.9 show that, for a fixed x, f is a continuous function of y.

10.3.22. Give an example of a function f defined on a set $A \subset \mathbb{R}^3$, and a point $(a,b,c) \in A$, so that f is continuous at (a,b,c) with respect to each of the 3 variables, but fails to be continuous at (a,b,c).

10.4 Boundedness of Continuous Functions

In this section we will revisit Theorem 3.9.8. (A continuous function on $[a,b]$ is bounded and it attains its minimum and maximum values.) We would like to generalize it to the case when the function f depends on more than one variable. We have seen in the case of one variable that the continuity of f is essential, so we will assume it here as well. The real challenge is to determine what kind of domain can replace $[a,b]$.

   An inspection of the proof of Theorem 3.9.8 reveals that a very important role was played by the Bolzano–Weierstrass Theorem. The fact that $[a,b]$ is a bounded set guaranteed that any sequence in it must be bounded. Therefore, our first priority is to extend the concept of a bounded set to subsets of $\mathbb{R}^n$. In $\mathbb{R}$, a set A is bounded if there exists M such that, for any $a \in A$, $|a| \le M$. In $\mathbb{R}^n$, we will replace the absolute values by the norm.

Definition 10.4.1. A set $A \subset \mathbb{R}^n$ is **bounded** (by M) if there is a real number M such that $\|\mathbf{a}\| \le M$, for all $\mathbf{a} \in A$.

   In addition to boundedness, there is a more subtle requirement. When the convergent subsequence is located its limit has to be in the same set. By definition, the limit of a sequence in a set A is a cluster point of A, so we are really asking that A contains its cluster points. Such sets are called *closed* sets.

Definition 10.4.2. A set $A \subset \mathbb{R}^n$ is **closed** if it contains all of its cluster points. The **closure** of a set $A \subset \mathbb{R}^n$ is the smallest closed set that contains A. It is denoted by $\overline{A}$.

Example 10.4.3. A closed set.

   Let $(a,b) \in \mathbb{R}^2$. Prove that the set $A = \{(x,y) \in \mathbb{R}^2 : (x-a)^2 + (y-b)^2 = r^2\}$ is closed.

Solution. We will show that (x,y) is a cluster point of A if and only if it belongs to A.

   Indeed, if $(x_1, y_1) \in A$, and $\delta > 0$, the disk with center (x_1, y_1) and radius δ contains a portion of A, so (x_1, y_1) is a cluster point of A. (See Figure 10.10.)

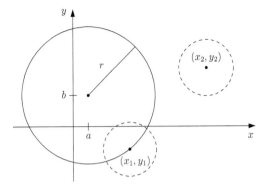

Figure 10.10: Cluster points of the circle.

On the other hand, if $(x_2, y_2) \notin A$, then its distance from the center (a, b) is a number different from r, say r'. Now the disk with center (x_2, y_2) and radius $|r - r'|/2$ does not intersect A, so (x_2, y_2) is not a cluster point of A. ♦

> The closed sets ("abgeschlossen" in German) were introduced by Cantor in 1884 in [15] through his work on the Continuum Hypothesis. Namely, he proved that no closed set can have a cardinal number strictly between the cardinal numbers of $\mathbb{N}$ and $\mathbb{R}$.

Our argument in the previous example was intuitive, appealing to the visual. We leave a formal proof as an exercise. Later, we will return to the general problem of determining when a set is closed. Right now, the important thing is that, for a set in $\mathbb{R}^n$ that is *closed* and bounded, the Bolzano–Weierstrass theorem holds.

Theorem 10.4.4 (Bolzano–Weierstrass Theorem). *Suppose that A is a closed and bounded set in $\mathbb{R}^n$. Every sequence of points in A has a convergent subsequence that converges to a point in A.*

Proof. The assumption that a set is bounded means that there exists a rectangle $[a_1, b_1] \times [a_2, b_2] \times \cdots \times [a_n, b_n]$ that contains A. Let

$$P_k = (x_1^{(k)}, x_2^{(k)}, \ldots, x_n^{(k)}), \quad k \in \mathbb{N},$$

be a sequence of points in A. Notice that the sequence of real numbers $x_1^{(1)}, x_1^{(2)}, x_1^{(3)}, \ldots$ lies in $[a_1, b_1]$. By Theorem 2.7.11, there exists a convergent subsequence $\{x_1^{(n_k)}\}$. We will now focus only at the points $P_{n_1}, P_{n_2}, P_{n_3}, \ldots$ and we will show that this sequence (a subsequence of $\{P_k\}$) has a convergent subsequence itself that converges to a point in A. To avoid multiple subscripts, we will denote $Q_k = P_{n_k}$, and

$$Q_k = (y_1^{(k)}, y_2^{(k)}, \ldots, y_n^{(k)}),$$

bearing in mind that the sequence $\{y_1^{(k)}\}$ converges.

Next, we consider the sequence $\{y_2^{(k)}\}$ that lies in $[a_2, b_2]$. Using Theorem 2.7.11, we can extract a convergent subsequence $\{y_2^{(m_k)}\}$. Then we will further reduce our sequence $\{Q_k\}$ to a subsequence

$$R_k = Q_{m_k} = (z_1^{(k)}, z_2^{(k)}, \ldots, z_n^{(k)})$$

which has the advantage that both $\{z_1^{(k)}\}$ and $\{z_2^{(k)}\}$ are convergent sequences. Continuing this process, after n iterations, we obtain a subsequence

$$T_k = (w_1^{(k)}, w_2^{(k)}, \ldots, w_n^{(k)})$$

in which $w_1^{(k)} \to w_1, w_2^{(k)} \to w_2, \ldots, w_n^{(k)} \to w_n$. By definition, the sequence $\{T_k\}$ converges to $T = (w_1, w_2, \ldots, w_n)$, and T is a cluster point of A or it belongs to A. Since A is closed it contains T. $\qquad\square$

Now we can establish a generalization of Theorem 3.9.8.

Theorem 10.4.5 (The Extreme Value Theorem). *Let f be a continuous function defined on a closed and bounded set A in $\mathbb{R}^n$. Then f is bounded and it attains both its minimum and its maximum value.*

Proof. First we will establish that f is bounded on A. Suppose, to the contrary, that it is not. Then, for each $k \in \mathbb{N}$, there exists $\mathbf{x}_k \in A$ such that

$$|f(\mathbf{x}_k)| > k.$$

By the Bolzano–Weierstrass Theorem, the sequence $\{\mathbf{x}_k\}$ has a convergent subsequence $\{\mathbf{x}_{n_k}\}$ converging to a limit $\mathbf{x} \in A$. Further, the continuity of f implies that $f(\mathbf{x}) = \lim f(\mathbf{x}_{n_k})$. However, the inequality $|f(\mathbf{x}_{n_k})| > n_k$ shows that $\{f(\mathbf{x}_{n_k})\}$ is not a bounded sequence of real numbers, so it cannot be convergent. Therefore, f is a bounded function on A.

Since f is bounded, its range B is a bounded subset of $\mathbb{R}$. Let $M = \sup B$ and $m = \inf B$. We will show that there exists a point $\mathbf{c} \in A$ such that $f(\mathbf{c}) = M$. Let k be a positive integer. Since M is the least upper bound of B, the number $M - 1/k$ cannot be an upper bound of B. Consequently, there exists $\mathbf{x}_k \in A$ such that

$$M - \frac{1}{k} < f(\mathbf{x}_k) \le M. \tag{10.4}$$

The Bolzano–Weierstrass Theorem (Theorem 10.4.4) guarantees the existence of a subsequence $\{\mathbf{x}_{k_j}\}$ and a point $\mathbf{c} \in A$ to which the subsequence converges. Further, the Squeeze Theorem applied to (10.4) implies that $\lim f(\mathbf{x}_{k_j}) = M$. Now the continuity of f shows that $M = f(\lim \mathbf{x}_{k_j}) = f(\mathbf{c})$. So, f attains its maximum value.

The fact that f attains its minimum value can be established by considering the function $g = -f$. $\qquad\square$

Theorem 10.4.5 shows that closed and bounded sets in $\mathbb{R}^n$ play an important role in calculus. Part of it is based on Theorem 10.4.4 which shows that in such sets every sequence has a convergent subsequence and that the limit remains in the set.

Definition 10.4.6. A set $A \in \mathbb{R}^n$ is **sequentially compact** if every sequence of points in A has a convergent subsequence that converges to a point in A.

Using this terminology, Theorem 10.4.4 asserts that if a set is closed and bounded then it is sequentially compact. In fact, this implication is reversible.

Theorem 10.4.7. *A set $A \in \mathbb{R}^n$ is sequentially compact if and only if it is closed and bounded.*

Proof. The "if" part is Theorem 10.4.4, so we need to establish its converse. Let A be a sequentially compact set in $\mathbb{R}^n$. First we will show that it is closed. Let $\mathbf{a}$ be a cluster point of A. Then there exists a sequence $\{\mathbf{a}_k\} \subset A$ that converges to $\mathbf{a}$. Since A is sequentially compact, the sequence $\{\mathbf{a}_k\}$ has a subsequence $\{\mathbf{a}_{k_j}\}$ that converges to a point $\mathbf{a}' \in A$. But $\{\mathbf{a}_{k_j}\}$ also converges to $\mathbf{a}$ so $\mathbf{a} = \mathbf{a}' \in A$. Thus, A contains all of its cluster points and it must be closed.

In order to show that A is bounded, we will argue by contradiction. So, suppose that

A is not bounded. Then, for every $k \in \mathbb{N}$, there exists a point $\mathbf{a}_k \in A$ such that $\|\mathbf{a}_k\| \geq k$. Since A is sequentially compact, the sequence $\{\mathbf{a}_k\}$ has a subsequence $\{\mathbf{a}_{k_j}\}$ that converges to a point $\mathbf{a} \in A$. Let $\varepsilon = 1$. The convergence of $\{\mathbf{a}_{k_j}\}$ implies that there exists $N \in \mathbb{N}$ such that

$$\|\mathbf{a}_{k_j} - \mathbf{a}\| < 1, \quad \text{for} \quad j \geq N.$$

Now, if $j \geq N$,

$$\|\mathbf{a}\| = \|\mathbf{a} - \mathbf{a}_{k_j} + \mathbf{a}_{k_j}\| \geq \|\mathbf{a}_{k_j}\| - \|\mathbf{a} - \mathbf{a}_{k_j}\| > k_j - 1 \to \infty, \quad j \to \infty.$$

This contradiction shows that A is bounded. $\qquad\qquad\qquad\qquad\qquad\qquad\qquad\square$

The concept of a sequentially compact set is due to Fréchet. By 1906, when he introduced the term "compact", it was known that this property of the interval $[a, b]$ was shared by some other objects. For example, the Arzelà–Ascoli theorem states that if A is a set of functions on $[a, b]$ that are uniformly bounded and equicontinuous (given $\varepsilon > 0$ the same δ can be used for all functions), then every sequence $\{f_n\} \subset A$ has a uniformly convergent subsequence. Later, a topological definition of a compact set emerged (see Section 10.7), so the older one got a prefix "sequential". A nice overview of the development of the concept can be found in [100]. Interestingly, towards the end of his life, Fréchet was asked what made him choose the word *compact*, and he could not remember.

MAURICE FRÉCHET (1878–1973) was a French mathematician. His father was the principal of a Protestant school. In the 1880s education in France became secularized, so he lost his job and the family survived by setting up a boarding house. Fréchet attended high school in Paris where one of the teachers was Hadamard, who recognized his talent and worked with him, even after moving to Bordeaux. Fréchet enrolled in the Ecole Normale Supérieure in Paris, still undecided whether to study mathematics or physics. He chose the former in order to avoid chemistry classes. As an undergraduate he published more than 20 papers. His dissertation was supervised by Hadamard and it introduced the concept of a metric space, although the name is due to Hausdorff (page 318), and the functionals on metric spaces. He started a teaching career in high schools in Besançon and Nantes, then as a professor of mechanics at the Faculty of Science in Poitiers (1910–19). He was planning to spend a year in the United States at the University of Illinois when WWI broke out and he was drafted, spending two and a half years at the front, yet continuing to publish research. After the war, he was a professor of higher analysis at the University of Strasbourg and Director of the Mathematics Institute (1919–27). Once again, he continued a high volume output, in spite of time consuming duties that involved setting up and organizing the International Congress of Mathematicians in Strasbourg in 1920. At that time Fréchet added statistics to topology and analysis as his research interests. In 1928 he moved to Paris where he held several different positions until 1948 when he retired. Despite his major achievements, he was not overly appreciated in France. As an illustration, while being nominated numerous times, he was not elected a member of the Academy of Sciences until the age of 78. On the other hand he was a member of the Polish Academy of Science and Arts, the Royal Netherlands Academy of Arts and Sciences, and the Royal Society of Edinburgh. Fréchet was a supporter of Esperanto, publishing some papers and articles in that constructed language.

Another result whose proof uses the Bolzano–Weierstrass theorem concerns the uniform continuity of a function. In Section 3.8 we have defined a uniformly continuous function f on a set $A \subset \mathbb{R}$. Now, we extend this concept to functions defined on subsets of $\mathbb{R}^n$.

Definition 10.4.8. Let f be a function with a domain $A \subset \mathbb{R}^n$. We say that f is **uniformly continuous** on A if for any $\varepsilon > 0$ there exists $\delta > 0$ such that, for any $\mathbf{x}, \mathbf{a} \in A$,

$$|f(\mathbf{x}) - f(\mathbf{a})| < \varepsilon \quad \text{whenever} \quad |x_k - a_k| < \delta, \quad 1 \leq k \leq n.$$

Example 10.4.9. A uniformly continuous function.

Prove that the function $f(x, y) = x^2 + y^2$ is uniformly continuous on $[0, 1] \times [0, 1]$.

Solution. Let $\varepsilon > 0$. If we examine the inequality $|f(x, y) - f(a, b)| < \varepsilon$, we see that

$$\begin{aligned}
|f(x, y) - f(a, b)| &= |x^2 + y^2 - a^2 - b^2| = |(x - a)(x + a) + (y - b)(y + b)| \\
&\leq |x - a|(|x| + |a|) + |y - b|(|y| + |b|).
\end{aligned} \tag{10.5}$$

Since $A = [0,1] \times [0,1]$, it follows that $|x|, |y|, |a|, |b| \leq 1$. Therefore,

$$|f(x,y) - f(a,b)| \leq 2|x - a| + 2|y - b|,$$

and if we take $\delta = \varepsilon/4$ it will work regardless of x, y, a, b. In other words, f is uniformly continuous on $[0,1]$.

Proof. Let $\varepsilon > 0$, and take $\delta = \varepsilon/4$. If $(x,y), (a,b) \in [0,1] \times [0,1]$, and if $|x - a|, |y - b| < \delta$ then, using (10.5),

$$\begin{aligned}
|f(x,y) - f(a,b)| &\leq |x - a|(|x| + |a|) + |y - b|(|y| + |b|) \\
&\leq 2|x - a| + 2|y - b| \\
&< 2\delta + 2\delta = 4\frac{\varepsilon}{4} = \varepsilon.
\end{aligned}$$ ◆

Example 10.4.10. A function that is not uniformly continuous.

Prove that the function $f(x,y) = x^2 + y^2$ is not uniformly continuous on $A = \mathbb{R}^2$.

Solution. Once again we will analyze the inequality $|f(x,y) - f(a,b)| < \varepsilon$. The left hand side can be written as $|(x-a)(x+a) + (y-b)(y+b)|$. Suppose that $|x-a|, |y-b| < \delta$. Then the factors $x - a$ and $y - b$ are small, but we have no control over the size of the factors $x + a$ and $y + b$. (Whatever δ we have optimistically selected, should work for all a, b, even if they go to ∞.) It looks as if f is not uniformly continuous on $\mathbb{R}^2$ and we will prove that.

Proof. Suppose, to the contrary, that f is uniformly continuous on $\mathbb{R}^2$. Let $\varepsilon > 0$. By definition, there exists $\delta > 0$ such that, regardless of the choice of $(x,y), (a,b) \in \mathbb{R}^2$, as soon as $|x - a| < \delta$ and $|y - b| < \delta$, we should have $|f(x,y) - f(a,b)| < \varepsilon$. It is easy to see that $x = a + \delta/2$ and $y = b$ satisfy $|x - a| < \delta$ and $|y - b| < \delta$. Therefore, we should have $|f(a + \delta/2, b) - f(a,b)| < \varepsilon$, and this should be true for any $(a,b) \in \mathbb{R}^2$. However,

$$\left| f\left(a + \frac{\delta}{2}, b\right) - f(a,b) \right| = \left| \left(a + \frac{\delta}{2}\right)^2 + b^2 - a^2 - b^2 \right| = \left| \left(2a + \frac{\delta}{2}\right)\frac{\delta}{2} \right|$$

and the last expression can be made arbitrarily large when a increases without a bound. For example, if $(a,b) = (\varepsilon/\delta, 0)$, then

$$\left| f\left(a + \frac{\delta}{2}, b\right) - f(a,b) \right| = \left(2\frac{\varepsilon}{\delta} + \frac{\delta}{2}\right)\frac{\delta}{2} > 2\frac{\varepsilon}{\delta} \cdot \frac{\delta}{2} = \varepsilon.$$

Thus, we cannot have $|f(a + \delta/2, b) - f(a,b)| < \varepsilon$ for all $(a,b) \in \mathbb{R}^2$, and f is not uniformly continuous on $\mathbb{R}^2$. ◆

Theorem 3.8.7 established that if a function is continuous on a closed and bounded subset of $\mathbb{R}$, it must be uniformly continuous. It turns out that exactly the same is true in $\mathbb{R}^n$.

Theorem 10.4.11. *A continuous function on a sequentially compact set is uniformly continuous.*

Proof. Let f be a function that is defined and continuous on a sequentially compact set A and suppose, to the contrary, that it is not uniformly continuous. Taking the negative in Definition 10.4.8 we obtain

$$(\exists \varepsilon_0)(\forall \delta)(\exists \mathbf{x})(\exists \mathbf{a})(\forall i)|x_i - a_i| < \delta \text{ and } |f(\mathbf{x}) - f(\mathbf{a})| \geq \varepsilon_0.$$

Let ε_0 be as above and, for every $k \in \mathbb{N}$, let $\delta = 1/k$. We obtain sequences $\{\mathbf{x}_k\}$ and $\{\mathbf{a}_k\}$ in A, $\mathbf{x}_k = (x_1^{(k)}, x_2^{(k)}, \ldots, x_n^{(k)})$ and $\mathbf{a}_k = (a_1^{(k)}, a_2^{(k)}, \ldots, a_n^{(k)})$, such that

$$|x_i^{(k)} - a_i^{(k)}| < \frac{1}{k}, \quad 1 \leq i \leq n, \quad \text{and} \quad |f(\mathbf{x}_k) - f(\mathbf{a}_k)| \geq \varepsilon_0. \tag{10.6}$$

For each $k \in \mathbb{N}$, $\mathbf{x}_k \in A$, so $\{\mathbf{x}_k\}$ has a convergent subsequence $\{\mathbf{x}_{k_j}\}$, converging to $\mathbf{x} = (x_1, x_2, \ldots, x_n) \in A$. Since (10.6) holds for every $k \in \mathbb{N}$, it holds for k_j, so we have

$$|x_i^{(k_j)} - a_i^{(k_j)}| < \frac{1}{k_j}, \quad 1 \leq i \leq n, \quad \text{and} \quad |f(\mathbf{x}_{k_j}) - f(\mathbf{a}_{k_j})| \geq \varepsilon_0. \tag{10.7}$$

Further, the fact that $x_i^{(k_j)} \to x_i$ implies that $a_i^{(k_j)} \to x_i$, as $j \to \infty$. Since f is continuous on A and $\mathbf{x} \in A$, Corollary 10.3.2 allows us to conclude that both $\lim f(\mathbf{x}_{k_j}) = f(\mathbf{x})$ and $\lim f(\mathbf{a}_{k_j}) = f(\mathbf{x})$. However, this contradicts the inequality $|f(\mathbf{x}_{k_j}) - f(\mathbf{a}_{k_j})| \geq \varepsilon_0$ in (10.7). $\qquad\square$

Exercises

In Exercises 10.4.1–10.4.16 determine whether the set is closed:

10.4.1. $\{(x, y) : x + y = 1\}$.

10.4.2. $\{(x, y) : x + y > 1\}$.

10.4.3. $\{(x, y) : x + y \leq 1\}$.

10.4.4. $\{(x, y) : x \geq 0, y \geq 0\}$.

10.4.5. $\{(x, y) : x \geq 0, y > 0\}$.

10.4.6. $\{(x, y) : x \geq 0, y \leq 0\}$.

10.4.7. $\{(x, y) : x \in \mathbb{Q}\}$.

10.4.8. $\{(x, y) : x \notin \mathbb{Q}\}$.

10.4.9. $\{(x, y) : x, y \in \mathbb{N}\}$.

10.4.10. $\{(x, y) : x \in \mathbb{N}\}$.

10.4.11. $\{(1, 2), (2, 3), (0, 0)\}$.

10.4.12. Any finite set in $\mathbb{R}^n$.

10.4.13. A closed n-ball in $\mathbb{R}^n$.

10.4.14. A closed rectangle in $\mathbb{R}^n$.

10.4.15. The whole $\mathbb{R}^n$.

10.4.16. The empty set.

10.4.17. Suppose that A and B are both closed subsets of $\mathbb{R}^n$. Prove that the set $A \cup B$ is closed.

10.4.18. Suppose that sets A_k, $k \in \mathbb{N}$, are all closed. Prove that the set $\bigcap_{k=1}^{\infty} A_k$ is closed.

10.4.19. Suppose that sets A_k, $k \in \mathbb{N}$, are all closed. Give an example to show that the set $\bigcup_{k=1}^{\infty} A_k$ need not be closed.

10.4.20. Prove that every convergent sequence in $\mathbb{R}^n$ is bounded.

10.4.21. Let $\{F_n\}$ be a sequence of non-empty, closed, bounded, nested sets: $F_{n+1} \subset F_n$ for all $n \in \mathbb{N}$. Prove that there exists a point $\mathbf{a}$ that belongs to $\cap_{n \in \mathbb{N}} F_n$.

10.4.22. Let $\{F_n\}$ be a sequence of non-empty, closed, bounded, nested sets: $F_{n+1} \subset F_n$ for all $n \in \mathbb{N}$. Further, let $d(F) = \sup\{\|\mathbf{x} - \mathbf{y}\| : \mathbf{x}, \mathbf{y} \in F\}$. Suppose that $d(F_n) \to 0$, as $n \to \infty$. Prove that there exists a unique point $\mathbf{a}$ that belongs to $\cap_{n \in \mathbb{N}} F_n$.

10.4.23.* Suppose that A and B are disjoint sets in $\mathbb{R}^n$ such that A is sequentially compact and B is closed. Prove that there exists $\varepsilon > 0$ such that $\|\mathbf{a} - \mathbf{b}\| > \varepsilon$ for all $\mathbf{a} \in A$ and $\mathbf{b} \in B$.

10.4.24.* Suppose that A and B are disjoint closed sets in $\mathbb{R}^n$. Prove or disprove: there exists $\varepsilon > 0$ such that $\|\mathbf{a} - \mathbf{b}\| > \varepsilon$ for all $\mathbf{a} \in A$ and $\mathbf{b} \in B$.

10.4.25.* Let A be any set in $\mathbb{R}^n$. Prove that the set of its cluster points is a closed set.

A point $\mathbf{a}$ is a **boundary point** of a set A if every open ball $B_r(\mathbf{a})$ contains points from both A and A^c. The **boundary** of A is the set of all of its boundary points and it is denoted by ∂A.

10.4.26. Prove that a set is closed if and only if it contains its boundary.

10.4.27.* Let A be any set in $\mathbb{R}^n$. Prove that its boundary is a closed set.

10.4.28. Prove or disprove: If $\mathbf{a}$ is a boundary point of $A \subset \mathbb{R}^n$ then $\mathbf{a}$ is a cluster point of A.

10.4.29. Prove or disprove: If $A \subset B \subset \mathbb{R}^n$ then the boundary of A is a subset of the boundary of B.

10.4.30. Give an example of a set $A \subset \mathbb{R}^n$ that is not sequentially compact and a continuous function $f : A \to \mathbb{R}$ that does not attain a largest value.

10.4.31. Suppose that A and B are sequentially compact sets in $\mathbb{R}$. Prove that the set $\{(x, y) : x \in A, y \in B\}$ is sequentially compact.

10.4.32. Suppose that $A \subset \mathbb{R}^n$ is sequentially compact and let $\mathbf{b} \notin A$. Prove that there exists a point $\mathbf{a} \in A$ such that $\|\mathbf{b} - \mathbf{a}\| \leq \|\mathbf{b} - \mathbf{x}\|$, for all $\mathbf{x} \in A$.

10.4.33. Prove or disprove: if A and B are sequentially compact sets in $\mathbb{R}^n$, then $A + B$ is sequentially compact.

10.4.34. A sequence $\{\mathbf{a}^{(k)}\} \subset \mathbb{R}^n$ is a **Cauchy sequence** if, for any $\varepsilon > 0$, there exists a positive integer N such that, if $m \geq k \geq N$ then $|\mathbf{a}^{(m)} - \mathbf{a}^{(k)}| < \varepsilon$. Prove that $\{\mathbf{a}^{(k)}\}$ is a Cauchy sequence in $\mathbb{R}^n$ if and only if it is convergent.

10.4.35. Suppose that $A \subset \mathbb{R}^n$ and that $f : A \to \mathbb{R}$ satisfies Lipschitz condition: there exists $C > 0$ such that $|f(\mathbf{x}) - f(\mathbf{y})| \leq C\|\mathbf{x} - \mathbf{y}\|$, for all $\mathbf{x}, \mathbf{y} \in A$. Prove that f is uniformly continuous on A.

10.5 Open Sets in $\mathbb{R}^n$

We have seen that sequentially compact sets represent a very useful generalization of closed intervals in $\mathbb{R}$. What about open intervals? What sets in $\mathbb{R}^n$ will play their role? In Section 10.1 we have introduced some natural candidates: open n-balls and open rectangles. Are there any others? In order to answer that question, we have to decide what feature of an open interval is important to us.

Let us consider the open interval $I = (0, 1)$ and $a = 0.3$. Notice that the interval $J = (0.2, 0.4)$ contains a and it is itself contained in I. Speaking loosely, a short move away

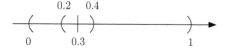

Figure 10.11: $0.3 \in J = (0.2, 0.4) \subset I = (0, 1)$.

from $a = 0.3$ in any direction results in a point that still belongs to I. This serves as the motivation for the following definition.

Definition 10.5.1. A point $\mathbf{a}$ is an **interior point** of a set $A \subset \mathbb{R}^n$ if there exists $\delta > 0$ and an open n-ball $B_\delta(\mathbf{a})$ that is completely contained in A.

Example 10.5.2. Interior points of a set.

Let $A = (0, 1) \cup [2, 3]$, $c_1 = 0.4$, $c_2 = 2.7$ and $c_3 = 3$. Then c_1 and c_2 are interior points of A, but $c_3 = 3$ is not. Figure 10.12 shows that from c_1 we can go at least 0.1 in each

Figure 10.12: $A = (0, 1) \cup [2, 3]$, $c_1 = 0.4$, $c_2 = 2.7$, $c_3 = 3$.

direction without leaving the set A, and the same works for c_2. Not for c_3! No matter how little we move to the right, the new location is not any more in A. ♦

Example 10.5.3. Interior points of a set in $\mathbb{R}^2$**.**

Let $A = [0, 1) \times [0, 1)$. Then $\mathbf{c}_1 = (0.7, 0.5)$ is an interior point of A and $\mathbf{c}_2 = (0, 0.5)$ is not.

Again, going from $\mathbf{c}_1$ in any direction no more than 0.1 keeps us within A. However, going from $\mathbf{c}_2$ to the left, no matter how short a distance, results in a point that is not in A. ♦

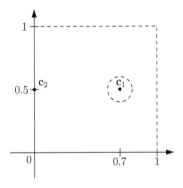

Figure 10.13: $A = [0, 1) \times [0, 1)$, $\mathbf{c}_1 = (0.7, 0.5)$, $\mathbf{c}_2 = (0, 0.5)$.

Notice that a disk centered at $\mathbf{c}_1$ with radius 0.1 lies completely inside A, while there is no such disk centered at $\mathbf{c}_2$.

In an open interval every point is an interior point. It is easy to see that the same is true for an open disk, and it is an exercise to prove it for any open n-ball. Notice that we use the adjective *open* in front of each of these sets. It is hardly a surprise that we will use it for any set that shares the same trait.

Definition 10.5.4. A set $A \subset \mathbb{R}^n$ is **open** if each of its points is an interior point.

Example 10.5.5. An open set.

Prove that the set $A = \{(x, y) \in \mathbb{R}^2 : x > 0, y > 0\}$ is open.

Solution. Let $(x_0, y_0) \in A$, and take $\delta = \min\{x_0/2, y_0/2)$. We will show that the disk with

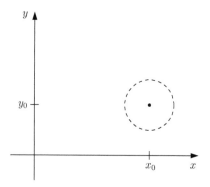

Figure 10.14: (x_0, y_0) is an interior point of A.

center (x_0, y_0) and radius δ is contained in A. So let (x, y) be a point in this disk, so that $(x - x_0)^2 + (y - y_0)^2 < \delta^2$. Then $(x - x_0)^2 < \delta^2$ which implies that

$$x_0 - x \leq |x - x_0| < \delta \leq \frac{x_0}{2}.$$

It follows that $x > x_0 - x_0/2 = x_0/2 > 0$. In a similar way we can prove that $y > 0$, so $(x, y) \in A$.   ♦

Open sets were introduced in 1914 in the book [61] by Hausdorff. Almost everything in this section can be found in this book.

FELIX HAUSDORFF (1868–1942) was a German mathematician. He came from a wealthy Jewish family in Leipzig and never needed to work to support himself. He studied at Leipzig University and received a doctorate in applications of mathematics to astronomy. After a successful Habilitation, he started teaching at Leipzig University. He wrote several papers on astronomy and optics, but he also published a literary work of 378 pages, a book on philosophy, a book of poems, and a number of articles on philosophy and literature. His 1904 comedy was performed more than 300 times between 1904 and 1912. Starting with 1904, he became interested in topology and set theory. He introduced the concept of a partially ordered set and proved many results about them. In 1910 he moved to Bonn, then 3 years later to Greifswalf. Hausdorff returned to Bonn in 1921 and worked there until 1935 when the Nazi regime forced him to retire. Faced with the prospect of being sent to a concentration camp, he and his wife committed suicide. His masterpiece on set theory and topology [61] is considered to be one of the most important mathematical books of the twentieth century. It created the theory of topological and metric spaces. In addition, he worked in measure theory (Hausdorff dimension), probability (the first correct proof of the strong law of large numbers), and functional analysis (Hausdorff-Young inequality).

There is an important relationship between open and closed sets.

Theorem 10.5.6. *A set is open if and only if its complement is closed.*

Proof. Suppose that the set A is open and let us show that its complement A^c is closed. Let $\mathbf{a}$ be a cluster point of A^c. In order to show that $\mathbf{a} \in A^c$ we will argue by contradiction. Suppose that $\mathbf{a} \notin A^c$ so $\mathbf{a} \in A$. Since A is open, there exists an n-ball B with center $\mathbf{a}$ such that $B \subset A$. Therefore B contains no points of A^c contradicting the assumption that $\mathbf{a}$ is a cluster point of A^c.

Suppose now that A^c is a closed set, and let us show that A must be open. Let $\mathbf{a} \in A$. We need to establish that $\mathbf{a}$ is an interior point of A. Suppose that this is not true. That would mean that every n-ball with center $\mathbf{a}$ contains at least one point $\mathbf{b} \in A^c$. Clearly, $\mathbf{b} \neq \mathbf{a}$, because $\mathbf{a} \in A$. By definition, $\mathbf{a}$ is a cluster point of A^c and, by assumption, A^c is closed which would imply that $\mathbf{a} \in A^c$. Yet, $\mathbf{a} \in A$. This contradiction shows that $\mathbf{a}$ is an interior point of A, so A is an open set.   □

Theorem 10.5.6 can be very useful when proving that a set is closed, because it is often easier to prove that its complement is open.

Example 10.5.7. A closed set.

Let $(a, b) \in \mathbb{R}^2$. Prove that the circle $A = \{(x, y) \in \mathbb{R}^2 : (x - a)^2 + (y - b)^2 = r^2\}$ is a closed set.

Solution. We will prove this by showing that the complement of A is open. So, let $(x_0, y_0) \in A^c$. Then

$$(x_0 - a)^2 + (y_0 - b)^2 = r_0^2 \tag{10.8}$$

for some number $r_0 \neq r$. Let B be a disk with center (x_0, y_0) and radius $|r - r_0|/2$. We will

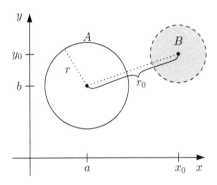

Figure 10.15: A^c is an open set.

show that $B \subset A^c$ when $r_0 > r$, and we will leave the case $r_0 < r$ as an exercise.

Suppose that $r_0 > r$ and that there exists a point $P = (x, y)$ that belongs to B and A. Since P belongs to A, its distance from (a, b) is r, and the fact that P belongs to B implies that its distance from (x_0, y_0) is less than $(r_0 - r)/2$. In other words,

$$\|(x, y) - (a, b)\| = r, \quad \|(x, y) - (x_0, y_0)\| < \frac{r_0 - r}{2}.$$

By the Triangle Inequality,

$$\|(a, b) - (x_0, y_0)\| < r + \frac{r_0 - r}{2} = \frac{r_0 + r}{2} < r_0,$$

which contradicts (10.8). ◆

Open sets are a very useful tool in calculus. Next, we present their connection with continuous functions. By definition, the *inverse image* (or the *preimage*) of a set A is the set

$$f^{-1}(A) = \{x \in \mathbb{R} : f(x) \in A\}.$$

Example 10.5.8. The inverse image of a set.

Let $f(x) = x^2$, $A = (1, 3)$, $B = (-1, 1) \cup (2, 3)$, $C = [3, 4)$. Determine the inverse image of each of these sets.

Solution. Notice that $y = x^2$ is not injective, so there is no inverse function f^{-1}. Nevertheless, it is not hard to see that $f^{-1}(A) = (-\sqrt{3}, -1) \cup (1, \sqrt{3})$, because if $x \in (-\sqrt{3}, -1) \cup (1, \sqrt{3})$ then $x^2 \in A$, and if $x \notin (-\sqrt{3}, -1) \cup (1, \sqrt{3})$ then $x^2 \notin A$.

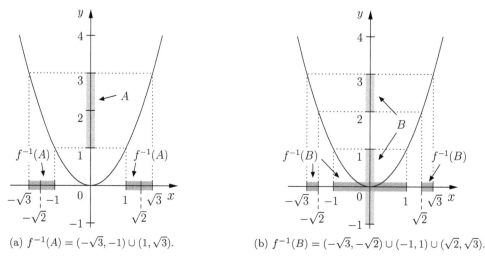

(a) $f^{-1}(A) = (-\sqrt{3}, -1) \cup (1, \sqrt{3})$.   (b) $f^{-1}(B) = (-\sqrt{3}, -\sqrt{2}) \cup (-1, 1) \cup (\sqrt{2}, \sqrt{3})$.

Figure 10.16: The inverse image.

Similarly, $f^{-1}(B) = (-\sqrt{3}, -\sqrt{2}) \cup (-1, 1) \cup (\sqrt{2}, \sqrt{3})$. Here, we mention that we are not concerned that B contains negative numbers, and x^2 cannot be negative. Finding the inverse image means splitting all real numbers in the *domain* of f into two sets: those that f maps to B, and those that it maps into B^c. Finally, $f^{-1}(C) = (-2, -\sqrt{3}] \cup [\sqrt{3}, 2)$. ♦

We make a quick observation that the sets A, B are open sets, and so are $f^{-1}(A), f^{-1}(B)$. On the other hand, C is not open, and neither is $f^{-1}(C)$. It turns out that this is always the case for continuous functions (such as $f(x) = x^2$).

Theorem 10.5.9. *Let f be a function defined on $\mathbb{R}^n$. Then f is continuous if and only if, for every open set $A \subset \mathbb{R}$, $f^{-1}(A)$ is an open set in $\mathbb{R}^n$.*

Proof. Let f be a continuous function and let A be an open set in $\mathbb{R}$. We will show that $f^{-1}(A)$ is open.

Let $\mathbf{a} \in f^{-1}(A)$. By definition, $b = f(\mathbf{a}) \in A$. Since A is open, there exists $\varepsilon > 0$ such that $B_\varepsilon(b) \subset A$. The continuity of f at $\mathbf{a}$ implies that $\lim_{\mathbf{x} \to \mathbf{a}} f(\mathbf{x}) = f(\mathbf{a}) = b$. This means that there exists $\delta > 0$ such that

$$|x_k - a_k| < \delta, \quad 1 \le k \le n, \quad \Rightarrow \quad |f(\mathbf{x}) - b| < \varepsilon. \tag{10.9}$$

We will show that $B_\delta(\mathbf{a}) \subset f^{-1}(A)$.

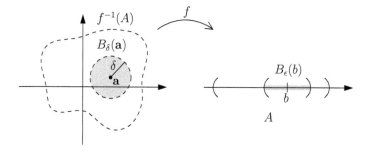

Figure 10.17: $f^{-1}(A)$ is open.

So, let $\mathbf{x} \in B_\delta(\mathbf{a})$. This means that $\|\mathbf{x} - \mathbf{a}\| < \delta$ and, all the more, $|x_k - a_k| < \delta$ for all

$1 \le k \le n$. Consequently, $|f(\mathbf{x}) - b| < \varepsilon$ which is just a way of saying that $f(\mathbf{x}) \in B_\varepsilon(b)$, so $f(\mathbf{x}) \in A$. Thus $\mathbf{x} \in f^{-1}(A)$.

Suppose now that the preimage of every open set is open, and let us prove that f is continuous. Let $\mathbf{a} \in \mathbb{R}^n$, let $b = f(\mathbf{a})$, and let $\varepsilon > 0$. We want to show that there exists $\delta > 0$ such that (10.9) holds. The set $B_\varepsilon(b)$ is open and, by assumption, so is $f^{-1}(B_\varepsilon(b))$. Since $\mathbf{a}$ belongs to this open set, it is an interior point and there exists $\delta_1 > 0$ such that $B_{\delta_1}(\mathbf{a}) \subset f^{-1}(B_\varepsilon(b))$.

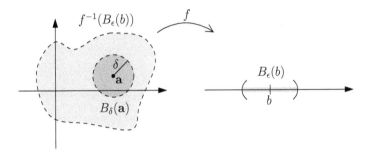

Figure 10.18: f is continuous at $\mathbf{a}$.

Let $\delta = \delta_1/\sqrt{n}$, and let $\mathbf{x} \in \mathbb{R}^n$ so that $|x_k - a_k| < \delta$, $1 \le k \le n$. Then

$$\|\mathbf{x} - \mathbf{a}\|^2 = \sum_{k=1}^{n} |x_k - a_k|^2 < \sum_{k=1}^{n} \delta^2 = n\delta^2$$

so $\|\mathbf{x} - \mathbf{a}\| < \delta\sqrt{n} = \delta_1$. It follows that $\mathbf{x} \in B_{\delta_1}(\mathbf{a}) \subset f^{-1}(B_\varepsilon(b))$ so $f(\mathbf{x}) \in B_\varepsilon(b)$ or, equivalently, $|f(\mathbf{x}) - b| < \varepsilon$. $\qquad\square$

Remark 10.5.10. The theorem does not hold for (direct) images. If f is a continuous function and A is an open set, it does not follow that its image $f(A)$ is open. Example: $f(x) = x^2$, $A = (-1, 1)$.

Theorem 10.5.9 is true when the domain of f is a proper subset of $\mathbb{R}^n$, but the formulation is not as crisp.

Example 10.5.11. The inverse image of an open set need not be open.

Let $f(x) = \sqrt{x}$, $A = (-1, 1)$. Prove that $f^{-1}(A)$ is not an open set even though f is continuous.

Solution. By definition, $x \in f^{-1}(A)$ if and only if $f(x) \in A$ and here it means if and only if $\sqrt{x} \in (-1, 1)$. It is not hard to see that $f^{-1}((-1, 1)) = [0, 1)$ which is not an open set. $\blacklozenge$

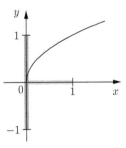

Figure 10.19: $f^{-1}((-1, 1)) = [0, 1)$.

Remark 10.5.12. The set $f^{-1}(A)$ in Example 10.5.11 is not open but it is a *relatively* open subset of the domain of f, meaning that it is the intersection of an open set and the domain of f. Namely, $[0,1) = (-\infty, 1) \cap [0, +\infty)$. (The domain of f is $[0, +\infty)$.)

Example 10.5.11 and Remark 10.5.12 suggest the formulation of the theorem.

Theorem 10.5.13. *Let f be a function defined on a set $A \subset \mathbb{R}^n$. Then f is continuous if and only if, for every open set $G \subset \mathbb{R}$, $f^{-1}(G)$ is a relatively open subset of A.*

The proof requires a modification of the proof of Theorem 10.5.9, and we leave it as an exercise.

If we look once again at Example 10.5.8 we may notice the following interesting phenomenon. The complements of the sets A, B, C are

$$A^c = (-\infty, 1] \cup [3, +\infty), \quad B^c = (-\infty, -1] \cup [1, 2] \cup [3, +\infty),$$
$$C^c = (-\infty, 3) \cup [4, +\infty).$$

Their inverse images are

$$f^{-1}(A^c) = (-\infty, -\sqrt{3}] \cup [-1, 1] \cup [\sqrt{3}, +\infty),$$
$$f^{-1}(B^c) = (-\infty, -\sqrt{3}] \cup [-\sqrt{2}, -1] \cup [1, \sqrt{2}] \cup [\sqrt{3}, +\infty),$$
$$f^{-1}(C^c) = (-\infty, -2] \cup (-\sqrt{3}, \sqrt{3}) \cup [2, +\infty).$$

The sets A^c, B^c are closed and so are $f^{-1}(A^c), f^{-1}(B^c)$. The set C^c is not closed, and neither is $f^{-1}(C^c)$. Coincidence?

Theorem 10.5.14. *Let f be a function defined on $\mathbb{R}^n$. Then f is continuous if and only if, for every closed set $A \subset \mathbb{R}$, $f^{-1}(A)$ is a closed set in $\mathbb{R}^n$.*

We leave the proof as an exercise, as well as the formulation and the proof of the case when f is defined on a proper subset of $\mathbb{R}^n$.

We make a caveat analogous to Remark 10.5.10: Theorem 10.5.14 does not hold for direct images. However, if a set is both closed and bounded, then its direct image (under a continuous function) is closed and bounded.

Theorem 10.5.15 (Preservation of Compactness). *Let f be a continuous function. If a set $A \subset \mathbb{R}^n$ is sequentially compact, then so is $f(A)$.*

Proof. By Theorem 10.4.5, $f(A)$ is a bounded set. Thus, we need to prove that it is closed. So let b be a cluster point of $f(A)$, and let us show that $b \in f(A)$. By definition, for any $\delta > 0$, the ball $B_\delta(b)$ contains a point $y \in f(A)$. By taking $\delta_k = 1/k$, we obtain a sequence $\{y_k\} \subset f(A)$ satisfying $|y_k - b| < 1/k$. Since $y_k \in f(A)$, there exists $\mathbf{x}_k \in A$ such that $y_k = f(\mathbf{x}_k)$.

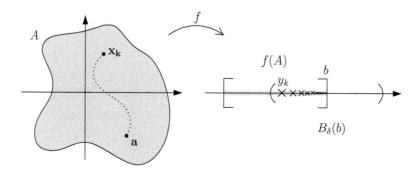

Figure 10.20: As $\mathbf{x_k} \to \mathbf{a}$, $y_k = f(\mathbf{x_k}) \to f(\mathbf{a}) = b$.

The fact that A is sequentially compact implies that the sequence $\{\mathbf{x}_k\}$ has a subsequence $\{\mathbf{x}_{k_j}\}$ converging to a point $\mathbf{a} \in A$. The assumption that f is continuous now implies that $f(\mathbf{x}_{k_j}) \to f(\mathbf{a})$, $j \to \infty$. On the other hand, $f(\mathbf{x}_{k_j}) = y_{k_j} \to b$, $j \to \infty$, so $b = f(\mathbf{a})$ and, hence, $b \in f(A)$. $\qquad\square$

Exercises

In Exercises 10.5.1–10.5.12 determine whether the set is open and find its interior:

10.5.1. $\{(x, y) : x + y = 1\}$.

10.5.2. $\{(x, y) : x + y > 1\}$.

10.5.3. $\{(x, y) : x \geq 0, y > 0\}$.

10.5.4. $\{(x, y) : x > 0, y < 0\}$.

10.5.5. $\{(x, y) : 0 < x < 1, y = 0\}$.

10.5.6. $\{(x, y) : 0 < x < 1\}$.

10.5.7. $\{(x, y) : x \in \mathbb{Q}\}$.

10.5.8. $\{(x, y) : x \notin \mathbb{Q}\}$.

10.5.9. An open n-ball in $\mathbb{R}^n$.

10.5.10. An open rectangle in $\mathbb{R}^n$.

10.5.11. The whole $\mathbb{R}^n$.

10.5.12. The empty set.

10.5.13. In Example 10.5.7 prove the case $r_0 < r$.

10.5.14. Suppose that sets A and B are both open. Prove that the set $A \cup B$ is open.

10.5.15. Suppose that sets A_k, $k \in \mathbb{N}$, are all open. Prove that the set $\bigcup_{k=1}^{\infty} A_k$ is open.

10.5.16. Suppose that sets A_k, $k \in \mathbb{N}$, are all open. Give an example to show that the set $\bigcap_{k=1}^{\infty} A_k$ need not be open.

10.5.17. Suppose that A and B are open sets in $\mathbb{R}$. Prove that the set $\{(x, y) : x \in A, y \in B\}$ is open.

10.5.18.* Suppose that A and B are sets in $\mathbb{R}$ and that the set $\{(x, y) : x \in A, y \in B\}$ is open. Prove or disprove: A and B are both open.

10.5.19.* Suppose that A is an open set in $\mathbb{R}^n$. Prove that the interior of A is open.

10.5.20. Prove or disprove: If $A \subset B \subset \mathbb{R}^n$ then the interior of A is a subset of the interior of B.

10.5.21. Prove Theorem 10.5.13.

10.5.22. Prove Theorem 10.5.14.

10.5.23. Formulate and prove a result analogous to Theorem 10.5.13 with closed sets instead of open.

10.5.24. Show by example that if f is a continuous function and A is a closed set, $f(A)$ need not be closed.

10.5.25. Suppose that G is an open set in $\mathbb{R}^n$, and that $m < n$. Prove that the set

$$\{(x_1, x_2, \ldots, x_m) \in \mathbb{R}^m : (x_1, x_2, \ldots, x_m, 0, 0, \ldots, 0) \in G\}$$

is an open set in $\mathbb{R}^m$.

10.5.26. Let $f : \mathbb{R}^n \to \mathbb{R}$. Prove that the set of the points where f is continuous is a G_δ set, i.e., it is an infinite intersection of open sets. Prove that the set of the points where f is not continuous is a F_σ set, i.e., it is an infinite union of closed sets.

10.5.27.* For any set $A \subset \mathbb{R}^n$, prove that the set of all cluster points of A is closed by showing that its complement is open.

10.5.28.* For any open set $A \subset \mathbb{R}^n$, prove that the set of all cluster points of A equals the closure of A.

10.5.29. Let $A \subset \mathbb{R}^n$ and suppose that the closure of A equals $\mathbb{R}^n$. Prove that each point of $\mathbb{R}^n$ is a cluster point of A.

10.5.30.* Let A be any set in $\mathbb{R}^n$. Prove that the closure of A is the union of A and the set of the cluster points of A.

10.5.31.* Let A be any set in $\mathbb{R}^n$. Prove that A is open if and only if it has an empty intersection with its boundary.

10.5.32. Let A be any set in $\mathbb{R}^n$. Prove that A is closed if and only if it contains its boundary.

10.5.33.* Let A be any set in $\mathbb{R}^n$, let $\mathbf{c} \in A$, and let $A + \mathbf{c} = \{\mathbf{a} + \mathbf{c} : \mathbf{a} \in A\}$. Prove that: (a) A is closed if and only if $A + \mathbf{c}$ is closed; (b) A is open if and only if $A + \mathbf{c}$ is open.

10.5.34.* Suppose that A and B are open sets in $\mathbb{R}^n$. Prove or disprove: the set $A + B$ is open. ($A + B = \{a + b : a \in A, b \in B\}$.)

10.5.35.* Suppose that A and B are open sets in $\mathbb{R}^n$. Prove or disprove: the set AB is open. ($AB = \{ab : a \in A, b \in B\}$.)

10.5.36. Let $f : \mathbb{R}^n \to \mathbb{R}$ be continuous. Prove that the set $\{\mathbf{x} \in \mathbb{R}^n : f(\mathbf{x}) = 0\}$ is closed.

10.5.37. Let $f : \mathbb{R}^n \to \mathbb{R}$ and $g : \mathbb{R}^n \to \mathbb{R}$ be both continuous. Prove that the set $\{\mathbf{x} \in \mathbb{R}^n : f(\mathbf{x}) = g(\mathbf{x}) = 0\}$ is closed.

10.5.38. Let $f : \mathbb{R}^n \to \mathbb{R}$ be continuous, and let $a, b \in \mathbb{R}$. Prove that the set $\{\mathbf{x} \in \mathbb{R}^n : a < f(\mathbf{x}) < b\}$ is open.

10.5.39. Let A be a sequentially compact set in $\mathbb{R}^2$ and define $B = \{x \in \mathbb{R} : \text{there exists } y \in \mathbb{R} \text{ such that } (x, y) \in A\}$. Prove that B is sequentially compact.

10.5.40. Give an example of a continuous function $f : \mathbb{R}^n \to \mathbb{R}$ and a sequentially compact set A in $\mathbb{R}$ such that $f^{-1}(A)$ is not sequentially compact.

10.6   Intermediate Value Theorem

In this section we will consider the Intermediate Value Theorem (Theorem 3.9.1, page 95), and we will explore its generalization in the multivariable setting. Let us assume that f is a function of two variables. If $f(a_1, a_2) < 0$ then the point $(a_1, a_2, f(a_1, a_2))$ lies below the xy-plane. Similarly, the condition $f(b_1, b_2) > 0$ means that the point $(b_1, b_2, f(b_1, b_2))$ lies above the xy-plane. The graph of f is a surface, and both $(a_1, a_2, f(a_1, a_2))$ and $(b_1, b_2, f(b_1, b_2))$ lie in this surface. If f is continuous, this surface should have some common point with the xy-plane, which would mean that there exists a point (c_1, c_2) such that $f(c_1, c_2) = 0$.

Although we cannot visualize the situation when f is a function of more than two variables, the result should still be true. However, Remark 3.9.4 reminds us that we have to worry about the domain.

Example 10.6.1. A continuous function that does not have the intermediate value property.

Prove that the function $f(x, y) = \begin{cases} -1, & \text{if } x^2 + y^2 < 1 \\ 2, & \text{if } x^2 + y^2 > 4 \end{cases}$ is continuous but there is no point (c_1, c_2) such that $f(c_1, c_2) = 0$.

Solution. The domain of f consists of the open unit disk and the outside of the disk centered at the origin and of radius 2 (see Figure 10.22). It is obvious that f is continuous at every point of its domain. Further, $f(0, 0) = -1$ and $f(2, 3) = 2$ (because $2^2 + 3^2 = 13 > 4$). Yet, there is no point (c_1, c_2) such that $f(c_1, c_2) = 0$. ♦

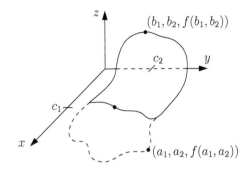

Figure 10.21: There exists a point (c_1, c_2) such that $f(c_1, c_2) = 0$.

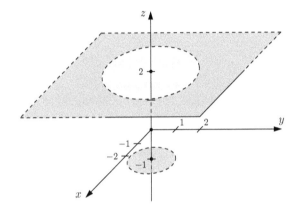

Figure 10.22: There is no point (c_1, c_2) such that $f(c_1, c_2) = 0$.

The problem here is that the domain consists of two separate sets. Therefore, in order to ensure the intermediate value property, we will require that the domain be *polygonally connected*.

Definition 10.6.2. A set $A \subset \mathbb{R}^n$ is **polygonally connected** if, for any two points $P, Q \in A$ there exists a polygonal line that connects them. More precisely, there exists a positive integer n and points $P_0 = P, P_1, \ldots, P_n = Q$ all in A, such that each line segment $P_i P_{i+1}$, $0 \le i \le n-1$, completely lies within A.

Example 10.6.3. A polygonally connected set.

Prove that the set $A = [a, b] \subset \mathbb{R}$ is polygonally connected.

Solution. It is easy to see that any 2 points in the interval $[a, b]$ can be connected with a straight line segment that lies entirely in $[a, b]$. Also, the same is true if the interval does not include one or both endpoints, or if it is an infinite interval. It is not hard to prove that these are the only polygonally connected sets in $\mathbb{R}$. ◆

Example 10.6.4. A polygonally connected set.

Prove that the set $A = [a, b] \times [c, d] \subset \mathbb{R}^2$ is polygonally connected.

Solution. This is a rectangle in $\mathbb{R}^2$ and it is quite obvious that any two points in A can be connected with a straight line segment that lies entirely in A. ◆

Example 10.6.5. A polygonally connected set.

Prove that the set $A = \{(x, y) \in \mathbb{R}^2 : 1 \le x^2 + y^2 \le 4\}$ is polygonally connected.

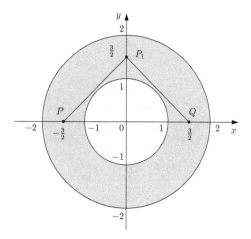

Figure 10.23: The annulus $A(\mathbf{0}; 1, 2)$ is polygonally connected.

Solution. This is an *annulus* with the *inner radius* 1 and the *outer radius* 2 (Figure 10.23). It is a polygonally connected set but it may not be possible to connect every two points with a single line segment. For example, if $P = (3/2, 0)$ and $Q = (-3/2, 0)$, the straight line segment PQ goes through the origin, which does not belong to A. Nevertheless, we can define $P_1 = (0, 3/2)$ and notice that $P_1 \in A$ and that both line segments PP_1 and P_1Q lie entirely in A. A similar argument can be used for any 2 points in A. ◆

Example 10.6.5 may be geometrically clear, but we need to be able to verify that a set is polygonally connected even in higher dimensions. The idea is to represent line segments by equations and then verify algebraically that each point on the segment belongs to the set. If $P = (a_1, a_2, \ldots, a_n)$ and $Q = (b_1, b_2, \ldots, b_n)$ then the **line segment** PQ is the set of points

$$\{(x_1, x_2, \ldots, x_n) : x_i = a_i + t(b_i - a_i), \ 1 \le i \le n, \ 0 \le t \le 1\}.$$

Example 10.6.6. Two points can be connected by a polygonal line.

We will show that the points P, Q in Example 10.6.5 can be connected by a polygonal line.

Solution. The equations of the line segment PP_1 are

$$x = \frac{3}{2}t - \frac{3}{2}, \quad y = \frac{3}{2}t, \quad 0 \le t \le 1.$$

Let us verify the claim that every point of PP_1 lies in the annulus A. We need to show that, for every $t \in [0, 1]$, the point $(\frac{3}{2}t - \frac{3}{2}, \frac{3}{2}t)$ lies in A, i.e., satisfies the inequalities

$$1 \le \left(\frac{3}{2}t - \frac{3}{2}\right)^2 + \left(\frac{3}{2}t\right)^2 \le 4. \tag{10.10}$$

Since

$$\left(\frac{3}{2}t - \frac{3}{2}\right)^2 + \left(\frac{3}{2}t\right)^2 = \frac{9}{4}t^2 - \frac{9}{2}t + \frac{9}{4} + \frac{9}{4}t^2 = \frac{1}{4}\left(18t^2 - 18t + 9\right),$$

inequalities (10.10) can be written as $4 \leq 18t^2 - 18t + 9 \leq 16$ or a pair of inequalities

$$18t^2 - 18t + 5 \geq 0, \quad 18t^2 - 18t - 7 \leq 0.$$

Both quadratic functions have a minimum at $t = 1/2$. For the first one, that minimum equals $18/4 - 18/2 + 5 = 1/2 > 0$. Since the minimum is positive, so is any other value of that function. For the second, the maximum is attained at the endpoints, and it equals -7 at both $t = 0$ and $t = 1$. Thus, it is less than -7 for any $t \in [0, 1]$. ◆

Example 10.6.7. A proof that a set is polygonally connected.

We will show that the set $A = \{(x, y, z, w) \in \mathbb{R}^4 : x^2 + y^2 + z^2 + w^2 \leq 1\}$ is polygonally connected.

Solution. This is the closed unit 4-ball, and we will show that it is polygonally connected. Let $P = (a_1, a_2, a_3, a_4)$, $Q = (b_1, b_2, b_3, b_4)$ be in A. The equations of the line segment PQ are

$$x = a_1 + t(b_1 - a_1), \; y = a_2 + t(b_2 - a_2), \; z = a_3 + t(b_3 - a_3), \; w = a_4 + t(b_4 - a_4), \; 0 \leq t \leq 1.$$

We need to show that, for every $t \in [0, 1]$,

$$(a_1 + t(b_1 - a_1))^2 + (a_2 + t(b_2 - a_2))^2 + (a_3 + t(b_3 - a_3))^2 + (a_4 + t(b_4 - a_4))^2 \leq 1.$$

This inequality can be written as $at^2 + bt + c \leq 0$ where

$$a = (b_1 - a_1)^2 + (b_2 - a_2)^2 + (b_3 - a_3)^2 + (b_4 - a_4)^2,$$
$$b = 2a_1(b_1 - a_1) + 2a_2(b_2 - a_2) + 2a_3(b_3 - a_3) + 2a_4(b_4 - a_4),$$
$$c = a_1^2 + a_2^2 + a_3^2 + a_4^2 - 1.$$

Notice that the values $t = 0$ and $t = 1$ correspond to the points P and Q which belong to A, so the quadratic function $at^2 + bt + c$ attains negative values at $t = 0$ and $t = 1$. If it attained a positive value for some $t \in (0, 1)$ then it would have a local maximum which is impossible because $a > 0$. ◆

Now we can state the n-dimensional version of the Intermediate Value Theorem.

Theorem 10.6.8. *Let f be a continuous function on a polygonally connected domain $A \subset \mathbb{R}^n$, and let $P, Q \in A$. If $f(P) < 0$ and $f(Q) > 0$ then there exists $M \in A$ such that $f(M) = 0$.*

Proof. Taking advantage of the fact that A is polygonally connected, we construct points $P_0 = P, P_1, \ldots, P_n = Q$ in A, such that each line segment $P_i P_{i+1}$, $0 \leq i \leq n - 1$, completely lies in A. If f vanishes at any of the vertices, we will take it as M. Otherwise, there has to be a segment $P_k P_{k+1}$ such that $f(P_k) < 0$ and $f(P_{k+1}) > 0$. Let $P_k = (a_1, a_2, \ldots, a_n)$ and $P_{k+1} = (b_1, b_2, \ldots, b_n)$. The equations of the segment $P_k P_{k+1}$ are then

$$x_i = a_i + t(b_i - a_i), \quad 1 \leq i \leq n, \quad 0 \leq t \leq 1. \tag{10.11}$$

Consider the function

$$F(t) = f(a_1 + t(b_1 - a_1), a_2 + t(b_2 - a_2), \ldots, a_n + t(b_n - a_n)), \quad 0 \leq t \leq 1. \tag{10.12}$$

By Theorem 10.3.6, F is a continuous function on $[0, 1]$. Further,

$$F(0) = f(a_1, a_2, \ldots, a_n) = f(P_k) < 0, \quad \text{and} \quad F(1) = f(b_1, b_2, \ldots, b_n) = f(P_{k+1}) > 0.$$

By Theorem 3.9.1, there exists $t_0 \in (0, 1)$ such that $F(t_0) = 0$. If

$$M = a_1 + t_0(b_1 - a_1), a_2 + t_0(b_2 - a_2), \ldots, a_n + t_0(b_n - a_n),$$

then M belongs to the segment $P_k P_{k+1}$ (and, hence, to A) and $f(M) = 0$. □

We have proved a result that can be further improved. For example, the unit circle is not polygonally connected. Yet, Theorem 10.6.8 holds for this set. Indeed, the difference in the proof would appear in (10.11) where we would need equations of the circle $x_i = a_i + r\cos t$. In turn, that would affect the definition of the function F in (10.12). Since we only need F to be continuous, the modified formula would still work. In fact, any path defined by a continuous parametrization $x_i = \varphi_i(t)$ would do.

Definition 10.6.9. A set $A \subset \mathbb{R}^n$ is **path connected** if, for any two points $P, Q \in A$ there exists a continuous curve that connects them. More precisely, there exists a continuous function $\gamma : [0,1] \to A$ such that $\gamma(0) = P$ and $\gamma(1) = Q$.

Can we do even better? Our motivation is to select a set A in such a fashion that it has the **Intermediate Value Property**. By that we mean that if f is a continuous function on A, if $\mathbf{a}, \mathbf{b} \in A$, and if $f(\mathbf{a}) < c < f(\mathbf{b})$, then c belongs to $f(A)$. Instead of describing such sets A, it is more profitable to consider the sets for which the theorem fails. So, suppose that $c \notin f(A)$. What does that say about A? Let

$$A_1 = f^{-1}((-\infty, c)), \quad A_2 = f^{-1}((c, +\infty)).$$

By Theorem 10.5.13, the sets A_1 and A_2 are relatively open subsets of A, i.e., there exist open sets B, C such that $A_1 = A \cap B$ and $A_2 = A \cap C$. Further, the sets A_1 and A_2 are disjoint and they cover A (meaning that $A \subset A_1 \cup A_2$).

Definition 10.6.10. A set $A \subset \mathbb{R}^n$ is **disconnected** if there exist open sets B, C such that $A \cap B$ and $A \cap C$ are non-empty disjoint sets and $A \subset B \cup C$. A set is **connected** if it is not disconnected.

Example 10.6.11. A connected set.

Prove that the set $A = [0,1] \cup [2,3]$ is disconnected.

Solution. Let $B = (-\infty, \frac{3}{2})$, $C = (\frac{3}{2}, +\infty)$. Then B, C are open sets and $A \subset B \cup C$. Further, the sets $A_1 = A \cap B = [0,1]$ and $A_2 = A \cap C = [2,3]$ are both non-empty and they

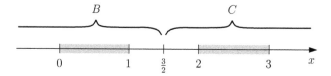

Figure 10.24: The set $A = [0,1] \cup [2,3]$ is disconnected.

are disjoint. Consequently, A is a disconnected set.   ♦

Connected sets are precisely the sets with the Intermediate Value Property.

Theorem 10.6.12. *A set $A \subset \mathbb{R}^n$ is connected if and only if it has the Intermediate Value Property.*

Proof. The implication "connected $\Rightarrow$ IVP" has been outlined above (albeit in the contrapositive and for $\mathbb{R}$), so we leave it as an exercise. In order to prove the converse, we will again use the contrapositive. Namely, we will assume that A is disconnected and we will show that it does not have the Intermediate Value Property. By definition, there exist open sets B, C such that $A \cap B$ and $A \cap C$ are non-empty disjoint sets and $A \subset B \cup C$. Let f be a function defined on A by

$$f(\mathbf{x}) = \begin{cases} 1, & \text{if } \mathbf{x} \in A \cap B \\ 0, & \text{if } \mathbf{x} \in A \cap C. \end{cases}$$

Clearly, $\frac{1}{2} \notin f(A)$, so it suffices to show that f is continuous. By Theorem 10.5.13, we will accomplish this goal if, for every open set $G \subset \mathbb{R}$, $f^{-1}(G)$ is relatively open in A. However, for any open set G, the set $f^{-1}(G)$ can be just one of the four possibilities: A, $A \cap B$, $A \cap C$, or $\emptyset$, depending on whether G contains both 0 and 1, or only one of them, or neither one. Since each of these sets is relatively open in A, f is continuous, and the proof is complete. $\qquad\square$

So far we have seen three different definitions of connectedness. We have started with polygonally connected sets, and then briefly mentioned *path connected* sets, where a polygonal line is replaced by any path. Since a polygonal line is a path, every polygonally connected set is path connected as well. The example of a circle shows that these two classes are not the same. Next, we introduced connected sets. How do they compare to path connected sets?

Theorem 10.6.13. *If a set is path connected then it is connected.*

We leave the proof as an exercise. Instead, we will show that the converse is not true.

Example 10.6.14. A set that is connected but not path connected.

Let $A_1 = \{(0, y) \in \mathbb{R}^2 : -1 \leq y \leq 1\}$, $A_2 = \{(x, \sin \frac{1}{x}) \in \mathbb{R}^2 : x > 0\}$, and $A = A_1 \cup A_2$. Prove that the set A is connected but not path connected.

Solution. First we will show that A is not path connected. Suppose, to the contrary, that it is. Then there exists a continuous function $\varphi : [0, 1] \to A$ such that $\varphi(0) = (\frac{1}{\pi}, 0)$, $\varphi(1) = (0, 0)$, and $\varphi(t) \in A$ for $0 < t < 1$. We define two sequences of real numbers by

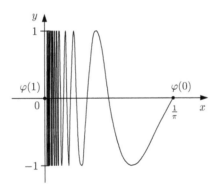

Figure 10.25: $\varphi : [0, 1] \to A$.

$x_n = \frac{2}{(4n+1)\pi}$, $y_n = \frac{2}{(4n+3)\pi}$, and two sequences of points in $\mathbb{R}^2$ by:

$$P_n = (x_n, 1), \quad Q_n = (y_n, -1).$$

All these points lie on the path between $\varphi(0)$ and $\varphi(1)$, so there exist two sequences $\{t_n\}, \{r_n\} \subset [0, 1]$ such that

$$P_n = \varphi(t_n), \quad Q_n = \varphi(r_n).$$

Let

$$c = \inf\{t \in [0, 1] : \varphi(t) \in A_1\}.$$

Since $P_n, Q_n \in A_2$, we see that $t_n, r_n \in [0, c]$. By the Bolzano–Weierstrass Theorem, there exist convergent subsequences $t_{n_k} \to t$ and $r_{n_k} \to r$. Since $t_n, r_n \in [0, c]$, we have that $t, r \in [0, c]$. However, they cannot be less than c. Indeed, φ is continuous, so

$$\varphi(t) = \lim \varphi(t_{n_k}) = \lim P_{n_k} = \lim(x_{n_k}, 1) = (0, 1), \quad \text{and}$$
$$\varphi(r) = \lim \varphi(r_{n_k}) = \lim Q_{n_k} = \lim(y_{n_k}, -1) = (0, -1),$$

whence φ maps t and r to A_1. It follows that $t = r = c$ which contradicts the fact that $\varphi(t) \neq \varphi(r)$. Thus, A is not path connected.

Now, we will show that A is connected. Suppose that it is disconnected, and that the open sets B, C are as in Definition 10.6.10, i.e., $A \cap B$ and $A \cap C$ are non-empty disjoint sets and $A \subset B \cup C$. Consider now $A_1 \cap B$ and $A_1 \cap C$. Clearly, $A_1 \subset B \cup C$, and the sets $A_1 \cap B$ and $A_1 \cap C$ are disjoint. If they were both non-empty, it would follow that A_1 is disconnected, which is obviously not true. Thus, one of the sets $A_1 \cap B$ and $A_1 \cap C$ is empty, which means that A_1 is completely contained in either B or C. We will assume that $A_1 \subset B$. A similar consideration shows that A_2 must be completely contained in either B or C. It cannot be B, because then we would have $A \subset B$ and $A \cap C$ would be empty. Thus, we have that

$$A_1 \subset B, \quad A_2 \subset C.$$

What is wrong with this picture? Since B is open, its complement is closed. Further, no points of A_2 belong to B, so they all belong to B^c. In particular, $P_n = (\frac{2}{(4n+1)\pi}, 1) \in B^c$ for all $n \in \mathbb{N}$. The sequence P_n converges to $(0, 1)$, so $(0, 1)$ is a cluster point of B^c, and it follows that $(0, 1) \in B^c$. This is a contradiction, since $(0, 1) \in A_1 \subset B$. Thus, A is a connected set.

$\blacklozenge$

The concept of connectedness has its roots in the attempts of ancient mathematicians to explain Zeno's paradoxes. In modern times, it can be found in the 1817 Bolzano's proof of the Intermediate Value Theorem (Theorem 3.9.1), in Cantor's 1883 paper [14] (see Exercise 10.6.30), Jordan's *Cours d'Analyse* (second edition) in 1893 (see Exercise 10.6.31), and, in 1904, in a paper by Arthur Moritz Schoenflies (1853–1928), a German mathematician and a student of Weierstrass (see Exercise 10.6.32). All these definitions, from Bolzano to Schoenflies, were restricted to closed sets. Finally, in the span of two months (December 1905–January 1906) two definitions emerged that were applicable to general sets. Nels Johann Lennes (1874–1951) was a Norwegian mathematician who graduated at the University of Chicago, and spent much of his life as Professor of Mathematics and Chairman of the Department at Montana State University. Frigyes Riesz (1880–1956) was a Hungarian mathematician, who did some of the fundamental work in developing functional analysis. Their definitions state: A set of points A is a connected set if at least one of any two complementary subsets contains a cluster point of the other set. Unaware of their results, Hausdorff published his book [61] in 1914 and presented the same definition. Only in the 3rd edition (published in 1944) did Hausdorff call attention to Lennes' definition. More on the evolution of this concept can be found in the excellent article [111].

Exercises

In Exercises 10.6.1–10.6.9 determine whether the set is connected, and if so whether it is path connected or polygonally connected:

10.6.1. $\{(x, y) : x + y = 1\}$.   10.6.2. $\{(x, y) : x + y > 1\}$.

10.6.3. $\{(x, y) : x^2/4 + y^2/9 = 1\}$.   10.6.4. $\{(x, y) : x^2 - y^2 = 1\}$.

10.6.5. $\{(x, y) : x \in \mathbb{Q} \text{ or } y \in \mathbb{Q}\}$.   10.6.6. $\{x : x \notin \mathbb{Q}\}$.

10.6.7. $\{x : x \in \mathbb{Z}\}$.   10.6.8. $\{(x, y) : |x| + |y| = 1\}$.

10.6.9. $\{(x, y) : xy \geq 1 \text{ and } x > 1\} \cup \{(x, y) : xy \leq 1 \text{ and } x \leq 1\}$.

10.6.10. Prove Theorem 10.6.13.

10.6.11.* Prove that an open set $A \subset \mathbb{R}^n$ is connected if and only if it is polygonally connected.

10.6.12. Prove the Preservation of Connectedness Theorem: if a set $A \subset \mathbb{R}^n$ is connected, and f is a continuous function defined on A, then $f(A)$ is connected. Does the result remain true if "connected" is replaced with "path connected"? With "polygonally connected"?

10.6.13. Let $f : \mathbb{R}^n \to \mathbb{R}$ be a continuous function and $A \subset \mathbb{R}$ a connected set. Prove or disprove: $f^{-1}(A)$ is connected.

10.6.14. Suppose that $f : [0,1] \times [0,1] \to [0,1]$ is a bijection. Prove that it cannot be continuous.

10.6.15. Let $A \subset \mathbb{R}^n$ and let f map A onto $\{1,2\}$. Prove that A is connected if and only if f is not continuous on A.

Two sets are **separated** if they are disjoint and neither one contains a cluster point of the other one.

10.6.16. Prove that a set $A \subset \mathbb{R}^n$ is connected if and only if it cannot be written as a union of two non-empty separated sets.

10.6.17. Let $A, B \subset \mathbb{R}^n$ be separated sets and suppose that C is a connected set contained in $A \cup B$. Prove that $C \subset A$ or $C \subset B$.

10.6.18. Let A be a set in $\mathbb{R}^n$ and suppose that for any $P, Q \in A$ there exists a connected set B such that $P, Q \in B \subset A$. Prove that A is connected.

10.6.19. Suppose that the sets $A, B \subset \mathbb{R}^n$ are connected and that they are not disjoint. Prove that $A \cup B$ is connected.

10.6.20. Suppose that the sets $A, B \subset \mathbb{R}^n$ are connected and that they are not disjoint. Prove or disprove: $A \cap B$ is connected.

10.6.21. Suppose that the set $A \subset \mathbb{R}^n$ is connected. Prove or disprove: The interior of A is connected.

10.6.22. Suppose that the set $A \subset \mathbb{R}^n$ is connected. Prove or disprove: The closure of A is connected.

10.6.23. Suppose that the sets $A_1, A_2 \subset \mathbb{R}^n$ are closed and that both $A_1 \cup A_2$ and $A_1 \cap A_2$ are connected. Prove that A_1 and A_2 are connected. Show that the assumption that A_1 and A_2 are closed cannot be omitted.

10.6.24. Let A be a connected set in $\mathbb{R}^n$ and suppose that $A \subset B \subset \overline{A}$. Prove that B is connected.

10.6.25. Suppose that the sets $A, B \subset \mathbb{R}^n$ are connected. Prove that their Cartesian product $A \times B$ is connected. Does the result remain true if "connected" is replaced with "path connected"? With "polygonally connected"?

10.6.26.* Suppose that the sets $A \subset \mathbb{R}^m$, $B \subset \mathbb{R}^n$ and that their Cartesian product $A \times B$ is connected. Prove that A is connected. Does the result remain true if "connected" is replaced with "path connected"? With "polygonally connected"?

10.6.27. Let $A \subset \mathbb{R}^n$ be a connected set with more than one point. Prove that every point of A is an accumulation point of A.

10.6.28. Prove or disprove: if f is defined on $A \subset \mathbb{R}^n$ and if its graph is connected then f is continuous.

10.6.29.* Prove or disprove: if f is defined on $A \subset \mathbb{R}^n$ and if its graph is path connected then f is continuous.

10.6.30. Cantor defined a connected set $A \in \mathbb{R}^n$ as having the following property:
   For any $\mathbf{a}, \mathbf{b} \in A$, and any $\varepsilon > 0$, there exists a finite number of points
   $\mathbf{a} = \mathbf{t}_0, \mathbf{t}_1, \mathbf{t}_2, \ldots, \mathbf{t}_n = \mathbf{b} \in A$ such that $\|\mathbf{t}_k - \mathbf{t}_{k-1}\| < \varepsilon$, $k = 1, 2, \ldots, n$.

(a) Prove that, if A is a closed and bounded set with Cantor's property, then it is connected.

(b) Show that neither the word "closed" nor "bounded" can be omitted in (a).

10.6.31.* Jordan (page 441) defined the distance between closed sets $A, B \in \mathbb{R}^n$ as $d(A, B) = \inf\{\|\mathbf{a} - \mathbf{b}\| : \mathbf{a} \in A, \mathbf{b} \in B\}$. Then, he defined *separated* sets as those closed sets A, B such that $d(A, B) > 0$. Finally, he defined a "single component" set $A \in \mathbb{R}^n$ as one that cannot be decomposed in 2 closed separated sets.

(a) Prove that, if A is a closed and bounded set with the "single component" property, then it is connected.

(b) Show that neither "closed" nor "bounded" can be omitted in (a).

10.6.32. A set A in $\mathbb{R}^n$ is **perfect** if it is a closed set with no isolated points. Schoenflies defined a perfect connected set $A \in \mathbb{R}^n$ as having the following property:

The set A cannot be decomposed into two non-empty disjoint subsets of A each of which is perfect.

Prove that, if A is a bounded perfect set, then it has the Schoenflies property if and only if it has Jordan's single component property.

Remark 10.6.15. The definition of Schoenflies does not include the concept of distance.

10.7 Compact Sets

In Section 3.8 we have proved Theorem 3.8.7: a continuous function on an interval $[a, b]$ is uniformly continuous. We have used a proof by contradiction and such proofs, although effective, frequently lack transparency. In this section we will try to shed more light on the phenomenon exhibited in this theorem.

Let us look again at the definition of a uniformly continuous function. If $\varepsilon > 0$ is given, we are looking for $\delta > 0$ such that $|x - y| < \delta$ implies $|f(x) - f(y)| < \varepsilon$. Since f is continuous at every point $c \in [a, b]$, we can certainly select $\delta = \delta(c)$ such that $|x - c| < \delta$ implies $|f(x) - f(c)| < \varepsilon/2$. Now all is good within each interval $(c - \delta, c + \delta)$: if x, y belong to it, then
$$|f(x) - f(y)| \le |f(x) - f(c)| + |f(y) - f(c)| < \varepsilon.$$
Let $\delta = \inf\{\delta(c) : c \in [a, b]\}$. Suppose now that $\delta > 0$ and let $|x - y| < \delta$. Then x, y belong to the interval $(z - \delta, z + \delta)$, where $z = \frac{x+y}{2}$. All the more, $x, y \in (z - \delta(z), z + \delta(z))$, so $|f(x) - f(y)| < \varepsilon$.

Trouble is, δ could very well be equal to 0, because we are taking the infimum over an infinite set of positive numbers. Wouldn't it be nice if it turned out that we did not need this set to be infinite? More precisely, we can ask if it is possible to keep only a finite number of points $c_1, c_2, \ldots, c_n$ so that, as soon as $|x - y| < \delta$ then x, y belong to some $(c_k - \delta(c_k), c_k + \delta(c_k))$, with $1 \le k \le n$? More generally, if K is a set, and $\mathcal{G}$ is an *open covering* of K, i.e., a collection of open sets such that
$$K \subset \bigcup_{G \in \mathcal{G}} G,$$
can it be reduced to a finite covering? It turns out that the answer is in the affirmative, and the crucial property is the sequential compactness.

Theorem 10.7.1. *Let K be a sequentially compact set in $\mathbb{R}^n$, and let $\mathcal{G}$ be a collection of open sets such that $K \subset \bigcup_{G \in \mathcal{G}} G$. Then there exists a finite collection of sets $G_i \in \mathcal{G}$, $1 \leq i \leq m$, such that $K \subset \bigcup_{i=1}^m G_k$.*

Proof. Suppose that there is no finite collection that would cover K. By Theorem 10.4.7, K is a bounded set, so it is contained in a closed rectangle R_1. Let L be the length of the diagonal of R_1. If we split each side of R_1 into two equal segments, then R_1 is split into 2^n rectangles, and $R_1 \cap K$ into up to 2^n nonempty pieces. At least one of these cannot be covered by a finite subcollection of $\mathcal{G}$, and let R_2 be a rectangle that contains it. It is easy to see that the diagonal of R_2 has length $L/2$. By splitting R_2 in the same manner, we get 2^n rectangles, and at least one has a non-empty intersection with K that cannot be covered by a finite subcollection of $\mathcal{G}$. We will denote that one by R_3, notice that its diagonal has length $L/2^2$, and continue the process. We obtain an infinite sequence of non-empty nested closed sets

$$R_1 \cap K \supset R_2 \cap K \supset R_3 \cap K \supset \dots,$$

so by Exercise 10.4.21 there exists a point $\mathbf{a}$ that belongs to all of them. Since $\mathbf{a} \in K$, there exists $G \in \mathcal{G}$ that contains $\mathbf{a}$. The set G is open, so there exists an open ball $B_r(\mathbf{a}) \subset G$. Let

$$k_0 = 1 + \left\lfloor \frac{\ln \frac{L}{r}}{\ln 2} \right\rfloor.$$

Then $k_0 \ln 2 > \ln \frac{L}{r}$ so $2^{k_0} > \frac{L}{r}$, and it follows that the diagonal of R_{k_0} has length

$$\frac{L}{2^{k_0}} < r.$$

Further R_{k_0} contains $\mathbf{a}$, so $R_{k_0} \subset B_r(\mathbf{a}) \subset G$, which means that R_{k_0} is covered by a single set from $\mathcal{G}$. Of course, this contradicts the choice of R_{k_0} and the theorem is proved. $\qquad\square$

Remark 10.7.2. The property of a set K stated in Theorem 10.7.1, namely that every open covering $\mathcal{G}$ of K, can be reduced to a finite subcovering is known as the Heine–Borel property.

Now we return to our earlier question: Is there a finite number of points $c_1, c_2, \dots, c_n$ so that, as soon as $|x - y| < \delta$ then x, y belong to some $(c_k - \delta(c_k), c_k + \delta(c_k))$, with $1 \leq k \leq n$? The answer is yes, and the proof is not very hard.

Corollary 10.7.3. *Let f be a continuous function on $[a, b]$, and let $\varepsilon > 0$. For each $c \in [a, b]$, let $\delta(c)$ be a positive number such that $|x - c| < \delta(c)$ implies $|f(x) - f(c)| < \varepsilon/2$. Then there exists a finite number of points $c_1, c_2, \dots, c_n \in [a, b]$ such that,*

$$\delta = \frac{1}{2} \min\{\delta(c_k) : 1 \leq k \leq n\}, \quad \text{and} \quad |x - y| < \delta \Rightarrow |f(x) - f(y)| < \varepsilon.$$

In other words, f is uniformly continuous on $[a, b]$.

Proof. By Theorem 10.7.1, there exists a finite number of points $c_1, c_2, \dots, c_n \in [a, b]$ and positive numbers $\delta(c_1), \delta(c_2), \dots, \delta(c_n)$, such that

$$[a, b] \subset \bigcup_{k=1}^n \left(c_k - \frac{1}{2}\delta(c_k), c_k + \frac{1}{2}\delta(c_k) \right),$$

and $|f(x) - f(c_k)| < \varepsilon/2$ when $|x - c_k| < \delta(c_k)$, $1 \leq k \leq n$. Let $\delta = \frac{1}{2} \min\{\delta(c_k) : 1 \leq k \leq n\}$ and suppose that $|x - y| < \delta$. (See Figure 10.26.) Since $x \in [a, b]$, there exists i, $1 \leq i \leq n$,

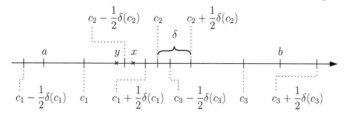

Figure 10.26: f is uniformly continuous on $[a, b]$.

such that $x \in (c_i - \frac{1}{2}\delta(c_i), c_i + \frac{1}{2}\delta(c_i))$. Now

$$|y - c_i| \leq |y - x| + |x - c_i| < \delta + \frac{1}{2}\delta(c_i) \leq \frac{1}{2}\delta(c_i) + \frac{1}{2}\delta(c_i) = \delta(c_i).$$

It follows that

$$|f(x) - f(y)| \leq |f(x) - f(c_i)| + |f(y) - f(c_i)| < \frac{\varepsilon}{2} + \frac{\varepsilon}{2} = \varepsilon,$$

and the proof is complete.   □

In the 1920s Russian topologists Pavel Alexandrov (1896–1982) and Pavel Urysohn (1898–1924) introduced the following definition:

Definition 10.7.4. A set K is **compact** if it has the Heine–Borel property, i.e., every open covering $\mathcal{G}$ of K can be reduced to a finite subcovering.

This definition gained popularity because it can be used in a general topological space. In a metric space, such as $\mathbb{R}^n$, it is equivalent to sequential compactness.

Theorem 10.7.5 (Heine–Borel Theorem). *A set $K \subset \mathbb{R}^n$ is compact if and only if it is sequentially compact.*

Proof. The "if" part is Theorem 10.7.1, so we assume that K is a compact set. By Theorem 10.4.7, it suffices to prove that K is closed and bounded.

Let $\mathbf{a}$ be a point that does not belong to K. We will show that there exists an open n-ball $B_r(\mathbf{a}) \subset K^c$ which will imply that the complement of K is open, so that K is closed. Let

$$G_k = \left\{\mathbf{x} \in \mathbb{R}^n : \|\mathbf{x} - \mathbf{a}\| > \frac{1}{k}\right\}, \quad k = 1, 2, 3, \ldots.$$

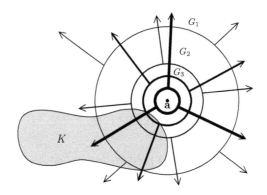

Figure 10.27: The union of sets G_k covers K.

Since the complement of G_k is the closed ball of radius $1/k$, with center at $\mathbf{a}$, each set G_k is open. Further, the union of all G_k contains every point in $\mathbb{R}^n$ with the exception of $\mathbf{a}$, so it covers K. By assumption, there is a finite collection $\{G_{k_i}\}$, $1 \le i \le m$, such that its union covers K. However, the sets $\{G_k\}$ satisfy the relation $G_k \subset G_{k+1}$, for all $k \in \mathbb{N}$, so

$$\bigcup_{i=1}^{m} G_{k_i} = G_{k_m}.$$

Since $K \subset G_{k_m}$, it follows that if $\mathbf{x} \in K$ then $\|\mathbf{x} - \mathbf{a}\| > 1/k_m$. Consequently, the open ball $B_{1/k_m}(\mathbf{a})$ contains no points of K and it lies completely in K^c. Thus K^c is open and K is closed.

In order to prove that K is bounded, let $B_k = B_k(\mathbf{a})$. Clearly, their union covers the whole of $\mathbb{R}^n$, and all the more the set K. By assumption,

$$K \subset \bigcup_{i=1}^{s} B_{k_i} = B_{k_s},$$

so K is bounded. $\qquad\square$

The idea of compactness is that in such sets we can go from a local property to a global property. The continuity (a local property) and uniform continuity (a global property) are a good illustration. Another one is left as an exercise (Exercise 10.7.10).

In the course of the proof of Theorem 3.8.7, Heine did prove that a specific collection of intervals covering $[a, b]$ can be reduced to a finite set, but he did not comment on it. In an earlier proof, Dirichlet used the same technique, and he did explicitly state that reducing the particular collection was a challenge. However, his lecture notes were not published until much later. In his dissertation in 1895, Borel proved a stronger result: whenever $[a, b]$ is contained in a *countable* union of open sets G_n, then there is a finite collection of these sets whose union contains $[a, b]$. Borel's result was generalized in [21] to two dimensional sets and the case when the collection $\mathcal{G}$ is not necessarily countable by Pierre Cousin (1867–1933), a student of Henri Poincaré (page 525). Lebesgue published the same result in 1904. The collection of open sets $\mathcal{G}$ is nowadays called an open covering of the set and the property that an arbitrary open covering can be reduced to a finite one is often referred to as the Borel–Lebesgue Property or the Heine–Borel Property. In spite of the minimal contribution of Heine, the theorem carries his name. Arthur Schoenflies is perhaps responsible, because in [92] he referred to "a theorem of Borel that extends a result of Heine". This was then quoted by an English mathematician William Henry Young (1863–1942), who is credited for the use of the word "covering". Further, in 1904 an American mathematician Oswald Veblen (1880–1960) published an article titled *The Heine–Borel Theorem*. Veblen was a reputable mathematician with many significant contributions in topology, differential geometry, and physics. Thus, in spite of much protest from the French, the name caught on. An exhaustive article on the subject is [38].

EMILE BOREL (1871–1956) was a French mathematician. He had two much older sisters, and when he was 11, he was sent to live with one of them in Montauben, where he attended high school. He showed a significant potential for mathematics and he went to Paris to continue his education at the Collège Sainte-Barbe and Lycée Louis-le-Grand. In the latter, one of his schoolmates was a son of Darboux. Through him he met leading mathematicians and Picard was a big influence on Borel. He took entrance exams for both the Ecole Normale Supérieure and the Ecole Polytechnique and ranked first on each one. He chose the former because it offered a better prospect for a career in science. His doctoral dissertation was written with Darboux as the advisor. It contained many important results in function theory, including a special case of Theorem 10.7.1. He went to teach at the University of Lille, during which time he published 22 research papers. He returned to the Ecole Normale in 1897, and was appointed to the chair of theory of function in 1909, which he held until 1941. He received numerous awards and by 1905 he was the president of the French Mathematical Society. His sister and her husband passed away so Borel and his wife adopted the nephew. During WWI he was the Secretary-General in the Cabinet Office, receiving the Croix de Guerre in 1918. After the war he became active in politics. He served in the French Chamber of Deputies (1924-36) and as Minister of the Navy (1925-40). During WWII, he was a member of the French Resistance, receiving the Grand Croix Légion d'Honneur in 1950. He became the president of the Science Committee of UNESCO in 1948. Together with Lebesgue, Borel is one of the creators of the Measure Theory. He pioneered its application to probability theory. He was the first to develop a systematic theory for a divergent series in 1899. Between 1921 and 1927, he published a series of papers on game theory and became the first to define games of strategy. In [8] he introduced the amusing thought experiment that entered popular culture under the name *infinite monkey theorem*.

Exercises

In Exercises 10.7.1–10.7.9 determine whether the set is compact. If not, find an open covering that cannot be reduced to a finite subcovering:

10.7.1. $\{(x, y) : x + y = 1\}$.

10.7.2. $\{(x, y) : 0 \leq x \leq 1\}$.

10.7.3. $\{\mathbf{x} \in \mathbb{R}^n : 1 \leq \|\mathbf{x}\| \leq 3\}$.

10.7.4. $\{\mathbf{x} \in \mathbb{R}^n : \|\mathbf{x}\| \leq 2\}$.

10.7.5. $\{\mathbf{x} \in \mathbb{R}^n : \|\mathbf{x}\| < 1\}$.

10.7.6. $\{\mathbf{x} \in \mathbb{R}^n : \|\mathbf{x}\| \geq 2\}$.

10.7.7. $\{(x, y) : 2x^2 - y^2 \leq 1\}$.

10.7.8. $\{(x, y) : xy \geq 1 \text{ and } x^2 + y^2 < 5\}$.

10.7.9. $\{(e^{-x} \cos \theta, e^{-x} \sin \theta) : x \geq 0,\, 0 \leq \theta \leq 2\pi\}$.

10.7.10. Let f_n be a sequence of functions defined on $[a, b]$. Suppose that, for every $c \in [a, b]$, there exists an interval around c in which f_n converges uniformly. Prove that f_n converges uniformly on $[a, b]$.

10.7.11. Without using the Heine–Borel Theorem prove that the interval $[0, 1]$ is compact.

10.7.12. Let a_n be a convergent sequence with limit L. Without using the Heine–Borel Theorem prove that the set $\{L, a_1, a_2, a_3, \dots\}$ is compact.

10.7.13. We say that a set A has the *finite intersection property* if, whenever $\mathcal{F}$ is a collection of closed sets such that $\cap \{F : F \in \mathcal{F}\} \cap A = \emptyset$ then there exists a finite set $F_1, F_2, \dots, F_m \in \mathcal{F}$ such that $\cap_{i=1}^m F_i \cap A = \emptyset$. Prove that a set $A \subset \mathbb{R}^n$ has the finite intersection property if and only if A is compact.

10.7.14. Prove that the intersection of two open sets is compact if and only if they are disjoint.

10.7.15. Let $\{G_k\}$ be a sequence of open bounded sets in $\mathbb{R}^n$. Prove or disprove: $\cap_{k=1}^{\infty} G_k^c$ is compact.

10.7.16. Let A be a compact set in $\mathbb{R}^n$ and $f : A \to B$ a continuous bijection. Prove that f^{-1} is continuous.

10.7.17. Let A be a set in $\mathbb{R}^n$ and let $\mathcal{G}$ be an open covering of A. Prove that $\mathcal{G}$ can be reduced to a countable subcovering.

10.7.18. Let A be a set in $\mathbb{R}^n$. Prove that the set of isolated points of A is countable.

10.7.19.* Let A be a set in $\mathbb{R}^n$ and suppose that for every $\mathbf{x} \in A$ there exists an n-ball $B_r(\mathbf{x})$ such that $B_r(\mathbf{x}) \cap A$ is countable. Prove that A is countable.

10.7.20. Let $\mathcal{G}$ be a collection of disjoint open subsets of $\mathbb{R}^n$. Prove that $\mathcal{G}$ is countable. What happens if "open" is replaced by "closed"?

10.7.21.* Let A be a non-empty perfect set in $\mathbb{R}^n$. Prove that A is uncountable.

10.7.22.* (Lebesgue) Let K be a compact subset of $\mathbb{R}^n$ and $\mathcal{G}$ an open covering of K. Prove that there exists a number $\delta > 0$ such that every subset of K having diameter less than δ is contained in some member of the covering.

10.7.23. Let $G_0 = (-1/5, 1/5)$, let $G_a = (a/2, 2)$, for $0 < a \leq 1$, and let $\mathcal{G} = G_0 \cup \{G_a : 0 < a \leq 1\}$. For the covering $\mathcal{G}$ of $A = [0, 1]$ find a number δ defined in Exercise 10.7.22.

10.7.24.* The purpose of this problem is to present the Cantor ternary set, and some of its properties. Let $I_0 = [0, 1]$. If we remove the middle third $(1/3, 2/3)$, we obtain the set $I_1 = [0, 1/3] \cup [2/3, 1]$. Next, we remove the middle thirds from each of the two parts, and we obtain $I_2 = [0, 1/9] \cup [2/9, 1/3] \cup [2/3, 7/9] \cup [8/9, 1]$. By removing the middle thirds from each of the four parts of I_2 we obtain I_3. Continuing in this fashion, we obtain a sequence I_n of sets. The Cantor set C is defined as $C = \cap_{n=0}^{\infty} I_n$.

(a) Prove that C is compact.

(b) Prove that C is a perfect set (a closed set with no isolated points).

(c) Prove that C is *nowhere dense* (its closure has no interior points).

This set appears in Cantor's paper [14] in 1883. However, Henry Smith had already constructed such sets in 1875 in [95]. More on the topic can be found in [45].

10.7.25.* The Principle of Continuity Induction is:
Let ϕ be a formula such that, for any real number x, $\phi(x)$ is either true or false. Suppose that

(a) $\phi(a)$ holds;

(b) if $\phi(y)$ holds for all $y \in [a, x]$, then there exists $\delta > 0$ such that $\phi(y)$ holds for all $y \in [a, x + \delta)$;

(c) if $\phi(y)$ holds for all $y \in [a, x)$, then $\phi(x)$ is true.

Then $\phi(x)$ is true for all $x \in [a, b]$.
Prove the Heine–Borel Theorem using the Principle of Continuity Induction.

The proof of the Principle of Continuity Induction can be found in [60].

11

Derivatives of Functions of Several Variables

In this chapter we will study functions of several variables, the concept of differentiability in this framework, and properties of differentiable functions. Although some of the material can be traced all the way to Newton and Leibniz, most of it was rigorously developed at the end of the nineteenth century, and the beginning of the twentieth century.

11.1 Computing Derivatives: Review

Example 11.1.1. Find the partial derivatives of $f(x, y) = 3x^2y^3 + 7x^4y + 5$.

Solution. To find $\partial f/\partial x$ we treat x as a variable and y as a constant. Therefore,

$$\frac{\partial f}{\partial x} = 6xy^3 + 28x^3y.$$

Similarly, when computing $\partial f/\partial y$ we treat y as a variable and x as a constant. We have that

$$\frac{\partial f}{\partial y} = 9x^2y^2 + 7x^4.$$
♦

Example 11.1.2. Find the partial derivatives of $f(x, y) = x^y$.

Solution. When y is a constant, f is a power function, so $\partial f/\partial x = yx^{y-1}$. When x is a constant, we have an exponential function. Therefore, $\partial f/\partial y = x^y \ln x$.
♦

Example 11.1.3. Find the partial derivatives of $f(x, y) = \begin{cases} xy\dfrac{x^2 - y^2}{x^2 + y^2}, & \text{if } x^2 + y^2 \neq 0, \\ 0, & \text{if } x = y = 0. \end{cases}$

Solution. First we will calculate $\partial f/\partial x$. If $(x, y) \neq (0, 0)$, then

$$\frac{\partial f}{\partial x} = y\frac{x^2 - y^2}{x^2 + y^2} + xy\frac{2x(x^2 + y^2) - 2x(x^2 - y^2)}{(x^2 + y^2)^2}$$

$$= \frac{y(x^2 - y^2)(x^2 + y^2) + 2x^2y(x^2 + y^2) - 2x^2y(x^2 - y^2)}{(x^2 + y^2)^2}$$

$$= \frac{yx^4 - y^5 + 4x^2y^3}{(x^2 + y^2)^2}.$$

If $(x, y) = (0, 0)$, then

$$\frac{\partial f}{\partial x}(0, 0) = \lim_{h \to 0} \frac{f(h, 0) - f(0, 0)}{h} = \lim_{h \to 0} \frac{0 - 0}{h} = 0.$$

Next, if $(x, y) \neq (0, 0)$, then

$$\frac{\partial f}{\partial y} = x \frac{x^2 - y^2}{x^2 + y^2} + xy \frac{-2y(x^2 + y^2) - 2y(x^2 - y^2)}{(x^2 + y^2)^2}$$

$$= \frac{x(x^2 - y^2)(x^2 + y^2) - 2xy^2(x^2 + y^2) - 2xy^2(x^2 - y^2)}{(x^2 + y^2)^2}$$

$$= \frac{x^5 - xy^4 - 4x^3 y^2}{(x^2 + y^2)^2}.$$

If $(x, y) = (0, 0)$, then

$$\frac{\partial f}{\partial y}(0, 0) = \lim_{h \to 0} \frac{f(0, h) - f(0, 0)}{h} = \lim_{h \to 0} \frac{0 - 0}{h} = 0. \qquad \blacklozenge$$

Example 11.1.4. Find the second-order partial derivatives of $f(x, y) = x^2 y + xy^2$.

Solution. Calculating partial derivatives is pretty straightforward:

$$\frac{\partial f}{\partial x} = f'_x = 2xy + y^2, \quad \frac{\partial f}{\partial y} = f'_y = x^2 + 2xy,$$

$$\frac{\partial^2 f}{\partial x^2} = f''_{xx} = 2x, \quad \frac{\partial}{\partial y}\left(\frac{\partial f}{\partial x}\right) = \frac{\partial^2 f}{\partial y \partial x} = f''_{xy} = (f'_x)'_y = 2x + 2y,$$

$$\frac{\partial^2 f}{\partial y^2} = f''_{yy} = 2x, \quad \frac{\partial}{\partial x}\left(\frac{\partial f}{\partial y}\right) = \frac{\partial^2 f}{\partial x \partial y} = f''_{yx} = (f'_y)'_x = 2x + 2y. \qquad \blacklozenge$$

Example 11.1.5. Find the second-order partial derivatives of $f(x, y) = 3x^2 + 5xy - 7y^2 + 8x + 4y - 11$.

Solution. It is not hard to compute $\partial f/\partial x = 6x + 5y + 8$ and $\partial f/\partial y = 5x - 14y + 4$. If we now take the partial derivative of $\partial f/\partial x$ with respect to x we obtain

$$\frac{\partial^2 f}{\partial x^2} = f''_{xx} = 6.$$

Similarly,

$$\frac{\partial^2 f}{\partial y^2} = f''_{yy} = -14, \quad \frac{\partial^2 f}{\partial x \partial y} = f''_{yx} = 5, \quad \frac{\partial^2 f}{\partial y \partial x} = f''_{xy} = 5. \qquad \blacklozenge$$

Example 11.1.6. Find the second-order mixed partial derivatives of the function f as in Example 11.1.3 and show that they are not continuous at $(0, 0)$.

Solution. We have calculated that

$$\frac{\partial f}{\partial x} = \begin{cases} \dfrac{yx^4 - y^5 + 4x^2 y^3}{(x^2 + y^2)^2}, & \text{if } x^2 + y^2 \neq 0 \\ 0, & \text{if } x = y = 0, \end{cases}$$

$$\frac{\partial f}{\partial y} = \begin{cases} \dfrac{x^5 - xy^4 - 4x^3 y^2}{(x^2 + y^2)^2}, & \text{if } x^2 + y^2 \neq 0 \\ 0, & \text{if } x = y = 0. \end{cases}$$

If $(x, y) \neq (0, 0)$, then

$$f''_{xy} = \frac{(x^4 - 5y^4 + 12x^2 y^2)(x^2 + y^2)^2 - (yx^4 - y^5 + 4x^2 y^3)2(x^2 + y^2)2y}{(x^2 + y^2)^4}$$

$$= \frac{(x^4 - 5y^4 + 12x^2y^2)(x^2 + y^2) - (yx^4 - y^5 + 4x^2y^3)4y}{(x^2 + y^2)^3}$$

$$= \frac{x^6 + 9x^4y^2 - 9x^2y^4 - y^6}{(x^2 + y^2)^3}, \text{ and}$$

$$f''_{yx} = \frac{(5x^4 - y^4 - 12x^2y^2)(x^2 + y^2)^2 - (x^5 - xy^4 - 4x^3y^2)2(x^2 + y^2)2x}{(x^2 + y^2)^4}$$

$$= \frac{(5x^4 - y^4 - 12x^2y^2)(x^2 + y^2) - (x^5 - xy^4 - 4x^3y^2)4x}{(x^2 + y^2)^3}$$

$$= \frac{x^6 + 9x^4y^2 - 9x^2y^4 - y^6}{(x^2 + y^2)^3}.$$

Next, we calculate the mixed second-order partial derivatives at $(0,0)$. By definition,

$$\frac{\partial}{\partial y}\left(\frac{\partial f}{\partial x}\right)(0,0) = \lim_{h \to 0} \frac{\frac{\partial f}{\partial x}(0,h) - \frac{\partial f}{\partial x}(0,0)}{h} = \lim_{h \to 0} \frac{\frac{-h^5}{h^4} - 0}{h} = -1,$$

$$\frac{\partial}{\partial x}\left(\frac{\partial f}{\partial y}\right)(0,0) = \lim_{h \to 0} \frac{\frac{\partial f}{\partial y}(h,0) - \frac{\partial f}{\partial y}(0,0)}{h} = \lim_{h \to 0} \frac{\frac{h^5}{h^4} - 0}{h} = 1.$$

Finally, f''_{xy} is not continuous at $(0,0)$. Indeed, $f''_{xy}(\frac{1}{n},0) = 1$ and $f''_{xy}(0,\frac{1}{n}) = -1$, so $\lim_{(x,y)\to(0,0)} f''_{xy}$ does not exist. ◆

We see that the mixed second-order partial derivatives at $(0,0)$ need not be equal. It is exactly the fact that they are not continuous that is to blame. Notice that in Examples 11.1.4 and 11.1.5 the mixed second-order partial derivatives were continuous at every point and we had $f''_{xy} = f''_{yx}$. Later we will see that the continuity of these derivatives guarantees that they are equal.

Example 11.1.7. Find the differential of $f(x,y) = \sin(x^2 + y^2)$.

Solution. By definition, the differential of f is $df = \frac{\partial f}{\partial x}\,dx + \frac{\partial f}{\partial y}\,dy$. Here,

$$\frac{\partial f}{\partial x} = 2x\cos(x^2 + y^2), \quad \frac{\partial f}{\partial y} = 2y\cos(x^2 + y^2), \tag{11.1}$$

so

$$df = 2x\cos(x^2 + y^2)\,dx + 2y\cos(x^2 + y^2)\,dy. \qquad ◆$$

Example 11.1.8. Find the gradient of $f(x,y) = x^3 + 3x^2y - y^2$.

Solution. The gradient ∇f is the vector $\left(\frac{\partial f}{\partial x}, \frac{\partial f}{\partial y}\right)$. Here $f'_x = 3x^2 + 6xy$ and $f'_y = 3x^2 - 2y$, so $\nabla f = (3x^2 - 6xy, 3x^2 - 2y)$. ◆

Example 11.1.9. Find the second-order differential of $f(x,y) = \sin(x^2 + y^2)$.

Solution. We have seen computed first-order partial derivatives in (11.1). Now,

$$f''_{xx} = 2\cos(x^2 + y^2) - 4x^2\sin(x^2 + y^2), \quad f''_{xy} = -4xy\sin(x^2 + y^2),$$
$$f''_{yy} = 2\cos(x^2 + y^2) - 4y^2\sin(x^2 + y^2).$$

Therefore,

$$d^2 f = \left(2\cos(x^2 + y^2) - 4x^2\sin(x^2 + y^2)\right)dx^2 - 8xy\sin(x^2 + y^2)\,dxdy$$
$$+ \left(2\cos(x^2 + y^2) - 4y^2\sin(x^2 + y^2)\right)dy^2. \qquad ◆$$

Exercises

In Exercises 11.1.1–11.1.4 find the required partial derivatives of f.

11.1.1. $f(x,y) = x\ln(xy)$, $\dfrac{\partial^3 f}{\partial x^2\, \partial y}$.   11.1.2. $f(x,y) = x^3\sin y + y^3\sin x$, $\dfrac{\partial^6 f}{\partial x^3\, \partial y^3}$.

11.1.3. $f(x,y,z) = e^{xyz}$, $\dfrac{\partial^3 f}{\partial x\, \partial y\, \partial z}$.   11.1.4. $f(x,y) = \dfrac{x+y}{x-y}$, $\dfrac{\partial^{m+n} f}{\partial x^m\, \partial y^n}$, $m,n \in \mathbb{N}$.

In Exercises 11.1.5–11.1.8 find the required differential of f.

11.1.5. $f(x,y) = \ln(x+y)$, $d^{10}f$.   11.1.6. $f(x,y) = x^3 + y^3 - 3xy(x-y)$, $d^3 f$.

11.1.7. $f(x,y,z) = xyz$, $d^3 f$.   11.1.8. $f(x,y,z) = e^{ax+by+cz}$, $d^n f$, $n \in \mathbb{N}$.

In Exercises 11.1.9–11.1.10 find the gradient of f.

11.1.9. $f(x,y) = \dfrac{x-2y}{x^2+y^2+1}$.   11.1.10. $f(x,y,z) = \dfrac{x+y}{e^z}$.

11.2   Derivatives and Differentiability

One of the unifying themes of calculus is the idea of approximating complicated functions by simpler ones. In Chapter 4, the desire to calculate $\sqrt[3]{9}$ (Example 4.2.1) has led us to the definition of the derivative. Using the derivative, we obtained the formula

$$\sqrt[3]{x} \approx 2 + \frac{1}{12}(x-8) = \frac{1}{12}x + \frac{4}{3}$$

which holds when x is close to 8. More generally, if f is a differentiable function we have that

$$f(x) \approx f(x_0) + f'(x_0)(x-x_0)$$

when $x - x_0$ is sufficiently small. Another way of saying the same thing, along the lines of Taylor's Formula, is that

$$f(x) = f(x_0) + f'(x_0)(x-x_0) + r(x) \tag{11.2}$$

where $r(x)/(x-x_0) \to 0$, as $x \to x_0$.

Now that we are studying functions of several variables we will look for a similar formula. Given a function $f(x,y)$, we would like to approximate it with a linear function of the form

$$f(x_0, y_0) + A(x-x_0) + B(y-y_0).$$

Let us make it more precise.

Definition 11.2.1. Let f be a function defined in an open disk D. If $(x_0, y_0) \in D$, and if there exist real numbers A, B such that, for all $(x,y) \in D$,

$$f(x,y) = f(x_0, y_0) + A(x-x_0) + B(y-y_0) + r(x,y), \quad \text{and}$$

$$\lim_{(x,y)\to(x_0,y_0)} \frac{r(x,y)}{\sqrt{(x-x_0)^2 + (y-y_0)^2}} = 0, \tag{11.3}$$

we say that f is **differentiable** at (x_0, y_0). If it is differentiable at every point of D, we say that it is differentiable on D.

Assuming that a function f is differentiable, our first task is to find the numbers A and B. Let us see what happens when we replace y by y_0 in (11.3):

$$f(x, y_0) = f(x_0, y_0) + A(x - x_0) + r(x, y_0), \quad \lim_{x \to x_0} \frac{r(x, y_0)}{|x - x_0|} = 0. \qquad (11.4)$$

Since y_0 is just a constant, we see that (11.4) is of the same form as (11.2), and the number A plays the role of $f'(x_0)$. The function f depends on two variables, so there is no such thing as $f'(x_0)$. What we really have is that one variable is being kept constant, and the derivative is taken using the other variable as in the one-variable case. In other words, A is a *partial derivative* of f.

Definition 11.2.2. Let f be a function defined on an open disk A and let $(x_0, y_0) \in A$. If the limit

$$\lim_{h \to 0} \frac{f(x_0 + h, y_0) - f(x_0, y_0)}{h}$$

exists, we say that f has the **partial derivative** with respect to x at (x_0, y_0) and we denote this limit by $(\partial f / \partial x)(x_0, y_0)$ or $f'_x(x_0, y_0)$. Similarly, if the limit

$$\lim_{k \to 0} \frac{f(x_0, y_0 + k) - f(x_0, y_0)}{k}$$

exists, we say that f has the **partial derivative** with respect to y at (x_0, y_0), and we denote it by $(\partial f / \partial y)(x_0, y_0)$ or $f'_y(x_0, y_0)$.

We see that the mysterious quantities A, B in (11.3) are $A = (\partial f / \partial x)(x_0, y_0)$ and $B = (\partial f / \partial y)(x_0, y_0)$. Consequently, formula (11.3) can be written as

$$f(x, y) = f(x_0, y_0) + \frac{\partial f}{\partial x}(x_0, y_0)(x - x_0) + \frac{\partial f}{\partial y}(x_0, y_0)(y - y_0) + r(x, y), \quad \text{and}$$
$$\lim_{(x,y) \to (x_0, y_0)} \frac{r(x, y)}{\sqrt{(x - x_0)^2 + (y - y_0)^2}} = 0. \qquad (11.5)$$

Example 11.2.3. A linear approximation.

Calculate $\sqrt{1.02^3 + 1.97^3}$.

Solution. We will consider the function $f(x, y) = \sqrt{x^3 + y^3}$ and $(x_0, y_0) = (1, 2)$. The partial derivatives of f are:

$$\frac{\partial f}{\partial x} = \frac{3x^2}{2\sqrt{x^3 + y^3}}, \quad \frac{\partial f}{\partial y} = \frac{3y^2}{2\sqrt{x^3 + y^3}},$$

so $(\partial f / \partial x)(1, 2) = 1/2$ and $(\partial f / \partial y)(1, 2) = 2$. Also, $f(1, 2) = 3$. Therefore,

$$f(x, y) \approx 3 + \frac{1}{2}(x - 1) + 2(y - 2).$$

In particular, $f(1.02, 1.97) \approx 3 + \frac{1}{2} 0.02 + 2(-0.03) = 2.95$. Not bad in view of the fact that the exact result (rounded off to 6 decimal places) is 2.950692. ♦

Partial derivatives were used by Newton, Leibniz, and Bernoullis in the seventeenth century. The symbol ∂ likely appears for the first time in an article by Marquis de Condorcet, in 1772. Legendre (page 437) was the first to use the modern notation in 1786, but he later abandoned it. A consistent use starts with the papers of Jacobi (page 357) in 1841, and his countrymen.

In the context of one variable, the differentiability of a function means that there exists a derivative. Where is it now? The differentiability in one variable means that, given a point $x_0 \in \mathbb{R}$, we can associate to a function f a real number $f'(x_0)$. Here, given a point $(x_0, y_0) \in \mathbb{R}^2$, we have obtained a pair of numbers (A, B), and we know that $A = \partial f / \partial x \, (x_0, y_0)$, $B = \partial f / \partial y \, (x_0, y_0)$. If we take a matricial point of view, we have a mapping

$$f \mapsto \left[\frac{\partial f}{\partial x}(x_0, y_0) \quad \frac{\partial f}{\partial y}(x_0, y_0) \right].$$

Each $m \times n$ matrix can be identified with a linear transformation from $\mathbb{R}^n$ to $\mathbb{R}^m$. In the present situation, our matrix is 1×2, so we have a linear transformation from $\mathbb{R}^2$ to $\mathbb{R}$. This transformation is **the derivative** of f at (x_0, y_0) and it is denoted by $\mathbf{D}f(x_0, y_0)$. It is often called the **total** derivative. We remark that in the single variable scenario, n equals 1, so the derivative is a linear transformation from $\mathbb{R}$ to $\mathbb{R}$: $y = f'(x_0)x$.

Next we turn our attention to the relationship between the differentiability and the existence of the derivative. When f is a function of one variable, these two are equivalent. That is, if $f : (a, b) \to \mathbb{R}$, and $x_0 \in (a, b)$, the derivative $f'(x_0)$ exists if and only if there exists $A \in \mathbb{R}$ such that, for all $x \in (a, b)$,

$$f(x) = f(x_0) + A(x - x_0) + r(x), \quad \text{and} \quad \lim_{x \to x_0} \frac{r(x)}{x - x_0} = 0. \tag{11.6}$$

Indeed, if (11.6) holds, then

$$\frac{f(x) - f(x_0)}{x - x_0} = A + \frac{r(x)}{x - x_0} \to A, \quad x \to x_0,$$

i.e., $f'(x_0)$ exists and equals A. Conversely, if $f'(x_0)$ exists, then (11.6) holds with $A = f'(x_0)$ and $r(x) = f(x) - f(x_0) - f'(x_0)(x - x_0)$. The fact that $r(x)/(x - x_0) \to 0$, as $x \to x_0$, follows from the definition of the derivative.

When f is a function of more than one variable, things are very different. If $f : A \to \mathbb{R}$, and if f is differentiable at $(x_0, y_0) \in A$, the partial derivatives of f exist, so $\mathbf{D}f(x_0, y_0)$ exists. However, the existence of partial derivatives does not imply (11.5).

Example 11.2.4. A function that has partial derivatives at a point but is not differentiable there.
    Prove that the function $f(x, y) = \begin{cases} x + y, & \text{if } x = 0 \text{ or } y = 0 \\ 1, & \text{otherwise} \end{cases}$ has partial derivatives at $(0, 0)$ but (11.5) does not hold.

Solution. By definition,

$$\frac{\partial f}{\partial x}(0, 0) = \lim_{h \to 0} \frac{f(h, 0) - f(0, 0)}{h} = \lim_{h \to 0} \frac{h}{h} = 1,$$

and similarly, $(\partial f / \partial y)(0, 0) = 1$. However, f is not continuous at $(0, 0)$. Indeed, $f(0, 0) = 0$, but for any $n \in \mathbb{N}$, $f(1/n, 1/n) = 1$ so $\lim f(1/n, 1/n) = 1$. This is bad news because, if we substitute $(x_0, y_0) = (0, 0)$ and $(x, y) = (1/n, 1/n)$ in (11.5), we would get

$$f\left(\frac{1}{n}, \frac{1}{n}\right) = 0 + 1\left(\frac{1}{n} - 0\right) + 1\left(\frac{1}{n} - 0\right) + r\left(\frac{1}{n}, \frac{1}{n}\right).$$

The left side equals 1, but the right side converges to 0. Thus, (11.5) does not hold.   ◆

Another important distinction that Example 11.2.4 brings forward is that the existence of partial derivatives does not imply continuity. What if, in addition to the existence of partial derivatives, we assume that f is continuous?

Example 11.2.5. A continuous function that has partial derivatives at a point but is not differentiable there.

Prove that the function $f(x,y) = \begin{cases} \dfrac{x^2 y}{x^2 + y^2}, & \text{if } x^2 + y^2 > 0 \\ 0, & \text{if } (x,y) = (0,0) \end{cases}$ has partial derivatives

at $(0,0)$, and it is continuous at $(0,0)$, but (11.5) does not hold.

Solution. This time f is continuous. It is easy to see that this is true at any point different from the origin. For the continuity at the origin, we will show that

$$\lim_{(x,y)\to(0,0)} \frac{x^2 y}{x^2 + y^2} = 0. \tag{11.7}$$

Using the arithmetic-geometric mean inequality $|2xy| \le x^2 + y^2$ we obtain that

$$0 \le \left| \frac{x^2 y}{x^2 + y^2} \right| \le \left| \frac{x^2 y}{2xy} \right| = \frac{|x|}{2}$$

which implies (11.7), via the Squeeze Theorem. So, f is continuous.

Also, the partial derivatives at $(0,0)$ exist:

$$\frac{\partial f}{\partial x}(0,0) = \lim_{h\to 0} \frac{f(h,0) - f(0,0)}{h} = \lim_{h\to 0} \frac{0}{h} = 0,$$

and similarly, $(\partial f/\partial y)(0,0) = 0$.

However, (11.5) does not hold. Otherwise, we would have $f(x,y) = r(x,y)$, and it would follow that

$$\lim_{(x,y)\to(0,0)} \frac{f(x,y)}{\sqrt{(x)^2 + (y)^2}} = 0.$$

In particular, taking once again $(x,y) = (1/n, 1/n)$, we would obtain that

$$\lim_{n\to\infty} \frac{\left(\frac{1}{n}\right)^2 \frac{1}{n}}{\left(\left(\frac{1}{n}\right)^2 + \left(\frac{1}{n}\right)^2\right)^{3/2}} = 0,$$

which is incorrect because the limit on the left side is $1/(2\sqrt{2})$. ◆

Thus, even the additional assumption about the continuity of f is not sufficient for (11.5) to hold. It turns out that the continuity of partial derivatives is.

Theorem 11.2.6. *Let f be a function with a domain an open disk D and suppose that its partial derivatives $\partial f/\partial x, \partial f/\partial y$ exist in D and that they are continuous at $(x_0, y_0) \in D$. Then f is differentiable at (x_0, y_0).*

Proof. We will start with the equality

$$f(x,y) - f(x_0, y_0) = [f(x,y) - f(x_0, y)] + [f(x_0, y) - f(x_0, y_0)].$$

The existence of partial derivatives allows us to apply the Mean Value Theorem to each pair above. We obtain that

$$f(x,y) - f(x_0, y_0) = \frac{\partial f}{\partial x}(z,y)(x - x_0) + \frac{\partial f}{\partial y}(x_0, w)(y - y_0),$$

for some real numbers z (between x and x_0) and w (between y and y_0). We will write

$$\frac{\partial f}{\partial x}(z,y) = \frac{\partial f}{\partial x}(x_0,y_0) + \alpha, \quad \frac{\partial f}{\partial y}(x_0,w) = \frac{\partial f}{\partial y}(x_0,y_0) + \beta,$$

and the continuity of partial derivatives at (x_0, y_0) implies that, when $(x,y) \to (x_0,y_0)$, $\alpha, \beta \to 0$. Therefore,

$$f(x,y) - f(x_0,y_0) = \left(\frac{\partial f}{\partial x}(x_0,y_0) + \alpha\right)(x - x_0) + \left(\frac{\partial f}{\partial y}(x_0,y_0) + \beta\right)(y - y_0)$$

$$= \frac{\partial f}{\partial x}(x_0,y_0)(x - x_0) + \frac{\partial f}{\partial y}(x_0,y_0)(y - y_0) + \alpha(x - x_0) + \beta(y - y_0).$$

Thus, it remains to show that

$$\lim_{(x,y)\to(x_0,y_0)} \frac{\alpha(x - x_0) + \beta(y - y_0)}{\sqrt{(x - x_0)^2 + (y - y_0)^2}} = 0.$$

Notice that

$$\frac{|x - x_0|}{\sqrt{(x - x_0)^2 + (y - y_0)^2}}, \frac{|y - y_0|}{\sqrt{(x - x_0)^2 + (y - y_0)^2}} \le 1.$$

It follows that

$$0 \le \left|\frac{\alpha(x - x_0) + \beta(y - y_0)}{\sqrt{(x - x_0)^2 + (y - y_0)^2}}\right| \le |\alpha| + |\beta| \to 0, \quad (x,y) \to (x_0,y_0). \qquad \square$$

Remark 11.2.7. The conditions of Theorem 11.2.6 are merely sufficient. The differentiability of f does not imply that any of its partial derivatives need to be continuous (Exercise 11.2.11). For the same conclusion as in Theorem 11.2.6 under weaker assumptions see Exercise 11.2.25.

> Definition 11.2.1 appears for the first time in 1893, in [99] by Stolz (see next page). In this book he gives an example (see Exercise 11.2.4) of a function that has partial derivatives at $(0,0)$ but that is not differentiable there. He also notes that a similar example can be found in Thomae's book [102] published in 1873.

We will close this section with a few words about functions that depend on more then two variables. First, we will state the definition of partial derivatives.

Definition 11.2.8. Let f be a function defined on an open set $A \subset \mathbb{R}^n$ and let $\mathbf{a} \in A$. Let i be a fixed integer, $1 \le i \le n$. If the limit

$$\lim_{h\to 0} \frac{f(a_1, a_2, \ldots, a_{i-1}, a_i + h, a_{i+1}, \ldots, a_n) - f(a_1, \ldots, a_{i-1}, a_i, a_{i+1}, \ldots, a_n)}{h}$$

exists, we say that f has the **partial derivative** with respect to x_i at $\mathbf{a}$ and we denote this limit by $(\partial f/\partial x_i)(\mathbf{a})$.

Next we define the differentiability.

Definition 11.2.9. Let f be a function defined in an open n-ball B. If $\mathbf{a} \in B$, and if there exist real numbers $A_1, A_2, \ldots, A_n$ such that, for all $\mathbf{x} \in B$,

$$f(\mathbf{x}) = f(\mathbf{a}) + A_1(x_1 - a_1) + A_2(x_2 - a_2) + \cdots + A_n(x_n - a_n) + r(\mathbf{x}), \quad \text{and}$$

$$\lim_{\mathbf{x}\to\mathbf{a}} \frac{r(\mathbf{x})}{\sqrt{(x_1 - a_1)^2 + (x_2 - a_2)^2 + \cdots + (x_n - a_n)^2}} = 0 \qquad (11.8)$$

we say that f is **differentiable** at $\mathbf{a}$. If it is differentiable at every point of a set A, we say that it is differentiable on A.

Naturally, the numbers A_k, $1 \le k \le n$, in (11.8) are the partial derivatives of f.

Definition 11.2.10. A function f that has continuous partial derivatives in an open n-ball B is said to be **continuously differentiable** in B. We write $f \in C^1(B)$, or $f \in C^1(A)$ if every point **a** of the open set A is contained in a ball $B(\mathbf{a}) \subset A$, and $f \in C^1(B)$. When the set A is tacitly understood, we will write $f \in C^1$.

Remark 11.2.11. Theorem 11.2.6 shows that, if $f \in C^1(B)$, then f is differentiable in B.

Whenever $n \ge 2$, the differentiability implies the existence of partial derivatives, but the converse is not true. It is not hard to see that Theorem 11.2.6 remains valid when $n \ge 3$.

Finally, the derivative of f at **a** is the linear mapping $Df(\mathbf{a})$ with the matrix

$$\left[\frac{\partial f}{\partial x_1}(\mathbf{a}) \quad \frac{\partial f}{\partial x_2}(\mathbf{a}) \quad \cdots \quad \frac{\partial f}{\partial x_n}(\mathbf{a}) \right].$$

In the next section we will explore the properties of the derivative.

 OTTO STOLZ (1842–1905) was an Austrian mathematician. He attended a high school in Innsbruck and studied at the University of Vienna. His 1864 Ph.D. was in geometry. After the habilitation in 1867, he taught at the university for 2 years. Before settling in Innsbruck, he spent 2 years in Germany. In Berlin, he attended the lectures by Weierstrass, Kummer, and Kronecker, and Weierstrass influenced Stolz towards analysis. In 1872 he became a professor at the University of Innsbruck, and remained there for the rest of his life, in spite of several offers from the University of Vienna. He served as the Dean of the Faculty of Arts (1877-78 and 1888-89) and as the Rector (1890-91). Stolz received many awards and he was a member of the Austrian Academy of Sciences and the Bavarian Academy of Sciences. He wrote several textbooks which were well received and an article about Bolzano's contributions, helping save him from oblivion. His name lives on in the Stolz-Cesàro theorem.

Exercises

In Exercises 11.2.1–11.2.6 determine whether the function f is differentiable at $(0,0)$ and, if so, find $\mathbf{D}f(0,0)$.

11.2.1. $f(x,y) = \sqrt[3]{xy}$.

11.2.2. $f(x,y) = \sqrt[3]{x^3 + y^3}$.

11.2.3. $f(x,y) = \begin{cases} e^{-\frac{1}{x^2+y^2}}, & \text{if } x^2 + y^2 \ne 0 \\ 0, & \text{if } x = y = 0. \end{cases}$

11.2.4. $f(x,y) = \sqrt{|xy|}$.

11.2.5. $f(x,y) = \begin{cases} \dfrac{xy}{\sqrt{x^2+y^2}}, & \text{if } x^2 + y^2 \ne 0 \\ 0, & \text{if } x = y = 0. \end{cases}$

11.2.6. $f(x,y) = \begin{cases} \dfrac{x^3 y}{x^6 + y^2}, & \text{if } x^6 + y^2 \ne 0 \\ 0, & \text{if } x = y = 0. \end{cases}$

In Exercises 11.2.7–11.2.10 find $\mathbf{D}f(\mathbf{a})$.

11.2.7. $f(x,y) = xy^2 + x^2 y$, $\mathbf{a} = (1,0)$.

11.2.8. $f(x,y) = e^{x^2 + y^2}$, $\mathbf{a} = (-1,2)$.

11.2.9. $f(x,y) = \ln(x^2 + y^2)$, $\mathbf{a} = (3,2)$.

11.2.10. $f(x,y,z) = x^2 + x \ln yz$, $\mathbf{a} = (-3, e, e)$.

11.2.11. Let $f(x,y) = \begin{cases} (x^2 + y^2) \sin \dfrac{1}{x^2 + y^2}, & \text{if } x^2 + y^2 \ne 0 \\ 0, & \text{if } x = y = 0. \end{cases}$ Prove that f is differentiable but its partial derivatives are not continuous at $(0,0)$.

11.2.12. Let φ be the Weierstrass function (continuous everywhere, differentiable nowhere), and let $f(x,y) = \sqrt{x^2 + y^2} \left[\varphi(\sqrt{x^2 + y^2}) - \varphi(0) \right]$. Prove that f is differentiable at $(0,0)$, but the partial derivatives do not exist at any point (x,y) such that $x \ne 0$ and $y \ne 0$.

11.2.13. Use the linear approximation to calculate: (a) $1.002 \cdot 2.003^2 \cdot 3.004^3$; (b) $0.97^{1.05}$; (c) $\sin 29° \cdot \tan 46°$; (d) $\dfrac{1.03^2}{\sqrt[3]{0.98}\,\sqrt[4]{1.05^3}}$.

Let $A \subset \mathbb{R}^n$, let $\mathbf{a} \in A$, let $f : A \to \mathbb{R}$, and let $\mathbf{u}$ be a unit vector in $\mathbb{R}^n$. The **directional derivative of f at a in the direction u** is

$$\mathbf{D_u}f(\mathbf{a}) = \lim_{h \to 0} \frac{f(\mathbf{a} + h\mathbf{u}) - f(\mathbf{a})}{h}$$

when the limit exists.

In Exercises 11.2.14–11.2.17 find the directional derivative of f at $\mathbf{a}$ in the direction $\mathbf{u}$.

11.2.14. $f(x,y) = 3x^2 - 4xy + 7y^2$, $\mathbf{a} = (1,-2)$, $\mathbf{u} = (\sqrt{3}/2, 1/2)$.

11.2.15. $f(x,y) = \ln(x + e^y)$, $\mathbf{a} = (e,1)$, $\mathbf{u} = (e/\sqrt{e^2+1}, 1/\sqrt{e^2+1})$.

11.2.16. $f(x,y) = \arcsin xy$, $\mathbf{a} = (1/5,3)$, $\mathbf{u} = (-1/\sqrt{2}, 1/\sqrt{2})$.

11.2.17. $f(x,y,z) = (x+y)/z$, $\mathbf{a} = (1,2,3)$, $\mathbf{u} = (1/\sqrt{3}, 1/\sqrt{3}, -1/\sqrt{3})$.

11.2.18.* Suppose that f has all directional derivatives at a point $\mathbf{a}$. Prove or disprove: f is continuous at $\mathbf{a}$.

Let $f : \mathbb{R}^2 \to \mathbb{R}$ and $(x_0, y_0) \in \mathbb{R}^2$. The **tangent plane** at $(x_0, y_0, f(x_0, y_0))$ is given by

$$z = f(x_0, y_0) + \frac{\partial f}{\partial x}(x_0, y_0)(x - x_0) + \frac{\partial f}{\partial y}(x_0, y_0)(y - y_0).$$

In Exercises 11.2.19–11.2.22 find an equation of the tangent plane to the graph of f at $\mathbf{a}$:

11.2.19. $z = x^3 - 7xy + e^y$, $(-1,0)$.   **11.2.20.** $z = 4\cos xy$, $(\pi/3, 1/2)$.

11.2.21. $z = e^{x+y}\cos xy$, $(0,1)$.   **11.2.22.** $z = x^2 + y^2$, $(0,1)$.

11.2.23. A set $A \subset \mathbb{R}^n$ is convex if for any two points $\mathbf{x}, \mathbf{y} \in \mathbb{R}^n$ and any $t \in [0,1]$, $t\mathbf{x} + (1-t)\mathbf{y} \in A$. Prove that if f is a function that is differentiable in a convex set $A \subset \mathbb{R}^2$, and has bounded partial derivatives in A, then f is uniformly continuous in A.

11.2.24. Suppose that f is a continuous function of x (for each fixed y) on a set $A \subset \mathbb{R}^2$, and that $\partial f/\partial y$ is bounded on A. Prove that f is a continuous function on A.

11.2.25.* Let f be a function with a domain an open disk $A \subset \mathbb{R}^2$ and suppose that both of its partial derivatives exist at $\mathbf{a} \in A$ and that $\partial f/\partial x$ is continuous in A. Prove that f is differentiable at $\mathbf{a}$.

11.2.26.* Let f be a function with a domain an open disk $A \subset \mathbb{R}^2$ and suppose that both of its partial derivatives exist at $\mathbf{a} \in A$ and that they are bounded in A. Prove that f is continuous in A.

11.3   Properties of the Derivative

In this section we will look at some basic properties of the derivative of functions of n variables.

Theorem 11.3.1. *Let f, g be two functions with a domain an open n-ball A and let $\mathbf{a} \in A$. Also, let α be a real number. If f and g are differentiable at $\mathbf{a}$ then the same is true for $f + g$ and αf and:*

(a) $\mathbf{D}(\alpha f)(\mathbf{a}) = \alpha \mathbf{D} f(\mathbf{a})$;

(b) $\mathbf{D}(f + g)(\mathbf{a}) = \mathbf{D} f(\mathbf{a}) + \mathbf{D} g(\mathbf{a})$.

Proof. By definition, there exist real numbers A_i, $1 \le i \le n$ such that

$$f(\mathbf{x}) = f(\mathbf{a}) + \sum_{i=1}^{n} A_i(x_i - a_i) + r_f(\mathbf{x}), \quad \frac{r_f(\mathbf{x})}{\sqrt{\sum_{i=1}^{n}(x_i - a_i)^2}} \to 0. \tag{11.9}$$

Now the function $\tilde{f} = \alpha f$ satisfies

$$\tilde{f}(\mathbf{x}) = \tilde{f}(\mathbf{a}) + \sum_{i=1}^{n} \tilde{A}_i(x_i - a_i) + \tilde{r}_f(\mathbf{x}),$$

with $\tilde{A}_i = \alpha A_i$, and $\tilde{r}_f(\mathbf{x}) = \alpha r_f(\mathbf{x})$. Therefore, $\tilde{f}$ is differentiable at $\mathbf{a}$, and

$$\mathbf{D}\tilde{f}(\mathbf{a}) = \left[\frac{\partial \tilde{f}}{\partial x_1}(\mathbf{a}) \quad \frac{\partial \tilde{f}}{\partial x_2}(\mathbf{a}) \quad \cdots \quad \frac{\partial \tilde{f}}{\partial x_n}(\mathbf{a}) \right]$$

$$= \alpha \left[\frac{\partial f}{\partial x_1}(\mathbf{a}) \quad \frac{\partial f}{\partial x_2}(\mathbf{a}) \quad \cdots \quad \frac{\partial f}{\partial x_n}(\mathbf{a}) \right] = \alpha \mathbf{D} f(\mathbf{a}).$$

That settles (a). We leave (b) as an exercise. $\qquad \square$

In the single variable case, if f is differentiable at $x = a$, then it is continuous at $x = a$.

Theorem 11.3.2. *If a function f is defined on an open set $A \subset \mathbb{R}^n$ and if it is differentiable at $\mathbf{a} \in A$, then it is continuous at $\mathbf{a}$.*

Proof. It follows immediately from (11.8) that, as $\mathbf{x} \to \mathbf{a}$, $f(\mathbf{x}) \to f(\mathbf{a})$. $\qquad \square$

Next we will establish the Product Rule and the Quotient Rule.

Theorem 11.3.3. *Let f, g be two functions with a domain an open n-ball A and let $\mathbf{a} \in A$. If f and g are differentiable at $\mathbf{a}$ then the same is true for $f \cdot g$ and*

$$\mathbf{D}(f \cdot g)(\mathbf{a}) = \mathbf{D} f(\mathbf{a}) g(\mathbf{a}) + f(\mathbf{a}) \mathbf{D} g(\mathbf{a}).$$

If, in addition, $g(\mathbf{a}) \ne 0$, then the function f/g is differentiable at $\mathbf{a}$ and

$$\mathbf{D}\left(\frac{f}{g} \right)(\mathbf{a}) = \frac{\mathbf{D} f(\mathbf{a}) g(\mathbf{a}) - f(\mathbf{a}) \mathbf{D} g(\mathbf{a})}{g(\mathbf{a})^2}.$$

Proof. If we multiply equations (11.9) we obtain that

$$f(\mathbf{x}) g(\mathbf{x}) = f(\mathbf{a}) g(\mathbf{a}) + \sum_{i=1}^{n} (A_i g(\mathbf{a}) + B_i f(\mathbf{a})) (x_i - a_i) + \sum_{i=1}^{n} \sum_{j=1}^{n} A_i B_j (x_i - a_i)(x_j - a_j)$$

$$+ r_f(\mathbf{x}) g(\mathbf{x}) + r_g(\mathbf{x}) f(\mathbf{x}) - r_f(\mathbf{x}) r_g(\mathbf{x}). \tag{11.10}$$

Further, it is obvious that the last three terms go to 0, as $\mathbf{x} \to \mathbf{a}$, faster then $\|\mathbf{x} - \mathbf{a}\|$. The same is true for the double sum because each of its n^2 terms satisfies

$$0 \le \frac{|x_i - a_i||x_j - a_j|}{\|\mathbf{x} - \mathbf{a}\|} \le |x_i - a_i| \to 0.$$

Finally, it follows from (11.10) that

$$\mathbf{D}(f \cdot g)(\mathbf{a}) = \begin{bmatrix} A_1 g(\mathbf{a}) + B_1 f(\mathbf{a}) & A_2 g(\mathbf{a}) + B_2 f(\mathbf{a}) & \cdots & A_n g(\mathbf{a}) + B_n f(\mathbf{a}) \end{bmatrix}$$
$$= g(\mathbf{a}) \begin{bmatrix} A_1 & A_2 & \cdots & A_n \end{bmatrix} + f(\mathbf{a}) \begin{bmatrix} B_1 & B_2 & \cdots & B_n \end{bmatrix}$$
$$= g(\mathbf{a})\mathbf{D}f(\mathbf{a}) + f(\mathbf{a})\mathbf{D}g(\mathbf{a}).$$

That settles the Product Rule.

In order to establish the Quotient Rule we multiply the first equation in (11.9) by $g(\mathbf{a})$, and the second by $f(\mathbf{a})$, and we subtract the resulting equations. This leads to

$$f(\mathbf{x})g(\mathbf{a}) - g(\mathbf{x})f(\mathbf{a}) = \sum_{i=1}^{n} (g(\mathbf{a})A_i - f(\mathbf{a})B_i)(x_i - a_i) + r_f(\mathbf{x})g(\mathbf{a}) - r_g(\mathbf{x})f(\mathbf{a}).$$

After dividing by $g(\mathbf{x})g(\mathbf{a})$ we obtain

$$\frac{f(\mathbf{x})}{g(\mathbf{x})} - \frac{f(\mathbf{a})}{g(\mathbf{a})} = \sum_{i=1}^{n} \frac{g(\mathbf{a})A_i - f(\mathbf{a})B_i}{g(\mathbf{x})g(\mathbf{a})}(x_i - a_i) + \frac{r_f(\mathbf{x})g(\mathbf{a}) - r_g(\mathbf{x})f(\mathbf{a})}{g(\mathbf{x})g(\mathbf{a})}$$
$$= \sum_{i=1}^{n} \frac{g(\mathbf{a})A_i - f(\mathbf{a})B_i}{g(\mathbf{a})^2}(x_i - a_i)$$
$$- \left(1 - \frac{g(\mathbf{a})}{g(\mathbf{x})}\right) \sum_{i=1}^{n} \frac{g(\mathbf{a})A_i - f(\mathbf{a})B_i}{g(\mathbf{a})^2}(x_i - a_i)$$
$$+ \frac{r_f(\mathbf{x})g(\mathbf{a}) - r_g(\mathbf{x})f(\mathbf{a})}{g(\mathbf{x})g(\mathbf{a})}.$$

It is not hard to see that, as $\mathbf{x} \to \mathbf{a}$, the last two expressions go to 0 faster than $\|\mathbf{x} - \mathbf{a}\|$. Thus, f/g is differentiable at $\mathbf{a}$ and

$$\mathbf{D}\left(\frac{f}{g}\right)(\mathbf{a}) = \begin{bmatrix} \dfrac{g(\mathbf{a})A_1 - f(\mathbf{a})B_1}{g(\mathbf{a})^2} & \dfrac{g(\mathbf{a})A_2 - f(\mathbf{a})B_2}{g(\mathbf{a})^2} & \cdots & \dfrac{g(\mathbf{a})A_n - f(\mathbf{a})B_n}{g(\mathbf{a})^2} \end{bmatrix}$$
$$= \frac{g(\mathbf{a}) \begin{bmatrix} A_1 & A_2 & \cdots & A_n \end{bmatrix} - f(\mathbf{a}) \begin{bmatrix} B_1 & B_2 & \cdots & B_n \end{bmatrix}}{g(\mathbf{a})^2}$$
$$= \frac{\mathbf{D}f(\mathbf{a})g(\mathbf{a}) - f(\mathbf{a})\mathbf{D}g(\mathbf{a})}{g(\mathbf{a})^2}. \qquad \square$$

When f is a function of one variable, higher derivatives were often very useful (Second Derivative Test, Taylor's Formula, etc.). What if f is a function of several variables? Is there such a thing as the second (total) derivative of f? If $f(x) = 3 - 5x + 7x^2$ then

$$7x^2 = \frac{f''(0)}{2!} x^2.$$

If $f(x, y) = 3x^2 + 5xy - 7y^2 + 8x + 4y - 11$ then we expect the second derivative to correspond to the *quadratic form*

$$Q(x, y) = 3x^2 + 5xy - 7y^2. \tag{11.11}$$

As we have seen in Example 11.1.5, the coefficients can be obtained by evaluating the second-order partial derivatives of f at $(0,0)$. Namely, $f''_{xx}(0,0) = 6$, $f''_{yy}(0,0) = -14$, $f''_{xy}(0,0) = 5$, and $f''_{yx}(0,0) = 5$. Therefore,

$$Q(x,y) = \frac{1}{2!}\left(f''_{xx}(0,0)x^2 + f''_{yy}(0,0)y^2 + f''_{xy}(0,0)yx + f''_{yx}(0,0)xy\right). \tag{11.12}$$

This serves as a motivation to define the **second derivative** of f, when f depends on n variables, as a *quadratic form*

$$\mathbf{D}^2 f(\mathbf{a})(\mathbf{u})^2 = \sum_{i=1}^{n}\sum_{j=1}^{n} f''_{x_j x_i}(\mathbf{a})\, u_i u_j. \tag{11.13}$$

Another way to represent the same thing is to use the **Hessian matrix**

$$H(\mathbf{a}) = \begin{bmatrix} f''_{x_1 x_1}(\mathbf{a}) & f''_{x_1 x_2}(\mathbf{a}) & \cdots & f''_{x_1 x_n}(\mathbf{a}) \\ f''_{x_2 x_1}(\mathbf{a}) & f''_{x_2 x_2}(\mathbf{a}) & \cdots & f''_{x_2 x_n}(\mathbf{a}) \\ \cdots\cdots\cdots\cdots \\ f''_{x_n x_1}(\mathbf{a}) & f''_{x_n x_2}(\mathbf{a}) & \cdots & f''_{x_n x_n}(\mathbf{a}) \end{bmatrix} \tag{11.14}$$

and notice that $\mathbf{D}^2 f(\mathbf{a})(\mathbf{u})^2 = H(\mathbf{a})\mathbf{u}\cdot\mathbf{u}$. When $\mathbf{u} = (dx_1, dx_2, \ldots, dx_n)$ we obtain the **total differential** df, and the second-order total differential $d^2 f$ of a function f. For example, when $n = 2$,

$$df = \frac{\partial f}{\partial x}\, dx + \frac{\partial f}{\partial y}\, dy,$$
$$d^2 f = f''_{xx}\, dx^2 + f''_{xy}\, dydx + f''_{yx}\, dxdy + f''_{yy}\, dy^2. \tag{11.15}$$

Example 11.3.4. The derivative and the second derivative of a function.

For $f(x,y) = x^3 + y^3 - 3x^2 y + 3xy^2$ and $(a,b) = (2,3)$, find $\mathbf{D}f(2,3)$ and $\mathbf{D}^2 f(2,3)$.

Solution. First we need the partial derivatives:

$$f'_x = 3x^2 - 6xy + 3y^2, \quad f'_y = 3y^2 - 3x^2 + 6xy,$$
$$f''_{xx} = 6x - 6y, \quad f''_{xy} = -6x + 6y, \quad f''_{yx} = -6x + 6y, \quad f''_{yy} = 6y + 6x.$$

Then we evaluate them at $(2,3)$ and we obtain that $f'_x(2,3) = 3$, $f'_y(2,3) = 51$, $f''_{xx}(2,3) = -6$, $f''_{xy}(2,3) = 6$, $f''_{yy}(2,3) = 30$. Thus,

$$\mathbf{D}f(2,3)(\mathbf{u}) = 3u_1 + 51u_2, \quad \mathbf{D}^2 f(2,3)(\mathbf{u})^2 = -6u_1^2 + 12u_1 u_2 + 30u_2^2.$$

In particular, when $u_1 = dx$ and $u_2 = dy$,

$$d^2 f(2,3) = -6dx^2 + 12dxdy + 30dy^2. \qquad \blacklozenge$$

The matrix $H(\mathbf{a})$ in (11.14) is called the Hessian matrix, by German mathematician Otto Hesse (1811–1874). Hesse introduced the "Hessian determinant" in a paper in 1842 as a part of the investigation of quadratic and cubic curves.

In Example 11.3.4 the mixed partial derivatives were equal but, as we have seen in Example 11.1.6, it can happen that $f''_{xy} \neq f''_{yx}$. Is there a way to predict whether changing the order of differentiation will give the same result or not? This is a question that can be asked when f depends on any number of variables. Yet, since we are exchanging the order with respect to *two* specific variables, we may as well assume that the function f depends precisely on these two variables.

Theorem 11.3.5. *Let f be a function defined on an open disk A in $\mathbb{R}^2$ and let $(x_0, y_0) \in A$. Suppose that the partial derivatives f'_x, f'_y, and f''_{xy} exist in A, and that f''_{xy} is continuous at (x_0, y_0). Then f''_{yx} exists at (x_0, y_0) and $f''_{yx}(x_0, y_0) = f''_{xy}(x_0, y_0)$.*

Proof. Let h and k be non-zero real numbers such that $(x_0 + h, y_0 + k)$, $(x_0 + h, y_0)$, and $(x_0, y_0 + k)$ belong to A, and let

$$W(h, k) = \frac{f(x_0 + h, y_0 + k) - f(x_0 + h, y_0) - f(x_0, y_0 + k) + f(x_0, y_0)}{hk}.$$

If we define

$$\varphi(x) = \frac{f(x, y_0 + k) - f(x, y_0)}{k},$$

then φ is differentiable in A. Using the Mean Value Theorem twice,

$$W(h, k) = \frac{\varphi(x_0 + h) - \varphi(x_0)}{h} = \varphi'(c_1) = \frac{f'_x(c_1, y_0 + k) - f'_x(c_1, y_0)}{k} = f''_{xy}(c_1, c_2),$$

where c_1 is between x_0 and $x_0 + h$, and c_2 is between y_0 and $y_0 + k$. Since f''_{xy} is continuous at (x_0, y_0), it follows that, when $h, k \to 0$, $W(h, k) \to f''_{xy}(x_0, y_0)$.

Let $\varepsilon > 0$. Now that we know that $\lim_{(h,k) \to (0,0)} W(h, k)$ exists, it follows that there exists $\delta > 0$ such that

$$0 < |h|, |k| < \delta \quad \Rightarrow \quad \left| W(h, k) - f''_{xy}(x_0, y_0) \right| < \frac{\varepsilon}{2}.$$

If we let $k \to 0$, we obtain that, for $0 < |h| < \delta$,

$$\left| \frac{f'_y(x_0 + h, y_0) - f'_y(x_0, y_0)}{h} - f''_{xy}(x_0, y_0) \right| \leq \frac{\varepsilon}{2}.$$

This implies that

$$\lim_{h \to 0} \frac{f'_y(x_0 + h, y_0) - f'_y(x_0, y_0)}{h}$$

exists and equals $f''_{xy}(x_0, y_0)$. Since this limit is, by definition, $f''_{yx}(x_0, y_0)$, the theorem is proved. $\qquad \square$

Several "proofs" of the equality of the mixed partial derivatives appeared in the eighteenth and the nineteenth century. Finally, in 1873 Schwarz produced the first counterexample in [93] (see Exercise 11.3.15). In the same text he gave a sufficient condition for the equality and the first correct proof. Peano made improvements on both results. In 1884 he came up with (simpler) Example 11.1.6, and in 1889 with (stronger) Theorem 11.3.5.

HERMANN SCHWARZ (1843–1921) was a German mathematician. He studied chemistry in Berlin, before Kummer and Weierstrass persuaded him to change to mathematics. He received his Ph.D. from the University of Berlin, with these two as advisors. He held positions at the University of Halle, ETH in Zürich, and the University of Göttingen. By the time he returned to the University of Berlin his best days were behind him. In the words of German mathematician Ludwig Bieberbach (1886–1982), "Schwarz retired to Berlin". It is possible that he did not have time for research because he was a dedicated teacher (with 20 Ph.D. students in 25 years), the captain of the local Voluntary Fire Brigade and, more surprisingly, he assisted the stationmaster at the local railway station by closing the doors of the trains. In research, his ideas were usually inspired by geometry. He excelled in Complex Analysis, especially in the theory of *conformal mappings*. Among other things, Schwarz improved the proof of the Riemann mapping theorem, developed a special case of the Cauchy–Schwarz inequality, and gave a proof that the ball has less surface area than any other body of equal volume. His work on the latter inspired Picard's proof of the existence of solutions of differential equations

Exercises

In Exercises 11.3.1–11.3.7 find total differentials df and d^2f:

11.3.1. $f(x,y) = x^4 + y^4 - 4x^2y^2$.   **11.3.2.** $f(x,y) = \dfrac{x}{\sqrt{x^2 + y^2}}$.

11.3.3. $f(x,y) = x\sin(x+y)$.   **11.3.4.** $f(x,y) = x^y$.

11.3.5. $f(x,y) = \ln(x+y^2)$.   **11.3.6.** $f(x,y) = \arctan\dfrac{x+y}{1-xy}$.

11.3.7. $f(x,y,z) = \dfrac{z}{x^2 + y^2}$.

In Exercises 11.3.8–11.3.14 find $\mathbf{D}f(\mathbf{a})$ and $\mathbf{D}^2f(\mathbf{a})$:

11.3.8. $f(x,y) = \dfrac{x-y}{x+y}$, $\mathbf{a} = (\frac{1}{2}, \frac{3}{2})$.   **11.3.9.** $f(r,\theta) = r\sin\theta$, $\mathbf{a} = (5, \frac{\pi}{6})$.

11.3.10. $f(x,y) = \ln(x+y)$, $\mathbf{a} = (1,2)$.   **11.3.11.** $f(x,y,z) = \arctan x + yz$, $\mathbf{a} = (0,3,1)$.

11.3.12. $f(x,y,z) = x + ye^z$, $\mathbf{a} = (1,1,0)$. **11.3.13.** $f(\mathbf{x}) = \|\mathbf{x}\|^2$, $\mathbf{a} = (1,1,\ldots,1)$.

11.3.14. $f(x_1, x_2, x_3, x_4) = \dfrac{x_1 + x_2}{x_3 + x_4}$, $\mathbf{a} = (1,1,1,1)$.

11.3.15. Let

$$f(x,y) = \begin{cases} x^2 \arctan\frac{y}{x} - y^2 \arctan\frac{x}{y}, & \text{if } xy \neq 0 \\ 0, & \text{if } xy = 0. \end{cases}$$

Prove that the second-order mixed partial derivatives at $(0,0)$ are not equal.

11.3.16.* Let f be a function defined on an open disk A in $\mathbb{R}^2$ and let $(x_0, y_0) \in A$. Suppose that the partial derivatives f'_x and f'_y exist in A, and that they are both differentiable at (x_0, y_0). Prove that $f''_{yx}(x_0, y_0) = f''_{xy}(x_0, y_0)$.

11.4 Functions from $\mathbb{R}^n$ to $\mathbb{R}^m$

It is often important to consider functions of n variables with values not real numbers but m-tuples of real numbers. For example,

$$\mathbf{f}(x,y) = (xe^y, \; 3x^2 - y, \; \sin x \cos y)$$

is a function that is defined on $\mathbb{R}^2$ and its values lie in $\mathbb{R}^3$. This function can be also viewed as a triple of functions:

$$f_1(x,y) = xe^y, \quad f_2(x,y) = 3x^2 - y, \quad f_3(x,y) = \sin x \cos y.$$

In general, if $\mathbf{f} : \mathbb{R}^n \to \mathbb{R}^m$, we can identify it with an ordered m-tuple of functions $(f_1, f_2, \ldots, f_m)$, where the **component functions** $f_i : \mathbb{R}^n \to \mathbb{R}$, $1 \leq i \leq m$. We will use this approach to generalize some of the results that hold when $m = 1$, especially regarding the continuity and differentiability of such functions. We start with the definition of the limit.

Definition 11.4.1. Let $\mathbf{f} = (f_1, f_2, \ldots, f_m)$ be a function defined in a domain $A \subset \mathbb{R}^n$, with values in $\mathbb{R}^m$, and let $\mathbf{a}$ be a cluster point of A. We say that $\mathbf{L} = (L_1, L_2, \ldots, L_m)$ is the limit of $\mathbf{f}$ as $\mathbf{x}$ approaches to $\mathbf{a}$, and we write $\lim_{\mathbf{x}\to\mathbf{a}} \mathbf{f}(\mathbf{x}) = \mathbf{L}$, if $\lim_{\mathbf{x}\to\mathbf{a}} f_i(\mathbf{x}) = L_i$, $1 \leq i \leq m$.

Example 11.4.2. The limit of a function.

Let $\mathbf{f}(x,y) = (3x^2 + y, e^x \cos y)$. Find $\lim_{(x,y)\to(0,0)} \mathbf{f}(x,y)$.

Solution. Since $f_1(x,y) = 3x^2 + y$ and $f_2(x,y) = e^x \cos y$, we calculate

$$\lim_{(x,y)\to(0,0)} f_1(x,y) = 0, \quad \text{and} \quad \lim_{(x,y)\to(0,0)} f_2(x,y) = 1,$$

so $\lim_{(x,y)\to(0,0)} \mathbf{f}(x,y) = (0,1)$.   ♦

With the definition of the limit, we can define continuity.

Definition 11.4.3. Let $\mathbf{f} = (f_1, f_2, \ldots, f_m)$ be a function defined in a domain $A \subset \mathbb{R}^n$, with values in $\mathbb{R}^m$, and let $\mathbf{a} \in A$. Then $\mathbf{f}$ is **continuous** at $\mathbf{a}$ if $\lim_{\mathbf{x}\to\mathbf{a}} \mathbf{f}(\mathbf{x})$ exists and equals $\mathbf{f}(\mathbf{a})$. If $\mathbf{f}$ is continuous at every point of A, then $\mathbf{f}$ is continuous on A.

The following result is a direct consequence of the definition of the limit.

Theorem 11.4.4. *Let* $\mathbf{f} = (f_1, f_2, \ldots, f_m)$ *be a function defined in a domain* $A \subset \mathbb{R}^n$, *with values in* $\mathbb{R}^m$, *and let* $\mathbf{a} \in A$. *Then* $\mathbf{f}$ *is continuous at* $\mathbf{a}$ *if and only if* f_i *is continuous at* $\mathbf{a}$ *for all* $1 \le i \le m$.

We stop to notice a lack of symmetry when it comes to continuity. Namely, Theorem 11.4.4 asserts that the continuity of $\mathbf{f}$ is equivalent to the continuity of its components $f_1, f_2, \ldots, f_m$. On the other hand, Example 10.3.9 shows that the continuity of $\mathbf{f}$ is *not* equivalent to the continuity in each variable x_i.

It is fairly easy to see that theorems about combinations of continuous functions still hold. Of course, we cannot consider the quotient of functions, because division is not defined in $\mathbb{R}^m$. Also, the product is really the inner product.

Theorem 11.4.5. *Let* $\mathbf{f}, \mathbf{g}$ *be two functions with a domain* $A \subset \mathbb{R}^n$ *and values in* $\mathbb{R}^m$, *and let* $\mathbf{a} \in A$. *Also, let* α *be a real number. If* $\lim_{\mathbf{x}\to\mathbf{a}} \mathbf{f}(\mathbf{x}) = \mathbf{L}$ *and* $\lim_{\mathbf{x}\to\mathbf{a}} \mathbf{g}(\mathbf{x}) = \mathbf{K}$ *then:*

(a) $\lim_{\mathbf{x}\to\mathbf{a}} [\alpha\mathbf{f}(\mathbf{x})] = \alpha \lim_{\mathbf{x}\to\mathbf{a}} \mathbf{f}(\mathbf{x})$;

(b) $\lim_{\mathbf{x}\to\mathbf{a}} [\mathbf{f}(\mathbf{x}) + \mathbf{g}(\mathbf{x})] = \lim_{\mathbf{x}\to\mathbf{a}} \mathbf{f}(\mathbf{x}) + \lim_{\mathbf{x}\to\mathbf{a}} \mathbf{g}(\mathbf{x})$;

(c) $\lim_{\mathbf{x}\to\mathbf{a}} [\mathbf{f}(\mathbf{x}) \cdot \mathbf{g}(\mathbf{x})] = [\lim_{\mathbf{x}\to\mathbf{a}} \mathbf{f}(\mathbf{x})] \cdot [\lim_{\mathbf{x}\to\mathbf{a}} \mathbf{g}(\mathbf{x})]$.

Proof. We will prove only assertion (a), and leave the rest as an exercise. Since $\mathbf{f} = (f_1, f_2, \ldots, f_m)$, then $\alpha\mathbf{f} = (\alpha f_1, \alpha f_2, \ldots, \alpha f_m)$. Therefore, using Definition 11.4.1,

$$\begin{aligned}
\lim_{\mathbf{x}\to\mathbf{a}} [\alpha\mathbf{f}(\mathbf{x})] &= \lim_{\mathbf{x}\to\mathbf{a}} (\alpha f_1(\mathbf{x}), \alpha f_2(\mathbf{x}), \ldots, \alpha f_m(\mathbf{x})) \\
&= \left(\lim_{\mathbf{x}\to\mathbf{a}} \alpha f_1(\mathbf{x}), \lim_{\mathbf{x}\to\mathbf{a}} \alpha f_2(\mathbf{x}), \ldots, \lim_{\mathbf{x}\to\mathbf{a}} \alpha f_m(\mathbf{x}) \right) \\
&= \alpha \left(\lim_{\mathbf{x}\to\mathbf{a}} f_1(\mathbf{x}), \lim_{\mathbf{x}\to\mathbf{a}} f_2(\mathbf{x}), \ldots, \lim_{\mathbf{x}\to\mathbf{a}} f_m(\mathbf{x}) \right) \\
&= \alpha \lim_{\mathbf{x}\to\mathbf{a}} (f_1(\mathbf{x}), f_2(\mathbf{x}), \ldots, f_m(\mathbf{x})) \\
&= \alpha \lim_{\mathbf{x}\to\mathbf{a}} \mathbf{f}(\mathbf{x}). \qquad \square
\end{aligned}$$

It follows immediately that all of these procedures preserve continuity.

Corollary 11.4.6. *Let* $\mathbf{f}, \mathbf{g}$ *be two functions with a domain* $A \subset \mathbb{R}^n$ *and values in* $\mathbb{R}^m$, *and let* $\mathbf{a} \in A$. *Also, let* α *be a real number and let* φ *be a real-valued function defined on* A. *If* $\mathbf{f}$, $\mathbf{g}$, *and* φ *are continuous at* $\mathbf{a}$ *then the same is true for: (a)* $\alpha\mathbf{f}$; *(b)* $\mathbf{f} + \mathbf{g}$; *(c)* $\varphi\mathbf{f}$; *(d)* $\mathbf{f} \cdot \mathbf{g}$.

Further, the composition of functions preserve continuity. That was proved in Theorem 10.3.6.

This brings us to the topic of derivatives. We have defined the derivative $\mathbf{D}f(\mathbf{a})$, in the case $m = 1$, as an $1 \times n$ matrix that satisfies (11.8):

$$f(\mathbf{x}) = f(\mathbf{a}) + \mathbf{D}f(\mathbf{a})(\mathbf{x} - \mathbf{a}) + r(\mathbf{x}), \tag{11.16}$$

where $r \to 0$. What if $m \neq 1$? Can we have a formula of the form (11.16)?

Let us consider the case when $m = n = 2$. Then $\mathbf{f} = (f_1, f_2)$ and, assuming that f_1 and f_2 are differentiable (so that (11.16) holds),

$\mathbf{f}(x, y) = (f_1(x, y), f_2(x, y))$

$$= \left(f_1(a, b) + \mathbf{D}f_1(a, b)(x - a, y - b) + r_1, \ f_2(a, b) + \mathbf{D}f_2(a, b)(x - a, y - b) + r_2 \right)$$

$$= \left(f_1(a, b), f_2(a, b) \right) + \left(\mathbf{D}f_1(a, b)(x - a, y - b), \mathbf{D}f_2(a, b)(x - a, y - b) \right) + (r_1, r_2).$$

Does this look like (11.16)? The first term is $\mathbf{f}(a, b)$, the last goes to 0, so it comes down to the middle one. Let us write $\mathbf{D}f_1(a, b)$ as the matrix $\begin{bmatrix} A_1 & B_1 \end{bmatrix}$, and $\mathbf{D}f_2(a, b)$ as the matrix $\begin{bmatrix} A_2 & B_2 \end{bmatrix}$. Then,

$$\left(\mathbf{D}f_1(a, b)(x - a, y - b), \mathbf{D}f_2(a, b)(x - a, y - b) \right) = \begin{bmatrix} A_1 & B_1 \\ A_2 & B_2 \end{bmatrix} (x - a, y - b)^T.$$

Thus, in this example, we have an expression like (11.16), with the role of $\mathbf{D}f$ played by this 2×2 matrix. What determines its dimensions? The reason that it has 2 rows is that there are 2 component functions f_1 and f_2. On the other hand, the number of columns was determined by the matrices for $\mathbf{D}f_1(a, b)$ and $\mathbf{D}f_2(a, b)$, and their number of columns equals the number of variables $x_1, x_2, \ldots, x_n$. In general, if $\mathbf{f}$ is a function defined in a domain $A \subset \mathbb{R}^n$, with values in $\mathbb{R}^m$, the matrix has m rows and n columns. This encourages us to use the following definition:

Definition 11.4.7. Let $\mathbf{f} = (f_1, f_2, \ldots, f_m)$ be a function defined in an open ball $A \subset \mathbb{R}^n$, with values in $\mathbb{R}^m$, and let $\mathbf{a} \in A$. Then $\mathbf{f}$ is **differentiable** at $\mathbf{a}$ if and only if there exists an $m \times n$ matrix $\mathbf{Df}(\mathbf{a})$, called the (**total**) **derivative** of $\mathbf{f}$ at $\mathbf{a}$, such that

$$\mathbf{f}(\mathbf{x}) = \mathbf{f}(\mathbf{a}) + \mathbf{Df}(\mathbf{a})(\mathbf{x} - \mathbf{a}) + \mathbf{r}(\mathbf{x}), \quad \text{and}$$

$$\lim_{\mathbf{x} \to \mathbf{a}} \frac{\mathbf{r}(\mathbf{x})}{\|\mathbf{x} - \mathbf{a}\|} = \mathbf{0}. \tag{11.17}$$

If $\mathbf{f}$ is differentiable at every point of a set A, we say that it is **differentiable** on A.

We are making a standard identification between elements of the Euclidean space of dimension n, and $n \times 1$ matrices. For example, $\mathbf{f}(\mathbf{x}) = (f_1(\mathbf{x}), f_2(\mathbf{x}), \ldots, f_m(\mathbf{x}))$ can be viewed as a column matrix of dimension $m \times 1$. That means that the first equation in (11.17) states an equality between matrices. If we read it row by row, we can conclude several things. First, the rows of $\mathbf{Df}$ are precisely the partial derivatives of the functions $f_1, f_2, \ldots, f_m$. It follows that

$$\mathbf{Df}(\mathbf{a}) = \begin{bmatrix} \dfrac{\partial f_1}{\partial x_1} & \dfrac{\partial f_1}{\partial x_2} & \cdots & \dfrac{\partial f_1}{\partial x_n} \\[2mm] \dfrac{\partial f_2}{\partial x_1} & \dfrac{\partial f_2}{\partial x_2} & \cdots & \dfrac{\partial f_2}{\partial x_n} \\[2mm] \vdots & \vdots & \ddots & \vdots \\[2mm] \dfrac{\partial f_m}{\partial x_1} & \dfrac{\partial f_m}{\partial x_2} & \cdots & \dfrac{\partial f_m}{\partial x_n} \end{bmatrix}.$$

Second, $\mathbf{f}$ is differentiable at $\mathbf{a}$ if and only if f_i is differentiable at $\mathbf{a}$ for all $1 \le i \le m$. Finally, all the rules for derivatives hold, simply because they hold for each of the component functions.

Theorem 11.4.8. *Let $\mathbf{f}, \mathbf{g}$ be two functions with a domain an open ball $A \subset \mathbb{R}^n$ and values in $\mathbb{R}^m$, and let $\mathbf{a} \in A$. Also, let α be a real number and let φ be a real-valued function defined on A. If $\mathbf{f}$, $\mathbf{g}$, and φ are differentiable at $\mathbf{a}$ then the same is true for $\mathbf{f} + \mathbf{g}$, $\alpha \mathbf{f}$, $\varphi \mathbf{f}$, and $\mathbf{f} \cdot \mathbf{g}$:*

(a) $\mathbf{D}(\alpha \mathbf{f})(\mathbf{a}) = \alpha \mathbf{Df}(\mathbf{a})$;

(b) $\mathbf{D}(\mathbf{f} + \mathbf{g})(\mathbf{a}) = \mathbf{Df}(\mathbf{a}) + \mathbf{Dg}(\mathbf{a})$;

(c) $\mathbf{D}(\varphi \mathbf{f})(\mathbf{a}) = \mathbf{f}(\mathbf{a})\mathbf{D}\varphi(\mathbf{a}) + \varphi(\mathbf{a})\mathbf{Df}(\mathbf{a})$;

(d) $\mathbf{D}(\mathbf{f} \cdot \mathbf{g})(\mathbf{a}) = \mathbf{g}(\mathbf{a})\mathbf{Df}(\mathbf{a}) + \mathbf{f}(\mathbf{a})\mathbf{Dg}(\mathbf{a})$.

Remark 11.4.9. In order to make assertion (d) as close to the "usual" product rule, we had to modify the meaning of the objects on the right hand side. Namely, $\mathbf{g}(\mathbf{a})$ and $\mathbf{f}(\mathbf{a})$ should be understood as $1 \times m$ matrices.

> The matrix for $\mathbf{Df}(\mathbf{a})$ is called the Jacobian matrix after Jacobi (see next page). However, the modern notion of the derivative came much later. The case $m = 1$ brew for a while until Stolz came up with a precise version in 1893. Strangely, the case $m > 1$ was first formulated and studied by Fréchet, in a much more abstract situation. In a 1925 paper [47] he developed the theory of functions between two abstract normed vector spaces. Since the Euclidean spaces $\mathbb{R}^n$ are prime examples of such spaces, his definition applies to them, and coincides with Definition 11.4.7.

Next, we will address the Chain Rule.

Theorem 11.4.10. *Let A be an open ball in $\mathbb{R}^n$, and let $\mathbf{f} : A \to \mathbb{R}^m$. Further, let B be an open set in $\mathbb{R}^m$ that contains the range of $\mathbf{f}$, and let $\mathbf{g} : B \to \mathbb{R}^p$. If $\mathbf{f}$ is differentiable at $\mathbf{a} \in A$, and if $\mathbf{g}$ is differentiable at $\mathbf{f}(\mathbf{a})$ then the composition $\mathbf{g} \circ \mathbf{f}$ is differentiable at $\mathbf{a}$, and*

$$\mathbf{D}(\mathbf{g} \circ \mathbf{f})(\mathbf{a}) = \mathbf{Dg}(\mathbf{f}(\mathbf{a}))\mathbf{Df}(\mathbf{a}).$$

Notice that the right side represents a product of matrices.

Proof. Since $\mathbf{f}$ is differentiable at $\mathbf{a}$ we have

$$\mathbf{f}(\mathbf{x}) = \mathbf{f}(\mathbf{a}) + \mathbf{Df}(\mathbf{a})(\mathbf{x} - \mathbf{a}) + \mathbf{r_f}, \tag{11.18}$$

where $\mathbf{r_f}/\|\mathbf{x} - \mathbf{a}\| \to 0$, $\mathbf{x} \to \mathbf{a}$. If we denote $\mathbf{b} = \mathbf{f}(\mathbf{a})$, then the fact that $\mathbf{g}$ is differentiable at $\mathbf{f}(\mathbf{a})$ can be written as

$$\mathbf{g}(\mathbf{y}) = \mathbf{g}(\mathbf{b}) + \mathbf{Dg}(\mathbf{b})(\mathbf{y} - \mathbf{b}) + \mathbf{r_g}, \tag{11.19}$$

where $\mathbf{r_g}/\|\mathbf{y} - \mathbf{b}\| \to 0$, $\mathbf{y} \to \mathbf{b}$. It follows from (11.18) that

$$\mathbf{g}\big(\mathbf{f}(\mathbf{x})\big) = \mathbf{g}\big(\mathbf{f}(\mathbf{a}) + \mathbf{Df}(\mathbf{a})(\mathbf{x} - \mathbf{a}) + \mathbf{r_f}\big)$$

so, with $\mathbf{y} = \mathbf{f}(\mathbf{a}) + \mathbf{Df}(\mathbf{a})(\mathbf{x} - \mathbf{a}) + \mathbf{r_f}$, (11.19) implies that

$$\begin{aligned}
\mathbf{g}\big(\mathbf{f}(\mathbf{x})\big) &= \mathbf{g}\big(\mathbf{f}(\mathbf{a})\big) + \mathbf{Dg}(\mathbf{f}(\mathbf{a}))\big(\mathbf{Df}(\mathbf{a})(\mathbf{x} - \mathbf{a}) + \mathbf{r_f}\big) + \mathbf{r_g} \\
&= \mathbf{g}\big(\mathbf{f}(\mathbf{a})\big) + \mathbf{Dg}(\mathbf{f}(\mathbf{a}))\mathbf{Df}(\mathbf{a})(\mathbf{x} - \mathbf{a}) + \mathbf{Dg}(\mathbf{f}(\mathbf{a}))\mathbf{r_f} + \mathbf{r_g}.
\end{aligned}$$

Thus, it remains to show that

$$\frac{\mathbf{Dg}(\mathbf{f}(\mathbf{a}))\mathbf{r_f} + \mathbf{r_g}}{\|\mathbf{x} - \mathbf{a}\|} \to 0, \quad \mathbf{x} \to \mathbf{a}.$$

By definition of $\mathbf{r_f}$, $\mathbf{r_f}/\|\mathbf{x} - \mathbf{a}\| \to \mathbf{0}$, so

$$\frac{\mathbf{Dg(f(a))r_f}}{\|\mathbf{x} - \mathbf{a}\|} \to \mathbf{0}.$$

Further, if $\mathbf{x} \to \mathbf{a}$, then

$$\mathbf{y} = \mathbf{f(a)} + \mathbf{Df(a)(x - a)} + \mathbf{r_f} \to \mathbf{f(a)} = \mathbf{b}.$$

Therefore, when $\mathbf{x} \to \mathbf{a}$, we have that $\mathbf{r_g}/\|\mathbf{y} - \mathbf{b}\| \to \mathbf{0}$. In other words,

$$\frac{\mathbf{r_g}}{\|\mathbf{Df(a)(x - a)} + \mathbf{r_f}\|} \to \mathbf{0}.$$

Now,

$$\frac{\mathbf{r_g}}{\|\mathbf{x} - \mathbf{a}\|} = \frac{\mathbf{r_g}}{\|\mathbf{Df(a)(x - a)} + \mathbf{r_f}\|} \frac{\|\mathbf{Df(a)(x - a)} + \mathbf{r_f}\|}{\|\mathbf{x} - \mathbf{a}\|} \to 0,$$

because the second fraction is bounded:

$$\frac{\|\mathbf{Df(a)(x - a)} + \mathbf{r_f}\|}{\|\mathbf{x} - \mathbf{a}\|} \le \|\mathbf{Df(a)}\| + \frac{\|\mathbf{r_f}\|}{\|\mathbf{x} - \mathbf{a}\|}.$$

This completes the proof. $\qquad\square$

CARL GUSTAV JACOBI (1804–1851) was a German mathematician. He came from a Jewish family in Potsdam and was given the name Jacques Simon. He was home schooled until the age of 12. He was so well prepared and talented that he was at that time put in the final year of high school. After that, he had to wait to be 16 in order to be admitted to the university, so he studied on his own. For example, he read Euler's [41]. When he enrolled at the University of Berlin he continued this practice, because the level of classes was not very high. After graduating, he taught in the best high school in Berlin. In 1825 he converted to Christianity making it possible to teach at a university. Indeed, he started at the University of Berlin, but moved to Königsburg in 1826. Jacobi's reputation as an excellent teacher attracted many students. He introduced the seminar method to teach the latest advances in mathematics. In 1843 he was diagnosed with diabetes and advised to go to Italy. Dirichlet urged von Humboldt, a scientist with strong ties to the court, to obtain support for Jacobi's trip. He received the money as well as the permission to relocate to Berlin and a stipend to offset the higher cost of living in the capital. During the upheaval of 1848 he did not endear himself to the government so he lost his stipend. Only after he accepted a position in Vienna, an arrangement was made for him to stay in Berlin. He died from smallpox in 1851. One of Jacobi's greatest accomplishments was his theory of elliptic functions which were used both in Mathematical Physics as well as in Number Theory. At the time he was the world's leading expert. He was one of the founders of the use of determinants. The Jacobian determinant bears his name, although it appeared in a Cauchy's paper from 1815.

Exercises

In Exercises 11.4.1–11.4.4 calculate the limits:

11.4.1. $\displaystyle\lim_{(x,y)\to(1,3)} \left(x + 3y - 1, x^2 - y^2, \sin \pi x\right)$. 11.4.2. $\displaystyle\lim_{(x,y)\to(-2,1)} \left(\frac{x}{y}, \ln(y - 3x), e^y\right)$.

11.4.3. $\displaystyle\lim_{(x,y)\to(0,0)} \left(\frac{\sin xy}{xy}, \frac{\ln(1 + xy^2)}{3xy^2}\right)$. 11.4.4. $\displaystyle\lim_{(x,y)\to(0,0)} \left(\frac{e^{x+y} - 1}{x + y}, \frac{\sqrt{x} - \sqrt{y}}{x - y}\right)$.

11.4.5. Prove Theorem 11.4.5 (b) and (c).

In Exercises 11.4.6–11.4.9 determine if $\mathbf{f}$ is continuous in its domain:

11.4.6. $\mathbf{f}(x, y) = \left(\dfrac{1}{x + y}, x + y\right)$. 11.4.7. $\mathbf{f}(x, y) = \left(\dfrac{e^x e^y}{2x^2 + y^4 + 3}, \sin\left(\dfrac{xy}{y^2 - 1}\right)\right)$.

11.4.8. $\mathbf{f}(x,y) = \left(\dfrac{e^x}{y}, \, xe^{-y} \right).$ **11.4.9.** $\mathbf{f}(x,y) = \left(\sqrt{x^2 + y^2}, \, \dfrac{1}{\sqrt{x^2 + y^2}} \right).$

In Exercises 11.4.10–11.4.14 find $\mathbf{Df}(\mathbf{a})$:

11.4.10. $\mathbf{f}(x,y) = (x^2 - y^2, \, 2xy)$, $\mathbf{a} = (3,1)$.

11.4.11. $\mathbf{f}(x,y) = \left(\arctan \dfrac{y}{x}, \, \dfrac{x}{x^2 + y^2}, \, \dfrac{y}{x^2 + y^2} \right)$, $\mathbf{a} = (1,1)$.

11.4.12. $\mathbf{f}(x,y,z) = (z^2 - y^2, \, x^2 - z^2, \, y^2 - x^2)$, $\mathbf{a} = (\frac{1}{2}, \frac{1}{3}, \frac{1}{6})$.

11.4.13. $\mathbf{f}(x_1, x_2, x_3, x_4, x_5) = (x_1 e^{x_2}, \, x_3 e^{-x_4}, \, x_5 e^{x_1})$, $\mathbf{a} = (0, -\ln 6, 4, \ln 2, 11)$.

11.4.14. $\mathbf{f}(x) = (5x + 1, 2\cos 3x, 3\sin 3x)$, $a = (\frac{\pi}{6})$.

11.4.15. Prove Theorem 11.4.8.

In Exercises 11.4.16–11.4.19 use the given data to find an approximation for $\mathbf{f}(\mathbf{x})$.

11.4.16. $\mathbf{f}(3,-1) = (2,6)$, $\mathbf{Df}(3,-1) = \begin{bmatrix} 2 & \frac{8}{3} \\ -3 & 2 \end{bmatrix}$, $\mathbf{x} = (3.1, -0.8)$.

11.4.17. $\mathbf{f}(0,0,0) = (1,-1)$, $\mathbf{Df}(0,0,0) = \begin{bmatrix} 0 & 2 & 1 \\ 4 & 3 & 5 \end{bmatrix}$, $\mathbf{x} = (0.0125, -0.1, 0.067)$.

11.4.18. $f(0,0,1,0) = 14$, $\mathbf{Df}(0,0,1,0) = \begin{bmatrix} -\frac{1}{3} & -\frac{1}{2} & \frac{2}{3} & 1 \end{bmatrix}$, $\mathbf{x} = (-0.01, -0.03, 1.02, 0.04)$.

11.4.19. $\mathbf{f}(4) = (3.21, -5.05, 4.8)$, $\mathbf{Df}(4) = \begin{bmatrix} 12 \\ 0 \\ -1 \end{bmatrix}$, $x = 4.13$.

11.4.20. Prove that if $\mathbf{f} : \mathbb{R}^n \to \mathbb{R}^n$ is a linear transformation, then $\mathbf{Df}(\mathbf{a})(\mathbf{x}) = \mathbf{f}(\mathbf{x})$.

A function $\mathbf{f} : \mathbb{R}^n \times \mathbb{R}^p \to \mathbb{R}^m$ is **bilinear** if it is linear in each component. That is, $\mathbf{f}(\alpha \mathbf{x} + \beta \mathbf{y}, \mathbf{z}) = \alpha \mathbf{f}(\mathbf{x}, \mathbf{z}) + \beta \mathbf{f}(\mathbf{y}, \mathbf{z})$, for all $\mathbf{x}, \mathbf{y} \in \mathbb{R}^n$ and $\mathbf{z} \in \mathbb{R}^p$, and all $\alpha, \beta \in \mathbb{R}$, with a similar formula for the second argument.

11.4.21.* Let $\mathbf{f} : \mathbb{R}^n \times \mathbb{R}^p \to \mathbb{R}^m$ be a bilinear map and $\mathbf{a} \in \mathbb{R}^n$, $\mathbf{b} \in \mathbb{R}^p$. Prove that $\mathbf{f}$ is differentiable and that $\mathbf{Df}(\mathbf{a}, \mathbf{b})(\mathbf{x}, \mathbf{y}) = \mathbf{f}(\mathbf{a}, \mathbf{y}) + \mathbf{f}(\mathbf{x}, \mathbf{b})$, for all $\mathbf{x} \in \mathbb{R}^n$, $\mathbf{y} \in \mathbb{R}^p$.

11.5   Taylor's Formula

In this section we will generalize Taylor's Formula (Theorem 4.5.1) to the case when f depends on more than one variable. Given $n \in \mathbb{N}$, the task is to come up with a polynomial of degree n that approximates f, and to estimate the error of the approximation.

For example, let $f(x,y) = 3x^2 + 5xy - 7y^2 + 8x + 4y - 11$. How can we express the coefficients in terms of the derivatives of f?

We have already seen in (11.11) and (11.12) that

$$3x^2 + 5xy - 7y^2 = \frac{1}{2!} \left(f''_{xx}(0,0)\, x^2 + f''_{yy}(0,0)\, y^2 + f''_{xy}(0,0)\, yx + f''_{yx}(0,0)\, xy \right).$$

Also, formula (11.3) shows that

$$8x + 4y - 11 = f(0,0) + f'_x(0,0)\, x + f'_y(0,0)\, y.$$

Together, we have that

$$f(x, y) = f(0, 0) + \mathbf{D}f(0, 0)(x, y) + \frac{1}{2!}\mathbf{D}^2 f(0, 0)(x, y)^2.$$

This gives us hope that we can use the same formula to *approximate* a function that is not a second-degree polynomial. What if we want to improve such an approximation? How do we define the third derivative $\mathbf{D}^3 f$? Once again, we can use the same strategy which led to formulas (11.11) and (11.12): take a polynomial (of degree at least 3) and find the relationship between the coefficients of the polynomial and its partial derivatives. A more sophisticated approach is to use the second formula in (11.15) and find the total differential:

$$
\begin{aligned}
d^3 f = d(d^2 f) &= d \left(f''_{xx} \, dx^2 + f''_{xy} \, dydx + f''_{yx} \, dxdy + f''_{yy} \, dy^2 \right) \\
&= \left(f'''_{xxx} \, dx^2 + f'''_{xyx} \, dydx + f'''_{yxx} \, dxdy + f'''_{yyx} \, dy^2 \right) dx \\
&\quad + \left(f'''_{xxy} \, dx^2 + f'''_{xyy} \, dydx + f'''_{yxy} \, dxdy + f'''_{yyy} \, dy^2 \right) dy \qquad (11.20) \\
&= f'''_{xxx} \, dx^3 + \left(f'''_{xyx} + f'''_{yxx} + f'''_{xxy} \right) dx^2 dy \\
&\quad + \left(f'''_{yyx} + f'''_{xyy} + f'''_{yxy} \right) dx dy^2 + f'''_{yyy} \, dy^3,
\end{aligned}
$$

which is a cubic expression. Therefore, we define

$$\mathbf{D}^3 f(\mathbf{a})(\mathbf{u})^3 = \sum_{i=1}^{n} \sum_{j=1}^{n} \sum_{k=1}^{n} f'''_{x_i x_j x_k}(\mathbf{a}) \, u_i u_j u_k. \qquad (11.21)$$

If we take $n = 2$ and $\mathbf{u} = (dx, dy)$ in (11.21), we obtain (11.20). Following along the same lines, we define $\mathbf{D}^m f$, with f a function of n variables, for any $m, n \in \mathbb{N}$, as

$$\mathbf{D}^m f(\mathbf{a})(\mathbf{u})^m = \sum_{i_1=1}^{n} \sum_{i_2=1}^{n} \cdots \sum_{i_m=1}^{n} f^{(m)}_{x_{i_1} x_{i_2} \ldots x_{i_m}}(\mathbf{a}) \, u_{i_1} u_{i_2} \ldots u_{i_m}.$$

Now we have all we need to state and prove the Taylor's Formula in several variables.

Theorem 11.5.1 (Taylor's Formula). *Let $m \in \mathbb{N}_0$ and suppose that a function f and all of its partial derivatives of order up to m are differentiable in an n-ball A. If $\mathbf{a} \in A$, then for any $\mathbf{x} \in A$, $\mathbf{x} \neq \mathbf{a}$, there exists a point $\mathbf{b}$ on the line segment connecting $\mathbf{a}$ and $\mathbf{x}$, such that*

$$f(\mathbf{x}) = f(\mathbf{a}) + \mathbf{D}f(\mathbf{a})(\mathbf{x} - \mathbf{a}) + \frac{\mathbf{D}^2 f(\mathbf{a})}{2!}(\mathbf{x} - \mathbf{a})^2 + \cdots$$

$$\cdots + \frac{\mathbf{D}^m f(\mathbf{a})}{m!}(\mathbf{x} - \mathbf{a})^m + \frac{\mathbf{D}^{m+1} f(\mathbf{b})}{(m+1)!}(\mathbf{x} - \mathbf{a})^{m+1}.$$

Proof. For a real number $t \in [0, 1]$ we define $F(t) = f(\mathbf{a} + t(\mathbf{x} - \mathbf{a}))$. This is a real-valued function of one variable, and we would like to apply the Taylor Theorem in one variable (Theorem 4.5.1). Therefore, we need to verify that F is differentiable enough times.

Notice that $F(t) = f(\mathbf{g}(t))$, where $\mathbf{g}$ is a vector valued function $\mathbf{g} : \mathbb{R} \to \mathbb{R}^n$, defined by $\mathbf{g}(t) = \mathbf{a} + t(\mathbf{x} - \mathbf{a})$. As we have remarked on page 356, $\mathbf{g}$ is differentiable if and only if each of its component functions $g_i(t) = a_i + t(x_i - a_i)$ is differentiable (which is, clearly, the case here), and $\mathbf{Dg}(t)$ is an $n \times 1$ matrix with entries $g'_i(t) = x_i - a_i$, $1 \leq i \leq n$. In other words, $\mathbf{Dg}(t) = \mathbf{x} - \mathbf{a}$. Thus, F is a differentiable function on $[0, 1]$, and

$$F'(t) = \mathbf{D}f(\mathbf{g}(t))\mathbf{Dg}(t) = \mathbf{D}f(\mathbf{a} + t(\mathbf{x} - \mathbf{a}))(\mathbf{x} - \mathbf{a}).$$

In fact, if the second-order partial derivatives of f are continuous, then we can conclude that F' is differentiable. Indeed,

$$F'(t) = \sum_{i=1}^{n} \frac{\partial f}{\partial x_i}(\mathbf{a} + t(\mathbf{x} - \mathbf{a}))(x_i - a_i)$$

and, for each $1 \le i \le n$, $(\partial f/\partial x_i)(\mathbf{a} + t(\mathbf{x} - \mathbf{a}))$ is differentiable by the same argument that established the differentiability of F. Further,

$$F''(t) = \sum_{i=1}^{n} \sum_{j=1}^{n} \frac{\partial^2 f}{\partial x_j \partial x_i}(\mathbf{a} + t(\mathbf{x} - \mathbf{a}))(x_j - a_j)(x_i - a_i) = \mathbf{D}^2 f(\mathbf{a} + t(\mathbf{x} - \mathbf{a}))(\mathbf{x} - \mathbf{a})^2.$$

Continuing along the same lines we can see that F is differentiable $m + 1$ times, and that

$$F^{(k)}(t) = \mathbf{D}^k f(\mathbf{a} + t(\mathbf{x} - \mathbf{a}))(\mathbf{x} - \mathbf{a})^k, \quad 1 \le k \le m + 1. \tag{11.22}$$

If we now apply the Taylor Theorem in one variable we obtain that there exists $\theta \in [0, 1]$ such that

$$F(1) = F(0) + F'(0) + \frac{F''(0)}{2!} + \cdots + \frac{F^{(m)}(0)}{m!} + \frac{F^{(m+1)}(\theta)}{(m+1)!}$$

Clearly, $F(1) = f(\mathbf{x})$ and $F(0) = f(\mathbf{a})$, so the theorem follows from (11.22) and the observation that, if $\theta \in [0, 1]$ then $\mathbf{a} + \theta(\mathbf{x} - \mathbf{a})$ is on the line segment connecting $\mathbf{a}$ and $\mathbf{x}$. $\qquad \square$

Example 11.5.2. A Taylor polynomial.

Write the Taylor polynomial of degree 3 for $f(x, y) = xe^{x+y}$ at $\mathbf{a} = (1, -1)$.

Solution. The constant term is $f(1, -1) = 1$. The partial derivatives are

$$f'_x = (x + 1)e^{x+y}, \quad f'_y = xe^{x+y}$$
$$f''_{xx} = (x + 2)e^{x+y}, \quad f''_{xy} = (x + 1)e^{x+y}, \quad f''_{yy} = xe^{x+y}$$
$$f'''_{xxx} = (x + 3)e^{x+y}, \; f'''_{xxy} = (x + 2)e^{x+y}, \; f'''_{xyy} = (x + 1)e^{x+y}, \; f'''_{yyy} = xe^{x+y}.$$

Evaluating at $(1, -1)$ yields

$$f'_x(1, -1) = 2, \; f'_y(1, -1) = 1,$$
$$f''_{xx}(1, -1) = 3, \; f''_{xy}(1, -1) = 2, \; f''_{yy}(1, -1) = 1,$$
$$f'''_{xxx}(1, -1) = 4, \; f'''_{xxy}(1, -1) = 3, \; f'''_{xyy}(1, -1) = 2, \; f'''_{yyy}(1, -1) = 1.$$

Therefore, the Taylor polynomial of degree 3 for f is

$$P_3(x, y) = 1 + 2(x - 1) + (y + 1) + \frac{1}{2!}\left[3(x - 1)^2 + 4(x - 1)(y + 1) + (y + 1)^2\right]$$
$$+ \frac{1}{3!}\left[4(x - 1)^3 + 9(x - 1)^2(y + 1) + 6(x - 1)(y + 1)^2 + (y + 1)^3\right]. \qquad \blacklozenge$$

> The idea to use a function of one variable to obtain a Taylor's Formula for a multivariable function can be found in the writings of Lagrange (page 398), but it is Cauchy who made it more precise in 1829.

The special case $m = 0$ in Taylor's Formula gives us the Mean Value Theorem for a function of several variables.

Corollary 11.5.3 (Mean Value Theorem). *Let f be a differentiable function in an n-ball A, and let $\mathbf{a} \in A$. Then, for any $\mathbf{x} \in A$, $\mathbf{x} \neq \mathbf{a}$, there exists a point $\mathbf{b}$ on the line segment connecting $\mathbf{a}$ and $\mathbf{x}$, such that*

$$f(\mathbf{x}) = f(\mathbf{a}) + \mathbf{D}f(\mathbf{b})(\mathbf{x} - \mathbf{a}).$$

We see that the Mean Value Theorem remains true for functions of more than one variable. How about functions from $\mathbb{R}^n$ to $\mathbb{R}^m$?

Example 11.5.4. A function that does not satisfy the mean value theorem.

Let $\mathbf{f}(x, y) = (x^2, x^3)$, $P = (1,0)$, $Q = (0,0)$. Is it true that $\mathbf{f}(P) - \mathbf{f}(Q) = \mathbf{Df}(C)(P - Q)$, for some point C on the line segment PQ.

Solution. The left side is

$$\mathbf{f}(1,0) - \mathbf{f}(0,0) = (1,1) - (0,0) = (1,1).$$

The derivative

$$\mathbf{Df}(x, y) = \begin{bmatrix} 2x & 0 \\ 3x^2 & 0 \end{bmatrix},$$

and a point C on the line segment PQ has coordinates $(c, 0)$, with $0 \leq c \leq 1$. Then

$$\mathbf{Df}(c, 0) = \begin{bmatrix} 2c & 0 \\ 3c^2 & 0 \end{bmatrix},$$

and $P - Q = (1,0) - (0,0) = (1,0)$ so

$$\mathbf{Df}(C)(P - Q) = \begin{bmatrix} 2c & 0 \\ 3c^2 & 0 \end{bmatrix} \begin{bmatrix} 1 \\ 0 \end{bmatrix} = \begin{bmatrix} 2c \\ 3c^2 \end{bmatrix}.$$

If the Mean Value Theorem were true, c would have to satisfy $2c = 1$ and $3c^2 = 1$ which is impossible. $\blacklozenge$

This example shows that there can be no straightforward generalization of the Mean Value Theorem. However, we can establish an inequality.

Theorem 11.5.5. *Let $\mathbf{f}$ be a differentiable function in an n-ball A with values in $\mathbb{R}^m$. If $\mathbf{a} \in A$, then for any $\mathbf{x} \in A$, $\mathbf{x} \neq \mathbf{a}$, there exists a point $\mathbf{b}$ on the line segment connecting $\mathbf{a}$ and $\mathbf{x}$, such that*

$$\|\mathbf{f}(\mathbf{x}) - \mathbf{f}(\mathbf{a})\| \leq \|\mathbf{Df}(\mathbf{b})(\mathbf{x} - \mathbf{a})\|.$$

Proof. For $\mathbf{x}$ and $\mathbf{a}$ fixed, and $\mathbf{u} \in A$, let

$$g(\mathbf{u}) = \mathbf{f}(\mathbf{u}) \cdot (\mathbf{f}(\mathbf{x}) - \mathbf{f}(\mathbf{a})).$$

If we use Theorem 11.4.8 and apply it according to Remark 11.4.9, we obtain that g is differentiable and $\mathbf{D}g(\mathbf{u}) = (\mathbf{f}(\mathbf{x}) - \mathbf{f}(\mathbf{a})) \mathbf{Df}(\mathbf{u})$. By Corollary 11.5.3, $g(\mathbf{x}) = g(\mathbf{a}) + \mathbf{D}g(\mathbf{b})(\mathbf{x} - \mathbf{a})$ which implies that

$$\mathbf{f}(\mathbf{x}) \cdot (\mathbf{f}(\mathbf{x}) - \mathbf{f}(\mathbf{a})) = \mathbf{f}(\mathbf{a}) \cdot (\mathbf{f}(\mathbf{x}) - \mathbf{f}(\mathbf{a})) + (\mathbf{f}(\mathbf{x}) - \mathbf{f}(\mathbf{a})) \mathbf{Df}(\mathbf{b})(\mathbf{x} - \mathbf{a}).$$

Therefore,

$$\|\mathbf{f}(\mathbf{x}) - \mathbf{f}(\mathbf{a})\|^2 = (\mathbf{f}(\mathbf{x}) - \mathbf{f}(\mathbf{a})) \mathbf{Df}(\mathbf{b})(\mathbf{x} - \mathbf{a}) \leq \|\mathbf{f}(\mathbf{x}) - \mathbf{f}(\mathbf{a})\| \|\mathbf{Df}(\mathbf{b})(\mathbf{x} - \mathbf{a})\|,$$

from which the result follows. $\square$

Exercises

In Exercises 11.5.1–11.5.5 write the Taylor polynomial of degree m for a function f at $\mathbf{a}$:

11.5.1. $f(x,y) = \sin(x^2 + y^2)$, $\mathbf{a} = (0,0)$, $m = 2$.

11.5.2. $f(x,y) = \dfrac{1}{1+xy}$, $\mathbf{a} = (0,0)$, $m = 3$.

11.5.3. $f(x,y) = e^{xy}\ln(1+x)$, $\mathbf{a} = (0,1)$, $m = 3$.

11.5.4. $f(x,y,z) = xy^2 z^3$, $\mathbf{a} = (1,0,-1)$, $m = 3$.

11.5.5. $f(x,y,z) = x\arctan yz$, $\mathbf{a} = (0,1,1)$, $m = 2$.

In Exercises 11.5.6–11.5.8 find $\mathbf{b}$ whose existence is asserted by Mean Value Theorem (Corollary 11.5.3):

11.5.6. $f(x,y) = x^2 + xy$, $\mathbf{a} = (3,-5)$, $\mathbf{x} = (2,-3)$.

11.5.7. $f(x,y) = x^2 y$, $\mathbf{a} = (1,1)$, $\mathbf{x} = (2,3)$.

11.5.8. $f(x,y,z) = xyz + x^2 + y^2$, $\mathbf{a} = (0,0,0)$, $\mathbf{x} = (1,1,1)$.

11.5.9. Let A be an n-ball, $\mathbf{a} \in A$, and $\mathbf{f} : \mathbb{R}^n \to \mathbb{R}^m$ a differentiable function in A. Prove that for any $\mathbf{x} \in A$, $\mathbf{x} \neq \mathbf{a}$, and any $1 \leq k \leq m$, there exists a point $\mathbf{b}_k$ on the line segment connecting $\mathbf{a}$ and $\mathbf{x}$, such that

$$f_k(\mathbf{x}) = f_k(\mathbf{a}) + \mathbf{D}f_k(\mathbf{b}_k)(\mathbf{x} - \mathbf{a}).$$

11.5.10. Prove or disprove: if $\mathbf{f} : \mathbb{R}^m \to \mathbb{R}^n$ is a differentiable function and $\mathbf{Df} = \mathbf{0}$, then $\mathbf{f}$ is a constant function.

11.5.11.* Let $\mathbf{f} : \mathbb{R}^n \to \mathbb{R}^m$ be differentiable and suppose that $\mathbf{Df}$ is constant. Prove that $\mathbf{f}$ is a sum of a linear term and a constant. What is the linear part of $\mathbf{f}$?

11.5.12. Let $f : \mathbb{R}^2 \to \mathbb{R}$, let $f(0,0) = 1$ and suppose that $\frac{\partial f}{\partial x}(x,y) = 2$, $\frac{\partial f}{\partial y}(x,y) = 3$, for all $(x,y) \in \mathbb{R}^2$. Prove that $f(x,y) = 1 + 2x + 3y$, for all $(x,y) \in \mathbb{R}^2$.

11.5.13.* What can you say about a function $f(x,y)$ all of whose second-order partial derivatives vanish in a disk $A \subset \mathbb{R}^2$?

11.5.14.* Let A be an open connected set in $\mathbb{R}^2$ and let $f : A \to \mathbb{R}$ satisfy $\frac{\partial f}{\partial x}(x,y) = 0$, for all $(x,y) \in A$. Prove or disprove: $f(a_1,b) = f(a_2,b)$, for any $(a_1,b),(a_2,b) \in A$.

In Exercises 11.5.15–11.5.17, A is a disk in $\mathbb{R}^2$ and $f : A \to \mathbb{R}$. What can you say about f if, for all $(x,y) \in A$, it satisfies:

11.5.15.* $\frac{\partial^2 f}{\partial x \partial y}(x,y) = \frac{\partial^2 f}{\partial y \partial x}(x,y) = 0$. $\qquad$ 11.5.16.* $\frac{\partial^2 f}{\partial x^2}(x,y) = \frac{\partial^2 f}{\partial y^2}(x,y) = 0$.

11.5.17.* $\frac{\partial f}{\partial x}(x,y) = \frac{\partial f}{\partial y}(x,y) = x$.

11.6   Extreme Values

A major application of the differential calculus is in the field of Optimization, which among other things, deals with the maximum and minimum values of a function. When a function depends on one variable, we say that f has a **relative maximum** (or a **local maximum**) at a point c of its domain A, if there exists an interval $(a,b) \subset A$ that contains c and such that $f(c) \geq f(x)$ for all $x \in (a,b)$. When f is a function of n variables, instead of an interval we require an n-ball.

Definition 11.6.1. Let f be defined on a domain $A \subset \mathbb{R}^n$. We say that f has a **relative maximum** at $\mathbf{a} \in A$ if there exists an n-ball $U \subset A$ such that $f(\mathbf{a}) \geq f(\mathbf{x})$ for all $\mathbf{x} \in U$.

A similar definition can be articulated for the **relative minimum**.

Example 11.6.2. A relative minimum.

Let $f(x,y) = x^2 + y^2$, $A = \mathbb{R}^2$. Does f have a relative minimum?

Solution. A look at the surface (a paraboloid) reveals that it has a minimum at $(x,y) = (0,0)$. For example, we can take U to be the open unit disk in the xy-plane. Then $f(0,0) = 0$

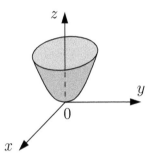

Figure 11.1: $f(x,y) = x^2 + y^2$ has a minimum at $(0,0)$.

and $f(x,y) \geq 0$ for all $(x,y) \in U$. ◆

How do we find where the relative maximum or the relative minimum occurs? In the single-variable calculus the procedure is: find critical points and test them. It turns out that this is the same when f depends on several variables. Of course, we have to define a critical point first. According to Fermat's Theorem (Theorem 4.4.2), if f attains its greatest/smallest value at a point c, and if f is differentiable at c, then $f'(c) = 0$. In the multivariable situation we have:

Theorem 11.6.3. *Let f be defined on a domain $A \subset \mathbb{R}^n$ and suppose that it attains its greatest/smallest value at an interior point $\mathbf{a} \in A$. If f has finite partial derivatives $f'_{x_i}(\mathbf{a})$, $1 \leq i \leq n$, then they all have to be equal to 0.*

Proof. Let $r > 0$ so that $B_r(\mathbf{a}) \subset A$. Let F be a function of one variable defined by $F(x) = f(x, a_2, a_3, \ldots, a_n)$, with x satisfying $(x, a_2, a_3, \ldots, a_n) \in B_r(\mathbf{a})$. More precisely,

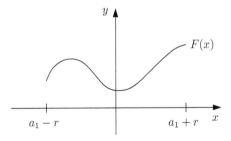

Figure 11.2: F is defined on $(a_1 - r, a_1 + r)$.

$x \in (a_1 - r, a_1 + r)$. Clearly, F is differentiable in this interval, and it attains its extreme value at $x = a_1$, so Fermat's Theorem implies that $F'(a_1) = 0$. It is not hard to see that this is the same as $f'_{x_1}(\mathbf{a}) = 0$, and that a similar argument can be used to establish that each of the partial derivatives of f vanishes at $\mathbf{a}$. □

Remark 11.6.4. A point **a**, where each of the partial derivatives of f equals 0, is called a **critical point** of f.

Example 11.6.5. Critical points of a function.

Let $f(x, y, z) = x^2 + y^2 + z^2 + 2x + 4y - 6z$. Find the critical points of f.

Solution. In order to find critical points we calculate $f'_x = 2x + 2$, $f'_y = 2y + 4$, $f'_z = 2z - 6$. Then we solve the system: $2x + 2 = 0$, $2y + 4 = 0$, $2z - 6 = 0$. This yields $x = -1$, $y = -2$, $z = 3$, so the only critical point is $(-1, -2, 3)$.   ♦

The next question is how to determine whether the critical point **a** is a point of a relative extremum or not. In the latter case we say that f has a **saddle point** at **a**. The case when f depends on a single variable is covered by Theorem 4.4.9: assuming that f is twice differentiable, if $f'(c) = 0$ and $f''(c) < 0$, then f has a local maximum at $x = c$. The situation is quite similar when f depends on several variables, although the hypothesis we make is somewhat stronger.

Theorem 11.6.6 (Second Derivative Test). *Let f be defined on an n-ball $A \subset \mathbb{R}^n$ and let $\mathbf{a} \in A$. Suppose that f has continuous second-order partial derivatives in A, and that $\mathbf{D}(f)(\mathbf{a}) = \mathbf{0}$.*

(a) If $\mathbf{D}^2(f)(\mathbf{a})(\mathbf{u})^2 > 0$ for all $\mathbf{u} \neq \mathbf{0}$, then f has a relative minimum at $\mathbf{a}$;

(b) If $\mathbf{D}^2(f)(\mathbf{a})(\mathbf{u})^2 < 0$ for all $\mathbf{u} \neq \mathbf{0}$, then f has a relative maximum at $\mathbf{a}$;

(c) If $\mathbf{D}^2(f)(\mathbf{a})(\mathbf{u})^2$ has values of both signs, then f has a saddle point at $\mathbf{a}$.

The quadratic form $\mathbf{D}^2(f)(\mathbf{a})(\mathbf{u})^2$ is said to be **positive definite** in (a), **negative definite** in (b), and **indefinite** in (c) (assuming that it equals 0 only when $\mathbf{u} = \mathbf{0}$).

Proof. We will apply Taylor's Formula (Theorem 11.5.1) to the function f, with $m = 1$. Taking into account that $\mathbf{D}(f)(\mathbf{a}) = \mathbf{0}$, we obtain that, for any $\mathbf{x} \in A$, $\mathbf{x} \neq \mathbf{a}$, there exists a point $\mathbf{b}$ on the line segment connecting $\mathbf{a}$ and $\mathbf{x}$, such that

$$f(\mathbf{x}) = f(\mathbf{a}) + \frac{\mathbf{D}^2 f(\mathbf{b})}{2!}(\mathbf{x} - \mathbf{a})^2. \tag{11.23}$$

In order to prove (a), we will show that $\mathbf{D}^2 f(\mathbf{b})(\mathbf{u})^2 > 0$ for $\mathbf{u} = \mathbf{x} - \mathbf{a} \neq \mathbf{0}$. This will imply that $f(\mathbf{x}) - f(\mathbf{a}) > 0$ for all $\mathbf{x} \neq \mathbf{a}$, so f will have a minimum at **a**. Let us write

$$\mathbf{D}^2 f(\mathbf{b})(\mathbf{x} - \mathbf{a})^2 = \mathbf{D}^2 f(\mathbf{a})(\mathbf{x} - \mathbf{a})^2 + \left(\mathbf{D}^2 f(\mathbf{b}) - \mathbf{D}^2 f(\mathbf{a})\right)(\mathbf{x} - \mathbf{a})^2. \tag{11.24}$$

By Theorem 10.4.5, the continuous function $\mathbf{v} \mapsto \mathbf{D}^2 f(\mathbf{a})(\mathbf{v})^2$ on a closed and bounded set $\{\mathbf{v} \in \mathbb{R}^n : \|\mathbf{v}\| = 1\}$ attains its minimum value m. The assumption that $\mathbf{D}^2(f)(\mathbf{a})(\mathbf{v})^2 > 0$ implies that $m > 0$. Therefore,

$$\mathbf{D}^2 f(\mathbf{a})(\mathbf{v})^2 \geq m > 0, \quad \|\mathbf{v}\| = 1.$$

This estimate can be extended for an arbitrary $\mathbf{u} \neq \mathbf{0}$ in the following way. Since $\mathbf{u}/\|\mathbf{u}\|$ has norm one,

$$\mathbf{D}^2 f(\mathbf{a}) \left(\frac{\mathbf{u}}{\|\mathbf{u}\|} \right)^2 \geq m > 0, \quad \mathbf{u} \neq \mathbf{0}.$$

The expression on the left hand side is a quadratic form, and quadratic forms have the property that $q(\alpha \mathbf{u}) = \alpha^2 q(\mathbf{u})$, for any $\alpha \in \mathbb{R}$. Consequently, we have that

$$\mathbf{D}^2 f(\mathbf{a})(\mathbf{u})^2 \geq m\|\mathbf{u}\|^2, \quad \mathbf{u} \neq \mathbf{0}.$$

In particular,
$$\mathbf{D}^2 f(\mathbf{a})(\mathbf{x} - \mathbf{a})^2 \geq m\|\mathbf{x} - \mathbf{a}\|^2, \quad \mathbf{x} \neq \mathbf{a}. \tag{11.25}$$

On the other hand,

$$\left| \left(\mathbf{D}^2 f(\mathbf{b}) - \mathbf{D}^2 f(\mathbf{a}) \right) (\mathbf{x} - \mathbf{a})^2 \right|$$

$$= \left| \sum_{i=1}^{n} \sum_{j=1}^{n} \left(\frac{\partial^2 f}{\partial x_i \partial x_j}(\mathbf{b}) - \frac{\partial^2 f}{\partial x_i \partial x_j}(\mathbf{a}) \right) (x_i - a_i)(x_j - a_j) \right|$$

$$\leq \sum_{i=1}^{n} \sum_{j=1}^{n} \left| \frac{\partial^2 f}{\partial x_i \partial x_j}(\mathbf{b}) - \frac{\partial^2 f}{\partial x_i \partial x_j}(\mathbf{a}) \right| |x_i - a_i||x_j - a_j|$$

$$\leq \sum_{i=1}^{n} \sum_{j=1}^{n} \left| \frac{\partial^2 f}{\partial x_i \partial x_j}(\mathbf{b}) - \frac{\partial^2 f}{\partial x_i \partial x_j}(\mathbf{a}) \right| \|\mathbf{x} - \mathbf{a}\|^2$$

The assumption that the second-order partial derivatives of f are continuous in A shows that each of the n^2 terms in the last sum goes to 0, when $\mathbf{x} \to \mathbf{a}$. Thus, when $\mathbf{x}$ is close enough to $\mathbf{a}$,
$$\left| \left(\mathbf{D}^2 f(\mathbf{b}) - \mathbf{D}^2 f(\mathbf{a}) \right) (\mathbf{x} - \mathbf{a})^2 \right| \leq \frac{m}{2} \|\mathbf{x} - \mathbf{a}\|^2.$$

Consequently,
$$\left(\mathbf{D}^2 f(\mathbf{b}) - \mathbf{D}^2 f(\mathbf{a}) \right) (\mathbf{x} - \mathbf{a})^2 \geq -\frac{m}{2} \|\mathbf{x} - \mathbf{a}\|^2. \tag{11.26}$$

If we now use equation (11.24), together with estimates (11.25) and (11.26), we obtain that

$$\mathbf{D}^2 f(\mathbf{b})(\mathbf{x} - \mathbf{a})^2 \geq \frac{m}{2} \|\mathbf{x} - \mathbf{a}\|^2 > 0,$$

and (a) is proved.

The proof of (b) is simple: just define $g = -f$, and conclude from (a) that g has a relative minimum at $\mathbf{a}$. This implies that f has a relative maximum at $\mathbf{a}$.

In order to prove (c), let us assume that there exist non-zero vectors $\mathbf{v}$ and $\mathbf{w}$ such that $\mathbf{D}^2(f)(\mathbf{a})(\mathbf{v})^2 > 0$ and $\mathbf{D}^2(f)(\mathbf{a})(\mathbf{w})^2 < 0$. If we replace $\mathbf{x}$ by $\mathbf{a} + t\mathbf{v}$ in (11.23), where $0 < t < 1$, we obtain
$$f(\mathbf{a} + t\mathbf{v}) = f(\mathbf{a}) + \frac{\mathbf{D}^2 f(\mathbf{a} + \theta t\mathbf{v})}{2!}(t\mathbf{v})^2, \tag{11.27}$$

for some $0 < \theta < 1$. We will show that $\mathbf{D}^2 f(\mathbf{a} + \theta t\mathbf{v})(\mathbf{v})^2 > 0$ for t small enough. Just like in (11.24) we write

$$\mathbf{D}^2 f(\mathbf{a} + \theta t\mathbf{v})(\mathbf{v})^2 = \mathbf{D}^2 f(\mathbf{a})(\mathbf{v})^2 + \left(\mathbf{D}^2 f(\mathbf{a} + \theta t\mathbf{v}) - \mathbf{D}^2 f(\mathbf{a}) \right) (\mathbf{v})^2.$$

Once again, the continuity of the second-order partial derivatives implies that, as $t \to 0$, the last expression on the right hand side approaches 0. Consequently, there exists $\delta > 0$ such that, if $|t| < \delta$,

$$\left| \left(\mathbf{D}^2 f(\mathbf{a} + \theta t\mathbf{v}) - \mathbf{D}^2 f(\mathbf{a}) \right) (\mathbf{v})^2 \right| < \frac{1}{2} \mathbf{D}^2 f(\mathbf{a})(\mathbf{v})^2.$$

Therefore, for such t, $\mathbf{D}^2 f(\mathbf{a} + \theta t\mathbf{v})(\mathbf{v})^2 > 0$, and (11.27) implies that $f(\mathbf{a} + t\mathbf{v}) > f(\mathbf{a})$. A similar argument shows that there exists $\eta > 0$ such that, if $|t| < \eta$, $f(\mathbf{a} + t\mathbf{w}) < f(\mathbf{a})$. Thus, f has a saddle point at $\mathbf{a}$. $\qquad\square$

Theorem 11.6.6 translated the problem of finding a relative extremum of a function of several variables into an algebraic question about symmetric quadratic forms. A quadratic form

$$q(x_1, x_2, \ldots, x_n) = \sum_{i=1}^{n} \sum_{j=1}^{n} q_{ij} x_i x_j$$

is **symmetric** if $q_{ij} = q_{ji}$ for all i, j, $1 \le i, j \le n$. This is clearly the case with $\mathbf{D}^2(f)(\mathbf{a})(\mathbf{u})^2$ when the second-order partial derivatives of f are continuous, because of Theorem 11.3.5. Given a symmetric quadratic form $q(x_1, x_2, \ldots, x_n)$ we associate to it a symmetric matrix

$$Q = \begin{bmatrix} q_{11} & q_{12} & \cdots & q_{1n} \\ q_{21} & q_{22} & \cdots & q_{2n} \\ \vdots & \vdots & \ddots & \vdots \\ q_{n1} & q_{n2} & \cdots & q_{nn} \end{bmatrix}$$

and the sequence $\{D_k\}$, $1 \le k \le n$, of determinants

$$D_k = \begin{vmatrix} q_{11} & q_{12} & \cdots & q_{1k} \\ q_{21} & q_{22} & \cdots & q_{2k} \\ \vdots & \vdots & \ddots & \vdots \\ q_{k1} & q_{k2} & \cdots & q_{kk} \end{vmatrix}.$$

The following result will establish the relationship between the properties of a quadratic form and the associated symmetric matrix. For a proof we recommend [67].

Theorem 11.6.7 (Sylvester's Rule). *Let q be a symmetric quadratic form in n variables, and let Q be the associated symmetric matrix. For each $1 \le k \le n$, let D_k be the $k \times k$ determinant as above.*

(a) If $D_k > 0$ for all $1 \le k \le n$, then q is positive definite;

(b) if $D_k > 0$ for k even, and $D_k < 0$, for k odd, $1 \le k \le n$, then q is negative definite.

Remark 11.6.8. The case of interest for us is when $q_{ij} = f''_{x_i x_j}$, i.e., when the matrix Q is the Hessian matrix of f. Being a symmetric matrix, it is a diagonal matrix $\operatorname{diag}(a_1, a_2, \ldots, a_n)$ in a suitably selected basis. That means that the quadratic form q can be written as $\sum_{i=1}^{n} a_i y_i^2$, with each y_i a linear combination of the x_j's. Now the positive (resp., negative) definiteness occurs if and only if all the numbers a_i, $1 \le i \le n$, are positive (resp., negative).

Let us consider the simplest case $n = 2$, i.e., when f is a function of 2 variables. Then,

$$\mathbf{D}^2 f(a, b)(x, y) = \frac{\partial^2 f}{\partial x^2}(a, b)\, x^2 + 2 \frac{\partial^2 f}{\partial x \partial y}(a, b)\, xy + \frac{\partial^2 f}{\partial y^2}(a, b)\, y^2.$$

In order to apply Theorem 11.6.6 we need to investigate the sign of this quadratic form. Let $a_{11} = f''_{xx}(a, b)$, $a_{12} = f''_{xy}(a, b)$, $a_{22} = f''_{yy}(a, b)$. Then, assuming that $a_{11} \ne 0$,

$$\mathbf{D}^2 f(a, b)(x, y) = a_{11} x^2 + 2a_{12} xy + a_{22} y^2$$

$$= a_{11} \left(x^2 + \frac{2a_{12}}{a_{11}} xy + \frac{a_{22}}{a_{11}} y^2 \right)$$

$$= a_{11} \left[\left(x + \frac{a_{12}}{a_{11}} y \right)^2 + \frac{a_{22}}{a_{11}} y^2 - \frac{a_{12}^2}{a_{11}^2} y^2 \right]$$

$$= a_{11}\left[\left(x + \frac{a_{12}}{a_{11}}y\right)^2 + \frac{a_{11}a_{22} - a_{12}^2}{a_{11}^2}y^2\right].$$

This leads us to the following conclusion, which goes back to the article [75] by Lagrange from 1759.

Theorem 11.6.9. *Let f be defined on an open disk $A \subset \mathbb{R}^2$ and let $(a, b) \in A$. Suppose that f has continuous second-order partial derivatives in A, and $f'_x(a, b) = f'_y(a, b) = 0$. Let $a_{11} = f''_{xx}(a, b)$, $a_{12} = f''_{xy}(a, b)$, $a_{22} = f''_{yy}(a, b)$.*

(a) *If $a_{11}a_{22} - a_{12}^2 > 0$ and $a_{11} > 0$, then f has a relative minimum at (a, b);*

(b) *if $a_{11}a_{22} - a_{12}^2 > 0$ and $a_{11} < 0$, then f has a relative maximum at (a, b);*

(c) *if $a_{11}a_{22} - a_{12}^2 < 0$, f has a saddle point at (a, b).*

Example 11.6.10. Finding and analyzing critical points of a function.

Let $f(x, y) = x^3 + y^3 - 3x - 12y + 20$. Find and analyze the critical points of f.

Solution. The partial derivatives are $f'_x = 3x^2 - 3$ and $f'_y = 3y^2 - 12$. The critical points are solutions of the system $3x^2 - 3 = 0$, $3y^2 - 12 = 0$. Since $3x^2 - 3 = 3(x - 1)(x + 1)$ we get $x_1 = 1$, $x_2 = -1$. Similarly, $3y^2 - 12 = 3(y - 2)(y + 2)$, so $y_1 = 2$, $y_2 = -2$. The critical points are $P_1 = (1, 2)$, $P_2 = (-1, -2)$, $P_3 = (1, -2)$, $P_4 = (-1, 2)$. The second-order partial derivatives are $f''_{xx} = 6x$, $f''_{xy} = 0$, $f''_{yy} = 6y$.

(i) At P_1, $a_{11} = 6 > 0$, and $a_{11}a_{22} - a_{12}^2 = 72 > 0$. Thus, f has a relative minimum at P_1.

(ii) At P_2, $a_{11} = -6 < 0$, and $a_{11}a_{22} - a_{12}^2 = 72 > 0$, so f has a relative maximum at P_2.

(iii) At P_3, $a_{11} = 6 > 0$, and $a_{11}a_{22} - a_{12}^2 = -72 < 0$, so f has a saddle at P_3.

(iv) At P_4, $a_{11} = -6 > 0$, and $a_{11}a_{22} - a_{12}^2 = -72 < 0$, so f has a saddle at P_4. ◆

Example 11.6.11. Finding and analyzing critical points of a function.

Let $f(x, y) = x^3y - x^3 + xy$. Find and analyze the critical points of f.

Solution. The partial derivatives are $f'_x = 3x^2y - 3x^2 + y$ and $f'_y = x^3 + x$. The critical points are solutions of the system $3x^2y - 3x^2 + y = 0$, $x^3 + x = 0$. Since $x^3 + x = x(x^2 + 1)$ we get $x = 0$. Substituting in the first equation yields $y = 0$. The only critical point is $P = (0, 0)$. The second-order partial derivatives are $f''_{xx} = 6xy - 6x$, $f''_{xy} = 3x^2 + 1$, $f''_{yy} = 0$, so $a_{11} = 0$, $a_{12} = 1$, $a_{22} = 0$. Since $a_{11} = 0$, we cannot apply Theorem 11.6.9. Nevertheless, $\mathbf{D}^2 f(0, 0)(x, y) = xy$ and it is easy to see that in the vicinity of $(0, 0)$ it is not of the same sign. For example, in the first quadrant $(x > 0, y > 0)$ it is positive, but in the second quadrant $(x < 0, y > 0)$ it is negative. Therefore, f has a saddle at $(0, 0)$.

Alternatively, one may find the eigenvalues of the Hessian matrix $\left(\begin{smallmatrix} 0 & 1 \\ 1 & 0 \end{smallmatrix}\right)$, which are 1 and -1. Since they are not of the same sign, we see that f has a saddle at $(0, 0)$. ◆

The last example is a reminder that Theorem 11.6.6 does not provide an answer when the associated quadratic form is **positive semi-definite** (meaning $\mathbf{D}^2(f)(\mathbf{a})(\mathbf{u})^2 \geq 0$ for all $\mathbf{u} \neq \mathbf{0}$) or **negative semi-definite** (i.e., $\mathbf{D}^2(f)(\mathbf{a})(\mathbf{u})^2 \leq 0$ for all $\mathbf{u} \neq \mathbf{0}$). For example, the functions $f(x, y) = x^2 + y^3$ and $g(x, y) = x^2 + y^4$ both have the critical point $(0, 0)$, and the associated quadratic form is the same: $q(x, y) = 2x^2$. Yet, at $(0, 0)$, f has a saddle while g has a minimum. Theorem 11.6.6 is not applicable because q is only semi-definite: $q(0, 1) = 0$. However, see Exercise 11.6.19.

JAMES JOSEPH SYLVESTER (1814–1897) was an English mathematician. He was born James Joseph, and added Sylvester when planning to immigrate to the United States. After finishing high school in London, he went to study in Cambridge. He was Jewish and refused to take oath to the Church of England so, although he passed the final examination and placed second overall, he did not get a diploma. Without it, he had few options and he taught physics at the University of London. He did research in mathematics and between 1838 and 1841 produced 15 papers on fluid dynamics and algebraic equations. At the age of 27 he went to teach mathematics at the University of Virginia, but resigned after a few months following a violent encounter with two students he had disciplined. After spending some time in New York trying to get a position there, he returned to England where he worked as an actuary and taught mathematics at the Royal Military Academy at Woolwich. During that period he was working on Matrix Theory and received several honors including being the president of the London Mathematical Society, the membership in the Paris Academy of Sciences and the Royal Society of London, which awarded Sylvester the Copley Medal, its highest award for scientific achievement. In 1901, it instituted the Sylvester Medal. The military rules required that he retire at the age of 55. Seven years later, following a period of low activity, he went to Johns Hopkins University. He founded the *American Journal of Mathematics*, the first mathematical journal in the United States and one of the most prestigious journals nowadays. He started doing research again. After 7 years he resigned feeling unable to follow "the tide of mathematical progress". His final post was the Savilian Chair of Geometry in Oxford. He is credited for inventing many mathematical terms such as "matrix", "graph" (in combinatorics), and "discriminant". In Discrete geometry he is remembered for *Sylvester's Problem* and a result on *the orchard problem*. He was fond of puzzles and here is one: *I have a large number of 5 cent stamps and 17 cent stamps. What is the largest denomination which I cannot make up with a combination of these two different values?*

Exercises

In Exercises 11.6.1–11.6.7 find the critical points of f.

11.6.1. $f(x,y) = 2x^4 + y^4 - x^2 - 2y^2$.   11.6.2. $f(x,y) = x^3 + y^3 - 3xy$.

11.6.3. $f(x,y) = xy\sqrt{1 - \dfrac{x^2}{a^2} - \dfrac{y^2}{b^2}}$, $a,b > 0$.   11.6.4. $f(x,y) = 1 - \sqrt{x^2 + y^2}$.

11.6.5. $f(x,y) = (x^2 + y^2)e^{-(x^2+y^2)}$.   11.6.6. $f(x,y) = xy\ln(x^2 + y^2)$.

11.6.7. $f(x,y,z) = x^3 + y^2 + z^2 + 12xy + 2z$.

In Exercises 11.6.8–11.6.14 find the extreme values of f.

11.6.8. $f(x,y) = x^2 + (y-1)^2$.   11.6.9. $f(x,y) = x^4 + y^4 - x^2 - 2xy - y^2$.

11.6.10. $f(x,y) = 2x - 3y + \ln xy$.   11.6.11. $f(x,y) = e^{2x+3y}(8x^2 - 6xy + 3y^2)$.

11.6.12. $f(x,y) = (x+y)(1-xy)$.   11.6.13. $f(x,y,z) = x^2 + y^2 + z^2 + 2x + 4y - 6z$.

11.6.14. $f(x_1, x_2, \ldots, x_n) = x_1 x_2^2 \ldots x_n^n (1 - x_1 - 2x_2 - \cdots - nx_n)$, if $x_1 > 0$, $x_2 > 0$, …, $x_n > 0$.

11.6.15. Let A be an open disk in $\mathbb{R}^n$ and $f : A \to \mathbb{R}$ a differentiable function that has no critical points. Prove that the set $f^{-1}(0)$ has no interior points.

11.6.16.* Let f be defined on an n-ball $A \subset \mathbb{R}^n$ and let $\mathbf{a} \in A$. Suppose that f has continuous second-order partial derivatives in A, and that it has a relative maximum at $\mathbf{a}$. Prove that $\mathbf{D}^2(f)(\mathbf{a})(\mathbf{u})^2 \leq 0$ for all $\mathbf{u} \neq \mathbf{0}$.

11.6.17. Let f be defined on an n-ball $A \subset \mathbb{R}^n$ and let $\mathbf{a} \in A$. Suppose that f has continuous partial derivatives of order p in A, that $\mathbf{D}(f)(\mathbf{a}) = \mathbf{D}^2(f)(\mathbf{a}) = \cdots = \mathbf{D}^{p-1}(f)(\mathbf{a}) = \mathbf{0}$, and $\mathbf{D}^p(f)(\mathbf{a})(\mathbf{u})^2 > 0$ for all $\mathbf{u} \neq \mathbf{0}$. Prove that f has a relative minimum at $\mathbf{a}$.

11.6.18. Let $f : \mathbb{R}^n \to \mathbb{R}$ be differentiable and let $f(\mathbf{x}) = 0$ whenever $\|\mathbf{x}\| = 1$. Prove that there exists $\mathbf{a} \in \mathbb{R}^n$ such that $\|\mathbf{a}\| < 1$ and $\mathbf{D}(f)(\mathbf{a}) = \mathbf{0}$.

11.6.19. Let f be defined on an n-ball $A \subset \mathbb{R}^n$ and let $\mathbf{a} \in A$. Suppose that f has continuous second-order partial derivatives in A, and that $\mathbf{D}(f)(\mathbf{a}) = \mathbf{0}$.

(a) Prove that if $\mathbf{D}^2(f)(\mathbf{x})(\mathbf{u})^2 \geq 0$ for all $\mathbf{x} \neq \mathbf{a}$, then f has a relative minimum at $\mathbf{a}$;

(b) Prove that if $\mathbf{D}^2(f)(\mathbf{x})(\mathbf{u})^2 \leq 0$ for all $\mathbf{x} \neq \mathbf{a}$, then f has a relative maximum at $\mathbf{a}$.

11.6.20. The purpose of this problem from [35] is to establish the Contraction Mapping Lemma. Let $A \subset \mathbb{R}^n$. A mapping $T : A \to A$ is a contraction if there exists a constant α, $0 < \alpha < 1$, such that

$$\|T(p) - T(q)\| \leq \alpha \|p - q\|, \quad \text{for all } p, q \in A.$$

A point p is fixed point for T if $T(p) = p$. The Contraction Mapping Lemma asserts that if A is a non-empty, closed set, and if $T : A \to A$ is a contraction, then T has a unique fixed point in A.

(a) Suppose that A is compact, define $f(x) = \|x - T(x)\|$. Prove that f attains its minimum at some point $p \in A$, and that $T(p) = p$.

(b) If A is not bounded, choose $q \in A$, and define $A_1 = \{x \in A : f(x) \leq f(q)\}$. Prove that A_1 is compact, and apply (a) to conclude that T has a fixed point in A.

(c) Show that the assumption that T has 2 fixed points leads to a contradiction.

12

Implicit Functions and Optimization

In this chapter we will pursue some centuries old questions: when does an equation $F(x, y) = 0$ determine a function $y = f(x)$? Is f differentiable and, if so, how to calculate its derivative? These questions can be asked when the variables x, y take values in $\mathbb{R}$, or in any Euclidean spaces. A path to the answers will lead us through a special case when $\mathbf{x}, \mathbf{y}$ belong to spaces of the same dimension, say to $\mathbb{R}^n$, and $\mathbf{F}(\mathbf{x}, \mathbf{y}) = \mathbf{x} - \mathbf{g}(\mathbf{y})$, for some function $\mathbf{g} : \mathbb{R}^n \to \mathbb{R}^n$. In other words, we will establish the existence and properties of the inverse function. Both of these can be best understood through the study of the derivative of $\mathbf{g}$, and the Jacobian matrix $J\mathbf{g}$. All these results will then be used in the field of the so-called Constrained Optimization.

12.1 Implicit Functions

In this section we will consider the problem: solve the equation $F(x, y) = 0$ for y. Let us start with a few examples.

Example 12.1.1. Solve $x - e^y = 0$ for y.

This one is fairly straightforward: $e^y = x$ so $y = \ln x$.   ◆

Example 12.1.2. Solve $x - ye^y = 0$ for y.

This time $ye^y = x$ but we cannot solve for y.   ◆

Example 12.1.3. Solve $x - y^2 = 0$ for y.

The equation has two solutions $y_1 = \sqrt{x}$ and $y_2 = -\sqrt{x}$. Although we have solved, we did not get a function $y = f(x)$ as a result: when $x = 4$, y is both 2 and -2, which is not acceptable for a function.   ◆

Examples 12.1.1–12.1.3 show that the success is not guaranteed. In Examples 12.1.1 and 12.1.3 there was no ambiguity: the function $y = f(x)$ either exists or it does not. Example 12.1.2 is much less clear. We admitted that we could not solve for y, but what does that really mean? On one hand, it is possible that the function $y = f(x)$ exists but we do not have a name for it. In a world where no one has heard of logarithms, we would not be able to solve the equation $e^y = x$ (Example 12.1.1), in spite of the fact that y is a perfectly good function of x. When a function $y = f(x)$ exists and it is given by an equation that cannot be solved, we say that it is *implicit*. On the other hand, it could be that equation $x - ye^y = 0$ does not define a function $y = f(x)$ (as in Example 12.1.3). Our first task will be to find a way to distinguish between these two possibilities.

Example 12.1.3 can be considered from a geometric viewpoint. The graph of $x - y^2 = 0$ is a parabola. The reason why this equation does not define a function is that its graph does not pass the "vertical line test", i.e., a vertical line intersects the parabola at 2 points. There is a way around this predicament. We will illustrate this on the point $P_1 = (4, 2)$ on the parabola. We will use a rectangle $[3,5] \times [1,3]$ that contains P_1. If we zoom in on this rectangle and ignore anything outside of it, then the graph of $x - y^2 = 0$ passes the vertical line test, so we obtain a function $y = \sqrt{x}$.

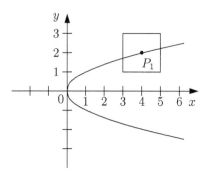

Figure 12.1: In the rectangle, $x - y^2 = 0$ passes the "vertical line test."

Next, let us consider the point $P_2 = (1/4, -1/2)$ and a rectangle $[-1, 1] \times [-1, 1]$. A quick look at the picture reveals that the rectangle is too big. If we zoom in further to the rectangle

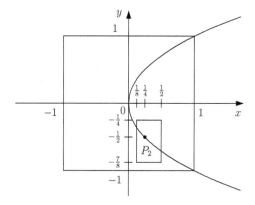

Figure 12.2: A smaller rectangle is needed.

$[1/8, 1/2] \times [-7/8, -1/4]$, we now obtain a function $y = -\sqrt{x}$.

However, if $P_3 = (0,0)$, no amount of zooming in will help (Figure 12.3). Any rectangle that contains P_3 in its interior must contain a portion of each branch of the parabola. In other words, equation $x - y^2 = 0$ does not define an implicit function in the vicinity of $(0,0)$.

We say that the equation $x - y^2 = 0$ defines an implicit function in the vicinity of P_1 and in the vicinity of P_2, but not P_3. Our investigation was geometric, and it would be useful to develop a criterion that can be generalized to the situations with more variables. One way to attack the problem is to make some approximations. In the equation $x - y^2 = 0$ we will use the linear approximation for the function $g(y) = y^2$. In the vicinity of a point y_0, we have that $g(y) \approx g(y_0) + g'(y_0)(y - y_0)$, i.e., $y^2 \approx y_0^2 + 2y_0(y - y_0)$. If we substitute this in the given equation, we obtain

$$x - y_0^2 - 2y_0(y - y_0) \approx 0.$$

Solving this equation for y will give us a linear approximation for the unknown function

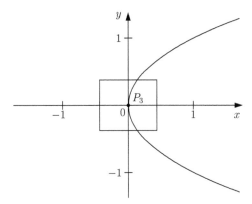

Figure 12.3: $x - y^2 = 0$ does not define an implicit function around $(0,0)$.

$f(x)$:

$$y \approx y_0 + \frac{x - y_0^2}{2y_0}.$$

Unfortunately, this is possible only if the denominator is different from 0. That explains why in the vicinity of P_1 and P_2 the equation $x - y^2 = 0$ defined an implicit function, but not at P_3. A second look at the denominator shows that it is $g'(y_0)$. What do we look for in general?

Let us consider another example: can we solve for y the equation $xy - y^2 + 1 = 0$? If we again approximate y^2 by $y_0^2 + 2y_0(y - y_0)$, we obtain

$$xy - y_0^2 - 2y_0(y - y_0) + 1 \approx 0.$$

This time the left side is $y(x - 2y_0) + y_0^2 + 1$, so the "bad" choice for y_0 is one that makes $x - 2y_0 = 0$. Of course, any choice of y_0 is really a choice of the point on the graph, so x is not arbitrary: (x_0, y_0) must satisfy the equation $x_0 y_0 - y_0^2 + 1 = 0$. In other words, when y is close to y_0, x is close to x_0, and we should try to avoid those points for which $x_0 - 2y_0 = 0$. In the last example, we asked that $g'(y_0) \neq 0$. Where does $x_0 - 2y_0$ come from? It is not hard to see that, if $F(x, y) = xy - y^2 + 1$, then $x_0 - 2y_0$ is really $F_y'(x_0, y_0)$. It looks as if the condition that is required for the existence of the implicit function is that $F_y'(x_0, y_0) \neq 0$.

Let us revisit Example 12.1.3 and assume that we have restricted our attention to a rectangle that does not include $(0,0)$. Since there is a function $y = f(x)$ we will try to find its derivative. One way to do this is to calculate the derivative of $F(x, y) = x - y^2$. If we denote $\mathbf{g}(x) = (x, y(x))$ then $F(x, y) = F(\mathbf{g}(x))$. By the Chain Rule,

$$\mathbf{D}(F \circ \mathbf{g})(x) = \mathbf{D}F(\mathbf{g}(x))\mathbf{D}\mathbf{g}(x).$$

The partial derivatives of F are $F_x' = 1$ and $F_y' = -2y$, so

$$\mathbf{D}F(\mathbf{g}(x)) = \begin{bmatrix} 1 & -2y(x) \end{bmatrix}.$$

On the other hand, $\mathbf{g} : \mathbb{R} \to \mathbb{R}^2$, so $\mathbf{D}\mathbf{g}(x)$ is a 2×1 matrix

$$\mathbf{D}\mathbf{g}(x) = \begin{bmatrix} 1 \\ y'(x) \end{bmatrix}.$$

It follows that $\mathbf{D}(F \circ \mathbf{g})(x) = 1 - 2y(x)y'(x)$. Of course, we are considering equation $F(x, y) = 0$, so $\mathbf{D}(F \circ \mathbf{g})(x) = 0$. Thus, we obtain

$$1 - 2y(x)y'(x) = 0,$$

and the assumption that $y(x) \neq 0$ allows us to conclude that the derivative of $y = f(x)$ exists and equals $y'(x) = 1/[2y(x)]$. Thus, the differentiability of F carries over to f. The following theorem confirms all this.

Theorem 12.1.4 (Implicit Function Theorem). *Let F be a function of 2 variables defined on a rectangle R, and let (x_0, y_0) be an interior point of R, such that $F(x_0, y_0) = 0$. Suppose that F has continuous partial derivatives in R, and that $F'_y(x_0, y_0) \neq 0$. Then:*

(a) there exists a rectangle $R' = (x_0 - \delta, x_0 + \delta) \times (y_0 - \eta, y_0 + \eta)$ such that the equation $F(x, y) = 0$ defines a function $y = f(x)$ in R';

(b) $f(x_0) = y_0$;

(c) f is continuous in $(x_0 - \delta, x_0 + \delta)$;

(d) f has a continuous derivative in $(x_0 - \delta, x_0 + \delta)$.

Equations of the form $f(x, y) = C$ were the central topic of Descartes' text [30] in 1637. While it was evident that these equations described curves in the xy-plane, in the nineteenth century it became important to tell when such an equation defined a function. According to [91], a modern presentation (including Theorem 12.1.4) can be found in Dini's 1878 lecture notes [31]. The second printing from 1907 became very influential inasmuch so that the Implicit Function Theorem is called *Dini's Theorem* in Italy. An excellent overview of the history of the subject can be found in the aforementioned [91], as well as in [74].

ULISSE DINI (1845–1918), was an Italian mathematician and politician from Pisa. He attended a teacher's college *Scuola Normale Superiore*, attached to the University of Pisa. A scholarship allowed him to continue his studies in Paris. During 1865-66 his research blossomed and he published 7 papers. In 1866 he returned to the University of Pisa to teach advanced algebra and the theory of geodesy. In 1871 he was promoted to a professor of analysis and higher geometry. These were turbulent years in Italy which was undergoing the process of unification. Dini was elected to the Pisa City Council and in 1880 to the National Parliament, as a representative from Pisa. In 1890 he became a senator. At the professional level, he served as the rector of the University of Pisa 1888-90 and as the director of the *Scuola Normale Superiore*, from 1908 until his death. Dini did some significant research in the field of analysis during a time when a major goal was to determine precisely when the theorems (which had earlier been stated and proved in an imprecise way) were valid. To achieve this aim mathematicians tried to see how far results could be generalized and they needed to find pathological counterexamples to show the limits to which generalization was possible. Dini was one of the greatest masters of generalization and constructing counterexamples. For example, he improved Kummer's Test (see page 207). In addition to one surface, several results were named after Dini: Dini's criterion for the pointwise convergence of Fourier series (Exercise 9.2.20), Dini's Theorem on uniform convergence (Exercise 8.1.17), and the Implicit Function Theorem is known as Dini's Theorem in Italy.

We will not prove Theorem 12.1.4 now. Instead, we will later prove a more general result. The straightforward proof from [31] is left as an exercise (Exercise 12.1.27). Right now, we will carefully analyze another example.

Example 12.1.5. The existence of implicit functions.

At what points does the system $x + y + z = 2$, $x^2 + y^2 + z^2 = 2$ determine y and z as functions of x?

Solution. If we solve the first equation for z and substitute it in the second equation we obtain

$$x^2 + y^2 + (2 - x - y)^2 = 2.$$

In order to apply Theorem 12.1.4, we take

$$F(x, y) = x^2 + y^2 + (2 - x - y)^2 - 2$$

and calculate
$$F'_y = 2y + 2(2 - x - y)(-1) = 2x + 4y - 4.$$
So the key is to have $2x + 4y - 4 \neq 0$. For example, the triple $(1, 1, 0)$ satisfies both equations, and $F'_y(1, 1, 0) = 2 \neq 0$. On the other hand, $(0, 1, 1)$ also satisfies both equations, but $F'_y(0, 1, 1) = 0$. Thus, the given equations define functions $y(x)$ and $z(x)$ in the vicinity of $(1, 1, 0)$, but not $(0, 1, 1)$. ◆

In the example above we took advantage of the opportunity to solve the first equation for z. What if this is impossible? It would not preclude the existence of the implicit functions y and z. It would be nice to have a general test, just like in Theorem 12.1.4. If we again use the linear approximations for y^2 and z^2, the second equation becomes
$$x^2 + y_0^2 + 2y_0(y - y_0) + z_0^2 + 2z_0(z - z_0) = 2$$
which can be simplified to $2y_0 y + 2z_0 z = 2 - x^2 + y_0^2 + z_0^2$. Thus, our system of equations can be written as
$$y + z = 2 - x$$
$$2y_0 y + 2z_0 z = 2 - x^2 + y_0^2 + z_0^2.$$
If we write this system in a matricial form
$$\begin{bmatrix} 1 & 1 \\ 2y_0 & 2z_0 \end{bmatrix} \begin{bmatrix} y \\ z \end{bmatrix} = \begin{bmatrix} 2 - x \\ 2 - x^2 + y_0^2 + z_0^2 \end{bmatrix},$$
then it is obvious that the existence of y and z will be guaranteed if the 2×2 matrix on the left is invertible. The last statement can also be rephrased using determinants: the condition is that the determinant of the 2×2 matrix on the left is not equal to 0. In the case of a function of one variable, Theorem 12.1.4 established the significance of the condition that $F'_y(x_0, y_0) \neq 0$. In the present example, if we denote
$$G(x, y, z) = x + y + z - 2, \quad H(x, y, z) = x^2 + y^2 + z^2 - 2,$$
then the 2×2 matrix is really
$$\begin{bmatrix} G'_y(x_0, y_0, z_0) & G'_z(x_0, y_0, z_0) \\ H'_y(x_0, y_0, z_0) & H'_z(x_0, y_0, z_0) \end{bmatrix}. \tag{12.1}$$
Just like the condition $F'_y(x_0, y_0) \neq 0$ in Theorem 12.1.4 required us to consider (temporarily) x as a constant, we can say the same thing here. If we denote
$$\mathbf{F}(y, z) = (G(x_0, y, z), H(x_0, y, z)),$$
then the matrix (12.1) is precisely the derivative $\mathbf{DF}(y_0, z_0)$. Thus the invertibility of this linear map plays an important role in the study of implicit functions. Incidentally, if we assume that it is invertible, and take the derivative of both given equations, we obtain
$$1 + y'(x) + z'(x) = 0, \quad 2x + 2y(x)y'(x) + 2z(x)z'(x) = 0.$$
The matricial form of this system
$$\begin{bmatrix} 1 & 1 \\ 2y(x) & 2z(x) \end{bmatrix} \begin{bmatrix} y'(x) \\ z'(x) \end{bmatrix} = \begin{bmatrix} -1 \\ -2x \end{bmatrix}$$
reveals that y' and z' exist, so we obtain that the implicit functions y and z not only exist but that they are differentiable.

Our goal is to understand the connection between the existence of the implicit functions, and the properties of the derivative as a linear map. We will start that program in the next section.

Exercises

In Exercises 12.1.1–12.1.3 find dy/dx and d^2y/dx^2.

12.1.1. $x \sin y + y \sin x = 1$.   12.1.2. $x = y + 2e^{-3y}$.   12.1.3. $x^{3/2} + y^{3/2} = 1$.

12.1.4. Find y', y'', y''' if $x^2 + xy + y^2 = 3$.

12.1.5. Find y', y'', y''' at $x = 0$, $y = 1$, if $x^2 - xy + 2y^2 + x - y - 1 = 0$.

In Exercises 12.1.6–12.1.13 find all first and second-order partial derivatives of $z(x,y)$.

12.1.6. $z^3 - 3xyz = 1$.   12.1.7. $x = ze^{yz}$.   12.1.8. $x^3 + y^3 + z^3 - 3xyz = 4$.

12.1.9. $xe^z = y \sin z$.   12.1.10. $e^z = x^2 + y^2 + z^2$.   12.1.11. $x + y = z + \cos(xyz)$.

12.1.12. $z^3 + xz + 2yz = 1$. 12.1.13. $z = \sqrt{x^2 - y^2} \tan(z/\sqrt{x^2 - y^2})$.

In Exercises 12.1.14–12.1.16 find dz and d^2z:

12.1.14. $\dfrac{x}{z} = \ln \dfrac{z}{y} + 1$.   12.1.15. $xyz = x + y + z$.   12.1.16. $z = x + \arctan \dfrac{y}{z - x}$.

In Exercises 12.1.17–12.1.18 the equations determine $x = x(z)$ and $y = y(z)$. Find dx/dz and dy/dz:

12.1.17. $x + y + z = 0$, $x^2 + y^2 + z^2 = 1$.   12.1.18. $z^2 + xy = 1$, $z^2 + x^2 = y^2 + 2$.

12.1.19. Find $\dfrac{dx}{dz}, \dfrac{dy}{dz}, \dfrac{d^2x}{dz^2}$, and $\dfrac{d^2y}{dz^2}$, for $x = 1$, $y = -1$, $z = 2$, if $x^2 + y^2 = \dfrac{1}{2}z^2$, $x + y + z = 2$.

In Exercises 12.1.20–12.1.23 the equations determine $z = z(x,y)$ and $u = u(x,y)$. Find $\partial u/\partial x$ and $\partial u/\partial y$:

12.1.20. $x^2 + y^2 + z^2 = 1$, $x^2 + y^2 + z^2 + u^2 = 1$.

12.1.21. $x^2 + y^2 + z^2 = z$, $x^2 + y^2 + z^2 + u^2 = u$.

12.1.22. $x + y + \sin z = z$, $x + y + z + e^u = u$.

12.1.23. $x^2 - y^2 + zu - u^2 = -3$, $x + y^2 + z^2 + uz = 2$.

12.1.24. Find $\dfrac{\partial z}{\partial x}$ and $\dfrac{\partial z}{\partial y}$ at $u = 1$, $v = 1$, if $x = u + \ln v$, $y = v - \ln u$, $z = 2u + v$.

12.1.25. Find $\dfrac{\partial^2 z}{\partial x \partial y}$ at $u = 2$, $v = 1$, if $x = u + v^2$, $y = u^2 - v^2$, $z = 2uv$.

12.1.26. Let $\mathbf{F} : \mathbb{R}^4 \to \mathbb{R}^2$, $\mathbf{F} = (F_1, F_2)$, let $\mathbf{a} \in \mathbb{R}^4$ be such that equations $F_1(x,y,z,u) = 0$ and $F_2(x,y,z,u) = 0$ are satisfied at $\mathbf{a}$, and suppose that the Jacobian matrix of $\mathbf{F}(\mathbf{a})$ is

$$\begin{bmatrix} 3 & 1 & 0 & 2 \\ 5 & 1 & -1 & 4 \end{bmatrix}.$$

If $x = x(z,u)$ and $y = y(z,u)$, find $\partial x/\partial z$ and $\partial y/\partial z$ at $\mathbf{a}$.

12.1.27. The purpose of this problem is to recreate Dini's original proof of Theorem 12.1.4.

(a) Prove that there exist $\delta, \eta, A, B > 0$ such that, if (x,y) is a point in the rectangle $R' = (x_0 - \delta, x_0 + \delta) \times (y_0 - \eta, y_0 + \eta)$, then $|F'_x(x,y)| < A$, $|F'_y(x,y)| > B$, and $A\delta < B\eta$.

(b) Use Taylor's Theorem to establish that there exists $\theta \in (0,1)$ such that

$$F(x_0 + h, y_0 + k) = hF'_x(x_0 + \theta h, y_0 + \theta k) + kF'_y(x_0 + \theta h, y_0 + \theta k).$$

(c) Prove that, for a fixed $h \le \delta$, the function $g(k) = F(x_0 + h, y_0 + k)$ must have precisely one zero between $k = -\eta$ and $k = \eta$.

(d) Conclude that, for each $x \in (x_0 - \delta, x_0 + \delta)$ there exists exactly one value of y in $(y_0 - \eta, y_0 + \eta)$ that satisfies $F(x, y) = 0$. Thus we have a function $y = f(x)$.

(e) Prove that $y_0 = f(x_0)$ and that f is continuous in $(x_0 - \delta, x_0 + \delta)$.

(f) Let h, k have the relationship established in (c), i.e., $F(x_0 + h, y_0 + k) = 0$. Prove that the limit of $\frac{h}{k}$, as $k \to 0$, exists.

(g) Conclude that f is in $(x_0 - \delta, x_0 + \delta)$ and that f' is continuous.

12.2   Derivative as a Linear Map

Our long-term goal is to prove Theorem 12.1.4. Our approach will be based on the following example. Let $F(x, y) = y - x + 1$. Clearly, we do not need any complicated theorems to see that the equation $F(x, y) = 0$ defines a function $f(x) = x - 1$. Let us, however, define

$$H(x, y) = (x, F(x, y)) = (x, y - x + 1)$$

and ask whether the function H has an inverse function. In other words, can we solve the system

$$u = x, \quad v = y - x + 1$$

for x and y? It is easy to see that the answer is yes, and that $x = u$, $y = u + v - 1$. If we set $v = 0$, which is another way of saying $F(x, y) = 0$, we obtain $x = u$, $y = u - 1$, which yields $y = x - 1$, our implicit function f.

From this example we see that, because of our interest in the implicit function, it might be helpful to study the existence and properties of the inverse function. In this section we will start working in this direction.

As a linear map, the derivative $\mathbf{Df}$ may have some properties like injectivity or surjectivity. How does that reflect on $\mathbf{f}$?

Example 12.2.1. A function is injective if and only if its derivative is injective.

Let $f(x) = x^2$. We will show that f is injective if and only if Df is injective.

Solution. When $\mathbf{f} : \mathbb{R}^n \to \mathbb{R}^m$, the derivative $\mathbf{Df}$ is a linear transformation from $\mathbb{R}^n$ to $\mathbb{R}^m$. Here, $m = n = 1$, and $f'(x) = 2x$ so $\mathbf{Df}$ is a multiplication by $2x$ on $\mathbb{R}$: $a \mapsto 2xa$. This map is invertible if and only if $2x \neq 0$. (If $2x = 0$, then each a is mapped to 0, so $\mathbf{Df}$ is not injective; if $2x \neq 0$, then the linear transformation $b \mapsto b/(2x)$ is the inverse of $\mathbf{Df}$). How does that affect f? It is clear from the graph that if $x \neq 0$ there exists an interval containing x in which f is monotone, hence injective. When $x = 0$, no such interval exists. ◆

Example 12.2.2. The relationship between the injectivity of a function and the injectivity of its derivative.

Let $\mathbf{f}(t) = (t^2, t^3)$. We will show that the injectivity of the derivative implies the injectivity of $\mathbf{f}$ (but not the other way round).

Solution. The derivative of $\mathbf{f}$ is a 2×1 matrix $\left[\begin{smallmatrix} 2t \\ 3t^2 \end{smallmatrix} \right]$, so $\mathbf{Df}(a) = \left[\begin{smallmatrix} 2a \\ 3a^2 \end{smallmatrix} \right] x$. When $a \neq 0$, $\mathbf{Df}(a)(x_1) = \mathbf{Df}(a)(x_2)$ if and only if $x_1 = x_2$, so for $a \neq 0$, $\mathbf{Df}(a)$ is injective. It is not hard to see that, if $a = 0$, $\mathbf{Df}(a) = \left[\begin{smallmatrix} 0 \\ 0 \end{smallmatrix} \right]$ is not injective. What about $\mathbf{f}$? It is easy to see that $\mathbf{f}(t_1) = \mathbf{f}(t_2)$ if and only if $t_1 = t_2$, so $\mathbf{f}$ is injective. Thus, if $\mathbf{Df}(a)$ is injective, then $\mathbf{f}$ is injective. ◆

Based on these examples, we might expect that, if the derivative is injective, then so is **f**. This is indeed true, and we will prove it below. First, we need a result from Linear Algebra.

Proposition 12.2.3. *Let $T : \mathbb{R}^m \to \mathbb{R}^n$ be a linear map, and let $[t_{ij}]$ be its matrix in the standard basis. Then:*

(a) $\|T(\mathbf{u})\| \leq M\|\mathbf{u}\|$, *for any* $\mathbf{u} \in \mathbb{R}^m$, *where* $M = (\sum_{i=1}^{n} \sum_{j=1}^{m} |t_{ij}|^2)^{1/2}$;

(b) T is continuous;

(c) T is injective if and only if there exists $\gamma > 0$ such that, for any $\mathbf{u} \in \mathbb{R}^m$, $\|T(\mathbf{u})\| \geq \gamma\|\mathbf{u}\|$.

Proof. Let $T(\mathbf{u}) = \mathbf{v}$. Then

$$v_i = \sum_{j=1}^{m} t_{ij} u_j, \quad 1 \leq i \leq n.$$

By the Cauchy-Schwarz inequality

$$|v_i|^2 \leq \sum_{j=1}^{m} |t_{ij}|^2 \sum_{j=1}^{m} |u_j|^2 = \|\mathbf{u}\|^2 \sum_{j=1}^{m} |t_{ij}|^2, \quad 1 \leq i \leq n,$$

and

$$\|T\mathbf{u}\|^2 = \|\mathbf{v}\|^2 = \sum_{i=1}^{n} |v_i|^2 \leq \|\mathbf{u}\|^2 \sum_{i=1}^{n} \sum_{j=1}^{m} |t_{ij}|^2,$$

so the assertion (a) follows by taking square roots.

If $\mathbf{u}, \mathbf{v} \in \mathbb{R}^m$, then (a) implies that

$$\|T(\mathbf{u}) - T(\mathbf{v})\| = \|T(\mathbf{u} - \mathbf{v})\| \leq M\|\mathbf{u} - \mathbf{v}\|,$$

so T is, in fact, uniformly continuous.

In order to prove (c), suppose first that

$$\|T(\mathbf{u})\| \geq \gamma\|\mathbf{u}\|, \quad \text{for all} \quad \mathbf{u} \in \mathbb{R}^m. \tag{12.2}$$

If T is not injective then there exist $\mathbf{u}, \mathbf{v} \in \mathbb{R}^m$ such that $T(\mathbf{u}) = T(\mathbf{v})$. This implies that $T(\mathbf{u} - \mathbf{v}) = \mathbf{0}$ so it follows from (12.2) that $\mathbf{u} - \mathbf{v} = \mathbf{0}$. Thus, T is injective.

In the other direction, suppose that T is injective and let

$$S = \{\mathbf{u} \in \mathbb{R}^m : \|\mathbf{u}\| = 1\}.$$

This is a closed and bounded set, so the continuous function $\|T(\mathbf{u})\|$ attains its minimum, say $\|T(\mathbf{w})\| = \gamma$. If $\gamma = 0$, then we would have $T(\mathbf{w}) = \mathbf{0}$, contradicting the injectivity of T. Thus $\gamma > 0$, and

$$\|T(\mathbf{u})\| \geq \gamma, \quad \text{for all} \quad \mathbf{u} \in S.$$

Now we can prove (12.2). It holds for $\mathbf{u} = \mathbf{0}$ because $T(\mathbf{0}) = \mathbf{0}$. If $\mathbf{u} \neq \mathbf{0}$, then $\mathbf{u}/\|\mathbf{u}\| \in S$, so $T(\mathbf{u}/\|\mathbf{u}\|) \geq \gamma$. The linearity of T now implies (12.2). $\qquad \square$

In part (a) we have made use of the quantity $(\sum_{i=1}^{n} \sum_{j=1}^{m} |t_{ij}|^2)^{1/2}$ that we have assigned to the linear transformation T with the matrix $[t_{ij}]$. This number is called the *Hilbert–Schmidt norm* of T, and it is denoted by $\|T\|_2$. We leave it as an exercise to verify that the Hilbert–Schmidt norm has the properties listed in Theorem 10.1.2, so it makes sense to call it a norm.

ERHARD SCHMIDT (1876–1959) was a German mathematician, born in what is today Estonia. He studied in Berlin with Schwarz, and his doctoral advisor was Hilbert at the University of Göttingen. The topic of his dissertation was integral equations. After defending his Habilitation thesis at the University of Bonn, he taught in Zürich, Erlangen, and Breslau. Finally he settled at the University of Berlin in 1917. He assumed the leadership position reviving a somewhat stagnant situation. He was instrumental in founding the Institute of Applied Mathematics and served as its director. His administrative skills brought him the position of Dean in 1921 and Vice Chancelor in 1929. During WWII Schmidt had to carry out various Nazi resolutions against the Jews—a job that he apparently did not do well, since he was criticized at one point for not understanding the "Jewish question". After the war he served as Director of the Mathematics Research Institute of the German Academy of Science until 1958. He was a co-founder and the first editor of *Mathematische Nachrichten*. His most important research came between 1905 and 1908. In a series of papers he built on the ideas of Hilbert to create the concept of abstract Hilbert space and established many of its geometric properties. The orthogonalization technique that he used is now known as the Gram–Schmidt process, although it was introduced by Laplace. In the later part of his career, Schmidt became interested in topology, finding a new proof of the Jordan curve theorem.

Recall that, if $\mathbf{x}$ is close to $\mathbf{y}$, then

$$f(\mathbf{x}) \approx f(\mathbf{y}) + \mathbf{D}f(\mathbf{y})(\mathbf{x} - \mathbf{y}). \tag{12.3}$$

Suppose now that both $\mathbf{x}$ and $\mathbf{y}$ are close to $\mathbf{a}$. Can we replace $\mathbf{D}f(\mathbf{y})$ by $\mathbf{D}f(\mathbf{a})$ in (12.3)? It is reasonable to expect that the answer is in the affirmative, but as usual, we need to make it precise.

Proposition 12.2.4. *Suppose that $\mathbf{f}$ is a function defined on an open n-ball A, with values in $\mathbb{R}^m$, and that its partial derivatives are continuous in A. Let $\mathbf{c} \in A$ and let $\varepsilon > 0$. Then there exists an n-ball $B \subset A$ such that, if $\mathbf{u}, \mathbf{v} \in B$, then*

$$\|\mathbf{f}(\mathbf{u}) - \mathbf{f}(\mathbf{v}) - \mathbf{D}\mathbf{f}(\mathbf{c})(\mathbf{u} - \mathbf{v})\| \leq \varepsilon \|\mathbf{u} - \mathbf{v}\|.$$

Proof. Let $\mathbf{g}$ be a function defined on A by $\mathbf{g}(\mathbf{u}) = \mathbf{f}(\mathbf{u}) - \mathbf{D}\mathbf{f}(\mathbf{c})(\mathbf{u})$. Then,

$$\mathbf{f}(\mathbf{u}) - \mathbf{f}(\mathbf{v}) - \mathbf{D}\mathbf{f}(\mathbf{c})(\mathbf{u} - \mathbf{v}) = \mathbf{g}(\mathbf{u}) - \mathbf{g}(\mathbf{v}). \tag{12.4}$$

The function $\mathbf{f}$ is differentiable by assumption, and $\mathbf{D}\mathbf{f}(\mathbf{c})$ because it is linear, so $\mathbf{g}$ is differentiable in A and $\mathbf{D}\mathbf{g}(\mathbf{u}) = \mathbf{D}\mathbf{f}(\mathbf{u}) - \mathbf{D}\mathbf{f}(\mathbf{c})$. Thus, $\mathbf{D}\mathbf{g}(\mathbf{u})$ is an $m \times n$ matrix with entries

$$\frac{\partial f_i}{\partial x_j}(\mathbf{u}) - \frac{\partial f_i}{\partial x_j}(\mathbf{c}), \quad 1 \leq i \leq m, \, 1 \leq j \leq n.$$

Each of these mn entries is continuous at $\mathbf{c}$ (as a function of $\mathbf{u}$) so, given $\varepsilon > 0$, there exists $\delta > 0$ and an open ball $B = B_\delta(\mathbf{c})$ such that, if $\mathbf{x} \in B$, then

$$\left| \frac{\partial f_i}{\partial x_j}(\mathbf{x}) - \frac{\partial f_i}{\partial x_j}(\mathbf{c}) \right| < \frac{\varepsilon}{\sqrt{mn}}, \quad 1 \leq i \leq m, \, 1 \leq j \leq n.$$

It follows that, for $\mathbf{x} \in B$,

$$\sum_{i=1}^{m} \sum_{j=1}^{n} \left| \frac{\partial f_i}{\partial x_j}(\mathbf{x}) - \frac{\partial f_i}{\partial x_j}(\mathbf{c}) \right|^2 < \varepsilon^2.$$

By Proposition 12.2.3 (a), if $\mathbf{x} \in B$, and $\mathbf{u} \in \mathbb{R}^n$,

$$\|\mathbf{D}\mathbf{g}(\mathbf{x})(\mathbf{u})\| < \varepsilon \|\mathbf{u}\|. \tag{12.5}$$

The rest is easy. Let $\mathbf{u}, \mathbf{v} \in B$. By Theorem 11.5.5, there exists a point $\mathbf{w}$ on the line segment from $\mathbf{u}$ to $\mathbf{v}$, such that

$$\|\mathbf{g}(\mathbf{u}) - \mathbf{g}(\mathbf{v})\| \leq \|\mathbf{D}\mathbf{g}(\mathbf{w})(\mathbf{u} - \mathbf{v})\|.$$

Since $\mathbf{w} \in B$, the result follows from (12.4) and (12.5). $\qquad\square$

Now we can establish the announced result about injectivity.

Theorem 12.2.5 (Injective Mapping Theorem). *Suppose that* $\mathbf{f}$ *is a function defined on an open n-ball* A*, with values in* $\mathbb{R}^m$*, and that its partial derivatives are continuous in* A*. Let* $\mathbf{c} \in A$ *and suppose that* $\mathbf{Df}(\mathbf{c})$ *is injective. Then there exists an n-ball* B *and* $\gamma > 0$ *such that:*

(a) $\|\mathbf{f}(\mathbf{u}) - \mathbf{f}(\mathbf{v})\| \geq \gamma \|\mathbf{u} - \mathbf{v}\|$*, for any* $\mathbf{u}, \mathbf{v} \in B$*;*

(b) the restriction $\mathbf{f}|B$ *is injective;*

(c) the inverse $(\mathbf{f}|B)^{-1}$ *is uniformly continuous on* $\mathbf{f}(B)$*.*

Proof. Since $\mathbf{Df}(\mathbf{c})$ is injective, Proposition 12.2.3 implies that there exists $\gamma > 0$, such that

$$\|\mathbf{Df}(\mathbf{c})(\mathbf{u})\| \geq 2\gamma \|\mathbf{u}\|, \quad \text{for any} \quad \mathbf{u} \in \mathbb{R}^m. \tag{12.6}$$

By Proposition 12.2.4, there exists $\delta > 0$ and an open ball $B = B_\delta(\mathbf{c})$ such that

$$\|\mathbf{f}(\mathbf{u}) - \mathbf{f}(\mathbf{v}) - \mathbf{Df}(\mathbf{c})(\mathbf{u} - \mathbf{v})\| < \gamma \|\mathbf{u} - \mathbf{v}\|, \quad \text{for any} \quad \mathbf{u}, \mathbf{v} \in B. \tag{12.7}$$

Combining (12.6) and (12.7) we obtain

$$\begin{aligned}\|\mathbf{f}(\mathbf{u}) - \mathbf{f}(\mathbf{v})\| &\geq \|\mathbf{Df}(\mathbf{c})(\mathbf{u} - \mathbf{v})\| - \|\mathbf{f}(\mathbf{u}) - \mathbf{f}(\mathbf{v}) - \mathbf{Df}(\mathbf{c})(\mathbf{u} - \mathbf{v})\| \\ &\geq 2\gamma \|\mathbf{u} - \mathbf{v}\| - \gamma \|\mathbf{u} - \mathbf{v}\| = \gamma \|\mathbf{u} - \mathbf{v}\|,\end{aligned} \tag{12.8}$$

so (a) is proved. Now (b) follows immediately: if $\mathbf{f}(\mathbf{u}) = \mathbf{f}(\mathbf{v})$ then $\mathbf{u} = \mathbf{v}$. Consequently, $\mathbf{f}$ is injective in B, and its restriction $\mathbf{f}|B$ is invertible.

Finally, let $\mathbf{h} = (\mathbf{f}|B)^{-1}$. It remains to prove that $\mathbf{h}$ is uniformly continuous on $\mathbf{f}(B)$. Let $\mathbf{y}, \mathbf{z} \in \mathbf{f}(B)$. Then there exist $\mathbf{u}, \mathbf{v} \in B$ such that $\mathbf{h}(\mathbf{y}) = \mathbf{u}$, $\mathbf{h}(\mathbf{z}) = \mathbf{v}$. Equivalently, $\mathbf{f}(\mathbf{u}) = \mathbf{y}$ and $\mathbf{f}(\mathbf{v}) = \mathbf{z}$. Now, (12.8) implies that

$$\|\mathbf{y} - \mathbf{z}\| \geq \frac{\gamma}{2} \|\mathbf{h}(\mathbf{y}) - \mathbf{h}(\mathbf{z})\|$$

so $\mathbf{h}$ is uniformly continuous on $\mathbf{f}(B)$. $\qquad\qquad\square$

Remark 12.2.6. The theorem makes no claim about the *global invertibility* of $\mathbf{f}$. It only establishes that $\mathbf{f}$ is *locally invertible*. Both of these statements have to do with the domain of $\mathbf{f}$. For example, if $f(x) = x^2$ and $c \neq 0$, Example 12.2.1 shows that there exists an interval B containing c, such that f is invertible on B. On the other hand, the function f with domain $\mathbb{R}$ is not invertible.

> Theorem 12.2.5 is almost present in Goursat's 1903 article [53] (see page 126). He actually proved a stronger theorem (see Exercise 12.2.13 for the case $m = n = 1$) which implies Theorem 12.2.5.

Injective Mapping Theorem uses a hypothesis that $\mathbf{f}$ has continuous partial derivatives in a ball around $\mathbf{c}$. What can we say about its inverse $\mathbf{h}$? By the definition of the derivative, differentiability is considered only at interior points. However, we do not know whether $\mathbf{f}(\mathbf{c})$ is an interior point of $\mathbf{f}(B)$.

Example 12.2.7. The continuous image of an interior point is not an interior point.

Let $f(x) = x^2$, $c = 0$, $A = (-1, 1)$. Then $f(A) = [0, 1)$ and $f(c) = 0$ is not an interior point of $f(A)$. $\qquad\qquad\blacklozenge$

Clearly, if we want to guarantee that $f(c)$ is an interior point of $f(A)$, we need to impose some additional conditions on f. It turns out that such a condition is the surjectivity of the derivative. We will establish this result in the next section.

Exercises

12.2.1. Verify that the Hilbert–Schmidt norm has the properties listed in Theorem 10.1.2.

12.2.2. Suppose that $\mathbf{f}$ has continuous partial derivatives at $\mathbf{c}$ and that the derivative $\mathbf{Df}(\mathbf{c})$ is a surjective linear transformation. Prove that there exists $r > 0$ such that $\mathbf{Df}(\mathbf{x})$ is surjective for all $\mathbf{x} \in B_r(\mathbf{c})$.

12.2.3. Let $\{e_i\}_{i=1}^n$ and $\{f_i\}_{i=1}^n$ be two orthonormal bases of $\mathbb{R}^n$, let $T : \mathbb{R}^n \to \mathbb{R}^n$ be a linear transformation. If $[t_{ij}]$ and $[r_{ij}]$ are the matrices of T in these two bases, prove that $\sum_{i,j=1}^n t_{ij}^2 = \sum_{i,j=1}^n r_{ij}^2$.

12.2.4. Prove or disprove: the Hilbert–Schmidt norm is independent of the choice of a basis of $\mathbb{R}^n$.

12.2.5. Let $A, B : \mathbb{R}^n \to \mathbb{R}^n$ be linear transformations. Prove that $\|AB\|_2 \leq \|A\|_2 \|B\|_2$.

12.2.6. Let $A : \mathbb{R}^n \to \mathbb{R}^m$ be an injective linear transformation. Prove that there exists $r > 0$ such that, if $B : \mathbb{R}^n \to \mathbb{R}^m$ is a linear transformation and $\|A - B\|_2 < r$, then B is also injective.

12.2.7. Let $A : \mathbb{R}^n \to \mathbb{R}^n$ be an invertible linear transformation. Prove that there exists $r > 0$ such that, if $B : \mathbb{R}^n \to \mathbb{R}^n$ is a linear transformation and $\|A - B\|_2 < r$, then B is also invertible.

12.2.8. Let A be a non-empty open set in $\mathbb{R}^n$, let $m > n$ and let $\mathbf{f} : A \to \mathbb{R}^m$ be a C^1 function. Suppose that $\operatorname{rank}(\mathbf{D}(\mathbf{f}(\mathbf{x}))) = n$ for all $\mathbf{x} \in A$. Prove that for every $x \in A$ there exists $r > 0$ so that $\mathbf{f}$ is injective in $B_r(\mathbf{x})$.

12.2.9. Let A be an n-ball and $\mathbf{F} : A \to \mathbb{R}^n$ a C^1 function. Suppose that the matrix B of $\mathbf{D}(\mathbf{F})(\mathbf{x})$ has the property that $B + B^T$ is positive definite. Prove that $\mathbf{F}$ is injective on A.

12.2.10. Let A be an n-ball and $\mathbf{f} : A \to \mathbb{R}^m$ a C^1 function. Suppose that there exists $\gamma > 0$ such that $\|\mathbf{f}(\mathbf{u}) - \mathbf{f}(\mathbf{v})\| \geq \gamma \|\mathbf{u} - \mathbf{v}\|$, for any $\mathbf{u}, \mathbf{v} \in A$. Prove that for any $\mathbf{x} \in A$, $\mathbf{D}(\mathbf{f})(\mathbf{x})$ is injective.

12.2.11. Let A be an n-ball, let $\mathbf{a} \in A$ and let $\mathbf{f} : A \to \mathbb{R}^m$ be a C^1 function. Suppose that $\mathbf{D}(\mathbf{f})(\mathbf{a})$ is not injective. Prove or disprove: $\mathbf{f}$ is not injective in A.

12.2.12. Give an example to show that Injective Mapping Theorem (Theorem 12.2.5) becomes false if the assumption that partial derivatives of $\mathbf{f}$ are continuous is replaced by the differentiability of $\mathbf{f}$.

12.2.13. The purpose of this problem is to recreate Goursat's proof of a weaker version of Theorem 12.1.4 from [53]. Let F be a function of 2 variables defined on a rectangle $R = (a, b) \times (c, d)$, and let (x_0, y_0) be an interior point of R, such that $F(x_0, y_0) = 0$. Suppose that: (i) F is continuous on R; (ii) for any $x \in (a, b)$, $F(x, y)$ is a continuously differentiable in (c, d); (iii) $F_y'(x_0, y_0) \neq 0$. Then:

(A) there exists a rectangle $R' = (x_0 - \delta, x_0 + \delta) \times (y_0 - \eta, y_0 + \eta)$ such that the equation $F(x, y) = 0$ defines a function $y = f(x)$ in R';

(B) $f(x_0) = y_0$;

(C) f is continuous in $(x_0 - \delta, x_0 + \delta)$.

(a) Show that $G(x, y) = y - y_0 - \gamma F(x, y)$, where $\gamma = 1/F_y'(x_0, y_0)$, satisfies $G(x_0, y_0) = G_y'(x_0, y_0) = 0$.

(b) Prove that, for any $x \in (a, b)$, $G(x, y)$ is continuously differentiable in (c, d).

(c) Define a sequence of functions $y_1 = y_0 + G(x, y_0)$, $y_2 = y_0 + G(x, y_1)$, etc. Prove that there exists a rectangle $R' = (x_0 - \delta, x_0 + \delta) \times (y_0 - \eta, y_0 + \eta)$ such that, for any $x \in (x_0 - \delta, x_0 + \delta)$ and any $n \in \mathbb{N}$, $y_n(x) \in (y_0 - \eta, y_0 + \eta)$.

(d) Prove that there exists $K > 0$ such that, for any $n \in \mathbb{N}$, $|y_{n+1} - y_n| < K|y_n - y_{n-1}|$. Deduce that the sequence y_n converges for $x \in (x_0 - \delta, x_0 + \delta)$ to a function $y = f(x)$.

(e) Prove that f satisfies $f(x_0) = y_0$ and that f is continuous for $x \in (x_0 - \delta, x_0 + \delta)$.

(f) Prove that f is a unique function that satisfies conditions in (e).

12.3   Open Mapping Theorem

In this section we will complete the proof of the Inverse Function Theorem: If $\mathbf{Df}(\mathbf{c})$ is invertible, then $\mathbf{f}$ is locally invertible, and the inverse function is differentiable. So far we have seen that the mere injectivity of the derivative guarantees that the restriction of $\mathbf{f}$ to an open ball B is invertible and that the inverse function is uniformly continuous on $\mathbf{f}(B)$. Thus, it remains to show that it is differentiable. There is a subtle difficulty here: we have defined differentiability only at interior points (page 346), so it would be useful to know that every point of $\mathbf{f}(B)$ is an interior point (i.e., that $\mathbf{f}(B)$ is an open set). The following theorem shows that this is indeed so, and the price we have to pay is not very steep.

Theorem 12.3.1 (Open Mapping Theorem). *Suppose that $\mathbf{f}$ is a function defined on an open n-ball A, with values in $\mathbb{R}^m$, and that its partial derivatives are continuous in A. Suppose that $\mathbf{Df}(\mathbf{x})$ is surjective for each $\mathbf{x} \in A$. Then $\mathbf{f}(A)$ is an open set in $\mathbb{R}^m$.*

Proof. Let $\mathbf{b} \in \mathbf{f}(A)$. We will show that there exists $\varepsilon > 0$ such that the open ball $B_\varepsilon(\mathbf{b})$ is contained in $\mathbf{f}(A)$.

Let $\mathbf{c} \in A$ so that $\mathbf{b} = \mathbf{f}(\mathbf{c})$. By assumption, $\mathbf{Df}(\mathbf{c})$ is a surjective map from $\mathbb{R}^n$ to $\mathbb{R}^m$, so it possesses the right inverse $T : \mathbb{R}^m \to \mathbb{R}^n$. By Proposition 12.2.4, there exists $\delta > 0$ and an open ball $B = B_\delta(\mathbf{c})$ such that, if $\mathbf{u}, \mathbf{v} \in B$, then

$$\|\mathbf{f}(\mathbf{u}) - \mathbf{f}(\mathbf{v}) - \mathbf{Df}(\mathbf{c})(\mathbf{u} - \mathbf{v})\| < \frac{1}{3\|T\|_2}\|\mathbf{u} - \mathbf{v}\|.$$

Let $\varepsilon = \delta/(3\|T\|_2)$. We will show that $B_\varepsilon(\mathbf{b}) \subset \mathbf{f}(A)$.

Let $\mathbf{y} \in B_\varepsilon(\mathbf{b})$. In order to show that $\mathbf{y} \in \mathbf{f}(A)$ we will exhibit $\mathbf{x} \in A$ such that $\mathbf{f}(\mathbf{x}) = \mathbf{y}$. We will obtain $\mathbf{x}$ as a limit of a sequence defined by

$$\mathbf{x}_0 = \mathbf{c}, \quad \mathbf{x}_1 = \mathbf{x}_0 + T(\mathbf{y} - \mathbf{b}),$$
$$\mathbf{x}_{k+1} = \mathbf{x}_k - T\left[\mathbf{f}(\mathbf{x}_k) - \mathbf{f}(\mathbf{x}_{k-1}) - \mathbf{Df}(\mathbf{c})(\mathbf{x}_k - \mathbf{x}_{k-1})\right]. \tag{12.9}$$

The crucial properties of this sequence will follow from the estimates

$$\|\mathbf{x}_{k+1} - \mathbf{c}\| < \frac{\delta}{2}, \quad \|\mathbf{x}_{k+1} - \mathbf{x}_k\| \leq \frac{\delta}{3^{k+1}}, \quad k \in \mathbb{N}_0 \tag{12.10}$$

which we will prove by induction. If $k = 0$ then $\mathbf{x}_0 = \mathbf{c} \in B$ and

$$\|\mathbf{x}_1 - \mathbf{x}_0\| = \|T(\mathbf{y} - \mathbf{b})\| \leq \|T\|_2 \|\mathbf{y} - \mathbf{b}\| \leq \|T\|_2 \varepsilon = \frac{\delta}{3},$$

so (12.10) holds for $k = 0$ and, in addition, $\mathbf{x}_1 \in B$. Suppose that (12.10) holds for $0, 1, 2, \ldots, k$. Then $\mathbf{x}_{k-1}, \mathbf{x}_k \in B$ and $\|\mathbf{x}_k - \mathbf{x}_{k-1}\| \le \delta/3^k$. Therefore,

$$
\begin{aligned}
\|\mathbf{x}_{k+1} - \mathbf{x}_k\| &= \|T\left[\mathbf{f}(\mathbf{x}_k) - \mathbf{f}(\mathbf{x}_{k-1}) - \mathbf{Df}(\mathbf{c})(\mathbf{x}_k - \mathbf{x}_{k-1})\right]\| \\
&\le \|T\|_2 \|\mathbf{f}(\mathbf{x}_k) - \mathbf{f}(\mathbf{x}_{k-1}) - \mathbf{Df}(\mathbf{c})(\mathbf{x}_k - \mathbf{x}_{k-1})\| \\
&< \|T\|_2 \frac{1}{3\|T\|_2} \|\mathbf{x}_k - \mathbf{x}_{k-1}\| \\
&\le \frac{\delta}{3^{k+1}}.
\end{aligned}
$$

Moreover,

$$
\|\mathbf{x}_{k+1} - \mathbf{c}\| \le \sum_{i=0}^{k} \|\mathbf{x}_{i+1} - \mathbf{x}_i\| \le \delta \sum_{i=0}^{k} \frac{1}{3^{i+1}} < \delta \sum_{i=0}^{\infty} \frac{1}{3^{i+1}} = \frac{\delta}{2},
$$

so we have established (12.10).

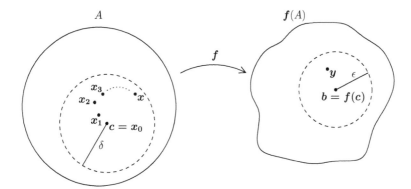

Figure 12.4: The sequence $\{\mathbf{x}_n\}$ converges to $\mathbf{x}$ such that $\mathbf{y} = \mathbf{f}(\mathbf{x})$.

Using (12.10) we will show that $\{\mathbf{x}_k\}$ is a Cauchy sequence. Let $\eta > 0$, and let

$$
N = \max\{\lfloor \ln(2\delta/\eta)/\ln 3 \rfloor, 1\}.
$$

Then, if $k \ge N$, $k + 1 > \ln(2\delta/\eta)/\ln 3$, which implies that $3^{k+1} > 2\delta/\eta$ and $2\delta/3^{k+1} < \eta$. It follows that, if $l \ge k \ge N$,

$$
\|\mathbf{x}_l - \mathbf{x}_k\| \le \sum_{i=k}^{l-1} \|\mathbf{x}_{i+1} - \mathbf{x}_i\| \le \delta \sum_{i=k}^{l-1} \frac{1}{3^{i+1}} < \sum_{i=k}^{\infty} \frac{1}{3^{i+1}} \delta = \frac{1}{3^{k+1}} \frac{3}{2} \delta < \eta.
$$

Thus, $\{\mathbf{x}_k\}$ is a Cauchy sequence and $\|\mathbf{x}_k - \mathbf{c}\| < \delta/2$, $k \in \mathbb{N}$. The limit $\mathbf{x}$ of this sequence satisfies $\|\mathbf{x} - \mathbf{c}\| \le \delta/2$, so $\mathbf{x} \in B$.

Finally, we will prove that $\mathbf{f}(\mathbf{x}) = \mathbf{y}$. In order to do that, we will establish by induction the equality $\mathbf{Df}(\mathbf{c})(\mathbf{x}_{k+1} - \mathbf{x}_k) = \mathbf{y} - \mathbf{f}(\mathbf{x}_k)$. The case $k = 0$ is easy:

$$
\mathbf{Df}(\mathbf{c})(\mathbf{x}_1 - \mathbf{x}_0) = \mathbf{Df}(\mathbf{c})T(\mathbf{y} - \mathbf{b}) = \mathbf{y} - \mathbf{b} = \mathbf{y} - \mathbf{f}(\mathbf{x}_0).
$$

Suppose that $\mathbf{Df}(\mathbf{c})(\mathbf{x}_{k+1} - \mathbf{x}_k) = \mathbf{y} - \mathbf{f}(\mathbf{x}_k)$. Then,

$$
\begin{aligned}
\mathbf{Df}(\mathbf{c})(\mathbf{x}_{k+2} - \mathbf{x}_{k+1}) &= -\mathbf{Df}(\mathbf{c})T\left[\mathbf{f}(\mathbf{x}_{k+1}) - \mathbf{f}(\mathbf{x}_k) - \mathbf{Df}(\mathbf{c})(\mathbf{x}_{k+1} - \mathbf{x}_k)\right] \\
&= -\mathbf{f}(\mathbf{x}_{k+1}) + \mathbf{f}(\mathbf{x}_k) + \mathbf{Df}(\mathbf{c})(\mathbf{x}_{k+1} - \mathbf{x}_k)
\end{aligned}
$$

$$= -\mathbf{f}(\mathbf{x}_{k+1}) + \mathbf{f}(\mathbf{x}_k) + (\mathbf{y} - \mathbf{f}(\mathbf{x}_k))$$
$$= \mathbf{y} - \mathbf{f}(\mathbf{x}_{k+1}).$$

Thus, $\mathbf{Df}(\mathbf{c})(\mathbf{x}_{k+1} - \mathbf{x}_k) = \mathbf{y} - \mathbf{f}(\mathbf{x}_k)$, for all $k \in \mathbb{N}_0$. If we now pass to the limit as $k \to \infty$, using the fact that both $\mathbf{Df}(\mathbf{c})$ and $\mathbf{f}$ are continuous, we obtain

$$\mathbf{y} - \mathbf{f}(\mathbf{x}) = \mathbf{Df}(\mathbf{c})(\mathbf{x} - \mathbf{x}) = \mathbf{0}. \qquad \square$$

Example 12.3.2. The continuous image of an open set is open.

Let $f(x,y) = x + \sin(x+y)$, $A = \{(x,y) \in \mathbb{R}^2 : x^2 + y^2 < \pi^2\}$. We will show that $f(A)$ is an open set.

Solution. Since

$$\mathbf{D}f(x,y) = \begin{bmatrix} 1 + \cos(x+y) & \cos(x+y) \end{bmatrix}$$

and the matrix defines a surjective linear transformation from $\mathbb{R}^2$ to $\mathbb{R}$, the Open Mapping Theorem implies that $f(A)$ is an open set. $\blacklozenge$

Theorem 12.3.1 appears in the 1950 article [56] by a University of Chicago professor Lawrence Graves (1896–1973).

As we have announced, our goal is to prove the Inverse Function Theorem. We will make a (rather natural) assumption that the dimensions of the domain and the codomain are the same. If $\mathbf{f} : \mathbb{R}^n \to \mathbb{R}^n$, and if $\mathbf{Df}$ is bijective, then both the Injective Map Theorem and the Open Mapping Theorem apply, and we get that $\mathbf{f}$ is locally invertible, with a continuous inverse $\mathbf{h}$ defined on the open set $\mathbf{f}(B)$. All that remains is to prove that $\mathbf{h}$ is differentiable.

Theorem 12.3.3 (Inverse Function Theorem). *Suppose that $\mathbf{f}$ is a function defined on an open n-ball A, with values in $\mathbb{R}^n$, and that its partial derivatives are continuous in A. Let $\mathbf{c} \in A$ and suppose that $\mathbf{Df}(\mathbf{c})$ is bijective. Then there exists an open n-ball B with center $\mathbf{c}$, such that:*

(a) the restriction $\mathbf{f}|B$ is a bijection between B and $\mathbf{f}(B)$;

(b) the set $V = \mathbf{f}(B)$ is open;

(c) the inverse $\mathbf{h} = (\mathbf{f}|B)^{-1}$ is uniformly continuous on V;

(d) $\mathbf{h}$ has continuous partial derivatives;

(e) $\mathbf{Dh}(\mathbf{v}) = (\mathbf{Df}(\mathbf{h}(\mathbf{v})))^{-1}$, for $\mathbf{v} \in V$.

Proof. Although $\mathbf{Df}$ is assumed to be bijective only at $\mathbf{c}$, this is actually true in an open ball with center $\mathbf{c}$. Let us prove this first. By assumption, $\mathbf{Df}(\mathbf{c})$ is injective, so there exists $\gamma > 0$ such that

$$\|\mathbf{Df}(\mathbf{c})(\mathbf{u})\| \geq 2\gamma\|\mathbf{u}\|, \quad \text{for any } \mathbf{u} \in \mathbb{R}^n.$$

Since partial derivatives of $\mathbf{f}$ are continuous, there exists a ball B_1 with center $\mathbf{c}$ so that,

$$\|\mathbf{Df}(\mathbf{x}) - \mathbf{Df}(\mathbf{c})\|_2 < \gamma, \quad \text{for any } \mathbf{x} \in B_1.$$

For such $\mathbf{x}$ and $\mathbf{u} \in \mathbb{R}^n$, $\|(\mathbf{Df}(\mathbf{x}) - \mathbf{Df}(\mathbf{c}))(\mathbf{u})\| \leq \gamma\|\mathbf{u}\|$, so we obtain that

$$\|\mathbf{Df}(\mathbf{x})(\mathbf{u})\| \geq \|\mathbf{Df}(\mathbf{c})(\mathbf{u})\| - \|(\mathbf{Df}(\mathbf{x}) - \mathbf{Df}(\mathbf{c}))(\mathbf{u})\| \geq 2\gamma\|\mathbf{u}\| - \gamma\|\mathbf{u}\| = \gamma\|\mathbf{u}\|.$$

This implies that $\mathbf{Df}(\mathbf{x})$ is injective, hence invertible, for all $\mathbf{x} \in B_1$. By the Injective Map

Theorem, there exists an open ball $B \subset B_1$ with center $\mathbf{c}$, so that $\mathbf{f}|B$ is injective, and the inverse function $\mathbf{h} = (\mathbf{f}|B)^{-1}$ is uniformly continuous on $\mathbf{f}(B)$. This establishes (a) and (c), while (b) follows from the Open Mapping Theorem.

In the remaining portion of the proof we will focus on the differentiability of $\mathbf{h}$ on the set $V = \mathbf{f}(B)$. Let $\mathbf{v} \in V$, and let $\mathbf{u} = \mathbf{h}(\mathbf{v})$. The mapping $\mathbf{Df}(\mathbf{u})$ is invertible since $\mathbf{u} \in B$. Let T be the inverse of $\mathbf{Df}(\mathbf{u})$. We will show that, for any $\mathbf{y} \in V$,

$$\|\mathbf{h}(\mathbf{y}) - \mathbf{h}(\mathbf{v}) - T(\mathbf{y} - \mathbf{v})\| \le \|T\|_2 \|\mathbf{r_h}(\mathbf{y})\|, \tag{12.11}$$

where

$$\frac{\|\mathbf{r_h}(\mathbf{y})\|}{\|\mathbf{y} - \mathbf{v}\|} \to 0, \quad \mathbf{y} \to \mathbf{v}.$$

Both (d) and (e) then follow immediately. Indeed, from (12.11) we conclude that $\mathbf{h}$ is

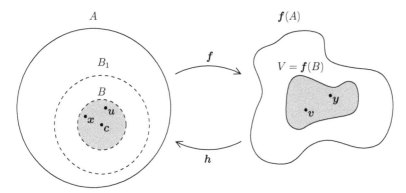

Figure 12.5: $\mathbf{h} = (\mathbf{f}|B)^{-1}$ is differentiable at $\mathbf{v}$.

differentiable at $\mathbf{v}$, with $\mathbf{Dh}(\mathbf{v}) = T = (\mathbf{Df}(\mathbf{u}))^{-1} = (\mathbf{Df}(\mathbf{h}(\mathbf{v})))^{-1}$, which is (e). The map $\mathbf{v} \mapsto \mathbf{Dh}(\mathbf{v})$ is continuous as a composition of three continuous maps:

$$\mathbf{v} \mapsto \mathbf{h}(\mathbf{v}) \mapsto \mathbf{Df}(\mathbf{h}(\mathbf{v})) \mapsto (\mathbf{Df}(\mathbf{h}(\mathbf{v})))^{-1},$$

so the entries of the matrix $\mathbf{Dh}(\mathbf{v})$ are continuous functions of $\mathbf{v}$ which settles (d).

Let $\mathbf{y} \in V$. Then $\mathbf{x} = \mathbf{h}(\mathbf{y}) \in B$, so we can write

$$\begin{aligned}
\|\mathbf{h}(\mathbf{y}) - \mathbf{h}(\mathbf{v}) - T(\mathbf{y} - \mathbf{v})\| &= \|\mathbf{x} - \mathbf{u} - T(\mathbf{y} - \mathbf{v})\| \\
&= \|T\mathbf{Df}(\mathbf{u})(\mathbf{x} - \mathbf{u}) - T[\mathbf{f}(\mathbf{x}) - \mathbf{f}(\mathbf{u})]\| \\
&\le \|T\|_2 \|\mathbf{Df}(\mathbf{u})(\mathbf{x} - \mathbf{u}) - (\mathbf{f}(\mathbf{x}) - \mathbf{f}(\mathbf{u}))\| \\
&= \|T\|_2 \|\mathbf{r_f}(\mathbf{x})\|,
\end{aligned}$$

where

$$\frac{\|\mathbf{r_f}(\mathbf{x})\|}{\|\mathbf{x} - \mathbf{u}\|} \to 0, \quad \mathbf{x} \to \mathbf{u}.$$

Thus, it remains to show that

$$\frac{\|\mathbf{r_f}(\mathbf{x})\|}{\|\mathbf{y} - \mathbf{v}\|} \to 0, \quad \mathbf{y} \to \mathbf{v}.$$

By the Injective Map Theorem (a),

$$\|\mathbf{y} - \mathbf{v}\| = \|\mathbf{f}(\mathbf{x}) - \mathbf{f}(\mathbf{u})\| \ge \gamma \|\mathbf{x} - \mathbf{u}\|, \quad \text{for any } \mathbf{x}, \mathbf{u} \in B.$$

That way, $\mathbf{y} \to \mathbf{v}$ implies that $\mathbf{x} \to \mathbf{u}$, and if we take $\mathbf{r_h}(\mathbf{y}) = \mathbf{r_f}(\mathbf{x})$ we have that

$$\frac{\|\mathbf{r_h}(\mathbf{y})\|}{\|\mathbf{y} - \mathbf{v}\|} \leq \frac{\|\mathbf{r_f}(\mathbf{x})\|}{\gamma \|\mathbf{x} - \mathbf{u}\|} \to 0. \qquad \square$$

Remark 12.3.4. The Inverse Function Theorem guarantees only that a function $\mathbf{f}$ is locally invertible, even if $\mathbf{Df}(\mathbf{c})$ is invertible at every point $\mathbf{c} \in \mathbb{R}^n$. For example, the function $\mathbf{f}(x, y) = (e^x \cos y, e^x \sin y)$ is differentiable at every point of $\mathbb{R}^2$, and

$$\mathbf{Df}(x, y) = \begin{bmatrix} e^x \cos y & -e^x \sin y \\ e^x \sin y & e^x \cos y \end{bmatrix}.$$

Since the last matrix has determinant $e^x \neq 0$, the derivative $\mathbf{Df}(x, y)$ is bijective for any $(x, y) \in \mathbb{R}^2$. Nevertheless, $\mathbf{f}$ is not globally invertible on $\mathbb{R}^2$, because it is not injective: $\mathbf{f}(x, y) = \mathbf{f}(x, y + 2\pi)$, for all $(x, y) \in \mathbb{R}^2$. This is in sharp contrast with the one-dimensional case. Namely, if f is a function of one variable that is differentiable for each $x \in \mathbb{R}$, and if $f'(x) \neq 0$ for each $x \in \mathbb{R}$, then f is injective, hence globally invertible. (When f is not injective there are $a, b \in \mathbb{R}$ such that $f(a) = f(b)$, so by Rolle's Theorem there would exist $c \in (a, b)$ such that $f'(c) = 0$.)

Example 12.3.5. Evaluating the inverse function at a point.

Let $\mathbf{f}(x, y) = (x^2 - y^2, xy)$, $\mathbf{c} = (1, 0)$. Show that there exists an open disk A containing $\mathbf{c}$ such that $\mathbf{f}$ is invertible in A and its inverse function $\mathbf{h}$ is differentiable in $\mathbf{f}(A)$. Calculate $\mathbf{h}(1.1, 0.1)$.

Solution. The derivative is a 2×2 matrix

$$\mathbf{Df}(x, y) = \begin{bmatrix} 2x & -2y \\ y & x \end{bmatrix}.$$

Its determinant equals $2x^2 + 2y^2$ and it is non-zero if and only if $(x, y) \neq (0, 0)$. In particular,

$$\mathbf{Df}(1, 0) = \begin{bmatrix} 2 & 0 \\ 0 & 1 \end{bmatrix}$$

is an invertible linear transformation with inverse $T = \begin{bmatrix} \frac{1}{2} & 0 \\ 0 & 1 \end{bmatrix}$. By the Inverse Function Theorem, f is locally invertible and its inverse function $\mathbf{h}$ is differentiable in some open disk containing $\mathbf{f}(\mathbf{c}) = \mathbf{f}((1, 0)) = (1, 0)$. Let us calculate $\mathbf{h}(1.1, 0.1)$.

Formulas (12.9) will allow us to find $\mathbf{h}(1.1, 0.1)$ approximately. The initial approximation $(x_0, y_0) = \mathbf{c} = (1, 0)$. Next,

$$\begin{bmatrix} x_1 \\ y_1 \end{bmatrix} = \begin{bmatrix} x_0 \\ y_0 \end{bmatrix} + T \left(\begin{bmatrix} 1.1 \\ 0.1 \end{bmatrix} - \begin{bmatrix} 1 \\ 0 \end{bmatrix} \right) = \begin{bmatrix} 1 \\ 0 \end{bmatrix} + \begin{bmatrix} 0.5 & 0 \\ 0 & 1 \end{bmatrix} \begin{bmatrix} 0.1 \\ 0.1 \end{bmatrix} = \begin{bmatrix} 1.05 \\ 0.1 \end{bmatrix}.$$

Now things get a little more tedious. Since $\mathbf{f}(x_0, y_0) = \mathbf{f}(1, 0) = (1, 0)$ and $\mathbf{f}(x_1, y_1) = \mathbf{f}(1.05, 0.1) = (1.0925, 0.105)$, we obtain

$$\begin{bmatrix} x_2 \\ y_2 \end{bmatrix} = \begin{bmatrix} x_1 \\ y_1 \end{bmatrix} - T \left(\mathbf{f}(x_1, y_1) - \mathbf{f}(x_0, y_0) \right) + \begin{bmatrix} x_1 \\ y_1 \end{bmatrix} - \begin{bmatrix} x_0 \\ y_0 \end{bmatrix}$$

$$= \begin{bmatrix} 1.05 \\ 0.1 \end{bmatrix} - \begin{bmatrix} 0.5 & 0 \\ 0 & 1 \end{bmatrix} \begin{bmatrix} 0.0925 \\ 0.105 \end{bmatrix} + \begin{bmatrix} 0.05 \\ 0.1 \end{bmatrix}$$

$$= \begin{bmatrix} 1.05375 \\ 0.095 \end{bmatrix}.$$

If we continue, we get $x_3 = 1.053067969$, $y_3 = 0.1389375$, $x_4 = 1.058243711$, $y_4 = 0.1366706191$, etc. The convergence is rather slow, but at least we have a method to approximate $\mathbf{h}(1.1, 0.1)$. $\blacklozenge$

An early version of the Inverse Function Theorem can be found in Jordan's *Cours d'Analyse*. The direct function is assumed to be only continuous, and the continuity of the inverse mapping is deduced.

PIERRE-SIMON LAPLACE (1749–1827) was a French scientist who made enormous contributions to engineering, mathematics, statistics, physics, astronomy, and philosophy. He came from an affluent family in Normandy. As a student at the University of Caen, Laplace wrote his first paper which attracted the attention of Lagrange. In 1771 he went to Paris where he impressed d'Alembert who helped him get a teaching position at the Ecole Militaire. His rise continued: already in 1773 he was elected to Académie des Sciences. In the following 10-15 years he established himself as one of the best mathematicians in France. The fact that he considered himself to be the best did not endear him to his colleagues. During that period he established a program of research in probability and celestial mechanics, in which he worked mathematically for the rest of his life. His masterpiece is a five-volume *Mécanique Céleste* (Celestial Mechanics) that appeared between 1799 and 1825. This work translated the geometric study of classical mechanics to one based on calculus. The stability of the Solar System baffled Newton so much that he believed that every now and then a divine intervention is needed. Dispensing with this hypothesis was a major contribution of Laplace's life work, although the problem is still open nowadays. In addition, he is recognized as one of the founders of the theory of probability and mathematical physics. He restated and developed the nebular hypothesis of the origin of the Solar System and was one of the first scientists to postulate the existence of black holes. In addition to being arrogant, Laplace changed his political views to fit in with the frequent changes during his lifetime, so he had few friends. Nevertheless, when he died the Academy did a very rare thing: a meeting was canceled as a sign of respect for the "French Newton".

Exercises

In Exercises 12.3.1–12.3.10 determine the points $\mathbf{a}$ such that the given function is invertible in the vicinity of $\mathbf{a}$.

12.3.1. $\mathbf{f}(x, y) = \left(\dfrac{x}{x^2 + y^2}, \dfrac{y}{x^2 + y^2} \right).$

12.3.2. $\mathbf{f}(x, y) = (e^x + e^y, \, e^x - e^y).$

12.3.3. $\mathbf{f}(x, y) = \left(x + y, \, x^2 + y^2\right).$

12.3.4. $u = x + e^y, \, v = y + e^z, \, w = z + e^x.$

12.3.5. $\mathbf{f}(x, y) = (\sin x \cos y + \cos x \sin y, \, \cos x \cos y - \sin x \sin y).$

12.3.6. $\mathbf{f}(r, \theta) = (r \cos \theta, \, r \sin \theta).$

12.3.7. $\mathbf{f}(x, y, z) = (yz, \, xz, \, xy).$

12.3.8. $\mathbf{f}(x, y) = \left(\dfrac{x}{1 + x + y}, \dfrac{y}{1 + x + y} \right).$

12.3.9. $\mathbf{f}(x, y) = (x \cos(\pi y/2), \, x \sin(\pi y/2)).$

12.3.10. $\mathbf{f}(x, y, z) = (e^y \cos x, \, e^y \sin x, \, 2 - \cos z).$

In Exercises 12.3.11–12.3.14 use formulas (12.9) to find $\mathbf{x}_4$:

12.3.11. $\mathbf{f}(x, y) = \left(x + y, \, x^2 + y^2\right)$, $\mathbf{c} = (0, 1)$, $\mathbf{y} = (0.9, 1.1)$.

12.3.12. $\mathbf{f}(x, y) = \left(2x - y, \, xy^2\right)$, $\mathbf{c} = (1, -1)$, $\mathbf{y} = (3.1, 0.9)$.

12.3.13. $\mathbf{f}(x, y) = \left(x^2 - y, \, x + y\right)$, $\mathbf{c} = (0, 2)$, $\mathbf{y} = (-1.9, 1.9)$.

12.3.14. $\mathbf{f}(x, y) = \left(x^2 + x - y, \, y^2 + x\right)$, $\mathbf{c} = (0, 1)$, $\mathbf{y} = (-1, 1.1)$.

In Exercises 12.3.15–12.3.18 $\mathbf{f}(x, y) = (u, v)$. Find $\partial u/\partial x$, $\partial u/\partial y$, $\partial v/\partial x$ and $\partial v/\partial y$ at $\mathbf{f}(\mathbf{a})$.

12.3.15. $\mathbf{f}(x, y) = \left(x + y, \, x^2 + y^2\right)$, $\mathbf{a} = (0, 1)$.

12.3.16. $\mathbf{f}(x, y) = \left(2x - y, \, xy^2\right)$, $\mathbf{a} = (1, -1)$.

12.3.17. $\mathbf{f}(x, y) = \left(x^2 - y, \, x + y\right)$, $\mathbf{a} = (0, 2)$.

12.3.18. $\mathbf{f}(x, y) = \left(x^2 + x - y, \, y^2 + x\right)$, $\mathbf{a} = (0, 1)$.

12.3.19. Give an example of a function $\mathbf{f} : \mathbb{R}^2 \to \mathbb{R}^2$ that has continuous partial derivatives, but $\mathbf{f}(A)$ is not an open set for any open set $A \subset \mathbb{R}^2$.

12.3.20. Give an example of a function $f : [a, b] \to [a, b]$ that is bijective, infinitely differentiable, has a continuous inverse function g, but g is not differentiable at some point of $[a, b]$.

12.3.21. Let A be an n-ball, let $\mathbf{f} : A \to \mathbb{R}^n$ and let $\mathbf{g} : \mathbf{f}(A) \to \mathbb{R}^n$ be its inverse function. Suppose that $\mathbf{f}$ is differentiable at $\mathbf{a} \in A$ and $\mathbf{g}$ is differentiable at $\mathbf{b} = \mathbf{f}(\mathbf{a})$. If $\mathbf{D}(\mathbf{f})(\mathbf{a})$ is not invertible, prove that $\mathbf{D}(\mathbf{g})(\mathbf{b})$ is not invertible.

12.3.22. Let $f : \mathbb{R}^3 \to \mathbb{R}$ and $g : \mathbb{R}^3 \to \mathbb{R}$ be C^1 functions and let $\mathbf{F} : \mathbb{R}^3 \to \mathbb{R}^3$ be defined by $\mathbf{F}(x, y, z) = (f(x, y, z), g(x, y, z), (f(x, y, z))^2 + (g(x, y, z))^2)$. Find all points (x, y, z) in $\mathbb{R}^3$ where the assumptions of the Inverse Function Theorem hold.

12.3.23. Let C be an $n \times n$ matrix, and let $\mathbf{a}$ and $\mathbf{b}$ be points in $\mathbb{R}^n$. Define $\mathbf{F}(\mathbf{x}) = \mathbf{a} + C(\mathbf{x} - \mathbf{b})$, for $\mathbf{x} \in \mathbb{R}^n$. Prove that $\mathbf{F}$ is bijective if and only if the matrix C is invertible.

12.3.24.* Let $\mathbf{f} : \mathbb{R}^n \to \mathbb{R}^n$ be a C^1 function and let $c > 0$. Suppose that $\|\mathbf{f}(\mathbf{u}) - \mathbf{f}(\mathbf{v})\| \geq c\|\mathbf{u} - \mathbf{v}\|$ for all $\mathbf{u}, \mathbf{v} \in \mathbb{R}^n$. Prove that $\mathbf{f}$ is invertible.

12.3.25. Suppose that, in addition to the hypotheses of the Inverse Function Theorem, $\mathbf{f} \in C^k(A)$ (all partial derivatives of order up to k are continuous). Prove that there exists an open n-ball B such that the inverse function $\mathbf{h} \in C^k(B)$.

12.3.26. The purpose of this problem is to show that the assumption about the continuity of the derivative in the Inverse Function Theorem cannot be deleted. Let $0 < \alpha < 1$ and let
$$f(x) = \begin{cases} \alpha x + x^2 \sin \frac{1}{x}, & \text{if } x \neq 0 \\ 0, & \text{if } x = 0. \end{cases}$$

(a) Prove that f is differentiable on $\mathbb{R}$.

(b) Prove that f' is not continuous at $x = 0$.

(c) Prove that f satisfies all other hypotheses of the Inverse Function Theorem.

(d) Prove that, for any $\varepsilon > 0$, the interval $(-\varepsilon, \varepsilon)$ contains infinitely many zeros of f'.

(e) Prove that there does not exist a point where both f' and f'' are equal to 0.

(f) Conclude that, for any $\varepsilon > 0$, f cannot be invertible in $(-\varepsilon, \varepsilon)$.

12.4   Implicit Function Theorem

In this section we return to the task of establishing the existence of an implicit function and we will look at some unexpected applications. A germ of the idea how to prove this result has already appeared at the beginning of Section 12.2. Let us take a better look.

Suppose that $F(x, y) = x - e^y$. Does $F(x, y) = 0$ define a function $y = f(x)$? Of course, we know that $y = \ln x$, but we want to develop a general strategy. We define

$$\mathbf{H}(x, y) = (x, F(x, y)) = (x, x - e^y).$$

Now $\mathbf{H} : \mathbb{R}^2 \to \mathbb{R}^2$. Is it invertible? In other words, if $\mathbf{H}(x, y) = (u, v)$, is there a function $\mathbf{G} : \mathbb{R}^2 \to \mathbb{R}^2$ such that $\mathbf{G}(u, v) = (x, y)$? In this example the answer is in the affirmative. The system

$$u = x, \quad v = x - e^y$$

can be solved for x, y. From the second equation (substituting $x = u$) we obtain $e^y = u - v$ and, solving for y, gives

$$x = u, \quad y = \ln(u - v).$$

That way, we have two functions $g_1, g_2 : \mathbb{R}^2 \to \mathbb{R}$:

$$g_1(u, v) = u, \quad g_2(u, v) = \ln(u - v).$$

How does this help? How can we get our hands on the desired function f?

The answer is that we can define $f(u) = g_2(u, 0)$. Will that work? First, the formula $g_1(u, v) = x$ really is $x = u$. Second, the assumption is that $F(x, y) = 0$, so $v = 0$. Thus, we obtain that $g_2(u, 0) = \ln u$.

In this example it was actually possible to solve for y. Even when this is impossible, we can guarantee the existence of the function $y = y(x)$ if $\mathbf{H}(x, y)$ is invertible. Notice that in our example the derivative of $\mathbf{H}(x, y)$ is a 2×2 matrix $\left[\begin{smallmatrix} 1 & 0 \\ 1 & -e^y \end{smallmatrix}\right]$. In the general case, the first row is still the same, so the matrix is invertible if and only if the entry in the lower right corner is not 0. Clearly, this entry is $F_y'(x, y)$.

We have already seen this condition in Theorem 12.1.4 on page 374 for the case of one independent variable and one function. Therefore, the function F is defined on a subset of $\mathbb{R}^2 = \mathbb{R}^{1+1}$. In the general case, these two numbers need not be 1 and 1.

Theorem 12.4.1 (Implicit Function Theorem). *Let $\mathbf{F}$ be a function defined on an open ball A in $\mathbb{R}^{n+m}$, with values in $\mathbb{R}^m$, and let $(\mathbf{x_0}, \mathbf{y_0}) \in A$ such that $\mathbf{F}(\mathbf{x_0}, \mathbf{y_0}) = \mathbf{0}$. Suppose that $\mathbf{F}$ has continuous partial derivatives in A, and that the map $\mathbf{v} \mapsto D\mathbf{F}(\mathbf{x_0}, \mathbf{y_0})(\mathbf{0}, \mathbf{v})$ is a bijection on $\mathbb{R}^m$.*

(a) There exists an open set $W \subset \mathbb{R}^n$ containing $\mathbf{x_0}$, and a unique function $\mathbf{f} : W \to \mathbb{R}^m$ such that:

(i) $\mathbf{y_0} = \mathbf{f}(\mathbf{x_0})$;

(ii) $\mathbf{F}(\mathbf{x}, \mathbf{f}(\mathbf{x})) = \mathbf{0}$, for all $\mathbf{x} \in W$;

(iii) $\mathbf{f}$ has continuous partial derivatives in W.

(b) There exists an open ball $B \subset \mathbb{R}^{n+m}$, with center $(\mathbf{x_0}, \mathbf{y_0})$ such that, if $(\mathbf{x}, \mathbf{y}) \in B$, $\mathbf{F}(\mathbf{x}, \mathbf{y}) = \mathbf{0}$ if and only if $\mathbf{y} = \mathbf{f}(\mathbf{x})$.

Proof. Following the strategy set forth at the beginning of this section, we define

$$\mathbf{H}(\mathbf{x}, \mathbf{y}) = (\mathbf{x}, \mathbf{F}(\mathbf{x}, \mathbf{y})).$$

The difference is that here $\mathbf{x} \in \mathbb{R}^n$ and $\mathbf{y} \in \mathbb{R}^m$. Thus, $\mathbf{H}$ is a function on A, with values in $\mathbb{R}^{n+m}$. In other words, we have

$$\mathbf{u} = \mathbf{x}, \quad \mathbf{v} = \mathbf{F}(\mathbf{x}, \mathbf{y}),$$

so we can obtain the function $\mathbf{y} = \mathbf{f}(\mathbf{x})$ if we are allowed to apply the Inverse Function Theorem to $\mathbf{H}$. More precisely, we need to show that $\mathbf{H}$ has continuous partial derivatives, and that its derivative $D\mathbf{H}(\mathbf{x_0}, \mathbf{y_0})$ is invertible. The first condition is easy to verify because the first n component functions of $\mathbf{H}$ are $h_i(\mathbf{x}, \mathbf{y}) = x_i$ and the next m are the component functions of $\mathbf{F}$. In order to verify that $D\mathbf{H}(\mathbf{x_0}, \mathbf{y_0})$ is invertible we will consider its $(m + n) \times (m + n)$ matrix T, and find the inverse matrix T^{-1}. We will present T as a 2×2 block matrix by grouping together the first n rows and columns, as well as the last m rows and columns:

$$T = \begin{bmatrix} I & 0 \\ P & Q \end{bmatrix}. \tag{12.12}$$

Here, I is the $n \times n$ identity matrix, and Q is an $m \times m$ matrix. In fact, $\begin{bmatrix} P & Q \end{bmatrix}$ is the matrix for $\mathbf{DF}(\mathbf{x_0}, \mathbf{y_0})$, and $Q(\mathbf{v}) = \mathbf{DF}(\mathbf{x_0}, \mathbf{y_0})(\mathbf{0}, \mathbf{v})$, so Q is invertible by assumption. It is not hard to verify that

$$T^{-1} = \begin{bmatrix} I & 0 \\ -Q^{-1}P & Q^{-1} \end{bmatrix},$$

so $\mathbf{DH}(\mathbf{x_0}, \mathbf{y_0})$ is invertible, and we can apply the Inverse Function Theorem. It gives us an open ball B with center $(\mathbf{x_0}, \mathbf{y_0})$, such that the restriction $\mathbf{H}|B$ is a bijection between B and the open set $V = \mathbf{H}(B)$. Let $\mathbf{G} : V \to \mathbf{R}^{m+n}$ be its inverse function, and let $\mathbf{G} = (\mathbf{g_1}, \mathbf{g_2})$, with $\mathbf{g_1} : V \to \mathbf{R}^n$, $\mathbf{g_2} : V \to \mathbf{R}^m$. We define

$$W = \{\mathbf{x} \in \mathbf{R}^n : (\mathbf{x}, \mathbf{0}) \in V\}, \quad \text{and} \quad \mathbf{f}(\mathbf{x}) = \mathbf{g_2}(\mathbf{x}, \mathbf{0}), \quad \text{for} \ \ \mathbf{x} \in W.$$

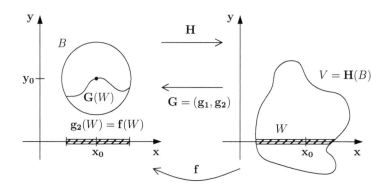

Figure 12.6: $\mathbf{f}(\mathbf{x}) = \mathbf{g_2}(\mathbf{x}, \mathbf{0})$.

Since V is an open set containing $\mathbf{H}(\mathbf{x_0}, \mathbf{y_0}) = (\mathbf{x_0}, \mathbf{F}(\mathbf{x_0}, \mathbf{y_0})) = (\mathbf{x_0}, \mathbf{0})$, it follows that W is an open set containing $\mathbf{x_0}$. Also, if $\mathbf{x} \in W$ then $(\mathbf{x}, \mathbf{0}) \in V$, so $\mathbf{f}$ is defined on W.

Now we can prove that $\mathbf{f}$ has desired properties. The equality $\mathbf{H}(\mathbf{x_0}, \mathbf{y_0}) = (\mathbf{x_0}, \mathbf{0})$ implies that $\mathbf{G}(\mathbf{x_0}, \mathbf{0}) = (\mathbf{x_0}, \mathbf{y_0})$, so $\mathbf{g_2}(\mathbf{x_0}, \mathbf{0}) = \mathbf{y_0}$. By the definition of $\mathbf{f}$, $\mathbf{f}(\mathbf{x_0}) = \mathbf{y_0}$, so (i) holds.

When $\mathbf{x} \in W$, then $(\mathbf{x}, \mathbf{0}) \in V$, so $\mathbf{G}(\mathbf{x}, \mathbf{0})$ is defined and

$$\mathbf{G}(\mathbf{x}, \mathbf{0}) = (\mathbf{g_1}(\mathbf{x}, \mathbf{0}), \mathbf{g_2}(\mathbf{x}, \mathbf{0})) = (\mathbf{x}, \mathbf{f}(\mathbf{x})).$$

Therefore,

$$(\mathbf{x}, \mathbf{0}) = \mathbf{H}(\mathbf{x}, \mathbf{f}(\mathbf{x})) = (\mathbf{x}, \mathbf{F}(\mathbf{x}, \mathbf{f}(\mathbf{x})))$$

and (ii) follows.

Finally, by the Inverse Function Theorem, $\mathbf{G}$ has continuous partial derivatives in V, so the same is true of $\mathbf{g_2}$ and, hence, $\mathbf{f}$ has continuous partial derivatives in W. This completes the proof of (a) except for the uniqueness of $\mathbf{f}$.

In order to prove (b), we will fix $(\mathbf{x}, \mathbf{y}) \in B$. Suppose first that $\mathbf{F}(\mathbf{x}, \mathbf{y}) = \mathbf{0}$. Then $(\mathbf{x}, \mathbf{0}) = (\mathbf{x}, \mathbf{F}(\mathbf{x}, \mathbf{y})) = \mathbf{H}(\mathbf{x}, \mathbf{y}) \in V$, so $\mathbf{x} \in W$. Also, applying $\mathbf{G}$ to both sides of the last equality,

$$(\mathbf{x}, \mathbf{y}) = \mathbf{G}(\mathbf{x}, \mathbf{0}) = (\mathbf{g_1}(\mathbf{x}, \mathbf{0}), \mathbf{g_2}(\mathbf{x}, \mathbf{0})) = (\mathbf{x}, \mathbf{f}(\mathbf{x})),$$

so $\mathbf{y} = \mathbf{f}(\mathbf{x})$. Incidentally, this settles the uniqueness of $\mathbf{f}$. In the other direction, let $\mathbf{y} = \mathbf{f}(\mathbf{x})$. Then

$$(\mathbf{x}, \mathbf{y}) = (\mathbf{x}, \mathbf{f}(\mathbf{x})) = (\mathbf{x}, \mathbf{g_2}(\mathbf{x}, \mathbf{0})) = \mathbf{G}(\mathbf{x}, \mathbf{0}).$$

If we apply $\mathbf{H}$ to both sides, it follows that

$$(\mathbf{x}, \mathbf{0}) = \mathbf{H}(\mathbf{x}, \mathbf{y}) = (\mathbf{x}, \mathbf{F}(\mathbf{x}, \mathbf{y}))$$

so $\mathbf{F}(\mathbf{x}, \mathbf{y}) = \mathbf{0}$, and the proof is complete.   □

Example 12.4.2. Implicitly defined functions.

Do the equations $x^2 - 2y^2 + z^3 - z^2 - 2 = 0$, $x^3 - y^3 - 3y + z - 4 = 0$ define y and z as functions of x, in the vicinity of $x = 2$? If so, write their second-order Taylor polynomials at $x = 2$.

Solution. It is easy to verify that the triple $(2, 1, 0)$ satisfies both equations. We would like to apply the Implicit Function Theorem. Our desired function $\mathbf{f}$ should map x to (y, z), so $n = 1$ and $m = 2$. Further, $\mathbf{x}_0 = 2$, $\mathbf{y}_0 = (1, 0)$, and $\mathbf{F}(x, y, z) = (x^2 - 2y^2 + z^3 - z^2 - 2, \ x^3 - y^3 - 3y + z - 4)$. Clearly, $\mathbf{F}$ has continuous partial derivatives, and the derivative

$$\mathbf{DF}(x, y, z) = \begin{bmatrix} 2x & -4y & 3z^2 - 2z \\ 3x^2 & -3y^2 - 3 & 1 \end{bmatrix},$$

whence

$$\mathbf{DF}(2, 1, 0) = \begin{bmatrix} 4 & -4 & 0 \\ 12 & -6 & 1 \end{bmatrix}.$$

It follows that

$$\mathbf{DF}(2, 1, 0) \begin{bmatrix} 0 \\ y \\ z \end{bmatrix} = \begin{bmatrix} -4y \\ -6y + z \end{bmatrix}.$$

The mapping $(y, z) \mapsto (-4y, -6y + z)$ is a linear map, and its matrix is $\begin{bmatrix} -4 & 0 \\ -6 & 1 \end{bmatrix}$, so the determinant equals $-4 \neq 0$. Therefore, it is a bijection on $\mathbb{R}^2$ and the Implicit Function Theorem guarantees that there exists a function $\mathbf{f} : \mathbb{R} \to \mathbb{R}^2$ such that $\mathbf{F}(x, \mathbf{f}(x)) = (0, \mathbf{0})$. If we denote the component functions of $\mathbf{f}$ by f_1, f_2, then $y = f_1(x)$ and $z = f_2(x)$.

It may not be possible to write these functions explicitly, but we can approximate them using Taylor polynomials. At $x = 2$, we have that $y(2) = 1$, $z(2) = 0$. Next we calculate $y'(2)$ and $z'(2)$. Taking the derivatives of the given equations, we have

$$2x - 4yy' + 3z^2 z' - 2zz' = 0, \quad 3x^2 - 3y^2 y' - 3y' + z' = 0.$$

When evaluated at $(2, 1, 0)$, we obtain that $y'(2) = 1$ and $z'(2) = -6$. If we take derivative with respect to x once again, we obtain

$$2 - 4y'^2 - 4yy'' + 6zz'^2 + 3z^2 z'' - 2z'^2 - 2zz'' = 0, \quad 6x - 6yy'^2 - 3y^2 y'' - 3y'' + z'' = 0$$

so, evaluating at $(2, 1, 0)$ yields $y''(2) = -37/2$ and $z''(2) = -123$. Thus, the second-order Taylor approximations are

$$y \approx 1 + (x - 2) - \frac{37}{2} \frac{(x - 2)^2}{2!}, \quad z \approx -6(x - 2) - 123 \frac{(x - 2)^2}{2!}.$$

Similar calculations can be used to obtain further terms in each polynomial. ◆

Example 12.4.3. Implicitly defined functions.

Do the equations $xu - yv = 0$, $yu + xv = 1$ define two functions $u = u(x, y)$, $v = v(x, y)$, in the vicinity of $(1, 0, 0, 1)$? If so, write their second-order Taylor polynomials at $(1, 0)$.

Solution. The function $\mathbf{F}(x, y, u, v) = (xu - yv, \ yu + xv - 1)$ has continuous partial derivatives, and

$$\mathbf{DF}(1, 0, 0, 1) = \begin{bmatrix} 0 & -1 & 1 & 0 \\ 1 & 0 & 0 & 1 \end{bmatrix}$$

so

$$\mathbf{DF}(1, 0, 0, 1) \begin{bmatrix} 0 \\ 0 \\ u \\ v \end{bmatrix} = \begin{bmatrix} u \\ v \end{bmatrix}.$$

which shows that the map $\mathbf{v} \mapsto \mathbf{DF}(\mathbf{x}_0, \mathbf{y}_0)(\mathbf{0}, \mathbf{v})$ in the Implicit Function Theorem is the identity map, hence a bijection on $\mathbb{R}^2$. It follows that there is a ball with center at $(1,0)$ so that u, v are functions of x, y in that ball. In order to write approximations for these functions, we differentiate both equations with respect to x, and we obtain

$$u + xu'_x - yv'_x = 0, \quad yu'_x + v + xv'_x = 0. \tag{12.13}$$

Evaluating at $(1,0,0,1)$ yields $u'_x = 0$ and $v'_x = -1$. Similarly, taking partial derivatives with respect to y,

$$xu'_y - v - yv'_y = 0, \quad u + yu'_y + xv'_y = 0, \tag{12.14}$$

so $u'_y = 1$, $v'_y = 0$. Differentiating (12.13) with respect to x and with respect to y yields

$$2u'_x + xu''_{xx} - yv''_{xx} = 0, \quad yu''_{xx} + 2v'_x + xv''_{xx} = 0$$
$$u'_y + xu''_{xy} - v'_x - yv''_{xy} = 0, \quad u'_x + yu''_{xy} + v'_y + xv''_{xy} = 0$$

so $u''_{xx} = 0$, $v''_{xx} = 2$, $u''_{xy} = -2$, $v''_{xy} = 0$. Similarly, differentiating (12.14) with respect to y, we obtain

$$xu''_{yy} - 2v'_y - yv''_{yy} = 0, \quad 2u'_y + yu''_{yy} + xv''_{yy} = 0$$

which leads to $u''_{yy} = 0$ and $v''_{yy} = -2$. Now we can write the second-degree Taylor polynomials for u, v in the vicinity of $(1,0)$:

$$u(x,y) \approx y - 2(x-1)y,$$
$$v(x,y) \approx 1 - (x-1) + (x-1)^2 - y^2. \qquad \blacklozenge$$

Exercises

In Exercises 12.4.1–12.4.8 determine whether given equations determine the implicit function in the vicinity of $\mathbf{a}$:

12.4.1. $xy - y \ln z + \sin xz = 0$; $z = z(x,y)$, $\mathbf{a} = (0,2,1)$.

12.4.2. $\sin(x+z) + \ln yz^2 = 0$, $e^{x+z} + yz = 0$; $y = y(x)$, $z = z(x)$; $\mathbf{a} = (1,1,-1)$.

12.4.3. $xu^2 + yzv + x^2z = 3$, $xyv^3 + 2zu - u^2v^2 = 2$; $u = u(x,y,z)$, $v = v(x,y,z)$; $\mathbf{a} = (1,1,1,1,1)$.

12.4.4. $(x^2 + y^2 + z^2)^3 - x + z = 0$, $\cos(x^2 + y^4) + e^z = 2$; $x = x(z)$, $y = y(z)$; $\mathbf{a} = (0,0,0)$.

12.4.5. $x^3 + x^2y + \sin(x+y+z) = 0$, $\ln(1+x^2) + 2x + (yz)^4 = 0$; $x = x(y)$, $z = z(y)$; $\mathbf{a} = (0,0,0)$.

12.4.6. $(uv)^4 + (u+s)^3 + t = 0$, $\sin uv + e^{v+t^2} = 1$; $u = u(s,t)$, $v = v(s,t)$; $\mathbf{a} = (0,0,0,0)$.

12.4.7. $x^3(y^3 + z^3) = 0$, $(x-y)^3 - z^2 = 7$; $y = y(x)$, $z = z(x)$; $\mathbf{a} = (1,-1,1)$.

12.4.8. $x + 2y + x^2 + (yz)^2 + w^3 = 0$, $-x + z + \sin(y^2 + z^2 + w^3) = 0$; $z = z(x,y)$, $w = w(x,y)$; $\mathbf{a} = (0,0,0,0)$.

12.4.9. Find for which points in the xy-plane does the system $x = u + v$, $y = u^2 + v^2$, $z = u^3 + 2v^3$ determine z as a differentiable function of x and y.

In Exercises 12.4.10–12.4.14 z is a function of x, y. Approximate this function by a second-degree polynomial at the indicated point $\mathbf{a} = (x,y,z)$:

12.4.10. $z^3 + 3xyz^2 - 5x^2y^2z + 14 = 0$, $\mathbf{a} = (1,-1,2)$.

12.4.11. $x + y + z = e^z$, $\mathbf{a} = (e,-1,1)$.

12.4.12. $x^2 + 2y^2 + 3z^2 + xy = z + 9$, $\mathbf{a} = (1, -2, 1)$.

12.4.13. $xyz = x + y + z$, $\mathbf{a} = (1, 0, -1)$.

12.4.14. $z^3 - xz + y = 0$, $\mathbf{a} = (3, -2, 2)$.

In Exercises 12.4.15–12.4.20 u and v are functions of the remaining variables. Approximate each of them by a second-degree polynomial at the indicated point $\mathbf{a}$:

12.4.15. $x^2 + y^2 + z^2 - u^2 - v^2 = 1$, $x^2 - y^2 + z^2 + u^2 + 2v^2 = 21$, $\mathbf{a} = (1, 1, 2, -1, 2)$.

12.4.16. $xy^2 + xzu + yv^2 = 3$, $u^3yz + 2xv - u^2v^2 = 2$, $\mathbf{a} = (1, 1, 1, 1, 1)$.

12.4.17. $x^2 - y^2 - u^3 + v^2 + 4 = 0$, $2xy + y^2 - 2u^2 + 3v^4 + 8 = 0$, $\mathbf{a} = (2, -1, 2, -1)$.

12.4.18. $2x - 3y + u - v = 0$, $x + 2y + u + 2v = 0$, $\mathbf{a} = (0, 0, 0, 0)$.

12.4.19. $x - 2y + u + v = 8$, $x^2 - 2y^2 - u^2 + v^2 = 4$, $\mathbf{a} = (3, -1, 2, 1)$.

12.4.20. $x^2 - y^2 + uv - v^2 = -3$, $x + y^2 + u^2 + uv = 2$, $\mathbf{a} = (2, 1, -1, 2)$.

12.4.21. Suppose that the equation $F(x, y, z) = 0$ can be solved for each of the variables as a differentiable function of the other two. Prove that

$$\frac{\partial x}{\partial y} \frac{\partial y}{\partial z} \frac{\partial z}{\partial z} = -1.$$

12.5 Constrained Optimization

In this section we will continue to work on optimization problems that we have started in Section 11.6. However, we will look at the problems where the extreme values are sought in a specific set.

Example 12.5.1. An optimization problem with a constraint.

Find the extreme values of $f(x, y) = x + y$, subject to $x^2 + y^2 = 1$.

Solution. It is not good enough to find critical points of f. If we use that approach we get $f'_x = 1$, $f'_y = 1$, so f has no critical points. Instead, we will use a method that is due to Lagrange. Let us denote $g(x, y) = x^2 + y^2 - 1$ and $F = f + \lambda g$, i.e.,

$$F(x, y, \lambda) = x + y + \lambda(x^2 + y^2 - 1).$$

Next, we will find all critical points of F. The partial derivatives are

$$F'_x = 1 + 2\lambda x, \quad F'_y = 1 + 2\lambda y, \quad F'_\lambda = x^2 + y^2 - 1.$$

We will solve the system

$$1 + 2\lambda x = 0, \quad 1 + 2\lambda y = 0, \quad x^2 + y^2 - 1 = 0. \tag{12.15}$$

Clearly, $\lambda \neq 0$, so the first two equations imply that $x = y = -\frac{1}{2\lambda}$. Substituting in the last equation we obtain that $x = y = \pm\frac{\sqrt{2}}{2}$. The corresponding values of f are: $f(\frac{\sqrt{2}}{2}, \frac{\sqrt{2}}{2}) = \sqrt{2}$ and $f(-\frac{\sqrt{2}}{2}, -\frac{\sqrt{2}}{2}) = -\sqrt{2}$, from which we deduce that f has a maximum at $P_1 = (\frac{\sqrt{2}}{2}, \frac{\sqrt{2}}{2})$ and a minimum at $P_2 = (-\frac{\sqrt{2}}{2}, -\frac{\sqrt{2}}{2})$. ♦

Later, we will present a proof that will justify the Lagrange's idea. Right now, let us look at the geometry of Example 12.5.1 (Figure 12.7). It will help us understand *why* the method works. The level curves of f are straight lines of the form $x + y = C$. Maximizing f amounts to selecting the largest possible C. Geometrically, that means choosing the rightmost among the parallel lines. Without constraints, we could pick any real number for C, hence any dotted line, so f would have no maximum. However, the constraint $x^2 + y^2 = 1$ requires that we restrict our attention to the unit circle. Now Figure 12.7 shows that both the maximum and the minimum are attained when the line $x + y = C$ is tangent to the circle. It is a well known fact that the gradient is always perpendicular to the tangent line

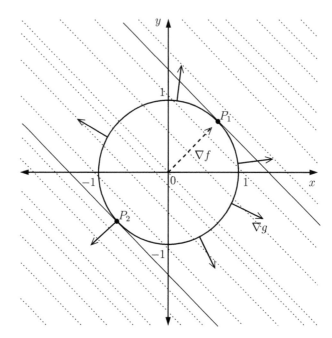

Figure 12.7: The extreme values of C correspond to tangent lines.

at the level curve. Thus, we see that the equalities (12.15) can be viewed as $\nabla f + \lambda \nabla g = \mathbf{0}$, expressing the geometric fact that ∇f and ∇g have the same direction (perpendicular to the tangent line) at P_1 and at P_2.

Example 12.5.2. An optimization problem with a constraint.

Find the extreme values of $f(x, y, z) = (x + 1)^2 + (y + 2)^2 + (z - 2)^2$, subject to $x^2 + y^2 + z^2 - 36 = 0$.

Solution. This time the Lagrange function is

$$F(x, y, z, \lambda) = (x + 1)^2 + (y + 2)^2 + (z - 2)^2 + \lambda(x^2 + y^2 + z^2 - 36),$$

and the partial derivatives are

$$F'_x = 2(x + 1) + 2\lambda x, \quad F'_y = 2(y + 2) + 2\lambda y,$$
$$F'_z = 2(z - 2) + 2\lambda z, \quad F'_\lambda = x^2 + y^2 + z^2 - 36.$$

The system we need to solve is

$$2(x + 1) + 2\lambda x = 0, \quad 2(y + 2) + 2\lambda y = 0,$$
$$2(z - 2) + 2\lambda z = 0, \quad x^2 + y^2 + z^2 - 36 = 0.$$

Solving the first three equations for x, y, and z, we obtain $x = -1/(\lambda+1)$, $y = -2/(\lambda+1)$, $z = 2/(\lambda+1)$. This requires that $\lambda \neq -1$, but it is easy to see that, with $\lambda = -1$, the first equation would yield $2 = 0$. Now we substitute the obtained expressions for x, y, z in the last equation:

$$\left(\frac{-1}{\lambda+1}\right)^2 + \left(\frac{-2}{\lambda+1}\right)^2 + \left(\frac{2}{\lambda+1}\right)^2 = 36.$$

Simplifying the left side yields $9/(\lambda+1)^2 = 36$ so $(\lambda+1)^2 = 1/4$. Thus, $\lambda+1 = \pm 1/2$, whence $\lambda_1 = -1/2$, $\lambda_2 = -3/2$. Consequently, the critical points are $P_1 = (-2, -4, 4)$ and $P_2 = (2, 4, -4)$. Since $f(-2, -4, 4) = 9$ and $f(2, 4, -4) = 81$ we conclude that f has a minimum at $(-2, -4, 4)$ and a maximum at $(2, 4, -4)$.   ◆

We remark that the level sets of f are spheres

$$(x+1)^2 + (y+2)^2 + (z-2)^2 = C \tag{12.16}$$

with center $(-1, -2, 2)$ and radius $\sqrt{C}$. The constraint represents a sphere with center at the origin and radius 6. Once again, the maximum is obtained when C is selected so that the sphere, given by (12.16), and the sphere $x^2 + y^2 + z^2 = 36$, have precisely one common point.

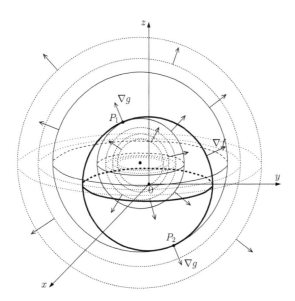

Figure 12.8: The spheres are touching at P_1 and P_2.

In other words, they share a common tangent plane at that point. Consequently, their gradients have the same direction, both being perpendicular to the tangent plane.

In the previous examples the point at which the extreme value was attained had to satisfy one condition. Sometimes, there is more than one constraint to be satisfied.

Example 12.5.3. An optimization problem with two constraints.

Find the extreme values of $f(x, y, z) = 3x^2 + y^2 + 3z^2$, subject to $x^2 + y^2 + z^2 = 1$ and $x - y + 5z = 0$.

Solution. Here we have to be a little more careful, because there are two constraints: $g_1(x, y, z) = x^2 + y^2 + z^2 - 1$ and $g_2(x, y, z) = x - y + 5z$ both have to be 0. We will use the Lagrange function $F = \mu f + \lambda_1 g_1 + \lambda_2 g_2$, and equate its partial derivatives with 0.

$$\mu(6x) + \lambda_1(2x) + \lambda_2 = 0, \quad \mu(2y) + \lambda_1(2y) - \lambda_2 = 0, \quad \mu(6z) + \lambda_1(2z) + 5\lambda_2 = 0,$$
$$x^2 + y^2 + z^2 - 1 = 0, \quad x - y + 5z = 0. \tag{12.17}$$

The first equation can be written as $2x(3\mu + \lambda_1) = -\lambda_2$. We want to solve it for x, but we need to explore the possibility that $3\mu + \lambda_1 = 0$. If that is the case, then $\lambda_2 = 0$, and the second equation becomes $2y(\mu + \lambda_1) = 0$. If $\mu + \lambda_1 = 0$, together with $3\mu + \lambda_1 = 0$, we would have that $\mu = \lambda_1 = 0$, which is impossible. (Cannot have all the parameters equal to 0.) So, $y = 0$, and the last 2 equations become

$$x^2 + z^2 = 1, \quad x + 5z = 0.$$

Solving them yields two critical points: $P_1 = (-\frac{5}{\sqrt{26}}, 0, \frac{1}{\sqrt{26}})$ and $P_2 = (\frac{5}{\sqrt{26}}, 0, -\frac{1}{\sqrt{26}})$.

For the rest, we will assume that $3\mu + \lambda_1 \neq 0$. The first and the third equation in (12.17) can be solved for x and z:

$$x = \frac{-\lambda_2}{6\mu + 2\lambda_1}, \quad z = \frac{-5\lambda_2}{6\mu + 2\lambda_1}. \tag{12.18}$$

We would like to solve the second equation in (12.17) for y, but we have that $2y(\mu + \lambda_1) = \lambda_2$, and it is possible that $\mu + \lambda_1 = 0$. However, that would make $\lambda_2 = 0$, so (12.18) would imply that $x = z = 0$, and the last 2 equations in (12.17) would become $y^2 = 1$ and $y = 0$, contradicting each other. Thus, $\mu + \lambda_1 \neq 0$, and we can solve for y:

$$y = \frac{\lambda_2}{2\mu + 2\lambda_1}. \tag{12.19}$$

If we substitute (12.18) and (12.19) in the last 2 equations in (12.17), we obtain

$$\frac{\lambda_2^2}{(6\mu + 2\lambda_1)^2} + \frac{\lambda_2^2}{(2\mu + 2\lambda_1)^2} + \frac{25\lambda_2^2}{(6\mu + 2\lambda_1)^2} = 1, \tag{12.20}$$

$$-\frac{\lambda_2}{6\mu + 2\lambda_1} - \frac{\lambda_2}{2\mu + 2\lambda_1} - \frac{25\lambda_2}{6\mu + 2\lambda_1} = 0. \tag{12.21}$$

Equation (12.21) can be multiplied by $(6\mu + 2\lambda_1)(2\mu + 2\lambda_1)$ to yield

$$\lambda_2 \left[26(2\mu + 2\lambda_1) + 6\mu + 2\lambda_1 \right] = 0.$$

Equation (12.20) shows that $\lambda_2 \neq 0$, so $6\mu + 2\lambda_1 = -26(2\mu + 2\lambda_1)$. Substituting this in (12.20), we obtain

$$\frac{\lambda_2^2}{26^2(2\mu + 2\lambda_1)^2} + \frac{\lambda_2^2}{(2\mu + 2\lambda_1)^2} + \frac{25\lambda_2^2}{26^2(2\mu + 2\lambda_1)^2} = 1.$$

Combining the fractions, we get the equation

$$\frac{702\lambda_2^2}{26^2(2\mu + 2\lambda_1)^2} = 1,$$

so

$$\frac{\lambda_2}{2\mu + 2\lambda_1} = \pm\frac{26}{\sqrt{702}}, \quad \text{and} \quad \frac{\lambda_2}{6\mu + 2\lambda_1} = \mp\frac{1}{\sqrt{702}}.$$

Using (12.18) and (12.19) we obtain two additional critical points $P_3 = (\frac{1}{\sqrt{702}}, \frac{26}{\sqrt{702}}, \frac{5}{\sqrt{702}})$, $P_4 = (-\frac{1}{\sqrt{702}}, -\frac{26}{\sqrt{702}}, -\frac{5}{\sqrt{702}})$. If we now evaluate f at each of the 4 critical points we obtain

$$f(P_1) = f(P_2) = 3, \quad f(P_3) = f(P_4) = \frac{29}{27},$$

so f attains its (constrained) minimum at P_3 and P_4, and its (constrained) maximum at P_1 and P_2. ♦

With two constraints the picture becomes more complicated. The level sets of f are ellipsoids $3x^2 + y^2 + 3z^2 = C$ (Figure 12.9(a)), and the constraints represent the sphere α (equation: $x^2 + y^2 + z^2 = 1$) and the plane β (equation: $x - y + 5z = 0$) which intersect along the circle S (Figure 12.9(b)). The maximum value of f will be obtained by choosing

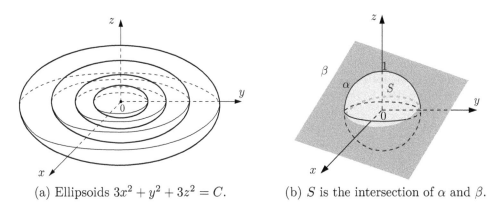

(a) Ellipsoids $3x^2 + y^2 + 3z^2 = C$.   (b) S is the intersection of α and β.

Figure 12.9: The level sets and the constraint.

the ellipsoid γ so that S is tangent to it (Figure 12.10(a)). That means that, at the common

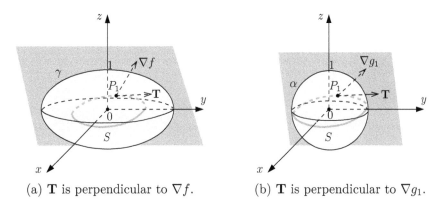

(a) $\mathbf{T}$ is perpendicular to ∇f.   (b) $\mathbf{T}$ is perpendicular to ∇g_1.

Figure 12.10: The ellipsoid γ is tangent to the circle S.

point P_1 of S and γ, the tangent vector $\mathbf{T}$ to the circle lies in the tangent plane to the ellipsoid. It follows that $\mathbf{T}$ is perpendicular to ∇f. On the other hand, S lies in the sphere α, so $\mathbf{T}$ is in the tangent plane to α at P_1 (Figure 12.10(b)). This implies that $\mathbf{T}$ is perpendicular to ∇g_1. The same argument shows that $\mathbf{T}$ must be perpendicular to ∇g_2. Since 3 vectors ∇f, ∇g_1, ∇g_2 are perpendicular to $\mathbf{T}$, they must be in the same plane. Consequently, there are scalars $\mu, \lambda_1, \lambda_2$ such that $\mu \nabla f + \lambda_1 \nabla g_1 + \lambda_2 \nabla g_2 = \mathbf{0}$.

When f depends on more than 3 variables, we cannot rely any more on the visual arguments. We must give a formal proof that the method that we have applied in these examples is sound.

Theorem 12.5.4. *Let $f, g_1, g_2, \ldots, g_m$ be functions with a domain an open ball $A \subset \mathbb{R}^n$, let $A_0 = \{\mathbf{x} \in A : g_i(\mathbf{x}) = 0, 1 \leq i \leq m\}$, and let $\mathbf{a} \in A_0$. Suppose that all these functions have continuous partial derivatives in A, and that $f(\mathbf{x}) \leq f(\mathbf{a})$ for all $\mathbf{x} \in A_0$. Then there exist real numbers $\mu, \lambda_1, \lambda_1, \ldots, \lambda_m$ not all zero such that*

$$\mu \mathbf{D} f(\mathbf{a}) = \lambda_1 \mathbf{D} g_1(\mathbf{a}) + \lambda_2 \mathbf{D} g_2(\mathbf{a}) + \cdots + \lambda_m \mathbf{D} g_m(\mathbf{a}). \tag{12.22}$$

Proof. Let $\mathbf{F}$ be a function on A with values in $\mathbb{R}^{m+1}$ defined by

$$\mathbf{F}(\mathbf{x}) = (f(\mathbf{x}), g_1(\mathbf{x}), g_2(\mathbf{x}), \ldots, g_m(\mathbf{x})). \tag{12.23}$$

Clearly the partial derivatives of $\mathbf{F}$ are continuous in A, and its derivative $\mathbf{DF}(\mathbf{x})$ is an $(m+1) \times n$ matrix with first row $\mathbf{D} f(\mathbf{x})$ and the remaining rows $\mathbf{D} g_1(\mathbf{x}), \mathbf{D} g_2(\mathbf{x})$, etc. We will show that $\mathbf{DF}(\mathbf{a})$ is not surjective.

Indeed, if $\mathbf{DF}(\mathbf{a})$ were surjective, then there would exist a ball B with center at $\mathbf{a}$ such that, for any $\mathbf{x} \in B$, $\mathbf{DF}(\mathbf{x})$ is surjective (Exercise 12.2.2). The Open Mapping Theorem would then imply that $\mathbf{F}(B)$ is an open set in $\mathbb{R}^{m+1}$. However, $\mathbf{a} \in B$, so $\mathbf{F}(\mathbf{a}) \in \mathbf{F}(B)$, and $\mathbf{F}(\mathbf{a}) = (f(\mathbf{a}), 0, 0, \ldots, 0)$ cannot be an interior point of $\mathbf{F}(B)$. This is because, for any $\varepsilon > 0$, the point $(f(\mathbf{a}) + \varepsilon, 0, 0, \ldots, 0)$ is not in $\mathbf{F}(B)$. Thus, $\mathbf{DF}(\mathbf{a})$ is not surjective, which means that its rows are linearly dependent, whence (12.22) follows. $\square$

Remark 12.5.5. Although Theorem 12.5.4 was formulated with f having a local maximum at $\mathbf{a}$, the analogous result is true if f has a local minimum.

The method established in Theorem 12.5.4 is known as the **method of Lagrange Multipliers**. Lagrange used it for the first time in *Analytical Mechanics* in 1788, in the analysis of equilibria for systems of particles. The application to the optimization appeared in his [79] in 1797.

JOSEPH-LOUIS LAGRANGE (1736–1813) was probably the best French mathematician of the eighteenth century. Italy also claims him because he was born in Turin, Italy, as Giuseppe Lodovico Lagrangia. He had French blood through his great-grandfather, and he always favored this line of ancestry, signing his name as Lodovico LaGrange or Luigi Lagrange. He studied at the College of Turin but did not show an interest in mathematics until he read a book on the use of algebra in optics. He was mostly self-taught which explains the blunder he made: after his first paper was published he discovered that the result appears in correspondence between Johann Bernoulli and Leibniz. His next work was on tautochrone (the curve on which a weighted particle will always arrive at a fixed point in the same time independent of its initial position). On the strength of this paper he became a professor of mathematics at the Royal Artillery School in Turin. In 1756 he generalized some results of Euler who was so impressed that he arranged for Lagrange to be offered a position in Prussia, but he declined. During the next 10 years he worked on various topics including the calculus of variations, theory of vibrating strings, and systems of differential equations. In 1766, on d'Alembert's initiative, Lagrange was offered a position in the Berlin Academy. Initially he declined, but after Euler left for Saint Petersburg he accepted to succeed him as Director of Mathematics at the Berlin Academy. He remained in Berlin for 20 years and during that period his work included astronomy, mechanics, and what can be considered as first steps towards group theory. His wife died in 1783 and his patron, King Frederick II, followed two years later. These events prompted Lagrange to move to Paris to become a member of the Académie des Sciences in Paris, where he remained for the rest of his career. In 1788 he completed a long work on a book *Mécanique Analytique* (*The Analytical Mechanics*), which offered a revolutionary view of mechanics as the four dimensional geometry. Lagrange himself was reported to have said that "mechanics was really a branch of pure mathematics" and that his book "does not contain a single diagram". Sir William Rowan Hamilton said the work could be described only as a scientific poem. During the French revolution and the following years he served on the committee to standardise weights and measures, the only branch of the Académie des Sciences that was allowed to operate. In 1794 he became a professor of analysis at the newly created Ecole

Polytechnique. He also taught at Ecole Normale which was founded to train school teachers. By many accounts he was not a good lecturer. Napoleon named Lagrange to the Legion of Honour and Count of the Empire in 1808. He died in 1813.

In Example 12.5.2 we have assumed that $\mu = 1$, and one less unknown is always a welcome sight. Are we allowed to do that? When $\mu \neq 0$, we can divide (12.22) by μ, which will result in an equation where the parameter with f equals 1. So, the real question is: can we assume that $\mu \neq 0$? The following result gives a sufficient condition for this.

Corollary 12.5.6. *In addition to the hypotheses of Theorem 12.5.4, suppose that $m \leq n$ and that $\mathbf{G}$ is a function on A with values in $\mathbb{R}^m$ defined by*

$$\mathbf{G}(\mathbf{x}) = (g_1(\mathbf{x}), g_2(\mathbf{x}), \ldots, g_m(\mathbf{x})).$$

If the rank of $\mathbf{DG}(\mathbf{a})$ equals m, then we can take $\mu = 1$ in (12.22).

Proof. If the $m \times n$ matrix $\mathbf{DG}(\mathbf{a})$ has rank $m \leq n$, then its rows are linearly independent. Thus, if $\mu = 0$, equation (12.22) would imply that $\lambda_i = 0$, for all i, $1 \leq i \leq m$. $\qquad \square$

Example 12.5.7. Applying Corollary 12.5.6.

Find the extreme values of $f(x, y) = 3x^2 + y^3$, subject to $x^2 + y^2 = 1$.

Solution. If we denote $g(x, y) = x^2 + y^2 - 1$, Equation (12.22) becomes $\mu(6x, 3y^2) = \lambda(2x, 2y)$. Further, the rank of $\mathbf{D}g(x, y) = \begin{bmatrix} 2x & 2y \end{bmatrix}$ is 1 unless $x = y = 0$. Since $g(0, 0) \neq 0$, we see that the rank is always 1, and Corollary 12.5.6 allows us to take $\mu = 1$. Therefore, we need to solve the system

$$6x = \lambda 2x, \quad 3y^2 = \lambda 2y, \quad x^2 + y^2 = 1.$$

The first equation shows that either $\lambda = 3$ or $x = 0$. If $x = 0$, the last equation yields $y = \pm 1$, so we get critical points $(0, 1)$ and $(0, -1)$. If $x \neq 0$ we must have $\lambda = 3$. Now the second equation becomes $3y^2 = 6y$ and it has solutions $y = 0$ and $y = 2$. Substituting $y = 0$ in the last equation gives $x = \pm 1$, while $y = 2$ is impossible. Thus we have 2 more critical points: $(1, 0)$ and $(-1, 0)$. Since $f(1, 0) = f(-1, 0) = 3$, $f(0, 1) = 1$, and $f(0, -1) = -1$ we conclude that f has a minimum at $(0, -1)$ and maxima at $(1, 0)$ and $(-1, 0)$. $\qquad \blacklozenge$

In all examples so far, we have been able to tell whether a particular critical point yields a minimum or a maximum. It is a consequence of Theorem 10.4.5 that, as long as the constraints define a compact set, the objective function f must attain both its minimum and the maximum values. In Example 12.5.2, the function $g(x, y) = x^2 + y^2 + z^2 - 36$ is continuous, and

$$G = \{(x, y) : x^2 + y^2 + z^2 - 36 = 0\} = g^{-1}(0)$$

so G is a closed set. It is bounded (because it is a sphere), so it is compact. Thus, the continuous function $f(x, y, z) = (x - 1)^2 + (y - 2)^2 + (z - 2)^2$ must attain its minimum and its maximum on G. However, when the set is not compact, all bets are off.

Example 12.5.8. The constraints define an unbounded set.

Find the extreme values of $f(x, y, z) = z + x(x^2y - x - 1)^2 + 2x^2$, subject to $z - \frac{x^5}{5} + \frac{3x^4}{4} = 0$.

Solution. We define $g(x, y, z) = z - \frac{x^5}{5} + \frac{3x^4}{4}$. Since $\mathbf{D}g(x, y, z) = \begin{bmatrix} -x^4 + 3x^3 & 0 & 1 \end{bmatrix}$, its rank is 1. Thus, we define $F(x, y, z) = f(x, y, z) + \lambda g(x, y, z)$ and equate its partial derivatives to 0:

$$(x^2y - x - 1)^2 + 2x(x^2y - x - 1)(2xy - 1) + 4x + \lambda(-x^4 + 3x^3) = 0,$$

$$2x(x^2y - x - 1)x^2 = 0, \quad 1 + \lambda = 0, \quad z - \frac{x^5}{5} + \frac{3x^4}{4} = 0. \tag{12.24}$$

The second equation implies that either $x = 0$ or $x^2 y - x - 1 = 0$. However, if $x = 0$, then the first equation becomes $1 = 0$. Thus, $x \neq 0$ and

$$x^2 y - x - 1 = 0. \qquad (12.25)$$

Now the third equation in (12.24) shows that $\lambda = -1$, and the first equation becomes $4x + x^4 - 3x^3 = 0$. It is an exercise in high-school algebra to write this as $x(x+1)(x-2)^2 = 0$, so we conclude that $x_1 = -1$ and $x_2 = 2$. (We have already seen that $x \neq 0$.) From (12.25) we obtain that $y_1 = 0$, $y_2 = 3/4$, and from the last equation in (12.24) that $z_1 = -19/20$, $z_2 = -28/5$. Thus, we have two critical points: $P_1 = (-1, 0, -\frac{19}{20})$ and $P_2 = (2, \frac{3}{4}, -\frac{28}{5})$, and $f(-1, 0, -\frac{19}{20}) = \frac{21}{20}$, $f(2, \frac{3}{4}, -\frac{28}{5}) = \frac{12}{5}$. Since $\frac{21}{20} < \frac{12}{5}$, it looks like $\frac{21}{20}$ is the minimum and $\frac{12}{5}$ is the maximum.

Unfortunately, this is wrong. At P_1, f has a local *maximum*! If $x = -1$ and $z = -19/20$, then the constraint $z - \frac{x^5}{5} + \frac{3x^4}{4} = 0$ is satisfied, regardless of what we choose for y. At the same time,

$$f\left(-1, y, -\frac{19}{20}\right) = \frac{21}{20} - y^2$$

so $f(-1, 0, -\frac{19}{20}) = \frac{21}{20}$ and $f(-1, y, -\frac{19}{20}) < \frac{21}{20}$ for any $y \neq 0$. In other words, the value of $21/20$ is not a minimum. ♦

Although we are sure that f does not have a local minimum at $(-1, 0, -\frac{19}{20})$, it is not clear whether it is a maximum. We will take a closer look at this issue in the next section.

Exercises

In Exercises 12.5.1–12.5.10 find the extreme values of f, subject to the given constraints:

12.5.1. $f(x, y) = x^2 + 12xy + 2y^2$ if $4x^2 + y^2 = 25$.

12.5.2. $f(x, y, z) = x - 2y + 2z$ if $x^2 + y^2 + z^2 = 1$.

12.5.3. $f(x, y, z) = xyz$ if $x^2 + y^2 + z^2 = 3$.

12.5.4. $f(x, y, z) = x^2 + y^2 + z^2$ if $\dfrac{x^2}{a^2} + \dfrac{y^2}{b^2} + \dfrac{z^2}{c^2} = 1$, $a > b > c > 0$.

12.5.5. $f(x, y, z) = x^6 + y^6 + z^6$ if $x^2 + y^2 + z^2 = 6$.

12.5.6. $f(x, y, z) = x^2 + y^2 + z^2$ if $x^4 + y^4 + z^4 = 1$.

12.5.7. $f(x, y, z) = xyz$ if $x^2 + y^2 + z^2 = 1$, $x + 2y + z = 0$.

12.5.8. $f(x, y, z) = x + y + z$ if $x^2 + y^2 + z^2 = 1$, $x + y + z = 0$.

12.5.9. $f(x, y, z) = 4x + y - z$ if $x^2 + y^2 = z^2$, $y + z = 1$.

12.5.10. $f(x, y, z) = xy + yz$ if $x^2 + y^2 = 2$, $y + z = 2$.

12.6 The Second Derivative Test for Constrained Optimization

In Example 12.5.8 we have seen that, just because there are two critical points, it does not mean that either one needs to be a minimum or a maximum. In the case without constraints, Theorem 11.6.6 shows that, when there are no constraints, it is the second derivative that provides additional insight. The same is true here.

Theorem 12.6.1. *In addition to the hypotheses of Corollary 12.5.6, suppose that:*

(i) the functions f and g_i, $1 \le i \le m$, have continuous second-order partial derivatives in A;

(ii) $\mu = 1$ and λ_i, $1 \le i \le m$ satisfy (12.22);

(iii) the $m \times m$ matrix consisting of the first m columns of $\mathbf{DG}(\mathbf{a})$ is invertible.

Finally, let $Q(u_{m+1}, u_{m+2}, \dots, u_n)$ be the quadratic form obtained by solving the system $\mathbf{DG}(\mathbf{a})(\mathbf{u}) = \mathbf{0}$ for u_i, $1 \le i \le m$, and by substituting the solutions in $\mathbf{D}^2 H(\mathbf{a})(\mathbf{u})^2$, where $H(\mathbf{u}) = f(\mathbf{u}) - \sum_{i=1}^{m} \lambda_i g_i(\mathbf{u})$.

(a) If Q is positive definite, then f has a relative minimum at $\mathbf{a}$;

(b) if Q is negative definite, then f has a relative maximum at $\mathbf{a}$.

Proof. Let $\mathbf{u} \in \mathbb{R}^n$ so that $\mathbf{a} + \mathbf{u} \in A_0$. Then $g_i(\mathbf{a} + \mathbf{u}) = g_i(\mathbf{a}) = 0$. Therefore,

$$f(\mathbf{a} + \mathbf{u}) - f(\mathbf{a}) = H(\mathbf{a} + \mathbf{u}) - H(\mathbf{a}).$$

Further, (12.22) implies that $\mathbf{D}H(\mathbf{a}) = \mathbf{0}$. By Taylor's Formula,

$$f(\mathbf{a} + \mathbf{u}) - f(\mathbf{a}) = H(\mathbf{a} + \mathbf{u}) - H(\mathbf{a}) = \frac{1}{2} \sum_{i=1}^{m} \sum_{j=1}^{m} \frac{\partial^2 H}{\partial x_i \partial x_j}(\mathbf{a}) u_i u_j + r(\mathbf{u}). \qquad (12.26)$$

Further, Taylor's Formula applied to $\mathbf{G}$ gives

$$\mathbf{G}(\mathbf{a} + \mathbf{u}) - \mathbf{G}(\mathbf{a}) = \mathbf{DG}(\mathbf{a})(\mathbf{u}) + \|\mathbf{u}\| E(\mathbf{u}),$$

where $E(\mathbf{u}) \to 0$, as $\|\mathbf{u}\| \to 0$. If we choose $\mathbf{u}$ so that $\mathbf{a} + \mathbf{u} \in A_0$, then the left side equals zero. On the other hand, condition (iii) implies that the matrix $\mathbf{DG}(\mathbf{a})$ can be viewed as a block matrix $\begin{bmatrix} C & T \end{bmatrix}$, where C is the invertible $m \times m$ block. If we denote $\mathbf{u}' = (u_1, u_2, \dots, u_m)$ and $\mathbf{u}'' = (u_{m+1}, u_{m+2}, \dots, u_n)$, we obtain that $C\mathbf{u}' + T\mathbf{u}'' + \|\mathbf{u}\| E(\mathbf{u}) = \mathbf{0}$. It follows that

$$\mathbf{u}' = -C^{-1} T \mathbf{u}'' - \|\mathbf{u}\| C^{-1} E(\mathbf{u})$$

If we now substitute this in the quadratic form in (12.26) we obtain

$$f(\mathbf{a} + \mathbf{u}) - f(\mathbf{a}) = Q(u_{m+1}, u_{m+2}, \dots, u_n) + r(\mathbf{u}) + R(\mathbf{u}),$$

where the additional term on the right hand side comes from $\|\mathbf{u}\| C^{-1} E(\mathbf{u})$. It is not hard to see that both r and R go to zero faster than $\|\mathbf{u}\|^2$. Therefore, the sign of $f(\mathbf{a} + \mathbf{u}) - f(\mathbf{a})$ is determined by the sign of $Q(u_{m+1}, u_{m+2}, \dots, u_n)$ and the theorem is proved. $\qquad \square$

Example 12.6.2. Analyzing a critical point.

Let us return to Example 12.5.7. We will show that f has a constrained minimum at $(0, 1)$.

Solution. Since $y = 1$ implies that $\lambda = 3/2$ we consider the function

$$H(x, y) = 3x^2 + y^3 - \frac{3}{2}(x^2 + y^2 - 1) = y^3 + \frac{3}{2}x^2 - \frac{3}{2}y^2 + \frac{3}{2}$$

and its second derivative

$$\mathbf{D}^2 H(x, y) = \begin{bmatrix} 3 & 0 \\ 0 & 6y - 3 \end{bmatrix}.$$

Since $\mathbf{D}g(0,1) = \begin{bmatrix} 0 & 2 \end{bmatrix}$ we have the relation $2v = 0$. Also,

$$\mathbf{D}^2 H(0,1) = \begin{bmatrix} 3 & 0 \\ 0 & 3 \end{bmatrix},$$

so together they yield the quadratic form $Q(u) = 3u^2$. This is a positive definite form, so f has a relative minimum at $(0,1)$. ♦

Example 12.6.3. A constrained optimization problem.

Find the extreme values of $f(x,y,z) = 8x + y + z^2$ subject to $x^2 - y^2 + z^2 = 0$, $y + z = 1$.

Solution. First we apply Corollary 12.5.6 to $\mathbf{G}(x,y,z) = (x^2 - y^2 + z^2, y + z - 1)$. The derivative

$$\mathbf{D}\mathbf{G}(x,y,z) = \begin{bmatrix} 2x & -2y & 2z \\ 0 & 1 & 1 \end{bmatrix},$$

so its rank is 2, unless $x = 0$ and $-2y - 2z = 0$. However, the latter equation is inconsistent with the constraint $y + z = 1$, so the rank of $\mathbf{D}\mathbf{G}(x,y,z)$ is 2. Thus, we are allowed to use $\mu = 1$. Accordingly, we define

$$F(x,y,z) = 8x + y + z^2 + \lambda_1(x^2 - y^2 + z^2) + \lambda_2(y + z - 1),$$

and we equate its partial derivatives with 0. We obtain the system

$$8 + 2\lambda_1 x = 0, \quad 1 - 2\lambda_1 y + \lambda_2 = 0, \quad 2z + 2\lambda_1 z + \lambda_2 = 0,$$
$$x^2 - y^2 + z^2 = 0, \quad y + z - 1 = 0. \tag{12.27}$$

The first equation implies that $\lambda_1 \neq 0$ and that

$$x = -\frac{4}{\lambda_1}. \tag{12.28}$$

From the second equation we obtain

$$y = \frac{1 + \lambda_2}{2\lambda_1}, \tag{12.29}$$

and from the third

$$z = -\frac{\lambda_2}{2(1 + \lambda_1)}, \tag{12.30}$$

assuming that $\lambda_1 \neq -1$. This assumption is justified, because, if $\lambda_1 = -1$ then it would follow from the third equation in (12.27) that $\lambda_2 = 0$, (12.28) and (12.29) would yield $x = 4$ and $y = -1/2$, so the last two equations in (12.27) would be $4 - \frac{1}{4} = z^2$ and $\frac{1}{2} + z = 1$, contradicting each other. Thus, equation (12.30) is valid. By substituting (12.28)–(12.30) in the last two equations in (12.27), we have

$$\frac{1 + \lambda_2}{2\lambda_1} - \frac{\lambda_2}{2(1 + \lambda_1)} = 1, \quad \frac{16}{\lambda_1^2} + \frac{\lambda_2^2}{4(1 + \lambda_1)^2} = \frac{(1 + \lambda_2)^2}{4\lambda_1^2}. \tag{12.31}$$

The first equation can be simplified to give

$$(1 + \lambda_2)(1 + \lambda_1) - \lambda_1\lambda_2 = 2\lambda_1(1 + \lambda_1)$$

or, after further simplification,

$$1 + \lambda_1 + \lambda_2 = 2\lambda_1 + 2\lambda_1^2.$$

From here we obtain two useful identities:

$$1 + \lambda_2 = \lambda_1(2\lambda_1 + 1), \quad \text{and} \quad \lambda_2 = (1 + \lambda_1)(2\lambda_1 - 1).$$

They can be used to simplify the second equation in (12.31):

$$\frac{16}{\lambda_1^2} = \frac{(2\lambda_1 + 1)^2}{4} - \frac{(2\lambda_1 - 1)^2}{4} = 2\lambda_1$$

so $\lambda_1 = 2$ and $\lambda_2 = 9$. It follows from (12.28)–(12.30) that $x = -2$, $y = 5/2$, $z = -3/2$.

Next, we will determine whether f has an extreme value at $(-2, \frac{5}{2}, -\frac{3}{2})$. Using the notation of Theorem 12.6.1, we have that

$$H(x, y, z) = 8x + y + z^2 + 2(x^2 - y^2 + z^2) + 9(y + z - 1) = 8x + 10y + 9z - 9 + 2x^2 - 2y^2 + 3z^2.$$

Therefore,

$$\mathbf{D}^2 H(x, y, z) = \begin{bmatrix} 4 & 0 & 0 \\ 0 & -4 & 0 \\ 0 & 0 & 6 \end{bmatrix},$$

and the associated quadratic form is $4u^2 - 4v^2 + 6w^2$. This form is indefinite but, thankfully, it is not the one we need. First we need to eliminate some variables. The system $\mathbf{DG}(-2, \frac{5}{2}, -\frac{3}{2})(u, v, w) = \mathbf{0}$ is

$$-4u - 5v - 3w = 0, \quad v + w = 0$$

and it can be solved to yield $v = -2u$, $w = 2u$. If we substitute these in the quadratic form above we have $Q(u) = 12u^2$ which is clearly a positive definite form, so f has a local (constrained) minimum at $(-2, \frac{5}{2}, -\frac{3}{2})$. ◆

Example 12.6.4. Analyzing a critical point.

Let us return to Example 12.5.8. We will show that f has a a constrained maximum at $(-1, 0, -\frac{19}{20})$ and a saddle point at $(2, \frac{3}{4}, -\frac{28}{5})$.

Solution. Both f and g have continuous second-order partial derivatives in $\mathbb{R}^3$ so we can apply Theorem 12.6.1. In order to test $(-1, 0, -\frac{19}{20})$ we will calculate $\mathbf{D}g(x, y, z) = \begin{bmatrix} -x^4 + 3x^3 & 0 & 1 \end{bmatrix}$ and evaluate it at $(-1, 0, -\frac{19}{20})$. We obtain that $\mathbf{D}g(-1, 0, -\frac{19}{20}) = \begin{bmatrix} -4 & 0 & 1 \end{bmatrix}$ and the equation $-4u_1 + u_3 = 0$. Now we have two (legitimate) choices: either eliminate u_1 or u_3. We will opt for the latter, and we will consider a quadratic form $Q(u_1, u_2)$. For that, we need the function H defined by $H(x, y, z) = f(x, y, z) + \lambda g(x, y, z)$. We know that $\lambda = -1$, so

$$H(x, y, z) = z + x(x^2y - x - 1)^2 + 2x^2 + (-1)\left(z - \frac{x^5}{5} + \frac{3x^4}{4}\right)$$

$$= x(x^2y - x - 1)^2 + 2x^2 + \frac{x^5}{5} - \frac{3x^4}{4}.$$

A calculation shows that $\mathbf{D}^2 H(-1, 0, -\frac{19}{20})$ is a 3×3 matrix

$$\mathbf{D}^2 H\left(-1, 0, -\frac{19}{20}\right) = \begin{bmatrix} -11 & 2 & 0 \\ 2 & -2 & 0 \\ 0 & 0 & 0 \end{bmatrix}.$$

The associated quadratic form (without the elimination of u_3) is $-11u_1^2 + 4u_1u_2 - 2u_2^2$, and

$$Q(u_1, u_2) = -11u_1^2 + 4u_1u_2 - 2u_2^2.$$

It follows easily from the Sylvester's Rule that this form is negative definite, so f has a constrained maximum at $(-1, 0, -\frac{19}{20})$.

What about P_2? Unfortunately,

$$\mathbf{D}^2 H\left(2, \frac{3}{4}, -\frac{28}{5}\right) = \begin{bmatrix} 16 & 32 & 0 \\ 32 & 64 & 0 \\ 0 & 0 & 0 \end{bmatrix}.$$

and the Second Derivative Test is inconclusive, because

$$\begin{vmatrix} 16 & 32 \\ 32 & 64 \end{vmatrix} = 0.$$

In a situation like this, we need to study the behavior of f in the vicinity of $(2, \frac{3}{4}, -\frac{28}{5})$. The constraint $z = \frac{x^5}{5} - \frac{3x^4}{4}$ requires that we restrict our attention to the function $f_1(x, y) = x(x^2 y - x - 1)^2 + 2x^2 + \frac{x^5}{5} - \frac{3x^4}{4}$ and its behavior in the vicinity of $(2, \frac{3}{4})$. Since we have terms both positive and negative, it could be that we have a saddle point. Notice that the term $x(x^2 y - x - 1)^2$ vanishes at $(2, \frac{3}{4})$, so a further simplification can be achieved if we choose only the values of x and y that satisfy $x(x^2 y - x - 1)^2 = 0$. Then, we have a function $f_2(x) = 2x^2 + \frac{x^5}{5} - \frac{3x^4}{4}$. Its derivative equals

$$f_2'(x) = 4x + x^4 - 3x^3 = x(x+1)(x-2)^2$$

so it is positive for $x > 0$. It follows that f_2 is increasing in the vicinity of $x = 2$, and $f_2(x) < f_2(2) = 12/5$ if $x < 2$, while $f_2(x) > 12/5$ if $x > 2$. Therefore, $f(2, \frac{3}{4}, -\frac{28}{5}) = \frac{12}{5}$, and in the vicinity of $(2, \frac{3}{4}, -\frac{28}{5})$ it takes on values both less and bigger than $12/5$, so it has a saddle point at $(2, \frac{3}{4}, -\frac{28}{5})$. ♦

Exercises

In Exercises 12.6.1–12.6.11 find the extreme values of f, subject to the given constraints:

12.6.1. $f(x, y) = xy$ if $x + y = 1$.

12.6.2. $f(x, y) = x^2 + y^2$ if $\dfrac{x}{a} + \dfrac{y}{b} = 1$.

12.6.3. $f(x, y, z) = xy^2 z^3$ if $x + 2y + 3z = 1$ and $x, y, z > 0$.

12.6.4. $f(x, y, z) = \sin x \sin y \sin z$ if $x + y + z = \dfrac{\pi}{2}$.

12.6.5. $f(x_1, x_2, \ldots, x_n) = x_1^m + x_2^m + \cdots + x_n^m$ if $x_1 + x_2 + \cdots + x_n = na$, $a > 0$, $m > 1$.

12.6.6. $f(x, y, z) = 2x^2 + y^2 + 4z^2$ if $3x + y - 2xz = 1$.

12.6.7. $f(x, y, z) = 3xy - 4z$ if $x + y + z = 1$.

12.6.8. $f(x, y, z) = x + y + z$ if $z = x^2 + y^2$.

12.6.9. $f(x, y, z) = (x + z)^2 + (y - z)^2 + x - z$ if $x + y + z = 1$, $2x + 4y + 3z = 0$.

12.6.10. $f(x, y, z) = 2x + y^2 - z^2$ if $x - 2y = 0$, $x + z = 0$.

12.6.11. $f(x, y, z) = xy + yz$ if $x^2 + y^2 = 1$, $yz = 1$.

12.6.1 Absolute Extrema

Example 12.6.5. Finding the absolute extrema: a review.

Find the absolute extrema of $f(x) = x^2$ on $A = [-1, 2]$.

Solution. It is easy to see that the minimum value of f on A is attained at $x = 0$. Since $f'(x) = 2x$, $x = 0$ is a critical point. However, f also attains its maximum value, because A is closed and bounded. This maximum is attained at $x = 2$. This point is not a critical point, but it is a boundary point of A. ◆

This example illustrates the principle: an extreme value of a function occurs either at a critical point or at a boundary point. This follows from Theorem 11.6.3 which asserts that, if a function attains an extreme value at an interior point P, then P must be a critical point. Therefore, if P is not a critical point, it cannot be an interior point, so it has to belong to the boundary.

Example 12.6.6. Finding absolute extrema of a function.

Find the absolute minimum and the absolute maximum of $f(x, y) = x^2 + y^2 - 12x + 16y$ if $x^2 + y^2 \leq 25$.

Solution. First we look for critical points: $f'_x = 2x - 12$, $f'_y = 2y + 16$. Solving the system

$$2x - 12 = 0, \quad 2y + 16 = 0$$

gives as the only critical point $(6, -8)$. However, it does not satisfy the condition $x^2 + y^2 \leq 25$. Thus we look at the boundary. These are the points that satisfy $x^2 + y^2 = 25$. Now we apply the Lagrange Theorem with $g(x, y) = x^2 + y^2 - 25$. Equation (12.22) becomes

$$\mu(2x - 12, 2y + 16) = \lambda(2x, 2y).$$

Further, the rank of $\mathbf{D}g(x, y) = \begin{bmatrix} 2x & 2y \end{bmatrix}$ is 1 unless $x = y = 0$. Since $g(0, 0) \neq 0$, we see that the rank is always 1, and Corollary 12.5.6 allows us to take $\mu = 1$. Therefore, we need to solve the system

$$2x - 12 = \lambda 2x, \quad 2y + 16 = \lambda 2y, \quad x^2 + y^2 = 25.$$

It is easy to see that $\lambda \neq 1$, and from the first two equations we obtain $x = 6/(1 - \lambda)$ and $y = -8/(1 - \lambda)$. When substituted in the last equation, we obtain

$$\frac{36}{(1 - \lambda)^2} + \frac{64}{(1 - \lambda)^2} = 25$$

which implies that $(1 - \lambda)^2 = 4$ and consequently $\lambda_1 = -1$, $\lambda_2 = 3$. We obtain two boundary points $(3, -4)$ and $(-3, 4)$. All that remains is to calculate $f(3, -4) = -75$ and $f(-3, 4) = 125$, so f has the minimum at $(3, -4)$ and the maximum at $(-3, 4)$. ◆

Example 12.6.7. Finding absolute extrema of a function.

Find the absolute minimum and the absolute maximum of $f(x, y, z) = x^2 + 2y^2 + 3z^2$ if $x^2 + y^2 + z^2 \leq 100$.

Solution. The critical points are solutions of the system

$$2x = 0, \quad 4y = 0, \quad 6z = 0,$$

so the only such point is $(0, 0, 0)$. Since it satisfies the condition $x^2 + y^2 + z^2 \leq 100$, we have our first point of interest $P_1 = (0,0,0)$. Next we consider the constrained optimization problem:

$$f(x,y,z) = x^2 + 2y^2 + 3z^2, \quad g(x,y,z) = x^2 + y^2 + z^2 - 100.$$

The rank of $\mathbf{D}g(x,y,z) = \begin{bmatrix} 2x & 2y & 2z \end{bmatrix}$ is 1, unless $x = y = z = 0$, which is impossible since $g(0,0,0) \neq 0$. Thus, we can use $\mu = 1$, and we define $F = f + \lambda g$. Taking partial derivatives of F leads to the system

$$2x + 2\lambda x = 0, \quad 4y + 2\lambda y = 0, \quad 6z + 2\lambda z = 0,$$

or, equivalently,

$$x(\lambda + 1) = 0, \quad y(\lambda + 2) = 0, \quad z(\lambda + 3) = 0.$$

Since we cannot have $x = y = z = 0$, we must have $\lambda \in \{-1, -2, -3\}$. If $\lambda = -1$, then $y = 0$, $z = 0$, and $x = \pm 10$. Thus, we obtain two points $P_2 = (10,0,0)$, $P_3 = (-10,0,0)$. Similarly, $\lambda = -2$ implies that $x = z = 0$ and $y = \pm 10$, so we get $P_4 = (0,10,0)$, $P_5 = (0,-10,0)$. Finally, $\lambda = -3$ yields $P_6 = (0,0,10)$, $P_7 = (0,0,-10)$. It is not hard to calculate that $f(P_2) = f(P_3) = 100$, $f(P_4) = f(P_5) = 200$, $f(P_6) = f(P_7) = 300$, so f attains its minimum at P_1 (because $f(P_1) = 0$) and its maximum at P_6 and P_7. $\quad\blacklozenge$

Exercises

In Exercises 12.6.12–12.6.16 find the absolute minimum and the absolute maximum of f in the given set:

12.6.12. $f(x,y) = xy$ if $x + y = 1$.

12.6.13. $f(x,y) = 3(x + 2y - \frac{3}{2})^2 + 4x^3 + 12y^2$ if $|x| \leq 1$, $0 \leq y \leq 1$.

12.6.14. $f(x,y) = x^2 - xy + y^2$ if $|x| + |y| \leq 1$.

12.6.15. $f(x,y,z) = \sin x + \sin y + \sin z - \sin(x+y+z)$ if $0 \leq x \leq \pi$, $0 \leq y \leq \pi$, $0 \leq z \leq \pi$.

12.6.16. $f(x,y,z) = x + y + z$ if $x^2 + y^2 \leq z \leq 1$.

13

Integrals Depending on a Parameter

Some important functions are defined by integrals, and many of these integrals are improper or infinite. For such a definition to be useful, we need to be able to perform the usual operations (derivative, antiderivative, etc.). In this chapter we will learn both how to do that, and when we are allowed to use a specific rule. Perhaps not surprisingly, the uniform convergence will play an essential role.

13.1   Uniform Convergence

Let us consider the following example:

$$f(t) = \int_0^1 \frac{dx}{1 + x^2 t^2}.$$

Clearly, for each $t \in \mathbb{R}$, the integrand is a continuous function on $[0, 1]$, so the integral is well defined, and so is the function f. The substitution $u = xt$ yields

$$\int_0^t \frac{1}{1 + u^2} \frac{1}{t} \, du = \frac{1}{t} \arctan u \Big|_0^t = \frac{1}{t}(\arctan t - \arctan 0) = \frac{\arctan t}{t}.$$

We see that $f(t) = \frac{\arctan t}{t}$, which is a rather nice function (continuous, differentiable), except at $t = 0$, where it is not defined. Is there a way to obtain this information without actually calculating the integral? After all, it is likely that in many examples there will be no elementary antiderivative.

We will start with the simplest of calculus procedures—taking the limit. In the equality

$$\int_0^1 \frac{dx}{1 + x^2 t^2} = \frac{\arctan t}{t}, \tag{13.1}$$

we will take the limit as $t \to 0$. On the right-hand side, using for example the L'Hôpital's Rule, we obtain 1. On the left side, we have

$$\lim_{t \to 0} \int_0^1 \frac{dx}{1 + x^2 t^2}.$$

If the integral and the limit were to trade places, we would get

$$\int_0^1 \lim_{t \to 0} \frac{dx}{1 + x^2 t^2} = \int_0^1 dx = 1.$$

In this example, taking the limit inside the integral gave us the correct result. Will that always work?

Example 13.1.1. Interchanging the limit and the integral yields a wrong answer.

Show that $\lim\limits_{t\to 0}\int_0^1 \dfrac{x}{t^2} e^{-\frac{x^2}{t^2}}\, dx \neq \int_0^1 \lim\limits_{t\to 0}\dfrac{x}{t^2} e^{-\frac{x^2}{t^2}}\, dx.$

Solution. The substitution $u = -x^2/t^2$ yields

$$\int_0^1 \frac{x}{t^2} e^{-\frac{x^2}{t^2}}\, dx = \int_0^{-1/t^2} e^u \left(-\frac{1}{2}\right) du = -\frac{1}{2} e^u \Big|_0^{-1/t^2} = \frac{1}{2}\left(1 - e^{-\frac{1}{t^2}}\right).$$

When $t \to 0$, the rightmost expression has the limit $1/2$. On the other hand, if we take the limit inside the leftmost integral, we have to evaluate

$$\lim_{t\to 0} \frac{x}{t^2} e^{-\frac{x^2}{t^2}}.$$

This can be done, for example, by substituting $t^2 = 1/u$ and finding the limit as $u \to +\infty$. We obtain

$$\lim_{u\to+\infty} xu e^{-x^2 u} = \lim_{u\to+\infty} \frac{xu}{e^{x^2 u}} = 0.$$

Thus, bringing the limit inside the integral would result in the equality $0 = 1/2$. ◆

This example shows that it is not always legitimate for the integral and the limit to trade places. We have already encountered this type of difficulty in Chapter 8, and we have seen (Theorem 8.2.3) that, if a sequence of functions $\{f_n\}$ converges uniformly, then we have equality (8.8). Perhaps the uniform convergence will be the key here as well? In that direction, the first task is to define precisely what we mean by the uniform convergence.

Definition 13.1.2. Let F be a function of 2 variables x, t defined on a rectangle $R = [a, b] \times [c, d]$ and let $t_0 \in [c, d]$. We say that F converges **uniformly** to a function f on $[a, b]$, as $t \to t_0$, if for any $\varepsilon > 0$ there exists $\delta > 0$ such that

$$|F(x, t) - f(x)| < \varepsilon, \quad \text{for } |t - t_0| < \delta, \quad \text{and } (x, t) \in R.$$

Example 13.1.3. A function converging uniformly.

Prove that $F(x, t) = 1/(1 + x^2 t^2)$ converges uniformly on $R = [-1, 1] \times [0, 1]$ as $t \to 0$.

Solution. When $t \to 0$, $F(x, t) \to 1$, so $f(x) = 1$. Is the convergence of F to f uniform?

$$|F(x, t) - f(x)| = \left|\frac{1}{1 + x^2 t^2} - 1\right| = \frac{x^2 t^2}{1 + x^2 t^2} \leq x^2 t^2 \leq t^2 \leq |t|. \qquad (13.2)$$

So, it suffices to take $\delta = \varepsilon$.

Proof. Let $\varepsilon > 0$. We define $\delta = \varepsilon$. Suppose that $|t| < \delta$, and $(x, t) \in R$. Then, calculations as in (13.2) show that $|F(x, t) - f(x)| \leq |t| < \delta = \varepsilon$. We conclude that the convergence is uniform. ◆

Example 13.1.4. A function not converging uniformly.

Show that $F(x, t) = \dfrac{x}{t^2} e^{-\frac{x^2}{t^2}}$ does not converge uniformly on $R = [-1, 1] \times [0, 1]$ as $t \to t_0 = 0$.

Solution. We have seen that, when $t \to 0$, $F(x, t) \to 0$, so $f(x) = 0$. This time, though, the convergence is not uniform. Taking the negative in Definition 13.1.2, we see that the assertion to prove is

$$(\exists \varepsilon)(\forall \delta)(\exists x, t) \text{ such that } |t - t_0| < \delta, \quad x \in [-1, 1], \quad \text{and} \quad |F(x, t) - f(x)| \geq \varepsilon. \quad (13.3)$$

Proof. Let $\varepsilon = 1$, and suppose that $\delta > 0$. We will show that there exists $(x, t) \in R$ such that $|t| < \delta$ but $|F(x, t) - f(x)| \geq 1$. Let $t = \min\{\delta/2, 1/e\}$ and $x = t$. It is obvious that $(x, t) \in R$ and $|t| < \delta$. On the other hand,

$$|F(x, t) - f(x)| = \left| \frac{x}{t^2} e^{-\frac{x^2}{t^2}} \right| = \left| \frac{1}{t} e^{-1} \right| = \frac{1}{te} \geq \frac{1}{\frac{1}{e} e} = 1.$$

Thus, the convergence is not uniform. ◆

We have seen throughout the text that the existence of a limit of a function can be described in terms of sequences. The present situation is no exception.

Theorem 13.1.5. *Let F be a function defined on a rectangle $R = [a, b] \times [c, d]$ and let $t_0 \in [c, d]$. Then $F(x, t) \to f(x)$ uniformly, as $t \to t_0$, if and only if $\{F(x, a_n)\}$ converges uniformly to $f(x)$, for every sequence $a_n \in [c, d]$ that converges to t_0.*

Proof. Suppose first that $F(x, t) \to f(x)$ uniformly. Let $\varepsilon > 0$. By definition, there exists $\delta > 0$ such that

$$|t - t_0| < \delta, \quad (x, t) \in R \quad \Rightarrow \quad |F(x, t) - f(x)| < \varepsilon. \tag{13.4}$$

Let $\{a_n\}$ be a sequence in $[c, d]$ that converges to t_0. Then there exists $N \in \mathbb{N}$ such that

$$n \geq N \quad \Rightarrow \quad |a_n - t_0| < \delta.$$

If $n \geq N$, (13.4) implies that $|F(x, a_n) - f(x)| < \varepsilon$, for all $x \in [a, b]$. Thus, the sequence $\{F(x, a_n)\}$ converges uniformly to $f(x)$ on $[a, b]$.

In the opposite direction, let $\varepsilon > 0$, and suppose that the convergence $F(x, t) \to f(x)$ is *not* uniform. Let $\varepsilon_0 > 0$ be the positive number guaranteed by (13.3), and let $\delta_n = 1/n$. Then there exist x_n, t_n such that

$$|t_n - t_0| < 1/n \quad \text{and} \quad |F(x_n, t_n) - f(x_n)| \geq \varepsilon_0. \tag{13.5}$$

Clearly, the sequence $\{t_n\}$ converges to t_0 and, by assumption, $F(x, t_n) \to f(x)$ uniformly. This implies that, for every $\varepsilon > 0$ (and, in particular, for ε_0) there exists δ_0 such that

$$|t - t_0| < \delta_0, \quad (x, t) \in R \quad \Rightarrow \quad |F(x, t) - f(x)| < \varepsilon_0.$$

If we choose $n > 1/\delta_0$, then $|t_n - t_0| < 1/n < \delta_0$, and it follows that, for any $x \in [a, b]$, $|F(x, t_n) - f(x)| < \varepsilon_0$. When $x = x_n$, we obtain a contradiction with (13.5). Consequently, $F(x, t) \to f(x)$ uniformly. □

Theorem 13.1.5 allows us to apply results from Chapter 8. In particular, it follows directly from Theorems 8.2.1 and 8.2.3 that:

Theorem 13.1.6. *Let F be a function defined on a rectangle $R = [a, b] \times [c, d]$ and let $t_0 \in [c, d]$. Suppose that, for each fixed $t \in [c, d]$, F is a continuous (integrable) function of x, and that $F(x, t) \to f(x)$ uniformly, as $t \to t_0$. Then f is a continuous (integrable) function on $[a, b]$.*

Examples 13.1.3 and 13.1.4 seem to point out that, in order to interchange the integral and the limit, the convergence should be uniform. The following result establishes that this is indeed true.

Theorem 13.1.7. *Let F be a function defined on a rectangle $R = [a, b] \times [c, d]$ and let $t_0 \in [c, d]$. Suppose that, for each fixed $t \in [c, d]$, F is a continuous (integrable) function of x, and that $F(x, t) \to f(x)$ uniformly, as $t \to t_0$. Then*

$$\lim_{t \to t_0} \int_a^b F(x, t)\, dx = \int_a^b f(x)\, dx.$$

Proof. By Theorem 13.1.6, f is continuous on $[a, b]$, hence it is integrable. Let $\varepsilon > 0$. By assumption, there exists $\delta > 0$ such that

$$|t - t_0| < \delta, \quad (x, t) \in R \quad \Rightarrow \quad |F(x, t) - f(x)| < \frac{\varepsilon}{b - a}.$$

Therefore, for such x, t,

$$\left| \int_a^b F(x, t)\, dx - \int_a^b f(x)\, dx \right| = \left| \int_a^b [F(x, t) - f(x)]\, dx \right|$$

$$\leq \int_a^b |F(x, t) - f(x)|\, dx$$

$$< \int_a^b \frac{\varepsilon}{b - a}\, dx = \frac{\varepsilon}{b - a}(b - a) = \varepsilon. \qquad \square$$

Example 13.1.8. Computing a limit.

Find $\lim\limits_{\alpha \to 0} \int_0^2 \cos(\alpha x^2)\, dx$.

Solution. There is no hope to evaluate the integral using the Fundamental Theorem of Calculus. Nevertheless, if $F(x, \alpha) = \cos(\alpha x^2)$, then $f(x) = \lim\limits_{\alpha \to 0} F(x, \alpha) = 1$. This convergence is uniform for $x \in [0, 2]$.

Proof. Let $\varepsilon > 0$ and let $\delta = \sqrt{\varepsilon}/4$. If $|\alpha| < \delta$ and $0 \leq x \leq 2$, then

$$|1 - \cos(\alpha x^2)| = 2 \sin^2 \frac{\alpha x^2}{2} \leq 2 \left(\frac{\alpha x^2}{2} \right)^2 = \frac{\alpha^2 x^4}{2} \leq 8\alpha^2 < 8\delta^2 = 8 \frac{\varepsilon}{16} < \varepsilon. \qquad \square$$

Since the convergence is uniform,

$$\lim_{\alpha \to 0} \int_0^2 \cos(\alpha x^2)\, dx = \int_0^2 \lim_{\alpha \to 0} \cos(\alpha x^2)\, dx = \int_0^2 1\, dx = 2. \qquad \blacklozenge$$

Remark 13.1.9. The uniform convergence is a sufficient but not a necessary condition in Theorem 13.1.7 (see Exercise 13.1.20).

Definition 13.1.2 and Theorem 13.1.7 can be found in Dini's 1878 [32].

Theorem 13.1.7 will allow us to conclude about some important properties of the function $I(t) = \int_a^b F(x, t)\, dx$. We will do that in the next section.

Exercises

In Exercises 13.1.1–13.1.6 determine whether the function $F(x,t)$ converges uniformly on a set A when $t \to 0$.

13.1.1. $F(x,t) = \dfrac{2xt}{x^2 + t^2}$, $A = [0,1]$. $\qquad$ 13.1.2. $F(x,t) = \sqrt{x^2 + t^2}$, $A = \mathbb{R}$.

13.1.3. $F(x,t) = \dfrac{\sqrt{x+t} - \sqrt{x}}{t}$, $A = (0,+\infty)$. $\qquad$ 13.1.4. $F(x,t) = t \sin \frac{x}{t}$, $A = \mathbb{R}$.

13.1.5. $F(x,t) = e^{\frac{x-1}{t}}$, $A = (0,1)$. $\qquad$ 13.1.6. $F(x,t) = xt \ln(xt)$, $A = (0,1)$.

In Exercises 13.1.7–13.1.10 determine whether it is correct to bring the limit inside of the integral:

13.1.7. $\displaystyle\lim_{\alpha \to 0} \int_{-1}^{1} \sqrt{x^2 + \alpha^2}\, dx$. $\qquad$ 13.1.8. $\displaystyle\lim_{t \to 0} \int_0^{\pi/2} \frac{\sin x}{1 + tx}\, dx$.

13.1.9. $\displaystyle\lim_{n \to \infty} \int_0^1 \frac{dx}{1 + \left(1 + \frac{x}{n}\right)^n}$. $\qquad$ 13.1.10. $\displaystyle\lim_{t \to \infty} \int_1^2 \frac{\ln(x + |t|)}{\ln(x^2 + t^2)}\, dx$.

13.1.11. Determine whether $\displaystyle\lim_{\alpha \to 0} \int_\alpha^{1+\alpha} \frac{dx}{1 + e^x + \alpha^2} = \int_0^1 \frac{dx}{1 + e^x}$.

In Exercises 13.1.12–13.1.15 find the limit:

13.1.12. $\displaystyle\lim_{t \to 0} \int_0^1 (x+t)(1+tx)^{1/t}\, dx$. $\qquad$ 13.1.13.* $\displaystyle\lim_{R \to \infty} \int_0^{\pi/2} e^{-R \sin \theta}\, d\theta$.

13.1.14. $\displaystyle\lim_{t \to 0} \int_0^2 x^2 \cos tx\, dx$. $\qquad$ 13.1.15.* $\displaystyle\lim_{t \to 0} \int_0^\pi x^t \sin x\, dx$.

13.1.16.* Determine whether the function $F(t) = \int_0^1 \dfrac{t f(x)}{x^2 + t^2}\, dx$ is continuous on $[0,1]$, if f is strictly positive and continuous on $[0,1]$.

13.1.17. Suppose that, in addition to the hypotheses of Theorem 13.1.7, g is an absolutely integrable function on $[a,b]$. Prove that

$$\lim_{t \to t_0} \int_a^b F(x,t)g(x)\, dx = \int_a^b f(x)g(x)\, dx.$$

13.1.18.* Let $f : [A, B] \to \mathbb{R}$ be continuous and let $A < a < x < B$. Prove that

$$\lim_{h \to 0} \int_a^x \left(f(t+h) - f(t)\right) dt = f(x) - f(a).$$

13.1.19. Let $F(x,t) = \begin{cases} 1, & \text{if } x > t \\ 0, & \text{if } x = t \\ -1, & \text{if } x < t. \end{cases}$ Show that $f(t) = \int_0^1 F(x,t)\, dx$ is a continuous function of t, even though F has a jump.

13.1.20.* Give an example of a function F such that

$$\lim_{t \to t_0} \int_a^b F(x,t)\, dx = \int_a^b f(x)\, dx,$$

but $F(x,t)$ does *not* converge to $f(x)$ uniformly on $[a,b]$.

13.2   Integral as a Function

Let F be a *bounded* function on a *finite* rectangle $[a, b] \times [c, d]$. In this section we will consider the function $I(t)$ defined by the integral

$$I(t) = \int_a^b F(x, t) \, dx \qquad (13.6)$$

and we will work on answering the questions such as: is I continuous? Differentiable? Our first result in this section provides the answer to the former question.

Theorem 13.2.1. *Let F be a function defined and continuous on a rectangle $R = [a, b] \times [c, d]$. Then I is continuous on $[c, d]$.*

Proof. Let $t_0 \in [c, d]$. We will show that I is continuous at t_0. In view of Theorem 13.1.7, it suffices to prove that, as $t \to t_0$, $F(x, t)$ converges uniformly to $F(x, t_0)$.

Let $\varepsilon > 0$. By Theorem 10.4.11, F is uniformly continuous, so there exists $\delta > 0$ such that

$$|x_1 - x_2| < \delta, \; |t_1 - t_2| < \delta, \; (x_1, t_1), (x_2, t_2) \in R \quad \Rightarrow \quad |F(x_1, t_1) - F(x_2, t_2)| < \varepsilon.$$

In particular, if we take $x_1 = x_2 = x$, $t_1 = t$, and $t_2 = t_0$, we obtain that

$$|t - t_0| < \delta, \; (x, t), (x, t_0) \in R \quad \Rightarrow \quad |F(x, t) - F(x, t_0)| < \varepsilon.$$

Thus, $F(x, t)$ converges uniformly to $F(x, t_0)$. By Theorem 13.1.7,

$$\lim_{t \to t_0} I(t) = \lim_{t \to t_0} \int_a^b F(x, t) \, dx = \int_a^b \lim_{t \to t_0} F(x, t) \, dx = \int_a^b F(x, t_0) \, dx = I(t_0).$$

Therefore, I is continuous at $t = t_0$. Since t_0 was arbitrary, it follows that I is continuous on $[c, d]$. □

Next we turn to the question whether the function I defined by (13.6) is differentiable. And, if so, is the derivative of the integral equal to the integral of the derivative?

Theorem 13.2.2. *Let F be a function defined on a rectangle $R = [a, b] \times [c, d]$. Suppose that, for any fixed $t \in [c, d]$, F is a continuous function of x. Also, suppose that the partial derivative F_t' exists and is continuous in R. Then I is differentiable on $[c, d]$ and*

$$I'(t) = \int_a^b F_t'(x, t) \, dx.$$

Proof. We will consider the case when $t_0 \in (c, d)$. The proof when $t_0 = c$ or $t_0 = d$ is left as an exercise. Let $\delta > 0$ be small enough so that $[t_0 - \delta, t_0 + \delta] \subset (c, d)$. Then, if $|h| < \delta$,

$$\lim_{h \to 0} \frac{I(t_0 + h) - I(t_0)}{h} = \lim_{h \to 0} \frac{\int_a^b F(x, t_0 + h) \, dx - \int_a^b F(x, t_0) \, dx}{h}$$

$$= \lim_{h \to 0} \int_a^b \frac{F(x, t_0 + h) - F(x, t_0)}{h} \, dx.$$

If we are permitted to bring the limit inside the integral, the result will follow. Thus, we would like to apply Theorem 13.2.1, and we need the integrand to be a continuous function of h. Clearly, it is continuous for $h \neq 0$. Since

$$\lim_{h \to 0} \frac{F(x, t_0 + h) - F(x, t_0)}{h} = F'_t(x, t_0),$$

the function

$$G(x, h) = \begin{cases} \dfrac{F(x, t_0 + h) - F(x, t_0)}{h}, & \text{if } h \neq 0 \\ F'_t(x, t_0), & \text{if } h = 0 \end{cases}$$

is continuous on $[a, b] \times [-\delta, \delta]$ and the result follows from Theorem 13.2.1. $\square$

Example 13.2.3. Computing an integral.

Find $I(t) = \int_0^{\pi/2} \ln \dfrac{1 + t \cos x}{1 - t \cos x} \dfrac{dx}{\cos x}$, $|t| < 1$.

Solution. The integrand $F(x, t)$ is not defined at $x = \pi/2$. However, using the L'Hôpital's Rule,

$$\lim_{x \to \pi/2} F(x, t) = \lim_{x \to \pi/2} \frac{\ln(1 + t \cos x) - \ln(1 - t \cos x)}{\cos x}$$

$$= \lim_{x \to \pi/2} \frac{\frac{-t \sin x}{1 + t \cos x} - \frac{t \sin x}{1 - t \cos x}}{- \sin x}$$

$$= 2t,$$

so we define

$$F(x, t) = \begin{cases} \ln \dfrac{1 + t \cos x}{1 - t \cos x} \dfrac{1}{\cos x}, & \text{if } 0 \le x < \dfrac{\pi}{2} \\ 2t, & \text{if } x = \dfrac{\pi}{2}. \end{cases}$$

Now, F is defined on $R = [0, \frac{\pi}{2}] \times [-c, c]$, where $c < 1$, and it is continuous. Further, it is not hard to see that the partial derivative

$$F'_t(x, t) = \begin{cases} \dfrac{2}{1 - t^2 \cos^2 x}, & \text{if } 0 \le x < \dfrac{\pi}{2} \\ 2, & \text{if } x = \dfrac{\pi}{2} \end{cases} = \frac{2}{1 - t^2 \cos^2 x}$$

is continuous in R, so Theorem 13.2.2 implies that

$$I'(t) = \int_0^{\pi/2} \frac{2}{1 - t^2 \cos^2 x} \, dx.$$

Using the substitution $u = \tan x$ (and the formulas on page 148) we obtain

$$I'(t) = \int_0^{\infty} \frac{2}{1 - t^2 \frac{1}{1 + u^2}} \frac{du}{1 + u^2} = \int_0^{\infty} \frac{2}{1 + u^2 - t^2} \, du = \frac{2}{\sqrt{1 - t^2}} \arctan \frac{u}{\sqrt{1 - t^2}} \Big|_{u=0}^{u=\infty}$$

$$= \frac{\pi}{\sqrt{1 - t^2}}.$$

It follows that $I(t) = \pi \arcsin t + C$. Since $I(0) = 0$, we have that $C = 0$, so $I(t) = \pi \arcsin t$. This formula holds for $|t| \le c < 1$. Since c is arbitrary, it holds for $|t| < 1$. $\blacklozenge$

It can happen that, in addition to the integrand, the limits of the integral also depend on the parameter t.

Theorem 13.2.4. *Let F be a function defined and continuous on a rectangle $R = [a, b] \times [c, d]$, and let α and β be continuous functions on $[c, d]$ such that, if $t \in [c, d]$, $\alpha(t), \beta(t) \in [a, b]$. Then*

$$I(t) = \int_{\alpha(t)}^{\beta(t)} F(x, t)\, dx$$

is a continuous function on $[c, d]$.

Proof. Let $t_0 \in [c, d]$. We will show that $\lim_{t \to t_0} I(t) = I(t_0)$. We write

$$I(t) = \int_{\alpha(t_0)}^{\beta(t_0)} F(x, t)\, dx + \int_{\beta(t_0)}^{\beta(t)} F(x, t)\, dx - \int_{\alpha(t_0)}^{\alpha(t)} F(x, t)\, dx. \tag{13.7}$$

When $t \to t_0$, the first integral has constant limits, so Theorem 13.2.1 applies:

$$\lim_{t \to t_0} \int_{\alpha(t_0)}^{\beta(t_0)} F(x, t)\, dx = \int_{\alpha(t_0)}^{\beta(t_0)} F(x, t_0)\, dx = I(t_0).$$

Therefore, it suffices to demonstrate that the other two integrals in (13.7) have the limit 0. Since F is continuous on a closed rectangle R, it is bounded: $|F(x, t)| \leq M$, for all $(x, t) \in R$. It follows that

$$\left| \int_{\beta(t_0)}^{\beta(t)} F(x, t)\, dx \right| \leq \left| \int_{\beta(t_0)}^{\beta(t)} |F(x, t)|\, dx \right| \leq \left| \int_{\beta(t_0)}^{\beta(t)} M\, dx \right| = M\, |\beta(t) - \beta(t_0)| \to 0$$

as $t \to t_0$. The same argument shows that the last integral in (13.7) also has the limit 0. $\square$

Next we consider the differentiability of $I(t)$.

Theorem 13.2.5 (Leibniz Rule). *Let F be a function defined and continuous on a rectangle $R = [a, b] \times [c, d]$, and let α and β be continuous functions on $[c, d]$ such that, if $t \in [c, d]$, $\alpha(t), \beta(t) \in [a, b]$. If, in addition, F_t' is continuous in R and α, β are differentiable, then so is $I(t)$ and*

$$I'(t) = \int_{\alpha(t)}^{\beta(t)} F_t'(x, t)\, dx + \beta'(t) F(\beta(t), t) - \alpha'(t) F(\alpha(t), t). \tag{13.8}$$

Proof. Let $t_0 \in [c, d]$. We will show that $I(t)$ has the derivative at $t = t_0$ and that it can be calculated using the formula above. We will again use (13.7). The first integral has constant limits, so Theorem 13.2.2 shows that its derivative is $\int_{\alpha(t_0)}^{\beta(t_0)} F_t'(x, t)\, dx$. In particular, at $t = t_0$, we obtain

$$\int_{\alpha(t_0)}^{\beta(t_0)} F_t'(x, t_0)\, dx.$$

Let us denote the second integral in (13.7) by $G(t)$. Notice that $G(t_0) = 0$, so

$$\frac{G(t) - G(t_0)}{t - t_0} = \frac{G(t)}{t - t_0} = \frac{1}{t - t_0} \int_{\beta(t_0)}^{\beta(t)} F(x, t)\, dx.$$

Further, Corollary 6.6.2 implies that

$$\frac{G(t) - G(t_0)}{t - t_0} = \frac{1}{t - t_0} \left(\beta(t) - \beta(t_0)\right) F(c, t)$$

where c is a real number between $\beta(t)$ and $\beta(t_0)$. If we now take the limit as $t \to t_0$, the continuity of F implies that

$$G'(t_0) = \beta'(t_0) F(\beta(t_0), t_0).$$

An analogous argument shows that the derivative of the last integral in (13.7) equals $\alpha'(t_0) F(\alpha(t_0), t_0)$. Thus, (13.8) holds at $t = t_0$. Since t_0 was arbitrary, the proof is complete. $\qquad\square$

> Differentiating under the integral sign (without justification) was first done by Leibniz. That is why Theorem 13.2.5 is referred to as the Leibniz Rule. In the nineteenth century, sufficient conditions were sought to justify it. One of the first publications dealing with these is Jordan's *Cours d'Analyse* (second edition from 1894 has it, but it is a challenge to see whether the first edition does too). In Germany, Axel Harnack (1851–1888) published the textbook [59] in 1881. He gives credit to Thomae and his 1875 book [103]. However, Thomae's text shows a serious lack of rigor, and he does not explicitly state what the hypotheses on the function f are.

Now we know that the derivative and the integral can trade places. What about two integrals? Is it true that

$$\int_c^d \left(\int_a^b F(x, t) \, dx \right) dt = \int_a^b \left(\int_c^d F(x, t) \, dt \right) dx? \tag{13.9}$$

Theorem 13.2.6. *Let F be a function defined and continuous on a rectangle $R = [a, b] \times [c, d]$. Then (13.9) holds.*

Proof. Let $y \in [c, d]$, and consider integrals

$$A(y) = \int_c^y \left(\int_a^b F(x, t) \, dx \right) dt \quad \text{and} \quad B(y) = \int_a^b \left(\int_c^y F(x, t) \, dt \right) dx.$$

They are both functions of y, and we will focus on their derivatives with respect to y. The first integral has the integrand $\int_a^b F(x, t) \, dx$ which is a continuous function of t by Theorem 13.2.1. By the Fundamental Theorem of Calculus, $A'(y) = \int_a^b F(x, y) \, dx$. As for the second integral, let

$$H(x, y) = \int_c^y F(x, t) \, dt.$$

For a fixed y, this is a continuous function of x. (Theorem 13.2.1 but x is the parameter now.) Further, the Fundamental Theorem of Calculus implies that $H'_y(x, y) = F(x, y)$ which is a continuous function in R. Thus, Theorem 13.2.2 shows that B is a differentiable function of y and that

$$B'(y) = \int_a^b H'_y(x, y) \, dx = \int_a^b F(x, y) \, dx.$$

Thus, $A'(y) = B'(y)$. Since $A(c) = B(c) = 0$, it follows that $A(y) = B(y)$. $\qquad\square$

Remark 13.2.7. It is a common practice to write the repeated integrals, as in the formula (13.9), by omitting the large parentheses, and placing the differential element (dt on the left, dx on the right side) before the second integral. Thus, (13.9) can be written as

$$\int\limits_c^d dt \int\limits_a^b F(x,t)\, dx = \int\limits_a^b dx \int\limits_c^d F(x,t)\, dt.$$

Example 13.2.8. Computing an integral.

Find $\int_0^1 \dfrac{x^b - x^a}{\ln x}\, dx$, if $0 < a < b$.

Solution. The antiderivative is not an elementary function, so the Fundamental Theorem of Calculus is out of the question. Let us consider the function $F(x,t) = x^t$ defined on $R = [0,1] \times [a,b]$. It is not hard to see that F is continuous on R so, by Theorem 13.2.6,

$$\int\limits_a^b \int\limits_0^1 x^t\, dx\, dt = \int\limits_0^1 \int\limits_a^b x^t\, dt\, dx.$$

On the left side we have a power function, so we obtain

$$\int\limits_a^b \frac{x^{t+1}}{t+1}\Big|_{x=0}^{x=1}\, dt = \int\limits_a^b \frac{1}{t+1}\, dt = \ln|t+1|\big|_a^b = \ln(b+1) - \ln(a+1) = \ln\frac{b+1}{a+1}.$$

On the right side, we are dealing with an exponential function, so we have

$$\int\limits_0^1 \frac{x^t}{\ln x}\Big|_{t=a}^{t=b}\, dx = \int\limits_0^1 \frac{x^b - x^a}{\ln x}\, dx.$$

Thus, the desired result is

$$\int\limits_0^1 \frac{x^b - x^a}{\ln x}\, dx = \ln\frac{b+1}{a+1}. \qquad \blacklozenge$$

> The question as to whether the order of integration can be changed was one of the important questions in the theory of double integrals (which we will study in Chapter 14). An example that the equality (13.9) need not be true can be found in Cauchy's 1814 article. In the last 25 years of the nineteenth century many more counterexamples surfaced, but the first result in the positive direction is due to Stolz in 1886 in [98]. We will present it in the next chapter as Theorem 14.3.2.

Exercises

In Exercises 13.2.1–13.2.4 find the derivative of $I(t)$ and justify your conclusion:

13.2.1. $I(t) = \int_{\pi/2}^{\pi} \dfrac{\cos xt}{x}\, dx.$

13.2.2. $I(t) = \int_0^t \dfrac{\ln(1+tx)}{x}\, dx.$

13.2.3. $I(t) = \int_{a+t}^{b+t} \dfrac{\sin tx}{x}\, dx.$

13.2.4. $I(t) = \int_0^{t^2} dx \int_{x-t}^{x+t} \sin(x^2 + y^2 - t^2)\, dy.$

In Exercises 13.2.5–13.2.10 use the differentiation with respect to the parameter to find $I(t)$ (and justify your conclusion):

13.2.5. $I(t) = \int_0^{\pi/2} \dfrac{\ln(1 + \cos t \cos x)}{\cos x}\, dx,\ t \in (0, \pi).$

13.2.6.* $I(t) = \int_0^{\pi/2} \dfrac{\arctan(t\tan x)}{\tan x}\, dx.$

13.2.7.* $I(t) = \int_0^{\pi/2} \ln(t^2 - \sin^2 x)\, dx,\ t > 1.$ 13.2.8. $I(t) = \int_0^{\pi} \ln(1 - t\cos x)\, dx.$

13.2.9.* $I(t) = \int_0^{\pi} \ln(1 - 2t\cos x + t^2)\, dx,\ |t| < 1.$

13.2.10. $I(a) = \int_0^{\pi/2} \ln(a^2 \sin^2 x + b^2 \cos^2 x)\, dx,$ if $a, b > 0.$

In Exercises 13.2.11–13.2.13 use the integration with respect to the parameter to find I (and justify your conclusion):

13.2.11. $I = \int_0^1 \sin\left(\ln\dfrac{1}{x}\right) \dfrac{x^b - x^a}{\ln x}\, dx,\ 0 < a < b.$

13.2.12. $I = \int_0^1 \cos\left(\ln\dfrac{1}{x}\right) \dfrac{x^b - x^a}{\ln x}\, dx,\ 0 < a < b.$

13.2.13. $I = \int_0^{\pi/2} \ln\dfrac{a + b\sin x}{a - b\sin x} \cdot \dfrac{dx}{\sin x},\ a > b > 0.$

13.2.14. In Theorem 13.2.2 prove that I is differentiable at $t = c$ and $t = d.$

13.2.15. Suppose that, in addition to the hypotheses of Theorem 13.2.2, g is an absolutely integrable function on $[a, b]$. Prove that

$$\frac{d}{dt}\int_a^b F(x,t)g(x)\, dx = \int_a^b F_t'(x,t)g(x)\, dx.$$

13.2.16. Suppose that, in addition to the hypotheses of Theorem 13.2.6, g is an absolutely integrable function on $[a, b]$. Prove that

$$\int_c^d dt \int_a^b F(x,t)g(x)\, dx = \int_a^b dx \int_c^d F(x,t)g(x)\, dt.$$

13.3 Uniform Convergence of Improper Integrals

So far, the study of the function $I(t)$, defined by (13.6), was done under the assumption that the function F was defined and bounded on a finite rectangle. In this section we will expand our investigations to include the cases when the domain of integration is unbounded, or F is not a bounded function. Either way, we will be dealing with improper integrals.

We will start with integrals of the form

$$I(t) = \int_a^\infty F(x,t)\, dx. \qquad (13.10)$$

Example 13.3.1. $I(t)$ is not continuous.

Let $I(t) = \int_0^\infty te^{-xt}\, dx,$ for $t \geq 0$. Prove that $I(t)$ is not continuous at $t = 0.$

Solution. We calculate $\int_0^b te^{-xt}\, dx$ using the substitution $u = -xt$:

$$\int\limits_0^b te^{-xt}\, dx = \int\limits_0^{-bt} e^u\,(-du) = -e^u\Big|_0^{-bt} = 1 - e^{-bt}.$$

As $b \to +\infty$, assuming that $t > 0$, the integral converges to 1. If we denote the integral by $I(t)$, then

$$I(t) = \begin{cases} 1, & \text{if } 0 < t \le 1 \\ 0, & \text{if } t = 0. \end{cases}$$

Thus, $I(t)$ has a jump at $t = 0$.   ♦

Notice that I is not continuous at $t = 0$ even though $F(x,t) = te^{-xt}$ is continuous on the infinite rectangle $[0, +\infty) \times [0, 1]$. This is in sharp contrast with the result of Theorem 13.2.1, where the continuity of F was sufficient to guarantee that the function I, defined by (13.6), is continuous. Of course, the integral in (13.3.1) is infinite, so it is essential that it converges (which is the case here). However, just like in the case of infinite series, the uniform convergence of the integral is the one that preserves important properties (such as the continuity). Thus, we will start by defining this concept here.

Definition 13.3.2. Let F be a function defined for $x \ge a$ and $t \in [c, d]$, and suppose that $\int_a^b F(x,t)\, dx$ exists for every $b > a$ and each $t \in [c, d]$. The integral (13.10) **converges uniformly** for $t \in [c, d]$ if, for every $\varepsilon > 0$, there exists $B > a$ such that, for any $b_2 \ge b_1 \ge B$ and any $t \in [c, d]$,

$$\left| \int\limits_a^{b_2} F(x,t)\, dx - \int\limits_a^{b_1} F(x,t)\, dx \right| < \varepsilon.$$

Example 13.3.3. Uniform convergence of an infinite integral.

Let us return to Example 13.3.1. We will show that $I(t)$ converges uniformly for $t \in [c, d]$, with $c > 0$.

Solution. Indeed, if $b_1 \le b_2$ and if we use the substitution $u = -xt$,

$$\int\limits_0^{b_2} te^{-xt}\, dx - \int\limits_0^{b_1} te^{-xt}\, dx = \int\limits_{b_1}^{b_2} te^{-xt}\, dx = \int\limits_{-tb_1}^{-tb_2} e^u\,(-du) = -e^u\Big|_{-tb_1}^{-tb_2} = e^{-tb_1} - e^{-tb_2}.$$

Therefore,

$$\left| \int\limits_0^{b_2} te^{-xt}\, dx - \int\limits_0^{b_1} te^{-xt}\, dx \right| = e^{-tb_1} - e^{-tb_2} \le e^{-tb_1}$$

and $e^{-tb_1} < \varepsilon$ is equivalent to $-tb_1 < \ln \varepsilon$, and thus to $b_1 > -\ln \varepsilon / t$. If we take $B = 1 - \ln \varepsilon / c$ (or $B = 1$ if $\varepsilon \ge 1$) then $b \ge B$ implies $b > -\ln \varepsilon / t$, and the integral converges uniformly.♦

Notice that we have made the assumption that $c > 0$. What if $c = 0$?

Example 13.3.4. Non-uniform convergence of an infinite integral.

We will prove that, if $c = 0$ in Example 13.3.3, the convergence is not uniform.

Solution. We need to demonstrate that

$$(\exists \varepsilon)(\forall B)(\exists b_1, b_2)(\exists t) \quad b_2 \ge b_1 \ge B, \quad t \in [0, d], \quad e^{-tb_1} - e^{-tb_2} \ge \varepsilon.$$

Let $\varepsilon = (e-1)/e^2$, and let $B > 0$. Without loss of generality we may assume that $B \geq 1/d$. We can take $b_1 = B$, $b_2 = 2B$, and $t = 1/B$. Then

$$e^{-tb_1} - e^{-tb_2} = \frac{1}{e} - \frac{1}{e^2} = \varepsilon,$$

so the convergence is not uniform. ◆

As usual, it is helpful to have an equivalent definition in terms of sequences. We will state the theorem and leave its proof as an exercise.

Theorem 13.3.5. *Let F be a function defined for $x \geq a$ and $t \in [c,d]$, and suppose that $\int_a^b F(x,t)\,dx$ exists for every $b > a$ and each $t \in [c,d]$. The integral (13.10) converges uniformly for $t \in [c,d]$ if and only if, for any sequence $\{b_n\}$ that converges to infinity, the sequence $f_n(t) = \int_a^{b_n} F(x,t)\,dx$ converges uniformly on $[c,d]$.*

Another type of improper integrals are those where the integrand is defined on a finite (or infinite) rectangle, but it is not necessarily bounded.

Example 13.3.6. Evaluating an integral.

Evaluate $\int_0^1 \frac{t}{x^2 + t^2}\,dx$, $t \in [0,1]$.

Solution. This is an improper integral, because the function $F(x,t) = t/(x^2 + t^2)$ is not defined when $(x,t) = (0,0)$. In fact, F is unbounded in $R = [0,1] \times [0,1]$. For example, $F(0,1/n) = n$. Nevertheless, we can calculate the integral. If $t = 0$, the integral is 0. When $t \neq 0$, the antiderivative of $1/(x^2 + t^2)$ is $\frac{1}{t}\arctan\frac{x}{t}$. Therefore,

$$\int_0^1 \frac{t}{x^2 + t^2}\,dx = \begin{cases} \arctan 1/t, & \text{if } 0 < t \leq 1 \\ 0, & \text{if } t = 0. \end{cases}$$ ◆

Example 13.3.7. Evaluating an integral.

Evaluate $\int_0^1 e^{-t/x}\frac{t}{x^2}\,dx$, $t \in [0,1]$.

Solution. When $t = 0$ the integrand is 0. If $t \neq 0$, then

$$\int_0^1 e^{-t/x}\frac{t}{x^2}\,dx = \lim_{a \to 0^+} \int_a^1 e^{-t/x}\frac{t}{x^2}\,dx = \lim_{a \to 0^+} e^{-t/x}\Big|_a^1 = \lim_{a \to 0^+}\left(e^{-t} - e^{-t/a}\right) = e^{-t}.$$ ◆

Just like in the case of infinite integrals, we will define the uniform convergence of improper integrals. To emphasize the analogy with the infinite integrals (for which we have always dealt with the domain being unbounded on the *right*), we are going to assume that the integrand is unbounded as x approaches the right endpoint of its domain.

Definition 13.3.8. The integral $\int_a^b F(x,t)\,dx$ **converges uniformly** for $t \in [c,d]$ if, for every $\varepsilon > 0$, there exists $\delta > 0$ such that, for any $b - \delta < b_1 \leq b_2 < b$ and any $t \in [c,d]$,

$$\left| \int_a^{b_1} F(x,t)\,dx - \int_a^{b_2} F(x,t)\,dx \right| < \varepsilon.$$

Once again, we can characterize the uniform convergence of the integral using sequences. We leave the proof as an exercise.

Theorem 13.3.9. *Let F be a function defined for $(x,t) \in [a,b) \times [c,d]$, and suppose that $\int_a^{b'} F(x,t)\,dx$ exists for every $a \leq b' < b$ and each $t \in [c,d]$. The integral $\int_a^b F(x,t)\,dx$ converges uniformly for $t \in [c,d]$ if and only if, for any sequence $\{b_n\} \subset [a,b)$ that converges to b, the sequence $f_n(t) = \int_a^{b_n} F(x,t)\,dx$ converges uniformly on $[c,d]$.*

> The concept of the uniform convergence of integrals (as in Definitions 13.3.2 and 13.3.8) was introduced by Dini in [32].

There is a strong connection between the infinite and improper integrals. When it comes to the uniform convergence, it can be stated in terms of sequences, and the only difference is that in one case the sequence diverges to ∞ and in the other converges to a finite number. In what follows we will talk about integrals of the form

$$I(t) = \int_a^b F(x,t)\,dx \tag{13.11}$$

where b could be $+\infty$. Thus, we will be able to prove results simultaneously for both types of integrals. The first such result is a variation on the Weierstrass M-Test for series. It was established by de la Vallée-Poussin in his 1892 article [28].

Theorem 13.3.10 (De la Vallée-Poussin's Test). *Let F be a function defined and integrable on the rectangle $R = [a,b) \times [c,d]$ and suppose that there exists a function $\varphi(x)$ integrable on $[a,b)$ such that $|F(x,t)| \leq \varphi(x)$ in R. Then the integral (13.11) converges uniformly.*

Proof. Let $\varepsilon > 0$ and let $\{b_n\}$ be a sequence of real numbers satisfying $a \leq b_n < b$, for all $n \in \mathbb{N}$, and $b_n \to b$. Since $\varphi(x)$ is integrable on $[a,b)$, the sequence $\int_a^{b_n} \varphi(x)\,dx$ is a Cauchy sequence. Therefore, there exists $N \in \mathbb{N}$, such that

$$m \geq n \geq N \quad \Rightarrow \quad \left| \int_{b_n}^{b_m} \varphi(x)\,dx \right| < \varepsilon.$$

For such m, n

$$\left| \int_{b_n}^{b_m} F(x,t)\,dx \right| \leq \left| \int_{b_n}^{b_m} |F(x,t)|\,dx \right| \leq \left| \int_{b_n}^{b_m} \varphi(x)\,dx \right| < \varepsilon.$$

By Theorem 8.1.10, the sequence $f_n(t) = \int_a^{b_n} F(x,t)\,dx$ converges uniformly, and the result follows from Theorem 13.3.5 or Theorem 13.3.9. $\square$

Example 13.3.11. A uniformly convergent integral.

Prove that the integral $\int_0^1 \frac{x^t}{\sqrt{1-x^2}}\, dx$ converges uniformly for $t \in [0,d]$, $d \geq 0$.

Solution. The integrand $F(x,t)$ is defined on $R = [0,1) \times [0,d]$, for any $d \geq 0$. Further,

$$|F(x,t)| = \frac{x^t}{\sqrt{1-x^2}} \leq \frac{1}{\sqrt{1-x^2}} = \varphi(x)$$

and $\varphi(x)$ is integrable on $[0,1)$. ($\int_0^1 \varphi(x)\, dx = \arcsin 1 - \arcsin 0 = \pi/2$.) By de la Vallée-Poussin's Test, the integral converges uniformly. ♦

Remark 13.3.12. Theorem 13.3.10 is sometimes referred to as the de la Vallée-Poussin's μ-Test, because of its similarity to the Weierstrass M-test.

Another useful test is a version of Dirichlet's Test for series (Theorem 8.3.6).

Theorem 13.3.13 (Dirichlet's Test). *Let F be a function defined and continuous on the rectangle $R = [a,b) \times [c,d]$ and suppose that there exists $M > 0$ such that*

$$\left| \int_a^{b'} F(x,t)\, dx \right| \leq M,$$

for all $b' \in [a,b)$ and $t \in [c,d]$. Let $\varphi(x,t)$ be a differentiable function defined on R and suppose that $\varphi(x,t)$ is monotone decreasing for each $t \in [c,d]$ and that $\varphi(x,t)$ converges uniformly to 0 as $x \to b$. Then the integral

$$I(t) = \int_a^b F(x,t)\varphi(x,t)\, dx \tag{13.12}$$

converges uniformly on $[c,d]$.

Proof. We will use Theorem 8.1.10. Let $\varepsilon > 0$. Since $\varphi(x,t)$ converges uniformly to 0, there exists $B \in [a,b)$ such that

$$(x,t) \in [B,b) \times [c,d] \quad \Rightarrow \quad |\varphi(x,t)| < \frac{\varepsilon}{4M}.$$

Let $B \leq b' \leq b'' < b$ and $G(x,t) = \int_a^x F(s,t)\, ds$. If we use integration by parts, we obtain

$$\left| \int_{b'}^{b''} F(x,t)\varphi(x,t)\, dx \right| = \left| G(x,t)\varphi(x,t) \Big|_{b'}^{b''} - \int_{b'}^{b''} G(x,t)\varphi'_x(x,t)\, dx \right|$$

$$\leq |G(b'',t)\varphi(b'',t)| + |G(b',t)\varphi(b',t)| + \int_{b'}^{b''} |G(x,t)\varphi'_x(x,t)|\, dx$$

$$< M\frac{\varepsilon}{4M} + M\frac{\varepsilon}{4M} + M\int_{b'}^{b''} -\varphi'_x(x,t)\, dx$$

$$= \frac{\varepsilon}{2} + M\left[\varphi(b',t) - \varphi(b'',t)\right]$$

$$< \frac{\varepsilon}{2} + M\left(\frac{\varepsilon}{4M} + \frac{\varepsilon}{4M}\right) = \varepsilon. \qquad \square$$

Remark 13.3.14. In a particular case, when neither F nor φ depends on t, we recover Dirichlet's Test for convergence of infinite integrals (Exercise 6.7.23) and improper integrals (Exercise 6.7.46).

Once we know that the integral (13.10) converges uniformly, we can derive some properties of the function I. We will do that in the next section.

Exercises

In Exercises 13.3.1–13.3.9 determine whether the given integral converges uniformly:

13.3.1. $\int_0^\infty e^{-tx^2}\,dx,\ t \geq t_0 > 0.$ **13.3.2.** $\int_0^\infty e^{-tx}x^a \cos x\,dx,\ t \geq t_0 > 0,\ a \geq 0.$

13.3.3. $\int_0^\infty x\sin x^3 \sin tx\,dx,\ t \in [c,d].$ **13.3.4.** $\int_0^\infty \dfrac{\sin tx}{x}\,dx.$ (a) $t \geq t_0 > 0$; (b) $t \geq 0.$

13.3.5. $\int_0^1 x^{t-1}\,dx.$ **13.3.6.** $\int_0^1 \dfrac{\sin x}{x^t}\,dx.$ (a) $t \leq t_0 < 2$; (b) $t \leq 2.$

13.3.7. $\int_0^1 x^{p-1}(1-x)^{q-1}\,dx,\ p \geq p_0 > 0,\ q \geq q_0 > 0.$

13.3.8. $\int_0^1 x^{t-1}\ln^m x\,dx,\ m \in \mathbb{N}.$ **13.3.9.** $\int_0^\infty \dfrac{\cos xt}{x^a}\,dx,\ t \geq t_0 > 0,\ 0 < a < 1.$

In Exercises 13.3.10–13.3.13 determine whether the given function is continuous:

13.3.10. $\int_0^\infty \dfrac{x\,dx}{2+x^t},\ t > 2.$ **13.3.11.** $\int_1^\infty \dfrac{\cos x}{x^t}\,dx,\ t > 0.$

13.3.12. $\int_0^\pi \dfrac{\sin x}{x^t(\pi - x^t)},\ t \in (0,2).$ **13.3.13.** $\int_0^\infty \dfrac{e^{-x}}{|\sin x|^t}\,dx,\ t \in (0,1).$

13.3.14. Use Equation (13.1) to find the integral $I = \int_0^1 \dfrac{\arctan x}{x\sqrt{1-x^2}}\,dx.$

13.3.15. Suppose that
$$\int_a^\infty F(x,t)\,dx = \varphi(t,a) + \int_a^\infty \psi(x,t)\,dx,$$
and that, as $a \to \infty$, $\varphi(t,a)$ converges uniformly to 0. Then $\int_a^\infty F(x,t)\,dx$ converges uniformly if and only if $\int_a^\infty \psi(x,t)\,dx$ does.

13.3.16. Let $\varphi(x,y)$ be a bounded function that is monotone in x. Suppose that the integral $\int_a^\infty f(x,y)\,dx$ converges uniformly for $y \in (c,d)$. Prove that the same is true of $\int_a^\infty f(x,y)\varphi(x,y)\,dx.$

13.3.17. Let $\varphi(x,y)$ be a function that is monotone in x, for $x > a$, and $\varphi(x,y)$ converges uniformly to 0, as $x \to \infty$, for $y \in (c,d)$. Suppose that $|\int_a^x f(t,y)\,dt| \leq M$ for all $y \in (c,d)$. Prove that the integral $\int_a^\infty f(x,y)\varphi(x,y)\,dx$ converges uniformly for $y \in (c,d)$.

13.3.18. Let F be a function defined for $x \geq a$ and $t \in [c,d]$. Prove that the integral (13.10) converges uniformly for $t \in [c,d]$ if and only if there exists a function $I(t)$ defined on $[c,d]$ with the following property: for every $\varepsilon > 0$, there exists $B > 0$ such that, for any $b \geq B$ and any $t \in [c,d]$,
$$\left|\int_a^b F(x,t)\,dx - I(t)\right| < \varepsilon.$$

13.3.19. Let F be a function defined for $a \leq x < b$ and $t \in [c,d]$. Prove that the integral $\int_a^b F(x,t)\,dx$ converges uniformly for $t \in [c,d]$ if and only if there exists a function $I(t)$ defined on $[c,d]$ with the following property: for every $\varepsilon > 0$, there exists $\delta > 0$ such that, for any $b - \delta < b' < b$ and any $t \in [c,d]$,
$$\left|\int_a^{b'} F(x,t)\,dx - I(t)\right| < \varepsilon.$$

13.3.20. Prove the Abel's test: Let F be a function defined and continuous on the rectangle $R = [a, b) \times [c, d]$ and suppose that $\int_a^b F(x, t)\, dx$ converges uniformly for $t \in [c, d]$. Also, suppose that, for each $t \in [c, d]$, $\varphi(x, t)$ is a monotone decreasing function of x, and there exists $M > 0$ such that $|\varphi(x, t)| \leq M$ for all $x \in [a, b)$ and $t \in [c, d]$. Then the integral (13.12) converges uniformly on $[c, d]$.

13.4 Improper Integral as a Function

Integral (13.11) defines a function $I(t)$. What can we say about this function? Is it continuous? Is it differentiable, and if so, is $I' = \int F_t'$? Let us start with the continuity.

Theorem 13.4.1. *Let F be a function defined and continuous on the rectangle $R = [a, b) \times [c, d]$ and suppose that the integral (13.11) converges uniformly for $t \in [c, d]$. Then the function $I(t)$ is continuous on $[c, d]$.*

Proof. Let b_n be a sequence of real numbers in $[a, b)$ that converges to b, and let $f_n(t) = \int_a^{b_n} F(x, t)\, dx$. By Theorem 13.3.5 or Theorem 13.3.9, the sequence $f_n(t)$ converges uniformly to $I(t)$ on $[c, d]$. Further, Theorem 13.2.1 implies that f_n is a sequence of continuous functions, so the result follows from Theorem 8.2.1. $\qquad\square$

Remark 13.4.2. It is now clear that neither of the integrals in Examples 13.3.6 and 13.3.7 converges uniformly on $[0, 1]$, because $I(t)$ is not continuous at $t = 0$.

In Section 13.2 we have seen (Theorem 13.2.6) that, when F is continuous on $R = [a, b] \times [c, d]$,

$$\int_c^d I(t)\, dt = \int_c^d dt \int_a^b F(x, t)\, dx = \int_a^b dx \int_c^d F(x, t)\, dt. \qquad (13.13)$$

Is it still true when the integral defining $I(t)$ is improper?

Example 13.4.3. Interchanging integrals gives an incorrect result.

Let $F(x, t) = \dfrac{x^2 t^2 - 1}{(x^2 t^2 + 1)^2}$. Prove that the iterated integrals in (13.13) exist, but they are not equal.

Solution. The function F is continuous on the rectangle $[1, \infty) \times [0, 1]$ so all integrals in (13.13) exist. Since

$$\frac{\partial}{\partial t}\left(\frac{-t}{x^2 t^2 + 1}\right) = F(x, t),$$

it follows that

$$\int_0^1 F(x, t)\, dt = \left.\frac{-t}{x^2 t^2 + 1}\right|_0^1 = -\frac{1}{x^2 + 1}, \text{ so}$$

$$\int_1^\infty dx \int_0^1 F(x, t)\, dt = \left.-\arctan x\right|_1^\infty = -\frac{\pi}{4}.$$

On the other hand,

$$\frac{\partial}{\partial x}\left(\frac{-x}{x^2 t^2 + 1}\right) = F(x, t), \text{ so}$$

$$I(t) = \int_1^\infty F(x, t)\, dx = \left.\frac{-x}{x^2 t^2 + 1}\right|_1^\infty = \frac{1}{t^2 + 1}, \text{ hence}$$

$$\int_0^1 dt \int_1^\infty F(x, t)\, dx = \frac{\pi}{4}. \qquad\qquad \blacklozenge$$

Integrals Depending on a Parameter

Once again, uniform convergence of the improper integral was missing. If we include it, the following result is obtained.

Theorem 13.4.4. *Under the hypotheses of Theorem 13.4.1 the formula (13.13) is valid.*

Proof. We will use the same notation as in the proof of Theorem 13.4.1. By Theorem 13.2.6,

$$\int\limits_c^d f_n(t)\,dt = \int\limits_c^d dt \int\limits_a^{b_n} F(x,t)\,dx = \int\limits_a^{b_n} dx \int\limits_c^d F(x,t)\,dt.$$

The assumption is that the sequence f_n converges uniformly to I, so Theorem 8.2.3 implies that the leftmost integral converges to $\int_c^d I(t)\,dt$. Therefore, the rightmost integral also converges, and its limit is $\int_a^b dx \int_c^d F(x,t)\,dt$. $\qquad\qquad\square$

Example 13.4.5. Evaluating an integral.

Evaluate $\int_0^\infty \dfrac{e^{-ax} - e^{-bx}}{x}\,dx$, $a, b > 0$.

Solution. We will use the formula

$$\int_0^\infty e^{-tx}\,dx = \frac{1}{t} \tag{13.14}$$

which holds for $t > 0$. The idea is to take the integral of both sides:

$$\int_a^b dt \int_0^\infty e^{-tx}\,dx = \int_a^b \frac{1}{t}\,dt = \ln\frac{b}{a}.$$

On the left side, we can interchange the integrals: the function $F(x,t) = e^{-tx}$ is continuous, and the integral (13.14) converges uniformly by the de la Vallée-Poussin's Test because $e^{-tx} \le e^{-ax}$. Theorem 13.4.4 now implies that

$$\ln\frac{b}{a} = \int_a^b dt \int_0^\infty e^{-tx}\,dx = \int_0^\infty dx \int_a^b e^{-tx}\,dt = \int_0^\infty \frac{e^{-ax} - e^{-bx}}{x}\,dx$$

so the value of the integral is $\ln\frac{b}{a}$. $\qquad\qquad\blacklozenge$

Theorem 13.4.4 is due to Dini, and it can be found in [32].

In the previous theorem we have used the assumption that t belongs to a finite interval $[c, d]$. What happens if d is replaced by $+\infty$?

Example 13.4.6. Interchanging integrals gives an incorrect result.

Let $F(x,t) = \dfrac{t^2 - x^2}{(x^2 + t^2)^2}$. Show that $\int_1^\infty dx \int_1^\infty F(x,t)\,dt \ne \int_1^\infty dt \int_1^\infty F(x,t)\,dx$.

Solution. The hypotheses of Theorem 13.4.4 are satisfied. It is obvious that F is defined and continuous on $R = [1, +\infty) \times [1, +\infty)$. The integral $\int_1^\infty F(x,t)\,dt$ converges uniformly for $x \ge 1$. Indeed, if we define $\varphi(t) = 1/(1 + t^2)$ then, for any $x \ge 1$,

$$|F(x,t)| = \left| \frac{t^2 - x^2}{(x^2 + t^2)^2} \right| \le \frac{t^2 + x^2}{(x^2 + t^2)^2} = \frac{1}{x^2 + t^2} \le \frac{1}{1 + t^2},$$

so the uniform convergence follows by the de la Vallée-Poussin's Test. In a similar fashion, we can show that the integral $\int_1^\infty F(x,t)\,dx$ converges uniformly for $t \ge 1$.

Now we compute the iterated integrals. Since

$$\frac{\partial}{\partial t}\left(\frac{-t}{x^2+t^2}\right) = F(x,t),$$

we have that

$$\int_1^\infty F(x,t)\,dt = \left.\frac{-t}{x^2+t^2}\right|_1^\infty = \frac{1}{x^2+1}.$$

Therefore,

$$\int_1^\infty dx \int_1^\infty F(x,t)\,dt = \int_1^\infty \frac{1}{x^2+1}\,dx = \left.\arctan x\right|_1^\infty = \frac{\pi}{2} - \frac{\pi}{4} = \frac{\pi}{4}.$$

On the other hand, $F(x,t) = -F(t,x)$, so

$$\int_1^\infty dt \int_1^\infty F(x,t)\,dx = -\frac{\pi}{4}. \qquad \blacklozenge$$

Examples 13.4.3 and 13.4.6 are modifications of an example given by Cauchy in his 1827 essay [17]. Namely, Cauchy had considered

$$\int_0^1 \int_0^1 \frac{y^2-z^2}{(y^2+z^2)^2}\,dy\,dz$$

and a change of variables $y = t$, $z = 1/x$, (resp., $y = 1/x$, $z = 1/t$) leads to the integral in Example 13.4.3 (resp., Example 13.4.6).

We see that justifying the change in the order of integration requires stronger hypotheses. The following theorem, coming from de la Vallée Poussin's 1892 article [27], gives a sufficient condition.

Theorem 13.4.7. *Let F be a non-negative function, defined and continuous on the infinite rectangle $R = [a,b] \times [c,+\infty)$ and suppose that both $I(t) = \int_a^b F(x,t)\,dx$ and $J(x) = \int_c^\infty F(x,t)\,dt$ converge uniformly. Then*

$$\int_a^b dx \int_c^\infty F(x,t)\,dt = \int_c^\infty dt \int_a^b F(x,t)\,dx \qquad (13.15)$$

as soon as one of the iterated integrals exists.

Proof. We will assume that the integral on the right side exists, and leave the other case as an exercise. Let $b' \in [a,b)$. Considering x as the parameter, it follows from Theorem 13.4.4 that

$$\int_a^{b'} dx \int_c^\infty F(x,t)\,dt = \int_c^\infty dt \int_a^{b'} F(x,t)\,dx$$

and the fact that $F \geq 0$ implies that

$$\int_c^\infty dt \int_a^{b'} F(x,t)\,dx \leq \int_c^\infty dt \int_a^b F(x,t)\,dx,$$

the last integral being convergent by our earlier assumption. The Monotone Convergence now implies that the integral on the left side of (13.15) converges and that an inequality holds in (13.15).

To prove the inequality in the other direction, let $c' > c$. By Theorem 13.4.4, the non-negativity of F, and the newly established convergence of the integral on the left side of (13.15),

$$\int_c^{c'} dt \int_a^b F(x,t)\,dx = \int_a^b dx \int_c^{c'} F(x,t)\,dt \leq \int_a^b dx \int_c^\infty F(x,t)\,dt.$$

Passing to the limit as $c' \to \infty$ completes the proof. $\qquad \square$

Example 13.4.8. Evaluating an integral.

Evaluate $W = \int_0^\infty e^{-x^2}\, dx$.

Solution. The fact that $\int_0^\infty e^{-x^2}\, dx$ exists is Exercise 6.7.10. We will now evaluate it. Let $c > 0$. First we will prove that

$$\int_c^\infty dt \int_0^\infty te^{-(x^2+1)t^2}\, dx = \int_0^\infty dx \int_c^\infty te^{-(x^2+1)t^2}\, dt. \qquad (13.16)$$

Clearly, $F(x,t) = te^{-(x^2+1)t^2}$ is a non-negative and continuous function. It is not hard to see that

$$J(x) = \int_c^\infty te^{-(x^2+1)t^2}\, dt = e^{-(x^2+1)t^2}\left.\frac{-1}{2(x^2+1)}\right|_c^\infty = \frac{e^{-(x^2+1)c^2}}{2(x^2+1)},$$

and the convergence is uniform for $x \geq 0$ because $te^{-(x^2+1)t^2} \leq te^{-t^2}$ and

$$\int_c^\infty te^{-t^2}\, dt = -\frac{1}{2}e^{-t^2}\Big|_c^\infty = \frac{1}{2}e^{-c^2}.$$

On the other hand, the substitution $u = xt$ yields

$$I(t) = \int_0^\infty te^{-(x^2+1)t^2}\, dx = e^{-t^2}\int_0^\infty te^{-x^2t^2}\, dx = e^{-t^2}\int_0^\infty e^{-u^2}\, du$$

and the convergence is uniform for $t \geq c$. Indeed,

$$\frac{\partial}{\partial t}\left(te^{-(x^2+1)t^2}\right) = e^{-(x^2+1)t^2}\left(1 - 2t^2(x^2+1)\right) \leq 1 - 2c^2(x^2+1) \leq 0$$

if $x \geq \sqrt{(1-2c^2)/(2c^2)}$. For such x, the function $te^{-(x^2+1)t^2}$ is a decreasing function of t, so it attains its maximum for $t = c$. Thus,

$$te^{-(x^2+1)t^2} \leq ce^{-(x^2+1)c^2}$$

and the de la Vallée Poussin's Test guarantees the uniform convergence. Notice that if $c = 0$, the convergence would not be uniform. This can be seen by the fact that I is not continuous at $t = 0$: $I(0) = 0$ but $e^{-t^2}W \to W$, as $t \to 0$.

Finally, $\int_c^\infty I(t)\, dt \leq W^2$, so the iterated integral on the left side of (13.16) exists. By Theorem 13.4.7, they both exist and the equality (13.16) holds.

Next, we will take the limit as $c \to 0$ in (13.16). On the left side,

$$\int_c^\infty dt \int_0^\infty te^{-(x^2+1)t^2}\, dx = \int_c^\infty e^{-t^2}\, dt \int_0^\infty e^{-u^2}\, du \to W^2, \quad c \to 0.$$

The right hand side is

$$\int_0^\infty dx \int_c^\infty te^{-(x^2+1)t^2}\, dt = \int_0^\infty \frac{e^{-(x^2+1)c^2}}{2(x^2+1)}\, dx \qquad (13.17)$$

and we will show that it is a continuous function of c. This will follow from Theorem 13.4.1 applied to $J(x,c) = e^{-(x^2+1)c^2}/(2(x^2+1))$. It is easy to see that J is continuous on $[0,\infty) \times [0,\infty)$ and that

$$J(x,c) = \frac{e^{-(x^2+1)c^2}}{2(x^2+1)} \leq \frac{1}{2(x^2+1)}.$$

By the de la Vallée Poussin's Test, the integral in (13.17) converges uniformly. Thus, Theorem 13.4.1 implies that we can bring the limit as $c \to 0$ inside this integral, and we obtain that

$$W^2 = \int_0^\infty \lim_{c\to 0}\frac{e^{-(x^2+1)c^2}}{2(x^2+1)}\, dx = \int_0^\infty \frac{1}{2(x^2+1)}\, dx = \frac{1}{2}\arctan x\Big|_0^\infty = \frac{\pi}{4}.$$

Consequently, $W = \sqrt{\pi}/2$. (It is not $-\sqrt{\pi}/2$ because $W > 0$.)   ◆

We close this section with a result about the derivative of the infinite integral, once again from de la Vallée Poussin's 1892 paper [27].

Theorem 13.4.9. *Let F and F'_t be defined and continuous on the rectangle $R = [a, b) \times [c, d]$ and suppose that, for each fixed $t \in [c, d]$ the integral (13.11) converges. If the integral*

$$J(t) = \int_a^b F'_t(x, t)\, dx$$

converges uniformly for $t \in [c, d]$, then the function $I(t)$ is differentiable on $[c, d]$ and $I'(t) = J(t)$.

Proof. Let $\{b_n\}$ be a sequence of real numbers in $[a, b)$ that converges to b, and let $f_n(t) = \int_a^{b_n} F(x, t)\, dx$. By Theorem 13.2.2, each function $f_n(t)$ is differentiable with the derivative $f'_n(t) = \int_a^{b_n} F'_t(x, t)\, dx$. The hypotheses of the theorem imply that the sequence $\{f'_n\}$ converges uniformly to $J(t)$ on $[c, d]$. On the other hand, $\{f_n(t)\}$ converges to $I(t)$ for each $t \in [c, d]$. Now the result follows from Theorem 8.2.5. $\qquad\square$

Example 13.4.10. Evaluating an integral.

Evaluate $\int_0^\infty \dfrac{dx}{(x^2 + a)^n}$, $a > 0$.

Solution. We denote the integral by $I_n(a)$, and notice that $I_1(a) = \pi/(2\sqrt{a})$. We will calculate $I'_n(a)$ using Theorem 13.4.9. Let $d > c > 0$. The function $F(x, a) = 1/(x^2 + a)^n$ is continuous on $R = [0, +\infty) \times [c, d]$. For each $a \in [c, d]$ the integral I_n converges because I_1 converges and

$$I_n(a) = \int_0^\infty \frac{dx}{(x^2 + a)^n} \le \int_0^\infty \frac{dx}{(x^2 + a)a^{n-1}} = \frac{1}{a^{n-1}} I_1(a).$$

Finally, $F'_a = -n/(x^2 + a)^{n+1}$, so we want to establish that the integral

$$\int_0^\infty \frac{-n}{(x^2 + a)^{n+1}}\, dx$$

converges uniformly on $[c, d]$. This follows from the de la Vallée-Poussin's Test and the inequality $1/(x^2 + a)^{n+1} \le 1/(x^2 + c)^{n+1}$. Now Theorem 13.4.9 implies that

$$I'_n(a) = \int_0^\infty \frac{-n}{(x^2 + a)^{n+1}}\, dx = -nI_{n+1}(a)$$

so $I_{n+1}(a) = -I'_n(a)/n$ and, inductively,

$$I_{n+1}(a) = (-1)^n \frac{I_1^{(n)}(a)}{n!}.$$

Since $I_1(a) = \pi/(2\sqrt{a})$, we obtain that

$$I_{n+1}(a) = \frac{(-1)^n}{n!} \frac{\pi}{2} \frac{(-1)^n (2n-1)!!}{2^n a^{\frac{2n+1}{2}}} = \frac{\pi}{2} \frac{(2n-1)!!}{(2n)!!} \frac{1}{(\sqrt{a})^{2n+1}}. \qquad\blacklozenge$$

Exercises

In Exercises 13.4.1–13.4.6 use the differentiation with respect to the parameter to find $I(t)$ (and justify your conclusion):

13.4.1. $I(a) = \displaystyle\int_0^\infty \frac{e^{-ax^2} - e^{-bx^2}}{x}\, dx,\ a,b > 0.$

13.4.2. $I(t) = \displaystyle\int_0^\infty \frac{e^{-ax} - e^{-bx}}{x}\, \sin tx\, dx,\ a,b > 0.$

13.4.3. $I(t) = \displaystyle\int_0^1 \frac{\ln(1 - t^2 x^2)}{x^2\sqrt{1 - x^2}}\, dx,\ |t| \le 1.$

13.4.4.* $I(t) = \displaystyle\int_0^1 \frac{\ln(1 - t^2 x^2)}{\sqrt{1 - x^2}}\, dx,\ |t| \le 1.$

13.4.5. $I(t) = \displaystyle\int_0^1 \frac{\arctan(tx)}{x\sqrt{1 - x^2}}\, dx.$

13.4.6.* $I(t) = \displaystyle\int_0^\infty \frac{\ln(t^2 + x^2)}{1 + x^2}\, dx.$

13.4.7. Prove that the function $I(t) = \displaystyle\int_0^\infty \frac{\cos x}{1 + (x + t)^2}\, dx$ is differentiable and has a continuous derivative for $t \in \mathbb{R}$.

13.4.8. Use the result of Example 13.4.8 to find $\displaystyle\int_{-\infty}^{+\infty} (At^2 + 2Bt + C)e^{-t^2}\, dt.$

In Exercises 13.4.9–13.4.13 use the result of Exercise 13.4.8 and the differentiation with respect to the parameter to find the integral (and justify your conclusion):

13.4.9.* $\displaystyle\int_0^\infty e^{-x^2 - \frac{t^2}{x^2}}\, dx.$ **13.4.10.** $\displaystyle\int_0^\infty e^{-ax^2}\cos tx\, dx,\ a > 0.$

13.4.11. $\displaystyle\int_0^\infty e^{-tx}\frac{\cos x - 1}{x}\, dx.$ **13.4.12.*** $\displaystyle\int_0^1 \frac{x^2 - 1}{\ln x}\, dx.$ **13.4.13.** $\displaystyle\int_{-\infty}^\infty x^n e^{-x^2}\, dx.$

13.4.14.* Let $I(t) = \displaystyle\int_0^\infty e^{-x^2}\cos(tx^2)\, dx$ and $J(t) = \displaystyle\int_0^\infty e^{-x^2}\sin(tx^2)\, dx$. Prove that I and J satisfy differential equations

$$I'(t) = -\frac{1}{2(1 + t^2)}\left[tI(t) + J(t)\right], \quad \text{and} \quad J'(t) = \frac{1}{2(1 + t^2)}\left[I(t) - tJ(t)\right].$$

In Exercises 13.4.15–13.4.16 use the integration with respect to the parameter to find $I(\alpha, \beta)$ (and justify your conclusion):

13.4.15.* $\displaystyle\int_0^\infty \frac{\tanh \beta x - \tanh \alpha x}{x}\, dx.$

13.4.16.* $\displaystyle\int_{-\infty}^\infty \frac{1}{\sqrt{x^2 + \alpha^2}} \arctan\left(\frac{\beta}{\sqrt{x^2 + \alpha^2}}\right)\, dx,\ a \ne 0.$

13.5   Some Important Integrals

In this section we will compute some frequently encountered integrals. In particular, we will define the so-called Euler integrals B (Beta) and Γ (Gamma), and we will establish some basic properties of the functions that they define.

Example 13.5.1. Evaluating an integral.

Evaluate $\int_0^\infty \frac{\sin x}{x}\,dx$.

Solution. One way to get rid of the denominator might be to start with $\sin(tx)/x$ and use differentiation with respect to the parameter. Unfortunately, that would yield $\int_0^\infty \cos(tx)\,dx$ which is divergent. We will introduce an additional factor that will make the integral converge. Let $k > 0$ and consider

$$I(t) = \int_0^\infty e^{-kx}\frac{\sin tx}{x}\,dx,$$

for $t \ge 0$. Now, we want to apply Theorem 13.4.9. The integrand is continuous in the rectangle $x \ge 0$, $t \ge 0$. Its partial derivative (with respect to t) is $e^{-kx}\cos(tx)$ which is also continuous. The integral $I(t)$ converges for any $t \ge 0$, because $|\sin(tx)/x| \le t$ and e^{-kx} is integrable. Finally,

$$\int_0^\infty e^{-kx}\cos(tx)\,dx$$

converges uniformly because $|\cos(tx)| \le 1$ and Theorem 13.3.10 applies. Further, the same method as in Example 5.1.10 (or any text on Differential Equations covering the Laplace Transform) can be used to determine that

$$\int_0^\infty e^{-kx}\cos(tx)\,dx = \lim_{b\to+\infty}\int_0^b e^{-kx}\cos(tx)\,dx$$

$$= \lim_{b\to+\infty}\frac{t\sin(tx) - k\cos(tx)}{k^2 + t^2}e^{-kx}\Big|_0^b$$

$$= \lim_{b\to+\infty}\left(\frac{t\sin(tb) - k\cos(tb)}{k^2 + t^2}e^{-kb} + \frac{k}{k^2 + t^2}\right) = \frac{k}{k^2 + t^2}.$$

Thus, $I'(t) = k/(k^2 + t^2)$ so $I(t) = \arctan(t/k) + C$. By the definition of $I(t)$, $I(0) = 0$, which implies that $C = 0$. Consequently,

$$I(t) = \int_0^\infty e^{-kx}\frac{\sin tx}{x}\,dx = \arctan(t/k).$$

The calculations so far were done with the assumption that $k > 0$. We would like to replace k with 0, so we need to verify that

$$\lim_{k\to 0^+}\int_0^\infty e^{-kx}\frac{\sin tx}{x}\,dx = \int_0^\infty \frac{\sin tx}{x}\,dx.$$

In other words, we need to show that

$$J(k) = \int_0^\infty e^{-kx}\frac{\sin tx}{x}\,dx$$

is a continuous function of k. We will keep t fixed, and apply Theorem 13.4.1. We need to establish that the function $e^{-kx}\sin tx/x$ is integrable for any $k \ge 0$. Since we already have that for $k > 0$, we only need to consider the case $k = 0$. The function $\sin tx/x$ is integrable by the Dirichlet's Test (Theorem 13.3.13), because $1/x$ is monotone decreasing to 0, and

$$\left|\int_a^b \sin tx\,dx\right| = \left|\frac{\cos(ta) - \cos(tb)}{t}\right| \le \frac{2}{t},$$

if $t \neq 0$. (If $t = 0$ then the integral equals 0.) Also, we need to prove that the integral $J(k)$ converges uniformly. Again, the Dirichlet's Test can be used, this time e^{-kx}/x is monotone decreasing converging to 0. Therefore,

$$\int_0^\infty \frac{\sin tx}{x}\,dx = \lim_{k \to 0+} \arctan \frac{t}{k} = \frac{\pi}{2}. \tag{13.18}$$

This is true for any $t \geq 0$, hence for $t = 1$. So,

$$\int_0^\infty \frac{\sin x}{x}\,dx = \frac{\pi}{2}. \tag{13.19}$$

$\blacklozenge$

Although the equality (13.19) was known in the eighteenth century, it is often associated with the name of Dirichlet. The reason is that it is a special case of the identity

$$\frac{2}{\pi}\int_0^\infty \frac{\sin x \cos tx}{x}\,dx = \begin{cases} 1, & \text{if } t < 1 \\ \dfrac{1}{2}, & \text{if } t = 1 \\ 0, & \text{if } t > 1, \end{cases} \tag{13.20}$$

which Dirichlet has used in the so-called Method of Reduction that also bears his name. (See Exercise 13.5.16.)

Example 13.5.2. Gamma Function.

The Euler's *Gamma Function* is defined by $\Gamma(t) = \int_0^\infty x^{t-1}e^{-x}\,dx$, $t > 0$. Let us derive some of its properties.

The integral $\int_1^\infty x^{t-1}e^{-x}\,dx$ converges by Example 6.7.9 and $\int_0^1 x^{t-1}e^{-x}\,dx$ converges by Example 6.7.17. Thus, $\Gamma(t)$ is defined for every $t > 0$. Using integration by parts,

$$\Gamma(t+1) = \int_0^\infty x^t e^{-x}\,dx = -e^{-x}x^t\Big|_0^\infty - \int_0^\infty -e^{-x}tx^{t-1}\,dx = t\Gamma(t).$$

It follows that $\Gamma(t+2) = (t+1)t\,\Gamma(t)$ and, inductively,

$$\Gamma(t+n) = (t+n-1)(t+n-2)\ldots t\,\Gamma(t).$$

In particular, for $t = 1$ we obtain

$$\Gamma(n+1) = n(n-1)\ldots 1\,\Gamma(1) = n!\int_0^\infty e^{-x}\,dx = n!(-e^{-x})\Big|_0^\infty = n!.$$

We see that the Gamma Function is a generalization of the factorial function to non-integers. As a consequence, we can now talk about its derivative. We will demonstrate that $\Gamma(t)$ is infinitely differentiable at each $t > 0$. Again, we will consider separately the intervals $(0, 1]$ and $[1, +\infty)$. Let $t > 0$ be fixed, and let $t \in [c, d] \subset (0, +\infty)$. We will apply Theorem 13.4.9 to

$$F(x, t) = x^{t-1}(\ln x)^m\,e^{-x} \quad \text{and} \quad R_1 = (0, 1] \times [c, d],$$

with m an arbitrary non-negative integer. Both F and $F_t' = x^{t-1}(\ln x)^{m+1}e^{-x}$ are continuous on R_1 and

$$|x^{t-1}(\ln x)^{m+1}e^{-x}| \leq x^{t-1}|\ln x|^{m+1},$$

so the uniform convergence of $\int_0^1 F_t'$ follows from Exercise 6.7.41. On the other hand, if we consider the rectangle $R_2 = [1, +\infty) \times [c, d]$, F and F_t' are continuous and $\ln x \leq x$ so

$$|x^{t-1}(\ln x)^{m+1}e^{-x}| \leq x^{t+m}e^{-x}.$$

Since $\int_1^\infty x^{t+m}e^{-x}\,dx$ converges (it is smaller than $\Gamma(t+m+1)$), we have that, for every $m \in \mathbb{N}_0$,

$$\frac{d}{dt}\int_0^\infty x^{t-1}(\ln x)^m\,e^{-x}\,dx = \int_0^\infty x^{t-1}(\ln x)^{m+1}\,e^{-x}\,dx.$$

Setting $m = 0$ yields

$$\Gamma'(t) = \int_0^\infty x^{t-1}\ln x\,e^{-x}\,dx.$$

When $m = 1$ we obtain $\Gamma''(t) = \int_0^\infty x^{t-1}(\ln x)^2\,e^{-x}\,dx$ and, inductively,

$$\Gamma^{(n)}(t) = \int_0^\infty x^{t-1}(\ln x)^n\,e^{-x}\,dx.$$

It follows that $\Gamma''(t) > 0$ for $t > 0$, so $\Gamma(t)$ is a convex function (see Exercise 4.4.26). Further, by Exercise 10.1.3,

$$\Gamma'(t) = \int_0^\infty x^{t-1}\ln x\,e^{-x}\,dx = \int_0^\infty \left(\sqrt{x^{t-1}e^{-x}}\right)\left(\sqrt{x^{t-1}e^{-x}}\ln x\right)\,dx$$

$$\leq \left(\int_0^\infty x^{t-1}e^{-x}\,dx\right)^{1/2}\left(\int_0^\infty x^{t-1}e^{-x}\ln^2 x\,dx\right)^{1/2} = \sqrt{\Gamma(t)\Gamma''(t)}.$$

This inequality is equivalent to $(\Gamma'(t))^2 \leq \Gamma(t)\Gamma''(t)$. Since the Gamma Function grows very fast, it is interesting to consider the function $L(t) = \ln\Gamma(t)$. Notice that

$$L'(t) = \frac{1}{\Gamma(t)}\Gamma'(t), \quad \text{and} \quad L''(t) = \frac{\Gamma''(t)\Gamma(t) - (\Gamma'(t))^2}{(\Gamma(t))^2} \geq 0,$$

so L is convex. We say that the Gamma Function is *logarithmically convex*. ♦

> The problem of extending the factorial to non-integers was considered by Daniel Bernoulli and Christian Goldbach in the 1720s. In 1730, in a letter to Goldbach, Euler presented the formula $t! = \int_0^1(-\ln s)^t\,ds$, for $t > 0$. In an article from 1781 (published only in 1794) he used the change of variables $x = -\ln s$ to obtain the formula given at the beginning of Example 13.5.2. The name "gamma function" and the symbol Γ were introduced by Legendre (page 437) around 1811.

Example 13.5.3. Evaluating two integrals.

Evaluate $I(t) = \int_0^\infty \frac{\cos tx}{a^2 + x^2}\,dx$ and $J(t) = \int_0^\infty \frac{x\sin tx}{a^2 + x^2}\,dx$, $a > 0$

Solution. First we will take the derivative of I, so we want to apply Theorem 13.4.9. Let us verify its hypotheses. The function $F(x, t) = \cos tx/(a^2 + x^2)$ is continuous on the rectangle $R = [0, +\infty) \times [c, d]$, where $c > 0$. Also,

$$F_t'(x, t) = -\frac{x\sin tx}{a^2 + x^2}$$

is continuous on R. The integral $I(t)$ converges uniformly by the de la Vallée-Poussin's Test:

$$\left|\frac{\cos tx}{a^2 + x^2}\right| \leq \frac{1}{a^2 + x^2} \tag{13.21}$$

and

$$\int_0^\infty \frac{1}{a^2 + x^2} \, dx = \frac{1}{a} \arctan \frac{x}{a} \Big|_0^\infty = \frac{\pi}{2a}. \tag{13.22}$$

Finally, $\int_0^\infty F_t'(x, t) \, dx = -J(t)$, and we need to show that this integral converges uniformly. Here we can use the Dirichlet's Test. For any $A > 0$,

$$\left| \int_0^A -\sin tx \, dx \right| = \left| \frac{\cos tx}{t} \Big|_0^A \right| = \left| \frac{\cos tA - 1}{t} \right| \le \frac{2}{c}.$$

Also, $x/(a^2 + x^2)$ converges monotonically to 0 (at least for $x > a$). Thus, by the Dirichlet's Test, $J(t)$ converges uniformly, and Theorem 13.4.9 implies that

$$I'(t) = -J(t).$$

Next, we will use (13.18):

$$I'(t) + \frac{\pi}{2} = \int_0^\infty \frac{\sin tx}{x} \, dx - \int_0^\infty \frac{x \sin tx}{a^2 + x^2} \, dx = \int_0^\infty \sin tx \, \frac{a^2 + x^2 - x^2}{x(a^2 + x^2)} \, dx$$

$$= a^2 \int_0^\infty \frac{\sin tx}{x(a^2 + x^2)} \, dx.$$

Now, we want to take the derivative again. The function $G(x, t) = \frac{\sin tx}{x(a^2 + x^2)}$ is defined for $x > 0$, and $\lim_{x \to 0} G(x, t) = t/a^2$, so we can extend G to a continuous function on R. Further, $G_t'(x, t) = \frac{\cos tx}{a^2 + x^2}$, also continuous on R. The uniform convergence of $\int_0^\infty G_t'(x, t) \, dt$ has been already established, and the estimate

$$|G(x, t)| \le \frac{|t|}{a^2 + x^2} \le \frac{d}{a^2 + x^2}$$

together with (13.22) show that $\int_0^\infty G(x, t) \, dt$ converges. Thus, we can use Theorem 13.4.9 and we obtain that

$$I''(t) = a^2 \int_0^\infty \frac{\cos tx}{a^2 + x^2} = a^2 I(t).$$

This is a linear differential equation of second-order, and its general solution is

$$I(t) = C_1 e^{at} + C_2 e^{-at}.$$

This formula holds for $t \in [c, d]$. However, d is arbitrary, so we can take it to be arbitrarily large. On the other hand, (13.21) and (13.22) imply that $I(t) \le \pi/(2a)$, so $C_1 = 0$ and $I(t) = C_2 e^{-at}$. Further, $c > 0$ is arbitrary, so this formula holds for $t > 0$. Our final move is to take the limit as $t \to 0$. Since F is continuous on $[0, \infty) \times [0, d]$ and $\int_0^\infty F(x, t) \, dt$ converges uniformly, Theorem 13.4.1 allows us to bring the limit inside the integral. Thus,

$$C_2 = \lim_{t \to 0} I(t) = \int_0^\infty \lim_{t \to 0} \frac{\cos tx}{a^2 + x^2} \, dx = \int_0^\infty \frac{1}{a^2 + x^2} \, dx = \frac{\pi}{2a}.$$

We conclude that $I(t) = \frac{\pi}{2a} e^{-at}$ and $J(t) = -I'(t) = \frac{\pi}{2} e^{-at}$.   ◆

The complete Example 13.5.3 is taken from the the de la Vallée-Poussin's [27].

Example 13.5.4. Evaluating an integral.

Evaluate $I(a) = \int_0^\infty \frac{x^{a-1}}{1+x} \, dx$, $0 < a < 1$.

Solution. We will write $I(a) = I_1(a) + I_2(a)$, where

$$I_1(a) = \int_0^1 \frac{x^{a-1}}{1+x} \, dx, \quad I_2(a) = \int_1^\infty \frac{x^{a-1}}{1+x} \, dx.$$

Both integrals converge: the integrand $x^{a-1}/(1+x)$ can be compared with x^{a-1} when $x \to 0$, and with x^{a-2} when $x \to \infty$. The integrals

$$\int_0^1 x^{a-1} \, dx \quad \text{and} \quad \int_1^\infty x^{a-2} \, dx$$

converge by Example 6.7.13 and Exercise 6.7.7. Let $0 < c < 1$. Then

$$\int_0^c \frac{x^{a-1}}{1+x} \, dx = \int_0^c x^{a-1} \sum_{n=0}^\infty (-1)^n x^n \, dx.$$

Since the geometric series converges uniformly for $x \in [0, c]$, we can integrate it term by term to obtain

$$\sum_{n=0}^\infty (-1)^n \int_0^c x^{n+a-1} \, dx = \sum_{n=0}^\infty (-1)^n \frac{x^{n+a}}{n+a} \Big|_0^c = \sum_{n=0}^\infty (-1)^n \frac{c^{n+a}}{n+a}.$$

We conclude that the equality

$$\int_0^c \frac{x^{a-1}}{1+x} \, dx = \sum_{n=0}^\infty (-1)^n \frac{c^{n+a}}{n+a}$$

holds for any $0 < c < 1$. The series on the right side converges for $c = 1$ by the Alternating Series Test, so by the Abel's Theorem we obtain that the equality holds for $c = 1$. In other words,

$$I_1(a) = \sum_{n=0}^\infty (-1)^n \frac{1}{n+a}.$$

The integral $I_2(a)$ can be transformed using the substitution $x = 1/u$:

$$I_2(a) = \int_0^1 \frac{u^{1-a}}{(1+\frac{1}{u})} \frac{du}{u^2} = \int_0^1 \frac{u^{-a}}{1+u} \, du$$

$$= I_1(1-a) = \sum_{n=0}^\infty (-1)^n \frac{1}{n+1-a} = \sum_{n=1}^\infty (-1)^n \frac{1}{a-n}.$$

It follows that

$$I(a) = \sum_{n=0}^\infty (-1)^n \frac{1}{n+a} + \sum_{n=1}^\infty (-1)^n \frac{1}{a-n}$$

$$= \frac{1}{a} + \sum_{n=1}^\infty (-1)^n \left(\frac{1}{n+a} + \frac{1}{a-n} \right)$$

$$= \frac{1}{a} + \sum_{n=1}^{\infty} (-1)^n \frac{2a}{a^2 - n^2}.$$

By Exercise 9.1.19, this equals $\pi/(2 \sin a\pi)$.   ♦

Example 13.5.5. Beta Function.

Euler's *Beta Function* is defined by $B(a,b) = \int_0^1 x^{a-1}(1-x)^{b-1}\, dx$, $a, b > 0$. Let us derive some of its properties.

By Example 6.7.13 the integral converges in spite of the fact that the integrand can be undefined at either endpoint (when $a < 1$ and $b < 1$).

Suppose that $b > 1$. Then, using integration by parts, with $u = (1-x)^{b-1}$ and $dv = x^{a-1}\, dx$, we obtain

$$B(a,b) = (1-x)^{b-1} \frac{x^a}{a} \Big|_0^1 + \int_0^1 (b-1)(1-x)^{b-2} \frac{x^a}{a}\, dx$$

$$= \frac{b-1}{a} \int_0^1 \left(x^{a-1}(1-x)^{b-2} - x^{a-1}(1-x)^{b-1} \right)\, dx$$

$$= \frac{b-1}{a} B(a, b-1) - \frac{b-1}{a} B(a,b).$$

Solving for $B(a,b)$ yields

$$B(a,b) = \frac{b-1}{a+b-1} B(a, b-1). \tag{13.23}$$

In particular, for $n \in \mathbb{N}$,

$$B(a,n) = \frac{n-1}{a+n-1} B(a, n-1) = \frac{n-1}{a+n-1} \frac{n-2}{a+n-2} B(a, n-2) = \dots$$

$$\dots = \frac{(n-1)!}{(a+n-1)(a+n-2)\dots(a+1)} B(a,1).$$

The definition of the Beta Function reveals that it is symmetric: $B(a,b) = B(b,a)$. (Use the substitution $u = 1 - x$.) If $m \in \mathbb{N}$, it follows that

$$B(m,n) = \frac{(n-1)!}{(m+n-1)(m+n-2)\dots(m+1)} B(m,1)$$

$$= \frac{(n-1)!}{(m+n-1)(m+n-2)\dots(m+1)} B(1,m)$$

$$= \frac{(n-1)!}{(m+n-1)(m+n-2)\dots(m+1)} \frac{(m-1)!}{m(m-1)\dots 2} B(1,1)$$

$$= \frac{(n-1)!(m-1)!}{(m+n-1)!},$$

because $B(1,1) = 1$.

The last relation can be written as $B(m,n) = \Gamma(n)\Gamma(m)/\Gamma(m+n)$. We will now demonstrate that the analogous formula holds for $B(a,b)$, where a, b are not necessarily integers. We start with the observation that if, in the definition of Gamma Function, we introduce a substitution $x = yu$ where u is a new variable, and y is a positive constant, we obtain

$$\Gamma(a) = \int_0^{\infty} u^{a-1} y^{a-1} e^{-uy} y\, du = y^a \int_0^{\infty} u^{a-1} e^{-uy}\, du.$$

Another observation concerns the Beta Function. Let us use the substitution $x = u/(1 + u)$. It is not hard to see that $1 - x = 1/(1 + u)$, $u = x/(1 - x)$, and that u varies from 0 to $+\infty$. Therefore,

$$B(a, b) = \int_0^\infty \frac{u^{a-1}}{(1 + u)^{a-1}} \frac{1}{(1 + u)^{b-1}} \frac{du}{(1 + u)^2} = \int_0^\infty \frac{u^{a-1}}{(1 + u)^{a+b}} du. \qquad (13.24)$$

When $a + b = 1$ (assuming that $0 < a, b < 1$), using Example 13.5.4, we have that

$$B(a, 1 - a) = \int_0^\infty \frac{u^{a-1}}{1 + u} du = \frac{\pi}{\sin a\pi}.$$

For example, $B(\frac{1}{2}, \frac{1}{2}) = \pi$.

Let $a, b > 1$. Then

$$\Gamma(a)\Gamma(b) = \Gamma(a) \int_0^\infty y^{b-1} e^{-y} dy$$

$$= \int_0^\infty y^{a+b-1} e^{-y} \frac{\Gamma(a)}{y^a} dy$$

$$= \int_0^\infty y^{a+b-1} e^{-y} \left(\int_0^\infty u^{a-1} e^{-uy} du \right) dy \qquad (13.25)$$

$$= \int_0^\infty dy \int_0^\infty y^{a+b-1} e^{-y} u^{a-1} e^{-uy} du.$$

At this point, we would like to interchange the order of integration. Theorem 13.4.7 requires that the integrand $F(u, y) = y^{a+b-1} e^{-y} u^{a-1} e^{-uy}$ be continuous on $R = [0, \infty) \times [0, \infty)$, and it is easy to see that this is the case here. It is an exercise in calculus to show that the function $f(u) = u^{a-1} e^{-uy} y^{a-1}$ attains its maximum at $u = (a - 1)/y$, when $y \neq 0$, and the said maximum equals $M_1 = ((a - 1)/e)^{a-1}$. (When $y = 0$, the maximum is 0.) Thus, $F(u, y) \leq M_1 y^b e^{-y}$ and the integral $\int_0^\infty F(u, y)\, dy$ converges uniformly by de la Vallée-Poussin's Test, because

$$\int_0^\infty M_1 y^b e^{-y} dy = M_1 \Gamma(b + 1).$$

Similarly, $g(y) = y^{a+b-1} e^{-y(u+1)}$ attains its maximum at $y = (a + b - 1)/(u + 1)$ and that maximum equals

$$\left(\frac{a + b - 1}{e(u + 1)} \right)^{a+b-1} = M_2 \frac{1}{(u + 1)^{a+b-1}}.$$

Thus, $F(u, y) \leq M_2 u^{a-1}/(u + 1)^{a+b-1}$ and the integral $\int_0^\infty F(u, y)\, du$ converges uniformly by de la Vallée-Poussin's Test, because

$$\int_0^\infty M_2 \frac{u^{a-1}}{(u + 1)^{a+b-1}} du = M_2 B(a, b - 1).$$

Since all the hypotheses of Theorem 13.4.7 are satisfied, we can change the order of integration in (13.25). We obtain that

$$\Gamma(a)\Gamma(b) = \int_0^\infty du \int_0^\infty y^{a+b-1} e^{-y} u^{a-1} e^{-uy} dy$$

$$= \int_0^\infty u^{a-1}\, du \int_0^\infty y^{a+b-1} e^{-y(u+1)}\, dy$$

$$= \int_0^\infty u^{a-1} \frac{\Gamma(a+b)}{(u+1)^{a+b}}\, du$$

$$= \Gamma(a+b) \int_0^\infty \frac{u^{a-1}}{(u+1)^{a+b}}\, du$$

$$= \Gamma(a+b)\mathrm{B}(a,b).$$

However, we have established this formula under the assumption that $a, b > 1$. What if at least one of them is between 0 and 1? In that case $a + 1, b + 1 > 1$ so

$$\Gamma(a+1)\Gamma(b+1) = \Gamma(a+b+2)\mathrm{B}(a+1, b+1). \tag{13.26}$$

The left hand side equals $a\Gamma(a)b\Gamma(b)$, and $\Gamma(a+b+2) = (a+b+1)(a+b)\Gamma(a+b)$. Finally, using (13.23),

$$\mathrm{B}(a+1, b+1) = \frac{b}{a+b+1}\,\mathrm{B}(a+1, b)$$

$$= \frac{b}{a+b+1}\,\mathrm{B}(b, a+1)$$

$$= \frac{b}{a+b+1}\,\frac{a}{a+b}\,\mathrm{B}(a, b),$$

so (13.26) becomes

$$a\Gamma(a)b\Gamma(b) = (a+b+1)(a+b)\Gamma(a+b)\frac{b}{a+b+1}\,\frac{a}{a+b}\,\mathrm{B}(a, b)$$

$$= ab\Gamma(a+b)\mathrm{B}(a, b)$$

which, after dividing both sides by ab, yields

$$\Gamma(a)\Gamma(b) = \Gamma(a+b)\mathrm{B}(a, b), \tag{13.27}$$

for any $a, b > 0$. This gives a nice description of the Beta Function in terms of the Gamma Function, which explains why the latter is much better known. A consequence of this formula is that, when $a + b = 1$, we have

$$\Gamma(a)\Gamma(1-a) = \Gamma(1)\mathrm{B}(a, 1-a) = \frac{\pi}{\sin a\pi}. \tag{13.28}$$

♦

Around 1655, 20 years before Newton's *Principia*, the English mathematician John Wallis (1616–1703) worked on evaluating $\int_0^1 x^e(1-x)^n\, dx$. Euler attacked this problem, but for him e and n were arbitrary numbers, and n was not necessarily an integer. By developing $(1-x)^n$ into a binomial series, he obtained that

$$\int_0^1 x^e(1-x)^n\, dx = \frac{n!}{(e+1)(e+2)\ldots(e+n+1)}.$$

The trouble was that $n!$ was meaningless unless n was an integer. So, Euler rolled up his sleeves, and came up with Gamma Function. The integral above is, clearly, $\mathrm{B}(e+1, n+1)$. The name "Beta function" and the symbol B were introduced by French mathematician Jacques Binet (1786–1856) in 1839, in [6]. There is a rumor that he has used the letter B because it is his initial, and that Legendre had selected Γ because the symbol looks like an inverted L. Binet is recognized as the first to describe the rule for multiplying matrices in 1812, and Binet's formula expressing Fibonacci numbers in closed form is named in his honor, although the same result was known to de Moivre a century earlier. Wallis is credited with introducing the symbol ∞ for infinity in his 1655 [108]. A year later he published [109] which contains what we now call the Wallis Product (see Exercise 13.5.17).

ADRIEN-MARIE LEGENDRE (1752–1833) was a French mathematician. He grew up in a wealthy family in Paris, studied at the Collège Mazarin, and defended his thesis in physics and mathematics in 1770. Although he did not need a job, he taught at the Ecole Militaire in Paris (1775–80) and at the Ecole Normale from 1795. His treatise on projectiles won the prize of the Berlin Academy in 1782. A year later his research on celestial mechanics earned him a membership in the Académie des Sciences. In 1787 he took part in the geodetic measurements jointly conducted with the Royal Greenwich Observatory in London. At this time he also became a member of the Royal Society of London. In 1791 he was a member of the committee to develop the metric system and, in particular, to conduct the necessary measurements to determine the standard meter. Between 1792 and 1801 he also worked on projects to produce logarithmic and trigonometric tables. During the French Revolution Legendre lost his family wealth. When the Académie des Sciences reopened as the Institut Nationale des Sciences et des Arts, Legendre was installed in the mathematics section. In 1824 he refused to endorse the government's candidate for the Institut and lost his pension. In 1831 he was made an officer of the Légion d'Honneur. Legendre has done impressive work in many areas, but much of it served as a springboard for the others. His work on roots of polynomials inspired Galois theory. Abel's work on elliptic functions was built on Legendre's. In number theory, he conjectured the quadratic reciprocity law, subsequently proved by Gauss; his 1798 conjecture of the Prime Number Theorem was rigorously proved by Hadamard and de la Vallée-Poussin in 1896. In 1806 he produced a book on the orbits of comets that contains the first comprehensive treatment of the method of least squares, although priority for its discovery is shared with Gauss. He named the Gamma function and introduced the symbol Γ. In 1830 he proved Fermat's last theorem for exponent $n = 5$, which was also done by Dirichlet in 1828. Legendre is best known as the author of *Eléments de géométrie* (*Foundations of Geometry*), which was published in 1794 and was the leading elementary text on the topic for around 100 years. This text greatly rearranged and simplified many of the propositions from Euclid's Elements to create a more effective textbook. In 2009 it was discovered that what had passed for his portrait for almost 200 years, has actually represented the politician Louis Legendre.

Exercises

In Exercises 13.5.1–13.5.6 evaluate the integrals:

13.5.1. $\displaystyle\int_0^{+\infty} \frac{\sin^3 tx}{x}\, dx.$ **13.5.2.*** $\displaystyle\int_0^{+\infty} \left(\frac{\sin tx}{x}\right)^2 dx.$ **13.5.3.*** $\displaystyle\int_0^{+\infty} \left(\frac{\sin tx}{x}\right)^3 dx.$

13.5.4.* $\displaystyle\int_0^{+\infty} \frac{\sin^4 x}{x^2}\, dx.$ **13.5.5.** $\displaystyle\int_0^{+\infty} \frac{\sin^2 x}{1+x^2}\, dx.$ **13.5.6.*** $\displaystyle\int_0^{+\infty} \frac{\cos tx}{\left(1+x^2\right)^2}\, dx.$

In Exercises 13.5.7–13.5.15 use Beta and Gamma functions to evaluate the integrals:

13.5.7. $\displaystyle\int_0^a x^2 \sqrt{a^2 - x^2}\, dx,\ a > 0.$ **13.5.8.** $\displaystyle\int_0^{+\infty} \frac{\sqrt[4]{x}}{(1+x)^2}\, dx.$ **13.5.9.** $\displaystyle\int_0^{+\infty} \frac{dx}{1+x^3}.$

13.5.10. $\displaystyle\int_0^1 \frac{1}{\sqrt[n]{1-x^n}}\, dx,\ n > 1.$ **13.5.11.** $\displaystyle\lim_{n\to\infty} \int_0^{+\infty} e^{-x^n}\, dx.$

13.5.12. $\displaystyle\int_0^{\pi/2} \tan^\alpha x\, dx,\ |\alpha| < 1.$ **13.5.13.** $\displaystyle\int_0^{+\infty} \frac{x^{p-1}\ln x}{1+x}\, dx,\ 0 < p < 1.$

13.5.14. $\displaystyle\int_0^1 \ln \Gamma(x)\, dx.$ **13.5.15.*** $\displaystyle\int_0^{+\infty} \frac{x^{p-1} - x^{q-1}}{(1+x)\ln x}\, dx,\ 0 < p, q < 1.$

13.5.16. The purpose of this problem is to illustrate the use of the Dirichlet's "discontinuous factor" (13.20), albeit in a slightly more general form.

(a) Let $k > 0$. Prove the identity

$$\frac{2}{\pi} \int_0^\infty \frac{\sin tx \cos kx}{x} \, dx = \begin{cases} 1, & \text{if } t > k \\ \dfrac{1}{2}, & \text{if } t = k \\ 0, & \text{if } t < k. \end{cases} \tag{13.29}$$

(b) Conclude that

$$\int_0^\infty e^{-t} \, dt \int_0^\infty \frac{\sin tx \cos kx}{x} \, dx = \frac{\pi}{2} e^{-k}. \tag{13.30}$$

(c) Justify the change of the order of integration in (13.30):

$$\int_0^\infty \frac{\cos kx}{x} \, dx \int_0^\infty e^{-t} \sin tx \, dt = \frac{\pi}{2} e^{-k}.$$

(d) Verify that

$$\int_0^\infty e^{-t} \sin tx \, dt = \frac{x}{1 + x^2}.$$

(e) Conclude that

$$\int_0^\infty \frac{\cos kx}{1 + x^2} \, dx = \frac{\pi}{2} e^{-k}.$$

13.5.17.* The purpose of this problem is to use the Beta function to derive the Wallis Product

$$\prod_{n=1}^\infty \left(\frac{2n}{2n - 1} \cdot \frac{2n}{2n + 1} \right) = \frac{2}{1} \cdot \frac{2}{3} \cdot \frac{4}{3} \cdot \frac{4}{5} \cdot \frac{6}{5} \cdot \frac{6}{7} \cdot \frac{8}{7} \cdot \frac{8}{9} \cdots = \frac{\pi}{2}. \tag{13.31}$$

(a) Prove that, if $n \in \mathbb{N}$,
$$B\left(\frac{1}{2}, n + \frac{1}{2}\right) = \pi \frac{(2n - 1)!!}{(2n)!!}.$$

(b) Prove that, if $n \in \mathbb{N}$,
$$B\left(\frac{1}{2}, n\right) = 2 \frac{(2n - 2)!!}{(2n - 1)!!}.$$

(c) Prove that, for a fixed a, $B(a, b)$ is a decreasing function of b.
(d) Use the Squeeze Theorem to prove that the sequence

$$\frac{B(\frac{1}{2}, n + \frac{1}{2})}{B(\frac{1}{2}, n)}$$

converges to 1.
(e) Conclude the validity of (13.31).

14

Integration in $\mathbb{R}^n$

A substantial amount of Integration theory of functions of one variable carries over to the case of functions defined on a subset of $\mathbb{R}^n$. Nevertheless, there are some important differences, and the proper theory of the multiple integrals was fully developed only at the end of the nineteenth century, in the work of Thomae, Peano, and Jordan. Much of the effort focused on the attempts to "measure" sets in $\mathbb{R}^2$. This laid the foundation for the work of Borel and Lebesgue in the twentieth century, which resulted in the modern theory of measure.

14.1 Double Integrals over Rectangles

When defining the double integral, we follow the pattern used for the definite integral of a single variable function (Section 6.2). Suppose that a bounded function f is defined on the rectangle $R = [a, b] \times [c, d]$, and we want to calculate the volume "under the graph" of f (as in Figure 14.1). First we create a **partition** P of R, by partitioning $[a, b]$ and $[c, d]$. More

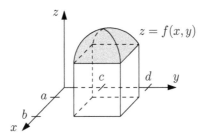

Figure 14.1: The volume under the graph of f.

precisely, we select positive integers m and n and **partition points**

$$x_0 = a < x_1 < x_2 < \cdots < x_{m-1} < x_m = b,$$
$$y_0 = c < y_1 < y_2 < \cdots < y_{n-1} < y_n = d. \tag{14.1}$$

That way, we have mn rectangles $R_{ij} = [x_{i-1}, x_i] \times [y_{j-1}, y_j]$, $1 \le i \le m$, $1 \le j \le n$. We write $P = \{x_1, x_2, \ldots, x_m; y_1, y_2, \ldots, y_n\}$, $\Delta x_i = x_{i+1} - x_i$, and $\Delta y_j = y_{j+1} - y_j$. Next we define $M_{ij} = \sup\{f(x, y) : (x, y) \in R_{ij}\}$ and $m_{ij} = \inf\{f(x, y) : (x, y) \in R_{ij}\}$. Just like in the case of one variable

$$L(f, P) = \sum_{i=1}^{m} \sum_{j=1}^{n} m_{ij} \Delta x_i \Delta y_j, \quad \text{and} \quad U(f, P) = \sum_{i=1}^{m} \sum_{j=1}^{n} M_{ij} \Delta x_i \Delta y_j,$$

are the **lower** and the **upper** Darboux sums, and $L = \sup L(f, P)$, $U = \inf U(f, P)$, taken over all possible partitions of R, are the **upper (Darboux) integral** and the **lower**

439

(Darboux) integral of f on R. Then f is **Darboux integrable** on the rectangle R if $L = U$. In that case, their common value is called the **double integral** of f over R, and it is denoted by

$$\iint\limits_{R} f(x,y)\, dA.$$

When establishing the integrability of f, it is often convenient to use the following analogue of Proposition 6.2.7.

Theorem 14.1.1. *A function f is integrable on $R = [a,b] \times [c,d]$ if and only if for every $\varepsilon > 0$ there exists a partition P of R such that $U(f,P) - L(f,P) < \varepsilon$.*

Another result that can be obtained by copying the single variable argument is the following theorem.

Theorem 14.1.2. *Every continuous function on $[a,b] \times [c,d]$ is integrable.*

The situation becomes more subtle, if f is not necessarily continuous.

Example 14.1.3. A discontinuous integrable function.

We will prove that the function $f(x,y) = \begin{cases} -1, & \text{if } 0 \le x < y \le 1 \\ 1, & \text{if } 0 \le y \le x \le 1 \end{cases}$ is integrable.

Solution. It is not hard to see that f is discontinuous precisely on the diagonal of the unit square $I = [0,1] \times [0,1]$, i.e. on the set $\{(x,y) \in I : y = x\}$. We will show that f is integrable

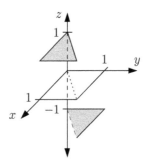

Figure 14.2: f is not continuous on the diagonal.

on I.

Let $\varepsilon > 0$, let $n = \lfloor 6/\varepsilon \rfloor + 1$, let P be a partition of I into n^2 squares of the same size. Then, both $L(f,P)$ and $U(f,P)$ consist of n^2 terms, and these terms fall into 2 groups: if $|i-j| \ge 2$, the rectangle R_{ij} has no common points with the diagonal; if $|i-j| < 2$, it does. In the former group, $m_{ij} = M_{ij}$ so these terms in $L(f,P)$ are exactly the same as those in $U(f,P)$ and they make no contribution in $U(f,P) - L(f,P)$. In the rectangles of the latter group, $m_{ij} = -1$, $M_{ij} = 1$, and there are

$$3n - 2 = n + (n-1) + (n-1)$$

of these, so

$$U(f,P) - L(f,P) = \sum_{i=1}^{n} \sum_{|i-j| \le 1} 1 \Delta x_i \Delta y_i - \sum_{i=1}^{n} \sum_{|i-j| \le 1} (-1) \Delta x_i \Delta y_i$$

$$= \sum_{i=1}^{n} \sum_{|i-j| \leq 1} 2 \frac{1}{n} \frac{1}{n}$$

$$= \frac{2}{n^2} (3n - 2) < \frac{6}{n} < \varepsilon.$$

By Theorem 14.1.1, f is integrable. ♦

Using a similar argument, one can prove the following result.

Theorem 14.1.4. *Let f be a bounded function on $R = [a, b] \times [c, d]$ and suppose that it is continuous on $(a, b) \times (c, d)$. Then f is integrable on R.*

Multiple integrals were used by Newton in his *Principia*. The first formal definition of the double integral was given by Thomae in [103] in 1875.

Example 14.1.3 shows that a function can be discontinuous, yet integrable. The crucial property of the set of discontinuity (the diagonal $y = x$) was that we were able to cover it with rectangles whose total area was less than $6/n$, and thus could be made arbitrarily small. Such sets deserve to be recognized.

Definition 14.1.5. A set $D \subset \mathbb{R}^2$ has (Jordan) **content** 0 if for every $\varepsilon > 0$ there exists a finite collection of rectangles R_k, $1 \leq k \leq n$, whose union covers D and the sum of their areas is less than ε.

CAMILLE JORDAN (1838–1922) was a French mathematician, born and raised in Lyon, in a prominent family. He studied at Ecole Polytechnique where he graduated in 1861, with a doctoral dissertation which had algebra in the first part and analysis in the second part. His education prepared him for a career in engineering, and he spent 15 years on jobs around France, doing mathematical research in his spare time. In 1876 he became a professor of analysis at Ecole Polytechnique. Surprisingly, he did more research as an engineer. Jordan's contributions to mathematics are numerous. He introduced the concepts of homotopy of paths and of groups. On his own he developed the theory of finite groups. In that direction, his best work between 1860 and 1870 is contained in his book *Traité des substitutions et des équations algebraique* (*Treatise on Substitutions and Algebraic Equations*). Among other results it contains the *Jordan form* of a matrix. (However, he is not the Jordan of the Gauss–Jordan Elimination.) His study of transformation groups led to the theory of Lie groups. His masterpiece is a 3-volume book *Cours d'Analyse*, the first edition coming from 1882 to 1887. In this text he introduced functions of bounded variations and he proved the *Jordan curve theorem* (a simple closed curve divides the plane into exactly 2 regions). He was the editor of the *Journal de Mathématiques Pure et Appliquées* for 35 years until his death. He retired from the university positions in 1912. In WWI he lost 3 sons. He was in high esteem: a member of the "Académie des Sciences" and an officer of the "Légion d'Honneur", he was also the Honorary President of the International Congress of Mathematicians at Strasbourg in 1920.

Example 14.1.6. A set of content 0.

Let $y = f(x)$ be a continuous function on $[a, b]$. Prove that its graph $D = \{(x, f(x)) : x \in [a, b]\}$ has content 0.

Proof. Let $\varepsilon > 0$. Since $[a, b]$ is a compact set, f is uniformly continuous, so there exists $\delta > 0$ such that

$$|x' - x''| < \delta \quad \Rightarrow \quad |f(x') - f(x'')| < \frac{\varepsilon}{b - a}.$$

Let $n = \lfloor (b - a)/\delta \rfloor + 1$, and let us partition $[a, b]$ into n intervals of equal length:

$$x_0 = a, \, x_1 = a + \frac{b - a}{n}, \, x_2 = a + 2\frac{b - a}{n}, \ldots, x_n = b.$$

Let $m_i = \inf\{f(x) : x_{i-1} \leq x \leq x_i\}$ and $M_i = \sup\{f(x) : x_{i-1} \leq x \leq x_i\}$, $1 \leq i \leq n$.

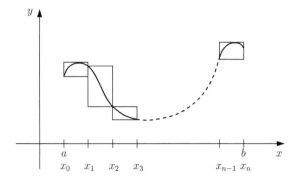

Figure 14.3: The graph of a continuous function has content 0.

It is easy to see that, for each i, $x_i - x_{i-1} = (b-a)/n < \delta$, so $M_i - m_i < \varepsilon/(b-a)$. Let $R_i = [x_{i-1}, x_i] \times [m_i, M_i]$. Then the union of these rectangles covers D, and their total area equals

$$\sum_{i=1}^{n} (x_i - x_{i-1})(M_i - m_i) < \sum_{i=1}^{n} \frac{b-a}{n} \frac{\varepsilon}{b-a} = \varepsilon. \qquad \blacklozenge$$

The property of having a "small" set of discontinuity guarantees the integrability.

Theorem 14.1.7. *Let f be a bounded function defined on $R = [a,b] \times [c,d]$, and let $D \subset R$ be the set of all the points at which f fails to be continuous. If D is of content 0, then f is integrable on R.*

Proof. Let $\varepsilon > 0$. We plan to use Theorem 14.1.1, so we need to select an appropriate partition P of the rectangle R. By assumption, f is bounded, so there exists $M > 0$ such that $|f(x,y)| \le M$, for all $(x,y) \in R$. The content of D is 0, so there exist rectangles $C_i = [p_i, q_i] \times [r_i, s_i]$, $1 \le i \le n$, whose total area does not exceed $\varepsilon/(4M)$ and $D \subset \cup_{i=1}^{n} [p_i, q_i] \times [r_i, s_i]$. It will be convenient to order the sets $\{p_1, p_2, \ldots, p_n, q_1, q_2, \ldots, q_n, a, b\}$

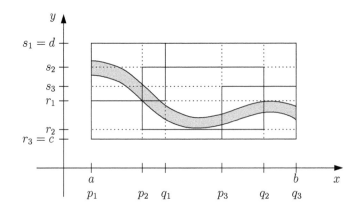

Figure 14.4: The set of discontinuities has content 0.

and $\{r_1, r_2, \ldots \ldots, r_n, s_1, s_2, \ldots, s_n, c, d\}$ as

$$a = x_0 \le x_1 \le \cdots \le x_{2n+1} = b, \quad \text{and} \quad c = y_0 \le y_1 \le \cdots \le y_{2n+1} = d.$$

Let J_1 be the set of pairs (i,j) such that the rectangle $R_{ij} = [x_{i-1}, x_i] \times [y_{j-1}, y_j]$ is completely contained in one of the rectangles C_k, $1 \le k \le n$. Also, let J_2 denote the set of pairs (i, j) such that R_{ij} has no common points with D except possibly at its boundary ∂R_{ij}. It is clear that, if $1 \le i, j \le 2n + 1$, the pair (i, j) must belong to either J_1 or J_2. If $(i, j) \in J_2$, the restriction f_{ij} of f to R_{ij} is continuous in the interior of R_{ij} hence, by Theorem 14.1.4, integrable on R_{ij}. Therefore, there exists a partition P_{ij} of R_{ij} such that

$$U(f_{ij}, P_{ij}) - L(f_{ij}, P_{ij}) < \frac{\varepsilon}{2(2n+1)^2}.$$

On the other hand, if $(i, j) \in J_1$, then R_{ij} is completely contained in some C_k, $1 \le k \le n$, and we will set $P_{ij} = \{x_{i-1}, x_i; y_{j-1}, y_j\}$. For such (i, j), denoting as usual by M_{ij} and m_{ij} the supremum and the infimum of f_{ij} on R_{ij}, and by $A(R_{ij})$ the area of R_{ij},

$$U(f_{ij}, P_{ij}) - L(f_{ij}, P_{ij}) = (M_{ij} - m_{ij})\, \Delta x_i \Delta y_j \le 2M \cdot A(R_{ij}).$$

Let P be a partition of R that contains all partition points of P_{ij}, $1 \le i, j \le 2n+1$. Then

$$
\begin{aligned}
U(f, P) - L(f, P) &= \sum_{(i,j) \in J_1} [U(f_{ij}, P_{ij}) - L(f_{ij}, P_{ij})] + \sum_{(i,j) \in J_2} [U(f_{ij}, P_{ij}) - L(f_{ij}, P_{ij})] \\
&< \sum_{(i,j) \in J_1} 2M \cdot A(R_{ij}) + \sum_{(i,j) \in J_2} \frac{\varepsilon}{2(2n+1)^2} \\
&\le 2M \sum_{i=1}^{n} A(C_i) + \sum_{i=1}^{n} \sum_{j=1}^{n} \frac{\varepsilon}{2(2n+1)^2} \\
&\le 2M \frac{\varepsilon}{4M} + \frac{\varepsilon}{2(2n+1)^2}(2n+1)(2n+1) = \varepsilon.
\end{aligned}
$$

$\square$

Definition 14.1.5 is associated with Jordan because Jordan content (or Jordan measure) appears in his work [69]. It should be mentioned that at about the same time Peano published [84] where the same ideas are used. Sadly, the significance of this concept is mostly historical. Some 20 years later, Lebesgue presented a new (improved) theory, in which the definition of a set of measure zero allows for an *infinite* number of rectangles. A modified Theorem 14.1.7 remains true, with almost the same proof (Exercise 14.1.29).

Just like in the case of functions of one variable, there is an equivalent definition of the double integral.

Definition 14.1.8. Let P be a partition of $[a, b] \times [c, d]$ given by (14.1), and let ξ be the collection of the **intermediate points** $(\xi_i, \eta_j) \in R_{ij} = [x_{i-1}, x_i] \times [y_{j-1}, y_j]$, $1 \le i \le m$, $1 \le j \le n$. A **Riemann Sum** for f on $[a, b] \times [c, d]$ is

$$\sum_{i=1}^{m} \sum_{j=1}^{n} f(\xi_i, \eta_j) \Delta x_i \Delta y_j,$$

and it is denoted by $S(f, P, \xi)$. The **norm of a partition** is defined to be the length of the largest diagonal of any rectangle R_{ij}. A function f is **Riemann integrable** on $R = [a, b] \times [c, d]$ if there exists a real number I with the property that, for any $\varepsilon > 0$ there exists a positive number δ, such that if P is any partition of R and $\|P\| < \delta$, then $|S(f, P, \xi) - I| < \varepsilon$.

The following is the 2-dimensional analogue of Theorem 6.4.5.

Theorem 14.1.9. *A function f is Riemann integrable on $R = [a, b] \times [c, d]$ if and only if it is Darboux integrable on R.*

Once again, the proof follows the same path as in the one-variable case. The same can be said for the following property of the integrals.

Theorem 14.1.10. *Let* f, g *be two functions that are integrable on* $R = [a, b] \times [c, d]$, *and let* $\alpha \in \mathbb{R}$. *Then the functions* αf *and* $f + g$ *are integrable on* R *as well and:*

(a) $\iint_R \alpha f(x, y) \, dA = \alpha \iint_R f(x, y) \, dA;$

(b) $\iint_R (f(x, y) + g(x, y)) \, dA = \iint_R f(x, y) \, dA + \iint_R g(x, y) \, dA.$

As we have seen, sets of content 0 do not affect the integrability of a function. They do not affect the value of the double integral either.

Theorem 14.1.11. *Let* f, g *be two functions that are integrable on* $R = [a, b] \times [c, d]$, *and let* $D \subset [a, b] \times [c, d]$ *be the set of all the points* (x, y) *at which* $f(x, y) \neq g(x, y)$. *If* D *has content 0, then* $\iint_R f(x, y) \, dA = \iint_R g(x, y) \, dA.$

Proof. Let $h = g - f$. Then h is integrable on R, $h = 0$ except on D, and the assertion is that $\iint_R h(x, y) \, dA = 0$. Let $\varepsilon > 0$. We will show that

$$-\varepsilon < L \leq U < \varepsilon. \tag{14.2}$$

Since h is integrable, it is bounded, so there exists $M > 0$ such that $|h(x, y)| \leq M$, for all $(x, y) \in R$. Using the same construction as in the proof of Theorem 14.1.7, we obtain a partition $P = \{x_0, x_1, \ldots, x_m; y_0, y_1, \ldots, y_n\}$ of R into rectangles $R_{ij} = [x_{i-1}, x_i] \times [y_{j-1}, y_j]$. This ensures that if J_1 (resp., J_2) denotes the set of all pairs (i, j) such that the rectangle R_{ij} has nonempty (resp., empty) intersection with D, then $\sum_{(i,j) \in J_1} A(R_{ij}) < \varepsilon/(2M)$. Let $h_{ij} = h|R_{ij}$ and $M_{ij} = \sup\{h_{ij}(x, y) : (x, y) \in R_{ij}\}$. Then

$$U(h, P) = \sum_{(i,j) \in J_1} M_{ij} \Delta x_i \Delta y_j + \sum_{(i,j) \in J_2} M_{ij} \Delta x_i \Delta y_j$$

$$\leq M \sum_{(i,j) \in J_1} A(R_{ij}) + \sum_{(i,j) \in J_2} 0 \cdot \Delta x_i \Delta y_j$$

$$\leq M \frac{\varepsilon}{2M} = \frac{\varepsilon}{2} < \varepsilon.$$

Similarly, $L(h, P) > -\varepsilon$. This implies (14.2) and, since ε is arbitrary, $L = U = 0$. $\square$

The next property is also very important and represents a generalization of Theorem 6.5.2.

Theorem 14.1.12. *Let* f *be a function that is integrable on* $R = [a, b] \times [c, d]$, *and let* P *be a partition of* R *into rectangles* R_{ij}. *Then* f *is integrable on* R *if and only if it is integrable on each* R_{ij}, *and in that case*

$$\iint_R f(x, y) \, dA = \sum_{i=1}^{m} \sum_{j=1}^{n} \iint_{R_{ij}} f(x, y) \, dA.$$

Exercises

14.1.1. Prove that the boundary of a rectangle $[a, b] \times [c, d]$ has content 0.

14.1.2. Prove that the unit circle $x^2 + y^2 = 1$ in $\mathbb{R}^2$ has content 0.

14.1.3. Prove that a bounded set in $\mathbb{R}^2$ with a finite number of accumulation points has content 0.

14.1.4. If D is a set with content 0, and if $A \subset D$, prove that the set A has content 0.

14.1.5. Let D be a set with content 0. Prove that D is a bounded set.

14.1.6. Let D be a set with content 0. Prove that the boundary of D has content 0.

14.1.7. Let f be a function defined on $I = [0, 1] \times [0, 1]$ by

$$f(x, y) = \begin{cases} \dfrac{1}{q}, & \text{if } y \in \mathbb{Q} \text{ and } x = \dfrac{p}{q}, \, p \in \mathbb{Z}, \, q \in \mathbb{N}, \text{ and } p, q \text{ are mutually prime} \\ 0, & \text{if either } x \text{ or } y \text{ is irrational, or } x = 0. \end{cases}$$

Prove that f is integrable on $R = [0, 1] \times [0, 1]$ and that $\iint_R f(x, y) \, dA = 0$.

14.1.8. Prove Theorem 14.1.1.

14.1.9. Prove Theorem 14.1.2.

14.1.10. Prove Theorem 14.1.4.

14.1.11.* Prove Theorem 14.1.9.

14.1.12. Prove Theorem 14.1.10.

14.1.13.* Prove Theorem 14.1.12.

14.1.14. Let f, g be two functions that are integrable on $R = [a, b] \times [c, d]$, and suppose that $f(x, y) \leq g(x, y)$, for all $(x, y) \in R$. Prove that $\iint_R f(x, y) \, dA \leq \iint_R g(x, y) \, dA$.

14.1.15. Let f be a bounded function on $R = [a, b] \times [c, d]$ and suppose that $f(x, y) = 0$ for all (x, y) in the interior of R. Prove that f is integrable on R and that $\iint_R f(x, y) \, dA = 0$.

14.1.16. Let f be an integrable function on $R = [a, b] \times [c, d]$ and suppose that $f(x, y) \geq 0$ if at least one of x, y is rational. Prove that $\iint_R f(x, y) \, dA \geq 0$.

14.1.17. Let f be an integrable function on $R = [a, b] \times [c, d]$, let $M > 0$ and suppose that $|f(x, y)| \leq M$, for all $(x, y) \in R$. Prove that $\left| \iint_R f(x, y) \, dA \right| \leq M(b - a)(d - c)$.

14.1.18. Let f be a bounded function on $R = [a, b] \times [c, d]$ and let $I \in \mathbb{R}$. Prove that f is integrable on R with $\iint_R f(x, y) \, dA = I$ if and only if $\lim_n L(f, P_n) = \lim U(f, P_n) = I$ for any sequence of partitions $\{P_n\}$ of R such that $\lim_n \|P_n\| = 0$.

In Exercises 14.1.19–14.1.21 use Exercise 14.1.18 to evaluate $\iint_{[0,1] \times [0,1]} f(x, y) \, dA$:

14.1.19. $f(x, y) = 2x - y$.   **14.1.20.** $f(x, y) = xy$.   **14.1.21.** $f(x, y) = x^2 y$.

14.1.22. Prove that a set $D \in \mathbb{R}^2$ has content 0 if and only if for each $\varepsilon > 0$ there exist $n \in \mathbb{N}$ and squares C_i, $1 \leq i \leq n$, such that $D \subset \cup C_i$ and $\sum A(C_i) < \varepsilon$.

14.1.23.* Let $D = \{(\frac{i}{p}, \frac{j}{p}) : p \text{ is a prime number}, i, j = 1, 2, \ldots, p - 1\}$. Determine whether the set D has content 0.

14.1.24.* Prove that a bounded function is integrable on R if and only if for each $\varepsilon > 0$ there exists $\delta > 0$ so that, if P, Q are two partitions of R satisfying $\|P\|, \|Q\| < \delta$, then $|S(f, P) - S(f, Q)| < \varepsilon$.

14.1.25. Let f and g be two integrable functions on a rectangle R. Prove that fg is integrable on R.

14.1.26. Without using Theorem 14.1.9 prove that every Riemann integrable function is bounded.

14.1.27.* Let f be a bounded function on $R = [a, b] \times [c, d]$. Suppose that for each fixed $y \in [c, d]$, $f(x, y)$ is an increasing function of x, and that for each fixed $x \in [a, b]$, $f(x, y)$ is an increasing function of y. Prove that f is integrable on R.

A set D has Lebesgue measure 0 if for every $\varepsilon > 0$ there exists an infinite collection of rectangles R_k, $k \in \mathbb{N}$, whose union covers D and the sum of their areas is less than ε.

14.1.28. Prove that the set of rational numbers has Lebesgue measure 0.

14.1.29.* Prove the Lebesgue theorem: Let f be a bounded function defined on $R = [a, b] \times [c, d]$, and let $D \subset R$ be the set of all the points at which f fails to be continuous. Then f is integrable on R if and only if D has Lebesgue measure 0.

14.2 Double Integrals over Jordan Sets

In the case of a single variable, most often an interval (finite or infinite) is a natural domain for a function. Unfortunately, there is no such a set for functions of two variables. In particular, rectangles cannot shoulder this responsibility. For example, $f(x, y) = \sqrt{1 - x^2 - y^2}$ has the domain the closed unit disk. Thus, it is essential to define the double integral over a region that is not necessarily a rectangle.

Definition 14.2.1. Let D be a bounded set in $\mathbb{R}^2$, and let R be a rectangle that contains D. If f is a function defined on D, we define its extension $\hat{f}$ to R by

$$\hat{f}(x, y) = \begin{cases} f(x, y), & \text{if } (x, y) \in D \\ 0, & \text{if } (x, y) \notin D. \end{cases} \tag{14.3}$$

We say that f is integrable on D if $\hat{f}$ is integrable on R, and we define

$$\iint\limits_{D} f(x, y) \, dA = \iint\limits_{R} \hat{f}(x, y) \, dA.$$

Whenever a definition involves the use of an object (like the rectangle R here) that is not uniquely determined, it raises a flag: would it matter if two different rectangles R_1 and R_2 were used?

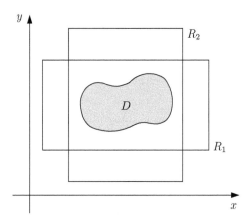

Figure 14.5: Extensions of f to different rectangles.

Proposition 14.2.2. *Let D be a bounded set in $\mathbb{R}^2$, and let R_1 and R_2 be rectangles that contain D. If f is a function defined on D, and if f_1 and f_2 are extensions of f to these rectangles as in (14.3), then*

$$\iint_{R_1} f_1(x,y)\, dA = \iint_{R_2} f_2(x,y)\, dA. \qquad (14.4)$$

Proof. Let $R_3 = R_1 \cap R_2$. Then R_3 is also a rectangle that contains D. Let f_3 be the extension of f to R_3 as in (14.3). Clearly, it suffices to show that $\iint_{R_1} f_1(x,y)\, dA = \iint_{R_3} f_3(x,y)\, dA$. In other words, it suffices to establish (14.4) in the case when $R_1 \subset R_2$. Let P be a partition of R_2 that includes R_1 as one of the rectangles. Theorem 14.1.12 makes it clear that we should prove that $\iint_R f(x,y)\, dA = 0$ for any rectangle $R \neq R_1$ in P. Since the restriction of f_2 on the interior of R is 0, and the boundary of R has content 0, Theorem 14.1.11 implies that $\iint_R f(x,y)\, dA = 0$, and the proof is complete. $\qquad\square$

Example 14.2.3. A function integrable on the unit disk.

Let $f(x,y) = 1$ for (x,y) in the unit disk D. Prove that f is integrable on D.

Solution. By definition, we need a rectangle R that contains D, say $R = [-2,2] \times [-2,2]$. We define the extension $\hat{f}$ of f to R as in (14.3):

$$\hat{f}(x,y) = \begin{cases} 1, & \text{if } (x,y) \in D \\ 0, & \text{if } (x,y) \notin D. \end{cases}$$

Since we are interested in the integrability of f, we must consider the integrability of $\hat{f}$. To that end, we notice that $\hat{f}$ is discontinuous at (x,y) if and only if (x,y) lies on the unit circle. By Exercise 14.1.2, this set has content zero, so Theorem 14.1.7 shows that f is integrable. $\qquad\blacklozenge$

Example 14.2.3 shows that it is possible that $\hat{f}$ has a discontinuity at every point of the boundary of D. (Of course, this can also happen when D is not the unit disk.) Thus, it is convenient to deal only with the sets that have the boundary of content 0. Recall that a point $\mathbf{a} \in \mathbb{R}^2$ is a boundary point of a set D if every open ball $B_r(\mathbf{a})$ has a nonempty intersection with D.

Definition 14.2.4. A bounded set D in $\mathbb{R}^2$ is a **Jordan domain** (or a **Jordan set**) if its boundary has Jordan content 0.

Exercises 14.1.1 and 14.1.2 show that a rectangle and a disk are Jordan sets. The following example shows that not every set is a Jordan domain.

Example 14.2.5. A set that is not a Jordan set.

Prove that the set $D = \{(x,y) \in [0,1] \times [0,1] : x, y \in \mathbb{Q}\}$ is not a Jordan set.

Solution. It is not hard to see that the boundary of D coincides with the square $[0,1] \times [0,1]$. Thus, it does not have content 0, hence D is not a Jordan set. $\qquad\blacklozenge$

Our next goal is to establish the rules of integration for functions defined on Jordan domains. By Theorem 14.1.10, the double integral is linear, when the domain of integration is a rectangle. Now we can show that this is true for any Jordan domain.

Theorem 14.2.6. *Let f, g be two functions that are integrable on a Jordan domain D, and let $\alpha \in \mathbb{R}$. Then the functions αf and $f + g$ are integrable on D as well and:*

(a) $\iint_D \alpha f(x,y)\, dA = \alpha \iint_D f(x,y)\, dA$;

(b) $\iint_D (f(x,y) + g(x,y))\, dA = \iint_D f(x,y)\, dA + \iint_D g(x,y)\, dA.$

Proof. Let R be a rectangle that contains D, and let $\hat{f}, \hat{g}$ be the usual extensions from D to R. Then $\widehat{\alpha f} = \alpha \hat{f}$ is the extension of αf, and $\widehat{f + g} = \hat{f} + \hat{g}$ is the extension of $f + g$. Therefore,

$$\iint_D \alpha f(x,y)\, dA = \iint_R \alpha \hat{f}(x,y)\, dA = \alpha \iint_R \hat{f}(x,y)\, dA = \alpha \iint_D f(x,y)\, dA$$

which establishes (a). The assertion (b) is proved the same way. $\qquad\square$

The fact that the double integral is additive with respect to the domain was established in Theorem 14.1.12, when the domain is a rectangle. Now we will show that it is true for any Jordan domain.

Theorem 14.2.7. *Let D be a Jordan set, and suppose that D is a disjoint union of Jordan sets D_1 and D_2. A function f is integrable on D if and only if it is integrable on D_1 and D_2, and in that case*

$$\iint_D f(x,y)\, dA = \iint_{D_1} f(x,y)\, dA + \iint_{D_2} f(x,y)\, dA. \tag{14.5}$$

Proof. Suppose first that f is integrable on D_1 and D_2, and let R be a rectangle that contains D. Let us denote by f_1 and f_2 restrictions of f to D_1 and D_2, and by $\hat{f}_1$ and $\hat{f}_2$ the usual extensions of these two functions to R. Since $\hat{f}_1$ and $\hat{f}_2$ are integrable, so is their sum $\hat{f}_1 + \hat{f}_2$, and

$$\iint_R (\hat{f}_1 + \hat{f}_2)\, dA = \iint_R \hat{f}_1\, dA + \iint_R \hat{f}_2\, dA.$$

It is not hard to see that $\hat{f}_1 + \hat{f}_2 = \hat{f}$, so f is integrable on D and satisfies (14.5).

In order to prove the converse, we need to verify that f_1 and f_2 are integrable functions. Clearly, it suffices to show that f_1 has this property. Let $\varepsilon > 0$. Since f is a bounded function, there exists $M > 0$ such that $|f(x,y)| \le M$, for $(x,y) \in D$. By assumption, D_1 is a Jordan set, so there exists a finite collection of rectangles C_i whose interiors cover the boundary of D_1, with the total area less than $\varepsilon/(4M)$. Let R be a rectangle that contains D and all the C_i. We repeat once again the argument of Theorem 14.1.7 to obtain a partition P of R such that each C_i is the union of rectangles in P. If necessary, P can be refined so

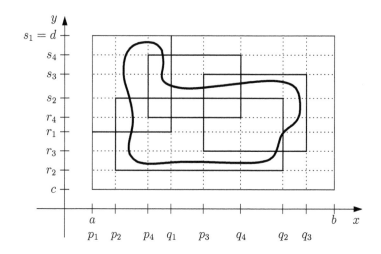

Figure 14.6: Proving that f_1 is integrable.

that, in addition, $U(\hat{f}, P) - L(\hat{f}, P) < \varepsilon/2$. Now all rectangles in P fall in three groups. Those that are entirely in D_1, those that are entirely outside of D_1, and those that contain a part of the boundary of D_1, hence are contained in some C_i. The upper and lower Darboux sums of $\hat{f}$ and $\hat{f}_1$ thus split into three sums. In the first, these functions are equal, so

$$U'(\hat{f}_1, P) - L'(\hat{f}_1, P) = U'(\hat{f}, P) - L'(\hat{f}, P). \tag{14.6}$$

In the second, $\hat{f}_1 = 0$, so

$$U''(\hat{f}_1, P) - L''(\hat{f}_1, P) = 0 \le U''(\hat{f}, P) - L''(\hat{f}, P). \tag{14.7}$$

Finally, in the third,

$$
\begin{aligned}
U'''(\hat{f}_1, P) - L'''(\hat{f}_1, P) &= \sum (M_i - m_i) A(R_i) \\
&\le 2M \frac{\varepsilon}{4M} = \frac{\varepsilon}{2} \\
&\le U'''(\hat{f}, P) - L'''(\hat{f}, P) + \frac{\varepsilon}{2}.
\end{aligned}
\tag{14.8}
$$

If we add (14.6)–(14.8), we obtain that

$$U(\hat{f}_1, P) - L(\hat{f}_1, P) \le U(\hat{f}, P) - L(\hat{f}, P) + \frac{\varepsilon}{2} < \frac{\varepsilon}{2} + \frac{\varepsilon}{2} = \varepsilon.$$

Thus, both f_1 and f_2 are integrable, and $\hat{f}_1 + \hat{f}_2 = \hat{f}$. It follows that

$$
\iint_D f(x,y)\, dA = \iint_R \hat{f}\, dA = \iint_R (\hat{f}_1 + \hat{f}_2)\, dA = \iint_R \hat{f}_1\, dA + \iint_R \hat{f}_2\, dA
$$
$$
= \iint_{D_1} f(x,y)\, dA + \iint_{D_2} f(x,y)\, dA. \qquad \square
$$

FERDINAND GEORG FROBENIUS (1849–1917) was a German mathematician, born and raised in Charlottenburg, a suburb of Berlin. After a semester at the University of Göttingen he transferred to the University of Berlin, where he received a doctorate in 1870. His advisor Weierstrass considered him one of the most gifted students. Frobenius was a high school teacher for 4 years, then a professor at the University of Berlin for one year, and then at ETF Zürich for 17 years. When Kronecker died, he returned to be the main mathematician at Berlin for 25 years. He tried to preserve the traditional style of German universities (in which there was no place for applied mathematics), and high academic standards, creating a sometimes unpleasant atmosphere. All this led to the decline of Berlin and Göttingen took over as the leading German university. Frobenius is best known for his work in the area of differential equations and group theory.

Exercises

14.2.1. Prove that the union of two Jordan sets is a Jordan set.

14.2.2. Prove that the intersection of two Jordan sets is a Jordan set.

14.2.3. Prove that the complement of a Jordan set is a Jordan set.

14.2.4. Let A and B be sets such that their closures $\overline{A}, \overline{B}$ are Jordan sets. Prove that $\overline{A \cap B}$ is a Jordan set.

14.2.5. Prove that the set of points with both rational coordinates in $[0,1] \times [0,1]$ is not a Jordan set.

14.2.6.* Prove or disprove: Every open set in $\mathbb{R}^2$ is a Jordan set.

14.2.7.* Prove or disprove: Every closed set in $\mathbb{R}^2$ is a Jordan set.

14.2.8. Let f be a bounded function on a Jordan set D, let $A \subset D$ be the set of all the points at which f fails to be continuous, and suppose that A is of content 0. Prove that f is integrable on D.

14.2.9.* Prove that a set $D \subset \mathbb{R}^2$ is of Jordan content 0 if and only if the characteristic function χ_D is integrable and $\iint \chi_D(x,y)\, dA = 0$.

14.2.10.* Prove that a set $D \subset \mathbb{R}^2$ is a Jordan set if and only if the characteristic function χ_D is integrable.

14.2.11. Suppose that f, g are two integrable functions on a Jordan set D, such that $f(x,y) \le g(x,y)$, for all $(x,y) \in D$. Prove that $\iint_D f(x,y)\, dA \le \iint_D g(x,y)\, dA$.

14.2.12.* Let f be a bounded function on a Jordan set D, and let $D \subset R = [a,b] \times [c,d]$. For each set $A \subset \mathbb{R}^2$ let $\omega_A(f) = \sup\{f(x,y) : (x,y) \in A\} - \inf\{f(x,y) : (x,y) \in A\}$. Prove that f is integrable on D if and only if for every $\varepsilon > 0$ there exists a partition P of R such that
$$\sum \text{Area}\{R_{ij} : \omega_{R_{ij}}(f) > \varepsilon\} < \varepsilon.$$

14.2.13. Let f be a bounded function on a Jordan set D and define the functions $f^+(x,y) = \max\{f(x,y), 0\}$ and $f^-(x,y) = \max\{-f(x,y), 0\}$. Prove that f is integrable on D if and only if both f^+ and f^- are integrable on D.

14.2.14. Let f and g be integrable functions on a Jordan set D. Prove that $\sup(f,g)$ and $\inf(f,g)$ are integrable on D.

14.2.15. Prove the Mean Value Theorem for double integrals: If f is continuous on a compact and connected Jordan set D, then there exists a point $(x_0, y_0) \in D$ such that $\iint_D f(x,y)\, dA = f(x_0, y_0) \iint_D dA$.

14.2.16. Let f and g be two integrable functions on a Jordan set D. Prove that fg is integrable on D.

14.2.17. Let f be an integrable function on a Jordan set D. Prove that $|f|$ is integrable on D and that $\left| \iint_D f(x,y)\, dA \right| \le \iint_D |f(x,y)|\, dA$.

14.2.18. Let f be an integrable function on a Jordan set D such that $f(x,y) \ge C > 0$ for $(x,y) \in D$. Prove that the function $1/f$ is integrable on D.

14.2.19. Let $\{f_n\}$ be a sequence of functions defined and integrable on a set $D \subset \mathbb{R}^2$, and suppose that $\{f_n\}$ converges uniformly on D to a function f. Prove that f is integrable on D and that $\iint_D f_n(x,y)\, dA \to \iint_D f(x,y)\, dA$, as $n \to \infty$.

In Exercises 14.2.20–14.2.22 determine whether the integral is positive or negative:

14.2.20. $\iint_D \ln(x^2 + y^2)\, dA(x,y)$, $D = \{(x,y) \in \mathbb{R}^2 : |x| + |y| \le 1\}$.

14.2.21. $\iint_D \arcsin(x+y)\, dA(x,y)$, $D = \{(x,y) \in \mathbb{R}^2 : 0 \le x \le 1, -1 \le y \le 1-x\}$.

14.2.22.* $\iint_D \sqrt[3]{1 - x^2 - y^2}\, dA(x,y)$, $D = \{(x,y) \in \mathbb{R}^2 : x^2 + y^2 \le 4\}$.

14.3 Double Integrals as Iterated Integrals

The definition of a double integral is not of much help when we want to calculate one. This is similar to the situation with a definite integral, except that we had the Fundamental

Theorem of Calculus there. Here, it is the ability to replace the double integral with an iterated integral.

Example 14.3.1. Evaluation of a double integral: a review.

Evaluate $\iint\limits_{R}(x+2y)\,dA$ if $R=[-1,2]\times[0,1]$.

Solution.

$$\iint\limits_{R}(x+2y)\,dA = \int\limits_{-1}^{2}\int\limits_{0}^{1}(x+2y)\,dydx$$

$$= \int\limits_{-1}^{2}(xy+y^2)\Big|_{y=0}^{y=1}\,dx$$

$$= \int\limits_{-1}^{2}(x+1)\,dx$$

$$= \left(\frac{x^2}{2}+x\right)\Big|_{-1}^{2} = \frac{9}{2}. \qquad \blacklozenge$$

Our first task is to justify the technique used in this example.

Theorem 14.3.2. *Let $R=[a,b]\times[c,d]$, and let f be a function integrable on R. For each $x\in[a,b]$, let F_x be a function on $[c,d]$ defined by $F_x(y)=f(x,y)$. Also, suppose that F_x is integrable and define $A(x)=\int_c^d F_x(y)\,dy$. Then A is integrable on $[a,b]$ and*

$$\iint\limits_{R}f(x,y)\,dA = \int\limits_a^b A(x)\,dx = \int\limits_a^b dx\int\limits_c^d f(x,y)\,dy.$$

Proof. We will demonstrate that, for any partition P of R, and the induced partition P_1 of $[a,b]$,
$$L(f,P) \le L(A,P_1) \le U(A,P_1) \le U(f,P). \qquad (14.9)$$
Once this is proved, it will follow that $U(A,P_1)-L(A,P_1) \le U(f,P)-L(f,P)$, so the integrability of f implies the integrability of A. Moreover, the integral of A lies between the lower and the upper Darboux sums, so

$$L(f,P) \le \int\limits_a^b A(x)\,dx \le U(f,P).$$

Since this is true for every P, we obtain that $\int_a^b A(x)\,dx = \iint_R f(x,y)\,dA$.

Thus, we concentrate on (14.9). Let $P_1=\{x_0,x_1,\dots,x_m\}$ be a partition of $[a,b]$ and let $P_2=\{y_0,y_1,\dots,y_n\}$ be a partition of $[c,d]$. We fix i, $1\le i\le m$, and $x\in[x_{i-1},x_i]$. If m_{ij} and M_{ij} are the infimum and the supremum of $f(x,y)$ in $R_{ij}=[x_{i-1},x_i]\times[y_{j-1},y_j]$, then

$$m_{ij} \le f(x,y) \le M_{ij}$$

for all $y\in[y_{j-1},y_j]$. Integrating over $[y_{j-1},y_j]$ and then summing over $1\le j\le n$, yields

$$\sum_{j=1}^{n}m_{ij}\Delta y_j \le \sum_{j=1}^{n}\int\limits_{y_{j-1}}^{y_j}f(x,y)\,dy = \int\limits_c^d f(x,y)\,dy = A(x) \le \sum_{j=1}^{n}M_{ij}\Delta y_j.$$

Since this is true for any $x \in [x_{i-1}, x_i]$, it is true for $m_i = \inf\{A(x) : x \in [x_{i-1}, x_i]\}$ and $M_i = \sup\{A(x) : x \in [x_{i-1}, x_i]\}$. In other words,

$$\sum_{j=1}^{n} m_{ij} \Delta y_j \leq m_i \leq M_i \leq \sum_{j=1}^{n} M_{ij} \Delta y_j.$$

All that remains now is to multiply through by Δx_i and sum over $1 \leq i \leq m$, and we get (14.9). □

Theorem 14.3.2 has the analogue in which the roles of x and y are reversed. If we define $G_y(x) = f(x, y)$ and assume that G_y is integrable then we get

$$\iint\limits_{R} f(x, y) \, dA = \int_{c}^{d} dy \int_{a}^{b} f(x, y) \, dx.$$

When f is continuous on R we get both conclusions.

Theorem 14.3.3. *Let $R = [a, b] \times [c, d]$, and let f be a function continuous on R. Then*

$$\iint\limits_{R} f(x, y) \, dA = \int_{a}^{b} dx \int_{c}^{d} f(x, y) \, dy = \int_{c}^{d} dy \int_{a}^{b} f(x, y) \, dx.$$

It is important to make sure that, whichever of the theorems we use, *all* hypotheses are satisfied.

Example 14.3.4. The iterated integrals are not equal.

Let $f(x, y) = \dfrac{x^2 - y^2}{(x^2 + y^2)^2}$ and $R = [0, 1] \times [0, 1]$. Show that $\int_0^1 \int_0^1 f(x, y) \, dy \, dx \neq \int_0^1 \int_0^1 f(x, y) \, dx \, dy$.

Solution. The equality $x^2 - y^2 = (x^2 + y^2) - 2y^2$ yields

$$\int_{0}^{1} \frac{x^2 - y^2}{(x^2 + y^2)^2} \, dy = \int_{0}^{1} \frac{1}{x^2 + y^2} \, dy + \int_{0}^{1} \frac{-2y^2}{(x^2 + y^2)^2} \, dy.$$

If we use Integration by Parts in the second integral, with $u = y$ and $dv = -2y \, dy/(x^2 + y^2)^2$, we obtain

$$\int_{0}^{1} \frac{x^2 - y^2}{(x^2 + y^2)^2} \, dy = \int_{0}^{1} \frac{1}{x^2 + y^2} \, dy + \frac{y}{x^2 + y^2} \Big|_{y=0}^{y=1} - \int_{0}^{1} \frac{1}{x^2 + y^2} \, dy = \frac{1}{x^2 + 1}.$$

Therefore,

$$\int_{0}^{1} dx \int_{0}^{1} \frac{x^2 - y^2}{(x^2 + y^2)^2} \, dy = \int_{0}^{1} \frac{1}{x^2 + 1} \, dx = \arctan x \Big|_{0}^{1} = \frac{\pi}{4}.$$

However, a similar calculation, or the observation that $f(x, y) = -f(y, x)$, shows that

$$\int_{0}^{1} dy \int_{0}^{1} \frac{x^2 - y^2}{(x^2 + y^2)^2} \, dx = -\frac{\pi}{4}.$$

The problem is that f is not bounded on R. Indeed, $f(\frac{1}{n}, 0) = n^2 \to \infty$, so f is not bounded and, hence, not integrable on R. ♦

The next level of difficulty arises when the domain of integration is not a rectangle.

Example 14.3.5. Evaluation of a double integral: a review.

Evaluate $\iint\limits_{D} (4x^2 - y^2) \, dA$, if D is the triangle formed by the lines $y = 0$, $x = 1$, $y = x$.

Solution.

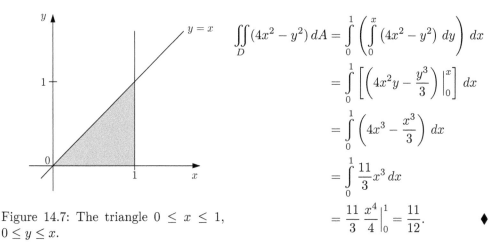

$$\iint_D (4x^2 - y^2)\, dA = \int_0^1 \left(\int_0^x (4x^2 - y^2)\, dy \right) dx$$

$$= \int_0^1 \left[\left(4x^2 y - \frac{y^3}{3} \right) \Big|_0^x \right] dx$$

$$= \int_0^1 \left(4x^3 - \frac{x^3}{3} \right) dx$$

$$= \int_0^1 \frac{11}{3} x^3\, dx$$

$$= \frac{11}{3} \frac{x^4}{4} \Big|_0^1 = \frac{11}{12}. \qquad \blacklozenge$$

Figure 14.7: The triangle $0 \le x \le 1$, $0 \le y \le x$.

The method that we used needs to be justified.

Theorem 14.3.6. *Let α, β be two functions defined and continuous on $[a, b]$ and suppose that, for all $x \in [a, b]$, $\alpha(x) \le \beta(x)$. Let $D = \{(x, y) : a \le x \le b,\ \alpha(x) \le y \le \beta(x)\}$, and let f be a function continuous on D. Then*

$$\iint_D f(x, y)\, dA = \int_a^b dx \int_{\alpha(x)}^{\beta(x)} f(x, y)\, dy.$$

Proof. We want to apply Theorem 14.3.2, so we extend f to a function $\hat{f}$ on a rectangle $R = [a, b] \times [c, d]$, in the usual manner. The function $\hat{f}$ is integrable because it is continuous,

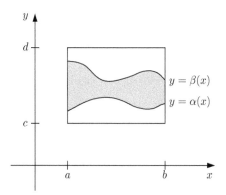

Figure 14.8: The region $a \le x \le b$, $\alpha(x) \le y \le \beta(x)$.

except possibly along the graphs of α and β, which have content 0. If we define $F_x(y) = \hat{f}(x, y)$, for x fixed, then this function is continuous with possible exception of two points: $y = \alpha(x)$, and $y = \beta(x)$. Thus, F_x is integrable, so Theorem 14.3.2 yields

$$\iint_D f(x, y)\, dA = \iint_R \hat{f}(x, y)\, dA = \int_a^b dx \int_c^d \hat{f}(x, y)\, dy. \qquad (14.10)$$

Finally, for any $x \in [a, b]$,

$$\int_c^d \hat{f}(x, y)\, dy = \int_c^{\alpha(x)} \hat{f}(x, y)\, dy + \int_{\alpha(x)}^{\beta(x)} \hat{f}(x, y)\, dy + \int_{\beta(x)}^d \hat{f}(x, y)\, dy$$

$$= 0 + \int_{\alpha(x)}^{\beta(x)} f(x, y)\, dy + 0 \qquad (14.11)$$

by the definition of $\hat{f}$. Combining (14.10) and (14.11) gives the desired result. □

Remark 14.3.7. We say that a region D in the xy-plane is *y-simple* if it can be described by inequalities $a \leq x \leq b$, $\alpha(x) \leq y \leq \beta(x)$, where α, β are continuous functions on $[a, b]$. It is *x-simple* if it can be described by inequalities $c \leq y \leq d$, $\gamma(y) \leq x \leq \delta(y)$, with γ, δ continuous functions on $[c, d]$.

> The fact that the integral of a continuous function over a rectangle $[a, b] \times [c, d]$ can be reduced to two successive integrations—first over $[c, d]$, then over $[a, b]$, was known to Cauchy. Much later, in the context of the Lebesgue Integral, the Italian mathematician Guido Fubini (1879–1943) proved the theorem under more general hypotheses. Since most of the theorems about the equality between a double *Riemann* integral and the iterated integrals (including Theorems 14.3.2 and 14.3.3) can be derived from the Fubini's Theorem, it is not uncommon to attach his name to any such result.

Exercises

In Exercises 14.3.1–14.3.6 evaluate the integral:

14.3.1. $\int_0^{\frac{\sqrt{\pi}}{2}} dx \int_{3x}^{\frac{3\sqrt{\pi}}{2}} \cos(y^2)\, dy.$

14.3.2. $\int_0^\pi dx \int_x^\pi \dfrac{\sin y}{y}\, dy.$

14.3.3. $\int_0^2 dy \int_{\frac{y}{2}}^1 e^{-x^2}\, dx.$

14.3.4. $\int_0^1 dy \int_0^{\arcsin y} e^{\cos x}\, dx.$

14.3.5. $\int_0^3 dx \int_0^{9-x^2} \dfrac{xe^{3y}}{9-y}\, dy.$

14.3.6. $\int_0^2 dy \int_{-\sqrt{4-y^2}}^{\sqrt{4-y^2}} y\, dx.$

In Exercises 14.3.7–14.3.12 evaluate $\iint_R f(x, y)\, dA$:

14.3.7. $f(x, y) = (x - 2y)^2$, R is the rectangle with sides parallel to the axes and opposite corners $(-1, 1)$ and $(3, 2)$.

14.3.8. $f(x, y) = x/(y + 1)^2$, R is the region bounded by $y = x^2$ and $y = 4x$.

14.3.9. $f(x, y) = ye^x$, R is the region bounded by $x = y^2$ and $x = 5y$.

14.3.10. $f(x, y) = x \cos y \sin y$, R is the region bounded by $x = 1$, $x = -1$, the x-axis and $y = \arctan x$.

14.3.11. $f(x, y) = \cos x - y$, R is the region bounded by $y = \sin x$, $y = 0$, $x = 0$ and $x = \pi$.

14.3.12. $f(x, y) = x$, R is the region in the first quadrant bounded by $y = 0$, $y = 2$, $x = 0$ and $x = 1 + y^2$.

In Exercises 14.3.13–14.3.19 change the order of integration in the iterated integral:

14.3.13. $\int_0^2 dx \int_x^{2x} f(x, y)\, dy.$

14.3.14. $\int_0^1 dx \int_{x^3}^{x^2} f(x, y)\, dy.$

14.3.15. $\int_{-1}^1 dx \int_{-\sqrt{1-x^2}}^{1-x^2} f(x, y)\, dy.$

14.3.16. $\int_1^e dx \int_0^{\ln x} f(x, y)\, dy.$

14.3.17.* $\int_0^{2\pi} dx \int_0^{\sin x} f(x, y)\, dy.$

14.3.18. $\int_1^2 dx \int_{2-x}^{\sqrt{2x-x^2}} f(x, y)\, dy.$

14.3.19. $\int_0^2 dx \int_{\sqrt{2x}}^{\sqrt{2x-x^2}} f(x,y)\, dy.$

In Exercises 14.3.20–14.3.21 transform the sum into one iterated integral:

14.3.20. $\int_0^1 dy \int_{y^2/9}^{y} f(x,y)\, dx + \int_1^2 dy \int_{y^2/9}^{1} f(x,y)\, dx.$

14.3.21. $\int_3^7 dy \int_{9/y}^{3} f(x,y)\, dx + \int_7^9 dy \int_{9/y}^{10-y} f(x,y)\, dx.$

14.3.22. Let $D = \{(\frac{i}{p}, \frac{j}{p}) : p \text{ is a prime number}, i,j = 1,2,\ldots,p-1\}$, and let f be the characteristic function of D: $f(x,y) = 1$ if $(x,y) \in D$, $f(x,y) = 0$ if $(x,y) \notin D$. Prove that both iterated integrals exist and satisfy

$$\int_0^1 dx \int_0^1 f(x,y)\, dy = \int_0^1 dy \int_0^1 f(x,y)\, dx$$

but $\iint_D f(x,y)\, dA(x,y)$ does not exist.

14.3.23. Let f be a function defined in Exercise 14.1.7. Prove that $\iint_I f(x,y)\, dA(x,y) = \int_0^1 dy \int_0^1 f(x,y)\, dx$ but $\int_0^1 f(x,y)\, dy$ does not exist if $x \in \mathbb{Q}$.

14.3.24. For a rational number x, let $x = p_x/q_x$ where p_x, q_x are mutually prime and $q_x > 0$. Let f be a function defined on $I^2 = [0,1] \times [0,1]$ by

$$f(x,y) = \begin{cases} \dfrac{1}{q_x} + \dfrac{1}{q_y}, & \text{if both } x,y \in \mathbb{Q} \\ 0, & \text{otherwise} \end{cases}$$

Prove that $\iint_I f(x,y)\, dA(x,y) = 0$ but the iterated integrals do not exist.

14.3.25.* (a) Let f be a function defined on $R = [a,b] \times [c,d]$. Prove that

$$\underline{\iint_R} f(x,y)\, dA \le \int_a^b dx \,\overline{\int_c^d} f(x,y)\, dy \le \overline{\int_a^b} dx \,\overline{\int_c^d} f(x,y)\, dy \le \overline{\iint_R} f(x,y)\, dA.$$

(b) Let f be a function integrable on $R = [a,b] \times [c,d]$. Prove that

$$\iint_R f(x,y)\, dA = \int_a^b dx \,\underline{\int_c^d} f(x,y)\, dy = \int_a^b dx \,\overline{\int_c^d} f(x,y)\, dy$$
$$= \int_c^d dy \,\underline{\int_a^b} f(x,y)\, dx = \int_c^d dy \,\overline{\int_a^b} f(x,y)\, dx.$$

14.3.26.* Let p be a positive integrable function on $[a,b]$, and let f,g be increasing functions on $[a,b]$. Prove that

$$\int_a^b p(x)f(x)\, dx \cdot \int_a^b p(x)g(x)\, dx \le \int_a^b p(x)\, dx \int_a^b p(x)f(x)g(x)\, dx.$$

14.4   Transformations of Jordan Sets in $\mathbb{R}^2$

In order to establish the formula for a change of variables in double integrals, we will have to acquire more information about Jordan sets. For example, if A is a Jordan set, what conditions on a mapping $\varphi : \mathbb{R}^2 \to \mathbb{R}^2$ guarantee that $\varphi(A)$ is also a Jordan set? If R is a rectangle, we can calculate its area. What about $\varphi(R)$? Is there a way to assign a number to it, and how?

We will start with the following lemma.

Lemma 14.4.1. *Let R be a rectangle, let $\varphi : R \to \mathbb{R}^2$ have continuous partial derivatives, and let $D \subset R$. Then there exists $M > 0$ such that:*

(a) if $C \subset R$ is a square with area α, then $\varphi(C)$ is contained in a square with area no more than $M^2\alpha$;

(b) if D is contained in a finite union of squares C_i, $1 \le i \le m$, with total area α, then $\varphi(D)$ is contained in a finite union of squares C_i', $1 \le i \le m$, with total area no more than $M^2\alpha$.

Proof. As usual, if $(u, v) \in R$, we will write $\varphi(u, v) = (\varphi_1(u, v), \varphi_2(u, v))$. Let us denote

$$\|\mathbf{D}\varphi(u, v)\| = \max_{i=1,2} \left\{ \left| \frac{\partial \varphi_i}{\partial u}(u, v) \right| + \left| \frac{\partial \varphi_i}{\partial v}(u, v) \right| \right\}, \tag{14.12}$$

and let $M = \sup\{\|\mathbf{D}\varphi(u, v)\| : (u, v) \in R\}$. (The existence of M is guaranteed by the continuity of $\|\mathbf{D}\varphi(u, v)\|$ and the compactness of R. See Exercise 14.4.12.) Let $C \subset R$ be a square with side of length $2s$, and center (p, q). For $i = 1, 2$, there exists $\theta_i \in [0, 1]$ and $(z_i, w_i) = (p + \theta_i(u - p), q + \theta_i(v - q))$, such that

$$\varphi_i(u, v) - \varphi_i(p, q) = \mathbf{D}\varphi_i(z_i, w_i)(u - p, v - q)$$
$$= \frac{\partial \varphi_i}{\partial u}(z_i, w_i)(u - p) + \frac{\partial \varphi_i}{\partial v}(z_i, w_i)(v - q).$$

Suppose that $(u, v) \in C$. Then $|u - p| \le s$ and $|v - q| \le s$, so

$$\sup_{(u,v)\in C} |\varphi_i(u, v) - \varphi_i(p, q)| \le s \sup_{(u,v)\in R} \left(\left| \frac{\partial \varphi_i}{\partial u}(u, v) \right| + \left| \frac{\partial \varphi_i}{\partial v}(u, v) \right| \right) \le sM,$$

for $i = 1, 2$. It follows that the set $\varphi(C)$ lies inside a square C' with center $(\varphi_1(p, q), \varphi_2(p, q))$ and side $2sM$, hence of area $(2s)^2 M^2 = \alpha M^2$. This establishes (a), and (b) follows directly from (a). $\square$

An important consequence of this lemma is that φ as above preserves the zero content.

Corollary 14.4.2. *Let R be a rectangle, let $\varphi : R \to \mathbb{R}^2$ have continuous partial derivatives, and let $D \subset R$. If D has content zero then $\varphi(D)$ has content zero.*

The following result is equally important but we need to work a little harder to prove it.

Theorem 14.4.3. *Let R be a rectangle, let $\varphi : R \to \mathbb{R}^2$ have continuous partial derivatives, and let $D \subset R$. If D is a Jordan set, and if $\mathbf{D}\varphi(x, y)$ is invertible on D, then $\varphi(D)$ is a Jordan set.*

Proof. We need to show that the boundary of $\varphi(D)$ has content zero. We will accomplish this goal by establishing the inclusion

$$\partial\varphi(D) \subset \varphi(\partial D). \tag{14.13}$$

Once this is proved, we will use the following argument. The boundary of D has content 0 (because D is a Jordan set) so Corollary 14.4.2 implies that $\varphi(\partial D)$ has content 0, whence the desired conclusion follows from Exercise 14.1.4.

Suppose that y belongs to the boundary of $\varphi(D)$. Then we can find a sequence $\{y_n\}$ in $\varphi(D)$ that converges to y. Therefore, there exists a sequence $\{x_n\} \subset D$ such that $\{\varphi(x_n)\}$ converges to y. By the Bolzano–Weierstrass Theorem, $\{x_n\}$ has a convergent subsequence $\{x_{n_k}\}$. Let $x = \lim x_{n_k}$. Since φ is continuous, $\varphi(x_{n_k}) \to \varphi(x)$, so $y = \varphi(x)$. The sequence $\{x_{n_k}\} \subset D$ so x belongs to D or its boundary. If x were an interior point of D, then the Open Mapping Principle would imply that $y = \varphi(x)$ is an interior point of $\varphi(D)$. Thus, x belongs to the boundary of D, and $y \in \varphi(\partial D)$. This establishes (14.13) and the theorem is proved. □

Theorem 14.4.3 points out a subtle but important difference between the multivariable and the one variable case. If $D = (a,b)$ and $\varphi : D \to \mathbb{R}$ is continuous, then $\varphi(D)$ is connected, hence an interval, hence a Jordan set. Notice that this argument does not require the existence of φ', let alone its invertibility. However, if $D = (0,1) \times (0,1)$, then both of these assumptions are essential. In order to justify this claim, we need a lemma.

Lemma 14.4.4. *Let $a, b, c, d, \beta, \gamma$ be real numbers such that $a < b < c < d$. Then there exists a function f that is differentiable on $\mathbb{R}$ and satisfies $f(x) = \beta$ if $x \in (a,b)$ and $f(x) = \gamma$ if $x \in (c,d)$.*

Proof. The idea is to define a function

$$f(x) = \begin{cases} \beta, & \text{if } a < x < b \\ \psi(x), & \text{if } b \le x \le c \\ \gamma, & \text{if } c < x < d. \end{cases}$$

In order for f to be continuous, we must have $\psi(b) = \beta$ and $\psi(c) = \gamma$. The differentiability

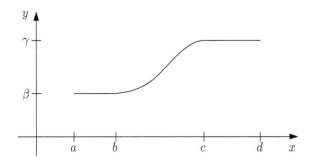

Figure 14.9: $f(x) = \beta$ on $[a,b]$, $f(x) = \gamma$ on $[c,d]$ and f is differentiable on $[a,d]$.

of f requires that $\psi'(b) = \psi'(c) = 0$. It is an exercise in elementary calculus to verify that the function

$$\psi(x) = \frac{2(\beta - \gamma)}{(c-b)^3}x^3 - \frac{3(\beta - \gamma)}{(c-b)^3}x^2 + \frac{6(\beta - \gamma)}{(c-b)^3}x + \frac{\beta c^2(c - 3b) - \gamma b^2(b - 3c)}{(c-b)^3}$$

has all the listed properties. Therefore, f is differentiable and it has values β and γ in the prescribed intervals. □

It is not hard to extend the lemma to an infinite number of intervals.

Corollary 14.4.5. *Let $\{a_n\}, \{b_n\}, \{\beta_n\}$ be infinite sequences of real numbers such that $a_1 > b_1 > a_2 > b_2 > \dots$. Then there exists a function f that is differentiable on $\mathbb{R}$ and satisfies $f(x) = \beta_n$ if $b_n < x < a_n$, for all $n \in \mathbb{N}$.*

Now we can showcase the announced difference between the multivariable and the single variable case.

Example 14.4.6. D **is a Jordan set and φ is differentiable on D but $\varphi(D)$ is not a Jordan set.**

Let $D = (0,1) \times (0,1)$, $a_n = \frac{1}{2n-1}$, $b_n = \frac{1}{2n}$ and let $\{\beta_n\}$ be the sequence of all rational numbers between 0 and 1. Let $\alpha(x)$ be a differentiable function such that $\alpha(x) = \beta_n$ if $x \in (b_n, a_n)$, for all $n \in \mathbb{N}$. Finally, let $\varphi(x,y) = (-\sin\frac{\pi}{x}, \alpha(x))$. We will show that $\varphi(D)$ is not a Jordan set, in spite of the fact that φ is a differentiable function on D. (Truth be told, its derivative is a matrix

$$\begin{bmatrix} \dfrac{\pi}{x^2}\cos\dfrac{\pi}{x} & 0 \\ \alpha'(x) & 0 \end{bmatrix}$$

which is clearly not invertible.)

Solution. The image of D under φ is a curve

$$\{(-\sin\frac{\pi}{x}, \alpha(x)) : 0 < x < 1\}.$$

Let us take a closer look at this curve (Figure 14.10). When x changes from 1 to $1/2$ we have that $-\sin\frac{\pi}{x}$ changes from 0 to 1 and back to 0. At the same time, $\alpha(x)$ equals β_1, so the curve starts at the point $(0, \beta_1)$, goes horizontally to $(1, \beta_1)$ and back to $(0, \beta_1)$. Next, we let x change from $1/2$ to $1/3$, so that $-\sin\frac{\pi}{x}$ changes from 0 to -1 and back to 0, and $\alpha(x)$ goes from β_1 to β_2. This means that the curve connects $(0, \beta_1)$ with $(0, \beta_2)$, while staying on the left of the y-axis. In phase two, x goes from $1/3$ to $1/4$, $-\sin\frac{\pi}{x}$ changes from 0 to 1 and back to 0, and $\alpha(x)$ equals β_2, so the curve starts at the point $(0, \beta_2)$, goes horizontally to $(1, \beta_2)$ and back to $(0, \beta_2)$. When x goes from $1/4$ to $1/5$, the curve connects $(0, \beta_2)$ with $(0, \beta_3)$, while staying on the left of the y-axis. As the process continues we see

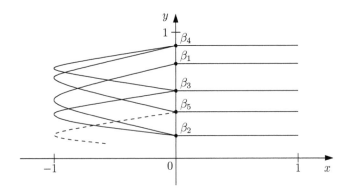

Figure 14.10: The curve $\{(-\sin\frac{\pi}{x}, \alpha(x)) : 0 < x < 1\}$.

that the portion of this (infinitely long) curve on the right of the y-axis traces precisely every horizontal line segment with the rational height. Thus, the boundary of the set $\varphi(D)$ contains every single point of $[0,1] \times [0,1]$ and it cannot have content 0. It follows that $\varphi(D)$ is not a Jordan set. ◆

Our next goal is to make sense of the concept of the area of a Jordan set. When D is a Jordan set, we will say that the area of D is

$$\mu(D) = \iint_D dA.$$

Considering the way we have defined a double integral over a Jordan set, we are really looking at the integral of the *characteristic function* of the set D. The function

$$\chi_D(x, y) = \begin{cases} 1, & \text{if } (x, y) \in D \\ 0, & \text{if } (x, y) \notin D \end{cases}$$

is called the characteristic function of the set D. Since χ_D is discontinuous precisely at the boundary of D, and ∂D has content 0, we see that χ_D is integrable, and $\mu(D)$ is well defined. When $D = [a, b] \times [c, d]$, then it is not hard to see that $\mu(D) = (b - a)(d - c)$, so the definition above gives us the expected result in a familiar situation.

We have seen in Lemma 14.4.1 that it is possible to compare the areas of D and $\varphi(D)$. The result can be made much more precise when φ is linear. It will follow from this theorem.

Theorem 14.4.7. *Suppose that φ is an invertible linear transformation on $\mathbb{R}^2$, with matrix Φ, and let D be a Jordan set in $\mathbb{R}^2$. If f is a continuous function on $\varphi(D)$, then*

$$\iint_{\varphi(D)} f(x, y)\, dA(x, y) = \iint_D f(\varphi(u, v))|\det(\Phi)|\, dA(u, v). \tag{14.14}$$

Proof. Every linear transformation $T : \mathbb{R}^2 \to \mathbb{R}^2$ is a composition of three types of linear transformations:

$$T_1(x, y) = (\lambda x, y), \quad T_2(x, y) = (x + y, y), \quad T_3(x, y) = (y, x),$$

with determinants of the respective matrices λ, 1, and -1. Let us consider the transformation T_1 first. Let $R = [a, b] \times [c, d]$ be a rectangle that contains D, and suppose that f is zero outside $T_1(D)$. If $\lambda > 0$,

$$\iint_D f \circ T_1(u, v)|\lambda|\, dA(u, v) = \iint_R f \circ T_1(u, v)|\lambda|\, dA(u, v)$$

$$= \int_c^d dv \int_a^b f(\lambda u, v)\lambda\, du$$

$$= \int_c^d dv \int_{\lambda a}^{\lambda b} f(t, v)\, dt$$

$$= \iint_{T_1(R)} f(t, v)\, dA(t, v)$$

$$= \iint_{T_1(D)} f(t, v)\, dA(t, v),$$

with the aid of the substitution $t = \lambda u$. The proof for the case $\lambda < 0$ is almost the same, and we leave it as an exercise, as well as the proofs for T_2 and T_3. It remains to show that, if the theorem is true for two transformations U and V, then it is true for UV.

$$\iint_D f \circ (UV)(u, v)|\det(UV)|\, dA(u, v)$$

$$= \iint\limits_{D} (f \circ U) \circ V(u,v) |\det(U)| |\det(V)| \, dA(u,v)$$

$$= \iint\limits_{V(D)} (f \circ U)(z,w) |\det(U)| \, dA(z,w)$$

$$= \iint\limits_{UV(D)} f(x,y) \, dA(x,y). \qquad \square$$

From here we derive an easy consequence.

Corollary 14.4.8. *Suppose that φ is an invertible linear transformation on $\mathbb{R}^2$, with matrix Φ, and let A be a Jordan set in $\mathbb{R}^2$. Then $\mu(\varphi(A)) = |\det(\Phi)|\mu(A)$.*

Proof. The set $\varphi(A)$ is a Jordan set by Theorem 14.4.3. Thus,

$$\mu(\varphi(A)) = \iint\limits_{\varphi(A)} dA = \iint\limits_{A} 1 \circ \varphi \, |\det \Phi| \, dA = |\det \Phi| \iint\limits_{A} dA = |\det \Phi| \, \mu(A). \qquad \square$$

Theorem 14.4.7 shows that we can introduce a linear change of variables in a double integral. Most of the time, however, such a change is not very useful. In the next section we will establish a much more general result.

Exercises

In Exercises 14.4.1–14.4.3 R is a rectangle and $\varphi : R \to \mathbb{R}^2$ is **Lipschitz continuous** on R, i.e., there exists $K > 0$ such that, for any $\mathbf{a}, \mathbf{b} \in R$, $\|\varphi(\mathbf{a}) - \varphi(\mathbf{b})\| \le K\|\mathbf{a} - \mathbf{b}\|$.

14.4.1. Let $C \subset R$ be a square with area α. Prove that $\varphi(C)$ is contained in a square with area no more than $K\sqrt{2\alpha}$.

14.4.2. Let $D \subset R$ be a compact set with content 0. Prove that $\varphi(D)$ has content 0.

14.4.3. Let $D \subset R$ be a Jordan set. Prove that $\varphi(D)$ is a Jordan set.

In Exercises 14.4.4–14.4.10 use a linear change of variables to compute $\iint_R f(x,y) \, dA$:

14.4.4. $f(x,y) = 1$, $R = \{(x,y) : \frac{x^2}{4} + \frac{y^2}{9} \le 1\}$.

14.4.5.* $f(x,y) = 1$, $R = \{(x,y) : 2x^2 + 2xy + 5y^2 \le 1\}$.

14.4.6. $f(x,y) = (y^2 - x^2)e^{(x^2+y^2)/2}$, $R = \{(x,y) : 0 \le x + y \le 2, 0 \le y - x \le 2\}$.

14.4.7. $f(x,y) = e^{(x-y)/(x+y)}$, R is the region in the first quadrant bounded by the coordinate axes and the line $x + y = 1$.

14.4.8. $f(x,y) = x + y$, R is the parallelogram with vertices at $(0,0)$, $(2,1)$, $(5,0)$ and $(3,-1)$.

14.4.9. $f(x,y) = xy$, R is the region bounded by the lines $y = \frac{1}{2}x$, $y = \frac{1}{2}x + 2$, $y = 3x$ and $y = 3x - 4$.

14.4.10. $f(x,y) = x$, R is the parallelogram with vertices at $(0,0)$, $(1,3)$, $(2,5)$ and $(3,8)$.

14.4.11.* Let $A = \begin{bmatrix} a & b \\ c & d \end{bmatrix}$ and $\|A\| = \max\{|a| + |b|, |c| + |d|\}$.
(a) Prove that this defines a norm on 2×2 matrices, i.e., it satisfies conditions (a)–(d) in Theorem 10.1.2.
(b) Prove that, if A, B are 2×2 matrices, $\|AB\| \le \|A\|\|B\|$.

14.4.12. Let R and $\varphi = (\varphi_1, \varphi_2)$ be as in Lemma 14.4.1 and, for $(u,v) \in R$, let $\|\mathbf{D}\varphi(u,v)\|$ be as in (14.12). Prove that $\|\mathbf{D}\varphi(u,v)\|$ is continuous on R.

14.4.13.* Suppose that D is a Jordan set such that $\mu(D) > 0$. Prove that there exists a rectangle $R \subset D$ such that $\mu(R) > 0$.

14.4.14. Suppose that D is a Jordan set and let D^o be its interior, i.e., the largest open set contained in D. Prove that D^o is a Jordan set and that $\mu(D^o) = \mu(D)$.

14.4.15. Suppose that D is a Jordan set and let $\overline{D}$ be its closure, i.e., the smallest closed set that contains D. Prove that $\overline{D}$ is a Jordan set and that $\mu(\overline{D}) = \mu(D)$.

14.4.16. Let $S, T : \mathbb{R}^2 \to \mathbb{R}^2$ be linear transformations and $I = [0,1] \times [0,1]$. Prove that $\mu((S \circ T)(I)) = \mu(S(I))\mu(T(I))$.

14.4.17. Let $T(x,y) = (3x - 2y, -2x + y)$ and let R be the region in the first quadrant bounded by the coordinate axes and the graph of $y = \frac{\sin x}{x}$. Find $\mu(T(R))/\mu(R)$.

14.4.18. Let R be a rectangle, let $\varphi : R \to \mathbb{R}^2$ be injective and have continuous partial derivatives, and let $D \subset R$. If D is a Jordan set, and if $\mathbf{D}\varphi(x,y)$ is invertible on D, prove that $\partial\varphi(D) = \varphi(\partial D)$.

14.4.19. Suppose that D is a Jordan set and that $\iint_D dA = \alpha$. Prove that, for each $\varepsilon > 0$, there exist rectangles R_i, $1 \le i \le n$, such that $D \subset \cup R_i$ and the total area of these rectangles is less than $\alpha + \varepsilon$.

14.4.20. Suppose that D is a Jordan set and that $\iint_D dA = \alpha$. Prove that, for each $\varepsilon > 0$, there exist rectangles R_i, $1 \le i \le n$, such that $\cup R_i \subset D$ and the total area of these rectangles is bigger than $\alpha - \varepsilon$.

14.4.21. Suppose that D is a Jordan set contained in the rectangle R. If f is integrable on R, prove that its restriction to D is integrable.

14.4.22. Complete the proof of Theorem 14.4.7 by showing that formula (14.14) holds for transformations $\varphi(x,y) = (x+y, y)$ and $\varphi(x,y) = (y,x)$.

Let $\mathcal{F}$ denote the collection of sets that can be represented as finite unions of rectangles. If $D \subset \mathbb{R}^2$, we define the **inner Jordan measure** $\mu_*(D) = \sup\{\mu(A) : A \in \mathcal{F}, A \subset D\}$, and the **outer Jordan measure** $\mu^*(D) = \inf\{\mu(A) : A \in \mathcal{F}, A \supset D\}$.

14.4.23. Prove that, for any D, $\mu_*(D) \le \mu^*(D)$.

14.4.24.* Prove that a set D is a Jordan set if and only if $\mu_*(D) = \mu^*(D)$.

14.4.25. Prove that $\mu_*(D) = \mu_*(\overline{D})$ for any $D \subset \mathbb{R}^2$.

14.4.26. If D_1, D_2 are disjoint sets, prove that $\mu^*(D_1 \cup D_2) \le \mu^*(D_1) + \mu^*(D_2)$.

14.4.27. Prove or disprove: there exist disjoint sets $D_1, D_2 \in \mathbb{R}^2$ such that $\mu^*(D_1) = \mu^*(D_2) = \mu^*(D_1 \cup D_2) \ne 0$.

14.5   Change of Variables in Double Integrals

Just like the substitution method for integrals in a single variable setting, the change of variables is an extremely effective tool to compute a double integral.

Example 14.5.1. Evaluation of a double integral: a review.

Let $f(x,y) = e^{x^2 + y^2}$, and let D be the disk $\{(x,y) : x^2 + y^2 \le 9\}$. Evaluate $\iint_D f(x,y)\, dA$.

Solution. We will use the polar coordinates $x = r\cos\theta$, $y = r\sin\theta$. The disk D can be described in polar coordinates as $\tilde{D} = \{(r, \theta) : 0 \leq r \leq 3, 0 \leq \theta \leq 2\pi\}$. Also, the Jacobian determinant for the polar coordinates has absolute value r. Therefore,

$$\iint_{D} e^{x^2+y^2}\, dxdy = \iint_{\tilde{D}} e^{r^2}\, r\, drd\theta = \int_{0}^{3} dr \int_{0}^{2\pi} e^{r^2}\, r\, d\theta = \int_{0}^{3} 2\pi e^{r^2}\, r\, dr = \pi e^{r^2}\big|_{0}^{3} = (e^9 - 1)\pi. \quad \blacklozenge$$

The first writer to use polar coordinates was Cavalieri in [18] (page 117) in order to calculate the area within an Archimedian spiral. Pascal used them to calculate the length of a parabolic arc. Newton was the first to consider them as a means of fixing any point in the plane but only to show how the tangent could be determined when the equation of the curve was given in polar form. A deeper interest was shown by Jacob Bernoulli, who went so far as to write the expression for the radius of curvature in polar coordinates.

The main result in this section will be to establish the validity of the change of coordinates in a double integral.

Lemma 14.4.1 shows that, if C is a square and φ has continuous partial derivatives,

$$\mu(\varphi(C)) \leq \sup_{(u,v) \in C} \|\mathbf{D}\varphi(u,v)\|^2 \mu(C). \tag{14.15}$$

We will now make another estimate.

Lemma 14.5.2. *Let K be a compact set in $\mathbb{R}^2$, let φ be a function defined on K, and suppose that the partial derivatives of φ are continuous on K. In addition, suppose that $\mathbf{D}\varphi(u,v)$ is invertible for $(u,v) \in K$. For any $\eta > 0$ there exists $\delta > 0$ such that, if $C \subset K$ is a square of side less than δ, and if $(a,b) \in C$, then*

$$\mu(\varphi(C)) \leq |\det(\mathbf{D}\varphi(a,b))|(1+\eta)\,\mu(C). \tag{14.16}$$

Proof. Let $T : \mathbb{R}^2 \to \mathbb{R}^2$ be an invertible linear transformation, and let $C \subset K$ be a square. By (14.15), applied to $T \circ \varphi$,

$$\mu\left(T \circ \varphi(C)\right) \leq \sup_{(u,v) \in C} \|\mathbf{D}(T \circ \varphi)(u,v)\|^2 \mu(C).$$

Let Δ be the determinant of the matrix for T. Corollary 14.4.8 implies that $\mu\left(T \circ \varphi(C)\right) = |\Delta|\mu(\varphi(C))$. It is not hard to verify that $\mathbf{D}(T \circ \varphi) = T\mathbf{D}\varphi$ so,

$$\mu(\varphi(C)) \leq \frac{1}{|\Delta|} \sup_{(u,v) \in C} \|T\mathbf{D}\varphi(u,v)\|^2 \mu(C). \tag{14.17}$$

Let $\eta > 0$, and let $\eta_1 = \min\{1, \eta/3\}$. The function $\|\mathbf{D}\varphi(u,v)^{-1}\|$ is continuous on K, so it is bounded. Let $M = \sup\{\|\mathbf{D}\varphi(u,v)^{-1}\| : (u,v) \in K\}$. Also, $\|\mathbf{D}\varphi(u,v)\|$ is continuous on K, so it is uniformly continuous. Therefore, there exists $\delta > 0$ such that, if $|u - x|, |v - y| < \delta$ then

$$\|\mathbf{D}\varphi(x,y) - \mathbf{D}\varphi(u,v)\| < \frac{\eta_1}{M}. \tag{14.18}$$

Suppose now that the side of the square $C \subset K$ is less than δ. If I is the 2×2 identity matrix and $(x,y),(u,v) \in C$, using Exercise 14.4.11,

$$
\begin{aligned}
\|I - [\mathbf{D}\varphi(x,y)]^{-1}\mathbf{D}\varphi(u,v)\| &= \|[\mathbf{D}\varphi(x,y)]^{-1}\,[\mathbf{D}\varphi(x,y) - \mathbf{D}\varphi(u,v)]\,\| \\
&\leq \|[\mathbf{D}\varphi(x,y)]^{-1}\|\,\|\mathbf{D}\varphi(x,y) - \mathbf{D}\varphi(u,v)\| \\
&\leq M\,\frac{\eta_1}{M} = \eta_1.
\end{aligned}
\tag{14.19}
$$

Another application of Exercise 14.4.11 shows that $\|[\mathbf{D}\varphi(x,y)]^{-1}\mathbf{D}\varphi(u,v)\| \leq 1 + \eta_1$. Let

$(a, b) \in C$, let $T = [\mathbf{D}\varphi(a,b)]^{-1}$, and let Δ be the determinant of the matrix for T. Then, (14.17) yields

$$\mu(\varphi(C)) \leq |\det(\mathbf{D}\varphi(a,b))| \sup_{(u,v) \in C} \|[\mathbf{D}\varphi(a,b)]^{-1}\mathbf{D}\varphi(u,v)\|^2 \mu(C)$$

$$\leq |\det(\mathbf{D}\varphi(a,b))|(1 + \eta_1)^2 \mu(C).$$

Since $(1 + \eta_1)^2 = 1 + \eta_1(2 + \eta_1) \leq 1 + 3\eta_1 \leq 1 + \eta$ we obtain the estimate (14.16). □

Finally, we have gathered enough material so that we can prove the main result.

Theorem 14.5.3 (Change of Variables Theorem). *Let A be an open set in $\mathbb{R}^2$, and let A_0 be an open Jordan set in A such that $K = A_0 \cup \partial A_0 \subset A$. Let φ be a function defined on A so that φ and its partial derivatives are continuous on A, and suppose that φ is a bijection from A_0 onto $\varphi(A_0)$ and that $\mathbf{D}\varphi$ is invertible on A_0. Then $\varphi(K)$ is a Jordan set in $\mathbb{R}^2$. Further, if f is a function defined and continuous on $\varphi(K)$, then*

$$\iint_{\varphi(K)} f(x,y) \, dA(x,y) = \iint_K f(\varphi(u,v))|\det(\mathbf{D}\varphi(u,v))| \, dA(u,v). \qquad (14.20)$$

Proof. It suffices to prove the result under the additional assumption that $f \geq 0$ which we make from here on. Indeed, we can always write $f = f^+ - f^-$, where

$$f^+ = \frac{f + |f|}{2}, \quad f^- = \frac{|f| - f}{2} \qquad (14.21)$$

are non-negative functions. Once the theorem is proved for non-negative functions, the linearity of double integrals shows that it holds for arbitrary real-valued functions. Also, f is defined only on $\varphi(K)$, so we may extend it to all of $\varphi(A)$ by setting it equal to zero outside of $\varphi(K)$.

Next, we make another reduction. Since A_0 is a Jordan set, its boundary ∂A_0 has content 0 and, by Corollary 14.4.2, so does the boundary of $\varphi(A_0)$. Consequently, the integrals in (14.20) are equal to the integrals over $\varphi(A_0)$ and A_0. We will prove only that

$$\iint_{\varphi(A_0)} f(x,y) \, dA(x,y) \leq \iint_{A_0} f(\varphi(u,v))|\det(\mathbf{D}\varphi(u,v))| \, dA(u,v). \qquad (14.22)$$

Indeed, once this is done, we can take advantage of the invertibility of φ on A_0 and apply the same estimate to $(f \circ \varphi)|\det(\mathbf{D}\varphi)|$ instead of f, $\varphi(A_0)$ instead of A_0, and φ^{-1} instead of φ, to obtain

$$\iint_{A_0} (f \circ \varphi)(u,v)|\det(\mathbf{D}\varphi(u,v))| \, dA(u,v)$$

$$\leq \iint_{\varphi(A_0)} (f \circ \varphi) \circ \varphi^{-1}(x,y)|\det(\mathbf{D}\varphi(\varphi^{-1}(x,y)))||\det(\mathbf{D}\varphi^{-1}(x,y))| \, dA(x,y)$$

$$= \iint_{\varphi(A_0)} f(x,y)|\det(\mathbf{D}\varphi(\varphi^{-1}(x,y)))(\mathbf{D}\varphi^{-1}(x,y))| \, dA(x,y)$$

$$= \iint_{\varphi(A_0)} f(x,y) \, dA(x,y).$$

The last equality follows from the Chain Rule:

$$(x,y) = \mathbf{D}I(x,y) = \mathbf{D}(\varphi \circ \varphi^{-1})(x,y) = \mathbf{D}\varphi(\varphi^{-1}(x,y)) \, \mathbf{D}\varphi^{-1}(x,y).$$

That way, we need to prove only (14.22).

Let $\varepsilon > 0$, and let P be a partition of A into *squares*. We will require that P is sufficiently fine to satisfy four conditions.

(H1) Let A_1 be an open Jordan set such that $K \subset A_1$ and $K_1 = A_1 \cup \partial A_1 \subset A$. If C is a

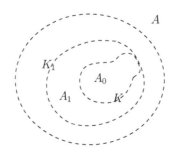

Figure 14.11: $K = A_0 \cup \partial A_0 \subset A_1$ and $K_1 = A_1 \cup \partial A_1 \subset A$.

square (formed by the partition P) that intersects K, then it has to be contained in A_1. As a consequence, φ will be defined on C.

(H2) Let J be the double integral on the right side of (14.20), and define

$$\eta = \min\left\{1, \frac{\varepsilon}{2(J+2)}\right\}. \tag{14.23}$$

We will require that the upper Darboux sum for this integral (corresponding to P) does not exceed J by more than η.

(H3) Let δ be the number corresponding to η as in Lemma 14.5.2. We will choose P so that each square defined by this partition has the side less than δ.

(H4) The boundary of A_0 has content 0, and we will ask that the total area of the squares that intersect ∂A_0 does not exceed $\varepsilon/(2M_1 M_2)$, where $M_1 = \sup\{f \circ \varphi(u,v) : (u,v) \in K_1\}$, and $M_2 = \sup\{\|\mathbf{D}\varphi(u,v)\|^2 : (u,v) \in K_1\}$.

The squares formed by P fall into three groups. Let $\{C_\alpha : \alpha \in \mathcal{A}\}$ be the squares that are completely contained in A_0, let $\{C_\beta : \beta \in \mathcal{B}\}$ be the squares that intersect ∂A_0, and let $\{C_\gamma : \gamma \in \mathcal{G}\}$ be the squares that completely miss K. We will denote $C_1 = \cup\{C_\alpha : \alpha \in \mathcal{A}\}$, $C_2 = \cup\{C_\beta : \beta \in \mathcal{B}\}$, and $C_3 = \cup\{C_\gamma : \gamma \in \mathcal{G}\}$. Since f vanishes outside $\varphi(K)$, the function $f \circ \varphi$ vanishes outside K, so

$$\iint\limits_{\varphi(C_3)} f(u,v)\, dA(u,v) = 0.$$

Further, if $\beta \in \mathcal{B}$, then C_β intersects ∂A_0, so it intersects K. By (H1), $C_\beta \subset A_1 \subset K_1$. Thus, using (14.15),

$$\iint\limits_{\varphi(C_\beta)} f(u,v)\, dA(u,v) \leq \iint\limits_{\varphi(C_\beta)} M_1\, dA(u,v) = M_1 \mu(\varphi(C_\beta))$$

$$\leq M_1 \sup_{(u,v)\in C_\beta} \|\mathbf{D}\varphi(u,v)\|^2 \mu(C_\beta) \tag{14.24}$$

$$\leq M_1 M_2 \mu(C_\beta).$$

Adding the inequalities (14.24) over all $\beta \in \mathcal{B}$, and using (H4), we obtain that

$$\iint\limits_{\varphi(C_2)} f(u,v) \, dA(u,v) \le M_1 M_2 \mu(C_2) \le M_1 M_2 \frac{\varepsilon}{2M_1 M_2} = \frac{\varepsilon}{2}.$$

Finally, we consider squares C_α. The function $f \circ \varphi$ is continuous on a compact set C_α, so it attains its extreme values. Let (a_α, b_α) be a point in C_α such that $f \circ \varphi(a_\alpha, b_\alpha) = \sup\{f \circ \varphi(u,v) : (u,v) \in C_\alpha\}$. By (H3), we are allowed to use (14.16) and we have that

$$\mu(\varphi(C_\alpha)) \le |\det(\mathbf{D}\varphi(a_\alpha, b_\alpha))|(1+\eta)\mu(C_\alpha).$$

Therefore,

$$\iint\limits_{\varphi(C_\alpha)} f \, dA \le \iint\limits_{\varphi(C_\alpha)} \sup_{(x,y)\in\varphi(C_\alpha)} f(x,y) \, dA = \sup_{(u,v)\in C_\alpha} f \circ \varphi(u,v) \mu(\varphi(C_\alpha))$$

$$\le f \circ \varphi(a_\alpha, b_\alpha) |\det(\mathbf{D}\varphi(a_\alpha, b_\alpha))|(1+\eta)\mu(C_\alpha)$$

$$\le \sup_{(u,v)\in C_\alpha} f \circ \varphi(u,v) |\det(\mathbf{D}\varphi(u,v))|(1+\eta)\mu(C_\alpha).$$

Summing up over all $\alpha \in \mathcal{A}$, we obtain

$$\iint\limits_{\varphi(C_1)} f \, dA \le (1+\eta) \left(\sum_{\alpha\in\mathcal{A}} \sup_{(u,v)\in C_\alpha} f \circ \varphi(u,v) |\det(\mathbf{D}\varphi(u,v))| \mu(C_\alpha) \right)$$

$$\le (1+\eta) \left(\iint\limits_{A_0} f(\varphi(u,v)) |\det(\mathbf{D}\varphi(u,v))| \, dA(u,v) + \eta \right)$$

because the sum above is a part of the upper Darboux sum for the integral on the right side of (14.22) taken over $C_1 \subset A_0 \subset K$, and the integrand is non-negative. By (H2),

$$\iint\limits_{\varphi(C_1)} f \, dA \le (1+\eta)(J+\eta)$$

and, using (14.23),

$$(1+\eta)(J+\eta) = J + \eta(1+J+\eta) < J + \eta(J+2) < J + \frac{\varepsilon}{2}.$$

It follows that

$$\iint\limits_{\varphi(A_0)} f \, dA = \iint\limits_{\varphi(C_1)} f(u,v) \, dA(u,v) + \iint\limits_{\varphi(C_2)} f(u,v) \, dA(u,v) + \iint\limits_{\varphi(C_3)} f(u,v) \, dA(u,v)$$

$$\le J + \frac{\varepsilon}{2} + \frac{\varepsilon}{2} = \iint\limits_{A_0} f(\varphi(u,v)) |\det(\mathbf{D}\varphi(u,v))| \, dA(u,v) + \varepsilon.$$

Since ε is arbitrary, we have established (14.22) and the proof is complete. $\qquad\square$

As an illustration, we return to Example 14.5.1. Using the notation of the theorem, we define $A = \mathbb{R}^2$, $\varphi(r,\theta) = (r\cos\theta, r\sin\theta)$, and $A_0 = (0,3) \times (0,2\pi)$ which implies that $K = [0,3] \times [0,2\pi]$. The derivative of φ is

$$\mathbf{D}\varphi(r,\theta) = \begin{bmatrix} \cos\theta & -r\sin\theta \\ \sin\theta & r\cos\theta \end{bmatrix}.$$

Clearly, the partial derivatives of φ are continuous in $\mathbb{R}^2$. Also, it is not hard to see that $\mathbf{D}\varphi(r,\theta)$ has determinant r, so it is invertible on A_0. Since $f(x,y) = e^{x^2+y^2}$ is continuous on $\varphi(K)$ which is the disk $\{(x,y) : x^2 + y^2 \le 9\}$, Change of Variables Theorem applies and the procedure that we have applied is legitimate.

> A very interesting essay on the history of this theorem is [71].

Exercises

In Exercises 14.5.1–14.5.10 use polar coordinates to evaluate the integrals:

14.5.1. $\displaystyle\iint_{x^2+y^2\leq 4}\sqrt{x^2+y^2}\,dA(x,y)$.   14.5.2. $\displaystyle\iint_{\pi^2\leq x^2+y^2\leq 4\pi^2}\sin\sqrt{x^2+y^2}\,dA(x,y)$.

14.5.3. $\displaystyle\iint_{x^2+y^2\leq 4x}(x^2+y^2)\,dA(x,y)$.   14.5.4. $\displaystyle\iint_{a^2\leq x^2+y^2\leq b^2}\ln(x^2+y^2)\,dA(x,y)$.

14.5.5. $\displaystyle\iint_D\frac{y}{x}\,dA(x,y)$, if D is the region in the first quadrant bounded by the lines $y=0$,

$y=x$, and the circles $x^2+y^2=9$, $x^2+y^2=25$.

14.5.6. $\displaystyle\iint_D(x^2+y^2)^2(x^2-3xy^2)\,dA(x,y)$, if D is the region in the first quadrant bounded

by $x=y\sqrt{3}$, $y=x\sqrt{3}$, $x^2+y^2=1$.

14.5.7. $\displaystyle\int_0^4 dx\int_x^4\frac{1}{\sqrt{x^2+y^2}}\,dy$.   14.5.8. $\displaystyle\int_1^2 dx\int_0^x\frac{y}{x}\sqrt{x^2+y^2}\,dy$.

14.5.9.* $\displaystyle\iint_{x^4+y^4\leq 1}(x^2+y^2)\,dA(x,y)$.   14.5.10.* $\displaystyle\iint_{x^2+y^2\leq 1}\left|\frac{x+y}{\sqrt{2}}-x^2-y^2\right|\,dA(x,y)$.

In Exercises 14.5.11–14.5.14 find the area of the region bounded by the given curve:

14.5.11. $r=\sin\frac{\theta}{2}$, $0\leq\theta\leq 2\pi$.   14.5.12. $(x^2+y^2)^2=2a^2(x^2-y^2)$.

14.5.13. $(x^2+y^2)^3=a^2(x^4+y^4)$.   14.5.14. $(x^2+y^2)^2=2ax^3$.

In Exercises 14.5.15–14.5.19 verify that $\varphi(x,y)=(u,v)$ is injective on $[0,1]\times[0,1]$:

14.5.15. $u=x+x^2$, $v=e^y$.

14.5.16. $u=xe^y$, $v=e^y$.

14.5.17.* $u=2x-y^2$, $v=y+xy$.

14.5.18. $u=e^x\cos y$, $v=e^x\sin y$.

14.5.19.* $u=5x-x^2+y^2$, $v=5y+10xy$.

14.5.20.* Give an example of a function φ that satisfies all hypotheses of Theorem 14.5.3 except for injectivity, and for which (14.20) does not hold.

In Exercises 14.5.21–14.5.27 use (14.20) to evaluate $\iint_D f(x,y)\,dA$, and verify that all hypotheses of Theorem 14.5.3 are satisfied. D is the region in the first quadrant bounded by the given curves.

14.5.21. $f(x,y)=x^2+y^2$; $xy=1$, $xy=3$, $x^2-y^2=1$, $x^2-y^2=4$.

14.5.22. $f(x,y)=(x^2\sin xy)/y$; $x^2=\frac{\pi y}{2}$, $x^2=\pi y$, $y^2=\frac{x}{2}$, $y^2=x$.

14.5.23.* $f(x,y)=(x^2+y^2)^{-2}$; $x^2+y^2=4x$, $x^2+y^2=6x$, $x^2+y^2=2y$, $x^2+y^2=8y$.

14.5.24. $f(x,y)=xy$; $x^2-y^2=a$, $x^2-y^2=b$, $x^2+y^2=c$, $x^2+y^2=d$.

14.5.25. $f(x,y)=x/(4x^2+y^2)$; $4x^2+y^2=16$, $4x^2+y^2=1$, $x=0$, $y=0$.

14.5.26. $f(x,y)=x^2+y^2$; $xy=1$, $xy=3$, $x^2-y^2=1$, $x^2-y^2=4$.

14.5.27.* $f(x,y)=(\sqrt{x}+\sqrt{y})^{1/2}$, $\sqrt{x}+\sqrt{y}=1$, $x=0$, $y=0$.

14.5.28.* The purpose of this problem is to evaluate the so-called Fresnel Integrals

$$F_0 = \int_0^\infty \cos x^2 \, dx \quad \text{and} \quad G_0 = \int_0^\infty \sin x^2 \, dx.$$

(a) Use the Dirichlet's Test to show that both integrals converge.

(b) Write

$$G_0 = \sum_{n=0}^\infty \int_{\sqrt{n\pi}}^{\sqrt{(n+1)\pi}} \sin x^2 \, dx,$$

and prove that the series is an alternating series $\sum(-1)^n c_n$, and that c_n is a strictly decreasing sequence of positive numbers. Conclude that $G_0 \geq 0$.
(c) Prove that $F_0 \geq 0$.
(d) Let $M > 0$, $t > 0$, and define

$$F(t) = \int_0^M e^{-tx^2} \cos x^2 \, dx \quad \text{and} \quad G(t) = \int_0^M e^{-tx^2} \sin x^2 \, dx.$$

Use polar coordinates to prove that

$$F(t)^2 - G(t)^2 = \int_0^{\pi/4} d\theta \int_0^{M^2/\cos^2\theta} e^{-tu} \cos u \, du.$$

(e) Show that, when $M \to \infty$, the right hand side has limit $\frac{t\pi}{4(1+t^2)}$.
(f) Prove that the functions $F(t)$ and $G(t)$ are continuous for $t \geq 0$. Conclude that

$$F_0^2 - G_0^2 = 0.$$

(g) Use the strategy similar to the one in (d)–(f) to prove that

$$2F_0 G_0 = \frac{\pi}{4}.$$

(h) Use (b), (c), (f), and (g) to conclude that $F_0 = G_0 = \sqrt{2\pi}/4$.

These integrals appear in Fresnel's work from 1798. As it might be expected, they were known earlier, and they were used by Euler. The equations $x(t) = \int_0^t \cos x^2 \, dx$, $y(t) = \int_0^t \sin x^2 \, dx$ are parametric equations of the Cornu spiral (or Euler's spiral), which was studied by Johann Bernoulli around 1696.

AUGUSTIN-JEAN FRESNEL (1788–1827), was a French engineer who studied the behavior of light both theoretically and experimentally. He was educated by his parents until the age of 12. By some accounts he could not read at the age of 8. He became interested in mathematics as a high school student in Caen. He studied for a civil engineer at the Ecole Polytechnique and the Ecole des Ponts et Chaussées (School of Bridges and Roads). While working on a road through France connecting Spain with Italy he developed a fascination with light and started conducting experiments. At that time he was unaware of the latest theories and results of other scientists. He became convinced in the wave nature of light, contrary to the mainstream science of the day. His mathematical study of the problem was instrumental in removing many objections to the wave theory. For his work on diffraction he received the Grand Prix of the Académie des Sciences in 1819. Next he introduced the hypothesis that light is a transverse wave (its oscillating being perpendicular to the direction of its propagation). Fresnel was elected to the Académie des Sciences in 1823 and the Royal Society of London in 1825, receiving its Rumford Medal in 1827. He died from tuberculosis in 1827 at the age of 39. He is perhaps best known as the inventor of the Fresnel lens, first adopted in lighthouses.

14.6 Improper Integrals

In Section 6.7 we have studied infinite and improper integrals of functions of a single variable. Further, we have noticed in Section 13.3 that both types of integrals can be treated simultaneously, and we will continue to do this. Further, we will focus on the situation when the integrand is a non-negative function. In other words, we will study the *absolute* convergence of these integrals. Towards the end of the section we will explain the reason for this attitude.

We start with the definition of a monotone covering.

Definition 14.6.1. Let $D \subset \mathbb{R}^2$ be an open connected set (not necessarily bounded). We say that the collection $\mathcal{D} = \{D_n : n \in \mathbb{N}\}$ is a **monotone covering** of D if:

(a) each D_n is a finite union of open, bounded and connected Jordan sets;

(b) $\cup_{n=1}^{\infty} D_n = D$;

(c) $D_n \subset D_{n+1}$, for all $n \in \mathbb{N}$.

Example 14.6.2. A monotone covering of a set in $\mathbb{R}^2$.

Let D denote the first quadrant $\{(x, y) \in \mathbb{R}^2 : x > 0,\, y > 0\}$. For every $n \in \mathbb{N}$, we define a square $D_n = \{(x, y) \in \mathbb{R}^2 : 0 < x < n,\, 0 < y < n\}$. It is easy to see that each square D_n is an open connected Jordan set, and that it is contained in D_{n+1}. Also, the union of all these squares covers D. Therefore, $\{D_n\}$ is a monotone covering of D. ◆

Although Definition 14.6.1 is about sets in $\mathbb{R}^2$, the same concept can be considered in $\mathbb{R}$.

Example 14.6.3. A monotone covering of a set in $\mathbb{R}$.

Let $D = (0, +\infty) \subset \mathbb{R}$, let a_n be a strictly increasing sequence of positive numbers, and let $D_n = (0, a_n)$. Then $\{D_n\}$ is a monotone covering of D. ◆

Now we can define improper integrals.

Definition 14.6.4. Let f be a function defined on an open connected set $D \subset \mathbb{R}^2$, and suppose that f is integrable on every bounded Jordan subset of D. We say that f is **integrable** on D if there exists a real number I such that, for any monotone covering $\{D_n\}$ of D, $\lim \iint_{D_n} f(x, y)\, dA(x, y) = I$. In that case we call I the **improper integral** of f over D, and we write $I = \iint_D f(x, y)\, dA(x, y)$.

As we have said, we will assume that $f \geq 0$. In that case, the requirements of Definition 14.6.4 can be considerably loosened.

Theorem 14.6.5. *Let f be a non-negative function on an open set $D \subset \mathbb{R}^2$. Then f is integrable on D if and only if there exists a monotone covering $\{D_n\}$ of D, such that the sequence $a_n = \iint_{D_n} f(x, y)\, dA(x, y)$ is bounded. In that case $\iint_D f(x, y)\, dA = \lim a_n$.*

Proof. By definition, if f is integrable, then $\{a_n\}$ is a convergent, hence bounded, sequence. This takes care of the "only if" part of the proof.

Suppose now, that the sequence a_n is bounded. Since $f \geq 0$ and $D_n \subset D_{n+1}$, we see that $\{a_n\}$ is an increasing sequence. Therefore, $\{a_n\}$ is convergent. Let $I = \lim a_n$. In order to show that $I = \iint_D f(x, y)\, dA(x, y)$, we need to demonstrate that, if $\{E_n\}$ is another monotone covering of D, if $b_n = \iint_{E_n} f(x,y)\, dA(x,y)$ and $\lim b_n = I'$, then $I = I'$. For each $k \in \mathbb{N}$, the collection $\{D_k \cap E_n\}_{n=1}^{\infty}$ is a monotone covering of D_k. By Exercise 14.6.24,

$$\iint_{D_k} f(x,y)\, dA(x,y) = \lim_{n\to\infty} \iint_{D_k \cap E_n} f(x,y)\, dA(x,y) \le \lim_{n\to\infty} \iint_{E_n} f(x,y)\, dA(x,y) = I'.$$

Taking the limit as $k \to \infty$, we obtain that $I \le I'$. Exactly the same argument can be used to establish the opposite inequality, so $I' = I$ and the theorem is proved. $\qquad\square$

Definitions 14.6.1 and 14.6.4 are due to Jordan and they can be found in his *Cours d'Analyse.*

Example 14.6.6. Evaluation of an improper integral.

Evaluate $\displaystyle \iint_{\substack{x+y>1 \\ 0<x<1}} \frac{1}{(x+y)^p}\, dA(x,y)$, for $p > 1$.

Solution. Let us denote the domain of integration by $D = \{(x,y) \in \mathbb{R}^2 : 0 < x < 1,\, x+y > 1\}$, and notice that D is an open set. Let $D_n = \{(x,y) \in \mathbb{R}^2 : 0 < x < 1,\, 1 < x+y < n\}$, for each $n \in \mathbb{N}$. Then $\{D_n : n \in \mathbb{N}\}$ is a monotone covering of D and

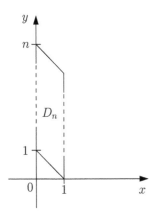

Figure 14.12: The set D_n.

$$\iint_{D_n} \frac{1}{(x+y)^p}\, dA(x,y) = \int_0^1 dx \int_{1-x}^{n-x} \frac{1}{(x+y)^p}\, dy$$

$$= \int_0^1 dx \int_1^n \frac{1}{t^p}\, dt$$

$$= \int_0^1 dx \left. \frac{t^{1-p}}{1-p}\right|_1^n$$

$$= \int_0^1 dx\, \frac{n^{1-p}-1}{1-p}$$

$$= \frac{n^{1-p}-1}{1-p} \to \frac{1}{p-1}.$$

Therefore, $\displaystyle \iint_D \frac{1}{(x+y)^p}\, dA(x,y) = \frac{1}{p-1}$. $\qquad\blacklozenge$

We have seen in Section 6.7 that a very useful tool when trying to establish the convergence of an improper integral is the Comparison Test. The same is true in the multivariable situation. We will leave the proof as an exercise.

Theorem 14.6.7 (Comparison Test). *Let f and g be two functions defined on an open connected set $D \subset \mathbb{R}^2$, and integrable on any Jordan subset of D. Suppose that $0 \le f(x,y) \le g(x,y)$ for all $(x,y) \in D$. If $\iint_D g(x,y)\, dA(x,y)$ converges then so does $\iint_D f(x,y)\, dA(x,y)$.*

Example 14.6.8. Convergence of an improper integral.

Prove that $\displaystyle \iint_{1<x<y} \frac{1}{(x+y)^3 + xy}\, dA(x,y)$ converges.

Solution. Let

$$f(x,y) = \frac{1}{(x+y)^3 + xy}, \quad g(x,y) = \frac{1}{2^3(\sqrt{xy})^3}, \quad D = \{(x,y) \in \mathbb{R}^2 : 1 < x < y\}.$$

Then $0 \le f(x,y) \le g(x,y)$ for all $(x,y) \in D$. (Reason: $xy \ge 0$ and $x + y \ge 2\sqrt{xy}$.) Further, g is integrable on D. If we define $D_n = \{(x,y) \in \mathbb{R}^2 : 1 < x < n,\ x < y < n\}$, it follows that

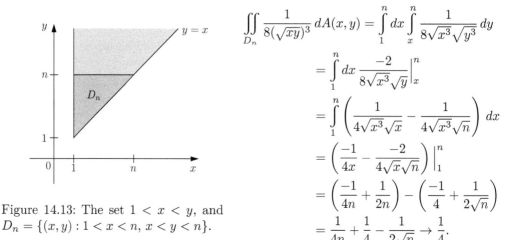

$$\iint_{D_n} \frac{1}{8(\sqrt{xy})^3} \, dA(x,y) = \int_1^n dx \int_x^n \frac{1}{8\sqrt{x^3}\sqrt{y^3}} \, dy$$

$$= \int_1^n dx \, \frac{-2}{8\sqrt{x^3}\sqrt{y}} \Big|_x^n$$

$$= \int_1^n \left(\frac{1}{4\sqrt{x^3}\sqrt{x}} - \frac{1}{4\sqrt{x^3}\sqrt{n}} \right) dx$$

$$= \left(\frac{-1}{4x} - \frac{-2}{4\sqrt{x}\sqrt{n}} \right) \Big|_1^n$$

$$= \left(\frac{-1}{4n} + \frac{1}{2n} \right) - \left(\frac{-1}{4} + \frac{1}{2\sqrt{n}} \right)$$

$$= \frac{1}{4n} + \frac{1}{4} - \frac{1}{2\sqrt{n}} \to \frac{1}{4}.$$

Figure 14.13: The set $1 < x < y$, and $D_n = \{(x,y) : 1 < x < n,\ x < y < n\}$.

Thus, g is integrable on D, and the Comparison Test implies that so is f. ◆

We started the section with a caveat about non-negative functions. The multivariable case is, in that regard, dramatically different from what we have learned about functions of a single variable. For example, we have seen in Example 13.5.1 that the integral

$$\int_0^\infty \frac{\sin x}{x} \, dx$$

converges. However, it does not converge absolutely. Let $n \in \mathbb{N}$. Then

$$\int_0^{(2n+1)\pi} \frac{|\sin x|}{x} \, dx = \sum_{k=0}^{2n} \int_{k\pi}^{(k+1)\pi} \frac{|\sin x|}{x} \, dx$$

$$\ge \sum_{j=0}^{n} \int_{2j\pi}^{(2j+1)\pi} \frac{\sin x}{x} \, dx$$

$$\ge \sum_{j=0}^{n} \int_{2j\pi}^{(2j+1)\pi} \frac{\sin x}{(2j+1)\pi} \, dx$$

$$= \sum_{j=0}^{n} \frac{1}{(2j+1)\pi} (-\cos x) \Big|_{2j\pi}^{(2j+1)\pi}$$

$$= \sum_{j=0}^{n} \frac{1}{(2j+1)\pi} \left(\cos(2j\pi) - \cos((2j+1)\pi) \right)$$

$$= \sum_{j=0}^{n} \frac{2}{(2j+1)\pi}$$

$$\ge \frac{1}{\pi} \sum_{j=0}^{n} \frac{1}{j+1}.$$

The last sum is just a partial sum of the Harmonic series, so the integral $\int_0^\infty \frac{\sin x}{x} \, dx$ converges conditionally but not absolutely. Such a phenomenon is impossible in the case of a double (improper) integral.

Theorem 14.6.9. *An improper integral $\iint_D f(x, y) \, dA(x, y)$ converges if and only if it converges absolutely.*

Proof. Suppose first that the integral converges absolutely, i.e., suppose that the integral $\iint_D |f(x, y)| \, dA(x, y)$ converges. If f^+ and f^- are as in (14.21), then $0 \le f^+, f^- \le |f|$ so, by the Comparison Test, both f^+ and f^- are integrable on D. Consequently, $f = f^+ - f^-$ is also integrable on D.

Let us now prove the converse. We will, actually, prove the contrapositive, i.e., we will assume that $|f|$ is not integrable on D, and we will establish that neither is f. Since $\iint_D |f(x, y)| \, dA(x, y)$ diverges, there is a monotone covering $\{D_n\}$ of D such that the sequence $\iint_{D_n} |f(x, y)| \, dA(x, y)$ is unbounded. By passing to a subsequence, if necessary, we may assume that

$$\iint_{D_{n+1}} |f(x, y)| \, dA(x, y) > 3 \iint_{D_n} |f(x, y)| \, dA(x, y) + 2n \tag{14.25}$$

for all $n \in \mathbb{N}$. If we define $E_n = D_{n+1} \setminus D_n$, then E_n is a Jordan set (see Exercises 14.2.1–

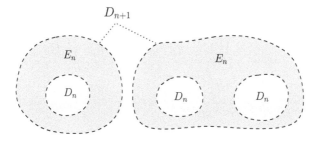

Figure 14.14: $E_n = D_{n+1} \setminus \overline{D}_n$.

14.2.3), and D_{n+1} is the union of Jordan sets D_n and E_n. It follows that

$$\iint_{D_{n+1}} |f(x, y)| \, dA(x, y) \le \iint_{D_n} |f(x, y)| \, dA(x, y) + \iint_{E_n} |f(x, y)| \, dA(x, y). \tag{14.26}$$

Combining (14.25) and (14.26) yields

$$\iint_{E_n} |f(x, y)| \, dA(x, y) > 2 \iint_{D_n} |f(x, y)| \, dA(x, y) + 2n. \tag{14.27}$$

Let us compare the non-negative numbers

$$I_1 = \iint_{E_n} f^+(x, y) \, dA(x, y) \quad \text{and} \quad I_2 = \iint_{E_n} f^-(x, y) \, dA(x, y),$$

and let us assume that $I_1 \ge I_2$. (The other option could be considered in exactly the same way as the proof that follows.) Since $|f| = f^+ + f^-$ we have that

$$\iint_{E_n} |f(x, y)| \, dA(x, y) = I_1 + I_2 \le 2I_1. \tag{14.28}$$

Now, (14.27) and (14.28) imply that

$$\iint_{E_n} f^+(x,y)\,dA(x,y) > \iint_{D_n} |f(x,y)|\,dA(x,y) + n.$$

Let P_n be a partition of E_n such that the lower Darboux sum

$$L(f^+, P_n) > \iint_{D_n} |f(x,y)|\,dA(x,y) + n.$$

The function $f^+ \geq 0$, so if R is a rectangle in P_n, $\inf\{f^+(x,y) : (x,y) \in R\} \geq 0$. Let $\mathcal{R}'_n$ be the collection of those rectangles in P_n where the infimum is strictly positive. (This implies that, on each such a rectangle, $f = f^+$.) For every $n \in \mathbb{N}$, we define $G_n = D_n \cup R'_n$, where R'_n is the union of the interiors of all rectangles in $\mathcal{R}'_n$. It is not hard to see that the collection $\{G_n\}$ is a monotone covering of D (Exercise 14.6.26). Further,

$$\iint_{G_n} f(x,y)\,dA(x,y) = \iint_{R'_n} f(x,y)\,dA(x,y) + \iint_{D_n} f(x,y)\,dA(x,y)$$

$$= \iint_{R'_n} f^+(x,y)\,dA(x,y) + \iint_{D_n} f(x,y)\,dA(x,y)$$

$$\geq L(f^+, P_n) + \iint_{D_n} f(x,y)\,dA(x,y)$$

$$\geq \iint_{D_n} |f(x,y)|\,dA(x,y) + n + \iint_{D_n} f(x,y)\,dA(x,y)$$

$$\geq n,$$

which shows that f is not integrable on D.   □

Remark 14.6.10. The example of $\int_0^\infty \frac{\sin x}{x}\,dx$ shows that Theorem 14.6.9 fails in the one dimensional case. The reason is that, for functions defined on a subset of $\mathbb{R}$, the definition of the convergence of an improper integral is different from Definition 14.6.4. Namely, in Chapter 6 we have also used the monotone covering (without calling it that way), but we allowed only intervals. As we have said before, there are no such dominant sets in $\mathbb{R}^2$, and it is quite possible that rectangles and disks yield a different result (see Exercise 14.6.22).

Exercises

In Exercises 14.6.1–14.6.8 determine whether the integral converges.

14.6.1. $\displaystyle\iint_D \frac{1}{x^2+y^2}\,dA(x,y)$, if D is the region defined by $|y| \leq x^2$, $x^2 + y^2 \leq 1$.

14.6.2. $\displaystyle\iint_{-1<x,y<1} \frac{1}{x^2+y^2}\,dA(x,y).$ 14.6.3. $\displaystyle\iint_{x^2+y^2\leq1} \frac{\ln(x^2+y^2)}{\sqrt{x^2+y^2}}\,dA(x,y).$

14.6.4. $\displaystyle\iint_{x^2+y^2\geq1} \ln(x^2+y^2)\,dA(x,y).$ 14.6.5. $\displaystyle\iint_{x^2+y^2\leq1} \frac{\sqrt{x^2+xy+y^2}}{x^2+y^2}\,dA(x,y).$

14.6.6. $\displaystyle\iint_D \frac{x^2-y^2}{(x^2+y^2)^2}\,dA(x,y)$ if D is: (a) the triangle with vertices $(0,0)$, $(0,1)$, $(1,1)$; (b) the strip $\{(x,y) : 0 \leq y \leq 1, x \geq 1\}$; (c) the region in the first quadrant bounded by $y = 1$ and $y = x$.

14.6.7.* $f(x, y) = 1/x$ if $x \neq 0$ and $f(0, y) = 0$, and R is the triangle with vertices $(0, 0)$, $(1, 1)$ and $(-1, 1)$.

14.6.8.* $\iint\limits_{D} x^{-3/2} e^{y-x} \, dA(x, y)$, where D is the region in the 1st quadrant bounded by $y = 0$ and $y = x$.

In Exercises 14.6.9–14.6.12 determine for what values of p, q the integral converges:

14.6.9. $\iint\limits_{x^2+y^2 \leq 1} \dfrac{1}{(x^2 + y^2)^p} \, dA(x, y)$. 　　14.6.10. $\iint\limits_{\mathbb{R}^2} (1 + x^2 + y^2)^p \, dA(x, y)$.

14.6.11.* $\iint\limits_{x+y \geq 1} \dfrac{\sin x \sin y}{(x + y)^p} \, dA(x, y)$. 　　14.6.12.* $\iint\limits_{|x|+|y| \leq 1} \dfrac{1}{|x|^p + |y|^q} \, dA(x, y), \ p, q > 0.$

In Exercises 14.6.13–14.6.21 prove that the integral converges then evaluate it:

14.6.13. $\iint\limits_{\mathbb{R}^2} e^{-x^2 - y^2} \cos(x^2 + y^2) \, dA(x, y)$. 　14.6.14. $\iint\limits_{0 \leq x, y \leq 1} \dfrac{1}{x + y} \, dA(x, y)$.

14.6.15. $\iint\limits_{x^2+y^2 \leq 1} \ln(x^2 + y^2) \, dA(x, y)$. 　　14.6.16. $\iint\limits_{x^2+y^2 \leq 1} \dfrac{1}{\sqrt{1 - x^2 - y^2}} \, dA(x, y)$.

14.6.17. $\iint\limits_{0 \leq x, y \leq 1} x^{-1/2} \, dA(x, y)$. 　　14.6.18. $\iint\limits_{0 \leq x, y \leq 1} y^2 / \sqrt{(x^2 + y^2)^3} \, dA(x, y)$.

14.6.19. $\iint\limits_{D} \dfrac{1}{1 + (x^2 + y^2)^2} \, dA(x, y)$, if D is the region in the first quadrant bounded by the x-axis and the line $y = x$.

14.6.20. $\iint\limits_{R} xy e^{-(x^2 + y^2)} \, dA(x, y)$, where R is the region in the first quadrant between $y = 0$ and $y = 1$.

14.6.21. $\iint\limits_{R} x^{-2}(1 - x)^{-1} \, dA(x, y)$, where R is the region between $y = x$ and $y = -x + 2$, to the right of $(1, 1)$.

14.6.22. Show that

$$\lim_{n \to \infty} \iint\limits_{|x|, |y| \leq n} \sin(x^2 + y^2) \, dA(x, y) = \pi, \quad \lim_{n \to \infty} \iint\limits_{x^2+y^2 \leq 2n\pi} \sin(x^2 + y^2) \, dA(x, y) = 0.$$

What does that say about $\iint\limits_{\mathbb{R}^2} \sin(x^2 + y^2) \, dA(x, y)$?

14.6.23. Let $\{D_n : n \in \mathbb{N}\}$ and $\{E_n : n \in \mathbb{N}\}$ be monotone coverings of a set D. Prove that the same is true of $\{D_n \cup E_n : n \in \mathbb{N}\}$. Is it true for $\{D_n \cap E_n : n \in \mathbb{N}\}$?

14.6.24. Prove that Definitions 14.6.4 and 14.2.1 are equivalent when D is a bounded set.

14.6.25. Prove Comparison Test (Theorem 14.6.7).

14.6.26. Prove that the collection $\{G_n\}$, as defined in the proof of Theorem 14.6.9, is a monotone covering of D.

14.6.27. Let D be an open set in $\mathbb{R}^2$. Prove that there exists an increasing sequence $\{D_n\}$ of compact Jordan sets such that $D = \cup_{n=1}^{\infty} D_n^o$, where D_n^o denotes the interior of D_n.

14.6.28. Prove that Theorem 14.5.3 holds if the integrals in (14.20) are assumed to be improper.

14.6.29. Let R be a region in the xy-plane and f a non-negative function on R. Suppose that $\iint_R f(x,y)\,dA$ exists, either as an ordinary double integral or as an improper double integral. Prove that, for any positive number ε, there is a positive number δ such that if $D \subset R$ and the area of D is less than δ then $\iint_D f(x,y)\,dA < \varepsilon$.

14.6.30. Prove that Theorem 14.2.6 remains true for improper integrals, i.e. if the integrals of f and g are assumed to exist as improper integrals.

14.6.31. Let D_1, D_2 be Jordan sets in the xy-plane such that $D_1 \subset D_2$. Suppose that $\iint_{D_2} f(x,y)\,dA$ exists, either as an ordinary double integral or as an improper double integral. Prove that the same is true of $\iint_{D_1} f(x,y)\,dA$.

14.6.32.* Let D_1, D_2 be Jordan sets in the xy-plane and suppose that both $\iint_{D_1} f(x,y)\,dA$ and $\iint_{D_2} f(x,y)\,dA$ exist. Prove that the same is true of $\iint_{D_1 \cup D_2} f(x,y)\,dA$. Further, show that if $D_1 \cap D_2 = \emptyset$ then $\iint_{D_1 \cup D_2} f(x,y)\,dA = \iint_{D_1} f(x,y)\,dA + \iint_{D_2} f(x,y)\,dA$.

14.6.33.* Let K be a compact Jordan set in $\mathbb{R}^2$ and suppose that f is non-negative and continuous on the interior of K. Prove that f is integrable on K if and only if for every $(x,y) \in K$ there exists an open disk D containing (x,y) so that f is integrable on $K \cap D$.

14.6.34.* Let f be a non-negative continuous function on $\mathbb{R}^2$ and let $I_1 = \iint_{\mathbb{R}^2} f(x,y)\,dA$, $I_2 = \int_{-\infty}^{\infty} dy \int_{-\infty}^{\infty} f(x,y)\,dx$. Suppose that for every $c \in \mathbb{R}$ there exists a non-negative function $F_c : \mathbb{R} \to \mathbb{R}$ that is integrable on $\mathbb{R}$ and such that $f(x,y) \le F_c(x)$, for all $(x,y) \in \mathbb{R} \times [-c,c]$. Also, suppose that $\int_{-\infty}^{\infty} f(x,y)\,dx$ is continuous on $[-c,c]$. Prove that the integral I_1 converges if and only if I_2 does.

14.7 Multiple Integrals

So far in this chapter we have dealt exclusively with functions defined on subsets of $\mathbb{R}^2$. This has simplified the exposition, yet the vast majority of the results and proofs can be easily modified to $\mathbb{R}^n$, $n \ge 3$. We will leave this task to the reader, and in this section we will focus on some distinct features of multiple integrals.

Let us quickly summarize the integration theory in $\mathbb{R}^n$. First we define integrals over *generalized rectangles*, i.e., sets of the form

$$[a_1, b_1] \times [a_2, b_2] \times \cdots \times [a_n, b_n].$$

Such sets are also called *hyperrectangles* or *boxes*. If f is a function defined on a generalized rectangle R, and if P is a partition of R, we define the upper and the lower Darboux sums $U(f, P)$ and $L(f, P)$. Then f is integrable if $L = \sup L(f, P)$ and $U = \inf U(f, P)$ are equal, in which case the common value is the **multiple integral** of f over R, denoted by

$$\int_R \cdots \int f(x_1, x_2, \ldots, x_n)\,dV_n \quad \text{or} \quad \int_R \cdots \int f(x_1, x_2, \ldots, x_n)\,dV.$$

When D is a Jordan set (meaning that its boundary has content 0), we enclose it by a generalized rectangle, and define the extension $\hat{f}$ to be zero outside of D. All the theorems from Sections 14.1 and 14.2 remain valid.

That brings us to Section 14.3 and the evaluation of multiple integrals as iterated integrals. Let us start with an example.

Example 14.7.1. Evaluation of a triple integral: a review.

Evaluate $\iiint_R \left(\dfrac{1}{x} + y + z^2 \right) dV$, if $R = [1,3] \times [0,5] \times [-1,0]$.

Solution.

$$
\begin{aligned}
\iiint_R \left(\frac{1}{x} + y + z^2 \right) dV &= \int_1^3 dx \int_0^5 dy \int_{-1}^0 \left(\frac{1}{x} + y + z^2 \right) dz \\
&= \int_1^3 dx \int_0^5 dy \left(\frac{z}{x} + yz + \frac{z^3}{3} \right) \Big|_{z=-1}^{z=0} \\
&= \int_1^3 dx \int_0^5 \left(\frac{1}{x} + y + \frac{1}{3} \right) dy \\
&= \int_1^3 dx \left(\frac{y}{x} + \frac{y^2}{2} + \frac{y}{3} \right) \Big|_{y=0}^{y=5} \\
&= \int_1^3 \left(\frac{5}{x} + \frac{25}{2} + \frac{5}{3} \right) dx \\
&= \left(5 \ln x + \frac{25x}{2} + \frac{5x}{3} \right) \Big|_1^3 \\
&= \left(5 \ln 3 + \frac{75}{2} + 5 \right) - \left(\frac{25}{2} + \frac{5}{3} \right) \\
&= 5 \ln 3 + 25 + \frac{10}{3} = 5 \ln 3 + \frac{85}{3}. \qquad \blacklozenge
\end{aligned}
$$

It is worth noticing that we have started with a triple integral and, only two steps later, we were left with a double integral (written as an iterated integral). In other words, we have evaluated a triple integral by reducing it to a double integral. This strategy is useful for any $n \in \mathbb{N}$ (not just $n = 3$).

Theorem 14.7.2. *Let* $R = [a_1, b_1] \times [a_2, b_2] \times \cdots \times [a_n, b_n]$, *and let* f *be a function integrable on* R. *Let* $R' = [a_1, b_1] \times [a_2, b_2] \times \cdots \times [a_{n-1}, b_{n-1}]$ *and, for each* $\mathbf{x} = (x_1, x_2, \dots, x_{n-1}) \in R'$, *let* $F_{\mathbf{x}}$ *be a function on* $[a_n, b_n]$ *defined by* $F_{\mathbf{x}}(x_n) = f(x_1, x_2, \dots, x_n)$. *Also, suppose that* $F_{\mathbf{x}}$ *is integrable and define* $A(\mathbf{x}) = \int_{a_n}^{b_n} F_{\mathbf{x}}(x_n) \, dx_n$. *Then* A *is integrable on* R' *and*

$$
\int_R \cdots \int f(x_1, x_2, \dots, x_n) \, dV_n = \int_{R'} \cdots \int A(\mathbf{x}) \, dV_{n-1}. \tag{14.29}
$$

Proof. The proof is based on the inequality

$$
L(f, P) \le L(A, P_1) \le U(A, P_1) \le U(f, P). \tag{14.30}
$$

where P is a partition of R, and P_1 is the induced partition of R'. Both the proof of (14.30) as well as the theorem follow the same path as in Theorem 14.3.2. $\qquad \square$

Theorem 14.7.2 allows us to use an inductive argument, which leads to the representation of a multiple integral as an iterated integral.

Theorem 14.7.3. *Let $R = [a_1, b_1] \times [a_2, b_2] \times \cdots \times [a_n, b_n]$, and let f be a function continuous on R. Then*

$$\int \cdots \int_R f(x_1, x_2, \ldots, x_n)\, dV_n = \int_{a_1}^{b_1} dx_1 \int_{a_2}^{b_2} dx_2 \cdots \int_{a_n}^{b_n} f(x_1, x_2, \ldots, x_n)\, dx_n.$$

Example 14.7.4. Evaluation of a multiple integral.

Let $n \in \mathbb{N}$, and $R = \{(x_1, x_2, \ldots, x_n) \in \mathbb{R}^n : 0 \leq x_i \leq 1, 1 \leq i \leq n\}$. Evaluate $I_n = \int \cdots \int_R (x_1 + x_2 + \cdots + x_n)\, dV_n$.

Solution. By Theorem 14.7.2,

$$I_n = \int \cdots \int_{R'} dV_{n-1} \int_0^1 (x_1 + x_2 + \cdots + x_n)\, dx_n$$

$$= \int \cdots \int_{R'} dV_{n-1} \left(x_n(x_1 + x_2 + \cdots + x_{n-1}) + \frac{x_n^2}{2} \right) \Bigg|_{x_n=0}^{x_n=1}$$

$$= \int \cdots \int_{R'} dV_{n-1} \left(x_1 + x_2 + \cdots + x_{n-1} + \frac{1}{2} \right)$$

$$= I_{n-1} + \frac{1}{2} \int \cdots \int_{R'} dV_{n-1}$$

$$= I_{n-1} + \frac{1}{2}.$$

Since $I_1 = 1/2$, it follows that $I_n = n/2$.   ◆

When the domain of integration D is not a rectangle, a formula similar to (14.29) is available, assuming that D is x_n-simple. We omit the proof because of the similarity with Theorem 14.3.6.

Theorem 14.7.5. *Let α, β be two functions defined and continuous on $D' \subset \mathbb{R}^{n-1}$ and suppose that, for all $\mathbf{x} = (x_1, x_2, \ldots, x_{n-1}) \in D'$, $\alpha(\mathbf{x}) \leq \beta(\mathbf{x})$. Let $D = \{(\mathbf{x}, x_n) : \mathbf{x} \in D', \alpha(\mathbf{x}) \leq x_n \leq \beta(\mathbf{x})\}$, and let f be a function continuous on D. Then*

$$\int \cdots \int_D f(x_1, x_2, \ldots, x_n)\, dV_n = \int \cdots \int_{D'} dV_{n-1} \int_{\alpha(x)}^{\beta(x)} f(x_1, x_2, \ldots, x_n)\, dx_n.$$

The material of Sections 14.4 and 14.5, including the change of variable formula for double integrals, allows a straightforward generalization to triple integrals and beyond. We will focus on some specific examples of the change of variables, that are often very useful.

Example 14.7.6. Using cylindrical coordinates to compute a triple integral.

Calculate $\iiint_S \sqrt{x^2 + y^2}\, dV$, if S is the solid bounded by $z^2 = x^2 + y^2$ and $z = 1$.

Solution. We will use *cylindrical coordinates*

$$x = r\cos\theta, \quad y = r\sin\theta, \quad z = w. \tag{14.31}$$

It is not hard to see that $0 \leq \theta \leq 2\pi$, and $0 \leq r \leq 1$. Also, $\sqrt{x^2 + y^2} \leq z \leq 1$, so $r \leq w \leq 1$. Consequently, S^* is determined by the inequalities

$$0 \leq \theta \leq 2\pi, \quad 0 \leq r \leq 1, \quad r \leq w \leq 1.$$

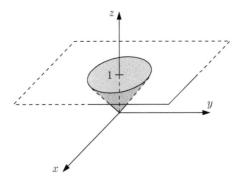

Figure 14.15: S is bounded by $z^2 = x^2 + y^2$ and $z = 1$.

Further, the integrand $f(x, y, z) = \sqrt{x^2 + y^2}$ becomes $f^*(r, \theta, w) = r$. Finally, the Jacobian determinant for cylindrical coordinates (meaning the Jacobian determinant for the function $\varphi(r, \theta, w) = (r \cos \theta, r \sin \theta, w)$) equals r (Exercise 14.7.1). According to Theorem 14.7.5, we obtain

$$\int\limits_0^{2\pi} d\theta \int\limits_0^1 dr \int\limits_r^1 r^2 \, dw = \int\limits_0^{2\pi} d\theta \int\limits_0^1 r^2 dr \, w \Big|_{w=r}^{w=1}$$

$$= \int\limits_0^{2\pi} d\theta \int\limits_0^1 r^2 (1 - r) dr$$

$$= \int\limits_0^{2\pi} d\theta \left(\frac{r^3}{3} - \frac{r^4}{4} \right) \Big|_{r=0}^{r=1}$$

$$= \int\limits_0^{2\pi} \left(\frac{1}{3} - \frac{1}{4} \right) d\theta$$

$$= \frac{1}{12} 2\pi = \frac{\pi}{6}. \qquad \blacklozenge$$

> Euler used coordinates given by (14.31) in 1776. The name was coined in the second half of the nineteenth century. One of the first occurrences is in *Solutions of the Cambridge Senate-house Problems and Riders for the Year 1875*. It also appears in the third edition of *Solid Geometry* by Percival Frost in 1886, but not in the second edition from 1875.

Example 14.7.7. Using spherical coordinates to compute a triple integral.
Calculate $\iiint_S \sqrt{x^2 + y^2 + z^2} \, dV$, if S is the solid enclosed by $x^2 + y^2 + z^2 = z$.
Solution. We will use *spherical coordinates*

$$x = \rho \sin \varphi \cos \theta, \quad y = \rho \sin \varphi \sin \theta, \quad z = \rho \cos \varphi. \qquad (14.32)$$

The equation $x^2 + y^2 + z^2 = z$ can be written as $x^2 + y^2 + (z - \frac{1}{2})^2 = \frac{1}{4}$, so it is a sphere with center $(0, 0, \frac{1}{2})$ and radius $\frac{1}{2}$ (Figure 14.16). Further, in spherical coordinates its equation is $\rho^2 = \rho \cos \varphi$. Thus, the region S^* is determined by

$$0 \leq \theta \leq 2\pi, \quad 0 \leq \varphi \leq \frac{\pi}{2}, \quad 0 \leq \rho \leq \cos \varphi.$$

Also, the integrand (in spherical coordinates) equals ρ. Finally, the Jacobian determinant

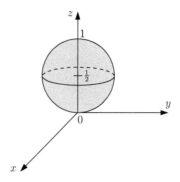

Figure 14.16: S is defined by $x^2 + y^2 + z^2 \leq z$.

for spherical coordinates equals $-\rho^2 \sin \varphi$, so its absolute value is $\rho^2 \sin \varphi$ (Exercise 14.7.2).
Thus, the triple integral equals

$$
\int_0^{2\pi} d\theta \int_0^{\frac{\pi}{2}} d\varphi \int_0^{\cos \varphi} \rho^3 \sin \varphi \, d\rho = \int_0^{2\pi} d\theta \int_0^{\frac{\pi}{2}} \sin \varphi \, d\varphi \left. \frac{\rho^4}{4} \right|_{\rho=0}^{\rho=\cos\varphi}
$$

$$
= \int_0^{2\pi} d\theta \int_0^{\frac{\pi}{2}} \sin \varphi \, \frac{\cos^4 \varphi}{4} \, d\varphi
$$

$$
= \int_0^{2\pi} d\theta \left(\left. -\frac{\cos^5 \varphi}{20} \right|_{\varphi=0}^{\varphi=\frac{\pi}{2}} \right)
$$

$$
= 2\pi \frac{1}{20} = \frac{\pi}{10}.
$$

♦

Formulas (14.32) can be found in a work of Euler from 1771, although the angles (labeled ξ and θ) appear already in [41]. As a method of computing an integral, spherical coordinates were introduced by Lagrange in 1773, in [78]. In fact he did more: in a somewhat unclear way, he gave the general method of the change of variables.

The integrals beyond triple are typically very hard, unless we can use induction.

Example 14.7.8. The volume of an n-dimensional ball.

Find the volume of the n-dimensional ball of radius R.

Solution. We will use induction to prove that the formula is

$$
V_n(R) = \frac{\pi^{\frac{n}{2}} R^n}{\Gamma(\frac{n}{2} + 1)}. \tag{14.33}
$$

Let us start with $n = 1$. The left side is just $2R$. On the right side we have $R\sqrt{\pi}/\Gamma(\frac{3}{2})$.
By (13.28), $\Gamma(\frac{1}{2}) = \sqrt{\pi}$, and the basic identity $\Gamma(t+1) = t\Gamma(t)$ implies that $\Gamma(\frac{3}{2}) = \sqrt{\pi}/2$.
Consequently, the right side of (14.33) is

$$
\frac{R\sqrt{\pi}}{\frac{\sqrt{\pi}}{2}} = 2R
$$

and (14.33) holds for $n = 1$. When $n = 2$, the right side of (14.33) equals $\pi R^2/\Gamma(2) = \pi R^2$, the area of the disk of radius R.

Suppose now that (14.33) is valid for $1, 2, \ldots, n-1$, and let us prove it for n. By definition, the volume of an n-ball B is the integral $\int \cdots \int_B dV_n$. Let us assume that $n \geq 3$, and let us introduce the change of variables where $x_{n-1} = r \cos \theta$, and $x_n = r \sin \theta$, while the variables x_i, $1 \leq i \leq n-2$, remain the same. The equation of the sphere $x_1^2 + x_2^2 + \cdots + x_n^2 = R^2$ becomes $x_1^2 + x_2^2 + \cdots + x_{n-2}^2 = R^2 - r^2$, and the Jacobian determinant of this transformation equals r (Exercise 14.7.3). We obtain that

$$V_n(R) = \int_0^{2\pi} d\theta \int_0^R r \, dr \int \cdots \int_{B'} dx_1 dx_2 \ldots dx_{n-2}$$

where B' is enclosed by the surface $x_1^2 + x_2^2 + \cdots + x_{n-2}^2 = R^2 - r^2$. In other words, B' is a ball of radius $\sqrt{R^2 - r^2}$ in $\mathbb{R}^{n-2}$. By induction hypothesis, the volume of B' equals

$$\frac{\pi^{\frac{n-2}{2}} (R^2 - r^2)^{\frac{n-2}{2}}}{\Gamma(\frac{n-2}{2} + 1)} = \frac{\pi^{\frac{n-2}{2}} (R^2 - r^2)^{\frac{n-2}{2}}}{\Gamma(\frac{n}{2})}.$$

It follows that

$$V_n(R) = \frac{\pi^{\frac{n-2}{2}}}{\Gamma(\frac{n}{2})} \int_0^{2\pi} d\theta \int_0^R (R^2 - r^2)^{\frac{n-2}{2}} r \, dr$$

$$= \frac{\pi^{\frac{n-2}{2}}}{\Gamma(\frac{n}{2})} \int_0^{2\pi} d\theta \left(-\frac{1}{n} \right) (R^2 - r^2)^{\frac{n}{2}} \Big|_{r=0}^{r=R}$$

$$= \frac{\pi^{\frac{n-2}{2}}}{\Gamma(\frac{n}{2})} \int_0^{2\pi} \frac{R^n}{n} d\theta$$

$$= \frac{\pi^{\frac{n-2}{2}}}{\Gamma(\frac{n}{2})} 2\pi \frac{R^n}{n}$$

$$= \frac{\pi^{\frac{n}{2}} R^n}{\Gamma(\frac{n}{2} + 1)}. \qquad \blacklozenge$$

Exercises

14.7.1. Prove that the Jacobian determinant for cylindrical coordinates equals r.

14.7.2. Prove that the Jacobian determinant for spherical coordinates equals $-\rho^2 \sin \varphi$.

14.7.3. Let $n \in \mathbb{N}$, $n \geq 3$, and suppose that the formulas

$$x_1 = y_1, \quad x_2 = y_2, \quad \ldots \quad x_{n-2} = y_{n-2}, \quad x_{n-1} = r \cos \theta, \quad x_n = r \sin \theta$$

determine a change of variables. Show that the Jacobian determinant of this transformation equals r.

14.7.4. Let $n \in \mathbb{N}$, $n \geq 3$, and suppose that the formulas

$$x_1 = \rho \cos \varphi_1, \quad x_2 = \rho \sin \varphi_1 \cos \varphi_2, \quad \ldots$$
$$\ldots \quad x_{n-1} = \rho \sin \varphi_1 \sin \varphi_2 \cdots \sin \varphi_{n-2} \cos \varphi_{n-1}, \quad x_n = \rho \sin \varphi_1 \sin \varphi_2 \cdots \sin \varphi_{n-2} \sin \varphi_{n-1}$$

determine a change of variables. Show that the Jacobian determinant of this transformation equals $\rho \sin^{n-2} \varphi_1 \sin^{n-3} \varphi_2 \cdots \sin \varphi_{n-2}$.

In Exercises 14.7.5–14.7.11 evaluate $\iiint_S f(x, y, z) \, dV$:

14.7.5. $f(x, y, z) = xy^2z^3$, S is bounded by $z = xy$, $y = x$, $x = 1$, $z = 0$.

14.7.6. $f(x, y, z) = (1 + x + y + z)^{-2}$, S is bounded by $x + y + z = 1$, $x = 0$, $y = 0$, $z = 0$.

14.7.7. $f(x, y, z) = x^2/a^2 + y^2/b^2 + z^2/c^2$, S is enclosed by $x^2/a^2 + y^2/b^2 + z^2/c^2 = 1$.

14.7.8. $f(x, y, z) = 1 + x(x^2 + y^2)^{-1/2}$, S is bounded by $z = x^2 + y^2$, $z = 1 - x^2 - y^2$.

14.7.9. $f(x, y, z) = x + y$, S is bounded above by $x^2 + y^2 + z^2 = 16$ and below by $z = \sqrt{3x^2 + 3y^2}$.

14.7.10. $f(x, y, z) = z$, $S = \{(x, y, z) : -1 \le x \le 0, |y| \le \sqrt{1 - x^2}, -1 \le z \le 3 - x^2 - y^2\}$.

14.7.11. $f(x, y, z) = 1/\left(1 + (x^2 + y^2 + z^2)^{3/2}\right)$, $S = \{(x, y, z) : 0 \le y \le 1/2, y \le x \le \sqrt{\frac{1}{2} - y^2}, \sqrt{x^2 + y^2} \le z \le \sqrt{1 - x^2 - y^2}\}$.

In Exercises 14.7.12–14.7.19 find $\int \cdots \int\limits_{R} f(x_1, x_2, \ldots, x_n)\, dV_n$:

14.7.12. $f(\mathbf{x}) = x_1^2 + x_2^2 + \cdots + x_n^2$, $R = [0, 1] \times [0, 1] \times \cdots \times [0, 1]$.

14.7.13. $f(\mathbf{x}) = x_n^2$, $R = \{\mathbf{x} \in \mathbb{R}^n : x_1^2 + x_2^2 + \cdots + x_{n-1}^2 \le a^2, -\frac{h}{2} \le x_n \le \frac{h}{2}\}$.

14.7.14.* $f(\mathbf{x}) = \min\{x_1, x_2, \ldots, x_n\}$, $R = [0, 1] \times [0, 1] \times \cdots \times [0, 1]$.

14.7.15.* $f(\mathbf{x}) = \min\{1, a/x_1, a/x_2, \ldots, a/x_n\}$, $0 < a < 1$, $R = [0, 1] \times [0, 1] \times \cdots \times [0, 1]$.

14.7.16.* $f(\mathbf{x}) = (x_1 + x_2 + \cdots + x_n)^2$, $R = [0, 1] \times [0, 1] \times \cdots \times [0, 1]$.

14.7.17.* $f(\mathbf{x}) = \sqrt{x_1 + x_2 + \cdots + x_n}$, $R = \{\mathbf{x} \in \mathbb{R}^n : x_1 + x_2 + \cdots + x_n \le 1, x_i \ge 0, 1 \le i \le n\}$.

14.7.18.* $f(\mathbf{x}) = \left(1 - x_1^2 - x_2^2 - \cdots - x_n^2\right)^{-1/2}$, $R = \{\mathbf{x} \in \mathbb{R}^n : x_1^2 + x_2^2 + \cdots + x_n^2 \le 1\}$.

14.7.19.* $f(\mathbf{x}) = x_1 x_2 \cdots x_n$, $R = \{\mathbf{x} \in \mathbb{R}^n : 0 \le x_1 \le 1, 0 \le x_{i+1} \le x_i, 1 \le i \le n - 1\}$.

In Exercises 14.7.20–14.7.23 find the volume of the n-dimensional region R:

14.7.20. $R = \{\mathbf{x} \in \mathbb{R}^n : \frac{x_1}{a_1} + \frac{x_2}{a_2} + \cdots + \frac{x_n}{a_n} \le 1, x_i \ge 0, a_i > 0, 1 \le i \le n\}$.

14.7.21. R is enclosed by $a_{i1}x_1 + a_{i2}x_2 + \cdots + a_{in}x_n = \pm h_i$, $1 \le i \le n$, and $\det(a_{ij}) \ne 0$.

14.7.22.* R is enclosed by $\frac{x_1^2}{a_1^2} + \frac{x_2^2}{a_2^2} + \cdots + \frac{x_{n-1}^2}{a_{n-1}^2} = \frac{x_n^2}{a_n^2}$ and $x_n = a_n$.

14.7.23. $R = \{\mathbf{x} \in \mathbb{R}^5 : x_1^2 + x_2^2 + x_3^2 \le 1, x_4^2 + x_5^2 \le 1\}$.

14.7.24.** Let $V(r, n)$ be the volume of the set $\{\mathbf{x} \in \mathbb{R}^n : \|\mathbf{x}\| \le r\}$, let $\varepsilon > 0$ and let $\{\mathbf{u}_i : 1 \le i \le n\}$ be an orthonormal set in $\mathbb{R}^n$. For any $1 \le k \le n$, define $V(r, n, k, \varepsilon)$ to be the volume of $\{\mathbf{x} \in \mathbb{R}^n : \|\mathbf{x}\| \le r, |\mathbf{x} \cdot \mathbf{u}_i| \le \varepsilon, 1 \le i \le k\}$. Find $\lim_{n \to \infty} V(r, n, k, \varepsilon)/V(r, n)$.

In Exercises 14.7.25–14.7.28 determine for what values of p, q, r, $\iiint\limits_{R} f(x, y, z)\, dV(x, y, z)$ converges:

14.7.25. $f(x, y, z) = (x^2 + y^2 + z^2)^p$, $R = \{(x, y, z) : x^2 + y^2 + z^2 \ge 1\}$.

14.7.26. $f(x, y, z) = (|x|^p + |y|^q + |z|^r)^{-1}$, $R = \{(x, y, z) : |x| + |y| + |z| \ge 1\}$, $p, q, r > 0$.

14.7.27. $f(x, y, z) = |x + y - z|^p$, $R = [-1, 1] \times [-1, 1] \times [-1, 1]$.

14.7.28. $f(x, y, z) = (1 - x^2 - y^2 - z^2)^p$, $R = \{(x, y, z) : x^2 + y^2 + z^2 \le 1\}$.

14.7.29. Let $U \subset \mathbb{R}^n$ and $V \subset \mathbb{R}^m$ be Jordan sets, and define $W = U \times V$. Prove that W is a Jordan set and that the volume (Jordan content) of W equals the product of volumes of U and V.

14.7.30.* Let R be a generalized rectangle in $\mathbb{R}^n$ and $g : R \to \mathbb{R}$ an integrable function. Prove that the set $\{(\mathbf{x}, g(\mathbf{x})) : \mathbf{x} \in R\}$ in $\mathbb{R}^{n+1}$ has Jordan content 0.

15

Fundamental Theorems of Multivariable Calculus

The Fundamental Theorem of Calculus expresses a relationship between the derivative and the definite integral. When a function depends on more than one variable, the connection is still there, although it is less transparent. In this chapter we will look at some multivariable generalizations of the Fundamental Theorem of Calculus.

15.1 Curves in $\mathbb{R}^n$

Let us start with the following question. Suppose that C is a curve in the xy-plane, and that we are interested in its length. The first order of business is to define what we mean by a *curve* and by its *length*.

Let $\mathbf{f} : [a, b] \to \mathbb{R}^n$. When $\mathbf{f}$ is continuous, we call it a **path** in $\mathbb{R}^n$. The image of $\mathbf{f}$ is called a **curve** in $\mathbb{R}^n$. The function $\mathbf{f}$ is a **parametrization** of the curve C.

Example 15.1.1. A parametrization of the unit circle.

The function $\mathbf{f}(t) = (\cos t, \sin t)$, defined for $0 \le t \le 2\pi$, is a parametrization of the unit circle in $\mathbb{R}^2$. ◆

Example 15.1.2. Another parametrization of the unit circle.

The path $\mathbf{g}(t) = (\cos 2t, \sin 2t)$, defined for $0 \le t \le \pi$, traces the same curve as in Example 15.1.1. ◆

We say that two paths $\mathbf{f} : [a, b] \to \mathbb{R}^n$ and $\mathbf{g} : [c, d] \to \mathbb{R}^n$ are **equivalent** if there exists a C^1 bijection $\varphi : [a, b] \to [c, d]$ such that $\varphi'(t) > 0$ for all $t \in [a, b]$ and

$$\mathbf{f} = \mathbf{g} \circ \varphi.$$

The paths $\mathbf{f}$ and $\mathbf{g}$ in the examples above are equivalent. The bijection $\varphi : [0, 2\pi] \to [0, \pi]$ is given by $\varphi(t) = t/2$. The relation of equivalence of paths is an equivalence relation (Exercise 15.1.1).

Example 15.1.3. A parametrization of the unit circle not equivalent to the previous.

The image of $\mathbf{h}(t) = (\cos t, \sin t)$, when $0 \le t \le 4\pi$, is again the unit circle, but the path is not equivalent to those in Examples 15.1.1 and 15.1.2 (Exercise 15.1.2). The obvious difference is that $\mathbf{f}$ traces the curve only once, whereas $\mathbf{h}$ loops twice around the origin. ◆

Throughout this chapter we will assume that the paths are **simple**, meaning that the function $\mathbf{f}$ is injective, with a possible exception at the endpoints. Namely, we can have $\mathbf{f}(a) = \mathbf{f}(b)$, in which case we say that a path is **closed**.

Example 15.1.4. Another parametrization of the unit circle not equivalent to the previous.

Let $\mathbf{k}(t) = (\sin t, \cos t)$, $\pi/2 \le t \le 5\pi/2$. Although the path traces the unit circle there is an essential differences between this one and the paths $\mathbf{f}$ and $\mathbf{g}$ in Examples 15.1.1 and 15.1.2. Namely, the circle is traversed *clockwise* whereas in the earlier examples it was always *counterclockwise*. Because of that the parametrization $\mathbf{k}$ is equivalent to neither $\mathbf{f}$ nor $\mathbf{g}$ (Exercise 15.1.3).   ♦

The observation that the unit circle can be oriented in 2 opposite ways extends to all simple closed curves in $\mathbb{R}^2$. Namely, such a curve is **positively oriented** if, when traveling along it one always has the curve interior to the left. Thus, the counterclockwise orientation, as in Examples 15.1.1 and 15.1.2, is called **positive**, while the path $\mathbf{k}$ is **oriented negatively**. We often write $-C$ to denote the curve C with the opposite orientation.

Remark 15.1.5. While intuitively obvious, the definition of the orientation relies on the fact that every simple closed curve in $\mathbb{R}^2$ divides the plane into two disjoint regions: the "interior" (bounded by the curve) and the "exterior" (on the other side). This fact is known as the **Jordan curve theorem**, and it is an extremely hard result to prove. Jordan's original proof has been disputed, and the known proofs often rely on the machinery of Algebraic Topology.

So, let $\mathbf{f}$ be a simple path. How can we compute its length? One way is to try to approximate it by the length of a polygonal line.

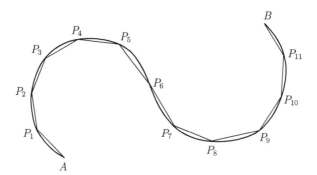

Figure 15.1: Approximating the curve by a polygonal line.

We expect that, as the partitions of $[a, b]$ get finer, the approximation gets better. If $P = \{t_0, t_1, \ldots, t_n\}$ is a partition of $[a, b]$, then the length of the polygonal line is

$$L(\mathbf{f}, P) = \sum_{k=1}^{n} \|\mathbf{f}(t_k) - \mathbf{f}(t_{k-1})\|, \tag{15.1}$$

and we define the **length of a curve** as

$$L = \sup\{L(\mathbf{f}, P) : P \text{ is a partition of } [a, b]\}.$$

When the supremum is finite, we say that the curve is **rectifiable**.

Clearly, an interesting question is to determine conditions that will guarantee the finiteness of the supremum above and, in the case when it is finite, to compute it. Our experience with integrals might suggest continuity. Unfortunately, this is not sufficient. In 1890, Peano was the first to give an example of a continuous function $\mathbf{f} : [0, 1] \to [0, 1] \times [0, 1]$ that is surjective! It is often referred to as a "space filling curve". A year later Hilbert gave a more geometric example (see Exercise 15.1.33). Therefore, we will restrict our attention to C^1 paths, i.e., the case when $\mathbf{f} \in C^1$.

Theorem 15.1.6. *Let* $\mathbf{f} : [a, b] \to \mathbb{R}^n$ *be a simple* C^1 *path. Then its length equals*

$$L = \int_a^b \|\mathbf{f}'(t)\| \, dt.$$

Proof. We will prove the case when $\mathbf{f}$ is a path in $\mathbb{R}^2$, and leave the general case for the reader. The formula (15.1) can be simplified. Namely, we write $\mathbf{f} = (f_1, f_2)$. Then, with the use of the Mean Value Theorem,

$$L(\mathbf{f}, P) = \sum_{k=1}^n \|(f_1(t_k), f_2(t_k)) - (f_1(t_{k-1}), f_2(t_{k-1}))\|$$

$$= \sum_{k=1}^n \|(f_1(t_k) - f_1(t_{k-1}), f_2(t_k) - f_2(t_{k-1}))\|$$

$$= \sum_{k=1}^n \|(f_1'(\xi_{1,k})(t_k - t_{k-1}), f_2'(\xi_{2,k})(t_k - t_{k-1}))\|$$

$$= \sum_{k=1}^n \|(t_k - t_{k-1})(f_1'(\xi_{1,k}), f_2'(\xi_{2,k}))\|$$

$$= \sum_{k=1}^n |t_k - t_{k-1}| \, \|(f_1'(\xi_{1,k}), f_2'(\xi_{2,k}))\|$$

$$= \sum_{k=1}^n \|(f_1'(\xi_{1,k}), f_2'(\xi_{2,k}))\| \, \Delta t_k.$$

The last sum is "almost" a Riemann sum for the function

$$g(t) = \|\mathbf{f}'(t)\| = \|(f_1'(t), f_2'(t))\|.$$

However, it is not a Riemann sum because the intermediate points $\xi_{1,k}$ and $\xi_{2,k}$ may be different. Nevertheless, we will show that these sums converge to $\int_a^b g(t) \, dt$.

Let $\varepsilon > 0$, and let $G(x, y) = \|(f_1'(x), f_2'(y))\|$, for $(x, y) \in [a, b] \times [a, b]$. Then G is a continuous function on a compact set, hence uniformly continuous. Therefore, there exists $\delta_1 > 0$ such that

$$|G(x, y) - G(z, w)| < \frac{\varepsilon}{2(b - a)}, \quad \text{whenever } |(x, y) - (z, w)| < \delta_1.$$

This implies that, if P is a partition of $[a, b]$ and $\|P\| < \delta_1$, then

$$|L(\mathbf{f}, P) - S(g, P)| = \left| \sum_{k=1}^n \|(f_1'(\xi_{1,k}), f_2'(\xi_{2,k}))\| \, \Delta t_k - \sum_{k=1}^n g(\xi_k) \, \Delta t_k \right|$$

$$= \left| \sum_{k=1}^n [G(\xi_{1,k}, \xi_{2,k}) - G(\xi_k, \xi_k)] \Delta t_k \right|$$

$$\leq \sum_{k=1}^n |G(\xi_{1,k}, \xi_{2,k}) - G(\xi_k, \xi_k)| \, \Delta t_k$$

$$< \sum_{k=1}^{n} \frac{\varepsilon}{2(b-a)} \, \Delta t_k$$

$$= \frac{\varepsilon}{2(b-a)} \, (b-a) = \frac{\varepsilon}{2}. \tag{15.2}$$

On the other hand, the function g is integrable, so there exists $\delta_2 > 0$ such that, if P is a partition of $[a,b]$ and $\|P\| < \delta_2$, then

$$\left| S(g,P) - \int_a^b g(t)\,dt \right| < \frac{\varepsilon}{2}. \tag{15.3}$$

Let $\delta = \min\{\delta_1, \delta_2\}$. If $\|P\| < \delta$, then both (15.2) and (15.3) hold and, combining them, we obtain that

$$\left| L(\mathbf{f},P) - \int_a^b g(t)\,dt \right| < \varepsilon.$$

Since ε was arbitrary, the theorem is proved. $\qquad\qquad\qquad\qquad\qquad\qquad\qquad\square$

Remark 15.1.7. Theorem 15.1.6 gives the length in terms of a specific parametrization. Exercise 15.1.6 states that choosing another equivalent parametrization yields the same result.

Example 15.1.8. Computing the length of a part of a parabola.

Let $f(x) = x^2$, defined for $x \in [0,1]$. Find the length of the graph of f.

Solution. Whenever a curve is given by an equation of the form $y = h(x)$, we can use the parametrization $\mathbf{f}(x) = (x, h(x))$. Here, $\mathbf{f}(x) = (x, x^2)$, so $\mathbf{f}'(x) = (1, 2x)$. By Theorem 15.1.6,

$$L = \int_0^1 \sqrt{1 + 4x^2}\,dx = 2\int_0^1 \sqrt{x^2 + \frac{1}{4}}\,dx. \tag{15.4}$$

Using Example 5.1.8 with $a = 1/2$, we obtain that

$$L = 2\left[\frac{x}{2}\sqrt{x^2 + \frac{1}{4}} + \frac{1}{8}\ln 2\left(x + \sqrt{x^2 + \frac{1}{4}}\right)\right]\Big|_0^1$$

$$= 2\left[\frac{1}{2}\sqrt{\frac{5}{4}} + \frac{1}{8}\ln 2\left(1 + \sqrt{\frac{5}{4}}\right)\right]$$

$$= \frac{\sqrt{5}}{2} + \frac{1}{4}\ln(2 + \sqrt{5}).$$

Consequently, the length of the parabolic arc equals $L \approx 1.48$. $\qquad\qquad\qquad\qquad\blacklozenge$

Finding the length of a curve intrigued mathematicians in the ancient world, but became a really hot topic in the seventeenth century. For example, it was important to find the distance traveled by a planet. Even before calculus made its appearance, there were isolated accomplishments. English architect Christopher Wren (1632–1723), famous for St.Paul's Cathedral in London, found the length of a cycloid (see Exercise 15.1.17). With calculus, there was a powerful tool, although one was often left with an integral without an elementary antiderivative. Such is, e.g., the case when computing the length of an arc of an ellipse.

Exercises

15.1.1. Prove that the equivalence of paths is an equivalence relation.

15.1.2. Prove that the paths $\mathbf{f}(t) = (\cos t, \sin t)$, $0 \le t \le 2\pi$, and $\mathbf{h}(t) = (\cos t, \sin t)$, $0 \le t \le 4\pi$ are not equivalent.

15.1.3. Prove that the path $\mathbf{k}(t) = (\sin t, \cos t)$, $0 \le t \le 2\pi$ is equivalent to neither $\mathbf{f}(t) = (\cos t, \sin t)$, $0 \le t \le 2\pi$, nor $\mathbf{g}(t) = (\cos 2t, \sin 2t)$, $0 \le t \le \pi$.

15.1.4. Let $\mathbf{f}(t) = (\cos t, \sin t)$, $0 \le t \le 2\pi$, and $\mathbf{g}(t) = (\cos t, \sin t)$, $-\pi \le t \le \pi$. Are these parametrizations of the unit circle equivalent?

15.1.5. Let $\mathbf{f} : [a, b] \to \mathbb{R}^2$ and $\mathbf{g} : [c, d] \to \mathbb{R}^2$ be two parametrizations of the unit circle, so that $\mathbf{f}$ is oriented positively and $\mathbf{g}$ is oriented negatively. Prove that they are not equivalent.

15.1.6. Let $\mathbf{f} : [a, b] \to \mathbb{R}^n$ and $\mathbf{g} : [c, d] \to \mathbb{R}^n$ be two equivalent C^1 paths. Prove that $\int_a^b \|\mathbf{f}'(t)\| \, dt = \int_c^d \|\mathbf{g}'(t)\| \, dt$.

15.1.7. Let $\mathbf{f} : [a, b] \to \mathbb{R}^n$ and $\mathbf{g} : [c, d] \to \mathbb{R}^n$ be two paths, each of which is one-to-one and such that $\mathbf{f}(a) = \mathbf{g}(c)$. Prove that $\mathbf{f}$ and $\mathbf{g}$ are equivalent if and only if they have the same image.

15.1.8. Let C be the curve in which the cylinders $x^2 + z^2 = 1$ and $y^2 + z^2 = 1$ intersect. Find a parametrization for the part of C in the first octant.

15.1.9. Prove Theorem 15.1.6 in the case when $n > 2$.

15.1.10. Let f be a continuous function and $r = f(\theta)$, $\alpha \le \theta \le \beta$, an equation of a curve C in polar coordinates. Prove that the length of C is $\int_\alpha^\beta \sqrt{f'(\theta)^2 + f(\theta)^2} \, d\theta$.

15.1.11. Derive a formula for the length of a curve C, if C is given in polar coordinates by $r = f(t)$, $\theta = g(t)$, $a \le t \le b$.

In Exercises 15.1.12–15.1.14 the set $A \subset \mathbb{R}^2$ is shaped like a letter E, consisting of segments joining the following pairs of points: $(0, 4)$ and $(3, 4)$; $(0, 2)$ and $(2, 2)$; $(0, 0)$ and $(3, 0)$; and $(0, 0)$ and $(0, 4)$.

15.1.12. Find a function $\mathbf{f} : [0, 1] \to \mathbb{R}^2$ that parametrizes A in such a way that $\mathbf{f}(0) = (3, 0)$ and $\mathbf{f}(1) = (3, 4)$. Determine the length of A under this parametrization.

15.1.13. Find a parametrization that yields A as a closed curve and the length of A under this parametrization.

15.1.14. Prove that A cannot be parametrized as a simple path.

In Exercises 15.1.15–15.1.19 find the length of the curve in $\mathbb{R}^2$:

15.1.15. $\mathbf{f}(t) = (\cos^3 t, \sin^3 t)$, $0 \le t \le \pi/2$.

15.1.16. $\mathbf{f}(t) = (\cos^4 t, \sin^4 t)$, $0 \le t \le \pi/2$.

15.1.17. $\mathbf{f}(t) = (a(t - \sin t), a(1 - \cos t))$, $0 \le t \le 2\pi$.

15.1.18. $\mathbf{f}(t) = (a(t \sin t + \cos t), a(\sin t - t \cos t))$, $0 \le t \le 2\pi$.

15.1.19. $y = e^x$, $0 \le x \le 1$.

15.1.20.* Let

$$g(x) = \begin{cases} x \cos(\pi/x) & \text{if } 0 < x \le 1 \\ 0, & \text{if } x = 0, \end{cases}$$

and define $\mathbf{f}(t) = (t, g(t))$, $0 \le t \le 1$. Prove that $\mathbf{f}$ is continuously differentiable in $(0, 1)$ but that the curve it traces is not rectifiable.

In Exercises 15.1.21–15.1.24 a curve in $\mathbb{R}^2$ is given in polar coordinates. Find its length.

15.1.21. $r = a\theta$, $0 \le \theta \le 2\pi$.

15.1.22. $r = a\cos\theta$, $-\pi/2 \le \theta \le \pi/2$.

15.1.23. $r = ae^{k\theta}$, $0 < r < a$.

15.1.24. $\theta = \sqrt{r}$, $0 \le r \le 5$.

In Exercises 15.1.25–15.1.31 find the length of the curve in $\mathbb{R}^3$:

15.1.25. $y = 2\arcsin\frac{x}{2}$, $z = \frac{1}{2}\ln\frac{2-x}{2+x}$, from $(0,0,0)$ to (a,b,c).

15.1.26. $(x-y)^2 = 3(x+y)$, $x^2 - y^2 = \frac{9}{8}z^2$, from $(0,0,0)$ to (a,b,c).

15.1.27. $\mathbf{f}(t) = (3\cos t, 3\sin t, 2t)$, $0 \le t \le 2\pi$.

15.1.28. $\mathbf{f}(t) = (3t, 3t^2, 2t^3)$, $(0,0,0)$ to $(3,3,2)$.

15.1.29. $\mathbf{f}(t) = (e^{-t}\cos t, e^{-t}\sin t, e^{-t})$, $(1,0,1)$ to $(0,0,0)$.

15.1.30. $\mathbf{f}(t) = (t, t^2, 1)$, $0 \le t \le 4$.

15.1.31. $x^2 + y^2 = 2z$, $\frac{y}{x} = \tan\frac{z}{2}$, from $(0,0,0)$ to (a,b,c).

15.1.32.* The purpose of this problem is to establish a sharp inequality involving the **width** w and length L of a closed curve C.

(a) Let $\mathbf{u} = (\cos\theta, \sin\theta)$, let $\mathbf{f} : [0,1] \to \mathbb{R}^2$ be a continuous parametrization of C, and define a projection $\mathbf{f_u} = (\mathbf{f} \cdot \mathbf{u})\mathbf{u}$. Find the formula for the length $L = L(\theta)$ of $\mathbf{f_u}(t)$, $0 \le t \le 1$.

(b) Prove that the formula

$$\frac{1}{2\pi}\int_0^{2\pi} L(\theta)\, d\theta = \frac{2L}{\pi}, \tag{15.5}$$

holds in the case when C is a line segment.

(c) Prove (15.5) in the general case.

(d) For a fixed θ, let w_θ denote the width of C in the direction $\mathbf{u}$, i.e., the distance between a pair of parallel tangent lines of C perpendicular to $\mathbf{u}$. Prove that

$$w_\theta \le \frac{L}{\pi}.$$

(e) Let w denote $\max\{w_\theta : 0 \le \theta \le 2\pi\}$. Conclude that

$$w \le \frac{L}{\pi}. \tag{15.6}$$

(f) Give an example to show that the inequality (15.6) is sharp.

15.1.33.* The purpose of this problem is to present Hilbert's example of a "space filling" curve published in [65] in 1891 (one year after Peano gave the first, more complicated, example). We will construct a function $\mathbf{f} : I \to A$, where $I = [0,1]$ and $A = [0,1] \times [0,1]$.

 The construction is done in stages. At stage 1, split the interval I into 4 equal subintervals, and the square A into 4 equal squares, labeling them by $1, 2, 3, 4$, as in Figure 15.2. At stage 2, split each of the 4 subintervals into 4 equal parts, and do the same to each of the 4 squares, labeling the intervals and squares by $1, 2, 3, \ldots, 15, 16$. Continue this process, by splitting each of the new intervals and the associated squares into 4 equal parts. By carefully labeling the intervals and the squares, we obtain a one-to-one correspondence between intervals and squares. Make sure that, at any stage, if an interval I' is contained in an interval I'', then the associated squares A' and A'' satisfy $A' \subset A''$. Also, if intervals

I' and I'' have a common point, then the corresponding squares A' and A'' have a common edge.

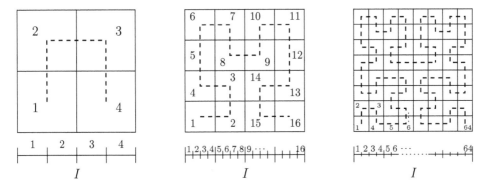

Figure 15.2: Hilbert's space filling curve.

Next we define a function $\mathbf{f} : I \to A$. Let $x \in I$. At every stage x belongs to at least one of the intervals. (It may belong to two intervals if it is a boundary point.) That way, we can associate to x a sequence of nested intervals and, by the construction above, a sequence of nested (closed) squares. By Exercise 10.4.22 there exists a unique point (a, b) that belongs to the intersection of these squares. We define $\mathbf{f}(x) = (a, b)$.

(a) Prove that $\mathbf{f}$ is well defined, i.e., that $\mathbf{f}(x)$ is independent of the choice of intervals that contain x.

(b) Prove that the range of $\mathbf{f}$ is A.

(c) Prove that $\mathbf{f}$ is continuous.

15.2   Line Integrals

In Chapter 6 we have studied the definite integral: a function f was defined on a line segment $[a, b]$ and we were interested in $\int_a^b f(x)\, dx$. In this section, the segment $[a, b]$ will be replaced by a curve in $\mathbb{R}^2$ or, more generally, in $\mathbb{R}^n$. Consequently, the integrand will be a function of n variables. The most interesting cases are when it is scalar valued or when it is a **vector field**, i.e., with values in $\mathbb{R}^n$. (The emphasis is that the dimension of the domain and the codomain are equal.)

We will start with the following question of practical importance. Let $\mathbf{f}$ be a path in $\mathbb{R}^2$, and let C be the curve traced by $\mathbf{f}$. Suppose that we are interested in the mass of a thin wire in the shape of C. We can use the formula $m = V\rho$ ("mass = volume × density"), except that we know neither its volume nor its density. If we disregard the units, the volume can be identified with the length. Thus, if we are given the density function u, we can approximate the curve by a polygonal line, and the mass by the sum

$$\sum_{k=1}^{n} u(\mathbf{f}(\xi_k))\|\mathbf{f}(x_k) - \mathbf{f}(x_{k-1})\|.$$

It is not hard to see that, as $\|P\| \to 0$, the sums converge to $\int_a^b u(\mathbf{f}(t))\|\mathbf{f}'(t)\|\, dt$. Motivated by this, we introduce the following definition.

Definition 15.2.1. Let $\mathbf{f} : [a, b] \to \mathbb{R}^n$ be a simple C^1 path, and let u be an integrable function defined on $C = \mathbf{f}([a, b])$. The **line integral** of u along the curve C is denoted by $\int_C u\, dL$ and it is defined by

$$\int_C u\, dL = \int_a^b u(\mathbf{f}(t))\|\mathbf{f}'(t)\|\, dt.$$

When the curve C is closed, we often write $\oint_C u\, dL$.

Example 15.2.2. Computing a line integral.

Calculate $\int_C (3x - 2y)\, dL$, if C is parametrized by $\mathbf{f}(t) = (4t^2 + 3t, 6t^2 - 2t)$, $0 \le t \le 1$.

Solution. Here $u(x, y) = 3x - 2y$, so

$$u(\mathbf{f}(t)) = 3(4t^2 + 3t) - 2(6t^2 - 2t) = 13t.$$

Since $\mathbf{f}'(t) = (8t + 3, 12t - 2)$, we have that

$$\|\mathbf{f}'(t)\| = \sqrt{(8t + 3)^2 + (12t - 2)^2} = \sqrt{64t^2 + 48t + 9 + 144t^2 - 48t + 4} = \sqrt{208t^2 + 13}.$$

Thus,

$$\int_C u\, dL = \int_0^1 13t\sqrt{208t^2 + 13}\, dt.$$

The substitution $w = 208t^2 + 13$ yields $dw = 416t\, dt = 32 \cdot 13t\, dt$ so

$$\int_C u\, dL = \int_{13}^{221} \frac{1}{32}\sqrt{w}\, dw = \frac{1}{32}\frac{2}{3} w^{3/2}\Big|_{13}^{221} = \frac{1}{48}\left(221\sqrt{221} - 13\sqrt{13}\right) \approx 67.47. \qquad \blacklozenge$$

Definition 15.2.1 involves the use of a parametrization $\mathbf{f}$. Since the same curve C can have more than one parametrization, it is important to establish that the value of the line integral does not depend on the choice of a parametrization.

Theorem 15.2.3. *If $\mathbf{f} : [a, b] \to \mathbb{R}^n$ and $\mathbf{g} : [c, d] \to \mathbb{R}^n$ are two equivalent C^1 parametrizations of a curve C in $\mathbb{R}^n$, and if u is an integrable function on C, then*

$$\int_a^b u(\mathbf{f}(t))\|\mathbf{f}'(t)\|\, dt = \int_c^d u(\mathbf{g}(t))\|\mathbf{g}'(t)\|\, dt.$$

Consequently, the line integral is independent of a parametrization.

We leave the proof as an exercise, just like the proofs of the following results.

Theorem 15.2.4. *Let u_1, u_2 be two functions defined and integrable on the curve C, and let $\alpha_1, \alpha_2 \in \mathbb{R}$. Then*

$$\int_C (\alpha_1 u_1 + \alpha_2 u_2)\, dL = \alpha_1 \int_C u_1\, dL + \alpha_2 \int_C u_2\, dL.$$

Theorem 15.2.5. *Let C_1 be a curve in $\mathbb{R}^2$ connecting points A_1, A_2, let C_2 be a curve in $\mathbb{R}^2$ connecting points A_2, A_3, and let $C = C_1 \cup C_2$. If $u : C \to \mathbb{R}$ is a continuous function, then*

$$\int_C u\, dL = \int_{C_1} u\, dL + \int_{C_2} u\, dL.$$

Theorem 15.2.6. *Let C_1 be a curve in $\mathbb{R}^2$ and let C_2 be the same curve with the opposite orientation. If $u : C_1 \to \mathbb{R}$ is a continuous function, then*

$$\int_{C_1} u \, dL = \int_{C_2} u \, dL.$$

Line integrals can also be used to calculate the work done by moving a particle along a curve. The formula $W = \mathbf{F} \cdot \mathbf{x}$ ("work is the dot product of the force and the distance") is useful when the force is constant. When the force is variable, the curve can be approximated by a polygonal line, so that $\mathbf{F}$ is "almost" constant on each piece of the polygonal line. This is the motivation behind the following definition.

Definition 15.2.7. Let $\mathbf{f} : [a, b] \to \mathbb{R}^n$ be a simple C^1 path, and let $\mathbf{F}$ be an integrable vector field defined on $C = \mathbf{f}([a, b])$. The **line integral** of $\mathbf{F}$ along the curve C is denoted by $\int_C \mathbf{F} \cdot d\mathbf{x}$ and it is defined by

$$\int_C \mathbf{F} \cdot d\mathbf{x} = \int_a^b \mathbf{F}(\mathbf{f}(t)) \cdot \mathbf{f}'(t) \, dt.$$

When the curve C is closed, we often write $\oint_C \mathbf{F} \cdot d\mathbf{x}$.

Example 15.2.8. Computing a line integral of a vector field.

Let $\mathbf{F}(x, y, z) = (z, x, y)$, $\mathbf{f}(t) = (\sin t, 3 \sin t, \sin^2 t)$, $0 \le t \le \pi/2$. Calculate $\int_C \mathbf{F} \cdot d\mathbf{x}$.
Solution. First, $\mathbf{F}(\mathbf{f}(t)) = (\sin^2 t, \sin t, 3 \sin t)$. Next, $\mathbf{f}'(t) = (\cos t, 3 \cos t, 2 \sin t \cos t)$, so

$$\int_C \mathbf{F} \cdot d\mathbf{x} = \int_0^{\pi/2} (\sin^2 t, \sin t, 3 \sin t) \cdot (\cos t, 3 \cos t, 2 \sin t \cos t) \, dt$$

$$= \int_0^{\pi/2} \left(\sin^2 t \cos t + 3 \sin t \cos t + 6 \sin^2 t \cos t\right) dt$$

$$= \int_0^{\pi/2} \left(7 \sin^2 t + 3 \sin t\right) \cos t \, dt$$

$$= \left(7 \frac{1}{3} \sin^3 t + 3 \frac{1}{2} \sin^2 t\right) \Big|_0^{\pi/2}$$

$$= \frac{7}{3} + \frac{3}{2} = \frac{23}{6}. \qquad \blacklozenge$$

Line integrals of vector fields share many properties with the case when the integrand is a real-valued function (Exercises 15.2.7–15.2.10).

Although line integrals were used by physicists in the eighteenth century, the mathematical development came through the use of complex numbers, and the study of paths in the complex plane. Gauss was the first to suggest that, if the integration was along a path in the complex plane, the Fundamental Theorem of Calculus might not be applicable. In 1820, Poisson (see below) published [88] in which he demonstrated an example where Gauss's concern was justified. This was one of the major reasons for Cauchy to define the integral not as an antiderivative (which was the popular view in the eighteenth century), but rather as a limit of sums. The term "line integral" was first used in 1873 by James Maxwell (page 280).

Let us denote the component functions of $\mathbf{F}$ by $F_1, F_2, \ldots, F_n$, so that for $\mathbf{x} \in \mathbb{R}^n$, $\mathbf{F}(\mathbf{x}) = (F_1(\mathbf{x}), F_2(\mathbf{x}), \ldots, F_n(\mathbf{x}))$. Also, let us use the notation $\mathbf{f}(t) = (f_1(t), f_2(t), \ldots, f_n(t))$. Then

$$\mathbf{F}(\mathbf{f}(t)) \cdot \mathbf{f}'(t) = F_1(\mathbf{f}(t)) \cdot f_1'(t) + F_2(\mathbf{f}(t)) \cdot f_2'(t) + \cdots + F_n(\mathbf{f}(t)) \cdot f_n'(t).$$

Finally, let $dx_k = f'_k(t)\,dt$, $1 \le k \le n$. Then

$$\int_a^b \mathbf{F}(\mathbf{f}(t)) \cdot \mathbf{f}'(t)\,dt = \int_a^b F_1(\mathbf{f}(t))\,dx_1 + F_2(\mathbf{f}(t))\,dx_2 + \cdots + F_n(\mathbf{f}(t))\,dx_n$$

and we ofen write $\int_C F_1\,dx_1 + F_2\,dx_2 + \cdots + F_n\,dx_n$.

Example 15.2.9. Computing a line integral.

Calculate $\oint_C (x+y)\,dx + (x-y)\,dy$, if C is the ellipse $\frac{x^2}{4} + \frac{y^2}{9} = 1$, oriented counter-clockwise.

Solution. We will use the parametrization $\mathbf{f} = (2\cos t, 3\sin t)$, $0 \le t \le 2\pi$. Then $dx = -2\sin t\,dt$, $dy = 3\cos t\,dt$, so we obtain

$$\int_0^{2\pi} \left[(2\cos t + 3\sin t)(-2\sin t) + (2\cos t - 3\sin t)(3\cos t) \right]\,dt$$

$$= \int_0^{2\pi} \left(6\cos^2 t - 6\sin^2 t - 13\sin t \cos t \right)\,dt$$

$$= \int_0^{2\pi} \left(6\cos 2t - \frac{13}{2}\sin 2t \right)\,dt = \left(3\sin 2t + \frac{13}{4}\cos 2t \right)\Big|_0^{2\pi} = 0. \qquad \blacklozenge$$

When the curve C is a graph of a function $y = f(x)$ in $\mathbb{R}^2$, the situation is even simpler.

Example 15.2.10. Computing a line integral.

Calculate $\int_C (x^2 - 2xy)\,dx + (y^2 - 2xy)\,dy$, if C is the parabola $y = x^2$, $-1 \le x \le 1$.

Solution. We will use x as a parameter, so $dy = 2x\,dx$. Consequently, we obtain

$$\int_{-1}^1 \left[x^2 - 2x(x^2) + ((x^2)^2 - 2x(x^2))\,2x \right]\,dx = \int_{-1}^1 \left(x^2 - 2x^3 - 4x^4 + 2x^5 \right)\,dx = -\frac{14}{15}. \qquad \blacklozenge$$

SIMÉON DENIS POISSON (1781–1840) was a French mathematician and physicist. At the age of 14 he was sent to train for a surgeon in Fontaine-bleau, but his lack of manual dexterity was an unsurmountable obstacle. Reportedly, his first patient died. So, he went to the Ecole Centrale and excelled. He placed first on the entrance exam for the Ecole Polytechnique in Paris. As a student he did very well except when drawing diagrams. He wrote his first paper when he was only 18, and it was so well received that he graduated next year without having to take the final examination. At the recommendation of Laplace he was offered a job in Paris, contrary to the standard practice to start away from the capital. He received many appoint-ments: a professor at the Ecole Polytechnique, an astronomer at Bureau des Longitudes, a professor of mechanics in the newly opened Faculté des Sciences, an examiner for the Ecole Militaire, an examiner for the final ex-aminations at the Ecole Polytechnique. Nevertheless, he authored between 300 and 400 publications. He made many significant contributions in applied mathematics, especially in the theory of electricity and magnetism, which virtually created a new branch of mathematical physics. In celestial mechanics, he worked on the stability of the planetary orbits and extended the results of his mentors Lagrange and Laplace. In probability he introduced the "Poisson" distribution and the "law of large numbers". In pure mathematics, his most im-portant works were his series of memoirs on definite integrals and his discussion of Fourier series, his work laying a foundation for the later work of Dirichlet and Riemann. Perhaps because he was a contemporary of Cauchy and Fourier, his reputation was bigger outside of France. He is quoted to say *Life is good for only two things, discovering mathematics and teaching mathematics.*

Exercises

15.2.1. Show that, if C is the graph of $y = f(x)$, $a \le x \le b$, and if F is an integrable function of 2 variables defined on C, then

$$\int_C F(x, y)\, dx = \int_a^b F(x, f(x))\, dx.$$

15.2.2. Show that, if C is a vertical line segment $c \le y \le d$, and if F is an integrable function of 2 variables defined on C, then

$$\int_C F(x, y)\, dx = 0.$$

15.2.3. Prove Theorem 15.2.3.

15.2.4. Prove Theorem 15.2.4.

15.2.5. Prove Theorem 15.2.5.

15.2.6. Prove Theorem 15.2.6

15.2.7. State and prove the analog of Theorem 15.2.4 for line integrals of vector fields.

15.2.8. Let C_1 be a curve in $\mathbb{R}^2$ connecting points A_1, A_2, let C_2 be a curve in $\mathbb{R}^2$ connecting points A_2, A_3, and let $C = C_1 \cup C_2$. If $\mathbf{F} : C \to \mathbb{R}^2$ is a continuous function, then

$$\int_C \mathbf{F} \cdot d\mathbf{x} = \int_{C_1} \mathbf{F} \cdot d\mathbf{x} + \int_{C_2} \mathbf{F} \cdot d\mathbf{x}.$$

15.2.9. Let C_1 be a curve in $\mathbb{R}^2$ and let C_2 be the same curve with the opposite orientation. If $\mathbf{F} : C_1 \to \mathbb{R}^2$ is a continuous function, then

$$\int_{C_1} \mathbf{F} \cdot d\mathbf{x} = \int_{C_2} \mathbf{F} \cdot d\mathbf{x}.$$

15.2.10. If $\mathbf{f} : [a, b] \to \mathbb{R}^n$ and $\mathbf{g} : [c, d] \to \mathbb{R}^n$ are two equivalent C^1 parametrizations of a curve C in $\mathbb{R}^n$, and if $\mathbf{F}$ is an integrable vector field on C, then

$$\int_a^b \mathbf{F}(\mathbf{f}(t)) \cdot \mathbf{f}'(t)\, dt = \int_c^d \mathbf{F}(\mathbf{g}(t)) \cdot \mathbf{g}'(t)\, dt.$$

Consequently, the line integral is independent of a parametrization.

In Exercises 15.2.11–15.2.26 find the line integrals:

15.2.11. $\int_C (x + y)\, dL$, C is the boundary of the triangle with vertices $(0, 0)$, $(1, 0)$, and $(0, 1)$.

15.2.12. $\int_C \left(x^{4/3} + y^{4/3}\right) dL$, C is the astroid $x^{2/3} + y^{2/3} = a^{2/3}$, $a > 0$.

15.2.13. $\int_C |y|\, dL$, C is the lemniscate $(x^2 + y^2)^2 = 4(x^2 - y^2)$.

15.2.14. $\int_C \sqrt{x^2 + y^2}\, dL$, C is the circle $x^2 + y^2 = 2x$.

15.2.15. $\int_C (x^2 + y^2)\, dx + (x^2 - y^2)\, dy$, C is given by $y = 1 - |1 - x|$, $0 \le x \le 2$.

15.2.16. $\int_C (2a - y)\, dx + x\, dy$, C is the arc of a cycloid $x = a(t - \sin t)$, $y = a(1 - \cos t)$, $0 \le t \le 2\pi$.

15.2.17. $\int_C \sin y \, dx + \sin x \, dy$, C is the line segment from $(0, \pi)$ to $(\pi, 0)$.

15.2.18. $\oint_C \arctan(y/x) \, dy - dx$, C is oriented counterclockwise and consists of the arc of parabola $y = x^2$ and a segment of the line $y = x$.

15.2.19. $\int_C (x^2 + y^2 + z^2) \, dL$, C is given by $x = a \cos t$, $y = a \sin t$, $z = bt$, $0 \leq t \leq 2\pi$.

15.2.20.* $\int_C x^2 \, dL$, C is the circle $x^2 + y^2 + z^2 = 9$, $x + y + z = 0$.

15.2.21. $\int_C z \, dL$, C is the curve given by $x^2 + y^2 = z^2$, $y^2 = 2x$, from $(0, 0, 0)$ to $(2, 2, 2\sqrt{2})$.

15.2.22. $\int_C z \, dL$, C is the curve given by $x = t \cos t$, $y = t \sin t$, $z = t$, $0 \leq t \leq \pi$.

15.2.23. $\int_C (y^2 - z^2) \, dx + 2yz \, dy - x^2 \, dz$, C is given by $x = t$, $y = t^2$, $z = t^3$, $0 \leq t \leq 1$.

15.2.24. $\int_C (y - z) \, dx + (z - x) \, dy + (x - y) \, dz$, C is the circle $x^2 + y^2 + z^2 = 4$, $y = x\sqrt{3}$, oriented counterclockwise when viewed from the positive x-axis.

15.2.25. $\int_C y \, dx + z \, dy + x \, dz$, C is the C is the curve given by $x = t \cos t$, $y = t \sin t$, $z = t$, as t increases from 0 to π.

15.2.26. $\int_C (y^2 - z^2) \, dx + (z^2 - x^2) \, dy + (x^2 - y^2) \, dz$, C is the boundary of the portion of the sphere $x^2 + y^2 + z^2 = 1$, $x \geq 0$, $y \geq 0$, $z \geq 0$, oriented so that the inner side of the sphere remains on the left.

15.2.27.* Find

$$\lim_{R \to \infty} \oint_{x^2 + y^2 = R^2} \frac{y \, dx - x \, dy}{(x^2 + xy + y^2)^2}.$$

15.3   Green's Theorem

The Fundamental Theorem of Calculus (Corollary 6.6.4) states that, if F is a differentiable function on $[a, b]$, then

$$\int_a^b F'(t) \, dt = F(b) - F(a). \tag{15.7}$$

Among other things, it shows that the integral of F' over the *whole* domain can be evaluated by considering the values of F on the *boundary* of the domain. In this section we will look at a generalization of (15.7) in the case when $\mathbf{F} : D \to \mathbb{R}^2$, and D is a 2 dimensional region. We will establish the following formula:

$$\iint_D [Q_x(x, y) - P_y(x, y)] \, dA = \oint_{\partial D} P(x, y) \, dx + Q(x, y) \, dy. \tag{15.8}$$

We will refer to it as Green's Theorem.

If we write $\mathbf{F} = (P, Q)$, then the right hand side is $\oint_{\partial D} \mathbf{F} \cdot d\mathbf{x}$, which depends only on the values of $\mathbf{F}$ on the boundary ∂D. On the left side we have the integral over the whole (2 dimensional) domain, hence a double integral. It is, perhaps, unexpected that the analogue of $F'(t)$ is $Q_x(x, y) - P_y(x, y)$. So, let us prove (15.8).

We will make the standard assumptions about $\mathbf{F}$. Namely, we want all 4 functions in (15.8) to be continuous: P, Q, P_y, and Q_x. To that end, we will assume that the region D, together with its boundary ∂D lies in an open set A, and that $\mathbf{F} \in C^1(A)$. Regarding D, we will start with the case when it is y-simple:

$$D = \{(x, y) : a \leq x \leq b, \, \alpha(x) \leq y \leq \beta(x)\}, \tag{15.9}$$

as in Figure 15.3.

Theorem 15.3.1. *Let α, β be two functions defined and continuous on $[a, b]$ and suppose that, for all $x \in [a, b]$, $\alpha(x) \leq \beta(x)$. Let D be as in (15.9), and let ∂D denote the boundary of D oriented counterclockwise. Finally, let $D \cup \partial D$ belong to an open set A, and let P be a function in $C^1(A)$. Then*

$$\iint_D P_y(x, y) \, dA = - \int_{\partial D} P(x, y) \, dx. \tag{15.10}$$

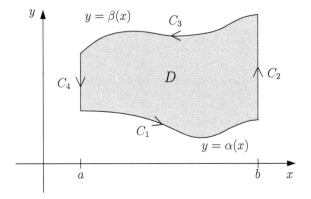

Figure 15.3: Region $D = \{(x, y) : a \leq x \leq b, \alpha(x) \leq y \leq \beta(x)\}$.

Proof. By Theorem 14.3.6,

$$\iint_D P_y(x, y) \, dA = \int_a^b dx \int_{\alpha(x)}^{\beta(x)} P_y(x, y) \, dy$$

$$= \int_a^b P(x, y) \Big|_{y=\alpha(x)}^{y=\beta(x)} dx$$

$$= \int_a^b (P(x, \beta(x)) - P(x, \alpha(x))) \, dx.$$

Using Exercise 15.2.1, and denoting the graphs of α and β by C_1 and C_3, respectively, oriented as in Figure 15.3, we obtain that

$$\iint_D P_y(x, y) \, dA = \int_{-C_3} P(x, y) \, dx - \int_{C_1} P(x, y) \, dx = - \int_{C_3} P(x, y) \, dx - \int_{C_1} P(x, y) \, dx$$

where we have used Exercise 15.2.9 to conclude that the integrals over C_3 and $-C_3$ are opposite numbers. Further, Exercise 15.2.2 shows that

$$\int_{C_2} P(x, y) \, dx = \int_{C_4} P(x, y) \, dx = 0.$$

Therefore,

$$\iint_D P_y(x, y) \, dA = - \int_{C_3} P(x, y) \, dx - \int_{C_1} P(x, y) \, dx - \int_{C_2} P(x, y) \, dx - \int_{C_4} P(x, y) \, dx$$

$$= - \int_{\partial D} P(x,y)\, dx,$$

and the theorem is proved.   □

Next, we turn our attention to *y-simple* regions in the *xy*-plane:

$$D = \{(x,y) : c \le y \le d,\ \gamma(y) \le x \le \delta(y)\}. \tag{15.11}$$

Theorem 15.3.2. *Let* γ, δ *be two functions defined and continuous on* $[c,d]$ *and suppose that, for all* $y \in [c,d]$, $\gamma(y) \le \delta(y)$. *Let* D *be as in (15.11), and let* ∂D *denote the boundary of* D *oriented counterclockwise. Finally, let* $D \cup \partial D$ *belong to an open set* A, *and let* Q *be a function in* $C^1(A)$. *Then*

$$\iint_D Q_x(x,y)\, dA = \int_{\partial D} Q(x,y)\, dy. \tag{15.12}$$

We will leave the proof to the reader, and we will now state a direct consequence of Theorems 15.3.1 and 15.3.2.

Theorem 15.3.3. *Suppose that* D *is a region in* $\mathbb{R}^2$ *that is both x-simple and y-simple, and let* ∂D *denote the boundary of* D *oriented counterclockwise. If* $D \cup \partial D$ *belongs to an open set* A, *and if* P, Q *are functions in* $C^1(A)$, *then (15.8) holds.*

Example 15.3.4. Using Green's Theorem when D is y-simple.

Calculate $\oint_C (2xy - x^2)\, dx + (x + y^2)\, dy$, if C is the boundary of the region between the parabolas $y = x^2$, $y^2 = x$.

Solution. Notice that the region D is both x-simple and y-simple (Figure 15.4). We will use Green's Theorem, with $P(x,y) = 2xy - x^2$ and $Q(x,y) = x + y^2$. The partial derivatives $P_y(x,y) = 2x$ and $Q_x(x,y) = 1$ are continuous on $\mathbb{R}$ just like the functions P and Q. Therefore,

$$\oint_C (2xy - x^2)\, dx + (x + y^2)\, dy = \iint_D (1 - 2x)\, dA$$

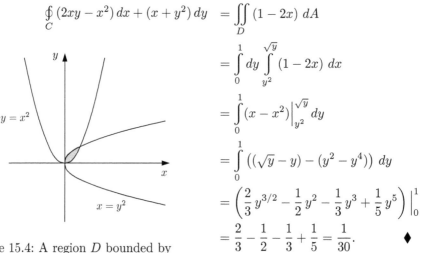

$$= \int_0^1 dy \int_{y^2}^{\sqrt{y}} (1 - 2x)\, dx$$

$$= \int_0^1 (x - x^2)\Big|_{y^2}^{\sqrt{y}}\, dy$$

$$= \int_0^1 \left((\sqrt{y} - y) - (y^2 - y^4) \right) dy$$

$$= \left(\frac{2}{3} y^{3/2} - \frac{1}{2} y^2 - \frac{1}{3} y^3 + \frac{1}{5} y^5 \right)\Big|_0^1$$

$$= \frac{2}{3} - \frac{1}{2} - \frac{1}{3} + \frac{1}{5} = \frac{1}{30}. \qquad \blacklozenge$$

Figure 15.4: A region D bounded by $y = x^2$, $y^2 = x$.

What if D does not satisfy the hypotheses of Theorem 15.3.3? Let us look at an example.

Example 15.3.5. Using Green's Theorem when D is not y-simple.

Let D be a region in Figure 15.5(a). We will show that, although D is not y-simple, (15.8) holds for D.

Solution. Notice that each of the 3 triangles D_1, D_2, D_3 in Figure 15.5(b) is y-simple.

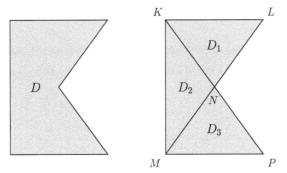

(a) D is not y-simple.   (b) D_1, D_2, D_3 are y-simple.

Figure 15.5: Region is not y-simple.

Therefore, (15.8) is true for every one of them:

$$\iint_{D_k} (Q_x(x,y) - P_y(x,y)) \, dA = \oint_{\partial D_k} P(x,y) \, dx + Q(x,y) \, dy, \quad 1 \le k \le 3. \tag{15.13}$$

If we add all 3 equations, Theorem 14.2.7 shows that the left sides add up to the double integral over D. On the other hand, Exercise 15.2.8 shows that each of the 3 integrals on the right side of (15.13) can be written as the sum of 3 line integrals over the edges of the corresponding triangle. When we add them all up we come up with 9 line integrals altogether. It is not hard to see that 5 of these add up to the line integral over the boundary of D (oriented counterclockwise). It remains to notice that all other integrals cancel. For example, the line segment MN between D_2 and D_3 features as a part of ∂D_2 (oriented from M to N) as well as a part of ∂D_3 (oriented from N to M). By Exercise 15.2.9, being of opposite orientation, two line integrals over this line segment cancel out. The same is true for the line segment KN. So, we obtain that (15.8) holds for D. ♦

Example 15.3.5 gives a blueprint how to deal with a more general region D. Recall that a curve with a parametrization $\mathbf{f} : [a, b] \to \mathbb{R}^2$ is **smooth** if $\mathbf{f} \in C^1$ and $\mathbf{f}'(t) \ne \mathbf{0}$ for all $t \in [a, b]$. It is **piecewise smooth** if there exists a partition $P = \{a = x_0, x_1, \dots, x_n = b\}$ of $[a, b]$ such that the restriction of $\mathbf{f}$ to $[x_{k-1}, x_k]$ is smooth, for $1 \le k \le n$.

Theorem 15.3.6 (Green's Theorem). *Let C be a positively oriented, piecewise smooth, simple closed curve in $\mathbb{R}^2$, and let D be the region bounded by C. If A is an open region containing D and C, and if P and Q are functions in $C^1(A)$, then*

$$\iint_D (Q_x(x,y) - P_y(x,y)) \, dA = \oint_{\partial D} P(x,y) \, dx + Q(x,y) \, dy.$$

We will not prove the theorem. When D can be decomposed into a finite number of regions that are both x-simple and y-simple, the result follows from Theorem 15.3.3, using the strategy as set forth in Example 15.3.5. For practical purposes, that is all that is needed. A complete proof of Theorem 15.3.6 is based on two assertions. The first has content and requires techniques that fall out of the scope of this book. It guarantees that every region D that satisfies the hypotheses of Green's Theorem can be decomposed into an *infinite* union of regions D_k, each being both x-simple and y-simple. Since Green's Theorem applies to each D_k, putting them together requires a limit version of the process we have used in Example 15.3.5. Such an argument can be found in [72].

The assumptions of Green's Theorem do not allow any holes in the region D. What if there are some?

Example 15.3.7. Using Green's Theorem when D has a hole.

Let D be an annulus, as in Figure 15.6(a). We will show that Green's Theorem is true for D.

Solution.

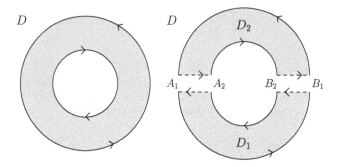

(a) The boundary consists of 2 curves.   (b) The boundary is a single curve.

Figure 15.6: Region D has a hole.

We introduce line segments $A_1 A_2$ and $A_2 A_1$, as in Figure 15.6(b). This is the same segment with two different orientations, but to make the illustration more obvious, they are drawn separately. Similarly, we have $B_1 B_2$ and $B_2 B_1$. The region D is now decomposed into 2 regions D_1 and D_2, each having a boundary that satisfies the hypotheses of Green's Theorem. ◆

> In 1845 William Thomson (a.k.a. Lord Kelvin) started referring to Theorem 15.3.6 as Green's Theorem. In the context of complex variables, Cauchy used it without proof in 1846, and the first proof (again in the complex domain) was given by Riemann in his doctoral dissertation. An excellent overview of the history of the theorem can be found in [70].

We close this section with an application of Green's Theorem. If D is a region, then $\iint_D dA$ is the area of D. Thus, if P, Q are functions such that $Q_x - P_y = 1$, then the area can be calculated using the line integral $\int_{\partial D} P\, dx + Q\, dy$.

Theorem 15.3.8. *Let D be a region in $\mathbb{R}^2$ with a positively oriented, piecewise smooth boundary C. If A is an open region containing D and C, then*

$$Area(D) = \oint_C x\, dy = \oint_C -y\, dx = \frac{1}{2} \oint_C -y\, dx + x\, dy. \tag{15.14}$$

Example 15.3.9. Computing the area using Green's theorem.

Find the area of the region bounded by the ellipse $\frac{x^2}{a^2} + \frac{y^2}{b^2} = 1$.

Solution. We will use a parametrization $x = a\cos t$, $y = b\sin t$, $0 \le t \le 2\pi$. The last expression in (15.14) becomes

$$\frac{1}{2} \int_0^{2\pi} [-b\sin t\, d(a\cos t) + a\cos t\, d(b\sin t)] = \frac{1}{2} \int_0^{2\pi} [-b\sin t(-a\sin t) + a\cos t(b\cos t)]\, dt$$

$$= \frac{1}{2} \int_0^{2\pi} \left(ab\sin^2 t + ab\cos^2 t\right)\, dt$$

$$= \frac{ab}{2} \int_0^{2\pi} dt = \pi ab.$$

Thus, the area of the region is πab. ◆

GEORGE GREEN (1793–1841) was born in Nottingham, England. His father was a baker and Green spent much of his life working in the family mill. He attended school for a year only between the ages of 8 and 9. It is conjectured that he was tutored by John Yoplis, a graduate of Queens' College in Cambridge who was a headmaster of a school in Nottingham, only a street away from where Green lived until 1817. It is known that, starting with 1817, he was a member of a local library, where he had access to the Transactions of the Royal Society of London. In 1828 he published himself an article which introduced several important concepts, among them the idea of potential functions as currently used in physics, the concept of what are now called Green's functions, and a result from which Theorem 15.3.6 can be deduced, although there are no indications that Green ever did. The article was sold through the library by subscription in 51 copies. One of the subscribers was Sir Edward Bromhead who had studied mathematics at Cambridge. He and Green met for three years and during that period Green wrote three further papers, which appeared in the Cambridge Philosophical Society and the Royal Society of Edinburgh. Bromhead convinced Green to study mathematics at Cambridge. Upon graduation in 1837 he remained at Cambridge to do research. Poor health forced him to return to Nottingham in 1840, and he died in 1841. In 1845 (four years after his death), his work was rediscovered and popularized by William Thomson, better known as Lord Kelvin (1824–1907), a Scots-Irish mathematical physicist and engineer.

Exercises

In Exercises 15.3.1–15.3.5 use Green's Theorem to find the line integrals:

15.3.1. $\oint_C xy^2\, dy - x^2 y\, dx$, C is the circle $x^2 + y^2 = a^2$.

15.3.2. $\oint_C e^x \left((1 - \cos y)\, dx - (y - \sin y)\, dy\right)$, C is the counterclockwise oriented boundary of the region $0 < x < \pi$, $0 < y < \sin x$.

15.3.3. $\oint_C (x + y)^2\, dx - (x^2 + y^2)\, dy$, C is the counterclockwise oriented boundary of the triangle with vertices $(1, 1)$, $(3, 2)$, $(2, 5)$.

15.3.4. $\oint_C e^{-(x^2 + y^2)} \left(\cos 2xy\, dx + \sin 2xy\, dy\right)$, C is the circle $x^2 + y^2 = R^2$.

15.3.5. $\int_C (e^x \sin y - 3y)\, dx + (e^x \cos y - 3)\, dy$, C is the upper semicircle $x^2 + y^2 = 2x$, going from $(2, 0)$ to $(0, 0)$.

15.3.6.* Let $P, Q : \mathbb{R}^2 \to \mathbb{R}$ be two functions in C^2. Prove that the integral

$$\oint_C P(x + a, y + b)\, dx + Q(x + a, y + b)\, dy$$

has the same value for all $a, b \in \mathbb{R}$ and all closed curves C if and only if $\partial Q/\partial x - \partial P/\partial y$ is constant.

15.3.7. Let C be a closed curve in the xy-plane, let $\mathbf{n}(x,y)$ denote the outer normal to C at (x,y), let $\mathbf{u}$ be a vector in the xy-plane and let $\alpha(x,y)$ be the angle between $\mathbf{u}$ and $\mathbf{n}(x,y)$. Prove that $\oint_C \cos\alpha(x,y)\,dL = 0$.

15.3.8. Let C be a simple smooth boundary of the region D in the xy-plane and let $\mathbf{n}(x,y)$ denote the outer normal to C at (x,y). Let (a,b) be a point in the xy-plane, let $\mathbf{r}(x,y)$ denote the vector with initial point at (a,b) and endpoint at (x,y) and let $\alpha(x,y)$ denote the angle between $\mathbf{r}(x,y)$ and $\mathbf{n}(x,y)$. Find

$$\oint_C \frac{\cos\alpha(x,y)}{\|\mathbf{r}\|}\,dL$$

if (a)* (a,b) is in D; (b)* (a,b) is outside D; (c)** (a,b) is on C.

15.3.9.* Let (a,b) be a point in the xy-plane, D an open Jordan set containing (a,b), S the area of D and d the diameter of D (i.e., $\sup\{\|\mathbf{u}-\mathbf{v}\| : \mathbf{u},\mathbf{v} \in D\}$). Let C be the boundary of D and let $\mathbf{n}(x,y)$ denote the unit vector in the direction of the outer normal to C at (x,y). Suppose that $\mathbf{F}(x,y) = (e^{-y^2}, y)$. Find

$$\lim_{d\to 0} \frac{1}{S}\oint_C \mathbf{F}\cdot\mathbf{n}\,dL.$$

In Exercises 15.3.10–15.3.15 use (15.14) to find the area of the region enclosed by:

15.3.10. The astroid $x = a\cos^3 t$, $y = b\sin^3 t$, $0 \le t \le 2\pi$.

15.3.11. The parabola $(x+y)^2 = ax$, $(a > 0)$, and the x-axis.

15.3.12.* The loop of the Folium of Descartes $x^3 + y^3 - 3axy = 0$, $a > 0$.

15.3.13.* The lemniscate $(x^2 + y^2)^2 = a^2(x^2 - y^2)$.

15.3.14. The ellipse $x = a\cos(t+\alpha) + b$, $y = c\sin t + d$.

15.3.15.* The curve $\left(\frac{x}{a}\right)^n + \left(\frac{y}{b}\right)^n = \left(\frac{x}{a}\right)^{n-1} + \left(\frac{y}{b}\right)^{n-1}$, $(a,b > 0, n \in \mathbb{N})$, and the coordinate axes.

15.4   Surface Integrals

In this section we will look at integrals of functions defined on a surface in $\mathbb{R}^3$. This will require a precise definition of a surface. Let us start with some examples.

Example 15.4.1. A surface in $\mathbb{R}^3$.

Let $z = x^2 + y^2$ for $-1 \le x, y \le 1$. The graph of this function is a surface consisting of all points $(x, y, x^2 + y^2) \in \mathbb{R}^3$, such that $-1 \le x \le 1$ and $-1 \le y \le 1$ (Figure 15.7).   ♦

We see that in our description we have used two independent parameters (x and y), and all three coordinates are functions of these parameters. Also, a (two dimensional) domain for the parameters is specified.

Example 15.4.2. A surface in $\mathbb{R}^3$.

Let $x = r\cos\theta$, $y = r\sin\theta$ and $z = r^2$, for $0 \le r \le 1$ and $0 \le \theta \le 2\pi$. Once again we have the description of a surface through 2 parameters (r and θ). It is not hard to see that, just as in Example 15.4.1, $z = x^2 + y^2$. However, in the previous example we had a portion of the paraboloid above the square $-1 \le x, y \le 1$. Here, it is above the unit disk (Figure 15.8).   ♦

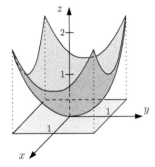

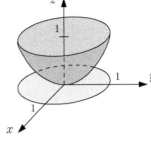

Figure 15.7: The surface $z = x^2 + y^2$, $-1 \le x, y \le 1$.

Figure 15.8: The surface $x = r \cos \theta$, $y = r \sin \theta$, $z = r^2$, $0 \le r \le 1$, $0 \le \theta \le 2\pi$.

Example 15.4.3. A surface in $\mathbb{R}^3$.

Let $x = 3 \cos t$, $y = s$ and $z = 3 \sin t$, for $0 \le t \le \pi$ and $0 \le s \le 1$. Since $x^2 + z^2 = 9$, we see that our surface is a portion of the cylindrical surface, that has the y-axis as an axis of symmetry (Figure 15.9). The cross sections are circles with center on the y-axis, and

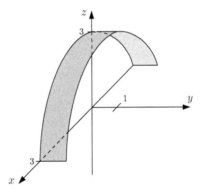

Figure 15.9: The surface $x = 3 \cos t$, $y = s$ and $z = 3 \sin t$, for $0 \le t \le \pi$ and $0 \le s \le 1$.

radius 3. The fact that $y = s$ and $0 \le s \le 1$, shows that the centers of these cross-sections lie between 0 and 1. Finally, $0 \le t \le \pi$ implies that the cross sections are not full circles but only the upper semi-circles. (Reason: $z = 3 \sin t \ge 0$ when $0 \le t \le \pi$.) ♦

These examples can be viewed as a motivation for the following definition.

Definition 15.4.4. Let A be an open set in $\mathbb{R}^2$, and let A_0 be an open Jordan set in A such that $D = A_0 \cup \partial A_0 \subset A$. Let $\mathbf{f} : A \to \mathbb{R}^3$ be a bijection from A_0 onto $\mathbf{f}(A_0)$. We say that a **surface** M is a set of points $\{\mathbf{f}(s, t) : (s, t) \in D\}$, and the function $\mathbf{f}$ is a **parametrization** of M. The parametrization is **smooth** when the partial derivatives $\mathbf{f}_s$ and $\mathbf{f}_t$ are continuous in A_0 and $\mathbf{f}_s(s, t) \times \mathbf{f}_t(s, t) \ne \mathbf{0}$, for any $(s, t) \in A_0$. In that case we denote the area of M by $\sigma(M)$ and we define it as

$$\sigma(M) = \iint_D \left\| \frac{\partial \mathbf{f}}{\partial s} \times \frac{\partial \mathbf{f}}{\partial t} \right\| \, dA. \tag{15.15}$$

Let us see if we can calculate the area of the surface in Example 15.4.2.

Example 15.4.5. Computing the area of a surface.

Compute the area of the surface in Example 15.4.2.

Solution. We are using the parametrization $\mathbf{f}(r, \theta) = (r \cos \theta, r \sin \theta, r^2)$, with $A_0 = \{(r, \theta) : 0 < r \leq 1, 0 \leq \theta < 2\pi\}$. Then, $\mathbf{f}_r = (\cos \theta, \sin \theta, 2r)$, $\mathbf{f}_\theta = (-r \sin \theta, r \cos \theta, 0)$, and $\mathbf{f}_r \times \mathbf{f}_\theta = (-2r^2 \cos \theta, 2r^2 \sin \theta, r)$. Thus, $\|\mathbf{f}_r \times \mathbf{f}_\theta\| = \sqrt{4r^4 + r^2} = r\sqrt{4r^2 + 1}$, and

$$
\begin{aligned}
\sigma(M) &= \int_0^{2\pi} d\theta \int_0^1 r\sqrt{4r^2 + 1}\, dr \qquad \text{[substitution } u = 4r^2 + 1] \\
&= \int_0^{2\pi} d\theta \int_1^5 \sqrt{u}\, \frac{1}{8}\, du \\
&= \int_0^{2\pi} \left(\frac{2}{3} u\sqrt{u} \right) \Big|_1^5 \frac{1}{8}\, d\theta \\
&= \int_0^{2\pi} \frac{1}{12} \left(5\sqrt{5} - 1 \right) d\theta = 2\pi \frac{5\sqrt{5} - 1}{12} \approx 5.33.
\end{aligned}
$$

Therefore, the area of M is approximately 5.33. $\quad\blacklozenge$

Remark 15.4.6. In Example 15.4.5 the boundary of the set A_0 had content 0, so the integrals over A_0 and $D = \{(r, \theta) : 0 \leq r \leq 1, 0 \leq \theta \leq 2\pi\}$ are equal. In situations like that it is often convenient to blur the distinction between A_0 and D.

Since we have defined the surface area in terms of a parametrization, it is important to establish that it is independent of the choice of a specific one. A special case, when the region D in Definition 15.4.4 is the unit square $[0, 1] \times [0, 1]$ is left as an exercise (Exercise 15.4.1). The general case is a consequence of the fact that every smooth compact surface can be decomposed into a finite number of surfaces that satisfy the assumptions of Exercise 15.4.1. Once again, pursuing this line of investigation is beyond the scope of this book.

> The definition of the area of a surface is very different from the definition of the length of a curve (the limit of the lengths of inscribed polygonal lines). Initially, an analogous definition was stated for the area, with inscribed polyhedral surfaces playing the role of polygonal lines. (A polyhedral surface consists of polygons so that just two faces join along any common edge.) For example, such a definition can be found in a 1880 calculus textbook by Serret (page 117). However, in 1880 Schwarz showed that the formula is inconsistent, even for simple surfaces such as a cylinder. Namely, he demonstrated that it is possible to get different limits, by selecting different sequences of inscribed polyhedral surfaces. (See [113] for the details.) In 1882, Peano independently showed that the surface area cannot be defined using inscribed polyhedral surfaces. Formula (15.15) is nowadays a standard way of defining the area of a surface. It was developed by W. H. Young (page 335) in a 1920 article [112].

Next, we will define *surface integrals*. We will assume that the domain of integration is a surface M. Just like in the case of the line integral, we will distinguish between the case when the integrand is a real-valued function and when its values are in $\mathbb{R}^3$. Let us start with the former.

Definition 15.4.7. Let M be a surface in $\mathbb{R}^3$, let $\mathbf{f} : D \to \mathbb{R}^3$ be a smooth parametrization of M and let u be an integrable function on M. The **surface integral** of u over the surface M is denoted by $\iint_M u\, d\sigma$ and it is defined by

$$
\iint_M u\, d\sigma = \iint_D u(\mathbf{f}(s, t)) \|\mathbf{f}_s(s, t) \times \mathbf{f}_t(s, t)\|\, dA. \tag{15.16}
$$

When the surface M is closed, we often write $\oiint_M u\, d\sigma$.

Example 15.4.8. Computing a surface integral.

Let $u(x, y, z) = x^2 y$. Find $\iint_M u \, d\sigma$ if M is the portion of $x^2 + z^2 = 1$ between $y = 0$ and $y = 1$, above the xy-plane.

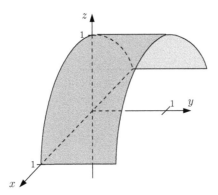

Figure 15.10: The surface of $x^2 + z^2 = 1$ between $y = 0$ and $y = 1$, above the xy-plane.

Solution. First we must write a parametrization for M. The equation $x^2 + z^2 = 1$ suggests $x = \cos t$, $z = \sin t$. Since the surface lies between the planes $y = 0$ and $y = 1$, we define $y = s$, and set $0 \leq s \leq 1$. Finally, the surface is above the xy-plane, which means that $z \geq 0$, hence $\sin t \geq 0$, so $0 \leq t \leq \pi$. Conclusion:

$$\mathbf{f}(s, t) = (\cos t, s, \sin t), \quad 0 \leq s \leq 1, 0 \leq t \leq \pi.$$

In order to calculate the surface integral, we find

$$\mathbf{f}_s(s, t) = (0, 1, 0), \ \mathbf{f}_t(s, t) = (-\sin t, 0, \cos t), \ \mathbf{f}_s(s, t) \times \mathbf{f}_t(s, t) = (\cos t, 0, \sin t),$$

$$\|\mathbf{f}_s(s, t) \times \mathbf{f}_t(s, t)\| = \sqrt{\cos^2 t + \sin^2 t} = 1.$$

Further, $u(\mathbf{f}(s, t)) = u(\cos t, s, \sin t) = (\cos t)^2 s$. Therefore,

$$\iint_M u \, d\sigma = \iint_D u(\cos t, s, \sin t) \cdot 1 \, dA$$

$$= \int_0^\pi dt \int_0^1 \cos^2 t \ s \, ds$$

$$= \int_0^\pi \cos^2 t \, dt \ \left(\frac{1}{2} s^2 \Big|_0^1 \right)$$

$$= \int_0^\pi \cos^2 t \, dt \ \frac{1}{2}$$

$$= \frac{1}{2} \int_0^\pi \frac{1 + \cos 2t}{2} \, dt$$

$$= \frac{1}{4} \left(t + \frac{\sin 2t}{2} \right) \Big|_0^\pi = \frac{\pi}{4}.$$

♦

> Surface integrals were developed by Gauss in the early years of the nineteenth century. He used them in his 1813 article [51].

Remark 15.4.9. The surface integral can be interpreted as the mass of a thin sheet in the shape of M, with variable density u.

A different situation arises when the integrand is a vector field $\mathbf{F} : M \to \mathbb{R}^3$.

Definition 15.4.10. Let M be a surface in $\mathbb{R}^3$, let $\mathbf{f} : D \to \mathbb{R}^3$ be a smooth parametrization of M and let $\mathbf{F} : M \to \mathbb{R}^3$ be an integrable vector field. The **surface integral** of $\mathbf{F}$ over the surface M is denoted by $\iint_M \mathbf{F} \cdot \mathbf{n} \, d\sigma$ and it is defined by

$$\iint_M \mathbf{F} \cdot \mathbf{n} \, d\sigma = \iint_D \mathbf{F}(\mathbf{f}(s,t)) \cdot (\mathbf{f}_s(s,t) \times \mathbf{f}_t(s,t)) \, dA. \tag{15.17}$$

When the surface M is closed, we often write $\oiint_M \mathbf{F} \cdot \mathbf{n} \, d\sigma$.

Remark 15.4.11. The vector $\mathbf{n}$ is the **unit normal vector** of the surface,

$$\mathbf{n} = \frac{\mathbf{f}_s(s,t) \times \mathbf{f}_t(s,t)}{\|\mathbf{f}_s(s,t) \times \mathbf{f}_t(s,t)\|}. \tag{15.18}$$

so formula (15.17) follows from (15.16) by taking u to be the scalar function $\mathbf{F} \cdot \mathbf{n}$. It is clear that $\|\mathbf{n}\| = 1$, and the modifier "normal" is most obvious when $x = s$ and $y = t$. In such a situation, the vectors $\mathbf{f}_s$ and $\mathbf{f}_t$ span a tangent plane to the surface, and the vector $\mathbf{n}$ is perpendicular (normal) to that plane.

Example 15.4.12. Computing the surface integral of a vector field.

Let $\mathbf{F}(x,y,z) = (x,y,2z)$. Find $\iint_M \mathbf{F} \cdot \mathbf{n} \, d\sigma$, if M is the portion of $z = 1 - x^2 - y^2$ above the xy-plane.

Solution. Whenever the equation of the surface can be solved for z, it is possible to use x and y as parameters. It is not a bad idea to call them s and t: $\mathbf{f}(s,t) = (s, t, 1 - s^2 - t^2)$. In order to find the region D, we notice that $z \geq 0$, so $1 - x^2 - y^2 \geq 0$, which means that $x^2 + y^2 \leq 1$. In other words, D is the closed unit disk (see Figure 15.12). Next,

$$\mathbf{f}_s(s,t) = (1, 0, -2s), \quad \mathbf{f}_t(s,t) = (0, 1, -2t), \tag{15.19}$$
$$\mathbf{f}_s(s,t) \times \mathbf{f}_t(s,t) = (2s, 2t, 1). \tag{15.20}$$

Also, $\mathbf{F}(s, t, 1 - s^2 - t^2) = (s, t, 2(1 - s^2 - t^2))$ and

$$(s, t, 2(1 - s^2 - t^2)) \cdot (2s, 2t, 1) = 2s^2 + 2t^2 + 2(1 - s^2 - t^2) = 2.$$

Therefore,

$$\iint_M \mathbf{F} \cdot \mathbf{n} \, d\sigma = \iint_D 2 \, dA = 2 \iint_D dA = 2\pi. \qquad \blacklozenge$$

Before we discuss the physical meaning of the surface integral, let us do the last example using a different parametrization.

Example 15.4.13. A different parametrization yields a different result.

We will show that if we use a different parametrization in Example 15.4.12, we will not get the same result.

Solution. This time we will set $x = t$ and $y = s$. Then $\mathbf{f}(s,t) = (t, s, 1 - s^2 - t^2)$, and D is again the closed unit disk. However,

$$\mathbf{f}_s(s,t) = (0, 1, -2s), \ \mathbf{f}_t(s,t) = (1, 0, -2t), \tag{15.21}$$

$$\mathbf{f}_s(s,t) \times \mathbf{f}_t(s,t) = (-2t, -2s, -1). \tag{15.22}$$

Since $\mathbf{F}(t, s, 1 - s^2 - t^2) = (t, s, 2(1 - s^2 - t^2))$, we obtain that

$$(t, s, 2(1 - s^2 - t^2)) \cdot (-2t, -2s, -1) = -2t^2 - 2s^2 - 2(1 - s^2 - t^2) = -2.$$

It follows that

$$\iint_M \mathbf{F} \cdot \mathbf{n} \, d\sigma = \iint_D -2 \, dA = -2 \iint_D dA = -2\pi. \qquad \blacklozenge$$

This is bad news, because the value of the surface integral should not depend on the choice of a parametrization. The good news is that the only change occurred in the sign of the result: depending on the parametrization it is $\pm 2\pi$. It is helpful to think of the function $\mathbf{F}$ as the velocity of a moving liquid. At each point $(x, y, z) \in \mathbb{R}^3$, the vector $\mathbf{F}(x, y, z)$ is the velocity vector of a particle passing through (x, y, z). (We are making an assumption that the flow is *steady*, i.e., that each particle passing through (x, y, z) has the same velocity.) In this scenario, the surface integral represents the *flux*—the rate at which the liquid flows through the surface. Two different results ($\pm 2\pi$) correspond to two different sides of the surface.

In order to make this meaningful, our surface M must have 2 sides. We say that such a surface is **orientable**. Geometrically, this means that if the normal vector $\mathbf{n}$ is moved along a closed curve in M, upon returning to the same point it will not have the opposite direction. A typical example of a *non-orientable* surface is the Möbius strip (Figure 15.11). Most surfaces that we encounter are orientable.

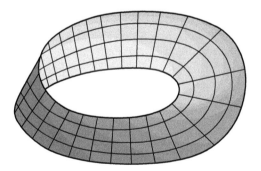

Figure 15.11: The Möbius strip is a non-orientable surface.

August Möbius (1790–1868) was a German mathematician and astronomer. He is best known for his discovery of the Möbius strip in 1858. In fact, the strip had been independently discovered and studied slightly earlier by another German mathematician Johann Listing (1808–1882). Listing was also the first to use the word *topology*.

Thus, when using the surface integral, we need to specify the side of the surface. For example, in Example 15.4.12 the surface of the paraboloid $z = 1 - x^2 - y^2$ has two sides. We can refer to them as the lower and the upper side, but it is usually more precise to talk about them in terms of the normal vector. The upper side has the normal vector pointing upward, the lower side has it pointing downward.

Incidentally, the direction of the normal vector is exactly the direction of the flow. This means that if we want to measure the flux of a liquid coming through the lower side and exiting on the upper side, we will select the upper side of the surface, and the vector **n** will be pointing up. How do we make this choice when computing the flux?

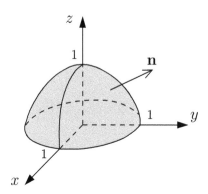

Figure 15.12: The normal vector for the upper side.

Example 15.4.14. Computing the surface integral of a vector field.

This is a restatement of Example 15.4.12. Let $\mathbf{F}(x, y, z) = (x, y, 2z)$. Find $\iint_M \mathbf{F} \cdot \mathbf{n} \, d\sigma$, if M is the portion of $z = 1 - x^2 - y^2$ above the xy-plane, oriented upward.

Solution. Since **n** satisfies (15.18), formulas (15.20) and (15.22) show that

$$\mathbf{n}(s, t) = \frac{(2s, 2t, 1)}{\sqrt{4s^2 + 4t^2 + 1}}, \quad \text{if } \mathbf{f}(s, t) = (s, t, 1 - s^2 - t^2), \text{ and}$$

$$\mathbf{n}(s, t) = \frac{(-2t, -2s, -1)}{\sqrt{4s^2 + 4t^2 + 1}}, \quad \text{if } \mathbf{f}(s, t) = (t, s, 1 - s^2 - t^2).$$

Which one is pointing up? It suffices to pick a point $(s, t) \in D$, and test $\mathbf{n}(s, t)$. For example, if $(s, t) = (0, 0)$ then we get $\mathbf{n} = (0, 0, 1)$ in the first case, and $\mathbf{n} = (0, 0, -1)$ in the second one. It is clear that the one pointing up is $(0, 0, 1)$, so we conclude that the flux is 2π. ◆

Exercises

15.4.1. Suppose that $\mathbf{f}, \mathbf{g} : [0, 1] \times [0, 1] \to \mathbb{R}^3$ are two smooth parametrizations of a surface M. Prove that $\iint_D \|\mathbf{f}_s \times \mathbf{f}_t\| \, dA = \iint_D \|\mathbf{g}_s \times \mathbf{g}_t\| \, dA$.

15.4.2. Prove that if a surface M is the graph of a C^1 function $z = f(x, y)$ for $(x, y) \in D$, then the area of M equals $\iint_D \sqrt{1 + (f'_x)^2 + (f'_y)^2} \, dA$.

15.4.3. Let $f : [a, b] \to \mathbb{R}$ be a non-negative C^1 function and let M be the surface obtained by rotating the graph of f around the x-axis. Prove that the area of M is given by

$$\int_a^b 2\pi f(x) \sqrt{1 + f'(x)^2} \, dx.$$

In Exercises 15.4.4–15.4.13 find a parametrization of the given surface. Find the sets A_0 and D and verify that they have properties as set in Definition 15.4.4.

15.4.4. A triangle with vertices $(2, 0, 0)$, $(0, 1, 0)$, $(0, 0, 3)$.

15.4.5. The portion of $x^2 + y^2 = 4$ between $z = 0$ and $z = y + 3$.

15.4.6. The portion of $x^2 + y^2 + z^2 = 1$ satisfying $z \geq \sqrt{x^2 + y^2}$.

15.4.7. The portion of $x = z$ satisfying $x^2 + y^2 \leq 1$.

15.4.8. The portion of $z = x^2/2$ cut out by $x = 2$, $y = 0$ and $x = y$.

15.4.9. The portion of $x^2 + y^2 - z^2 = 1$ between $z = 0$ and $z = 1$.

15.4.10. The portion of $x^2 - 2x + y^2 = 0$ cut off by $z^2 = x^2 + y^2$.

15.4.11. The portion of $y = \sin x + \cos z$ cut off by $x^2 + z^2 = 6$.

15.4.12. The portion of $r = \theta$ (in cylindrical coordinates), $0 \leq \theta \leq \pi/2$, between $z = 0$ and $z = \sqrt{x^2 + y^2}$.

15.4.13. The portion of $\rho = \varphi$ (in spherical coordinates), $0 \leq \varphi \leq \pi/2$, $0 \leq \theta \leq \pi$.

In Exercises 15.4.14–15.4.19 find the area of the given surface:

15.4.14. The portion of $az = xy$, $a > 0$, cut off by $x^2 + y^2 = a^2$.

15.4.15. The surface of the solid bounded by $x^2 + z^2 = a^2$, $y^2 + z^2 = a^2$.

15.4.16. The portion of $z = \frac{1}{2}(x^2 - y^2)$, cut off by $x - y = \pm 1$, $x + y = \pm 1$.

15.4.17. The portion of $(x^2 + y^2)^{3/2} + z = 1$, cut off by $z = 0$.

15.4.18. The portion of $2z = x^2 + y^2$, cut off by $z = \sqrt{x^2 + y^2}$.

15.4.19. The outer portion of the ring that is obtained when a cylindrical hole of radius r is drilled through the center of a ball of radius R.

In Exercises 15.4.20–15.4.30 find the surface integrals:

15.4.20. $\iint_M z \, d\sigma$, M is the portion of the surface $x^2 + z^2 = 2az$, $(a > 0)$, cut off by $z = \sqrt{x^2 + y^2}$.

15.4.21. $\iint_M (x + y + z) \, d\sigma$, M is the surface $x^2 + y^2 + z^2 = 4$, $z \geq 0$.

15.4.22. $\iint_M (x^2 + y^2) \, d\sigma$, M is the boundary of the solid $\sqrt{x^2 + y^2} \leq z \leq 1$.

15.4.23. $\iint_M z \, d\sigma$, M is a portion of the helicoid $x = u \cos v$, $y = u \sin v$, $z = v$, $0 < u < a$, $0 < v < 2\pi$.

15.4.24. $\iint_M z^2 \, d\sigma$, M is a portion of the cone $x = \rho \cos \varphi \sin \alpha$, $y = \rho \sin \varphi \sin \alpha$, $z = \rho \cos \alpha$, $0 \leq \rho \leq a$, $0 \leq \varphi \leq 2\pi$, and α is a constant $0 < \alpha < \pi/2$.

15.4.25. $\iint_M (xy + yz + zx) \, d\sigma$, M is a portion of the cone $z = \sqrt{x^2 + y^2}$, cut off by $x^2 + y^2 = 2x$.

15.4.26. $\iint_M \mathbf{F} \cdot \mathbf{n} \, d\sigma$, $\mathbf{F}(x, y, z) = (y^3 z, -xy, x + y + z)$, M is a portion of the surface $z = ye^x$ above the unit square in the xy-plane, oriented upward.

15.4.27. $\iint_M \mathbf{F} \cdot \mathbf{n} \, d\sigma$, $\mathbf{F}(x, y, z) = (x^2, 4z, y - x)$, M is the boundary of the tetrahedron with vertices $(0, 0, 0)$, $(1, 0, 0)$, $(0, 2, 0)$, and $(0, 0, 3)$, oriented towards the outside.

15.4.28. $\iint_M \mathbf{F} \cdot \mathbf{n} \, d\sigma$, $\mathbf{F}(x, y, z) = (x^2, 0, 0)$, M is the boundary of the cylinder $x^2 + y^2 = 9$, $0 \leq z \leq 4$, oriented outward.

15.4.29. $\iint_M \mathbf{F} \cdot \mathbf{n} \, d\sigma$, $\mathbf{F}(x, y, z) = (y - z, z - x, x - y)$, M is the outer side of the cone $x^2 + y^2 = z^2$, $0 \leq z \leq 3$.

15.4.30. $\iint_M \mathbf{F} \cdot \mathbf{n} \, d\sigma$, $\mathbf{F}(x, y, z) = (x, y, z)$, M is the outer side of the sphere $x^2 + y^2 + z^2 = a^2$.

15.5   Divergence Theorem

Green's Theorem represents a generalization of the formula (15.7) in the case when $\mathbf{F} : D \to \mathbb{R}^2$, and D is a 2 dimensional region. In this section we will explore a further generalization to the 3 dimensional case: we will assume that $\mathbf{F} : S \to \mathbb{R}^3$, and S is a solid in $\mathbb{R}^3$.

In order to formulate the result we need to introduce the concept of *divergence* first. Given a vector field $\mathbf{F} : \mathbb{R}^3 \to \mathbb{R}^3$, $\mathbf{F} = (F_1, F_2, F_3)$, we denote the **divergence of F** by div $\mathbf{F}$, and we define it by

$$\operatorname{div} \mathbf{F} = \frac{\partial F_1}{\partial x} + \frac{\partial F_2}{\partial y} + \frac{\partial F_3}{\partial z}.$$

Remark 15.5.1. If $\mathbf{F}$ represents the velocity of a moving liquid, the divergence measures the rate of the outward flow at each point.

Example 15.5.2. Computing the divergence of a vector field.

Find the divergence of $\mathbf{F}(x, y, z) = (x^2 y, 2y^3 z, 3z)$.

Solution. Clearly $F_1(x, y, z) = x^2 y$, so $(F_1)_x = 2xy$. Similarly, $(F_2)_y = 6y^2 z$ and $(F_3)_z = 3$. Thus, div $\mathbf{F} = 2xy + 6y^2 z + 3$.   ♦

Before we state the Divergence Theorem, we make a comment. Just like in the case of Green's Theorem, we will first prove the result for sufficiently simple solids. A solid $S \subset \mathbb{R}^3$ is *z-simple* if there exists a 2 dimensional region D and continuous functions α, β on D, such that

$$S = \{(x, y, z) \in \mathbb{R}^3 : (x, y) \in D, \ \alpha(x, y) \le z \le \beta(x, y)\}. \tag{15.23}$$

Similar definitions can be written for the x-simple and y-simple solids in $\mathbb{R}^3$.

Theorem 15.5.3 (Divergence Theorem). *Let S be a solid in $\mathbb{R}^3$ that is compact and has a piecewise smooth, outward oriented boundary ∂S. If A is an open region containing S and ∂S, and if $\mathbf{F}$ is a vector field in $C^1(A)$, then*

$$\iiint_S \operatorname{div} \mathbf{F} \, dV = \oiint_{\partial S} \mathbf{F} \cdot \mathbf{n} \, d\sigma, \tag{15.24}$$

Proof. We can write (15.24) as

$$\iiint_S (F_1)_x \, dV + \iiint_S (F_2)_y \, dV + \iiint_S (F_3)_z \, dV = \oiint_{\partial S} F_1 n_1 \, d\sigma + \oiint_{\partial S} F_2 n_2 \, d\sigma + \oiint_{\partial S} F_3 n_3 \, d\sigma.$$

It suffices to prove the equality between the appropriate triple integral and the surface integral. We will prove that

$$\iiint_S (F_3)_z \, dV = \oiint_{\partial S} F_3 n_3 \, d\sigma, \tag{15.25}$$

and we will leave the other two as an exercise.

Let us assume that S is a z-simple region, as in (15.23). Notice that, in this situation, the boundary of S consists of up to 3 surfaces. The "top" is $z = \beta(x, y)$ and the "bottom" is $z = \alpha(x, y)$. There is possibly a side surface. However, the normal vector to it is parallel to the xy-plane, so n_3 equals 0 at these points. Thus, it remains to focus on two surfaces. Let M_1 denote the graph of $z = \beta(x, y)$. Since M_1 is parametrized by $\mathbf{f}(s, t) = (s, t, \beta(s, t))$, we have that

$$\mathbf{f}_s(s, t) = (1, 0, \beta_s(s, t)), \quad \mathbf{f}_t(s, t) = (0, 1, \beta_t(s, t)),$$

$$\mathbf{f}_s(s,t) \times \mathbf{f}_t(s,t) = (-\beta_s(s,t), -\beta_t(s,t), 1),$$

$$n_3(s,t) = \frac{1}{\sqrt{(\beta_s(s,t))^2 + (\beta_t(s,t))^2 + 1}}.$$

Therefore,

$$\iint_{M_1} F_3 n_3 \, d\sigma = \iint_D F_3(s,t,\beta(s,t)) \, n_3(s,t,) \|\mathbf{f}_s(s,t) \times \mathbf{f}_t(s,t)\| \, dA$$

$$= \iint_D F_3(s,t,\beta(s,t)) \, dA. \tag{15.26}$$

In a similar way we can show that, if M_2 is the graph of $z = \alpha(x,y)$ (the bottom of S), then

$$\iint_{M_2} F_3 n_3 \, d\sigma = - \iint_D F_3(s,t,\alpha(s,t)) \, dA. \tag{15.27}$$

Combining (15.26) and (15.27) we obtain

$$\oiint_{\partial S} F_3 n_3 \, d\sigma = \iint_D F_3(s,t,\beta(s,t)) \, dA - \iint_D F_3(s,t,\alpha(s,t)) \, dA$$

$$= \iint_D \left(F_3(s,t,\beta(s,t)) - F_3(s,t,\alpha(s,t)) \right) \, dA$$

$$= \iint_D F_3(s,t,z) \Big|_{z=\alpha(s,t)}^{z=\beta(s,t)} \, dA$$

$$= \iint_D \int_{\alpha(s,t)}^{\beta(s,t)} (F_3)_z(s,t,z) \, dz \, dA$$

$$= \iiint_S (F_3)_z \, dV.$$

Thus, (15.24) holds for a solid that is, at the same time, z-simple, y-simple, and x-simple. Just like in the proof of Green's Theorem, we can now deduce that the theorem is true whenever S can be decomposed into a finite number of regions that have all 3 properties (see Exercise 15.5.2). Finally, we will not discuss the case when S is an infinite union of simpler solids, but we will again direct the reader to [72]. □

Special cases of Theorem 15.5.3 have appeared in the very diverse body of work: Lagrange (gravitation, 1764), Gauss (magnetism, 1813), Sarrus (floating bodies, 1828), Poisson (elasticity, 1829). The first general statement was due to Ostrogradsky (see next page) who presented his work to the Paris Academy of Sciences in 1826. The paper did not appear until 1831, and there is some evidence that Poisson and Sarrus (page 176) were aware of it. As we have said earlier, Green published his essay in 1828, and it contained a result that is equivalent to the Divergence Theorem, although nothing close to the formulation of Theorem 15.5.3 appears in his work. Nowadays, the theorem is often called the Gauss Theorem or, less often, the Gauss-Ostrogradsky Theorem. The article [70] offers many more details about the theorem from the historic viewpoint.

Example 15.5.4. Computing a surface integral using Divergence Theorem.

Find the outward flux of $\mathbf{F}(x,y,z) = (x^3, y^3, z^2)$ across the surface of the solid S enclosed by $x^2 + y^2 = 9$, $z = 0$, $z = 2$.

Solution. By definition, the flux equals $\iint_{\partial S} \mathbf{F} \cdot \mathbf{n} \, d\sigma$, and by the Divergence Theorem it can be calculated as

$$\iiint_S \text{div} \, \mathbf{F} \, dV = \iiint_S (3x^2 + 3y^2 + 2z) \, dV.$$

In view of the shape of S, we will use cylindrical coordinates. Then $3x^2 + 3y^2 + 2z$ becomes $3r^2 \cos^2 \theta + 3r^2 \sin^2 \theta + 2w = 3r^2 + 2w$. It is not hard to see that $0 \le r \le 3$, $0 \le \theta \le 2\pi$, and $0 \le w \le 2$. Since the Jacobian determinant for the cylindrical coordinates equals r, we obtain

$$\int_0^3 dr \int_0^{2\pi} d\theta \int_0^2 \left(3r^2\cos^2\theta + 3r^2\sin^2\theta + 2w\right) r\, dw = \int_0^3 dr \int_0^{2\pi} d\theta \int_0^2 \left(3r^2 + 2w\right) r\, dw$$

$$= \int_0^3 dr \int_0^{2\pi} d\theta\, r \left(3r^2 w + w^2\right) \Big|_{w=0}^{w=2}$$

$$= \int_0^3 dr \int_0^{2\pi} r \left(6r^2 + 4\right) d\theta$$

$$= \int_0^3 dr \left(6r^3 + 4r\right) \theta \Big|_{\theta=0}^{\theta=2\pi}$$

$$= \int_0^3 \left(6r^3 + 4r\right) 2\pi\, dr$$

$$= 2\pi \left(\frac{3}{2}r^4 + 2r^2\right) \Big|_0^3 = 279\pi. \quad \blacklozenge$$

Figure 15.13: The solid enclosed by $z = 0$, $z = 2$, and $x^2 + y^2 = 9$.

MIKHAIL OSTROGRADSKY (1801–1862) was a Russian mathematician, born and raised on the territory of today's Ukraine. As a child he enjoyed measuring objects and always carried a rock with a long string tied around it so that he could measure a depth of any well that he came across. He did not shine in high school and needed additional study to be ready for college. He started as a student of physics and mathematics at the University of Kharkov. In 1820, his mentor Osipovsky was accused of not teaching from a truly Christian viewpoint and suspended, while all of his students were required to retake their exams. Ostrogradsky refused and left for Paris where he stayed from 1822 to 1827, attending lectures by leading mathematicians. He started publishing papers and in 1826 proved the Divergence Theorem. When his father stopped sending him money he accumulated debts and was even taken to court, but Cauchy paid off all his financial obligations and found him a teaching position at a high school. Upon returning to Russia he settled in Saint Petersburg, where he was elected a member of the Academy of Sciences and where he remained for the rest of his life. He started lecturing at the Naval Academy, but added many other duties. Ostrogradsky was a big tall man with a loud voice. His appearance was quite formidable, especially after having lost his right eye during the 1830 visit to Paris, but had a cheerful character. He passionately loved his native land, its people, and its culture. He loved classical French and Russian literature although his language of choice when at home was always Ukrainian. By his request, he was buried in Ukraine. His most significant work was in applied mathematics. His goal was to provide a combined theory of hydrodynamics, elasticity, heat, and electricity. He is the founder of the Russian school of theoretical mechanics. Together with Bunyakovsky (page 295) he calculated the pensions for the Naval Department when they had to send into retirement a large number of office workers. He did not appreciate the work on non-Euclidean geometry of Lobachevsky and he rejected it when it was submitted for publication in the Saint Petersburg Academy of Sciences. Some of his work was ignored in Western Europe, such as the change of variables theorem in n variables (see [71] for more). He wrote numerous textbooks, and had many administrative duties. Chebyshev expressed the opinion that Ostrogradsky was "of brilliant mind, he did not accomplish even half of what he could have done if he were not 'bogged down' with tiresome permanent pedagogic work".

Exercises

15.5.1. Prove the remaining equalities in (15.24):

$$\iiint_S (F_1)_x\, dV = \iint_{\partial S} F_1 n_1\, d\sigma \quad \text{and} \quad \iiint_S (F_2)_y\, dV = \iint_{\partial S} F_2 n_2\, d\sigma.$$

15.5.2. Prove that Divergence Theorem is true when a compact solid S, with a piecewise smooth, outward oriented boundary, can be written as a finite union of solids S_n, each of them being z-simple, y-simple, and x-simple.

In Exercises 15.5.3–15.5.7 S, A and $\mathbf{F}$ are as in Theorem 15.5.3, $f : A \to \mathbb{R}$ is a C^1 function, V is the volume of S, $(a, b, c) \in S$ and V_r is the volume of the ball with center (a, b, c) and radius r. Prove the identities:

15.5.3. $\iiint_S f \operatorname{div} \mathbf{F}\, dV = \oiint_{\partial S} f\mathbf{F} \cdot \mathbf{n}\, d\sigma - \iiint_S \nabla f \cdot \mathbf{F}\, dV.$

15.5.4. $\oiint_{\partial S} f\mathbf{n}\, d\sigma = \iiint_S \nabla f\, dV.$

15.5.5. $\operatorname{div} \mathbf{F}(a, b, c) = \lim_{r \to 0} \frac{1}{V_r} \oiint_{\partial S} \mathbf{F} \cdot \mathbf{n}\, d\sigma.$

15.5.6. $\oiint_{\partial S} f\mathbf{n} \cdot \mathbf{i}\, d\sigma = \iiint_S \frac{\partial f}{\partial x}\, dV.$

15.5.7. $V = \oiint_{\partial S} x\mathbf{n} \cdot \mathbf{i}\, d\sigma = \oiint_{\partial S} y\mathbf{n} \cdot \mathbf{j}\, d\sigma = \oiint_{\partial S} z\mathbf{n} \cdot \mathbf{k}\, d\sigma = \frac{1}{3} \oiint_{\partial S} (x\mathbf{n} \cdot \mathbf{i} + y\mathbf{n} \cdot \mathbf{j} + z\mathbf{n} \cdot \mathbf{k})\, d\sigma.$

In Exercises 15.5.8–15.5.9 S and A are as in Theorem 15.5.3, $\mathbf{r}$ is the vector with initial point $(x, y, z) \in S$ and terminal point $(a, b, c) \in \partial S$ and α is the angle between $\mathbf{r}$ and $\mathbf{n}$.

15.5.8. Prove the equality: $\frac{1}{2} \oiint_{\partial S} \cos \alpha\, d\sigma = \iiint_S \frac{1}{\|\mathbf{r}\|}\, dV.$

15.5.9. Find $I(x, y, z) = \oiint_{\partial S} \frac{\cos \alpha}{\|\mathbf{r}\|^2}\, d\sigma$, if:

(a) (x, y, z) does not belong to $S \cup \partial S$;   (b)* (x, y, z) belongs to $S \setminus \partial S$.

In Exercises 15.5.10–15.5.19 find the surface integral $\iint_M \mathbf{F} \cdot \mathbf{n}\, d\sigma$ using the Divergence Theorem:

15.5.10. $\mathbf{F}(x, y, z) = (x^2, y^2, z^2)$, M is the outer surface of the cube $0 \le x \le a$, $0 \le y \le a$, $0 \le z \le a$.

15.5.11. $\mathbf{F}(x, y, z) = (x^3, y^3, z^3)$, M is the outer surface of the sphere $x^2 + y^2 + z^2 = a^2$.

15.5.12. $\mathbf{F}(x, y, z) = (x - y + z, y - z + x, z - x + y)$, M is the outer side of the surface $|x - y + z| + |y - z + x| + |z - x + y| = 1$.

15.5.13. $\mathbf{F}(x, y, z) = (x^2 + y^2, y^2 + x^2 z, z^2 + x^2 y)$, M is the outward oriented boundary of the solid bounded by $x^2 + y^2 = 1$, $z = x + 2$ and $z = 0$.

15.5.14. $\mathbf{F}(x, y, z) = (xz^2, x^2 y - z^3, 2xy + y^2 z)$, M is the outward oriented boundary of the solid bounded by $z = \sqrt{a^2 - x^2 - y^2}$ and $z = 0$.

15.5.15. $\mathbf{F}(x, y, z) = (z^2 - x, -xy, 3z)$, M is the outward oriented boundary of the solid bounded by $z = 4 - y^2$, $x = 0$, $x = 3$, and $z = 0$.

15.5.16. $\mathbf{F}(x, y, z) = (x, y, z)$, M is the outward oriented boundary of the solid defined by $x^2 + y^2 > 1$ and $x^2 + y^2 + z^2 < 4$.

15.5.17. $\mathbf{F}(x, y, z) = (x, y, z)$, M is the outward oriented boundary of the solid defined by $x^2 + y^2 < 4$ and $x^2 + y^2 - z^2 > 1$.

15.5.18. $\mathbf{F}(x, y, z) = (x^2, y^2, z^2)$, M is the outward oriented boundary of the tetrahedron in the first octant, between the coordinate planes and $6x + 3y + 2z = 6$.

15.5.19. $\mathbf{F}(x, y, z) = (e^y \cos z, \sin x\sqrt{1 + z}, x^2 + y^3)$, M is the outward oriented graph of $z = (1 - x^2 - y^2)e^{1 - x^2 - 3y^2}$ for $z \ge 0$.

In Exercises 15.5.20–15.5.23 use the results of Exercise 15.5.7 to calculate the volume of the solid:

15.5.20. A ball of radius R.

15.5.21. The solid $x^2 + y^2 \le r^2$, $0 \le z \le h$.

15.5.22. The solid bounded by $z = 3x^2 + 3y^2 - 16$ and $z = 9 - x^2 y^2$.

15.5.23.* The solid bounded by the torus $x = (b + a \cos \psi) \cos \varphi$, $y = (b + a \cos \psi) \sin \varphi$, $z = a \sin \psi$, where $0 < a < b$.

15.6   Stokes' Theorem

Green's Theorem establishes a relation between two integrals: a double integral over a region D in $\mathbb{R}^2$ and the line integral along the boundary of D. What happens when we replace D by a surface M in $\mathbb{R}^3$? We can expect that the double integral over D will be replaced by the surface integral over M, and that we would still have a line integral, this time along the boundary of M. What about the integrands?

Before we can state the theorem, there is the issue of orientation to address. Assuming that the surface M is orientable, it can have one of the two orientations. The same holds for the curve ∂M. In order to get the quantities of the same sign on both sides of the anticipated equation, we need to match these orientations. Let us see how this works in Green's Theorem. The boundary ∂D is positively oriented: this means that if we travel along the boundary (but stay inside the region, very close to the boundary), then the region will always remain on our left (and the boundary on our right). We will use this as the definition of the orientation of ∂M. Namely, if M is an orientable surface M, we will say that the boundary ∂M is **positively oriented** if, when you walk near the edge on the positive side of the surface in the direction corresponding to the orientation of boundary, then the surface must be to your left and the edge must be to your right.

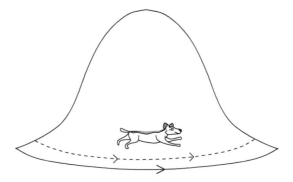

Figure 15.14: The surface remains on the left.

There remains the question of the integrand in the surface integral. In Green's Theorem it was $Q_x(x, y) - P_y(x, y)$. If $\mathbf{F} : \mathbb{R}^3 \to \mathbb{R}^3$ is a differentiable function, we define the **curl** of $\mathbf{F}$ by

$$\operatorname{curl} \mathbf{F} = ((F_3)_y - (F_2)_z, \ (F_1)_z - (F_3)_x, \ (F_2)_x - (F_1)_y).$$

A popular memorization trick is

$$\operatorname{curl} \mathbf{F} = \begin{vmatrix} \mathbf{i} & \mathbf{j} & \mathbf{k} \\ \dfrac{\partial}{\partial x} & \dfrac{\partial}{\partial y} & \dfrac{\partial}{\partial z} \\ F_1 & F_2 & F_3 \end{vmatrix}.$$

Example 15.6.1. Computing the curl of a vector field.

 Find the curl of $\mathbf{F}(x, y, z) = (x^2 y, 2y^3 z, 3z)$.

Solution. First, $F_1(x, y, z) = x^2 y$, so $(F_1)_y = x^2$ and $(F_1)_z = 0$. Next, $F_2(x, y, z) = 2y^3 z$, so $(F_2)_x = 0$ and $(F_2)_z = 2y^3$. Finally, $F_3(x, y, z) = 3z$, so $(F_3)_x = 0$ and $(F_3)_y = 0$. It follows that $\operatorname{curl} \mathbf{F} = (-2y^3, 0, -x^2)$.   ◆

Remark 15.6.2. Assuming that $\mathbf{F}$ is the velocity vector of a moving liquid, the curl of $\mathbf{F}$ describes the angular velocity at each point.

Now we can formulate the main result of this section.

Theorem 15.6.3 (Stokes' Theorem). *Let M be a piecewise smooth, orientable surface, and let its boundary ∂M be a closed, simple, piecewise smooth, positively oriented curve. If A is an open set in $\mathbb{R}^3$ containing M and ∂M, and if $\mathbf{F} : A \to \mathbb{R}^3$ is a function in $C^1(A)$, then*

$$\iint_M \operatorname{curl} \mathbf{F} \cdot \mathbf{n} \, d\sigma = \oint_{\partial M} \mathbf{F} \cdot d\mathbf{x}. \tag{15.28}$$

Proof. First we will make a simplification. Namely, we will assume that there exists a region $D \in \mathbb{R}^2$ and a function

$$\mathbf{f} : D \to \mathbb{R}^3, \text{ such that } \mathbf{f}(D) = M \text{ and } \mathbf{f}(\partial D) = \partial M. \tag{15.29}$$

This is certainly possible when, for example, M is given by an equation $z = h(x, y)$. (Take $\mathbf{f}(x, y) = (x, y, h(x, y))$.) Once again, we will merely state that more complicated surfaces can be decomposed into simpler ones, and focus on a parametrization $\mathbf{f}$ as in (15.29).

The idea of the proof is to write the left side of (15.28) as a double integral over D, the right side as a line integral over ∂D, and then use Green's Theorem. Let us start with the left side of (15.28):

$$\iint_M \operatorname{curl} \mathbf{F} \cdot \mathbf{n} \, d\sigma = \iint_D \operatorname{curl} \mathbf{F} \left(\mathbf{f}(s, t) \right) \cdot \left(\mathbf{f}_s(s, t) \times \mathbf{f}_t(s, t) \right) \, dA.$$

By the definition of the curl and the cross product, the integrand in this double integral is the dot product of

$$\Big((F_3)_y - (F_2)_z, (F_1)_z - (F_3)_x, (F_2)_x - (F_1)_y \Big), \text{ and}$$

$$\Big((f_2)_s(f_3)_t - (f_2)_t(f_3)_s, (f_3)_s(f_1)_t - (f_3)_t(f_1)_s, (f_1)_s(f_2)_t - (f_1)_t(f_2)_s \Big).$$

This leads to

$$(F_3)_y \big[(f_2)_s(f_3)_t - (f_2)_t(f_3)_s \big] - (F_2)_z \big[(f_2)_s(f_3)_t - (f_2)_t(f_3)_s \big]$$
$$+ (F_1)_z \big[(f_3)_s(f_1)_t - (f_3)_t(f_1)_s \big] - (F_3)_x \big[(f_3)_s(f_1)_t - (f_3)_t(f_1)_s \big]$$
$$+ (F_2)_x \big[(f_1)_s(f_2)_t - (f_1)_t(f_2)_s \big] - (F_1)_y \big[(f_1)_s(f_2)_t - (f_1)_t(f_2)_s \big]$$

Let us focus on the terms containing F_1. We have

$$(F_1)_z \big[(f_3)_s(f_1)_t - (f_3)_t(f_1)_s \big] - (F_1)_y \big[(f_1)_s(f_2)_t - (f_1)_t(f_2)_s \big]$$
$$= \big[(F_1)_z(f_3)_s + (F_1)_y(f_2)_s \big](f_1)_t - \big[(F_1)_z(f_3)_t + (F_1)_y(f_2)_t \big](f_1)_s$$
$$= \big[(F_1)_z(f_3)_s + (F_1)_y(f_2)_s + (F_1)_x(f_1)_s \big](f_1)_t$$
$$\quad - \big[(F_1)_z(f_3)_t + (F_1)_y(f_2)_t + (F_1)_x(f_1)_t \big](f_1)_s$$
$$= (F_1 \circ \mathbf{f})_s(f_1)_t - (F_1 \circ \mathbf{f})_t(f_1)_s.$$

In a similar fashion, it can be shown that the terms containing F_2 add up to $(F_2 \circ \mathbf{f})_s(f_2)_t - (F_2 \circ \mathbf{f})_t(f_2)_s$, and those with F_3 to $(F_3 \circ \mathbf{f})_s(f_3)_t - (F_3 \circ \mathbf{f})_t(f_3)_s$. Combining these 3 expressions yields

$$(F_1 \circ \mathbf{f})_s(f_1)_t - (F_1 \circ \mathbf{f})_t(f_1)_s + (F_2 \circ \mathbf{f})_s(f_2)_t - (F_2 \circ \mathbf{f})_t(f_2)_s$$

$$+ (F_3 \circ \mathbf{f})_s (f_3)_t - (F_3 \circ \mathbf{f})_t (f_3)_s$$
$$= \big[(F_1 \circ \mathbf{f})_s (f_1)_t + (F_2 \circ \mathbf{f})_s (f_2)_t + (F_3 \circ \mathbf{f})_s (f_3)_t \big]$$
$$- \big[(F_1 \circ \mathbf{f})_t (f_1)_s + (F_2 \circ \mathbf{f})_t (f_2)_s + (F_3 \circ \mathbf{f})_t (f_3)_s \big]$$
$$= (\mathbf{F} \circ \mathbf{f})_s \cdot \mathbf{f}_t - (\mathbf{F} \circ \mathbf{f})_t \cdot \mathbf{f}_s.$$

Let $G_1 = (\mathbf{F} \circ \mathbf{f}) \cdot \mathbf{f}_s$ and $G_2 = (\mathbf{F} \circ \mathbf{f}) \cdot \mathbf{f}_t$. Using Theorem 11.4.8 (d),

$$(G_2)_s - (G_1)_t = \big[(\mathbf{F} \circ \mathbf{f})_s \cdot \mathbf{f}_t + (\mathbf{F} \circ \mathbf{f}) \cdot \mathbf{f}_{st} \big] - \big[(\mathbf{F} \circ \mathbf{f})_t \cdot \mathbf{f}_s + (\mathbf{F} \circ \mathbf{f}) \cdot \mathbf{f}_{st} \big]$$
$$= (\mathbf{F} \circ \mathbf{f})_s \cdot \mathbf{f}_t - (\mathbf{F} \circ \mathbf{f})_t \cdot \mathbf{f}_s.$$

It follows that

$$\iint\limits_{M} \operatorname{curl} \mathbf{F} \cdot \mathbf{n} \, d\sigma = \iint\limits_{D} \big[(G_2)_s - (G_1)_t \big] \, dA$$
$$= \oint_{\partial D} G_1 \, ds + G_2 \, dt,$$

with the last equality coming from Green's Theorem. Let $\mathbf{g} : [a, b] \to \mathbb{R}^2$ be a parametrization of ∂D. Then the last integral can be written as

$$\int_a^b \Big(G_1(\mathbf{g}(u)), G_2(\mathbf{g}(u)) \Big) \cdot \mathbf{g}'(u) \, du.$$

Notice that, if $\mathbf{g} = (g_1, g_2)$,

$$\big(G_1(\mathbf{g}(u)), G_2(\mathbf{g}(u)) \big) \cdot (g_1'(u), g_2'(u))$$
$$= \Big((\mathbf{F} \circ \mathbf{f})(\mathbf{g}(u)) \cdot \mathbf{f}_s(\mathbf{g}(u)), (\mathbf{F} \circ \mathbf{f})(\mathbf{g}(u)) \cdot \mathbf{f}_t(\mathbf{g}(u)) \Big) \cdot (g_1'(u), g_2'(u))$$
$$= \big[(\mathbf{F} \circ \mathbf{f})(\mathbf{g}(u)) \cdot \mathbf{f}_s(\mathbf{g}(u)) \big] g_1'(u) + \big[(\mathbf{F} \circ \mathbf{f})(\mathbf{g}(u)) \cdot \mathbf{f}_t(\mathbf{g}(u)) \big] g_2'(u)$$
$$= (\mathbf{F} \circ \mathbf{f})(\mathbf{g}(u)) \cdot \big[\mathbf{f}_s(\mathbf{g}(u)) g_1'(u) + \mathbf{f}_t(\mathbf{g}(u)) g_2'(u) \big]$$
$$= \mathbf{F}(\mathbf{f} \circ \mathbf{g}(u)) (\mathbf{f} \circ \mathbf{g})'(u).$$

Thanks to (15.29), $\mathbf{f} \circ \mathbf{g}$ is a parametrization of ∂M, so

$$\oint_{\partial D} G_1 \, ds + G_2 \, dt = \int_a^b \mathbf{F}(\mathbf{f} \circ \mathbf{g}(u)) (\mathbf{f} \circ \mathbf{g})'(u) \, du = \oint_{\partial M} \mathbf{F} \cdot d\mathbf{x}$$

and the theorem is proved. □

Theorem 15.6.3 appears for the first time in a letter from Lord Kelvin to Stokes in 1850, although the surface integral on the left side of (15.28) may have appeared in some earlier work of Stokes. In 1854 Stokes included this question in the prestigious Smith's Prize Exam (awarded annually to two research students in theoretical physics, mathematics, and applied mathematics at the University of Cambridge). The first published proof is in the 1861 monograph [58] on the motion of the fluids. In this work Hankel (page 291) proved the result in a way that is very similar to ours (assuming that the surface is given by $z = h(x, y)$). More details can be found in the excellent survey article [70].

Example 15.6.4. Computing the work done by vector field.

Find the work performed by the vector field $\mathbf{F}(x, y, z) = (x^2, 4xy^3, y^2 x)$ on a particle that goes along the rectangle with vertices $(1, 0, 0)$, $(0, 0, 0)$, $(0, 3, 3)$, $(1, 3, 3)$.

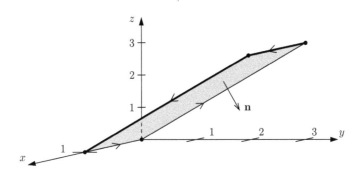

Figure 15.15: The particle being moved around the rectangle.

Solution. If we denote the rectangle by R, and its boundary by ∂R, then Stokes' Theorem shows that the work W equals

$$W = \oint_{\partial R} \mathbf{F} \cdot d\mathbf{x} = \iint_R \operatorname{curl} \mathbf{F} \cdot \mathbf{n}\, d\sigma.$$

A straightforward calculation shows that $\operatorname{curl} \mathbf{F}(x, y, z) = (2xy, -y^2, 4y^3)$. The rectangle R lies in the plane $y - z = 0$, so we can use the parametrization $\mathbf{f}(s, t) = (s, t, t)$, with $0 \le s \le 1$, $0 \le t \le 3$. Then

$$\mathbf{f}_s = (1, 0, 0), \mathbf{f}_t = (0, 1, 1), \mathbf{f}_s \times \mathbf{f}_t = (0, -1, 1).$$

Figure 15.15 reveals that the normal vector should be pointing downward, so we must replace $(0, -1, 1)$ by $(0, 1, -1)$. It follows that

$$W = \int_0^1 ds \int_0^3 (2st, -t^2, 4t^3) \cdot (0, 1, -1)\, dt$$

$$= \int_0^1 ds \int_0^3 (-t^2 - 4t^3)\, dt$$

$$= \int_0^1 ds \left(-\frac{1}{3} t^3 - t^4 \right) \Big|_0^3$$

$$= \int_0^1 -90\, ds = -90. \qquad \blacklozenge$$

he was appointed secretary of the Society in 1854. He remained the secretary until 1885, and served as the president from 1885-1890. However, in 1857 he got married and became much more involved in the administrative duties than research. He remained a big influence on subsequent generations of Cambridge scientists, including Maxwell (page 280). Together with Maxwell and Lord Kelvin, he contributed to the fame of the Cambridge school of mathematical physics in the middle of the nineteenth century. His main contributions are in fluid dynamics (the Navier–Stokes equation), fluorescence (the Stokes line), polarization of light, and a study of divergent series.

Exercises

In Exercises 15.6.1–15.6.8 use Stokes' Theorem to find the line integrals:

15.6.1. $\oint_C y\,dx + z\,dy + x\,dz$, C is the circle $x^2 + y^2 + z^2 = a^2$, $x + y + z = 0$, oriented counterclockwise when viewed from the positive x-axis.

15.6.2. $\oint_C (y - z)\,dx + (z - x)\,dy + (x - y)\,dz$, C is the curve $x^2 + y^2 = a^2$, $\frac{x}{a} + \frac{z}{b} = 1$, $(a, b > 0)$, oriented counterclockwise when viewed from the positive x-axis.

15.6.3.* $\oint_C (y^2 + z^2)\,dx + (x^2 + z^2)\,dy + (x^2 + y^2)\,dz$, C is the curve $x^2 + y^2 + z^2 = 2Rx$, $x^2 + y^2 = 2rx$, $(0 < r < R, \ z > 0)$, oriented so that the portion of the outer surface of the hemisphere inside the cylinder remains on the left.

15.6.4. $\oint_C y^2 z^2\,dx + x^2 z^2\,dy + x^2 y^2\,dz$, C is the curve $x = a\cos t$, $y = a\cos 2t$, $z = a\cos 3t$, oriented in the direction of the increase of t.

15.6.5. $\oint_C z^2\,dx + x^2\,dy + y^2\,dz$, C is the curve $\mathbf{x} = (\sin 2t, \cos t, \sin t)$, $0 \le t \le 2\pi$.

15.6.6. $\oint_C x^2\,dx + y^2\,dy + z^2\,dz$, C is the curve

$$\mathbf{x} = \begin{cases} (6t, 0, 5t), & \text{if } 0 \le t \le 1 \\ (12 - 6t, 0, 3t + 2), & \text{if } 1 \le t \le 2 \\ (0, 0, 24 - 8t), & \text{if } 2 \le t \le 3. \end{cases}$$

15.6.7. $\oint_C \mathbf{F} \cdot d\mathbf{x}$, $\mathbf{F} = (-y/2 + z, x/2 + 3z/2, -x - 3y/2)$, C is the curve in which the surface $z = x^2 - y^2$ intersects the cylinder $x^2 + y^2 = 1$, oriented counterclockwise when viewed from the positive z-axis.

15.6.8. $\oint_C \mathbf{F} \cdot d\mathbf{x}$, $\mathbf{F} = (xy, xz, yz)$, C is the boundary of the triangle with vertices $(1, 0, 0)$, $(0, 2, 0)$, $(0, 0, 3)$, oriented by this ordering of the points.

In Exercises 15.6.9–15.6.11 use Stokes' Theorem to find the $\iint_M \operatorname{curl} \mathbf{F} \cdot \mathbf{n}\,d\sigma$:

15.6.9. $\mathbf{F} = (3y, -xz, yz^2)$, and M is the outer surface of $x^2 + y^2 = 2z$, bounded by $z = 2$.

15.6.10. $\mathbf{F} = (x^3 + xz + yz^2, xyz^3 + y^7, x^2 z^5)$, $M = M_1 \cup M_2$, M_1 is the outer surface of $x^2 + y^2 = 9$, $0 \le z \le 8$, and M_2 is the upper surface of $z = 11 - \sqrt{x^2 + y^2}$, $z \ge 8$.

15.6.11.* $\mathbf{F}(x, y, z) = (x^3, e^{y^2}, ze^{xy})$, and M is defined by $z = 4 - 4x^2 - y^2$ and $z \ge 0$, oriented by a normal with non-negative $\mathbf{k}$-component.

In Exercises 15.6.12–15.6.18 the surface M is as in Stokes' Theorem, $\mathbf{a}$ is a constant vector, $\mathbf{F} : \mathbb{R}^3 \to \mathbb{R}^3$ and $f, g : \mathbb{R}^3 \to \mathbb{R}$ are C^2 functions. Prove the identities:

15.6.12. $\iint_M \mathbf{a} \cdot \mathbf{n}\,d\sigma = \dfrac{1}{2} \oint_{\partial M} (\mathbf{a} \times \mathbf{x}) \cdot d\mathbf{x}$.   **15.6.13.** $\iint_M (\nabla f \times \nabla g) \cdot \mathbf{n}\,d\sigma = \oint_{\partial M} [f(\nabla g)] \cdot d\mathbf{x}$.

15.6.14. $\iint_M (\mathbf{n} \times \nabla f) \cdot \mathbf{i}\,d\sigma = \oint_{\partial M} f\mathbf{i} \cdot d\mathbf{x}$.   **15.6.15.** $\iint_M \mathbf{n} \times \nabla f\,d\sigma = \oint_{\partial M} f\,d\mathbf{x}$.

15.6.16. $\operatorname{curl}(f\mathbf{F}) = f\operatorname{curl}\mathbf{F} + \nabla f \times \mathbf{F}$.   **15.6.17.** $\operatorname{curl}(\nabla f) = \mathbf{0}$.   **15.6.18.** $\operatorname{div}\operatorname{curl}\mathbf{F} = \mathbf{0}$.

15.6.19. Let $\mathbf{F} : \mathbb{R}^3 \to \mathbb{R}^3$ be a C^1 function. Prove that $\oint_C \mathbf{F} \cdot d\mathbf{x} = 0$ for every closed curve C if and only if curl $\mathbf{F} = \mathbf{0}$ identically.

15.6.20. Let $\mathbf{F} : \mathbb{R}^3 \to \mathbb{R}^3$ be a C^1 function. Prove that curl $\mathbf{F} = \mathbf{0}$ identically if and only if $\mathbf{F} = \nabla f$ for some $f : \mathbb{R}^3 \to \mathbb{R}$.

15.6.21. Let M and $\mathbf{F}$ be as in Stokes' Theorem, let $(a, b, c) \in M$, let M_r denote the portion of the surface M that lies in the ball with center (a, b, c) and radius r and let A_r be the area of M_r. Prove that

$$\text{curl } \mathbf{F}(a, b, c) \cdot \mathbf{n}(a, b, c) = \lim_{r \to 0} \frac{1}{A_r} \oint_{\partial M_r} \mathbf{F} \cdot d\mathbf{x}.$$

15.7 Differential Forms on $\mathbb{R}^n$

Let us look at the formulas in Green's Theorem, Stokes' Theorem and Divergence Theorem

$$\iint_D (Q_x(x, y) - P_y(x, y))\, dA = \oint_{\partial D} P(x, y)\, dx + Q(x, y)\, dy,$$

$$\iint_M \text{curl } \mathbf{F} \cdot \mathbf{n}\, d\sigma = \oint_{\partial M} \mathbf{F} \cdot d\mathbf{x},$$

$$\iiint_S \text{div } \mathbf{F}\, dV = \oiint_{\partial S} \mathbf{F} \cdot \mathbf{n}\, d\sigma.$$

Notice that they all follow the pattern similar to the Fundamental Theorem of Calculus. The right hand side is an integral over the boundary of a region, and the left sides feature the integral over the whole region, the integrand being obtained from the one on the right side using derivatives.

In this section we will present a theory that unifies these four theorems. It is associated with the name of French mathematician Elie Cartan (1869–1951). We will start with expressions like $P\, dx + Q\, dy$ or $F_1\, dx + F_2\, dy + F_3\, dz$.

Definition 15.7.1. Let $a_1, a_2, \ldots, a_n$ be real-valued functions, defined on an n dimensional domain D. We say that

$$\omega(\mathbf{x}) = a_1(\mathbf{x})\, dx_1 + a_2(\mathbf{x})\, dx_2 + \cdots + a_n(\mathbf{x})\, dx_n \qquad (15.30)$$

is a **differential form of degree 1** on D, or a **1-form** on D.

Thus, $P(x, y)\, dx + Q(x, y)\, dy$ is a 1-form in 2 variables, and $F_1(x, y, z)\, dx + F_2(x, y, z)\, dy + F_3(x, y, z)\, dz$ is a 1-form in 3 variables. Another very important example is the total differential du of a differentiable function u (page 351). We make a note that *not* every 1-form is a differential of some function. Later we will talk more about this issue.

Assuming that the region D in Green's Formula is parametrized by x and y, the symbol dA gets replaced by $dxdy$. Then the integrand on the left side becomes $(Q_x - P_y)\, dxdy$, and it does not follow the pattern as laid out in (15.30). Instead of individual differentials dx_i, it features a product of *two* such factors.

Definition 15.7.2. Let $\{a_{ij}\}$, $1 \le i < j \le n$, be real-valued functions, defined on an

n dimensional domain D. We say that

$$\omega(\mathbf{x}) = a_{12}dx_1 dx_2 + a_{13}dx_1 dx_3 + \cdots + a_{1n}dx_1 dx_n$$
$$+ a_{23}dx_2 dx_3 + \cdots + a_{2n}dx_2 dx_n$$
$$\cdots \cdots$$
$$+ a_{n-1,n}dx_{n-1}dx_n$$

is a **differential form of degree 2** on D, or a **2-form** on D.

Further, div $\mathbf{F}\, dxdydz$ (featuring in Divergence Theorem, with dV replaced by $dxdydz$) is neither a 1-form, nor a 2-form, because it contains a product of 3 factors dx, dy, dz, so we say that it is a **differential form of degree 3**, or a **3-form**. A general 3-form is a sum of all terms that can be written as $a_{ijk}\, dx_i dx_j dx_k$, with $1 \le i < j < k \le n$. It is easy to see that these definitions can be extended to differential forms of any degree.

Before we continue, we recall Examples 15.4.12 and 15.4.13. A change of roles of s and t resulted in a change of sign of the result. This motivated Cartan to introduce the following rule for the "product" of forms:

$$dx_i dx_j = -dx_j dx_i, \text{ if } i \ne j, \quad \text{and} \quad dx_i dx_i = 0. \tag{15.31}$$

We call this operation the **external product** of differential forms, and very often it is denoted by $\omega_1 \wedge \omega_2$. Since differential forms are not numbers, there is no "usual" product defined on them, and there is no confusion if the symbol $\wedge$ is omitted.

Example 15.7.3. The external product of differential forms.

Find the product of the differential forms $\omega_1(x, y, z) = xy\, dx + yz\, dy$ and $\omega_2(x, y, z) = dx - xy\, dy + x\, dz$.

Solution. Using (15.31), the product $\omega_1\omega_2$ equals

$$(xy\, dx + yz\, dy)\,(dx - xy\, dy + x\, dz)$$
$$= xy\, dxdx - (xy)^2\, dxdy + x^2 y\, dxdz + yz\, dydx - xy^2 z\, dydy + xyz\, dydz$$
$$= -(xy)^2\, dxdy + x^2 y\, dxdz - yz\, dxdy + xyz\, dydz$$
$$= -(x^2 y^2 + yz)\, dxdy + x^2 y\, dxdz + xyz\, dydz.$$

Notice that ω_1 and ω_2 are 1-forms, but their product is a 2-form. ◆

Next, we define the "derivative" of a differential form.

Definition 15.7.4. Let $\omega(\mathbf{x}) = \sum_{k=1}^{n} a_k(\mathbf{x})\, dx_k$ be a 1-form in n variables. The **external derivative** $d\omega$ is a 2-form

$$d\omega(\mathbf{x}) = \sum_{k=1}^{n} da_k(\mathbf{x})\, dx_k = \sum_{k=1}^{n} \left(\sum_{j=1}^{n} \frac{\partial a_k}{\partial x_j}\, dx_j \right) dx_k.$$

Example 15.7.5. The external derivative of a 1-form.

$\omega(x, y) = x^3 y\, dx + xy^2\, dy$. Find $d\omega$.

Solution. Let $P(x, y) = x^3 y$, $Q(x, y) = xy^2$. Then $P_x = 3x^2 y$, $P_y = x^3$, $Q_x = y^2$, $Q_y = 2xy$, so $dP = 3x^2 y\, dx + x^3\, dy$, and $dQ = y^2\, dx + 2xy\, dy$. Therefore,

$$d\omega(x, y) = \left(3x^2 y\, dx + x^3\, dy\right) dx + \left(y^2\, dx + 2xy\, dy\right) dy$$
$$= 3x^2 y\, dxdx + x^3\, dydx + y^2\, dxdy + 2xy\, dydy$$
$$= (y^2 - x^3)\, dxdy. \qquad ◆$$

Notice that $d(P\,dx + Q\,dy)$ turned out to be $(Q_x - P_y)\,dxdy$. Exercise 15.7.15 asserts that this is generally true. Thus, using the language of differential forms, if $\omega(x,y) = P(x,y)\,dx + Q(x,y)\,dy$, Green's Theorem states that

$$\iint_D d\omega = \oint_{\partial D} \omega. \tag{15.32}$$

It is not hard to see that, if ω is a 1-form in one variable and D is a line segment, equality (15.32) yields the Fundamental Theorem of Calculus. We will show that it also represents a generalization of the formulas in Stokes' Theorem and Divergence Theorem.

> The three fundamental theorems appear as one general result in 1889, in the work [106] of Volterra (page 158). Poincaré (page 525) did the same thing in his book [87] in 1899, but improved the notation. Cartan (see next page) then developed the theory of differential forms and expressed their results in this new language. The article [70] shows all this in more detail.

Example 15.7.6. The external derivative of a 1-form in 3 variables.

Let $\omega(x,y,z) = x^3y^2z\,dx - xy^2z^3\,dy + xyz\,dz$. Find $d\omega$.

Solution. Let $F_1(x,y,z) = x^3y^2z$, $F_2(x,y,z) = -xy^2z^3$, $F_3(x,y,z) = xyz$. Their partial derivatives are

$$\begin{aligned}
(F_1)_x &= 3x^2y^2z, & (F_1)_y &= 2x^3yz, & (F_1)_z &= x^3y^2, \\
(F_2)_x &= -y^2z^3, & (F_2)_y &= -2xyz^3, & (F_2)_z &= -3xy^2z^2, \\
(F_3)_x &= yz, & (F_3)_y &= xz, & (F_3)_z &= xy.
\end{aligned}$$

Therefore,

$$\begin{aligned}
d\omega(x,y,z) &= \left(3x^2y^2z\,dx + 2x^3yz\,dy + x^3y^2\,dz\right)\,dx \\
&\quad + \left(-y^2z^3\,dx - 2xyz^3\,dy - 3xy^2z^2\,dz\right)\,dy \\
&\quad + \left(yz\,dx + xz\,dy + xy\,dz\right)\,dz \\
&= \left(-2x^3yz - y^2z^3\right)\,dxdy + \left(-x^3y^2 + yz\right)\,dxdz + \left(3xy^2z^2 + xz\right)\,dydz. \quad \blacklozenge
\end{aligned}$$

We see that, in this example,

$$d\omega = \left(-(F_1)_y + (F_2)_x\right)\,dxdy + \left(-(F_1)_z + (F_3)_x\right)\,dxdz + \left(-(F_2)_z + (F_3)_y\right)\,dydz \tag{15.33}$$

which looks like the dot product of curl $\mathbf{F}$ and a vector with components $dydz$, $-dxdz$, and $dxdy$. Exercise 15.7.16 states that (15.33) is universally true. Since our goal is to establish the equivalence of the Stokes' Formula (15.28) and the general formula (15.32), we need to prove the equality of forms

$$n_1\,d\sigma = dydz, \quad n_2\,d\sigma = -dxdz, \quad n_3\,d\sigma = dxdy. \tag{15.34}$$

If we write $\mathbf{x} = \mathbf{f}(s,t)$, then $x = f_1(s,t)$, $y = f_2(s,t)$, $z = f_3(s,t)$, and

$$\begin{aligned}
dydz &= ((f_2)_s\,ds + (f_2)_t\,dt)\,((f_3)_s\,ds + (f_3)_t\,dt) \\
&= ((f_2)_s(f_3)_t - (f_2)_t(f_3)_s)\,dsdt.
\end{aligned}$$

On the other hand, $\mathbf{n}\,d\sigma = \mathbf{f}_s \times \mathbf{f}_t\,dsdt$, and the first component of $\mathbf{f}_s \times \mathbf{f}_t$ is precisely $((f_2)_s(f_3)_t - (f_2)_t(f_3)_s)$, hence $n_1\,d\sigma = dydz$. A similar calculation can be used to verify the remaining equalities in (15.34), so it follows that Stokes' Theorem is also of the form (15.32). (Except that now D is a surface, and ∂D is its boundary.)

Definition 15.7.4 can be easily extended to forms of any degree. We will formulate it only for 2-forms.

Definition 15.7.7. Let $\omega(\mathbf{x}) = \sum_{1 \leq i < k \leq n} a_{ik}(\mathbf{x})\, dx_i dx_k$ be a 2-form in n variables. The **external derivative** $d\omega$ is a 3-form

$$d\omega(\mathbf{x}) = \sum_{1 \leq i < k \leq n} da_{ik}(\mathbf{x})\, dx_i dx_k = \sum_{1 \leq i < k \leq n} \left(\sum_{j=1}^{n} \frac{\partial a_{ik}}{\partial x_j}\, dx_j \right) dx_i dx_k.$$

Example 15.7.8. The external derivative of a 2-form in 3 variables.

Let $\omega(x, y, z) = xyz\, dxdy + x^2\, dxdz - y^2 z\, dydz$. Find $d\omega$.

Solution. Let $F_1(x, y, z) = xyz$, $F_2(x, y, z) = x^2$, $F_3(x, y, z) = -y^2 z$. Their partial derivatives are

$$
\begin{aligned}
(F_1)_x &= yz, & (F_1)_y &= xz, & (F_1)_z &= xy, \\
(F_2)_x &= 2x, & (F_2)_y &= 0, & (F_2)_z &= 0, \\
(F_3)_x &= 0, & (F_3)_y &= -2yz, & (F_3)_z &= -y^2.
\end{aligned}
$$

Therefore,

$$
\begin{aligned}
d\omega(x, y, z) &= (yz\, dx + xz\, dy + xy\, dz)\, dxdy + (2x\, dx)\, dxdz + \left(-2yz\, dy - y^2\, dz\right) dydz \\
&= xy\, dzdxdy = xy\, dxdydz. \qquad \blacklozenge
\end{aligned}
$$

Finally, we turn our attention to Divergence Theorem. Here, the integrand on the right side is the form $\mathbf{F} \cdot \mathbf{n}\, d\sigma$. As (15.34) shows, this integrand is a 2-form $\omega = F_1\, dydz - F_2\, dxdz + F_3\, dxdy$. Then

$$
\begin{aligned}
d\omega &= dF_1\, dydz - dF_2\, dxdz + dF_3\, dxdy \\
&= [(F_1)_x\, dx + (F_1)_y\, dy + (F_1)_z\, dz]\, dydz \\
&\quad - [(F_2)_x\, dx + (F_2)_y\, dy + (F_2)_z\, dz]\, dxdz \\
&\quad + [(F_3)_x\, dx + (F_3)_y\, dy + (F_3)_z\, dz]\, dxdy \\
&= [(F_1)_x + (F_2)_y + (F_3)_z]\, dxdydz \\
&= \operatorname{div} \mathbf{F}\, dxdydz,
\end{aligned}
$$

which implies that Divergence Theorem is another special case of (15.32).

ELIE CARTAN (1869–1951) was a French mathematician, one of the greatest of the twentieth century. His father was a blacksmith and his family was not affluent enough to pay for the education of the children. Cartan was discovered by a school inspector Antonin Dubost, later the president of the French Senate. He helped Cartan secure the funding to attend the high school in Vienne, Grenoble, and Paris. He became a student at the Ecole Normale Supérieure in 1888, where he attended courses by the leading mathematicians of the day including Poincaré (page 525). Cartan graduated in 1891 and then served for a year in the army before returning for a doctorate. He received it at Sorbonne in 1894 on the topic of Lie algebras. He lectured at Montpellier, Lyon, Nancy, before he was appointed at Sorbonne in 1909, where he remained until he retired in 1940. He worked on continuous groups, Lie algebras, differential equations, and geometry. His work achieved a synthesis between these areas. In 1913 he discovered the theory of spinors which later played a fundamental role in quantum mechanics. Because of his extraordinary originality, the significance of his work was not fully recognized until the 1930s. He received honorary degrees from several universities, including Harvard University in 1936, the Free University of Berlin and the Catholic University of Louvain in 1947. He was elected a Fellow of the Royal Society of London, to the French Academy of Sciences in 1931, where he served as a vice-president in 1945 and President in 1946. A crater on the moon is named Cartan in his honor. His son Henri, another distinguished mathematician, was one of the seven founding members of the Bourbaki group.

Exercises

In Exercises 15.7.1–15.7.4 calculate the product of the given differential forms:

15.7.1. $(7\,dx + 3xy\,dy)(x^2y\,dx - 5x\,dy)$.   15.7.2. $(y\,dx + x\,dy)(x^2\,dx + y^2\,dy - 3\,dz)$.

15.7.3. $(x^3\,dxdy + y^2\,dxdz)(z\,dx - x\,dz)$.   15.7.4. $(xy\,dxdydz)(3z\,dx - 2x\,dy + 5y\,dz)$.

15.7.5. Prove that $(A\,dx + B\,dy + C\,dz)(a\,dx + b\,dy + c\,dz)$

$$= \begin{vmatrix} B & C \\ b & c \end{vmatrix} dydz + \begin{vmatrix} C & A \\ c & a \end{vmatrix} dzdx + \begin{vmatrix} A & B \\ a & b \end{vmatrix} dxdy.$$

15.7.6. Prove that

$$(A\,dx + B\,dy + C\,dz)(a\,dydz + b\,dzdx + c\,dxdy) = (aA + bB + cC)\,dxdydz.$$

15.7.7.* Let $\omega_1, \omega_2, \ldots, \omega_n$ be 1-forms on $\mathbb{R}^n$, let $\omega_i = \sum_{j=1}^{n} a_{ij}\,dx_j$, $1 \le i \le n$, and let A be the $n \times n$ matrix (a_{ij}). Prove that $\omega_1\omega_2\cdots\omega_n = (\det A)dx_1dx_2\cdots dx_n$.

In Exercises 15.7.8–15.7.14 calculate the derivatives of the given differential forms:

15.7.8. $4y\,dx - 3xy\,dy$.   15.7.9. $(xy+z)\,dx + (x^2-z)\,dy + (x+z^2)\,dz$.

15.7.10. $\sin(xy)\,dx + \cos(xy)\,dy$.   15.7.11. $x^3\,dxdy + y^2\,dxdz - z\,dydz$.

15.7.12. $yx\,dydx + zx\,dzdx + yz\,dydz$.

15.7.13. $\sum_{i=1}^{n}(-1)^{i-1}x_idx_1\cdots dx_{i-1}dx_{i+1}\cdots dx_n$.

15.7.14. $(x_1^2 + x_2^2 + \cdots + x_n^2)^{-n/2}\sum_{i=1}^{n}(-1)^{i-1}dx_1\cdots dx_{i-1}dx_{i+1}\cdots dx_n$.

15.7.15. Let $\omega(x, y) = P(x, y)\,dx + Q(x, y)\,dy$, where P, Q are C_1 functions. Prove that $d\omega = (Q_x - P_y)\,dxdy$.

15.7.16. Let $\omega(x, y, z) = F_1(x, y, z)\,dx + F_2(x, y, z)\,dy + F_3(x, y, z)\,dz$, where $\mathbf{F} = (F_1, F_2, F_3)$ is a C^1 function. Prove that (15.33) is valid.

15.7.17. Prove that the external derivative is a linear operator on differential forms, i.e., if ω_1, ω_2 are differential forms of the same order, and $\alpha, \beta \in \mathbb{R}$, then $d(\alpha\omega_1 + \beta\omega_2) = \alpha d\omega_1 + \beta d\omega_2$.

15.7.18. Let $\mathbf{F} : \mathbb{R}^3 \to \mathbb{R}^3$ be a C^1 function. Prove that

$$dF_1dF_2dF_3 = \mathbf{DF}(x, y, z)\,dxdydz.$$

15.7.19. Let $\omega = \sum_{1 \le i < j \le n} a_{ij}\,dx_idx_j$ be a differential 2-form on $\mathbb{R}^n$. Prove that

$$d\omega = \sum_{1 \le i < j < k \le n} \left(\frac{\partial a_{ij}}{\partial x_k} - \frac{\partial a_{ik}}{\partial x_j} + \frac{\partial a_{jk}}{\partial x_i} \right) dx_idx_jdx_k.$$

In Exercises 15.7.20–15.7.23 $\mathbf{F} : \mathbb{R}^m \to \mathbb{R}^n$ is an injective C^1 function, ω is a k-form on $\mathbb{R}^n$ and $\mathbf{F}^*\omega$ is the **pullback** of ω. It is defined as $\mathbf{F}^*f = f \circ \mathbf{F}$ for a 0-form f and $\mathbf{F}^*(\sum_{i=1}^{n} a_i\,dx_i) = \sum_{i=1}^{n} a_i \circ \mathbf{F}\,d\mathbf{F}_i$. The definition is then extended to arbitrary k-forms by requiring that $\mathbf{F}^*$ is additive and multiplicative.

15.7.20. Prove that $\mathbf{F}^*(d\omega) = d(\mathbf{F}^*\omega)$ when ω is a 0-form on $\mathbb{R}^3$.

15.7.21. Prove that $\mathbf{F}^*(d\omega) = d(\mathbf{F}^*\omega)$ when ω is a 1-form on $\mathbb{R}^3$.

15.7.22. Let $\omega = \dfrac{-y}{x^2 + y^2}\,dx + \dfrac{x}{x^2 + y^2}\,dy$, $A = \{(r, \theta) : r > 0 \text{ and } 0 < \theta < 2\pi\}$ and let $\mathbf{F} : A \to \mathbb{R}^2$ be defined by $\mathbf{F}(r, \theta) = (r\cos\theta, r\sin\theta)$. Find $\mathbf{F}^*\omega$.

15.7.23. Let R be a k-dimensional rectangle in $\mathbb{R}^n$, let $\mathbf{F} : R \to \mathbb{R}^n$ be a C^1 parametrization of a surface M and let ω be a k-form on $\mathbb{R}^n$. Prove that $\int_M \omega = \int_R \mathbf{F}^*\omega$.

15.8   Exact Differential Forms on $\mathbb{R}^n$

Among all differential forms ω, those that are of the form $\omega = du$ have a special status.

Theorem 15.8.1. *Let $u : \mathbb{R}^n \to \mathbb{R}$ be a C^1 function on an open set A, and let C be a curve in $\mathbb{R}^n$ parametrized by a C^1 path $\mathbf{f} : [a,b] \to A$. Then*

$$\int_C du = u(\mathbf{f}(b)) - u(\mathbf{f}(a)).$$

Proof. It is helpful to define $g = u \circ \mathbf{f}$. Clearly, g is a C^1 function mapping $[a,b]$ to $\mathbb{R}$. By the Chain Rule, $g'(t) = \mathbf{D}u(\mathbf{f}(t)) \cdot \mathbf{f}'(t)$. Further, if we denote $d\mathbf{x} = (dx_1, dx_2, \ldots, dx_n)$, then

$$du = \sum_{k=1}^{n} \frac{\partial u}{\partial x_k} dx_k = \mathbf{D}u \cdot d\mathbf{x}.$$

It follows that

$$\int_C du = \int_C \mathbf{D}u \cdot d\mathbf{x}$$

$$= \int_a^b \mathbf{D}u(\mathbf{f}(t)) \cdot \mathbf{f}'(t)\, dt$$

$$= \int_a^b g'(t)\, dt$$

$$= g(b) - g(a) = u(\mathbf{f}(b)) - u(\mathbf{f}(a)),$$

and the theorem is proved. $\square$

Theorem 15.8.1 shows that, when the differential form is a total differential of a function, the line integral along a curve depends only on the endpoints of the curve.

Corollary 15.8.2. *Let $u : \mathbb{R}^n \to \mathbb{R}$ be a C^1 function on an open set A, and let C_1, C_2 be two curves in $\mathbb{R}^n$ parametrized by C^1 paths $\mathbf{f}, \mathbf{g} : [a,b] \to A$. If $\mathbf{f}(a) = \mathbf{g}(a)$ and $\mathbf{f}(b) = \mathbf{g}(b)$, and if $\omega = du$ then*

$$\int_{C_1} \omega = \int_{C_2} \omega.$$

Another easy consequence of Theorem 15.8.1 concerns the case when C is a closed curve.

Corollary 15.8.3. *Let $u : \mathbb{R}^n \to \mathbb{R}$ be a C^1 function on an open set A, and let C be a closed curve in A. Then*

$$\int_C du = 0.$$

Remark 15.8.4. A very similar argument to the one in the proof of Theorem 15.8.1 shows that the first three results of this section can be generalized to the case when u is not just a function (i.e., a 0-form) but a k-form. The difference is that in the case $k = 1$ we would use Green's Theorem and in the case $k = 2$ Stokes' Theorem. The result is true beyond $k = 2$, but it requires a generalization of the mentioned theorems, namely an equality of the form (15.32). For a proof of the latter we refer the reader to [96].

We see that differential forms that are total differentials of C^1 functions deserve special attention. Thus, it is natural to look for a way to identify them. The following "Product Rule" is a first step in this direction.

Theorem 15.8.5. *Suppose that ω_1 is a k-form, that ω_2 is a m-form, and that both are C^1 forms. Then*

$$d(\omega_1\omega_2) = (d\omega_1)\omega_2 + (-1)^k\omega_1(d\omega_2).$$

Proof. Since the external derivative is linear (Exercise 15.7.17), it suffices to consider the case when $\omega_1 = a\,dx_{i_1}\ldots dx_{i_k}$ and $\omega_2 = b\,dx_{j_1}\ldots dx_{j_m}$. Then

$$
\begin{aligned}
d(\omega_1\omega_2) &= d\left(ab\,dx_{i_1}\ldots dx_{i_k}\,dx_{j_1}\ldots dx_{j_m}\right) \\
&= d(ab)\,dx_{i_1}\ldots dx_{i_k}\,dx_{j_1}\ldots dx_{j_m} \\
&= (da)b\,dx_{i_1}\ldots dx_{i_k}\,dx_{j_1}\ldots dx_{j_m} + a(db)\,dx_{i_1}\ldots dx_{i_k}\,dx_{j_1}\ldots dx_{j_m} \\
&= (da\,dx_{i_1}\ldots dx_{i_k})(b\,dx_{j_1}\ldots dx_{j_m}) + (-1)^k(a\,dx_{i_1}\ldots dx_{i_k})(db\,dx_{j_1}\ldots dx_{j_m}) \\
&= (d\omega_1)\omega_2 + (-1)^k\omega_1(d\omega_2),
\end{aligned}
$$

and the theorem is proved. $\qquad\square$

While the formula in Theorem 15.8.5 bears resemblance to the "usual" product rule, the next result is quite surprising. Since it concerns the "second derivative", we will assume that the differential form belongs to C^2, the class of functions with continuous second-order partial derivatives.

Theorem 15.8.6. *If ω is a k-form in C^2, then $d(d\omega) = 0$.*

Proof. We will prove the result in the case when

$$\omega = a\,dx_{i_1}\ldots dx_{i_k}. \qquad (15.35)$$

Since every k-form is a linear combination of such forms, the general case will then follow from the fact that the external derivative is linear (Exercise 15.7.17).

By definition, $d\omega = da\,dx_{i_1}\ldots dx_{i_k}$. In order to calculate $d(d\omega)$ we will use Theorem 15.8.5. Of course, for any j, $d(dx_{i_j}) = d(1dx_{i_j}) = 0$, so $d(d\omega) = d(da)\,dx_{i_1}\ldots dx_{i_k}$, and it remains to prove that $d(da) = 0$. If we denote

$$a_i = \frac{\partial a}{\partial x_i}, \quad a_{ij} = \frac{\partial a_i}{\partial x_j} = \frac{\partial^2 a}{\partial x_j\partial x_i},$$

then $da = \sum_{i=1}^{n} a_i dx_i$ and

$$d(da) = \sum_{j=1}^{n} \frac{\partial}{\partial x_j}\left(\sum_{i=1}^{n} a_i dx_i\right)dx_j = \sum_{j=1}^{n}\sum_{i=1}^{n} a_{ij}\,dx_i dx_j.$$

In the last sum there are terms where $i = j$. They are each equal to 0, because $dx_i dx_i = 0$. The remaining ones, when $i \neq j$ can be grouped in pairs which have the same indices but in the opposite order (such as, for example, the pairs $(2,3)$ and $(3,2)$). Now

$$a_{ij}\,dx_i dx_j + a_{ji}\,dx_j dx_i = a_{ij}\,dx_i dx_j - a_{ji}\,dx_i dx_j = 0$$

because $a \in C^2$, so that $a_{ij} = a_{ji}$ (Theorem 11.3.5). Thus $d(da) = 0$ and the proof is complete. $\qquad\square$

A k-form ω is **closed** if $d\omega = 0$, and it is **exact** if there exists a $(k-1)$-form ω_1 such that $\omega = d\omega_1$. Corollaries 15.8.2 and 15.8.3 show that exact forms are in high demand. Theorem 15.8.6 gives a necessary condition for a form to be exact: if a form is exact then it is closed. In the remainder of the section we will consider the converse of this statement.

Our first observation is that, without additional assumptions, such an assertion is false.

Example 15.8.7. A differential form that is closed but not exact.

Prove that the form $\omega = \dfrac{-y}{x^2 + y^2}\, dx + \dfrac{x}{x^2 + y^2}\, dy$ is closed but not exact in $A = \mathbb{R}^2 \setminus \{(0,0)\}$.

Solution. It is not hard to verify that $d\omega = 0$ (Exercise 15.8.1). However, if ω were exact in A, Corollary 15.8.3 would imply that $\int_C \omega = 0$, where C is the unit circle. If we parametrize C by the polar coordinates $x = \cos t$, $y = \sin t$, $0 \le t \le 2\pi$, then

$$\int_C \omega = \int_0^{2\pi} (-\sin t(-\sin t)\, dt + \cos t(\cos t)\, dt) = \int_0^{2\pi} dt = 2\pi.$$

This implies that ω is not exact in A.   ♦

Thus, in order to establish a converse to Theorem 15.8.6 we will need to strengthen our hypotheses. One way to do that is to require that the set A is *starlike with respect to* $\mathbf{a} \in A$. This means that, for any $\mathbf{x} \in A$, the line segment connecting $\mathbf{a}$ and $\mathbf{x}$ lies entirely in A. To simplify the calculations, we will consider the case $\mathbf{a} = \mathbf{0}$, and leave the general case as an exercise (Exercise 15.8.7).

We will start by defining an operator I (the "integral") that will associate to a k-form $\omega = dx_{i_1} dx_{i_2} \ldots dx_{i_k}$ a $(k-1)$-form $I(\omega)$, such that $d(I(\omega)) = \omega$. We will often write $I\omega$ instead of $I(\omega)$. When ω is a 1-form dx, it is easy to see that $I\omega$ should be x. If $\omega = dxdy$, it is not hard to verify that

$$\omega = d\left(\frac{x}{2}\, dy - \frac{y}{2}\, dx\right),$$

so we define $I\omega = \frac{1}{2}(x\, dy - y\, dx)$. When $\omega = dxdydz$, we define $I\omega = \frac{1}{3}(x\, dydz - y\, dxdz + z\, dxdy)$. In general, if $d\mathbf{x} = dx_{i_1} dx_{i_2} \ldots dx_{i_k}$ and if $d\mathbf{x}^{-j}$ denotes the $(k-1)$-form obtained from $d\mathbf{x}$ by deleting dx_{i_j}, we define

$$I\omega = \frac{1}{k}\left(\sum_{j=1}^{k}(-1)^{j-1} x_{i_j}\, d\mathbf{x}^{-j}\right).$$

Next, we will consider the forms ω as in (15.35). For example, what if $\omega(x,y) = e^{xy}\, dxdy$? It turns out that

$$I\omega = \frac{e^{xy} - 1}{2y}\, dy - \frac{e^{xy} - 1}{2x}\, dx.$$

While the verification is simple, the question is what formula leads to it. Notice that

$$\frac{e^{xy} - 1}{2y}\, dy - \frac{e^{xy} - 1}{2x}\, dx = \frac{e^{xy} - 1}{xy} \cdot \frac{1}{2}(x\, dy - y\, dx) = \frac{e^{xy} - 1}{xy}\, I(dxdy), \text{ and}$$

$$\frac{e^{xy} - 1}{xy} = \int_0^1 2te^{t^2 xy}\, dt.$$

Further, $e^{t^2 xy} = e^{(tx)(ty)}$ so, if $f(x,y) = e^{xy}$, then $e^{t^2 xy} = f(tx, ty)$, and for the integrand above to exist it is essential that f be defined for all $t \in [0,1]$, meaning for points on the line segment connecting $(0,0)$ and (x,y). So, assuming that the domain A is starlike with respect to the origin, and writing $\mathbf{x} = (x_1, x_2, \ldots, x_n)$, $t\mathbf{x} = (tx_1, tx_2, \ldots, tx_n)$, we define

$$I\left(a(\mathbf{x})\, dx_{i_1} dx_{i_2} \ldots dx_{i_k}\right) = \int\limits_0^1 kt^{k-1} a(t\mathbf{x})\, dt\, I(dx_{i_1} dx_{i_2} \ldots dx_{i_k}), \qquad (15.36)$$

and we extend I as a linear operator to all k forms. This means that every k-form ω can be written in a unique way as a sum $\omega = \sum_{m=1}^p \omega_m$, where each ω_m is as in (15.35), and we define $I\omega = \sum_{m=1}^p I\omega_m$. We will demonstrate that, when ω is a closed form, then $d(I\omega) = 0$. The following lemma is an important step in this direction.

Lemma 15.8.8. *Let A be a set that is starlike with respect to the origin, and let $\omega = a(\mathbf{x})\, dx_{i_1} dx_{i_2} \ldots dx_{i_k}$ be a k-form in $C^1(A)$. Then, assuming that $I\omega$ is defined by (15.36) and extended to all k forms as a linear operator,*

$$I(d\omega) = \sum_{j=1}^k \left(\int\limits_0^1 t^k \frac{\partial a}{\partial x_j}(t\mathbf{x})\, dt \right) x_j\, d\mathbf{x} - \sum_{j=1}^k \frac{\partial}{\partial x_j}\left(\int\limits_0^1 kt^{k-1} a(t\mathbf{x})\, dt \right) dx_j\, I(d\mathbf{x}). \quad (15.37)$$

Proof. By assumption I is linear, so

$$I(d\omega) = I\left(\sum_{j=1}^k \frac{\partial a}{\partial x_j}\, dx_j\, d\mathbf{x} \right) = \sum_{j=1}^k I\left(\frac{\partial a}{\partial x_j}\, dx_j\, d\mathbf{x} \right).$$

The form $dx_j\, d\mathbf{x}$ is a $(k+1)$-form so, for each j, $1 \le j \le k$,

$$\begin{aligned}
I\left(\frac{\partial a}{\partial x_j}\, dx_j\, d\mathbf{x} \right) &= \left(\int\limits_0^1 (k+1)t^k \frac{\partial a}{\partial x_j}(t\mathbf{x})\, dt \right) I(dx_j d\mathbf{x}) \\
&= \left(\int\limits_0^1 (k+1)t^k \frac{\partial a}{\partial x_j}(t\mathbf{x})\, dt \right) \frac{x_j\, d\mathbf{x} - dx_j\, kI(d\mathbf{x})}{k+1}.
\end{aligned}$$

It follows that

$$I(d\omega) = \sum_{j=1}^k \left(\int\limits_0^1 t^k \frac{\partial a}{\partial x_j}(t\mathbf{x})\, dt \right) x_j\, d\mathbf{x} - \sum_{j=1}^k \left(\int\limits_0^1 kt^k \frac{\partial a}{\partial x_j}(t\mathbf{x})\, dt \right) dx_j\, I(d\mathbf{x}). \quad (15.38)$$

Since the partial derivatives of a are continuous in A, Theorem 13.2.2 shows that

$$\frac{\partial}{\partial x_j}\left(\int\limits_0^1 t^{k-1} a(t\mathbf{x})\, dt \right) = \int\limits_0^1 t^k \frac{\partial a}{\partial x_j}(t\mathbf{x})\, dt. \quad (15.39)$$

Together, (15.38) and (15.39) imply (15.37). $\qquad\square$

Now we can prove the promised converse of Theorem 15.8.6.

Theorem 15.8.9. *If ω is a closed k-form in $C^1(A)$, and A is an open set that is starlike with respect to the origin, then ω is exact.*

Proof. We will prove that, if ω is a k-form in $C^1(A)$, then

$$d(I\omega) + I(d\omega) = \omega. \quad (15.40)$$

Since $d\omega = 0$, the result will follow.

We will write $\omega = \sum_{m=1}^{p} \omega_m$, where each ω_m, $1 \le m \le p$, is of the form (15.35). Both d and I are linear operators, so (15.40) becomes $\sum_{m=1}^{p} d(I\omega_m) + \sum_{m=1}^{p} I\,(d\omega_m) = \sum_{m=1}^{p} \omega_m$, and it suffices to prove (15.40) for $\omega = a(\mathbf{x})\,dx_{i_1}dx_{i_2}\ldots dx_{i_k}$. Using product rule for d and Lemma 15.8.8,

$$d\,(I\omega) = d\left(\int_0^1 kt^{k-1}a(t\mathbf{x})\,dt\,I(dx_{i_1}dx_{i_2}\ldots dx_{i_k})\right)$$

$$= d\left(\int_0^1 kt^{k-1}a(t\mathbf{x})\,dt\,\frac{1}{k}\left(\sum_{j=1}^{k}(-1)^{j-1}x_{i_j}\,d\mathbf{x}^{-j}\right)\right)$$

$$= \sum_{j=1}^{k}(-1)^{j-1}d\left(\int_0^1 t^{k-1}a(t\mathbf{x})\,dt\,x_{i_j}\right)\,d\mathbf{x}^{-j}$$

$$= \sum_{j=1}^{k}(-1)^{j-1}\left[d\left(\int_0^1 t^{k-1}a(t\mathbf{x})\,dt\right)x_{i_j} + \int_0^1 t^{k-1}a(t\mathbf{x})\,dt\,dx_{i_j}\right]d\mathbf{x}^{-j}$$

$$= d\left(\int_0^1 t^{k-1}a(t\mathbf{x})\,dt\right)\sum_{j=1}^{k}(-1)^{j-1}x_{i_j}\,d\mathbf{x}^{-j} + \int_0^1 t^{k-1}a(t\mathbf{x})\,dt\,\sum_{j=1}^{k}(-1)^{j-1}dx_{i_j}d\mathbf{x}^{-j}$$

$$= d\left(\int_0^1 t^{k-1}a(t\mathbf{x})\,dt\right)kI(d\mathbf{x}) + \int_0^1 t^{k-1}a(t\mathbf{x})\,dt\,k\,d\mathbf{x}$$

$$= \sum_{j=1}^{k}\frac{\partial}{\partial x_j}\left(\int_0^1 kt^{k-1}a(t\mathbf{x})\,dt\right)dx_j\,I(d\mathbf{x}) + \int_0^1 kt^{k-1}a(t\mathbf{x})\,dt\,d\mathbf{x}$$

$$= -I(d\omega) + \sum_{j=1}^{k}\left(\int_0^1 t^k\frac{\partial a}{\partial x_j}(t\mathbf{x})\,dt\right)x_j\,d\mathbf{x} + \int_0^1 kt^{k-1}a(t\mathbf{x})\,dt\,d\mathbf{x} \qquad \text{[Lemma 15.8.8]}$$

$$= -I(d\omega) + \int_0^1\left(\sum_{j=1}^{k}t^k\frac{\partial a}{\partial x_j}(t\mathbf{x})\,x_j + kt^{k-1}a(t\mathbf{x})\right)dt\,d\mathbf{x}$$

$$= -I(d\omega) + \int_0^1\left(t^k\frac{d}{dt}(a(t\mathbf{x})) + \frac{d}{dt}(t^k)\,a(t\mathbf{x})\right)dt\,d\mathbf{x}$$

$$= -I(d\omega) + \int_0^1\frac{d}{dt}\left(t^k a(t\mathbf{x})\right)dt\,d\mathbf{x}$$

$$= -I(d\omega) + t^k a(t\mathbf{x})\Big|_{t=0}^{t=1}\,d\mathbf{x}$$

$$= -I(d\omega) + a(\mathbf{x})\,d\mathbf{x}$$

$$= -I(d\omega) + \omega,$$

and the theorem is proved. $\qquad\square$

Poincaré coined the terms "closed" and "exact" for the properties of differential forms, and proved Theorem 15.8.9 in [87]. Today, it is known as the *Poincaré's Lemma*. Ten years earlier, Volterra (page 158) proved a more general result in [106].

Remark 15.8.10. In the case when A is a starlike set with respect to $\mathbf{a} \neq \mathbf{0}$, a linear change of variables can be used to translate the domain, so that it is starlike with respect to the origin. (See Exercise 15.8.7.)

Example 15.8.11. Finding a potential function for an exact differential form.

Let $\omega = (6xy^2 - y^3)\,dx + (6x^2y - 3xy^2)\,dy$. Show that ω is exact and find the potential function $I\omega$.

Solution. The partial derivatives of $P(x,y) = 6xy^2 - y^3$ and $Q(x,y) = 6x^2y - 3xy^2$ are $P_x = 6y^2$, $P_y = 12xy - 3y^2$, $Q_x = 12xy - 3y^2$, $Q_y = 6x^2 - 6xy$. Therefore,

$$d\omega = \left(6y^2\,dx + (12xy - 3y^2)\,dy\right)\,dx + \left((12xy - 3y^2)\,dx + (6x^2 - 6xy)\,dy\right)\,dy$$
$$= (12xy - 3y^2)\,dydx + (12xy - 3y^2)\,dxdy = 0,$$

so ω is exact. To find $I\omega$ we will do it separately for $P\,dx$ and $Q\,dy$. Both are 1-forms so $k = 1$. Using (15.36),

$$I(P\,dx) = \left(\int_0^1 t^0 P(tx, ty)\,dt\right) x = \left(\int_0^1 t^3(6xy^2 - y^3)\,dt\right) x = \frac{x}{4}(6xy^2 - y^3),$$

$$I(Q\,dy) = \left(\int_0^1 t^0 Q(tx, ty)\,dt\right) y = \left(\int_0^1 t^3(6x^2y - 3xy^2)\,dt\right) y = \frac{y}{4}(6x^2y - 3xy^2),$$

$$I(\omega) = \frac{x}{4}(6xy^2 - y^3) + \frac{y}{4}(6x^2y - 3xy^2) = 3x^2y^2 - xy^3. \qquad \blacklozenge$$

JULES HENRI POINCARE (1854–1912) was a French mathematician, theoretical physicist, engineer, and philosopher of science. His family was a distinguished one—his cousin Raymond serve as President of France from 1913 to 1920. Henri was an excellent student, both in high school in Nancy (now renamed the Lycée Henri-Poincaré in his honor) and later at Ecole Polytechnique where he was a student of Charles Hermitte. During that time he published his first paper and wrote a doctoral dissertation in the field of differential equations. After receiving his degree, he taught at the University of Caen for 2 years, then at the Faculty of Sciences of the University of Paris (the Sorbonne) for the rest of his life. He is often described as "The Last Universalist," with contributions in mathematics, celestial mechanics, special theory of relativity, and the philosophy of science. He created algebraic topology and initiated the study of complex functions of several variables. In applied mathematics he studied optics, electricity, telegraphy, capillarity, elasticity, thermodynamics, potential theory, quantum theory, theory of relativity and cosmology. In the field of celestial mechanics he studied the three-body problem, and the theories of light and of electromagnetic waves. He is acknowledged as a co-discoverer, with Albert Einstein and Hendrik Lorentz, of the special theory of relativity. The *Poincaré Conjecture* was one of the most famous open problems in mathematics until it was proved in 2003 by Grigori Perelman. During his lifetime Poincaré received many awards and recognitions including membership in all five sections of the Académie des Sciences (geometry, mechanics, physics, geography, and navigation). He was a strong believer in intuition. He wrote: *It is by logic we prove, it is by intuition that we invent.*

Exercises

15.8.1. Prove that the form $\omega = \dfrac{-y}{x^2 + y^2}\,dx + \dfrac{x}{x^2 + y^2}\,dy$ is closed in $A = \mathbb{R}^2 \setminus \{(0,0)\}$.

In Exercises 15.8.2–15.8.6 verify that the integrand is an exact form and calculate the line integral along a curve with given endpoints:

15.8.2. $\displaystyle\int_{(-1,2)}^{(2,3)} x\,dy + y\,dx$.

15.8.3. $\displaystyle\int_{(0,1)}^{(3,-4)} x\,dx + y\,dy$.

15.8.4. $\displaystyle\int_{(0,1)}^{(2,3)} (x + y)\,dx + (x - y)\,dy$.

15.8.5. $\displaystyle\int_{(0,0)}^{(a,b)} e^x\,(\cos y\,dx - \sin y\,dy)$.

15.8.6. $\int_{(1,2,3)}^{(0,1,1)} yz\,dx + xz\,dy + xy\,dz.$

15.8.7. Prove that Theorem 15.8.9 remains true if A is starlike with respect to $\mathbf{a} \neq \mathbf{0}$.

15.8.8. Suppose that $P(x,y)\,dx + Q(x,y)\,dy$ is an exact form in a rectangle A, and that C is a curve in A with endpoints (x_1, y_1) and (x_2, y_2). Prove that

$$\int_C P(x,y)\,dx + Q(x,y)\,dy = \int_{x_1}^{x_2} P(x, y_1)\,dx + \int_{y_1}^{y_2} Q(x_2, y)\,dy.$$

In Exercises 15.8.9–15.8.16 determine whether ω is exact and, if so, find the potential function $I\omega$.

15.8.9. $\omega = (x^2 + 2xy + y^2)\,dx + (x^2 - 2xy + y^2)\,dy$.

15.8.10. $\omega = \dfrac{y\,dx - x\,dy}{3x^2 - 2xy + 3y^2}$.

15.8.11. $\omega = e^x\left(e^y(x - y + 2) + y\right)\,dx + e^x\left(e^y(x - y) + 1\right)\,dy$.

15.8.12. $\omega = (x^2 - 2yz)\,dx + (y^2 - 2xz)\,dy + (z^2 - 2xy)\,dz$.

15.8.13. $\omega = \left(1 - \dfrac{1}{y} + \dfrac{y}{z}\right)\,dx + \left(\dfrac{x}{z} + \dfrac{x}{y^2}\right)\,dy - \dfrac{xy}{z^2}\,dz$.

15.8.14. $\omega = \dfrac{(x + y - z)\,dx + (x + y - z)\,dy + (x + y + z)\,dz}{x^2 + y^2 + z^2 + 2xy}$.

15.8.15. $\omega = (2y - 4)\,dydz + (y^2 - 2x)\,dzdx + (3 - x - 2yz)\,dxdy$.

15.8.16. $\omega = (xy^2 + yz^2 + zx^2)\,dxdydz$.

15.8.17.* Let ω be a closed 1-form on $A = \mathbb{R}^2 \setminus \{(0,0)\}$, let C be the unit circle, and suppose that $\int_C \omega = 0$. Prove that ω is exact on A.

15.8.18.* Let ω be a closed 1-form on $A = \mathbb{R}^3 \setminus \{(0,0,0)\}$. Prove that ω is exact on A.

15.8.19. Let $\mathbf{F}$ be a C^1 vector field on a starlike open set $A \subset \mathbb{R}^3$. Prove that $\operatorname{curl}\mathbf{F} = \mathbf{0}$ if and only if there exists $f : A \to \mathbb{R}$ such that $\mathbf{F} = \nabla f$.

15.8.20. Let $\mathbf{F}$ be a C^1 vector field on a starlike open set $A \subset \mathbb{R}^3$. Prove that $\operatorname{div}\mathbf{F} = 0$ if and only if there exists $\mathbf{G} : A \to \mathbb{R}^3$ such that $\mathbf{F} = \operatorname{curl}\mathbf{G}$.

A function $f : \mathbb{R}^n \to \mathbb{R}$ is called an **integrating factor** for a k-form ω if $f(\mathbf{x})$ is never 0 and $d(f\omega) = 0$. In Exercises 15.8.21–15.8.24 find the integrating factor for the given form or show that it does not exist:

15.8.21. $(x^2 + 2y)\,dx - x\,dy$.   15.8.22. $3yz^2\,dx + xz^2\,dy + 2xyz\,dz$.

15.8.23. $xy\,dx + xy\,dy + yz\,dz$.   15.8.24. $dx + z\,dy$.

15.8.25. Let ω be a 1-form on $\mathbb{R}^n$ that has an integrating factor f. Prove that $\omega\,d\omega = 0$.

Solutions and Answers to Selected Exercises

1.1.1. Suppose that $x = z$. Then $z = x$ (Axiom 3). By Axiom 4, $z = x$ and $x = y$ imply that $z = y$, so by Axiom 3, $y = z$ which contradicts the assumption that $y \neq z$.

1.1.3. By definition, $2 = S(1)$. Using Exercise 1.1.9, $2 + 2 = 2 + S(1)$. By Rule 2 for addition, $2 + S(1) = S(2 + 1)$. By definition, $2 + 1 = S(2) = 3$, and $S(3) = 4$. Therefore, $2 + 2 = 4$.

1.1.5. Let x, y be arbitrary elements of $\mathbb{N}$ and define $K = \{z \in \mathbb{N} : x(yz) = (xy)z\}$. Using Rule 1 for multiplication, $x \cdot (y \cdot 1) = x \cdot y = (x \cdot y) \cdot 1$ so $1 \in K$. Suppose that $n \in K$, i.e., $x \cdot (y \cdot n) = (x \cdot y) \cdot n$. Then, by Theorem 1.1.3, $x \cdot (y \cdot (n + 1)) = x \cdot (y \cdot n + y \cdot 1) = x \cdot (y \cdot n) + x \cdot (y \cdot 1)$, whence the assertions for n and 1 imply that $x \cdot (y \cdot (n + 1)) = (x \cdot y) \cdot n + (x \cdot y) \cdot 1$. Finally, applying Theorem 1.1.3 once again, $x \cdot (y \cdot (n + 1)) = (x \cdot y) \cdot (n + 1)$.

1.1.7. Let $K = \{x \in \mathbb{N} : x \neq x + 1\}$. By Axiom 8, the statement $1 = 1 + 1$ is false, so $1 \in K$. Suppose that $n \neq n + 1$ and that $S(n) = S(n) + 1$. By Exercise 1.1.6, $S(n) + 1 = 1 + S(n)$ and by Rule 2 for addition $1 + S(n) = S(1 + n)$. Finally, using Exercise 1.1.6 once again, we get that $S(1 + n) = S(n + 1)$. Combining, we get $S(n) = S(n + 1)$, whence Axiom 7 implies that $n = n + 1$, contradicting the induction hypothesis.

1.1.9. Let $x, z \in \mathbb{N}$ satisfy $x = z$, and let $K = \{w \in \mathbb{N} : y = w \Rightarrow x + y = z + w\}$. If $w = 1$ then $y = 1$ and the relation $x + 1 = z + 1$ follows from Axiom 7, whence $1 \in K$. Suppose that $n \in K$, meaning that $y = n$ implies $x + n = z + n$. Axiom 7 yields $S(x + n) = S(z + n)$, i.e., $(x + n) + 1 = (z + n) + 1$. By Theorem 1.1.2, $x + (n + 1) = z + (n + 1)$. In other words, that $S(n) \in K$.

1.1.11. First we will prove that, for every $x \in \mathbb{N}$, $1 \cdot x = x$. Let $L = \{x \in \mathbb{N} : 1 \cdot x = x\}$. By Rule 1 for multiplication, $1 \cdot 1 = 1$, so $1 \in L$. Suppose that $1 \cdot n = n$. Then, using Theorem 1.1.3, $1 \cdot (n + 1) = 1 \cdot n + 1 \cdot 1$ and by the induction hypothesis, $1 \cdot (n + 1) = n \cdot 1 + 1 \cdot 1$. Finally, Theorem 1.1.3 and Rule 1 for multiplication imply that $1 \cdot (n + 1) = (n + 1) \cdot 1 = n + 1$. Thus, $S(n) \in L$.

Now we will prove the identity $x \cdot y = y \cdot x$. Let $x \in \mathbb{N}$ and $K = \{y \in \mathbb{N} : x \cdot y = y \cdot x\}$. The fact that $1 \in K$ follows from the established assertion and Rule 1 for multiplication. If $n \cdot 1 = 1 \cdot n$ then, by Theorem 1.1.3, $(n + 1) \cdot 1 = n \cdot 1 + 1 \cdot 1 = 1 \cdot n + 1 \cdot 1$ and, again by Theorem 1.1.3, $(n + 1) \cdot 1 = 1 \cdot (n + 1)$.

1.2.1. Let $c = \sup B$. Then, for any $b \in B$, $b \leq c$. If $a \in A$, then $a \in B$, so $a \leq c$. Thus, c is an upper bound for A, whence $\sup A \leq c$.

1.2.3. The inequality $x^2 < 3x$ is equivalent to $x(x - 3) < 0$, which means that either $x > 0$ and $x - 3 < 0$, or $x < 0$ and $x - 3 > 0$. Since there is no number satisfying the second pair of inequalities, we see that $0 < x < 3$. Thus, $\sup A = 3$ and $\inf A = 0$.

1.2.5. The set A is not bounded above. (Set $m = n$ and let $n \to \infty$.) Thus, $\sup A = +\infty$. Notice that $mn \geq 1$, $mn \geq m$, and $mn \geq n$. Therefore, $3mn \geq 1 + m + n$, so $\inf A \geq 1/3$. To see that it is $1/3$, take $m = n = 1$.

1.2.7. Let $a = \sup A$ and $b = \sup B$. If $c \in C$, then there exist $x \in A$ and $y \in B$ such that $c = x + y$. By definition of supremum, $x \leq a$ and $y \leq b$, whence $c \leq a + b$. That way, $a + b$ is an upper bound for the set C. Suppose that it is not the least upper bound. Then there exists $d < a + b$ such that $c \leq d$, for all $c \in C$. It follows that, for any $x \in A$ and $y \in B$, $x + y \leq d$. Let $y \in B$ be fixed. Then, for any $x \in A$, $x \leq d - y$. In other words, $d - y$ is an upper bound for A, whence $a \leq d - y$. Thus, for any $y \in B$, $y \leq d - a$. Since b is the least upper bound for B, we get that $b \leq d - a$, hence $a + b \leq d$. This contradiction implies that $a + b$ is the least upper bound for C.

1.2.9. If there are two suprema a and b, one of them has to be smaller. Let $a < b$. Then b is not the least upper bound, contradicting the assumption that $= \sup A$.

1.2.11. Let $x \in A$. By definition, $f(x) \leq \sup\{f(x) : x \in A\}$ and $g(x) \leq \sup\{g(x) : x \in A\}$. It follows that $f(x) + g(x) \leq \sup\{f(x) : x \in A\} + \sup\{g(x) : x \in A\}$. Since this is true for all $x \in A$, we get that $\sup\{f(x) + g(x) : x \in A\} \leq \sup\{f(x) : x \in A\} + \sup\{g(x) : x \in A\}$.

1.3.1. Clearly, a is an upper bound of S, so it remains to prove that it is the least upper bound. Suppose it is not. Then, there exists $b < a$, such that b is an upper bound of S. By Theorem 1.3.7, there exists a rational number r between b and a. Since $r < a$ it belongs to S. Thus, b cannot be an upper bound of S.

1.3.3. Suppose that $\sqrt{n - 1} + \sqrt{n + 1}$ is a rational number r. By raising both sides to the second power, we get $n - 1 + 2\sqrt{n^2 - 1} + n + 1 = r^2$, and it follows that $\sqrt{n^2 - 1}$ is also a rational number p/q, with p, q mutually prime. Squaring again, we get $n^2 - p^2/q^2 = 1$, which leads to $q^2 n^2 - p^2 = q^2$, hence $(qn - p)(qn + p) = q^2$. The right-hand side is divisible by q^2, so at least one of the factors on the left side must be divisible by q. If $qn - p$ is divisible by q, it would follow that so is p, contrary to the assumption that p, q are mutually prime. A similar argument shows that $qn + p$ cannot be divisible by q.

1.3.5. By Trichotomy, either $i < 0$ or $i = 0$ or $i > 0$. If $i > 0$, we apply the last axiom in Table 1.2 with $a = 0$ and $b = c = i$, and we obtain that $0 < i^2 = -1$. We leave to the reader to show that the remaining cases are impossible as well.

1.3.6. Since α is irrational, it is not 0. Consider the real numbers a/α and b/α. By Theorem 1.3.7, there exists a rational number r between them, and multiplying through by α yields the desired result.

1.3.8. Hint: simplify $x^{1/\log_a x}$.

1.3.10. Let $n \in \mathbb{N}$ be fixed. Consider $n + 1$ numbers

$$0, \; \alpha - \lfloor \alpha \rfloor, \; 2\alpha - \lfloor 2\alpha \rfloor, \; \ldots, \; n\alpha - \lfloor n\alpha \rfloor$$

that all belong to $[0, 1)$ and n intervals

$$\left[0, \frac{1}{n}\right), \; \left[\frac{1}{n}, \frac{2}{n}\right), \; \ldots, \; \left[\frac{n-1}{n}, 1\right)$$

that cover $[0, 1)$. Then, one of these intervals contains 2 numbers $k\alpha - \lfloor k\alpha \rfloor$ and $m\alpha - \lfloor m\alpha \rfloor$, with $k < m$. Let $p_n = \lfloor m\alpha \rfloor - \lfloor k\alpha \rfloor$, and $q_n = m - k$. First,

$$|\alpha q_n - p_n| = |\alpha m - \alpha k - \lfloor m\alpha \rfloor + \lfloor k\alpha \rfloor| = |(m\alpha - \lfloor m\alpha \rfloor) - (k\alpha - \lfloor k\alpha \rfloor)| < \frac{1}{n}.$$

It follows that

$$\left| \alpha - \frac{p_n}{q_n} \right| = \left| \frac{\alpha q_n - p_n}{q_n} \right| = \frac{|\alpha q_n - p_n|}{q_n} < \frac{1}{nq_n}.$$

Since $0 \leq m, k \leq n$, we have that $q_n \leq n$, so

$$\left| \alpha - \frac{p_n}{q_n} \right| < \frac{1}{nq_n} \leq \frac{1}{q_n^2}.$$

1.4.1. We will show that Dedekind's Axiom implies Completeness Axiom and leave the converse to the reader. Let A be a set that is bounded above. For each $a \in A$, let $A(a) = \{x \in \mathbb{F} : x \leq a\}$, and let $\tilde{A} = \cup_{a \in A} A(a)$. Then $\tilde{A}$ is bounded above so it is not $\mathbb{F}$ and the complement $\tilde{B}$ of $\tilde{A}$ is non-empty. The sets $\tilde{A}$ and $\tilde{B}$ satisfy the assumptions of Dedekind's Axiom. Indeed, if $b_1 < a < b_2$, then $b_1 \in A(a) \subset \tilde{A}$; if $a_1 < b < a_2$, then $b \in A(a_2) \subset \tilde{A}$. Thus, there exists $x \in \mathbb{F}$ that lies between $\tilde{A}$ and $\tilde{B}$, i.e., if $a \in \tilde{A}$ and $b \in \tilde{B}$, then $a \leq x \leq b$. Clearly, x is an upper bound of $\tilde{A}$ and therefore of A as well. It remains to show that it is the least upper bound. Suppose that there is $c < x$ that is an upper bound of A. Then $(c + x)/2 > a$, for any $a \in \tilde{A}$, so $(c + x)/2 \in \tilde{B}$. Now, $a < (c + x)/2 < x$ contradicting the conclusion of Dedekind's Axiom.

2.1.1. Answer: $1/3$.

2.1.3. Answer: $2/3$.

2.1.5. Answer: $-\infty$.

2.1.7. Answer: $1/2$.

2.1.9. Answer: 0. Hint: $a^3 - b^3 = (a - b)(a^2 + ab + b^2)$.

2.1.11. Answer: 2. Hint: use Theorem 2.6.3.

2.1.13. First, $\sqrt{n^2 + n} = n + \left(\sqrt{n^2 + n} - n\right)$, and $\sin(n\pi + x) = (-1)^n \sin x$, so

$$\sin^2(\pi \sqrt{n^2 + n}) = \sin^2\left[n\pi + \pi\left(\sqrt{n^2 + n} - n\right)\right] = \sin^2\left[\pi\left(\sqrt{n^2 + n} - n\right)\right].$$

Also,

$$\sqrt{n^2 + n} = \frac{n}{\sqrt{n^2 + n} + n} \to \frac{1}{2}$$

and it follows that $\lim \sin^2(\pi \sqrt{n^2 + n}) = \sin^2(\pi/2) = 1$.

2.1.14. Answer: $\sin x / x$. Hint: multiply (and divide) by $\sin(x/2^n)$.

2.1.15. Hint: use mathematical induction to prove that the formula $1 + 2 + 3 + \cdots + n = n(n + 1)/2$ holds for every $n \in \mathbb{N}$. Answer: $1/2$.

2.1.17. Hint: use mathematical induction to prove that the inequality $\frac{1}{2} \cdot \frac{3}{4} \cdots \cdots \frac{2n-1}{2n} \leq \frac{1}{\sqrt{n+1}}$ holds for every $n \in \mathbb{N}$. Answer: 0.

2.2.1. Let $\varepsilon > 0$ and $N = \lfloor 1/3 + 4/(3\varepsilon) \rfloor + 1$. If $n \geq N$ then $n > 1/3 + 4/(3\varepsilon)$ so $3n > 1 + 4/\varepsilon$ and it follows that $4/(3n - 1) < \varepsilon$. Now, $|(n + 1)/(3n - 1) - 1/3| = 4/(3n - 1) < \varepsilon$, so the limit is $1/3$.

2.2.3. Let $\varepsilon > 0$ and $N = \lfloor 7/(3\varepsilon) \rfloor + 1$. If $n \geq N$ then $3n^2 + 2 > n > 7/(3\varepsilon)$ and it follows that $7/(3(3n^2 + 2)) < \varepsilon$. Now, $|(2n^2 - 1)/(3n^2 + 2) - 2/3| = 7/(3(3n^2 + 2)) < \varepsilon$, so the limit is $2/3$.

2.2.5. Let $M > 0$ and $N = \lfloor M \rfloor + 1$. If $n \geq N$ then $n > M$ and it follows that $(2n^4 - 1)/(n^3 - n^2) > (2n^4 - n^4)/n^3 = n > M$, so the limit is $+\infty$.

2.2.7. Let $\varepsilon > 0$ and $N = \lfloor 1/\varepsilon \rfloor + 1$. If $n \geq N$ then $n > 1/\varepsilon$ and it follows that $1/n < \varepsilon$. Now, $|n^2 \sin n/(n^3 + 1)| < n^2/(n^3 + 1) < n^2/n^3 = 1/n < \varepsilon$, so the limit is 0.

2.2.9. Let $\varepsilon > 0$ and $N = 1$. If $n \geq N$ then $|a_n - a| = 0 < \varepsilon$, so $\lim a_n = a$.

2.2.11. Suppose that $\lim a_n = -\infty$ and let $M > 0$. Since $-M < 0$, by definition there exists $n \in \mathbb{N}$ such that, for all $n \geq N$, $a_n < -M$. It follows that, for $n \geq N$, $-a_n > M$ so $\lim(-a_n) = +\infty$. We leave the other direction to the reader.

2.2.13. Suppose that $\lim \sin n = L$ and that $L \leq 0$. Let $\varepsilon = 1/2$. By definition, there exists $N \in \mathbb{N}$ such that, for all $n \geq N$, $|\sin n - L| < 1/2$. Notice that the distance between $\pi/4$ and $3\pi/4$ is $\pi/2 > 1$, which means that there exists $\alpha \in (\pi/4, 3\pi/4)$ and $n \geq N$ such that $n \equiv \alpha \pmod{2\pi}$. Thus $|\sin n - L| = |\sin \alpha - L| = \sin \alpha - L > \sqrt{2}/2 > \varepsilon$. We leave to the reader to come up with the proof in case $L > 0$.

2.3.1. Let $\varepsilon > 0$. Since $\lim a_n = a$ there exists $n \in \mathbb{N}$ such that, for any $n \geq N$, $|a_n - a| < \varepsilon$. By Theorem 2.3.1, $||a_n| - |a|| \leq |a_n - a|$ so, for all $n \geq N$, $||a_n| - |a|| < \varepsilon$. Therefore, $\lim |a_n| = |a|$. The converse is not true: consider $a_n = 1 - 1/n$ and $a = -1$.

2.3.3. Hint: write $a_{n+1} - a_n$ as $(a_{n+1} - L) - (a_n - L)$ and use Exercise 2.2.10. Consider $a_n = \sqrt{n}$.

2.3.5. Let $\varepsilon = L + \frac{1}{2}(1 - L)$. Then, there exists $N \in \mathbb{N}$ such that, if $n \geq N$, $a_{n+1}/a_n \leq L + \varepsilon < 1$. By induction, prove that $0 \leq a_{N+k} \leq a_N (L + \varepsilon)^k$. Squeeze Theorem and Exercise 2.2.10 imply that $\lim a_n = 0$.

2.3.7. Since $\lim a_n = 0$, there exists $N \in \mathbb{N}$ such that, if $n \geq N$, $|a_n| < 1$. For such n, $|a_n^n| < |a_n|$. By Squeeze Theorem, $\lim a_n^n = 0$.

2.3.9. Suppose that $a > b$, so $\max\{a, b\} = a$. Let $\varepsilon = (a - b)/2$. Since $\lim a_n = a$ there exists $N_1 \in \mathbb{N}$ such that, for all $n \geq N_1$, $|a_n - a| < \varepsilon$. Also, $\lim b_n = b$ so there exists $N_2 \in \mathbb{N}$ such that, for all $n \geq N_2$, $|b_n - b| < \varepsilon$. Let $N = \max\{N_1, N_2\}$ and $n \geq N$. By Theorem 2.3.1, $a - a_n < |a_n - a| < (a - b)/2$ so $a_n > a - (a - b)/2 = (a + b)/2$. Similarly, $b_n - b < |b_n - b| < (a - b)/2$ yields $b_n < b + (a - b)/2 = (a + b)/2$. Thus, $a_n > b_n$, and $\max\{a_n, b_n\} = a_n \to a = \max\{a, b\}$. We leave the remaining cases to the reader.

2.4.1. Notice that $a_{n+1} = (n+1)/((n+1)^2 + 1)$ so

$$\frac{a_{n+1}}{a_n} = \frac{-n^2 - n + 1}{((n+1)^2 + 1)(n^2 + 1)} < 0,$$

and $\{a_n\}$ is decreasing.

2.4.3. Notice that $a_{n+1} = (n+1)!/(2n+3)!!$ so $a_{n+1}/a_n = (n+1)/(2n+3) < (n+1)/(2n+2) = 1/2 < 1$, and $\{a_n\}$ is decreasing.

2.4.5. A calculation shows that $a_{n+1}/a_n = ((n+1)/2^{n+1})/(n/2^n) = (n+1)/2n \leq (n+n)/2n = 1$. We see that $a_1 = a_2 > a_3 > a_4 > \ldots$. Following Remark 2.4.4, $\{a_n\}$ is decreasing.

2.4.7. Answer: increasing.

2.4.9. First we prove by induction that $\{a_n\}$ is increasing. More precisely, we will prove that, for all $n \in \mathbb{N}$, $a_{n+1} > a_n > 0$. It is easy to see that $a_2 > a_1 > 0$. Suppose that $a_{n+1} > a_n > 0$, and let us show that $a_{n+2} > a_{n+1} > 0$:

$$a_{n+2} - a_{n+1} = \frac{1}{2}(c + a_{n+1}^2) - \frac{1}{2}(c + a_n^2) = \frac{1}{2}(a_{n+1} - a_n)(a_{n+1} + a_n) > 0,$$

so $a_{n+2} > a_{n+1} > 0$.

Suppose now that the sequence $\{a_n\}$ is bounded above. Then it is convergent, and let $L = \lim a_n$. Passing to the limit in $a_{n+1} = \frac{1}{2}(c + a_n^2)$ we obtain $L = \frac{1}{2}(c + L^2)$, which leads to $L^2 - 2L + c = 0$. The last equation can be written as $(L - 1)^2 = 1 - c$. From here we see that $c \leq 1$.

Suppose, to the contrary, that $0 < c \leq 1$. We will show by induction, that $\{a_n\}$ is bounded above by $1 - \sqrt{1-c}$. When $n = 1$,

$$1 - \sqrt{1 - c} = 1 - \sqrt{1 - c} \frac{1 + \sqrt{1 - c}}{1 + \sqrt{1 - c}} = \frac{c}{1 + \sqrt{1 - c}} > \frac{c}{2} = a_1.$$

So, we assume that $a_n \leq 1 - \sqrt{1 - c}$. Then

$$a_{n+1} = \frac{1}{2}(c + a_n^2) \leq \frac{1}{2}\left(c + (1 - \sqrt{1 - c})^2\right) = \frac{1}{2}\left(c + 1 - 2\sqrt{1 - c} + 1 - c\right) = 1 - \sqrt{1 - c}.$$

Thus, for any $0 < c \leq 1$, the sequence $\{a_n\}$ is increasing and bounded above by $1 - \sqrt{1 - c}$, so it is convergent. We have seen that its limit satisfies the quadratic equation $L^2 - 2L + c = 0$, so $L = 1 \pm \sqrt{1 - c}$. Since $a_n \leq 1 - \sqrt{1 - c}$ for all $n \in \mathbb{N}$, it follows that $L = 1 - \sqrt{1 - c}$. On the other hand, if $c > 1$, the equation $L^2 - 2L + c = 0$ has no solution, so $\{a_n\}$ diverges to $+\infty$.

2.4.11. Hint: prove that $2 \leq a_n \leq \frac{\sqrt{15}}{3} + 1$ and that $\{a_n\}$ is increasing. Answer: $\lim a_n = \frac{\sqrt{15}}{3} + 1$.

2.4.13. Hint: prove that $0 \leq a_n \leq (\sqrt{5} - 1)/2$ and that $\{a_n\}$ is increasing. Answer: $\lim a_n = (\sqrt{5} - 1)/2$.

2.5.1. Notice that

$$b_{n+1} \leq b_n \Leftrightarrow \left(1 + \frac{1}{n+1}\right)^{n+2} \leq \left(1 + \frac{1}{n}\right)^{n+1} \Leftrightarrow \sqrt[n+2]{\left(\frac{n}{n+1}\right)^{n+1}} \leq \frac{n+1}{n+2}.$$

The last inequality is a consequence of the arithmetic-geometric mean inequality

$$\sqrt[m]{x_1 x_2 \ldots x_m} \leq \frac{x_1 + x_2 + \cdots + x_m}{m},$$

which holds for all $m \in \mathbb{N}$ and all $x_1, x_2, \ldots, x_m \geq 0$. Using $m = n + 2$ and $x_1 = 1$, $x_2 = x_3 = \cdots = x_{n+2} = n/(n+1)$ yields the desired result.

2.5.3. It was shown in the proof of Theorem 2.5.2 that, for all $n \in \mathbb{N}$,

$$\left(1 + \frac{1}{n}\right)^n < e < \left(1 + \frac{1}{n}\right)^{n+1}.$$

Taking the nth root and subtracting 1 yields

$$\frac{1}{n} < \sqrt[n]{e} - 1 < \left(1 + \frac{1}{n}\right)^{1 + \frac{1}{n}} - 1, \text{ so } 1 < n\left(\sqrt[n]{e} - 1\right) < n\left[\left(1 + \frac{1}{n}\right)^{1 + \frac{1}{n}} - 1\right].$$

Using Bernoulli's inequality, $\left(1 + \frac{1}{n^2}\right)^n \geq 1 + \frac{1}{n}$, so

$$\left(1 + \frac{1}{n}\right)^{\frac{1}{n}} \leq 1 + \frac{1}{n^2}.$$

It follows that

$$n\left(\sqrt[n]{e} - 1\right) < n\left[\left(1 + \frac{1}{n}\right)\left(1 + \frac{1}{n}\right)^{\frac{1}{n}} - 1\right] \leq n\left[\left(1 + \frac{1}{n}\right)\left(1 + \frac{1}{n^2}\right) - 1\right]$$

$$= n\left(1 + \frac{1}{n} + \frac{1}{n^2} + \frac{1}{n^3} - 1\right) = 1 + \frac{1}{n} + \frac{1}{n^2}.$$

Thus,

$$1 < n\left(\sqrt[n]{e} - 1\right) < 1 + \frac{1}{n} + \frac{1}{n^2}$$

and Squeeze Theorem implies that $\lim n\left(\sqrt[n]{e} - 1\right) = 1$.

2.5.5. Hint: Show that $a_{n+1} - a_n = 1/n!$.

2.5.7. Notice that

$$a_{n+1} \geq a_n \Leftrightarrow \frac{1}{n+1} \geq \ln\frac{n+1}{n} \Leftrightarrow \ln\left(\frac{n+1}{n}\right)^{n+1} \geq 1 \Leftrightarrow \left(1 + \frac{1}{n}\right)^{n+1} \geq e,$$

which was established in the proof of Theorem 2.5.2. Thus, $\{a_n\}$ is increasing. To prove that it is bounded above we will use the inequality $(1 + 1/n)^n \leq e$, which implies that $\ln(1 + 1/k) \leq 1/k$, for all $k \in \mathbb{N}$. Now,

$$a_n \geq \ln(1 + 1) + \ln\left(1 + \frac{1}{2}\right) + \cdots + \left(1 + \frac{1}{n}\right) - \ln n = \ln\left(2 \cdot \frac{3}{2} \cdot \frac{4}{3} \cdots \cdot \frac{n+1}{n}\right) - \ln n$$

$$= \ln(n+1) - \ln n \geq 0.$$

2.5.9. Let m and n be positive integers. Then

$$c_{m+n} - c_n = \frac{1}{(n+1)!} + \frac{1}{(n+2)!} + \cdots + \frac{1}{(n+m)!}$$

$$= \frac{1}{(n+1)!}\left[1 + \frac{1}{n+2} + \frac{1}{(n+2)(n+3)} + \cdots + \frac{1}{(n+2)(n+3)\dots(n+m)}\right]$$

$$< \frac{1}{(n+1)!}\left[1 + \frac{1}{n+2} + \frac{1}{(n+2)^2} + \cdots + \frac{1}{(n+2)^{m-1}}\right]$$

$$= \frac{1}{(n+1)!} \cdot \frac{1 - \left(\frac{1}{n+2}\right)^m}{1 - \frac{1}{n+2}}$$

$$< \frac{1}{(n+1)!} \cdot \frac{1}{1 - \frac{1}{n+2}} = \frac{1}{(n+1)!} \cdot \frac{n+2}{n+1} = \frac{1}{n!} \cdot \frac{n+2}{(n+1)^2}$$

$$= \frac{1}{n!}\left(\frac{1}{n+1} + \frac{1}{(n+1)^2}\right) < \frac{1}{n!}\left(\frac{3}{4}\right).$$

When n is fixed and $m \to \infty$ we obtain that $e - c_n \leq 3/4n! < 1/n!$. As we have already shown, $c_n < e$ so $0 < e - c_n < 1/n!$ and $0 < \theta_n < 1$.

2.5.11. Hint: Assume that $e = m/n$ and use Lemma 2.5.6 to show that θ_n must be an integer.

2.6.1. Let $\varepsilon > 0$ and let $N = \lfloor -\ln\varepsilon/\ln 2 \rfloor + 1$. Suppose that $m > n \geq N$. By induction one can prove that $n^2 \leq 2^n$, for $n \geq 4$. Also, $n > -\ln\varepsilon/\ln 2$, which implies that $1/2^n < \varepsilon$. Now,

$$|a_m - a_n| = \frac{(n+1)^2}{4^{n+1}} + \frac{(n+2)^2}{4^{n+2}} + \cdots + \frac{m^2}{4^m} \leq \frac{2^{n+1}}{4^{n+1}} + \frac{2^{n+2}}{4^{n+2}} + \cdots + \frac{2^m}{4^m} = \frac{1}{2^{n+1}} \cdot \frac{1 - 1/2^{m-n}}{1 - 1/2} \leq \frac{1}{2^n} < \varepsilon.$$

2.6.3. Hint: Prove that $a_{2n} - a_n \geq n \cdot \frac{2n}{(2n+1)^2} \geq \frac{2}{9}$.

2.6.5. Let $\varepsilon > 0$ and let $N = \lfloor 1/\varepsilon \rfloor + 1$. Suppose that $m > n \geq N$. Then, $n > 1/\varepsilon$ so $1/n < \varepsilon$. Now,

$$|a_m - a_n| = \left|\frac{m+1}{m} - \frac{n+1}{n}\right| = \frac{m-n}{mn} = \frac{1}{n} - \frac{1}{m} < \frac{1}{n} < \varepsilon.$$

2.6.7. Hint: Consider $a_n = \sqrt{n}$.

2.6.9. The assumption that $|a_n| < 2$ implies that $|a_{n+1} + a_n| < 4$, so

$$|a_{n+2} - a_{n+1}| \leq \frac{1}{8}|a_{n+1}^2 - a_n^2| = \frac{1}{8}|a_{n+1} + a_n||a_{n+1} - a_n| < \frac{1}{2}|a_{n+1} - a_n|.$$

By induction,

$$|a_{n+1} - a_n| \leq \frac{1}{2^{n-1}}|a_2 - a_1|,$$

and the result follows from Exercise 2.6.8.

2.7.1. Let $\varepsilon > 0$. Since $a_n \to L$, there exists $N \in \mathbb{N}$ such that, if $n \geq N$, then $|a_n - L| < \varepsilon$. Next, $n_k \to \infty$, as $k \to \infty$, so there exists $K \in \mathbb{N}$ such that, if $k \geq K$, then $n_k \geq N$. For such k, $n_k \geq N$ so $|a_{n_k} - L| < \varepsilon$.

2.7.3. Suppose that $\{a_n\}$ does not converge to L. Then, there exists $\varepsilon_0 > 0$ and a subsequence $\{a_{n_k}\}$ such that $|a_{n_k} - L| \geq \varepsilon_0$, for all $k \in \mathbb{N}$. The subsequence $\{a_{n_k}\}$ is bounded because $\{a_n\}$ is. By Bolzano–Weierstrass theorem, there is a convergent subsequence $\{b_n\} \subset \{a_{n_k}\}$, and by assumption it converges to L. This contradicts the inequality $|b_n - L| \geq \varepsilon_0$.

2.7.5. We will look for those positive integers n such that, if $k \geq n$ then $a_k \geq a_n$. (For example, if $\{a_n\}$ is $1, 0, 2, 0, 3, 0, 4, 0, \ldots$, then every even integer has this property, but no odd integer does.) If there are infinitely many such integers, we will denote them by $n_1 < n_2 < n_3 < \ldots$. Then $a_{n_1} \leq a_{n_2} \leq a_{n_3} \leq \ldots$, so we have a monotone increasing subsequence. If there are only finitely many such integers, let N be the largest one among them, and let $m_1 = N + 1$. Since m_1 does not have the property under consideration, there exists $m_2 > m_1$ such that $a_{m_2} < a_{m_1}$. Also $m_2 > N$, so it does not have the said property, and there exists $m_3 > m_2$ such that $a_{m_3} < a_{m_2}$. Continuing, we get a subsequence $\{a_{m_i}\}$ that is monotone decreasing.

2.7.7. By definition, for each $M > 0$ there exists $n \in \mathbb{N}$ such that $a_n \geq M$. Let $M = k$. Then, there exists n_k such that $a_{n_k} \geq k$, and the sequence $\{a_{n_k}\}$ converges to $+\infty$.

2.7.9. The implication (a) $\Rightarrow$ (b) is the Monotone Convergence Theorem and (b) $\Rightarrow$ (c) is Theorem 2.7.5, although the proof needs to be slightly modified to avoid the use of suprema and infima. Similarly, (c) $\Rightarrow$ (d) is Bolzano–Weierstrass theorem, again with a slightly modified proof. Cauchy's test implies that (d) $\Rightarrow$ (e) so it remains to prove (e) $\Rightarrow$ (a). Let A be a non-empty set that is bounded above. We define $B = \{n \in \mathbb{N} : n \text{ is an upper bound of } A\}$. By the Archimedean property, the set B is non-empty. Let a_1 be the smallest element of B. Then a_1 is an upper bound of A but $a_1 - 1$ is not. Define $a_2 = a_1 - 1/2$ if $a_1 - 1/2$ is an upper bound of A, or $a_2 = a_1$ if that is not the case. Notice that a_2 is an upper bound of A but $a_2 - 1/2$ is not. Also $a_2 \leq a_1$ and $|a_2 - a_1| \leq 1/2$. We continue the process, by defining $a_{n+1} = a_n - 1/2^n$ if $a_n - 1/2^n$ is an upper bound of A, or $a_{n+1} = a_n$ if that is not the case. Again, $a_{n+1} \geq a_n$ and $|a_{n+1} - a_n| \leq 1/2^n$. That way we obtain a decreasing sequence $\{a_n\}$ and, by Exercise 2.6.8, it is a Cauchy sequence. By Cauchy's test it is convergent. Let $L = \lim a_n$. We will show that L is the least upper bound of A. Suppose, to the contrary, that x is an upper bound of A and $x < L$. Let $\varepsilon = L - x$. Since $\lim 1/2^n = 0$, there exists $n \in \mathbb{N}$ such that $1/2^n < \varepsilon$. By the construction above, $a_{n+1} - 1/2^n$ is not an upper bound of A so there exists $a \in A$ such that $a > a_{n+1} - 1/2^n \geq L - 1/2^n > L - \varepsilon = x$.

2.8.1. By definition, $L = \limsup a_n = \liminf a_n$ is the only accumulation point of $\{a_n\}$. Let $\varepsilon > 0$. The members of $\{a_n\}$ outside of $(L - \varepsilon, L + \varepsilon)$ are a bounded sequence. If it were infinite, Bolzano–Weierstrass Theorem would imply that there is an accumulation point of $\{a_n\}$ different from L. Thus, there are only finitely many members of $\{a_n\}$ in $(L - \varepsilon, L + \varepsilon)$. In other words, there exists $n \in \mathbb{N}$ such that, for all $n \geq N$, $a_n \in (L - \varepsilon, L + \varepsilon)$.

2.8.3. ($\Rightarrow$) Let $\varepsilon > 0$. Since L is the biggest accumulation point there can be only finitely many members of $\{a_n\}$ that are bigger than $L + \varepsilon$ so (a) is proved. Let $N \in \mathbb{N}$. If we delete the first N members of $\{a_n\}$, the L is still an accumulation point, so there exists $n \geq N$ such that $a_n > L - \varepsilon$.

($\Leftarrow$) Let $k \in \mathbb{N}$. By (a), there exists $N \in \mathbb{N}$ such that, for all $n \geq N$, $a_n < L + 1/k$. By (b), there exists $n > N$, depending on k, such that $a_n > L - 1/k$. If we denote that particular n by n_k, we have proved that $L - 1/n_k < a_{n_k} < L + 1/k$, so the subsequence $\{a_{n_k}\}$ converges to L, whence L is an accumulation point of $\{a_n\}$. Suppose that it is not the greatest, i.e., that there exists an accumulation point $L_1 > L$. Let $\varepsilon = (L_1 - L)/2$. By (a), there exists $N \in \mathbb{N}$ such that, for all $n \geq N$, $a_n < L + \varepsilon = (L_1 + L)/2 < L_1$, so L_1 cannot be an accumulation point of $\{a_n\}$.

2.8.5. Hint: Write the first 10 members of the sequence.

2.8.7. Hint: Prove that $|a_{2n+1} - \frac{1}{3}| = \frac{1}{2}|a_{2n} - \frac{2}{3}| = \frac{1}{4}|a_{2n-1} - \frac{1}{3}|$.

2.8.9. We prove part (a) and leave the rest to the reader. Let $\limsup a_n = L_1$ and $\limsup b_n = L_2$, and let $\varepsilon > 0$. By Exercise 2.8.3 (a), there exists $N_1 \in \mathbb{N}$ such that, for all $n \geq N_1$, $a_n < L_1 + \varepsilon L_1$. Also, there exists $N_2 \in \mathbb{N}$ such that, for all $n \geq N_2$, $b_n < L_2 + \varepsilon L_2$. It follows that, for $n \geq N \equiv \max\{N_1, N_2\}$, $a_n b_n < L_1 L_2 (1 + \varepsilon)^2$, whence $\limsup(a_n b_n) \leq L_1 L_2 (1 + \varepsilon)^2$. Since this is true for any $\varepsilon > 0$, we obtain that $\limsup(a_n b_n) \leq L_1 L_2$. The inequality is strict when $\{a_n\}$ is the sequence $1, 0, 1, 0, 1, 0, \ldots$ and $\{b_n\}$ is the sequence $0, 1, 0, 1, 0, 1, \ldots$. If $\{a_n\}$ and $\{b_n\}$ are allowed to be negative we could use $a_n = b_n = -1$. Can you find an example, where $a_n \geq 0$ and $b_n < 0$?

2.8.11. The middle inequality $\liminf \sqrt[n]{a_n} \leq \limsup \sqrt[n]{a_n}$ is obvious. We will prove the rightmost inequality, and the leftmost will then follow from Exercise 2.8.2. Further, if $\limsup a_{n+1}/a_n$ is infinite, then there is nothing to prove, so we will assume that $\limsup a_{n+1}/a_n$ is finite and denote it by L.

Let $\varepsilon > 0$. Then there exists $N \in \mathbb{N}$ such that, if $n \geq N$, then $a_{n+1}/a_n < L + \varepsilon$. If we write m inequalities for $n = N, N + 1, \ldots, N + m - 1$, and multiply them all, we obtain

$$\frac{a_m}{a_N} = \frac{a_m}{a_{m-1}} \cdot \frac{a_{m-1}}{a_{m-2}} \cdot \ldots \cdot \frac{a_{N+1}}{a_N} < (L + \varepsilon)^m.$$

It follows that

$$\sqrt[m]{a_m} < \sqrt[m]{a_N}\,(L + \varepsilon), \quad \text{for any } n \in \mathbb{N}.$$

By Exercise 2.8.8,

$$\limsup \sqrt[m]{a_m} \leq \limsup \sqrt[m]{a_N}\,(L + \varepsilon) = \lim \sqrt[m]{a_N}\,(L + \varepsilon) = L + \varepsilon.$$

Since ε is arbitrary, we have that $\limsup \sqrt[m]{a_m} \le L$.

2.8.13. Hint: In one direction use Exercise 2.8.9. In the other, try $b_n = 1/a_n$.

2.9.1. $1 \le \sqrt[n]{1^7 + 2^7 + \cdots + n^6} \le \sqrt[n]{n \cdot n^7} = (\sqrt[n]{n})^8 \to 1$, by Example 2.9.4.

2.9.3. The Binomial Formula can be used to write

$$(2 + \sqrt{3})^n = \sum_{k=0}^{n} \binom{n}{k} 2^{n-k} (\sqrt{3})^k = A_n + B_n \sqrt{3}$$

where A_n and B_n are both positive integers. It is not hard to see that $(2 - \sqrt{3})^n = A_n - B_n \sqrt{3}$. Since $2 + \sqrt{3} > 1$ and $0 < 2 - \sqrt{3} < 1$, we have that

$$\lim \left(A_n + B_n \sqrt{3} \right) = +\infty, \quad \text{and} \quad \lim \left(A_n - B_n \sqrt{3} \right) = 0.$$

Adding these two equalities shows that $\lim A_n = \infty$, and the equality $A_n - B_n \sqrt{3} = A_n (1 - B_n \sqrt{3}/A_n)$ implies that

$$\lim \frac{B_n \sqrt{3}}{A_n} = 1.$$

Moreover, there exists $N_1 \in \mathbb{N}$ such that, if $n \ge N_1$, then $B_n \sqrt{3}/A_n < 1$, hence $B_n \sqrt{3} < A_n$. On the other hand, $A_n - B_n \sqrt{3} \to 0$, so there exists $N_2 \in \mathbb{N}$ such that, if $n \ge N_2$, then $A_n - B_n \sqrt{3} < 1$, hence $A_n - 1 < B_n \sqrt{3}$. Let $N = \max\{N_1, N_2\}$, and let $n \ge N$. Then $A_n - 1 < B_n \sqrt{3} < A_n$ so $\lfloor B_n \sqrt{3} \rfloor = A_n - 1$. It follows that

$$\{A_n + B_n \sqrt{3}\} = \{B_n \sqrt{3}\} = B_n \sqrt{3} - \lfloor B_n \sqrt{3} \rfloor = B_n \sqrt{3} - A_n + 1 \to 1.$$

2.9.5. $1 \le (1 + 3n)^{1/n} \le (4n)^{1/n} \to 1$, by Example 2.9.4.

2.9.7. We will prove the inequality

$$x^2 \left(1 - \frac{x^2}{n (\sqrt[n]{x})^2} \right) \le \left(2 \sqrt[n]{x} - 1 \right)^n \le x^2.$$

The result will then follow from Squeeze Theorem.

The right-hand inequality follows from

$$2 \sqrt[n]{x} - 1 = \left(\sqrt[n]{x} \right)^2 - \left(\sqrt[n]{x} - 1 \right)^2 < \left(\sqrt[n]{x} \right)^2$$

by raising both sides to the nth power. For the other inequality, notice that

$$\lim_{n \to \infty} \frac{x - 1}{n \sqrt[n]{x}} = 0$$

so there exists $N \in \mathbb{N}$ such that, if $n \ge N$, $(x - 1)/(n \sqrt[n]{x}) < 1$. For such n, Bernoulli's Inequality yields the estimate

$$\left[1 - \left(\frac{x - 1}{n \sqrt[n]{x}} \right)^2 \right]^n \ge 1 - \frac{x^2}{n (\sqrt[n]{x})^2}.$$

Thus, it remains to show that

$$\left(\sqrt[n]{x} \right)^2 \left[1 - \left(\frac{x - 1}{n \sqrt[n]{x}} \right)^2 \right] \le 2 \sqrt[n]{x} - 1.$$

However, the last inequality can be written as

$$\left(\sqrt[n]{x} \right)^2 - \left(\frac{x - 1}{n} \right)^2 \le 2 \sqrt[n]{x} - 1$$

which can be obtained from (2.17).

2.9.9. Notice that $a_2 \le a_1 + a_1 = 2a_1$, $a_3 \le a_2 + a_1 \le 3a_1$, ... By induction, $a_n \le na_1$, so the sequence $\{a_n/n\}$ is bounded above. Let $L = \limsup a_n$. There exists a subsequence $\{a_{n_k}/n_k\}$ that converges to L. Fix $n \in \mathbb{N}$ and write $n_k = nq_k + r_k$, where $0 \le r_k \le n - 1$. Now, $a_{n_k} \le a_{nq_k} + a_{r_k} \le q_k a_n + a_{r_k}$. Thus,

$$\frac{a_{n_k}}{n_k} \le \frac{q_k a_n}{nq_k + r_k} + \frac{a_{r_k}}{n_k}.$$

When $k \to \infty$ so does q_k, but r_k remains bounded by n. We obtain that $L \le a_n/n$. Since this is true for any $n \in \mathbb{N}$, it follows that $L \le \liminf a_n/n$. By Exercise 2.8.1, the sequence $\{a_n/n\}$ is convergent.

2.9.11. Take $a_n = 1 + 1/\sqrt{2} + \cdots + 1/\sqrt{n}$, $b_n = \sqrt{n}$ and apply Exercise 2.9.10.

2.9.13. Let $a_n = 1^k + 2^k + \cdots + n^k$ and $b_n = n^{k+1}$. Then

$$\frac{a_{n+1} - a_n}{b_{n+1} - b_n} = \frac{(n+1)^k}{(n+1)^{k+1} - n^{k+1}}$$

$$= \frac{(n+1)^k}{(n+1)^k + (n+1)^{k-1}n + (n+1)^{k-2}n^2 + \cdots + (n+1)n^{k-1} + n^k} \to \frac{1}{k+1}.$$

2.9.15. Take $a_n = k! + (k+1)!/1! + \cdots + (k+n)!/n!$, $b_n = n^{k+1}$ and apply Exercise 2.9.10.

2.9.17. Let $a_n = (k+1)(1^k + 2^k + \cdots + n^k) - n^{k+1}$ and $b_n = (k+1)n^k$. Then

$$a_{n+1} - a_n = (k+1)(n+1)^k - (n+1)^{k+1} + n^{k+1}$$

$$= (k+1)(n+1)^k - [(n+1)^k + (n+1)^{k-1}n + \cdots + (n+1)n^{k-1} + n^k]$$

$$= [(n+1)^k - (n+1)^{k-1}n] + [(n+1)^k - (n+1)^{k-2}n^2] + \cdots + [(n+1)^k - n^k]$$

$$= (n+1)^{k-1} + (n+1)^{k-2}[(n+1)^2 - n^2] + \cdots + [(n+1)^k - n^k]$$

$$= (n+1)^{k-1} + (n+1)^{k-2}[(n+1) + n] + \cdots + [(n+1)^{k-1} + (n+1)^{k-2}n + \cdots + n^{k-1}].$$

Also, $b_{n+1} - b_n = (k+1)[(n+1)^k - n^k] = (k+1)[(n+1)^{k-1} + (n+1)^{k-2}n + \cdots + n^{k-1}]$. Therefore, $\lim(a_{n+1} - a_n)/(b_{n+1} - b_n) = (1 + 2 + \cdots + k)/((k+1)k) = 1/2$.

2.9.19. Take $b_n = n$ and apply Exercise 2.9.10.

3.1.1. $x^3 - 2x^2 - 4x + 8 = (x-2)^2(x+2)$ and $x^4 - 8x^2 + 16 = (x-2)^2(x+2)^2$. Answer: 1/4.

3.1.3. $(\sqrt[4]{x} - 2)\left[(\sqrt[4]{x})^3 + 2(\sqrt[4]{x})^2 + 4(\sqrt[4]{x}) + 8\right] = x - 16$, and $(\sqrt{x} - 4)(\sqrt{x} + 4) = x - 16$. Answer: 1/4.

3.1.5. Let $x = 1 - t$. then $t \to 0$ and we have

$$\lim_{t \to 0} \frac{1 - \sqrt{1-t}}{t} \cdot \frac{1 - \sqrt[3]{1-t}}{t} \cdots \cdots \frac{1 - \sqrt[n]{1-t}}{t} = \frac{1}{2} \cdot \frac{1}{3} \cdots \cdots \frac{1}{n} = \frac{1}{n!},$$

by Example 3.1.16.

3.1.7. $1 - \cos x = 2\sin^2\left(\frac{x}{2}\right)$, so

$$\frac{1 - \cos x}{x^2} = \frac{2\sin^2\left(\frac{x}{2}\right)}{4\left(\frac{x}{2}\right)^2}.$$

Answer: 1/2.

3.1.9. Write $1 - \cos x \cos 2x = 1 - \cos 2x + \cos 2x(1 - \cos x)$ and use Exercise 3.1.7. Answer: 3.

3.1.11. Write

$$\frac{\ln \cos 3x}{\ln \cos 2x} = \frac{\ln \cos 3x}{\cos 3x - 1} \cdot \frac{\cos 2x - 1}{\ln \cos 2x} \cdot \frac{\cos 3x - 1}{\cos 2x - 1},$$

and use Example 3.1.13 and Exercise 3.1.7.

3.1.13. Write $a^{x^2} - b^{x^2} = (a^{x^2} - 1) - (b^{x^2} - 1)$ and $a^x - b^x = (a^x - 1) - (b^x - 1)$ and use Example 3.1.14. Answer: $(\ln a/b)^{-1}$.

3.2.1. $\sin(\sqrt{x}) \geq 0$ if and only if $\sqrt{x} \in [2k\pi, (2k+1)\pi]$ for some $k \in \mathbb{Z}$, hence if and only if $x \in [4k^2\pi^2, (2k+1)^2\pi^2]$ for some $k \in \mathbb{Z}$.

3.2.3. $2\sin x \in [-1, 1]$ if and only if $\sin x \in [-1/2, 1/2]$, hence if and only if $x \in [k\pi - \pi/6, k\pi + \pi/6]$, for some $k \in \mathbb{Z}$.

3.2.5. $2 + x - x^2 \geq 0$ if and only if $-1 \leq x \leq 2$. Since $f(-1) = f(2) = 0$ and the function $2 + x - x^2$ attains its maximum at $x = 1/2$ which is 9/4m the range of f is $[0, 3/2]$.

3.2.7. $-1 \leq \frac{2x}{1+x^2} \leq 1$ if and only if $-(1 + x^2) \leq 2x \leq 1 + x^2$, hence if and only if $x^2 + 2x + 1 \geq 0$ and $x^2 - 2x + 1 \geq 0$. Since both of the last two inequalities hold for all real numbers, we get that the domain is $\mathbb{R}$. The range is $[0, \pi]$.

3.2.9. Answer: $E_y = [0, 4]$.

3.2.11. Answer: $E_y = (1, +\infty)$.

3.2.13. Let $x + 1 = t$. Then $f(t) = (t-1)^2 - 3(t-1) + 2 = t^2 - 5t + 6$, so $f(x) = x^2 - 5x + 6$.

3.2.15. Take $1/x = t$. Answer: $f(x) = (1 + \sqrt{1 + x^2})/x$.

3.2.17. Hint: $\ln a/b = -\ln b/a$. Answer: odd.

3.2.19. Answer: even.

3.2.21. Answer: even.

3.3.1. f is not defined at $x = 1$ and $x = 2$. At $x = 1$, $\lim_{x \to 1} f(x) = -2$ and at $x = 2$, $\lim_{x \to 2} f(x)$ is infinite. Therefore, f has a removable discontinuity at $x = 1$, and an essential discontinuity at $x = 2$.

3.3.3. f is not defined at $x = 0$, $x = -1$ and $x = 1$. At $x = 0$, $\lim_{x \to 0} f(x) = -1$, at $x = -1$ the limit is infinite and at $x = 1$, $\lim_{x \to 1} f(x) = 0$. Therefore, f has a removable discontinuity at $x = 0$ and $x = 1$, and an essential discontinuity at $x = -1$.

3.3.5. f is not defined at $x = 0$. The limits at $x = 0$ are: $\lim_{x \to 0+} f(x) = \pi/2$ and $\lim_{x \to 0-} f(x) = -\pi/2$, so f has a jump discontinuity at $x = 0$.

3.3.7. f is not defined at $x = 0$. The limits at $x = 0$ are: $\lim_{x \to 0+} f(x) = +\infty$ and $\lim_{x \to 0-} f(x) = 0$, so f has an essential discontinuity at $x = 0$.

3.3.9. Both $2x + c$ and $x^2 + 3$ are continuous, so we need to check only at $x = 1$, where $\lim_{x \to 1^-} f(x) = 2 + c$ and $\lim_{x \to 1^+} f(x) = 2^2 + 3$. Thus, $2 + c = 9$, so $c = 7$.

3.3.11. $\lim_{x \to 1^-} f(x) = \sin(x + 1)$ and $\lim_{x \to 1^+} f(x) = 0$. Thus, $\sin(1 + c) = 0$, whence $c = k\pi - 1$ for any $k \in \mathbb{Z}$.

3.3.13. $\lim_{x \to 4^-} f(x) = 8 - c$ and $\lim_{x \to 4^+} f(x) = (4 - 2c)^2$. Thus, $(4 - 2c)^2 = 8 - c$, whence $c = (15 \pm \sqrt{97})/8$.

3.3.15. Answer: $a = 1$, $b = 2$.

3.4.1. The limit is 28. Proof: Let $\varepsilon > 0$. Define $\delta = \min\{1/3, \varepsilon/21\}$. If $0 < |x - 2| < \delta$ then $|3x^2 + 8x - 28| = |3(x - 2)(x + 14/3)| < 3\delta|x - 2 + 20/3| < 3\delta(\delta + 20/3) < 21\delta < \varepsilon$.

3.4.3. The limit is $1/3$. Proof: Let $\varepsilon > 0$. Define $\delta = \min\{1, 3\varepsilon\}$. If $0 < |x - 1| < \delta$ then $x + 2 = (x - 1) + 3 > -\delta + 3 \geq 2$ and

$$\left| \frac{x}{x + 2} - \frac{1}{3} \right| = \frac{2|x - 1|}{3(x + 2)} < \frac{2\delta}{3 \cdot 2} \leq \varepsilon.$$

3.4.5. The limit is 2. Proof: Let $\varepsilon > 0$. Define $\delta = \varepsilon$. If $0 < |x - 1| < \delta$ then $x \neq 1$ so $(x^2 - 1)/(x - 1) = x + 1$. Therefore, $|(x^2 - 1)/(x - 1) - 2| = |(x + 1) - 2| = |x - 1| < \delta = \varepsilon$.

3.4.7. Let $a_n = 1/n$, $b_n = -1/n$, for all $n \in \mathbb{N}$. Then $\lim f(a_n) = 1 \neq -1 = \lim f(b_n)$.

3.4.9. Answer: $B = [0, 1]$.

3.4.11. Answer: $B = \mathbb{R}$.

3.4.13. Let $\{a_n\}$ be a sequence that converges to a, so that $\lim f(a_n) = L$. Then $\lim |f(a_n)| = |L|$ by Exercise 2.3.1.

3.4.15. Let $\{a_n\}$ be a sequence that converges to $1/a$ and, for all $n \in \mathbb{N}$, $a_n \neq 1/a$. We will show that $\lim f(1/a_n) = L$. Since $a_n \to 1/a \neq 0$, there exists $N \in \mathbb{N}$ such that, if $n \geq N$, $a_n \neq 0$. Let $\{b_n\}$ be the sequence $1/a_N$, $1/a_{N+1}$, $1/a_{N+2}$, $\ldots$. Then $\lim b_n = c$ (because $a_n \to 1/c$), so $\lim f(b_n) = L$. Therefore, $\lim f(1/a_n) = L$.

3.4.17. Suppose that there exists $\delta > 0$ such that the interval $(a - \delta, a + \delta)$ contains only finitely many points of A. Let b be the closest to a among them, and let $d = |b - a|$. Then the interval $(a - d, a + d)$ contains no point of A (except possibly a).

3.4.19. Hint: use Exercise 1.2.10.

3.5.1. The limit is 0. Proof: let $\varepsilon > 0$ and let $\delta = 1/\ln(1/\varepsilon - 1)$ if $\varepsilon < 1$ or $\delta = 1$ if $\varepsilon \geq 1$. Suppose that $0 < x < \delta$. When $\varepsilon < 1$, we have that $1/x > \ln(1/\varepsilon - 1)$ so $e^{1/x} > 1/\varepsilon - 1$. Of course, when $\varepsilon \geq 1$ the last inequality holds for any x. It follows that $1 + e^{1/x} > 1/\varepsilon$, so $|1/(1 + e^{1/x})| < \varepsilon$.

3.5.3. Let $\varepsilon > 0$, and choose $\delta \in (0, 1)$. Then, if $8 < x < 8 + \delta$,

$$\left| \left\lfloor \frac{x}{2} \right\rfloor - 4 \right| = \left| \left\lfloor \frac{x - 8}{2} \right\rfloor \right| = 0 < \varepsilon.$$

3.5.5. The limit is 1. Let $\varepsilon > 0$, and choose $\delta = 1$. If $0 < x < \delta$ then $|x| = x$, so $||x|/x - 1| = 0 < \varepsilon$.

3.5.7. The limit is 0. Let $\varepsilon > 0$, and choose $\delta = \varepsilon$. If $-\delta < x < 0$ then $|x| = -x$ and $-x < \delta$. Thus, $|x^2/|x|| = |x^2/(-x)| = |-x| = -x < \varepsilon$.

3.5.9. The limit is 0. Let $\varepsilon > 0$, and choose $\delta = 1$. If $0 < x < \delta$ then $\lfloor x \rfloor = 0$, so $|x\lfloor x \rfloor| = 0 < \varepsilon$.

3.5.11. The limit is 1. Let $\varepsilon > 0$, and choose $\delta = \min\{1, \varepsilon\}$. If $1 < x < 1 + \delta$ then $\lfloor x \rfloor = 1$, so $|x\lfloor x \rfloor - 1| = |x - 1| < \varepsilon$.

3.5.13. The limit is $\pi/2$. Let $\varepsilon > 0$, and choose $\delta = \varepsilon$. Using $|x| < |\tan x|$, we have that

$$\left| \arctan \frac{1}{1 - x} - \frac{\pi}{2} \right| < \left| \tan \left(\arctan \frac{1}{1 - x} - \frac{\pi}{2} \right) \right| = \left| \cot \left(\arctan \frac{1}{1 - x} \right) \right| = |1 - x| < \varepsilon.$$

3.5.15. Suppose first that $\lim_{x \to a^-} f(x) = L$ and let $\{a_n\} \subset A$ be a sequence converging to a from the left. Let $\varepsilon > 0$. Then, there exists $\delta > 0$ such that, for any $a - \delta < x < a$, $|f(x) - L| < \varepsilon$. Also, there exists $N \in \mathbb{N}$ such that, for all $n \in \mathbb{N}$, $0 < a - a_n < \delta$. It follows that, for all $n \in \mathbb{N}$, $a - \delta < a_n < a$ whence $|f(a_n) - L| < \varepsilon$. In the other direction, suppose that $\lim_{x \to a^-} f(x) \neq L$. That means that there exists $\varepsilon_0 > 0$ such that, for any $n \in \mathbb{N}$, there exists a_n satisfying $a - 1/n < a_n < a$ and $|f(a_n) - L| \geq \varepsilon_0$. By Squeeze Theorem, the sequence $\{a_n\}$ converges to a from the left, $a_n \neq a$, so by assumption $\lim_{n \to \infty} f(a_n) = L$, contradicting the inequality $|f(a_n) - L| \geq \varepsilon_0$.

3.5.17. Let $c \in (a, b)$ and let $\varepsilon > 0$. The number $m(c) + \varepsilon$ is not a lower bound for $\{f(t) : t \in [a, c)\}$ (it is bigger than $m(c)$), so there exists $t \in [a, c)$ such that $f(t) < m(c) + \varepsilon$. Let $\delta = c - t$ and let $c - \delta < x < c$. Then $a \leq t < x < c$, so using the fact that the function m is decreasing, $m(c) \leq m(x)$. Also, the definition of m shows that $m(x) < f(t)$. Therefore,

$$|m(x) - m(c)| = m(x) - m(c) < f(t) - m(c) < \varepsilon,$$

and m is continuous from the left at c.

3.5.19. Let $\varepsilon > 0$ and choose $M = 4 + 11/\varepsilon$. Now, if $x > M$ then $x > 4$ so

$$\left| \frac{2x + 3}{x - 4} - 2 \right| = \frac{11}{|x - 4|} = \frac{11}{x - 4} < \frac{11}{M - 4} < \varepsilon.$$

3.5.21. Let $\varepsilon > 0$ and choose $M = \max\{4, 20/(3\varepsilon)\}$. Now, if $x > M$ then

$$\left| \frac{2x + 3}{x^2 - 4} \right| = \frac{2x + 3}{x^2 - 4} < \frac{5x}{x^2 - x^2/4} = \frac{20}{3x} < \frac{20}{3M} < \varepsilon.$$

3.5.23. The limit is 0. Proof: Let $\varepsilon > 0$ and take $M = \ln(1 + 1/\varepsilon)$. If $x \geq M$, then $e^x > 1$, so

$$\left| \frac{1}{e^x - 1} \right| = \frac{1}{e^x - 1} < \frac{1}{e^M - 1} = \varepsilon.$$

3.5.25. The limit is 1. Proof: Let $\varepsilon > 0$ and take $M = \max\{1, 2/\varepsilon^2\}$. If $x \geq M$, then

$$\left| \frac{\sqrt{x + \sqrt{x + \sqrt{x}}}}{\sqrt{x + 1}} - 1 \right| = \left| \frac{\sqrt{x + \sqrt{x + \sqrt{x}}} - \sqrt{x + 1}}{\sqrt{x + 1}} \right| \cdot \left| \frac{\sqrt{x + \sqrt{x + \sqrt{x}}} + \sqrt{x + 1}}{\sqrt{x + \sqrt{x + \sqrt{x}}} + \sqrt{x + 1}} \right|$$

$$= \frac{\sqrt{x + \sqrt{x}} - 1}{\sqrt{x + 1}\left(\sqrt{x + \sqrt{x + \sqrt{x}}} + \sqrt{x + 1} \right)} < \frac{\sqrt{2x}}{\sqrt{x + 1}\sqrt{x + 1}}$$

$$= \sqrt{\frac{2x}{x + 1}} \cdot \frac{1}{\sqrt{x + 1}} < \sqrt{2} \cdot \frac{1}{\sqrt{x}} < \sqrt{\frac{2}{M}} = \varepsilon.$$

3.5.27. Answer: 3/2. Proof: Let $\varepsilon > 0$ and take $M = \max\{1, 9^{3/\varepsilon}\}$. If $x \geq M$, then

$$\left| \frac{\ln(1 + \sqrt{x} + \sqrt[3]{x})}{\ln(1 + \sqrt[3]{x} + \sqrt[4]{x})} - \frac{3}{2} \right| = \left| \frac{\ln(1 + \sqrt{x} + \sqrt[3]{x})^2 - \ln(1 + \sqrt[3]{x} + \sqrt[4]{x})^3}{\ln(1 + \sqrt[3]{x} + \sqrt[4]{x})} \right|$$

$$\leq \frac{1}{\ln \sqrt[3]{M}} \ln \frac{(1 + \sqrt{x} + \sqrt[3]{x})^2}{(1 + \sqrt[3]{x} + \sqrt[4]{x})^3} < \frac{3}{\ln M} \ln \frac{(3\sqrt{x})^2}{(\sqrt[3]{x})^3} = \frac{3 \ln 9}{\ln M} < \varepsilon,$$

3.5.29. Since $(x^2 + 1)/(x + 1) - ax - b = [x^2(1 - a) - x(a + b) + (1 - b)]/(x + 1)$ we see that $1 - a = 0$ and $a + b = 0$, so $a = 1$ and $b = -1$.

3.5.31. Notice that

$$\left(\frac{x^2 - 1}{x^2 + 1} \right)^{x^2} = \left(1 + \frac{2}{x^2 + 1} \right)^{\frac{x^2 + 1}{2} \cdot \frac{2x^2}{x^2 + 1}} = \left(\left(1 + \frac{2}{x^2 + 1} \right)^{\frac{x^2 + 1}{2}} \right)^{\frac{2x^2}{x^2 + 1}} \to e^2,$$

3.5.33. First,

$$\ln(x^2 - x + 1) = \ln\left[x^2 \left(1 - \frac{1}{x} + \frac{1}{x^2} \right) \right] = 2\ln x + \ln\left(1 - \frac{1}{x} + \frac{1}{x^2} \right)$$

$$= \ln x \left[2 + \frac{\ln\left(1 - \frac{1}{x} + \frac{1}{x^2} \right)}{\ln x} \right], \quad \text{and similarly}$$

$$\ln(x^{10} + x + 1) = \ln x \left[10 + \frac{\ln\left(1 + \frac{1}{x^9} + \frac{1}{x^{10}} \right)}{\ln x} \right].$$

Finally,

$$\lim_{x \to \infty} \frac{\ln\left(1 - \frac{1}{x} + \frac{1}{x^2} \right)}{\ln x} = \lim_{x \to \infty} \frac{\ln\left(1 + \frac{1}{x^9} + \frac{1}{x^{10}} \right)}{\ln x} = 0, \quad \text{so} \quad \lim_{x \to \infty} \frac{\ln(x^2 - x + 1)}{\ln(x^{10} + x + 1)} = \frac{2}{10} = \frac{1}{5}.$$

3.5.35. Answer: $\ln(3/2)$.

3.5.37. Let $\varepsilon > 0$. There exists $M_0 > 0$ such that, if $x > M_0$ then $|xf(x) - L| < 1$. Let $M = \max\{M_0, (1 + |L|)/\varepsilon\}$. If $x > M$ then $|f(x)| = |xf(x) - L + L|/|x| \leq (1 + |L|)/M < \varepsilon$.

3.5.39. Let $M > 0$ and choose $\delta = 5/M$. If $2 < x < 2 + \delta$ then $x + 3 > 5$ and $0 < x - 2 < \delta$, so $(x + 3)/(x - 2) > 5/\delta = M$.

3.5.41. Let $M > 0$ and select $\delta = \frac{\pi}{2} - \arctan M$. If $\frac{\pi}{2} - \delta < x < \frac{\pi}{2}$, then $\arctan M = \frac{\pi}{2} - \delta < x < \frac{\pi}{2}$ so $\tan x > M$.

3.5.43. Let $M > 0$ and select $\delta = \min\{1/2, 1/(4M)\}$. If $1 - \delta < x < 1$ then $x^2 - 1 < 0$, so $x/(x^2 - 1) < (\delta - 1)/(2\delta) < -1/(4\delta) \leq -M$.

3.5.45. Let $M > 0$ and select $K = M^2$. If $x > K$ then $x > M^2$ so $\sqrt{x} > M$.

3.5.47. Let $M > 0$. Without loss of generality, we may assume that $M > \ln 2$, so that $e^M - 1 > 1$. Select $K = \ln(e^M - 1)$. If $x > K$ then $\ln(1 + e^x) > \ln(1 + e^K) = M$.

3.5.49. Hint: see the solution to Exercise 3.5.15.

3.5.51. Suppose that $\lim_{x \to a} f(x) = +\infty$. Let $\varepsilon > 0$ and select $M = 1/\varepsilon$. There exists $\delta > 0$ such that, if $0 < |x - a| < \delta$ then $f(x) > M$. It follows that $|1/f(x)| = 1/f(x) < 1/M = \varepsilon$. In the other direction, let $M > 0$ and select $\varepsilon = 1/M$. By assumption, there is $\delta > 0$ such that, if $0 < |x - a| < \delta$ then $0 < 1/f(x) < \varepsilon$, whence $f(x) > 1/\varepsilon = m$.

3.6.1. Let $\varepsilon > 0$ and take $\delta = 2\varepsilon$. If $0 < |x - 4| < \varepsilon$ then $|\sqrt{x} - 2| = |(x - 4)/(\sqrt{x} + 2)| < \delta/(\sqrt{x} + 2) \leq \delta/2 = \varepsilon$.

3.6.3. Let $\varepsilon > 0$ and take $\delta = \min\{3e^\varepsilon - 3, 3 - 3e^{-\varepsilon}\}$. If $|x - 3| < \delta$, then $3 - \delta < x < 3 + \delta$, so

$$x > 3 - (3 - 3e^{-\varepsilon}) = 3e^{-\varepsilon} \quad \text{and} \quad x < 3 + (3e^\varepsilon - 3) = 3e^\varepsilon.$$

It follows that $e^{-\varepsilon} < \frac{x}{3} < e^{\varepsilon}$, hence $-\varepsilon < \ln \frac{x}{3} < \varepsilon$, and we obtain that $|\ln x - \ln 3| < \varepsilon$.

3.6.5. Let $\varepsilon > 0$ and take $\delta = \varepsilon/3$. If $0 < |x - a| < \varepsilon$ then $|(3x - 2) - (3a - 2)| = 3|x - a| < 3\delta = \varepsilon$.

3.6.7. Hint: use Triangle Inequality.

3.6.9. Hint: (a) Consider $g = -f$. (b) $g = (f + g) - f$.

3.6.11. Hint: modify Dirichlet function.

3.6.13. Suppose that a is an irrational number and denote by f Dirichlet function. By Exercise 2.2.12, there exists a sequence $\{a_n\}$ of rational numbers converging to a. Thus, $f(a_n) = 1$, for all $n \in \mathbb{N}$, but $f(a) = 0$. By Theorem 3.4.9, f is not continuous at a.

3.6.15. Hint: prove that $\min\{x, y\} = (x + y - |x - y|)/2$.

3.6.17. f need not be continuous at a. Example: $f(x) = x - 1$ if $x < a$, $f(a) = a$ and $f(x) = x + 1$ if $x > a$. Given $\varepsilon > 0$ choose $\delta = \varepsilon$. If $x < a$ then $|a - x| = a - x < a + 1 - x = |x - 1 + a| = |f(x) - f(a)|$. If $x > a$ then $|x - a| = x - a < x + 1 - a = |x + 1 - a| = |f(x) - f(a)|$. Prove that the given condition holds if and only if for any $b \in \mathbb{R}$, the set $\{x \in \mathbb{R} : f(x) = b\}$ is a bounded set.

3.6.19. Let a be an irrational number. By Exercise 2.2.12, there exists a sequence $\{a_n\}$ of rational numbers converging to a. Thus, $f(a_n) = g(a_n)$, for all $n \in \mathbb{N}$. Since f and g are continuous, Theorem 3.4.9 implies that $f(a) = g(a)$.

3.6.21. Notice that

$$s = \frac{t - s}{t - r} r + \left(1 - \frac{t - s}{t - r}\right) t, \quad \text{so,}$$

$$f(s) \le \frac{t - s}{t - r} f(r) + \left(1 - \frac{t - s}{t - r}\right) f(t) = \frac{t - s}{t - r} f(r) + \frac{s - r}{t - r} f(t), \quad \text{and thus}$$

$$\frac{f(s) - f(r)}{s - r} \le \frac{1}{s - r}\left[\frac{t - s}{t - r} f(r) + \frac{s - r}{t - r} f(t) - f(r)\right] = \frac{f(t) - f(r)}{t - r}.$$

This proves (a), and we leave (b) to the reader. In (c), we will prove that f is continuous from the right, and leave the continuity from the left to the reader. Let $c \in \mathbb{R}$, and select $a, b \in \mathbb{R}$ so that $a < c < b$. Let $\{c_n\}$ be a sequence that converges to c, and suppose that $c_n > c$ for any $n \in \mathbb{N}$. Using parts (a) and (b),

$$\frac{f(c) - f(a)}{c - a} \le \frac{f(c_n) - f(c)}{c_n - c} \le \frac{f(b) - f(c)}{b - c}.$$

Let $M = \max\left\{\left|\frac{f(c) - f(a)}{c - a}\right|, \left|\frac{f(b) - f(c)}{b - c}\right|\right\}$. Then

$$\left|\frac{f(c_n) - f(c)}{c_n - c}\right| \le M,$$

so $|f(c_n) - f(c)| \le M|c_n - c|$. The result now follows from Squeeze Theorem.

3.7.1. Hint: $|x| = \sqrt{x^2}$.

3.7.3. By Theorem 3.6.4, a rational function f/g is continuous at a unless $g(a) = 0$. If $f(a) = 0$ the fraction f/g can be reduced by a power of $x - a$, leading to a function u/v where at most one of $u(a)$ and $v(a)$ can be zero. If $v(a) \neq 0$ then $\lim_{x \to a} u(x)/v(x) = u(a)/v(a)$ so the discontinuity is removable. If $v(a) = 0$ then $\lim_{x \to a} u(x)/v(x) = \infty$ so the discontinuity is essential.

3.7.5. Answer: $\frac{\pi}{2} - 2x$.

3.7.7. First we prove that $f(x) = \arcsin x$ is continuous at $x = 0$. Let $\{a_n\}$ be a sequence that converges to 0. Without loss of generality, we may assume that $a_n \in (-1, 1)$, for all $n \in \mathbb{N}$. Inequality (3.1) implies that $|\sin x| \ge |x| \cos x$, for $x \in (-\pi/2, \pi/2)$. If we use a substitution $u = \sin x$, then $x = \arcsin u$ and $\cos x = \sqrt{1 - u^2}$. That way, we obtain the inequality $|u| \ge |\arcsin u|\sqrt{1 - u^2}$. It follows that

$$|\arcsin a_n| \le \frac{|a_n|}{\sqrt{1 - a_n^2}}.$$

Since the expression on the right side goes to 0, as $n \to \infty$, Squeeze Theorem implies that $\lim \arcsin a_n = 0$. Thus, f is continuous at $x = 0$.

Next, let $c \in [-1, 1]$, and let $\{c_n\}$ be a sequence in $[-1, 1]$ that converges to c. We will show that $\lim f(c_n) = f(c)$. One knows from trigonometry that $\sin(x - y) = \sin x \cos y - \cos x \sin y$. Substituting $u = \sin x$ and $v = \sin y$, and assuming that $x, y \in (-\pi/2, \pi/2)$, we obtain

$$\arcsin u - \arcsin v = \arcsin\left(u\sqrt{1 - v^2} - v\sqrt{1 - u^2}\right).$$

In particular,

$$\arcsin c - \arcsin c_n = \arcsin\left(c\sqrt{1 - c_n^2} - c_n\sqrt{1 - c^2}\right).$$

The expression inside the parentheses has limit 0, so by the established continuity of f at $x = 0$, the right side has limit $\arcsin 0 = 0$. Thus, $\lim \arcsin c_n = \arcsin c$, and f is continuous at $x = c$.

3.7.9. By definition, $a^x = \sup\{a^r : r \in \mathbb{Q}, r < x\}$. Define a sequence $\{r_n\}$ of rational numbers inductively. Let

$r_1 < x$. Given $n \geq 2$, there exists a rational number r_n such that $r_{n-1} < r_n < x$ and $a^x - 1/n < a^{r_n} < a^x$. Now, $\{r_n\}$ has the desired properties. We leave the construction of $\{s_n\}$ as an exercise.

3.7.11. Let $\{r_n\}$ and $\{s_n\}$ be the sequences of rational numbers provided by Exercise 3.7.9. Let $\varepsilon > 0$. There exists $k \in \mathbb{N}$ such that $a^x - \varepsilon < a^{r_k} < a^x < a^{s_k} < a^x + \varepsilon$. Since $r_k < x < s_k$, there exists $N \in \mathbb{N}$ such that, if $n \geq N$, $r_k < x_n < s_k$. Using Exercise 3.7.10, we have that $a^x - \varepsilon < a^{r_k} < a^{x_n} < a^{s_k} < a^x + \varepsilon$, whence $|a^{x_n} - a^x| < \varepsilon$.

3.7.13. Suppose first that $x \in \mathbb{Q}$. Let $\{r_n\}$ be the sequence of rational numbers provided by Exercise 3.7.9 such that $\lim r_n = y$ and $\lim (a^x)^{r_n} = (a^x)^y$. Then, $(a^x)^{r_n} = a^{x r_n}$ and, by Exercise 3.7.11, $\lim a^{x r_n} = a^{xy}$. If $x \notin \mathbb{Q}$, let $\{t_n\}$ be a sequence of rational numbers converging to x. By the first part, $(a^{t_n})^y = a^{t_n y}$, for all $n \in \mathbb{N}$. Taking the limit when $n \to \infty$ and using Theorem 3.7.6 and Exercise 3.7.11 yields the result.

3.7.13. The area of the triangle that has the same vertices as the shaded region equals $a\sqrt{a^2 - 1}$. The area of the shaded region is the difference between the area of the triangle and

$$2 \int_1^a \sqrt{x^2 - 1}\, dx = \left(x\sqrt{x^2 - 1} - \ln|x + \sqrt{x^2 - 1}| \right) \Big|_1^a = a\sqrt{a^2 - 1} - \ln(a + \sqrt{a^2 - 1}).$$

The result follows from Exercise 3.7.14.

3.7.17. Hint: $\log_a x = \ln x / \ln a$ and $y = \ln x$ is continuous (see Exercise 3.6.3).

3.8.1. Let $\varepsilon > 0$ and choose $\delta = \varepsilon$. If $x, y \in [1, +\infty)$ and $|x - y| < \delta$ then $|1/x - 1/y| = |x - a|/|xa| \leq |x - a| < \delta = \varepsilon$.

3.8.3. The function is not uniformly continuous on $(0, 1)$. We will take $\varepsilon = 1$, and we will show that, for every $\delta > 0$, there exists $x, y \in (0, 1)$ such that $|x - y| < \delta$ and $|f(x) - f(y)| \geq 1$. So, let $\delta > 0$. We take

$$n = \left\lfloor \frac{1}{\delta \pi} \right\rfloor, \quad x = \frac{1}{2n\pi}, \quad y = \frac{1}{3n\pi}.$$

Then

$$|x - y| = \left| \frac{1}{2n\pi} - \frac{1}{3n\pi} \right| = \frac{1}{6n\pi} < \frac{1}{n\pi} \leq \delta, \quad \text{and}$$

$$|f(x) - f(y)| = \left| e^{\frac{1}{2n\pi}} \cos 2n\pi - e^{\frac{1}{3n\pi}} \cos 3n\pi \right| = e^{\frac{1}{2n\pi}} + e^{\frac{1}{3n\pi}} > 1.$$

3.8.5. The function is uniformly continuous on $[1, +\infty)$. Let $\varepsilon > 0$, and take $\delta = \varepsilon$. If $|x - y| < \delta$ then

$$|\sqrt{x} - \sqrt{y}| = |\sqrt{x} - \sqrt{y}| \frac{\sqrt{x} + \sqrt{y}}{\sqrt{x} + \sqrt{y}} = \frac{|x - y|}{\sqrt{x} + \sqrt{y}} \leq \frac{|x - y|}{2} < |x - y| < \delta = \varepsilon.$$

3.8.7. The function is not uniformly continuous on $(-\pi/2, \pi/2)$. Take $\varepsilon_0 = 1$ and let $\delta > 0$. Choose $n \in \mathbb{N}$ such that $1/(2n) < \delta$ and $x = \pi/2 - 1/n$, $y = \pi/2 - 1/(2n)$. Then $|x - y| = 1/(2n) < \delta$. On the other hand, using the identity $\sin(u - v) = \sin u \cos v - \cos u \sin v$,

$$\tan\left(\frac{\pi}{2} - \frac{1}{n} \right) - \tan\left(\frac{\pi}{2} - \frac{1}{2n} \right) = \cot\left(\frac{1}{n} \right) - \cot\left(\frac{1}{2n} \right) = \frac{1}{\sin(1/n)} \geq \varepsilon_0.$$

3.8.9. Extend f to $[0, 1]$ by $f(0) = 1$ and $f(1) = |\sin 1|$. Then f is continuous on $[0, 1]$, hence uniformly continuous on $[0, 1]$, and all the more uniformly continuous on $(0, 1)$. A similar argument shows that f is uniformly continuous on $(-1, 0)$. However, $\lim_{x \to 0^-} f(x) = -1$ and $\lim_{x \to 0^+} f(x) = 1$ so f cannot be extended to a continuous function on $[-1, 1]$.

3.8.11. Let $\varepsilon_0 = 1$. By assumption, there exists $\delta_0 > 0$ such that, if $x, y \in A$ and $|x - y| < \delta_0$ then $|f(x) - f(y)| < 1$. Since A is bounded, there exist real numbers a, b such that $A \subset [a, b]$. Let

$$c \in A, \quad n_0 = \left\lfloor \frac{b - a}{\delta_0} \right\rfloor, \quad \text{and} \quad M = |f(c)| + n.$$

We will show that $|f(x)| \leq M$, for all $x \in A$.

Let $x \in A$. Then $|x - c| < b - a < n\delta_0$, so there exist numbers $x_0 = x, x_1, x_2, \ldots, x_{n_0} = c$ such that $|x_k - x_{k-1}| < \delta_0$, for $1 \leq k \leq n_0$. It follows that $|f(x_k) - f(x_{k-1})| < 1$, and by Triangle Inequality, $|f(x) - f(c)| < n_0$. This implies that $|f(x)| \leq |f(c)| + |f(x) - f(c)| < |f(c)| + n_0 = M$.

3.8.13. Let $\varepsilon > 0$. By assumption $\lim_{x \to \infty} f(x) = L$ so there exists $M > 0$ such that, if $x > M$ then $|f(x) - L| < \varepsilon/2$. Since f is continuous on $[0, M + 1]$, it is uniformly continuous on $[0, M + 1]$. Thus, there exists δ_1 such that, if $x, y \leq M + 1$ and $|x - y| < \delta_1$, then $|f(x) - f(y)| < \varepsilon$. If at least one of x, y is bigger than $M + 1$ and $|x - y| < 1$, then $x, y > M$, so $|f(x) - f(y)| < |f(x) - L| + |f(y) - L| < \varepsilon$. Therefore, taking $\delta = \min\{\delta_1, 1\}$ guarantees that if $|x - y| < \delta$, then $|f(x) - f(y)| < \varepsilon$.

3.8.15. In one direction it is trivial. In the other, suppose that f is uniformly continuous on (a, b) and let $\varepsilon > 0$. There exists $\delta_1 > 0$ such that, if $x, y \in (a, b)$ and $|x - y| < \delta_1$ then $|f(x) - f(y)| < \varepsilon$. Also, f is continuous at a and b so there exists $\delta_2 > 0$ such that if $x \in (a, b)$ and $|x - a| < \delta_2$, (resp., $|x - b| < \delta_2$) then $|f(x) - f(a)| < \varepsilon$ (resp., $|f(x) - f(b)| < \varepsilon$). Now, let $\delta = \min\{\delta_1, \delta_2, b - a\}$. If $x, y \in [a, b]$ and $|x - y| < \delta$ then either both $x, y \in (a, b)$ or at most one of them is an endpoint. In either case, $|f(x) - f(y)| < \varepsilon$.

3.8.17. Let $\varepsilon > 0$. Since f is uniformly continuous on (a, b) there exists δ_0 such that, if $x, y \in (a, b)$ and $|x - y| < \delta$, then $|f(x) - f(y)| < \varepsilon$. Since $\{x_n\}$ is a Cauchy sequence in (a, b), there exists $N \in \mathbb{N}$ such that if $m \geq n \geq N$ then $|x_m - x_n| < \delta$. It follows that if $m \geq n \geq N$, then $|f(x_m) - f(x_n)| < \varepsilon$.

3.8.19. Let $\varepsilon > 0$ and choose $\delta = \varepsilon/M$. If $x, y \in (a, b)$ and $|x - y| < \delta$ then $|f(x) - f(y)| \leq M|x - y| < M\delta = \varepsilon$. In the other direction consider $f(x) = \sqrt{x}$ on $(0, 1)$.

3.9.1. Let $g(x) = f(x + 1) - f(x)$, defined on $[0, 1]$. The function g is continuous and

$$g(0) = f(1) - f(0), \quad \text{and} \quad g(1) = f(2) - f(1) = f(0) - f(1) = -g(0).$$

If $g(0) = 0$, then $f(1) = f(0)$ and the solution is to take $x_1 = 0$ and $x_2 = 1$. If $g(0) \neq 0$, then $g(1)$ and $g(0)$ are non-zero numbers of the opposite signs. By the Intermediate Value Theorem, there exists $c \in [0, 1]$ such that $g(c) = 0$. Now, the solution is to define $x_1 = c$ and $x_2 = c + 1$.

3.9.3. Let

$$g_1(x) = \begin{cases} x & \text{if } 0 \leq x \leq 1, \\ 1 - x & \text{if } 1 \leq x \leq 2, \\ x - 2 & \text{if } 2 \leq x \leq 3. \end{cases}$$

It is not hard to see that g_1 attains each value between 0 and 1 exactly 3 times, while 0 and 1 are attained twice. Next we extend it to $[0, 6]$ by defining

$$g_2(x) = \begin{cases} x - 2 & \text{if } 3 \leq x \leq 4, \\ 6 - x & \text{if } 4 \leq x \leq 5, \\ x - 4 & \text{if } 5 \leq x \leq 6. \end{cases}$$

Now we have a function on $[0, 6]$ that attains each value between 0 and 2 exactly 3 times, while 0 and 2 are attained twice. We continue by defining, for each $n \in \mathbb{Z}$,

$$g_n(x) = \begin{cases} x - 2n + 2 & \text{if } 3n - 3 \leq x \leq 3n - 2, \\ 4n - 2 - x & \text{if } 3n - 2 \leq x \leq 3n - 1, \\ x - 2n & \text{if } 3n - 1 \leq x \leq 3n. \end{cases}$$

The function f which equals g_n on $[3n - 3, 3n]$ attains each real number exactly 3 times.

It is impossible for a continuous function f on $\mathbb{R}$ to attain each of its values exactly twice. Let $a < b$ be real numbers such that $f(a) = f(b) = C$. By the Intermediate Value Theorem, in the interval (a, b) we have either that $f(x) > C$ or $f(x) < C$, for all x. Without loss of generality, we will assume that $f(x) > C$, for all $x \in (a, b)$. By the Extreme Value Theorem, there exists $c \in (a, b)$ where f attains its maximum M. Further, the maximum can be attained only at one point. (If it were attained at two different points $c_1 < c_2$, and if p is a point between them, then $f(p)$ would have to be attained in each of the intervals (a, c_1) and (c_2, b), so it would be attained 3 times.) Therefore, there exists d outside of $[a, b]$, such that $f(d) = M$. Without loss of generality, we will assume that $a < b < d$. Now, each value between C and M is attained in (a, c), in (c, b), and in (b, d). This contradiction shows that it is impossible for f to attain each of its values exactly twice.

3.9.5. The function $g(x) = f(x) - x$ is continuous on $[0, 1]$ so it attains its minimum at some point $a \in [0, 1]$ and a maximum at some point $b \in [0, 1]$. If $g(a) \geq 0$, then $g(1) \geq 0$ so $f(1) = 1$. Similarly, if $g(b) \leq 0$ then $f(0) = 0$. Otherwise, we have that $g(a) < 0 < g(b)$ so Theorem 3.9.3 implies that there exists $c \in [0, 1]$ such that $g(c) = 0$.

3.9.7. By Example 3.4.8, $\lim_{x \to 0+} f(x)$ does not exist, so f is not continuous on $[0, 1]$. Let $u, v \in [0, 1]$. If both $u, v \in (0, 1)$, the fact that f is continuous on $(0, 1)$ implies the intermediate value property. Thus, it suffices to consider the case when $u = 0$ or $v = 1$. We will prove the former. Since $0 < v \leq 1$, there exists $k \in \mathbb{Z}$ such that $1/[2(k + 1)\pi] < v \leq 1/[2k\pi]$. Notice that if $1/[2(k + 2)\pi] < x \leq 1/[2(k + 1)\pi]$, then $2(k + 1)\pi \leq 1/x < 2(k + 2)\pi$ so $\sin(1/x)$ attains all values in $[-1, 1]$. Consequently, regardless of the choice of y between 0 and $f(v)$, we have that $y \in [-1, 1]$ so there exists $x \in [0, v]$ such that $f(x) = y$.

3.9.9. Hint: modify Example 3.9.6.

3.9.11. Let T be the period of f, and let $M = \sup\{f(x) : 0 \leq x \leq T\}$. Then $M = \sup\{f(x) : x \in \mathbb{R}\}$. Indeed, suppose that there exists $c \in \mathbb{R}$ such that $f(c) > M$. Let $k = \lfloor c/T \rfloor$. Then $k \leq c/T < k + 1$ so $kT \leq c < (k + 1)T$, hence $0 \leq c - kT < T$. Since f is periodic, with period T, $f(c - kT) = f(c) > M$ which contradicts the assumption that M is the maximum of f on $[0, T]$. Thus, $M = \sup\{f(x) : x \in \mathbb{R}\}$. Since M is also $\sup\{f(x) : 0 \leq x \leq T\}$, and f is continuous, the Extreme Value Theorem implies that this maximum is attained. A similar proof can be used to show that f attains its minimum as well.

3.9.13. Let $c \in [a, b]$ and let $\varepsilon > 0$. We will prove that the function M is continuous at $x = c$, and leave the function m to the reader.

First we will show that M is right continuous at $x = c$. Let $\{a_n\}$ be a sequence in $[c, b]$ that converges to c. Suppose that $M(a_n)$ does not converge to $M(c)$. Then there exists $K > M(c)$ and a subsequence $\{b_n\}$ of $\{a_n\}$ such that, for all $n \in \mathbb{N}$, $M(b_n) \geq K$. By the Extreme Value Theorem, for each $n \in \mathbb{N}$, the function f attains its supremum on $[a, b_n]$. Let y_n be a real number such that $f(y_n) = M(b_n)$. Since $M(b_n) > M(c)$, we see that $y_n \notin [a, c]$. Thus $c < y_n \leq b_n$. Since $\{b_n\}$ converges to c, it follows by Squeeze Theorem that $\lim y_n = c$. On the other hand,

$$f(y_n) = M(b_n) \geq K > M(c) \geq f(c)$$

so $\lim f(y_n) \geq K > f(c)$, contradicting the continuity of f at c.

Next, we will show that M is left continuous at $x = c$. Let $\{a_n\}$ be a sequence in $[a, c]$ that converges to c, and suppose that $M(a_n)$ does not converge to $M(c)$. Then there exists $L < M(c)$ and a subsequence $\{b_n\}$ of $\{a_n\}$ such that, for all $n \in \mathbb{N}$, $M(b_n) \leq L$. By the Extreme Value Theorem, the function f attains its supremum on $[a, c]$. Let y be a real number such that $f(y) = M(c)$. Since $M(b_n) < f(y)$, we see that $y \notin [a, b_n]$, so $y \in [b_n, c]$. However, $\lim b_n = c$, so $y = c$. Let $\varepsilon = \frac{1}{2}[M(c) - L]$. By assumption, f is continuous at c, so $\lim f(b_n) = f(c)$. Therefore, there exists $n \in \mathbb{N}$ such that $f(c) - f(b_n) < \varepsilon$. Now,

$$M(c) = f(c) < f(b_n) + \varepsilon \leq M(b_n) + \varepsilon \leq L + \varepsilon = L + \frac{1}{2}[M(c) - L] < M(c),$$

and this contradiction shows that M is left continuous at $x = c$.

4.1.1. Answer: $2(1 - 2x)/(1 - x + x^2)^2$.

4.1.3. Answer: $1/(1 - x^2)^{3/2}$.

4.1.5. Answer: $\cos(\sin(\sin x)) \cos(\sin x) \cos x$.

4.1.7. Answer: $x^{a^a-1} a^a + a^{x^a+1} x^{a-1} \ln a + a^{a^x+x} \ln^2 a$.

4.1.9. $4x/(x^4 - 1)$.

4.1.11. Answer: $1/(x^2 + 1)$.

4.2.1. $f(x+h) - f(x) = [(x+h)^2 - 2(x+h) + 3] - (x^2 - 2x + 3) = h^2 + 2xh - 2h$, so $[f(x+h) - f(x)]/h = h + 2x - 3$. Thus, $f'(x) = \lim_{h \to 0}(h + 2x - 3) = 2x - 3$

4.2.3. $f(x + h) - f(x) = \sqrt{3(x + h) - 2} - \sqrt{3x - 2}$

$$= \frac{[3(x + h) - 2] - (3x - 2)}{\sqrt{3(x + h) - 2} + \sqrt{3x - 2}} = \frac{3h}{\sqrt{3(x + h) - 2} + \sqrt{3x - 2}},$$

so $[f(x + h) - f(x)]/h = 3/(\sqrt{3(x + h) - 2} + \sqrt{3x - 2})$. Thus, $f'(x) = 3/(2\sqrt{3x - 2})$.

4.2.5. Using (4.3) with $f(x) = \ln x$, $x_0 = e$ and $x = 2$, we obtain that $\ln 2 \approx 1 + (1/e)(2 - e) = 2/e \approx 0.736$.

4.2.7. (c) $\arctan x \approx \arctan 1 + f'(1)(x - 1)$. Since $(\arctan x)' = 1/(1 + x^2)$, $f'(1) = 1/2$. Also, $\arctan 1 = \pi/4$, so we obtain

$$\arctan x \approx \frac{\pi}{4} + \frac{1}{2}(x - 1).$$

Using this formula, $\arctan 0.9 \approx \frac{\pi}{4} + \frac{1}{2}(0.9 - 1) \approx 0.735$.

4.2.9. For $h \neq 0$, $f(h) - f(0) = he^{-1/h^2}$ so $f'(0) = \lim_{h \to 0} e^{-1/h^2} = 0$

4.2.11. Using the definition of the derivative,

$$f'(0) = \lim_{x \to 0} \frac{f(x) - f(0)}{x - 0} = \lim_{x \to 0} \frac{f(x)}{x} = \lim_{x \to 0} \begin{cases} 1 + x & \text{if } x \in \mathbb{Q}, \\ 1 & \text{if } x \notin \mathbb{Q} \end{cases} = 1,$$

so f is differentiable at $x = 0$ and $f'(0) = 1$.

4.2.13. $xf(a) - af(x) = f(a)(x - a) - a(f(x) - f(a))$ so $[xf(a) - af(x)]/(x - a) = f(a) + a[f(x) - f(a)]/(x - a)$. Thus the limit equals $f(a) + af'(a)$.

4.2.15. Notice that $g(0) = 0$. It follows that for $h \neq 0$, $g(0+h) - g(0) = g(h) = hf(h)$. Thus $g'(0) = \lim_{h \to 0} f(h) = f(0)$, because f is continuous at $x = 0$.

4.2.17. Hint: consider f defined by $|x|/x$ when $x \neq 0$ and $f(0) = 0$.

4.2.19. Let $\varepsilon > 0$. Since f is right differentiable at $x = a$, there exists $\delta_1 > 0$ such that, if $a < x < a + \delta_1$ then

$$\left| \frac{f(x) - f(a)}{x - a} - f'_+(a) \right| < 1.$$

Using the same technique as in the proof of Theorem 4.2.3, this implies that $|f(x) - f(a)| \leq (1 + |f'_+(a)|)|x - a|$. Let $\delta = \min\{\delta_1, \dfrac{\varepsilon/2}{1 + |f'_+(a)|}\}$. Now, if $a < x < a + \delta$, then

$$|f(x) - f(a)| \leq (1 + |f'_+(a)|)\delta \leq \frac{\varepsilon}{2} < \varepsilon.$$

4.2.21. By Exercise 4.2.19, f is continuous from the left at $x = c$. Using a similar argument we get that f is continuous from the right at $x = c$. By definition this means that $\lim_{x \to c^-} f(x) = f(c)$ and $\lim_{x \to c^+} f(x) = f(c)$. It follows that $\lim_{x \to c} f(x)$ exists and equals $f(c)$.

4.3.1. $f + g$ cannot be differentiable at $x = a$. Reason: $g = (f + g) - f$.

4.3.3. If the limit exists then, in particular, taking first the limit as $y \to c^-$, and using the continuity of f, we have that

$$\lim_{x \to c^+} \frac{f(x) - f(c)}{x - c}$$

exists. Similarly, if we first let $x \to c^+$, we must obtain the same limit

$$\lim_{y \to c^-} \frac{f(x) - f(c)}{x - c}.$$

Since this means that both $f'_+(c)$ and $f'_-(c)$ exist and are equal, we obtain that f is differentiable at $x = c$.

In the other direction, suppose that f is differentiable at $x = c$. Let $\varepsilon > 0$. There exists $\delta > 0$ such that

$$0 < |x - c| < \delta \quad \Rightarrow \quad \left| \frac{f(x) - f(c)}{x - c} - f'(c) \right| < \frac{\varepsilon}{2}.$$

We will show that, for $c - \delta < y < c < x < c + \delta$,

$$\left| \frac{f(x) - f(y)}{x - y} - f'(c) \right| < \varepsilon.$$

To that end, we notice that

$$\frac{f(x) - f(y)}{x - y} - f'(c) = \frac{f(x) - f(y) - f'(c)(x - y)}{x - y} = \frac{[f(x) - f(c) - f'(c)(x - c)] - [f(y) - f(c) - f'(c)(y - c)]}{x - y}.$$

Our next observation is that, if $a, b, c, d > 0$, $\frac{a+b}{c+d} \leq \frac{a}{c} + \frac{b}{d}$. Therefore,

$$\left| \frac{f(x) - f(y)}{x - y} - f'(c) \right| \leq \frac{|f(x) - f(c) - f'(c)(x - c)| + |f(y) - f(c) - f'(c)(y - c)|}{[x - c] + [c - y]}$$

$$\leq \frac{|f(x) - f(c) - f'(c)(x - c)|}{x - c} + \frac{|f(y) - f(c) - f'(c)(y - c)|}{c - y} < \frac{\varepsilon}{2} + \frac{\varepsilon}{2} = \varepsilon.$$

4.3.5. If we divide the numerator by $x - 0$, it has the limit

$$\lim_{x \to 0} \frac{e^x f(x) - f(0)}{x - 0} = \lim_{x \to 0} \left(e^x \frac{f(x) - f(0)}{x - 0} + f(0) \frac{e^x - e^0}{x - 0} \right) = f'(0) + f(0).$$

Similarly, dividing the denominator by $x - 0$, we obtain

$$\lim_{x \to 0} \frac{\cos x f(x) - f(0)}{x - 0} = \lim_{x \to 0} \left(\cos x \frac{f(x) - f(0)}{x - 0} + f(0) \frac{\cos x - \cos 0}{x - 0} \right) = f'(0).$$

Therefore, if we divide both the numerator and the denominator by $x - 0$, we obtain $1 + f(0)/f'(0)$.
4.3.7. If $x > 0$ then $f(x) = x^{4/3}$ so $f'(x) = 4x^{1/3}/3$. If $x < 0$ then $f(x) = -x^{4/3}$ so $f'(x) = -4x^{1/3}/3$. For $x = 0$,

$$f'(0) = \lim_{h \to 0} \frac{\sqrt[3]{h}|h|}{h} = 0.$$

Thus, $f'(x) = 4\sqrt[3]{|x|}/3$.
4.3.9. For any $x \neq 0$, the function is differentiable and $f'(x) = 2x \sin(1/x) - \cos(1/x)$. When $x = 0$,

$$f'(0) = \lim_{h \to 0} \frac{h^2 \sin(1/h)}{h} = 0.$$

Therefore, f is differentiable and

$$f'(x) = \begin{cases} 2x \sin(1/x) - \cos(1/x), & \text{if } x \neq 0 \\ 0, & \text{if } x = 0. \end{cases}$$

To see that f' is not continuous at $x = 0$ use Example 3.4.8 and the fundamental trigonometric identity.
4.3.11. Since $F(x) = \tan x$ is differentiable, Theorem 4.3.4 shows that $f(x) = F^{-1}(x)$ is differentiable and, if $F(c) = d$,

$$f'(d) = (F^{-1})'(d) = \frac{1}{F'(c)} = \frac{1}{\sec^2 c} = \cos^2 c = \cos^2(\arctan d) = \frac{1}{1 + d^2}.$$

4.3.13. By Theorem 4.3.4, if $F(x) = \sinh x$ and $F(c) = d$, then

$$(F^{-1})'(d) = \frac{1}{F'(c)} = \frac{1}{\cosh c}.$$

Further, $\cosh c = \cosh(\text{arsinh } d) = \sqrt{1 + \sinh^2(\text{arsinh } d)} = \sqrt{1 + d^2}$. Therefore,

$$(\text{arsinh } x)' = \frac{1}{\sqrt{1 + x^2}}.$$

4.3.15. We will use induction on n. The case $n = 0$ is obvious, so we assume that the statement is true for some $n \in \mathbb{N}_0$, and we will prove that it holds for $n + 1$.

$$f^{(n+1)}(x) = \frac{d}{dx} f^{(n)}(x) = \frac{d}{dx} \sin\left(x + \frac{n\pi}{2} \right) = \cos\left(x + \frac{n\pi}{2} \right) = \cos\left[\left(x + \frac{(n+1)\pi}{2} \right) - \frac{\pi}{2} \right]$$

$$= \cos\left(x + \frac{(n+1)\pi}{2} \right) \cos \frac{\pi}{2} + \sin\left(x + \frac{(n+1)\pi}{2} \right) \sin \frac{\pi}{2} = \sin\left(x + \frac{(n+1)\pi}{2} \right).$$

4.3.17. Let $g(x) = \ln f(x)$. Since $f(a) > 0$ and f is continuous at $x = a$, there exists $\varepsilon > 0$ such that $f(x) > 0$, for $x \in (a - \varepsilon, a + \varepsilon)$. If $1/n < \varepsilon$ then

$$\ln\left(\frac{f(a + 1/n)}{f(a)}\right)^{1/n} = \frac{1}{n}\left(g\left(a + 1/n\right) - g(a)\right) = \frac{1}{n^2}\frac{\left(g\left(a + 1/n\right) - g(a)\right)}{\frac{1}{n}} \to 0,$$

as $n \to \infty$. Reason: g is differentiable at $x = a$ so the second factor goes to $g'(a)$.

4.3.19. Use induction. For $n = 0$ it is obvious. Suppose that the formula holds for some positive integer n. Differentiating both sides with respect to n we get

$$(fg)^{(n+1)} = \sum_{k=0}^{n}\binom{n}{k}\left(f^{(k+1)}g^{(n-k)} + f^{(k)}g^{(n-k+1)}\right)$$

$$= \sum_{k=0}^{n}\binom{n}{k}f^{(k+1)}g^{(n-k)} + \sum_{k=0}^{n}\binom{n}{k}f^{(k)}g^{(n-k+1)}$$

$$= \sum_{k=1}^{n+1}\binom{n}{k-1}f^{(k)}g^{(n-k+1)} + \sum_{k=0}^{n}\binom{n}{k}f^{(k)}g^{(n-k+1)}$$

$$= g^{(n+1)} + \sum_{k=1}^{n}\left(\binom{n}{k-1} + \binom{n}{k}\right)f^{(k)}g^{(n-k+1)} + f^{(n+1)}$$

$$= \sum_{k=0}^{n+1}\binom{n+1}{k}f^{(k)}g^{(n-k+1)}.$$

4.4.1. $f'(x) = 1 - 2x > 0 \Leftrightarrow x < 1/2$. Answer: f is increasing on $(-\infty, 1/2)$ and decreasing on $(1/2, +\infty)$.

4.4.3. $f'(x) = (100 - x)/(2\sqrt{x}(x + 100)^2) > 0 \Leftrightarrow x < 100$. Answer: f is increasing on $(0, 100)$ and decreasing on $(100, +\infty)$.

4.4.5. $f'(x) = 3x^2 - 12x + 9 = 3(x - 1)(x - 3) = 0 \Leftrightarrow x = 1$ or $x = 3$. $f''(x) = 6x - 12 \Rightarrow f''(1) = -6$ and $f''(3) = 6$. By Second Derivative Test, f has a local maximum at $x = 1$ and a local minimum at $x = 3$.

4.4.7. $f'(x) = (1 - x)/\sqrt{2x - x^2} = 0 \Leftrightarrow x = 1$. $f''(x) = -1/(2x - x^2)^{3/2} \Rightarrow f''(1) = -1$. By Second Derivative Test, f has a local maximum at $x = 1$

4.4.9. No. Example: $y = x^3$, $c = 0$.

4.4.11. Let $f(x) = e^x$ and $g(x) = x + 1$. Notice that $f(0) = g(0)$ and that $f'(x) = e^x > 1 = g'(x)$, for all $x > 0$. By Exercise 4.4.10, $f(x) > g(x)$, for all $x > 0$. If $x < 0$, let $x = -t$, $t > 0$, and consider $\tilde{f}(t) = e^{-t}$, $\tilde{g}(t) = 1 - t$.

4.4.13. Let $f(x) = x^\alpha - 1$ and $g(x) = \alpha(x - 1)$. Notice that $f(1) = g(1)$, $f'(1) = g'(1)$, and that $f''(x) = \alpha(\alpha - 1)x^{\alpha - 2} > 0 = g''(x)$, for all $x > 1$. By Exercise 4.4.10, $f(x) > g(x)$, for all $x > 1$.

4.4.15. Let $x, y \in \mathbb{R}$. By Mean Value Theorem there exists c between x and y such that $f(x) - f(y) = (x - y)f'(c) = 0$. Thus, $f(x) = f(y)$.

4.4.17. Hint: Apply Rolle's Theorem to $h(x) = f(x) - g(x)\dfrac{f(b) - f(a)}{g(b) - g(a)}$.

4.4.19. Let $f(x) = 3x^4 - 8x^3 + 7x^2 - 45$. Then $f(-2) = 95$, $f(-1) = -27$ and $f(3) = 55$. By Intermediate Value Theorem there exist $x_1 \in (-2, -2)$ and $x_2 \in (-2, 3)$ such that $f(x_1) = f(x_2) = 0$. If there were any other real roots of the given equation, the derivative f' would have at least 2 real roots. However, $f'(x) = 12x^3 - 24x^2 + 14x = 2x[6(x - 1)^2 + 1]$ so f' has the only zero at $x = 0$.

4.4.21. Let $g(x) = f(x) - Cx$. Then $g'(x) = f'(x) - C$, for $a < x < b$, and g has one-sided derivatives at the endpoints. In fact, $g'_+(a)$ and $g'_-(b)$ are of the opposite sign, and without loss of generality we will assume that $g'_+(a) < 0 < g'_-(b)$. By Exercise 4.2.19 and Theorem 4.2.3, g is continuous on $[a, b]$, so it attains its minimum at $z \in [a, b]$. If $z \in (a, b)$, by Fermat's Theorem $g'(z) = 0$, which implies that $f'(z) = C$. In order to complete the proof, it suffices to show $z \neq a$ and $z \neq b$.

Let $\varepsilon > 0$. Since $\lim_{x \to a^+} \frac{g(x) - g(a)}{x - a} < 0$, there exists $\delta > 0$ such that, if $a < x < a + \delta$, $\frac{g(x) - g(a)}{x - a} < L + \varepsilon < 0$. Consequently, for such x, $g(x) < g(a)$, and the minimum is not at $x = a$. A similar argument shows that the minimum is not at $x = b$.

4.4.23. If $h \neq 0$, $f(0 + h) - f(0) = f(h) = h + h^2\sin(2/h)$, so $f'(0) = \lim_{h \to 0} f(h)/h = \lim_{h \to 0} 1 + h\sin(2/h) = 1$. Thus, f is differentiable at 0 and $f'(0) = 1 > 0$. Notice that, for $x \neq 0$, $f'(x) = 1 + 2x\sin(2/x) - 2\cos(2/x)$. Let $a > 0$. Take n large enough so that $x_1 = 1/(n\pi - \pi/4) < a$. Then $f'(x_1) = -1$. Also, f' is continuous on $(0, a)$ so, by Remark 3.6.13, there exists $\delta > 0$ such that $f'(x) < 0$ for all $x \in (x_1 - \delta, x_1 + \delta)$. By Remark 4.4.7, f is decreasing on $(x_1 - \delta, x_1 + \delta)$.

4.4.26. Since $f''(x) > 0$ for $x \in (a, b)$, Theorem 4.4.6 implies that f' is an increasing function on (a, b). Let $a \leq x < c < y \leq b$. By the Mean Value Theorem, there exist points $c_1 \in (x, c)$ and $c_2 \in (c, y)$ such that

$$\frac{f(c) - f(x)}{c - x} = f'(c_1) \quad \text{and} \quad \frac{f(y) - f(c)}{y - c} = f'(c_2).$$

The fact that f' is increasing implies that $f'(c_1) < f'(c_2)$, so

$$\frac{f(c) - f(x)}{c - x} < \frac{f(y) - f(c)}{y - c}.$$

If we multiply both sides by $(c - x)(y - c)$, and solve the inequality for $f(c)$, we obtain

$$f(c) < \frac{f(y)(c - x) + f(x)(y - c)}{y - x}.$$

Let $t \in [0, 1]$ and let $c = tx + (1-t)y$. Then $c - x = tx + (1-t)y - x = (1-t)(y-x)$, and $y - c = y - tx - (1-t)y = t(y - x)$, so

$$f(tx + (1-t)y) < \frac{f(y)(1-t)(y-x) + f(x)t(y-x)}{y-x} = tf(x) + (1-t)f(y).$$

4.5.1. Multiplying both the numerator and denominator by $1 + x$ and simplifying further leads to $f(x) = 1 + (2x + 2x^2)/(1 + x^3)$. Next, let $g(x) = (1 + x)^\alpha$. Notice that $g(0) = 1$ and $g'(0) = \alpha$, so $(1 + x)^\alpha = 1 + \alpha x + r_1(x)$. Here $\alpha = -1$, so $(1 + x^3)^{-1} = 1 - x^3 + r_1(x^3)$. It follows that $f(x) = 1 + (2x + 2x^2)(1 - x^3 + r_1(x^3))$, whence $p_4(x) = 1 + 2x + 2x^2 - 2x^4$.

4.5.3. Write $f(x) = \ln\sqrt{1 - \sin^2 x} = \frac{1}{2}\ln(1 - \sin^2 x)$. Next, let $g(x) = \ln(1+x)$. A calculation shows that $g(0) = 0$, $g'(0) = 1$, $g''(0) = -1$, and $g'''(0) = 2$, so $\ln(1 + x) = x - x^2/2 + x^3/3 + r_3(x)$. Combining with the formula $\sin x = x - x^3/3! + x^5/5! + r_5(x)$, we obtain that

$$f(x) = \frac{1}{2}\left(-\sin^2 x - \frac{\sin^4 x}{2} - \frac{\sin^6 x}{3} + r_3(-\sin^2 x)\right)$$

$$= -\frac{1}{2}\left[\left(x - \frac{x^3}{3!} + \frac{x^5}{5!} - r_5(x)\right)^2 + \frac{1}{2}\left(x - \frac{x^3}{3!} + \frac{x^5}{5!} - r_5(x)\right)^4 + \frac{1}{3}\left(x - \frac{x^3}{3!} + \frac{x^5}{5!} - r_5(x)\right)^6 + r_3(-\sin^2 x)\right].$$

Thus, $p_6(x) = -x^2/2 - x^4/12 - x^6/45$.

4.5.5. Use $t = x - 1$ and consider $g(t) = \sqrt{t+1}$. Then $g(0) = 1$, $g'(0) = 1/2$, $g''(0) = -1/4$ and $g'''(0) = 3/8$. It follows that $p_3(t) = 1 + t/2 - t^2/8 + t^3/16$ and $p_3(x) = 1 + (x-1)/2 - (x-1)^2/8 + (x-1)^3/16$.

4.5.7. Write $f(c + h) = f(c) + hf'(c) + (h^2/2)f''(c) + r_3(h)$ and $f(c - h) = f(c) - hf'(c) + (h^2/2)f''(c) + r_3(-h)$. Then $f(c + h) - 2f(c) + f(c - h) = h^2 f''(c) + r_3(h) + r_3(-h)$. Dividing by h^2 and letting h go to 0, we obtain that the given limit equals $f''(c)$.

4.5.9. If $f(x) = \tan x$, then $f'(x) = \sec^2 x$, $f''(x) = 2\tan x \sec^2 x$, $f'''(x) = 2\sec^4 x + 4\tan^2 x \sec^2 x$, $f^{(4)}(x) = 16\sec^4 x \tan x + 8\tan^3 x \sec^2 x$, and $f^{(5)}(x) = 88\sec^4 x \tan^2 x + 16\sec^6 x + 16\tan^4 x \sec^2 x$. Thus, $f(0) = f''(0) = f^{(4)}(0) = 0$, $f'(0) = 1$, and $f'''(0) = 2$. It follows that $\tan x = x + x^3/3 + r_4(x)$, where $r_4(x) = (f^{(5)}(x_0)/5!)x^5$, and $|x_0| \le 0.1$. Now,

$$f^{(5)}(x_0) = 8\frac{11\sin^2 x_0 + 2 + 2\sin^4 x_0}{\cos^6 x_0} \le 8\frac{11\sin^2 0.1 + 2 + 2\sin^4 0.1}{\cos^6 0.1} \approx 17.39356365, \text{ so}$$

$$|r_4(x)| \le \frac{17.39356365}{5!}(0.1)^5 \approx 1.45 \times 10^{-6}.$$

4.5.11. The difference between $\cos x$ and the given polynomial is exactly the remainder in Taylor's Theorem, $r_5(x)$. In Lagrange's form it is $(-\sin^6 x_0/6!)x^6$, so $|r_5(x)| \le x^{12}/6! \le (0.5)^{12}/6! < 3.4 \times 10^{-7}$.

4.5.13. In radians, $1° = \pi/180$, so $|r_n(x)| \le \pi^{n+1}/((n+1)!180^{n+1})$. Substituting $n = 1$ in the last expression yields approximately 1.5×10^{-4} which is too big. Similarly, for $n = 2$ we get 8.8×10^{-7}. Finally, for $n = 3$ we have $3.8 \times 10^{-9} \le 10^{-8}$, so we will use $n = 3$. Now, $\sin 1° = \sin(\pi/180) \approx \pi/180 - \pi^3/(180^3\, 3!) \approx 0.01745244$.

4.5.15. Write $\log_{10} 11 = \log_{10}(10 + 1) = \log_{10}(10(1 + 1/10)) = \log_{10} 10 + \log_{10}(1 + 1/10) = 1 + \ln(1 + 1/10)/\ln 10$. For the function $\ln(1 + x)$ the remainder in Lagrange's form is

$$r_n(x) = \frac{(-1)^n n!}{(1 + x_0)^{n+1}(n + 1)!}x^{n+1}.$$

Since $x = 0.1$ and $x_0 \ge 0$ we have that $|r_n(x)| \le (0.1)^{n+1}/n$. Testing various choices for n, we get that $n = 4$. Then, $\log_{10} 11 \approx 1 + (0.1 - 0.1^2/2 + 0.1^3/3 - 0.1^4/4)/\ln 10 \approx 1.04139$.

4.5.17. The function $\sin(\sin x)$ can be written as

$$\sin x - \frac{\sin^3 x}{3!} + \frac{\sin^5 x}{5!} + r_6(\sin x),$$

and $r_6(\sin x)$ is a multiple of $\sin^7 x$, so $r_6(\sin x)/x^5 \to 0$, as $x \to 0$. Similarly,

$$x\sqrt[3]{1 - x^2} = x\left(1 - \frac{x^2}{3} - \frac{x^4}{9} + r_3(x^2)\right)$$

and $r_3(x^2)$ is a multiple of x^6, so $r_3(x^2)/x^5 \to 0$, as $x \to 0$. Consequently, our problem is reduced to

$$\lim_{x \to 0} \frac{\sin x - \frac{1}{6}\sin^3 x + \frac{1}{120}\sin^5 x - x + \frac{1}{3}x^3 + \frac{1}{9}x^5}{x^5}.$$

Further, $\sin x = x - x^3/3! + x^5/5! + r_6(x)$, and $r_6(x)/x^5 \to 0$, as $x \to 0$, so we obtain

$$\lim_{x \to 0} \frac{\left[x - \frac{x^3}{3!} + \frac{x^5}{5!}\right] - \frac{1}{6}\left[x - \frac{x^3}{3!} + \frac{x^5}{5!}\right]^3 + \frac{1}{120}\left[x - \frac{x^3}{3!} + \frac{x^5}{5!}\right]^5 - x + \frac{1}{3}x^3 + \frac{1}{9}x^5}{x^5}$$

$$= \lim_{x \to 0} \frac{\frac{1}{6}x^3 + \frac{43}{360}x^5 - \frac{1}{6}x^3\left[1 - \frac{x^2}{3!} + \frac{x^4}{5!}\right]^3 + \frac{1}{120}x^5\left[1 - \frac{x^2}{3!} + \frac{x^4}{5!}\right]^5}{x^5}$$

$$= \lim_{x \to 0} \frac{\frac{1}{6} + \frac{43}{360}x^2 - \frac{1}{6}\left[1 - \frac{x^2}{3!} + \frac{x^4}{5!}\right]^3 + \frac{1}{120}x^2\left[1 - \frac{x^2}{3!} + \frac{x^4}{5!}\right]^5}{x^2}.$$

Notice,

$$1 - \left[1 - \frac{x^2}{3!} + \frac{x^4}{5!}\right]^3 = \left(\frac{x^2}{6} - \frac{x^4}{120}\right)\left[1 + \left(1 - \frac{x^2}{6} + \frac{x^4}{120}\right) + \left(1 - \frac{x^2}{6} + \frac{x^4}{120}\right)^2\right]$$

$$= x^2\left(\frac{1}{6} - \frac{x^2}{120}\right)\left[1 + \left(1 - \frac{x^2}{6} + \frac{x^4}{120}\right) + \left(1 - \frac{x^2}{6} + \frac{x^4}{120}\right)^2\right], \quad \text{so}$$

$$\lim_{x \to 0} \frac{\frac{1}{6} - \frac{1}{6}\left[1 - \frac{x^2}{3!} + \frac{x^4}{5!}\right]^3}{x^2} = \lim_{x \to 0} \frac{1}{6}\left(\frac{1}{6} - \frac{x^2}{120}\right)\left[1 + \left(1 - \frac{x^2}{6} + \frac{x^4}{120}\right) + \left(1 - \frac{x^2}{6} + \frac{x^4}{120}\right)^2\right] = \frac{1}{12}.$$

It follows that the desired limit equals $1/2 + 43/360 + 1/120 = 19/90$.

4.5.19. $x - x^2 \ln(1 + 1/x) = x - x^2(1/x - 1/(2x^2) + r_2(1/x)) = 1/2 + x^2 r_2(1/x) \to 1/2$, when $x \to \infty$.

4.6.1. Since $\lim_{x \to 0}(\tan x - x) = \lim_{x \to 0}(x - \sin x) = 0$, and $(\tan x - x)' = \sec^2 x - 1$, $(x - \sin x)' = 1 - \cos x$, we see that the given limit equals

$$\lim_{x \to 0} \frac{\sec^2 x - 1}{1 - \cos x} = \lim_{x \to 0} \frac{1 - \cos^2 x}{\cos^2 x(1 - \cos x)} = \lim_{x \to 0} \frac{1 + \cos x}{\cos^2 x} = 2.$$

4.6.3. First, $\lim_{x \to \pi/4}(\sqrt[3]{\tan x} - 1) = \lim_{x \to \pi/4}(2\sin^2 x - 1) = 0$. Next, $(\sqrt[3]{\tan x} - 1)' = (1/3)\tan^{-2/3} x \sec^2 x$ and $(2\sin^2 x - 1)' = 4\sin x \cos x$. Finally, $\lim_{x \to \pi/4}(1/3)\tan^{-2/3} x \sec^2 x = 2/3$ and $\lim_{x \to \pi/4} 4\sin x \cos x = 2$. Answer: $1/3$.

4.6.5. First, $\lim_{x \to +\infty} \ln x = +\infty$ and $\lim_{x \to +\infty} x^2 = +\infty$. Next, $(\ln x)' = 1/x$ and $(x^2)' = 2x$. Finally, $(1/x)/2x = 1/(2x^2) \to 0$, as $x \to +\infty$.

4.6.7. First, $\lim_{x \to \pi/2-}(3 + 4\sec x) = \lim_{x \to \pi/2-}(2 + \tan x) = \infty$. Next, $(3 + 4\sec x)' = 4\tan x \sec x$ and $(2 + \tan x)' = \sec^2 x$. Finally, $4\tan x \sec x/\sec^2 x = 4\sin x \to 4$, as $x \to \pi/2-$.

4.6.9. Write $(\cot x)^{\sin x} = e^{\sin x \ln \cot x}$. Now, $\lim_{x \to 0} \sin x = 0$ and $\lim_{x \to 0} \ln \cot x = \infty$. Therefore, we write $\sin x \ln \cot x = \ln \cot x/\csc x$. Next, $(\ln \cot x)' = -\tan x \csc^2 x$ and $(\csc x)' = -\cot x \csc x$. Finally, $(-\tan x \csc^2 x)/(-\cot x \csc x) = \tan x \sec x \to 0$, as $x \to 0$. Answer: 1.

4.6.11. Write $x^x = e^{x \ln x}$. Now, $\lim_{x \to 0+} x = 0$ and $\lim_{x \to 0+} \ln x = -\infty$. Therefore, we write $x \ln x = \ln x/(1/x)$. Next, $(\ln x)' = 1/x$ and $(1/x)' = -1/x^2$. Finally, $(1/x)/(-1/x^2) = -x \to 0$, as $x \to 0^+$. Answer: 1.

4.6.13. First, $\lim_{x \to 0}(1 + x)^{1/x} - e = \lim_{x \to 0} x = 0$. Next,

$$\left((1 + x)^{1/x}\right)' = \left(e^{\ln(1+x)/x}\right)' = (1 + x)^{1/x} \frac{\frac{x}{1+x} - \ln(1 + x)}{x^2}.$$

Further, $\lim_{x \to 0} x/(1 + x) - \ln(1 + x) = \lim_{x \to 0} x^2 = 0$. Taking derivatives we get

$$\frac{\frac{1}{(1+x)^2} - \frac{1}{1+x}}{2x} = \frac{-1}{2(1+x)^2} \to -\frac{1}{2}.$$

Answer: $-e/2$.

4.6.15. First,

$$(\cos x)^{1/x^2} = e^{\ln(\cos x)^{1/x^2}} = e^{\frac{\ln \cos x}{x^2}},$$

and $\lim_{x \to 0} \ln \cos x = \lim_{x \to 0} x^2 = 0$, so Theorem 4.6.1 can be used. We obtain

$$\lim_{x \to 0} \frac{\ln \cos x}{x^2} = \lim_{x \to 0} \frac{\frac{1}{\cos x}(-\sin x)}{2x} = \lim_{x \to 0} \frac{\sin x}{x} \frac{-1}{2\cos x} = -\frac{1}{2}.$$

Thus, $\lim_{x \to 0}(\cos x)^{1/x^2} = e^{-1/2}$.

4.6.17. If we take the natural logarithm of the given function, we have

$$\ln\left(\frac{x^{\ln x}}{(\ln x)^x}\right) = \ln\left(x^{\ln x}\right) - \ln\left((\ln x)^x\right) = (\ln x)^2 - x \ln(\ln x) = x\left[\frac{(\ln x)^2}{x} - \ln(\ln x)\right].$$

Next, we calculate $\lim_{x\to\infty} \frac{(\ln x)^2}{x}$. It is of the form $\left(\frac{\infty}{\infty}\right)$ so we take the derivatives of the numerator and the denominator. This leads to $\lim_{x\to\infty} \frac{2\ln x}{x}$ which is again of the form $\left(\frac{\infty}{\infty}\right)$. Taking derivatives again, we obtain $\lim_{x\to\infty} \frac{2}{x} = 0$. It follows that

$$\lim_{x\to\infty} x\left[\frac{(\ln x)^2}{x} - \ln(\ln x)\right] = -\infty, \quad \text{so} \quad \lim_{x\to\infty}\left(\frac{x^{\ln x}}{(\ln x)^x}\right) = \lim_{x\to\infty} e^{\left[\frac{(\ln x)^2}{x} - \ln(\ln x)\right]} = 0.$$

4.6.19. This is an indeterminate form 1^∞. Let $f(x) = (a^x - x\ln a)/(b^x - x\ln b)$. Then $f(x)^{1/x^2} = e^{\ln f(x)/x^2}$ and the exponent is now of the form $0/0$. Next, $(\ln f(x))' = f'(x)/f(x)$, so the task becomes to compute $\lim_{x\to 0} f'(x)/(2xf(x))$. Further, $\lim_{x\to 0} f(x) = 1$, so we focus on $\lim_{x\to 0} f'(x)/x$. We have that,

$$\frac{f'(x)}{x} = \frac{(a^x - 1)\ln a(b^x - x\ln b) - (b^x - 1)\ln b(a^x - x\ln a)}{x(a^x - x\ln a)(b^x - x\ln b)}$$

$$= \frac{1}{(a^x - x\ln a)(b^x - x\ln b)}\left(\frac{a^x - 1}{x}\ln a(b^x - x\ln b) - \frac{b^x - 1}{x}\ln b(a^x - x\ln a)\right).$$

Now use Example 3.1.14. Answer: $e^{(\ln^2 a - \ln^2 b)/2}$.

4.6.21. Write $1/\sin^2 x - 1/x^2 = (x^2 - \sin^2 x)/(\sin^2 x x^2)$. Now, $\lim_{x\to 0}(x^2 - \sin^2 x) = \lim_{x\to 0}(\sin^2 x x^2) = 0$. Next, $(x^2 - \sin^2 x)' = 2x - 2\sin x\cos x$ and $(\sin^2 x x^2)' = 2\sin x\cos x x^2 + \sin^2 x 2x$. Again, $\lim_{x\to 0}(2x - 2\sin x\cos x) = \lim_{x\to 0}(2\sin x\cos x x^2 + \sin^2 x 2x) = 0$. Calculate derivatives again: $(2x - 2\sin x\cos x)' = 4\sin^2 x$ and $(2\sin x\cos x x^2 + \sin^2 x 2x)' = 2x^2(1 - 2\sin^2 x) + 8x\sin x\cos x + 2\sin^2 x$. Now use $\lim_{x\to 0}\sin x/x = 1$. Answer: $1/3$.

4.6.23. We use algebra:

$$\frac{1}{\ln(x + \sqrt{1 + x^2})} - \frac{1}{\ln(1 + x)} = \frac{\ln(1 + x) - \ln(x + \sqrt{1 + x^2})}{\ln(x + \sqrt{1 + x^2})\ln(1 + x)}$$

which is of the form $\left(\frac{0}{0}\right)$, as $x \to 0$. Taking the derivative of the numerator and the denominator yields

$$\frac{\frac{1}{1+x} - \frac{1}{\sqrt{1+x^2}}}{\frac{1}{\sqrt{1+x^2}}\ln(1 + x) + \frac{1}{1+x}\ln(x + \sqrt{1 + x^2})} = \frac{\sqrt{1 + x^2} - (1 + x)}{(1 + x)\ln(1 + x) + \sqrt{1 + x^2}\ln(x + \sqrt{1 + x^2})}$$

which is again of the form $\left(\frac{0}{0}\right)$. Taking derivatives we obtain

$$\frac{\frac{x}{\sqrt{1+x^2}} - 1}{\ln(1 + x) + (1 + x)\frac{1}{1+x} + \sqrt{1 + x^2}\frac{1}{\sqrt{1+x^2}} + \frac{x}{\sqrt{1+x^2}}\ln(x + \sqrt{1 + x^2})}$$

which has the limit $-1/2$, as $x \to 0$.

5.1.1. Notice that $(x + 1)/\sqrt{x} = \sqrt{x} + 1/\sqrt{x} = x^{1/2} + x^{-1/2}$. Thus, $\int((x + 1)/\sqrt{x})\,dx = 2x^{3/2}/3 + 2x^{1/2} + C$.

5.1.3. Use the substitution $u = x\sqrt{3/2}$. Then $dx = du\sqrt{2/3}$, so we obtain $\int du\sqrt{2/3}/(2 + 2u^2) = 1/\sqrt{6}\arctan u + C = 1/\sqrt{6}\arctan(x\sqrt{3/2}) + C$.

5.1.5. Hint: Use $u = \sqrt{x^2 + 1}$. Answer: $\frac{1}{2}\ln\frac{\sqrt{x^2 + 1} + 1}{\sqrt{x^2 + 1} - 1} + C$.

5.1.7. Hint: Write as $\int \frac{\sec^2 x}{\tan^2 x + 2}\,dx$ and use $u = \tan x$. Answer: $\frac{1}{\sqrt{2}}\arctan\frac{\tan x}{\sqrt{2}} + C$.

5.1.9. Hint: Write as $\int \frac{\left(\frac{3}{2}\right)^x}{\left(\frac{9}{4}\right)^x - 1}\,dx$ and use $u = \left(\frac{3}{2}\right)^x$. Answer: $\frac{1}{2\ln\frac{3}{2}}\ln\frac{3^x - 2^x}{3^x + 2^x} + C$.

5.1.11. Hint: Use $\sin^2 x = (1 - \cos 2x)/2$. Answer: $(3x/2 - \sin 2x + \sin 4x/8)/4 + C$.

5.1.13. Hint: Use $u = e^x$. Answer: $\ln\frac{e^x}{1 + e^x} + C$.

5.1.15. Answer: $x/\sqrt{1 - x^2} + C$.

5.1.17. Answer: $\frac{1}{3}x^3\arccos x - \frac{1}{9}(x^2 + 2)\sqrt{1 - x^2} + C$.

5.1.19. Hint: Use Integration by Parts with $u = \ln(x + \sqrt{1 + x^2})$ and $dv = \frac{x}{\sqrt{1 + x^2}}\,dx$. Answer: $\sqrt{1 + x^2}\ln(x + \sqrt{1 + x^2}) - x + C$.

5.2.1. Use $u = \sqrt{\sin x}$. Then $du = (1/2\sqrt{\sin x})\cos x\,dx$, so $\cos x\,dx = 2u\,du$. Also, $\cos^4 x = (1 - u^4)^2$. Answer: $(2/3)\sin x\sqrt{\sin x} - (4/7)\sin^3 x\sqrt{\sin x} + (2/11)\sin^5 x\sqrt{\sin x} + C$.

5.2.3. Use $u = \arctan\sqrt{x}$. Then $du = 1/((1 + x)2\sqrt{x})\,dx$. Answer: $(\arctan\sqrt{x})^2 + C$.

5.2.5. Use $x = a\sin t$. Then $dx = a\cos t\, dt$ and $\sqrt{a^2 - x^2} = \sqrt{a^2 - a^2\sin^2 t} = a\cos t$. Answer: $\arcsin(x/a) + C$.

5.2.7. Use $x = a\cosh t$. Then $dx = a\sinh t\, dt$ and $\sqrt{x^2 - a^2} = \sqrt{a^2\cosh^2 t - a^2} = a\sinh t$. Now $\int \sqrt{x^2 - a^2}\, dx = a^2 \int \sinh^2 t\, dt = a^2 \int \frac{1 + \cosh 2t}{2}\, dt = \frac{a^2}{2}(t + \frac{\sinh 2t}{2})\, dt$. Finally, $\sinh 2t = 2\cosh t \sinh t = 2\cosh t\sqrt{\cosh^2 t - 1} = \frac{2x}{a}\sqrt{\frac{x^2}{a^2} - 1}$. Answer: $\frac{a^2}{2}[(\operatorname{arcosh} \frac{x}{a}) + \frac{2x}{a}\sqrt{\frac{x^2}{a^2} - 1}] = \frac{a^2}{2}[\ln(\frac{x}{a} + \sqrt{\frac{x^2}{a^2} - 1}) + \frac{2x}{a}\sqrt{\frac{x^2}{a^2} - 1}]$.

5.2.9. Use $u = \arcsin x$ and $dv = dx$. Then $du = dx/\sqrt{1 - x^2}$ and $v = x$, so we obtain $x\arcsin x - \int x\, dx/\sqrt{1 - x^2}$. The last integral can be calculated using the substitution $z = 1 - x^2$. Answer: $x\arcsin x + \sqrt{1 - x^2} + C$.

5.2.11. Use $u = x$ and $dv = \cos x\, dx$. Then $du = dx$ and $v = \sin x$, so we obtain $x\sin x - \int \sin x\, dx = x\sin x + \cos x + C$.

5.2.13. Use $u = x$ and $dv = f'(x)\, dx$. Then $du = dx$ and $v = f(x)$, so we obtain $xf'(x) - \int f'(x)\, dx = xf'(x) - f(x) + C$.

5.2.15. Use $u = x^2$. Then $f'(u) = 1/\sqrt{u}$, so $f(u) = 2\sqrt{u} + C$.

5.2.17. Let us denote the function f^{-1} by g. Then $x = f(g(x)) = F'(g(x))$. It follows that $\int xg'(x)\, dx = \int F'(g(x))g'(x)\, dx = F(g(x)) + C$. On the other hand, using integration by parts, $\int xg'(x)\, dx = xg(x) - \int g(x)\, dx$.

5.2.19. Write $x/[(x+1)(x+2)(x+3)] = A/(x+1) + B/(x+2) + C/(x+3)$, which leads to $A(x+2)(x+3) + B(x+1)(x+3) + C(x+1)(x+2) = x$. Substituting $x = -1$, $x = -2$ and $x = -3$ yields $A = -1/2$, $B = 2$ and $C = -3/2$. Answer: $-\frac{1}{2}\ln|x+1| + 2\ln|x+2| - \frac{3}{2}\ln|x+3| + C$.

5.2.21. $\dfrac{x^{10}}{x^2 + x - 2} = x^8 - x^7 + 3x^6 - 5x^5 + 11x^4 - 21x^3 + 47x^2 - 85x + 171 - \dfrac{348x - 342}{(x-1)(x+2)}$. Answer: $\frac{1}{9}x^9 - \frac{1}{8}x^8 + \frac{3}{7}x^7 - \frac{5}{6}x^6 + \frac{11}{5}x^5 - \frac{21}{4}x^4 + \frac{43}{3}x^3 - \frac{85}{2}x^2 + 171x - \frac{1024}{3}\ln(x+2) + \frac{1}{3}\ln(x-1) + C$.

5.2.23. Since $x^3 + 1 = (x+1)(x^2 - x + 1)$, write $1/(x^3 + 1) = A/(x+1) + (Bx + C)/(x^2 - x + 1)$. This leads to $A(x^2 - x + 1) + (Bx + C)(x + 1) = 1$. Substituting $x = -1$ yields $A = 1/3$, while substituting $x = 0$ and $x = 1$ yields $A + C = 1$ and $A + 2(B + C) = 1$. It follows that $B = -1/3$ and $C = 2/3$. Since $\int \frac{(1/3)}{x+1}\, dx = \frac{1}{3}\ln|x+1| + C$, we focus on the other fraction. Notice that $x^2 - x + 1 = (x - 1/2)^2 + 3/4$ and $(-1/3)x + 2/3 = (-1/3)(x - 1/2) + 1/2$. Answer: $\frac{1}{3}\ln|x+1| - \frac{1}{6}\ln(x^2 - x + 1) + \frac{1}{\sqrt{3}}\arctan\frac{2x - 1}{\sqrt{3}} + C$.

5.2.25. Hint: Use $u = x - 1/x$. Answer: $\frac{\sqrt{3}}{3}\arctan\frac{(2x+1)\sqrt{3}}{3} + \frac{\sqrt{3}}{3}\arctan\frac{(2x-1)\sqrt{3}}{3} + C$.

5.2.27. Write $x^2/(x^2 + 2x + 2)^2 = (Ax + B)/(x^2 + 2x + 2) + (Cx + D)/(x^2 + 2x + 2)^2$. This leads to $x^2 = (Ax + B)(x^2 + 2x + 2) + (Cx + D) = Ax^3 + (2A + B)x^2 + (2A + 2B + C)x + (2B + D)$. It follows that $A = 0$, $B = 1$, $2B + C = 0$ and $2B + D = 0$, so $C = D = -2$. Since $x^2 + 2x + 2 = (x + 1)^2 + 1$ we have that $\int dx/(x^2 + 2x + 2) = \arctan(x + 1) + C$. Also, $-2x - 2 = -2(x + 1)$, so $\int(-2x - 2)\, dx/(x^2 + 2x + 2)^2 = 1/(x^2 + 2x + 2)$.

5.2.29. $x^4 + 1 = x^4 + 2x^2 + 1 - (x\sqrt{2})^2 = (x^2 + x\sqrt{2} + 1)(x^2 - x\sqrt{2} + 1)$. Therefore, $1/(x^4 + 1) = (\frac{1}{2\sqrt{2}}x + \frac{1}{2})/(x^2 + x\sqrt{2} + 1) - (\frac{1}{2\sqrt{2}}x - \frac{1}{2})/(x^2 - x\sqrt{2} + 1)$. Answer: $\frac{\sqrt{2}}{8}\ln\frac{x^2 + x\sqrt{2} + 1}{x^2 - x\sqrt{2} + 1} + \frac{\sqrt{2}}{4}\arctan(x\sqrt{2} + 1) + \frac{\sqrt{2}}{4}\arctan(x\sqrt{2} - 1) + C$.

5.2.31. Write $(ax^2 + bx + c)/[x^3(x - 1)^2] = A/x + B/x^2 + C/x^3 + D/(x - 1) + E/(x - 1)^2$. This leads to $ax^2 + bx + c = Ax^2(x - 1)^2 + Bx(x - 1)^2 + C(x - 1)^2 + D(x - 1)x^3 + Ex^3 = (A + D)x^4 + (-2A + B - D + E)x^3 + (A - 2B + C)x^2 + (B - 2C)x + C$. Solving for A, B, C, D, E yields $A = a + 2b + 3c$, $B = b + 2c$, $C = c$, $D = -(a + 2b + 3c)$ and $E = a + b + c$. In order for the integral to be a rational function we must have $A = D = 0$, so we obtain the condition $a + 2b + 3c = 0$.

5.2.33. Use integration by parts with $u = (ax^2 + bx + c)^{-n}$ and $dv = dx$. Then $I_n = x(ax^2 + bx + c)^{-n} + n\int(2ax + b)x(ax^2 + bx + c)^{-n-1}\, dx$. Write $(2ax + b)x = 2(ax^2 + bx + c) - (bx + 2c)$, so the last integral equals $2I_n - \int(bx + 2c)(ax^2 + bx + c)^{-n-1}\, dx$. Next, write $bx + 2c = (b/(2a))(2ax + b) + (2c - b^2/(2a))$. Now, the latest integral becomes $(b/(2a))(-1/n)(ax^2 + bx + c)^{-n} + (2c - b^2/(2a))I_{n+1}$. Thus, we have the equation $I_n = x(ax^2 + bx + c)^{-n} + 2nI_n + (b/(2a))(ax^2 + bx + c)^{-n} - n(2c - b^2/(2a))I_{n+1}$. Solving for I_{n+1} yields $I_{n+1} = [(1 - 2n)/n]I_n[2a/(b^2 - 4ac)] - (2ax + b)/[(b^2 - 4ac)n(ax^2 + bx + c)^{-n}]$.

When $a = b = c = 1$ and $n = 2$, we have

$$\int \frac{dx}{(x^2 + x + 1)^3} = \frac{2x + 1}{6(x^2 + x + 1)^2} + \int \frac{dx}{(x^2 + x + 1)^2} = \frac{2x + 1}{6(x^2 + x + 1)^2} + \frac{2x + 1}{3(x^2 + x + 1)} + \frac{2}{3}\int \frac{dx}{x^2 + x + 1}$$

$$= \frac{2x + 1}{6(x^2 + x + 1)^2} + \frac{2x + 1}{3(x^2 + x + 1)} + \frac{4}{3\sqrt{3}}\arctan\frac{2x + 1}{\sqrt{3}} + C.$$

5.2.35. Hint: Use $x = t^6$. Answer:

$$\ln x - \frac{3}{2}\ln\left(\sqrt[6]{x} + 1\right) - \frac{9}{4}\ln\left(2\sqrt[3]{x} - \sqrt[6]{x} + 1\right) - \frac{3\sqrt{7}}{14}\arctan\left(\frac{\sqrt{7}}{7}\left(4\sqrt[6]{x} - 1\right)\right) + C.$$

5.2.37. Use $u = \sqrt[4]{x} + 1$. Then $x = (u - 1)^4$ and $dx = 4(u - 1)^2\, du$. Answer: $4\ln(1 + \sqrt[4]{x}) + 4/(1 + \sqrt[4]{x}) + C$.

5.2.39. Hint: Use $u = \sqrt[4]{\dfrac{x}{1 - x}}$. Answer:

$$\frac{\sqrt{2}}{8}\ln\frac{\sqrt{\frac{x}{1-x}} + \sqrt[4]{\frac{x}{1-x}}\sqrt{2} + 1}{\sqrt{\frac{x}{1-x}} - \sqrt[4]{\frac{x}{1-x}}\sqrt{2} + 1} + \frac{\sqrt{2}}{4}\arctan\left(\sqrt[4]{\frac{x}{1-x}}\sqrt{2} + 1\right)$$

$$+ \frac{\sqrt{2}}{4}\arctan\left(\sqrt[4]{\frac{x}{1-x}}\sqrt{2} - 1\right) + x\sqrt[4]{\frac{x}{1-x}} - \sqrt[4]{\frac{x}{1-x}} + C.$$

5.2.41. Since $1 - \frac{1}{x-2} = \frac{x-3}{x-2}$ we use $t = \sqrt[3]{\frac{x-3}{x-2}}$. Then $x = (2t^3 - 3)/(t^3 - 1)$ so $dx = 3t^2/(t^3 - 1)^2\, du$, and we obtain $\int 3t^3/(t^3 - 1)^2\, dt$. Partial fraction decomposition yields

$$\int \left(\frac{t+1}{(t^2+t+1)^2} + \frac{1}{3(t-1)^2} + \frac{1}{3(t-1)} - \frac{t+3}{3(t^2+t+1)} \right) dx.$$

Answer:

$$\frac{t-1}{3(t^2+t+1)} - \frac{1}{\sqrt{3}} \arctan \frac{2t+1}{\sqrt{3}} - \frac{1}{3(t-1)} + \frac{\ln(t-1)}{3} - \frac{\ln(t^2+t+1)}{6} + C,$$

where $t = \sqrt[3]{\frac{x-3}{x-2}}$.

5.2.43. Use $x = 1/t$. Answer: $-\sqrt{1 + 1/x^2} + C$.

5.2.45. Write $x^2/\sqrt{x^2+1} = \sqrt{x^2+1} - 1/\sqrt{x^2+1}$, and use Example 5.1.8 and Exercise 5.2.4. Answer:

$$\frac{1}{2}x\sqrt{x^2+1} - \frac{1}{2}\ln|x + \sqrt{x^2+1}| + C.$$

5.2.47. Write $x^2/(1-x^2)^{3/2} = -1/\sqrt{1-x^2} + 1/(1-x^2)^{3/2}$ and for the integral of the second fraction use substitution $x = 1/t$. Answer: $-\arcsin x + x/\sqrt{1-x^2} + C$.

5.2.49. Use $u = x + \sqrt{x^2 - x + 1}$. Then $x = (u^2 - 1)/(2u - 1)$ and $dx = (2u^2 - 2u + 2)/(2u - 1)^2\, du$. Answer: $2\ln|x + \sqrt{x^2 - x + 1}| - \frac{3}{2}\ln|2x - 1 + 2\sqrt{x^2 - x + 1}| - \frac{3}{2}(2x - 1 + 2\sqrt{x^2 - x + 1})^{-1} + C$.

5.2.51. Use $x = t^4$. Obtain $\int 4t^4\, dt/(1 + t^2)^3 = 4I_1 - 8I_2 + 4I_3$, where I_n is as in Example 5.1.11. Answer: $\sqrt[4]{x}/(\sqrt{x} + 1)^2 - (5/2)\sqrt[4]{x}/(\sqrt{x} + 1) + (3/2)\arctan\sqrt[4]{x} + C$.

5.2.53. Use $u = \sqrt[3]{1 + x^5}$. Obtain $\int 3u^2\, du/(u(u^3 - 1))$. Answer: $(3/5)\ln(\sqrt[3]{1+x^5} - 1) - (1/10)\ln((\sqrt[3]{1+x^5})^2 + \sqrt[3]{1+x^5} + 1) + (\sqrt{3}/5)\arctan(2\sqrt[3]{1+x^5} + 1)/\sqrt{3} + C$.

5.2.55. Use $u = \sqrt{1 + \sqrt[3]{x}}$. Obtain $3\int(u^2 - 1)^5\, du/u$. Answer: $(6/11)(1 + \sqrt[3]{x})^{11/2} - (10/3)(1 + \sqrt[3]{x})^{9/2} + (60/7)(1 + \sqrt[3]{x})^{7/2} - 12(1 + \sqrt[3]{x})^{3/2} - 6(1 + \sqrt[3]{x})^{1/2} + C$.

5.2.57. Use $x = 1/t$ and Exercise 5.2.4. Answer: $(\sqrt{2}/2)\ln|x\sqrt{2}/(\sqrt{2 + x^2} + \sqrt{2})| + C$.

5.2.59. Use $u = \sqrt{1 - x^2}$. Obtain $-\int(u^2 - 1)^2\, du$. Answer: $-(1 - x^2)^{5/2}/5 + 2(1 - x^2)^{3/2}/3 - \sqrt{1 - x^2} + C$.

5.2.61. Hint: Use $u = \sqrt{1 + \sqrt[3]{x^2}}$. Answer: $\frac{1}{5}\sqrt{1 + \sqrt[3]{x^2}}\left(3\sqrt[3]{x^4} - 4\sqrt[3]{x^2} + 8\right) + C$.

5.2.63. Hint: Use $u = \sqrt[3]{3 - x^2/x^2}$. Answer:

$$-\frac{1}{2}\ln\left(\sqrt[3]{\frac{3-x}{x^2}} + 1\right) + \frac{1}{2}\frac{\sqrt[3]{\frac{3-x}{x^2}} + 1}{\left(\sqrt[3]{\frac{3-x}{x^2}}\right)^2 - \sqrt[3]{\frac{3-x}{x^2}} + 1} - \frac{1}{2}\arctan\frac{\sqrt{3}}{3}\left(2\sqrt[3]{\frac{3-x}{x^2}} - 1\right)$$

$$+ \frac{1}{4}\ln\left(\left(\sqrt[3]{\frac{3-x}{x^2}}\right)^2 - \sqrt[3]{\frac{3-x}{x^2}} + 1\right) - \frac{1}{2\left(\sqrt[3]{\frac{3-x}{x^2}} + 1\right)} + C.$$

5.2.65. Hint: Use $u = \sqrt[3]{1 + x^6}/x$.

5.2.67. Hint: Use $u = \tan\frac{x}{2}$. Answer:

$$\frac{1}{4}\ln\frac{\tan^2\frac{x}{2} - \tan\frac{x}{2} + 1}{\tan^2\frac{x}{2} + \tan\frac{x}{2} + 1} + \frac{4\sqrt{3}}{3}\arctan\frac{2\tan\frac{x}{2} - 1}{\sqrt{3}} - \frac{4\sqrt{3}}{3}\arctan\frac{2\tan\frac{x}{2} + 1}{\sqrt{3}} + C.$$

5.2.69. Hint: Use $u = \tan x$. Answer: $\frac{ax + b\ln(a\cos x + b\sin x)}{a^2 + b^2} + C$, for $x \neq \frac{\pi}{2} + k\pi$, $k \in \mathbb{Z}$.

5.2.71. Use $u = \cos x$. Obtain $-\int(1 - u^2)^2\, du/u^4$. Answer: $1/(3\cos^3 x) - 2/\cos x - \cos x + C$.

5.2.73. Use $u = \sin x$. Obtain $\int u^2(1 - u^2)\, du$. Answer: $\sin^3 x/3 - \sin^5 x/5 + C$.

5.2.75. Use $u = \sin x$. Obtain $\int du/(1 - u^2)^3$. Answer: $(3/16)\ln|(1 + \sin x)/(1 - \sin x)| + (1/4)\sin x/\cos^4 x + (3/8)\sin x/\cos^2 x + C$.

5.2.77. Use $u = \tan x$. Obtain $\int u\, du/((a + bu)(1 + u^2))$. Answer: $\frac{a}{2(a^2 + b^2)}\ln|1 + \tan^2 x| - \frac{a}{a^2 + b^2}\ln|a + b\tan x| + \frac{bx}{a^2 + b^2} + C$.

5.2.79. Use $u = \tan(x/2)$. Obtain $\int 4u(1 - u^2)\, du/((1 - 2u + u^2)(1 + u^2)^2)$. Answer: $\cos(x/2)(\sin(x/2) - \cos(x/2)) + (\sqrt{2}/4)\ln|1 + \sqrt{2} - \tan(x/2)| - (\sqrt{2}/4)\ln|1 - \sqrt{2} - \tan(x/2)| + C$.

5.2.81. Answer: $2\arctan\left(\frac{1+r}{1-r}\tan\frac{x}{2}\right) + C$.

6.1.1. Use the substitution $u = x + \frac{1}{2}$. Obtain

$$\int_{-1/2}^{3/2} \frac{u - \frac{1}{2}}{u^2 + \frac{3}{4}} \, du = \left[\frac{1}{2} \ln\left(u^2 + \frac{3}{4}\right) - \frac{1}{2}\frac{2}{\sqrt{3}} \arctan \frac{2u}{\sqrt{3}} \right]\Bigg|_{-1/2}^{3/2}$$

$$= \left[\frac{1}{2} \ln 3 - \frac{1}{\sqrt{3}} \frac{\pi}{3} \right] - \left[-\frac{1}{\sqrt{3}}\left(-\frac{\pi}{6}\right) \right] = \frac{1}{2} \ln 3 - \frac{\pi}{2\sqrt{3}}.$$

6.1.3. Use Integration by Parts with $u = \arcsin\sqrt{\frac{x}{1-x}}$ and $dv = dx$. Obtain

$$x \arcsin\sqrt{\frac{x}{1-x}} \Big|_0^3 - \int_0^3 \frac{\sqrt{x}\,dx}{2(1+x)} = 3 \arcsin\frac{\sqrt{3}}{2} - \left(\sqrt{x} - \arctan\sqrt{x}\right)\Big|_0^3$$

$$= 3\left(\frac{\pi}{3}\right) - \left(\sqrt{3} - \arctan\sqrt{3}\right) = \frac{4\pi}{3} - \sqrt{3}.$$

6.1.5. $\int_1^{e^{2n\pi}} \frac{|\sin \ln x|}{x} \, dx = \int_0^{2n\pi} |\sin t| \, dt = n \int_0^{2\pi} |\sin t| \, dt = 2n \int_0^{\pi} \sin t \, dt = 4n.$

6.1.7. Using partial fraction decomposition, $1/(x^2 + x - 2) = (1/3)/(x-1) - (1/3)1/(x+2)$. Therefore, an antiderivative of $1/(x^2 + x - 2)$ is $(1/3)\ln|(x-1)/(x+2)|$. Finally,

$$\int_2^{\infty} \frac{dx}{x^2 + x - 2} = \lim_{b \to \infty} \frac{1}{3} \ln\left|\frac{x-1}{x+2}\right|\Big|_2^b = \frac{\ln 4}{3}.$$

6.1.9. Use the identity $\cos x \sin y = \frac{1}{2}(\sin(x+y) - \sin(x-y))$ and Example 6.1.5. Answer: 0.

6.1.11. Write

$$\int_{-1}^1 \frac{dx}{\sqrt{1-x^2}} = \lim_{a \to -1} \lim_{b \to 1} \int_a^b \frac{dx}{\sqrt{1-x^2}} = \lim_{a \to -1} \lim_{b \to 1} \arcsin x\Big|_a^b = \lim_{a \to -1} \lim_{b \to 1} (\arcsin b - \arcsin a)$$

$$= \arcsin 1 - \arcsin(-1) = \pi.$$

6.1.13. Use substitution $x = \pi - t$. Since $\sin(\pi - t) = \sin t$ and $\cos(\pi - t) = -\cos t$, the given integral I satisfies

$$I = \int_0^{\pi} \frac{(\pi - t)\sin t}{1 + \cos^2 t} \, dt = \pi \int_0^{\pi} \frac{\sin t}{1 + \cos^2 t} \, dt - \int_0^{\pi} \frac{t \sin t}{1 + \cos^2 t} \, dt = -\arctan(\cos t)|_0^{\pi} - I.$$

Answer: $\pi/4$.

6.1.15. Use integration by parts with $u = \sin^{n-1} x$ and $dv = \sin x \, dx$. Then $du = (n-1)\sin^{n-2} x \cos x \, dx$ and $v = -\cos x$. Therefore, the given integral I_n satisfies

$$I_n = -\cos x \sin^{n-1} x\Big|_0^{\pi/2} + \int_0^{\pi/2} (n-1)\sin^{n-2} x \cos^2 x \, dx = (n-1)\int_0^{\pi/2} \left(\sin^{n-2} x - \sin^n x\right) dx$$

$$= (n-1)I_{n-2} - (n-1)I_n,$$

so $I_n = \frac{n-1}{n} I_{n-2}$. Since $I_0 = \pi/2$ and $I_1 = 1$, we obtain that

$$I_{2k} = \frac{(2k-1)!!}{(2k)!!} \frac{\pi}{2}, \text{ and } I_{2k+1} = \frac{(2k)!!}{(2k+1)!!}.$$

6.2.1. The function f is decreasing for $x < 1/4$ and increasing for $x > 1/4$. Therefore, $m_1 = f(1/4) = -1/8$, $m_2 = f(1/3) = -1/9$, $m_3 = f(1) = 1$, $M_1 = f(0) = 0$, $M_2 = f(1) = 1$, and $M_3 = f(3/2) = 3$. Answer: $L(f, P) = 83/216$, $U(f, P) = 13/6$.

6.2.3. Answer: $L(f, P) = 97/12$, $U(f, P) = 41/4$.

6.2.5. Let $P_1 = \{x_0, x_1, \ldots, x_n\}$. It suffices to prove the assertion in the case when P_2 has exactly one additional partition point c. Suppose that $x_k \leq c \leq x_{k+1}$. Since $L(f, P_1)$ and $L(f, P_2)$ differ only in the terms corresponding to the interval $[x_k, x_{k+1}]$, we focus on them. In $L(f, P_1)$ there is only one term, and it equals $(b-a)\min\{f(x) : x_k \leq x \leq x_{k+1}\}$. In $L(f, P_2)$, these terms are $(c-a)\min\{f(x) : x_k \leq x \leq c\} + (b-c)\min\{f(x) : c \leq x \leq x_{k+1}\}$. Writing, $m_k = \min\{f(x) : x_k \leq x \leq x_{k+1}\}$, $m_k' = \min\{f(x) : x_k \leq x \leq c\}$ and $m_k'' = \min\{f(x) : c \leq x \leq x_{k+1}\}$, we have that

$$(b-a)m_k = (c-a)m_k + (b-c)m_k \leq (c-a)m_k' + (b-c)m_k''.$$

6.2.7. If $[c, d]$ is any subinterval of $[a, b]$, then

$$\sup_{c \leq x \leq d} [f(x) + g(x)] \leq \sup_{c \leq x \leq d} f(x) + \sup_{c \leq x \leq d} g(x), \text{ and } \inf_{c \leq x \leq d} [f(x) + g(x)] \geq \inf_{c \leq x \leq d} f(x) + \inf_{c \leq x \leq d} g(x).$$

Let us denote by M_k, M'_k and M''_k the suprema of $f + g$, f and g over $[x_{k-1}, x_k]$, and by m_k, m'_k and m''_k the infima of $f + g$, f and g over the same interval. Then

$$M_k \leq M'_k + M''_k, \quad \text{and} \quad m_k \geq m'_k + m''_k.$$

Subtracting these two inequalities and summing them (for $1 \leq k \leq n$) yields

$$U(f + g, P) - L(f + g, P) \leq U(f, P) - L(f, P) + U(g, P) - L(g, P).$$

If we only add the inequalities $M_k \leq M'_k + M''_k$ (for $1 \leq k \leq n$), we obtain that $U(f + g, P) \leq U(f, P) + U(g, P)$.

6.2.9. Let $P = \{0, \frac{1}{n}, \frac{2}{n}, \ldots, \frac{3n-1}{n}, 3\}$. Since f is increasing on $[0, 3]$, we get that

$$L(f, P) = \sum_{k=1}^{3n} \frac{1}{n} \left(\frac{k-1}{n}\right)^2 = \frac{1}{n^3} \sum_{k=1}^{3n} (k-1)^2 = \frac{1}{n^3} \frac{(3n-1)3n(6n-1)}{6}, \quad \text{and}$$

$$U(f, P) = \sum_{k=1}^{3n} \frac{1}{n} \left(\frac{k}{n}\right)^2 = \frac{1}{n^3} \sum_{k=1}^{3n} k^2 = \frac{1}{n^3} \frac{3n(3n+1)(6n+1)}{6}.$$

Therefore, $U(f, P) - L(f, P) = 9/n$. If we choose $n > 900$, we obtain that $U(f, P) - L(f, P) < 0.01$.

6.2.11. Let $P = \{0, \frac{1}{n}, \frac{2}{n}, \ldots, \frac{n-1}{n}, 1\}$. Since f is increasing on $[0, 1]$, we get that

$$L(f, P) = \sum_{k=1}^{n} \frac{1}{n} \left(\frac{k-1}{n}\right)^2 = \frac{1}{n^3} \sum_{k=1}^{n} (k-1)^2 = \frac{1}{n^3} \frac{(n-1)n(2n-1)}{6} \to \frac{1}{3}, \quad \text{and}$$

$$U(f, P) = \sum_{k=1}^{n} \frac{1}{n} \left(\frac{k}{n}\right)^2 = \frac{1}{n^3} \sum_{k=1}^{n} k^2 = \frac{1}{n^3} \frac{n(n+1)(2n+1)}{6} \to \frac{1}{3}.$$

By Exercise 6.2.10, f is integrable on $[0, 1]$ and $\int_0^1 x^2 \, dx = 1/3$.

6.2.13. Let $P = \{0, \frac{1}{n}, \frac{2}{n}, \ldots, \frac{n-1}{n}, 1\}$. Since f is increasing on $[0, 1]$, we get that

$$L(f, P) = \sum_{k=1}^{n} \frac{1}{n} f\left(\frac{k-1}{n}\right) = \frac{(n-1)(2n-1) + 6n(n-1)}{6n^2} \to \frac{4}{3}, \quad \text{and}$$

$$U(f, P) = \sum_{k=1}^{n} \frac{1}{n} f\left(\frac{k}{n}\right) = \frac{(n+1)(2n+1) + 6n(n+1)}{6n^2} \to \frac{4}{3}.$$

By Exercise 6.2.10, f is integrable on $[0, 1]$ and $\int_0^1 (x^2 + 2x) \, dx = 4/3$.

6.2.15. Let $P = \{3, 3 + \frac{1}{n}, 3 + \frac{2}{n}, \ldots, 3 + \frac{3n-1}{n}, 6\}$. Then, $L(f, P) = \frac{1}{n} \cdot 0 + \sum_{k=2}^{3n-1} \frac{1}{n} \cdot 1 + \frac{1}{n} \cdot 0 = (3n-2)/n \to 3$. Similarly, $U(f, P) = \frac{1}{n} \cdot 1 + \sum_{k=2}^{3n-1} \frac{1}{n} \cdot 1 + \frac{1}{n} \cdot 1 = 3$. By Exercise 6.2.10, f is integrable on $[3, 6]$ and $\int_3^6 \chi_{(3,6)} \, dx = 3$.

6.2.17. (a) It was shown in the proof of Lemma 6.2.3 that $U(f, Q_{i+1}) \leq U(f, Q_i)$, so the sequence $\{U(f, P_n)\}$ is decreasing. It is bounded below, so it must be convergent. Similarly, $\{L(f, P_n)\}$ is increasing and bounded above, hence convergent. Therefore, leting $n \to \infty$ in the inequality $L(f, P_n) \leq U(f, P_n)$, we obtain that $A \leq B$.

(b) Let $U = \inf\{U(f, P) : P \in \mathcal{P}\}$ and $L = \sup\{L(f, P) : P \in \mathcal{P}\}$. Then $A = \lim L(f, P_n) \leq L \leq U \leq \lim U(f, P_n) = B$. Thus, $A = B$ implies that $L = U$, so f is integrable and $\int_a^b f(x) \, dx = A$.

(c) Let $f(x) = x^2$ and, for each $n \in \mathbb{N}$, $P_n = \{0, 1\}$. Then f is integrable on $[0, 1]$ by Exercise 6.2.11. However, $L(f, P_n) = 0$ and $U(f, P_n) = 1$, so $\lim L(f, P_n) = 0 \neq 1 = \lim U(f, P_n)$.

6.2.19. Let $P = \{x_0, x_1, \ldots, x_n\}$ be a partition of $[a, b]$. Let $M_k(f) = \sup\{f(x) : x_{k-1} \leq x \leq x_k\}$ and $M_k(g) = \sup\{g(x) : x_{k-1} \leq x \leq x_k\}$. If $c \in [x_{k-1}, x_k]$, then $f(c) \leq g(c) \leq M_k(g)$. Therefore, $M_k(g)$ is an upper bound for the set $\{f(x) : x_{k-1} \leq x \leq x_k\}$, hence $M_k(f) \leq M_k(g)$. Since this is true for all k, $1 \leq k \leq n$, we have that $U(f, P) \leq U(g, P) \leq U_g$. It follows that $U(f, P) \leq U_g$ for any partition P, so U_g is an upper bound for the set $\{U(f, P) : P \in \mathcal{P}\}$. Since U_f is the least upper bound for this set, we conclude that $U_f \leq U_g$.

6.2.21. We will first show that the condition of the problem implies that f is Darboux integrable. Let $\varepsilon > 0$. By assumption, there exists $\delta > 0$ such that, if P is a partition of $[a, b]$ with $\|P\| < \delta$ then $U(f, P) - L(f, P) < \varepsilon$. Let $n = \lfloor \frac{b-a}{\delta} \rfloor + 1$, and let P be a partition of $[a, b]$ into n subintervals of equal length. Then $n > \frac{b-a}{\delta}$ so $\frac{b-a}{n} < \delta$, hence $\|P\| = \frac{b-a}{n} < \delta$. Thus, $U(f, P) - L(f, P) < \varepsilon$ and the result now follows from Proposition 6.2.7.

In the other direction, suppose that f is Darboux integrable, and let $\varepsilon > 0$. By Proposition 6.2.7 there exists a partition $P = \{x_0, x_1, \ldots, x_n\}$ of $[a, b]$, such that $U(f, P) - L(f, P) < \varepsilon/2$. Since f is bounded, there exists $M > 0$ such that $|f(x)| \leq M$ for all $x \in [a, b]$. Let $\delta = \varepsilon/(12nM)$, and let P_1 be a partition of $[a, b]$, such that $\|P_1\| < \delta$. We will show that $U(f, P_1) - L(f, P_1) < \varepsilon$. Let $P_2 = P \cup P_1$. Then P_2 has up to n more partition points than P_1. By Lemma 6.2.3,

$$U(f, P_1) \leq U(f, P_2) + 3nM\|P_1\| \quad \text{and} \quad L(f, P_1) \geq L(f, P_2) - 3nM\|P_1\|, \quad \text{hence}$$
$$U(f, P_1) - L(f, P_1) \leq U(f, P_2) - L(f, P_2) + 6nM\|P_1\|.$$

By the same Lemma, $U(f, P_2) \leq U(f, P)$ and $L(f, P_2) \geq L(f, P)$. Therefore,

$$U(f, P_1) - L(f, P_1) \leq U(f, P) - L(f, P) + 6nM\delta < \frac{\varepsilon}{2} + 6nM \frac{\varepsilon}{12nM} = \varepsilon.$$

6.3.8. Let $\varepsilon > 0$. Since f is bounded on $[a, b]$, there exists M such that $|f(x)| \leq M$ for all $x \in [a, b]$. By definition, there exist a positive integer n and intervals $[a_i, b_i]$, $1 \leq i \leq n$, such that

$$A \subset \bigcup_{i=1}^{n} [a_i, b_i], \quad \text{and} \quad \sum_{i=1}^{n} |b_i - a_i| < \frac{\varepsilon}{4M}.$$

The set $[a, b] \setminus \cup_{i=1}^{n}[a_i, b_i]$ is the union of open intervals (two of which may include the endpoint a or b). Let $\{(c_i, d_i)\}_{i=1}^{m}$ be these remaining intervals. For each i, $1 \leq i \leq m$, f is continuous on (c_i, d_i), so there exists a partition P_i of (c_i, d_i), such that $U(f, P_i) - L(f, P_i) < \varepsilon/(2m)$. Now, let P be a partition of $[a, b]$ that includes all partition points of each P_i, $1 \leq i \leq m$, and all a_i, b_i, $1 \leq i \leq n$. Let us denote $M_i = \sup\{f(x) : a_i \leq x \leq b_i\}$ and $m_i = \inf\{f(x) : a_i \leq x \leq b_i\}$. Then,

$$U(f, P) - L(f, P) = \sum_{i=1}^{m} [U(f, P_i) - L(f, P_i)] + \sum_{i=1}^{n} (M_i - m_i)(b_i - a_i)$$

$$< \sum_{i=1}^{m} \frac{\varepsilon}{2m} + \sum_{i=1}^{n} 2M|b_i - a_i| < m \cdot \frac{\varepsilon}{2m} + 2M \cdot \frac{\varepsilon}{4M} = \varepsilon.$$

6.3.17. Let f be the Thomae function and let g be the characteristic function of $(0, 1]$, both defined on $[0, 1]$.
6.3.20. It is useful to define the oscillation of a function first. The **oscillation of f on an interval** I in its domain is the difference between the supremum and infimum of f:

$$\omega_f(I) = \sup\{f(x) : x \in I\} - \inf\{f(x) : x \in I\}.$$

The **oscillation of f at a point** c is defined by

$$\omega_f(c) = \lim_{\epsilon \to 0} \omega_f(c - \epsilon, c + \epsilon).$$

We will use the fact that a function f is discontinuous at c if and only if $\omega_f(c) \neq 0$.

Suppose, to the contrary, that f is integrable on $[0, 1]$. Let $m \in \mathbb{N}$ and let $\varepsilon > 0$. Then, there exists a partition $P = \{x_0, x_1, \ldots, x_n\}$ of $[0, 1]$ such that $U(f, P) - L(f, P) < \varepsilon/m$. Let E_m be the set of all integers k, $1 \leq k \leq m$, such that $[x_{k-1}, x_k]$ contains a point c with $\omega_f(c) \geq 1/m$. If $k \in E_m$, then $M_k - m_k \geq 1/m$, where as usual, M_k and m_k are the supremum and the infimum of f on $[x_{k-1}, x_k]$. Therefore,

$$\frac{1}{m} \sum_{k \in E_m} \Delta x_k \leq \sum_{k \in E_m} (M_k - m_k) \Delta x_k \leq U(f, P) - L(f, P) < \frac{\varepsilon}{m},$$

so $\sum_{k \in E_m} \Delta x_k < \varepsilon$. Let $D_{1/m}$ be the set of points in $[0, 1]$ with the oscillation at least $1/m$. Since ε was arbitrary, we can take $\varepsilon = 1/3^m$, which means that $D_{1/m}$ is contained in the union of intervals $\{I_{1m}, I_{2m}, \ldots, I_{k_m, m}\}$ of total length not exceeding $1/3^m$.

By assumption, f is discontinuous at every point of $[0, 1]$, so

$$[0, 1] \subset D_1 \cup D_{1/2} \cup D_{1/3} \cup \cdots \subset \left(I_{11} \cup I_{21} \cup \cdots \cup I_{k_1, 1}\right) \cup \left(I_{12} \cup I_{22} \cup \cdots \cup I_{k_2, 2}\right) \cup \cdots.$$

Considering the lengths, we obtain a contradiction

$$1 \leq \frac{1}{3} + \frac{1}{3^2} + \frac{1}{3^3} + \cdots = \frac{1}{3} \frac{1}{1 - \frac{1}{3}} = \frac{1}{2}.$$

Thus, f cannot be integrable on $[0, 1]$.

6.4.4. Let $f(x) = 2^x$, $P = \{0, \frac{1}{n}, \frac{2}{n}, \ldots, 1\}$, and $\xi = \{0, \frac{1}{n}, \frac{2}{n}, \ldots, \frac{n-1}{n}\}$. Then

$$S(f, P, \xi) = \frac{1}{n} \sum_{k=1}^{n} 2^{\frac{k-1}{n}} = \frac{1}{n} \frac{(\sqrt[n]{2})^n - 1}{\sqrt[n]{2} - 1} = \frac{1}{n} \frac{1}{\sqrt[n]{2} - 1}.$$

When $n \to \infty$, the limit is $\ln 2$ by Example 3.1.14 (with $a = 2$ and $x = 1/n$).
6.4.14. This is a Riemann sum $S(f, P, \xi)$, where $f(x) = \sin \pi x$, $P = \{0, \frac{1}{n}, \frac{2}{n}, \ldots, 1\}$, and $\xi = \{\frac{1}{n}, \frac{2}{n}, \ldots, 1\}$. Since $\int_0^1 \sin \pi x \, dx = 2/\pi$, the limit equals $2/\pi$.

6.4.18. Let $\varepsilon > 0$ and let $P = \{x_0, x_1, \ldots, x_n\}$ be a partition of $[0, a]$ such that $U(f, P) - L(f, P) < \varepsilon/2$. Define P' to be the partition $\{-x_n, -x_{n-1}, \ldots, -x_1, -x_0 = 0\}$ of $[-a, 0]$. Then $\sup\{f(t) : -x_k \leq t \leq -x_{k-1}\} = \sup\{f(t) : x_{k-1} \leq t \leq x_k\}$, and the analogous equality holds for the infima. Thus, $L(f, P) = L(f, P')$ and $U(f, P) = U(f, P')$. Let $Q = P \cup P'$ be a partition of $[-a, a]$. Then

$$U(f, Q) - L(f, Q) = U(f, P) - L(f, P) + U(f, P') - L(f, P') < \frac{\varepsilon}{2} + \frac{\varepsilon}{2} = \varepsilon.$$

So, f is integrable on $[-a, a]$. Further, $L' = \sup L(f, P') = \sup L(f, P) = L$, and similarly $U' = U$. Since f is integrable on $[0, a]$, $U = L$, and it follows that $U' = L' = U = L$, meaning that $\int_{-a}^{0} f(x)\, dx = \int_{0}^{a} f(x)\, dx$.

6.5.6. Hint: $\max\{a, b\} = \dfrac{a + b}{2} + \dfrac{|a - b|}{2}$.

6.5.13. Since $f(x) \leq M$, for all $x \in [a, b]$, we have $f(x)^n \leq M^n$, hence $\int_a^b f(x)^n\, dx \leq \int_a^b M^n\, dx = M^n(b - a)$. It follows that

$$\lim_{n \to \infty} \left(\int_a^b f(x)^n\, dx \right)^{1/n} \leq \lim_{n \to \infty} M(b - a)^{1/n} = M.$$

To prove the reverse inequality, let $\varepsilon > 0$. By definition of M, there exists $x \in [a, b]$ such that $f(x) > M - \varepsilon$. Since f is continuous, there exists an interval $[\alpha, \beta]$ such that $f(x) > M - \varepsilon$ on $[\alpha, \beta]$. Then,

$$\int_a^b f(x)^n\, dx \geq \int_\alpha^\beta f(x)^n\, dx \geq \int_\alpha^\beta (M - \varepsilon)^n\, dx = (M - \varepsilon)^n(\beta - \alpha), \quad \text{so}$$

$$\lim_{n \to \infty} \left(\int_a^b f(x)^n\, dx \right)^{1/n} \geq \lim_{n \to \infty} (M - \varepsilon)(\beta - \alpha)^{1/n} = M - \varepsilon.$$

Since ε is arbitrary, we conclude that $\left(\int_a^b f(x)^n\, dx \right)^{1/n} \geq M$.

6.5.15. Let $\varepsilon > 0$. Since g is uniformly continuous, there exists $\delta_1 > 0$ such that

$$x, y \in f([a, b]) \quad \text{and} \quad |x - y| < \delta_1 \Rightarrow |g(x) - g(y)| < \frac{\varepsilon}{4(b - a)}.$$

Also, by Exercise 3.8.11, g is bounded so there exists $M > 0$ such that $|g(x)| \leq M$, for all $x \in f([a, b])$. Let

$$\delta = \min\{\delta_1, \frac{\varepsilon}{4M}\}.$$

By assumption, f is integrable, so Proposition 6.2.7 guarantees the existence of a partition $P = \{x_0, x_1, \ldots, x_n\}$ of $[a, b]$, such that $U(f, P) - L(f, P) < \delta^2$. Let $M_k = \sup\{f(x) : x \in [x_{k-1}, x_k]\}$, $m_k = \inf\{f(x) : x \in [x_{k-1}, x_k]\}$, and let M_k^* and m_k^* have the analogous meaning for the composition $g \circ f$. The set $\{1, 2, 3, \ldots, n\}$ can be written as a disjoint union $A \cup B$, where $k \in A$ if and only if $M_k - m_k < \delta$ and $k \in B$ if and only if $M_k - m_k \geq \delta$. If $k \in A$ and $x, y \in [x_{k-1}, x_k]$, then

$$|f(x) - f(y)| \leq M_k - m_k < \delta \leq \delta_1, \quad \text{so } |g(f(x)) - g(f(y))| < \frac{\varepsilon}{4(b - a)}, \quad \text{hence } M_k^* - m_k^* \leq \frac{\varepsilon}{4(b - a)}.$$

If $k \in B$, then

$$\delta \sum_{k \in B} \Delta x_k \leq \sum_{k \in B} (M_k - m_k)\, \Delta x_k \leq U(f, P) - L(f, P) < \delta^2, \quad \text{hence } \sum_{k \in B} \Delta x_k < \delta.$$

It follows that

$$U(g \circ f, P) - L(g \circ f, P) = \sum_{k \in A} \left(M_k^* - m_k^* \right) \Delta x_k + \sum_{k \in B} \left(M_k^* - m_k^* \right) \Delta x_k$$

$$\leq \frac{\varepsilon}{4(b - a)} \sum_{k \in A} \Delta x_k + 2M \sum_{k \in B} \Delta x_k \leq \frac{\varepsilon}{4(b - a)} (b - a) + 2M\delta \leq \frac{\varepsilon}{4} + 2M \frac{\varepsilon}{4M} < \varepsilon.$$

The integrability of $g \circ f$ is now a consequence of Proposition 6.2.7.

6.5.16. By assumption $1/f$ is bounded, so there exists $m > 0$ such that $1/f(x) \leq 1/m$ for all $x \in [a, b]$. It follows that, for all $x \in [a, b]$, $f(x) \geq m$. Further, f is integrable, so it is bounded, and there exists $M > 0$ such that, for all $x \in [a, b]$, $f(x) \leq M$. If $P = \{x_0, x_1, \ldots, x_n\}$ is a partition of $[a, b]$, and if M_k and m_k are the supremum and the infimum of f on $[x_{k-1}, x_k]$, then $M_k, m_k \in [m, M]$.

   Let $\varepsilon > 0$, and let P be a partition of $[a, b]$ such that $U(f, P) - L(f, P) < \varepsilon m^2$. Since $\sup(\frac{1}{f}) = \frac{1}{\inf f}$, if we denote

$$M_k' = \sup \left\{ \frac{1}{f(x)} : x \in [x_{k-1}, x_k] \right\}, \qquad m_k' = \inf \left\{ \frac{1}{f(x)} : x \in [x_{k-1}, x_k] \right\}$$

then $M'_k = 1/m_k$ and $m'_k = 1/M_k$. Therefore,

$$U\left(\frac{1}{f}, P\right) - L\left(\frac{1}{f}, P\right) = \sum_{k=1}^{n}(M'_k - m'_k)\Delta x_k = \sum_{k=1}^{n}\left(\frac{1}{m_k} - \frac{1}{M_k}\right)\Delta x_k$$

$$= \sum_{k=1}^{n}\frac{M_k - m_k}{M_k m_k}\Delta x_k < \frac{1}{m^2}\sum_{k=1}^{n}(M_k - m_k)\Delta x_k = \frac{1}{m^2}\left[U(f, P) - L(f, P)\right] < \frac{1}{m^2}\varepsilon m^2 = \varepsilon.$$

6.6.1. Answer: 0.

6.6.2. Answer: $-\sin x^2$.

6.6.3. Answer: $2x\sin x^4$.

6.6.8. Hint: Consider the function $g(x)\int_a^x f(t)\,dt$ for a suitable choice of g.

6.6.16. Hint: Write $F(b) - F(a) = \sum[F(x_i) - F(x_{i-1})]$ and apply Theorem 4.4.5.

6.6.18. Let $\varepsilon > 0$, and choose $\delta > 0$ so that $|u - v| < \delta \Rightarrow |f(u) - f(v)| < \varepsilon$. Let $P = \{x_0, x_1, \ldots, x_n\}$ be a partition of $[a, b]$ so that $\|P\| < \delta$. We write

$$\int_a^b f(x)g(x)\,dx = \sum_{k=0}^{n-1}\int_{x_k}^{x_{k+1}} f(x)g(x)\,dx = \sum_{k=0}^{n-1}f(x_k)\int_{x_k}^{x_{k+1}} g(x)\,dx + \sum_{k=0}^{n-1}\int_{x_k}^{x_{k+1}}[f(x) - f(x_k)]g(x)\,dx.$$

Since g is continuous it is bounded, $|g(x)| \leq L$. Then

$$\left|\sum_{k=0}^{n-1}\int_{x_k}^{x_{k+1}}[f(x) - f(x_k)]g(x)\,dx\right| \leq \sum_{k=0}^{n-1}\int_{x_k}^{x_{k+1}}|f(x) - f(x_k)||g(x)|\,dx \leq L\varepsilon(b - a).$$

If we write $G(x) = \int_a^x g(t)\,dt$, then

$$\sum_{k=0}^{n-1}f(x_k)\int_{x_k}^{x_{k+1}} g(x)\,dx = \sum_{k=0}^{n-1}f(x_k)[G(x_{k+1}) - G(x_k)] = \sum_{k=1}^{n-1}[f(x_{k-1}) - f(x_k)]G(x_k) + f(x_{n-1})G(x_n).$$

(See Lemma 7.5.8.) Since f is decreasing, $f(x_{k-1}) - f(x_k) \geq 0$ and since f is non-negative, $f(x_{n-1}) \geq 0$. Further, G is bounded $m \leq G(x) \leq M$. It follows that

$$mf(a) = m\sum_{k=0}^{n-1}f(x_k) \leq \sum_{k=0}^{n-1}f(x_k)\int_{x_k}^{x_{k+1}} g(x)\,dx \leq Mf(a).$$

Finally, G is continuous, so by the Intermediate Value Theorem there exists $c \in [a, b]$ such that $\sum_{k=0}^{n-1} f(x_k)\int_{x_k}^{x_{k+1}} g(x)\,dx = f(a)G(c)$.

6.6.19. Hint: Define $h(x) = f(b) - f(x)$ and apply Exercise 6.6.18.

6.6.21. Hint: $\cos x = \sin(\pi/2 - x)$.

6.6.30. Let $I_n = \int_0^{\pi/2}\sin^n x\,dx$. We use Integration by Parts with $u = \sin^{n-1} x$ and $dv = \sin x\,dx$. Then $du = (n-1)\sin^{n-2} x\cos x\,dx$ and $v = -\cos x$, so

$$I_n = -\cos x\sin^{n-1} x\Big|_0^{\pi/2} + \int_0^{\pi/2}(n-1)\sin^{n-2} x\cos^2 x\,dx$$

$$= (n-1)\int_0^{\pi/2}\left(\sin^{n-2} x - \sin^n x\right)dx = (n-1)(I_{n-2} - I_n).$$

It follows that $nI_n = (n-1)I_{n-2}$, hence

$$I_n = \frac{n-1}{n}I_{n-2}.$$

We obtain that

$$I_{2k} = \frac{2k-1}{2k}\cdot\frac{2k-3}{2k-2}\cdot\ldots\cdot\frac{3}{4}\cdot\frac{1}{2}I_0 = \frac{(2k-1)!!}{(2k)!!}\frac{\pi}{2}, \quad \text{and } I_{2k-1} = \frac{2k-2}{2k-1}\cdot\frac{2k-4}{2k-3}\cdot\ldots\cdot\frac{2}{3}I_1 = \frac{(2k-2)!!}{(2k-1)!!}.$$

6.7.1. Answer: $2/e$.

6.7.3. $x^4 + 4 = x^4 + 4x^2 + 4 - 4x^2 = (x^2 + 2)^2 - 4x^2 = (x^2 - 2x + 2)(x^2 + 2x + 2)$. Using Partial Fraction Decomposition,

$$\frac{1}{x^4 + 4} = \frac{1}{8} \frac{x + 2}{(x + 1)^2 + 1} - \frac{1}{8} \frac{x - 2}{(x - 1)^2 + 1}$$

$$= \frac{1}{8} \frac{x + 1}{(x + 1)^2 + 1} + \frac{1}{8} \frac{1}{(x + 1)^2 + 1} - \frac{1}{8} \frac{x - 1}{(x - 1)^2 + 1} + \frac{1}{8} \frac{1}{(x - 1)^2 + 1}.$$

For $b > 0$,

$$\int_0^b \frac{dx}{x^4 + 4} = \left(\frac{1}{16} \ln[(x + 1)^2 + 1] + \frac{1}{8} \arctan(x + 1) - \frac{1}{16} \ln[(x - 1)^2 + 1] + \frac{1}{8} \arctan(x - 1) \right) \Big|_0^b$$

$$= \frac{1}{16} \ln \left(\frac{(b + 1)^2 + 1}{(b - 1)^2 + 1} \right) + \frac{1}{8} \arctan(b - 1) + \frac{1}{8} \arctan(b + 1).$$

When we let $b \to +\infty$, we obtain that $\int_0^\infty \frac{dx}{x^4 + 4} = \frac{1}{8} \frac{\pi}{2} + \frac{1}{8} \frac{\pi}{2} = \frac{\pi}{8}$.

6.7.4. Answer: $\ln 2/4$.

6.7.8. Hint: $3x^4 + 5x^2 + 1 > x^4$.

6.7.10. Hint: $-x^2 \le -x$ for $x \ge 1$.

6.7.12. Use the limit comparison test and $g(x) = x^{-2/3}$.

6.7.16. Use the limit comparison test and Remark 6.7.6.

6.7.22. We may assume that $g \ge 0$. Otherwise, since g is bounded, there exists $\gamma > 0$ such that $\tilde{g} \equiv g + \gamma \ge 0$. If we show that the integral $\int_a^\infty f(x)\tilde{g}(x)\,dx$ converges, then

$$\int_a^b f(x)g(x)\,dx = \int_a^b f(x)\tilde{g}(x)\,dx - \gamma \int_a^b f(x)\,dx \to \int_a^\infty f(x)\tilde{g}(x)\,dx - \gamma \int_a^\infty f(x)\,dx.$$

So, let $g(x) \ge 0$ for all $x \ge a$.

Let $\varepsilon > 0$. Since g is bounded, there exists $M > 0$ such that $|g(x)| \le M$, for all $x \ge a$. Further, the infinite integral of f exists, so there exists $B > 0$ such that, if $b_1 \ge b_2 \ge B$, $\left| \int_{b_1}^{b_2} f(x)\,dx \right| < \varepsilon/M$. Using Exercise 6.6.18, there exists $c \in [b_1, b_2]$ such that

$$\left| \int_{b_1}^{b_2} f(x)g(x)\,dx \right| = \left| g(b_1) \int_{b_1}^c f(x)\,dx \right| < M \frac{\varepsilon}{M} = \varepsilon.$$

Thus the integral $\int_a^\infty f(x)g(x)\,dx$ converges by Exercise 6.7.20.

6.7.23. Hint: Use Exercises 6.6.18 and 6.7.20.

6.7.24. Since $\sin^2 x/x \ge 0$, the absolute and conditional convergence are the same. To see that the integral diverges, write $\sin^2 x = (1 - \cos 2x)/2$, and notice that $\int_1^\infty 1/(2x)\,dx$ diverges while $\int_1^\infty \cos 2x/(2x)\,dx$ converges by Dirichlet's test. Indeed, $|\int_1^b \cos 2x\,dx| = |(\sin 2b - \sin 2)/2| \le 1$, and $g(x) = 1/(2x)$ is monotone decreasing with $\lim_{x\to\infty} g(x) = 0$.

6.7.26. The integral converges (Dirichlet's Test with $f(x) = \sin x$ and $g(x) = 1/x$). It does not converge absolutely in view of the inequality $|\sin x| \ge \sin^2 x$ and Exercise 6.7.24.

6.7.29. Answer: $\pi/2$.

6.7.32. Start with $x = 2t$. Obtain

$$I = 2\int_0^{\pi/4} \ln \sin 2t\,dt = 2\int_0^{\pi/4} \ln 2\,dt + 2\int_0^{\pi/4} \ln \sin t\,dt + 2\int_0^{\pi/4} \ln \cos t\,dt.$$

In the last integral use $t = \pi/2 - u$. Then $\int_0^{\pi/4} \ln \cos t\,dt = \int_{\pi/4}^{\pi/2} \ln \sin u\,du$, so

$$I = \frac{\pi}{2} \ln 2 + 2\int_0^{\pi/4} \ln \sin t\,dt + 2\int_{\pi/4}^{\pi/2} \ln \sin u\,du = \frac{\pi}{2} \ln 2 + 2I.$$

Therefore, $I = -\frac{\pi}{2} \ln 2$.

6.7.35. Use Comparison Test: $\sin x/x \le 1$.

6.7.37. The integral $\int_0^{1/2} x^q\,dx/\sqrt{1 - x^4}$ converges if and only if $q > -1$. The integral $\int_{1/2}^1 x^q\,dx/\sqrt{1 - x^4}$ converges for all q (Limit Comparison Test with $g(x) = 1/\sqrt{1 - x}$).

6.7.39. Hint: Use Limit Comparison Test with $g(x) = x^{-3/4}$.

6.7.41. Let us use the substitution $u = -\ln x$. Then $x = e^{-u}$ and $dx = -e^{-u}\,du$. The limits of the integral also change to ∞ ("$-\ln 0$") and 0 ($= -\ln 1$). We obtain

$$(-1)^{m+1} \int_\infty^0 e^{-uq} u^m e^{-u}\,du = (-1)^m \int_0^\infty e^{-u(q+1)} u^m\,du = (-1)^m \left(\int_0^1 e^{-u(q+1)} u^m\,du + \int_1^\infty e^{-u(q+1)} u^m\,du \right).$$

To establish the convergence the first integral we will use Limit Comparison Test (Theorem 6.7.12), and compare $e^{-u(q+1)}u^m$ with u^m. Since $\lim_{u\to 0+} e^{-u(q+1)}u^m/u^m = 1$ and $\int_0^1 u^m\,du$ converges if and only if $m > -1$ (Example 6.7.13), we see that $\int_0^1 e^{-u(q+1)}u^m\,du$ converges if and only if $m > -1$.

For the second integral, we will use Limit Comparison Test (Theorem 6.7.5) and we will consider separately 3 cases: $q > -1$, $q = -1$, and $q < -1$. If $q > -1$, then we compare $e^{-u(q+1)}u^m$ with u^{-2}. Since $\lim_{u\to\infty} e^{-u(q+1)}u^m/u^{-2} = 0$ and $\int_0^1 u^{-2}\,du$ converges (see Remark 6.7.6), it follows that $\int_1^\infty e^{-u(q+1)}u^m\,du$ converges. When $q = -1$, we have the integral $\int_1^\infty u^m\,du$. In view of our previous assumption that $m > -1$, we see that the integral diverges. Finally, if $q < -1$ then we can compare $e^{-u(q+1)}u^m$ with $e^{-u(q+1)/2}$. Since $\lim_{u\to\infty} e^{-u(q+1)}u^m/e^{-u(q+1)/2} = \infty$ and $\int_1^\infty e^{-u(q+1)/2}\,du$ diverges, Remark 6.7.7 shows that the integral $\int_1^\infty e^{-u(q+1)}u^m\,du$ diverges. Thus, the second integral converges if and only if $q > -1$.

6.7.44. Suppose first that the integral $\int_a^b f(x)\,dx$ converges. In other words, there exists $\lim_{c\to a+}\int_c^b f(x)\,dx$. Let us denote this limit by I. Let $\varepsilon > 0$. Then there exists $\delta > 0$ such that, if $a < c < a+\delta$ then $|\int_c^b f(x)\,dx - I| < \varepsilon/2$. It follows that, if $c_1, c_2 \in (a, a+\delta)$,

$$\left|\int_{c_1}^{c_2} f(x)\,dx\right| = \left|\int_{c_1}^b f(x)\,dx - I + I - \int_{c_2}^b f(x)\,dx\right| < \frac{\varepsilon}{2} + \frac{\varepsilon}{2} = \varepsilon.$$

To prove the converse, let $\{c_n\}$ be a sequence in (a, b) that converges to a. The condition that is now the assumption, implies that $\{\int_{c_n}^b f(x)\,dx\}$ is a Cauchy sequence, hence a convergent one. Let I denote its limit. It remains to show that $\lim_{c\to a+}\int_c^b f(x)\,dx = I$. Let $\varepsilon > 0$. There exists $N \in \mathbb{N}$ such that, if $n \geq N$, $\left|\int_{c_n}^b f(x)\,dx - I\right| < \varepsilon/2$. By assumption, there exists $\delta_1 > 0$ such that, if $p, q \in (a, a+\delta)$, then $\left|\int_p^q f(x)\,dx\right| < \varepsilon/2$. Let $\delta = \min\{\delta_1, \frac{1}{2}(c_N - a)\}$, and let $c \in (a, a+\delta)$. Then

$$\left|\int_c^b f(x)\,dx - I\right| \leq \left|\int_{c_N}^b f(x)\,dx - I\right| + \left|\int_c^{c_N} f(x)\,dx\right| < \frac{\varepsilon}{2} + \frac{\varepsilon}{2} = \varepsilon.$$

7.1.2. Notice that the series under consideration can be written as $\sum_{n=1}^\infty \frac{2n-1}{2^n}$. Let

$$s_n = \sum_{k=1}^n \frac{2k-1}{2^k} = \sum_{k=1}^n \frac{k}{2^{k-1}} - \sum_{k=1}^n \frac{1}{2^k}.$$

For every $n \in \mathbb{N}$ define a function t_n by

$$t_n(x) = \sum_{k=1}^n \frac{kx^{k-1}}{2^{k-1}}.$$

Then

$$\int t_n(x)\,dx = \sum_{k=1}^n \frac{x^k}{2^{k-1}} + C = x\,\frac{1 - \left(\frac{x}{2}\right)^n}{1 - \frac{x}{2}} + C.$$

Taking the derivative now yields

$$t_n(x) = \left[x\,\frac{1 - \left(\frac{x}{2}\right)^n}{1 - \frac{x}{2}}\right]' = \frac{1 - \left(\frac{x}{2}\right)^n}{1 - \frac{x}{2}} + x\,\frac{-n\left(\frac{x}{2}\right)^{n-1}\cdot\frac{1}{2}\left(1 - \frac{x}{2}\right) - \left[1 - \left(\frac{x}{2}\right)^n\right]\left(-\frac{1}{2}\right)}{\left(1 - \frac{x}{2}\right)^2}.$$

It follows that

$$t_n(1) = \frac{1 - \left(\frac{1}{2}\right)^n}{1 - \frac{1}{2}} + \frac{-n\left(\frac{1}{2}\right)^{n-1}\cdot\frac{1}{2}\left(1 - \frac{1}{2}\right) - \left[1 - \left(\frac{1}{2}\right)^n\right]\left(-\frac{1}{2}\right)}{\left(1 - \frac{1}{2}\right)^2} \to 2 + 2 = 4$$

as $n \to \infty$. Since

$$\sum_{k=1}^n \frac{1}{2^k} = \frac{1}{2}\,\frac{1 - \left(\frac{1}{2}\right)^n}{1 - \frac{1}{2}} \to 1$$

we have that $\lim s_n = 3$.

7.1.3. The nth partial sum of the series can be written as

$$s_n = \sum_{k=1}^{n} \frac{1}{(3k-2)(3k+1)} = \sum_{k=1}^{n} \left[\frac{\frac{1}{3}}{3k-2} - \frac{\frac{1}{3}}{3k+1} \right]$$

$$= \frac{1}{3} \left[\left(1 - \frac{1}{4}\right) + \left(\frac{1}{4} - \frac{1}{7}\right) + \cdots + \left(\frac{1}{3n-2} - \frac{1}{3n+1}\right) \right] = \frac{1}{3} \left[1 - \frac{1}{3n+1} \right].$$

It follows that $\lim s_n = 1/3$, so the sum of the series is $1/3$.

7.1.6. If we denote $a_n = n!/n^n$ then

$$\frac{a_{n+1}}{a_n} = \frac{(n+1)!}{(n+1)^{n+1}} \cdot \frac{n^n}{n!} = \frac{n^n}{(n+1)^n} = \frac{1}{\left(1 + \frac{1}{n}\right)^n} \to \frac{1}{e} < 1,$$

so the series $\sum_{n=1}^{\infty} a_n$ converges by the Ratio Test.

7.2.5. Hint: $n^2 + 5n + 6 = (n+3)(n+2)$.

7.2.7. Answer: $1/4$.

7.2.9. Hint: $5040 = 7!$.

7.2.10. Hint: Write the numerator as $[(2n+1) - 1]/2$.

7.2.14. Let $s_n = \sum_{k=0}^{n} a_k$ and $t_n = \sum_{k=0}^{n} 2^k a_{2^k}$. We will show that the sequence $\{s_n\}$ is bounded if and only if the sequence $\{t_n\}$ is bounded. Since both are monotone, then one of them will converge if and only if the other one does.

Suppose first that $\{s_n\}$ is bounded, and let us use the fact that the sequence $\{a_n\}$ is decreasing to write

$$t_n = a_0 + 2a_2 + 2^2 a_{2^2} + \cdots + 2^n a_{2^n} = a_0 + 2 \left(a_2 + 2a_4 + 4a_8 + \cdots + 2^{n-1} a_{2^n} \right)$$

$$\leq a_0 + 2 \left[a_2 + (a_3 + a_4) + (a_5 + a_6 + a_7 + a_8) + \cdots + (a_{2^{n-1}+1} + a_{2^{n-1}+2} + \cdots + a_{2^n}) \right]$$

$$\leq a_0 + 2s_{2^n}.$$

Therefore, $\{t_n\}$ is bounded as well.

In the other direction,

$$s_{2^n - 1} = a_0 + a_1 + (a_2 + a_3) + (a_4 + a_5 + a_6 + a_7) + \cdots + (a_{2^{n-1}} + a_{2^{n-1}+1} + \cdots + a_{2^n-1})$$

$$\leq a_0 + a_1 + 2a_2 + 4a_4 + \cdots + 2^{n-1} a_{2^{n-1}} = a_1 + t_{n-1},$$

which shows that the subsequence $\{s_{2^n - 1}\}$ is bounded. Since the sequence $\{s_n\}$ is monotone, it follows that it is bounded, hence convergent.

7.2.18. Let $\varepsilon > 0$ and choose $N = \lfloor 1/\varepsilon \rfloor + 1$. If $m \geq n \geq N$, then $n > 1/\varepsilon$, and

$$|s_m - s_n| = \left| \sum_{k=n+1}^{m} \frac{\cos\left(x^k\right)}{k^2} \right| \leq \sum_{k=n+1}^{m} \left| \frac{\cos\left(x^k\right)}{k^2} \right| \leq \sum_{k=n+1}^{m} \frac{1}{k^2}$$

$$< \sum_{k=n+1}^{m} \frac{1}{(k-1)k} = \sum_{k=n+1}^{m} \left[\frac{1}{k-1} - \frac{1}{k} \right] = \frac{1}{n} - \frac{1}{m} < \frac{1}{n} < \varepsilon.$$

7.2.19. Hint: Prove that $s_{2n} - s_n > 1/4$.

7.3.2. By Example 3.1.14, $\lim\limits_{x \to 0} \dfrac{a^x - 1}{x} = \ln a$. If we substitute x by $1/n$, then $n \to \infty$, so we obtain

$$\lim_{n \to \infty} \frac{a^{1/n} - 1}{\frac{1}{n}} = \ln a.$$

By Limit Comparison Test (Theorem 7.3.4), the series $\sum_{n=0}^{\infty} a_n$ diverges.

7.3.6. Hint: Prove that $\lim \dfrac{e^{-n^2}}{\frac{1}{n^2}} = 0$.

7.3.16. If $\{a_n\}$ is not bounded above, then there exists a subsequence $\{a_{n_k}\}$ that diverges to $+\infty$. Therefore, $a_{n_k}/(1 + a_{n_k}) \to 1$ so $\sum_{n=1}^{\infty} a_n/(1 + a_n)$ diverges by Divergence Test. On the other hand, if there exists $M > 0$ so that $a_n \leq M$ for all $n \in \mathbb{N}$, then

$$\frac{a_n}{1 + a_n} \geq \frac{a_n}{1 + M},$$

so $\sum_{n=1}^{\infty} a_n/(1 + a_n)$ diverges by Comparison Test.

7.3.17. Let t_n and s_n denote the nth partial sum of the series $\sum_{n=1}^{\infty} \frac{1}{n^2}$ and $\sum_{n=1}^{\infty} a_n$, respectively. Then

$$s_{n^2} = 1 + \frac{1}{2^2} + \frac{1}{3^2} + \frac{1}{4} + \frac{1}{5^2} + \frac{1}{6^2} + \frac{1}{7^2} + \frac{1}{8^2} + \frac{1}{9} + \frac{1}{10^2} + \cdots + \frac{1}{(n^2-2)^2} + \frac{1}{(n^2-1)^2} + \frac{1}{n^2} < t_{n^2} + t_n.$$

Since $\sum_{n=1}^{\infty} \frac{1}{n^2}$ converges, the sequence of partial sums $\{t_n\}$ is bounded, and it follows that so is $\{s_n\}$. The fact that $\{s_n\}$ is an increasing sequence implies that the series $\sum_{n=1}^{\infty} a_n$ converges.

7.3.25. Hint: Prove that $\ln(n+1) \cdot \ln(1+n^n) \geq n \ln^2 n$ and use Integral Test.

7.4.3. Using Root Test and Example 2.9.8,

$$C_n = \left(\frac{n-1}{n+1}\right)^{n-1} = \frac{1}{\left(\frac{n+1}{n-1}\right)^{n-1}} = \frac{1}{\left(1 + \frac{2}{n-1}\right)^{n-1}} \to \frac{1}{e^2} < 1,$$

so the series converges.

7.4.15. Answer: The series converges.

7.4.21. Let $r < R < 1$. By Exercise 2.8.3, there exists $N \in \mathbb{N}$ such that, if $n \geq N$ then $C_n < R$. Now the convergence of $\sum_{n=1}^{\infty} a_n$ follows from Root Test. On the other hand, if $r > 1$, then there exists a subsequence $\{C_{n_k}\}$ that converges to r. Consequently, there exists $K \in \mathbb{N}$ such that, if $k \geq K$, $\sqrt[n_k]{a_{n_k}} > 1$. Thus, $\{a_n\}$ cannot converge to 0 and Divergence Test implies that $\sum_{n=1}^{\infty} a_n$ diverges.

7.4.28. $\mathcal{R}_n = n\left[\left(\frac{3n+3}{3n+1}\right)^2 - 1\right] = n\frac{12n+8}{(3n+1)^2} \to \frac{12}{9} > 1$, so the series converges by Raabe's Test.

7.4.31. We will use Raabe's Test:

$$\mathcal{R}_n = n\left[\frac{n^n e}{(n+1)^n} - 1\right] = \frac{1}{\left(1 + \frac{1}{n}\right)^n} n\left[e - \left(1 + \frac{1}{n}\right)^n\right].$$

In order to calculate $\lim \mathcal{R}_n$, we consider

$$\lim_{x \to 0} \frac{e - (1+x)^{1/x}}{x}.$$

Using L'Hôpital's Rule, we obtain

$$\lim_{x \to 0} \frac{-(1+x)^{1/x}\left[\frac{x}{1+x} - \ln(1+x)\right]}{x^2} = -e \lim_{x \to 0} \frac{\frac{x}{1+x} - \ln(1+x)}{x^2}$$

which is again of the form $\left(\frac{0}{0}\right)$. Another application of L'Hôpital's Rule yields

$$-e \lim_{x \to 0} \frac{\frac{1}{(1+x)^2} - \frac{1}{1+x}}{2x} = -e \lim_{x \to 0} \frac{\frac{-x}{(1+x)^2}}{2x} = \frac{e}{2}.$$

Therefore, $\lim \mathcal{R}_n = \frac{1}{e} \cdot \frac{e}{2} = \frac{1}{2} < 1$, so the series $\sum_{n=1}^{\infty} \frac{1}{n!}\left(\frac{n}{e}\right)^n$ diverges.

7.4.34. A calculation shows that

$$\frac{a_n}{a_{n+1}} = \frac{1}{e}\left(\frac{n+1}{n}\right)^{n+p} = e^{-1+(n+p)\ln(1+1/n)}.$$

Using Taylor's formula for $\ln(1+x)$ we have that $\ln(1+1/n) = 1/n - 1/(2n^2) + A_n$, where $\lim n^2 A_n = 0$. Thus. $-1 + (n+p)\ln(1+1/n) = -1 + (n+p)[1/n - 1/(2n^2) + A_n] = (p-1/2)/n + B_n$, where $\lim n B_n = 0$. Applying Taylor's formula to e^x we obtain that

$$\frac{a_n}{a_{n+1}} = e^{(p-1/2)/n+B_n} = 1 + \frac{p-1/2}{n} + C_n,$$

where $\lim n C_n = 0$. It follows that $\mathcal{R} = p - 1/2$, so the series converges for $p > 3/2$ and diverges for $p < 3/2$.

7.4.36. Answer: The series converges if and only if $p \geq 2$.

7.4.37. Answer: The series converges if $q > p$.

7.4.45. Hint: Make a suitable choice of $\{c_n\}$ in Kummer's Test.

7.5.21. The sequence $\{a_n\}$ converges to 0, so it must be bounded. Let $|a_n| \leq M$, for all $n \in \mathbb{N}$. Then $|a_n b_n| \leq M|b_n|$, for all $n \in \mathbb{N}$, and $\sum_{n=1}^{\infty} |a_n b_n|$ converges by the Comparison Test. It follows that $\sum_{n=1}^{\infty} a_n b_n$ is absolutely convergent.

7.5.22. Let $P_n = \sum_{k=1}^{n} p_k$, $Q_n = \sum_{k=1}^{n} q_k$, and $S_n = \sum_{k=1}^{n} a_k$. For each $n \in \mathbb{N}$, there exist positive integers

σ_n and π_n such that $-Q_n = S_{\sigma_n} - P_{\pi_n}$. It is not hard to see that both sequences $\{\sigma_n\}$ and $\{\pi_n\}$ are increasing and diverging to $+\infty$. By assumption, $S = \lim S_n$ exists. Suppose, to the contrary, that $\lim P_n$ also exists, and denote it by P. We will show that $\lim Q_n = P - S$.

Let $\varepsilon > 0$. First, $\lim S_n = S$ and $\lim P_n = P$ so there exists $N_1 \in \mathbb{N}$ such that

$$n \geq N_1 \Rightarrow |S_n - S| < \frac{\varepsilon}{2} \quad \text{and} \quad |P_n - P| < \frac{\varepsilon}{2}.$$

Further, $\lim \pi_n = \infty$, so there exists $N_2 \in \mathbb{N}$ such that

$$n \geq N_2 \Rightarrow \pi_n \geq N_1.$$

Let $N = \max\{N_1, N_2\}$, and suppose that $n \geq N$. Then $\sigma_n \geq n \geq N \geq N_1$ and $\pi_n \geq N_1$, so

$$|Q_n + (S - P)| = |-S_{\sigma_n} + P_{\pi_n} + S - P| \leq |S - S_{\sigma_n}| + |P - P_{\pi_n}| < \frac{\varepsilon}{2} + \frac{\varepsilon}{2} = \varepsilon.$$

Thus, if $\{P_n\}$ converges, so does $\{Q_n\}$. (The proof that the convergence of $\{Q_n\}$ implies the convergence of $\{P_n\}$ is similar.) However, they cannot both converge. Indeed, $\sum_{n=1}^{\infty} |a_k| = \sum_{n=1}^{\infty} p_k - \sum_{n=1}^{\infty} q_k$, so it would follow that $\sum_{n=1}^{\infty} a_n$ is absolutely convergent.

7.5.24. We are looking at the series

$$-1 - \frac{1}{2} - \frac{1}{3} + \frac{1}{4} + \frac{1}{5} + \frac{1}{6} + \frac{1}{7} + \frac{1}{8} - \frac{1}{9} - \frac{1}{10} - \cdots$$

Let $A_1 = 1 + \frac{1}{2} + \frac{1}{3}$, $A_2 = \frac{1}{4} + \frac{1}{5} + \frac{1}{6} + \frac{1}{7} + \frac{1}{8}$, etc. Our first step will be to show that the alternating series $\sum_{n=1}^{\infty} (-1)^n A_n$ converges. Notice that

$$0 < A_n = \frac{1}{n^2} + \frac{1}{n^2 + 1} + \cdots + \frac{1}{(n+1)^2 - 1} \leq \frac{2n+1}{n^2} \to 0, \quad \text{as } n \to \infty.$$

In order to apply the Alternating Series Test, it remains to prove that the sequence $\{A_n\}$ is decreasing. Indeed,

$$
\begin{aligned}
A_n - A_{n+1} &= \frac{1}{n^2} + \frac{1}{n^2+1} + \cdots + \frac{1}{(n+1)^2 - 1} - \frac{1}{(n+1)^2} - \frac{1}{(n+1)^2 + 1} - \cdots - \frac{1}{(n+2)^2 - 1} \\
&= \sum_{k=0}^{2n} \left(\frac{1}{n^2 + k} - \frac{1}{(n+1)^2 + k} \right) - \frac{1}{(n+2)^2 - 2} - \frac{1}{(n+2)^2 - 1} \\
&= \sum_{k=0}^{2n} \frac{2n+1}{(n^2+k)[(n+1)^2+k]} - \frac{1}{n^2 + 4n + 2} - \frac{1}{n^2 + 4n + 3} \\
&> (2n+1) \cdot \frac{2n+1}{(n^2+2n)[(n+1)^2+2n]} - \frac{1}{n^2+4n+2} - \frac{1}{n^2+4n+3} \\
&> \frac{(2n+1)^2}{(n^2+2n)(n^2+4n+1)} - \frac{1}{n^2+2n} - \frac{1}{n^2+4n+1} \\
&= \frac{(2n+1)^2 - (2n^2 + 6n + 1)}{(n^2+2n)(n^2+4n+1)} = \frac{2n^2 - 2n}{(n^2+2n)(n^2+4n+1)} \geq 0.
\end{aligned}
$$

By the Alternating Series Test, $\sum_{n=1}^{\infty} (-1)^n A_n$ converges.

Now we will show that $\sum_{n=1}^{\infty} \frac{(-1)^{\lfloor \sqrt{n} \rfloor}}{n}$ converges. Let $\{s_n\}$ be the sequence of its partial sums, and let $\{S_n\}$ be the sequence of partial sums of $\sum_{n=1}^{\infty} (-1)^n A_n$. The latter series converges, and let S denote its sum. We will prove that $\lim s_n = S$ as well. Let $\varepsilon > 0$. There exists $N \in \mathbb{N}$ such that

$$n \geq N \Rightarrow |S_n - S| < \frac{\varepsilon}{2} \quad \text{and} \quad |A_n| < \frac{\varepsilon}{2}.$$

Let $n \geq N^2$. Then there exists a positive integer m such that $m^2 \leq n < (m+1)^2$. This implies that $m + 1 > N$, so $m \geq N$. For such m and n,

$$
\begin{aligned}
|s_n - S| &= \left| S_m + (-1)^m \left(\frac{1}{m^2} + \frac{1}{m^2 + 1} + \cdots + \frac{1}{n} \right) - S \right| \\
&\leq |S_m - S| + \left| \frac{1}{m^2} + \frac{1}{m^2 + 1} + \cdots + \frac{1}{n} \right| < \frac{\varepsilon}{2} + A_m < \frac{\varepsilon}{2} + \frac{\varepsilon}{2} = \varepsilon.
\end{aligned}
$$

Thus, the series $\sum_{n=1}^{\infty} \frac{(-1)^{\lfloor\sqrt{n}\rfloor}}{n}$ converges.

7.5.26. The series converges by Dirichlet's Test. It does not converge absolutely. Let $S_n = \sum_{k=1}^{n} \left| \frac{\ln^{100} k}{k} \sin \frac{k\pi}{4} \right|$. When $k = 4m$, $\sin k\pi/4 = 0$, and for $k \neq 4m$, $|\sin k\pi/4| \geq \sqrt{2}/2$. Also, $\ln^{100} k \geq 1$, so

$$S_{4m} = \sum_{n=1}^{4m} \left| \frac{\ln^{100} n}{n} \sin \frac{n\pi}{4} \right| \geq 1 + \frac{1}{2} + \frac{1}{3} + \frac{1}{5} + \frac{1}{6} + \frac{1}{7} + \frac{1}{9} + \cdots + \frac{1}{4m-1}$$

$$\geq 1 + \frac{1}{3} + \frac{1}{5} + \frac{1}{7} + \frac{1}{9} + \cdots + \frac{1}{4m-1} \geq \frac{1}{2} + \frac{1}{4} + \frac{1}{6} + \cdots + + \frac{1}{4m-2} = \frac{1}{2} H_{2m-1} \to \infty,$$

where H_n denotes the nth partial sum of Harmonic Series. It follows that the series $\sum_{n=1}^{\infty} \frac{\ln^{100} n}{n} \sin \frac{n\pi}{4}$ is not absolutely convergent.

7.5.28. Hint: Prove that $\cos \frac{\pi n^2}{n+1} = (-1)^{n+1} \cos \frac{\pi}{n+1}$.

7.5.30. Since $\sin^2 n = (1 - \cos 2n)/2$, our series can be written as a sum of 2 convergent series:

$$\sum_{n=1}^{\infty} (-1)^n \frac{\sin^2 n}{n} = \sum_{n=1}^{\infty} (-1)^n \frac{1}{2n} + \sum_{n=1}^{\infty} (-1)^{n+1} \frac{\cos 2n}{2n}.$$

The first series on the right hand side converges by the Alternating Series Test, the second by Dirichlet's Test. However, the series $\sum_{n=1}^{\infty} (-1)^n \frac{\sin^2 n}{n}$ does not converge absolutely because $\sum_{n=1}^{\infty} \frac{\sin^2 n}{n}$ diverges. Indeed, if the last series were convergent, using once again the identity $\sin^2 n = (1 - \cos 2n)/2$, we would have that

$$\sum_{n=1}^{\infty} \frac{1}{2n} = \sum_{n=1}^{\infty} \frac{\sin^2 n}{n} + \sum_{n=1}^{\infty} \frac{\cos 2n}{2n}.$$

Now, the right hand side would be a sum of two convergent series (the second one being convergent by Dirichlet's Test), while the left side is a divergent series. This contradiction shows that $\sum_{n=1}^{\infty} (-1)^n \frac{\sin^2 n}{n}$ is not absolutely convergent.

7.5.33. Hint: Use Example 2.9.6.
7.5.34. Hint: Write

$$\frac{\sin \frac{n\pi}{4}}{n^p + \sin \frac{n\pi}{4}} = \frac{\sin \frac{n\pi}{4}}{n^p} - \frac{\sin^2 \frac{n\pi}{4}}{n^p \left(1 + \sin \frac{n\pi}{4} \right)}.$$

Answer: The series converges for $p > 1/2$ and diverges for $0 < p \leq 1/2$.
7.5.37. Hint: Use Lemma 7.5.8 with $a_n = 1/n$ and prove that $|B_n| \leq c\sqrt{n}$.
7.5.38. Answer: The series diverges.

7.5.45. Hint: Prove that $\frac{1}{\sqrt{4n-3}} + \frac{1}{\sqrt{4n-1}} - \frac{1}{\sqrt{2n}} > \left(1 - \frac{\sqrt{2}}{2} \right) \frac{1}{\sqrt{n}}$.
7.5.49. By Exercise 2.5.7, if $H_n = \sum_{k=1}^{n} 1/k$, then

$$H_n - \ln n = C + \gamma_n$$

where C is the Euler's constant, and $\lim \gamma_n = 0$. Notice that

$$t_m \equiv \frac{1}{2} + \frac{1}{4} + \cdots + \frac{1}{2m} = \frac{1}{2} H_m = \frac{1}{2} \ln m + \frac{1}{2} C + \frac{1}{2} \gamma_m, \quad \text{and}$$

$$r_m \equiv 1 + \frac{1}{3} + \frac{1}{5} + \cdots + \frac{1}{2m-1} = H_{2m} - \frac{1}{2} H_m$$

$$= \ln 2m + C + \gamma_{2m} - \frac{1}{2} \ln m - \frac{1}{2} C - \frac{1}{2} \gamma_m = \ln 2 + \frac{1}{2} \ln m + \frac{1}{2} C + \gamma_{2m} - \frac{1}{2} \gamma_m.$$

Therefore, if $\{s_n\}$ is the sequence of partial sums of the rearranged series, then

$$s_{n(p+q)} = r_{np} - t_{nq} = \left(\ln 2 + \frac{1}{2} \ln np + \frac{1}{2} C + \gamma_{2np} - \frac{1}{2} \gamma_{np} \right) - \left(\frac{1}{2} \ln nq + \frac{1}{2} C + \frac{1}{2} \gamma_{nq} \right)$$

$$= \ln 2 + \frac{1}{2} \ln \frac{p}{q} + \gamma_{2np} - \frac{1}{2} \gamma_{np} - \frac{1}{2} \gamma_{nq} \to \ln 2 + \frac{1}{2} \ln \frac{p}{q}.$$

If m is not a multiple of $p + q$, let $\varepsilon > 0$, and let $m > (p+q)(1 + \frac{2}{\varepsilon})$. There exists $n \in \mathbb{N}$ such that $n(p+q) < m < (n+1)(p+q)$. It follows that

$$n > \frac{m}{p+q} - 1 > \frac{2}{\varepsilon}, \quad \text{so}$$

$$\left| s_m - s_{n(p+q)} \right| \leq \frac{1}{n(p+q)+1} + \frac{1}{n(p+q)+2} + \cdots + \frac{1}{(n+1)(p+q)} < \frac{p+q}{n(p+q)+1} < \frac{1}{n} < \frac{\varepsilon}{2}.$$

Letting $n \to \infty$, we obtain that $\left| s_m - \left(\ln 2 + \frac{1}{2} \ln \frac{p}{q} \right) \right| \leq \frac{\varepsilon}{2} < \varepsilon$, which shows that $\lim s_m = \ln 2 + \frac{1}{2} \ln \frac{p}{q}$.

8.1.5. Notice that

$$f(x) \leq \frac{\lfloor n f(x) \rfloor}{n} \leq \frac{n f(x) + 1}{n}$$

so $\lim f_n(x) = f(x)$. The convergence is uniform because

$$|f_n(x) - f(x)| = \left| \frac{\lfloor n f(x) \rfloor}{n} - \frac{n f(x)}{n} \right| = \frac{|\lfloor n f(x) \rfloor - n f(x)|}{n} \leq \frac{1}{n} \to 0.$$

8.1.8. For each $x \in \mathbb{R}$, $\lim x/n = 0$ so $\lim \sin \frac{x}{n} = 0$, hence $f(x) = 0$. The convergence is not uniform on $\mathbb{R}$, because for each fixed $n \in \mathbb{N}$,

$$\sup\{|f_n(x) - f(x)| : x \in \mathbb{R}\} = \sup\left\{ \left| \sin \frac{x}{n} \right| : x \in \mathbb{R} \right\} = 1.$$

8.1.10. First, for any $x > 0$,

$$f_n(x) = n \left(\sqrt{x + \frac{1}{n}} - \sqrt{x} \right) \cdot \frac{\sqrt{x + \frac{1}{n}} + \sqrt{x}}{\sqrt{x + \frac{1}{n}} + \sqrt{x}} = \frac{1}{\sqrt{x + \frac{1}{n}} + \sqrt{x}} \to \frac{1}{2\sqrt{x}},$$

so $f(x) = \frac{1}{2\sqrt{x}}$. Further,

$$f_n(x) - f(x) = \frac{1}{\sqrt{x + \frac{1}{n}} + \sqrt{x}} - \frac{1}{2\sqrt{x}} = \frac{-\frac{1}{n}}{2\sqrt{x} \left(\sqrt{x + \frac{1}{n}} + \sqrt{x} \right)^2},$$

so, for each fixed $n \in \mathbb{N}$, $\sup\{|f_n(x) - f(x)| : x > 0\}$ is infinite. We conclude that the convergence of $\{f_n\}$ to f is not uniform on $(0, +\infty)$.

8.1.17. Suppose, to the contrary, that the convergence is not uniform on $[a, b]$. Then there exists $\gamma > 0$ and a subsequence $\{f_{n_k}\}$ such that

$$\sup\{|f_{n_k}(x) - f(x)| : x \in [a, b]\} > \gamma.$$

It follows that, for each $k \in \mathbb{N}$, there exists $x_k \in [a, b]$ such that $|f_{n_k}(x_k) - f(x_k)| > \gamma$. The assumption that the sequence $\{f_n\}$ is increasing implies that, for any $n \in \mathbb{N}$ and any $x \in [a, b]$, $f_n(x) \leq f(x)$. Therefore, we have that

$$f(x_k) - f_{n_k}(x_k) > \gamma.$$

Let $n \in \mathbb{N}$ be arbitrary, and suppose that $n_k \geq n$. Then $f_{n_k}(x_k) \geq f_n(x_k)$, so

$$f(x_k) - f_n(x_k) \geq f(x_k) - f_{n_k}(x_k) > \gamma.$$

The sequence $\{x_k\}$ belongs to $[a, b]$ so, by Bolzano–Weierstrass Theorem it has a convergent subsequence $\{x_{k_j}\}$ and its limit c belongs to $[a, b]$. Now $f(x_{k_j}) - f_n(x_{k_j}) > \gamma$ and, letting $j \to \infty$, we obtain that $f(c) - f_n(c) \geq \gamma$. Since n is arbitrary, this contradicts the assumption that $\lim f_n(c) = f(c)$.

8.1.20. Example: Let f be an unbounded function on $[a, b]$ and define $f_n(x) = f(x)$, $g_n(x) = 1/n$, for all $n \in \mathbb{N}$. Then $\{f_n\}$ converges to f uniformly on $[a, b]$, and $\{g_n\}$ converges to 0 uniformly on $[a, b]$, so $f_n(x)g_n(x) \to 0$, but the convergence is not uniform on $[a, b]$.

8.2.1. By assumption, for each $n \in \mathbb{N}$, there exists $M_n > 0$ such that $|f_n(x)| \leq M_n$, for all $x \in \mathbb{R}$. Our first step will be to prove that the sequence $\{f_n\}$ is uniformly bounded, i.e., that there exists $M > 0$ such that $|f_n(x)| \leq M$, for all $n \in \mathbb{N}$ and $x \in \mathbb{R}$.

The assumption that $\{f_n\}$ converges uniformly implies that, if $\varepsilon = 1$, there exists $N \in \mathbb{N}$ such that

$$m \geq n \geq N \quad \text{and} \quad x \in \mathbb{R} \Rightarrow |f_m(x) - f_n(x)| < 1.$$

In particular, if $n = N$, we have that for all $x \in \mathbb{R}$ and all $m \geq N$, $|f_m(x) - f_N(x)| < 1$. It follows that

$$|f_m(x)| \leq |f_N(x)| + |f_m(x) - f_N(x)| \leq M_N + 1.$$

Next, we define $M = \max\{M_1, M_2, \ldots, M_{N-1}, M_N + 1\}$, and it is not hard to see that, for all $n \in \mathbb{N}$ and $x \in \mathbb{R}$, $|f_n(x)| \leq M$. If we now let $n \to \infty$, we obtain that $|f(x)| \leq M$ for all $x \in \mathbb{R}$.

8.2.3. Hint: If p_m and p_n have different degrees, then $\sup |p_m(x) - p_n(x)| = \infty$.

8.2.7. Hint: Consider $f_n(x) = \frac{\sin nx}{\sqrt{n}}$.

8.2.14. Let $\varepsilon > 0$. The sequence $\{f_n\}$ is uniformly Cauchy on (a, b) so there exists $N \in \mathbb{N}$ such that, if $m \geq n \geq N$ and $x \in (a, b)$, then $|f_m(x) - f_n(x)| < \varepsilon/3$. It remains to show that $|f_m(a) - f_n(a)| < \varepsilon$ and $|f_m(b) - f_n(b)| < \varepsilon$.

Let $m \geq n \geq N$ be fixed. The function f_m is continuous at $x = a$, so there exists $\delta_1 > 0$ such that

$$0 < |x - a| < \delta_1 \Rightarrow |f_m(x) - f_m(a)| < \frac{\varepsilon}{3}.$$

Similarly, there exists $\delta_2 > 0$ such that

$$0 < |x - a| < \delta_2 \Rightarrow |f_n(x) - f_n(a)| < \frac{\varepsilon}{3}.$$

Now, if $\delta = \min\{\delta_1, \delta_2\}$ and $0 < |x - a| < \delta$ then

$$|f_m(a) - f_n(a)| \leq |f_m(a) - f_m(x)| + |f_m(x) - f_n(x)| + |f_n(x) - f_n(a)| < \frac{\varepsilon}{3} + \frac{\varepsilon}{3} + \frac{\varepsilon}{3} = \varepsilon.$$

The same argument can be used to show that $|f_m(b) - f_n(b)| < \varepsilon$.

8.2.23. First, $\lim f_n(x) = 0$, for all $x \in [0,1]$, so $f(x) = 0$. Further, using the substitution $u = 1 - x$,

$$\int_0^1 nx(1-x)^n \, dx = \int_1^0 n(1-u)u^n \, (-du) = n \int_0^1 \left(u^n - u^{n+1} \right) du$$

$$= n \left[\frac{u^{n+1}}{n+1} - \frac{u^n}{n} \right] \Big|_0^1 = n \left[\frac{1}{n+1} - \frac{1}{n+2} \right] = \frac{n}{(n+1)(n+2)} \to 0.$$

However, $\{f_n\}$ does not converge uniformly to 0 on $[0,1]$. Indeed,

$$\sup_{x \in [0,1]} |f_n(x) - 0| = \sup_{x \in [0,1]} nx(1-x)^n \geq n \left(\frac{1}{n} \right) \left(1 - \frac{1}{n} \right)^n \to \frac{1}{e} \neq 0.$$

8.2.24. Hint: Consider $f_n(x) = e^{-(x-n)^2}$.

8.3.27. For the terms to be defined, we must have $x \neq -k\pi$, $k \in \mathbb{N}$. Let

$$A = \mathbb{R} \setminus \{-k\pi : k \in \mathbb{N}\}.$$

For $x \in A$,

$$\sum_{k=1}^n \frac{1}{(x+k)(x+k+1)} = \frac{1}{(x+1)(x+2)} + \frac{1}{(x+2)(x+3)} + \cdots + \frac{1}{(x+n)(x+n+1)}$$

$$= \left(\frac{1}{x+1} - \frac{1}{x+2} \right) + \left(\frac{1}{x+2} - \frac{1}{x+3} \right) + \cdots + \left(\frac{1}{x+n} - \frac{1}{x+n+1} \right) = \frac{1}{x+1} - \frac{1}{x+n+1} \to \frac{1}{x+1}.$$

The convergence is not uniform on A. Indeed,

$$\left| s_n(x) - \frac{1}{x+1} \right| = \frac{1}{|x+n+1|},$$

and $\sup_{x \in A} \frac{1}{|x+n+1|}$ is infinite if x can be arbitrarily close to negative integers. One way to ensure the uniform convergence is to have $x \geq 0$, i.e., $B = [0, +\infty)$. Then

$$\sup_{x \geq 0} \frac{1}{|x+n+1|} = \frac{1}{n+1} \to 0,$$

so the convergence is uniform on B.

8.3.28. We will use Root Test:

$$\sqrt[n]{\frac{1}{n(1+x^2)^n}} = \frac{1}{1+x^2} \cdot \frac{1}{\sqrt[n]{n}} \to \frac{1}{1+x^2} \leq 1$$

so the series converges whenever $1/(1+x^2) \neq 1$, i.e., $x \neq 0$. Therefore, $A = \mathbb{R} \setminus \{0\}$. The series does not converge uniformly on A because if $x_n = 1/\sqrt{n}$, the sequence $\{s_n(x_n)\}$ is not convergent. ($\{s_n\}$ is the sequence of partial sums.) Indeed,

$$s_n(x_n) = \sum_{k=1}^n \frac{1}{k \left(1 + \frac{1}{k} \right)^k}$$

and the series $\sum_{k=1}^\infty \frac{1}{k \left(1 + \frac{1}{k} \right)^k}$ does not converge. This can be established using the comparison with Harmonic Series and the fact that the latter diverges.

If we take $B = [a, +\infty)$, where $a > 0$, then the series converges uniformly on B. This follows from the Weierstrass Test, with $M_n = 1/(1 + a^2)^n$.

8.4.3. $\sqrt[n]{\dfrac{3^n + (-2)^n}{n}} = \sqrt[n]{3^n}\,\sqrt[n]{\dfrac{1 + \left(-\frac{2}{3}\right)^n}{n}} \to 3$, so the radius of convergence is $R = 1/3$. The endpoints of the interval of convergence are $-1 - \frac{1}{3}$ and $-1 + \frac{1}{3}$. At the left endpoint the series converges because it is

$$\sum_{n=1}^{\infty} \frac{3^n + (-2)^n}{n} \left(-\frac{1}{3}\right)^n = \sum_{n=1}^{\infty} \left(\frac{(-1)^n}{n} + \frac{\left(\frac{2}{3}\right)^n}{n}\right),$$

hence the sum of the Alternating Harmonic Series and a series that is dominated by the geometric series with ratio $2/3$. At the right endpoint the series diverges because it is

$$\sum_{n=1}^{\infty} \frac{3^n + (-2)^n}{n} \left(\frac{1}{3}\right)^n = \sum_{n=1}^{\infty} \left(\frac{1}{n} + \frac{\left(-\frac{2}{3}\right)^n}{n}\right),$$

hence the sum of a convergent series and a divergent Harmonic Series.

8.4.5. The series can be written as

$$\frac{1}{2}\,x + \frac{1}{2^2}\,x^4 + \frac{1}{2^3}\,x^9 + \frac{1}{2^4}\,x^{16} + \dots .$$

This allows us to see that

$$a_n = \begin{cases} \frac{1}{2^k} & \text{if } n = k^2, \\ 0 & \text{otherwise.} \end{cases}$$

Consequently,

$$\sqrt[n]{a_n} = \begin{cases} \sqrt[k^2]{\frac{1}{2^k}} & \text{if } n = k^2, \\ 0 & \text{otherwise,} \end{cases}$$

so $\limsup \sqrt[n]{a_n} = 1$, hence the radius of convergence is $R = 1$. At both endpoints $x = \pm 1$, we get an absolutely convergent series. (Taking the absolute values of each term yields a geometric series with ratio $1/2$.)

8.4.6. $\lim \sqrt[n]{|a_n|} = 1$, so the radius of convergence is $R = 1$. At the endpoints $x = 1$ we obtain the series

$$\sum_{n=1}^{\infty} \frac{(-1)^{\lfloor \sqrt{n} \rfloor}}{n}$$

which converges by Exercise 7.5.24. At the other endpoint, $x = -1$, we obtain

$$\sum_{n=1}^{\infty} \frac{(-1)^{\lfloor \sqrt{n} \rfloor + n}}{n} = 1 - \frac{1}{2} + \frac{1}{3} + \frac{1}{4} - \frac{1}{5} + \frac{1}{6} - \frac{1}{7} + \frac{1}{8} + \frac{1}{9} - \dots, \tag{16.1}$$

which is "almost" alternating. Namely, the successive terms of the same (positive) sign are a_{k^2-1} and a_{k^2}, for $k \geq 2$. Indeed,

$$(-1)^{\lfloor \sqrt{k^2-1} \rfloor + k^2 - 1} = (-1)^{k-1+k^2-1} = (-1)^{k(k+1)} \text{ and } (-1)^{\lfloor \sqrt{k^2} \rfloor + k^2} = (-1)^{k+k^2} = (-1)^{k(k+1)},$$

and $k(k+1)$ is an even number. Thus, the series (16.1) is a sum of two series: one alternating and the other one that contains the "extra" terms $\frac{1}{4} + \frac{1}{9} + \frac{1}{16} + \dots$, i.e., the series $\sum_{n=1}^{\infty} \frac{1}{k^2}$ which converges.

8.4.12. We will use Ratio Test:

$$\left|\frac{a_{n+1}}{a_n}\right| |x| = \left[\frac{(n+1)!}{(2n+3)!!}\right]^2 \cdot \left[\frac{(2n+1)!!}{n!}\right]^2 |x| = \frac{(n+1)^2}{(2n+3)^2} |x| \to \frac{|x|}{4}$$

so the radius of convergence is $R = 4$.

At $x = -4$ the series diverges. We will prove this using Kummer's Test with $c_n = n \ln n$. The series $\sum_{n=1}^{\infty} \frac{1}{c_n}$ diverges (use Integral Test), so Kummer's Test can be used. Now

$$\mathcal{K}_n = c_n \frac{a_n}{a_{n+1}} - c_{n+1} = n \ln n \left(\frac{2n+3}{n+1}\right)^2 \cdot \frac{1}{4} - (n+1) \ln(n+1)$$

$$= n \ln n + n \ln n \frac{4n+5}{4n^2 + 8n + 4} - (n+1) \ln(n+1) = n \ln \frac{n}{n+1} - \ln(n+1) + \ln n \frac{4n^2 + 5n}{4n^2 + 8n + 4}$$

$$= \ln \frac{1}{\left(1 + \frac{1}{n}\right)^n} - \ln(n+1) + \ln n \left(\frac{4n^2 + 5n}{4n^2 + 8n + 4} - 1\right) + \ln n$$

$$= \ln \frac{1}{\left(1 + \frac{1}{n}\right)^n} - \ln\left(1 + \frac{1}{n}\right) + \ln n \left(\frac{-3n - 4}{4n^2 + 8n + 4}\right) \to \ln \frac{1}{e} = -1 < 0,$$

so the series diverges.

At $x = 4$, the series converges by Alternating Series Test. The series is

$$\sum_{n=1}^{\infty} (-1)^{n+1} \left(\frac{n!}{(2n+1)!!} \right)^2 4^n$$

and if we denote $b_n = \left(\frac{n!}{(2n+1)!!} \right)^2 4^n$, then

$$\frac{b_{n+1}}{b_n} = \left(\frac{(n+1)!}{(2n+3)!!} \right)^2 4^{n+1} \cdot \left(\frac{(2n+1)!!}{n!} \right)^2 \frac{1}{4^n} = \left(\frac{n+1}{2n+3} \right)^2 4 = \left(\frac{2n+2}{2n+3} \right)^2 < 1$$

so the sequence $\{b_n\}$ is a decreasing sequence of positive numbers. It remains to show that $\lim b_n = 0$. Notice that, for any $k \in \mathbb{N}$,

$$\frac{2k}{2k+1} = 1 - \frac{1}{2k+1} \leq 1 - \frac{1}{2k+2} = \frac{2k+1}{2k+2}.$$

Therefore,

$$0 \leq b_n = \left(\frac{(2n)!!}{(2n+1)!!} \right)^2 = \left(\frac{(2n)!!}{(2n+1)!!} \right) \left(\frac{(2n)!!}{(2n+1)!!} \right) \leq \left(\frac{(2n)!!}{(2n+1)!!} \right) \left(\frac{(2n+1)!!}{(2n+2)!!} \right) = \frac{1}{2n+2} \to 0,$$

so Squeeze Theorem implies that $\lim b_n = 0$.

8.5.3. Since $y' = \frac{1}{\sqrt{1+x^2}}$, we can use (8.37), with x^2 instead of x.

$$y' = \frac{1}{\sqrt{1+x^2}} = 1 - \frac{1}{2} x^2 + \frac{3!!}{2^2(2!)} x^4 - \frac{5!!}{2^3(3!)} x^6 + \frac{7!!}{2^4(4!)} x^8 - \cdots.$$

Integrating, we obtain

$$y = x - \frac{1}{2} \frac{x^3}{3} + \frac{3!!}{2^2(2!)} \frac{x^5}{5} - \frac{5!!}{2^3(3!)} \frac{x^7}{7} + \frac{7!!}{2^4(4!)} \frac{x^9}{9} - \cdots + C. \tag{16.2}$$

In fact, $y(0) = 0$ so $C = 0$. The series converges for $|x| < 1$ and diverges for $|x| > 1$. At the endpoints, the series converges absolutely. Let

$$a_n = \frac{(2n-1)!!}{2^n \, n!(2n+1)}.$$

The absolute convergence of the series means that $\sum_{n=1}^{\infty} a_n$ converges. In order to accomplish this, we will use Raabe's Test.

$$\mathcal{R}_n = n \left(\frac{a_n}{a_{n+1}} - 1 \right) = n \left[\left(\frac{(2n-1)!!}{2^n \, n!(2n+1)} \right) \left(\frac{2^{n+1}(n+1)!(2n+3)}{(2n+1)!!} \right) - 1 \right]$$

$$= n \left(\frac{2(n+1)(2n+3)}{(2n+1)^2} - 1 \right) = n \left(\frac{4n^2 + 10n + 6}{4n^2 + 4n + 1} - 1 \right) = n \cdot \frac{6n+5}{4n^2 + 4n + 1} \to \frac{3}{2} > 1.$$

Thus, the series (16.2) converges if and only if $|x| \leq 1$.

8.5.5. Hint: Use partial fraction decomposition.

8.5.6. Using some algebra and (8.44), we have

$$y = \ln(1 + x + x^2 + x^3) = \ln(1+x)(1+x^2) = \ln(1+x) + \ln(1+x^2)$$

$$= \left(x - \frac{x^2}{2} + \frac{x^3}{3} - \frac{x^4}{4} + \cdots \right) + \left(x^2 - \frac{x^4}{2} + \frac{x^6}{3} - \frac{x^8}{4} + \cdots \right)$$

$$= \sum_{n=1}^{\infty} \frac{x^n}{n} (-1)^{n+1} + \sum_{n=1}^{\infty} \frac{x^{2n}}{n} (-1)^{n+1} = \sum_{n=1}^{\infty} \frac{x^n}{n} \left[(-1)^{n+1} + 2\sin(n-1)\frac{\pi}{2} \right],$$

for $|x| < 1$. When $x = 1$, the series converges as a sum of two Alternating Harmonic Series. When $x = -1$, the series diverges because it is a sum of the Harmonic Series and Alternating Harmonic Series.

8.5.7. Using (8.44) and (8.38) we have

$$y = \frac{\ln(1+x)}{1+x} = \sum_{n=1}^{\infty} \frac{x^n}{n} (-1)^{n+1} \sum_{k=0}^{\infty} (-x)^k = \sum_{n=1}^{\infty} \sum_{k=0}^{\infty} \frac{x^{n+k}}{n} (-1)^{n+1+k}.$$

Using the substitution $m = n + k$ and interchanging the order of summation, we obtain

$$y = \sum_{n=1}^{\infty} \sum_{m=n}^{\infty} \frac{x^m}{n} (-1)^{m+1} = \sum_{m=1}^{\infty} \sum_{n=1}^{m} \frac{x^m}{n} (-1)^{m+1} = \sum_{m=1}^{\infty} (-1)^{m+1} x^m \sum_{n=1}^{m} \frac{1}{n}.$$

The series converges for $|x| < 1$. When $|x| = 1$, the series diverges by Divergence Test.

8.5.12. Hint: you will need to solve a second-order differential equation.

8.5.15. If we denote the sum of the series by $S(x)$ then

$$\int \frac{S(x)}{x} dx = x - 2x^2 + 3x^3 - 4x^4 + \cdots + C = (x^2 - x^3 + x^4 - \dots)' - x + x^2 - x^3 + \cdots + C = \frac{x}{1+x^2} + C.$$

Taking the derivatives we obtain $S(x) = x(1-x)/(1+x)^3$.

8.5.16. Answer: $2x/(1-x)^3$.

8.5.20. Let $f(x) = \sum_{n=0}^{\infty} \frac{(-1)^n}{3n+1} x^{3n+1}$. To determine when the series converges, we compute

$$\limsup \sqrt[n]{\left| \frac{(-1)^n}{3n+1} \right|} = 1$$

so the radius of convergence is $R = 1$. Further, when $x = 1$ we have $\sum_{n=0}^{\infty} \frac{(-1)^n}{3n+1}$ which converges by Alternating Series Test. Thus, the desired result is $f(1)$.

For any $|x| < 1$, using Corollary 8.5.12 and formula (8.38),

$$f'(x) = \sum_{n=0}^{\infty} (-1)^n x^{3n} = \frac{1}{1+x^3}.$$

It follows that $f(x) = \int \frac{1}{1+x^3} dx$. Using Partial Fraction Decomposition, we obtain that

$$\frac{1}{1+x^3} = \frac{1}{3} \frac{1}{x+1} - \frac{1}{3} \frac{x-2}{x^2-x+1} = \frac{1}{3} \frac{1}{x+1} - \frac{1}{3} \frac{x - \frac{1}{2}}{\left(x-\frac{1}{2}\right)^2 + \frac{3}{4}} - \frac{1}{3} \frac{-\frac{3}{2}}{\left(x-\frac{1}{2}\right)^2 + \frac{3}{4}}, \quad \text{so}$$

$$\int \frac{1}{1+x^3} dx = \frac{1}{3} \ln(x+1) - \frac{1}{3} \frac{1}{2} \ln\left(\left(x-\frac{1}{2}\right)^2 + \frac{3}{4}\right) + \frac{1}{2} \frac{2}{\sqrt{3}} \arctan\left[\left(x+\frac{1}{2}\right) \frac{2}{\sqrt{3}}\right].$$

It follows that $f(1) = \frac{1}{3} \ln 2 - \frac{1}{6} \ln 1 + \frac{1}{\sqrt{3}} \arctan \sqrt{3} = \frac{1}{3} \ln 2 + \frac{\pi}{3\sqrt{3}}$.

8.5.21. Answer: $5e - 1$.

8.5.27. By (8.40), we have that

$$\frac{\ln(1+x)}{x} dx = 1 - \frac{x}{2} + \frac{x^2}{3} - \frac{x^3}{4} + \dots.$$

It is not hard to prove that the last series has radius of convergence 1. Further, the alternating series test shows that it converges for $x = 1$. Thus, Abel's Theorem guarantees that it converges uniformly on $[0, 1]$. Now, Exercise 8.5.26 shows that

$$\int_0^1 \frac{\ln(1+x)}{x} dx = 1 - \frac{1}{2^2} + \frac{1}{3^2} - \frac{1}{4^2} + \dots.$$

By Exercise 9.2.10, $\sum_{n=1}^{\infty} 1/n^2 = \pi^2/6$. Therefore,

$$\int_0^1 \frac{\ln(1+x)}{x} dx = \left(1 + \frac{1}{2^2} + \frac{1}{3^2} + \frac{1}{4^2} + \dots\right) - 2\left(\frac{1}{2^2} + \frac{1}{4^2} + \frac{1}{6^2} + \dots\right) = \frac{\pi^2}{6} - 2\left(\frac{1}{2^2}\right)\left(\frac{\pi^2}{6}\right) = \frac{\pi^2}{12}.$$

8.5.29. Hint: Prove that the integrand equals $1 + 2\sum_{n=1}^{\infty} r^n \cos nx$.

9.1.6. Using the substitution $t = x - \pi$, we obtain the function $g(t) = -\frac{1}{2} t$, for $-\pi \le t \le \pi$. Formula (9.8) yields a Fourier series for g: $\sum_{n=1}^{\infty} \frac{1}{n} \cos n\pi \sin nt$. Since $t = x - \pi$ and

$$\sin nt = \sin n(x - \pi) = \sin nx \cos n\pi - \cos nx \sin n\pi = \sin nx \cos n\pi.$$

Therefore, a Fourier series for f is

$$\sum_{n=1}^{\infty} \frac{1}{n} \cos n\pi \sin nx \cos n\pi = \sum_{n=1}^{\infty} \frac{1}{n} \sin nx.$$

because $\cos n\pi = (-1)^n$.

9.1.9. The coefficients $b_n = 0$, $n \in \mathbb{N}$, because the function is even. Also,

$$a_n = \frac{1}{\pi} \int_{-\pi}^{\pi} |x| \cos nx \, dx = \frac{2}{\pi} \int_0^\pi |x| \cos nx \, dx = \frac{2}{\pi} \int_0^\pi x \cos nx \, dx.$$

When $n = 0$,

$$a_0 = \frac{2}{\pi} \int_0^\pi x \, dx = \frac{2}{\pi} \cdot \frac{\pi^2}{2} = \pi.$$

For $n \geq 1$, we use Integration by Parts with $u = x$ and $dv = \cos nx \, dx$. Then $du = dx$ and $v = \frac{1}{n} \sin nx$ so

$$a_n = \frac{2}{\pi} \left[\frac{x \sin nx}{n} \Big|_0^\pi - \int_0^\pi \frac{\sin nx}{n} \, dx \right] = \left(-\frac{2}{n\pi} \right) \frac{-\cos nx}{n} \Big|_0^\pi = \left(-\frac{2}{n\pi} \right) \left(\frac{-\cos n\pi}{n} + \frac{1}{n} \right) = \frac{-2}{n^2 \pi} (1 - \cos n\pi).$$

When n is an even number, $n = 2k$, we have $\cos n\pi = 1$ so $a_{2k} = 0$. When n is odd, $n = 2k - 1$, we have $\cos n\pi = -1$ so $a_{2k-1} = -\frac{4}{(2k-1)^2 \pi}$. A Fourier series for $y = |x|$ is

$$\pi - \frac{4}{\pi} \sum_{k=1}^{\infty} \frac{\cos(2k-1)x}{(2k-1)^2}.$$

9.1.10. Answer: $\dfrac{2}{\pi} - \dfrac{4}{\pi} \displaystyle\sum_{k=1}^{\infty} \dfrac{\cos 2kx}{4k^2 - 1}$.

9.1.19. The coefficients $b_n = 0$, $n \in \mathbb{N}$, because the function is even. Also, we will asume that a is not an integer. (Otherwise, the function is already in the form of a Fourier series.) Finally,

$$a_n = \frac{2}{\pi} \int_0^\pi \cos ax \cos nx \, dx$$

because the function is even. For $n = 0$,

$$a_0 = \frac{2}{\pi} \int_0^\pi \cos ax \, dx = \frac{2}{\pi} \frac{\sin ax}{a} \Big|_0^\pi = \frac{2}{\pi} \sin a\pi.$$

For $n \geq 1$, we will use a trigonometric identity $2 \cos \alpha \cos \beta = \cos(\alpha + \beta) + \cos(\alpha - \beta)$. Then

$$a_n = \frac{1}{\pi} \int_0^\pi [\cos(a + n)x + \cos(a - n)x] \, dx = \frac{1}{\pi} \left[\frac{\sin(a+n)x}{a+n} + \frac{\sin(a-n)x}{a-n} \right] \Big|_0^\pi$$

$$= \frac{1}{\pi} \left[\frac{\sin(a+n)\pi}{a+n} + \frac{\sin(a-n)\pi}{a-n} \right].$$

Next,

$$\sin(a+n)\pi = \sin a\pi \cos n\pi + \cos a\pi \sin n\pi = (-1)^n \sin a\pi, \quad \text{and}$$

$$\sin(a-n)\pi = \sin a\pi \cos n\pi - \cos a\pi \sin n\pi = (-1)^n \sin a\pi.$$

It follows that

$$a_n = \frac{(-1)^n \sin a\pi}{\pi} \left[\frac{1}{a+n} + \frac{1}{a-n} \right] = \frac{(-1)^n \sin a\pi}{\pi} \cdot \frac{2a}{(a+n)(a-n)}.$$

Therefore, a Fourier series for $\cos ax$ is

$$\cos ax \sim \frac{1}{a\pi} \sin a\pi + \frac{2a}{\pi} \sin a\pi \sum_{n=1}^{\infty} \frac{(-1)^n}{a^2 - n^2} \cos nx.$$

The function $y = \cos ax$ is continuous at $x = 0$, so the equality holds for $x = 0$:

$$1 = \frac{1}{a\pi} \sin a\pi + \frac{2a}{\pi} \sin a\pi \sum_{n=1}^{\infty} \frac{(-1)^n}{a^2 - n^2}.$$

Multiplying both sides by $\frac{\pi}{2 \sin a\pi}$, we obtain

$$\frac{\pi}{2 \sin a\pi} = \frac{1}{2a} + \sum_{n=1}^{\infty} (-1)^n \frac{a}{a^2 - n^2}.$$

9.2.8. Let $\varepsilon > 0$. Since f is integrable, it is bounded, $|f(x)| \le M$ for all $x \in (a, b)$. Let $N_1 = \lfloor \frac{8\pi M}{\varepsilon} \rfloor + 1$. Then

$$n \ge N_1 \Rightarrow n > \frac{8\pi M}{\varepsilon} \Rightarrow \frac{\pi M}{n} < \frac{\varepsilon}{8}.$$

Next, let $N_2 = \lfloor \frac{32}{\varepsilon} \rfloor + 1$. Then

$$n \ge N_2 \Rightarrow n > \frac{32}{\varepsilon} \Rightarrow \frac{4}{n} < \frac{\varepsilon}{8}.$$

Let $N_3 = \lfloor \frac{32M}{\varepsilon} \rfloor + 1$. Then

$$n \ge N_3 \Rightarrow n > \frac{32M}{\varepsilon} \Rightarrow \frac{4M}{n} < \frac{\varepsilon}{8}.$$

Finally, f is integrable on (a, b), so there exists $\delta > 0$ such that, if P is a partition of (a, b) with the property that $\|P\| < \delta$, and if ξ is any selection of intermediate points, then the Riemann sum $S(f, P, \xi)$ satisfies

$$\left| S(f, P, \xi) - \int_a^b f(x)\, dx \right| < \frac{\pi \varepsilon}{16}. \tag{16.3}$$

Let $N_4 = \lfloor \frac{\pi}{\delta} \rfloor + 1$. Now, suppose that $n \ge N_4$, so that $\pi/n < \delta$. There exist integers p, q such that

$$\frac{(p-1)\pi}{n} \le a < \frac{p\pi}{n} \quad \text{and} \quad \frac{q\pi}{n} < b \le \frac{(q-1)\pi}{n}.$$

If $P = \{x_0, x_1, \ldots, x_m\} = \{a, \frac{p\pi}{n}, \frac{(p+1)\pi}{n}, \ldots, \frac{q\pi}{n}, b\}$, then $\|P\| \le \pi/n < \delta$, and (16.3) holds for any $\xi = \{\xi_1, \xi_2, \ldots, \xi_m\}$.

Let $N = \max\{N_1, N_2, N_3, N_4\}$, and let $n \ge N$ be fixed. For P as above and $1 \le i \le m$, we consider integrals $\int_{x_{i-1}}^{x_i} f(x) |\sin nx|\, dx$. Using Exercise 6.6.17, there exist $\xi_i \in [x_{i-1}, x_i]$ such that

$$\int_{x_{i-1}}^{x_i} f(x) |\sin nx|\, dx = f(\xi_i) \int_{x_{i-1}}^{x_i} |\sin nx|\, dx.$$

Now, we have that

$$\left| \int_a^b f(x) \ |\sin nx|\, dx - \frac{2}{\pi} \int_a^b f(x)\, dx \right| \le \left| \int_a^{p\pi/n} f(x) |\sin nx|\, dx \right| + \left| \int_{q\pi/n}^b f(x) |\sin nx|\, dx \right|$$

$$+ \frac{2}{\pi} \left| S(f, P, \xi) - \int_a^b f(x)\, dx \right| + \left| \int_{p\pi/n}^{q\pi/n} f(x) |\sin nx|\, dx - \frac{2}{\pi} S(f, P, \xi) \right|$$

$$< \frac{\varepsilon}{8} + \frac{\varepsilon}{8} + \frac{2}{\pi} \frac{\pi \varepsilon}{16} + \left| \sum_{i=2}^{m-1} \int_{x_{i-1}}^{x_i} f(x) |\sin nx|\, dx - \frac{2}{\pi} \sum_{i=1}^{m} f(\xi_i) \Delta x_i \right|$$

$$\le \frac{3\varepsilon}{8} + |f(\xi_1)| \Delta x_1 + |f(\xi_m)| \Delta x_m + \sum_{i=2}^{m-1} \left| \int_{x_{i-1}}^{x_i} f(x) |\sin nx|\, dx - \frac{2}{\pi} f(\xi_i) \Delta x_i \right|$$

$$\le \frac{3\varepsilon}{8} + M \frac{\pi}{n} + M \frac{\pi}{n} + \sum_{i=2}^{m-1} \left| f(\xi_i) \int_{x_{i-1}}^{x_i} |\sin nx|\, dx - \frac{2}{\pi} f(\xi_i) \frac{\pi}{n} \right|$$

$$< \frac{3\varepsilon}{8} + 2\frac{\varepsilon}{8} + \sum_{i=2}^{m-1} |f(\xi_i)| \left| \int_{x_{i-1}}^{x_i} |\sin nx|\, dx - \frac{2}{n} \right| = \frac{5\varepsilon}{8}$$

because $\int_{x_{i-1}}^{x_i} |\sin nx|\, dx = 2/n$. Indeed, $\sin nx$ is of constant sign on $[x_{i-1}, x_i]$ and vanishes at the endpoints so the values of $\cos nx$ at these endpoints alternate between 1 and -1. Thus,

$$\int_{x_{i-1}}^{x_i} |\sin nx|\, dx = \left| \int_{x_{i-1}}^{x_i} \sin nx\, dx \right| = \left| -\frac{\cos nx}{n} \Big|_{x_{i-1}}^{x_i} \right| = \frac{2}{n}.$$

9.2.11. By Exercise 9.1.9, the Fourier series for $y = |x|$ on $-\pi \le x \le \pi$ is

$$\pi - \frac{4}{\pi} \sum_{k=1}^{\infty} \frac{\cos(2k-1)x}{(2k-1)^2}.$$

Further, since $y = |x|$ is continuous at $x = 0$, the series converges for $x = 0$ to $|0| = 0$. Thus,

$$0 = \pi - \frac{4}{\pi} \sum_{k=1}^{\infty} \frac{\cos(2k-1)0}{(2k-1)^2} = \pi - \frac{4}{\pi} \sum_{k=1}^{\infty} \frac{1}{(2k-1)^2}, \quad \text{so} \quad \sum_{k=1}^{\infty} \frac{1}{(2k-1)^2} = \frac{\pi^2}{4}.$$

9.2.12. Hint: Find the Fourier series for $f(x) = ax^3 + bx^2 + cx + d$, then choose the coefficients a, b, c, d.

9.2.18. Formula (9.11) shows that, if $c \in \mathbb{R}$,

$$S_n(c) = \frac{1}{\pi} \int_0^\pi \frac{f(c+t) + f(c-t)}{2} \frac{\sin\left(n + \frac{1}{2}\right)t}{\sin \frac{t}{2}}\, dt,$$

and Theorem 9.2.6 asserts that, if f is a piecewise differentiable function in $PC(2\pi)$, then $\{S_n(c)\}$ converges to $(f(c+) + f(c-))/2$. Let

$$F(t) = \begin{cases} \dfrac{1}{t} - \dfrac{1}{2\sin \frac{t}{2}}, & \text{if } 0 < t < \pi, \\ 0, & \text{if } t = 0. \end{cases}$$

Then F is a continuous function on $[0, \pi)$. If we assume that $f \in PC(2\pi)$, then $[f(c+t) - f(c-t)]\, F(t)$ is integrable on $[0, \pi)$, so

$$\lim_{n \to \infty} \int_0^\pi [f(c+t) - f(c-t)]\, F(t) \sin\left(n + \frac{1}{2}\right)t\, dt = 0.$$

This implies that $\{S_n(c)\}$ converges if and only if

$$\frac{1}{\pi} \int_0^\pi \frac{f(c+t) - f(c-t)}{t} \sin\left(n + \frac{1}{2}\right)t\, dt \tag{16.4}$$

converges. Further, if $0 < \delta < \pi$, we can write the integral in (16.4) as a sum of integrals over $[0, \delta]$ and $[\delta, \pi]$. On $[\delta, \pi]$, $(f(c+t) + f(c-t))/t$ is a piecewise continuous function, hence integrable, so Riemann–Lebesgue Lemma implies that

$$\frac{1}{\pi} \int_\delta^\pi \frac{f(c+t) - f(c-t)}{t} \sin\left(n + \frac{1}{2}\right)t\, dt \to 0.$$

Therefore, $\{S_n(c)\}$ converges if and only if

$$\frac{1}{\pi} \int_0^\delta \frac{f(c+t) - f(c-t)}{t} \sin\left(n + \frac{1}{2}\right)t\, dt$$

converges.

9.2.19. Hint: write $\int_0^\pi \frac{f(c+t) - f(c+)}{2\sin \frac{t}{2}} \sin\left(n + \frac{1}{2}\right)t\, dt$ as a sum of integrals over $[0, \delta]$ and $[\delta, \pi]$, then apply Exercise 6.6.18 and the Riemann–Lebesgue Lemma.

9.2.20. (a) We write

$$\int_0^\delta g(t) \frac{\sin \lambda t}{t}\, dt = \int_0^\delta \frac{g(t) - g(0+)}{t} \sin \lambda t\, dt + g(0+) \int_0^\delta \frac{\sin \lambda t}{t}\, dt.$$

By assumption, $(g(t) - g(0+))/t$ is integrable on $[0, \delta]$, so Riemann–Lebesgue Lemma implies that

$$\int_0^\delta \frac{g(t) - g(0+)}{t} \sin \lambda t\, dt \to 0, \quad \text{as} \quad \lambda \to \infty.$$

Further, using the substitution $u = \lambda t$,

$$\int_0^\delta \frac{\sin \lambda t}{t}\, dt = \int_0^{\lambda \delta} \frac{\sin u}{u}\, du \to \int_0^\infty \frac{\sin u}{u}\, du = \frac{\pi}{2}.$$

(See Example 13.5.1.) Thus,

$$\frac{2}{\pi} \int_0^\delta g(t) \frac{\sin \lambda t}{t}\, dt \to g(0+), \quad \text{as} \quad \lambda \to \infty.$$

(b) The integrability of $(g(t) - S(x))/t$ on $[0, \delta]$ implies (via Riemann–Lebesgue Lemma) that

$$\lim_{\lambda \to \infty} \int_0^\delta \frac{S(x)}{t} \sin \lambda t\, dt = \lim_{\lambda \to \infty} \int_0^\delta \frac{g(t)}{t} \sin \lambda t\, dt = \frac{\pi}{2} g(0+).$$

Since

$$\lim_{\lambda \to \infty} \int_0^\delta \frac{\sin \lambda t}{t}\, dt = \frac{\pi}{2},$$

we obtain that $S(x) = g(0+) = (f(x+) + f(x-))/2$.

9.3.2. Let $P = \{x_0, x_1, \ldots, x_n\}$ be a partition of $[-\pi, \pi]$ such that u and v are continuous on (x_{k-1}, x_k), $1 \le k \le n$. Now, the functions U, V are differentiable, and have continuous derivatives in (x_{k-1}, x_k), $1 \le k \le n$, so the "standard" Integration by Parts applies.

9.3.4. Since the function $y = x^3 - \pi^2 x$ is odd, we have that $a_n = $ for all $n \in \mathbb{N}_0$. For the same reason,

$$b_n = \frac{2}{\pi} \int_0^\pi (x^3 - \pi^2 x) \sin nx \, dx.$$

Using Integration by Parts with $u = x^3 - \pi^2 x$ and $dv = \sin nx \, dx$,

$$b_n = \frac{2}{\pi} \left[(x^3 - \pi^2 x) \left(\frac{-\cos nx}{n} \right) \Big|_0^\pi + \int_0^\pi (3x^2 - \pi^2) \frac{\cos nx}{n} \, dx \right] = \frac{2}{n\pi} \int_0^\pi (3x^2 - \pi^2) \cos nx \, dx.$$

Another Integration by Parts, using $u = 3x^2 - \pi^2$ and $dv = \cos nx \, dx$, yields

$$b_n = \frac{2}{n\pi} \left[(3x^2 - \pi^2) \frac{\sin nx}{n} \Big|_0^\pi - \int_0^\pi 6x \frac{\sin nx}{n} \, dx \right] = \frac{12}{n^2 \pi} \int_0^\pi x \sin nx \, dx.$$

The third (and final) Integration by Parts, with $u = x$ and $dv = \sin nx \, dx$, leads to

$$b_n = \frac{12}{n^2 \pi} \left[-\frac{x \cos nx}{n} \Big|_0^\pi + \int_0^\pi \frac{\cos nx}{n} \, dx \right] = \frac{12}{n^2 \pi} \left[-\frac{\pi(-1)^n}{n} + \frac{\sin nx}{n^2} \Big|_0^\pi \right] = \frac{-12(-1)^n}{n^3}.$$

Therefore,

$$x^3 - \pi^2 x \sim \sum_{n=1}^\infty \frac{-12(-1)^n}{n^3} \sin nx,$$

so term by term differentiation gives

$$\sum_{n=1}^\infty \frac{-12(-1)^n}{n^2} \cos nx.$$

On the other hand, $(x^3 - \pi^2 x)' = 3x^2 - \pi^2$, which is an even function. Therefore, $b_n' = 0$, for all $n \in \mathbb{N}$, and

$$a_0 = \frac{2}{\pi} \int_0^\pi (3x^2 - \pi^2) \, dx = \frac{2}{\pi} (x^3 - \pi^2 x) \Big|_0^\pi = 0, \quad a_n = \frac{2}{\pi} \int_0^\pi (3x^2 - \pi^2) \cos nx \, dx = \frac{-12(-1)^n}{n^2}.$$

9.4.13. Let c be a real number in $[-\pi, \pi]$ such that $s(c) = \lim_{t \to 0+} (f(c+t) + f(c-t))/2$ exists, and let $\varepsilon > 0$. As in the proof of Fejér's Theorem,

$$\sigma_n(c) - s(c) = \frac{1}{\pi} \int_{-\pi}^\pi [f(c-s) - s(c)] F_n(s) \, ds.$$

By assumption, f is integrable, hence bounded, so there exists $M > 0$ such that $|f(x)| \le M$, for all $x \in \mathbb{R}$. This also implies that $|s(c)| \le M$. If $\delta > 0$, we can define $N = \lfloor \frac{4\pi^2 M}{\delta^2 \varepsilon} \rfloor$ and conclude, just like in the proof of Fejér's Theorem, that

$$\frac{1}{\pi} \int_{-\pi}^{-\delta} |f(c-s) - s(c)| \, F_n(s) \, ds + \frac{1}{\pi} \int_\delta^\pi |f(c-s) - s(c)| \, F_n(s) \, ds \le \frac{\varepsilon}{2}.$$

It remains to consider the integral over $[-\delta, \delta]$, for a suitable choice of δ. We will choose $\delta > 0$ such that

$$\left| \frac{f(c+s) + f(c-s)}{2} - s(c) \right| < \frac{\varepsilon}{4}, \quad \text{if } |s| < \delta.$$

Now, with the aid of the substitution $w = -s$,

$$\frac{1}{\pi} \int_{-\delta}^\delta [f(c-s) - s(c)] \, F_n(s) \, ds = \frac{1}{\pi} \int_{-\delta}^0 [f(c-s) - s(c)] \, F_n(s) \, ds + \frac{1}{\pi} \int_0^\delta [f(c-s) - s(c)] \, F_n(s) \, ds$$

$$= \frac{1}{\pi} \int_0^\delta [f(c+w) - s(c)] \, F_n(w) \, dw + \frac{1}{\pi} \int_0^\delta [f(c-s) - s(c)] \, F_n(s) \, ds$$

$$= \frac{1}{\pi} \int_0^\delta [f(c+s) - s(c) + f(c-s) - s(c)] \, F_n(s) \, ds$$

$$= 2 \frac{1}{\pi} \int_0^\delta \left[\frac{f(c+s) + f(c-s)}{2} - s(c) \right] F_n(s) \, ds.$$

It follows that

$$|\sigma_n(c) - s(c)| \le \frac{\varepsilon}{2} + 2 \frac{1}{\pi} \int_0^\delta \left| \frac{f(c+s) + f(c-s)}{2} - s(c) \right| F_n(s) \, ds$$

$$< \frac{\varepsilon}{2} + 2 \frac{1}{\pi} \int_0^\delta \frac{\varepsilon}{4} F_n(s) \, ds \le \frac{\varepsilon}{2} + 2 \frac{1}{\pi} \int_0^\pi \frac{\varepsilon}{4} F_n(s) \, ds = \frac{\varepsilon}{2} + 2 \frac{\varepsilon}{4} = \varepsilon.$$

9.4.18. Hint: Follow the proof of Fejér's Theorem.

9.4.22. Let $\varepsilon > 0$. Since $na_n \to 0$, there exists $N_1 \in \mathbb{N}$ such that

$$|na_n| < \frac{\varepsilon}{3}, \quad \text{if } n \geq N_1.$$

Further, by Example 2.9.6, $\frac{1}{n}\sum_{k=1}^{n}|ka_k| \to 0$ so there exists $N_2 \in \mathbb{N}$ such that

$$\frac{1}{n}\sum_{k=1}^{n}|ka_k| < \frac{\varepsilon}{3}, \quad \text{if } n \geq N_2.$$

Finally, the series $\sum_{n=0}^{\infty}a_n$ is Abel summable, so the limit $\lim_{x\to 1^-}\sum_{n=0}^{\infty}a_n x^n$ exists. Consequently, there exists $\delta > 0$ and $L \in \mathbb{R}$ such that

$$\left|L - \sum_{k=0}^{\infty}a_k x^k\right| < \frac{\varepsilon}{3}, \quad \text{if } 1 - \delta < x < 1.$$

If $N_3 = \lfloor 1/\delta \rfloor$, then $N_3 > 1/\delta$. Furthermore, if $n \geq N_3$ then $n > 1/\delta$ so $1/n < \delta$, and it follows that $1 - \delta < 1 - \frac{1}{n}$. Therefore,

$$\left|L - \sum_{k=0}^{\infty}a_k x^k\right| < \frac{\varepsilon}{3}, \quad \text{if } 1 - \frac{1}{n} < x < 1 \text{ and } n \geq N_3.$$

Let $N = \max\{N_1, N_2, N_3\}$, and let $n \geq N$. If $1 - \frac{1}{n} < x < 1 - \frac{1}{n+1}$, we have

$$\left|L - \sum_{k=0}^{n}a_k\right| = \left|L - \sum_{k=0}^{\infty}a_k x^k + \sum_{k=0}^{\infty}a_k x^k - \sum_{k=0}^{n}a_k\right| = \left|L - \sum_{k=0}^{\infty}a_k x^k + \sum_{k=1}^{n}a_k(x^k - 1) + \sum_{k=n+1}^{\infty}a_k x^k\right|$$

$$\leq \left|L - \sum_{k=0}^{\infty}a_k x^k\right| + \sum_{k=1}^{n}|a_k|(1 - x^k) + \sum_{k=n+1}^{\infty}|a_k|x^k.$$

The first of the last three terms is less than $\varepsilon/3$ and we will show that the same is true for the other two sums.

Now, $1 - x^k = (1-x)(1 + x + x^2 + \ldots + x^{k-1}) \leq k(1-x)$, for any $k \in \mathbb{N}$. Hence, since $1 - x < 1/n$, we have that $1 - x^k < k/n$. Therefore,

$$\sum_{k=1}^{n}|a_k|(1 - x^k) < \sum_{k=1}^{n}|a_k|\frac{k}{n} = \frac{1}{n}\sum_{k=1}^{n}|ka_k| < \frac{\varepsilon}{3}.$$

Finally,

$$\sum_{k=n+1}^{\infty}|a_k|x^k = \sum_{k=n+1}^{\infty}|ka_k|\frac{x^k}{k} < \frac{\varepsilon}{3}\sum_{k=n+1}^{\infty}\frac{x^k}{k} \leq \frac{\varepsilon}{3(n+1)}\sum_{k=n+1}^{\infty}x^k \leq \frac{\varepsilon}{3(n+1)}\sum_{k=0}^{\infty}x^k = \frac{\varepsilon}{3(n+1)(1-x)}.$$

But $x < 1 - 1/(n+1)$ and so $1 - x > 1/(n+1)$. Thus $(n+1)(1-x) > 1$ and so

$$\sum_{k=n+1}^{\infty}|a_k|x^k < \frac{\varepsilon}{3}.$$

We conclude that, if $n \geq N$, then $\left|L - \sum_{k=0}^{n}a_k\right| < \varepsilon$, so the series $\sum_{k=0}^{n}a_k$ converges and its sum is L.

9.5.7. We will show that assertion (b) is false, which implies that all the more (a) is false. In order to do that, we will construct a sequence $\{f_n\}$ that converges to 0 in the mean, but for any $x \in [0,1]$ the sequence $\{f_n(x)\}$ does not converge to 0. We define $f_1(x) = 1 = \chi_{[0,1]}(x)$. Next we split $[0,1]$ into $[0,1/2]$ and $[1/2,1]$ and define $f_2(x) = \chi_{[0,1/2]}(x)$, $f_3(x) = \chi_{[1/2,1]}(x)$. For the next 4 functions we split $[0,1]$ into 4 intervals of equal length, and define $f_4(x) = \chi_{[0,1/4]}(x)$, $f_5(x) = \chi_{[1/4,1/2]}(x)$, $f_6(x) = \chi_{[1/2,3/4]}(x)$, $f_7(x) = \chi_{[3/4,1]}(x)$. The next 8 functions are obtained by splitting into 8 intervals of equal length, and defining functions f_8 through f_{15} as the characteristic functions of these intervals. Continuing this process we obtain the infinite sequence $\{f_n\}$. Since each f_n is taking only values 0 and 1, $|f_n - 0|^2 = f_n$, so

$$\int_0^1 |f_n(x) - 0|^2\, dx = \int_0^1 f_n(x)\, dx = \ell(I_n)$$

the length of the nth interval. Clearly, these lengths go to 0, so the sequence $\{f_n\}$ converges to 0 in the mean. On the other hand, for any $x \in [0, 1]$, the sequence $\{f_n(x)\}$ consists of infinitely many 0's and 1's, so it cannot be convergent. We see that $\{f_n\}$ does not converge pointwise for any $x \in [0, 1]$.

The final question of the problem also has the negative answer. Namely, even if the functions f_n and f are all continuous, and $\{f_n\}$ converges in the mean to f, it need not converge pointwise for any $x \in [0, 1]$. To see that, it suffices to make a small improvement of the example above. Namely, if f_n is the characteristic function of the interval

$$I_n = \left[\frac{j}{2^k}, \frac{j+1}{2^k}\right],$$

then we will define g_n to be

$$g_n(x) = \begin{cases} 0 & \text{if } x \in [0, \frac{j-1}{2^k}], \\ 2^k x - (j-1) & \text{if } x \in [\frac{j-1}{2^k}, \frac{j}{2^k}], \\ 1 & \text{if } x \in [\frac{j}{2^k}, \frac{j+1}{2^k}], \\ -2^k x + (j+2) & \text{if } x \in [\frac{j+1}{2^k}, \frac{j+2}{2^k}], \\ 0 & \text{if } x \in [\frac{j+2}{2^k}, 1]. \end{cases}$$

It is not hard to see that, for each $n \in \mathbb{N}$, the function g_n is continuous, $|g_n(x)|^2 \leq g_n(x)$ for all $x \in [0, 1]$, and $\int_0^1 g_n(x)\, dx \leq 2\ell(I_n)$. Therefore, the sequence $\{g_n\}$ converges to 0 in the mean, but not pointwise. Indeed, for any $x \in [0, 1]$, the sequence $\{g_n(x)\}$ contains infinitely many 0's and 1's.

9.5.9. Let $\varepsilon > 0$. By Exercise 9.5.8, there exists a continuous periodic function g with period 2π such that

$$\int_{-\pi}^{\pi} |f(t) - g(t)|^2 \, dt < \frac{\varepsilon}{4}.$$

Next, since g is continuous, the Cesàro means $\{\sigma_n[g]\}$ of the Fourier series of g converge uniformly to g. This means that there exists $N \in \mathbb{N}$ such that

$$|g(x) - \sigma_n[g](x)| < \sqrt{\frac{\varepsilon}{8\pi}} \quad \text{for all } x \in \mathbb{R} \text{ and all } n \geq N.$$

Therefore, for $n \geq N$,

$$|f(x) - \sigma_n[g](x)| \leq |f(x) - g(x)| + |g(x) - \sigma_n[g](x)| < |f(x) - g(x)| + \sqrt{\frac{\varepsilon}{8\pi}}, \quad \text{so}$$

$$|f(x) - \sigma_n[g](x)|^2 < \left(|f(x) - g(x)| + \sqrt{\frac{\varepsilon}{8\pi}}\right)^2 \leq 2\,|f(x) - g(x)|^2 + 2\frac{\varepsilon}{8\pi},$$

where we have used the inequality $(a+b)^2 \leq 2(a^2 + b^2)$. Finally, $\sigma_n[g]$ is a trigonometric polynomial. If we denote by $\{S_n\}$ the partial sums of the Fourier series of f, Lemma 9.5.2 implies that

$$\int_{-\pi}^{\pi} |f(t) - S_n(t)|^2 \, dt \leq \int_{-\pi}^{\pi} |f(t) - \sigma_n[g](t)|^2 \, dt < 2\int_{-\pi}^{\pi} |f(t) - g(t)|^2 \, dt + \int_{-\pi}^{\pi} \frac{\varepsilon}{4\pi} \, dt < 2\frac{\varepsilon}{4} + 2\pi \frac{\varepsilon}{4\pi} = \varepsilon.$$

9.5.20. Answer: $\dfrac{\alpha(\pi - \alpha)}{2}$ and $\dfrac{\pi^2 - 3\alpha\pi + 3\alpha^2}{6}$.

10.1.6. $r = \delta\sqrt{n}/2$.

10.1.10. The set of cluster points of A is $[0, 1] \times [0, 1]$.

If $(a, b) \notin [0, 1] \times [0, 1]$ then at least one of a, b is not in $[0, 1]$. E.g., if $a > 1$, then there exists $\delta > 0$ such that $a - \delta > 1$. Then $B_\delta(x, y) \cap [0, 1] \times [0, 1] = \emptyset$.

If $(a, b) \in [0, 1] \times [0, 1]$, then there exists $x, y \in [0, 1]$ such that $0 < |x - a| < \delta/2$ and $0 < |y - b| < \delta/2$. By Theorem 1.3.7 there exist a rational number r_1 between a and x, and a rational number r_2 between b and y. Then $(r_1, r_2) \in B_\delta(a, b) \cap A$.

10.2.1. We will use Theorem 10.2.5. Let $\{x_n\}, \{y_n\}$ be sequences such that $\lim x_n = 0$ and $\lim y_n = \pi/2$. Then

$$\lim \cos x_n \sin y_n = (\lim \cos x_n)(\lim \sin y_n) = \cos(\lim x_n) \sin(\lim y_n) = \cos 0 \sin \frac{\pi}{2} = 1.$$

Since $\{x_n\}, \{y_n\}$ are arbitrary, we conclude that $\lim_{(x,y) \to (0, \pi/2)} \cos x \sin y = 1$.

10.2.5. By Cauchy-Schwarz inequality, $|xy| \leq \frac{1}{2}(x^2 + y^2)$, which implies that $|xy| \leq \frac{1}{2}(x^2 + y^2 + z^2)$. Therefore,

$$0 \leq \left|\frac{xyz}{x^2 + y^2 + z^2}\right| \leq \frac{1}{2}z \to 0, \quad z \to 0,$$

and we see that $\lim_{(x,y,z) \to (0,0,0)} xyz/(x^2 + y^2 + z^2) = 0$.

10.2.9. Let $u = x + y^2$. Then $u \to 0$ as $(x, y) \to (0, 0)$. Thus, we can write the given limit as $\lim_{u \to 0} \sin u/u$ which equals 1 (see Example 3.1.17). Thus, $\lim_{(x,y) \to (0,0)} \sin(x + y^2)/(x + y^2) = 1$.

10.2.13. Let $\{x_n\}, \{y_n\}$ be sequences defined by $x_n = 1/n$, $y_n = 0$. Then

$$\lim \frac{x_n}{x_n^2 + y_n^2} = \lim \frac{\frac{1}{n}}{(\frac{1}{n})^2} = \lim n = +\infty.$$

On the other hand, if we define $x_n = 0$, $y_n = 1/n$ then

$$\lim \frac{x_n}{x_n^2 + y_n^2} = \lim \frac{0}{(\frac{1}{n})^2} = 0.$$

Thus, by Theorem 10.2.5, the limit does not exist.

10.2.15. Hint: Consider sequences of the form $(1/k, a/k^2)$.

10.3.1. Let (a_n, b_n) be a sequence in $\mathbb{R}^2$ such that $\lim a_n = -1$ and $\lim b_n = 2$. Then $\lim f(a_n, b_n) = \lim(a_n b_n + 6a_n) = -8$. Since $f(-1, 2) = -8$, the result follows from Corollary 10.3.2.

10.3.5. Let $(a, b) \in \mathbb{R}^2$ and let (a_n, b_n) be a sequence in $\mathbb{R}^2$ that converges to (a, b). Then $\lim f(a_n, b_n) = \lim(a_n b_n^2 - 3a_n) = ab^2 - 3a$. Since $f(a, b) = ab^2 - 3a$, the result follows from Corollary 10.3.2.

10.3.11. Let $(a, b) \in A$ and let $(a_n, b_n) \to (a, b)$. For any $\varepsilon > 0$, there exists $N \in \mathbb{N}$ such that, if $n \geq N$ then $|b_n - b| < \varepsilon/(2M)$ and $|f(a_n, b_n) - f(a, b)| < \varepsilon/2$. For $n \geq N$,

$$|f(a_n, b_n) - f(a, b)| \leq |f(a_n, b_n) - f(a_n, b)| + |f(a_n, b) - f(a, b)| \leq M|b_n - b| + \frac{\varepsilon}{2} < \frac{M\varepsilon}{2M} + \frac{\varepsilon}{2} = \varepsilon.$$

10.3.14. Hint: Try $\mathbf{a_k} = (1/k, 1/k)$ and $\mathbf{b_k} = (1/k^2, 1/k)$.

10.3.16. Answer: No.

10.3.17. Answer: No.

10.3.18. Hint: Use the inequality $xy \leq (x^2 + y^2)/2$ and the fact that $\lim_{t \to 0} t \ln t = 0$.

10.3.20. Let $a_n \to a$. Then $(a_n, b) \to (a, b)$ and $f(a_n, b) \to f(a, b)$, since f is continuous at (a, b). Therefore, $g(a_n) \to g(a)$, so g is continuous at $x = a$.

10.4.3. Let $A = \{(x, y) : x + y \leq 1\}$. We will show that A contains all of its cluster points. Suppose to the contrary that there exists a point in $(a, b) \in A^c$ that is a cluster point of A. Since $a + b > 1$, there exists $r \in (0, (a + b - 1)/2)$. We will show that the ball $B_r(a, b)$ contains no points of A. Indeed, let $(x, y) \in B_r(a, b)$. By Cauchy-Schwarz inequality $|x - a| + |y - b| \leq \sqrt{2}\sqrt{|x - a|^2 + |y - b|^2} < r\sqrt{2} < a + b - 1$. Further, by Triangle inequality $x > a - |x - a|$ and $y > b - |y - b|$. Thus, $x + y > a + b - |x - a| - |y - b| > a + b - (a + b - 1) = 1$, whence $(x, y) \in A^c$.

10.4.7. Let $A = \{(x, y) : x \in \mathbb{Q}\}$. We will show that $(\sqrt{2}, 0)$ is a cluster point of A, and it is obvious that it does not belong to A. To that end, let $r > 0$ and consider the disk $B_r(\sqrt{2}, 0)$. It suffices to show that there exists a point $(x, y) \in A \cap B_r(\sqrt{2}, 0)$. Since rational numbers are dense in $\mathbb{R}$ there exists $q \in \mathbb{Q}$ such that $|q - \sqrt{2}| < r$. Then $(q, 0) \in A$ and $\|(q, 0) - (\sqrt{2}, 0)\| = [(q - \sqrt{2})^2 + 0^2]^{1/2} = |q - \sqrt{2}| < r$, so $(q, 0) \in B_r(\sqrt{2}, 0)$.

10.4.17. Let $\mathbf{c}$ be a cluster point of $A \cup B$. By Exercise 10.2.26, for each $n \in \mathbb{N}$, there exists $\mathbf{x}_n \in A \cup B$ such that $\|\mathbf{x}_n - \mathbf{c}\| < 1/n$. Now A or B must contain infinitely many members of $\{\mathbf{x}_n\}$. Suppose that A does. Then $\mathbf{c}$ is a cluster point of A, so $\mathbf{c} \in A \subset A \cup B$.

10.4.20. See the proof of Theorem 2.3.7.

10.4.21. For each $n \in \mathbb{N}$, select $\mathbf{x}_n \in F_n$. By Bolzano–Weierstrass Theorem, there exists a convergent subsequence $\mathbf{x}_{n_k}$. Let $\mathbf{a}$ be its limit. Now, if n is a positive integer, then for any $k \geq n$, $\mathbf{x}_{n_k} \in F_{n_k} \subset F_n$. Consequently, $\mathbf{a} \in F_n$.

10.4.26. If A is closed and $\mathbf{c} \in \partial A$, then for each $n \in \mathbb{N}$, there exists $\mathbf{x}_n \in A$ such that $\|\mathbf{x}_n - \mathbf{c}\| < 1/n$. Therefore, $\mathbf{c}$ is a cluster point of A, so $\mathbf{c} \in A$.

If A contains its boundary and $\mathbf{c}$ is a cluster point of A, then $\mathbf{c} \in A$. Indeed, for each $n \in \mathbb{N}$, the ball $B_{1/n}(\mathbf{c})$ contains a point of A. If $\mathbf{c} \notin A$, then $B_{1/n}(\mathbf{c})$ contains also a point $\mathbf{c}$ in the complement of A, so $\mathbf{c} \in \partial A \subset A$.

10.5.1. The set is not open. Indeed, let $r > 0$. The disk $B_r((1, 0))$ contains the point $(1 + r/2, 0)$ because $\|(1 + r/2, 0) - (1, 0)\| = r/2 < r$. However, $(1 + r/2, 0)$ does not belong to the set $\{(x, y) : x + y = 1\}$.

10.5.7. The set $A = \{(x, y) : x \in \mathbb{Q}\}$ is not open. Indeed, let $r > 0$ and consider the point $(0, 0) \in A$. If $n > \sqrt{2}/r$ then the point $(\sqrt{2}/n, 0)$ belongs to the disk $B_r((0, 0))$ but not to A.

10.5.14. Let $\mathbf{x} \in A \cup B$. Without loss of generality, suppose that $\mathbf{x} \in A$. Then there exists $r > 0$ such that $B_r(\mathbf{x}) \subset A \subset A \cup B$. It follows that $\mathbf{x}$ is an interior point of $A \cup B$.

10.5.17. Let $(a, b) \in G = \{(x, y) : x \in A, y \in B\}$. Then $a \in A$, $b \in B$ so there exist $r_1, r_2 > 0$ such that $B_{r_1}(a) \subset A$ and $B_{r_1}(b) \subset B$. Let $r = \min\{r_1, r_2\}$. We will show that $B_r((a, b)) \subset G$. Indeed, let $(x, y) \in B_r((a, b))$. Then $\|(x, y) - (a, b)\| < r \leq r_1$ so it follows that $|x - a| < r_1$, whence $x \in A$. Similarly, $y \in B$ so $(x, y) \in G$.

10.5.21. Suppose first that f is continuous on A, and let G be an open set in $\mathbb{R}$. If $\mathbf{c}$ is an arbitrary point in $f^{-1}(G)$, and $d = f(\mathbf{c})$, then there exists $\varepsilon > 0$ so that $B_\varepsilon(d) \subset G$. In other words, if $|y - d| < \varepsilon$ then $y \in G$. The continuity of f at $\mathbf{c}$ implies that there exists $\delta > 0$ so that, if $\mathbf{x} \in A$ and $\|\mathbf{x} - \mathbf{c}\| < \delta$, then $|f(\mathbf{x}) - d| < \varepsilon$. The latter implies that $f(\mathbf{x}) \in G$ and, combined with $\mathbf{x} \in A$, that $\mathbf{x} \in f^{-1}(G)$. To summarize, for any $\mathbf{c} \in f^{-1}(G)$ there exists $\delta > 0$ such that $B_\delta(\mathbf{c}) \cap A \subset f^{-1}(G)$. Let G_0 be the union of all $B_\delta(\mathbf{c})$, as $\mathbf{c} \in f^{-1}(G)$. Then G_0 is an open set, and $G_0 \cap A \subset f^{-1}(G)$. The other inclusion is obvious since $\mathbf{c} \in f^{-1}(G)$ implies that $\mathbf{c} \in A$, and $\mathbf{c} \in B_\delta(\mathbf{c})$ for some δ.

To prove the converse, let $\mathbf{c} \in A$ and $d = f(\mathbf{c})$. Let $\varepsilon > 0$. The set $B_\varepsilon(d)$ is open so, by assumption, there exists an open set G_0 such that $f^{-1}(B_\varepsilon(d)) = G_0 \cap A$. Since $\mathbf{c} \in f^{-1}(B_\varepsilon(d))$, $\mathbf{c}$ belongs to the open set G_0, and there exists $\delta > 0$ such that $B_\delta(\mathbf{c}) \subset G_0$. Let $\mathbf{x} \in A$ and $\|\mathbf{x} - \mathbf{c}\| < \delta$. Then $\mathbf{x} \in A$ and $\mathbf{x} \in G_0$, so $\mathbf{x} \in f^{-1}(B_\varepsilon(d))$. Therefore, $f(\mathbf{x}) \in B_\varepsilon(d)$, which means that $|f(\mathbf{x}) - d| < \varepsilon$, so f is continuous at $\mathbf{c}$. Since $\mathbf{c}$ was an arbitrary point in A, f is continuous in A.

10.5.26. For any $k \in \mathbb{N}$, the set

$$A_k = \left\{ \mathbf{x} \in \mathbb{R}^n : \exists r = r(k, \mathbf{x}) > 0, \text{ such that } \mathbf{y}, \mathbf{z} \in B_r(\mathbf{x}) \Rightarrow |f(\mathbf{y}) - f(\mathbf{z})| < \frac{1}{k} \right\},$$

is open. Indeed, let $k \in \mathbb{N}$, let $\mathbf{c} \in A_k$, and let $r = r(k, \mathbf{c})$. We will show that $B_r(\mathbf{c}) \subset A_k$.

Let $\mathbf{a} \in B_r(\mathbf{c})$. Then $\|\mathbf{a} - \mathbf{c}\| < r$, and we define $r' = r - \|\mathbf{a} - \mathbf{c}\|$. If $\mathbf{y} \in B_{r'}(\mathbf{a})$, then

$$\|\mathbf{y} - \mathbf{c}\| \le \|\mathbf{y} - \mathbf{a}\| + \|\mathbf{a} - \mathbf{c}\| < r' + \|\mathbf{a} - \mathbf{c}\| = r,$$

so $\mathbf{y} \in B_r(\mathbf{c})$. It follows that, if $\mathbf{y}, \mathbf{z} \in B_{r'}(\mathbf{a})$ then $\mathbf{y}, \mathbf{z} \in B_r(\mathbf{c})$, so $|f(\mathbf{y}) - f(\mathbf{z})| < \frac{1}{k}$. We conclude that $\mathbf{a} \in A_k$, so $\mathbf{c}$ is an interior point of A_k. Since $\mathbf{c}$ was arbitrary, the set A_k is open, and the set $A = \cap_{k \in \mathbb{N}} A_k$ is a G_δ set.

It remains to show that $\mathbf{x} \in A$ if and only if f is continuous at $\mathbf{x}$. If f is continuous at $\mathbf{x}$ and $k \in \mathbb{N}$, then there exists $r > 0$ such that $\|\mathbf{a} - \mathbf{x}\| < r$ implies $|f(\mathbf{a}) - f(\mathbf{x})| < 1/(2k)$. By Triangle Inequality, if $\mathbf{y}, \mathbf{z} \in B_r(\mathbf{x})$, then $|f(\mathbf{y}) - f(\mathbf{z})| < \frac{1}{k}$, so $\mathbf{x} \in A_k$.

In the other direction, let $\mathbf{x} \in A$, and let $\varepsilon > 0$. We select $k \in \mathbb{N}$ so that $1/k \le \varepsilon$. Since $\mathbf{x} \in A_k$, there exists $r > 0$ such that, $\|\mathbf{a} - \mathbf{x}\| < r$ implies $|f(\mathbf{a}) - f(\mathbf{x})| < 1/k \le \varepsilon$. This means that f is continuous at $\mathbf{x}$.

The statement about the discontinuities of f can be proved by taking complements: a set A is G_δ if and only if A^c is F_σ.

10.5.32. Suppose first that A is closed and let $x \in \partial A$. For any $r > 0$, $B^r(x)$ has a nonempty intersection with both A and A', so it is either an isolated point of A, or it is in A', and hence in A (because A is closed). Thus, $\partial A \subset A$. In the other direction suppose that A is not closed. Then, there is a cluster point y of A that is not in A. By definition, for any $r > 0$, $B_r(y)$ has a non-empty intersection with A. It also has a non-empty intersection with A^c (it contains y) so $y \in \partial A \setminus A$. In other words, if A is not closed it does not contain its boundary.

10.6.1. The line $A = \{(x, y) : x + y = 1\}$ is polygonally connected, and therefore path connected and connected. Indeed, let P, Q be two points in A. The line segment PQ is a polygonal path connecting P and Q and it lies in A.

10.6.7. The set $\mathbb{Z}$ is not connected, and therefore neither path connected nor polygonally connected. Indeed, take $B = (-\infty, 1/2)$ and $C = (1/2, +\infty)$. The sets B and C are open, $\mathbb{Z} \subset B \cup C$, both $\mathbb{Z} \cap B$ and $\mathbb{Z} \cap C$ are non-empty, and $(\mathbb{Z} \cap B) \cap (\mathbb{Z} \cap C) = \emptyset$.

10.6.10. Modify the proof of Theorem 10.6.8 to show that if A is path connected then it has the Intermediate Value Property.

10.6.12. Suppose that A is connected but that $f(A)$ is not. By definition, there exist open set $\tilde{B}, \tilde{C}$ such that the sets $f(A) \cap \tilde{B}$ and $f(A) \cap \tilde{C}$ are non-empty, disjoint sets whose union contains $f(A)$. If we denote $B = f^{-1}(\tilde{B})$ and $C = f^{-1}(\tilde{C})$, then B, C are open sets, and $A \cap B, A \cap C$ are non-empty, disjoint sets whose union contains A. Indeed, if $\mathbf{x}$ belongs to both $A \cap B$ and $A \cap C$, then $f(\mathbf{x})$ belongs to $f(A) \cap \tilde{B}$ and $f(A) \cap \tilde{C}$ which is impossible. Also, if $\mathbf{x} \in A$, then $y = f(\mathbf{x})$ belongs to $\tilde{B} \cup \tilde{C}$, so $\mathbf{x} \in f^{-1}(\tilde{B} \cup \tilde{C}) = f^{-1}(\tilde{B}) \cup f^{-1}(\tilde{C}) = B \cup C$. Finally, if $A \cap B$ were empty, then it would follow that $A \subset C$, so $f(A) \subset f(C) \subset \tilde{C}$, which would imply that $f(A) \cap \tilde{B}$ is empty. Thus, the assumption that $f(A)$ is disconnected leads to the conclusion that A is disconnected.

If A is path connected then so is $f(A)$. Indeed, for any $c, d \in f(A)$ there exist $\mathbf{a}, \mathbf{b} \in A$ such that $f(\mathbf{a}) = c$ and $f(\mathbf{b}) = d$. Since A is path connected there exists a continuous function $\varphi : [0, 1] \to A$ with $\varphi(0) = \mathbf{a}$, $\varphi(1) = \mathbf{b}$. The function $f \circ \varphi$ is a continuous function from $[0, 1]$ to $f(A)$, and $f \circ \varphi(0) = c$, $f \circ \varphi(1) = d$, so $f(A)$ is path connected.

Finally, when A is polygonally connected and f is a continuous function with values in $\mathbb{R}^2$, $f(A)$ need not be polygonally connected. Example: $A = [0, 1]$, $f(t) = (\cos t, \sin t)$.

10.6.23. We will prove that A_1 must be connected. Suppose that this is not true. Then there exist open sets B, C so that $U = A_1 \cap B$ and $V = A_1 \cap C$ are disjoint, non-empty sets whose union is A_1.

First we will show that $\overline{U} \cap V = \emptyset$. Notice that

$$A_1 \cap B = A_1 \setminus (A_1 \cap C) = A_1 \cap (A_1 \cap C)^c = A_1 \cap (A_1^c \cup C^c) = (A_1 \cap A_1^c) \cup (A_1 \cap C^c) = A_1 \cap C^c.$$

Therefore, $\overline{A_1 \cap B} = \overline{A_1 \cap C^c} \subset \overline{A_1} \cap \overline{C^c} = A_1 \cap C^c$, because both A_1 and C^c are closed. It follows that

$$\overline{U} \cap V = \overline{A_1 \cap B} \cap V \subset (A_1 \cap C^c) \cap V = A_1 \cap C^c \cap A_1 \cap C = \emptyset,$$

and similarly, $U \cap \overline{V} = \emptyset$.

Next we notice that $A_1 \cap A_2$ is a connected subset of A, so it must be contained in U or it must be contained in V. Without loss of generality, suppose that $A_1 \cap A_2 \subset U$, and let $W = U \cup A_2$. Clearly, $V \cup W = V \cup U \cup A_2 = A_1 \cup A_2$. Also, $\overline{V} \cap A_2 = \emptyset$. Indeed,

$$\overline{V} \cap A_2 = \overline{V} \cap ((A_1 \cap A_2) \cup (A_2 \setminus A_1)) = \left(\overline{V} \cap A_1 \cap A_2\right) \cup \left(\overline{V} \cap (A_2 \setminus A_1)\right)$$

$$\subset \left(\overline{V} \cap U\right) \cup \left(\overline{V} \cap A_2 \cap A_1^c\right) = \overline{A_1 \cap C} \cap A_2 \cap A_1^c \subset \overline{A_1} \cap \overline{C} \cap A_2 \cap A_1^c = \emptyset.$$

It follows that both $V \cap \overline{W}$ and $\overline{V} \cap W$ are empty. First,

$$V \cap \overline{W} = V \cap \overline{U \cup A_2} = V \cap \left(\overline{U} \cup \overline{A_2}\right) = \left(V \cap \overline{U}\right) \cup \left(V \cap \overline{A_2}\right) = V \cap \overline{A_2} = \emptyset.$$

Also,

$$\overline{V} \cap W = \overline{V} \cap (U \cup A_2) = \left(\overline{V} \cap U\right) \cup \left(\overline{V} \cap A_2\right) = \overline{V} \cap A_2 = \emptyset.$$

Let $P = \left(\overline{V}\right)^c$ and $Q = \left(\overline{W}\right)^c$. Clearly, they are open sets, and we will show that they form a disconnection of $A_1 \cup A_2$. To begin with,

$$(A_1 \cup A_2) \cap P = (A_1 \cup A_2) \cap \left(\overline{V}\right)^c \supset (A_1 \cup A_2) \cap W = W,$$

$$(A_1 \cup A_2) \cap Q = (A_1 \cup A_2) \cap \left(\overline{W}\right)^c \supset (A_1 \cup A_2) \cap V = V,$$

so the sets $(A_1 \cup A_2) \cap P$ and $(A_1 \cup A_2) \cap Q$ are non-empty. They are also disjoint:

$$[(A_1 \cup A_2) \cap P] \cap [(A_1 \cup A_2) \cap Q] = [(A_1 \cup A_2) \cap \left(\overline{V}\right)^c] \cap [(A_1 \cup A_2) \cap \left(\overline{W}\right)^c] = (A_1 \cup A_2) \cap \left(\overline{V} \cup \overline{W}\right)^c$$
$$\subset (A_1 \cup A_2) \cap (V \cup W)^c = (A_1 \cup A_2) \cap (A_1 \cup A_2)^c = \emptyset.$$

Finally,
$$A_1 \cup A_2 \supset [(A_1 \cup A_2) \cap P] \cup [(A_1 \cup A_2) \cap Q] \supset W \cup V = A_1 \cup A_2,$$

which implies that $A_1 \cup A_2 = [(A_1 \cup A_2) \cap P] \cup [(A_1 \cup A_2) \cap Q]$, and the proof is complete.

If $A_1 = (0, 1]$ and $A_2 = \{0, 1\}$, then $A_1 \cap A_2 = \{1\}$ and $A_1 \cup A_2 = [0, 1]$ which are both connected, but A_2 is not. Thus, the assumption that the sets A_1, A_2 are closed cannot be omitted.

10.6.25. By Exercise 10.6.12 it suffices to show that if $F : A \times B \to \{1, 2\}$ is continuous then it must be constant. Let $(a, b), (a', b') \in A \times B$. The sets $\{a\} \times B$ and $A \times \{b'\}$ are connected, so F is constant on each of them. Since both sets contain (a, b'), $F(a, b) = f(a', b')$. For path connected the answer is "yes". Let $\gamma_1 : [0, 1] \to A$ and $\gamma_2 : [0, 1] \to B$ so that $\gamma_1(0) = a$, $\gamma_1(1) = a'$, $\gamma_2(0) = b$, $\gamma_2(0) = b'$. Take $\gamma(t) = (\gamma_1(t), \gamma_2(t))$. For polygonally connected use the same argument with γ_1, γ_2 piecewise linear.

10.6.30. Suppose that A is not connected and let B, C be open set such that $A \cap B$ and $A \cap C$ are non-empty disjoint sets, whose union is A. The fact that they are disjoint implies that $A \cap B = A \cap B \cap (A \cap C)^c$ and $A \cap C = A \cap C \cap (A \cap B)^c$. Further,

$$A \cap B \cap (A \cap C)^c = A \cap B \cap (A^c \cup C^c) = (A \cap B \cap A^c) \cup (A \cap B \cap C^c) \subset A \cap C^c,$$

and $A \cap C \cap (A \cap B)^c \subset A \cap B^c$. This implies that

$$A \subset (A \cap C^c) \cup (A \cap B^c) \subset A,$$

so (assuming that A is closed) A is a union of 2 closed sets $U = A \cap B^c$ and $V = A \cap C^c$.

Further, since $A = (A \cap B) \cup (A \cap C)$ we have that

$$U \cap V = (A \cap B^c) \cap (A \cap C^c) = [(A \cap B) \cup (A \cap C)] \cap B^c \cap C^c$$
$$= [(A \cap B) \cap B^c \cap C^c] \cup [(A \cap C) \cap B^c \cap C^c] = \emptyset.$$

Now, let $u \in U$, $v \in V$. Since the set A has Cantor's property, we can define inductively two sequences $\{u_n\} \subset U$ and $\{v_n\} \subset V$ such that, for every $n \in \mathbb{N}$, $|u_n - v_n| < 1/n$. The boundedness of U and V implies that there exist convergent subsequences $\{u_{n_k}\}, \{v_{n_k}\}$. The sets U and V are closed so $u = \lim u_{n_k} \in U$ and $v = \lim v_{n_k} \in V$. On the other hand, $|u_{n_k} - v_{n_k}| < 1/n_k$ shows that $u = v$ which would contradict the fact that U and V are disjoint. Thus, A must be connected.

To show that it is essential that A be closed, consider $A = [0, 1) \cup (1, 2]$. To show that it is essential that A be bounded, consider the set $A = \{(x, y) \in \mathbb{R}^2 : x \geq 1, x|y| \geq 1\}$.

10.7.1. The set is not compact because it is not bounded. Indeed, let $M > 0$. The point $(M + 1, -M)$ belongs to the set, but $\|(M + 1, -M)\| > M$. The covering $\{B_r(0, 0) : r > 0\}$ is an open covering of $\mathbb{R}^2$ and therefore of A. If there were a finite subcovering, one of the balls would have the largest radius R, so we would have that $A \subset B_R(0, 0)$. In other words, we would conclude that A is bounded which is false.

10.7.11. Let $\{G_\alpha : \alpha \in A\}$ be an open covering of $[0, 1]$. Let

$$C = \{x \in [0, 1] : \text{ the interval } [0, x] \text{ is covered by finitely many } G_\alpha\},$$

and let $a = \sup C$. The set C is non-empty, because it contains 0. It is bounded, because it is a subset of $[0, 1]$, so a is well-defined.

Notice that $a \in C$. Indeed, there exists $\alpha_0 \in A$ so that $a \in G_{\alpha_0}$. Since G_{α_0} is open, there exists $r > 0$ such that $(a - r, a + r) \subset G_{\alpha_0}$. Clearly, $a - r$ is not an upper bound of A, so $[0, a - r]$ is covered by finitely many sets G_{α_k}, $1 \leq k \leq n$. Then $[0, a]$ is covered by G_{α_k}, $0 \leq k \leq n$, so $a \in C$.

It remains to prove that $a = 1$. Suppose that $a < 1$. If r is as above, then $a + r/2$ would also belong to C, so $a = 1$.

10.7.14. If A and B are such sets, then $A \cap B$ is both open and closed. Therefore it is either $\mathbb{R}^n$ or the empty set. Since $\mathbb{R}^n$ is not compact, it follows that $A \cap B = \emptyset$.

10.7.17. Let $\mathbf{x} \in A$. Then there exists $G^{\mathbf{x}} \in \mathcal{G}$ such that $\mathbf{x} \in G^{\mathbf{x}}$. Since $G^{\mathbf{x}}$ is open, there exists a ball $B_r^{\mathbf{x}}(\mathbf{a})$ with center $\mathbf{a}$ having all coordinates rational and with radius $r \in \mathbb{Q}$, such that $\mathbf{x} \in B_r^{\mathbf{x}}(\mathbf{a}) \subset G^{\mathbf{x}}$. It is clear that $\{B_r^{\mathbf{x}}(\mathbf{a}) : \mathbf{x} \in A\}$ is a countable open covering of A. Further, for each $B_r^{\mathbf{x}}(\mathbf{a})$ there exists at least one set $G(\mathbf{a}) \in \mathcal{G}$ such that $B_r^{\mathbf{x}}(\mathbf{a}) \subset G(\mathbf{a})$. The collection of the sets $\{G(\mathbf{a})\}$ is a countable open covering of A.

11.1.2. $-6 \sin x = 6 \cos y$.

11.1.4. Since $f(x, y) = 1 + \dfrac{2y}{x - y}$, we have that

$$\frac{\partial^m f}{\partial x^m} = \frac{(-1)^m m! \, 2y}{(x - y)^{m+1}} = \frac{(-1)^{m+1} m! \, 2}{(x - y)^m} + \frac{(-1)^m m! \, 2x}{(x - y)^{m+1}}, \quad \text{so} \quad \frac{\partial^{m+n} f}{\partial x^m \partial y^n} = \frac{2(-1)^m (m + n - 1)!}{(x - y)^{m+n-1}} (nx + my).$$

11.1.6. $6dx^3 - 18dx^2 dy + 18dx dy^2 + 6dy^3$.

11.1.8. $e^{ax + by + cz} (a \, dx + b \, dy + c \, dz)^n$.

11.2.2. No. If f satisfies (11.8), then

$$A_1 = f'_x(0,0) = \lim_{h \to 0} \frac{\sqrt[3]{h^3}}{h} = 1,$$

and similarly, $A_2 = 1$. Then $r(x,y) = \sqrt[3]{x^3 + y^3} - (x+y)$ and, if f were differentiable, we would have

$$\lim_{(x,y)\to(0,0)} \frac{\sqrt[3]{x^3 + y^3} - (x+y)}{\sqrt{x^2 + y^2}} = 0.$$

However, this is not true. E.g., if $x = y = 1/n$, the limit is $(\sqrt[3]{2} - 2)/\sqrt{2}$.

11.2.4. No. The partial derivatives are $f'_x(0,0) = f'_y(0,0) = 0$, so $r(x,y) = \sqrt{|xy|}$. However, the limit $\lim_{(x,y)\to(0,0)} r(x,y)/\sqrt{x^2 + y^2} \neq 0$. (Try $x = y = 1/n$.)

11.2.12. By definition,

$$f'_x(0,0) = \lim_{h \to 0} \frac{f(0+h,0) - f(0,0)}{h} = \lim_{h \to 0} \frac{|h|(\varphi(h) - \varphi(0))}{h} = 0,$$

because φ is continuous. Similarly, $f'_y(0,0) = 0$, so $r(x,y) = f(x,y)$ and

$$\frac{r(x,y)}{\sqrt{x^2 + y^2}} = \varphi(\sqrt{x^2 + y^2}) - \varphi(0) \to 0, \text{ as } (x,y) \to (0,0).$$

Thus, f is differentiable at $(0,0)$. However, if $(a,b) \neq (0,0)$, f does not have partial derivatives. Indeed,

$$f(a+h,b) - f(a,b) = \sqrt{(a+h)^2 + b^2}\left[\varphi(\sqrt{(a+h)^2 + b^2}) - \varphi(0)\right] - \sqrt{a^2 + b^2}\left[\varphi\left(\sqrt{a^2 + b^2}\right) - \varphi(0)\right]$$

$$= \left(\sqrt{(a+h)^2 + b^2} - \sqrt{a^2 + b^2}\right)\left[\varphi(\sqrt{(a+h)^2 + b^2}) - \varphi(0)\right] + \sqrt{a^2 + b^2}\left[\varphi(\sqrt{(a+h)^2 + b^2}) - \varphi(\sqrt{a^2 + b^2})\right].$$

Now

$$\lim_{h \to 0} \frac{\sqrt{(a+h)^2 + b^2} - \sqrt{a^2 + b^2}}{h}$$

exists and equals $g'_x(a,b)$, where $g(x,y) = \sqrt{x^2 + y^2}$. Also,

$$\lim_{h \to 0} \left[\varphi(\sqrt{(a+h)^2 + b^2}) - \varphi(0)\right] = \varphi\left(\sqrt{a^2 + b^2}\right) - \varphi(0).$$

Thus,

$$\lim_{h \to 0} \frac{f(a+h,b) - f(a,b)}{h}$$

exists if and only if

$$\lim_{h \to 0} \frac{\varphi(\sqrt{(a+h)^2 + b^2}) - \varphi(\sqrt{a^2 + b^2})}{h}$$

does. However, the latter does not exist. To see that, let $c = \sqrt{a^2 + b^2}$ and $k = \sqrt{h^2 + 2ah}$. Then the limit in question equals

$$\lim_{k \to 0} \frac{\varphi(c+k) - \varphi(c)}{k}$$

which does not exist since it would have been the derivative of φ at c.

11.2.13. (c) The function $f(x,y) = \sin x \tan y$ has partial derivatives $f'_x(\pi/6, \pi/4) = \sqrt{3}/2$ and $f'_y(\pi/6, \pi/4) = 1$. Since $f(\pi/6, \pi/4) = 1/2$, and $1° = \pi/180$ we obtain the approximation

$$\frac{1}{2} + \frac{\sqrt{3}}{2}\left(-\frac{\pi}{180}\right) + 1\left(\frac{\pi}{180}\right) \approx 0.5023382978.$$

11.2.23. Since partial derivatives of f are bounded, there exists $M > 0$ such that $|f'_x(a,b)|, |f'_y(a,b)| \leq M$, for all $(a,b) \in A$. Let $\varepsilon > 0$ and choose $\delta = \varepsilon/(2M)$. Suppose now that (a,b) and (c,d) are any 2 points in A such that $\|(a,b) - (c,d)\| < \delta$. Define $g(t) = f(a + t(c-a), b + t(d-b))$ and notice that g is defined for $t \in [0,1]$, because

A is convex. Also, g is differentiable and, by Mean Value Theorem, $g(1) - g(0) = g'(z)$ for some $z \in (0,1)$. By the Chain Rule,

$$g'(z) = f'_x[a + z(c - a), \, b + z(d - b)](c - a) + f'_y[a + z(c - a), \, b + z(d - b)](d - b),$$

so

$$|f(c, d) - f(a,b)| = |g(1) - g(0)| = \left|g'(z)\right|$$
$$= \left|f'_x(a + z(c - a), \, b + z(d - b))(c - a) + f'_y(a + z(c - a), \, b + z(d - b))(d - b)\right|$$
$$\leq M|c - a| + M|d - b| \leq M\sqrt{2}\sqrt{(c - a)^2 + (d - b)^2} < M\sqrt{2}\delta = M\sqrt{2}\,\frac{\varepsilon}{2M} < \varepsilon.$$

11.2.24. Let $(a, b) \in A$, and let M be such that $|f'_y(a, b)| \leq M$. Let $\varepsilon > 0$ and take $\delta < \varepsilon/(2M)$ and such that whenever $|c - a| < \delta$, $|f(c, b) - f(a, b)| < \varepsilon/2$. If $\|(a, b) - (c, d)\| < \delta$, then

$$|f(c, d) - f(a, b)| \leq |f(c, d) - f(c, b)| + |f(c, b) - f(a, b)| \leq M\delta + \frac{\varepsilon}{2} < \varepsilon.$$

11.3.4. $df = yx^{y-1}\, dx + x^y \ln y \, dy$; $d^2f = y(y-1)x^{y-2}\, dx^2 + 2(x^{y-1} + yx^{y-1}\ln x)\, dx dy + x^y(\ln x)^2\, dy^2$.

11.3.6. $df = \dfrac{dx}{1 + x^2} + \dfrac{dy}{1 + y^2}$; $d^2f = \dfrac{-2x}{(1 + x^2)^2}\, dx - \dfrac{-2y}{(1 + y^2)^2}\, dy$.

11.3.9. $\mathbf{D}f(\mathbf{a})(\mathbf{u}) = \frac{1}{2}u_1 + \frac{5\sqrt{3}}{2}u_2$. $\mathbf{D}^2f(\mathbf{a})(\mathbf{u})^2 = \sqrt{3}u_1 u_2 - \frac{5}{2}u_2^2$.

11.3.11. $\mathbf{D}f(\mathbf{a})(\mathbf{u}) = u_1 + u_2 + 3u_3$. $\mathbf{D}^2f(\mathbf{a})(\mathbf{u})^2 = 2u_2 u_3$.

11.4.12. $\begin{bmatrix} 0 & -2/3 & 1/3 \\ 1 & 0 & -1/3 \\ -1 & 2/3 & 0 \end{bmatrix}$. **11.4.14.** $\begin{bmatrix} 5 \\ -6 \\ 0 \end{bmatrix}$. **11.4.17.** $(0.867, -0.915)$.

11.4.20. If $\mathbf{f}$ is linear, then $\forall \mathbf{x}, \mathbf{a} \in \mathbb{R}^n$, $\mathbf{f}(\mathbf{x}) = \mathbf{f}(\mathbf{a}) + \mathbf{f}(\mathbf{x} - \mathbf{a})$. Therefore, (11.17) holds with $\mathbf{D}\mathbf{f}(\mathbf{a}) = \mathbf{f}$ and $\mathbf{r} = \mathbf{0}$.

11.5.2. Hint: Use the formula $\frac{1}{1+x} = \sum_{n=0}^{\infty}(-x)^n$.

11.5.3. $x + \frac{1}{2}x^2 + \frac{1}{3}x^3 + x^2(y - 1)$.

11.5.5. $\frac{1}{4}\pi x + \frac{1}{2}x(z - 1) + \frac{1}{2}x(y - 1)$.

11.6.9. $f'_x = 4x^3 - 2x - 2y$ and $f'_y = 4y^3 - 2x - 2y$. By equating both to 0 and subtracting, we obtain that $4x^3 = 4y^3$, hence $x = y$. Substituting in $4x^3 - 2x - 2y = 0$ we obtain $4x^3 - 4x = 0$, so $x = 0$ or $x = -1$ or $x = 1$. It follows that the critical points of f are $(0, 0)$, $(-1, -1)$, and $(1, 1)$. The Hessian matrix is $Hf(x, y) = \begin{bmatrix} 12x^2 - 2 & -2 \\ -2 & 12y^2 - 2 \end{bmatrix}$, so $Hf(-1, -1) = Hf(1, 1) = \begin{bmatrix} 10 & -2 \\ -2 & 10 \end{bmatrix}$, and f has a relative minimum at both. Since $Hf(0, 0) = \begin{bmatrix} -2 & -2 \\ -2 & -2 \end{bmatrix}$, we cannot use Sylvester's Rule. However,

$$\Delta f = f(h, k) - f(0, 0) = h^4 + k^4 - h^2 - 2hk - k^2.$$

If $h = k$, then $\Delta f = 2h^4 - 4h^2 = 2h^2(h^2 - 2) < 0$, for small values of h. If $h = -k$, then $\Delta f = 2h^4 > 0$. We conclude that f has a saddle at $(0, 0)$.

11.6.11. A saddle at $(-1/4, -1/2)$ and a relative minimum at $(0, 0)$.

11.6.13. A saddle at $(-1, -2, 2)$.

11.6.14. Let $A = (1 - x_1 - 2x_2 - \cdots - nx_n)$. For any $1 \leq k \leq n$,

$$f'_{x_k} = kx_1 x_2^2 \ldots x_n^n \frac{A - x_k}{x_k},$$

so $f'_{x_k} = 0$ implies that $A = x_k$, and it follows that a critical point must satisfy

$$x_1 = x_2 = \cdots = x_k = A.$$

It is not hard to see that they are all equal to $c = \frac{2}{n^2 + n + 2}$. To test this only critical point, we compute the second-order derivatives:

$$f'_{x_k x_k}(c, c, \ldots, c) = -k(k + 1)c^{\frac{n^2 + n - 2}{2}}, \quad f'_{x_k x_j}(c, c, \ldots, c) = -kjc^{\frac{n^2 + n - 2}{2}}.$$

The Hessian matrix is negative definite, because the sequence of determinants $\{D_i\}_{i=1}^{n}$ satisfies

$$D_m = (-1)^m \, m! \left[c^{\frac{n^2 + n - 2}{2}}\right]^m \begin{vmatrix} 2 & 2 & 3 & 4 & \ldots & m \\ 1 & 3 & 3 & 4 & \ldots & m \\ 1 & 2 & 4 & 4 & \ldots & m \\ 1 & 2 & 3 & 5 & \ldots & m \\ \ldots & & & & & \\ 1 & 2 & 3 & 4 & \ldots & m+1 \end{vmatrix}.$$

If we subtract the row $(m-1)$ from row m, row $(m-2)$ from row $(m-1)$, etc., we obtain

$$C_m = \begin{vmatrix} 2 & 2 & 3 & 4 & \ldots & m \\ -1 & 1 & 0 & 0 & \ldots & 0 \\ 0 & -1 & 1 & 0 & \ldots & 0 \\ 0 & 0 & -1 & 1 & \ldots & 0 \\ \ldots \\ 0 & 0 & 0 & 0 & \ldots & 1 \end{vmatrix}.$$

Expanding along the last column yields the recursive formula $C_m = C_{m-1} + m$. Since $C_1 = 2$, we obtain that $C_m = 2 + 2 + 3 + \cdots + m = \frac{m^2 + m + 2}{2} > 0$. Consequently we have that $D_1 < 0$, $D_2 > 0$, $D_3 < 0$, etc., whence f has a local maximum at the critical point $(c, c, \ldots, c)$.

12.1.4. $y' = -\dfrac{2x+y}{x+2y}$, $y'' = -\dfrac{6(x^2+xy+y^2)}{(x+2y)^3}$, $y''' = -\dfrac{54x(x^2+xy+y^2)}{(x+2y)^5}$.

12.1.6. $z'_x = \dfrac{yz}{z^2-xy}$, $z'_y = \dfrac{xz}{z^2-xy}$, $z'_{xx} = -\dfrac{2xy^3z}{(z^2-xy)^3}$, $z'_{xx} = -\dfrac{2x^3yz}{(z^2-xy)^3}$, $z'_{xy} = \dfrac{z(z^4-2xyz^2-x^2y^2)}{(z^2-xy)^3}$.

12.1.15. $z'_x = \dfrac{yz-1}{1-xy}$, $z'_y = \dfrac{xz-1}{1-xy}$, so $dz = \dfrac{yz-1}{1-xy}dx + \dfrac{xz-1}{1-xy}dy$.

$z'_{xx} = \dfrac{2y(yz-1)}{(1-xy)^2}$, $z'_{yy} = \dfrac{2x(xz-1)}{(1-xy)^2}$, $z'_{xy} = \dfrac{z-x-y+xyz}{(1-xy)^2}$, so

$d^2z = \dfrac{2y(yz-1)}{(1-xy)^2}dx^2 + 2\dfrac{z-x-y+xyz}{(1-xy)^2}dxdy + \dfrac{2x(xz-1)}{(1-xy)^2}dy^2$.

12.1.19. $\dfrac{dx}{dz} = 0$, $\dfrac{dy}{dz} = -1$, $\dfrac{d^2x}{dz^2} = -\dfrac{1}{4}$, $\dfrac{d^2y}{dz^2} = \dfrac{1}{4}$.

12.1.25. Taking partial derivative with respect to x yields

$$1 = u'_x + 2vv'_x, \quad 0 = 2uu'_x - 2vv'_x, \quad z'_x = 2vu'_x + 2uv'_x.$$

From here, $u'_x(2,1) = 1/5$, $v'_x(2,1) = 2/5$. Similarly, taking partial derivatives with respect to y leads to $u'_y(2,1) = 1/5$, $v'_y(2,1) = -1/10$, $z'_y(2,1) = 0$. Finally, differentiating the three equations above with respect to y, we obtain $z''_{xy} = 7/25$.

12.2.3. The linear transformation $U : \mathbb{R}^n \to \mathbb{R}^n$, defined by $Ue_i = f_i$, $1 \le i \le n$, is orthogonal. This means that $U^TU = I$. Indeed, $U^Tf_i = \sum_{k=1}^n c_k e_k$ and $\langle U^Tf_i, e_j\rangle = c_j$. But $\langle U^Tf_i, e_j\rangle = \langle f_i, Ue_j\rangle = \langle f_i, f_j\rangle$. Since the last inner product is δ_{ij}, i.e., 0 unless $i = j$ in which case it is 1, we see that $U^Tf_i = e_i$. In other words, the linear transformation U^T is inverse to U.

In the basis $\{e_i\}_{i=1}^n$, T has the matrix

$$\left[t_{ij}\right] = \left[\langle Ae_j, e_i\rangle\right] = \left[\langle AU^Tf_j, U^Tf_i\rangle\right] = \left[\langle UAU^Tf_j, f_i\rangle\right],$$

which is the matrix of T in the basis $\{f_i\}_{i=1}^n$, hence $B = \left[r_{ij}\right]$. It follows that $B = UAU^T$. Further, $B^TB = (UAU^T)^T(UAU^T) = (UA^TU^T)(UAU^T) = UA^TAU^T$. Finally, for any $1 \le i \le n$,

$$\|Bf_i\|^2 = \langle Bf_i, Bf_i\rangle = \langle B^TBf_i, f_i\rangle = \langle UA^TAU^Tf_i, f_i\rangle = \langle A^TAe_i, e_i\rangle = \|Ae_i\|^2.$$

It remains to notice that $\sum_{i,j=1}^n t_{ij}^2 = \sum_{i=1}^n \|Ae_i\|^2$, and $\sum_{i,j=1}^n r_{ij}^2 = \sum_{i=1}^n \|Ae_i\|^2$.

12.2.6. Since A is injective, there exists $\gamma > 0$ such that, for any $\mathbf{u}$, $\|A\mathbf{u}\| \ge \gamma\|\mathbf{u}\|$. If $r < \gamma$ and $\|A - B\|_2 < r$ then, for any $\mathbf{u}$,

$$\|B\mathbf{u}\| \ge \|A\mathbf{u}\| - \|(A-B)\mathbf{u}\| \ge \gamma\|\mathbf{u}\| - \|(A-B)\|_2\|\mathbf{u}\| \ge \gamma\|\mathbf{u}\| - r\|\mathbf{u}\|$$

so B is injective.

12.3.3. $D\mathbf{f}(x,y) = \begin{bmatrix} 1 & 1 \\ 2x & 2y \end{bmatrix}$ is bijective if $y \ne x$. By the Inverse Function Theorem, $\mathbf{f}$ is locally invertible at any (x,y) where $x \ne y$. If $x = y$, at such a point $\mathbf{f}$ is not bijective. Reason: for any $\varepsilon, \eta > 0$, $\mathbf{f}(x+\varepsilon, x+\eta) = \mathbf{f}(x+\eta, x+\varepsilon)$, so $\mathbf{f}$ is not injective in any ball containing (x,x).
12.3.19. $\mathbf{f}(x,y) = (x,x)$.
12.3.20. $f(x) = x^3$.

12.4.1. $D\mathbf{f}(\mathbf{a}) = \begin{bmatrix} 3 & 0 & -2 \end{bmatrix}$ and the mapping

$$z \mapsto \begin{bmatrix} 3 & 0 & -2 \end{bmatrix}\begin{bmatrix} 0 \\ 0 \\ z \end{bmatrix} = -2z$$

is bijective. Therefore, z is determined as a function of x, y in a ball centered at $\mathbf{a}$ that does not contain any points with $z \le 0$.

12.4.3. Df(a) $= \begin{bmatrix} 3 & 1 & 2 & 2 & 1 \\ 1 & 1 & 2 & 0 & 1 \end{bmatrix}$ and the mapping

$$(u,v) \mapsto \mathbf{Df(a)} \begin{bmatrix} 0 \\ 0 \\ 0 \\ u \\ v \end{bmatrix} = \begin{bmatrix} 2u+v \\ v \end{bmatrix}$$

is bijective, so the answer is yes!

12.4.5. Yes.

12.4.9. Must have u,v functions of x,y: if $\mathbf{f}(u,v) = (u+v, u^2+v^2)$ then $\mathbf{Df}(u,v) = \begin{bmatrix} 1 & 1 \\ 2u & 2v \end{bmatrix}$ so it is invertible if and only if $u \neq v$. The equation $u = v$ translates to $x = 2u$, $y = 2u^2$, so $2y = x^2$. In addition, $2y - x^2 = 2(u^2+v^2) - (u+v)^2 \geq 0$ for all u,v. Thus, z is a function of x,y, in the part of the xy-plane determined by $2y - x^2 > 0$.

12.5.2. Take $F(x,y,z;\lambda) = x - 2y + 2z + \lambda(x^2+y^2+z^2-1)$. The equation $\nabla F = \mathbf{0}$ yields $1 + 2\lambda x = 0$, $-2 + 2\lambda y = 0$, $2 + 2\lambda z = 0$. Since $\lambda \neq 0$ (otherwise the first equation would be $1 = 0$), we get $x = -1/(2\lambda)$, $y = 1/\lambda$, $z = -1/\lambda$. Substituting in $x^2+y^2+z^2-1 = 0$ yields $\lambda = \pm 3/2$, so the critical points of F are $A(\frac{1}{3}, -\frac{2}{3}, \frac{2}{3})$ and $B(-\frac{1}{3}, \frac{2}{3}, -\frac{2}{3})$. The function f has a maximum at A and a minimum at B.

12.5.9. Dg$(x,y,z) = \begin{bmatrix} 2x & 2y & -2z \\ 0 & 1 & 1 \end{bmatrix}$, and $\begin{vmatrix} 2y & -2z \\ 1 & 1 \end{vmatrix} = 2(y+z) = 2 \cdot 1 \neq 0$, so we can take $\mu = 1$. The first three equations we obtain are

$$4 + 2\lambda_1 x = 0, \quad 1 + 2\lambda_1 y + \lambda_2 = 0, \text{ and } -1 - 2\lambda_1 z + \lambda_2 = 0.$$

The first one shows that $\lambda_1 \neq 0$, so we have $x = -2/\lambda_1$. If we subtract the third equation from the second, and if we take advantage of $y + z = 1$, we obtain that $\lambda_1 = -1$, so $x = 2$. Substituting $\lambda_1 = -1$ now leads to $y = (1+\lambda_2)/2$, $z = (1-\lambda_2)/2$. The equation $x^2 + y^2 = z^2$ becomes

$$4 + (1+\lambda_2)^2/4 = (1-\lambda_2)^2/4$$

which has the solution $\lambda_2 = -4$. Consequently $y = -3/2$ and $z = 5/2$. The critical point is $(2, -\frac{3}{2}, \frac{5}{2})$ and $f(2, -\frac{3}{2}, \frac{5}{2}) = 4$.

12.6.3. We consider, instead, $h(x,y,z) = \ln f(x,y,z) = \ln x + 2\ln y + 3\ln z$. Then $F(x,y,z;\lambda) = \ln x + 2\ln y + 3\ln z + \lambda(x + 2y + 3z - 1)$. The first three equations we obtain are

$$\frac{1}{x} + \lambda = 0, \quad \frac{2}{y} + 2\lambda = 0, \quad \frac{3}{z} + 3\lambda = 0.$$

Clearly, $\lambda \neq 0$ and $x = y = z$. It follows from $x + 2y + 3z - 1 = 0$ that $x = y = z = -1/6$ and $\lambda = -6$. Now $H(x,y,z) = \ln x + 2\ln y + 3\ln z - 6(x + 2y + 3z - 1)$ and a calculation shows that

$$\mathbf{D}^2 H(x,y,z) = \begin{bmatrix} -1/x^2 & 0 & 0 \\ 0 & -2/y^2 & 0 \\ 0 & 0 & -3/z^2 \end{bmatrix}$$

which is clearly negative definite. Thus, f has a constrained maximum at $(-\frac{1}{6}, -\frac{1}{6}, -\frac{1}{6})$.

12.6.7. Since $\mathbf{D}g(x,y,z) = \begin{bmatrix} 1 & 1 & 1 \end{bmatrix}$, its rank is 1, and we can use $\mu = 1$. Thus, $F(x,y,z;\lambda) = 3xy - 4z + \lambda(x+y+z-1)$, and we obtain the system

$$3y + \lambda = 0, \quad 3x + \lambda = 0, \quad -4 + \lambda = 0, \quad x + y + z = 1.$$

Its solution is $\lambda = 4$, $x = y = -4/3$, $z = 11/3$, so $H(x,y,z) = 3xy - 4z + 4(x+y+z-1)$. Then,

$$\mathbf{D}^2 H(x,y,z) = \begin{bmatrix} 0 & 3 & 0 \\ 3 & 0 & 0 \\ 0 & 0 & 0 \end{bmatrix}$$

which corresponds to the quadratic form $Q(u_1, u_2) = 6u_1 u_2$. This form takes both positive and negative values, so f has a saddle at $(-\frac{4}{3}, -\frac{4}{3}, \frac{11}{3})$.

12.6.11. Since $\mathbf{D}g(x,y,z) = \begin{bmatrix} 2x & 2y & 0 \\ 0 & z & y \end{bmatrix}$, and $y \neq 0$, the rank is 2. We take $\mu = 1$ and define

$$F(x,y,z;\lambda_1, \lambda_2) = xy + yz + \lambda_1(x^2 + y^2 - 1) + \lambda_2(yz - 1).$$

Taking partial derivatives with respect to x, y, z, we obtain equations

$$y + 2\lambda_1 x = 0, \quad x + z + 2\lambda_1 y + \lambda_2 z = 0, \quad y + \lambda_2 y = 0.$$

Since $y \neq 0$, the last of these implies that $\lambda_2 = -1$. The remaining two equations become $y + 2\lambda_1 x = 0$ and $x + 2\lambda_1 y = 0$ whence $y(1 - 4\lambda_1^2) = 0$. Since $y \neq 0$, $\lambda_1 = \pm\frac{1}{2}$.

When $\lambda_1 = 1/2$ we get $x + y = 0$ which, together with $x^2 + y^2 = 1$ and $yz = 1$ yields critical points $A(\frac{\sqrt{2}}{2}, -\frac{\sqrt{2}}{2}, -\sqrt{2})$ and $B(-\frac{\sqrt{2}}{2}, \frac{\sqrt{2}}{2}, \sqrt{2})$. In this case $H(x, y, z) = xy + \frac{x^2}{2} + \frac{y^2}{2} + \frac{1}{2}$ so

$$\mathbf{D}^2 H(x, y, z) = \begin{bmatrix} 1 & 1 & 0 \\ 1 & 1 & 0 \\ 0 & 0 & 0 \end{bmatrix}$$

which corresponds to the quadratic form $Q(u, v, w) = u^2 + 2uv + v^2$. Notice that

$$\mathbf{D}g\left(\frac{\sqrt{2}}{2}, -\frac{\sqrt{2}}{2}, -\sqrt{2}\right) = \begin{bmatrix} \sqrt{2} & -\sqrt{2} & 0 \\ 0 & -\sqrt{2} & -\frac{\sqrt{2}}{2} \end{bmatrix}$$

leads to equations $u - v = 0$ and $v + w/2 = 0$, so the quadratic form above reduces to $Q(u) = 4u^2$ which is positive definite. Similarly,

$$\mathbf{D}g\left(-\frac{\sqrt{2}}{2}, \frac{\sqrt{2}}{2}, \sqrt{2}\right) = \begin{bmatrix} -\sqrt{2} & \sqrt{2} & 0 \\ 0 & \sqrt{2} & \frac{\sqrt{2}}{2} \end{bmatrix}$$

leads to the same equations $u - v = 0$ and $v + w/2 = 0$, hence to the same quadratic form $Q(u) = 4u^2$. We conclude that f has a minimum at both A and B.

When $\lambda_1 = -1/2$ we get $x - y = 0$ which, together with $x^2 + y^2 = 1$ and $yz = 1$ yields critical points $C(\frac{\sqrt{2}}{2}, \frac{\sqrt{2}}{2}, \sqrt{2})$ and $D(-\frac{\sqrt{2}}{2}, -\frac{\sqrt{2}}{2}, -\sqrt{2})$. In this case $H(x, y, z) = xy - \frac{x^2}{2} - \frac{y^2}{2} + \frac{3}{2}$ so

$$\mathbf{D}^2 H(x, y, z) = \begin{bmatrix} -1 & 1 & 0 \\ 1 & -1 & 0 \\ 0 & 0 & 0 \end{bmatrix}$$

which corresponds to the quadratic form $Q(u, v, w) = -u^2 + 2uv - v^2$. This time

$$\mathbf{D}g\left(\frac{\sqrt{2}}{2}, \frac{\sqrt{2}}{2}, \sqrt{2}\right) = \begin{bmatrix} \sqrt{2} & \sqrt{2} & 0 \\ 0 & \sqrt{2} & \frac{\sqrt{2}}{2} \end{bmatrix}, \mathbf{D}g\left(-\frac{\sqrt{2}}{2}, -\frac{\sqrt{2}}{2}, -\sqrt{2}\right) = \begin{bmatrix} -\sqrt{2} & -\sqrt{2} & 0 \\ 0 & -\sqrt{2} & -\frac{\sqrt{2}}{2} \end{bmatrix},$$

both leading to equations $u + v = 0$ and $v + w/2 = 0$, so the quadratic form above reduces to $Q(u) = -4u^2$ which is negative definite. We conclude that f has a maximum at both C and D.

12.6.14. Since $\nabla f(x, y) = (2x - y, -x + 2y)$, the only critical point is $P_1(0, 0)$. The boundary consists of 4 lines. If $x + y = 1$, then $F(x, y; \lambda) = x^2 - xy + y^2 + \lambda(x + y - 1)$, so calculating ∇F leads to equations

$$2x - y + \lambda = 0, \quad -x + 2y + \lambda = 0, \text{ and } x + y = 1.$$

Solving this system yields $P_2(\frac{1}{2}, \frac{1}{2})$. Similarly, when $-x + y = 1$ we obtain $P_3(-\frac{1}{2}, \frac{1}{2})$, $-x - y = 1$ leads to $P_4(-\frac{1}{2}, -\frac{1}{2})$, and $x - y = 1$ to $P_5(-\frac{1}{2}, \frac{1}{2})$. Finally, $f(0, 0) = 0$, $f(\frac{1}{2}, \frac{1}{2}) = f(-\frac{1}{2}, -\frac{1}{2}) = \frac{1}{4}$, and $f(-\frac{1}{2}, \frac{1}{2}) = f(\frac{1}{2}, -\frac{1}{2}) = \frac{3}{4}$, so f has the absolute maximum at P_2 and P_3, and the absolute minimum at P_1.

13.1.2. Let $a_n \to 0$. Then

$$|F(x, a_n) - F(x, 0)| = |\sqrt{x^2 + a_n^2} - \sqrt{x^2}| = \frac{a_n^2}{\sqrt{x^2 + a_n^2} + \sqrt{x^2}} \leq \frac{a_n^2}{\sqrt{a_n^2}} = |a_n| \to 0.$$

By Theorem 13.1.5, the convergence is uniform.

13.1.6. Let $a_n \to 0$, and suppose that $a_n \in (0, 1/e)$. For a fixed n, the function $|F(x, a_n)|$ attains its maximum at $x = 1$. Indeed, $xa_n < 1/e$, so $\ln xa_n + 1 < 0$, and $F_x'(x, a_n) = a_n(\ln xa_n + 1) < 0$. Therefore, the negative function F is decreasing for $0 < x < 1$, so $|F(x, a_n)|$ has a maximum at $x = 1$, and that maximum is $a_n \ln a_n$. Since $\lim a_n \ln a_n = 0$, the convergence is uniform by Theorem 13.1.5.

13.1.9. We will prove that $f_n(x) = \dfrac{1}{1 + \left(1 + \frac{x}{n}\right)^n}$ converges uniformly to $f(x) = \dfrac{1}{1 + e^x}$.

$$\left| \frac{1}{1 + \left(1 + \frac{x}{n}\right)^n} - \frac{1}{1 + e^x} \right| = \left| \frac{e^x - \left(1 + \frac{x}{n}\right)^n}{\left[1 + \left(1 + \frac{x}{n}\right)^n\right](1 + e^x)} \right| < \left| e^x - \left(1 + \frac{x}{n}\right)^n \right| \leq \sup_{0 \leq x \leq 1} \left| e^x - \left(1 + \frac{x}{n}\right)^n \right|.$$

The function $g(x) = e^x - \left(1 + \frac{x}{n}\right)^n$ attains its maximum either at one of the endpoints or at a point x_0 where $g'(x_0) = 0$. It is obvious that $g(0) = 0$, and that $g(1) = e - \left(1 + \frac{1}{n}\right)^n \to 0$, as $n \to \infty$. Finally, $g'(x) = e^x - \left(1 + \frac{x}{n}\right)^{n-1}$, so $g'(x_0) = 0$ implies that $e^{x_0} = \left(1 + \frac{x_0}{n}\right)^{n-1}$. It follows that

$$|g(x_0)| = \left| e^{x_0} - \left(1 + \frac{x_0}{n}\right)^n \right| = \left| \left(1 + \frac{x_0}{n}\right)^{n-1} - \left(1 + \frac{x_0}{n}\right)^n \right| = \left(1 + \frac{x_0}{n}\right)^{n-1} \frac{x_0}{n} \leq \left(1 + \frac{1}{n}\right)^{n-1} \frac{1}{n} \to 0.$$

Thus, f_n converges uniformly to f on $[0,1]$, and bringing the limit inside the integral is justified.

13.1.10. Let $f(x) = \sin x - \frac{2}{\pi} x$. Notice that $f(0) = f(\frac{\pi}{2}) = 0$, and that the derivative $f'(x) = \cos x - \frac{2}{\pi}$ is positive for $0 \le x < \arccos \frac{2}{\pi}$ and negative for $\arccos \frac{2}{\pi} < x \le 1$, so f has a maximum at $\arccos \frac{2}{\pi}$. It follows that its minimum is at the endpoints and $f(x) \ge 0$ for $0 \le x \le \frac{\pi}{2}$.

Now we will use the inequality $\sin \theta \ge \frac{2}{\pi} \theta$ to obtain that

$$\int_0^{\pi/2} e^{-R\sin\theta}\, d\theta \le \int_0^{\pi/2} e^{-R\frac{2}{\pi}\theta}\, d\theta = -\frac{\pi}{2R} e^{-\frac{2R\theta}{\pi}} \Big|_0^{\pi/2} = -\frac{\pi}{2R}\left(e^{-R}-1\right) \to 0, \quad \text{as } R \to \infty.$$

13.2.2. Let the function F be defined by

$$F(x,t) = \begin{cases} \dfrac{\ln(1+tx)}{x} & \text{if } x \ne 0, \\ t & \text{if } x = 0. \end{cases}$$

This function is continuous on $[0,b] \times [0,b]$. Further, $\alpha(t) = 0$ and $\beta(t) = t$ are continuous on $[0,b]$ and differentiable on $(0,b)$. Finally, $F_t'(x,t) = \frac{1}{1+tx}$ if $x \ne 0$, and $F_t'(x,t) = 1$ if $x = 0$, and it is continuous on $[0,b] \times [0,b]$. By Leibniz Rule

$$I'(t) = \int_0^t \frac{1}{1+tx}\, dx + \frac{\ln(1+t^2)}{t} = \frac{1}{t}\ln(1+tx)\,|_0^t + \frac{\ln(1+t^2)}{t} = \frac{2\ln(1+t^2)}{t}.$$

13.2.10. The function $F(x,a,b) = \ln(a^2 \sin^2 x + b^2 \cos^2 x)$ is continuous and

$$F_a'(x,a,b) = \frac{2a\sin^2 x}{a^2 \sin^2 x + b^2 \cos^2 x}$$

is also continuous. Therefore,

$$I'(a) = \int_0^{\pi/2} \frac{2a\sin^2 x}{a^2 \sin^2 x + b^2 \cos^2 x}\, dx.$$

If $a \ne b$, the substitution $u = \cot x$ leads to

$$I'(a) = \int_0^\infty \frac{2a}{(a^2 + b^2 u^2)(1 + u^2)}\, du = \frac{2a}{a^2 - b^2}\int_0^\infty \left(\frac{1}{1+u^2} - \frac{b^2}{a^2 + b^2 u^2}\right) du = \frac{2a}{a^2 - b^2}\left(\frac{\pi}{2} - \frac{b}{a}\frac{\pi}{2}\right) = \frac{\pi}{a+b}.$$

If $a = b$, using Example 5.1.11,

$$I'(a) = \frac{2}{a}\int_0^\infty \frac{du}{(1+u^2)^2} = \frac{2}{a}\left[\frac{1}{2}\arctan u + \frac{u}{2(u^2+1)}\right]\Big|_0^\infty = \frac{\pi}{2a}.$$

It follows that $I'(a) = \frac{\pi}{a+b}$ (even if $a = b$), so $I(a) = \pi \ln(a+b) + C$. However,

$$I(b) = \int_0^{\pi/2} \ln b^2\, dx = \pi \ln b,$$

so $C = \pi \ln b - \pi \ln 2b = -\pi \ln 2$. We conclude that $I(a) = \pi \ln \dfrac{a+b}{2}$.

13.2.11. It is not hard to see that $\dfrac{x^b - x^a}{\ln x} = \int_a^b x^t\, dt$. Therefore,

$$I = \int_0^1 \sin\left(\ln \frac{1}{x}\right) \int_a^b x^t\, dt\, dx = \int_0^1 dx \int_a^b \sin\left(\ln \frac{1}{x}\right) x^t\, dt.$$

The function $F(x,t) = \sin\left(\ln \frac{1}{x}\right) x^t$ is continuous on $(0,1] \times [a,b]$, and it can be extended to a continuous function on $[0,1] \times [a,b]$, since $\lim_{x\to 0+} F(x,t) = 0$. Thus, we can reverse the order of integration. If we combine it with the substitution $x = e^{-u}$, we obtain

$$I = \int_a^b dt \int_0^1 \sin\left(\ln \frac{1}{x}\right) x^t\, dx = \int_a^b dt \int_0^\infty \sin u\, e^{-ut} e^{-u}\, du = \int_a^b dt \int_0^\infty \sin u\, e^{-u(t+1)}\, du$$

$$= \int_a^b dt \left[\frac{e^{-u(t+1)}}{1 + (t+1)^2}((t+1)\sin u - \cos u)\Big|_0^\infty\right] = \int_a^b \frac{dt}{1 + (t+1)^2}$$

$$= \arctan(t+1)\,|_a^b = \arctan(b+1) - \arctan(a+1).$$

13.3.3. If we apply the formula $2\sin a \sin b = \cos(a-b) - \cos(a+b)$, we obtain that

$$\sin x^3 \sin tx = \frac{1}{2}[\cos(x^3 - tx) - \cos(x^3 + tx)].$$

We will now show that the integral

$$\int_{\sqrt{d/3}}^{\infty} x \cos(x^3 + tx)\, dx$$

converges uniformly, and leave the integral with $\cos(x^3 - tx)$ to the reader. Let

$$F(x,t) = (3x^2 + t)\cos(x^3 + tx), \quad \text{and} \quad \varphi(x,t) = \frac{x}{3x^2 + t}.$$

We will apply the Dirichlet's Test. First,

$$\int_{0}^{b} F(x,t)\, dx = \sin(x^3 + tx)\,|_0^b = \sin(b^3 + tb)$$

so $|\int_0^b F(x,t)\, dx| \le 1$ for all $b > 0$. Also, for any $t \in [c,d]$, $\varphi(x,t)$ is a decreasing function: $\varphi'_x(x,t) = (t - 3x^2)/(3x^2 + t)^2 < 0$, for $x > \sqrt{d/3}$. Finally, $\varphi'_t(x,t) = -x/(3x^2 + t)^2 < 0$, so $\sup_t |\varphi(x,t)|$ is attained at an end point. Since both $\varphi(x,c) \to 0$ and $\varphi(x,d) \to 0$ when $x \to \infty$, we see that Dirichlet's Test can be applied, and the integral converges uniformly.

13.3.5. The integral converges uniformly for $t \ge t_0 > 0$. Indeed, if $0 < x \le 1$, $x^{t-1} \le x^{t_0-1}$ and $\int_0^1 x^{t_0-1}\, dx = 1/t_0$. However, the integral does not converge uniformly in $[0,d]$, for any $d > 0$. Indeed, if $\{a_n\} \subset [0,d]$ and $a_n \to 0$, then

$$\int_{a_n}^{1} x^{t-1}\, dx = \frac{1}{t} - \frac{a_n^t}{t},$$

and the sequence a_n^t/t does not converge to 0 uniformly because $\sup_t a_n^t/t = +\infty$.

13.3.14. Let

$$F(x,t) = \frac{1}{(1 + x^2 t^2)\sqrt{1 - x^2}}.$$

Using (13.1) we have that $I = \int_0^1 dx \int_0^1 F(x,t)\, dt$. We will start by calculating the integral $I(b) = \int_0^b dx \int_0^1 F(x,t)\, dt$, where $0 < b < 1$. Since $F(x,t)$ is continuous on $[0,b] \times [0,1]$ we can reverse the order of integration. The substitution $u = \arcsin x$ then leads to

$$I(b) = \int_0^1 dt \int_0^{\arcsin b} \frac{du}{1 + t^2 \sin^2 u}.$$

Let $w = \tan u$. Using formulas established on page 148, we obtain

$$I(b) = \int_0^1 dt \int_0^{\tan(\arcsin b)} \frac{dw}{1 + w^2(1 + t^2)} = \int_0^1 \frac{dt}{1 + t^2}\left(\sqrt{1 + t^2}\arctan(w\sqrt{1 + t^2})\right)\Big|_0^{\tan(\arcsin b)}$$

$$= \int_0^1 \frac{dt}{1 + t^2}\arctan\left(\sqrt{1 + t^2}\tan(\arcsin b)\right).$$

Now we will take the limit as $b \to 1^-$. The integrand is continuous for $(t,b) \in [0,1] \times [0,1)$, and it has the limit when $b \to 1^-$, which equals $\frac{\pi}{2(1+t^2)}$. Further, the convergence (as $b \to 1^-$) is uniform. Indeed, let $g(z) = \frac{1}{z}[\arctan(\sqrt{z}\tan(\arcsin b)) - \frac{\pi}{2}]$. Then

$$g'(z) = \frac{1}{z^2}\left[\frac{\sqrt{z}\tan(\arcsin b)}{2(1 + z\tan^2(\arcsin b))} + \frac{\pi}{2} - \arctan(\sqrt{z}\tan(\arcsin b))\right] \ge 0,$$

so g is a negative increasing function. It follows that $|g|$ is decreasing, so it attains its maximum at the left endpoint. We are interested in $z = 1 + t^2$ and $t \in [0,1]$, so the maximum ocurrs at $z = 1$, and $g(1) = \arcsin b - \frac{\pi}{2} \to 0$, as $b \to 1^-$.

Thus we can take the limit inside the integral, and we obtain that

$$I = \lim_{b \to 1^-} I(b) = \int_0^1 \lim_{b \to 1^-} \frac{dt}{1 + t^2}\arctan\left(\sqrt{1 + t^2}\tan(\arcsin b)\right) = \frac{\pi}{2}\int_0^1 \frac{dt}{1 + t^2} = \frac{\pi}{2}.$$

13.3.18. Suppose that there exists $I(t)$ with the stated properties, and let $\varepsilon > 0$. Then there exists $B > 0$ such that if $b \ge B$ and $t \in [c,d]$, $|\int_a^b F(x,t)\, dx - I(t)| < \varepsilon/2$. If $b_2 \ge b_1 \ge B$ and $t \in [c,d]$,

$$\left|\int_a^{b_2} F(x,t)\, dx - \int_a^{b_1} F(x,t)\, dx\right| \le \left|\int_a^{b_2} F(x,t)\, dx - I(t)\right| + \left|I(t) - \int_a^{b_1} F(x,t)\, dx\right| < \frac{\varepsilon}{2} + \frac{\varepsilon}{2} = \varepsilon.$$

Next, we will prove the converse. Let $t \in [c, d]$ be fixed. By Exercise 6.7.20, there exists $\lim_{b \to \infty} \int_a^b F(x, t) \, dx$, which we denote by $I(t)$. Thus, for every $t \in [c, d]$, we have a well defined real number $I(t)$. Let $\varepsilon > 0$. By assumption, there exists $B > 0$ such that, if $b_2 \geq b_1 \geq B$ and $t \in [c, d]$, $|\int_a^{b_2} F(x, t) \, dx - \int_a^{b_1} F(x, t) \, dx| < \varepsilon/2$. Now let $b \geq B$ and $t_0 \in [c, d]$ be arbitrary. Then, if $b_2 \geq b$,

$$\left| \int_a^{b_2} F(x, t_0) \, dx - \int_a^b F(x, t_0) \, dx \right| < \frac{\varepsilon}{2}.$$

Also, there exists B_0 such that, if $b \geq B_0$,

$$\left| \int_a^b F(x, t_0) \, dx - I(t_0) \right| < \frac{\varepsilon}{2}.$$

This, if $b_2 \geq B_0$,

$$\left| \int_a^{b_2} F(x, t_0) \, dx - I(t_0) \right| < \frac{\varepsilon}{2}.$$

It follows that

$$\left| \int_a^b F(x, t_0) \, dx - I(t_0) \right| < \frac{\varepsilon}{2} + \frac{\varepsilon}{2} = \varepsilon.$$

13.3.20. Use a version of Exercise 6.6.18.

13.4.3. Let

$$F(x, t) = \frac{\ln(1 - t^2 x^2)}{x^2 \sqrt{1 - x^2}}.$$

First, F is continuous on $(0, 1) \times [-1, 1]$, and $\lim_{x \to 0^+} F(x, t) = -t^2$, so F can be defined to be continuous on $[0, 1) \times [-1, 1]$. Next, the integral $\int_0^1 F(x, t) \, dx$ converges uniformly because it is a sum of two integrals: $\int_0^{1/2} F(x, t) \, dx$ is not improper and its integrand is bounded, while for $x \in [-\frac{1}{2}, 1]$, $|F(x, t)| \leq g(x) = 4 |\ln(1 - x^2)|/\sqrt{1 - x^2}$, and $\int_{1/2}^1 g(x) \, dx$ converges. [Substitution $u = \sqrt{1 - x^2}$.]

Let $0 < d < 1$. The derivative

$$F_t'(x, t) = \frac{-2t}{(1 - t^2 x^2)\sqrt{1 - x^2}}$$

is continuous on $[0, 1) \times [-d, d]$. Also, $\int_0^1 F_t'(x, t) \, dx$ converges uniformly for $t \in [-d, d]$, because $|F_t'(x, t)| \leq \frac{2}{(1 - d^2)\sqrt{1 - x^2}}$, and the integral $\int_0^1 \frac{1}{\sqrt{1 - x^2}} \, dx$ converges. By Theorem 13.4.9, using the substitution $x = \sin u$,

$$I'(t) = \int_0^1 \frac{-2t}{(1 - t^2 x^2)\sqrt{1 - x^2}} \, dx = \int_0^{\pi/2} \frac{-2t}{1 - t^2 \sin^2 u} \, du.$$

The substitution $w = \tan u$ now yields

$$I'(t) = -2t \int_0^\infty \frac{dw}{1 + w^2(1 - t^2)} = \frac{-2t}{1 - t^2} \sqrt{1 - t^2} \arctan(\sqrt{1 - t^2}\, w)\Big|_0^\infty = -\frac{\pi t}{\sqrt{1 - t^2}}.$$

It follows that $I(t) = \pi \sqrt{1 - t^2} + C$. Since $I(0) = 0$ we see that $C = -\pi$, so $I(t) = \pi \sqrt{1 - t^2} - \pi$.

This result holds for $t \in [-d, d]$, for any $0 < d < 1$. Thus, it holds for $t \in (-1, 1)$, and it remains to show that it is true in $[-1, 1]$. It suffices to establish that $I(t)$ is continuous at $t = 1$ and $t = -1$. As we have already seen, $F(x, t)$ is continuous on $[0, 1) \times [-1, 1]$ and $\int_0^1 F(x, t) \, dx$ converges uniformly for $|x| \leq 1$. By Theorem 13.4.1, $I(t)$ is continuous so the formula $I(t) = \pi \sqrt{1 - t^2} - \pi$ holds for $t = \pm 1$.

13.4.13. Both $F(x, t) = e^{-ax^2} \cos tx$ and $F_t'(x, t) = -x e^{-ax^2} \sin tx$ are continuous on $[0, \infty) \times (-\infty, \infty)$. Further, $|F(x, t)| \leq e^{-ax^2}$, and $\int_0^\infty e^{-ax^2} \, dx$ converges. [Use substitution $u = x\sqrt{a}$ and Example 13.4.8.] Also, $|F_t'(x, t)| \leq x e^{-ax^2}$, and $\int_0^\infty x e^{-ax^2} \, dx$ converges uniformly. [Substitution $u = -ax^2$.] Thus, Theorem 13.4.9 applies and

$$I'(t) = \int_0^\infty -x e^{-ax^2} \sin tx \, dx.$$

Using integration by parts with $u = \sin tx$ and $dv = -x e^{-ax^2} \, dx$, we obtain

$$I'(t) = \frac{1}{2a} e^{-ax^2} \sin tx \Big|_0^\infty - \int_0^\infty \frac{1}{2a} e^{-ax^2} t \cos tx \, dx = -\frac{t}{2a} \int_0^\infty e^{-ax^2} \cos tx \, dx = -\frac{t}{2a} I(t).$$

The separable differential equation $I'(t) = -tI(t)/2a$ has the general solution $I(t) = Ce^{-t^2/(4a)}$. It is easy to see that $C = I(0) = \frac{\sqrt{\pi}}{2\sqrt{a}}$, so $I(t) = \frac{\sqrt{\pi}}{2\sqrt{a}} e^{-t^2/(4a)}$.

13.5.8. Using (13.24), (13.27), and (13.28),

$$\frac{\sqrt[4]{x}}{(1+x)^2} = B\left(\frac{5}{4}, \frac{3}{4}\right) = \frac{\Gamma(\frac{5}{4})\Gamma(\frac{3}{4})}{\Gamma(2)} = \frac{1}{4}\frac{\Gamma(\frac{1}{4})\Gamma(\frac{3}{4})}{1!} = \frac{1}{4}\frac{\pi}{\sin\frac{\pi}{4}} = \frac{\pi\sqrt{2}}{4}.$$

13.5.10. Use $x = t^{1/n}$.

13.5.13. By (13.24),

$$B(p, 1-p) = \int_0^\infty \frac{x^{p-1}}{1+x}\,dx.$$

We will show that this is a differentiable function of p, for $0 < p < 1$, and that its derivative can be calculated by differentiating inside the integral. That way we will obtain that

$$\int_0^\infty \frac{x^{p-1}\ln x}{1+x}\,dx = \frac{d}{dp}B(p, 1-p) = \frac{d}{dp}\left(\frac{\pi}{\sin p\pi}\right) = -\frac{\pi^2 \cos p\pi}{\sin^2 p\pi}. \qquad (16.5)$$

Let $0 < c < d < 1$. We will consider two rectangles: $R_1 = (0, 1] \times [c, d]$ and $R_2 = [1, \infty) \times [c, d]$, and we will show that our claims are justified in both. Then (16.5) will hold for all $p \in [c, d]$, and since c, d are arbitrary, it will be true for $p \in (0, 1)$.

The function $F(x, p) = x^{p-1}/(1+x)$ is continuous in both R_1 and R_2, and so is $F'_p(x, p) = x^{p-1}\ln x/(1+x)$. The convergence of $\int_0^\infty F(x, p)\,dx$ was established in Example 13.5.5, so it remains to prove that $\int_0^\infty F'_p(x, p)\,dx$ converges uniformly for $p \in [c, d]$. If $x \geq 1$ then $\ln x \leq x + 1$, so $|F'_p(x, p)| \leq x^{p-1} \leq x^{d-1}$ and $\int_1^\infty x^{d-1}\,dx$ converges. If $0 < x \leq 1$, then $|\ln x| \leq 2x^{-c/2}/(ce)$. Indeed, if $u \geq 1$, the function $\ln u/u^{c/2}$ attains its maximum at $e^{2/c}$ and this maximum is $2/ce$. Therefore, $\ln u \leq \frac{2}{ce} u^{c/2}$. Replacing u by $1/x$, $0 < x \leq 1$, we obtain that $-\ln x \leq x^{-c/2}\frac{2}{ce}$, so $|x^{c/2}\ln x| \leq \frac{2}{ce}$. It follows that, for $0 < x \leq 1$,

$$\left|\frac{x^{p-1}\ln x}{1+x}\right| \leq |x^{c-1}\ln x| \leq \frac{2}{ce}x^{-1+c/2}.$$

Since $\int_0^1 x^{-1+c/2}\,dx$ converges, we conclude that $\int_0^1 F'_p(x, p)\,dx$ converges uniformly for $p \in [c, d]$.

13.5.14. If we denote this integral by I, then $\int_0^1 \ln\Gamma(1-x)\,dx = I$, so

$$I = \frac{1}{2}\int_0^1 \ln\Gamma(x)\,dx + \frac{1}{2}\int_0^1 \ln\Gamma(1-x)\,dx = \frac{1}{2}\int_0^1 \ln(\Gamma(x)\Gamma(1-x))\,dx$$

$$= \frac{1}{2}\int_0^1 \ln\frac{\pi}{\sin\pi x}\,dx = \frac{1}{2}\int_0^1 [\ln\pi - \ln\sin\pi x]\,dx = \frac{1}{2}\ln\pi - \frac{1}{2}\int_0^1 \ln\sin\pi x\,dx.$$

To calculate the last integral, we use the substitution $\pi x = 2t$ which leads to

$$\int_0^1 \ln\sin\pi x\,dx = \frac{2}{\pi}\int_0^{\pi/2} \ln\sin 2t\,dt = \frac{2}{\pi}\int_0^{\pi/2} \ln(2\sin t\cos t)\,dt$$

$$= \frac{2}{\pi}\int_0^{\pi/2} [\ln 2 + \ln\sin t + \ln\cos t]\,dt = \frac{2}{\pi}\ln 2\frac{\pi}{2} + \frac{2}{\pi}\int_0^{\pi/2} \ln\sin t\,dt + \frac{2}{\pi}\int_0^{\pi/2} \ln\cos t\,dt.$$

The last integral can be transformed using $t = \frac{\pi}{2} - u$ into $\int_0^{\pi/2} \ln\sin u\,du$, so

$$I = \frac{1}{2}\ln\pi - \frac{1}{2}\left(\ln 2 + \frac{4}{\pi}\int_0^{\pi/2} \ln\sin t\,dt\right).$$

Finally, Exercise 13.2.10 shows that $\int_0^{\pi/2} \ln(a^2\sin^2 x + b^2\cos^2 x) = \pi\ln\frac{a+b}{2}$. If we take $a = 1$ and $b = 0$, we obtain that $\int_0^{\pi/2} 2\ln\sin t\,dt = \pi\ln\frac{1}{2} = -\pi\ln 2$. It follows that

$$I = \frac{1}{2}\ln\pi - \frac{1}{2}\ln 2 - \frac{2}{\pi}\left(-\frac{\pi}{2}\ln 2\right) = \frac{1}{2}\ln\pi + \frac{1}{2}\ln 2 = \ln\sqrt{2\pi}.$$

14.1.2. It consists of the graphs of $y = \sqrt{1 - x^2}$ and $y = -\sqrt{1 - x^2}$.

14.1.22. Since squares are rectangles, the "if" part is clear. In the other direction, the assertion follows from the following lemma: If R is a rectangle of area A, and if $\delta > 0$, then there exist squares $C_1, C_2, \ldots, C_n$ such that $R \subset \cup_{i=1}^n C_i$ and $\sum_{i=1}^n A(C_i) < A + \delta$. Proof: Let a, b be the sides of R and extend a by δ/b to obtain the rectangle R'. By Theorem 1.3.7, there exists a rational number p/q between a/b and $a/b + \delta/b^2$. The pq squares of the side b/q now cover R: q along the side of length b make a perfect covering, p along the side of length a cover slightly more, because $a \le p(b/q) \le a + (\delta/b)$.

14.1.28. Let $\varepsilon > 0$. The set $\mathbb{Q}$ is countable which means that it can be arranged into a sequence $\{r_n\}$. For each $n \in \mathbb{N}$, let $\varepsilon_n = \varepsilon/4^n$, and notice that $r_n \in (r_n - \varepsilon_n, r_n + \varepsilon_n) = J_n$. That way the union $\cup_{i=1}^\infty J_n$ covers $\mathbb{Q}$, the length of J_n is $2\varepsilon_n$, so the sum of all these lengths is $\sum_{i=1}^\infty 2\varepsilon_n = \sum_{i=1}^\infty 2\varepsilon/4^n = 2\varepsilon/3 < \varepsilon$.

14.2.1. Prove that $\partial(A \cup B) \subset \partial(A) \cup \partial(B)$.

14.2.4. The inclusion $\overline{A \cap B} \subset \overline{A} \cap \overline{B}$ implies that $\partial(\overline{A \cap B}) \subset \partial(\overline{A} \cap \overline{B})$. Now the result follows from Exercise 14.2.2.

14.2.21. Let $D = D_1 \cup D_2 \cup D_3$, where $D_1 = \{(x, y) \in \mathbb{R}^2 : 0 \le x \le 1, 0 \le y \le 1 - x\}$, $D_2 = \{(x, y) \in \mathbb{R}^2 : 0 \le x \le 1, -1 \le y \le -x\}$, $D_3 = \{(x, y) \in \mathbb{R}^2 : 0 \le x \le 1, -x \le y \le 0\}$. Then $\iint_{D_1} \arcsin(x + y) \, dA(x, y) > 0$, because $x + y \ge 0$, so $\arcsin(x + y)$ is a non-negative function. On the other hand,

$$\iint_{D_2} \arcsin(x + y) \, dA(x, y) = -\iint_{D_3} \arcsin(x + y) \, dA(x, y),$$

because D_2 and D_3 are symmetric with respect to $x + y = 0$ and, at symetric points (a, b) and $(-b, -a)$, $\arcsin(p + q) = -\arcsin(-q - p)$.

14.3.2. Let $D = \{(x, y) \in \mathbb{R}^2 : 0 \le x \le \pi, x \le y \le \pi\}$, and let

$$F(x, y) = \begin{cases} \frac{\sin y}{y}, & \text{if } (x, y) \ne (0, 0) \\ 1, & \text{if } (x, y) = (0, 0). \end{cases}$$

Then F is continuous on D and Theorem 14.3.6 implies that

$$\int_0^\pi dx \int_x^\pi \frac{\sin y}{y} \, dy = \iint_D F(x, y) \, dA.$$

On the other hand, D can be represented as x-simple: $0 \le y \le \pi, 0 \le x \le y$. Another application of Theorem 14.3.6 yields

$$\int_0^\pi dx \int_x^\pi \frac{\sin y}{y} \, dx = \int_0^\pi dy \int_0^y \frac{\sin y}{y} \, dx = \int_0^\pi \frac{\sin y}{y} \, y \, dy = 2.$$

14.3.14.   **14.3.16.**

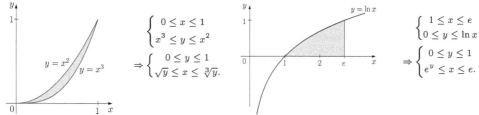

For 14.3.14:
$$\begin{cases} 0 \le x \le 1 \\ x^3 \le y \le x^2 \end{cases} \Rightarrow \begin{cases} 0 \le y \le 1 \\ \sqrt{y} \le x \le \sqrt[3]{y}. \end{cases}$$

with $y = x^2$, $y = x^3$

For 14.3.16:
$y = \ln x$
$$\begin{cases} 1 \le x \le e \\ 0 \le y \le \ln x \end{cases} \Rightarrow \begin{cases} 0 \le y \le 1 \\ e^y \le x \le e. \end{cases}$$

14.3.22. For a fixed $x \in [0, 1]$, $f(x, y) = 0$ if $x \notin \mathbb{Q}$ or $x = i/p$ but y is not one of the $p-1$ numbers j/p, $1 \le j \le p-1$. In other words, for any x, $f(x, y) = 0$ except for finitely many values of y. Consequently, $\int_0^1 f(x, y) \, dy = 0$ for all x, and $\int_0^1 dx \int_0^1 f(x, y) \, dy = 0$. Similarly, $\int_0^1 dy \int_0^1 f(x, y) \, dx = 0$. However, the double integral does not exist because for every partition P of $[0, 1] \times [0, 1]$, $L(f, P) = 0$ and $U(f, P) = 1$. Namely, in every rectangle there are points that belong to D and those that do not.

14.4.7. We will use the change of variables $u = x - y$, $v = x + y$. Clearly, the Jacobian determinant of (u, v) with respect to (x, y) is $\left| \begin{smallmatrix} 1 & -1 \\ 1 & 1 \end{smallmatrix} \right| = 2$, so the Jacobian of (x, y) with respect to (u, v) equals $1/2$. The domain of the integration is bounded by the lines $v = -u$, $v = u$, and $v = 1$. Thus, the integral is

$$\int_0^1 dv \int_{-v}^v e^{u/v} \frac{1}{2} \, du = \frac{1}{2} \int_0^1 dv \, (ve^{u/v}) \Big|_{u=-v}^{u=v} = \frac{1}{2} \int_0^1 \left(e - \frac{1}{e} \right) v \, dv = \sinh 1 \left(\frac{v^2}{2} \Big|_0^1 \right) = \frac{1}{2} \sinh 1.$$

14.4.11. Let $A = \left[\begin{smallmatrix} a_1 & a_2 \\ a_3 & a_4 \end{smallmatrix} \right]$ and $B = \left[\begin{smallmatrix} b_1 & b_2 \\ b_3 & b_4 \end{smallmatrix} \right]$. Notice that

$$|a_1 + b_1| + |a_2 + b_2| \le |a_1| + |a_2| + |b_1| + |b_2|, \quad \text{and} \quad |a_3 + b_3| + |a_4 + b_4| \le |a_3| + |a_4| + |b_3| + |b_4|, \quad \text{so}$$
$$\|A + B\| \le \max\{|a_1| + |a_2| + |b_1| + |b_2|, |a_3| + |a_4| + |b_3| + |b_4|\}.$$

Next,

$|a_1| + |a_2| + |b_1| + |b_2| \leq \max\{|a_1| + |a_2|, |a_3| + |a_4|\} + \max\{|b_1| + |b_2|, |b_3| + |b_4|\} = \|A\| + \|B\|$ and, similarly, $|a_3| + |a_4| + |b_3| + |b_4| \leq \|A\| + \|B\|$.

It follows that $\|A + B\| \leq \|A\| + \|B\|$. The other properties are easier to verify.

To prove assertion (b), we make the following estimates:

$$\|AB\| = \max\{|a_1 b_1 + a_2 b_3| + |a_1 b_2 + a_2 b_4|, |a_3 b_1 + a_4 b_3| + |a_3 b_2 + a_4 b_4|\}$$
$$\leq \max\{|a_1|(|b_1| + |b_2|) + |a_2|(|b_3| + |b_4|), |a_3|(|b_1| + |b_2|) + |a_4|(|b_3| + |b_4|)\}$$
$$\leq \max\{|a_1|\|B\| + |a_2|\|B\|, |a_3|\|B\| + |a_4|\|B\|\} = \|B\|\|A\|.$$

14.4.13. Suppose that for every rectangle $R \subset D$, $\mu(R) = 0$. Let R_0 be a rectangle that contains D, and let $\varepsilon > 0$. Since D is a Jordan set, its boundary can be covered by rectangles of total area not exceeding ε. Using these rectangles, like in the proof of Theorem 14.1.7, we obtain a partition $P = \{R_{ij}\}$ of R_0, such that each of the mentioned rectangles is R_{ij} for some i, j. Let $f = \chi_D$, and let ξ be any selection of intermediate points. The Riemann sum $S(f, P, \xi)$ contains three types of terms. They are equal to 0 for rectangles not intersecting D (because $f = 0$), and for those completely contained in D (because their measure is 0). For rectangles that intersect both D and $R_0 \setminus D$, their total area is less than ε, so the terms in the Riemann sum add up to no more than ε. Thus, $S(f, P, \xi) < \varepsilon$. Since ε is arbitrary, we obtain that $\mu(D) = \iint_D \chi(x, y)\, dA = 0$.

14.4.23. Suppose that $A', A'' \in \mathcal{F}$ and $A' \subset D \subset A''$. Then

$$A' = \bigcup_{i=1}^{n} R_i', \quad \text{and} \quad A'' = \bigcup_{j=1}^{m} R_j'', \quad \text{where } R_i' = [a_i', b_i'] \times [c_i', d_i'], \text{ and } R_j'' = [a_j'', b_j''] \times [c_j'', d_j''].$$

Let $\{p_k\}, \{q_k\}, \{r_k\}, \{s_k\}$ be the collection of all a_i' and a_j'', respectively b_i' and b_j'', c_i' and c_j'', and d_i' and d_j''. Among all rectangles $Q_k = [p_k, q_k] \times [r_k, s_k]$, let Q_k' be those contained in D, and let Q_k'' be those that have nonempty intersection with D. If we denote $B = \cup Q_k'$ and $C = \cup Q_k''$, then $B, C \in \mathcal{F}$, and $A' \subset B \subset D \subset C \subset A''$. It follows that $\mu(A') \leq \mu(A'')$. Since this is true for any $A' \subset D$, we see that $\mu_*(D) \leq \mu(A'')$. Now this is true for any $A'' \in \mathcal{F}$ that contains D, so $\mu^*(D) \geq \mu_*(D)$.

14.4.26. First notice that if $A_1, A_2 \in \mathcal{F}$, then $\mu(A_1 \cup A_2) \leq \mu(A_1) + \mu(A_2)$. Let $\varepsilon > 0$. Then there exist $A_1, A_2 \in \mathcal{F}$ such that $D_1 \subset A_1$, $D_2 \subset A_2$, and $\mu(A_1) < \mu^*(D_1) + \varepsilon/2$, $\mu(A_2) < \mu^*(D_2) + \varepsilon/2$. Thus, $D_1 \cup D_2 \subset A_1 \cup A_2$, so

$$\mu^*(D_1 \cup D_2) \leq \mu(A_1 \cup A_2) \leq \mu(A_1) + \mu(A_2) < \mu^*(D_1) + \mu^*(D_2) + \varepsilon.$$

Since ε is arbitrary, we obtain the inequality $\mu^*(D_1 \cup D_2) \leq \mu^*(D_1) + \mu^*(D_2)$.

14.5.2. The domain of integration is $\pi \leq r \leq 2\pi$, $0 \leq \theta \leq 2\pi$. Therefore, the integral equals

$$\int_0^{2\pi} d\theta \int_\pi^{2\pi} (\sin r)\, r\, dr = \int_0^{2\pi} d\theta\, [\sin r - r \cos r]\,\big|_\pi^{2\pi} = \int_0^{2\pi} d\theta(-3\pi) = -3\pi \cdot 2\pi = -6\pi^2.$$

14.5.3. The inequality $x^2 + y^2 \leq 4x$ is equivalent to $(x - 2)^2 + y^2 \leq 4$, so the domain is the disk with center $(2, 0)$ and radius 2. In polar coordinates, we have $-\pi/2 \leq \theta \leq \pi/2$ and, for each fixed θ, $r^2 \leq 4r \cos \theta$. Since $r \geq 0$, we obtain that $r \leq 4 \cos \theta$. Now, the integral is

$$\int_{-\pi/2}^{\pi/2} d\theta \int_0^{4\cos\theta} r^2 \cdot r\, dr = \int_{-\pi/2}^{\pi/2} \frac{r^4}{4}\,\big|_0^{4\cos\theta}\, d\theta = \int_{-\pi/2}^{\pi/2} \frac{(4\cos\theta)^4}{4}\, d\theta = 64 \int_{-\pi/2}^{\pi/2} \cos^4\theta\, d\theta.$$

The substitution $u = \tan\theta$ leads to $64 \int_{-\infty}^\infty \frac{1}{(1+u^2)^3}\, du$. Now the result follows from the recursive formula

$$I_{n+1} = \frac{2n-1}{2n} I_n + \frac{u}{2n(u^2+1)^n}.$$

established in Example 5.1.11. Here,

$$64 \int_{-\infty}^\infty \frac{1}{(1+u^2)^3}\, du = 64 \left[\frac{3}{2} \int_{-\infty}^\infty \frac{1}{(1+u^2)^2}\, du + \frac{u}{4(u^2+1)^2}\,\Big|_{-\infty}^\infty \right]$$

$$= 96 \left[\frac{1}{2} \int_{-\infty}^\infty \frac{1}{1+u^2}\, du + \frac{u}{2(u^2+1)}\,\Big|_{-\infty}^\infty \right] = 48 \left[\arctan u \,\Big|_{-\infty}^\infty \right] = 48\pi.$$

14.5.24. Use the change of variables $x = \frac{u}{u^2+v^2}$, $y = \frac{v}{u^2+v^2}$.

14.5.28. Hints: (a) $\sin x^2 = (2x \sin x^2)(1/(2x))$. (b) Use the substitution $x = \sqrt{t^2 + \pi}$.

14.6.10. Since the integrand is non-negative, we can apply Theorem 14.6.5. Let D_n be the disk $0 \le \theta \le 2\pi$, $0 \le r \le n$. Then the integral becomes, assuming that $p \ne -1$,

$$\int_0^{2\pi} d\theta \int_0^n (1+r^2)^p \, r \, dr = \int_0^{2\pi} d\theta \int_0^{n^2} \frac{1}{2}(1+u)^p \, du = \int_0^{2\pi} d\theta \frac{(1+u)^{p+1}}{p+1} \Big|_{u=0}^{u=n^2}$$

$$= \int_0^{2\pi} d\theta \frac{(1+n^2)^{p+1} - 1}{p+1} = 2\pi \frac{(1+n^2)^{p+1} - 1}{p+1}.$$

Clearly, this sequence is bounded if $p < -1$, and unbounded if $p > -1$. When $p = -1$, we get

$$\int_0^{2\pi} d\theta \ln(1+u) \Big|_0^{n^2} = \int_0^{2\pi} d\theta \ln(1+n^2) = 2\pi \ln(1+n^2)$$

which is also unbounded. Therefore, the integral converges if and only if $p < -1$.

14.6.14. Let D_n be the rectangle $[0,1] \times [\frac{1}{n}, 1]$. Then

$$\iint_{D_n} \frac{1}{x+y} \, dA = \int_0^1 dx \int_{1/n}^1 \frac{1}{x+y} \, dy = \int_0^1 dx \, \ln(x+y) \, |_{y=1/n}^{y=1} = \int_0^1 \left[\ln(x+1) - \ln\left(x+\frac{1}{n}\right) \right] dx$$

$$= [(x+1)\ln(x+1) - (x+1)] \, |_0^1 - \left[\left(x+\frac{1}{n}\right) \ln\left(x+\frac{1}{n}\right) - \left(x+\frac{1}{n}\right) \right] \Big|_0^1$$

$$= [(2\ln 2 - 2) - (-1)] - \left[\left(1+\frac{1}{n}\right) \ln\left(1+\frac{1}{n}\right) - \left(1+\frac{1}{n}\right) \right] + \left[\frac{1}{n}\ln\frac{1}{n} - \frac{1}{n} \right] \to 2\ln 2, \text{ as } n \to \infty.$$

14.6.22. Use Exercise 14.5.28.

14.7.6. The solid S is determined by the inequalities $0 \le x \le 1$, $0 \le y \le 1-x$, $0 \le z \le 1-x-y$, so we use the integral

$$\int_0^1 dx \int_0^{1-x} dy \int_0^{1-x-y} \frac{dz}{(1+x+y+z)^2} = \int_0^1 dx \int_0^{1-x} dy \frac{-1}{1+x+y+z} \Big|_{z=0}^{z=1-x-y}$$

$$= \int_0^1 dx \int_0^{1-x} dy \left(-\frac{1}{2} + \frac{1}{1+x+y} \right) = \int_0^1 dx \left(-\frac{1}{2}y + \ln(1+x+y) \right) \Big|_{y=0}^{y=1-x}$$

$$= \int_0^1 dx \left(-\frac{1}{2}(1-x) + \ln 2 - \ln(1+x) \right)$$

$$= \left[\frac{1}{4}(x-1)^2 + x\ln 2 - (x+1)\ln(x+1) + (x+1) \right] \Big|_0^1 = -\ln 2 + \frac{3}{4}.$$

14.7.9. We will use spherical coordinates: $0 \le \rho \le 4$, $0 \le \theta \le 2\pi$, $0 \le \varphi \le \frac{\pi}{6}$. (The cone is obtained by the revolution of the line $z = x\sqrt{3}$, with slope $\pi/3$, so $\varphi = \frac{\pi}{2} - \frac{\pi}{3} = \frac{\pi}{6}$.) Thus, we have

$$\int_0^4 d\rho \int_0^{2\pi} d\theta \int_0^{\pi/6} (\rho\cos\theta\sin\varphi + \rho\sin\theta\sin\varphi)(\rho^2\sin\varphi) \, d\varphi = \int_0^4 \rho^3 \, d\rho \int_0^{2\pi} (\cos\theta + \sin\theta) \, d\theta \int_0^{\pi/6} \sin^2\varphi \, d\varphi = 0,$$

because $\int_0^{2\pi} (\cos\theta + \sin\theta) \, d\theta = 0$.

14.7.11. Answer: $\dfrac{h^3 \pi^{\frac{n-1}{2}} a^{n-1}}{12\Gamma\left(\frac{n}{2} - 1\right)}$.

14.7.20. Let $I_n(h)$ denote the volume of the n-dimensional solid $S_n(h)$ determined by the constraints $x_k \ge 0$, $1 \le k \le n$, and $x_1 + x_2 + \cdots + x_n \le h$. Notice that

$$I_n(h) = \int_0^h dx_1 \int_0^{h-x_1} dx_2 \int_0^{h-x_1-x_2} dx_3 \cdots \int_0^{h-x_1-x_2-\cdots-x_{n-1}} dx_n.$$

The substitution $x_k = hy_k$, $1 \le k \le n$ yields $I_n(h) = h^n I_n(1)$. Further,

$$I_n(1) = \int_0^1 dx_n \int\cdots\int_{S_{n-1}(1-x_n)} dV_{n-1} = \int_0^1 I_{n-1}(1-x_n) dx_n = \int_0^1 (1-x_n)^{n-1} I_{n-1}(1) \, dx_n$$

$$= I_{n-1}(1) \frac{(1-x_n)^n}{-n} \, |_0^1 = \frac{1}{n} I_{n-1}(1).$$

Inductively, we obtain that $I_n(1) = I_1(1)\frac{1}{n!}$. Since I_1 is the length of the unit segment, we have that $I_n(1) = \frac{1}{n!}$, so $I_n(h) = \frac{h^n}{n!}$.

Finally, the volume V_n of the solid determined by $x_k \geq 0$, $1 \leq k \leq n$, and $\frac{x_1}{a_1} + \frac{x_2}{a_2} + \cdots + \frac{x_n}{a_n} \leq 1$ can be calculated using the substitution $x_k = a_k y_k$, $1 \leq k \leq n$. This yields $V_n = a_1 a_2 \ldots a_n I_n(1) = \frac{a_1 a_2 \ldots a_n}{n!}$.

14.7.26. It is not hard to see that the given integral converges if and only if its portion in the first octant $\mathbb{O}_1$ does. We will show that

$$\iiint_{\substack{x+y+z \geq 1 \\ x,y,z \geq 0}} \frac{1}{x^p + y^q + z^r}\, dV(x,y,z)$$

converges if and only if $\frac{1}{p} + \frac{1}{q} + \frac{1}{r} < 1$.

The domain of integration now consists of three parts: $D_1 = \{(x,y,z) \in \mathbb{O}_1 : x + y \leq 1, 1 - x - y \leq z \leq 1\}$, $D_2 = \{(x,y,z) \in \mathbb{O}_1 : x + y \leq 1, z \geq 1\}$, $D_3 = \{(x,y,z) \in \mathbb{O}_1 : x + y \geq 1, z \geq 0\}$. The integral over D_1 is not improper, because the integrand is a bounded function there. If $\frac{1}{p} + \frac{1}{q} + \frac{1}{r} < 1$, then $r > 1$, so $\int_1^\infty \frac{1}{z^r}\, dz$ converges. This implies that the integral over D_2 converges as well:

$$\iiint_{D_2} \frac{1}{x^p + y^q + z^r}\, dV(x,y,z) = \iint_{\substack{x+y \leq 1 \\ x,y \geq 0}} dA(x,y) \int_1^\infty \frac{1}{x^p + y^q + z^r}\, dz \leq \iint_{\substack{x+y \leq 1 \\ x,y \geq 0}} dA(x,y) \int_1^\infty \frac{1}{z^r}\, dz.$$

Thus, it remains to consider the integral over D_3. Notice that the substitution $z = t^{1/r} w$ transforms the integral $\int_0^\infty \frac{dz}{t + z^r}$ to $\int_0^\infty \frac{t^{1/r}\, dw}{t(1 + w^r)}$. When $r > 1$, $\int_0^\infty \frac{dw}{1 + w^r}$ converges to a positive number α. This shows that

$$\iiint_{D_3} \frac{1}{x^p + y^q + z^r}\, dV(x,y,z) = \iint_{\substack{x+y \geq 1 \\ x,y \geq 0}} dA(x,y) \int_0^\infty \frac{1}{x^p + y^q + z^r}\, dz = \alpha \iint_{\substack{x+y \geq 1 \\ x,y \geq 0}} \frac{1}{(x^p + y^q)^{1-1/r}}\, dA(x,y)$$

and we need to prove that the last integral converges.

Now we will split the domain of integration into 3 parts: $E_1 = \{(x,y) \in \mathbb{R}^2 : 0 \leq x \leq 1, 1 - x \leq y \leq 1\}$, $E_2 = \{(x,y) \in \mathbb{R}^2 : 0 \leq x \leq 1, y \geq 1\}$, $E_3 = \{(x,y) \in \mathbb{R}^2 : x \geq 1, y \geq 0\}$. On E_1, the integrand is a bounded function, and on E_2 it is dominated by $1/y^{q(1-1/r)}$. The assumption that $\frac{1}{p} + \frac{1}{q} + \frac{1}{r} < 1$ implies that $\frac{1}{q} + \frac{1}{r} < 1$, hence $q\left(1 - \frac{1}{r}\right) > 1$. Therefore, the integral over E_2 converges. Finally, the substitution $y = t^{1/q} v$ transforms the integral $\int_0^\infty \frac{dy}{(t + y^q)^{1-1/r}}$ to $\int_0^\infty \frac{t^{1/q}\, dv}{t^{1-1/r}(1 + v^q)^{1-1/r}}$. The last integral converges to a positive number β, because $q\left(1 - \frac{1}{r}\right) > 1$. Therefore,

$$\iint_{E_3} \frac{1}{(x^p + y^q)^{1-1/r}}\, dA(x,y) = \int_1^\infty dx \int_0^\infty \frac{1}{(x^p + y^q)^{1-1/r}}\, dy = \beta \int_1^\infty \frac{dx}{(x^p)^{1-1/r-1/q}},$$

and the convergence of the last integral is guaranteed because $p(1 - \frac{1}{r} - \frac{1}{q}) > 1$. This shows that, if $\frac{1}{p} + \frac{1}{q} + \frac{1}{r} < 1$, then the given integral converges.

In the other direction,

$$\iiint_{\substack{x+y+z \geq 1 \\ x,y,z \geq 0}} \frac{1}{x^p + y^q + z^r}\, dV(x,y,z) \geq \iiint_{D_3} \frac{1}{x^p + y^q + z^r}\, dV(x,y,z) \geq \iint_{E_3} dA(x,y) \int_0^\infty \frac{1}{x^p + y^q + z^r}\, dz$$

$$= \iint_{E_3} \frac{1}{(x^p + y^q)^{1-1/r}}\, dA(x,y) \int_0^\infty \frac{dw}{1 + w^r} = \int_1^\infty dx \int_0^\infty \frac{1}{(x^p + y^q)^{1-1/r}}\, dy \int_0^\infty \frac{dw}{1 + w^r}$$

$$= \int_1^\infty \frac{dx}{(x^p)^{1-1/r-1/q}} \int_0^\infty \frac{dv}{(1 + v^q)^{1-1/r}} \int_0^\infty \frac{dw}{1 + w^r},$$

and it is not hard to see that the convergence requires that $r > 1$, $q\left(1 - \frac{1}{r}\right) > 1$, and $p\left(1 - \frac{1}{q} - \frac{1}{r}\right) > 1$. The last one can be written as $\frac{1}{p} + \frac{1}{q} + \frac{1}{r} < 1$.

14.7.27. The integral converges if and only if $p < 1$.

15.1.2. Suppose that there exists $\varphi : [0, 2\pi] \to [0, 4\pi]$ such that φ is a C^1 bijection and $\mathbf{f} = \mathbf{h} \circ \varphi$. Then $\cos t = \cos \varphi(t)$, for all $t \in [0, 2\pi]$. Since φ is surjective, there exist $t_1, t_2 \in [0, 2\pi]$ such that $\varphi(t_1) = \pi$ and $\varphi(t_2) = 3\pi$. Then $\cos t_1 = \cos \varphi(t_1) = -1$ and $\cos t_2 = -1$. This implies that $t_1 = t_2 = \pi$ which is a contradiction.

15.1.26. If we use z as a parameter, and if we solve for x and y, we obtain a parametrization

$$\mathbf{f}(z) = \left(\frac{3z^{4/3} + 6z^{2/3}}{8}, \frac{3z^{4/3} - 6z^{2/3}}{8}, z\right), \text{ so } \mathbf{f}'(z) = \left(\frac{z^{1/3} + z^{-1/3}}{2}, \frac{z^{1/3} - z^{-1/3}}{2}, 1\right), \text{ and}$$

$$\|\mathbf{f}'(z)\| = \left(\frac{z^{1/3} + z^{-1/3}}{\sqrt{2}}\right)^2.$$

Now, the length of the curve is

$$L = \int_0^c \frac{z^{1/3} + z^{-1/3}}{\sqrt{2}} \, dz = \frac{1}{\sqrt{2}} \left(\frac{3}{4} x^{4/3} + \frac{3}{2} z^{2/3} \right) \Big|_0^c = \frac{3\sqrt{2}}{8} c^{4/3} + \frac{3\sqrt{2}}{4} c^{2/3}.$$

15.2.13. In polar coordinates the lemniscate is $r^4 = 4r^2(\cos^2 t - \sin^2 t) = 4r^2 \cos 2t$, so $r^2 = 4 \cos 2t$. This implies that $\cos 2t \geq 0$, so $2t \in [-\pi/2, \pi/2] \cup [3\pi/2, 5\pi/2]$, i.e.,

$$t \in \left[-\frac{\pi}{4}, \frac{\pi}{4} \right] \cup \left[\frac{3\pi}{4}, \frac{5\pi}{4} \right].$$

For such t, $r = 2\sqrt{\cos 2t}$, so $x = r \cos t = 2 \cos t \sqrt{\cos 2t}$ and $y = r \sin t = 2 \sin t \sqrt{\cos 2t}$. It follows that

$$dx = \left[\frac{2}{\sqrt{\cos 2t}} (-\sin 2t) \cos t - 2 \sin t \sqrt{\cos 2t} \right] dt = \frac{-2}{\sqrt{\cos 2t}} (\sin 2t \cos t + \cos 2t \sin t) \, dt = \frac{-2 \sin 3t}{\sqrt{\cos 2t}} \, dt, \text{ and}$$

$$dy = \left[\frac{2}{\sqrt{\cos 2t}} (-\sin 2t) \sin t + 2 \cos t \sqrt{\cos 2t} \right] dt = \frac{2}{\sqrt{\cos 2t}} (-\sin 2t \sin t + \cos 2t \cos t) \, dt = \frac{2 \cos 3t}{\sqrt{\cos 2t}} \, dt.$$

Then

$$\|\mathbf{f}'(t)\|^2 = \frac{4 \sin^2 3t}{\cos 2t} + \frac{4 \cos^2 3t}{\cos 2t} = \frac{4}{\cos 2t}$$

and $\|\mathbf{f}'(t)\| = \frac{2}{\sqrt{\cos 2t}}$. Now we have

$$\int_{-\pi/4}^{\pi/4} 2\sqrt{\cos 2t} \, |\sin t| \, \frac{2}{\sqrt{\cos 2t}} \, dt + \int_{3\pi/4}^{5\pi/4} 2\sqrt{\cos 2t} \, |\sin t| \, \frac{2}{\sqrt{\cos 2t}} \, dt$$

$$= 8 \int_0^{\pi/4} \sin t \, dt + 8 \int_{3\pi/4}^{\pi} \sin t \, dt = 8 \left(-\cos t \Big|_0^{\pi/4} - \cos t \Big|_{3\pi/4}^{\pi} \right) = 16 - 8\sqrt{2}.$$

15.2.17. The equation of the line is $y = \pi - x$, so we use x as a parameter, $0 \leq x \leq \pi$. Now $dy = -dx$ so we have

$$\int_0^{\pi} [\sin(\pi - x) + (\sin x)(-1)] \, dx = \int_0^{\pi} (\sin x - \sin x) \, dx = 0.$$

15.2.21. Use x as a parameter: $y = \sqrt{2x}$, $z = \sqrt{x^2 + 2x}$, $0 \leq x \leq 2$. Then $dy = \frac{\sqrt{2}}{2\sqrt{x}} \, dx$ and $dz = \frac{2x+2}{2\sqrt{x^2+2x}} \, dx$, so we obtain

$$\int_0^2 \sqrt{x^2 + 2x} \sqrt{1 + \frac{1}{2x} + \frac{(x+1)^2}{x^2 + 2x}} \, dx = \int_0^2 \sqrt{x^2 + 2x + \frac{1}{2}(x+2) + (x+1)^2} \, dx$$

$$= \int_0^2 \sqrt{2x^2 + \frac{9}{2} x + 2} \, dx = \int_0^2 \sqrt{2 \left(x + \frac{9}{8} \right)^2 - \frac{17}{32}} \, dx$$

$$= \sqrt{2} \left[\frac{1}{2} \left(x + \frac{9}{8} \right) \sqrt{\left(x + \frac{9}{8} \right)^2 - \frac{17}{64}} - \frac{1}{2} \frac{17}{64} \ln \left(x + \frac{9}{8} + \sqrt{\left(x + \frac{9}{8} \right)^2 - \frac{17}{64}} \right) \right] \Big|_0^2$$

$$= \frac{\sqrt{2}}{128} \left(100\sqrt{38} - 72 + 17 \ln(25 - \sqrt{608}) \right).$$

15.2.26. Let C_1, C_2, C_3 be as in the figure. First we compute the integral over C_1. Now $z = 0$ and $x^2 + y^2 = 1$, so $x = \cos t$, $y = \sin t$, $0 \leq t \leq \pi/2$. We have

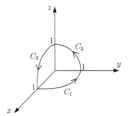

$$\int_0^{\pi/2} \left[(\sin^2 t - 0^2)(-\sin t) + (0^2 - \cos^2 t) \cos t + (\cos^2 t - \sin^2 t) \right] dt$$

$$= \int_0^{\pi/2} \left[(1 - \cos^2 t)(-\sin t) + (\sin^2 t - 1) \cos t + \cos 2t \right] dt$$

$$= \left[\cos t - \frac{1}{3} \cos^3 t + \frac{1}{3} \sin^3 t - \sin t + \frac{1}{2} \sin 2t \right] \Big|_0^{\pi/2} = -\frac{4}{3}.$$

Next, on C_2, $x = 0$, $y = \cos t$, $z = \sin t$, $0 \le t \le \pi/2$, so

$$\int\limits_0^{\pi/2} \left[(\cos^2 t - \sin^2 t) + (\sin^2 t - 0^2)(-\sin t) + (0^2 - \cos^2 t)\cos t \right] dt = -\frac{4}{3}.$$

Finally, on C_3, $y = 0$, $z = \cos t$, $x = \sin t$, $0 \le t \le \pi/2$, so

$$\int\limits_0^{\pi/2} \left[(0^2 - \cos^2 t)\cos t + (\cos^2 t - \sin^2 t) + (\sin^2 t - 0^2)(-\sin t) \right] dt = -\frac{4}{3}.$$

Therefore, we get that the integral over the whole curve C is -4.

15.3.3. Here $P(x,y) = (x+y)^2$, $Q(x,y) = -(x^2 + y^2)$, so $Q_x(x,y) = -2x$ and $P_y(x,y) = 2(x+y)$. Therefore, Green's Theorem leads to the integral $\iint_D [(-2x) - 2(x+y)]\, dA$, where D is the triangular region bounded by the lines $y = \frac{1}{2}x + \frac{1}{2}$, $y = -3x + 11$, and $y = 4x - 3$. As we see in the figure, D can be split into 2 y-simple regions, so we have to compute

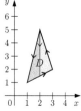

$$\int\limits_1^2 dx \int\limits_{\frac{1}{2}x+\frac{1}{2}}^{4x-3} (-4x - 2y)\, dy + \int\limits_2^3 dx \int\limits_{\frac{1}{2}x+\frac{1}{2}}^{-3x+11} (-4x - 2y)\, dy$$

$$= \int\limits_1^2 (-4xy - y^2)\Big|_{y=\frac{1}{2}x+\frac{1}{2}}^{y=4x-3}\, dx + \int\limits_2^3 (-4xy - y^2)\Big|_{y=\frac{1}{2}x+\frac{1}{2}}^{y=-3x+11}\, dx$$

$$= \int\limits_1^2 \left[-4x(4x-3) - (4x-3)^2 + 4x\left(\frac{1}{2}x + \frac{1}{2}\right) + \left(\frac{1}{2}x + \frac{1}{2}\right)^2 \right] dx$$

$$+ \int\limits_2^3 \left[-4x(-3x+11) - (-3x+11)^2 + 4x\left(\frac{1}{2}x + \frac{1}{2}\right) + \left(\frac{1}{2}x + \frac{1}{2}\right)^2 \right] dx$$

$$= \int\limits_1^2 \left[-\frac{119}{4}x^2 + \frac{77}{2}x - \frac{35}{4} \right] dx + \int\limits_2^3 \left[\frac{21}{4}x^2 + \frac{49}{2}x - \frac{483}{4} \right] dx = -\frac{140}{3}.$$

15.3.5. Let C_1 denote the line segment from $(0,0)$ to $(2,0)$. Now C and C_1 together represent the boundary of the upper half of a disk D with center at $(1,0)$ and radius 1. Also, $P(x,y) = e^x \sin y - 3y$ and $Q(x,y) = e^x \cos y - 3$, so $Q_x(x,y) = e^x \cos y$ and $P_y(x,y) = e^x \cos y - 3$, and it follows that $Q_x(x,y) - P_y(x,y) = 3$. Therefore,

$$\int\limits_C (e^x \sin y - 3y)\, dx + (e^x \cos y - 3)\, dy + \int\limits_{C_1} (e^x \sin y - 3y)\, dx + (e^x \cos y - 3)\, dy = \iint\limits_D 3\, dA.$$

On C_1 we have that $y = 0$ (and $dy = 0$), so $\int_{C_1} (e^x \sin y - 3y)\, dx + (e^x \cos y - 3)\, dy = 0$. Thus,

$$\int\limits_C (e^x \sin y - 3y)\, dx + (e^x \cos y - 3)\, dy = \iint\limits_D 3\, dA = 3\,\frac{\pi}{2}$$

because the area of D is $\pi/2$.

15.3.11. This parabola can be best understood using the change of variables $u = x$ and $v = x + y$, which transforms the equation into $v^2 = au$. So, we have a parabola that is symmetric with respect to the u-axis, i.e., the set $\{(u,v) : v = 0\} = \{(x,y) : y = -x\}$. Also, the parabola lies to the right of the v-axis, i.e., the set $\{(u,v) : u = 0\} = \{(x,y) : x = 0\}$, i.e., the y-axis. We will use the formula

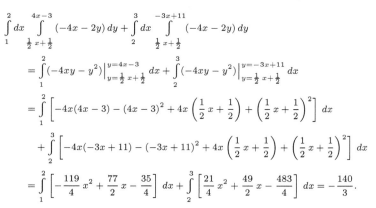

$$A = \oint\limits_C x\, dy.$$

To determine a parametrization for C, let $x = u$. Then $y = v - x = v - u = v - v^2/a$. The intersection of the parabola with the x-axis occur when $x^2 = ax$, hence at $x = 0$ and $x = a$, i.e., $u = 0$ and $u = a$. The region remains on the left if we orient that u flows from a to 0. However, this is only the parabolic portion of the curve. For it to be closed, we will include the line segment C_1 from $(0,0)$ to $(a,0)$. Since on this segment $y = 0$, we have that $\int_{C_1} x\, dy = 0$. Therefore, the area equals

$$\int\limits_a^0 \frac{v^2}{a}\left(1 - \frac{2v}{a}\right) dv = \int\limits_a^0 \left(\frac{v^2}{a} - \frac{2v^3}{a^2} \right) dv = \frac{a^2}{6}.$$

15.4.15. The surface consists of 16 pieces of equal area. One of them lies above the triangle $0 \le y \le a$, $0 \le x \le y$

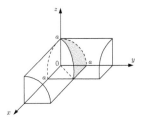

in the xy-plane. Therefore, the area equals

$$16 \int\limits_0^a \int\limits_0^y \sqrt{1 + (z_x')^2 + (z_y')^2}\, dxdy.$$

Here, $z = \sqrt{a^2 - y^2}$, so $z_x' = 0$ and $z_y' = -y/\sqrt{a^2 - y^2}$. It follows that

$$\sqrt{1 + (z_x')^2 + (z_y')^2} = \frac{a}{\sqrt{a^2 - y^2}},$$

and the area is

$$16a \int\limits_0^a \int\limits_0^y \frac{1}{\sqrt{a^2 - y^2}}\, dxdy = 16a \int\limits_0^a \frac{1}{\sqrt{a^2 - y^2}}\, x \Big|_{x=0}^{x=y}\, dy = 16a \int\limits_0^a \frac{y}{\sqrt{a^2 - y^2}}\, dy = 16a(-\sqrt{a^2 - y^2})\Big|_0^a = 16a^2.$$

15.4.17. We will use the parametrization $x = r\cos\theta$, $y = r\sin\theta$, which implies that $z = 1 - r^3$. It is straightforward to compute that $\mathbf{n} = (3r^3\cos\theta, 3r^3\sin\theta, r)$ and $\|\mathbf{n}\| = \sqrt{9r^6 + r^2} = r\sqrt{1 + 9r^4}$. When $z = 0$, we get $r = 1$, which is the unit circle, so $0 \le r \le 1$ and $0 \le \theta \le 2\pi$. Now, using the substitution $u = 3r^2$,

$$\int\limits_0^{2\pi} \int\limits_0^1 r\sqrt{1 + 9r^4}\, drd\theta = \int\limits_0^{2\pi} \int\limits_0^3 \frac{1}{6}\sqrt{1 + u^2}\, dud\theta = \int\limits_0^{2\pi} \frac{1}{12}\left(u\sqrt{1 + u^2} + \ln(u + \sqrt{1 + u^2})\right)\Big|_{u=0}^{u=3}\, d\theta$$

$$= \int\limits_0^{2\pi} \frac{1}{12}\left(3\sqrt{10} + \ln(3 + \sqrt{10})\right)\, d\theta = 2\pi\left(\frac{1}{12}3\sqrt{10} + \frac{1}{12}\ln(3 + \sqrt{10})\right).$$

15.4.21. Inspired by spherical coordinates we take $x = 2\sin\varphi\cos\theta$, $y = 2\sin\varphi\sin\theta$, $z = 2\cos\varphi$, and $0 \le \theta \le 2\pi$, $0 \le \varphi \le \frac{\pi}{2}$. Then $\mathbf{n} = (4\sin^2\varphi\cos\theta, 4\sin^2\varphi\sin\theta, 4\sin\varphi\cos\varphi)$, so $\|\mathbf{n}\| = 4|\sin\varphi| = 4\sin\varphi$, because $\varphi \in [0, \pi/2]$. We obtain

$$\int\limits_0^{2\pi} \int\limits_0^{\pi/2} 4\sin\varphi\, (2\sin\varphi\cos\theta + 2\sin\varphi\sin\theta + 2\cos\varphi)\, d\varphi d\theta = 8 \int\limits_0^{2\pi} \int\limits_0^{\pi/2} \left[\sin^2\varphi(\sin\theta + \cos\theta) + \sin\varphi\cos\varphi\right]\, d\varphi d\theta$$

$$= 8 \int\limits_0^{2\pi} \left[\frac{\pi}{4}(\sin\theta + \cos\theta) + \frac{1}{2}\right]\, d\theta = 8\left[\frac{\pi}{4}(\sin\theta - \cos\theta) + \frac{\theta}{2}\right]\Big|_0^{2\pi} = 8\pi.$$

15.4.23. It is straightforward to obtain that $\mathbf{n} = (\sin v, -\cos v, u)$ and $\|\mathbf{n}\| = \sqrt{1 + u^2}$. Thus, we have

$$\int\limits_0^{2\pi} \int\limits_0^a v\sqrt{1 + u^2}\, dudv = \int\limits_0^{2\pi} v\, dv\, \frac{1}{2}\left[u\sqrt{1 + u^2} + \ln(u + \sqrt{1 + u^2})\right]\Big|_0^a$$

$$= \frac{1}{2} \int\limits_0^{2\pi} v\, dv\left[a\sqrt{1 + a^2} + \ln(a + \sqrt{1 + a^2})\right] = \pi^2\left[a\sqrt{1 + a^2} + \ln(a + \sqrt{1 + a^2})\right].$$

15.4.28. Since we are dealing with a cylinder of radius 3, we use $x = 3\cos t$, $y = 3\sin t$, and $z = s$. Further, $0 \le s \le 4$, and $0 \le t \le 2\pi$. When $t = 0$, $\mathbf{n} = (3, 0, 0)$, pointing in the direction of the positive x-axis, when starting at $(3, 0, s)$. Thus, it is an outward normal to the surface of the cylinder. Therefore, the flux equals

$$\int\limits_0^{2\pi} \int\limits_0^4 (9\cos^2 t, 0, 0) \cdot (3\cos t, 3\sin t, 0)\, dsdt = \int\limits_0^{2\pi} \int\limits_0^4 27\cos^3 t\, dsdt = 108 \int\limits_0^{2\pi} \cos^3 t\, dt = 0.$$

15.5.12. It is easy to see that $\operatorname{div}\mathbf{F} = 3$, so by Divergence Theorem our problem is reduced to computing $\iiint_S 3\, dV$, where S is the solid defined by

$$|x - y + z| + |y - z + x| + |z - x + y| \le 1.$$

In other words we need to calculate the volume of S and multiply it by 3. A change of variables $u = x - y + z$, $v = y - z + x$, $w = z - x + y$ has the Jacobian determinant

$$\begin{vmatrix} 1 & -1 & 1 \\ 1 & 1 & -1 \\ -1 & 1 & 1 \end{vmatrix} = 4$$

so we need to compute

$$\frac{3}{4} \iiint\limits_{|u|+|v|+|w|\leq 1} dV(u,v,w).$$

The solid determined by $|u|+|v|+|w| \leq 1$ has in each octant a part of the same volume, so it suffices to compute the volume of S', its portion in the 1st octant and multiple by 8. Further, S' is a tetrahedron of volume $\frac{1}{3} \cdot \frac{1}{2} = \frac{1}{6}$. Thus, the result is $\frac{3}{4} \cdot 8 \cdot \frac{1}{6} = 1$.

15.5.15. The solid is defined by $0 \leq x \leq 3$, $-2 \leq y \leq 2$, and $0 \leq z \leq 4 - y^2$, and div $\mathbf{F} = 2 - x$. Using Divergence Theorem we obtain

$$\int_0^3 dx \int_{-2}^2 dy \int_0^{4-y^2} (2-x)\, dz = \int_0^3 dx \int_{-2}^2 (2-x)(4-y^2)\, dy$$

$$= \int_0^3 (2-x)\, dx \left(4y - \frac{1}{3}y^3\right) \Big|_{y=-2}^{y=2} = \int_0^3 \frac{32}{3}(2-x)\, dx = \frac{32}{3}\left(2x - \frac{1}{2}x^2\right) \Big|_0^3 = 16.$$

15.6.2. The surface M is a portion of the plane $\frac{x}{a} + \frac{z}{b} = 1$. We will use the parametrization

$$\mathbf{f}(r,\theta) = \left(r\cos\theta, r\sin\theta, b\left(1 - \frac{r\cos\theta}{a}\right)\right), \quad 0 \leq r \leq a, \quad 0 \leq \theta \leq 2\pi.$$

Then $\mathbf{n} = (\frac{b}{a}r, 0, r)$, which is pointing upward (the z-coordinate is $r > 0$ so it agrees with the counterclockwise flow of the curve. The curl equals curl $\mathbf{F}(x,y,z) = (-2,-2,-2)$ so Stokes' Theorem leads to

$$\int_0^{2\pi} d\theta \int_0^a (-2,-2,-2) \cdot \left(\frac{b}{a}r, 0, r\right) dr = \int_0^{2\pi} d\theta \int_0^a \left(-\frac{2b}{a}r - 2r\right) dr$$

$$= \int_0^{2\pi} \left(-\frac{b}{a}r^2 - r^2\right) \Big|_{r=0}^{r=a} d\theta = \int_0^{2\pi} (-ab - a^2)\, d\theta = 2\pi(-ab - a^2).$$

15.6.6. curl $\mathbf{F}(x,y,z) = \mathbf{0}$.
15.6.8. The plane determined by the vertices of the triangle has an equation $x + \frac{y}{2} + \frac{z}{3} = 1$, so we use the parametrization

$$\mathbf{f}(x,y) = \left(x, y, 3\left(1 - x - \frac{y}{2}\right)\right), \quad \text{with } 0 \leq x \leq 1, \text{ and } 0 \leq y \leq 2 - 2x.$$

Then, $\mathbf{n} = (3, \frac{3}{2}, 1)$ is pointing up, which agrees with the orientation of the triangle. Next, curl $\mathbf{F}(x,y,z) = (z - x, 0, z - x)$, so

$$\int_0^1 dx \int_0^{2-2x} \left(3 - 3x - \frac{3}{2}y - x, 0, 3 - 3x - \frac{3}{2}y - x\right) \cdot \left(3, \frac{3}{2}, 1\right) dy = \int_0^1 dx \int_0^{2-2x} (12 - 16x - 6y)\, dy$$

$$= \int_0^1 \left(12y - 16xy - 3y^2\right) \Big|_{y=0}^{y=2-2x} dx = \int_0^1 \left[12(2-2x) - 16x(2-2x) - 3(2-2x)^2\right] dx$$

$$= \int_0^1 \left(12 - 32x + 20x^2\right) dx = \left(12x - 16x^2 + \frac{20}{3}x^3\right) \Big|_0^1 = \frac{8}{3}.$$

15.6.9. The boundary of M is a circle $x^2 + y^2 = 4$, $z = 2$. Since M is the outer surface of the paraboloid, the circle should be travelled clockwise (when viewed from the positive z-axis). If we use $\mathbf{f}(t) = (2\cos t, 2\sin t, 2)$, then t should go from 2π to 0. It is easy to see that $\mathbf{f}'(t) = (-2\sin t, 2\cos t, 0)$, so we obtain the integral

$$\int_{2\pi}^0 (6\sin t, -4\cos t, 8\sin t) \cdot (-2\sin t, 2\cos t, 0)\, dt = \int_{2\pi}^0 (-12\sin^2 t - 8\cos^2 t)\, dt$$

$$= \int_{2\pi}^0 \left(-12\frac{1 - \cos 2t}{2} - 8\frac{1 + \cos 2t}{2}\right) dt = \int_{2\pi}^0 (-10 + 2\cos 2t)\, dt = 20\pi.$$

15.7.1. $(-35x - 3x^3 y^2)\, dx dy$.
15.7.8. By definition, $d\omega = d(4y)\, dx + d(-3xy)\, dy$. Since $d(4y) = 4\, dy$ and $d(-3xy) = -3y\, dx - 3x\, dy$, $d\omega = 4\, dydx - 3y\, dxdy - 3x\, dydy = (-3y - 4)\, dxdy$.
15.7.11. By definition, $d\omega = d(x^3)\, dxdy + d(y^2)\, dxdz + d(-z)\, dydz = 3x^2\, dxdxdy + 2y\, dydxdz - dzdydx = -2y\, dxdydz + dxdydz = (1 - 2y)\, dxdydz$.

15.7.17. We will prove the case when the forms ω_1, ω_2 are 2-forms. The general case can be proved using the same strategy. Let $\omega_1(\mathbf{x}) = \sum_{1 \leq i \leq k \leq n} a_{ik}(\mathbf{x})\, dx_i dx_k$, $\omega_2(\mathbf{x}) = \sum_{1 \leq i \leq k \leq n} b_{ik}(\mathbf{x})\, dx_i dx_k$. Then

$$\alpha\omega_1(\mathbf{x}) + \beta\omega_2(\mathbf{x}) = \sum_{1 \leq i \leq k \leq n} (\alpha a_{ik}(\mathbf{x}) + \beta b_{ik}(\mathbf{x}))\, dx_i dx_k, \text{ and}$$

$$d\left(\alpha\omega_1(\mathbf{x}) + \beta\omega_2(\mathbf{x})\right) = \sum_{1 \leq i \leq k \leq n} d\left(\alpha a_{ik}(\mathbf{x}) + \beta b_{ik}(\mathbf{x})\right)\, dx_i dx_k, \text{ while}$$

$$\alpha\, d\left(\omega_1(\mathbf{x})\right) + \beta\, d\left(\omega_2(\mathbf{x})\right) = \alpha \sum_{1 \leq i \leq k \leq n} d\left(a_{ik}(\mathbf{x})\right)\, dx_i dx_k + \beta \sum_{1 \leq i \leq k \leq n} d\left(b_{ik}(\mathbf{x})\right)\, dx_i dx_k$$

$$= \sum_{1 \leq i \leq k \leq n} (\alpha\, da_{ik}(\mathbf{x}) + \beta\, db_{ik}(\mathbf{x}))\, dx_i dx_k.$$

The result now follows from the fact that the derivative is a linear transformation when applied to functions.

15.8.2. $d(x\, dy + y\, dx) = dxdy + dydx = 0$ so ω is exact. Also, $\omega = d(xy)$, and the integral equals $xy\, |_{(-1,2)}^{(2,3)} = 6 - (-2) = 8$.

15.8.4. 4.

15.8.5. $e^a \cos b - 1$.

15.8.10. The form ω is not defined at $(0,0)$, so we introduce new variables $x = u$, $y = v - 1$. This gives a form

$$\gamma(u,v) = \frac{(v-1)\, du - u\, dv}{3u^2 - 2u(v-1) + 3(v-1)^2} = \frac{v-1}{3u^2 - 2u(v-1) + 3(v-1)^2}\, du + \frac{-u}{3u^2 - 2u(v-1) + 3(v-1)^2}\, dv$$

$$= a(u,v)\, du + b(u,v)\, dv,$$

and we will show that it is exact. Clearly,

$$d\gamma = (a_u'\, du + a_v'\, dv)\, du + (b_u'\, du + b_v'\, dv)\, dv = a_v'\, dvdu + b_u'\, dudv = (b_u' - a_v')\, dudv,$$

so it suffices to show that $a_v' = b_u'$. Now

$$a_v' = \frac{\partial}{\partial v}\left(\frac{v-1}{3u^2 - 2u(v-1) + 3(v-1)^2}\right) = \frac{(3u^2 - 2u(v-1) + 3(v-1)^2) - (v-1)(-2u + 6(v-1))}{(3u^2 - 2u(v-1) + 3(v-1)^2)^2}$$

$$= \frac{3u^2 - 3(v-1)^2}{(3u^2 - 2u(v-1) + 3(v-1)^2)^2}, \text{ and}$$

$$b_u' = \frac{\partial}{\partial u}\left(\frac{-u}{3u^2 - 2u(v-1) + 3(v-1)^2}\right) = \frac{-(3u^2 - 2u(v-1) + 3(v-1)^2) - (-u)(6u - 2(v-1))}{(3u^2 - 2u(v-1) + 3(v-1)^2)^2}$$

$$= \frac{3u^2 - 3(v-1)^2}{(3u^2 - 2u(v-1) + 3(v-1)^2)^2}.$$

and we see that γ is exact.

Next, we apply formula (15.36) to both $a(u,v)\, du$ and $b(u,v)\, dv$. Since $k = 1$, we obtain

$$\int_0^1 \frac{tv - 1}{3t^2u^2 - 2tu(tv-1) + 3(tv-1)^2}\, dt \cdot u + \int_0^1 \frac{-tu}{3t^2u^2 - 2tu(tv-1) + 3(tv-1)^2}\, dt \cdot v$$

$$= \int_0^1 \frac{-u}{3t^2u^2 - 2tu(tv-1) + 3(tv-1)^2}\, dt$$

The denominator of the last fraction can be written as

$$3t^2u^2 - 2tu(tv-1) + 3(tv-1)^2 = t^2\left(3u^2 - 2uv + 3v^2\right) + t\left(2u - 6v\right) + 3$$

$$= \left(3u^2 - 2uv + 3v^2\right)\left(t^2 + \frac{2u - 6v}{3u^2 - 2uv + 3v^2}\, t + \frac{3}{3u^2 - 2uv + 3v^2}\right)$$

$$= \left(3u^2 - 2uv + 3v^2\right)\left[\left(t + \frac{u - 3v}{3u^2 - 2uv + 3v^2}\right)^2 + \frac{3(3u^2 - 2uv + 3v^2) - (u - 3v)^2}{(3u^2 - 2uv + 3v^2)^2}\right]$$

$$= \left(3u^2 - 2uv + 3v^2\right)\left[\left(t + \frac{u - 3v}{3u^2 - 2uv + 3v^2}\right)^2 + \frac{8u^2}{(3u^2 - 2uv + 3v^2)^2}\right], \text{ so}$$

$$I\gamma = \int_0^1 \frac{-u}{3u^2 - 2uv + 3v^2} \frac{1}{\left(t + \dfrac{u - 3v}{3u^2 - 2uv + 3v^2}\right)^2 + \dfrac{8u^2}{(3u^2 - 2uv + 3v^2)^2}}\, dt$$

$$= \frac{-u}{3u^2 - 2uv + 3v^2} \frac{3u^2 - 2uv + 3v^2}{2|u|\sqrt{2}} \arctan \frac{3u^2 - 2uv + 3v^2}{2|u|\sqrt{2}} \left(t + \frac{u - 3v}{3u^2 - 2uv + 3v^2} \right) \Big|_0^1$$

$$= \frac{-u}{2|u|\sqrt{2}} \left[\arctan \frac{3u^2 - 2uv + 3v^2}{2|u|\sqrt{2}} \left(1 + \frac{u - 3v}{3u^2 - 2uv + 3v^2} \right) \right.$$

$$\left. - \arctan \frac{3u^2 - 2uv + 3v^2}{2|u|\sqrt{2}} \left(\frac{u - 3v}{3u^2 - 2uv + 3v^2} \right) \right]$$

$$= \frac{-u}{2|u|\sqrt{2}} \left[\arctan \frac{3u^2 - 2uv + 3v^2 + u - 3v}{2|u|\sqrt{2}} - \arctan \frac{u - 3v}{2|u|\sqrt{2}} \right]$$

$$= \frac{-1}{2\sqrt{2}} \left[\arctan \frac{3u^2 - 2uv + 3v^2 + u - 3v}{2u\sqrt{2}} - \arctan \frac{u - 3v}{2u\sqrt{2}} \right].$$

Next, we use the trigonometric formula $\tan(\alpha - \beta) = \frac{\tan \alpha - \tan \beta}{1 + \tan \alpha \tan \beta}$, and we apply it to $\alpha = \arctan a$, $\beta = \arctan b$. This leads to

$$\arctan a - \arctan b = \arctan \frac{a - b}{1 + ab}.$$

It follows that

$$I\gamma = \frac{-1}{2\sqrt{2}} \arctan \frac{\dfrac{3u^2 - 2uv + 3v^2 + u - 3v}{2u\sqrt{2}} - \dfrac{u - 3v}{2u\sqrt{2}}}{1 + \dfrac{3u^2 - 2uv + 3v^2 + u - 3v}{2u\sqrt{2}} \dfrac{u - 3v}{2u\sqrt{2}}}$$

$$= \frac{-1}{2\sqrt{2}} \arctan \frac{2u\sqrt{2}(3u^2 - 2uv + 3v^2)}{8u^2 + (3u^2 - 2uv + 3v^2 + u - 3v)(u - 3v)}$$

$$= \frac{-1}{2\sqrt{2}} \arctan \frac{2u\sqrt{2}(3u^2 - 2uv + 3v^2)}{8u^2 + (3u^2 - 2uv + 3v^2)(u - 3v) + (u - 3v)^2}$$

$$= \frac{-1}{2\sqrt{2}} \arctan \frac{2u\sqrt{2}(3u^2 - 2uv + 3v^2)}{3(3u^2 - 2uv + 3v^2) + (3u^2 - 2uv + 3v^2)(u - 3v)}$$

$$= \frac{-1}{2\sqrt{2}} \arctan \frac{2u\sqrt{2}}{3 + u - 3v} = \frac{-1}{2\sqrt{2}} \arctan \frac{2x\sqrt{2}}{3 + x - 3(y + 1)} = \frac{-1}{2\sqrt{2}} \arctan \frac{2x\sqrt{2}}{x - 3y}.$$

15.8.12. The form ω is closed because

$$d\omega = (2x\,dx - 2z\,dy - 2y\,dz)\,dx + (-2z\,dx + 2y\,dy - 2x\,dz)\,dy + (-2y\,dx - 2x\,dy + 2z\,dz)\,dz$$
$$= -2z\,dydx - 2y\,dzdx - 2z\,dxdy - 2x\,dzdy - 2y\,dxdz - 2x\,dydz = 0.$$

Here, $k = 1$, so

$$I\omega = \int_0^1 (t^2 x^2 - 2t^2 yz)\,dt \cdot x + \int_0^1 (t^2 y^2 - 2t^2 xz)\,dt \cdot y + \int_0^1 (t^2 z^2 - 2t^2 xy)\,dt \cdot z$$

$$= (x^2 - 2yz)x \left(\frac{1}{3} t^3 \right) \Big|_0^1 + (y^2 - 2xz)y \left(\frac{1}{3} t^3 \right) \Big|_0^1 + (x^2 - 2xy)z \left(\frac{1}{3} t^3 \right) \Big|_0^1$$

$$= \frac{1}{3} \left(x^3 - 2xyz + y^3 - 2xyz + z^3 - 2xyz \right) = \frac{1}{3} \left(x^3 + y^3 + z^3 \right) - 2xyz.$$

15.8.14. $I\omega = \frac{1}{2} \ln[(x + y)^2 + z^2] + \arctan \frac{z}{x + y}$.

15.8.17. Define $\gamma(x, y)$ to be the arc of C from $(1, 0)$ to $(x/|x|, y/|y|)$ followed by the line segment to (x, y); define $f(x, y) = \int_\gamma \omega$; prove that $\omega = df$.

Bibliography

[1] Niels Henrik Abel, *Untersuchungen über die Reihe:* $1 + (m/1)x + m \cdot (m-1)/(1 \cdot 2) \cdot x^2 + m \cdot (m-1) \cdot (m-2)/(1 \cdot 2 \cdot 3) \cdot x^3 + \ldots$, [Investigation of the series …]. Journal für die reine und angewandte Mathematik (Crelle's Journal) 1 (1826), 311–339.

[2] Isaac Barrow, *Lectiones Geometricae*, [The Lessons in Geometry]. London, 1670.

[3] Daniel Bernoulli, *Réflexions et éclaircissemens sur les nouvelles vibrations des cordes*, [Thoughts and clarifications on the new string vibrations]. Mémoire de l'Académie Royale des Sciences et Belles-Lettres de Berlin 9 (1753/1755), 147–172.

[4] Jacob Bernoulli, *Positiones arithmeticae de seriebus infinitis earumque summa finita*, [Treatise on Infinite Series]. Basel, 1689. Also in Basiliensis, opera. Vol. I. Genevae, 1744.

[5] Johann Bernoulli, *Principia calculi exponentialium seu percurrentium*, [Principles of the exponential calculus]. Leipzig, Grosse & Gleditsch, 1697.

[6] Jacques Binet, *Mémoire sur les intégrales définies Eulériennes*, [A memoir on Euler integrals]. Journ. de l'Ecole Roy. Polyt. 16 (1839), 123–343.

[7] Ralph P. Boas, *Can we make mathematics intelligible?*, American Mathematical Monthly 88 (1981), 727–731.

[8] Emile Borel, *Mécanique Statistique et Irréversibilité*, [Statistical mechanics and irreversibility]. J. Phys. 5e série 3: 189–196.

[9] Viktor Bunyakovskiĭ, *Sur quelques inégalités concernant les intégrales ordinaires et les intégrales aux différences finies*, [On some inequalities about integrals and integrals of finite differences]. Mémoires de l'Acad. de St. Pétersbourg (VII) 1, No. 9, (1859).

[10] Georg Cantor, *Beweis, dass eine für jeden reellen werth von x durch eine trigonometrische reihe gegebene function $f(x)$ sich nur auf eine einzige weise in dieser form darstellen lässt*, [Proof that a function can be represented by only one trigonometric series for every real value x.] Journal für die reine und angewandte Mathematik (Crelle's Journal) 72 (1870), 139–142.

[11] Georg Cantor, *Notiz zu dem Aufsatz: Beweiss, dass eine für jeden reellen Wert von x durch eine trigonometrische Reihe gegebene Function $f(x)$ sich nur auf eine einzige Weise in dieser Form darstellen lässt*, [A note to the article: Proof that a function can be represented by only one trigonometric series for every real value x.] Journal für die reine und angewandte Mathematik (Crelle's Journal) 73 (1871), 294–296.

[12] Georg Cantor, *Über die Ausdehnung eines Satzes aus der Theorie der trigonometrischen Reihen*, [On the extension of a result from the theory of trigonometric series.] Math. Ann. 5 (1872), no. 1, 123–132.

591

[13] Georg Cantor, *Über eine Eigenschaft des des Inbegriffs aller reelen algebraischen Zahlen*, [About a property of the set of all real algebraic numbers]. Journal für die reine und angewandte Mathematik (Crelle's Journal) 77 (1874), 258–262. English translation in *From Kant to Hilbert*. William Bragg Ewald, (ed.) A Source Book in the Foundations of Mathematics, vol. 2, 839–843. Oxford University Press, Oxford 1996.

[14] Georg Cantor, *Über unendliche, lineare Punktmannichfaltigkeiten, Part 5*, [On Infinite Linear Aggregates of Points]. Math. Ann. 21 (1883), 545–591. Reprinted in Gesammelte Abhandlungen mathematischen und philosophischen Inhalts, [Collected essays of mathematical and philosophical content] E. Zermelo (ed.), Springer-Verlag, New York, 1980, 165–209.

[15] Georg Cantor, *Über unendliche, lineare Punktmannichfaltigkeiten, Part 6*, [On Infinite Linear Aggregates of Points]. Math. Ann. 23 (1884), 453–488. Reprinted in Gesammelte Abhandlungen mathematischen und philosophischen Inhalts, [Collected essays of mathematical and philosophical content] E. Zermelo (ed.), Springer-Verlag, New York, 1980, 210–246.

[16] Augustin-Louis Cauchy, *Résumé des leçons données à l'Ecole Royale Polytechnique sur le calcul infinitésimal*, [Summary of the lectures at the Ecole Royale Polytechnique on the infinitesimal calculus]. Paris 1823. Reprinted in Oeuvres Complètes, sér. 2, vol.4; Gauthier-Villars, 1882.

[17] Augustin-Louis Cauchy, *Mémoire sur les intégrales definies*, [A memoire on definite integrals]. Mémoires présentés par divers savants à l'Académie des Sciences 1 (1827), 601–799; reprinted in Oeuvres Complètes I,1 (1882), 319–506.

[18] Bonaventura Cavalieri, *Geometria indivisibilibus continuorum nova quadam ratione promota*, [Geometry, developed by a new method through the indivisibles of the continua]. Bologna, 1635.

[19] Ernesto Cesàro, *Sur la multiplication des séries*, [On the product of series]. Bull. Sci. Math. (2), vol. 14 (1890), 114–120.

[20] John B. Conway, *Subnormal operators*, Research Notes in Math., 51, Pitman Advanced Publishing Program, Boston-London-Melbourne, 1981.

[21] Pierre Cousin, *Sur les Functions de n variables complexes*, [On the functions of n complex variables]. Acta Mathematica, 19, (1895) 1–61.

[22] Harold Scott MacDonald Coxeter, *Non-Euclidean geometry*. Sixth edition. MAA Spectrum. Mathematical Association of America, Washington, DC, 1998.

[23] Jean Le Rond d'Alembert, *Recherches sur la courbe que forme une corde tendue mise en vibration*, [Researche on the curve that a stretched cord forms when set into vibration]. Mem. Acad. Sci. Berlin, vol. 3, (1747) 214–219. *Suite des recherches sur la courbe que forme une corde tendue mise en vibration*, [Further researche on the curve that a stretched cord forms when set into vibration], Mem. Acad. Sci. Berlin, vol. 3, (1747) 220–249.

[24] Jean Le Rond d'Alembert, *Réflexions sur les suites et sur les racines imaginaires*, [Thoughts on sequences and on the imaginary roots]. Opuscules mathématiques, vol. 5, (1768), 171–215. Briasson, Paris.

[25] Gaston Darboux, *Mémoire sur les fonctions discontinues*, [A memoir on discontinuous functions]. Annales scientifiques de l'Ecole Normale Supérieure, Sér. 2, no. 4 (1875), 57–112.

[26] Richard Dedekind, *Stetigkeit und irrationale Zahlen*, [Continuity and irrational numbers]. Vieweg, Braunschweig 1872. English translation in *From Kant to Hilbert*. William Bragg Ewald, (ed.) A Source Book in the Foundations of Mathematics, vol. 2, 765–779. Oxford University Press, Oxford 1996.

[27] Charles de la Vallée-Poussin, *Etude des intégrales a limites infinies pour lesquelles la fonction sous le signe est continue*, [A study of integrals with infinite limits in which the integrand is a continuous function]. Ann. Soc. Sci. Bruxelles, 16 (1892), 150–180.

[28] Charles de la Vallée-Poussin, *Recherches sur la convergence des intégrales définies*, [Research on the convergence of definite integrals]. Journal de Mathématiques Pures et Appliquées, 8 (1892), 421–467.

[29] Charles de la Vallée-Poussin, *Sur quelques applications de l'intégrale de Poisson*, [On some applications of the Poisson's integral]. Ann. Soc. Sci. Bruxelles, 17 (1893), B, 18–34.

[30] René Descartes, *La Géométrie*, [Geometry]. Published as an appendix to *Discours de la méthode pour bien conduire sa raison et chercher la vérité dans les sciences*, [Discourse on the Method of Rightly Conducting One's Reason and of Seeking Truth in the Sciences]. Leyden, 1637.

[31] Ulisse Dini, *Analisi infinitesimale*, [Analysis of Infinitesimals]. Litografia Gorani, Pisa 1877-1878 (Lithographed version of manuscript lecture notes).

[32] Ulisse Dini, *Fondamenti per la teoria delle funzioni di variabili reali*, [Foundations of the theory of functions of real variables]. Pisa, 1878.

[33] J.P.G. Lejeune Dirichlet, *Sur la convergence des séries trigonométriques qui servent à réprésenter une fonction arbitraire entre des limites donées* [On the convergence of trigonometric series which serve to represent an arbitrary function between given limits], Journal für die reine und angewandte Mathematik (Crelle's Jounral) 4 (1829), 157–169.

[34] J.P.G. Lejeune Dirichlet, *Beweis des Satzes, dass jede unbegrenzte arithmetische Progression, deren erstes Glied und Differenz ganze Zahlen ohne gemeinschaftlichen Factor sing, unendlich viele Primzahlen erh'alt*, [Proof of the theorem that any infinite arithmetic progression whose first term and difference are integers with no common factor contains an infinite number of primes]. Abhand. Ak. Wiss. Berlin 48, (1837), 45–81.

[35] Lance D. Drager, Robert L. Foote, *The Contraction Mapping Lemma and the Inverse Function Theorem in Advanced Calculus*, Amer. Math. Monthly, 93 (1986) 52–54.

[36] Paul Du Bois-Reymond, *Versuch einer Klassification der willkürlichen Functionen reeller Argumente nach ihren Änderungen in den kleinsten Intervallen*, [An attempt to classify arbitrary functions of real argument by their oscillation in the small intervals]. Journal für die reine und angewandte Mathematik (Crelle's Journal) 79 (1875), 21–37.

[37] Paul Du Bois-Reymond, *Untersuchungen über die konvergenz und divergenz der Fourierschen Darstellungsformeln*, [Studies on the convergence and divergence of Fourier series]. Bayerischen Abhandlungen 12 (1876), 1–103.

[38] Pierre Dugac, *Sur la correspondance de Borel et le théorème de Dirichlet–Heine–Weierstrass–Borel–Schoenflies–Lebesgue*, [On the Borel's correspondence and the theorem of Dirichlet–Heine–Weierstrass–Borel–Schoenflies–Lebesgue]. Arch. Internat. Hist. Sci. 39 (1989), 69–110.

[39] Charles Henry Edwards, *The Historical Development of the Calculus*, Springer-Verlag. New York, 1979.

[40] Leonhard Euler, *Institutiones calculi differentialis*, [Foundations of differential calculus]. 1748. Reprinted in Opera Omnia: Series 1, Volume 10, B.G. Teubner, Leipzig and Berlin, 1911.

[41] Leonhard Euler, *Introductio in Analysin infinitorum*, [Introduction to the analysis of the infinite], 1748. Reprinted in Opera Omnia: Series 1, vol. 8–9, B.G. Teubner, Leipzig and Berlin, 1911.

[42] Leonhard Euler, *Sur la vibration des cordes*, [On the vibration of a string]. Mém. Acad. Sci. Berlin, vol. 4 (1748), 69–85.

[43] Pierre Fatou, *Séries trigonométriques et séries de Taylor*, [Trigonometric and Taylor series]. Acta Math. 30 (1906), 335–400.

[44] Leopold Fejér, *Untersuchungen über Fouriersche Reihen*, [Investigations of Fourier series]. Math. Ann. 58 (1904), 51–69.

[45] Julian F. Fleron, *A Note on the History of the Cantor Set and Cantor Function*, Mathematics Magazine, Vol. 67, No. 2. (1994), 136–140.

[46] Joseph Fourier, *Mémoire sur la propagation de la chaleur dans les corps solides*, [Memoir on the propagation of heat in solid bodies]. Manuscript, Paris, 1807.

[47] Maurice Fréchet, *La notion de différentielle dans l'analyse générale*, [The concept of the differential in general analysis]. Ann. Ec. Norm. Sup., 42 (1925), 293–323.

[48] Ferdinand Georg Frobenius, *Über die Leibnitzsche Reihe*, [On the Leibniz series]. Journal für die reine und angewandte Mathematik (Crelle's Journal) 89 (1880), 262–264.

[49] Carl Friedrich Gauss, *Theorematis arithmetici demonstratio nova*, [Demonstration of a new theorem of arithmetic]. Comm. Soc. Reg. Sci. Gott. 16 (1808). Reprinted in his Werke (Göttingen), vol. 2 (1860), 3–8;

[50] Carl Friedrich Gauss, *Disquisitiones generales circa seriem infinitam*, [General investigations of infinite series)]. Comm. Soc. Regia Sci. Göttingen Rec., 2 (1812). Reprinted in his Werke (Göttingen), vol. 3 (1860), 123–163.

[51] Carl Friedrich Gauss, *Theoria attractionis corporum sphaeroidicorum ellipticorum homogeneorum methodo novo tractata*, [A new method in the attraction theory of elliptic homogeneous spheroidal bodies]. Comm. Soc. Reg. Scient. Gottingensis, 2 (1813): Reprinted in his Werke (Göttingen), vol. 5 (1860), 1–22.

[52] Enrique A. Gonzalez-Velasco, *Connections in Mathematical Analysis: The Case of Fourier Series*. American Mathematical Monthly 99, no. 5 (1992), 427–441.

[53] Edouard Jean Baptiste Goursat, *Sur la théorie des fonctions implicites*, [On the theory of implicit functions]. Bulletin de la Société Mathématique de France, 31 (1903), 184–192.

[54] Hermann Grassmann, *Die Lineale Ausdehnungslehre, ein neuer Zweig der Mathematik*, [The Theory of Linear Extension, a New Branch of Mathematics], Leipzig 1844.

[55] Ivor Grattan-Guinness, *The Development of the Foundations of Mathematical Analysis from Euler to Riemann*, Cambridge and London, 1970.

[56] Lawrence M. Graves, *Some mapping theorems*, Duke Math. J. 17, No. 2 (1950), 111–114.

[57] Ernst Hairer, Gerhard Wanner, *Analysis by Its History*, Undergraduate Texts in Mathematics, Springer-Verlag. New York 1996.

[58] Hermann Hankel, *Zur allgemeinen Theorie der Bewegung der Flüssigkeiten*, [On the general theory of motion of liquids]. Dieterische Univ. Buchdruckerei. W.F. Kaestner, Göttingen, 1861.

[59] Axel Harnack, *Die Elemente der Differential- und Integralrechnung. Zur Einführung in das Studium*, [The elements of differential and integral calculus. An introduction to the study]. B.G. Teubner, Leipzig 1881.

[60] Dan Hathaway, *Using Continuity Induction*, College Math. Journal 42 (2011), 229–231.

[61] Felix Hausdorff, *Grundzüge der Mengenlehre*, [Elements of the Set Theory]. Veit and Company, Leipzig, 1914.

[62] Heinrich Eduard Heine, *Über trigonometrische Reihen*, [On trigonometric series]. Journal für die reine und angewandte Mathematik (Crelle's Journal) 71 (1870), 353–365.

[63] Heinrich Eduard Heine, *Die Elemente der Funktionenlehre*, [Elements of the Function Theory]. Journal für die reine und angewandte Mathematik (Crelle's Journal) 74 (1872), 172–188.

[64] Edwin Hewitt, Karl Stromberg, *Real and abstract analysis. A modern treatment of the theory of functions of a real variable.* Third printing. Graduate Texts in Mathematics, No. 25. Springer-Verlag, New York-Heidelberg, 1975.

[65] David Hilbert, *Über die stetige Abbildung einer Linie auf ein Flächenstück*, [About the continuous map of a line on a surface element]. Mathematische Annalen 38 (1891), 459–460.

[66] David Hilbert, *Über den Zahlbegriff*, [About the Concept of a Number]. Jahresbericht der Deutschen Mathematiker-Vereinigung, 8 (1900), 180–84. English translation in *From Kant to Hilbert*. William Bragg Ewald, (ed.) A Source Book in the Foundations of Mathematics, vol. 2, 1089–1096. Oxford University Press, Oxford, 1996.

[67] Roger A. Horn and Charles R. Johnson, *Matrix Analysis*, Cambridge University Press, Cambridge, 1985.

[68] Hans Niels Jahnke (ed.), *A History of Analysis.* Vol. 24 of the Series 'History of Mathematics'. American Mathematical Society & London Mathematical Society 2003.

[69] Camille Jordan, *Remarques sur les intégrales définies*, Journal de Mathématiques Pures et Appliquées 8, (1892) 69–99.

[70] Victor J. Katz, *The History of Stokes' Theorem*, Mathematics Magazine 52, no. 3 (1979) 146–156.

[71] Victor J. Katz, *Change of variables in multiple integrals: Euler to Cartan*, Mathematics Magazine 55, no. 1 (1982), 3–11.

[72] Oliver Dimon Kellogg, *Foundations of potential theory*. Reprint from the first edition of 1929. Die Grundlehren der Mathematischen Wissenschaften, Band 31 Springer-Verlag, Berlin-New York, 1967.

[73] Morris Kline, *Mathematical Thought From Ancient to Modern Times*, Oxford University Press, New York, 1972.

[74] Steven Krantz, Harold Parks, *The Implicit Function Theorem: History, Theory, and Applications*. Birkhäuser, Boston, 2002.

[75] Joseph Louis Lagrange, *Recherches sur la méthode de maximis et minimis*, [Research on the method of maxima and minima]. Misc. Taurinensia, Torino 1 (1759), 18–32. Reprinted in his Oeuvres, Vol. 1, Gauthier-Villars, Paris, 1858, 3–21.

[76] Joseph Louis Lagrange, *Nouvelle méthode pour résoudre les équations littérales par le moyen des séries*, [New method for solving explicit equations by means of series]. Mém. de l'Acad. Roy. des Sciences de Berlin 24 (1768), 251–326. Reprinted in Oeuvres de Lagrange, vol. III, Gauthier-Villars, Paris, 1858, 5–73.

[77] Joseph Louis Lagrange, *Sur une nouvelle espèce de calcul relatif à la différentiation et à l'integration des quantités variables*, [On a new type of calculation relating to the differentiation and the integration of variable quantities]. Nouveaux Memoires de l'Academie royale des Sciences et Belles-Lettres de Berlin, 1772. Reprinted in Oeuvres de Lagrange, vol. III, Gauthier-Villars, Paris, 1858, 451–478.

[78] Joseph Louis Lagrange, *L'attraction des sphéroides elliptiques*, [Attraction of elliptic spheroids]. Nouv. Mem. Acad, royale Berlin (1773), 121–148. Reprinted in Oeuvres de Lagrange, vol. III, Gauthier-Villars, Paris, 1858, 619–649.

[79] Joseph Louis Lagrange, *Théorie de Fonctions Analytiques*, [Theory of Analytic Functions]. Imprimerie de la République, Paris, 1797.

[80] Edmund Landau, *Foundations of Analysis: The Arithmetic of Whole, Rational, Irrational and Complex Numbers*. Translated by F. Steinhardt. Chelsea Publishing Company, New York, 1951. Translation of *Grundlagen der Analysis*, Akademische Verlagsgesellschaft, Leipzig, 1930.

[81] Guillaume de L'Hôpital, *Analyse des Infiniment Petits pour l'Intelligence des Lignes Courbes*, [Analysis of the Infinitely Small for the Understanding of Curved Lines]. Paris, 1696.

[82] John Napier, *Mirifici Logarithmorum Canonis Descriptio*, [Description of the Wonderful Rule of Logarithms]. Edinburgh, 1614.

[83] Blaise Pascal, *Traité du triangle arithmétique*, [Treatise on the Arithmetical Triangle], 1653.

[84] Giuseppe Peano, *Applicazioni geometriche del calcolo infinitesimale*, [Geometrical applications of calculus.] Fratelli Bocca, Torino, 1887.

[85] Giuseppe Peano, *Calcolo geometrico secondo l'Ausdehnungslehre di H. Grassmann preceduto dalle operazioni della logica deduttiva*, [Geometric calculations according to the *Extensions theory* of H. Grassmann preceded by the operations of deductive logic.] Fratelli Bocca Editori, Torino, 1888.

[86] Giuseppe Peano, *Arithmetices principia, nova methodo exposita*, [The principles of arithmetic, presented in a new manner]. Bocca, Turin, 1889.

[87] Henri Poincaré, *Les Méthodes Nouvelles de la Mécanique Céleste*, [New methods in celestial mechanics]. Gauthier-Villars et fils, 1892–99. Translated, edited and introduced by Daniel L. Goroff, as New Methods of Celestial Mechanics, American Institute of Physics, Woodbury, NY, 1993.

[88] Siméon Denis Poisson, *Sur les intégrales des fonctions, qui passent par l'infini entre les limites de l'intégration, et sur l'usage des imaginaires dans la détermination des intégrales définies*, [On integrals of functions that are unbounded between the limits of integration, and the use of imaginary numbers in determining definite integrals]. Journal de l'Ecole Polytechn. Cah. 18, (1820) 320–341.

[89] Bernhard Riemann, *Über die Darstellbarkeit einer Function durch eine trigonometrische Reihe*, [About the description of a function by a trigonometric series], Abhandlungen der Königlichen Gesellschaft der Wissenschaften zu Göttingen, XIII (1854). Reprinted in Gesammelte Math. Abhandlungen. New York: Dover, 227–264, 1957.

[90] Walter Rudin, *Principles of Mathematical Analysis*, Third edition. International Series in Pure and Applied Mathematics. McGraw-Hill Book Co., New York-Auckland-Düsseldorf, 1976.

[91] Giovanni Mingari Scarpello, Daniele Ritelli, *A historical outline of the theorem of implicit functions*. Divulg. Mat. 10 (2002), no. 2, 171–180.

[92] Arthur Moritz Schoenflies, *Die Entwicklung der Lehre von den Punktmannigfaltigkeiten I*, [The development of the theory of point sets]. Jahresbericht der Deutschen Mathematiker-Vereinigung, Bd. 8 (1900), 1–250.

[93] Hermann Amandus Schwarz, *Communication*. Archives des Sciences Physiques et Naturelles, 48 (1873), 38–44.

[94] Hermann Amandus Schwarz, *Ueber ein die Flächen kleinsten Flächeninhalts betreffendes Problem der Variationsrechnung*, [A problem of variational calculus: the surfaces with a minimum surface area]. Acta Soc. Scient. Fenn. 15 (1885), 315–362.

[95] Henry John Stephen Smith, *On the integration of discontinuous functions*, Proc. London Math. Soc. (1) 6 (1875), 140–153.

[96] Michael Spivak, *Calculus on Manifolds: A Modern Approach to Classical Theorems of Advanced Calculus*, W. A. Benjamin, Inc., New York-Amsterdam, 1965.

[97] Michael Stifel, *Arithmetica integra*, [The arithmetic of integers]. Nurnberg, 1544.

[98] Otto Stolz, *Die gleichmässige Konvergenz von Funktionen mehrerer Veränderlichen zu den dadurch sich ergebenden Grenzwerthen, dass einige derselben constanten Werthen sich nähern*, [The uniform convergence of functions of several variables to the resulting limits, so that some of them approach constant values]. Math. Ann. 26 (1886), 83–96.

[99] Otto Stolz, *Grundzüge der differential und integralrechnung*, [Fundamentals of differential and integral calculus]. B.G. Teubner, Leipzig, 1893.

[100] Manya Raman Sundström, *A pedagogical history of compactness*, (2010). Posted on http://arxiv.org/abs/1006.4131.

[101] Brook Taylor, *Methodus incrementorum directa et inversa*, [Direct and Indirect Methods of Incrementation]. London, 1715.

[102] Karl Johannes Thomae, *Abriss einer Theorie der complexen Functionen und der Thetafunctionen einer Veränderlichen*, [An outline of a theory of the complex functions of one variable and the Theta function]. Halle, 1873.

[103] Karl Johannes Thomae, *Einleitung in die Theorie der bestimmten Integrale*, [Introduction to the theory of definite integrals]. Halle, 1875.

[104] Edward Burr Van Vleck, *The influence of Fourier's series upon the development of mathematics*, Science 39, no. 995 (1914), 113–124.

[105] Vito Volterra, *Sui principii del calcolo integrale*, [On the principles of integral calculus]. Giornale di Matematiche 19 (1881), 333–372.

[106] Vito Volterra, *Delle Variabili Complesse Negli Iperspazi*, [Complex variables in hyperspaces]. Rend. Accad. dei Lincei 5 (1889) 158–165. Reprinted in *Opere matematiche; memorie e note*, [Mathematical Works: Memories and Notes] Vol. 1, 403–432.

[107] George Walker, *Elements of Algebraical Notation and Expansion*, Leeds 1828.

[108] John Wallis, *De sectionibus conicis, nova methodo expositis tractatus*, [Treatise on conic sections]. Oxford, 1655.

[109] John Wallis, *Arithmetica infinitorum*, [The arithmetic of infinitesimals]. Oxford, 1656.

[110] Karl Weierstrass, *Zur funktionenlehre*, [On the function theory]. Monatsber. Acad. Wiss. Berlin (1880), 719–743, 1880. Reprinted in *Mathematische Werke von Karl Weierstrass*, vol. 2, 201–230. Mayer & Mueller, Berlin, 1895.

[111] Raymond Louis Wilder, *Evolution of the topological concept of "connected"*, Amer. Math. Monthly, 85 (1978), 720–726.

[112] William Henry Young, *On the area of surfaces*, Proc. Royal Soc. London ser. A, vol. 96 (1920), 71–81.

[113] Frieda Zames, *Surface Area and the Cylinder Area Paradox*, The College Mathematics Journal, Vol. 8, No. 4, (1977), 207–211.

[114] Antoni Zygmund, *Trigonometric series*. Vol. I, II. Reprinting of the 1968 version of the second edition with Volumes I and II bound together. Cambridge University Press, Cambridge-New York-Melbourne, 1977.

Subject Index

Author Index